Y0-CMJ-761

ISBN 978-0-266-96728-6
PIBN 10916875

HANDBOOK OF INDIANA GEOLOGY

Indiana. Division of Geology,
William Newton Logan, ...

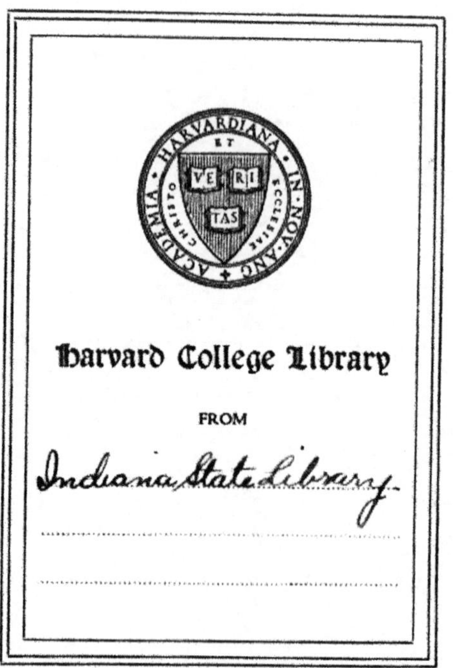

THE

DEPARTMENT OF CONSERVATION

STATE OF INDIANA

PUBLICATION No. 21

RICHARD LIEBER
DIRECTOR

1922

Handbook of Indiana Geology

By

W. N. Logan
E. R. Cumings
C. A. Malott
S. S. Visher
W. M. Tucker
J. R. Reeves

THE DEPARTMENT OF CONSERVATION.

Division of Geology.

W. N. LOGAN, State Geologist.

INDIANAPOLIS:
WM. B. BURFORD, CONTRACTOR FOR STATE PRINTING AND BINDING
1922

GEOLOGICAL CORPS.

W. N. Logan, Ph.D., State Geologist.................... Economic Geology
E. R. Cumings, Ph.D...................... Stratigraphy and Paleontology
C. A. Malott, Ph.D.................................. Physiography
S. S. Visher, Ph.D.................................. Geography
W. M. Tucker, Ph.D.................................. Hydrology
J. R. Reeves, A.B., A.M.............................. Economic Geology
H. W. Legge... Preparator

Office Division.

Theodore Kingsbury......................... Supervisor of Natural Gas
E. H. Shaw.................................. Curator of Museum
Adda Rinker .... Stenographer

Field Division, 1921.

W. N. Logan
E. R. Cumings
C. A. Malott
J. R. Reeves
R. E. Esarey
G. G. Bartle
M. A. Harrell
H. C. Barnett
E. L. Lucas
J. I. Moore
W. P. Rawles
K. W. Ray
P. B. Moore

TABLE OF CONTENTS

Part I. Page

Foreword

The Handbook of Indiana Geology is, in a large measure, the result of a co-operative plan which had its inception shortly after the establishment of the Department of Conservation of Indiana. The plan constitutes an agreement whereby the instructional staff of the Department of Geology of Indiana University conducts the field and laboratory investigations for the Department of Conservation, thus obviating the necessity for large salary expenditures. In this manner our slender geological funds may be used to defray other expenses incurred in the investigation of the mineral resources of Indiana. This plan of co-operation under which not even the State Geologist's salary is paid by the Division, provides the Department of Conservation a corps of trained workers in all the various phases of geological work, phases having a cultural, educational, economic and scientific bearing. Evidence of the success of the co-operative plan is to be found in the present volume.

There is a Latin proverb to the effect that nothing worth while is accomplished without great effort. It is well exemplified in this treatise. Some of the preliminary investigations necessary for the results embodied in this volume have been carried on by the collaborators for more than a score of years. During the past two years their efforts have been concentrated toward the completion of the work. Each spare moment has been employed without expectation of monetary reward. Thus we have here garnered the fruits of years of labor at an expense to the State of what amounts virtually to the cost of printing the volume. Sales from two of the chapters which were printed in advance will assist materially in defraying the expenses of printing the complete work.

To Dr. W. N. Logan, State Geologist, and Professor of Economic Geology in Indiana University, who is keenly alive at all times to the development of Indiana's mineral resources, is due in a large measure, credit for the successful completion of this volume.

The Handbook of Indiana Geology has been prepared by the members of the technical force of the Division of Geology with the hope that it will prove helpful to those who are interested in a scientific or commercial way in the geography, physiography, hydrology, stratigraphy and economic geology of Indiana. In its preparation the proper division of labor has been maintained in that each subject is presented by that member whose previous investigations and interests have been mostly directed along the line which he discusses. The preparation of the Handbook was undertaken in response to a rather insistent demand for a comprehensive discussion of the geology and mineral resources of Indiana. However, in this volume no attempt has been made to say the final word on the geology of Indiana. To make such an attempt would be to deny the possibility of any future progress of the science of geology. On the other hand the object of the publication is rather to record the progress of the science of geology as applied to Indiana in so far as the limitations of time and funds will permit.

Many interesting and complex problems still face the student of geology in Indiana and much information which will aid in the solution of these problems is being acquired daily. As the student of Indiana geology reads within the pages of this volume the history of past accomplishments let him turn his eyes hopefully toward the future which holds opportunities for greater achievements. To the encouragement of these students the present volume is dedicated.

RICHARD LIEBER,
Director, The Department of Conservation.

The Geography of Indiana

By

STEPHEN S. VISHER

CONTENTS

LIST OF TABLES

LIST OF ILLUSTRATIONS

The Geography of Indiana

CHAPTER I.

LOCATION.

Indiana is located at the center of the eastern half of the United States. The parallel of latitude passing through the center of the United States, which is in north-central Kansas, also passes through Indianapolis, near the center of Indiana. The north-south line bisecting the eastern half of the United States passes just east of Indianapolis. The eastern half of the United States is far more important in most respects than is the western half because it receives more rainfall, has more valuable natural resources and is otherwise favored. Most of the people of the United States live in the eastern half of the country; the crop production is vastly greater than that of the western half; nearly all of the country's manufacturing is done in the eastern half, most of it in the northeast quarter. Hence it is not surprising that the "center of population" of the United States should be near the center of the eastern half. It has been in Indiana since 1880. Since 1900 it has been a short distance south of Indianapolis, near Columbus in 1900, in Bloomington in 1910 and exactly ten miles due west of Bloomington in 1920. The "median point" of the population of the United States is a short distance northeast of Indianapolis, in Randolph County. The term "center of population" as used by the Census Bureau is the point to which all the people of the United States could come with the least possible travel, if they could travel in a straight line. The median point is the point in respect to which as many people live east as live west, and as many live north as live south. The center of the urban population of the United States is a few miles east of Indiana, northeast of Indianapolis. The center of the rural population is in Illinois only a few miles northwest of Evansville, Indiana. The center of the foreign-born population is near the northeastern corner of Indiana, just across the line in Ohio. Thus Indiana is at the center of the nation in several important respects, and Indianapolis, with its excellent railway facilities, is logically the great convention city which it is coming to be.

Indiana in the "heart" of the more important half of the United States is one of the rich states of the Union, and the bordering states all rank high. To the east is Ohio, the fourth state in population; on the west is Illinois, the third state in population; to the north is Michigan, the eighth state in population; to the south is Kentucky, the fourteenth state in population. Indiana, the smallest of the group and possessing no city as large as the largest in the neighboring states (except Kentucky) ranks ninth in population. In density of population per square mile, Indiana ranks next to Illinois among the states of the Union. (81.3 per square mile in 1920.)

Indiana is favored by being not far from several great industrial centers. The northwestern corner of the state is indeed a part of Greater Chicago (2nd city of the U. S. in 1920) and is increasing rapidly in population with the growth of that remarkable city. Louisville (29th city) is just across the Ohio River from Indiana. Cincinnati (16th city) is not much farther away. Detroit (4th city in 1920) and Toledo (26th city) are nearer the northeastern corner of the state than is the centrally located Indianapolis, and Cleveland (5th city) is about as near as Indianapolis. St. Louis (6th city) is nearer the southwestern corner of the state than is Indianapolis. The presence of these strong rivals, five of them much larger than Indianapolis (21st city), has hindered the growth of large cities in Indiana because they serve parts of Indiana. No Indiana city serves as the metropolis of any appreciable part of another state.

Because of the location of Indiana in respect to the rest of the nation it is crossed by many of the more important railroad lines. Its extent from the south end of Lake Michigan southward to the Ohio River, beyond the latitude in which there are great industrial cities, means that it is crossed by all great lines between the more western and the more eastern parts of the nation. Important north-south lines also pass through the state between Chicago or more eastern points and the south. Many of the lines might have crossed the northern part of the state to Chicago without benefiting the state very much if the state legislature during the years of rapid railroad extension westward, just before the Civil War, had not declined to grant a charter of right of way across Indiana except to companies which agreed to construct at least a spur of their line to Indianapolis, the State Capital. One result of this policy was that Indianapolis soon became well supplied with railways.

Not only is Indiana at present fortunately located in respect to trade routes but even before the building of railroads her location was fortunate. The Ohio River, a great highway in pre-railroad days, forms much of the southern and southeastern boundary. The southern part of the state was settled chiefly by frontiersmen who came west along the Ohio. The Wabash River furnished an outlet to the Ohio and Mississippi for many of the commodities shipped out of the southwestern part of the state in early years. For example, in 1831, fifty-five steamers ascended this river to get farm produce and several hundred flat boats descended it. Still earlier, during the period of exploration, the eighteenth century, many explorers were enabled to easily enter what is now Indiana by passing southward from the Great Lakes to one of the nearby tributaries of the Mississippi River. From Lake Michigan the St. Joseph was ascended to the present site of South Bend where a short portage was made to the Kankakee River. From Lake Erie, the Maumee River was ascended to the present site of Fort Wayne where a short portage was made to the Wabash River. The importance of the great rivers as highways was formerly so great that for decades nearly all important towns were on navigable streams. The oldest continuous settlement in Indiana was made at Vincennes, on the Wabash, in 1727. Before the railroad era commenced, in the decade beginning in 1850, it was not generally foreseen that many areas remote from navigable streams would soon have a far better highway and outlet than even the Ohio River furnished. As an illustration of this, when the commission delegated in 1820 to select a site near the center of the state on which to locate the State Capital made the selection, the present site of Indianapolis was chosen because these commissioners considered that the West Fork of White River was navigable to that point. No one now thinks of Indianapolis as being on a navigable stream. The development of railways has made unnecessary the use of such a poor waterway.

Although Lake Michigan forms nearly one-fourth of Indiana's northern boundary, it was a highway of subordinate local importance in the days of active settlement of the state. A broad and complex belt of sand dunes and lagoon-like marshes border the lake in Indiana. Inland from this belt there are wide marshes along the Kankakee and a rugged glacial moraine farther east. These physical features discouraged settlers from entering Indiana from Lake Michigan. Home-seekers who might otherwise have entered Indiana via Lake Michigan went on to Chicago, and westward from there; instead of settling in Indiana. Lake Michigan has, however, furnished an important outlet by boat for Michigan City on the one good natural harbor in Indiana. The recent vast development at Gary is related closely to its splendid artificial harbor and to its lake commerce.

CHAPTER II.

AREA.

Indiana contains 36,045 square miles of land area besides 280 square miles of rivers and small lakes and 230 square miles of Lake Michigan. Indiana is the thirty-seventh state in size. Ten of the eleven states smaller than Indiana were numbered among the original thirteen states. The other smaller state, West Virginia, is only one-fifth smaller than Indiana.

Indiana is a small state as compared with its neighbors. Illinois, Michigan, and Iowa are each about fifty per cent. larger than Indiana. Ohio and Kentucky are each about eleven per cent larger, and Missouri is nearly twice as large as Indiana.

The comparatively small area of Indiana often masks the richness of this state, when comparison between states is made. The area of Indiana is about as rich an area as any of like size in the nation. Many states surpass Indiana in the statistics of production, etc., but after allowance has been made for Indiana's smaller area, it is seen that Indiana really ranks with the leaders.

Though thirty-seventh in size, Indiana ranks eleventh in population. In spite of ranking sixteenth in proportion of urban population to rural population it ranks about tenth in density (81.3 persons per square mile). This is because, although slightly more than one-half of the people live in cities, (50.6%), many other states, with larger cities, have a larger proportion living in cities of over 2,500. Indiana, however, is nearly average in regard to proportion of rural and urban population. The United States as a whole had 51.4% of its population living in cities in 1920. Though often surpassed by five states in the area planted to corn, Indiana ranks with Illinois and Iowa as a corn state, when total area is considered. Indeed in average yield of corn per square mile, Indiana surpasses these states and the other three, Missouri, Nebraska and Kansas, which often have more acres planted to corn than Indiana has. In wheat production, Indiana ranks eighth among the states when area is ignored. However, in proportion to total area, Indiana usually ranks third, being exceeded only by North Dakota and Kansas in the average yield of wheat per square mile of total area. In oats production, Indiana ranks fourth, when area is considered, though seventh when area is ignored.

Industrially, Indiana is seen to rank much higher when her comparatively small area is considered than when comparative area is ignored. For example, though Indiana ranked eighth among the states in 1909 in the value added by manufacture and ninth in total value of manufactured products, when its smaller area is considered, Indiana is seen to clearly surpass its nearest rivals, Michigan and Wisconsin. Likewise, though Indiana was the eighth state (1909) in the value of products of flour mills and grist mills, Indiana is enough smaller than the seven states which ranked higher than it so as to surpass all but the leading state (Minnesota) in value per square mile of area. As a canning state, Indiana takes fourth place, instead of sixth, when size is considered.

One of the interesting correlations between area and a point in which Indiana is often compared unfavorably with her neighboring states is in respect to the attendance at her state university. The attendance at the universities of some of the neighboring states is far greater than at Indiana University. It should be recalled, however, that the state university and the schools of agriculture and mechanical arts are combined in several of the nearby states. When the combined attendance at Indiana University and at Purdue University are compared with corresponding figures for Illinois, Ohio, Wisconsin, Minnesota, Missouri, etc., it will be seen that the attendance at the state universities of Indiana is little if any less per square mile of area or per million of population than is the attendance in these other states. This is in spite of the presence, just beyond the borders of Indiana, of strong rival universities, more conveniently located for a considerable number of Indiana citizens than are Indiana's universities.

CHAPTER III.

THE QUALITY OF THE LAND.

The topography and natural resources of Indiana are considered at length in following chapters. Here, only a few facts need be mentioned.

Most of the state is distinctly smooth and no considerable part of it is really rugged. All was glaciated except about six thousand square miles in the southwestern part of the state. (See Plate III.) The unglaciated part is moderately rugged as are also some small morainic areas in the northeastern part of the state. All but a very small fraction of the state is smooth enough so that it can be tilled. Indiana ranks third among the states in the proportion of improved land, being surpassed by only Iowa and Illinois. Most of the state possesses a deep subsoil covered by a fertile soil. The hills in the unglaciated section form the chief exception to this statement.

The chief hindrance to successful cultivation was originally the poor natural drainage characteristic of wide areas in the glaciated part of the state. Many millions of acres needed to be artificially drained before they were available for high grade farming. The amount of draining already completed in Indiana is enormous and every year hundreds of miles of additional lines of drain-tiling are laid. Relatively few farms contain no tiling. Many of the best farms of today are in what not many decades ago was considered almost valueless swamp. Indeed many of the pioneer settlers located their homesteads on lands which are now much less valuable than lands which were long unoccupied. They could not anticipate future inventions and developments nor did they realize the abiding worth of the nearly level areas of deep soil. For example, many of the pioneers who homesteaded in the unglaciated part of the state were quite sure they had picked the best land in the state. It was well drained, relatively free from the troublesome mosquito and the dreaded ague (Malaria) and it possessed abundant game, fine woods, numerous streams, an attractive variety of forage for their stock, and beautiful scenery. Travel from place to place by the common mode of those days, horseback or with oxen, was less hindered by the ruggedness than was travel in the glaciated part by lakes and swamps. Not until the coming of railroads, which made possible the shipment of bulky farm products, could farming surpass stock-raising in importance. The rougher and more varied land was better adapted originally for stock-raising than were the "black swamps" and marshes of the long largely neglected northern half of the state. The development of labor-saving farm machinery aided farming more

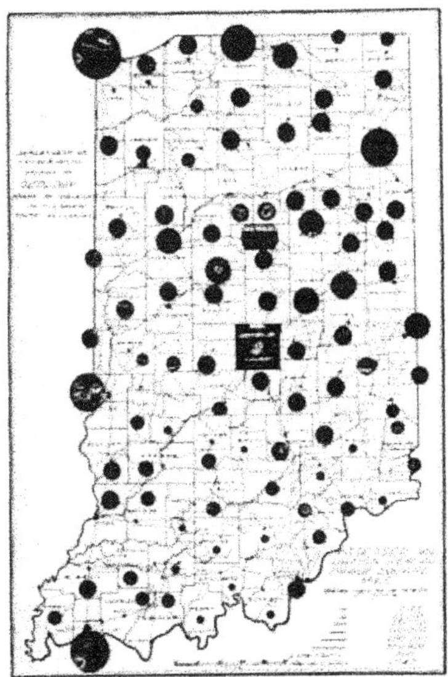

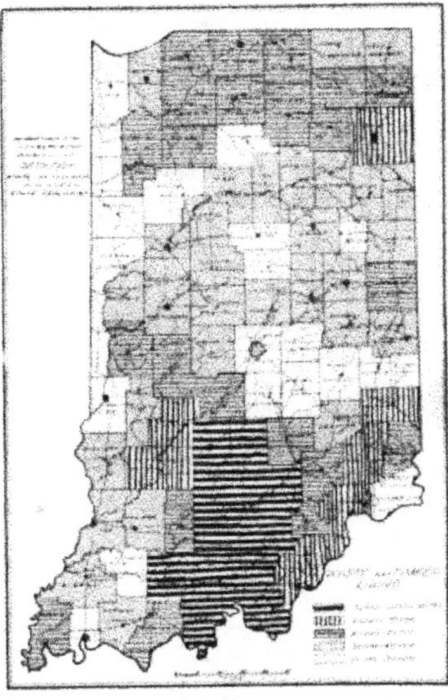

Fig. 1. Assessed valuation per average square mile by counties. Values are proportional to areas of circles from $800,000 in Lake County, to less than $15,000 in some of the poor counties in the unglaciated section. Marion County has such a large valuation, because of Indianapolis, that the circle is only outlined. It was $1,743,000 per square mile in 1920.

Fig. 2. Waste and timber land per county. Darkest shading represents 75,000-100,000 acres, white represents 11,000-30,000 acres.

in the smoother part of the state than in the rougher, because fields must be small in rough areas. The development of the art and science of tiling benefited the glaciated region greatly. After proper tiling, many formerly worthless areas became far better farm land than the rougher land with its thinner, less fertile soil. The heavy rainfalls, frequent in Indiana, produce enough runoff so that the soil on even fairly gentle slopes is eroded and carried away, if loosened. Many of the long-cultivated steeper slopes of Southern Indiana have lost so much of their soil as to have become distinctly infertile. The average value of farm land per acre in the non-glaciated part of the state is only a small fraction of that in the northern half of the state. (See Figs. 1 and 4.) In 1920, with most land in the north selling for more than three hundred dollars per acre, tens of thousands of acres in the hills and ridges in the unglaciated part could be bought for less than one-twentieth as much. Indeed much brush-covered land could be had for less than ten dollars per acre.

In forest resources Indiana was very wealthy a century ago. Ninety-five per cent of the area was formerly forested. Some of the finest lumber trees in the world were in Indiana. A large fraction of the rougher, unglaciated part of the state is still wooded, and a much larger fraction should be. If forestry is scientifically practiced, Indiana should continue to rank as a leading state in supplying hardwood lumber. Land which is of little value for the growing of cultivated crops may be excellent for the growing of timber. The rougher land of southern Indiana is certainly splendidly adapted to lumber production. The increased demand for lumber with the consequent clearing of level, tillable lands elsewhere, makes the growing of timber on the rougher land all the more profitable. The increased desirability of checking soil erosion on the slopes, and of reducing floods also contributes to make the extension of woodland increasingly important. Forestry should be encouraged in every way possible in the rough land of Indiana. Figure 2 shows the land not timbered, and the land reported as waste land by the tax assessors in 1918. It is based on statistics in the Indiana Year Book for 1919.

In mineral wealth, Indiana is also very rich. She has usually ranked fifth or sixth among the states in the production of coal; in clay products about fourth; in limestone for building purposes she now stands first; in oil and gas she has long stood high. For the decade 1900-1910 Indiana was seventh state in oil production, and about fifth in the production of natural gas. For the total value of the products of mines, quarries and wells, Indiana ranked seventh among the states in 1904, eleventh in 1909, and about twelfth in 1920, in spite of ranking 37th in size.

CHAPTER IV.

CLIMATE.

In climate Indiana closely resembles neighboring areas. A large region in eastern United States has strongly marked seasons, with hot summers and cold winters; frequent changes of weather, due to the passage eastward of numerous cyclonic storms; high humidity and considerable rainfall; moderate cloudiness and windiness. These characteristics are in response to Indiana's position in mid-latitudes (the belt of westerly winds, of cyclonic storms and of highly variable temperatures), in the interior of a great continent but not shut off from a great source of moisture, the Gulf of Mexico.

Temperature. Indiana is near the southern margin of the area having long, severe winters. The normal January temperature for the state is about 28° F. At the southern end of the state the normal January temperature is 33° F., while at the north it is 25° F. Even in the northern half of the state, two-thirds of the winter precipitation is rain. In the snowiest month, February, only 37.5% of the precipitation is snowfall in the northernmost part of the state.

Although all parts of the state have experienced temperatures as low as 20° below zero F., such cold snaps are rare and of short duration. Even in the extraordinarily cold winter of 1917-1918, when new minima records were established at many points, Indianapolis had only seven days when temperatures below zero were experienced and only one day when a temperature below —9° F. occurred. That day it reached 19° below, which was the coldest temperature experienced in fifty years, except one year, 1884, when —25° occurred.

Before 1916 the lowest official temperature recorded for the state was —33° (Lafayette) and the next lowest —28° (Jeffersonville and Paoli), near the southern margin of the state. Fort Wayne had, in twenty years, not experienced a temperature lower than —19° F.

Indiana has rather hot summers. The average July temperature is 75° F. The extreme southwestern part, the hottest area, averages 79° F. The extreme northeast, the coolest part, averages about 73° F. The mean maximum daily temperatures for June, July and August are more than 80°. for all but a very few of the forty-four well distributed Weather Bureau Stations. Many stations have a temperature of almost 90° each average day in July.[1] Night-time temperatures are normally quite high also. The lowest temperature normally experienced at night (or early in the morning) in June is about 58° for the northern half of the state and about 62° for the southern half. For July the figures are about 62° (northern half) and 65° (southern half). In August the mean minimum temperature for the northern half is about 60° and for the southern half about 63°. Frequently the temperature remains high until late in the night. Indeed commonly the coolest temperatures occur at sunrise.

The high average summer temperature of Indiana is favorable for the growth of corn. The hot nights are especially valuable in this respect. Corn makes practically no growth when the temperature is below 49° F. For rapid growth, the soil temperature must be 75° or above for at least part of the day. The longer the time when the soil temperature is between 75° and 85°, the more rapid growth, when sufficient moisture is available.

Temperatures of nearly 100° F. are not rare in summer, and all of the state but the northeast has experienced a temperature above 105°, though few points have had a temperature of 108° and none above 111°. The northeastern corner has never had an official temperature above 103° F. The state's extreme range (the difference between the lowest temperatures of winter and the highest of summer), hence is almost 130° F. The mean seasonal range (the difference between the normal January temperature of about 28° F. and the normal July temperature, of about 76° F.), is about 44°. A great seasonal range, such as this, is characteristic of continental climates in mid-latitudes. Indiana's normal seasonal range is however somewhat less than in states to the west or north.

The normal daily range, or difference between the coolest temperature at night and the warmest temperature by day is usually less than 25° F., and often much less than 15° F. In winter, the normal range from the highest by day to the lowest at night averages almost 15° F., and for the summer about 20° F. The daily range of temperature in Indiana is normally

[1] The climate statistics given in this section are the official figures as published in different reports of the United States Weather Bureau, especially the summary of the climatological data of the United States by sections, Sec. 67 (Southern Indiana), Sec. 68 (Northern Indiana), 1916.

rather small, in response (1) to the abundant moisture in the air, (2) the low altitude of most of Indiana, and (3) the latitude. Most of the state is between seven hundred and eight hundred feet above sea level. The more moisture there is in the air, the less effective is the day-time warming by the sun. Clouds reduce day-time temperature conspicuously but the haze which is common in moist climates even when clouds are not present has a similar effect. Water vapor in the air lessens the escape of heat at night. Clouds are particularly effective in this regard. The altitude is important in affecting the range (1) because atmospheric moisture decreases with altitude and (2) because air itself retards the passage of heat, though far less than water vapor does. The latitude is important because the length of day and night in summer and winter is related to the latitude. In Indiana's latitude the summer days are notably longer than the summer nights. There is not so much time for cooling during the summer nights as there is farther south where the summer days are shorter and the summer nights are longer.

The aspect of the range of temperature which is commonly considered of most importance is that involving freezing temperatures. The importance of the fluctuations of temperature whenever the lowest temperatures are below the freezing point is very great. During the cooler half year in Indiana freezing temperatures at night and thawing temperatures at the warmest time of the day are characteristic. Such temperature conditions are partly favorable and partly unfavorable for winter grains, clover, etc. The freezing of the surface layers of the soil causes them to expand and often warp upward and heave away from the unfrozen soil beneath. This heaving breaks many roots. A few inches of snow protects crops against such alternate freezing and thawing, and, therefore, is welcomed by farmers having fields of such crops. Moderately cool temperatures are favorable for the stooling of grains. Stooling is the process of multiplication of stalks. Wherever it is warm, wheat yields poorly because only a few, or possibly only one stalk developes from each seed. Where heavy stooling has occurred, a dozen stalks may grow from a single seed. The moderately cool temperature of the winter and early spring in Indiana is normally much more favorable for stooling than are the higher temperatures which are characteristic farther south. Not far north of Indiana winter temperatures are too cold for winter wheat, or else the snow lies too thick upon the ground. A thick layer of snow smothers wheat if it stays long, and hence is quite undesirable for winter wheat.

There is a second respect in which the long period during which night-time temperatures are about freezing and daytime temperatures about 50° F. is important. Ellsworth Huntington's studies of human efficiency, given in his "Civilization and Climate", indicate clearly that cool temperatures (30° to 50° outdoors and 50° to 60° indoors) are most favorable for intellectual work. If Huntington's conclusions are correct, and the evidence is quite convincing, the temperature conditions which normally prevail in Indiana during the cooler half year are one of Indiana's greatest assets.

The fact as to the temperature which is often considered most important by farmers is the length of the period in which killing frosts are absent. (See Fig. 3.) In Indiana the average last killing frost in spring is April 15 in the extreme southeast; April 20 in the southern; April 25 in the central, and May 1 in the northeastern part of the state. Killing frosts sometimes occur as late as May 10 in the southern half and as late as June 1 in the northern half. The average first killing frost in the fall is on October 5 for the northern half of the state and October 15 for the southern half, but October 20 for the southwestern part. Rarely, killing frosts occur as early as September 20 in all the state but the extreme southern part, where the earliest has been September 25.

The growing season averages about one hundred seventy days in length, averaging about one hundred sixty days for the northern half, and one hundred eighty days for the southern half. (See Fig. 3.) Some stations in the north have as short a growing season as one hundred fifty days, and the extreme southwest has one of nearly two hundred days. Latitude is the chief factor affecting the length of the growing season; on the average it decreases in length towards the north. Three other factors contribute, however—

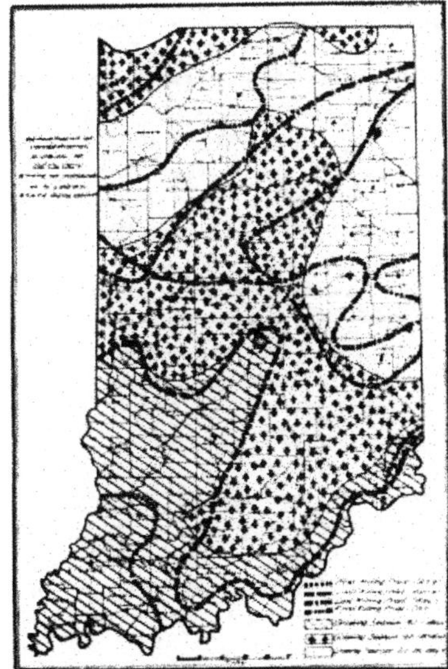

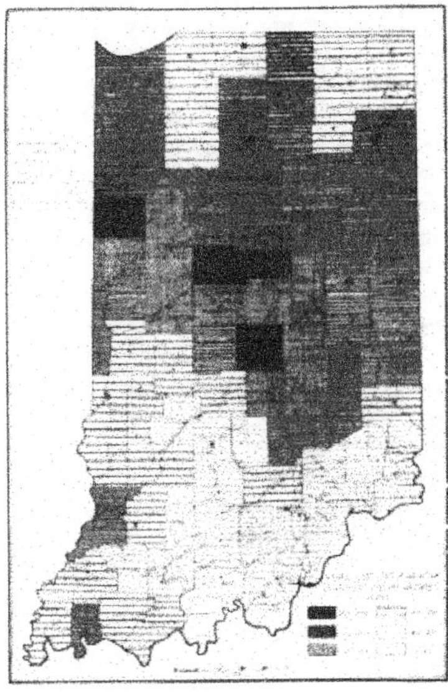

Fig. 3. Average length of growing season, and dates of last killing frost in spring and first in fall. Frost lines: April 21, May 1, October 11 and October 21. Growing season lengths: white, 150-160 days; checked, 160-180 days; lined, over 180 days.

Fig. 4. Value of farm land per acre, according to the 1920 Census, by counties. White=$17-25, black=$200-226.

altitude, storminess and nearness to Lake Michigan. The longest growing season at the south is along the lower Ohio and Wabash valleys, little more than three hundred feet above sea level. The parts of the state having an elevation of more than one thousand feet have a shorter growing season than do lower altitudes in the same latitude. The northern part of the state is affected more strongly by cyclonic storms than the southern part. Unseasonable frosts practically always occur only when an intense Low is closely followed by a well developed High. As the centers of more Lows and Highs pass northern Indiana than pass southern Indiana, this is another reason for the shorter growing season there. Lake Michigan affects the length of the growing season for a very small part of Indiana. The lake is relatively warm in the fall, and frosts do not occur so early near the lake as some distance south of the lake. For example, the earliest killing frost at Michigan City, and elsewhere near the lake, is a week or more later than in the latitude of Logansport. In the spring the lake is comparatively cold, and although frosts are not notably delayed, fruit near the lake is less likely to be injured by a late frost than farther south, as the cool, raw, cloudy springs induced by the cold lake, commonly delay the flowering of fruit trees and vines until the danger of injury from late frosts is relatively small.

Precipitation. The average annual precipitation received in Indiana is, for most of the state, about forty inches. In parts of the north, slightly less than thirty-five inches is the average, while in the south more than forty inches is normal, and at the extreme central south one small area usually receives as much as fifty inches, or about forty per cent. more than the driest part of the state. All but a small part of the state receives within three inches of thirty-eight inches rainfall. In this respect, Indiana is much more uniform than most of the states. Of the forty-eight states, more than forty normally receive in their rainiest part more than fifty per cent. more precipitation than their driest part receives, and in many states, the wettest part receives two or more times as much as the driest. The uniformity in Indiana is related to (a) its levelness, (b) its small extent, (c) its location in the great cyclonic storm belt.

The average decrease in precipitation northward in Indiana is due chiefly to increase in distance from the Gulf of Mexico, the great source of rainfall in eastern United States. The eastern margin of the state receives on the average more rainfall than the western margin, in keeping with the general increase in precipitation from the Rocky Mountains to the Atlantic Coast, because the cyclonic "Lows", the great agency which draws the moisture northward from the Gulf of Mexico, move eastward hour by hour. They thus draw more rain into the eastern than into the western parts of the state.

The precipitation is fairly evenly distributed throughout the year. Most months receive about three inches. However, a little more than half of Indiana's precipitation falls in the warmer six months, April to September inclusive, but less than sixty per cent. The southern margin receives exactly half of its total precipitation in the warmer half year, while the northwestern corner receives sixty per cent. then.

The great reasons why droughts are more severe in the southern half of the state than in the northern, in spite of the southern half's greater annual rainfall are (1) a larger share of the rainfall received at the south comes in winter when it does the crops little good, (2) the higher average temperature in the southern half than in the northern half results in greater evaporation there than in the north. The water requirement of plants depends largely upon the amount of evaporation. (3) A third reason why summer droughts are more frequent in the southern part of the state than farther north is that thunderstorms decrease in frequency and intensity towards the north. Thunderstorm rainfall comes quickly and is more likely to run off without benefiting the crops than is the case where the rain falls more quietly and has time to soak in.

The forty to fifty per cent. of the precipitation which falls in the cooler half year (October to March inclusive) is far less valuable per unit than the summer rainfall. Indeed Indiana would be better off perhaps, if she received less precipitation in autumn, winter and early spring. Soil erosion would certainly be less rapid.

Of Indiana's approximately thirty-nine inches of precipitation, about one-tenth is snow at the north and about one-twentieth is snow at the south. On the average almost thirty inches of snow falls at the north and twenty inches at the south, except in the extreme southwest, where an average of only about fifteen inches is received. Little snow remains long on the ground in the southern half of the state. The lesser snowfall, longer growing season, the higher temperature, and certain other climate peculiarities of the extreme southwestern part of the state is chiefly due to its low elevations along the Wabash and Ohio valleys. Watermelons are raised extensively in this lowland section. There is a slight increase in snow-fall near Lake Michigan due to the influence of the large body of cool water in early spring.

There is a very close relationship between the amount and distribution of rainfall and the yield of crops. Corn yields illustrated this clearly. Corn should receive an average of at least 2.5 inches of rainfall per month during the three months of its growth. When more rain falls at the right time, the yield is much increased. For example, on a plot of land at the University of Illinois, in eight years when a total of less than seven inches fell in the three months, June, July and August, the average yield was 25.3 bushels per acre. In nine years when between seven and ten inches occurred in these months, the average yield was 32.4 bushels per acre. In the eleven years when more than ten inches fell, the average yield was 40 bushels per acre. A hundred bushels per acre yield requires about sixteen inches of rain. In general, rainy weather in the last half of July and the first half of August results in a good corn yield, while a drought in this period is very costly.

Other Climatic Elements.—On the average about fifty inches of water evaporates from an exposed water surface in a year at the southern part of the state, and about thirty-five inches at the north. Thus a little more water evaporates from a lake than falls upon it. Lakes are maintained by streams. More water often falls on the land than can promptly evaporate. The excess forms streams. During dry spells, much more water could evaporate from the soil than is available for evaporation.

The air of Indiana normally is rather moist. For most of the state the relative humidity averages about seventy per cent. during the day. That is, the air possesses nearly three-fourths as much moisture as it could hold. At night, when dew is forming, for example, the lower air is supersaturated and is forced to give up part of its moisture. The relative humidity is important because, first, the rate of evaporation depends largely on it, and second, the sensible temperature is powerfully influenced by the humidity. The sensible temperature is the temperature as it feels to us. Moist air feels warmer than dry air when the temperature is high. Moist air feels colder than dry air at low temperatures. That is to say, a raw or chilly cold is more uncomfortable than a dry cold. Zero weather in Indiana is often as trying, while it lasts, as twenty degrees below zero in Dakota. Also a sultry heat of 95° F. is worse than a dry heat of 110°.

Indiana is comparatively sunny. There is least sunshine at the northeast, about 2,500 hours per year. There is most at the southwest, almost 2,850 hours per year. The average for the state is, however, nearly 2,800 hours. Most of Indiana receives almost fifty-five per cent. of the amount of sunshine which would be received if no cloudiness occurred. In winter, the sky is clear about forty per cent. of the day time. In summer, it is clear about seventy per cent. of the day time, on the average. In December the average day has only three and one-half hours of sunshine, while in July the average day receives sunshine slightly more than ten hours. On the average for the year about one-third of all the days are classed as "clear", almost one-third are "cloudy", and about a third are "partly cloudy".

Indiana is within the belt of frequent cyclonic storms (not tornadoes). Northern Indiana is crossed by the centers of about thirty "Lows" a year, while the southern part is crossed by about twenty. Changes of weather hence are somewhat more frequent and intense in the northern part than in the southern. In so far as such changes of weather are stimulating, the northern part of the state has a more invigorating climate than the southern part.

Tornadoes occur occasionally in Indiana, but they are notably less frequent than in Illinois and Missouri. Tornadoes damage so narrow a strip of country and so few occur, that there is slight likelihood that any given spot will be affected. Indeed, a United States Weather Bureau expert has calculated that even in Missouri, where they are most common, the chance that any given half-mile square of land (160 acres) will be crossed by a tornado in any century, is one in sixteen hundred. As most tornadoes demolish a strip only a few hundred feet wide, a farm might be crossed several times without the buildings being demolished. Many buildings are injured without the inhabitants being killed.

Indiana is not a windy state. The average wind velocity is about eight miles per hour. Gales (winds of forty miles per hour), are rare. In winter the direction of the wind is from the northwest usually. Southwest winds prevail on the average in summer.

CHAPTER V.

AGRICULTURE.

The natural advantages of climate, soil, topography ,and location make Indiana one of the leading agricultural states, in proportion to area, as we have seen in an earlier paragraph dealing with size. In 1919 the value of all crops in Indiana was nearly one-half billion dollars, or more than $150 for each man, woman, and child in the state.

A large share of the people of the state are engaged in agriculture, and farm products are the raw materials of several important manufacturing industries. The rural population, as that term is defined by the census, made up 49.4% of the population in 1920. However, the census bureau classes cities of less than 2,500 as rural, and hence many people are called rural who are not farmers. If only those who do not live in incorporated cities be classed as rural, 39% of the state's population was rural in 1920.

Indiana has 23 million acres of land, of which 91% is in farms, and most of the remainder is in roads, railroads, cities, villages, or lakes. Seventy-nine per cent. is classed by the census as improved land, and seventy-one per cent. as land in crops. Unimproved land includes wild pasturage, brush land, and any other raw land.

Value of Farm Land. The average value of all farm property for the state in 1920 was $144.44 per acre. The farm land alone was worth on the average $104.00 per acre. There is considerable variation among the counties and among the sections of the state as is shown by Fig. 4 and by table 1.

TABLE 1.

Physiographic Regions and Land Values.

Average value of land alone per acre, by counties; values from 1920 Census for counties solely in physiographic provinces as shown in Plate II of Dr. Malott's Chapter.

(1) Northern Moraine and Lake Region	
(Average of 18 counties)	$107.00 per acre
(a) Kankakee Lacustrian Section (4 counties)	102.00 per acre
(b) Steuben Morainal and Lake Section (8 counties)	96.00 per acre
(2) Tipton Till Plain Region (Average of 24 counties)	159.00 per acre
(3) Southern Hills and Valleys Region	
(Average of 30 counties)	47.00 per acre
(a) Wabash Lowland (10 counties)	77.00 per acre
(b) Crawford Upland (5 counties)	26.00 per acre
(c) Unglaciated Mitchell Plain and Norman Upland (5 counties)	34.00 per acre
(d) Muscatatuck Regional Slope (3 counties)	26.00 per acre
(e) Dearborn Upland (6 counties)	74.00 per acre

Marion County led with an average value of $226 per acre of land alone. The presence of Indianapolis increases the value of the nearby farm land. Benton County had an average value of $221.34; Tipton, $211.25; Clinton, $202.66, and Howard, $197.79 per acre. See Fig 1 for value of the land in all counties.

The most valuable farm land is in the north-central portion of the state on what is known as the Indiana prairie or Tipton Till Plain. The land is nearly level and is well adapted to the use of farm machinery, has good transportation facilities, excellent soil, and and has produced banner crops for more than a half century.

Brown County has the lowest average valuation of farm land, $17.24 per acre. In 1920, Crawford County on the Ohio River had an average valuation of $19.06, Perry $20.46, and Monroe $34.00 per acre. These counties contain much land which is not well suited to farming nor to most other industries. They are hilly, have barren, rocky soil and lack adequate transportation facilities. Their resources are harder to develop than those of the level prairies; and their people commonly lack the training and capital to make the most of the resources of this rougher part of the state. Dairying, sheep raising, and tree crop culture, all of which are promising here, will help very materially in developing this rougher region.

Number and Size of Farms. There were slightly over 200,000 farms at the 1920 census. The majority of the farms contain about 100 acres, of which about 80 acres are improved. Twenty-eight per cent. of the farms contain from 100 to 175 acres, and only 1,035 farms, or 0.5% of the total, have more than 500 acres.

Large farms have the largest percentage of unimproved land while those under twenty acres have the least. It is easier to improve a small farm than a large one, but it likewise takes more intensive farming to obtain a living on a small piece than on a large one. Almost one-half of the Indiana farms are worked by their owners, while nearly one-third are leased. The majority of the tenants are share-tenants, only about one-sixth being cash renters. Randolph County had the most tenant farmers, 1,333. Benton had the largest per cent., 59%, and Dubois had the smallest, 12%. The level northern counties are better adapted to the tenant system than the hilly portion of the south. One owner can easily oversee several farms, and the soil and topography make it easier to secure profitable tenants. Furthermore, rented farms are less likely to be permanently injured by mismanagement in the level portion of the state than in the rougher portion.

Leading Crops. Grain-farming was the first agricultural activity to be highly developed in Indiana and still remains the chief form of agriculture.

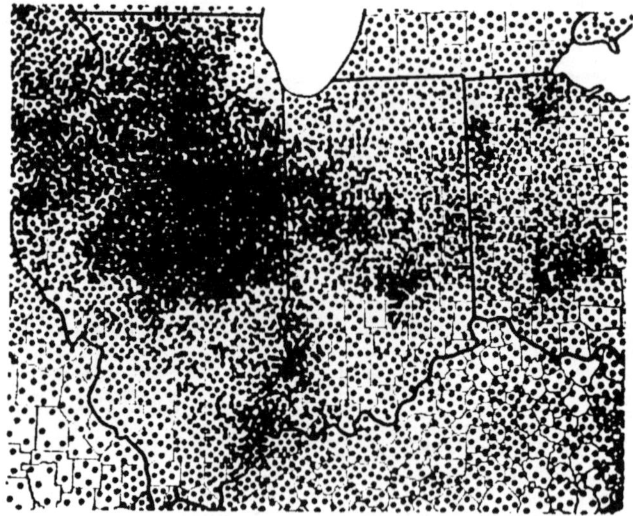

Fig. 5. The distribution of Corn. Each dot represents 5,000 acres. (From O. E. Baker, in Finch & Baker's Geography of World Agriculture, 1917.)

Corn is the most important crop. It is grown on practically every farm. In 1919 the total production was 158⅓ million bushels, coming from 4½ million acres. Benton County led in 1919 with over 4,200,000 bushels, followed by Tippecanoe, Rush, Allen and Randolph Counties. Brown County, one of the poorest counties, produces 366,000 bushels. See Fig. 5 for the average production of all parts of the state.

The average farm has about twenty-five acres in corn which normally yields about forty bushels per acre. With the exception of the state of Ohio, which has the same average yield, in no other corn states does corn normally yield as much per acre as in Indiana. At the last three censuses, 1900-10-20 there were about 4½ million acres, or 27% of the farm land, in corn each year. According to the Indiana Year Book, 80% of the corn produced in 1919 was used on the farm. The amount used on the farm varies among the counties from 95% in Brown County to 35% in Benton. About 15% of the corn is shipped out of the state, mostly to Chicago.

Corn is so important in Indiana agriculture because the climate, soil and topographic conditions are all favorable for its production. The climatic aspect has been discussed, it will be recalled, in the section on climate. In addition, corn responds to intensive cultivation, has many uses, and requires work at times of the year when the farmer is not occupied with the other chief crops grown.

Wheat is second in value among the cereals and also among all crops. In 1920 there were slightly more than 2½ million acres in wheat, producing over 45 million bushels, which gives an average of about seventeen bushels per acre.

The yield of wheat in Indiana is much affected by weather conditions, however, and since 1908 it has fluctuated from a little over 10 million bushels

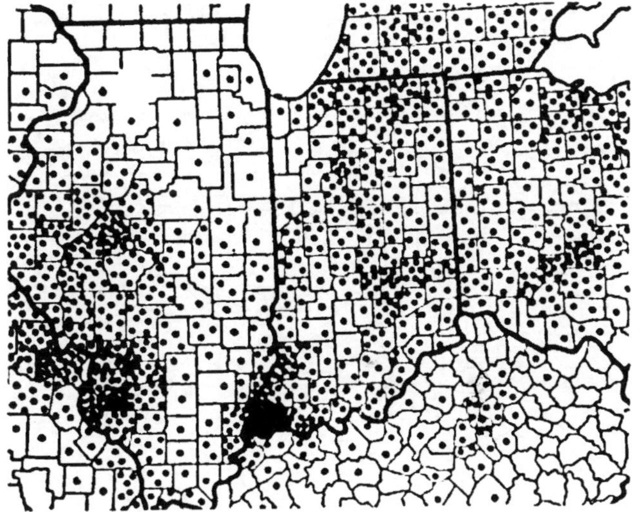

Fig. 6. The production of Wheat. Each dot represents 100,000 bushels. (From Baker.)

in 1912 to over 45 million bushels in several years and to 49 million bushels in 1918. This fluctuation is in sharp contrast to the conditions in respect to corn, Indiana's climate being much more favorable for corn than for wheat. In the period 1909-1920 instead of more than a four-fold fluctuation in the yield, as in wheat, the yield of corn varied only from 163 million to 199 million bushels, or less than 20%, according to estimates of the Department of Agriculture. This gives a normal fluctuation of less than 10% from the average yield. With wheat the normal fluctuation from the average yield is from 25 to 35%.

The most important wheat area is in the extreme southwest, in the Ohio and Wabash valley of Posey, Knox and Gibson Counties. (See Fig. 6). A second belt of considerable importance in wheat production extends northward from Jackson county. In 1909 twenty-five counties in this belt produced two-fifths of the state's wheat, nearly twice as much per county as the average for the state. Practically all the wheat of the state is marketed. The wheat acreage has been approximately the same for the last forty years, about 2¼ million acres, or 15% of the farm land of the state. Practically all wheat grown in Indiana is winter wheat. The soil is usually too wet early in the spring for the sowing of spring wheat in time for it to stool heavily before hot weather sets in. Even corn is often planted late on low fields because the ground does not dry off sufficiently for cultivation until the middle of May.

The third crop in value is hay and forage. It is fairly well distributed over the state (Fig. 7). In 1919, 3,300,000 acres were devoted to the growth of hay, from which over five million tons were obtained. Timothy and clover

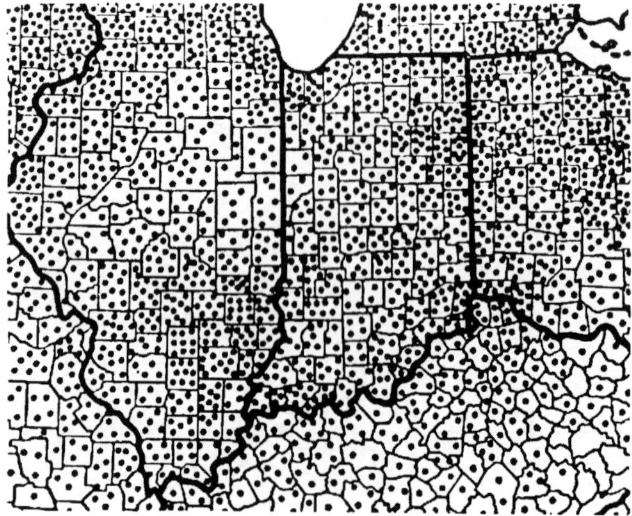

Fig. 7. The production of Hay and Forage. Each dot represents 5,000 acres. (From Baker.)

mixed, was the leader, while plain timothy was second. Seventy-six per cent of the hay produced, in 1919, was used on the farm. According to the Indiana Year Book (official) for 1920, the amount used on the farm varies among the counties from 91% in Pulaski and Union counties to 51% in Vermillion. For the last forty years there has been an approximate average of two and a half million acres, or 15% of the state's farmland, in hay and forage. The greater acreage, in 1919, was related to its high price in 1917-19. Allen County produced the most hay in 1919. This leadership is related to the large local market for hay and milk in the city of Ft. Wayne, and to the fact that it is in the northern part of the state, where more hay is required than in the southern part. The majority of the counties have between 20 and 40 acres per farm devoted to the growth of forage.

Oats rank fourth in value among the crops of the state. In 1920, there were nearly two million acres in oats, with a production of 52,539,000 bushels, an average yield of thirty bushels per acre. In 1919, Benton County led with nearly three million bushels. (See Fig. 8.) It will be recalled that Benton County is a leading corn producing county and one of the wheat counties. For the state as a whole, about seventy per cent of the oats produced, in 1919, was used on the farm. The percentage used on the farm varies among the counties from 96% in Orange County to 21% in Benton. The oats acreage increased from one million acres in 1879 to nearly two million in 1919.

Wheat and oats require work at somewhat different times of the year, which is advantageous for the farmer. The wheat is sown in the late summer, the oats in the spring before corn can be planted. The wheat ripens before the oats. The wheat also is a cash crop while oats are used mostly as horse feed.

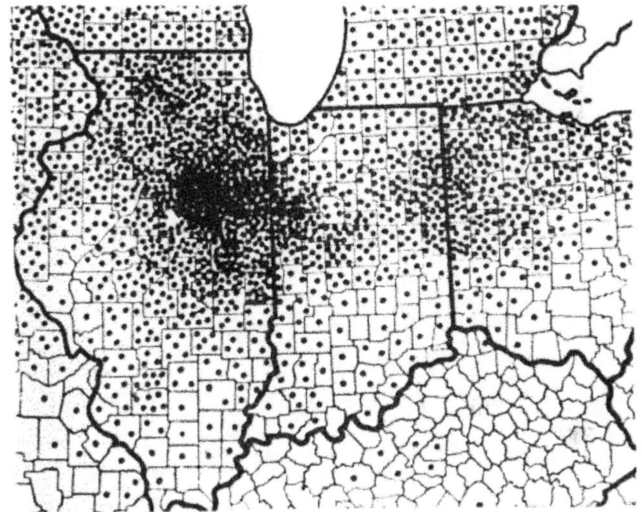

Fig. 8. The production of Oats. Each dot represents 5,000 acres. (From Baker.)

Barley and rye had, in 1919, an acreage of 74,000 and 350,000 acres respectively. Steuben County led in rye, followed by Lagrange, Porter, and Lake counties. Noble County led in barley, followed by Steuben, Allen and White. These counties are all in the northern part of the state, and have many areas of light sandy soil, which is well suited to the growth of rye and barley, while it is unfavorable for wheat or oats. The acreage of both rye and barley has increased greatly recently. The 1919 acreage of rye was nearly six times the average acreage for 1889-1909, and the barley acreage was nearly seven times as great.

Next to barley in acreage, in 1919, was potatoes, with 62,000 acres. As 1919 was a poor year for potatoes, beause of the drought at the critical time, the average yield was only forty bushels per acre, whereas the average in the United States is about 100 bushels. Indiana's average in 1920 was ninety bushels per acre, according to the 1920 Indiana Year Book (official). White, or Irish, potatoes are not a dependable crop in Indiana because of frequent dry spells in June when the potato needs rain. Such dry spells are more frequent in the southern part of the state than in the northern part. This is one reason why the extreme northern counties lead in potato raising. (See Fig. 9.) The sandy soil locally present in this region, also, is better adapted to their production than to the production of most crops.

Minor Crops. Both orchard and small fruits can be grown especially well in the southern, rougher part of the state, on soils that are not suited for grains. The limestone soils are suited for orchard fruit, and the knobstone soils produce fine berry crops. More young fruit trees are being set out in these rougher counties than elsewhere in the state.

The apple is by far the leading fruit grown in Indiana, as it also is in the United States as a whole. There were three and one-third million bearing apple trees in Indiana in 1919, yielding nearly one million bushels, valued at

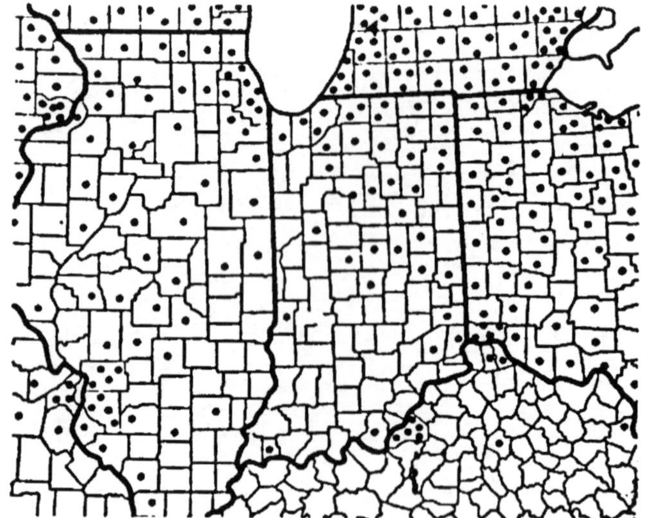

Fig. 9. The production of Potatoes. Each dot represents 2,000 acres, on farms reporting one acre or more at the Federal Census of 1910. (From Baker.)

two and a quarter million dollars. Clay County led with 39,000 bushels, while Vigo was a close second. Apples were grown, in 1919, on nearly two-thirds of all the farms. (See Fig. 10.)

Grapes are second to apples in value among the fruits in Indiana. However, to continue with the tree crops, the peach is next in order. There were 860,000 peach trees of bearing age yielding 82,000 bushels, valued at $255,000, in 1919. Twenty-one per cent of all farms reported peaches. Morgan, Harrison, Washington and Jackson counties have the most peach trees (31,000-45,000 each), though Brown County produced most peaches in 1919. Lawrence, Morgan and Brown counties have most peach trees not of bearing age (19,-000-30,000 each). In pears, Posey, Gibson and Marion counties lead with 11,000-14,000 bearing trees each. In younger trees, Pike County leads (7,000), with Laporte, Fayette and Lawrence next in order (1,600-2,000).

In 1919, 62,000 bushels of cherries, valued at $215,000 were harvested in Indiana. Forty per cent of all farms reported cherries in 1920. The extreme northern and the southern counties grow most cherries. Only one central county, Marion, had over 2,700 bushels in 1920. Elkhart County was the leader with 3,500 bushels, but Vigo, Clay and Franklin rank high.

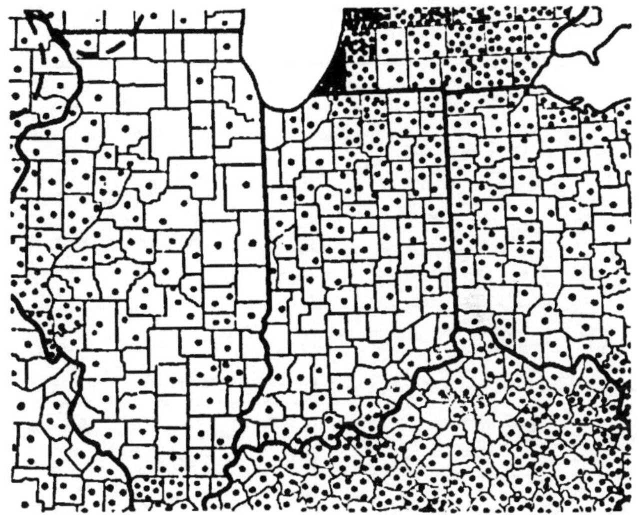

Fig. 10. The production of Apples. Each dot represents 20,000 bushels. (From Baker.)

TABLE 2.

Minor Crops.

	Year	Per cent of Farms	Acres	Quantity	Value
1. Cabbages	1919	1.1	1,953	....	$250,000
2. Muskmelons	1919	0.8	4,182	....	498,000
	1918	....	2,300	....	....
3. Sweet Corn	1919	1.5	10,101	....	488,000
4. Cucumbers	1919	1.1	7,878	....	218,033
5. Onions	1919	1.2	4,191	....	1,008,000
	1918	....	2,950		1,228,000
6. Tomatoes	1919	4.1	20,790	....	1,990,374
	1918	....	21,800	....	....
7. Watermelons	1919	1.0	4,850	....	446,902
	1918	....	2,990	....	....
8. Sorghum	1919	10.6	12,307	681,000 gal.	987,000
	1918	10.1	11,829	965,000 gal.	450,000
9. Sugar Beets	1919	0.3	4,119	44,450 gal.	444,550
10. Maple Syrup	1919	2.1	560,000 trees	170,000 gal.	443,000
	1909	2.5	740,000 trees	274,000 gal.	297,000
11. Strawberries	1919	6.4	3,400	4¼ mil. qts.	855,000
	1909	2.5	2,570	3¼ mil. qts.	310,000
12. Raspberries	1919	3.0	1,988	1¼ mil. qts.	287,883
13. Blackberries	1919	2.8	1,965	1,087,317 qts.	195,716
14. Currants	1919	.5	56	56,742 qts.	13,049
15. Apples	1919	61.7	....	925,624 bu.	2,221,498
16. Peaches	1919	21.2	....	82,266 bu.	255,027
17. Pears	1919	30.8	....	109,463 bu.	218,926
18. Plums	1919	19.7	....	35,536 bu.	88,877
19. Cherries	1919	40.3	....	63,365 bu.	215,161
20. Grapes	1919	34.5	....	6,612,804 bu.	462,899

The statistics for 1919 and 1909 are from the 1920 census. Those for 1918 are from the 1919 Indiana Year Book (official).

In 1919, 4.1% of all farms reported tomatoes, a total of 20,790 acres being reported. The crop was valued at nearly two million dollars. Seven counties, mostly in the central part of the state, had 900 acres or more in tomatoes in 1918. See Fig. 11 for distribution in 1909, as well as for the distribution of five other minor crops.

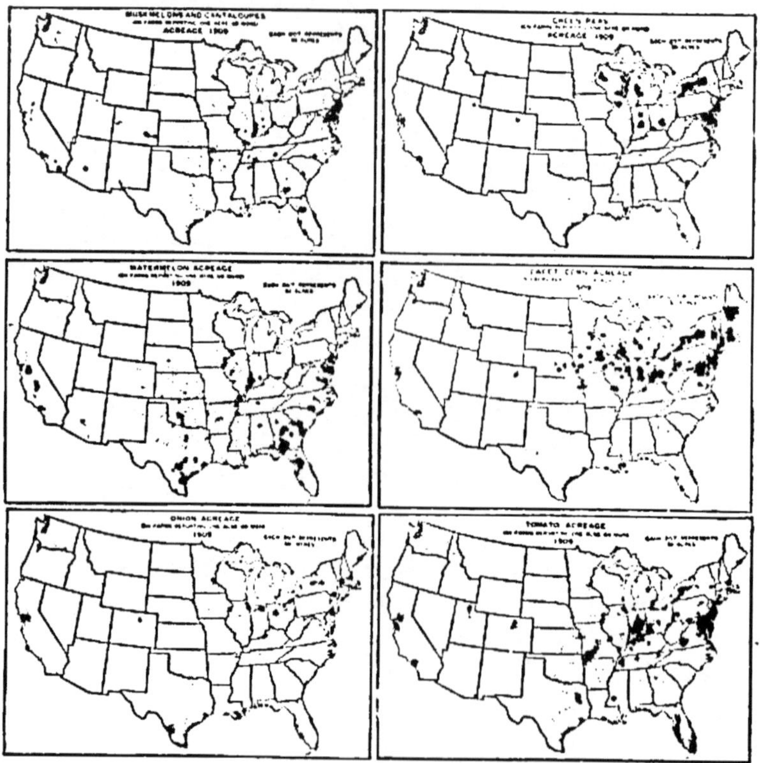

Fig. 11. Distribution of six minor crops in 1909. Each dot represents a total of 50 acres, on farms reporting an acre. Notice the concentration of production of melons in the southwestern part of Indiana, the warmest and sunniest part of the state; onions are grown chiefly in the coolest part of the state. Peas, sweet corn, and tomatoes are all chiefly grown in the central part of the state. (From O. E. Baker, in Finch & Baker's Geography of World Agriculture, U. S. Department of Agriculture, 1917.)

Domestic Animals. In 1920, 96% of all farms reported domestic animals, and their value was placed at about one-fourth billion dollars. Because of the extensive cereal and forage farming, horses are found in large numbers, and in 1920 were of greater value than any other live stock. There were 71,000 horses valued at sixty-seven million dollars in 1920. The distribution of horses is widespread, 91% of all farms reporting them. However, they are most numerous in the north-central farming belt (Fig. 12). The number is rapidly decreasing, having been reduced one-half between 1910 and 1920, according to a press-bulletin of the Census Bureau. This decrease is mostly due to the multiplication of autos, trucks and tractors, though partly due to the war demands for horses and subsequent slump, and also to the high cost of hay during 1916-20. Not only are horses and mules raised for local use, for tasks, like the working of wet ground, which the tractor cannot perform as well, but many are shipped to other regions, especially the south. This is partly a result of favorable climatic conditions for rearing colts and partly due to the relatively cheap fodder and forage available in Indiana. Mules are found chiefly in the southern portion of the state, Posey County, with 4,000, having the most. They are great pullers and sure of foot; there-

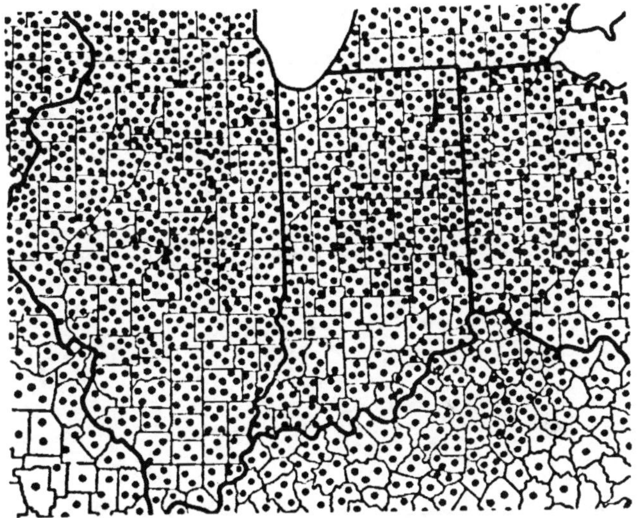

Fig. 12. The distribution of Horses in 1910. Each dot represents 2,000 horses. (From Baker.)

fore, they are better than horses in the hilly parts of the south; also, the chief market for mules is to the south of Indiana, in the cotton fields.

In Indiana, cattle rank next to horses in value. Indiana's small area prevents it from ranking high as a beef producing state. However, cattle were reported from 92% of all the farms, and had a total value of nearly 100 million dollars in 1920. (See Fig. 13.) About three-fifths of the cattle are dairy cows. Dairying is carried on very generally in the state and the production of milk reaches a large total, over 200 million dollars being reported for 1919 by the fourteenth census. Dairying is most extensively carried on in the northern counties, party because of the great industrial cities there or close at hand. Allen, Porter and Lake counties each produced between five and five and one-half million gallons per year, and Elkhart County produced four and one-third million gallons. The only other counties producing over three million gallons are Marion County, in which Indianapolis is located (four and two-thirds million gallons), and the adjacent Madison County (four and one-fifth million gallons).

The rich till plain of the central portion supports a large majority of the million and a half cattle. Forage and feed are especially abundant here and the market is close at hand in the nearby industrial cities.

Swine. In 1920, Indiana had nearly four million hogs valued at approximately sixty-three million dollars. When relative area is considered, Indiana is seen to be surpassed as a hog producing state only by Iowa. The importance of hogs here is related to the greatness of the corn crop, the large local or neighboring market, the accessibility of packing centers in Indianapolis, Chicago and Cincinnati, and to the exceptionally good railway facilities. The distribution of hogs over the state is shown by Fig. 14.

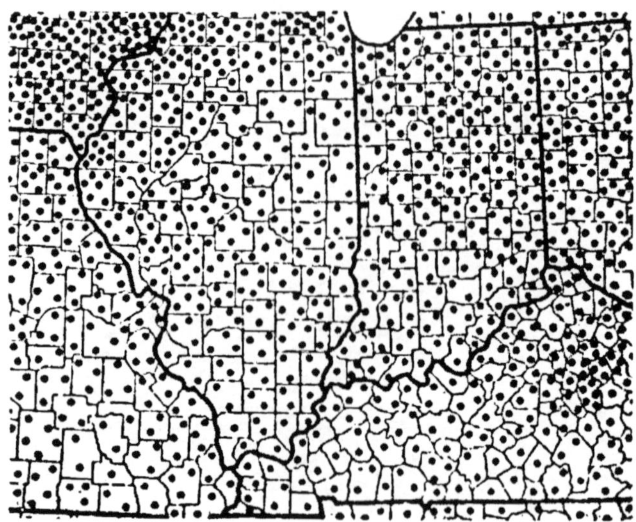

Fig. 13. The distribution of Cattle. Each dot represents 5,000 cattle. (From Baker.)

In 1920, hogs were reported from 84% of the farms. The leading counties were Rush, Boone and Henry, with from 86,000 to 102,000 hogs each. These counties are not far from Indianapolis, an important packing center. It will be recalled that Rush County was the third county in corn production in 1919. The two counties, Boone and Henry, which surpassed it in corn production are both nearer to Chicago than Rush. In the region near to Chicago, the market there for corn, for use in corn products and other human food, raises the price of corn sufficiently so that hog raising is less certainly profitable than in counties more remote from Chicago.

Poultry. The distribution of poultry is moderately dense throughout Indiana. They are reported from 95% of all farms. Poultry is of far greater value than the sheep of the state. Chickens number sixteen and one-half million, and all the poultry of the state is valued at $16,700,000. The northern counties led in 1920, especially Allen, the largest county in the state in area.

Sheep. Indiana has 643,000 sheep, valued at seven and one-half million dollars. The five counties in the northeast corner, Lagrange, Steuben, Noble, Dekalb and Allen, led in the number of sheep, having respectively from 34,000 to 16,000 each. These five counties produced nearly 700,000 pounds of wool. See Fig. 15 for the distribution of sheep in Indiana in 1909.

The hills of southern Indiana are only slightly less suitable to sheep raising than are the hills of northeastern Indiana. Many more sheep should be raised in the south than at present. The great drawback is the many sheep-killing dogs. Such dogs have been an important factor in causing the marked decrease, since 1870, in the number of sheep raised in Indiana. Indiana had only 40% as many sheep in 1920 as in 1900. Farmers on good land spend their time and energy raising crops and hogs, for which the market is certain,

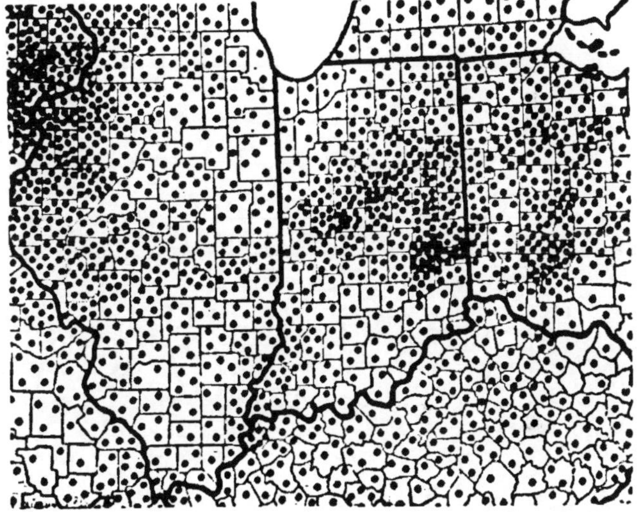

Fig. 14. The distribution of Hogs. Each dot represents 5,000 hogs. (From Baker.)

and the profit fairly sure, rather than bother with sheep, which may be killed by a dog any night. Since 1870, sheep have been raised in larger numbers in other regions, especially in the west and in newer countries, such as Australia, and Argentina, where, for one reason or another, they can be raised more cheaply than in Indiana.

Recent Changes. Although Indiana is one of the older states, and is undergoing less marked changes than many newer states, nevertheless notable changes have occurred in recent years. Some of the changes in respect to agriculture are the following: There has been a progressive decline in the percentage of the population living on the farms. In 1900, 65.7% of Indiana's people were "rural"; in 1910, 57.8%; and in 1920, 49.4%. It must not be thought, however, that agriculture has declined because a smaller fraction of the people are on the farm. With the help of labor-saving machinery, a farmer can work more land than formerly.

The increased importance of the farm is illustrated by the increased value of land—it doubled between 1910 and 1920, and has tripled since 1900—of farm buildings and equipment, and especially of crops. The crops of 1919 were worth two and a half times as much as the crops of 1909. The value of machinery increased fivefold between 1900 and 1920, and the value of buildings threefold. Many farms are well improved. For instance, 27,000 farms, or about one-eighth of the total number, have electric lights, and 37,000 farm houses, or nearly one-fifth of the total, have furnaces. The increased use of farm equipment is also illustrated by the fact that 2,700 farms have milking machines.*

The farm land itself is much improved. Millions of acres have been tile-drained since 1900, and hundreds of thousands of acres cleared of brush. There was a 4% increase in the area in crops between 1910 and 1920. Much

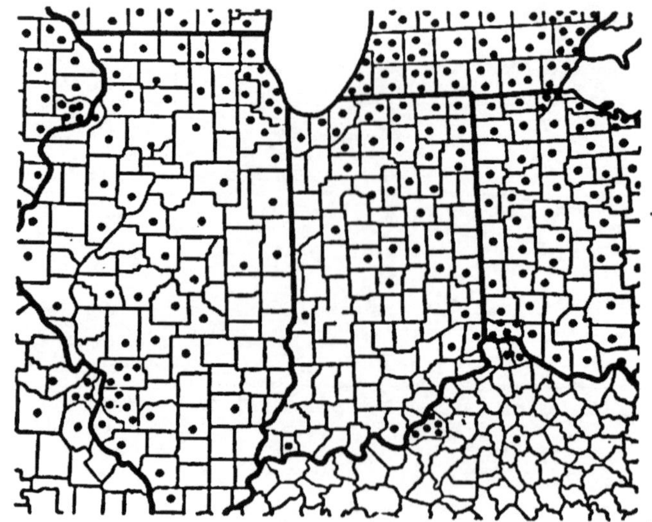

Fig. 15. The distribution of Sheep. Each dot represents 10,000 sheep. (From Baker.)

more fertilizer is also used, a total of $8,000,000 worth in 1919. The six-fold increase in rye acreage and the sevenfold increase in barley acreage has already been mentioned. On the other hand, there has been a decrease in the number of horses, sheep and potatoes. According to a Census Press Bulletin, there were only 49% as many horses in the state in 1920 as in 1910. Instead of horses there were about 15,000 tractors, 17,000 trucks, and 150,000 autos on the farms of Indiana in October, 1920, according to an official estimate (1920 Year Book). Ten years ago there were few of these substitutes for the horse and in 1900 practically none.

There has also been a slight decrease, 5%, in the number of farms, with a corresponding slight increase in the size of the farms.

The decrease in fruit trees is marked. Between 1910 and the 1920 Census the number of bearing trees fell off 40% for apples, 50% for peaches, 52%

* 1920 Year Book (official).

for pears, 62% for plums and 41% for cherries. Grapes also have decreased, though a lesser amount, 29%. This decline is chiefly due to the introduction of the San Jose scale, and other enemies of fruit trees. The increased competition of areas, in other states, which specialize in fruit growing is, however, another important factor.

The northern two-thirds of the state has continued to gain upon the southern one-third in importance. A century ago the northern two-thirds was altogether an unimportant agricultural region while the southern third, with its river highways and varied topography, was by far the most important part of the state. With the coming of numerous railroads between 1850 and 1870, the northern part of the state secured an outlet for its products which was far more satisfactory than the rivers of the south. With the draining

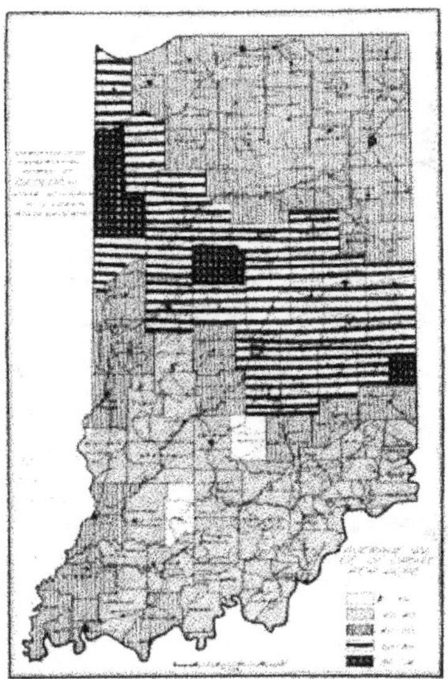

FIG. 16. Average value of crops per acre in 1919, according to the 1920 Census, by counties. White—less than $20, black—$80-114.

of much of the swamp land and with the growth of industrial cities the central and northern part has increased in importance until in 1919, the crops of the northern two-thirds were worth approximately five times as much as the crops of the southern one-third, or nearly twice as much in proportion to area. The average value per acre of the crops by counties, in 1919, is shown in Fig. 16.

CHAPTER VI.

TRANSPORTATION.

Transportation is highly important in Indiana as in almost all other well developed regions. Moreover, during Indiana's history there have been several important changes in chief modes of transportation. The earliest type was that of the Indian war parties and of French explorers, the use of the natural waterways. For example, many French explorers found the Maumee-Wabash route the most convenient way to go from the Great Lakes to the Ohio-Mississippi system. Others used the St. Joseph-Kankakee route. The earliest settlers also used the rivers, especially for flat-boats upon which much of their surplus produce was carried to New Orleans. Often when the crop was good and local demand was small, corn, for example, was worth more than three times as much at New Orleans as in Indiana. Hence, flat-boat transportation was encouraged.

The next stage in the use of the waterways was ushered in by the steamboat. The first steamboat on the Ohio appeared in 1811. The first one to ascend the river from New Orleans to Cincinnati did so in 1817. Indiana rivers, aside from the Ohio, were too shallow for extensive use of steamboats. However, after 1823, during the high water of spring, steamers ascended the Wabash to Lafayette and occasionally to Peru, and the White River to Spencer. The flat-boat era and the steamboat era overlapped. For example, in 1823, several steamers made regular trips on the Wabash while hundreds of flat-boats descended it. Often heavy loads of farm produce or lumber were taken down the rivers on flat-boats and lighter loads of manufactured products were brought up against the current on steamboats. In the first three decades of Indiana's settlement, from 1800-1830, the Ohio and Wabash rivers were immensely important. A large share of the incoming settlers came via the Ohio River. In 1830, five-sixths of the state's population lived in the southern counties (See Fig. 18). Most of the residents in southern counties still are descended from pioneers who came into the state from Pennsylvania and Virginia mostly by the Ohio.

The third great type of water transportation in Indiana was more pretentious, but less successful financially than the foregoing types. It seems hard to realize in this age of railroads and autos that Indiana ever was as much concerned with canals as was the case. However, from 1832 to 1849, and locally as late as 1853, many miles of canals were constructed in Indiana. About 453 miles of canals in Indiana were nearly completed. Such vast funds were appropriated by the young state in its ambitious program of canal construction—something like 1,200 miles of canals were planned—that Indiana was almost bankrupted by the expenditures. The two chief canals were the Whitewater and the Wabash-Maumee. The former was commenced in 1832 and the latter in 1834. Fig. 17 indicates the location of these canals. Neither was fully completed, and both were dismal failures financially. The state lost nearly three million dollars on the latter. The financial failure of the canals in Indiana was due to a combination of several factors. The chief of these was the competition of the railroad. Had the canals been completed, a few decades before the railroads came, they doubtless would have paid well, but the railroads appeared even before the canals were completed, and proved so superior to the canals that the latter promptly became almost useless. Reasons why the railroads are superior to canals include:

(1.) The railroads can be used the entire year, while in Indiana most sections of the canals were closed a considerable period in winter by ice and in summer by low water.

(2.) The railroads are located so as to bring their freight directly to its destination while the course of the canals is closely determined by physical features. It was impractical to construct canals where there was greatest

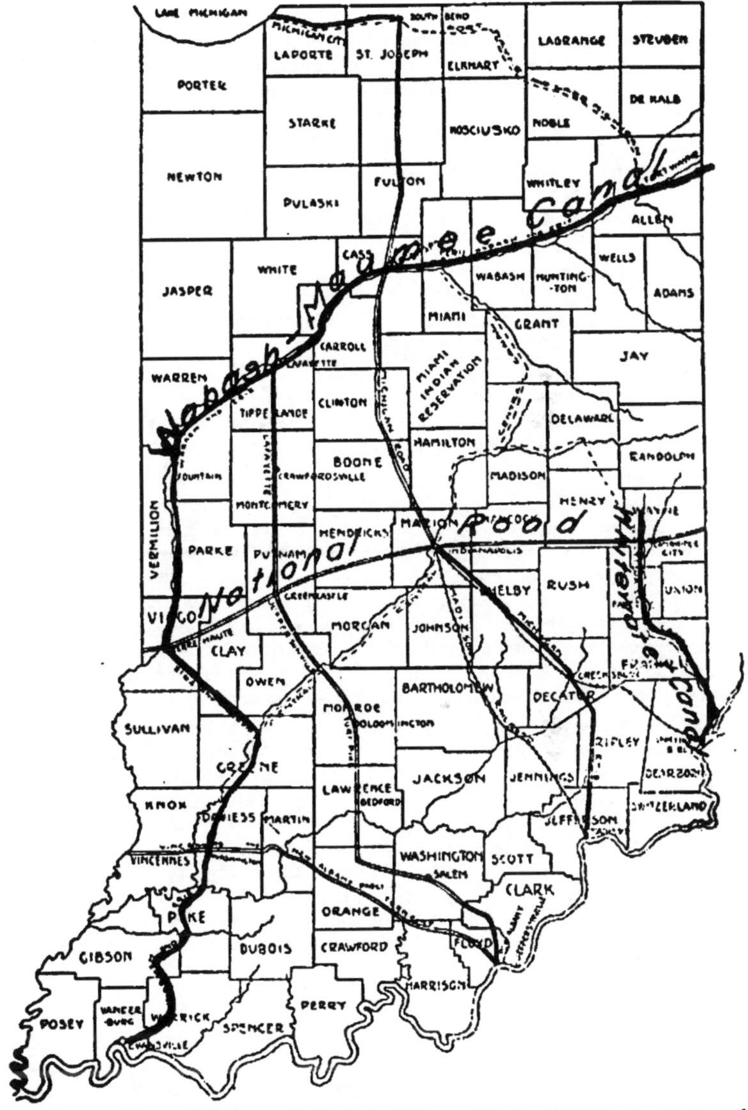

Fig. 17. Showing the location of the Wabash-Maumee Canal and Whitewater Canal, and of the National Road and other roads important in early years. The first railroad, that between Madison and Indianapolis, and the second, that between Lawrenceburg and Indianapolis, are also shown. The county boundaries are as of 1836. (Revised from Shockley's map in Esarey, History of Indiana, Vol. II, 1918.)

demand for transportation. For example, the chief cities of the state were not connected by canals.

(3.) While the railroads are almost straight and nearly always in usable condition, the canals were crooked, small, often inadequately supplied with water, and frequently clogged with vegetation, mud and gravel washed in by floods, etc.

(4.) The greater speed possible upon railroads than canals is one of their great advantages. In case boats were propelled or drawn through the canals at more than three or five miles per hour, the waves set up by the passing boat eroded the banks badly and caused landslides and other damaging consequences.

(5.) Still another advantage of railroads, is the ease with which they may be approached with a load and the load transferred to the railway cars. Many canals were in relatively inaccessible places at the base of steep slopes and it was difficult to transfer a load from wagon to barge.

(6.) Furthermore, once a consignment is loaded on a railway car it need not be unloaded until it reaches its destination or almost there, while very little freight shipped on the canals or rivers could be brought to its ultimate destination without repeated trans-shipment.

Railroads have similar advantages over river transportation, and as a result of the abandonment of canals in Indiana, closely coincided in point of time with a marked decline in the use of rivers, wherever railroads were available. The above mentioned respects in which transportation on canals is less satisfactory than railway transportation is supplemented by still other disadvantages of rivers, such as the great difficulty with the current, with floating logs, etc., and with muddy water causing boiler explosions. Indeed, the extensive use of Indiana's rivers for transportation is a thing of the past. No feasible amount of improvement of our rivers will lead to any permanent extensive renewal of their usefulness as highways.

Although the canal and river stage of transportation was thus logically brought to a close by the coming of the railroads, it must be recognized that in their day they were highly important. The Wabash-Maumee canal, opened from Toledo to Lafayette in 1843, and to Terre Haute in 1849, was especially important. The possibilities of this canal were early recognized. Indeed, during wet years the portage from the Wabash to the Maumee at Fort Wayne was very easy and short. From 1800 until 1832, when it was commenced, there was much discussion as to the desirability of the canal. It was pointed out that with the canal, there would be ready communication and cheap transportation between Indiana and the Great Lakes, and thus with the east, after the Erie Canal was opened in 1825. Furthermore, if the canal was constructed, Hoosiers would have a choice between the eastern markets and that at New Orleans. Several legislatures, after 1816, appealed to Congress for aid. A grant of $30,000 was made in 1824 for a survey and preliminary work. Digging was begun at Ft. Wayne in 1832, and slowly extended southwestward along the Wabash and eastward to Toledo. The canal was completed to Evansville in 1853. It had a depth of four feet between Toledo and Ft. Wayne, and a depth of three feet the rest of the way. It was 307 miles long. Locks admitted barges of about sixty tons burden.

Some of the effects of the canal were the following: *

I. It reversed most of the trade of Middle Indiana. It was able to do this because:

(a.) The faciltics for receiving and forwarding grain at Toledo were much better than at such Ohio River ports as Evansville.

(b.) The freight rates in the Great Lakes were lower than on the Ohio and Mississippi rivers.

(c.) Prices paid for farm products were higher at Toledo than at New Orleans while merchandise cost less.

(d.) There was less danger of grain being injured by the climate while en route on the northern route than on the southern. The warehouses at New Orleans had an especially bad reputation.

(e.) It took much longer to ship grain to market via the Mississippi to New Orleans than it did via the Great Lakes.

(f.) Finally, New York, the terminus of the northern route, was a better market than New Orleans, the terminus of the southern route. In brief, it freed much of Indiana from the economic domination of the south.

II. The second great effect of the canal was the fact that it raised the price of farm products. For example, at one point wheat went from forty cents a bushel, in 1840, to $1.00, in 1842, after the canal had been completed to that point. On the other hand, the price of salt, one of the heavy imports, was halved in the same interval at that place. As a consequence of higher prices for farm products, large farms tended to replace small clearings, and much wheat and corn were raised.

III. The canal promptly became the greatest highway for settlement in the northern part of the state. Many German and Irish immigrants came in by way of the canal. By 1840, the area tributary to the canal had a population of 270,000, while before the canal was built the region was almost unpopulated. See Fig. 18 for charts showing the spread of population, 1810-1880.

IV. The canal created many cities or gave them their first substantial growth. Huntington, Wabash, Peru, Logansport, Delphi, Lafayette, Covington, and Attica are examples. Older cities, such as Ft. Wayne, Terre Haute and Evansville, were much helped by the canal.

V. The surplus water power developed along the canal was used for manufacturing; and flour mills, saw mills, etc., sprang up.

VI. Partly because of the canal, the value of real estate and personal property increased rapidly, and hence taxation yielded larger funds for public enterprises. The assessed valuation in the state doubled between 1840 and 1850. During the next decade it increased two and one-half fold, but much of this last increase was due to the railroads.

The traffic on the canal increased until 1856 and then declined rather rapidly, as railroads became more plentiful. The canal was abandoned in 1874.

In respect to land transportation, Indiana has gone through several epochs also. The first was the use of trails made by wild animals and Indians. For a long time pack horses and saddle horses were far more

* Based on a lecture delivered July 21, 1911, by Harlan H. Barrows, Professor of Geography, University of Chicago.

common than wheeled vehicles. However, early in the nineteenth century, the federal government began to encourage road building. When Indiana was admitted as a state, in 1816, it was provided that five per cent of the receipts from the sale of public lands was to go into road building. The state expended most of this fund.. By 1820, twenty-six roads were projected to connect the older towns. Five radiated out from Indianapolis. By 1835 about two-thirds of the state was fairly well supplied with state-built highways. Upon several of the more important of these roads, stage lines were maintained until the coming of the railroad. For example, in 1820, a line between Louisville and St. Louis via Vincennes was put into operation. In 1826 a line between Madison, then an important river port, and Indianapolis was opened. The passenger fare was six and one-fourth cents per mile. In 1832, there was a stage line from Indianapolis to Dayton, Ohio. The location of several important roads is shown in Fig. 17.

Not only did the state build many highways, but the federal government built the famous National Road across Indiana. It had reached Wheeling, W. Va., in 1817; Richmond, on the eastern border of Indiana, in 1827; Indianapolis in 1834; Terre Haute in 1838-1843, and Vandalia, Ill., in 1852. This road was well built. The roadway was eighty feet wide; the road itself was 30-40 feet wide, and was macadamized with ten inches of crushed stone, the culverts and bridges were of masonry. The road was very important before the coming of the railroads. Although it has been graded to St. Louis, the National Road was not surfaced much beyond the Indiana boundary because railway extension overtook it when it had been completed to Vandalia. The federal government turned the road over to the several states crossed, in 1848. Indiana in turn gave the road to private companies which were to maintain it for the tolls received. However, most of these companies relinquished their claims by 1860.

Another important highway was the Michigan Road, begun in 1828 and extending from Madison to Indianapolis, Logansport, South Bend and thence to Michigan City. (Fig. 17.) It was paid for by the receipts from the sale of lands donated to the state for that purpose.

From 1845 until 1879, road building was largely a private or county proposition. Many turnpikes were built upon which a toll was charged. In 1879, however, the system of free roads, built or supervised by the county road commissions, was inaugurated. During recent years the state has taken increased interest in roads, paying for its share from the automobile license fees. Indiana now has about 73,000 miles of public road, of which 31,000 or 42.5% are surfaced. Only one state, Massachusetts, had as large a percentage of public roads surfaced, and Massachusetts being a much smaller state, had less than one-third as many miles as Indiana. The high rank of Indiana in surfaced roads is related in part to the numerous outcrops of limestone in Indiana, especially in the western and southern parts of the state. Glacial gravels, abundant in the valleys of rivers which flowed out from the melting ice, is also much used as a surfacing material in parts of the state.

The railroad era in Indiana was commenced in 1847, with the completion of the railway from Madison, on the Ohio River, to Indianapolis. By 1850, there were seven lines with a total of 228 miles of railroads connecting Indianapolis with Terre Haute and Lafayette as well as with Madison. The mileage at certain other dates is given in Table 3.

TABLE 3.

Railway mileage for Indiana in specified years, 1850 to 1918.

(From Statistical Abstract of U. S., 1920.)

Year	Mileage	Year	Mileage
1850	228 miles	1900	6,470.61 miles
1860	2,163 miles	1910	7,420.14 miles
1870	3,177 miles	1917	7,436.35 miles
1880	4,373 miles	1918	7,410.70 miles
1890	5,971.03 miles		

From 1850 to 1860, more than $34,000,000 were invested in Indiana railways, and a total of 2,163 miles was completed. Since that decade railroad building has been less rapid but has continued, as the foregoing table shows, until the present. In 1920, there were about thirty-four operating and eleven terminal companies having over 7,200 miles of main track and in addition enough secondary track and side track to bring the total to over 13,000 miles. The chief companies and their mileage of main track, according to the 1919 Official Yearbook, is given below.

Steam Railroad Mileage of Indiana for 1919.

Road	*Mileage of Main Track*
Chicago, Terre Haute & Southeastern	207.8
Central Ind. Ry. Co.	117.7
Cheapeake & Ohio of Ind.	227.6
Chicago & Erie R. R. Co	160.1
Chicago & Eastern Ill.	379.6
Chicago, Indianapolis & Louisville	608.1
Chicago, Terre Haute & Southeastern	207.8
Cincinnati, Indianapolis & Western	154.4
Evansville & Indianapolis	134.0
Illinois Central	175.5
New York Central	465.0
Big Four	803.8
Lake Erie & Western	445.0
New York, Chicago & St. Louis	204.1
Pittsburgh, Cincinnati, Chicago & St. Louis	1,477.8
Toledo, St. Louis & Western	171.1
Wabash, R. R. Co.	351.1
Lesser Lines	578.7
Total	7,195.7

Indiana is comparatively well supplied with railways. Only six states of the Union have more miles in proportion to area than has Indiana, and only two, New Jersey and Pennsylvania, have notably more.

Not only does Indiana stand high among the states in regard to steam railroads, but it is rich in regard to electric interurban lines. The first interurban in Indiana was that from Anderson to Alexandria, first operated on January 1, 1898. The first to reach Indianapolis did so in 1900. In the following decade many lines were built. By 1905 Indianapolis had nine interurban lines in operation and three more were under construction. Now she has thirteen lines. (See Fig. 19.)

Since 1911, the electric system has continued to grow, but at a much slower rate. In 1920, there were 2,420 miles of main track, ninety-one miles of second track and eighty-nine miles of side track. Upon these were operated 1,920 passenger cars, 500 freight and 440 other cars. The lines centering in Indianapolis are united into four companies, the Ft. Wayne and Northern Indiana Traction Company, the Interstate Public Service Co., the Terre Haute, Indianapolis & Eastern Traction Co., and the Union Traction Co.

Indiana is as well served by electric interurban railways as any other large state in the Union. Indeed, Indianapolis claims to be the world's greatest

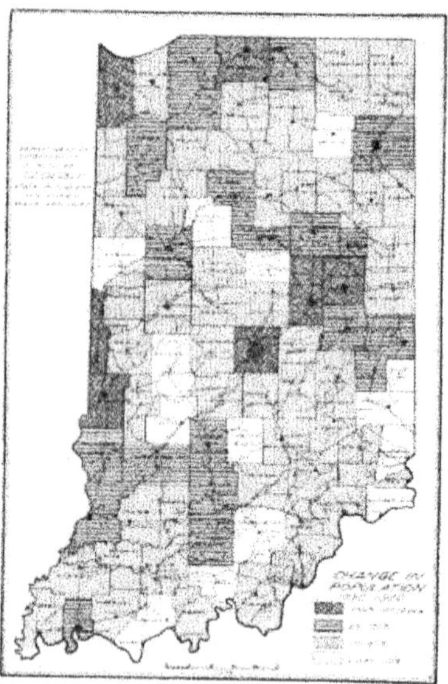

Fig. 19. Electric Interurbans of Indiana in 1920.

Fig. 20. Change in total population between 1890 and 1920, by counties. White counties lost population: darkest ones doubled.

interurban center. Although the interurbans are popular and important, they have not been financially very successful in recent years. They have earned fair returns on the actual investment, but the profits are hardly enough to attract capital into further expansion of the system.

The electric interurban has found that its business has been largely to carry passengers, especially those who desire to ride less than fifty miles; for beyond that limit the superior comfort and speed of the steam railway coaches cannot usually be overcome. It is generally conceded that the interurban has failed to solve the question of local transportation. Hauling to the road and delivering from the terminal station makes it necessary to handle freight at least four times. This gives a great advantage to the automobile.

Thus we are introduced to the last great stage of transportation in Indiana, the present type. The auto now supplements the railway, the interurban and the wagon. For local traffic it is the most important of all. There are now more than 400,000 autos in the state. Vast quantities of freight are carried in the common auto while an ordinary auto truck will haul three to five tons at a rate of fifteen to twenty miles per hour, picking up the freight in the field and delivering it to the final destination.

CHAPTER VII.

POPULATION.

The total population of Indiana was 2,930,390 in 1920. It has increased steadily, though at a declining rate since 1820. There were about twenty times as many people in Indiana in 1920 as in 1820, and about twice as many as at the close of the civil war. The increase of 1920 over 1910 was 8.5%. Table 4, below, gives the state's population by decades, according to the fourteenth census, together with the absolute and relative increases over the preceding decade.

TABLE 4.

Indiana's Population by Decades.

Total population and increase from 1800 to 1920.

Year	Population	Increase over Preceding Census		Per cent of Increase for U.S.
		Number	Per cent	
1920	2,930,390	229,514	8.5	14.9
1910	2,700,876	184,414	7.3	21.0
1900	2,516,462	324,058	14.8	20.7
1890	2,192,404	214,105	10.8	25.5
1880	1,978,301	297,664	17.7	30.1
1870	1,680,637	330,209	24.5	22.6
1860	1,350,428	362,012	36.6	35.6
1850	988,416	302,550	44.1	35.9
1840	685,866	342,835	99.9	32.7
1830	343,031	195,853	133.1	33.5
1820	147,178	122,658	500.2	33.1
1810	24,520	18,879	334.7	36.4
1800	5,641	....	....	....

It will be noted that, since 1870, Indiana's population increased more slowly than the average for the entire United States. In the decade, 1900-1910, it increased at only one-third the rate of the country as a whole. In the last decade, however, the increase was somewhat more rapid, being about 60% as rapid as that for the United States as a whole. The variations in the rate of increases are due to two sorts of conditions. In the first place, as any population increases in numbers its percentage increases tend to decline. For example, an increase of 122,000 between 1810 and 1820, when the state's population was small, meant a percentage increase of 500%, while an increase of more than twice as many people, now that the state's population is approximately three million, means a percentage increase of only about one-fiftieth as much, or 10%.

The second great reason why Indiana's increase has declined has been related to the greater development of other parts of the nation, newer states, or states which have developed more rapidly industrially than has Indiana. The greater increase during the last decade than during the preceding one, while the country as a whole increased far less rapidly, doubtless is related to the fact that Indiana's industrial development during the decade 1910-1920, was quite rapid. However, Indiana ranks lower among the states in population than since 1840, when it was tenth state. In 1850 it ranked seventh, from 1860 to 1880 it ranked sixth, in 1890 and 1900 it ranked eighth, and in 1910, ninth. Fig. 20 shows the counties which have gained or lost in population since 1890.

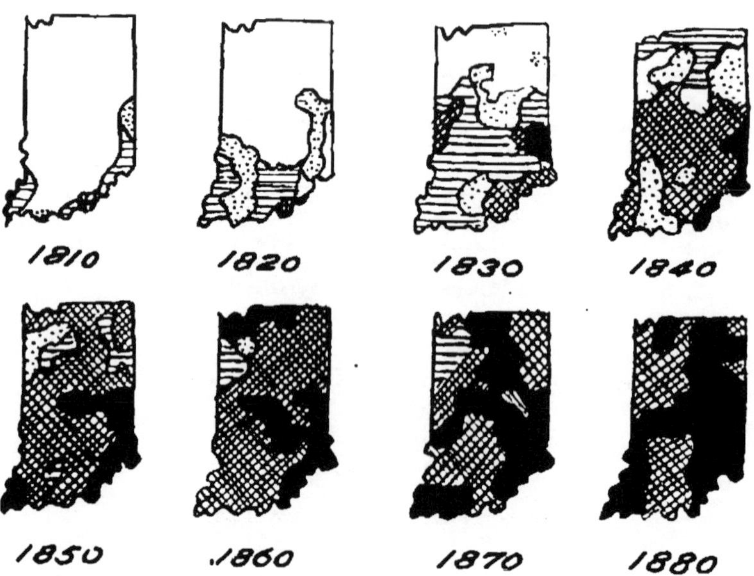

Fig. 18. Spread of population. Depth of shading indicates density of population.

Density of Population. In 1920, Indiana's population represented an average of 81.3 persons per square mile. In density Indiana ranks eleventh among the states. The average density here is slightly over twice the average for the United States as a whole, but only about one-eighth of that of Belgium and one-fifth of that of Britain. Table 5, below, shows the density by decades since 1880.

TABLE 5.

Density of Population by Decades Since 1880.

Year	Indiana	United States
1920	81.3	35.5
1910	74.9	30.9
1900	70.1	25.6
1890	61.1	21.2
1880	55.1	16.9

The density of population varies among the counties. Marion County, in which Indianapolis is located, had 876 persons per square mile in 1920, while Brown County, with no city and only a few miles of railroads, had only 21.7 persons per square mile.

Fig. 21 shows the average density in 1920 by counties. It will be noticed that the twenty-seven counties having fewer than forty-five persons per square mile occur chiefly in (a) the rougher southern half of the state; (b) the almost strictly rural flat land in the northwest quarter, where corn thrives, but where neither coal nor oil is present to encourage industrial development, and (c) the extreme northeast corner, in the morainal portion of the state, north of the main east-west railroads, and their great industrial cities.

On the other hand, sixteen counties have more than ninety persons per square mile. Ten of these are found on the borders of the state, partly near Chicago and partly in the coal region of the southwest. The other six counties are in the east-central portion, chiefly in or near the eastern oil and gas

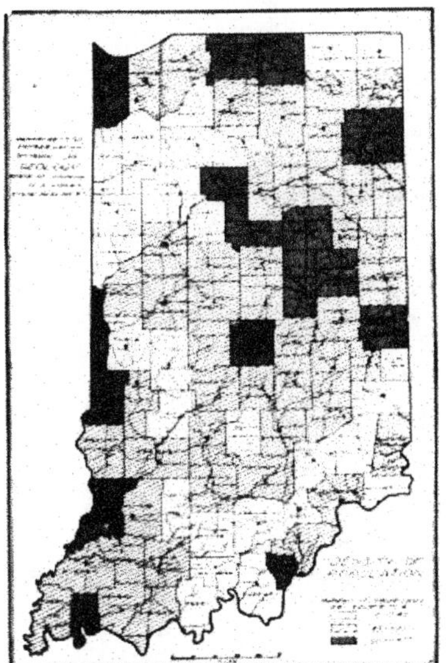

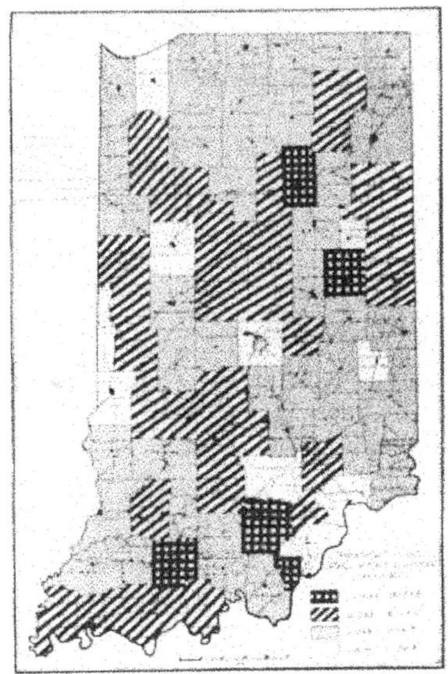

Fig. 21. Density of Population in 1920, by counties. White=21-45, darkest=90-877.

Fig. 22. Change in Rural Population, 1910-1920, by counties. All shaded counties lost, some as much as 29%.

region. They form a loop with Marion at one end and Cass at the other, with Howard, Grant, Delaware and Madison between.

In respect to density of population, many counties have lost since 1900. Of the ninety-two counties, sixty-four decreased between 1910 and 1920, while only twenty-eight increased. Between 1900 and 1910, fifty-six counties lost while thirty-six gained. Fig. 22 shows the gain and loss in rural population by counties, 1910-1920. Five counties have lost more than 19% each and three others more than 15%. On the other hand, only six counties have gained in rural population as much as 5%, and only two of the six have gained as much as 15% (Blackford and Vermillion).

Rural Population vs. Urban Population. More than 60% of Indiana's population now lives in cities, and towns; slightly over one-half living in cities of over 2,500, and nearly one-fourth living in cities of over 50,000. Table 6, below, shows the distribution of population in this respect for 1890-1920.

TABLE 6.

Urban and Rural Population.

Year	Total Pop.	Urban and Small Towns		Strictly Rural	
		Number	Per cent	Number	Per cent
1920	2,930,390	1,777,742	60.7	1,152,648	39.3
1910	2,700,876	1,444,323	53.5	1,256,553	46.5
1900	2,516,462	1,126,683	44.8	1,389,779	55.2
1890	2,192,404	815,362	37.2	1,377,042	62.8

This table reveals a rapid increase in urban population and a rapid decrease in rural. Indeed, the number of people on the farms has declined absolutely since 1890. This does not mean, of course, that the productivity of the farms has declined. Rather, it means that with the help of labor-saving machinery, a farmer can handle more land than formerly. While the rural population was 10% smaller in 1920 than in 1890, the urban population has nearly doubled.

Cities. In 1920, Indiana had ninety-eight cities of 2,500 or more. The thirty largest are given below, with the percentage of increase of 1920 over 1910.

TABLE 7.

The Larger Cities in 1920.

	Population	% Increase		Population	% Increase
1. Indianapolis	319,194	34.5	16. Lafayette	23,486	10.6
2. Fort Wayne	86,549	35.4	17. Logansport	21,626	11.9
3. Evansville	85,264	22.4	18. Michigan City	19,457	2.2
4. South Bend	70,983	32.2	19. Vincennes	17,210	13.4
5. Terre Haute	65,914	13.6	20. Mishawaka	15,195	21.1
6. Gary	55,344	229.6	21. Laporte	15,158	35.0
7. Muncie	36,524	52.2	22. Newcastle	14,458	34.5
8. Hammond	36,004	72.1	23. Huntington	14,000	36.3
9. East Chicago	35,967	88.3	24. Peru	12,561	15.1
10. Kokomo	30,067	76.8	25. Bloomington	11,595	23.7
11. Anderson	29,767	32.4	26. Frankfort	11,585	25.5
12. Richmond	26,728	19.9	27. Clinton	10,982	43.2
13. Elkhart	24,277	25.9	28. Whiting	10,140	35.0
14. Marion	23,747	18.8	29. Crawfordsville	10,139	7.5
15. New Albany	22,992	10.3	30. Jeffersonville	10,098	3.0

It will be noticed that the cities fall into five groups. (1) Indianapolis, the capital and chief city, nearly four times as large as the second city. (2) Cities in Greater Chicago (Gary, Hammond, East Chicago, Whiting). (3) Cities in the northeastern industrial area, on the railroads between Chicago and the east (Fort Wayne, South Bend, LaPorte, Michigan City). (4) Cities in the coal fields (Evansville, Terre Haute, Vincennes, Clinton). (5) Cities in the oil and gas fields (Muncie, Kokomo, Richmond).

In the decade 1910 to 1920 the cities of the northern half of the state grew more than those in the southern half. Ft. Wayne passed Evansville, South Bend passed Terre Haute, Gary grew notably and became the sixth

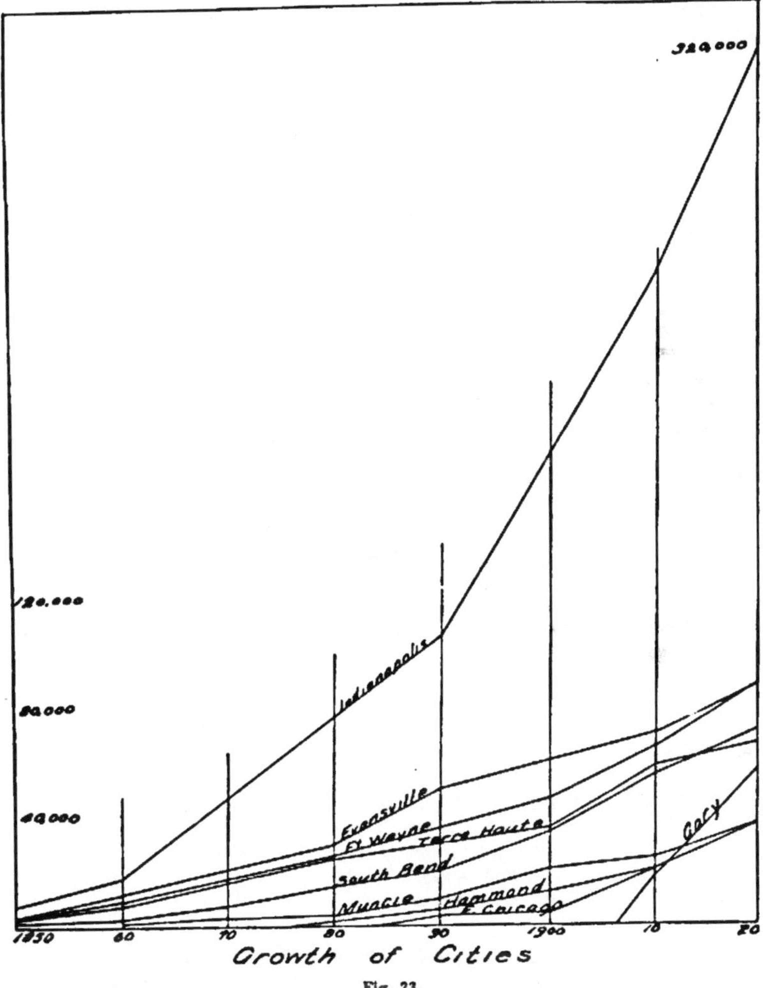

Fig. 23.

city, although it was laid out only in 1906. Fig. 23 is a graph showing by the steepness of the rise of the lines the rate of growth of several cities.

The development of Indiana's cities has been brought about by a number of interesting factors and so a brief sketch of each of the larger thirty cities, of 1920, will be given. The arrangement is alphabetical. All but eight are county seats, and four of these eight are parts of Greater Chicago (Gary, Hammond, Whiting and East Chicago). The other four exceptions are Elkhart, Michigan City and Mishawaka, on railroads not far east of Chicago, and finally, Clinton, a mining city in Vermillion County. The four cities near Chicago (Gary, Whiting, Hammond and East Chicago) are the only ones besides Laporte, which did not exist before the coming of railroads. Now, however, all are conspicuous as railroad centers. Several of the thirty owe their start or their early growth to the Wabash-Maumee Canal, for example, Ft. Wayne, Huntington, Peru, Logansport and Lafayette. Terre Haute and Evansville also were much affected by the canal. The Cumberland National Road was important in the early development of Richmond, Indianapolis and Terre Haute. The discovery of oil and gas, in 1886-87, in the eastern oil field, stimulated powerfully the growth of several cities, indeed converting some villages into important cities. Examples are Muncie, Kokomo, Richmond, Peru, Marion, Anderson and Newcastle.

ANDERSON. In 1836, the county seat of Madison County was moved from Pendleton to Anderson, the site of an old Indian village. It is situated in the old gas region (1886) to which it owes it early rapid growth. It manufactures glass, iron and steel, tin plate, paper, automobiles, tractors and wagon and carriage wheels.

BLOOMINGTON. In 1818, the first lots were sold for the site of Bloomington at the head of Clear Creek, in Monroe County. In 1824, the State University was opened with twenty students for the first session. Bloomington owes much of its growth to the University and to the stone quarries. However, in recent years it has become notable for its furniture factories. It is claimed that in 1920 Bloomington had one of the largest kitchen cabinet factories and furniture factories in the nation.

CLINTON. The town of Clinton was laid out in 1824, and named in honor of DeWitt Clinton, Governor of New York. Its first growth was very slow. Indeed many residents traded chiefly at Terre Haute, fifteen miles away. The early industries were pork packing, dealing in grain, coal and wood, and wagon-making. All of these closed down during the Civil War and never reopened. Present industrial establishments produce: Paving brick, overalls and shirts, canned vegetables, ice, hardwood, lumber, etc. Coal mining is the chief industry. Coal is found one mile from the city and has been worked successfully for many years.

CRAWFORDSVILLE. In 1823, Crawfordsville was located on the Wabash River in Montgomery County. It was an early railway and turnpike center. There are mineral springs of medicinal importance here. There is an abundance of limestone, sand and brown sandstone. It manufactures wire, nails, matches, pressed brick, boxes, sheet metal, castings, clay and tile products, and vehicles.

EAST CHICAGO. (Indiana Harbor), on the shore of Lake Michigan is near the center of the extensive industries covering the Calumet region of Indiana. It was founded in 1885. With the building of the Chicago and Calumet Terminal Railway, in 1888, manufacturers came here in search of cheap factory sites outside of the great city of Chicago, yet in ready communication with it. The harbor and canal constructed in 1901-03 have been important factors in the development of East Chicago. The city is noted for its iron and steel manufacturing.

ELKHART is beautifully located on the St. Joseph River in Elkhart County. Manufacturing is favored by the water power developed by both Elkhart and St. Joseph rivers. It manufactures wagons, musical instruments, starch and paper. It is a great commercial and railway center, being the terminus of three railway divisions. The machine shops of the Lake Shore and Michigan Southern Railroad are located here and give employment to about fifteen hundred railroad men.

EVANSVILLE was founded in 1812. The geographic conditions affecting the choice of this site are said to have been: (1) The recognition that it was favorably located to serve as a river port for Vincennes and other interior towns, as well as of a promising rural district near at hand. (2) Good mill sites were available on Pigeon Creek here, and at that time of costly transportation it was considered very desirable for a city to have water power enough to run a flour mill. The river commerce was considerable for several decades after the establishment of the city, but it grew slowly, and was not incorporated until 1847. However, with the arrival of the Wabash-Maumee Canal, in 1853, the city had a boom and increased three and one-half fold in population between 1850 and 1860. The growth of the city attracted railroads. Indeed, the first, the Crawfordsville and Evansville, arrived in 1850. Soon after, it had several lines and became the second city in the state in population. It held this rank until the 1920 census. It is the metropolis of the southern coal fields and also does considerable manufacturing of lumber, furniture, flour, tobacco and leather. It also has slaughtering and packing industries, foundries and machine shops.

FRANKFORT, county seat of Clinton County, was settled in 1830, and incorporated in 1846. It was named after the capital of Kentucky. The city grew slowly until 1870, when the first railroad entered. It is surrounded by rich agricultural country, and its manufacturing plants produced products valued, in 1919, at over six million dollars. The principal articles manufactured are kitchen cabinets, plumbing fixtures, enameled advertising signs, porcelain refrigerator and stove linings and table tops, ash and hickory handles and wire wheels for automobiles. The Clover Leaf Route has its principal shops here and three other railroads and two interurban lines serve the city.

FT. WAYNE. In 1794, the military post of Ft. Wayne was erected at the head of the Maumee River. In 1815, a fur trading station was established, and it was for many years one of the principal depots for the fur trade. Its early growth was due to the Wabash-Maumee Canal, which was commenced here in 1832. It was a commercial center for flat boats. With the coming of the railroads the "Summit City" became the great railroad center of northern Indiana. It became the state's third city in size in 1870, and the second in 1920. Ft. Wayne lies in a great agricultural region, is the center of the electric railways of northern Indiana, and manufactures car wheels, engines, bar iron, wagons, electric apparatus, hosiery, knit goods, cotton goods, pianos and Edison Mazda lamps.

GARY. The city of Gary at the southern end of Lake Michigan was made to order by the United States Steel Corporation. Until April, 1906, the region was a waste of sand dunes interspersed with marshes and ponds. Three years later there was a great steel plant, employing 10,000 men, a splendid artificial harbor and a modern city. By 1914, Gary's population was 40,000 and in 1920, 55,000. Gary has the five requirements for the establishment of an industry-capital, labor, raw material, transportation facilities and market. There is a good artificial harbor with a ship canal a mile long connected with it. Thus, iron ore from the Lake Superior region is accessible. It has many of Chicago's great railroad facilities for obtaining coal from the Indiana and Illinois coal fields, and the Chicago labor market. Gary is famous for iron and steel; however, other products, as cement made from furnace slag, and automobiles, are of importance. Its excellent school system is not less worthy of mention.

HAMMOND. The original town of Hammond was laid out in 1875. In 1888, shipping by boat was abandoned because of the competition of railroads. Now Hammond is a great railroad and commercial center, having eleven trunk lines, seven belt lines, and three electric lines, and many manufacturing establishments, producing chiefly manufactures of steel, however, other industries, such as canning, slaughtering and meat packing, are also of importance.

HUNTINGTON. In 1831, the first white settlement was made at Flint Springs, the site of Huntington. The supply of pure water and the absence of underbrush, because of the ground for many rods around the springs was white with flint rock, made a favorite camping place for traders, prospectors and adventurers. In 1834, it became the county seat of Huntington County. Upon the arrival of the Wabash-Maumee Canal, large warehouses were built, and there was a heavy boat trade in lumber, grain, etc. In 1848, the town was incorporated and in 1855 the Toledo, Wabash and Western Railroad was built. The second railroad, the Chicago and Erie, arrived in 1882 and created a mild boom, as the car shops, round house and division headquarters were located here. The chief manufacturing is lumber, staves, spokes, carriage materials, plow handles, barrels, hoops, lime, shoes, gas engines, bicycles and pianos. Limestone is also quarried in this region.

INDIANAPOLIS was selected by the General Assembly of 1820 to become the state capital and the sale of lots began in 1821. The capital was moved from Corydon in 1825. The city had an average growth of 50% each year from 1826 when it had a population of 600 until 1866. The Cumberland National Road reached it in 1835. Its growth is due to its central location and to its being the state's capital. The generally level surface facilitates railroads. The radiation of railroads, interurbans and roads in all directions makes it accessible from all directions. It is near the coal field, is near the center of the "pork" and hardwood region, is in a rich grain region, and is in or near the most direct line of communication between seaboard and the Mississippi Valley, the Great Lakes and the South. In 1914 it had 996 manufacturing establishments employing 31,000 persons, and manufacturing products having a value of 140 million dollars, according to the special Federal census of that year. Now it is claimed to have 700 separate industries. It is the great bakery center of the state, has flour mills, makes milling machinery, steel and iron implements, automobiles, clothing, tractors, copper products, tin plate, etc.

JEFFERSONVILLE was founded in 1802 and was promptly made the county seat of Clark County. It is said the city was planned by Thomas Jefferson. Between Jeffersonville and Louisville the Ohio River is relatively narrow and is confined in a gorge below the falls, hence it was relatively easy to bridge it there. The first bridge across the Ohio was made here in 1870.

There is manufacturing of railway cars, steamboats, and machinery, iron and oil stove factories, clothing and also a pork packing plant. The Jeffersonville, Madison and Indianapolis R. R. Co. shops and car works, two ship yards, Ohio Falls Car Co., the Southern Car Co., the State Prison, and the Depot of the Quartermaster Department of the U. S. Army are also located here.

KOKOMO was located in 1844 in a swamp covered with dense forests. Mosquitoes and malaria were the bane of the city. It grew slowly because the heavy timber, underbrush, and swampy soil retarded growth and in 1852 it had a population of only 152. A railroad arrived (the Indianapolis, Peru and Chicago) and the city was incorporated in 1855. Natural gas was found in 1887 and soon after Kokomo's industries included paper mills, bit works, lumber mills, and plants manufacturing rubber, pottery, stoves and ranges, plate glass, automobiles, woven wire, clothing, paper from straw, brass and stelite. The last is used for making points of cutting tools.

LAFAYETTE was founded in 1825 upon the terraces of the Wabash River in Tippecanoe County. Its early growth was due to the fact that it was at the head of navigation on the Wabash River. It was made a port of entry on the Wabash Canal which reached this point in 1843 and soon outstripped less favorably located neighboring towns such as Crawfordsville. Not long after this it became a railroad center. It manufactures wagons, farming implements, wire goods, bridges, hosiery, bicycles, hard-wood lumber, electric meters, steering gears, safes, and strawboard. It is also a meat packing center. The railroad repair shops located here give work to many men. Purdue University, the state college of agriculture and mechanic arts, is located here, as is a federal agricultural experiment station.

LAPORTE, laid out in 1833 and incorporated in 1854, is a manufacturing and railroad center twelve miles from Michigan City, in Laporte County. It manufactures farm machinery, vehicles, woolen goods and bicycles. Pine Lake Assembly is a popular summer resort near the city.

LOGANSPORT. In 1828 a fur trading post was established between the Wabash and Eel Rivers, and named in honor of Captain Logan, a Shawanee chief, who lost his life while attesting his fidelity to the white people. Its early growth was due to the coming of the Michigan Road, the Wabash Canal and the Richmond, Newcastle and Lake Erie, and Wabash and St. Louis Railroads, Logansport being at a point where the main line of travel crossed the Wabash. The water power derived from Eel River made it a manufacturing town. Here are located machine shops of the Pennsylvania Railroad, several automobile factories, and limestone quarries, the stone from which is used in manufacturing quicklime and also for building purposes.

MARION. The city of Marion was founded in 1831, and named in honor of the Revolutionary general, Francis Marion. The most important early industry was a ginseng factory. Later a tannery was built which supplied leather for the settlers' boots and shoes. In 1886 natural gas was discovered and Marion rapidly became a manufacturing town, making glass, iron, paper and furniture. More recently it has done much canning of vegetables and is also a center for automobile manufacture.

MICHIGAN CITY was laid out in 1831. It has a fair harbor and has lake commerce between Chicago and other lake cities. It is the terminus for two railways and is served by another. The early growth of Michigan City was due to its location on Lake Michigan. It manufactures clothing, furniture, and steam-railway cars. One of the State Prisons is located here.

MISHAWAKA is an Indian name meaning "the town of big rapids", which was applied to an Indian village, near the big rapids in the St. Joseph River. The city's growth has been largely dependent upon the water power. From 1833 until 1856 bog ore, obtained from a swamp south of the town, was smelted here. Upon the exhaustion of the bog ore in 1856 pig iron was shipped in by rail to a foundry in which plows and cultivators were made. This foundry was one of the effective forces in building up Mishawaka. Some of the present industries manufacture hydraulic cement, furniture, wagons, axles, refrigerators, woolens, windmills, axes, paper, wood pulp, pumps, flour, felt, putty, harness, furniture, carpets and stockyard products.

MUNCIE was located in 1825. The name is a corruption of that of an Indian chieftain of a neighboring tribe. In 1837 the roads from Ohio to Indianapolis, Richmond to Logansport, and Newcastle to Ft. Wayne, one to Pendleton, and one to Delphi, were built thru Muncie. The discovery of oil and gas near, in 1886, led to a rapid increase in wealth and changed Muncie from a commercial into a manufacturing city. At present it is a great interurban center as well. It manufactures automobile parts, wire and wire fencing, steel and iron goods, machinery tools, tin plate, paper, gloves, canvas and boxes. Its chief factories, however, manufacture glass fruit-jars. Indeed Muncie produces a large share of the Ball Mason Jars made in this section of the country. The Eastern Division of the State Normal School is located here.

NEW ALBANY was laid out in 1813 across the Ohio from Louisville. It has a beautiful situation. The nearby "knobs" are celebrated for their superior quality and abundance of peaches, pears, plums, apples, grapes, raspberries. New Albany has had the advantage of river transportation, and is a good railroad center. It has become a manufacturing city of some consequence and makes plate glass, furniture, woolen and cotton goods, clothing, and leather.

NEWCASTLE is in a good agricultural section, and is a good commercial and railroad center, having several railroads and interurbans. It produces furniture and automobiles and is also noted for rose culture.

PERU was started in 1827 and like several other towns along the Wabash-Maumee Canal it had a very rapid growth at first. In 1887 a small, but rich, oil field aided its development. There are large quarries of limestone, much of which is burned for quicklime and also is used as a flux in the blast furnaces of Chicago. Several railroads have their repair shops located here also.

RICHMOND was located on Whitewater River in 1816. It was settled largely by the Society of Friends. In 1822, it had factories, several stores, and two newspapers. Its early growth was partly due to the Whitewater Canal and the National Road. Richmond is surrounded by a rich agricultural section, has ample railroad facilities and is a commercial and manufacturing city. It is noted for the manufacture of farm implements, pianos, caskets, carriages, furniture, and the growing of greenhouse flowers. Earlham College is located here.

SOUTH BEND was established in 1823 as an outpost of the fur trading post at Ft. Wayne. Its early growth was stimulated by the Michigan Road. The St. Joseph River brought commerce from the Great Lakes, and furnished water power for mills. South Bend has considerable manufacturing. In

1852 Studebaker Brothers started a wagon-smith shop. In 1904 they produced more than one hundred thousand vehicles. Some of the present manufactures of South Bend are: harnesses, automobiles, iron products, plows, stoves, clover hullers, Singer sewing machines, varnish, woolen underwear, and folding paper boxes.

TERRE HAUTE was founded by pioneers from more southern portions of Indiana in 1817. The name signifies high land. Its early growth was stimulated by the National Road which reached it in 1838, and the Wabash-Maumee Canal which reached it in 1849. It is situated near the center of the Indiana coal field, and the cheapness of fuel along with the good railroad facilities have attracted manufacturing. The rapid growth of the city from 1900-1905 was partly due to the distilleries and breweries, the river bottom lands furnishing the necessary large supplies of corn. It is noted for the manufacture of sewer pipes and paving bricks from shale for glass, enameled ware, iron and steel implements, wheels, paper, flour, confections, and men's clothing. Mining of coal is a chief industry. Slaughtering and meat packing are also engaged in. The State Normal School was established here in 1867.

VINCENNES is the oldest continuous settlement in Indiana. In 1860 a French trading post and military station was established, in 1702 a permanent mission, and in 1705 a regular trading post. Vincennes has had a slow growth. The capital of Indiana Territory was located here from 1800-13, increasing its growth considerably, and it was a severe blow to the town when the capital was moved to Corydon. It is in the center of a fine timber, coal and agricultural region, and has several rolling mills, and glass factories. Vincennes is the business center of the Illinois and Indiana coal and southwestern oil field. The first school founded in the state, Vincennes University, is still flourishing.

WHITING. In 1889 the Standard Oil Company established the immense refinery and storage system at Whiting. It is located between Lakes Wolf, George, and Michigan. In 1874 it was established as a little station on the Michigan Southern Railroad. It has now become one of the greatest petroleum refineries in the world, if not actually the greatest. Pipe lines lead to it from Texas, Oklahoma, Kansas, and from Ohio and Pennsylvania, and also connect it with the Atlantic coast at Jersey City.

Racial Complexion of Population. Most of the people of Indiana are old-time Americans. According to the phraseology of the Census reports, nearly four-fifths of the people were "native whites of native parentage" in 1910 while 13% were "native whites of foreign or mixed parentage". Nearly 6% were foreign-born and 2% were negroes. The percentage of foreign born probably decreased between 1910 and 1920, tho the figures are not yet available. The percentage of negroes increased notably, there having been a northward migration of southern negroes during the World War. In only one county, Lake, in which Gary, Hammond, East Chicago, Indiana Harbor, etc., are located, is the percentage of foreign born high. In 1910 it made up between one-third and one-half of the county's population. Laporte County, adjacent, had between 16% and 25% of its population foreign born. There were only six counties in 1910 in which the percentage of those who were foreign born and native born having one or both parents born abroad together comprised as much as one-fourth of the total population.

Of the native American population four-fifths in 1910 were born in Indiana and one-fifth born outside of the state. More than half of the negroes residing in the state in 1910 were born in some other state. Of the state's foreign born in 1910, the country of nativity was as follows:

Germany...... 39%	Austria...... 7%	England.... 6%	Italy........ 4%	Sweden...... 2%
Hungary...... 9%	Ireland...... 7%	Russia...... 6%	Canada...... 4%	Scotland.... 2%

Table 8, below, taken from the 13th Census, sums up the color and nativity of the population by decades.

TABLE 8.

Color—Nativity—Parentage.

	Total Number			Per Cent of Total		
	1910	1900	1890	1910	1900	1890
Total	2,700,876	2,516,462	2,192,404	100	100	100
White	2,639,961	2,548,502	2,146,736	97.7	97.7	97.9
Negro	60,320	57,505	45,215	2.2	2.3	2.1
Indian	279	243	343	Less	than .1 of	1%
Chinese	276	207	92	Less	than .1 of	1%
Japanese	38	5	18	Less	than .1 of	1%
Total Native	2,541,213	2,374,341	2,046,199	94.1	94.4	93.3
Foreign Born	159,663	142,121	146,205	5.9	5.6	6.7
Total Native White	2,480,639	2,316,641	2,000,733	91.8	92.1	91.3
Native Parents	2,130,088	1,952,194	1,697,998	78.9	77.6	77.4
Foreign Parents	211,008	215,785	189,226	7.8	8.6	8.6
Mixed Parents	139,543	148,662	113,509	5.2	5.9	5.2
Foreign Born White	159,322	141,861	146,003	5.9	5.6	6.7
Urban						
Total	1,143,835	862,689	590,039	100	100	100
White	1,195,026	820,188	561,049	95.7	95.1	95.1
Negro	48,425	42,274	28,839	4.2	4.9	4.9
Indians, Chinese, etc	384	227	151	Less	than .1 of	1%
Rural						
Total	1,557,041	1,653,773	1,602,365	100	100	100
White	1,544,935	1,638,314	1,385,687	99.2	99.1	99.0
Negroes	11,895	15,231	16,376	.8	.9	1.0
Others	211	288	302	Less	than .1 of	1%

In respect to the percentage of her population born in foreign lands, Indiana ranked thirteenth among the states in 1910. The states that had a smaller percentage were nearly all southern states, where there has been only a slight industrial development. As to country of birth, Indiana ranks highest in regard to the number of Austro-Hungarians and Germans. Of both of these nationalities only eleven states have more. Fourteen states have more Irish, eighteen have more English and Scotch, nineteen more Italians, twenty-one more Russians and twenty-five states have more Scandinavians and Canadians. Table 9, below, gives the number of foreign born in Indiana by country of birth at the census of 1910.

TABLE 9.

Foreign Born Population. Distributed According to Country of Birth.

Country	*1910*	*Country*	*1910*
England	9,783	Spain	40
Ireland	11,206	Italy	6,911
Scotland	3,419	Russia	9,599
Wales	1,498	Finland	215
Norway	531	Austria	11,831
Sweden	5,081	Hungary	14,370
Denmark	900	Roumania	709
Netherlands	2,131	Bulgaria	434
Belgium	2,298	Serbia	99
Luxemburg	11	Montenegro	43
Germany	62,179	Turkey in Europe	2,274
Switzerland	2,765	Greece	1,370
France	2,388	Other Europeans	62
Portugal	6	French Canadian	789
Other Canadians	5,049	India	16
Newfoundland	19	Turkey, Asiatic	809
Cuba	48	Other Asians	30
Other West Indies	34	Africa	29
Mexico	47	Australia	45
Central America	7	Atlantic Islands	1
South America	36	Pacific Islands	5
Japan	41	Other Countries	52
China	196	Born at Sea	197
Total Foreign Born Population			159,663

CHAPTER VIII.

MANUFACTURING.

During recent decades manufacturing has become important in Indiana. Table 10, below, gives many facts from the Federal Censuses.

TABLE 10.

Summary of Manufactures in Indiana.

	1904	1909	1914	1919
Number of establishments	7,044	7,969	8,022	7,916
Proprietors, firm members	7,191	7,674	7,229	6,768
Salaried employees	14,862	23,605	28,538	45,797
Wage earners	154,174	186,894	197,503	277,580
Total employees	176,227	218,263	233,270	330,145
Primary horse-power	380,758	633,377	709,703	1,095,912
Capital invested, in $1,000	312,071	508,717	668,863	1,335,714
Salaries paid, in $1,000	15,029	26,305	36,596	85,117
Wages paid, in $1,000	72,058	95,510	119,258	317,043
Cost of Material, in $1,000	220,507	334,375	423,857	1,174,951
Value of products, in $1,000	393,954	576,075	730,795	1,898,753
Value added by manufacture, in $1,000	173,447	244,700	306,938	723,802
Per cent increase in value of products	46.9	47.0	26.2	159.8
Per cent increase wages	10.9	21.3	5.6	165.9

Indiana ranked ninth among the states in manufacturing at the 1920 Census. It had ranked 8th in 1910. The more important industries, the value of the products, and Indiana's rank among the states in respect to the industry are given in Table 11, below, which is based on the Statistical Abstract for 1920, the Statistical Atlas, and recent publications of the Census Bureau.

TABLE 11.

Leading Industries, Value of Products, Indiana's Rank.

1. Steel, Forgings, etc.—
   1919 $192 million, rank 3
   1914 54 million, rank 4
   1909 36 million, rank 5

2. Automobiles and parts—
   1919 $179 million, rank 2
   1914 29 million, rank
   1909 24 million, rank 4
   1904 3 million, rank 7

3. R. R. Cars, repairs, etc.—
   1919 $149 million, rank
   1914 43 million, rank
   1909 28 million, rank 6
   1899 10 million, rank 5

4. Slaughtering and Meat Packing—
   1919 $134 million, rank 9
   1914 51 million, rank 9
   1909 47 million, rank 9
   1899 45 million, rank 5

5. Foundry and Machine Shops—
   1919 $133 million, rank
   1914 48 million, rank
   1909 40 million, rank 10
   1899 18 million, rank 10

6. Lumber and Its Manufactures—
   1919 $90 million, rank
   1914 43 million, rank
   1909 42 million, rank 23
   1899 26 million, rank 9

7. Flour and Grist Mills—
   1919 $75 million, rank
   1914 37 million, rank
   1909 41 million, rank 8
   1899 28 million, rank

8. Electrical Machines—
   1919 $42 million, rank
   1014 9 million, rank
   1909 8 million, rank 8
   1899 2 million, rank 9

9. Butter, Condensed Milk and Cheese—
   1919 $33 million, rank
   1914 9 million, rank
   1909 4 million, rank 15
   1899 1 million, rank

10. Agricultural Implements—
   1919 $32 million, rank 3
   1914 13 million, rank 5
   1909 14 million, rank

11. Printing and Publishing—
   1919 $31 million
   1914 17 million
   1909 14 million

12. Glass—
   1919 $30 million
   1914 15 million
   1909 12 million

13. Bakeries—
   1919 $30 million, rank
   1914 12 million, rank
   1909 10 million, rank 10
   1899 4 million, rank

14. Canning and Preserving—
   1919 $28 million, rank
   1914 14 million, rank
   1909 9 million, rank 6
   1899 3 million, rank

15. Clothing, (Ready Made)—
   1919 $27 million, rank
   1914 10 million, rank 10
   1909 10 million, rank

16. Rubber, tires, etc.—
   1919 $21 million
   1914 6 million
   1909 4 million

17. Carriages and Wagons—
   1919 $16 million, rank
   1914 23 million, rank
   1909 23 million, rank 2
   1899 16 million, rank

18. Copper, Tin, Brass, etc.—
   1919 $16 million
   1914 10 million
   1909 7 million

19. Paper and Wood Pulp—
   1919 $15 million
   1914 6 million
   1909 5 million

20. Drugs, etc.—
   1919 $14 million
   1914 6 million
   1909 4 million

21. Cement—
   1919 $13 million
   1914 10 million
   1909 7 million

22. Brick and Tile—
   1919 $13 million, rank
   1914 9 million, rank
   1909 8 million, rank 5
   1899 3 million, rank 5

23. Gas (domestic and commercial)
   1919 $12 million, rank
   1914 6 million, rank
   1909 3 million, rank 12

24. Confections and Ice Cream—
   1919 $11 million, rank
   1914 5 million, rank
   1909 3 million, rank 13

25. Hosiery and Knit Goods—
   1919 $10 million, rank
   1914 4 million, rank
   1909 2 million, rank 16

26. Leather and Leather Goods—
   1919 $10 million
   1914 6 million
   1909 6 million

27. Tobacco Manufactures—
   1919 $10 million
   1914 6 million
   1909 4 million

TABLE 12.

Production of Lumber in Indiana in Billions of Board Feet.

Year		Year		Year		Year	
1880	916	1900	978	1910	423	1918	350
1890	707	1909	556	1915	356	1919	282

TABLE 13.

Production of Pig Iron in Indiana in Millions of Tons.

Year		Year		Year		Year	
1909	.6	1912	1.7	1916	2.0	1919	2.3
1910	1.0	1914	1.5	1918	3.0		

Indiana ranked third among the states in 1914 and fifth in 1919.

TABLE 14.

Production of Crude Petroleum in Thousands of Barrels.

Year		Year		Year	
1890	63	1916	769	1920	932
1900	4,874	1918	878		

TABLE 15.

Receipts at Indianapolis of Cattle, Hogs and Sheep in Thousands, Average of Five-Year Periods, and in 1920.

	Cattle	Hogs	Sheep
1896—1900	138	1,500	90
1901—1905	250	1,600	100
1906—1910	385	1,900	100
1911—1915	375	2,100	140
1916—1920	500	2,600	120
1920	597	2,900	135

CHAPTER IX.

CONCLUSION: THE RICHNESS OF INDIANA.

Since a state's wealth depends largely upon the geographic factors dominating it, a statement of the wealth is a summary of the geography. Hence a suitable conclusion to this treatise on the Geography of Indiana is a statement of the state's richness.

Indiana is very rich. The indications are that no other area of similar size in the nation is much better endowed. Indiana has great natural resources, a favorable location, a comparatively dependable climate, and much fairly smooth land. Because of this superior endowment, Indiana has been able to surpass most of the older states in the production of many commodities, and is able to hold a high rank among the states in total production. This is the more notable when Indiana's comparatively small size is recalled. Only eleven states are smaller. All of Indiana's neighbors are notably larger. For example, Illinois, Michigan, Iowa and Wisconsin are each about fifty per cent larger than Indiana, while Ohio and Kentucky are each about eleven per cent larger. Eighteen states are more than twice Indiana's size.

Although a considerable number of states surpass Indiana in total production, it is seen, when allowance is made for her smaller size, that she really ranks with the leaders.

The gross value of farm products affords one excellent basis for comparison between states. In this, Indiana ranked eighth state in 1919, according to a recent census report. However, only three states ranked ahead of Indiana, after comparative area is considered, namely Iowa, Illinois and Ohio. Indiana produced an average of $22,300 worth of agricultural products per square mile in 1919. Iowa, the best state, produced about ten per cent more.

The "gross value of farm products" is made up of crops, livestock products and animals sold or slaughtered. In respect to value of crops, although Indiana ranked thirteenth state in 1919, when allowance is made for comparative size, Indiana is seen to surpass all but Iowa, Illinois and Ohio. In livestock products, Indiana ranked eleventh state in 1919, but six of the ten states ahead of Indiana are enough larger than Indiana so as to fall behind her when comparative area is considered. Indiana stands still higher in the value of animals sold or slaughtered, ranking seventh state when size is ignored, but claiming third place when area is considered.

In respect to individual crops: Though often surpassed by five states in the area planted to corn, Indiana ranks with Illinois and Iowa as a corn state, when area is considered. Indeed, in average yield of corn per square mile, Indiana surpasses these states, and the other three, Missouri, Nebraska and Kansas, which often have more acres planted to corn than Indiana has. In wheat production Indiana normally ranks eighth among the states. However, Indiana usually is exceeded only by North Dakota and Kansas in the average yield of wheat per square mile of total area. In oats production Indiana ranks fourth when area is considered, though seventh when area is ignored.

Forest products: Few people realize how large a total income is obtained from Indiana's woods and woodlots. In 1919, according to the 1920 census, Indiana marketed eleven million dollars' worth of forest products, ranking nineteenth among the states. However, in proportion to area, Indiana ranked thirteenth state.

The value added by manufacturing (724 million dollars) was only a little less than the gross value of farm products (782 million dollars) in 1919, in Indiana, so it is logical to next consider Indiana's rank as a manufacturing state. Indiana ranked ninth among all the states in 1919, but the eighth state, California, which barely passed Indiana, is more than four times as large as Indiana. Indiana had more than two dozen manufactures yielding products worth over 10 million dollars in 1919.

The leading industries of Indiana, according to the 1920 census, are: (1) Steel, forgings, etc., producing products worth 192 million dollars and giving Indiana third rank among the states. (2) The Indiana products of the automobile industry had a value of 179 million dollars in 1919, and Indiana was surpassed only by Michigan in this respect. (3) The third industry in value of products was the manufacture and repair of railroad cars, etc. The total for 1919 was 149 million dollars and Indiana's rank about fifth state. Indiana ranked ninth state in the value of the slaughtering and meat packing industry, with a value of 134 million dollars. However, if comparative area is considered, Indiana would surpass half of the eight states which have greater totals. Indiana ranked as third state in the production of agricultural implements in 1919. In the value of products of the canning industry, Indiana should take fourth place instead of sixth. Likewise, though about eighth state in the value of the products of flour mills and grist mills, Indiana is enough smaller than the seven states which ranked higher than it to surpass all but the leading state, Minnesota, in value per square mile of area.

In mineral products also, Indiana ranks high, second in the production of cement, fifth or sixth in coal, fourth in coke and fifth in brick and tile. In the production of limestone for building purposes, Indiana stands far ahead of all others.

To sum up, although Indiana is thirty-seventh state in size, she ranks second in the production of automobiles and parts and of cement; third in the production of steel and forgings; fourth in coke; fifth in coal, brick and tile, and in railroad cars and repairs; sixth in corn acreage, and in value of canned products; seventh in value of animals sold or slaughtered, and in oats production; eighth in the gross value of farm products, in wheat production and in the products of flour and grist mills; ninth in the value added by manufacture, and in the products of the meat packing industry. If allowance is made for Indiana's comparatively small size, she ranks notably higher in several of these categories, ranking second in automobiles, cement and flour and grist mill products; third in steel, forgings, etc., in corn and wheat production and in animals sold or killed; fourth in the gross value of farm products, of crops, and of livestock, of oats, and of coke, and in the canning industry; fifth in meat packing and coal mining; eighth in the value added by manufacturing, and thirteenth in forest products.

PART II.

The Physiography of Indiana

By

Clyde A. Malott.

(81)

LIST OF ILLUSTRATIONS

LIST OF ILLUSTRATIONS—Continued

LIST OF PLATES

LIST OF TABLES

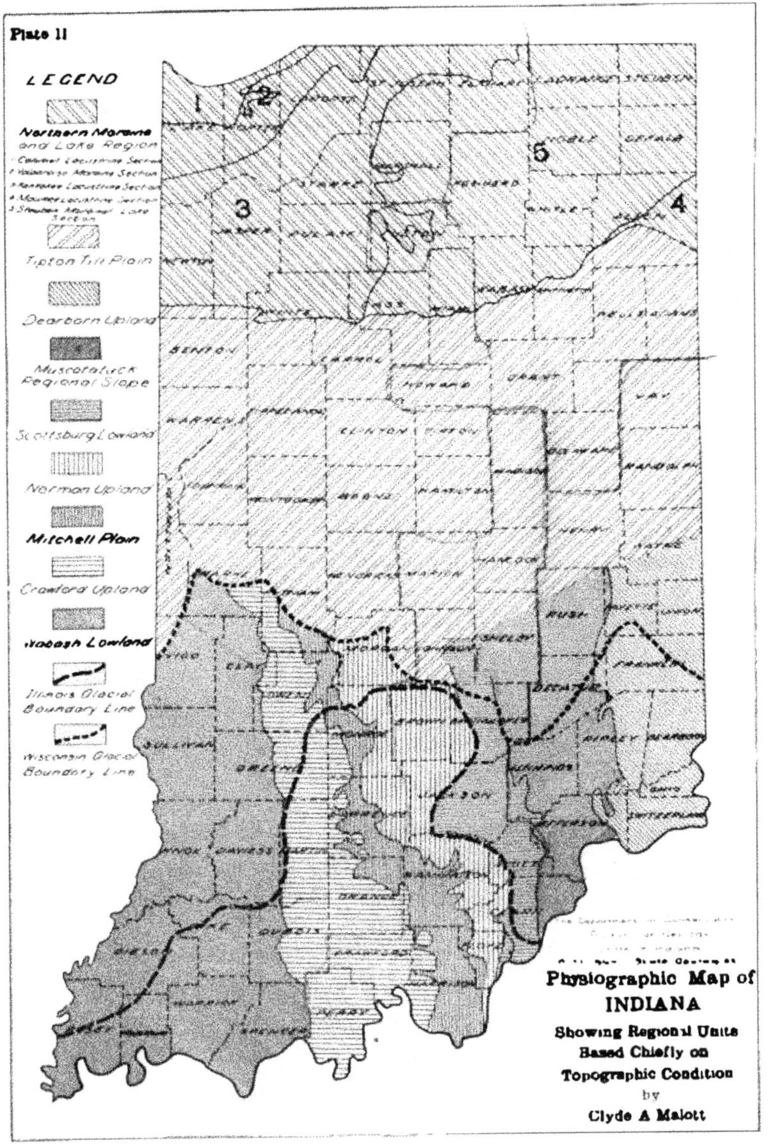
Plate II
LEGEND
Northern Moraine and Lake Region
Tipton Till Plain
Dearborn Upland
Muscatatuck Regional Slope
Scottsburg Lowland
Norman Upland
Mitchell Plain
Crawford Upland
Wabash Lowland
Illinois Glacial Boundary Line
Wisconsin Glacial Boundary Line
1
2
3
4
5
Physiographic Map of
INDIANA
Showing Regional Units
Based Chiefly on
Topographic Condition
by
Clyde A Malott

The Physiography of Indiana

By CLYDE A. MALOTT.

CHAPTER I.

INTRODUCTION.

Definition of Subject Matter.

PHYSIOGRAPHY is a study of the surface features of the lands of the earth. Inasmuch as physiography deals with the sizes, shapes and relationships in position and distance of land forms, it includes topography. This phase is largely descriptive and geographic, since it deals with dimensions and locations. Physiography also treats of the origin and development of land forms. It attempts to explain the landscape as it exists, and calls attention to the action of certain natural processes which are working on the materials of the earth under particular conditions, and is therefore explanatory as well as descriptive. Physiography as a science treats in detail of all those processes which give rise to land forms, calling attention to particular land forms or types which serve as illustrations of the work of the processes involved on earth materials under particular conditions. Physiography as applied to regions consist chiefly in the description and explanation of the existing land forms. Their development is important, and the conditions under which they have been developed are scarcely less important. It is regional physiography with respect to Indiana which will be discussed in this report.

Scope of Treatment.

A full description of the land forms of Indiana, a grouping of the related forms into natural regional units, and a comprehensive treatment of physiographic development, are attempted for the entire state, for the first time. Such an undertaking must of necessity draw from many sources of information, and recognition of the various sources of information is considered a part of this treatise. Acknowledgment is given in the text and in the accompanying footnotes. Special acknowledgment is due Dr. E. R. Cumings, head of the Department of Geology, Indiana University, for his careful and detailed criticism of the manuscript and his helpful suggestions at all times during the preparation of the manuscript. The writer, however, assumes full responsibility for much of the matter as presented here.

Regional physiography usually confines its treatment to natural units wherein the land forms are similar or have had similar development. Such units are not usually limited by civil boundaries. In the physiographic treatment of Indiana, attention will first be directed to the position of the Indiana region as it is embraced in one or more of the larger natural units of the United States. Drainage description will be treated in a chapter by itself. An appreciation of the drainage lines and drainage relationships is an important introductory step in an understanding of the physiography of a region.

The general topographic condition of any region is usually closely related to drainage, but such is not always the case. A considerable portion of Indiana is an exception to this general rule. The topographic map accompanying this report bears this out. Nevertheless drainage is an important factor, for the physiography of Indiana or any region is largely dependent upon the work performed by running water.

Probably the most important part of this treatise is concerned with the detailed presentation of the topographic condition of Indiana and the division of the state into natural units on the basis of altitude, relief and related type of topographic forms present. This presentation must be largely descriptive, but in addition to the text, topographic and physiographic maps have been prepared. The topographic map is designed to present the altitudes and reliefs and their relationships. The physiographic map shows the natural units as presented in the text and is as important as the details of the text itself.

No particular attempt is made towards a rational classification of the factors involved in the development of land forms, yet a brief consideration of the factors and their relative importance is offered. The sequence of events which entered into the physiographic development of Indiana and related regions is given an important place, and some of the problems connected with the larger portion of the Mississippi basin are touched upon. Special attention is given to the factors instrumental in bringing about the present topographic condition in Indiana. The close relationship of the Indiana region to the gulf embayment is emphasized, and the events that are known to have taken place in one region are correlated with those of the other region. The correlations are, however, of a tentative nature.

The treatment here must be more general than complete, yet it is desirable that a fair conception of the physiography of Indiana be presented. For this reason an attempt is made to show specifically the more important phases of development in the various physiographic units. Special attention is called to the factors which have acted to produce the land forms; and also those factors which have conditioned the physiographic processes. Such a consideration leads naturally into form development which is both descriptive and explanatory, and is perhaps the most comprehensive and important phase of physiography. When all the conditioning and active factors in form development have been properly determined in any region, it may be seen that inasmuch as this or that condition, or this or that active factor, has prevailed here and there, the region may be marked off into divisions or subdivisions, each with its particular array of topographic forms. Thus Indiana in a broad sense is divided at once into two divisions on the basis of glaciation alone, part of it having been subjected to that important agency while a smaller portion has never been glaciated. The unglaciated portion has long been subjected to weathering agencies and the destructive work of running water, and these active agencies have been conditioned to an important degree by great resistance of the rock strata in certain localities while in other localities nonresistant rocks permitted rapid destruction. These diverse conditions in the present erosion cycle have largely controlled the forms present in the different localities. The glaciated portion, on the other hand, is largely a plain of glacial construction, and the destructive agencies have only locally been of importance since its glaciation.

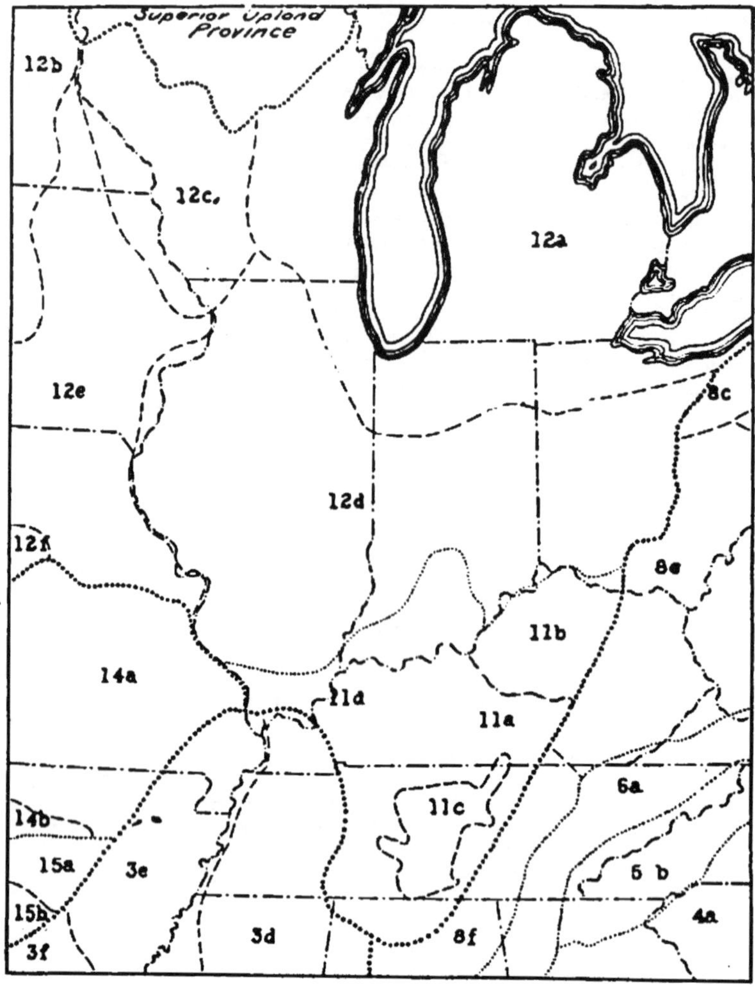

Fig. 1. Physiographic Situation of Indiana (after Fenneman).

3 COASTAL PLAIN PROVINCE
- 3d East Gulf Coastal Plain
- 3e Mississippi Alluvial Plain
- 3f West-Gulf Coastal Plain

4 PIEDMONT PROVINCE
- 4a Piedmont Upland

5 BLUE RIDGE PROVINCE
- 5b Southern Section

6 APPALACHIAN VALLEY PROVINCE
- 6a Tennessee Section

8 APPALACHIAN PLATEAU PROVINCE
- 8c Allegheny Plateau, Glaciated Section
- 8e Allegheny Plateau, Kanawha Section
- 8f Cumberland Section.

15 OUACHITA PROVINCE

11 INTERIOR LOW PLATEAUS PROVINCE
- 11a Highland Rim Plateau
- 11b Lexington Plain
- 11c Nashville Basin
- 11d Aggraded Valley Section

12 CENTRAL LOWLAND PROVINCE
- 12a Eastern Lake Section
- 12b Western Lake Section
- 12c Wisconsin Driftless Section
- 12d Till Plains Section
- 12e Dissected Till Plains Section
- 12f Osage Plains Section

14 OZARK PLATEAUS PROVINCE
- 14a Springfield-Salem Plateau
- 14b Boston Mountains

Geographic and Physiographic Location of Indiana.

Indiana is located in the middle-eastern part of the United States, approximately 600 miles from either the Gulf of Mexico on the south or the Atlantic Ocean on the east. Longitude 86° and latitude 40° cross near the geographic center of the state. The Great Lakes system is immediately north, but the drainage is chiefly to the Gulf of Mexico through the Ohio and Mississippi rivers. A small portion of the state discharges its drainage into the St. Lawrence system. The divide in the northeastern part of the state, between the St. Lawrence and the Mississippi drainage systems, is about equidistant from the sea-level head of the Gulf of St. Lawrence and the Gulf of Mexico. It is interesting to note that the distance is but little greater to the head of the Hudson Bay on the north, which, of course, is not related in any way to the physiography of Indiana.

Indiana is located towards the eastern edge of the great interior plains of the United States, which stretch between the Rocky Mountains and the Appalachian highlands. It is in the area of low plains, a great part of which are glacial till plains. A considerable portion of the southern part of the state constitutes a part of the well dissected low plateau area which stretches from the foot of the higher Appalachian plateaus westward to the axis of the Mississippi River. The extreme southwestern part of the state approaches within fifty miles of the Gulf embayment area of the Coastal plain. It is towards this area that the greater portion of the drainage of the state is discharged.

The accompanying map, Fig. 1, following Fenneman's scheme of physiographic classification, shows the physiographic divisions which the area of Indiana partly embraces.[1] Indiana, following Fenneman, lies in two physiographic provinces. The portion of the state belonging to the Interior Low Plateau province is essentially a highly dissected portion of the Highland Rim uplifted peneplain, but towards the western side other characters besides those inferred by dissection of an uplifted peneplain prevail. This is recognized by Fenneman, and is signified by his "Western (unnamed) Section." Approximately five-sixths of the state belongs to the great Central Lowland province where local relief is small "regardless of absolute altitude," and which is divided into several sections. The line of demarkation between the Low Plateau province and the Central Lowland province is the southern limit of glaciation. This is recognized as a natural boundary line of importance. Indiana embraces portions of two sections of the Central Lowland province. The northern one-fourth of the state lies within the "Eastern Lake Section," which is characterized by "broad cuestas and corresponding lowlands in which lie the basins of the great lakes and great lacustrine plains. It is characterized also by numerous moraine lakes in the younger glacial drift." This latter descriptive sentence in particular applies well to the Indiana portion of the Eastern Lake section. The remainder of the glaciated portion of the state lies in the "Till Plains section," chiefly characterized by a covering of glacial till of nearly level surface without lakes and not well dissected by streams. "This section consists of two subsections distinguished as older (Illinoian) and

[1] Fenneman, Physiographic Divisions of the United States: Annals of Assoc. of Amer. Geogr., Vol. VI, 1917, pp. 19-98, Pl. I.

younger (Wisconsin)." Thus, Indiana lies in and occupies portions of two physiographic provinces within the great Interior Plains, a major Physiographic division of the United States. Two physiographic sections of each province are represented, and these sections may be further divided into smaller units, as shown beyond under more detailed treatment.

CHAPTER II.

GENERAL DRAINAGE AND CHIEF DRAINAGE LINES.

The Mississippi-St. Lawrence Divide.

Slightly less than nine-tenths of the area of Indiana belongs to the Mississippi drainage basin, and a little more than one-tenth belongs to the St. Lawrence drainage basin. A part of the divide between these two major drainage basins of the North American continent passes across the northeastern and the northern part of Indiana. This divide may in a sense be considered a continental divide, as the waters discharged from each side go into remote water bodies. The divide enters Indiana from Ohio between the Wabash and St. Marys rivers some 25 miles south of Ft. Wayne, and bears in an approximately north direction, passing just west of Ft. Wayne and around the rather vague headwaters of Eel River in eastern Noble County, thence westward to northern central Kosciusko County north of Warsaw, where it turns in a northwesterly direction. It passes just south and west of South Bend and crosses the Indiana-Michigan line to the northwest of South Bend. It immediately re-enters Indiana and passes southwest and west parallel to Lake Michigan and on the average some 10 miles distant from the lake. This divide is nowhere high and is not sharply defined. In places it is so indefinite that the water near it at times goes either way, as in the old glacial water routes near Ft. Wayne and South Bend. During the flood of March, 1913, water from the St. Marys River passed over the broad, flat divide immediately west of Ft. Wayne in a stream several feet deep and nearly one-half mile wide. Under natural conditions extraordinarily high waters of the St. Joseph River would merge with the headwaters of the Kankakee River in the broad, marshy divide in the vicinity of South Bend. A rise of less than 10 feet of the waters of Lake Michigan would allow an overflow into the Mississippi basin at the great col south of Chicago, through which the glacial waters of the late Pleistocene discharged. The divide is almost entirely developed on glacial deposits, and represents one of the youngest major divides of the continent.

The St. Lawrence Drainage System in Indiana.

The drainage area in Indiana belonging to the St. Lawrence basin is divided into three main systems. Some 1,277 square miles[1] of the Maumee system lie in the northeastern edge of the state, and comprise portions of the St. Marys and St. Joseph rivers and a small part of Maumee River. St. Joseph River comes into Indiana from Ohio flowing southwest to Ft. Wayne, where it joins the St. Marys River which comes from the southeast. The

[1] The figures given for the areas of the various drainage basins in Indiana have been carefully worked out by W. M. Tucker, and these are incorporated in this work. Mr. Tucker's report on the "Hydrology of Indiana" forms Part III of this publication.

junction of these two streams forms the Maumee which flows east and northeast out of the state and across a portion of Ohio into Lake Erie at Toledo. The peculiar barbed pattern which the Maumee system presents will be discussed at some length below. The second and largest drainage system belonging to the St. Lawrence basin, which is represented in Indiana, is the St. Joseph system. This particular St. Joseph River enters Lake Michigan north of the Indiana line. It drains 1,668 square miles of territory in Indiana on the northeastern border. The St. Joseph River lies mostly in Michigan, but enters Indiana flowing southwest about midway between Lake Michigan and the northeastern corner of the state. It turns abruptly northwest at South Bend and flows out of the state some 12 miles to the northwest of South Bend. Its chief tributary in Indiana is Elkhart River which drains part of the inter-lobate moraine in Elkhart and Noble Counties. The Calumet River and minor Lake Michigan drainage lies immediately south of Lake Michigan. Calumet River makes up most of the 665 square miles in this system. It flows parallel to the southern end of Lake Michigan and only a few miles distant. After entering Illinois it makes a turn and comes back into Indiana, having formerly looped back upon itself for more than 20 miles before entering the lake east of the present site of Gary.[3] It now enters the lake near the Indiana-Illinois line through an artificial channel. The lower part now flows in a reversed direction due in part to artificial interference at Gary and to the artificial outlet in the Calumet Lake and Wolf Lake region. It is a very sluggish stream on the flat plain south of Lake Michigan, and very much resembles a canal.

It is interesting to note the perched condition of the upper St. Lawrence system. The Great Lakes are simply great expanded portions of the St. Lawrence system, and are practically level. There is very little descent from the south end of Lake Michigan (581 feet above sea level) to the lower end of Lake Erie (573 feet above sea level), though the distance is approximately 850 miles. But this great expanded portion of the St. Lawrence system is perched high above Lake Ontario (246 feet above sea level), and Lake Ontario is in turn somewhat perched with respect to the sea-level Gulf of St. Lawrence. The waters which descend from the Mississippi-St. Lawrence divide on either side have about an equal distance to go before reaching the sea. But the Great Lakes act as temporary base-levels toward which the entering streams work. Lake Michigan itself is but little lower than the divide near by, and the streams have little advantage from the standpoint of erosion in a downward direction. On account of the perched condition of the lakes, the exceedingly low fall in a great distance seaward is not advantageous to the streams on the St. Lawrence side of the divide. The perched condition of the upper St. Lawrence system thus offsets the extremely low fall seaward. Were the descent of the St. Lawrence drainage dependent upon a more gradual fall similar to the gradients of the streams of the Mississippi system, it is likely that the streams on the two sides of the divide would have very similar gradients, due to their equal distances from the sea.

The Kankakee Drainage Basin.[4]

A broad, flat area consisting of 3,101 square miles and 650 feet or more above sea level lying between the Calumet basin on the north and the

[3] Alden, The Chicago Folio, No. 81, U. S. Geol. Surv., 1902, p. 1.

[4] Leverett, U. S. Geol. Surv., Monogr. 38, 1899, pp. 505-508.

Tippecanoe and Wabash on the south, is rather imperfectly drained by the Kankakee River and its tributaries. A considerable part of this basin in Indiana was until recently a great marsh, but artificial channels now drain much of the former marsh. The Kankakee River crosses the State line into Illinois and enters the Illinois River which empties into the Mississippi. A chief tributary, the Iroquois River, lies to the south and joins the Kankakee after abruptly turning to the north in Illinois. The Kankakee River rises in a broad, marshy tract in the vicinity of South Bend, very near the abrupt northward turn of the St. Joseph River. Its principal headwater tributary is Yellow River which drains a portion of the higher moraine country in Marshall County. The Kankakee from its source to where it leaves the state has a fall of about 100 feet. The river over this 100-mile stretch in its natural state meandered to such an extent that the river length was not less than 300 miles. Its average fall per mile was then not more than 4 inches, very low, indeed, for a stream no larger than the Kankakee. The river as it now exists has a greater fall, as it has been artificially straightened, and little undrained area exists.

Wabash-White River Drainage Area

The chief trunk stream of Indiana is the Wabash River. It enters Indiana from Ohio just south of the Maumee basin. Its source is not far distant from the state line where it is a rather small stream. After flowing northwest some 35 miles it turns west and southwest, and on approaching to within a few miles of the west line of the state, turns southward. A short distance below Terre Haute it serves as the western boundary of the state to its junction with the Ohio River in the extreme southwestern corner of the state. The stream distance as a boundary line is approximately 200 miles and the air-line distance is approximately 110 miles. This trunk stream with its tributaries exclusive of White and Patoka rivers constitutes approximately one-third of the drainage area of the state, and with all its tributaries constitutes approximately two-thirds of the area of the state. According to Tucker's figures the area of its drainage in Indiana above the entrance of White River is 11,481 square miles. The chief tributaries on its westward sweep across the state are the Salamonie and Mississinewa Rivers and Wildcat Creek from the south, and Eel and Tippecanoe Rivers on the north. The Tippecanoe River is the largest of these, draining over 3,000 square miles. After the Wabash turns south it receives several small creeks from the east, chief of which are Coal, Sugar, Raccoon and Busseron Creeks. White River enters from the east some 90 miles by stream above the junction of the Wabash with the Ohio River. White River drainage is divided into two areas approximately equal, and constitutes 11,155 square miles or a little over 30 per cent. of the total area of the state.

The two forks of White River head near the middle eastern side of the state in Randolph County where the highest land in the state is found. This elevation at its maximum is approximately 1,285 feet above sea level. The Mississinewa and Whitewater Rivers also have their sources near this high area. "The West Fork flows in a westerly direction through Muncie and Anderson, to Noblesville, then almost due south to Indianapolis. From Indianapolis it takes a direct southwesterly course to Petersburg. The West Fork flows through the Wisconsin glacial drift from its source to Martinsville a distance of 125 miles, and in Illinois glacial drift from Martinsville to the forks, a distance of 180 miles by the river. The East Fork flows in a tortuous, winding manner, thus increasing its length and decreasing its fall by numerous meanders. The East Fork flows through the Wisconsin glacial drift from its source to Columbus about 155 miles; then in the Illinois drift from Columbus to Brownstown, a distance of 40 miles. From Brownstown it flows through the unglaciated part of the state for about 90 miles, and the last 20 miles are again in the Illinois drift."[6] The West Fork is a relatively narrow basin, receiving only a few important tributaries. These are Fall and Eagle creeks at Indianapolis, White Lick near Martinsville, Beanblossom near Gosport and Eel River at Worthington. A number of tributaries run parallel with the East Fork for a considerable distance before joining it in the headwater area. Muscatatuck River is the largest tributary of the East Fork. Above the mouth of the Muscatatuck, White River is locally known as Driftwood River, and is a relatively swift stream. It is more sluggish from the mouth of the Muscatatuck westward. East Fork in its westward course receives Salt Creek and Indian Creek on the north and Lost River on the south as its chief tributaries.

The Patoka River is the largest Indiana tributary to the Wabash south of the mouth of White River. Black Creek and Big Creek, south of the Patoka, drain a combined area of about 350 square miles directly into the Wabash. Patoka River rises in the rugged region of southeastern Orange County, and in a very tortuous route flows west to the Wabash, entering only 1 mile below the mouth of White River. The basin of the Patoka lies immediately south of East Fork of White River and White River proper. It is rather peculiar in that its tributaries are very short, and that the basin, while reaching half way across the state, does not have an average width of more than 12 miles. Throughout most of its course it is a sluggish, meandering stream in a flat, aggraded valley. A considerable part of its volume is lost as underflow in the lower part of its course.

Whitewater Drainage Basin

South of the high drift-covered area in Randolph County, from which radiate the headwater streams of the Mississinewa and White Rivers, lies the Whitewater drainage basin. It embraces an area of 1,319 square miles in Indiana. Only a very small portion of the basin laps over the state line into Ohio. Whitewater River in its headwater portion consists of two main forks which flow south and slightly west until near the southern margin of the basin, where the west fork turns abruptly eastward. The east fork enters the west fork near the center of Franklin County at Brookville, forming the Whitewater River proper, which flows southeast out of Indiana. After crossing the Indiana-Ohio line it joins the Big Miami River which enters the Ohio River where the Ohio River first becomes the state boundary line. The eastward extending portion of the Whitewater drainage basin southeast of Brookville is remarkably narrow. A small portion of the area of the state east of the Whitewater basin drains directly into the Big Miami by small streams heading on the Indiana side of the state line. Whitewater River as a stream is relatively swift and carries much sediment, which on being received by the more slowly-moving Ohio is dropped in the form of sand-bars which are so numerous below the mouth of the Big Miami.

---

[6] Bybee-Malott, The Flood of 1913: Ind. Univ. Studies No. 22, 1914, p. 123.

Ohio River Drainage

The Ohio River, a trunk-stream tributary coming into the Mississippi from the east, forms the entire boundary line between Indiana and Kentucky with a mileage of 352 miles.[6] This stream drains a large area from the east, but receives only rather small streams from Indiana before the entrance of the Wabash. With the exception of Big Blue River, all of the streams head within 30 miles of the Ohio. At one point a short distance west of Madison along an extended westward bend of the Ohio River, the divide is within 2 miles of the river. The water draining from the very brink of the high bluffs above the Ohio goes into the Muscatatuck River, thence into White River and the Wabash before reaching the Ohio, traversing a distance of 300 miles, and from the crest of the divide descending approximately 500 feet. The Ohio River flows in a generally southwest direction, but does so in many great nearly right angled turns alternately west and south, with the exception of two prominent turns and several lesser turns to the north at the west ends of a number of the westward stretches. The Ohio River has a descent of 115 feet from the mouth of the Big Miami to the mouth of the Wabash and a little more than 23 feet of this occurs in the falls and rapids between Jeffersonville and New Albany.

Laughery Creek in the southeastern part of the state, after flowing more than half its length southward, turns abruptly to the east-northeast and joins the Ohio a short distance below the mouth of Big Miami River. Indian-Kentuck, Fourteen Mile, Silver, Buck, and Indian creeks, Big Blue River, Little Blue River, Anderson, Little Pigeon and Big Pigeon creeks enter the Ohio River in succession as its chief tributaries from the Indiana side. The combined drainage area of these streams and other less important ones is approximately 4,400 square miles. Blue River, the largest stream of the minor Ohio drainage has approximately 475 square miles of drainage area. The drainage area of Blue River and Indian Creek in particular is almost impossible to determine, as these streams drain a high plateau-like area largely underlain by limestone which is especially characterized by underground drainage. These streams have few surface tributaries in their lower courses, but many large springs drain into them. With the exception of the Pigeon creeks, all of the minor Ohio tributaries have their sources on high rugged land and have rather steep gradients (Silver Creek in part may also be considered an exception). While these streams are short and rise on high, rugged land and descend into the deeply intrenched Ohio gorge-like valley, none of them, with the exception of a few very minor streams in the vicinity of Madison, possess waterfalls worthy of note. The streams entering the Ohio River in the lower one-third of the minor Ohio drainage are all aggraded streams in their lower courses. This condition is not particularly noticeable in the streams above Cannelton.

General Drainage Relations

Approximately five-sixths of Indiana is covered with glacial drift of greater or less depth. As noted above all of the drainage area belonging to the St. Lawrence basin in Indiana is almost wholly consequent upon glacial material. Likewise the areas drained by the Tippecanoe and Eel Rivers of the Wabash system are largely as they were left by the latest

[6] Leverett, Water Resources of Indiana and Ohio: U. S. Geol. Surv., Eighteenth Ann. Report, Part IV. 1898, p. 441.

glacial invasion. Many lakes and marshes prevail, and drainage is very imperfect. The headwaters of these streams north of the Wabash channel are the areas of greatest glacial deposition and are practically unmodified by running water. These streams at no place in their headwater regions come in contact with bed-rock. A similar condition prevails in the Kankakee basin. The Wabash channel itself is entrenched for the most part in glacial drift. Everywhere, even to the brink of the valley bluffs, the monotony of a constructed glacial plain prevails. The main Wabash valley from the vicinity of Huntington is largely the result of the drainage which existed during the run-off of glacial waters formed in the retreat of the continental glacial ice. This valley in numerous places comes in contact with the bed-rock. The valley, however, is narrow and the stream relatively rapid. The valley is almost wholly the result of stream action, with little evidence of weathering and beveling which are so characteristic of normally developed streams.

All drainage south of the Wabash in the northern two-thirds of the state, or more definitely the drainage developed on the area of Wisconsin drift south of the Wabash River (Plate III), is much better developed than the area north of the Wabash. No lakes exist. The streams below their headwaters are frequently entrenched rather deeply in the glacial drift and are rather swift. Locally bed-rock is exposed, though this condition is some what exceptional. Throughout the headwater portions of the tributaries of the Wabash from the south and the headwater portions of both forks of White River and their tributaries in their upper portions, there exists a great glacial plain of construction with low relief and sluggish drainage. This great central plain is excellently shown on the topographic map, Pl. I.

Wabash River below Lafayette or southward from the great bend follows an old pre-glacial valley. This valley must have been a large trunk stream before its partial destruction by the several glacial invasions from the north. Above the great south bend of the Wabash it was wholly obliterated. The Wabash drainage has partially cleared out the glacial drift in its southward course. The valley is wide and has notable flood plain extensions, yet the pre-glacial bed-rock floor of the old valley is deeply buried beneath glacial material. Great terraces of valley-train material flank the present valley-flat as far south as southern Vigo County. Below Terre Haute the Wabash channel meanders over a wide flood plain which is high above the old bed-rock floor. All the streams in the southwestern part of the state are mere channels in broad aggraded valleys which are extraordinarily extensive for the size of the streams which meander over them. Since the two forks of White River and certain of their tributaries were the outlet channels of glacial waters and their charge of diminuted rock material, their valleys are aggraded more deeply than any of the streams which were not so affected, and their gradients are accordingly steeper. All streams which enter these valleys are also aggraded accordantly, but usually have a relatively lower gradient in their aggraded condition.

The Ohio River in the upper two-thirds of its passage along the southern border of the state nowhere flows in a deeply filled valley. It is apparently filled no deeper than the stream is itself capable of scouring in times of high water. It is gorge-like, and its flood-plain is very narrow considering the size of the stream. Only very locally are its bluffs beveled in accordance with what should be expected of a stream of its size. High bluffs and steep, rocky slopes are almost everywhere in sight from the channel. Yet in only one place does the entire stream flow over uncovered bed-rock, and this occurs at the Falls of the Ohio over a stretch of a little more than 2 miles between Jeffersonville and New Albany. It is apparent here that a deeper buried channel is present at one side of the bed-rock course. The tributaries of the Ohio River above Cannelton are deeply trenched and flow over rocky floors nearly to the river itself. Blue River, the largest of the minor tributaries, is typical in this respect. It is flowing over bed-rock in a narrow, deeply entrenched valley within 5 miles of the Ohio and has a much higher gradient in its lower portion than is normally to be expected. The lower one-third of the Ohio River in its passage along the southern border is quite in contrast with the upper part. Below Cannelton the stream obviously flows over a filled valley with a bed-rock floor which is never scoured by stream action. The tributary streams which enter the river are aggraded accordingly, and flow in wide valleys as tiny channels and everywhere possess mud banks. The Ohio valley is miles in width and the slopes from the upland are gentle and deeply mantled with weathered rock material. This filled-valley condition of southwestern Indiana is the chief physiographic characteristic, and with its gentle slopes and wide, flat valleys and sluggish streams is in great contrast to the sharply cut condition and swift streams of the area in middle southern and southeastern Indiana.

The drainage of Indiana through the Wabash and the Ohio Rivers passes from the extreme southwestern corner of the state through the Ohio River into the Mississippi River, and thence into the Gulf of Mexico. The water in the lower part of the trunk streams flows rather slowly, as the streams are fairly well graded and are not being made deeper. The surface of the land is being lowered toward their level rather than the bed of the streams being cut farther below the upland surface. The water on leaving the area of Indiana must pass over a route, 1,170 miles in length before reaching the sea, and in that distance descends 313 feet or slightly more (low water at the mouth of the Wabash being 313 feet above sea level.) The drainage as a whole is in a southwesterly direction, and is in keeping with the position of the highest land in the state in the middle of the eastern edge, centering in Randolph County, and the lowest land of the state in the southwestern corner, where a great tract of more than 2,000 square miles are at or below the 500-foot contour line. This drainage is not everywhere in response to the present topographic condition of the surface, but the drainage has developed most of the surface features as they exist in the southern third of the state. A discussion of the topographic condition of Indiana follows.

CHAPTER III.

TOPOGRAPHIC CONDITION AND PHYSIOGRAPHIC DIVISIONS OF INDIANA.

Definition and Method of Presentation.

The topographic condition of an area is its state or condition with respect to relief, altitude, and the form, size and relationship of the physical features present. The simplest topographic condition is that of a level plain. The coastal plain is an example. The topographic condition of a region becomes more complex as relief or difference in elevation from place to place is produced by the physiographic processes, chief of which is the action of running water. Essentially, the topographic condition is expressed generally by stating the stage of topographic development in terms of the erosion cycle. But to say that a region is in topographic youth, maturity, or old age is usually insufficient, since most regions with any but the softest rocks have more than one cycle represented. If the region is limited in area or forms a very small natural unit, its topographic condition may be signified by a statement of its particular stage in the erosion cycle. But the terms youth, maturity and old age are only very general terms and must be supplemented when applied to any region of considerable areal extent. The detail of topographic condition is adequately depicted only when altitude, relief, and form, size and relationship of the physical features are expressed either in figures or comparisons. But such expression is burdensome and the effort is largely unappreciated. The only adequate way in which the full presentation of topographic condition may be placed before the student is by means of the detailed topographic map.

In the absence of the topographic map, resort is often had to the use of physiographic terms. In addition to descriptive detail, such as altitudes, reliefs, sizes of streams, proximity to major streams, presence and nature of divides, size, shape and relationship of topographic forms, etc., such terms as base-level, grade, peneplain, aggraded valleys, rejuvenation, interrupted cycle of erosion, etc., are used, and the whole is frequently illumined to a great degree by a discussion of the lithology and regional and local rock structure in their relation to the denudational agencies actively producing the topographic forms present. But such a treatment is explanatory as well as descriptive, and may be called a physiographic rather than a topographic treatment.

In the presentation of the topographic condition of Indiana, a plan is followed which involves both topographic and physiographic treatment. A detailed topographic map is not at present available, but a general one with a contour interval of 100 feet accompanies this report. Such a large vertical interval results in much generalization of the topographic forms. This map, however, presents the regional topographic condition of Indiana with a much greater degree of detail than any map which has heretofore been prepared.[1]

[1] The topographic map (Pl. I) is a product of all available data that the writer has been able to procure. The following topographic sheets of the U. S. Geol. Surv. furnished the basis for approximately 3,800 square miles within the State: Tell City, Owensboro, Newburg, Henderson, Mt. Vernon, New Harmony, Mt. Carmel, Princeton, Birds, Vincennes, Petersburg, Velpen, Degonia Springs, Haubstadt, Boonville, St. Meinrad, Danville, Calumet, Tolleston, Clay City, Bloomington, Winchester and Kosmosdale. The topographic sheets of the western Ohio state boundary line were the basis for the elevations along the eastern boundary line of the State. Approximately 3,300 listed elevations of the railway surveys as published by Barrett in the 36th Ann. Report of the State Geol. Surv. have been made use of. A number of these railway surveys have elevations marked not only at the stations but at the mile posts also. Thus the Monon Railway, extending from Chicago to Louisville, from Monon to Indianapolis, from Bedford to Switz City, from Wallace Junction to Shirley Hill and from Orleans to French Lick, lists over 500 mile post elevations along the routes. Various annual reports of the State Geol. Surv. have scattered topographic data in them and these have been made use of. The Coal Report of Ashley, 1898, in particular contains much topographic data for the southwestern part of the State. The writer has made many topographic sketches within the hilly, southern part of the State, and has during the last few years made thousands of barometric determinations. These contribute largely to the topographic detail of the rougher portion of Indiana. Where precise elevations are scanty, the geologic horizons as shown on the areal geologic map and the knowledge of stream gradients have been used in giving close approximations to altitudes.

It shows at once the areas of small relief contrasted with those of greater relief. Regional slopes, escarpments, trenched valleys, nature and character of both major and minor divides, and the altitudes and reliefs are depicted within a relatively narrow margin of error. Its inclusion here is expected to largely take the place of what would otherwise be a rather extended description of topographic form and condition, such as would probably prove burdensome to the student and much less effective as a presentation of the topographic condition of the state. The map will be supplemented by a brief presentation of the physical features present in each of the physiographic divisions, which in themselves present a unity of topographic form and condition. Since topographic condition is the chief basis of the physiographic divisions as here given, special attention is given to its presentation, and recourse to physiographic terms is frequent, though description itself is chiefly employed.

General Topographic Condition of Indiana.

Much of Indiana belongs to the glacial plains region of the Upper Mississippi basin, and the surface forms over large areas are those produced by the constructional phases of glaciation. Approximately 30,100 s uare miles of Indiana have been subjected to glaciation, while only about 6,250 square miles have not been touched by glacial ice. The unglaciated area in the middle-western portion of the southern third of the state is frequently spoken of as the driftless area as opposed to the glaciated area. Such a division of the state is often regarded as the most natural, and in the treatment which follows this distinction will frequently be made, though the limit of glacial drift is not used here as a basis of division into regional units.

The area north of the Wabash River still largely retains the original forms following continental glaciation in the zone of ice wastage. Little modification by running water has taken place, and such as has taken place has resulted in reduction of relief. Where obstructed drainage following glaciation gave rise to lakes, great tracts now exist as flat, monotonous lacustrine plains, such as those adjacent the Kankakee River and the middle Tippecanoe basin. Local areas of silted-up basins occur commonly along stream courses. Where the moraines were piled high near the headwaters of the streams in the northeastern part of the state, many lakes still exist, and the topography is often rugged, consisting of variable forms of glacial construction which are now being destroyed and levelled through drainage development.

The middle portion of Indiana is largely a great till plain which has been modified to only a slight degree by stream action. No lakes exist, though marshes and swamps were formerly common throughout most of the territory and have since been drained artificially. This great till plain of central Indiana is remarkably level, and is representative of a great ground moraine with only slight development of terminal moraine topography locally. Local relief is very slight even where the best developed terminal moraines are present. Farther south the glacial plain is older and the streams larger. Considerable dissection of the glacial plain has taken place along the Wabash River in the area between Lafayette and Terre Haute.

The southern third of the state as a whole is rugged, though notable exceptions occur. The acme of relief diversity is reached in the unglaciated or driftless portion of southern Indiana. The region bordering the Ohio River, though the portion above Jeffersonville was once subjected to glaciation, represents the severest relief of the entire state. The severity, however, dies out along the lower third of the Ohio, or from near the mouth of Anderson Creek, below Cannelton to the mouth of the Wabash River. All of the driftless area north of the Ohio River between New Albany and Troy at the mouth of Anderson Creek is very rugged. The relief diversity does not die out at the glacial drift margin, but continues northward in Morgan and Owen counties in particular. The much dissected and hilly upland portion of the state may be largely ascribed to the resistant rocks present and to the relative youth of the Ohio River as a trunk stream.

Altitude and Relief.

The altitudes within the state fall between the maximum of 1,285 feet in southern Randolph county in the middle-eastern part of the state, adjacent to the State of Ohio, and the minimum of 313 feet at the mouth of the Wabash in the extreme southwestern part of the state. Approximately 1,600 square miles in area have an altitude of 1,000 feet or more. The largest area comprising about 1,400 square miles centers in southern Randolph County and includes portions of Randolph, Jay, Henry, Hancock, Wayne, Union, Fayette, Rush, Franklin, Decatur, Ripley and Dearborn Counties. A second area of about 190 square miles comprises portions of Steuben, Lagrange, Noble and Dekalb counties in the high morainal northeastern portion of the state. Small areas and isolated hills reach 1,000 feet or slightly above in Floyd, Clark, Washington, Scott, Bartholomew, Brown and Monroe Counties within the driftless area of the state, and morainal material in Hendricks and Elhart counties also reach slightly above the 1,000-foot contour. The aggregate area of these smaller heights of 1,000 feet or more is approximately 10 square miles. A maximum elevation of 1,000 feet or more is attained in 26 counties, as above named. Approximately 2,250 square miles in area in Indiana lie below the 500-foot contour line. Nearly all of this area lies in the southwestern part of the state, though the 500-foot contour line passes beyond the state into Ohio up the Ohio River, and up the Wabash valley into Tippecanoe County. Thirty-three of the ninety-two counties are touched by the 500-foot contour line. Just as no single county has all of its area above the 1,000-foot contour, so no county is entirely below the 500-foot contour. Thus, approximately 4.4% of the area of Indiana is 1,000 feet or more in elevation, and approximately 6.2% is below 500 feet in elevation. An area of 32,500 square miles or 89.4% of the state lies between the 500-foot and the 1,000-foot contour lines. The average altitude of the state as based on the estimates of the average altitudes of the counties as shown in the accompanying table is 760 feet above sea level.

The accompanying table of elevations and reliefs reveals many interesting features relative to the topographic condition of Indiana. As noted above, twenty-six counties have a maximum elevation of 1,000 feet or more. Eleven counties, in the southwestern part of the state, have maximum elevation below 700 feet. Only one county, Posey, has a maximum elevation below 600 feet. Randolph and Steuben counties have minimum elevations above 900 feet. Thirty-three counties have minimum altitudes below 500 feet, and eleven counties have a minimum of 400 feet or less. Harrison, Floyd, Clark, Washington, Monroe and Brown counties of the driftless area, and Dearborn and Jefferson along the Ohio River and Franklin on the deeply dissected portion of the Whitewater basin, have maximum reliefs greater than 500 feet, and three of these, Clark, Floyd and Harrison, have immediate reliefs exceeding 500 feet.

TABLE OF ALTITUDES AND RELIEFS BY COUNTIES IN INDIANA.*

County	Maximum Altitude	Minimum Altitude	Average Altitude	Maximum Relief	Maximum Local Relief
Adams	895	760	840	135	60
Allen	925	705	780	220	100
Bartholomew	1,003	555	685	448	300
Benton	857	630	750	227	70
Blackford	940	825	875	115	60
Boone	950	775	850	175	80
Brown	1,050	515	810	535	400
Carroll	790	520	680	270	100
Cass	825	550	725	275	120
Clark	1,020	375	650	645	500
Clay	740	505	625	235	160
Clinton	975	775	910	200	100
Crawford	905	355	680	550	360
Daviess	690	400	540	290	175
Dearborn	1,010	425	775	585	400
Decatur	1,030	690	900	340	140
Dekalb	1,025	775	890	250	100
Delaware	1,075	865	945	210	80
Dubois	775	415	600	360	260
Elkhart	1,025	710	815	315	225
Fayette	1,110	695	910	415	250
Floyd	1,020	370	800	650	560
Fountain	780	470	650	310	150
Franklin	1,040	525	735	515	300
Fulton	900	715	760	185	80
Gibson	650	355	470	295	250
Grant	930	720	870	210	65
Greene	925	470	630	455	300
Hamilton	950	725	870	225	85
Hancock	1,025	790	890	235	70
Harrison	980	360	770	620	610
Hendricks	1,010	685	885	325	120
Henry	1,150	880	1,040	270	100
Howard	875	700	830	175	55
Huntington	860	680	775	180	90
Jackson	940	495	675	445	360
Jasper	770	640	670	130	55
Jay	1,120	840	945	280	100
Jefferson	925	410	740	515	460
Jennings	885	535	710	350	165
Johnson	930	610	780	320	225
Knox	610	376	480	234	90
Kosciusko	950	770	855	180	90
Lagrange	1,027	830	940	197	100
Lake	750	581	630	169	100
Laporte	885	581	730	304	150
Lawrence	900	470	700	430	350
Madison	975	780	855	195	70
Marion	850	625	775	225	125
Marshall	895	705	810	190	100
Martin	860	425	655	435	300
Miami	870	610	760	260	100
Monroe	1,006	490	760	516	400
Montgomery	900	550	750	350	180
Morgan	915	560	730	355	275
Newton	770	620	655	150	60
Noble	1,025	845	950	180	[illegible]
Ohio	875	425	740	450	400
Orange	925	455	700	470	[illegible]
Owen	950	500	703	[illegible]	[illegible]

*Figures are actual or estimated

TABLE OF ALTITUDES AND RELIEFS BY COUNTIES IN INDIANA.
(Continued.)

County	Maximum Altitude	Minimum Altitude	Average Altitude	Maximum Relief	Maximum Local Relief
Parke	800	455	650	345	200
Perry	825	348	640	477	325
Pike	660	395	510	265	180
Porter	830	581	680	249	190
Posey	585	313	420	272	160
Pulaski	750	660	700	90	45
Putnam	925	580	770	345	180
Randolph	1,285	915	1,110	370	140
Ripley	1,030	550	885	480	300
Rush	1,116	825	975	291	75
Scott	1,015	520	625	495	400
Shelby	920	650	730	270	75
Spencer	660	338	480	322	220
Starke	760	660	710	100	60
Steuben	1,200	920	990	280	230
St. Joseph	875	654	775	221	100
Sullivan	610	415	520	195	170
Switzerland	850	415	740	435	405
Tippecanoe	840	495	680	345	185
Tipton	930	815	830	115	40
Union	1,130	680	960	450	250
Vanderburgh	600	321	430	279	140
Vermillion	662	460	585	202	140
Vigo	673	440	550	233	100
Wabash	930	685	775	245	120
Warren	750	485	670	245	150
Warrick	660	330	450	330	260
Washington	1,015	495	800	520	400
Wayne	1,250	820	1,035	430	150
Wells	900	750	830	150	75
White	760	540	670	220	110
Whitley	960	775	855	185	80

The greatest immediate relief in the state is found in Harrison County on the Ohio River bluff, just south of the Floyd and Harrison county line, where a hill (Locust Point) near brow of the bluff, reaches an altitude of 980 feet, and scarcely one-half mile distant from the crest of the hill low water of the Ohio River is 370 feet above sea level, giving an immediate relief of 610 feet. Nine counties have maximum local reliefs of 400 feet or more; twenty-one counties have 300 feet or more, and thirty-three counties have 200 feet or more. Tipton County has the least local relief in the state, though Newton, Jasper and Pulaski counties with their great extent of lacustrine plain rival Tipton County in flatness, and Pulaski County exceeds it in levelness, having a maximum relief of only 90 feet. Twenty-two counties in all have an immediate relief of 90 feet or less. Only three counties, Randolph, with an average altitude of 1,110 feet, Henry with 1,040 feet and Wayne, with 1,035 feet, have an average altitude of 1,000 feet or more. Six counties consisting of Knox, Gibson, Posey, Vanderburgh, Warrick and Spencer, all bordering the Wabash or Ohio rivers in the southwestern part of the state, have an average altitude less than 500 feet. As stated above, the average altitude of the entire state is estimated to be 760 feet above sea level.[a]

[a] Gorby estimated the average altitude of Indiana to be 800 feet (Sixteenth Ann. Rept., Ind. Geol. Surv. for 1891, p. 217) and Gannett made an estimate of 700 feet (Thirteenth Ann. Report, U. S. Geol. Surv., 1893, p. 289).

Regional Units of Indiana and Their Topographic Condition.

Introductory

Dryer's physiographic divisions. Dryer[9] divides Indiana into three physiographic regions: 1. *Southern Indiana.* 2. *The Central Drift Plain.* 3. *The Northern Moraine and Lake Region.* Southern Indiana is further divided from east to west into "a succession of lowlands and uplands bounded by relatively steep slopes or escarpments formed by the outcropping edges of the harder strata." These divisions of Dryer's from east to west are: *The Eastern Lowland,* divided from the *Middle Lowland* by the *Limestone escarpment* which is described as extending from near Madison northward about 40 miles. The *Middle Upland* occupies the middle of Southern Indiana and is divided from the *Middle Lowland* by the *Knobstone escarpment* which passes northward from the Ohio River near New Albany to Johnson County. The *Western Lowland* is the westernmost of the divisions in Southern Indiana and is separated from the *Middle Upland* "by a descent almost as abrupt [as the Knobstone escarpment] but about half as high, formed by a bed of sandstone." Southern Indiana is limited to the area south of the Wisconsin glacial drift as represented by the rather low Shelbyville moraine. The *Central Drift Plain* is not divided. Its southern boundary is given as the Shelbyville moraine, and its northern boundary is designated as "the valleys of the St. Joseph and Wabash Rivers from northern Allen County to southern Warren." The *Northern Moraine and Lake Region* constitutes the remainder of the state and is divided into a number of divisions on the physiographic map, but no distinct limits are given to the minor divisions.

Basis of present physiographic divisions. Dryer's classification of the natural or physiographic units of Indiana is not used in this treatise, though a number of the units are approximately equivalent. His *Eastern lowland* is essentially an upland, though deeply dissected by the Whitewater and Ohio Rivers and tributaries. Part of his *Middle lowland* is a rather high upland area. No sandstone escarpment marks the western boundary of his *Middle upland.* The massive Mansfield sandstone is present there, but it gradually dips beneath the softer Pennsylvanian shales which overlie it. It may also be mentioned that the division line drawn across southern Indiana at the glacial drift margin by Fenneman in his "Physiographic Divisions of the United States"[10] is not used, since the characteristics of the Low Interior Plateaus, or more properly the Highland Rim Province, are still retained a considerable distance north of the glacial boundary line. The Indiana portion of the Illinoian drift-covered section only slightly modifies this northern portion of the Highland Rim Province, as most of the characteristic topographic forms and reliefs prevail as far north as the Wisconsin drift margin, where the glacial drift largely obscures the underlying bed-rock features. From the larger point of view it is probable that Fenneman's division line drawn at the oldest drift margin is the most natural position, but when smaller areas and divisions are considered the thin mantle of older drift, often largely removed by erosion, does not stand out topographically from the wholly driftless area adjacent to it. The scheme followed here is to allow the several separate divisions of the southern portion of Indiana to extend as far north as their topographic forms and the relief condition

[9] Geography of Indiana: Supplement to High School Geography, Amer. Book Co., 1912.
[10] Fenneman, Annals of Assoc. of Amer. Geogr., Vol. VI, 1917, pp. 19-98, pl. I.

are characteristic regardless of glaciation. These several units, however, may be said to possess a glaciated section as contrasted to their unglaciated section. The regional units as shown on the physiographic map accompanying this treatise, (Pl. II) are of equal rank, though the areas are far from equal, especially, when regarded as limited to Indiana.

Differences in the topographic condition of different parts of the southern older portion of Indiana (older both from the standpoint of geologic history and physiographic development) give rise to a number of regional units which are more or less sharply separated from each other. These regional units rather alternate from east to west as regions of great relief diversity and regions of much less relief. It may be said here that the unity of topographic forms characterizing each of the alternating units is even more striking than the unity of relief condition. The topographic condition of each regional unit is largely the result of the conditioning effect of the underlying rock in its control over the land features as sculptured by denudational agents, and as a result of such lithologic control the regional units correspond rather closely with the lithologic and geologic divisions.

The names of the physiographic divisions of Indiana as here presented are as follows: *Dearborn upland, Muscatatuck regional slope, Scottsburg lowland, Norman upland, Mitchell plain, Crawford upland, Wabash lowland, Tipton till plain, and the Northern Moraine and Lake region. The Northern Moraine and Lake region* is subdivided into five units which are given geographic names as follows: *Calumet lacustrine section, Valparaiso moraine section, Kankakee lacustrine section, Steuben morainal lake section and Maumee lacustrine section.*

The Dearborn Upland

Extent and boundaries. An area in southeastern Indiana extending southward from southern Wayne County to the Ohio River in a strip about 25 miles wide and 75 miles long is marked by considerable relief diversity. The area is an upland, the undissected surface and surface remnants of which vary from about 800 feet in elevation to about 1,100 feet. Streams have trenched this upland from 200 to 500 feet below the general upland surface, though much of the upland surface is still preserved. This regional strip in Indiana comprising about 1,925 square miles is a portion of a larger area which extends eastward into Ohio chiefly south of the Wisconsin drift limit, and also south into Kentucky where it makes up a portion of the Blue Grass Region. To this upland area the name of *Dearborn upland* is given from its typical development in and throughout Dearborn County, Indiana. Only a part of the boundary line of the Dearborn upland lies in Indiana. Beginning on the bluffs of the Ohio River west of Indian-Kentuck Creek some 5 or 6 miles east of Madison, the western boundary in Indiana trends almost north, not considering irregularities. The line passes around the West Fork of Indian-Kentuck Creek just beyond the Jefferson County line and then returns to within a short distance of the junction of the two forks of Indian-Kentuck Creek, but in the angle between them, and thence northward around the headwaters of the east fork somewhat east of the village of Cross Plains in southeastern Ripley County. From here it passes northward along the crest of a scarp just west of Laughery Creek formed by the rather resistant Niagara limestone. This escarpment is not everywhere distinct; but it is distinct enough to trace mainly west of Indian-Kentuck Creek and just west of Laughery Creek, where it is well developed, and as far north as southeastern Decatur County. It is here called the *Laughery escarpment* after its typical development just west of Laughery Creek through middle Ripley County. The boundary from southeastern Decatur County may be defined as the divide between Whitewater and the east fork of White River drainage, as the Laughery escarpment is mostly obscured by glacial drift. The line passes from Decatur County into southeastern Rush, and thence slightly east of north into Fayette to the northwestern corner of the county. The northern edge of the Dearborn upland merges into the rather undissected glacial plain to the north in southern Wayne County. It is rather arbitrarily drawn from northwestern

Fig. 2. Long, even-topped spurs extend as promontories between the narrow valleys in the dissected portion of the Dearborn upland. Steep, uniform slopes abound, with rarely ever a bluff or cliff. View along highway between Dillsboro and Aurora, in ravine leading into South Hogan Creek.

Fayette County eastward, bending around Richmond to the north and passing into Ohio southeast of Richmond.

Topographic Condition. The Dearborn upland in Indiana is composed of the drainage basin of the Whitewater River and a number of smaller basins to the south which discharge directly into the deeply entrenched Ohio River, in Dearborn, Ohio, and eastern Jefferson counties. The maximum altitudes vary from 850 feet to slightly over 1,100 feet, while the Ohio River has an elevation of about 425 feet. The Ohio River is the base toward which the streams draining the Dearborn upland are cutting. Since most of the streams are short, the fall from the upland headwaters is relatively great. They have trenched the upland plain deeply, and the local relief is usually measured in hundreds of feet. Along the main streams the slopes are most frequently quite steep, though bluffs are rare. Steep slopes are the rule, and long, evenly elevated tongues of upland lie between the deeply trenched valleys. Valley lands are frequently well developed, but never of great width. Terraces of glacial outwash flank the present flood plain of the Whitewater River. Areas of upland near the headwaters of the streams are notably flat and often quite extensive. Most of the divides between the streams rise to approximately the same elevation, so that a view from one of them presents an even sky-line of upland area. A rather extensive upland tract of marked flatness occurs in western Dearborn and eastern Ripley counties east of the headwater portion of Laughery Creek. This tract of upland rises slightly above the 1,000-foot contour, and is excellently shown on the topographic map. It is a portion of the upland plain which has not yet been severely acted upon by streams. The Dearborn upland at its northern edge merges into the glacial plain, but in southern Wayne County stream dissection becomes rather severe and tongues of upland begin to show between the entrenched valleys. The region as a whole presents a beautiful panorama of elevated, prolonged divides flanked by smooth, steep slopes and ending in rounded spurs overlooking relatively narrow valley lands often hundreds of feet below. (See Fig. 2.) Distant from the main valleys the elevated upland tongues merge into rather flat upland tracts of considerable extent.

The Muscatatuck Regional Slope.

Extent and boundaries. Just west of the Dearborn upland lies the most notable regional slope in the entire state. This great regional slope begins at the crest of the divide between the Whitewater, Laughery, and Indian-Kentuck basins and east fork of White River and Muscatatuck basins. This line was traced above as the western boundary of the Dearborn upland. It slopes west and south for twenty-five or thirty miles and merges into a broad lowland which lies east of a great escarpment. It embraces eastern Clark, a small portion of southeastern Scott, middle and western Jefferson, western Ripley, all but southwestern Jennings, nearly all of Decatur, northwestern Bartholomew, and the southern two-thirds of both Shelby and Rush counties, comprising about 1,875 square miles. Across the Ohio River in Kentucky it is well represented, and is usually considered a portion of the Blue Grass Region. It is here named the *Muscatatuck regional slope*, from its typical development as a well defined slope along the numerous tributaries of Muscatatuck River in Ripley, Jennings, and Jefferson counties. As a regional slope it merges rather indistinctly into the lowland area on the west and the glacial plain on the north. From the Ohio River north its western boundary may be given as the divide between Fourteen Mile and Silver creeks in Clark County, thence north to Lexington, where it bends to the east, but returns to its northward alignment again at the westernmost boundary line of Jefferson County. It again bends eastward to near Deputy and Paris Crossing, and then turns to the northwest and north past Weston to Hayden. Here it turns southwest nearly to the Jennings county-line, which it follows to the northern boundary, where near Elizabethtown it becomes identified with the Shelbyville moraine which it follows to near Clifford Station in northeastern Bartholomew County. From Clifford Station it continues north and slightly west to near the middle of the eastern line of Johnson County where the regional slope is practically lost in the glacial plain. The northern indistinct boundary is drawn broadly from the middle of the east line of Johnson County northeast to the northwestern corner of Fayette County where it joins the northern end of the eastern boundary, as described above."

Fig. 3. View of the Muscatatuck regional slope at North Madison. Here the plain is about 880 feet in altitude to the very bluffs of the deeply intrenched Ohio River. View shows the levelness of an interstream space. This plain is structural, having been developed on the Silurian-Devonian limestone unit. No beveling towards the Ohio River occurs in the vicinity of Madison. View taken looking toward Ohio River valley, which is distant about one mile.

Topographic Condition. The eastern edge of the Muscatatuck regional slope near the Ohio River has an elevation of 875 feet, but gradually rises to the northward where its elevation is about 1,100 feet. The western edge of the regional slope is about 500 feet near the Ohio River and about 725 feet at the north. At the north its surface is composed of the Wisconsin glacial drift which is rather thin. It merges broadly into the glacial plain to the north. South of the Wisconsin glacial drift it is much more dissected, and across the Ohio River in Kentucky it is very highly dissected by stream action. As a regional slope it descends to the southwest at the rate of about 15 feet to the mile, and the streams heading at the crest of the Laughery escarpment follow the slope rather precisely in many parallel rather shallow grooves with rather small tributaries which enter the main stream in accordance with the regional slope. In Jennings, Jefferson and Clark counties the streams are canyon-like in character, but are never entrenched deeper than 175 feet below the gently inclined plain. Where the greatest entrenchment occurs, the tributary streams have dissected the upland. Such dissection is typically and best illustrated along the Muscatatuck in central Jennings County, near Vernon. Between the streams the slope is typically quite flat, and the gentle southwestward slant is not perceptible to the eye. The thin veneer of Illinois glacial

[11] The *Muscatatuck regional slope* is approximately equivalent to the "*Eastern Plateau*" of Newsom. See A Geological Section Across Southern Indiana, from Hanover to Vincennes: Jour Geol., Vol. 6, 1898, pp. 250-256. Newsom only by implication includes the area here designated the *Dearborn upland* in his "*Eastern Plateau*" area.

drift obscures the bed-rock in the interstream areas, but its depth is never great. The streams passing down the slope are almost everywhere set in limestone and have precipitous biuffs and very narrow flood plains in the middle portion, but in the eastern part they flow over the surface with little entrenchment and with a fall very little greater than the westward descent of the regional slope itself. The Ohio River crosses the regional slope as a master stream, and at the eastern edge the exceedingly narrow valley is bordered by steep bluffs 400 to 450 feet high.[12] The height of the bluffs diminishes to the west in accordance with the descent of the slope itself, and at the west are less than 100 feet in height. The anomalous feature of the Ohio trough across the slope is the entire lack of beveling of the upland in the vicinity of the trough. The upland is just as highly elevated at the very edge of the bluff as it is miles back from it. The trench through the Dearborn upland to the east is a rather striking gorge, but it is not wholly without beveling. This feature is one of the strongest evidences of the youth of the Ohio River as a trunk stream.

The Scottsburg Lowland.

Extent and boundaries. The western edge of the Muscatatuck regional slope merges rather imperceptibly into a lowland of slight relief characterized by great expanses of valley land along the streams and a notable lack of bluffs or steep slopes. This lowland lies between the Muscatatuck regional slope on the east and a well defined escarpment on the west, and extends from southeastern Johnson County southward into Kentucky. The name of *Scottsburg lowland* is here proposed for this regional lowland. The name is taken from Scottsburg in Scott County, Indiana, where the lowland is excellently developed. It comprises an area of about 950 square miles in Indiana. Its western line lies at the foot of the *Knobstone escarpment*, the largest and most prominent topographic feature in Indiana, and is therefore well defined. Beginning at the Ohio River southwest of New Albany the line passes northward and slightly eastward to southern Scott County just west of the village of Underwood. This line is direct with the exception of a few minor valley irregularities and a considerable westward indenture where Muddy Fork of Silver Creek comes in from the adjacent upland area. From Underwood the line passes northwest along the foot of the ragged and dissected escarpment adjacent to the flood plain of the Muscatatuck River. At the junction of the Muscatatuck and White rivers the line is drawn across the valley to the foot of the 350-foot escarpment southwest of Medora; it passes just west of Medora, and to the northwest into southwestern Bartholomew County where it curves back to a northwest direction. The escarpment becomes obscured by glacial drift in southern Johnson County and the Scottsburg lowland itself merges with the glacial plain to the north in the vicinity of Franklin. The western line as described is also the boundary line between the Illinois drift covered area and the driftless area, except for the region south of the entrance of Muddy Fork of Silver Creek in Clark County and a small portion of the line north of northwestern Bartholomew County.[13]

[12] See "The Geology of Jefferson County" by Glenn Culbertson, 40th Ann. Rept., Ind. Dept. of Geol. and Nat. Res. for 1915.

[13] The Scottsburg lowland is described by Newsom as the "Eastern lowland." See Newsom, Jour. Geol., Vol. 6, 1898, p. 252.

Topographic Conditions. The Scottsburg lowland is throughout most of its extent a broad shallow concavity broken by two or three outlets to the west where the main drainage lines pass out of the region. The concavity first becomes apparent in southern Johnson County along Youngs Creek, and extends slightly east of south along the line of Blue River and White River through Bartholomew County where it becomes a prominent lowland hundreds of feet below the upland on the west. White River turns to the southwest at the southern line of Bartholomew County and the concavity follows the broad valley in conjunction with that of White Creek. South of Vallonia in Jackson County the Muscatatuck prong of the lowland merges with the White River portion, and the drainage discharges out of the lowland through the gorge-like valley of White River to the west. But the main concavity continues in a southeasterly direction from northeastern Jackson County, past Seymour and Crothersville to Scottsburg, crossing obliquely the several forks of the Musca-

Fig. 4. The Brownstown hills as viewed from near Brownstown, Jackson County. These hills represent an upland mass isolated from the Norman upland by stream erosion. They overlook the broad valley of White River on the north, while to the south lies the valley of the Muscatatuck. During the Illinois glacial stage the upper parts of these hills stood above the continental ice-sheet as a nunatak.

tatuck to Silver Creek. Silver Creek forms the axis of the lowland to the Ohio River. Along the axis of the lowland from southern Johnson County to the Ohio River, the elevation is everywhere below 700 feet, and from southern Bartholomew County to the Ohio River it scarcely exceeds 600 feet. At the divide north of Silver Creek near the north line of Clark County, the elevation reaches 625 feet, though here considerable roughness exists, and much greater elevations occur.

In the angle between Muscatatuck and White rivers northeast of their junction and west of the main axis of the lowland, an upland area of 25 or 30 square miles rises to a maximum height of 300 feet above the lowland. This upland mass is greatly dissected and is known as the "Brownstown Hills." On the southwest and south the slopes of this upland mass are quite steep, and from all sides it is in great contrast to the low relief of the lowland. On the north and west it is flanked by sand dunes with a hammocky topography of the knob and basin type, but with small local relief. The dunes do not exceed a height of more than 50 feet above the general lowland. Morainal material flanks the northeastern and eastern sides of the isolated upland which itself is composed of bed-rock, and is essentially an outlier detached by stream action from the main highland mass to the west. This same stream action is responsible for great V-shaped extension of the Scottsburg lowland on the west to the gorge-like opening in the highland area through which White River passes (see Pl. II). Beginning some three miles south of Seymour a prominent morainal ridge, one-half to one mile wide and rising from 50 to 170 feet above the lowlands adjacent, extends 8 miles in a southerly direction. This morainal mass is known as "Chestnut Ridge."[14]

The valleys over the Scottsburg lowland are exceedingly broad and flat, and the upland slopes adjacent the valleys are gentle and usually without cliffs. Local relief in the low upland areas between the streams is never great, seldom reaching 75 feet. Where the Ohio River crosses the Scottsburg lowland, the valley is open and several miles in width, and is quite in contrast to the narrow valley and high rocky bluffs either up or down the river from the lowland. Yet it is in this broad portion of the river valley across the lowland that the only stretch of bed-rock is to be seen entirely across the channel, and over which occurs the "Falls of the Ohio" where a descent of about 23 feet is made in a little over 2 miles.

The Norman Upland.

The Knobstone escarpment. Immediately west of the Scottsburg lowland a bold escarpment towers hundreds of feet above the lowland. This escarpment is known as the *Knobstone escarpment*, and extends from southern Johnson County southward across the Ohio River into Kentucky. It is called the Knobstone escarpment from the old formation name given to the rocks in which it is formed. The name has been long used for the escarpment and will likely stand. The Knobstone escarpment in Indiana extends from the Ohio River in southeastern Harrison County chiefly northward for about 150 miles including the western notch through which flows White River. Beginning at the Ohio River opposite the mouth of Salt Creek in Kentucky, the escarpment extends parallel to the river to near New Albany, and then continues northward and slightly eastward to southern Scott County. There is only one prominent indenture in the wall-like escarpment in this stretch of about 45 miles, and this is where Muddy Fork of Silver Creek comes from the west in a broad valley. It is in this stretch that the escarpment reaches its greatest perfection, as farther north it is somewhat ragged in appearance, due to the many small streams descending from the upland back of the crest. It is barely broken in Harrison and Floyd counties, and its crest rises 400 to 600 feet above the valleys and the Scottsburg lowland to the east. The greatest local relief in Indiana is the descent from a prominence on the crest of the escarpment down into the Ohio River channel, where in a horizontal distance of barely one-half mile in southeastern Harrison County a descent of 610 feet

[14] Leverett, Glacial Formations of Erie and Ohio Basins: U. S. Geol. Surv., Monogr. 41, 1901-2, pp. 255-256.

occurs.[13] Prominences on the crest in Floyd, Clark, Washington, and Scott counties reach an altitude of over 1,000 feet. From southwestern Scott County northwest and west to the notch through which White River passes out of the Scottsburg lowland, the Knobstone escarpment is much dissected, and broad, flat valleys open up between the prominent spurs which represent the escarpment. Northward from the White River opening the crest of the escarpment is not so high above the lowland to the east. Just west of Medora it is 360 feet above, and west of Columbus it is 250 to 350 feet above the lowland. Taylor Hill, 6 miles west and 2 miles south of Columbus is 1,003 feet in elevation and is 375 above the level of the railway station at Columbus. In southern Johnson County the Wisconsin glacial drift obscures the escarpment, largely on account of the till having deeply filled the adjacent lowland.

Fig. 5. Near view of a spur of the Knobstone escarpment, about five miles north of New Albany. The escarpment here rises approximately 500 feet above the Scottsburg lowland at its foot.

The Knobstone escarpment is the greatest relief feature and the most prominent topographic form in Indiana.[14] Everywhere it is a steep slope, but rarely if ever an unscalable bluff. As seen at a distance from the Scottsburg lowland at the south it appears wall-like and unbroken, and farther north where its face is much more broken by stream dissection, the prominent spurs appear as great knobs, and the outliers appear as conical hills and are spoken of as "The Knobs." The "Brownstown Hills" in the angle formed by White and Muscatatuck rivers is merely a more or less dissected outlier of the upland area to the west, and is a veritable monadnock-mass rising high above the

[13] See Kosmosdale Quadrangle (Kentucky-Indiana), U. S. Geol. Surv., 1912. The locality is near the small village of Locust Point.

[14] Bennett, Notes on the Eastern Escarpment of the Knobstone Formation in Indiana: Proc. Ind. Acad. Sci. for 1898, pp. 283-287. Malott, Some Special Features of the Knobstone Cuesta Region of Southern Indiana: Proc. Ind. Acad. Sci. for 1919, pp. 361-383.

much reduced Scottsburg lowland in which it lies. The Knobstone escarpment is present in Kentucky, but it is highly dissected and many detached masses exist, due to the stream work of Salt Creek and its tributaries.

Extent and boundaries of the Norman Upland. The knobstone escarpment as described above is the eastern edge of a notable upland area comprising some 2,075 square miles, and which is here named the *Norman Upland* from its typical development in the vicinity of Norman in western Jackson County, Indiana. The Norman upland embraces most of Floyd, eastern and northern Washington, western Clark, western Jackson, eastern Lawrence, eastern Monroe, all of Brown, western Bartholomew, and southern Morgan counties, and in the very small portion of Johnson and northwestern Bartholomew counties has been glaciated, and possesses a thin covering of Illinois drift. The northern line is drawn at the Shelbyville moraine of the Wisconsin glacial drift, since north of this line the glacial material has deeply buried much of

Fig. 6. View on Weedpatch Hill, south of Nashville, Brown County. Weedpatch Hill is not a hill in the ordinary sense, but is a rather extensive plateau-like eminence, reaching an altitude of approximately 1,050 feet. It is representative of the Highland Rim peneplain in southern Indiana, making up one of the largest remnants of this old and nearly destroyed uplifted peneplain.

the bed-rock and the characteristics of the Norman upland are almost wholly obscured.

The western boundary line of the Norman upland is very difficult to describe because of the general merging of the upland with the general limestone plain on the west. The divides composed of limestone carry the characteristics of the adjacent plain far into the Norman upland area while the valleys possess the characteristics of the valleys of the Norman upland, though they are well within the general area of the adjacent limestone plain. The western limit of the Norman upland may be said to extend along the line where the characteristics pass from the deeply dissected clastic rocks to that of a rolling limestone plain which possesses sinkholes as a distinct feature.

The boundary line as shown on the physiographic map and as here described is based on these characteristics chiefly. The western boundary line in Harrison County is the crest of the Knobstone escarpment, and the Norman upland there is very narrow. The line continues north to the tunnel on the Southern Railway some 4 miles west of New Albany. The upland here widens to the westward along Indian Creek and its tributaries, the boundary line passing west to the vicinity of Georgetown, and thence northward across Indian Creek, and along the divide west of Richland Fork of Indian Creek to Greenville, thence to a point a short distance east of the junction of Floyd, Clark and Washington counties, where it turns northwest to Blue River below Pekin, and thence generally north with much irregularity to middle northern Washington County. It here turns to a westerly direction around the headwaters of Rush, Twin, and Clifty creeks, and along the bluffs of White River, passing near Rivervale and crossing White River at the mouth of Guthrie Creek. From the mouth of Guthrie Creek it passes in a bend to the east and north and back west to the north of Bedford, crossing Salt Creek near Oolitic; thence north on the west bluff of Salt Creek to Harrodsburg, and then up Little Clear Creek to near Smithville, where it turns to the northeast and north to Unionville northeast of Bloomington. From Unionville the boundary line passes west and northwest along the upland west of Beanblossom to White River at Gosport. From Gosport the line passes northeast to near the northeastern corner of Owen County, where it turns abruptly northwest into the southeastern corner of Putnam County. Here it strikes the Shelbyville moraine which well serves as the northern boundary of the Norman upland. The Shelbyville moraine from southeastern Putnam County passes east and slightly north to near Mooresville where it turns abruptly south and east to near the southeastern corner of Morgan County; thence east a short distance north of the Johnson County line to the northwestern corner of Bartholomew and to the west line of the Scottsburg lowland.

Topographic condition. The Knobstone escarpment has already been described, and the eastern margin of the Norman upland was there contrasted with the Scottsburg lowland to the east. Brown County, Indiana, is known throughout the state for its rugged picturesqueness, and is much visited by artists and others who appreciate the wild wholesomeness of its rugged hills. The Norman upland reaches its greatest width through southern Brown and adjacent parts of Monroe and Bartholomew counties, where it is approximately 40 miles across the rugged, much dissected area. In Brown County also it reaches its greatest elevation, a tract of several square miles attaining an altitude of over 1,000 feet and reaching a climax in Weedpatch Hill southeast of Nashville where an elevation of 1,050 feet[37] is attained. Bearwallow Hill north of Nashville attains a height of 1,020 feet, and affords one of the most picturesque panoramic views in the entire region.

Aside from great local relief in the Norman upland the chief characteristic which gives unity to the entire area is the uniformity of its topographic condition. Everywhere it is maturely dissected by stream action. The long

[37] Weedpatch Hill is a plateau-like expanse of upland, and is not an isolated hill in the ordinary sense. Careful aneroid measurements by the writer show that the elevation of the highest portion of Weedpatch Hill is approximately 1,050 feet above sea level, and not 1,147 feet as given in a number of reports of the Indiana State Survey. Collett (Geology of Brown County: 6th Ann. Rept., Ind. Geol. Sur., 1874, p. 103) makes it clear that the figure 1,147 was determined by a single aneroid survey and not by precise leveling.

sharp ridges are the reverse of the deep stream trenches. The dissection has reached that degree where the amount of the material left in the divides and ridges is about equivalent to the amount that has been carried away by erosion in the carving out of the valleys between the ridges. Everywhere steep slopes exist with rarely a bluff. The valleys have the same general shape and the same depth where they are of equal size. The only diversity of relief forms is found in size; the shapes are monotonously alike. The valleys descend sharply and deeply to near their local base-level and then become flat-bottomed areas flanked by steep ascents, but never by bluffs. The valleys are always V-shaped where they have not reached their local base-levels. No waterfalls of any consequence exist. The rather uniform height of the uplands between the deeply set streams gives an even sky-line view when one looks across the area from any ridge in the upland. South of the Muscatatuck River in the head-water area of the forks of Blue River the depth of dissection is somewhat lessened, and the angularity so characteristic of the Brown County portion of the upland is considerably subdued. The topography between the streams is somewhat gently rolling, especially where the capping of Harrodsburg lime-stone is still present or has only lately been removed. The Norman upland area is especially known for the perfection and symmetry of its drainage lines. The perfection of development is largely due to the mature stage of the erosion cycle in an area of considerable relief, and the symmetry results from stream development in rocks of uniform structure and lithology.

The Mitchell Plain.

Extent and boundaries. A notable westward sloping limestone plain with a well developed karst topography extends in a southeasterly direction from near Greencastle in Putnam County to the Ohio River, and thence south-ward and westward into Kentucky where its greatest development is reached. This sinkhole plain is locally well developed north of east fork of White River, but is very irregular and rather narrow in its extension through southern Putnam, eastern Owen, northwestern and middle Monroe, and middle northern Lawrence counties. In southern Lawrence, northeastern Orange, western Washington and middle and eastern Harrison counties it is widely developed, and forms a regional unit of considerable importance. It has been named the *Mitchell plain* by Beede from its excellent development in the vicinity of Mitchell in southern Lawrence County, Indiana.[18] In Kentucky it reaches its maximum development, extending over an area of about 2,500 square miles. The area of the Mitchell plain in Indiana is approximately 1,125 square miles.

The boundary on the east has already been described in connection with the boundary of the Norman upland which lies to the east of the Mitchell plain. The western boundary is even more irregular than that on the east and may be defined much more easily than traced. It may be said to lie in the zone of passage from the limestone sinkhole plain to the upland on the west, which is usually much dissected and composed largely of clastic rocks except in the narrowed and deep-set valleys on the east, partially excavated in the limestone. This zone of passage aside from being marked by a decided change in topographic forms is characterized by considerable increase in alti-tude and by a great diversity of relief, the amount of which is usually meas-

[18] Beede, Proc. Ind. Acad. Sci. for 1910, p. 195.

ured in hundreds of feet. Inasmuch as the upland on the west begins in a very much dissected and broken scarp, there are great numbers of rather prominent outliers of clastic rock material surrounded entirely by portions of the limestone sinkhole plain. These isolated outliers belong to the Mitchell plain. But in so far as irregular upland spurs extend out from the adjacent upland into the lower sinkhole plain but are not enclosed by it, these spurs may be said to belong to the adjacent upland. The irregularity is so great that the line drawn on the physiographic map is only an attempt to show the average position of the actual boundary within the zone described above.

The western boundary line of the Mitchell plain in general passes from the Ohio River near Mauckport northwest to Indian Creek some 2 miles below Corydon, and thence in a rather irregular line westward to within a short distance of Blue River near White Cloud along the southern margin of a streamless valley. The line returns on the northern side of the sinkhole plain of

Fig. 7. View of sinkhole topography on the Mitchell plain 7 miles southwest of Bloomington (see south half of Sec. 21. T. 8 N., R. 2 W., Bloomington Quadrangle).

the streamless valley of irregular width to some 5 miles east of White Cloud and turns northward to near Ramsey, where it again passes to Blue River around the westward extending streamless Brushy Valley. The line then passes northward and westward, crossing Blue River in the northwestern part of Harrison County, and trends northward a short distance east of the western boundary of Washington County, passing by Hardinsburg to near Livonia. From near Livonia the line passes about the headwaters of Stampers Creek and other swallow-hole streams, in a northwesterly direction to Orangeville where occurs the "Rise of Lost River." From Orangeville the line is generally northeast to a point a short distance southwest of Mitchell, where it turns westward along the southern margin of Beaver Valley nearly to Huron. The line returns to the east on the north side of Beaver Valley to a short distance west of Mitchell, where it passes north and west along the foothills on the south side of White River to near Williams. From Williams the line is irregularly north to some 2 miles west of Harrodsburg in Monroe County, where it continues north and slightly westward to southeastern Owen County, and thence to Spencer on the west fork of White River. From Spencer, the line passes somewhat east of north to near Cataract, where it turns west and north to the Shelbyville moraine, marking the southern limits of the Wisconsin glaciation, a short distance west of Greencastle. The Shelbyville moraine marks the northern boundary line of the Mitchell plain, because the sinkhole topography becomes buried beneath glacial drift north of that line.

The Mitchell plain as a distinct plain is not co-terminous with the areal outcrop of the so-called Mitchell limestone, but is best developed near the top of the St. Louis limestone some 100 to 140 feet below the top of the Mitchell

Fig. 8. View of plugged sinkhole lake on the Mitchell plain near Leonard School, 5 miles southwest of Bloomington. Lakes of this sort are very numerous near the western margin of the Mitchell plain. The eastern margin of the Crawford upland shows in the background. The Crawford upland in this locality rises 150 feet above the sinkhole plain shown in the foreground. (See Bloomington Quadrangle, Sec. 23, T. 8 N., R. 2 W.)

limestone group. In so far as the clastic rocks have been stripped from the Mitchell limestone, sinkholes exist on the western edge of the plain, but the plain is not distinct as a sinkhole plain until the upper portion of the Mitchell limestone itself has been taken away. The eastern edge includes portions of the areal outcrop of the limestones below the Mitchell, including the rolling limestone lands which have some development of sinkholes and a slight relief, except where the major streams have trenched the general plain.

Topographic condition. The topographic unity of the Mitchell plain is expressed in the peculiar forms of a karst upland plain with its great array of minor irregularities composed of sinkhole depressions and knob-to-ridge-like divides. The smaller sinkholes are funnel-shaped, and the larger ones are more irregular and usually serve as the opening into which the surface water drains. In the latter case the openings themselves are frequently referred to as "swallow-holes." The more extensive basins of irregular outline, such as those near the western border of the Mitchell plain, frequently occupy as much as a square mile or more in area, and contain numerous small individual sinks with low rounded divides between. (See Fig. 7.) It is quite usual for the sinks to become choked, giving rise to numerous small lakes or ponds commonly round in outline and partly filling the depression. (See Fig. 8.) The amount of local relief is not great, only infrequently being measured in more than a few tens of feet, but is minutely rough where the typical development of sinks occurs. The rounded irregularities, knobs, and sag-like divides between the sinks are approximately of the same elevation in local areas. Where the sinkholes are incipiently developed, or little developed, notable flat areas exist, as in southern Lawrence and northeastern Orange counties in the immediate vicinity of Mitchell and immediately south of Leipsic, Saltillo, and Campbellsburg, along the headwaters of Lost River. Only a few streams which have their headwaters on the Mitchell plain cross it as surface streams. All streams crossing the Mitchell plain as surface streams (streams arising east of the Mitchell plain and flowing southwest across it) are entrenched below the surface of the sinkhole plain from 80 to 175 feet, and the sinkholes are the largest and best developed near these incised streams. Distant from the incised surface streams, the sinkholes are small and in an early stage of development. Such areas are exceptionally flat, as noted above.

Fig. 9. View looking south from the surface of the Mitchell plain above Shirley Spring across the intrenched valley headed by Shirley and Leonard springs, 5 miles southwest of Bloomington. The even surface of the Mitchell plain shows distinctly south of the valley. The valley is entrenched here 100 to 125 feet below the sinkhole plain above, and has been dammed to catch the water of Shirley and Leonard springs, which furnish most of the water supply for the city of Bloomington. These springs during the winter and spring seasons flow over a half million gallons per day. (See Bloomington Quadrangle, Secs. 23 and 24, T. 8 N., R. 2 W.)

The Mitchell plain is essentially a regional slope inclining westward and slightly south at the rate of approximately twenty feet to the mile, and as a plain it varies in height from less than 600 to more than 900 feet. South of east fork of White River the higher elevations are at the eastern margin and the lower ones on the western margin. North of east fork of White River the plain is largely destroyed by surface stream dissection. Salt Creek in Lawrence County and Clear Creek and the tributaries of Beanblossom Creek and the west fork of White River in Monroe County, have removed much of the plain, and relatively narrow remnants only remain, mostly to the west of these streams. In Owen County the Mitchell plain is very effectually effaced by stream erosion and glaciation, though its characteristics are locally prominent.[19] In Monroe County west of Bloomington on the south side of the divide between the two White Rivers the plain is well developed at an elevation of 800 to 900 feet.[20] (See Fig. 36.) Farther south and west of Salt Creek it is 650 to 750 feet in elevation, and as a sinkhole plain largely destroyed by surface dissection. (See Fig. 9.) The Mitchell plain is best developed and has its widest extension in the area between Salem and Orangeville in Washington and Orange counties. Lost River is within this area. The upper part of Lost River flows over the inclined plain with little entrenchment, and is perched rather high above its potential local base-level. In the vicinity southeast of Orleans the present stream disappears downward through the limestone at an elevation of 625 feet and comes out as a much enlarged stream some 8 miles distant to the west, near Orangeville, at an elevation of 490 feet.

The eastern edge of the Mitchell plain interlaces with the Norman upland; tongues of sinkhole upland rising higher and higher to the east, flanked on either side by deeply dissected surface streams which have all the characteristics of the Norman upland. The western portion of the Mitchell plain frequently extends as streamless, sinkhole valleys far to the west of the general edge, and in between these westward extensions occur spurs of upland rising high above the rather low-lying sinkhole plain. Outliers are characteristic of the western half of the Mitchell plain, especially the portion south of Lost River. Some of these hills with their capping of clastic rock rise 150 to 200 feet above the sinkhole plain upon which they stand as monadnock-like masses of unreduced upland.

The Crawford Upland.

Extent and boundaries. The maximum of relief diversity and variability of topographic form in Indiana is found in the *Crawford upland;* the name applied to the exceedingly rugged highland west of the Mitchell plain. The Crawford upland receives its name from Crawford County, Indiana, which is one of the roughest counties in the state. The Crawford upland is the second largest of the southern Indiana regional units, comprising an area of approximately 2,900 square miles. It is relatively narrow at the north. Its northern line is drawn at the limit of Wisconsin glacial drift in southeastern Parke and southwestern Putnam counties. It occupies southeastern Parke, southwestern Putnam, a small portion of eastern Clay, a large portion of Owen, the eastern half of Greene, southwestern Monroe, western Lawrence, practically all of Martin, all but northeastern one-fourth of Orange, the eastern half of Dubois, all of Perry and Crawford, and western Harrison counties in Indiana. It

[19] Malott, The Flatwoods Region of Owen and Monroe Counties, Indiana: Proc. Ind. Acad. Sci. for 1914, pp. 418-419.

[20] See Bloomington Quadrangle, U. S. Geol. Surv., 1910.

continues southward in Kentucky where it occupies an area of several thousand square miles.

The eastern boundary line has already been described in connection with the western boundary of the Mitchell plain. The western boundary line is somewhat arbitrarily drawn on the physiographic map. The topographic map (Pl. I) shows that the relief becomes greatly lessened at its western border. The decrease in relief is rather gradual over a few miles, and near this belt another characteristic becomes quite prominent. The valleys become rather wide and have all the appearances of being deeply filled with alluvium above their bed-rock floors. Wide, deeply aggraded valleys constitute the most characteristic feature of the adjacent region of rather low relief on the west. This condition is not wholly lacking in the Crawford upland, but is not characteristic, and is a real feature only in the western part. The line may be defined as passing along this belt of greatly decreasing relief associated with the beginning of aggraded valleys. From the Ohio River this line passes up Anderson Creek from its mouth at Troy to St. Meinrad,[21] and thence up Blackhawk Creek and over a low divide to Ferdinand and Hunley Creek and down Hunley Creek northward to Patoka River. From the mouth of Hunley Creek the line is drawn up Patoka River to about 3 miles northeast of Jasper, where a turn is made to the north and slightly west over a low flat divide to Mill Creek, and down Mill Creek to White River west of Haysville. After crossing the east fork of White River the line passes up Slate Creek in southeastern Daviess County to the limit of the Illinois glacial till. It follows the drift limit which passes north slightly east of the county-line between Daviess and Martin counties to the northwestern corner of Martin County. From here the western boundary of the Crawford upland passes north to Scotland, and thence down Doans and Bogard creeks to White River; thence up White River to the mouth of Eel River at Worthington; thence up Eel River to the mouth of Lick Creek. The line passes up Lick Creek to Denmark,[22] and is then drawn arbitrarily across country for 5 miles to Eel River at the mouth of Six Mile Creek, and thence up Eel River to the mouth of Croys Creek in the southwestern corner of Putnam County; thence up Groys Creek to the village of Lena, and thence over a flat divide and down Little Rocky Fork and Rocky Fork creeks to Big Raccoon Creek at Mansfield. From Mansfield the line is drawn arbitrarily northwest towards Rockville to the Shelbyville moraine just south of Rockville. As noted above, the Shelbyville moraine well serves as the northern boundary of the Crawford upland, because the glacial drift largely obscures the bed-rock relief to the north. Considerable relief occurs along Sugar Creek and its near tributaries in northern Parke and southwestern Montgomery counties. These features, however, are rather exceptional in the generally level glacial plain.

Topographic condition. Throughout the whole extent of the Crawford upland the variety of topographic form and diversity of relief is bewildering. High hills, low hills; great ridges, some sharp, some rounded, and others flat-topped, but none of even altitude for any great distance; trench-like valleys, flat-bottomed valleys with wall-like bluffs, especially at the western margin; rock benches and structural plains of greater or less extent; remnants of an uplifted peneplain of rolling character left here and there amidst the extreme

[21] See St. Meinrad Quadrangle, U. S. Geol. Surv.

[22] See Clay City Quadrangle, U. S. Geol. Surv.

of differential dissection of the present erosion cycle; sinkholes, and swallow-holes, some of which are deeply sunk in loose sandstones, and due to underlying limestone high above the local water table; monadnocks, outliers, local cuestas, escarpments, waterfalls, canyon-like gorges, natural bridges, caves, etc., all contribute their quota to the makeup of Indiana's most inaccessible area. The beautiful Marengo and the magnificent Wyandotte caves of eastern Crawford County; the precipitous bluffs of the exceedingly narrow valley of the Ohio River at the southern margin of Crawford and Perry counties; extremely deep-set and meandering valleys, such as lower Lost River, the Blue Rivers, east fork of White River, and Indian Creek, along all of which occur mineral springs some of which are widely known; "Jug Rock,"[23] an unusual erosional remnant, and the "Pinnacle," a narrow upright spur rising almost directly

Fig. 10. View of the mouth of Boone's Cave, southeastern Owen County, showing the recession of the open surface valley at the expense of the lower end of the subterranean stream. Large caves developed in the upper Mitchell limestone are common in the eastern part of the Crawford upland area.

180 feet above White River at Shoals in Martin County; the "American Bottoms"[24] of eastern Greene County, a lacustrine plain deeply set among rocky hills and which possesses a very unique subterranean drainage system; these constitute a few of the features of the Crawford upland which render it attractive from a scenic point of view. This maze of variety of land forms and the diversity of relief are due to topographic maturity in a region of considerable relief with respect to the major streams (minimum altitude of 348 feet at Troy on the Ohio River; maximum altitude of 980 feet on the divide between the White rivers in western Monroe County),[25] and to a great series of alternating non-resistant shales and resistant sandstones with many inter-

[23] Cox, 2nd Ann. Rept. of Ind. Geol. Surv., 1870, p. 83; Ashley, 23rd Ann. Rept., Ind. Geol. Surv., 1898, pp. 947-948; Shannon, Proc. Ind. Acad. Sci. for 1906, pp. 53-70.

[24] Malott, The American Bottoms Region: Ind. Univ. Studies No. 40, 1919.

[25] See Tell City and Bloomington Quadrangles, U. S. Geol. Surv.

calated beds of limestone. At the eastern margin these beds of rock, none of which attain a great thickness, overlie the great thickness of Mitchell limestone the upper surface of which is high above the deeply-set streams, and here occur deep caverns and many artesian springs of enormous size, and locally at this margin occur streamless valleys surrounded by high ridges, such as "Possum Valley" east of Indian Creek, northeast of Silverville in western Lawrence County. This valley along with many others is a small circumscribed sinkhole plain of the Mitchell plain type set within the eastern margin of the Crawford upland. At the western margin occurs the thick and resistant Mansfield sandstone which frequently caps hills almost to the eastern margin of the upland, especially in Lawrence, Monroe, and Putnam counties, where

Fig. 11. View of wall-like cliff formed by the Beech Creek limestone and the Cypress sandstone, Eagle Ranch Farm, southeast of French Lick, Orange County. Such cliffs are common in the area of the outcrop of the Chester series in the Crawford upland, where angularity of topographic forms is persistent. Frequently such wall-like cliffs occur far up the slopes along the streams, and are common at the head of gulches, where large springs come from the Beech Creek limestone.

the alternating Chester clastics and limestones are thin; and is responsible for the retention of high elevations in the early post-mature stage of the erosion cycle.[38]

Perry County has the distinction of being the roughest county in Indiana. Its condition is due both to its diversity of rocks, the most massive of which serve as ineffective cap rocks, but make excellent bluff formers; and to its nearness to the deep gorge of the Ohio River, which with the similarly deep-set Anderson Creek on the west nearly surrounds the county, making the area a huge south-extending spur of upland in a highly dissected state. The only level land is found in the valleys. Crawford County is very similar and barely

[38] The Influence of the outcropping rocks upon the topography is treated specifically in a local area of the Crawford upland by the writer, in "The American Bottoms Region," Ind. Univ. Studies No. 40, March, 1919, pp. 1-64.

escapes being as rough and angular as Perry County by having wider undulating divides, especially in the area between the deeply entrenched Blue rivers. Brown County of the Norman upland with its great fairly uniformly steep slopes and unusually perfected drainage, probably ranks third as a whole county in roughness, though it is exceeded by none in hillside area.

The Wabash Lowland.

Extent and boundaries. *Wabash lowland* is the name applied to a great lowland tract extending widely on both sides of the Wabash River south of the outermost Wisconsin or Shelbyville moraine. Its eastern border has been described as the western limit of the Crawford upland. It occupies a large area

Fig. 12. Filled valleys are everywhere present in the Wabash lowland area of southwestern Indiana. View from near the mouth of Splunge Creek adjacent Eel River, west of Clay City, Clay County. Many square miles here are near or slightly below the level of Eel River valley. The valley of Splunge Creek was formerly the site of a large reservoir for the old canal system, but the region is now being ditched for agricultural purposes.

in Kentucky and Illinois, and in the latter state is most typically developed as a lowland of wide extent. The Indiana portion occupies an area of approximately 4,900 square miles, and is the largest of the series of north and south extending physiographic units in southern Indiana.[27] The larger part of it in both Indiana and Illinois has been glaciated by the great lobe of ice which reached the farthest south of any of the glacial lobes. Much of the levelness of this extensive lowland plain is undoubtedly due, directly and indirectly, to glaciation, though other agencies and conditions enter largely into the development of the physiographic unit. The area as a whole is characterized by wide extent of alluvial lands some of which are actually lacustrine in origin.

[27] The *Wabash Lowland* is approximately equivalent to the "*Western Lowland*" of Newsom and Dryer.

This condition is characteristic of both the glaciated and unglaciated portion of the Wabash lowland.

Topographic condition. The unity of the Wabash lowland as a physiographic division lies chiefly in its topographic condition, which is the result of its physiographic development. The Wabash lowland area as a whole is composed of surface rocks which are non-resistant to denudational agencies, and this in connection with its position along the axis of a trunk stream, goes far toward explaining the generally low altitude of the area as compared with the Crawford upland on the east. Associated with the low altitude, almost wholly below 700 feet and on an average in Indiana about 500 feet, is the great extent of aggraded valley area.[28] Every stream which has reached any size runs in a flat, wide valley far too large for it, and very frequently the upland rises rather abruptly in rather steep slopes from the flat flood plain area. Many of the aggraded valleys are associated with the stream obstruction attending the Illinois glaciation, as along Patoka River and the headwaters of Pigeon Creek; but much of the aggraded valley condition cannot be directly associated with this cause. The treatment of this problem will be taken up below under physiographic development. The thing to be emphasized here is the prevalence over the entire Wabash lowland of the aggraded valley condition, which chiefly gives unity to the entire division. The general levelness of the glaciated area in Illinois and Indiana is very striking, but this levelness or flatness is not at once noticeable at glacial boundary limits. It must be said, however, that as a whole the driftless upland area is much more minutely dissected and is considerably rougher than the glaciated. This more general roughness of the driftless section of the Wabash lowland is not altogether due to lack of glaciation. A considerable part of this may be ascribed to the differences of the rocks present in the two sections. Southeastern Daviess County has been glaciated, yet where the altitude and rocks are similar it is nearly as rough as the driftless area. Knox and Sullivan counties as a whole are the levelest of the Indiana counties in the Wabash lowland, but their general bed-rock surface is somewhat lower than the rougher counties, and non-resistant shales are more predominant. These counties would probably possess lower relief and would be characterized by more gentle topography than the present somewhat rougher driftless counties regardless of glaciation.

One of the peculiarities associated with the aggraded valley condition of the Wabash lowland is the common occurrence of isolated bed-rock hills standing island-like in the midst of alluvium or the silt of a lacustrine plain. These range in size and shape from mere conical knolls a few feet in height to great detached uplands 100 or more feet in height and which may cover an area of a square mile or several square miles. Illustrations of these upland areas in the midst of extensive valley lands occur in the Mumford Hills,[29] which reach up 140 feet above the lowlands, and Foot Pond Hills near the line of Posey and Gibson counties. Gordon Hills and Claypool Hills near the mouth of White River are other examples. Such outstanding, half-buried uplands are prominent in Kentucky along Pond and Green rivers, and Shaw has suggested that these hills be called *island-hills* after the half-buried upland in the midst of

[28] The following topographic sheets of the U. S. Geol. Surv. show the topographic condition of much of the Wabash lowland: Owensboro, Newburg, Henderson, Mt. Vernon, New Haven, New Harmony, Mt. Carmel, Princeton, Birds, Vincennes, Petersburg, Velpen, Degonia Springs, Haubstadt, Boonville, St. Meinrad and Clay City.

[29] See Vincennes and New Harmony topographic sheets, U. S. Geol. Survey.

an extensive valley flat on which the town of Island, Kentucky, is situated.[30] The uplands commonly rise 150 feet above the flat, wide valleys, but most commonly immediate relief is not much over 100 feet. The uplands south of White River in both the glaciated and unglaciated sections are well dissected and rather rough. Sharp ravines descend rapidly and widen out as extensive valley flats on approaching the stream into which they discharge. This roughness is present in southern and eastern Daviess County to as great a degree as in any of the driftless sections. Perhaps a distinction should be made between the more rugged uplands which reach 600 feet or more in altitude south of White River and the more gently rolling uplands which rarely are higher than 550 feet.[31] More than one-fourth of the area of Knox County is valley land. This county is bounded by rivers except on the north, and possesses more river boundary than any other county in the state. Practically all of the upland area of Knox County is undulating and rolling. Very little flat land occurs although the relief is small. Sullivan and western Greene counties are relatively flat, with local relief of rarely more than 75 feet. The valleys are broad and much marsh land exists. The divides between the streams are low and generally rather flat. The maximum local relief of the Wabash lowland in Indiana occurs at Merom where a precipitous bluff composed mostly of massive sandstone rises 170 feet above the Wabash River. The slope of this eminence away from the river is relatively gentle. In the vicinity of the Wabash the adjacent uplands are frequently composed of or covered by sand dunes, which have probably been formed mostly of sand blown from the valley immediately following glaciation. The action is not noticeable at the present time. These dunes are rather characteristic north and east of Merom in Sullivan County and are present in a number of places east of the Wabash on the uplands adjacent to the valley.[32]

The Tipton Till Plain.

Extent and boundaries. We pass now to that greater portion of the state where the features of continental glaciation predominate, and instead of features produced mainly by running water upon various beds of rock, features of an entirely different origin characterize the landscape. Topographic forms resulting from destructive denudation here merely modify the surface of the great till plain, and over much of it are insignificant. A great area lying south of the Wabash valley from Fort Wayne to Logansport and west past Kentland, and north of the series of regional physiographic units of the southern part of the state, is characterized by a deep covering of glacial material, the surface of which is mainly flat or very gently rolling. This glacial plain extends eastward into Ohio where it covers a large area in the western middle portion of the state. It also extends westward into Illinois bearing the same characteristics as in its most typical representation in Indiana. It is a vast till plain of little relief and meager modification by dissecting streams. The name of *Tipton till plain* is proposed for it from its excellent representation in Tipton County, Indiana. Tipton County lies in the midst of the greatest expanse of its most featureless portion. The Tipton till plain in Indiana com-

[30] Shaw, New System of Quaternary Lakes in the Mississippi Basin: Jour. Geol., Vol. 19, 1911, pp. 481-491.

[31] Fuller, Ditney Folio No. 84, U. S. Geol. Surv., 1902, page 1, under heading "Topography."

[32] Shannon, The Sand Areas of Indiana: Proc. Ind. Acad. Sci. for 1915, pp. 185-188.

prises approximately 11,900 square miles, and is the largest physiographic division in the state, occupying nearly one-third of the entire area of the state.[33]

The northern boundary line in Indiana from east to west passes along the southern edge of the lacustrine plain of the former glacial lake Maumee, and is marked by a distinct beach line called the Van Wert beach line to Fort Wayne, at the western end of the old glacial lake. From Fort Wayne the boundary line passes southwest down a broad, flat-bottomed valley, occupied partly by the insignificant Abóite Creek, to Huntington where the Wabash River enters the broad valley from the southeast; thence down the Wabash Valley to near the Cass-Carroll County line some 10 miles below Logansport, and thence west to Monticello in White County, this latter portion of the boundary being arbitrarily drawn. Westward from Monticello the boundary passes along the edge of the till plain at its juction with a flat lacustrine plain into Illinois. A considerable portion of this line is marked by a distinct beach line which runs on an east and west line just north of Kentland.[34]

The boundary line on the south of the Tipton till plain has already been delimited in connection with the northern boundaries of the divisions present in the southern part of the state. The Shelbyville terminal moraine of the Wisconsin glaciation serves as the boundary for slightly more than half of the distance across the state, but from Nineveh in southern Johnson County the line passes northward to near Franklin and thence northeast to Richmond and the state-line in southern Wayne County. No natural boundary exists in the eastern part of the state at the south of the Tipton till plain. The Scottsburg lowland, the Muscatatuck regional slope, and the Dearborn upland preserve their characteristics as regional units, though they are tremendously modified at the north on account of the increasing thickness of drift. The Tipton till plain and these regional units have no distinct line of separation. Since the southern divisions were once present much farther north than now, it was deemed best to trace them as far as possible. Certainly the concavity of the Scottsburg lowland persists beyond the limits of the Shelbyville moraine. The Muscatatuck regional slope persists as a slope well within the Wisconsin glacial drift, though the drift deeply covers the bed-rock. South of Richmond the Whitewater drainage system has largely destroyed the till plain surface and bed-rock is commonly exposed, causing the features of the Dearborn upland to prevail. But the southern units are gradually overwhelmed by the increasing thickness of glacial material northward of a line from Nineveh to Rushville and running thence to the southern boundary of Wayne County. The zone between these lines may be considered as belonging to either the southern units or the till plain.

Topographic condition. The Tipton till plain is characteristically a slightly modified ground moraine plain, and over wide areas is monotonously flat. Poorly developed terminal moraine topography detracts little from its general levelness over wide areas, especially through several thousand square miles in its central and northeastern portion in Indiana. Dryer calls it the

[33] The *Tipton till plain* is nearly equivalent to the *Central Drift* Plain of Dryer.

[34] The northern boundary of the Tipton till plain across Indiana is identical with Fenneman's boundary between the "Eastern lake section" and "Till Plain section" of the "Central lowland province," as shown on Pl. I, Physiographic Divisions of the United States: Annals Assoc. Amer. Geogr., Vol. VI, pp. 19-98.

"Central Drift Plain"[35] and says of it: " . . . the topography is that of an almost featureless drift plain. It is traversed by numerous morainic ridges, but they are low and inconspicuous. The traveler may ride upon the railway train for hours without seeing a greater elevation than a haystack or a pile of sawdust. The divides are flat and sometimes swampy, the streams muddy and sluggish. The valleys begin on the uplands as scarcely perceptible grooves in the compact boulder clay, widen much more rapidly than they become deep, and seldom reach down to the rock floor."[36] Leverett in describing the most typical portion of the Tipton till plain says: "In north-central Indiana a tract of about 4,000 square miles, extending from the Bloomington morainic system northeastward to the Mississinewa morainic system and northward to Wabash River, is without strong or well defined moraines, though it has small tracts where the surface is undulatory and a few places where boulders are numerous. In general its surface is plane and its drift is much thinner than on the belt south of it. This tract shows a westward descent of 200 to 250 feet in 45 miles, dropping from 1,100 feet at the Ohio-Indiana state line to 850 to 900 feet on the meridian at Anderson. For 20 miles west from the Anderson meridian, in Madison, Hamilton, and Tipton counties, the altitude is remarkably uniform. The southern part of the tract rises westward into the great drift belt in western Hamilton and southwestern Tipton counties, ascending 60 to 100 feet in 8 or 10 miles. Farther north the tract very gradually descends to the northeast toward the Wabash Valley, its altitude at the border of that valley being about 700 feet. From the great drift belt in eastern Indiana northward about to White River the descent is abrupt, being 120 feet in 10 miles from Bloomingsport [near the highest point in Indiana] to Winchester, 150 feet in 9 miles from ridges near Luray [northern Henry County] to Muncie, and 140 feet in 14 miles from Warrington [northeastern Hancock County] to Anderson."[37]

In dealing with the topographic condition of a glacial plain there are two types of forms to be considered: first, those inherent on a glacial plain of construction, and second, those which have come into existence since the plain was formed, or features of subsequent development. Topographic forms inherent on the glacial plain are any forms made during the construction of the plain by glacial agencies. These are usually features which give differences in relief, or which detract from planeness. Such forms are the knolls, basins and ridges of terminal moraines, drumlins peculiar to certain gound moraines, and eskers, kames, and other fluvio-glacial features always directly associated with glaciation. The second type of forms consists of the forms produced by running water, etc., subsequent to the formation of the glacial plain. Such may only slightly modify the plain or they may completely destroy it, depending upon time and condition. Both of these types will be taken up with respect to the Tipton till plain.

The chief inherent irregularities of the Tipton till plain are the terminal moraines and their associated features. No drumlins exist in Indiana. A glacial map chiefly following Leverett accompanies this report, and this shows the position, extent, and relationship of the terminal moraines. (Pl. III.)

[35] Dryer, Geography of Indiana, Supplement to High School Geography: Amer. Book Co., 1912.

[36] Dryer, Studies in Indiana Geography: Inland Pub. Co, Terre Haute, Ind., 1908, p. 19.

[37] Leverett, The Pleistocene of Indiana and Michigan: U. S. Geol. Surv., Monogr. 53, 1915, p. 107.

These features are not revealed on the topographic map, and no conception may be formed from it of the unevenness of the terminal moraines. Only where they are massively piled up do they control the 100-foot contours. Thus in Randolph, Jay and Benton counties in particular, and in Henry, Adams, and Wayne counties to a less degree, the outlines of the higher morainic material are indicated by the position of the contours. The individual topography of the moraines, however, is not shown, the contours showing rather the general altitude of the morainic masses.

The Shelbyville morainic system[38] along the southern boundary of the Tipton till plain is traceable over a considerable portion of its extent by the change in soil condition rather than by morainic topography. In Vermillion and Parke counties it is quite distinct, though only a slight increase in altitude occurs at its outer margin. The knolls and sags have a variable relief though the amount is small, rarely being more than 30 feet. A similar condition occurs in Morgan County between Monrovia and Mooresville. East of Martinsville and southeast into southern Johnson County, the knob and basin type of moraine is quite distinct, some of the knolls reaching 50 or 60 feet above the marginal area. This moraine at the eastern margin of the Scottsburg lowland, east of Columbus, and across the Muscatatuck regional slope from Elizabethtown east and northeast past Greensburg to Connersville, presents hummocky topography of low relief over a width of from 1 to 4 miles. The knolls seldom rise more than 20 or 30 feet above the sag-like basins between them.

In Fountain, Montgomery, Warren, Tippecanoe, and Benton counties lies a tangled mass of moraines which belong to two or possibly three morainic systems. Those of the Champaign system[39] in Fountain and Montgomery counties stand above the flat inter-morainic tracts usually 25 or 30 feet, with occasional knolls rising 50 or 60 feet above their surroundings. Northward in Warren, Benton and Tippecanoe counties the Bloomington[40] morainic system presents a complex series of moraines which are difficult to correlate. "Clusters of sharp knolls covering a square mile or more, surrounded by gently undulating or nearly plane tracts, can be linked into two parallel chains leading eastward and then southeastward into Montgomery County. Some of the knolls in these clusters reach a height of 50 or 60 feet and many reach 30 feet or more. Shawnee Mound, in southwestern Tippecanoe County, rises 60 feet above the surrounding country and 80 feet above the marsh on its northwest border. About 5 miles southeast is Cemetery Hill, also 60 feet high. In Tippecanoe County a gently undulating surface, extending entirely across the southern townships, is about 100 feet higher than the next range of townships to the north. In places the descent to this northern plain is abrupt, but as a rule it extends over a distance of 2 miles or more."[41]

In the southern middle portion of the Tipton till plain, in Hendricks, Boone, Hamilton, Marion, Johnson, Hancock, and Shelby counties, the moraines are occasionally well developed and stand in knoll and basin form with occa-

---

[38] Leverett, Illinois Glacial Lobe: U. S. Geol. Surv., Monogr. 38, 1899, pp. 189-218; The Pleistocene of Indiana and Michigan: U. S. Geol. Surv., Monogr. 53, 1915, pp. 77-86.

[39] Leverett, The Illinois Glacial Lobe: U. S. Geol. Surv., Monogr. 38, 1899, pp. 223-210; The Pleistocene of Indiana and Michigan: U. S. Geol. Surv. Monogr., 53, 1915, pp. 87-93.

[40] Leverett, U. S. Geol. Surv., Monogr. 38, 1899, pp. 240-290; U. S. Geol. Surv., Monogr. 53, 1915, pp. 95-122.

[41] Leverett, U. S. Geol. Surv., Monogr. 53, 1915, pp. 95-96.

sional ridge-like prominences above the general plain. As a rule, the knolls are no more than 30 feet above the gently undulating to flat intermorainic tracts, though exceptions locally occur, heights of 50 or 60 feet being attained. The Champaign system from Waverly southeast in Johnson County to the south and east of Franklin presents a number of rather conspicuous knobs on its rather broad tract. Knolls of 40 feet or more rise above the adjacent plain. Doty Mound northwest of Bargersville rises sharpely 75 feet above its surroundings, and Donnel Mound east of Franklin rises 90 feet above the plain. The Bloomington morainic system in Marion and northern Johnson counties gives rise to considerable local relief. Crown Hill[42] in Indianapolis rises 125 feet above White River valley-train plain, and a sharp ridge south of Indianapolis near the Marion-Johnson line rises 75 to 125 feet above the plain. The Champaign and Bloomington morainic belts in Shelby, Rush, Hancock and southern Henry counties are rather low and are nowhere prominent.

In northeastern Henry, southeastern Delaware, Randolph, and northern Wayne counties, the Champaign and Bloomington morainic systems are quite prominent, and a great mass of morainic accumulation here gives rise to the highest area in the state. Over considerable of this great morainic pile the local relief is not great; the knolls rise as gentle eminences and the slopes to the streams are not abrupt.[43] The highest area in southern Randolph County is rather a morainic plateau of uneven surface. In northern Henry County in the vicinity of Mt. Summit a system of ridges and knolls rises rather abruptly 75 or 100 feet above the bordering marshy plain of the headwaters of Blue River. The morainic mass is thickly strewn with boulders. Leverett[44] says of this morainic mass: "In southern Randolph and northern Wayne counties, Ind., the moraine covers an uneven surface of uplands and lowlands, the lowlands being occupied by the headwater branches of Whitewater River. . . . The knolls of this region are closely aggregated and thickly strewn with boulders, but few of them reach a height exceeding 30 feet."

In the northeastern part of the Tipton till plain a series of similarly curving terminal moraines occurs. These moraines in connection with their correlatives north of the Erie-Wabash glacial waterway, furnish an excellent illustration of recessional moraines formed during the retreat of a great glacial ice-lobe.[45] The Union City moraine (see Pl. III) is the outermost in the series, but is poorly developed and only locally possesses the typical characteristics of a terminal moraine. The Mississinewa moraine is the second in the series and is well developed. It extends from the Wabash River just east of Wabash, southeast in a curve to Hartford City, Redkey, and eastward into Ohio. It is several miles broad and composed of a mass of morainic material which rises above the plain rather gently, but consists of numerous knolls and basins. None of the knolls rise high, though the entire mass is 20 to 100 feet above the flat ground moraine on which the hummocky terminal moraine appears to lie. The higher figure is toward the Indiana-Ohio line where the altitude reaches over 1,100 feet. The Mississinewa River flanks the moraine on the outside curve, and the moraine rises rather abruptly above it. The second of

[42] Dryer, The Physiography of Indianapolis: Proc. Ind. Acad. Sci. for 1917, pp. 55-57.

[43] See Winchester topographic sheet, U. S. Geol. Surv., 1918.

[44] Leverett, The Pleistocene of Indiana and Michigan: U. S. Geol. Surv., Monogr. 53, 1915, p. 98.

[45] Dryer, The Drift of the Wabash-Erie Region: Eighteenth Ann. Rept., Ind. Geol. Surv., 1894, pp. 83-90.

the series is the Salamonie moraine extending east from Portland in Jay County in one direction and northwest in the other direction. It lies immediately adjacent to the Salamonie River, and rises rather abruptly above it. It is a very narrow moraine of the knob and basin type, the knolls rising 20 or 30 feet above the flat plain to the east. The third of the series of recessional moraines lies alongside the Wabash River, and is a fairly well-developed moraine with its higher knolls rising 30 to 50 feet above the intermorainic flat plain. The fourth of the series south of the Erie-Wabash waterway is a broad moraine adjacent to the St. Mary's River, with its higher portion rising 40 to 60 feet above the flat intermorainal plain.

The Tipton till plain as a constructional feature in the zone of ice wastage is a plain without valleys; but as the ice melted from the region the older and first exposed portion of it was subject to overflow by the waters coming from the ice of the portion still under construction. Consequently there are numerous broad valleys leading southward and southwestward across the plain. Most of these glacial waterways are now occupied by streams, though in many cases the streams are very small and feeble as compared to their Pleistocene ancestors, which during the melting seasons occupied the whole valley and carved out the valley practically as it exists today. The master valley of the whole state is the great valley of the Wabash. The valley begins at the west end of the Maumee lacustrine plain (described below), but is not occupied by the Wabash River between Ft. Wayne and Huntington, as the present river enters this Pleistocene valley at Huntington. But most of the Pleistocene waterways were of a different type than the Pleistocene Wabash Valley. The Wabash Valley carried relatively clear water from glacial Lake Maumee, a vast ice-dammed lake stretching eastward from Fort Wayne.[**] Most of the ice-fed streams were more temporary in character, and must have carried great quantities of glacial material, so that aggradation was an important process in their descent towards the sea. The clear waters coming from Lake Maumee were capable of cutting downward, and thus gave rise to a trench cut below the general plain. The most notable valley of this sort is the Illinois Valley extending from Lake Michigan at Chicago southwestward to the Mississippi. This valley is broad and is deeply entrenched beneath the level till plain through which it flows. The Illinois Valley was maintained a very long time and was reduced to a very low gradient. The Wabash Valley from Ft. Wayne was of this type, though it lasted as the outlet of lake waters only a relatively short time.

The chief waterways across the Tipton till plain flowed either south or southwest. A number of these headed near the high morainic mass in southern Randolph County, and flowed southward down the present route of the Whitewater River. This stream was especially favorable for receiving the discharge of the ice-fed waters, because it was located between two rather prominent ice lobes, as evidenced by the positions of the moraines. Flatrock Valley, heading in northeastern Rush County, Blue River Valley heading north of Newcastle, Sugar Creek Valley south of Acton, west fork of White River Valley south of Noblesville, Eel River Valley from near Lebanon, are valleys which owe their existence largely to the work done by glacial waters. The streams which now occupy these valleys have modified them very little since the close of the glacial period. The present streams are but feeble representatives of

[**] Dryer, The Maumee-Wabash Water-way: Annals Assoc. Amer. Geogr., Vol. IX, 1920, pp. 41-51.

their mighty progenitors. Another rather remarkable valley which must have been a route of the glacial waters begins near Marion and extends southward. It crosses Pipe Creek at Alexandria, White River at Anderson, Fall Creek at Pendleton, and Sugar Creek in northern Hancock County, beyond which it is followed by Brandywine Creek to its junction with Blue River west of Shelbyville. This rather remarkable glacial valley is approximately 70 miles in length, and varies from one-eighth to one mile in width. It is nowhere conspicuous, but is easily traceable throughout its length. In a broad stretch from Anderson southward it is occupied by one of the best defined and longest eskers in the entire state.[47] The presence of the esker throws some light on the origin of this rather remarkable glacial waterway. It may be said here that other eskers in Indiana are also definitely associated with glacial waterways, though usually with the shorter and more inconspicuous routes which on account of their topographic position are not now occupied by any considerable stream. The more conspicuous eskers, and these are usually broad shallow deposits of water-laid sand and gravel, occur near Muncie in Delaware County, near Burney in Decatur County, near Franklin in Johnson, near Malta in Putnam County, near Greensboro in Henry County, and near South Raub in Tippecanoe.[48] The esker at South Raub is perhaps the most prominent and best defined esker in Indiana.[49]

Post-glacial stream development is rather feeble over the greater part of the Tipton till plain. The drainage lines over extensive areas of the plain are mere channels without noticeable valley formation. The tributary streams on their approach to the great valleys developed by the Pleistocene waters are entrenched somewhat in accordance with the main valley. Such entrenchment is best developed by the streams which enter the Wabash valley. The Wabash valley as developed by the glacial waters from glacial Lake Maumee and other ice-discharged waters which came into it, extending from Ft. Wayne west and southwest to near Delphi, is cut from 50 to 125 feet below the bordering plain, and is one-fourth to one mile in width. Over much of this course it is cut through the glacial drift, and occasional bed-rock bluffs rise sharply from the valley floor. From the vicinity of Delphi the Wabash Valley follows a broad pre-glacial valley from which the glacial material has been partly removed. It is apparent that glaciation completely filled this pre-glacial valley, and through Tippecanoe, Warren-Fountain, Vermillion-Fountain, and Vermillion-Parke counties to the Shelbyville moraine the glacial material has been re-excavated to a depth of approximately 200 feet below the plain. The valley is much wider in this portion than in the portion above Delphi. This portion of the valley is flanked by two sets of terraces, the higher one of which is approximately 150 feet above the river. Prominent remnants of this older glacial valley floor occur as broad, level gravel plains, such as the one on which Purdue University at West Lafayette stands, and the Wea Plains southwest of Lafayette on the south side of the valley.[50]

---

[47] Leverett, U. S. Geol. Surv., Monogr. 53, 1915, pp. 108-111. Reeves, The Anderson Esker: Amer. Jour. Sci., Vol. 50, 1920, pp. 65-68.

[48] Leverett, U S. Geol. Surv., Monogr. 53, 1915, pp. 83, 93 and 107-111.

[49] McBeth, An Esker in Tippecanoe County, Indiana: Proc. Ind. Acad. Sci. for 1904, pp. 45-46.

[50] The details of the topographic condition of the Wabash Valley and its glacial history, and in particular with respect to its terraces, may be found in the following references: *McBeth*, The Physical Geography of the Region of the Great Bend of the Wabash: Proc. Ind. Acad. Sci. for 1899, pp. 156-161; The Development of the Wabash Drainage System and the

The entrenchment of the tributary streams of the Wabash River above Delphi is inconspicuous a short distance from the river. The streams are swift and flow in rather shallow valleys. The Wabash River above where it enters the old glacial waterway at Huntington is in a rather shallow valley and nowhere has cut through the boulder clay of which the plain is made. The lower part of Wildcat Creek in Tippecanoe and western Carroll counties is in a considerable valley, for the most part a re-excavated pre-glacial valley. North and west of the Wabash River below Delphi the streams on approaching the river become deeply trenched in the till plain. The lower part of the Tippecanoe River is very crooked and entrenched approximately 100 feet.[31] Just back of the bluffs the plain is excellently preserved as a typical till plain. In Warren and Vermillion counties adjacent to the Wabash Valley the plain is quite rugged and much dissected by minor streams. The valley trench of Pine Creek in Warren County is conspicuous. Vermillion River in middle Vermillion County is also deeply entrenched, the till slopes being relatively steep and descending 150 feet or more below the general level of the plain.[32] In Fountain, Montgomery, and Parke counties considerable relief exists where the main streams have dissected the plain, though between the main streams the till plain is well preserved and is fairly representative. Coal, Sugar, and Raccoon creeks were probably partly excavated by outgoing glacial waters, but their deep-set tributaries near the main trenches are chiefly the work of post-glacial run-off. Sugar Creek in southwestern Montgomery and northern Parke counties is deeply entrenched in and below the massive, resistant Mansfield sandstone, and sheer cliffs of 100 feet or more are present. In the "Shades of Death" park[33] and the Turkey Run State Park, Sugar Creek and its tributaries exhibit wild and rugged scenery. A maximum entrenchment below the plain of over 200 feet occurs here. In general the western part of the Tipton till plain in Indiana has lost its chief characteristics, general flatness and small local relief due to the entrenchment of the Wabash Valley and the resulting dissection of the plain in its vicinity, yet notable portions of the typical till plain are preserved between the main streams, even close to the Wabash Valley itself. This condition is very well shown on the topographic map, Pl. I.

The headwater streams of Whitewater River are in broad, basin-like valleys considerably below the higher parts of the high moraine but they are not sharply trenched until they approach the southern part of Wayne County where they pass into the Dearborn upland area. The parallel streams which constitute the headwaters of east fork of White River head on the westward

---

Recession of the Ice Sheet in Indiana: Proc. Ind. Acad. Sci. for 1900, pp. 184-192; Wabash Terraces in Tippecanoe County, Indiana: Proc. Ind. Acad. Sci. for 1901, pp. 237-243; History of Wea Creek in Tippecanoe County, Indiana: Proc. Ind. Acad. Sci. for 1901, pp. 244-247; Volume of the Ancient Wabash River: Proc. Ind. Acad. Sci. for 1915, pp. 189-190. *Soovell*, Terraces of the Lower Wabash: Proc. Ind. Acad. Sci. for 1898, pp. 274-277. *Dryer*, A Physiographic Survey of the Terre Haute Area: Proc. Ind. Acad. Sci. for 1910, pp. 145-146; Wabash Studies: Proc. Ind. Acad. Sci. for 1912, pp. 199-206; The Maumee-Wabash Waterway: Annals Assoc. Amer. Geogr., Vol. IX, 1920, pp. 41-51. (Contains detailed maps of the Wabash terraces from Fort Wayne to the mouth of the Wabash.)

[31] Breeze, The Valley of the Lower Tippecanoe River: Proc. Ind. Acad. Sci. for 1901, pp. 215-216; Some Topographic Features in the Lower Tippecanoe Valley: Proc. Ind. Acad. Sci. for 1902, pp. 198-200.

[32] Danville (Illinois-Indiana) Quadrangle, U. S. Geol. Surv., 1902.

[33] Barrett, The Beautiful Shades: 41st Ann. Rept. Ind. Geol. Surv. for 1916, pp. 80-89.

slope of the till plain (which is here continuous with the northern part of the Muscatatuck regional slope) and are little below the general level of the till plain. The valleys of these streams are veritable valley-trains and are little modified by post-glacial stream action.

The west fork of White River occupies a broad valley but slightly below the till plain surface beyond eastern Hamilton County. In the vicinity of Noblesville the valley is somewhat below the plain and till bluffs become common. Southward from Noblesville the valley was clearly the passageway of large volumes of water during the retreat of the glacial ice. It is broad, and in southern Marion and Morgan counties is 175 feet below the till plain.

Northern Moraine and Lake Region.

General Character, Name and Subdivisions. North of the Tipton till plain lies a great region of variable relief character, the surface of which is composed of the younger drift of the Wisconsin glacial epoch. Fenneman[34] has called this region east of the Wisconsin driftless section the Eastern Lake section of the Central Lowland province. Dryer[35] has called the portion which lies in Indiana the Northern Moraine and Lake Region. It is certainly occupied by numerous lakes which constitute one of its chief features. Most of the lakes are relatively small and are nearly, if not quite, restricted to the terminal moraines. The Great Lakes themselves belong in part to this region, and represent lake basins of a different character from the smaller morainic lakes. Lacustrine plains of great extent occur, and these are marked by broad marshes, or areas which were formerly marshes, broken by low sand ridges or knolls. The Indiana portion of the region is typically a compound of massive rugged moraines occupied by lakes and broad lacustrine plains. Only locally occur small areas of level till plain, but it is likely that much of the lacustrine plains consists of areas of level till plain covered with a veneer of lacustrine deposit. Nowhere north of the Wabash Valley is bed-rock exposed at the surface, except in southeastern Jasper, southwestern Pulaski and northwestern White counties where the coating of till and lacustrine deposit is thin, and local outcrops occur. Practically everywhere else in the region the thickness of the glacial drift is measured in hundreds of feet.

No geographic name applicable to the region as a whole is available, and Dryer's name *Northern Moraine and Lake Region* is used here. It will be noticed that the lakes of the region are limited to the rugged moraines, and that most of the level areas are lacustrine plains. Massive, rugged moraines and broad lacustrine plains chiefly characterize the region. These features are used as the basis of subdivision of the region. Three areas of the Indiana portion are chiefly characterized by predominance of lacustrine plains, and two areas are principally uneven morainic masses of considerable altitude as compared with the lacustrine plains. The morainic masses are occupied by morainic lakes. Geographic names are applied to these subdivisions, as follows: *1. Calumet Lacustrine section. 2. Valparaiso Moraine section. 3. Kankakee Lacustrine section. 4. Steuben Morainal Lake section. 5. Maumee Lacustrine section.* All of these subdivisions or sections extend beyond the limits of the state. Their combined area in Indiana comprises approximately 8,700 square miles.

[34] Fenneman, Annals Assoc. Amer. Geogr., Vol. VI. 1917. pp. 19-98.

[35] Dryer, Indiana Supplement, Natural One-Book Geography: Amer. Book Co., 1911.

The Calumet Lacustrine Section. The Calumet lacustrine section is mainly the area south of the head of Lake Michigan which was once covered by lake waters. It extends northeastward a few miles along the lake shore into Michigan to near the mouth of Gilien River, and westward into Illinois to a point a few miles north and west of Chicago. During this expansion of the lake it had its outlet through the Illinois River to the Mississippi. Its present outlet was covered by glacial ice. The southern boundary of the lacustrine plain is marked by an old beach line known as the Glenwood beach.[56] This old beach line has an altitude of about 640 feet, and is about 60 feet above Lake Michigan. It extends from Dyer eastward to near Wheeler in Porter County, and thence north and northeast to within 2 or 3 miles of the lake shore, and parallel with it into Michigan. The lake plain passes a few miles up Salt Creek and up the Calumet River to near the east line of Porter County. The name is taken from the Calumet River which traverses the lacustrine plain in its best developed portion.[57] The area covered by the Calumet lacustrine section in Indiana is approximately 275 square miles. The entire section around the head of Lake Michigan has an area of 600 square miles or more.

The Calumet lacustrine section is by no means a featureless lake plain. It is marked by a number of low sand ridges which extend parallel to Lake Michigan. These low sandy ridges are old beach lines marking the successive stages in the withdrawal of the old lake from the area. The outer margin of the plain is somewhat trenched by Deep River and its small tributaries in the vicinity of Hobart.[58] East and northeast of Gary and adjacent to the lake occurs a remarkable belt of dunes.[59] This dune belt has in it some of the largest sand dunes in the United States. Mt. Tom, north of Chesterton in Porter County, rises 190 feet above the surface of Lake Michigan, and several other massive dunes in that vicinity nearly equal it in height. But except for the low ridges of the old beach lines, the belt of dunes between Gary and Michigan City, and the stream trenching at the outer and higher part of the lacustrine plain, the region is fairly featureless. The Calumet River flows sluggishly over the flat plain in a shallow channel which resembles a canal. The altitude of the general plain is below 640 feet. Lake Michigan has an altitude of 581 feet.

The Valparaiso Moraine Section. The Valparaiso Moraine[60] is a massive moraine curving about the head of Lake Michigan. It receives its name from

[56] In the following references the full details of the Calumet lacustrine region are taken up: *Leverett*, Pleistocene Features of the Chicago Area: Bull. Geol. and Nat. History, Chic. Acad. Science, No. 2, 1897, pp. 1-86; The Pleistocene of Indiana and Michigan: U. S. Geol. Surv., Monogr. 53, 1915, pp. 350-357. *Alden*, The Chicago Folio No. 81, U. S. Geol. Surv., 1902, pp. 1 and 7-12. *Blatchley*, Geology of Lake and Porter Counties, Indiana: 22nd Ann. Rept., Ind. Geol. Surv. for 1897, pp. 25-104.

[57] The Calumet lacustrine section is exactly equivalent to the "Chicago Plain" of Alden. See Chicago Folio No. 81, U. S. Geol. Surv., 1902.

[58] See, Tolleston topographic sheet, U. S. Geol. Surv., 1900.

[59] *Blatchley*, Geology of Lake and Porter Counties, Indiana: 22nd Ann. Rept., Ind. Geol. Surv. for 1897, pp. 29-41. *Alden*, Chicago Folio No. 81, U. S. Geol. Surv., 1902, p. 12. *Barrett*, The Dunes of Northwestern Indiana: 41st Ann. Rept., Ind. Geol. Surv. for 1916, pp. 11-27.

[60] *Leverett*, Pleistocene Features of the Chicago Area: Bull. Geol. and Nat. History, Chic. Acad. Science, No. 2, 1897; The Pleistocene of Indiana and Michigan: U. S. Geol. Surv., Monogr. 53, 1915, pp. 214-221. *Blatchley*, Geology of Lake and Porter Counties, Indiana: 22nd Ann. Rept., Ind. Geol. Surv. for 1897, pp. 43-55.

Valparaiso in Porter County, in the vicinity of which its knoll and sag type of topography is well represented. The Valparaiso moraine is the divide between the St. Lawrence and Mississippi basins in the northwestern part of the state. Its inner margin at the Illinois-Indiana line is about 15 miles from Lake Michigan and at the Indiana-Michigan line it is only 2 miles distant. It covers an area of approximately 600 square miles in Indiana. It is much broader and lower westward from Valparaiso. The westward portion is from 12 to 15 miles broad and its highest knolls in Lake County attain an altitude of about 750 feet. This portion is composed of three morainic ridges which nearly run together, leaving little or no level till plain between them. Few of the knolls rise more than 20 or 30 feet above the nearby sags. Northeast from Valparaiso the moraine has an average width of 8 miles and its highest knoll, which is in Laporte County, reaches an altitude of 885 feet. Its outer or southern margin is but little above the massive outwash plain which flanks it on the south. But on the inner margin it rises abruptly above the Calumet lacustrine plain, and within 2 miles has risen 150 feet. Its highest knolls are approximately 300 feet above the surface of Lake Michigan and approximately 10 miles distant. Its basins contain a number of small lakes. Hudson Lake in northeastern Laporte County and Cedar Lake in middle Lake County cover more than a square mile each, and are the largest lakes on the Indiana portion of the moraine. Both of these lakes are very shallow.

The Kankakee Lacustrine Section. The name Kankakee lacustrine section is applied to a great system of sandy lacustrine plains, outwash plains, valley trains, and local enclosed till plains associated with a great line of glacial drainage and ponding along the St. Joseph, Kankakee, Tippecanoe, and Iroquois rivers, embracing portions of southwestern Michigan, northwestern Indiana and a smaller portion of northeastern Illinois. The section is bounded on the west and north by the Valparaiso moraine in Indiana. The southern boundary in southern Newton, southern Jasper, and middle White counties has been already described in connection with the northern boundary of the Tipton till plain. The eastern boundary extends northward from the Wabash River in western Cass County along the morainal country to near South Bend; thence eastward south of the St. Joseph River to northeastern Elkhart County. The gravel plains continue up the Pigeon and Elkhart rivers and may be considered either as extensions of the Kankakee section or as local plain areas in the generally morainic country into which they extend. In Fulton County the Tippecanoe River and its tributary Mud Creek are bordered by wide marshy tracts which are apparently lacustrine in origin, and are therefore included in the Kankakee lacustrine section.

The section may be further divided into smaller units. The largest and most characteristic portion is the area adjacent the Kankakee River extending from near South Bend southwestward into Illinois, with a width of 15 or 20 miles. It comprises the Kankakee marsh[a] and adjacent sand and gravel plains. The second most prominent division occurs along the Tippecanoe River from near Rochester in Fulton County to near Ora in southeastern Starke County, and thence southward as a great imperceptible sag to Monticello, and west through northwestern White, southern Jasper, and southern Newton counties, comprising the areas drained by Monon Creek and Iroquois River.

[a] Blatchley, Geology of Lake and Porter Counties, Indiana: 22nd Ann. Rept., Ind. Geol. Surv. for 1897, pp. 55-65.

This Tippecanoe-Iroquois plain merges with that of the Kankakee near the line between Starke and Pulaski counties. These two divisions are separated in Jasper and Newton counties by a third division which is in no way characteristic of the region as a whole. The Marseilles moraine passes from Illinois through middle Newton and Jasper counties, just to the north of Rensselaer and thence northwest to northwestern Pulaski County where it dies out in the midst of the lacustrine sand plain. This narrow moraine with its hummocky topography is quite in contrast with the Kankakee and Iroquois plains to the north and south respectively. The fourth division of the Kankakee lacustrine section is only partly represented in Indiana. It is the great gravel plain of the St. Joseph, Kalamazoo, and Dowagiac rivers which comprises a large area in southwestern Michigan and laps over somewhat into Indiana. It stretches eastward from South Bend along the St. Joseph and Pigeon rivers. This great gravel plain was the course of glacial drainage which passed near the present site of South Bend and into the ponded drainage of the Kankakee area.[a] It slopes toward the great sag between the present head of the Kankakee River and the St. Joseph at South Bend, and is continuous with the Kankakee plain, though its slope is somewhat greater.

The greater part of the Kankakee lacustrine section is characterized by a thin deposit of somewhat ridged sand. Dunes are common. This sand was probably deposited in ponded or laked waters, and takes the place of the more commonly deposited silt so characteristic of lake plains. It was probably derived from the glacial drainage over areas associated with the outwash plains and valley trains, chief of which must have been the drainage leading down the glacial St. Joseph Valley. The general altitude of the sand plains of the Kankakee lacustrine section is very well shown on the topographic map. The gravel plains of the St. Joseph River area; the outwash gravel plain stretching southeast in St. Joseph County from the outer margin of the Valparaiso moraine; the Marseilles moraine and a broad area lying between the Kankakee and Tippecanoe rivers and a broad strip flanking the morainal country at the eastern margin of the section, rise above the 700-foot contour line. The great area of the Kankakee marsh, the great sag occupied by the Tippecanoe River, and the large area drained by Monon Creek and Iroquois River, lie at an altitude mainly between 650 feet and 700 feet above sea level. In general the slope is to the southwest, except for the sag along the Tippecanoe River north of Monticello.

"The greater part of the sand-covered area has a nearly plane surface. The Kankakee marsh, with an area of nearly 1,000 square miles, is very flat, while the portion north of the marsh has scarcely any ridges worthy of note. In southeastern Pulaski and northern White counties, Indiana, the surface is mainly level and level tracts are quite extensive in southern Jasper County. In Illinois the surface is mainly plane, except on the outer face of the Iroquois [Marseilles] moraine and in a belt a few miles wide which follows the south border of the Kankakee River, in both of which places there are ridges and dunes of some prominence.

"The ridged portions of the sand in Indiana occupy southeastern Starke, much of Pulaski, and the central portions of Jasper and Newton counties, all of which lie southeast of the Kankakee marsh, as well as a narrow strip on the east and south borders of the sand area in Fulton, Cass, White, and

[a] Montgomery, The Kankakee Valley: Proc. Ind. Acad. Sci. for 1898, pp. 277-282.

Jasper counties. There are also scattering ridges in the midst of the level portions of the sand area in that state.

"The most prominent ridges are 35 or 40 feet in height, but the majority are less than 20 feet, and many are only 5 or 10 feet. The individual ridges vary in breadth from 50 feet or less up to nearly one-eighth mile, but are usually about 200 to 300 feet. Among the ridges even where most prominent, there are narrow strips with nearly plane surfaces.

"The prevailing trend of the ridges is usually easy to determine, though in places the ridges wind about and interlock, forming an intricate network. Those on the east border, in Pulaski County, Indiana, show a tendency to a north-south trend, while those on the south border in Cass, White, and Jasper counties, trend nearly east to west. Those on the south border of the Kankakee trend about with the course of the stream, south of west in the Indiana portion and north of west in the Illinois portion. Between the ridges bordering the Kankakee in Indiana, and those on the south and east borders of the sand area, the trend is not so easily systematized. The ridges there are arranged in groups or strips, among which there are extensive plane tracts, often boulder strewn, and having only a thin sand coating."[63]

Flanking the outer margin of the Valparaiso moraine in middle eastern Porter County, and extending northwest through Laporte County into Michigan, is a great outwash apron which has an altitude of 775 to 800 feet at the margin of the moraine. It slopes southeast for 8 or 10 miles to the edge of the Kankakee marsh where it has an elevation of about 700 feet. This outwash apron constitutes an extensive gravel plain made by the drainage from the ice during the deposition of the Kalamazoo moraine. In places it is pitted. The drainage from the Valparaiso moraine passed through the broad valleys of Crooked and Mill creeks chiefly. These valleys are somewhat trenched below the gravel plain.

The Marseilles[64] moraine in Indiana has a "long gradual slope on its north or iceward side, and a rather abrupt descent on its south or landward side. The rise from the plain to the crest of the moraine on the north amounts to 20 to 40 feet in about 2 miles and on the south to 30 to 50 feet in less than 1 mile." The moraine dies out in the midst of the sandy plain in northwestern Pulaski County. It is highest and best developed north of Rensselaer where its highest knolls attain an altitude of 80 or 100 feet above the sand plain. But much of the moraine lies between 700 and 725 feet above sea level. "The surface of the ridge carries low knolls and ridges among which sloughs and shallow basins are enclosed."

The Steuben Morainal Lake Section. The Steuben morainal lake section receives its name as a section from Steuben County, Indiana, where the moraines of both the Saginaw and Erie ice-lobes are well represented, and among which numerous lakes abound. The section includes the moraines made by the differentiated Saginaw ice-lobe and those made by the north limb of the differentiated Erie ice-lobe north of the Wabash River. These ice-lobes were somewhat crowded together in an area extending from near Logansport on the Wabash River, northeast into Michigan near the northeast corner of the

[63] Leverett, U. S. Geol. Surv., Monogr. 38, 1899, pp. 332-333. For further interesting details of the region see *Leverett*, U. S. Geol. Surv., Monogr. 53, 1915, pp. 128-130; *McBeth*, The Tippecanoe, An Infantile Drainage System: *Proc.* Ind. Acad. Sci. for 1909, pp. 341-343; *Barrett*, Dunes of Northwestern Indiana: 41st Ann. Rept. Ind. Geol. Surv. for 1916, pp. 11-27.

[64] Leverett, U. S. Geol. Surv., Monogr. 53, 1915, pp. 126-128.

state. This crowding of the ice-lobes resulted in great massing of morainal material on this area, and particular in Kosciusko, Whitley, Noble, Dekalb, Lagrange, and Steuben counties. Chamberlin[65] has defined such a morainal accumulation an *interlobate moraine*, a designation very appropriate for the crowded assemblage of moraines between the positions of the two ice-lobes. He regards the eastern limb of the outer Saginaw morainic loop and the northern limb of the Maumee or Erie moraine "as a joint intermediate [interlobate] moraine." Dryer[66] definitely calls the morainic mass the Saginaw-Erie *interlobate moraine*. "This interlobate moraine," says Dryer,[67] "is the joint product of the glacial ice which passed from Saginaw Bay southwestward across Michigan and northern Indiana, and another lobe which entered Indiana from Lake Erie and covered nearly the whole state south of the Wabash River." This morainic mass is here given the name Packerton moraine, from the village of Packerton in southern Kosciusko County. Aside from this dominating interlobate morainal mass extending from near Logansport northeast through the state, several distinct moraines of both the Erie and Saginaw ice-lobes are prominent. The Maxinkuckee moraine extending northward from the Wabash River some 10 miles west of Logansport to the state line north of South Bend, marks the western limit of the Steuben morainal lake section. Several small moraines extend nearly at right angles between the broad Maxinkuckee moraine and the interlobate morainal mass. These smaller moraines mark the stands of the Saginaw ice-lobes in its retreat northward. Following the names suggested by Leverett,[68] the Bremen, New Paris, Middlebury, Lagrange, and Sturgis moraines are the most prominent. Other moraines belonging to the same series, but not distinguished as such by Leverett, are the Rochester, Burket, and Topeka moraines. These are shown in position on the glacial map, Pl. III. The moraines of the Erie ice-lobe north of the Wabash are the Mississinewa, Salamonie, Wabash and Ft. Wayne moraines named in order from the outer moraine inward, as represented in Indiana. The southern boundary of the Steuben morainal lake section in Indiana extends along the Wabash River to Huntington, and thence up the Wabash-Erie valley to Ft. Wayne, and thence northeast at the inner margin of the north limb of the Ft. Wayne moraine which is marked by an old lake beach line.

Genetically the topographic condition of the Steuben morainal lake section is the result of the Saginaw and Erie ice-lobes and the associated glacio-fluviatile features. This condition can be described more understandingly by taking the moraines produced by each of the ice-lobes in order of their age. Accordingly the moraines of the Saginaw ice-lobe will be discussed first, beginning with the outer moraine.

From the standpoint of the topographic features of the surface, the ice-lobes of the Wisconsin glacier gave rise to a series of curving terminal moraines concentric with the margin of the ice where the successive series of halts were made during the waning period of the last glacial stage. The Saginaw ice-lobe as differentiated from the Erie on the east and the Lake

---

[65] Chamberlin, *Preliminary Paper on the Terminal Moraine of the Second Glacial Epoch*: U. S. Geol. Surv., Third Ann. Rept., 1882-83, pp. 301-2 and 330.

[66] Dryer, Geology of Whitley County, Indiana: Ind. Dept. of Geol. and Nat. Resources for 1891, p. 168.

[67] Dryer, Geology of Noble County: Ind. Dept. of Geol. and Nat. Resources for 1893, page 18.

[68] Leverett, The *Pleistocene of Indiana and Michigan*: U. S. Geol. Surv., Monogr. 53, 1915, pp. 130-170.

Michigan on the west first became marked when the ice had receded to the vicinity west of Logansport in western Cass County. From this vicinity northward extends the Maxinkuckee moraine. The moraine is rather weak through western Cass and western Fulton counties. For the most part the boulder till seems to be of ground moraine origin with only a slightly undulating topography, but a weak terminal moraine some 2 miles wide with knolls seldom more than 25 feet in height and an altitude ranging from 760 to 800 feet has been defined south of the Tippecanoe River. North of the Tippecanoe River to the vicinity of South Bend occurs a belt of uneven-surfaced drift 5 to 15 miles wide, and attaining in its highest knolls an altitude of nearly 900 feet. This is the Maxinkuckee moraine. The moraine at its outer margin adjacent to the Kankakee plain is prominent, but its inner border is rather indistinguishable from the gently undulating drift that stands nearly as high as the moraine itself. For some distance south of South Bend the knolls are prominent, rising abruptly 50 or 60 feet above their bases. Sloughs and enclosed basins are numerous, and lakes are plentiful. Lake Maxinkuckee in southwestern Marshall County is the largest lake, having an area of approximately 3 square miles and a maximum depth of 89 feet. The outwash at the outer border is conspicuous in places and extends up the Yellow and Tippecanoe rivers.

Northeast from the apex of the differentiated Saginaw outer moraine in western Cass County extends a great morainic mass which was referred to above as an interlobate morainal mass between the Saginaw and Erie lobes. It is likely that much if not most of this moraine was made by the Saginaw lobe, but it may subsequently have been partly moved or over-ridden by the Erie ice-lobe which very probably persisted longer than the Saginaw ice-lobe. Leverett[9] regards this moraine as largely the result of the Erie ice-lobe. Dryer[10] regarded it as the product of the Saginaw ice-lobe. The writer is inclined to favor the view of Dryer. The moraine, here designated the Packerton moraine, may be logically regarded as the east limb of the Maxinkuckee moraine. It is no more massive than should be expected, considering the crowding of this ice-lobe by the north limb of the Erie ice-lobe. Further it is not likely that the western margin of the Saginaw ice-lobe would give rise to a moraine as prominent as the Maxinkuckee moraine and have no similar correlative on its eastern limb, a feature not fully considered by Leverett. Northeastward from near Albion in Noble County, the Mississinewa moraine of the Erie ice-lobe laps upon and over this moraine, giving rise to altitudes of nearly 1,200 feet in Steuben County, on the highest knolls of the moraine. It is likely that the Saginaw outer moraine north of Albion was feeble, in keeping with its condition north of South Bend on its western limb.

Leverett[11] in discussing the topography of this morainic mass calls attention to the altitudes as "below 700 feet near Delphi, but reach 775 feet north Logansport, 800 feet in northeastern Cass County, 900 feet in southeastern Fulton County, 1,000 feet in eastern Noble County, and 1,100 to 1,150 feet on the highest points in northern Steuben County.

"From data furnished by deep borings it appears that the altitude of the bed-rock surface does not rise to the northeast, but is remarkably uniform

[9] Leverett, The Pleistocene of Indiana and Michigan: U. S. Geol. Surv., Monogr. 53, 1915, pp. 158-159.

[10] Dryer, The Geology of Noble County: Ind. Dept. of Geol. and Nat. Res., 1893, pp. 22-28.

[11] Leverett, U. S. Geol. Surv., Monogr. 53, 1915, pp. 159-160.

at about 600 feet above sea level all along the morainic belt in northeastern Indiana. A few miles north of the state line, however, in Hillsdale County, Mich., bed-rock rises very rapidly and attains an altitude of over 1,100 feet. The great drift accumulations in northeast Indiana are laid down, as it were, in the lee of the prominent table-land of sandstone of the Marshall formation that has its southern terminus in Hillsdale County, Mich.

"In Whitley, Kosciusko, Fulton, Wabash, Miami, and Cass counties the moraines form the divide between Tippecanoe and Eel rivers, both of which are tributary to the Wabash. North of the head of the Tippecanoe the several headwaters of Elkhart River find their source in these moraines and lead northwestward to St. Joseph River and thence to Lake Michigan. The opposite or southeastern slope of the moraines is drained by the tributaries of the St. Joseph (of the Maumee), which belongs to the Lake Erie drainage. In Steuben County these moraines are traversed from west to east by three small streams, Turkey Creek, Pigeon River, and Crooked Creek, which have their sources near the eastern border on much lower land than the main crest of the moraine. Their discharge is into St. Joseph River and thence to Lake Michigan.

"In most places the drainage divide is at the crest of the main moraine, most of which is higher than the general level of the western edge, though some prominent points along the western edge are as high as the main crest. The relief of the main crest above the outer border district is generally between 50 and 75 feet and above the plain on the inner or eastern border 100 to 150 feet, the country to the southeast being lower than to the northwest.

"The topography of the southwestern portion of this series of moraines in Cass County and northern Miami County, Ind., is of a swell and sag type in which there are few basins. The swells are 10 to 40 feet in height and are closely aggregated, giving the moraines a strong expression compared with the nearly plane district to the northwest. Lake basins become conspicuous in southeastern Fulton and northwestern Wabash counties and continue numerous all along the morainal belt to the northwest. They are much more prominent on the northwestern . . . slope than on the southeastern . . . slope and are especially numerous where moraines of the Saginaw lobe connect these moraines. The largest lakes exceed a square mile in area, and one, James Lake, has an area of 2.6 square miles; the majority, however, cover but a small fraction of a square mile and some occupy but a few acres. Many of them have depths 50 to 75 feet or more below the level of the outlets and more than 100 feet below the surrounding parts of the moraine. James Lake, with a reported depth of 87 feet, has on its eastern border hills that rise more than 100 feet above the water surface, thus giving the basin a depth of about 200 feet below the neighboring hills. The highest knolls along the moraine are probably those in northern Steuben County, which rise more than 100 feet above bordering sags. Knolls 50 to 60 feet in height are found at short intervals as far southwest as northern Wabash County, but most knolls are only 20 to 40 feet high and many of them rise less than 20 feet above the bordering sags. . . . Marshes and nearly plain tracts are much less common on the inner [eastern] slope than on the outer [western]. In places on the outer [western] slope sharp ridges known as 'hogbacks,' some of which are to be classed as eskers, are not uncommon; they are, however, very short, none exceeding a mile in length having been observed. They commonly form the borders of the lake basins; one lake in Steuben County has received the name 'Hogback Lake' because of its close association with one of these ridges.

"On the outer [western] slope of the moraine deep indentations show where the lines of glacial drainage lead away from the moraine. In Steuben County these glacial channels have their heads on the inner [eastern] border of the moraine and thus completely traverse the moraine. One of them is occupied by Turkey Creek, another by Pigeon River, and a third by Concord Creek and a string of lakes tributary to Crooked Creek."

Dryer[11] in discussing the topographic condition of the interlobate morainal mass in Noble County emphasizes the action of glacio-fluviatile waters on the Saginaw morainic drift. "Much of this diversity," he says, "and irregularity is due to the fact already alluded to, that during the period of glacial retreat and rapid melting of ice, vast volumes of water escaped over the crest of the outer moraine, and flowing across the face of this country, cut and gashed it into irregular gorges, leaving corresponding ridges between. The greater part of the material left by the Saginaw glacier has been washed away or redistributed to such an extent as to leave few of the original features recognizable. Scattered patches of hills and the more prominent ridges are probably all that escaped the action of the flood of water. The valleys themselves, originally cut to a depth of one hundred feet below their present bottom levels, have since been filled up with silt and vegetable growth, forming large areas of marsh and much meadow, with frequent pools of open water. The area of marsh land has been estimated to comprise ten or fifteen per cent of the whole county, and in some townships it must be nearly fifty per cent."

Several broken moraines cross the inner space between the western limb of the Maxinkuckee moraine to its eastern correlative, the Packerton moraine, and the over-topping Mississinewa moraine. From southwest to the northeast Leverett has named these moraines the Bremen, New Paris, Middlebury, Lagrange, and Sturgis moraines. The Bremen moraine extends from the Maxinkuckee moraine near Lakeville in southern St. Joseph County southeast past Bremen to near Warsaw where it connects with the eastern limb of the Maxinkuckee moraine. This moraine is broken by a number of gaps. It is low as a morainic mass and the knolls rise 10 to 20 feet above their bases. The New Paris moraine, named from the town of New Paris in southern Elkhart County, arises south of the St. Joseph River southeast of Elkhart as a broad undulating mass of drift and passes southeast through northeastern Kosciusko County to the east limb of the Maxinkuckee moraine in the southwestern Noble County. Dryer has called this moraine the Turkey Creek moraine from Turkey Creek which flows parallel with it just to the north. The moraine rises only slightly above the outer margin. "It has low swells with gentle slopes, a common height being 20 feet, though more fall below than exceed that height. Some knolls north of Boydstown Lake are 30 to 40 feet high and have steep slopes. 'Waybright hill,' a conspicuous irregular-shaped mass of drift about 2½ miles south of New Paris, is by far the most prominent feature of the moraine; it reaches an altitude (barometric) of 1,025 feet, or about 215 feet above New Paris station. . . . The remainder of the moraine in Elkhart County, except perhaps a few knolls 3 to 4 miles southwest of Goshen, is below the 900-foot contour, and the part in Kosciusko County reaches 900 feet only at the extreme eastern edge of the county. The portion in Noble County rises a little

[11] Dryer, Geology of Noble County: 18th Ann. Rept., Ind. Geol. Surv. for 1893, pp. 22 and 29.

above 900 feet, with a few points near High Lake that approach 1,000 feet.'"[73] A number of lakes are along or near the New Paris moraine. Turkey Lake with an area of 5.66 square miles lies just to the north of it in northeastern Kosciusko County. Tippecanoe, Boydstown, Little Eagle, Bear, High, Ridinger, Dwart, and Milford are the principal lakes. Tippecanoe Lake with a depth of 121 feet is reported to be the deepest lake in Indiana.

The Middlebury moraine, named by Leverett from the town of Middlebury in northeastern Elkhart County, is about 20 miles long and extends from the St. Joseph River near Bristol in northern Elkhart County southeast into western Lagrange County. The relief of the southern border ranges from 10 to 120 feet. A prominent point south of Bristol rises over 1,000 feet above sea level, and more than 200 feet above the St. Joseph Valley to the northwest. This portion of the moraine has many sharp knolls, a few of them 75 to 100 feet high.

The Lagrange moraine, named from Lagrange in central Lagrange County, extends from the state line in northeastern Elkhart County southeast to a short distance east of Lagrange, where a prong turns southward and joins the Mississinewa moraine a short distance southeast of Wolcottville in northeastern Noble County. South of Lagrange another prong of the moraine or probably a portion of an older moraine extends southwest to Elkhart River near Ligonier, and after a turn to the west is cut by Elkhart River, but is continuous westward to near the point where Solomon Creek enters Elkhart River a short distance east of New Paris. Leverett believes this moraine belongs to the Erie ice-lobe, but it more likely belongs to the Saginaw system and is intermediate in age between the New Paris and Middlebury moraines. The Lagrange moraine is continued to the northwest in Michigan beyond the St. Joseph River. The Lagrange moraine is nearly everywhere more than 900 feet in altitude, and a point on the county line in extreme northeastern Elkhart County reaches more than 1,000 feet in altitude. It consists of numerous steep-sided knolls among which are marshes and undulating tracts. Several notable lakes are associated with the moraine in southern Lagrange County northwest of Wolcottville among which are Oliver, Witmer, Dallas, Clearspring, and Adams lakes. In northeastern Elkhart and northwestern Lagrange counties Dryer[74] describes it as quite rugged. The region "displays a tract of hills or knobs, about twenty square miles in area, which, for tumultuous ruggedness and irregularity, are scarcely surpassed in Indiana. The northern and eastern borders of the Pigeon River moraine [Lagrange] are bold and well defined, while its southern and western borders display an extremely irregular apron or fringe where differentiation and delimitation of ground moraine from terminal prove elusive and baffling."

The Sturgis moraine lies mostly in Michigan, being barely represented in Indiana in northeastern Lagrange County. This moraine is designated the Fawn River moraine by Dryer, and is described as "about two miles wide, distinct but feeble, except around Wall Lake."

The intermorainic spaces within the area covered by the Saginaw ice-lobe in the northern tier of counties is largely made up of outwash and gravel plains. These plains have been designated as principally belonging to the St. Joseph gravel plain and places as a division of the Kankakee lacustrine sec-

[73] Leverett, The Pleistocene of Indiana and Michigan: N. S. Geol. Surv., Monogr. 53, 1915, p. 129.

[74] Dryer, Geology of Lagrange County: 18th Ann. Rept., Ind. Geol. Surv. for 1893, p. 79.

tion. Through northeastern St. Joseph, Elkhart, Lagrange, northeastern Kosciusko, and northwestern Noble counties the gravel plains make up probably half of the area. Southward the intermorainic spaces are largely ground moraine usually of an undulating nature, though a considerable tract in St. Joseph and western Elkhart counties north of the Bremen moraine is quite flat. In the district extending from the apex of the differentiated Saginaw outer moraine in western Cass County and the Bremen moraine "swampy basins or depressions are numerous, probably one-fourth of the surface being occupied by them. Most of the dry land has a wavy or gently undulating surface. Most knolls are but 10 or 12 feet high and few of them reach 20 feet. The features are such as might result from a rapid receding ice border. The most prominent ridge noted is one just west of Atwood, in western Kosciusko County. It rises about 40 feet above the surrounding country, is a mile in length, and nearly one-half mile in width."[73]

The second division genetically of the Steuben morainal lake section consists of the moraines constructed by the northern limb of the Erie ice-lobe, lying north of Maumee-Wabash waterway, and occupying northern Wabash, Huntington, and Allen counties, practically all of Whitley, Dekalb, and Steuben counties, and eastern Noble and Lagrange counties in Indiana. The northern limbs of the Mississinewa, Salamonie, Wabash, and Ft. Wayne moraines are represented.

The Mississinewa moraine extending from the Wabash River at Lagro northeast through northeastern Wabash, Whitley, eastern Noble and western and northern Steuben counties into Michigan with its southern limb southeast and east from the Wabash at Lagro, is the largest and best defined moraine in Indiana, and is probably not surpassed in excellence of development elsewhere. It is from 3 to 4 miles wide from Lagro to Columbia City. Northeastward from Columbia City it is quite broad, attaining widths varying from 5 to 15 miles. Its elevation varies from 800 feet or more at the south to about 1,200 feet in "Hell's Point" in Steuben County. As a morainic mass it is best developed in Whitley, Noble, Lagrange, and Steuben counties. Dryer[74] says of it: "It is an irregular, variously undulating pile of clay, sand, gravel, and boulders with the coarse materials predominating. Its surface is 150 to 300 feet above the country on either side, and its total thickness down to bedrock from 200 to 485 feet. Its topography defies verbal description in detail, but may be included under a few general types. The greater part of the area may be designated as *crumpled*, resembling the surface of a sheet of paper which has been carelessly crushed in the hand and then spread out. The ridges have no particular direction, their tops are broad and slopes gentle, yet there is very little level ground. This type passes by insensible graduations into the *corrugated*, in which the ridges are steeper, sharper, and arranged in somewhat parallel lines. Similar features very much exaggerated produce what may be called *gouged* or *chasmed* country, found in perfection southwest of Columbia City. The surface is entirely occupied by deep, irregular, elongated valleys, with narrow, sharp, winding ridges between, all in indescribable confusion. The roads through it are very crooked in order to

[73] Leverett, The Pleistocene of Indiana and Michigan: U. S. Geol. Surv., Monogr., 53, 1915, p. 132.

[74] Dryer, Studies in Indiana Geography: Inland Pub. Co., Terre Haute, 1908, p. 46.

avoid the marshes, yet in every direction, they are a series of steep descents and ascents. The relief might be imitated by taking a block of plastic clay and gouging it with some blunt instrument in as irregular a manner as possible.

"Scarcely more extreme and peculiar is the topography usually regarded as typical of terminal moraines, 'the knob and basin.' It consists of confused groups of dome-shaped or conical hills, often as steep and sharp as the materials, usually sand and gravel, will lie, with hollows of corresponding shape between. The impression made is as if the material had been dumped from above and left as it fell, like gravel from a wagon. Some of the finest specimens in America occur south of Albion, in the Diamond lake hills near Ligonier, east of Lagrange, in the northwest corner of Lagrange County, and grandest of all, north of Angola, where the peaks rise to about 1,200 feet. Throughout this morainic region the hollows or 'kettle-holes' are occupied by marshes or lakes, . . . and the marshes, or extinct lakes, out-number the living ones."

The northern limb of the Salamonie moraine extends northeastward from Wabash a short distance west of Huntington, through northern Huntington, southeastern Whitley, northwestern Allen, southeastern Noble, northwestern Dekalb and eastern Steuben counties. Through Noble and most of Dekalb counties it coalesces with the more massive Mississinewa on the west. It has a rather mild topography of the swell and sag type, though many knobs and basins exist. A number of small lakes are present on the moraine. It is best developed in eastern Steuben County where it is 2 to 6 miles wide and has a considerable range in relief and possesses many massive, rounded and dome-shaped hills. The altitude of the moraine varies from somewhat over 800 feet at the south to 1,050 feet in Steuben County.

The northern limb of the Wabash moraine extends from the Erie-Wabash glacial col in western Allen County northeast into Ohio from northeastern Dekalb County. It is separated from the northern limb of the Ft. Wayne moraine on the south by the valley of the St. Joseph River. The moraine varies from 4 to 8 miles in width. The wider portion in Dekalb County is not of a distinctly terminal moraine character on its northern edge. The general moraine stands about 80 feet above the St. Joseph valley. It is a broad, rolling upland with few knolls exceeding 20 feet in height. It has only a few small lakes, though many hollows and small saucer-shaped marshes exist bespeaking the former existence of small lakes. The moraine is crossed by Fish Creek in northeastern Dekalb County in a gorge 50 or more feet in depth, and by Cedar Creek in northern Allen County, with a gorge 100 feet in depth.

The northern limb of the Ft. Wayne, or St. Marys-St. Joseph moraine of Dryer, extends from Ft. Wayne northeast out of the state just south of the St. Joseph River. It has an average width of about 4 miles, and rises 50 to 75 feet above the Maumee lacustrine plain to the south. Its surface topography is mild and of an undulating character. A number of kame-like hills lie in the eastern part of the glacial col west of Ft. Wayne.

Between the recessional moraines of the northern limb of the Erie ice-lobe occur notable areas of level to gently undulating ground moraine, lying considerably below the crests of the moraines. The strips of ground moraine are well represented in the south of Noble and Steuben counties. They are fairly featureless, and have their companion in the vast Tipton till plain south of the Wabash.

Aside from the system of recessional moraines which mark the area occupied by the north limb of the Erie ice-lobe and the alternating areas of level to undulating ground moraine, the glacial drainage lines are worthy of note. Those in Steuben and eastern Lagrange counties have already been mentioned. Probably next to the Maumee-Wabash glacial waterway, that of Cedar Creek and Eel River is the most striking in the northern portion of Indiana. This glacial channel extends along Cedar Creek in Dekalb County into northern Allen County, where Cedar Creek turns abruptly from it into a gorge through the Wabash moraine. The glacial channel continues more directly southwest, and crosses the Salamonie and the massive Mississinewa moraines without the slightest deviation from its course. It is quite evident that this route was made by a great glacial river and that Cedar Creek and Eel River simply inherited their present courses in this glacial channel. The route of Cedar Creek through the Wabash moraine was very probably made by a glacial tributary which flowed in the opposite direction from the present lower Cedar Creek through the gorge. The valley of the St. Joseph River was also the route of a large glacial stream, as evidenced by the gravel terrace of the valley and the presence of the gravel hills in the vicinity of the eastern portion of the glacial col southwest of Ft. Wayne.

The Maumee Lacustrine Section. The Maumee lacustrine section is the name applied to the lacustrine plain which was once occupied by an ice-dammed lake with its outlet near Ft. Wayne southwestward through a broad valley to the Wabash at Huntington. This lacustrine plain covers a large area in Ohio, and its western end or apex laps over into Indiana, where about 120 square miles of it is represented in Allen County. A distinct beach line extends along the inner margin of the Ft. Wayne moraine northeast from Ft. Wayne to the extreme northeastern corner of Allen County. The old beach line extending east and southeast on the southern border of the flat lacustrine plain is scarcely less distinct. The plain is almost completely featureless. The Maumee River, formed by the junction of the St. Mary's and St. Joseph rivers at Ft. Wayne flows in a tortuous channel sunk 25 to 40 feet below the plain, and is without flood plain or terrace. The plain has an elevation of about 750 feet above sea level. The opening to the Wabash is at about the same elevation or slightly higher. Near New Haven is a low sandy tract slightly above the general elevation of the plain. This sandy tract is evidently a delta deposit where the St. Mary's River debouched into the old lake during its existence. The lake which gave rise to the plain is known as Glacial Lake Maumee.[77]

[77] Dryer, Geology of Allen County: 16th Ann. Rept., Ind. Geol. Surv. for 1888, pp. 107-114.

CHAPTER IV.

GENERAL PHYSIOGRAPHIC DEVELOPMENT.

General Considerations.

Physiographic Factors and Their Results.

The physiographic development of a region involves a number of factors which in the course of time have left their impress directly or indirectly upon the landscape. Some of these factors are active while others are conditioning. Fundamentally, denudation, diastrophism, and vulcanism as geologic processes encompass any and all of the details of physiographical geology, yet these terms are seldom used. Forms directly due to diastrophism and vulcanism may be considered initial; they are surface expressions of readjustments taking place within the earth. Land forms due to denudation are subsequently developed upon the earth's surface, the position of which as a land mass may be due to diastrophism. Much of the detail of physiographic development, however, is given in terms which are classified under denudation. The active factors are the processes which act upon the land materials. The denudational agencies from the physiographic point of view may be placed in two convenient classes. First, the weathering agencies, or those processes which disintegrate or decompose the rock materials of the land *in situ*. Freezing and thawing, expansion and contraction, the work of organisms, oxidation, carbonation, solution and hydration are the chief destructional processes belonging to this class. These agencies rather prepare the rock materials of the lands for the erosional agencies, or those of the second class. Running water, waves, wind, and glacial ice are largely transporting agencies, yet they are also effective in getting possession of rock materials directly. Their final work is deposition, and usually more permanently where they cease to act because of the absence of the downward component under the influence of gravity.

Many of the land forms over the surface of the earth are the results of the temporary deposition of material by the erosive agencies, especially wind and glacial ice. But such forms as dunes and moraines are usually subject to removal by running water and eventually must succumb to its tendency to carry all the materials of the land to the sea, where they are permanently deposited.

The final result of the agencies of weathering and erosion is the reduction of the lands of the earth nearly to the level of the sea. Such reduction of an uplifted land mass is frequently referred to as the cycle of erosion. The various agencies acting during the development of the cycle of erosion are very unequal in importance, depending upon a number of conditions. Those factors which condition the relative amount of work done by the active agencies are of great importance in giving rise to land forms. Among the most important conditioning factors are climate, topographic conditions (including stage in the erosion cycle), geologic structure, and lithology.

In an arid climate wind is the agency which dominates, and its results are the most common. Other agencies are important, though they are overshadowed by the more obvious wind action. In a humid climate running water is the chief agency, and the forms produced by it largely predominate. Weathering in all cases is a silent destroyer of the surface rocks, but in the early and mature stages of the erosion cycle or in areas of considerable relief, mechanical agencies may predominate. In the later stages of the erosion cycle or on areas of slight relief chemical weathering is of prime importance, as water running over the surface has little erosive effect on account of the low gradient. Geologic structure is of importance in that it causes rocks of different degrees of resistance or rocks of different lithological characteristics to be presented at the surface. Land forms are frequently regionally dependent upon the lithology. Thus, limestone rocks often give rise to sinkholes; soft shales never permit angularity; sandstones usually furnish the roughest land forms; and granites give rise to rounded bosses.

The combined effect of all denudational processes upon a land mass produces a progressive change in topography. The materials of the land are wasted away, giving first a topographic condition known as the youthful stage of the erosion cycle, in which only a relatively small amount of the land mass has been removed. The stage of maturity in the erosion cycle is marked by extreme dissection and maximum relief. The amount of material which has been removed by the processes of land sculpture is approximately equal to the amount which is yet to be removed. The final stage in the cycle of erosion is that of the peneplain, or old age condition of topography, in which the entire area is cut down so low that running water has little effect. Regardless of the initial form of the uplifted land mass or the material of which it is constructed, peneplanation is the final result and marks the completion of the erosion cycle. All land masses are subject to this fate unless the erosion cycle is interrupted. Where the geologic structure, the lithology, and the altitude and distance from the sea are known the topographic condition in any particular stage of the erosion cycle may be reconstructed.

General Physiographic Development of the Eastern United States.

The topographic condition of the lands of the earth today is such that their physiographic histories may be read far back into the past. It is known that the peneplaned condition has prevailed over wide areas, and that many regions which have been peneplaned have since been uplifted and a new cycle or successive cycles of erosion have ensued. The eastern half of the United States reveals certainly two periods of wide peneplanation and two shorter periods of more localized peneplanation, each of which has been interrupted by uplift or warping. Indeed, the history of the last geological era is interpreted largely from the topographic condition of the lands. The broad diastrophic movements of the last geological era are clearly revealed in the present topography. The chronological order of the series of uplifts and partial peneplanations may be determined by the topographic forms present. But from the standpoint of geologic time these forms must be related to the fossils embedded in strata that were deposited contemporaneously. Thus a period of peneplanation of the land implies deposition of the removed material in the seas of the time. It remains to correlate these two phases of geologic activity in order to determine the geologic date of the period of peneplanation. By the application of such methods Davis, Willis, Hayes, and Campbell and others have shown that uplifted, deformed, and partly destroyed peneplains in the eastern part of the United States date back into the Mesozoic era. The oldest recognized peneplain was completed some time in the Cretaceous period, and is represented by the Cumberland plateau, the tops of the higher parallel ridges of the Appalachian mountains, and the mountainous plateau of New England. A later period of peneplanation less prolonged followed, and was ended about the middle of the Tertiary. The resulting peneplain was less wide-spread than the one completed in the Cretaceous period. A prominent wide-spread uplift over the United States marked the close of the Tertiary. Since the close of the Tertiary the great valley trenches which exist in the plains of the Mississippi basin have come into existence. The complete or partial filling of some of the valley trenches by glacial till shows that their development was well advanced at the time of the great spread of the continental ice sheets during the Pleistocene.

The surface of Indiana has been subjected in general to the same broad diastrophic impulses and intercalated periods of crustal stability accompanied by denudation, as the entire Mississippi basin. In company with the St. Lawrence and upper Mississippi basins it was subjected to a series of glacial invasions. The topographic forms present are such as to partly reveal the physiographic history. This history is in harmony with that of adjacent regions.

Relationship of the Indiana Region to the Gulf Embayment Area.

Apparently most of the decipherable history of the physiographic development of the Indiana region must be closely related to the Gulf embayment area. Drainage from the larger portion of Indiana has probably always been directed towards the Gulf embayment and the material carried by the streams from the region has been deposited within the seas of the Gulf of Mexico or its former embayment area extending to the northward. When the Gulf embayment reached as far north as southern Illinois, the Indiana region with its drainage directed towards the Gulf was not far removed from the margin of the sea, and at the close of any period of peneplanation must have been but little above sea level. As the land area of North America grew larger during the latter half of the Tertiary on account of the withdrawal of the embayment seas to the southward, the Indiana region must have become higher above sea level. This condition of greater elevation and increase in area was attained by uplift and by deposition of the waste material from the interior of the continent. It is likely that the Indiana region is farther from the sea at the present time than at any time in the past history of its topographic development. This conclusion is based on the fact that instead of an embayment extending towards the Indiana region there now exists a seaward extension of the land. This land extension has been built by the Mississippi River during Pleistocene and Recent times. The relationship of the various cycles or partial cycles of erosion in Indiana to the deposits of the Gulf embayment area is not entirely clear. Attempts have been made, however, to correlate the erosion cycles with those of other areas adjacent to the embayment deposits, where the relationships are more apparent. Lack of detailed topographic maps has hindered definite correlation, but the broader phases have been fairly well determined.

Formation and Influence of the Cincinnati Arch.

Towards the close of the Paleozoic era the continent of North America was uplifted and the long prevalent interior continental seas were withdrawn. Most of eastern North America has been above the sea since that time. It is very likely that the crest of the Cincinnati arch or geanticline was uplifted some time before the withdrawal of the sea from the basins on either side, and that it furnished the source of some of the sediments for the basins. At any rate the Cincinnati arch was uplifted to a much greater degree than the geosynclinal basin centering in southern Illinois and lying between the Cincinnati arch and the Ozark uplift of Missouri.

The crest of the Cincinnati arch is in Jessamine County, Kentucky. The axis of the arch descends northward and southward from the crest. The axis of the uplift passes the vicinity of Cincinnati, and thence north-northwest to Richmond, Kokomo, and Logansport. Near the latter place the strata commence to rise again along an axis that passes through the vicinity of Chicago and unites with the Wisconsin uplift. It may be said that the axes of the Wisconsin and Cincinnati uplifts merge in Indiana in the vicinity of Logansport. These axes of uplifts are of considerable importance physiographically, since they determine the distribution and position of the rocks which compose their flanks, especially as conditioned by their peneplaned and beveled edges which were later uplifted. The strata on the western flank of these uplifts dip 10 to 40 feet to the mile in a direction normal to their axes. This dip converges toward the synclinal basin in south-central Illinois. In northeastern Indiana on the eastern flank of the axes of the uplifts, the dip is somewhat lower and is in a northeastern direction toward the synclinal basin in central Michigan. The inclination of the strata from the axes of the uplifts was attained for the most part at the close of the Paleozoic era, as evidenced by the approximate equality of the dip of the strata of all the rock systems on the western flank of the geanticline.

The initial diastrophic arch was probably eroded as rapidly as it was formed, and the drainage was probably down its flanks. Great thicknesses of the strata of the later rock systems which were laid down over the area before the final bowing of the arch were removed from the crest and axis, though geologic evidence indicates that some of the strata deposited over it were much thinner than on the flanks or in the basins adjacent, and that during certain epochs of the Paleozoic following the Ordovician period no deposition at all occurred over the crest of the low initial arch.

Mesozoic Peneplanation.

The erosion of the Cincinnati arch and adjacent regions continued throughout the Mesozoic, and at its close the Cretaceous sea extended up the Gulf embayment area into southern Illinois. The Indiana region must have then been perfectly peneplaned and the surface little above the sea. It is probable that the Cretaceous sea actually invaded the southwestern part of Indiana and that the materials which were deposited there were subsequently removed. No known Cretaceous deposits occur at present nearer than southern Illinois and western Kentucky, in the vicinity of the mouth of the Tennessee River, approximately fifty miles distant from the southwestern corner of Indiana. In Kentucky and Tennessee outliers of the Tuscaloosa and Eutaw formations occur far east of the present general outcrop. In Lewis County, Tennessee, outliers occur as much as forty miles east of the main line of outcrop.[79] The

[79] Berry, Upper Cretaceous Floras of the Eastern Gulf Region in Tennessee, Mississippi, Alabama and Georgia: U. S. Geol. Surv., Prof. Paper 112, 1919, pp. 26-28 and Plate I.

Cretaceous sea floor upon which these outlying deposits rest is the resurrected Cretaceous peneplain, which is now designated as the Highland Rim peneplain, of early Tertiary age.

The Mesozoic era in the eastern part of the United States was largely given over to denudation of the lands. The Appalachian area which came into existence as a high mountainous mass at the close of the Paleozoic was peneplained. It is not known how many erosion cycles preceded the one which gave rise to an extensive peneplain at the close of the Cretaceous period. This peneplain is now represented by the Cumberland plateau surface[79] and other upland but more restricted surfaces in the Appalachian region. The geologic evidence indicates that conditions were such as to give at least two and possibly three erosion cycles. These erosion cycles probably formed peneplains, but it is likely that each erosion cycle in turn destroyed the peneplain made by its predecessor; or they may locally have coincided with each other or with the extensive peneplain present at the close of the Cretaceous period. It is not likely that any portion of the Cretaceous peneplain has been preserved in Indiana. It is possible, however, that some portion of the later developed Tertiary peneplain coincided with it.

Uplift at the Close of the Mesozoic.

The close of the Mesozoic era on the continent of North America was marked by wide crustal deformation over most of the continent. The well developed Cretaceous peneplain over the eastern part of the United States was uplifted and somewhat deformed.[80] Contrary to the generally elevated condition over the continent, the Gulf embayment area does not appear to have been disturbed. The Eocene Gulf embayment sea remained in approximately the same position as that occupied by the Cretaceous sea, and the sediments are such as to indicate that deposition was continuous.[81] The drainage of the Indiana region into the Gulf embayment area was probably unchanged. Without uplift of the Gulf embayment area there could have been no rejuvenation of the immediately contiguous areas, and the erosion cycle initiated by the uplifted continent elsewhere was not effective about the margin of the Gulf embayment area. Thus it is very likely that the Cretaceous peneplain immediately adjacent the embayment area was not destroyed, but was continued into the Tertiary. But uplift was effective at a distance from the Gulf embayment area, and the early Tertiary erosion cycle ensued.

Early Tertiary Peneplain.

During the Eocene Tertiary the erosion cycle initiated near the close of the Cretaceous was completed over areas of considerable expanse on either side of the Mississippi, though full peneplanation was largely restricted to areas of nonresistant rocks and adjacent to the more vigorous drainage lines. East of the Mississippi, the area immediately adjoining the Gulf embayment as a

[79] Not all geologists are agreed that the Cumberland plateau represents an up-lifted base-level plain of erosion. See Miller, A. M., Geology of Kentucky: Ky. Geol. Surv., Ser. V, Bull. II, 1919, pp. 195-196.

[80] Hayes and Campbell, Geomorphology of the Southern Appalachians: Nat. Geog. Mag., Vol. VI, 1894, pp. 63-126.

[81] Veatch, Geology and Underground Water Resources of Northern Louisiana and Southern Arkansas: U. S. Geol. Surv., *Prof. Paper* 46, 1906, p. 31.

peneplaned region was probably undifferentiated from the peneplain present at the close of the Cretaceous period, because the seas of the two periods were approximately in the same position. There is some evidence, however, that the Eocene sea spread much more widely over the lower Mississippi region than the late Cretaceous sea, though such extension was probably of short duration. It is likely that the greater part of Illinois lying in the geosynclinal basin was not elevated perceptibly at the close of the Mesozoic, and the peneplain which had been developed there was continued without erosion through the early Tertiary cycle.[52] The vertical distance between the early Tertiary peneplain and the older Cretaceous one undoubtedly developed in Indiana is unknown, though it has been inferred that the difference in the two levels could not have been great.[53]

The early Tertiary peneplain was widely developed in Tennessee where it is known as the Highland Rim peneplain.[54] Here it is about 1,000 feet below

Fig. 13. View of the Highland Rim peneplain near the crest of the Knobstone escarpment 2 miles west of Medora, Jackson County.

the Cumberland plateau which is supposed to represent the Cretaceous peneplain. In the region of the crest of the Cincinnati arch in central Kentucky, the underlying limestone has been reduced to a plain now standing 900 to 1,000 feet above sea level. This plain, known as the Lexington plain is correlated with the Highland Rim peneplain, though much of it has been reduced somewhat below the original uplifted level.[55] Much of the upland surface of the Dearborn upland represents remnants of this peneplain. Likewise the highest

---

[52] This idea is discussed somewhat by Hershey, Erosion Cycles in Northwestern Illinois: Amer. Geol., Vol. 18, 1896, pp. 72-100; and is apparently approved by Bain, Zinc and Lead Deposits of the Upper Mississippi Valley: U. S. Geol. Surv., Bull. 264, 1906, pp. 8-11. Hershey credits the suggestion to Leverett.

[53] Hayes and Campbell, Geomorphology of the Southern Appalachians: Nat. Geog. Mag., Vol. VI, 1894, pp. 63-126.

[54] Hayes, Physiography of the Chattanooga District, in Tennessee, Georgia and Alabama: 19th Ann. Rept. of the U. S. Geol. Surv., 1899, pp. 1-58.

[55] Campbell, U. S. Geol. Surv., The Richmond Folio No. 46, 1898. Also Latrobe Folio No. 110, 1904.

portion of the Norman upland, where the even-topped ridges reach an elevation of 900 to 1,000 feet (see Figures 6 and 13), and the highest upland masses of the Crawford upland where even-topped spurs attain heights of 800 to 900 feet or more, are probably remnants of this peneplain.

Eastward from the Crawford upland the uplifted peneplain remnants retain a remarkably even elevation, possessing an altitude from 900 to 1,000 feet over southeastern Indiana, southwestern Ohio, central Kentucky, and middle Tennessee. Westward from the Crawford upland there appears to be a distinct descent, since the tops of the hills and upland spurs which probably represent remnants or slightly reduced remnants of the peneplain become distinctly lower in elevation. This descent is probably associated with differential uplift and with later depression of the lower Mississippi Valley area. It is possible, however, that no remnants of the old peneplain exist at the western edge of the Crawford upland and westward, and that the tops of hills and upland spurs are representative of a lower and later peneplain. One interpretation and correlation of certain cherty gravels at elevations of from 700 to 750 feet in southern Indiana sustain this latter view.

It is presumed that the oldest recognized peneplain level in southern Indiana and associated regions was formed largely during the Eocene. It appears that the Gulf embayment area during the Eocene denudation period on the adjacent lands was slightly sinking, permitting the waters of the embayment to reach as far north as the mouth of the present Ohio River during the entire period. During this period, which represents the larger portion of Tertiary time, the lands adjacent to the embayment sea must have been reduced to a peneplain, as outlined above. Thus the Highland Rim peneplain has a geologic basis.

Middle Tertiary Uplift.

Near the beginning of the Oligocene, uplift occurred, causing the withdrawal of the embayment sea to near the vicinity of Vicksburg, and leaving but little indenture of the coast in the embayment area.[56] The deposits of the Oligocene are in part of an off-shore nature, and indicate erosion of the lands to the north.

Late Tertiary Peneplain.

Throughout the Oligocene, Miocene and early Pliocene erosion took place, and large areas in the Mississippi valley were reduced below the level of the uplifted early Tertiary peneplain. A partially base-leveled plain was formed on the Eocene deposits of the embayment area and adjacent areas of soft rocks were reduced for the most part to the same level.[57] Portions of this late Tertiary peneplain in the embayment area and adjacent regions are now

[56] Veatch, Geology and Underground Water Resources of Northern Louisiana and Southern Arkansas: U. S. Geol. Surv., Prof. Paper 46, 1906, p. 41 and Pl. II.

[57] Veatch, U. S. Geol. Surv., Prof. Paper 46, 1906, p. 44. Also see Shaw, The Pliocene History of Northern and Central Mississippi: U. S. Geol. Surv., Prof. Paper 108, 1918, pp. 148-163. Shaw has carefully analyzed the topographic condition of the Gulf embayment area in Mississippi and has carefully traced the Highland Rim peneplain, as identified in Tennessee, into the Gulf embayment area. He identifies an erosion plain below the Highland Rim level developed on Eocene deposits. This plain may be correlated with the late Tertiary peneplain developed in areas adjacent to the Gulf Embayment area. This work of Shaw in this critical area is highly commendable.

preserved at an elevation of 500 to 550 feet. This level in southwestern Indiana is well developed over the Wabash lowland. Over wide areas in southern Indiana, Kentucky, and Tennessee the erosion following the middle Tertiary uplift nearly destroyed the early Tertiary or Highland Rim peneplain. In Indiana only the interstream tracts where the rocks are resistant retain areas which reach approximately to the older peneplain level.

Late Tertiary Gravels and Related Physiographic Problems.

Lafayette Formation.

Near the close of the late Tertiary erosion cycle, terminated by the widespread uplift which brought the Tertiary to a close and initiated the Pleistocene, cherty gravels with associated quartz pebbles, geodes, and reddish sands

Fig. 11. View of late Tertiary peneplain northeast of Salem, Washington County. View taken from a ridge which reaches approximately 900 feet in altitude on the Vallonia-Salem road, about 3 miles north of Salem. The valley of Blue River is very wide and approximately 750 feet in altitude. Long slopes lead down to the broad valley level from the Highland Rim level.

were widely deposited over the Mississippi Valley region. These gravels away from the Gulf embayment area are apparently stream gravels and are almost always found adjacent to the present entrenched drainage lines. These gravels and sands generally known as the Lafayette formation, have been described by Hilgard, McGee, and others as a deposit forming a more or less complete and unconformable mantle over the surface of the late Tertiary peneplain of the Gulf embayment area.

The problem of the method of the deposition of the Lafayette formation has not yet been satisfactorily solved. Some geologists assign the gravels and sands to a Pliocene depression as wide-spread as the occurrence of the deposits themselves.[56] Others realizing that the topographic situations of the gravels

[56] McGee, The Lafayette Formation: U. S. Geol. Surv., Twelfth Ann. Rept., part 1, pp. 347-521. Hilgard, The Age and Origin of the Lafayette Formation: Amer. Jour. Sci., 3rd ser., Vol. 43, 1892, pp. 389-402. Veatch, Geology and Underground Water Resources of Northern Louisiana and Northern Arkansas: U. S. Geol. Surv., Prof. Paper 46, 1906, pp. 41-46.

will not admit of such a simple explanation, have assigned the formation to a period of differential and progressive warping. Under this latter interpretation it is thought that the areas along and adjacent to the widened valleys inland from the Gulf embayment region, were more or less covered with river deposits derived from the deeply accumulated residuals of the bordering lands, while oceanward these gravels merged into true marine deposits.[69] Recently members of the United States Geological Survey have called into question many deposits which have usually been assigned to the Lafayette. Shaw[70] attempts to show that the Lafayette formation consists largely of "a more or less weathered portion of the older and underlying formations." In summary he concludes: " . . . the material called 'Lafayette formation' in Mississippi is the product neither of Pleistocene icy floods from the north nor of a marine invasion; it is not a Pliocene blanket of waste from the Appalachians gradually spread over the state by streams; and it does not consist altogether of parts of pre-Pliocene formations with their surface of residuum. It is believed to be made up of unrelated or distantly related materials that have been erroneously grouped together and to consist in the main of more or less modified parts of the underlying formations, including some residuum and colluvium, and of terrace deposits of Pliocene and Quaternary age."

Cherty River Gravels.

Shaw's disposal of the Lafayette formation, while applying in particular to the deposits of Mississippi, cannot be regarded as conclusive. The formation name Lafayette has beyond doubt been greatly misapplied, but the denial of the existence of the formation inland from the Gulf margin merely complicates the problem. The deposits of southern Indiana, Kentucky, southern Illinois, Missouri, and Tennessee, so frequently occurring on the uplands adjacent to the chief valleys, are distinctly river gravels, and these deposits are commonly regarded as belonging to the Lafayette formation. The writer is convinced that they bear some close relation to the more wide-spread upland deposits of the Gulf embayment area known as the Lafayette formation and to the recently defined Citronelle formation of the Gulf Coastal plain.[71]

The topographic situation of the cherty gravels with admixture of geodes and quartz pebbles has not been worked out over wide areas. It would appear that much difficulty is encountered in distinguishing the original deposits from redepositions of the same. Future studies will probably reveal two or more sets of these gravels which have been in the past assigned to the Lafayette formation. Certainly these gravels are of utmost importance from the physiographic point of view, and when the relationships of the deposits have been worked out in detail and proper correlations have been made, the physiographic development of the Mississippi Valley area will be much better understood than at present.[72]

[69] Chamberlin and Salisbury, Geology, Vol. III, Earth History, 1906, pp. 301-308.

[70] Shaw, The Pliocene History of Northern and Central Mississippi: U. S. Geol. Surv., Prof. Paper 108, 1918, pp. 125-163.

[71] Recent work on the rather inclusive and elastic Lafayette deposits has shown the unsatisfactory condition of this hodge-podge formation in geological literature. An excellent piece of work by G. C. Matson of the U. S. Geol. Surv. shows that at least the Gulfward margin of the Lafayette formation may be given definite status, and his delimitation of the Citronelle formation marks a distinct step towards this end. See Matson, The Pliocene Citronelle Formation of the Gulf Coastal Plain: U. S. Geol. Surv., Prof. Paper 98, 1916, pp. 167-192.

[72] Little serious work has been attempted on these gravels. Such work as has been done has been of rather an incidental nature in connection with the prosecution of other work in regions where the gravels occur.

Such gravels assigned to the Lafayette formation or to the supposedly equivalent Irvine formation of Campbell occur widely over Kentucky and the driftless portion of southern Indiana. So far as known the gravels slope downward somewhat in accord with the present graded courses of the streams. They occupy the upland surfaces 200 to 300 feet above the present streams. In the vicinity of Irvine on the Kentucky River they are at an elevation of about 900 feet, and at Frankfort their elevation is about 800 feet. In Harrison County, Indiana, near the Ohio River, gravels of similar character are 735 feet in altitude; farther down the river at Stephensport, 700 feet, and near Cannelton and Hawsville somewhat less than 700 feet. Along Salt Creek one mile south of Harrodsburg, Indiana, the base of the gravels in place is at an altitude of 720 feet. Down stream along White River in the vicinity of Trinity Springs and Shoals in Martin County, the gravels have an elevation of 700 feet. In southern Illinois and western Kentucky correlated gravels of similar character occupy uplands at an elevation of 500 to 550 feet.

Ohio River Formation of Ashley.

The so-called Ohio River formation of Ashley[18] offers an interesting problem relative to the Tertiary gravels. This formation consists mainly of loose sand much cross-bedded and locally containing white quartz pebbles. In southeastern Harrison County it forms the upper part of Bearwallow Ridge, and in section 28, 1 mile east of Buena Vista (see Kosmosdale topographic sheet), a section was obtained by the writer showing a thickness of nearly 80 feet. The base of the sand here rests on the St. Louis limestone at an altitude of 765 feet. The sand contains pockets of clay and at the base water-deposited cherts and geodes are much in evidence. The base of this formation rises to the northward. White sand is present in the old road at an elevation of 935 feet on the west side of the prominent hill known as Locust Point adjacent to the Ohio River in Sec. 12, T. 4 S., R. 5 E. The hill reaches an altitude of 980 feet. No indication of the sand is found on top of the hill. Farther north relatively pure white sand 30 or more feet in thickness makes up the crest of the high land at the junction of Floyd, Clark, and Washington counties. The formation has an elevation of 900 feet or slightly more. This is the highest land in the locality, and it is not near any drainage line of importance. The drainage radiates from the locality. Likewise at an elevation of 900 feet or slightly more, the same formation occurs on a high prominence two miles southwest of Farabee, Washington County. Here the base is a cherty conglomerate, while the upper portion is mainly loose sand very much cross-bedded, and reaches a thickness of approximately 50 feet.

This formation in Washington and Harrison counties is an important geological formation, and deserves attention from that point of view. It has the appearance of a beach formation. A thickness of 50 to 80 feet is locally present on the high divides. Evidently the thickness has been much greater since its position and the topographic condition of the region are such that erosion is rather rapid. Only remnants are left. Certainly this formation cannot be correlated with the chert gravels which occur along the Ohio River at elevations of 700 feet. It is more likely that the formation as found

[18] Ashley, Geology of the Lower Carboniferous Area of Southern Indiana: 27th Ann. Rept. of Ind. Geol. Surv., 1902, pp. 68-70. Also Ellis, A Soil Survey of Clark, Floyd and Harrison Counties: 32nd Ann. Rept., Ind. Geol. Surv., 1907, pp. 294-295.

in Harrison and Washington counties is a source of the polished chert gravel with its accompanying geodes and quartz pebbles. It is certainly of importance as a geological formation in addition to its imperative need of consideration from the physiographic standpoint.

Possible Correlations and Interpretations.

The highest land of the Norman upland and the eastern edge of the Mitchell plain has been indicated as forming remnants of the early Tertiary peneplain and correlated with the Highland Rim peneplain. The so-called Ohio River formation of Ashley rests upon and forms these remnants, and should these sands prove to be of Eocene age, the age of the peneplain would be verified. Should the formation be demonstrated to be equivalent to the Pliocene Lafayette, not only would the highest land of southern Indiana prove to be remnants of the late Tertiary peneplain, but the Lafayette formation itself would certainly find strong proof of having been deposited during a submergence more wide-spread than is now accepted by geologists. The gravels which lie on the uplands adjacent to the chief drainage lines, as along the Ohio River, would then have to be brought forward into the Pleistocene. Further, an even more extreme hypothesis may well be entertained, that these sands, bearing much evidence of marine deposition on a former beach, are of Cretaceous age. Certain topographic phenomena could on that supposition be far more simply explained. Evidently detailed work on these sands is imperative, and until such work is done the interpretation of the physiographic development of southern Indiana and associated regions must remain uncertain and based on grounds largely insufficient.

The previously mentioned polished chert gravels with rounded quartz pebbles and numerous geodes, frequently of large size, which may be seen resting on the upland surfaces adjacent to the Ohio River and at many places along Salt Creek and the east fork of White River, are very probably associated with the uplift which brought the Tertiary to a close. They may have been deposited before the uplift was completed, and certainly before the present valley trenches were made. The topographic condition of the lands with respect to the streams which deposited the gravels and the position of the coastal waters towards which the streams carried their burden are not yet clearly known. These gravels may be seen to advantage along Salt Creek on Judah's Hill, one mile south of Harrodsburg, where a deposit 25 or 30 feet thick occurs. The base of this deposit is at an altitude of 720 feet. On the east side of Salt Creek in the same vicinity the gravels occur on the Harrodsburg limestone at elevations varying from 650 to 700 feet.[63a] These deposits have probably been redeposited on the Harrodsburg structural plain which is so prominent in the locality. Salt Creek Valley in the vicinity of Harrodsburg has an altitude of 500 feet. Again, along White River in the vicinity of Trinity Springs and some two miles west of Shoals in Martin County, the gravels are in place at an elevation of approximately 700 feet. Numerous patches of these same gravels, redeposited, occur at lower levels down to the valley of White River itself. White River here flows in an entrenched valley 250 feet lower than the gravels in place. Fuller and Clapp[64] have described a deposit of bronzed gravel and associated red sand on the hills north of

[63a] Malott. Planation Stream Piracy: Proc. Ind. Acad. Sci. for 1920, p. 253.
[64] Patoka Folio No. 84, 1902, U. S. Geol. Surv.

Princeton in Gibson County. The base of the gravel here is 610 feet A.T. On the basis of topographic relationships this gravel may be interpreted as equivalent to Ashley's Ohio River formation. It is interpreted by Fuller and Clapp as lying on a monadnock eminence which it is thought may reach up to the early Tertiary peneplain level. The writer is inclined to believe that this deposit should be correlated with the cherty river gravels of late Tertiary age, and that it is equivalent to the deposits near Shoals in Martin County. The monadnock on which the deposit rests, under such an interpretation reaches up to the late Tertiary peneplain level.

Tertiary-Pleistocene Uplift and Valley Trenching.

The most prominent common topographic feature of the unglaciated regions in the Mississippi Valley is seen in the valley trenches which are everywhere sunk below the uplands. The trenched condition of southern Indiana, unglaciated Ohio, Kentucky, and associated regions was brought about by the erosion following the wide-spread uplift which closed the Tertiary and initiated the Pleistocene. If the cherty gravels which lie on the uplands near the chief drainage lines were deposited on the late Tertiary valley lands and associated peneplained regions at the close of late Tertiary denudational period, the amount of the uplift at the close of the Tertiary would equal the difference in the elevation of the gravels *in situ* and the bottom of the valley trenches where the condition of the trench is such as to denote a graded condition. In Indiana the uplift amounted to from 200 to 300 feet. But it is apparent on close observation that this total uplift was not attained during one uninterrupted emergence. Two or more periods of uplift are thought to have taken place. Portions of the Mitchell plain and practically all of the upland portion of the Scottsburg lowland and local areas within the other physiographic divisions of southern Indiana are developed below the level of the gravels, yet they are high above the valleys entrenched below. These areas have all the topographic features of local peneplains developed by subaerial denudation. Such areas are usually located on nonresistant rocks, the weakness of which permitted rapid removal of their material. Frequently local areas of similar characteristics and at similar elevations may be definitely related to the structural and lithological characters of the rocks. It is thought, however, that the local upland plains developed below the cherty gravel level mark a period of crustal stability of post-Tertiary age and preceding the Illinois glacial stage. That such a period of local peneplanation took place has only recently been recognized.[**]

Early Glaciation and Formation of the Ohio River.

It is quite clear that the advance of the ice sheets which moved southward from the Canadian region into the United States destroyed or deranged the drainage systems previously developed. Streams which flowed northward had their lower courses obstructed and destroyed by the ice and its burden of till. The earliest advance of ice across the present position of the St.

[**] Malott, Valley Trenching and Gradation Plains in Southern Indiana and Associated Regions: Sci. N. S., Vol. XLIII, p. 398, Mar. 17, 1916. Galloway, Geol. and Nat. Res. of Rutherford Co., Tenn.: Tenn. Geol. Surv., Bull. 22, 1919, pp. 17-24. Cleland, A Pleistocene Peneplain in the Coastal Plain: Jour. Geol., Vol. XXVIII, pp. 702-706, 1920.

Lawrence into New York, Ohio, and Indiana undoubtedly caused violent stream changes. The total derangements brought about by the several ice advances down to the Illinois glacial stage were largely responsible for the present system of the Ohio. It has been shown that the Ohio River is essentially a stream skirting the outer limits of the advance of the several ice invasions.[96] A number of formerly northward flowing streams were turned westward over the low places in intervening divides on account of the blocking of the lower portions of their preglacial courses. In Pennsylvania and southern Ohio several of the preglacial courses are quite well marked, and the details have been described by Leverett, Tight, and others. It is evident that the trunk line of the present Ohio River was well established before the advent of the third or Illinois glacial stage.[97] The establishment of the largest tributary of the Mississippi River along the southern border of Indiana in the place of a much smaller stream has been of considerable importance in giving rise, as previously described, to the present topographic condition of southern Indiana.

There is some question as to the size of the preglacial Ohio. Considerable difference of opinion exists as to the position of the last divide over which the ponded drainage broke. The opinion of Leverett who has studied the question in its broader relationships is that the last divide was near Manchester, Ohio, some seventy miles up the present Ohio River from Cincinnati.[98] Such an interpretation would make the pre-glacial Ohio rise near this point. It would receive the Licking River and probably an important tributary along the course of the present Big Miami. After a detour by Hamilton some twenty miles north of the present route west from Cincinnati, the pre-glacial Ohio followed approximately its present course along the southern border of Indiana. Fenneman,[99] who has given the question some study, accepts this view of Leverett. The writer favors the view of Fowke,[100] who postulates the existence of a major divide near Madison, Indiana. Such an interpretation would place the head of the pre-glacial Ohio on the Muscatatuck regional slope between the Muscatatuck River on the north and the Salt Creek of Kentucky on the south. It presumes that the Kentucky River turned northeastward in the vicinity of Carrollton and flowed up the course of the present Ohio to the mouth of the Big Miami and thence northward and slightly eastward to near Hamilton where it was joined by the Licking. The route then continued northward and either entered the pre-glacial St. Lawrence or turned westward into Indiana where it connected with other streams to make the pre-glacial valley of the Wabash River which is distinctly discernible as a valley many times larger than the present one, in the vicinity of Lafayette.

Leverett and Fennoman object to Fowke's interpretation, chiefly because of the elevation of the bed-rock floors, as revealed by drillings, and the width and apparent age of the valleys at different points. The chief evidence of the existence of a former major divide near Madison is found in the extreme narrowness of the valley and the non-dissected land adjacent. Fenneman admits further that borings made at Madison and at Carrollton reveal a bed-rock slope to the northeast.

[96] Leverett, U. S. Geol. Surv., Monogr. 41, 1902, pp. 82-219.

[97] Leverett, U. S. Geol. Surv., Monogr. 41, 1902, pp. 83-84.

[98] Leverett, U. S. Geol. Surv., Monogr. 41, 1902, pp. 116-118.

[99] Fenneman, Pre-glacial Miami and Kentucky Rivers: Bull. Geol. Soc. Amer., Vol. 25, 1914, p. 85.

[100] Fowke, Bull. Denison Univ., Vol. XI, 1898, pp. 1-10; Ohio Acad. Sci., Special Papers No. 3, 1900, pp. 68-75.

But there are three lines of evidence which Leverett and Fenneman failed to consider, that tend to support the interpretation of Fowke. First, general beveling of the upland surface towards the Kentucky River and its supposed continuation from Carrollton northeastward, has taken place, whereas in the vicinity of Madison where the former divide is postulated absolutely no beveling towards the present Ohio is discernible. Further south on the small streams such as Corn, Barebone, Pattons, and Harrods creeks beveling is distinctly noticeable on the same rocks as those existing at Madison. Where the present valley of the Ohio passes across the Mitchell plain and the Crawford upland the beveling and broken down condition is noted by Leverett. Such a condition furnishes no argument in either case, except as it furnishes a contrast to the condition at Madison; and this argues in favor of the former divide at Madison. Second, cherty gravels are present on the bluffs above the Kentucky and Licking rivers, and similar gravels are present above the bluffs on either side of the Ohio River some distance below Louisville. So far as the writer is aware no gravels have been found along the present Ohio in the vicinity of Madison. The gravels are so abundant next to the Kentucky River that if this stream passed the vicinity of Madison gravels should have been deposited. So much of the upland is present that their subsequent removal would be less likely than along the more broken down and beveled area adjacent to the Kentucky River. Third, differential deformation is not given full consideration. The apparent rising of the bed-rock floor up the present Big Miami could be explained on the assumption of post-glacial uplift at the north where uplift was considerable.[109a] It is possible also that the Pleistocene uplift was somewhat greater on the Jessamine dome than on its flanks and adjacent geosynclines. The post-glacial uplift at the north combined with the greater uplift of the Jessamine dome at the south would have the effect of allowing the route of the present Ohio to occupy the position of a comparative syncline, a feature which is suggested in the broader structural conditions. A careful study of the Tertiary gravels should be made with this suggestion in mind, though the gravels may prove too meagre to give definite results.

Further it is tacitly granted by both Leverett and Fenneman that the conditions at Madison denote recent drainage changes. It is difficult to conceive of any drainage changes other than those postulated by Fowke. Certainly there is no evidence that the pre-glacial Kentucky River followed a now buried route either to the north or the south of the narrows at Madison. Again the full significance of the present upstream direction of a number of the tributaries is not appreciated by Leverett and Fenneman. The dip of the rock does not favor the eastward trend of any tributary. The westward dip in itself is in part sufficient to explain the direction of those streams which tend in that direction against the general law that tributary streams tend to enter the main stream in a downstream direction. Considerable adjustment has taken place since the postulated drainage changes. But it is a fact that a number of the tributaries of the present Ohio are conspicuously oriented in an upstream direction and against the structural conditions of the region. The question might well be asked why a single one of these tributaries has such a relationship. It is apparent to the writer that the evidence of a former divide at Madison is of the same nature and is just as trechant as the evidence at Manchester, New Martinsville, and other well established former divides.

[109a] See Fairchild's isobasic map of post-glacial uplift. Sci. N. S., Vol. 47, 1918, p. 166.

Valley Filling.

It is apparent that the trenching of the present valleys in the lower Mississippi region and for some distance up the tributary streams on the lower portion of the flanks of the Ozark and Cincinnati uplifts was somewhat deeper than is indicated by the surface of the present flood plains. The valleys of southwestern Indiana, western Kentucky, and southern Illinois are filled with gravels and sands to a depth of 100 to 200 feet. Most of the valley filling probably took place some time before the Illinois glacial stage, though no doubt some valley filling was associated with both the Illinois and Wisconsin glacial stages. Veatch[141] has shown that the lower Mississippi region was filled from 150 to 200 feet in depth over wide areas. The valley-fill deposits of the lower Mississippi valley have been named the Port Hudson deposits by Veatch. He infers that the Port Hudson deposits belong to the early half of the Pleistocene. These deposits have been partly removed, but are still 150 feet thick at Vicksburg and 175 feet at the head of the present delta near the mouth of Red River.

This period of valley filling may be definitely associated, first, with depression along the axis of the lower Mississippi, second, with the earlier glacial flood waters and their associated burden of material, and third, with coastal extension in the Gulf of Mexico. The relative importance of these three physiographic factors instrumental in valley filling is difficult to evaluate, but their importance as responsible factors in valley filling in the lower Mississippi and adjacent regions must be recognized.

Certainly depression has taken place in southwestern Indiana, western Kentucky, southern Illinois, and southward to the Gulf of Mexico. Valleys everywhere upon approach to this region are built up above their bed-rock floors with sands, silts, and gravels. Valley filling independent of glaciation becomes evident in the western half of the Crawford upland, and at its western margin becomes a distinct and characteristic topographic feature. Direct evidence of warping may be sought for in the comparative altitude of the cherty gravels assigned to the late Tertiary. As yet this line of evidence has received little study. It is known, however, that these gravels in the middle and eastern part of the Crawford upland have an elevation of 700 feet or slightly more. In the vicinity of Tell City and Troy they should be somewhat lower, as valley filling is quite evident. Anderson Creek, marking the boundary line between the Crawford upland and the Wabash lowland in its lower course, occupies a prominently filled valley. Its broad flat-bottomed valley with the half-buried, abrupt bluffs and slopes is conspicuous as far as the northern line of Perry County. Other streams in southwestern Indiana flowing through the western half of the Crawford upland into the prominently valley-filled Wabash lowland also show a similar condition.[142] It would appear that the warping must have begun in this zone and the region to the west has become distinctly depressed. The Tertiary gravels in southern Illinois and western Kentucky have an altitude of 500 to 550 feet, having descended from an altitude of about 700 feet in the vicinity of Rome, Perry County, where they cap the bluffs adjacent

[141] Veatch, U. S. Geol. Surv., *Prof. Paper* 46, 1906, pp. 46-52.

[142] The contrast between the valley and stream conditions of the Wabash lowland and the provinces on the east has been ascribed to depression on the west by Ashley, Geology of the Lower Carboniferous Area of Southern Indiana: 27th Ann. Rept., Ind. Geol. Surv., 1902, pp. 54-65.

to the river. Further study of the gravels should be made with reference to the warping of the surface upon which they are located.

Several lines of evidence indicate that glacial flood waters were an important factor in bringing about valley filling in the lower Mississippi and associated regions. First, the materials which make up the valley-fill are definitely related to the glacial debris discharged from the north. The Port Hudson deposits of Veatch in the Lower Mississippi are themselves highly calcareous silts wholly in keeping with the rock-flour discharged from the glacial mill at the north. The materials in the valleys adjacent to the drift covered areas contain pebbles and sands which have been carried by waters discharging from the drift deposited by the ice sheet. The chief water routes, such as the upper Mississippi, the Wabash, and the Ohio River valleys, were built up by valley train material deposited by the glacial floods. These valley trains caused ponding of the side streams in the unglaciated areas. This ponding, due to valley trains and the high waters of the glacial floods, gave rise to extensive lake systems, such as have been described by Shaw,[103] related in particular to the last glacial stage. Veatch[104] has called attention to the difference in height of the banks of the Mississippi and the Ouachita rivers in the latitude of the Arkansas-Louisiana line. The top of the Mississippi banks is 112 feet in altitude, while those of the Ouachita are but 63 feet. A small part of this is ascribed to a recent fault. The 100-foot contour line passes up the valley of the relatively small Ouachita some 70 or 80 miles beyond the northern line of Louisiana, and in an equal valley distance up the Mississippi the elevation of the valley of the Mississippi is 160 feet. This condition of greater stream gradient and much higher elevation of the Mississippi with respect to the rather insignificant Ouachita is evidently the result of the greater silt burden which the former stream has carried in the past and is still carrying. The Ouachita has never been subjected to glacial floods with their associated burden of silt, such as have repeatedly deluged the Mississippi. Thus, the glacial floods of silt and sand laden waters have been of considerable importance in valley filling.

Coastal extension in itself has been of some importance in increasing the depth of valley-fill material since the beginning of the Pleistocene. The Gulf coastal plain has grown notably since the beginning of the Pleistocene.[105] East of the mouth of the Mississippi the increased width of the coast plain is not great. In the vicinity of Mobile recent local depression has counterbalanced the effect of the Pleistocene deposits. A strip some 25 miles in width has been added to the coast in the vicinity of the mouth of Pearl River. The delta of the Mississippi has extended the coastal plain approximately 125 miles since the beginning of the Pleistocene. The river at present flows across this delta obliquely in a straight-line distance of approximately 200 miles and a river distance of over 300 miles. The Pleistocene coastal extension westward from the delta of the Mississippi amounts to approximately 70 miles in the vicinity of Houston, 30 miles at Corpus Christi, and 60 miles at the mouth of the Rio

---

[103] Shaw, Newly Discovered Beds of Extinct Lakes in Southern and Western Illinois and Adjacent States: Bull. 20, Ill. Stale Ceol. Surv., 1915.

[104] Veatch, U. S. Geol. Surv., *Prof. Paper* 46, 1906, pp. 14 and 55.

[105] See Geologic Map of North America by Balley Willis: U. S. Geol. Surv., *Prof. Paper* 71, 1912, Pl. I: Matson, The *Pliocene* Citronelle Formation of the Gulf Coastal *Plain*: U .S. Geol. Surv., *Prof. Paper* 98, 1916: and Shaw, The *Pliocene* History of Northern and Central Mississippi: U. S. Geol. Surv., *Prof. Paper* 108, 1918, p. 157.

Grande River. A large part of the Pleistocene material added to the coastal plain no doubt has been contributed by the Mississippi River.

The extension of the coast by the great growth of deposits at the mouth of the Mississippi is directly related to the problem of valley filling. The mouth of the Mississippi near the beginning of the Pleistocene was probably in the vicinity of the present mouth of Red River or near the present head of the delta. Here the river once debouched its waters directly into the Gulf. This same point is now built up approximately 50 feet above the level of the sea. This former sea level position was slowly built up as the delta advanced into the Gulf. While the upbuilding of the delta head was slowly taking place the graded portion of the valley upstream from the delta was correspondingly raised. The valleys were undoubtedly in a graded condition well into southwestern Indiana where valley filling is now so evident.[106] Such coastal extension as has occurred at the mouth of the Mississippi with an upbuilding of its delta head of 50 feet would bring about valley filling in the graded portion of valleys above by an amount approximately equal to 50 feet, provided other conditions have remained the same. Undoubtedly the delta extension of the Mississippi River has contributed materially to valley filling in southwestern Indiana and associated regions.[106a]

Illinois Glacial Stage.

General Influence of the Earliest Glaciation.

It is not known positively that either of the first two recognized ice advances of the Pleistocene period came into the Indiana region. Both the Nebraskan and Kansas ice sheets locally advanced much farther south than the latitude of northern Indiana, and it is quite likely that one or both of these earlier ice sheets advanced far enough south to enter the Indiana region. The possible existence of pre-Illinoian till has been discussed by Leverett[107] and Fuller.[108]

The earliest ice sheet no doubt advanced over a dissected upland surface very similar to that in the driftless area of southern Indiana. Such a condition is revealed by the local outcrops of bed-rock and by well records in middle and northern Indiana. Locally the glacial drift covers the bed-rock surface several hundred feet deep yet within a short distance the drift cover may be a few feet thick or bed-rock may be exposed. Where the drift cover is deep old pre-glacial drainage lines are indicated. Some of the old drainage lines are of considerable width, while others are quite narrow or gorge-like.

---

[106] Malott, The American Bottoms Region of Eastern Greene County, Indiana: Ind. Univ. Studies No. 40, 1919, pp. 26-34.

[106a] Since the above was written an article by Mr. Frank Leverett has appeared in the Jour. Geol., Vol. 29, 1921, pp. 615-626, entitled "Outline of Pleistocene History of Mississippi Valley," which ascribes valley filling in the Mississippi valley definitely to the lengthening of the lower course of the Mississippi River by delta building and to the clogged condition of the Mississippi and Missouri resulting from glacial outwash and to their present great silt burden. These two causes, Mr. Leverett says, are sufficient to account for the full difference between the level of the rock bed and the level of the present river, and that there is no need for postulating diastrophic movement. The ideas which Mr. Leverett uses were evidently suggested to him by the discussion of these two causes of valley filling by the writer in his paper on the "American Bottoms Region," Ind. Univ. Studies No. 40, 1919, pp. 26-34.

[107] Leverett, U. S. Geol. Surv., Monogr. 38, 1899, p. 109.

[108] Fuller, Ditney Folio No. 84, U. S. Geol. Surv., 1902, p. 3.

Not much is known as to the alignment and direction of former flow of the pre-glacial streams in the middle and northern part of the state where the drift is deep. No systematic study of well records has been made over wide areas in Indiana. Locally, however, some features of the pre-glacial drainage lines have been determined.[108]

It is likely that the Illinois ice sheet spread farther than any earlier ice sheet in the Indiana region, and was the first ice sheet to advance over certain parts of the previously unglaciated topography. Such an advance of any ice sheet must have destroyed and buried under its till practically all of the minor drainage lines over which it passed. The Illinois ice sheet may be regarded as having passed over a normally developed pre-glacial topography in southern Indiana.

The Illinois drift sheet beyond the Shelbyville moraine of the Wisconsin glacial stage covers an area of 7,200 square miles in southern Indiana. 4,100 square miles of the Illinois drift surface in southwestern Indiana belongs to the western section of the Illinois drift covered area, and 3,100 square miles in southeastern Indiana belongs to the eastern section. These two drift covered sections are separated by the driftless area of southern Indiana. The drift varies in depth over this surface from practically nothing where it has been removed by subsequent erosion to a depth of over 100 feet. On the average it is perhaps less than 25 feet in depth over the greater part of the Illinoian drift covered area.[109] The deeper drift occupies the pre-glacial valleys. Much of the drift covered portion of southern Indiana is exceptionally flat, still exhibiting the characteristics of a featureless ground moraine. Rarely does ridged drift occur. The margin of the Illinois drift sheet may be said to

[108] For the thickness of the glacial drift in Indiana and data bearing on pre-glacial valleys now filled with drift, see the following: *Leverett*, Monogr. 38, U. S. Geol. Surv., 1899, pp. 27, 67-70, 199-200, 237, 335 and 392; Monogr. 41, U. S. Geol. Surv., 1902, pp. 261-267, 322-325, 331-332, 484-487, 502-505, 516-522, 560-561 and 577-578; Monogr. 53, U. S. Geol. Surv., 1915, pp. 66-69, 81-82, 90-92, 100-105, 127-128, 133-135, 140, 145, 161-163, 170, 173 and 217: Natural Gas Borings in Indiana: Amer. Geologist, Vol. IV, 1899, pp. 6-21; Pre-glacial Valleys of the Mississippi and its Tributaries: Jour. Geol., Vol. III, 1895, pp. 744-745 and 756-757; Wells of Northern Indiana: U. S. Water Supply paper No. 21, 1899; Wells of Southern Indiana: U. S. Water Supply paper No. 26, 1899. *Beachler*, Erosion of Small Basins in Northwestern Indiana: Amer. Geologist, Vol. XII, 1893, pp. 51-53. *Bownocker*, A Deep Pre-glacial Channel in Western Ohio and Eastern Indiana: Amer. Geol., Vol. XXIII, 1899, pp. 178-182. *Moore*, Glacial and Preglacial Erosion in the Vicinity of Richmond, Indiana: Proc. Ind. Acad. Sci. for 1892, p. 27. *Clem*, The Preglacial Valleys of the Upper Mississippi and its Eastern Tributaries: Proc. Ind. Acad. Sci. for 1910, pp. 335-352. *Dryer*, Wabash Studies: Raccoon Valley, Parke County, Indiana: Proc. Ind. Acad. Sci. for 1912, pp. 206-212. Numerous references are made to the thickness of the glacial drift and the occurrence of pre-glacial valleys in Indiana in the various reports of the Ind. Geol. Surv., as follows: Fifth Ann. Rept., 1873, pp. 193-196; Seventh Ann. Rept., 1875, pp. 91-95, 129-130, 195-196; Tenth Ann. Rept., 1878, pp. 205-207; Eleventh Ann. Rept., 1881, pp. 63-68, 92-93; Twelfth Ann. Rept., 1882, pp. 51-57, 66-67, 91-93, 156-165; Thirteenth Ann. Rept., 1883, pp. 100-103; Fourteenth Ann. Rept., 1884, pp. 55-57, 170; Fifteenth Ann. Rept., 1886, pp. 76-78, 87-96, 165-175, 208, 209, 212-220; Sixteenth Ann. Rept., 1888, pp. 21, 32-37, 178, 233-266; Seventeenth Ann. Rept., 1891, pp. 238-239; Eighteenth Ann. Rept., 1893, pp. 18, 30-31, 222-255; Nineteenth Ann. Rept., 1894, pp. 31-39; Twenty-first Ann. Rept., 1896, pp. 555-557; Twenty-third Ann. Rept., 1898, pp. 82-84, 183-184, 195, 216-217, 236, 286, 300, 398, 473, 474, 519-520, 561, 687-688, 768-769, 895, 803, 818, 966, 1050, 1158; Thirtieth Ann. Rept., 1905, pp. 123, 213, 251, 359-360, 389, 440, 534, 571-573, 577-587, 599, 638, 642, 883, 894; Thirty-fourth Ann. Rept., 1909, pp. 17-18, 109; Thirty-sixth Ann. Rept., 1911, pp. 17, 37, 75, 85, 257-264, 267-268, 284, 294, 323; Thirty-seventh Ann. Rept., 1912, pp. 169, 171, 187, 189, 217, 254, 313; Thirty-ninth Ann. Rept., 1914, pp. 57, 158; Fortieth Ann. Rept., 1915, pp. 12-16, 54, 85, 123-124, 171, 256.

[109] See plate IV, Monogr. 53, U. S. Geol. Surv., 1915.

possess little or no evidence of a terminal moraine in Indiana. Much of the margin is composed of fluviatile material which is much more evident than the attenuated edge of the till sheet itself.

The Illinois Glacial Boundary.

The Illinois ice sheet covered all of Indiana with the exception of about 6,250 square miles in the southern middle portion of the state. The glacial limits were first mapped by Wright[111] and later and more accurately by Leverett.[112] The ice limits as shown on the physiographic and glacial maps accompanying this report differ from Leverett to a considerable extent locally. The glacial boundary as it appears on the maps is based on the actual presence of glacial till. Outwash and lake silts directly associated with the proximity of glacial ice extend locally beyond the boundary line as drawn. Such deposits are extra-glacial and areas covered by them without direct evidence of the presence of glacial ice, have been excluded from the glaciated area.

The configuration of the outer limits of the Illinois glacial materials indicate that two great ice-lobes reached into southern Indiana and adjacent regions. The great ice-lobe on the west extended further south than any previous or subsequent ice sheet, reaching the southern part of Williamson County, Illinois, some 10 or 15 miles south of Marion. This great lobe has been named and described in detail by Leverett as the Illinois glacial lobe.[113] The Illinois glacial stage has received its name from this most extensive and largest of the known Pleistocene ice-lobes.

From southern Williamson County, Illinois, the boundary of the Illinois ice sheet passes east and north, crossing the Wabash River near the middle of Posey County, Indiana. The boundary[114] continues east and north to near Princeton in Gibson County, and then more easterly to southeastern Daviess County; thence north along the western line of Martin County, through eastern Greene County to southeastern Owen and northwestern Monroe counties. In northern Monroe and Brown counties the Illinois ice-lobe becomes a portion of the main ice sheet, or overlaps or is overlapped by a probably contemporaneous ice-lobe whose western boundary extends southward along the Knobstone escarpment to the vicinity of Jeffersonville in Clark County. Just north of Jeffersonville the margin of this eastern division of the Illinois drift sheet turns northeastward and follows the Ohio River for about 25 miles to northeastern Clark County where it crosses the river into Kentucky. As mapped by Leverett the southern margin remains in Kentucky keeping a few miles south of the Ohio River to a point a short distance up the river above Cincinnati where it crosses and enters Ohio.

The glacial boundary as indicated on the maps accompanying the present report is notably different from Leverett's boundary line, especially in southeastern Indiana. In the broad valleys of White and Muscatatuck rivers drift masses are rare. The writer, however, on examining the gravels on the side of the bluff in the vicinity of Millport Knobs in northern Washington County

[111] Bull. 58, U. S. Geol. Surv., 1890, p. 65.

[112] Monogr. 38, U. S. Geol. Surv., 1899, plates VI, VIII and IX; Monogr. 43, U. S. Geol. Surv., 1902, plate II; Monogr. 53, U. S. Geol. Surv., pl. VI.

[113] Leverett, U. S. Geol. Surv., Monogr. 38, 1898.

[114] The till and associated glacio-fluviatile deposits have been mapped in detail south of White River in southwestern Indiana. See Ditney and Patoka Folios, U. S. Geol. Surv.

a short distance above the mouth of the Muscatatuck River, found that the gravel deposit was underlain by several feet of glacial till in a highly decayed state. Large granite and schist boulders were decayed to their centers. The till, while having the appearance of being much older than the Illinois till, has been correlated with the Illinoian stage. Its identification at this position on the north facing bluffs south of the Muscatatuck River furnishes the basis for drawing the glacial boundary approximately at the foot of the Knobstone escarpment in this locality. It is apparent that the tops of the Brownstown Hills south of Brownstown were not entirely covered by the glacial ice, but were probably surrounded by it for a time. It would appear that the Knobstone escarpment furnished a barrier over which the glacial ice did not pass; but this escarpment was an effective barrier most likely because the ice had already approximately reached its limit. That it would deploy down the wide opening of the White and Muscatatuck rivers in southern middle Jackson County should cause no surprise.

It is quite possible and even probable that the glacial ice for a short time advanced farther in northeastern Brown and western Bartholomew counties than is shown by the limits of the map, which here follows substantially Leverett's latest map of the glacial boundary.[115] Boulders of large size are frequently to be seen south and west of the line as drawn in that locality, though no typical till has been observed by the writer.

Drainage Derangements Near the Drift Border.

The drainage of the driftless area of southern Indiana was locally notably modified during the Illinois glacial stage. "The ice of the Illinois Glacial Epoch pushed in from the west and northwest upon the drainage systems leading toward it on the one hand, and on the other it pushed in from the east and north down the drainage systems leading away from it. Thus, there are only areas of inconsiderable size entirely immune from the influence of the Illinois Glacial Lobe in southern Indiana. Any particular area must be considered with respect to its relation to the particular glacial flank controlling its modification . . . .

"It may be noticed that the glacial ice advancing from the west and northwest vigorously over-rode the pre-glacial valley of the west fork of White River, having crossed it everywhere almost at right angles, and that it advanced several miles beyond, going somewhat farther to the south than to the north. In transgressing somewhat beyond the pre-glacial White River valley, it advanced up the drainage lines coming in from the east, blocking them very effectively at the time with ice and ice-borne material. This condition of blocked drainage, which in a number of instances persisted after the ablation of the ice, is everywhere in evidence along the former position of the ice front. Streams in a number of cases follow their pre-glacial courses to near the position of the ice front and then change their courses to some position that gave a favorable outlet to their waters previous to the melting of the ice. Others persist to the White River Valley, showing only the effects of the temporary blocking of their courses. The smaller pre-glacial streams or ravines back of the position of the ice-front were for the most part obliter-

[115] Leverett, U. S. Geol. Surv., Monogr. 53, 1915, Pl. VI. The limit of drift in Brown County as given on page 620 of the text diverges considerably from that shown on plate VI.

ated by the over-riding ice with its burden of till and the glacio-fluviatile deposits that characterized the ablational period. . . .

"In connection with the blocking of the drainage lines leading from the unglaciated area to the ice front, great quantities of material were carried locally by water coming from the ice margin itself, sweeping the ice-contributed debris farther inland than it could ever have been brought by the ice alone. The morainic material has everywhere been carried by the glacio-fluviatile streams beyond the actual position of the ice-front. Occasionally debris has been carried in this manner several miles beyond the position of the ice front. Locally great outwash aprons occur, covering several square miles. In the re-entrants between these expansions of outwash material there may be for a number of miles along the ice margin comparatively little material of this sort. These local expansions of outwash material have frequently nearly obliterated all the pre-glacial surface irregularities, and their fairly even, eastwardly sloping surfaces may still be seen descending from near the former position of the ice front. These fans are now [locally] deeply trenched by post-glacial erosion."[116]

In the above described manner stream obstruction and ponding gave rise to notable stream derangement and prominent lake deposits in southwestern Indiana south of White River. Big Creek of Posey County was considerably enlarged at the expense of Black River in Gibson County to the north, and extensive lake flats were produced in the vicinity of Poseyville and Cynthiana. The drainage basin of Pigeon Creek which enters the Ohio River at Evansville was greatly increased on the north at the expense of the once important but now minor drainage basin of Indian Camp Creek in western Gibson County, and the extensive flats of former glacial lakes McGary and Pigeon, described by Fuller and Clapp, were deposited.[117]

Probably the most striking and important modification in southwestern Indiana was in connection with Patoka River. It is apparent that the long narrow basin of this stream, reaching 70 or 80 miles eastward from the Wabash River, has resulted from the deflection of three important former tributaries of White River and their combination into one continuous stream in conjunction with a fourth stream which probably entered the Wabash. On account of the encroachment of the glacial ice south of White River, the Upper Patoka which formerly entered White River north of Jasper was obstructed and its waters forced over a divide in the vicinity of Jasper into Middle Patoka River. (See Fig. 51.) This position has been maintained since the time of deflection. The Middle Patoka formerly flowed northwest past Otwell and Algiers in northeastern Pike County where it entered White River. It was likewise obstructed, and, upon ponding, its waters escaped over the divide into the Lower Patoka a short distance east of Velpen. The obstruction of the former Middle Patoka was so effectual that the new route was permanently maintained. The Lower Patoka flowed west and slightly north and evidently entered White River through the now obstructed area between Patoka and Hazleton in northern Gibson County. The Patoka at present flows through a narrow gorge east of the village of Patoka, and thence due west to the Wa-

[116] Malott, The American Bottoms Region of Eastern Greene County, Indiana: Ind. Univ. Studies No. 40, 1919, pp. 36-38.

[117] Leverett, U. S. Geol. Surv., Monogr. 38, 1898, p. 97-98; Fuller and Clapp, *Patoka* Folio No. 105, 1904.

bash River. The lower portion of the present Patoka from the gorge a short distance east of the village of Patoka constitutes a portion of the fourth formerly independent stream which helps to make up the present Patoka River. Because of such notable stream derangement and stream ponding, lake deposits now locally trenched are prominent in northern Dubois, Pike, and Gibson counties along and near the Patoka River. It is evident that the waters which were discharged from the ice front and those which collected in the ponded drainage in the unglaciated area in front of the ice sheet while White and Wabash river valleys were still deeply covered with glacial ice, escaped to the southward over a divide near the village of Francisco, and thence passed by way of Pigeon Creek to the Ohio River.[118]

North of the east fork of White River the blocked and deflected drainage leading to the west fork of White River has been briefly described by Leverett and Siebenthal.[119] Furse Creek in northwestern Martin County formerly flowed into Doans Creek near Scotland, but was blocked and upon readjustment took a course southward into an adjacent stream. Plummers, Clifty, Beech, and Richland creeks of eastern Greene County were notably obstructed. Plummers and Beech creeks have retained their pre-glacial routes. Clifty Creek and a main tributary from the north were ponded and deranged largely by outwash material, and on account of the peculiar subsequent subterranean drainage developed, a prominent lake flat known as the "American Bottoms" is preserved high above the local base level of the region. This region has been mapped and describd in detail.[120] Richland Creek, formerly flowing west and slightly north past Newark to White River, was ponded just west of the Bloomington Quadrangle area and deflected southward over a divide into a tributary of Beech Creek, and now with the Beech Creek drainage enters White River some 20 miles below its former mouth. The ponded drainage gave rise to a lake named Lake Richland by Siebenthal. The present drainage has trenched the lake deposits and a beautiful set of terraces occurs along the stream on the Bloomington Quadrangle area. Raccoon Creek of southeastern Owen County was ponded, but does not appear to have been deflected into an adjacent stream. The terraces which flank the present stream coalesce with the surface of "Flatwoods," a wide lacustrine flat lying between Ellettsville and Spencer. It has been shown that the "Flatwoods" region was formerly occupied by a drainage channel deep below the present surface which discharged south of the present steep, rocky outlet through McCormicks Creek. The extensive lacustrine plain of "Flatwoods" lies approximately 200 feet above White River and is but slightly trenched by McCormicks Creek. McCormicks Creek is the chief drainage outlet and is entirely post-glacial. It descends 160 feet in its lower 2 miles through a deep rocky gorge in the St. Louis and Salem limestones. This lower portion of McCormicks Creek is now a State Park, and is noted for its rugged beauty.[121]

Bean Blossom Creek in northern Monroe and Brown counties presents a more complicated problem. It is evident that the valley was effectually ob-

---

[118] Leverett, U. S. Geol. Surv., Monogr. 38, 1898, pp. 98-102; Ashley, 23rd Ann. Rept., Ind. Geol. Surv., 1898, pp. 1099-1102; Ditney Folio No. 84, 1902; Patoka Folio No. 105, 1904.

[119] Mono. 38, U. S. Geol. Surv., 1898, pp. 102-104.

[120] Malott, The American Bottoms Region of Eastern Green County, Indiana: Ind. Univ. Studies No. 40, 1919, pp. 1-61.

[121] Siebenthal, 21st Ann. Rept., Ind. Geol. Surv., 1898, pp. 301-302. Malott, The Flatwoods Region of Owen and Monroe Counties, Indiana: *Proc. Ind. Acad. Sci. for 1914*, pp. 399-428; 39th Ann. Rept., Ind. Geol. Surv., 1914, pp. 217-222.

structed and ponded for a time, and there is evidence of derangement in the vicinity of Tabor Hill in the lower portion of its course in northwestern Monroe County. The headwaters of Bean Blossom Valley in Brown County also received glacial waters from both the Illinois and Wisconsin ice sheets. While considerable attention has been given to its well developed and prominent terraces, their explanation is still considered an unsolved problem by the writer. The problem of the terraces must be viewed from the standpoint of obstruction and ponding in the lower course and of the valley and valley-train deposition from both the Illinois and Wisconsin ice margins in the upper course. The explanation of the terraces given by Marsters, namely, that they are due to local deltas of incoming streams, is not in harmony with the form and materials of the terraces, nor is the principle as presented sound. It is evident to the writer that much detailed work remains to be done before the terraces of Bean Blossom Valley can be adequately explained.[122]

Stream derangement on the eastern margin of the driftless area was of little consequence. The pre-glacial drainage was largely away from the ice margin, and the glacial waters escaped down the regular drainage lines without becoming ponded. The stream alignments and topographical conditions, however, in southwestern Bartholomew and northwestern Jackson counties along White Creek and the south fork of Salt Creek suggest that some stream ponding and derangement have taken place; but the locality has not been carefully investigated.

The north fork of Salt Creek southeast of Spearsville in northeastern Brown County undoubtedly received a notable flood of glacial waters from the ice margin, and in connection with Muddy Fork in southern Brown County exhibits the results in the form of terraces flanking the present streams. Salt Creek was an important avenue for the discharge of glacial waters across the driftless area. The east fork of White River west of its junction with the Muscatatuck, was also an important sluice-way for the escape of glacial waters from the western margin of the ice sheet in southeastern Indiana. The effects of such glacial discharge have been largely obscured or effaced by the freer passage of the glacial waters and their burden of gravel and sand during the Wisconsin glacial stage. Likewise the Ohio River valley must have been an important outlet for the glacial floods of the Illinois glacial stage, and here again the glacio-fluviatile discharge from the Wisconsin ice sheet has largely covered the material deposited in the valley during the Illinois stage. Leverett,[123] however, recognizes and distinguishes the deposits of each stage.

The passage of glacial waters from the eastern section of the Illinois ice sheet across the driftless area, and the partial impounding of these same waters on the west, resulted in valley trains of gravel, sand, silt and clay, which caused local upbuilding of the tributary streams of the driftless area where they enter the main valleys. Such effects are seen in the alluvial terraces preserved along a number of the streams within the driftless area. One pronounced effect of such local upbuilding is seen in the numerous abandoned meanders enclosing bed-rock, island-like masses of upland, standing isolated in the midst of valley alluvium or terrace material. A number of such

[122] Siebenthal, 21st Ann. Rept., Ind. Geol. Surv., 1896, pp. 302-303. Marsters, Topography and Geography of Bean Blossom Valley, Monroe County, Ind.: *Proc.* Ind. Acad. Sci. for 1901, pp. 222-237. Reagen, Geol. of Monroe County, Ind.: *Proc.* Ind. Acad. Sci. for 1903, pp. 205-233.

[123] Monogr. 41, U. S. Geol. Surv., 1902, pp. 257, 258, 267, 290-291; Monogr. 53, U. S. Geol. Surv., 1915, pp. 71-72.

valley monadnocks are present along Salt Creek south of Harrodsburg; along the lower part of Indian Creek in Martin County; and along the east fork of White River, at Williams in western Lawrence County and at Shoals and Hindostan in Martin County. The many similar valley monadnocks occuring in the Wabash lowland may be definitely associated with the valley filling characteristic of that region.

Sangamon-Loessial-Peorian Inter-glacial Stage.

Following the Illinois glacial stage denudation of the lands continued. Leverett[124] has named the interval of time between the Illinois glacial stage and the period of loess deposition, the Sangamon interval. "The Sangamon soil and weathered zone, formed on the surface of the Illinoian drift, has received attention in Monographs XXXVIII and XLI of this survey, and little need be added concerning it. It separates the Illinoian drift by a very marked interval from the overlying loess and associated silt deposits. The writer has never observed the Illinoian drift to grade upward into the loess, either in Indiana or in the neighboring states, but has everywhere found it capped either with a deeply weathered zone or with a dark colored soil."[125]

"The weathered surface of the Illinoian drift and the outlying driftless territory in southern Indiana both bear a thin deposit of silt which is part of a practically continuous sheet that extends from Ohio westward beyond the Mississippi. In the vicinity of the main drainage line it is loose textured and is commonly termed 'loess.' On the interfluvial tracts it ranges from a deposit readily pervious to water to one slowly pervious."[126] The time and in particular the method of deposition of the loess is a question which has been much discussed by Mississippi Valley geologists. The consensus of opinion at present is that it was chiefly a wind-blown deposit, the material having been gathered from the dried and largely non-vegetated mud flats or similar surfaces succeeding the Illinois glacial stage; though locally the loessial deposits are undoubtedly water deposited. Loessial deposits are also associated with an apparently localized advance of the Keewatin ice sheet in Iowa which has been designated the Iowan glacial stage, intervening between the Illinoian and Wisconsin stages. "The thickness of the [loessial] deposit in Indiana in few places exceeds 40 feet and is greatest along the Ohio and Wabash valleys. Within a short distance back from the bluffs of these valleys it decreases to 10 or 12 feet and over the greater part of the region is between 5 and 10 feet. On slopes it is very thin, or wanting, because of erosion."

Over large areas in both southwestern and southeastern Indiana the Illinois glacial flats are covered with a light ash colored clay or structureless silt correlated with the loess of the Wabash and Mississippi bluffs. In the opinion of the writer this structureless clay overlying the Illinoian drift and grading indefinitely into it below need not be considered a loessial deposit. It more probably originated on the glacial or other flats, through continuous leaching and decay of the upper portion of the drift or soil materials.[127]

---

[124] Jour. Geol., Vol. VI, 1898, pp. 171-181, 238-243.

[125] Leverett, Monogr. 53, U. S. Geol. Surv., 1915, p. 73.

[126] Leverett, U. S. Geol. Surv., Monogr. 53, 1915, pp. 74-75.

[127] Mr. Leverett gives an excellent discussion of the so-called Iowan loess and its problems in Monograph XXXVIII, U. S. Geol. Surv., 1899, pp. 153-184. The distribution, thickness, composition, structure, fossils and possible modes of deposition are considered in some detail.

Following the deposition of the loess over the Illinoian drift came a period of erosjon which continued as a part of the Peorian inter-glacial interval in the upper Mississippi Valley. This Peorian interval is marked stratigraphically by locally weathered soils and evidences of erosion where the area of its occurrence was over-ridden by later glaciation.[128] This period of erosion is not distinct from the general post-Illinoian erosion seen beyond the limits of the Wisconsin glacial drift.

Wisconsin Glacial Stage.

General Characteristics and Lobation.

The last great ice sheet which came down from Canada into the United States left a series of terminal moraines separated by broad areas of flattish ground moraine. From the well preserved materials left by this ice sheet much has been deduced concerning continental glaciation. The Wisconsin glacial stage, like the other glacial stages, was named and defined by Chamberlin.[129] The maximum advance of the Wisconsin ice sheet fell far short of that of the Illinoian in Illinois and Indiana, but to the eastward it extended south as far or even farther. The limits of the Wisconsin drift as shown on both the physiographic and glacial maps accompanying this report have been carefully traced by Leverett.[130] All of Indiana north of the margin of the Wisconsin drift was covered by the glacial ice, and the various halts, retreats, and re-advances of the ice, so characteristic of the last glacial stage, have been partially deciphered from the topographic forms and their relationships as shown in Indiana and contiguous territory. The area in Indiana covered by the Wisconsin ice sheet is approximately 22,900 square miles.

The margin of the Wisconsin ice sheet, as indicated by the drift border, was distinctly lobate, a characteristic which became accentuated as the ice retreated to the northward. The well preserved terminal moraine systems indicate that the Wisconsin glacial stage was characterized by notable re-advances of the ice of the various lobes. Such re-advances probably also characterized the earlier ice sheets, but it is thought that lobation was best developed during the Wisconsin stage. Leverett[131] in his discussion of the lobation of the Wisconsin drift border, notes that the lobes of the Indiana region were local deployments down major valleys. These local lobes were off-shoots of the two major lobes occupying Lake Michigan and Huron-Erie basins. "In Illinois the Wisconsin [ice sheet] had but one pronounced lobe, the Lake Michigan, whose outline was rudely concentric with that of the Illinoian drift

and numerous references bearing on the loess problem are given. In Monograph XLI, U. S. Geol. Surv., 1902, pp. 295-301, also occurs a brief discussion of the loess more particularly with respect to its occurrence in the Ohio valley. Fuller in the Ditney Folio No. 84, U. S. Geol. Surv., 1902, and Fuller and Clapp in the Patoka Folio No. 105, U. S. Geol. Surv., 1904, and further in a special article, The Marl Loess of the Lower Wabash Valley: Bull. Geol. Soc. Amer., Vol. 14, 1903, pp. 213-227, discuss the distribution, composition and structure of the loessial deposits of southwestern Indiana, and distinguish between the water-laid and wind-deposited loessial deposits.

[128] See Leverett, U. S. Geol. Surv., Monogr. 53, 1915, pp. 27 and 75. Also Monogr. 38, 1899, pp. 185-190.

[129] Chamberlin, Glacial Phenomena of North America: Geikie's Great Ice Age, 3rd ed., 1895, pp. 724-775; Classification of American Glacial Deposits: Jour. Geol., Vol. III, pp. 270-277; Vol. IV, 1896, pp. 272-276.

[130] Monographs 38, 41 and 53, U. S. Geol. Surv.

[131] Monogr. 53, U. S. Geol. Surv., 1915, pp. 27-28.

border. This lobe was an extension from the Lake Michigan basin. It formed, in the course of its withdrawal from Illinois, a succession of bulky but rather smooth morainic ridges, which extend into western Indiana for a few miles to the head of the re-entrant between it and the portion of the Labrador ice field that passed through the Huron and Erie basins. This re-entrant is a few miles farther west than the great re-entrant between the Illinois and the Huron-Erie lobes of the Illinoian stage, the latter being in south-central rather than in southwestern Indiana."

Causes of Lobation in Indiana.

It appears from the configuration of the terminal moraines left by the lower reaches of the Huron-Erie lobe in Indiana and Ohio, that deployment of the ice was largely dependent upon the altitudes of the pre-glacial topographic sub-divisions. The development of the lobe of East White River southward from the latitude of Noblesville and Anderson was probably caused by the relatively low altitude of the pre-glacial Scottsburg lowland in Hamilton and Marion counties, as compared with the now buried Norman upland immediately west. The pre-glacial Knobstone escarpment in eastern Boone County and southward through western Marion and Johnson counties was over-ridden and over-lapped, but its ragged front in conjunction with the Norman upland retarded and discouraged ice movement, with the result of deployment southward down the Scottsburg lowland tract. Thus, lobation of the ice took place because of these major topographic differences in the vicinity of the glacial margin. The lobation of the moraines is noticeable for a considerable distance back from the glacial margin. In the East White River lobe, it begins in the latitude of Noblesville and Anderson, a distance of 50 to 60 miles from the farthest extension of the ice; though at the time of maximum advance the extension of the ice-lobes beyond the re-entrants was less than half this distance. The Miami lobe, chiefly in southwestern Ohio, deployed southward along the line of a major pre-glacial valley, east of the high pre-glacial upland in Randolph, Wayne, and other counties in Indiana. The main Erie lobe reaching southwestward across middle northern Indiana apparently did not deploy along the line of a continuous valley. That it did so, however, in Ohio is indicated by the depth of the drift over the bed-rock surface of the region.

Lobes and Moraines of the Late Wisconsin Substage.

When the ice sheet had retreated to the vicinity of Delphi or Logansport in northwestern Indiana the separation of the Saginaw lobe began. Further withdrawal of the ice northward and eastward was attended by the deposition of three distinct sets of roughly concentric terminal moraines. These moraines were formed during halts or temporary re-advances of the ice margin. In northwestern Indiana the moraines were built by the Lake Michigan lobe and were conformable to its spatulate margin. In central northeastern Indiana the morainic systems are in excellent harmony with the main Erie lobe; while in north-central Indiana the moraines conform to the somewhat crowded position of the enclosed Saginaw lobe. Such a systematic arrangement of the morainic systems is in keeping with the lobation of the ice, and barring minor deployment and irregularities, the separation of the lobes is well borne out by field evidence. Under this interpretation the outer moraine of the Saginaw lobe consists of the Maxinkuckee moraine, extending northward from a few miles west of Logansport, and the massive Packerton moraine leading northeast from the same position. The weak Union moraine of the main Erie lobe, running northwest from Muncie to near Logansport, supports this interpretation, since it seems quite unreasonable to assume that its correlative could be the unusually massive Packerton moraine. Leverett[132] assumes that the Saginaw lobe retreated rapidly from northern Indiana and southern Michigan, while the Erie and Lake Michigan lobes still extended deeply into northern Indiana, though he has some difficulty in explaining the rather prominent moraines in northern Indiana assigned to the Saginaw lobe. The interpretation here suggested does not assume a rapid withdrawal of the Saginaw lobe from northern Indiana. The evidence rather supports the idea of a rapid retreat of the Lake Michigan lobe from the Kankakee-Tippecanoe basin. The absence of moraines in this locality is rather striking. It is probable that the Saginaw lobe receded quite rapidly across southern Michigan. This is suggested by the dying out of the massive Maxinkuckee outer moraine in the vicinity of South Bend and the disappearance of the Packerton component in Noble County. Further, the relationships of the moraines in Steuben, Lagrange, and Noble counties indicate that the Erie lobe over-rode the earlier position of the outer Saginaw lobe after the recession of the latter from that locality.

Glacial Lakes Maumee and Chicago.

The recession of the Wisconsin ice sheet beyond the Mississippi-St. Lawrence divide initiated an interesting phase of drainage development. It was the beginning of the series of extensive ice-dammed lakes which were the precursors of the present Great Lakes system. Probably the first notable ice-dammed lake to appear was glacial Lake Maumee. This lake expanded within the drainage basin of the present Maumee River as the ice receded eastward, and uncovered more of the drainage basin. The outlet was over the lowest point in the divide lying between Fort Wayne and Huntington. The divide at this place was cut down nearly 100 feet by the water which escaped to the Mississippi system through this outlet. The St. Joseph and St. Marys rivers, the positions of which are controlled by terminal moraines, emptied into the glacial Maumee-Wabash River near the head of the lake. Only the western end of Lake Maumee reached into Indiana. Some 120 square miles were flooded in northeastern Allen County. When the ice had retreated far enough to uncover a lower spot in the divide, the lake waters were lowered and the outlet through the Maumee-Wabash channel abandoned. This gave rise to the silt covered area now known as the Maumee lacustrine plain.[133] Over this plain flowed the new-born Maumee River, which at first drained eastward into the lower lake, and later into the successor to Lake Maumee. The lowering of the lake waters and the initiation of Maumee River favored a discharge of the waters from the St. Joseph and St. Marys rivers to the eastward, thus giving rise to the barbed pattern of the present drainage of the Maumee.[134]

---

[132] Monogr. 53, U. S. Geol. Surv., 1915, pp. 123-124.

[133] See Geological map of Allen County by Price, Ind. Geol. Surv., 30th Ann. Rept., 1905, p. 275.

[134] For the details of the development of Glacial Lake Maumee see the following: Gilbert, Surface Geology of the Maumee Valley: Ohio Geol. Surv., Vol. I, 1873, pp. 535-590; Dryer,

Only one other of the series of ice-dammed lakes, connected with the Great Lakes system, invaded the Indiana region. This was the glacial Lake Chicago. It was the southern expanded portion of the present Lake Michigan, with an outlet southward through the Illinois River into the Mississippi. The area of the Calumet lacustrine plain marks the site of this glacial lake in Indiana. Lake Chicago endured for a much longer period than Lake Maumee. It is marked by several successive stages of lowering, but until the present stage was reached it retained its outlet through the Illinois River. The details of its connection with the stages of ice retreat, and its connection with the waters which came from other contemporary ice-dammed lakes to the east constitute a fascinating chapter in the history of the Great Lakes.[135]

Summary of the Wisconsin Substages.

The Wisconsin glacial stage is marked by the development of a series of successive terminal moraines. These moraines are not always single and well-defined, but each consists of a morainic system frequently comprising a number of related moraines, distinct from any other system of related moraines. The outermost morainic system of the Wisconsin glacial stage is known as the Shelbyville system, so named by Leverett from its typical development near Shelbyville, Illinois. It is poorly developed in Indiana. The second Wisconsin substage is marked by the moraines of the Champaign system, named from Champaign, Illinois. This system as a whole is not well developed in Indiana, but moraines included in this system are quite prominent locally. The Bloomington morainic system marks a very prominent substage in the development of massive moraines. This system is also best developed in the area of the great Michigan lobe, and receives its name from Bloomington, Illinois. In western Indiana this system is rather complex, owing to the tendency of the ice sheet as a whole to break up into lobes which advanced and retreated somewhat independently. Following the Bloomington substage is what the writer prefers to call the Late Wisconsin substage. It terminates the Wisconsin glacial stage, and is marked by the separation of the great ice sheet into distinct lobes which did not necessarily advance and retreat as a unit. The chronology of the morainic systems belonging to the Late Wisconsin substage is not satisfactorily worked out. The moraines of the different lobes take on a recessional character. In Indiana this feature is best developed in the area of recession of the Erie lobe.

An attempt has been made to show the Wisconsin substages as named above on the glacial map of Indiana, Pl. III.[136]

Geology of Allen Co., Ind., 16th Ann. Rept., Ind. Geol. Surv., 1888, pp. 107-114; Taylor, A Short History of the Great Lakes: Dryer's Studies in Indiana Geography, pp. 90-110, Inland Pub. Co., Terre Haute, 1897; Taylor, The Great Ice-Dams of Lakes Maumee, Whittlesley and Warren: Amer. Geologist, Vol. 24, 1899, pp. 6-38; Taylor, Glacial Lake Maumee: Monogr. 53, U. S. Ceol. Surv., 1915, pp. 334-349. Also pp. 318-322; Leverett, Monogr. 41, U. S. Geol. Surv., 1902, pp. 710-740.

[135] See the following: Leverett, *Pleistocene Features of the Chicago Area*: Bull. Geol. and Nat. Hist., No. 2, Chicago Acad. Sci., 1897, pp. 1-86; Leverett, The Chicago Outlet and the Beaches of Lake Chicago: Monogr. 38, U. S. Geol. Surv., 1899, pp. 418-459; Leverett, Glacial Lake Chicago: Monogr. 53, U. S. Ceol. Surv., 1915, pp. 350-357; Blatchley, Geology of Lake and *Porter* Counties, Indiana: 22nd Ann. Rept., Ind. Ceol. Surv. for 1897, pp. 25-104; Alden, The Chicago Folio No. 81, U. S. G. S., 1902, pp. 1, 7-12.

[136] For the details of the substages and morainic systems of the Wisconsin glacial stage, see Monographs of the U S. Ceol. Surv. 38, 1899, pp. 191-417; 41, 1902, pp. 304-580; 53, 1915, pp. 77-463.

Recent Physiographic Work.

The time since the Wisconsin ice sheet permanently withdrew from the Indiana region has been relatively short, and little physiographic work has been accomplished. The area covered by the Late Wisconsin ice is essentially as it was left by the glacier and its attendant waters. The materials deposited by the glacier look remarkably fresh. Minute scratches are perfectly preserved on soluble limestone pebbles in the upper part of the drift. The work of post-glacial time is manifest, however, in the destruction of the numerous small basins which characterize morainic topography. Thousands of miniature lakes have been filled by inwash and vegetal accumulations. The areas of these small shallow basins are now marked by level deposits of black earth. Many of the larger basins have become smaller, and much of their former area has become marsh or meadow of peaty soil surrounding what is left of the once more extensive lake. In the area of the earlier Wisconsin drift, as

Fig. 15. Man has brought about great changes in the topography of the lands. Head cut, Illinois Central Railway, in eastern Greene County, 2 miles west of Solsberry.

for example on the Tipton till plain, no lakes at all exist, though in the natural state great areas of marsh and tracts of level, black land bespeak their former existence. Nearly all such lands are now artificially drained and are highly productive agricultural areas. Much of the change within the area of older Wisconsin drift, such as the filling up of the shallow basins and the development of drainage, has taken place since the withdrawal of the ice from the region, but during the Pleistocene period. In southern Indiana erosion continued through the latter part of the Pleistocene, and the stream adjustments and drainage development following the Illinoian glaciation were practically complete before the close of the Pleistocene. The terraces which characterize the valleys where temporary or permanent drainage obstruction took place had reached approximately their present form by the close of the Pleistocene.

It should be noted, however, that the most marked change in the topography and general aspect of the lands since the Pleistocene period, has taken place since the advent of man. Man has been a potent factor in topographic change in the short time that he has occupied the lands. In Indiana as elsewhere, deep cuts have been made for railroads and other thoroughfares; dams across valleys have been constructed; mines with miles of subterranean passageways have been made, and the collapsed pits of abandoned mines and great piles of refuse many feet in height mark the mining areas in southwestern Indiana; great irregular ridges of earth material and deep trenches and irregular basins cover wide stretches of waste land where stripping of shallow coal seams has taken place; great quarry pits scar the hillsides where thousands of tons of stone have been removed; gravel pits are characteristic of the glaciated area; artificial drainage has greatly accelerated the natural drainage, and as much geologic work has been accomplished in a few decades as had previously been wrought during the tens of thousands of years since the close of the Pleistocene period. Perhaps one of the most notable changes in the aspect of the lands is due directly and indirectly to deforestation. Re-

Fig. 16. In procuring coal for industrial needs man has locally greatly changed the topographic aspect of the landscape. Coal IV. Primrose mine and strip pit, 1 mile north of Jasonville.

moval of the vegetation in the humid region has enormously increased the destructive power of the denudational agencies. Rainwash is directly effective, as it is in the arid regions; run-off is greatly increased, and gully erosion unnoticeable in the previously forested areas is now wide-spread in the deforested lands. Deforestation and the attendant tilling of the lands have given rise to rapid streams of turbid water where formerly clear sluggish streams flowed seaward. Thus, the present is a time of great destruction of the lands and marked topographic changes as compared with the post-glacial period previous to the advent of man. Such destruction and change are attained through the agency of man in gaining his needs and supplying his wants, in doing which he co-operates with the natural destructive agencies of the lands.

CHAPTER V.

SOME DETAILS OF PHYSIOGRAPHIC DEVELOPMENT IN INDIANA.

Influences in the Topographic Development of the Dearborn Upland.

General Statement.

It has been stated that the topographic condition of the Dearborn upland in Switzerland, eastern Jefferson, Ohio, Dearborn, eastern Ripley, and Franklin counties, and to a lesser degree in Union, Fayette, and southern Wayne counties, is essentially that of a deeply dissected upland surface. The main valleys lie from 300 to 500 feet below the remnants of the former extensive and nearly level surface. On the main interstream surfaces tracts of considerable area are as yet untrenched by the drainage lines. Even in the dissected parts adjacent to the main valleys, the long, even-topped ridges between the deeply set valley trenches retain to a remarkable degree the elevation of the formerly extensive upland surface. The upland surface with its plateau-like aspects, it will be recalled, has been correlated with the Highland Rim peneplain surface of Tennessee.

Several things enter into the explanation of the present topographic condition of the Dearborn upland area, in addition to the normal destructional agencies common to humid uplands well inland from the sea. First, the structure and lithological constitution of the rocks enter as primary conditioning factors; second, drainage modification attendant on the Pleistocene period, especially in connection with the Ohio River, has brought about considerable topographic change; third, the Illinoian glaciation greatly interfered with the normal erosional destruction of the uplifted peneplain surface; and fourth, the Wisconsin ice sheet greatly modified the northern part of the area. The influence of these several factors on the development of the present topography of the region will now be discussed.

Influence of Geologic Structure and Lithology.

The Dearborn upland of Indiana lies near the axis of the Cincinnati arch, and the rock layers vary but slightly from the horizontal. There is a slight dip westward and southwestward along the western side. The southwest dip along the Ohio River is perhaps more pronounced than elsewhere. The influence of the westerly dip is apparent in the course of Laughery Creek and perhaps also of Indian-Kentuck Creek. It is a well established law that streams flowing on rock tend to migrate down the dip of the rock. This shifting is true in the Dearborn upland area to a considerable degree, yet there are several notable exceptions to it. Lower Whitewater River, Tanners, Hogan, lower Laughery, and other smaller creeks in Ohio and Switzerland counties turn and flow against the dip, contrary to the well established law for minor streams. This feature is one of the very suggestive evidences of the former flow of the Kentucky River northeastward up the present Ohio and Big Miami rivers, as has been previously suggested.

With regard to structural influences on the course of the upper part of Laughery Creek, Ward[187] says: "As a consequence of the dip of the rock to

[187] Ward, Soil Survey of Ripley County: 32nd Ann. Rept., Ind. Geol. Surv., 1907, p. 211.

the west in this county [Ripley], Laughery has been forced always toward the western bank with greater pressure than toward the eastern. This fact, in connection with the presence of a resistant rock on the west, and soft rock on the east, has resulted in a peculiar valley form. The western valley side is practically everywhere steep—so steep that cultivation is entirely out of the question. It is an ideal cliff-and-talus slope for much of its course, averaging something more than 125 feet in height throughout the county. The eastern valley-side is much less precipitous, in some places arable from top to bottom, and nearly everywhere of gentle slope enough to allow an accumulation of soil over the rocks." Ward has here also described the Laughery escarpment which marks a portion of the western boundary of the Dearborn upland. The rather striking alignment of the valleys of the west fork of Whitewater River, Salt, Laughery, and Indian-Kentuck creeks has been explained by Newsom[135] as probably due to the influence of westward dip of the strata. Such is very probably the case.

The lithological constitution and succession of the rock strata which compose the Dearborn upland area have been important in controlling the topography of many of the land forms. The main streams have cut deeply below the uplifted peneplain surface, and the rock on the great rounded spurs and steep hillsides is very near the surface, but not so frequently exposed as should be expected where the relief is so extreme. The strata are in the aggregate more than one-half shales and clays of very non-resistant character, the remainder being rather resistant limestones of a flaggy nature intercalated with the soft shales. Towards the bases of the hills the shales are much more abundant than the limestones, but the upper parts of the hills and the upland surface are largely composed of limestones which at certain horizons take on a relatively massive character. The limestones are much more resistant to the destructional agencies than the shales, but the weathering of the shales allows the flaggy limestones to slump down the slopes. Bed-rock in place is rather infrequently seen, considering the sharpness of the relief. The presence of the more abundant and rather resistant limestone in the upper horizons has been an important factor in the preservation of much of uplifted peneplain surface which composes the relatively level upland tracts, reaching an elevation of 900 to 1,000 feet. This feature also is responsible for the steeper slopes of the lower parts of the valley sides. The presence of the intercalated shales throughout the whole thickness of the upland strata causes a general lack of resistance to the destructional agencies, and is responsible for the great rounded spurs, smooth, uniform slopes and general lack of angularity which characterize the dissected portion of the Dearborn upland. Cliffs are rare features throughout the area. The valleys are rather broad, though the streams are swift except in their lower courses. The breadth of the valley floors may be assigned for the most part to the predominance of non-resistant shales at the horizon of the valley floors.

Influence of Drainage Modifications.

The second factor of importance affecting the present topographic condition of the Dearborn upland area was the drainage changes due to stream obstruction by glacial ice during the earlier part of the Pleistocene period.

[135] Newsom, Geologic Section across Southern Indiana: 26th Ann. Rept., Ind. Geol. Surv., 1901, pp. 290-291.

The question of the size of the pre-glacial Ohio River along the southern border of Indiana has already been discussed at some length. Under either interpretation as to its former size and the position of its former headwaters, the present Ohio River is many times larger than its pre-glacial predecessor. The great increase in the size of the stream undoubtedly caused considerable deepening and widening of the valley formerly occupied by it, or other valleys which it supplanted. With the advent of a major trunk stream in such close proximity to an upland which possessed no major drainage lines, erosion was greatly increased. Tributary streams were greatly steepened and erosion along their courses was also speeded up. But the greatest accentuation of relief because of such drainage changes was in the vicinity of the Ohio. How great the increase in relief was is somewhat conjectural. A careful study of the terraces and benches along the Kentucky and Licking rivers of Kentucky would probably throw much light on the problem.

Other drainage changes of importance are more definitely related to the Illinois glacial stage. It is probable that a portion of the upper part of Laughery Creek owes its position to post-Illinoian stream adjustments. Certainly the head water areas of many of the minor streams began on the glacial plain developed at the level of the old peneplain surface, and flowed for some distance without reference to the pre-Illinoian valleys. Just what relation the Whitewater River drainage basin bears to its pre-Illinoian predecessor is not clear. The middle and lower parts of the present stream, however, occupy the pre-Illinoian valley, but at an elevation of 100 to 200 feet above the bed-rock floor of the old valley.

Influence of the Illinoian Glaciation.

The third factor mentioned above as entering into the full explanation of the present topographic features of the Dearborn upland area, is that of the Illinoian glaciation. Previous to the Illinoian glaciation of the region, dissection was probably greater than at present. The ice-sheet over-rode the entire upland area in Indiana, and on its retreat practically all of the smaller valleys were effaced and the larger ones were more or less deeply filled. The great Ohio River valley which had previously come into existence was probably at no time entirely obstructed[139] though the theory of an ice-dam in the vicinity of Cincinnati has been stoutly held by Wright[140] and evidence to support the existence of the ice-dam has been advanced. A general effect of the ice-sheet was the reconstruction of a plain somewhat simulating the old Tertiary peneplain surface and at its identical level. This effect is not generally appreciated.

The relatively wide areas between the deeply set valleys of the old Tertiary peneplain appear, when looked at casually, to have changed little since the development of the peneplain. Yet evidence is at hand which shows that the similarity of the upland surface to the old peneplain surface is more apparent than real. Not so much of it has actually been preserved intact as the present surface would seem to indicate. Wells on the plateau-like interstream tracts of the upland usually encounter bed-rock within ten or twenty feet, since the

[139] See Leverett, U. S. Geol. Surv., Monogr. 41, 1902, pp. 290-291.

[140] Wright, Theory of a Glacial Dam at Cincinnati and Verifications: Am. Naturalist, Vol. XVIII, 1884, pp. 563-567.

drift as a whole is quite thin. But occasionally drift is penetrated to depths of fifty and eighty feet, indicating the presence of pre-Illinoian valleys where now no vestige of them exists. The restoration of the surface of the old uplifted peneplain by deposition of glacial debris in the valleys was probably fairly complete over much of the upland surface, but the great relief combined with the proximity to the major streams has caused much of the reconstructed surface to be destroyed. The present valleys are nearly everywhere sunk in bed-rock. Little drift is noticeable, partly because much of it has been removed and partly because on the upland tracts where it has not been entirely removed it is covered by a fine, whitish clay soil termed loess. Evidence of glaciation is largely removed or obscured. Ward[141] in briefly discussing glaciation in the counties of southeastern Indiana, says: "But of the actual presence of ice within the limits of these counties there is little evidence, except in the extreme northern portion." Again Bigney[142] says: "Dearborn County seems not to have been invaded by the Glaciers, except along the eastern margin." The evidence obtained from wells, however, shows that glaciation was of more importance than is generally believed.

Influence of the Wisconsin Glaciation.

The Wisconsin glaciation changed the larger part of the present Dearborn upland area but little. The northern edge, however, was covered with drift material, and the former surface largely obscured. In the vicinity of Whitewater River and its chief tributaries the Wisconsin drift has been mostly removed and bed-rock conditions prevail. Here valleys with steep sides and great projecting rounded promontories between the valley trenches are quite as characteristic of the Dearborn upland as they are farther south. Towards the northern edge of the upland in southern Wayne County the drift mantle becomes quite thick and the bed-rock surface no longer gives expression to the landscape. The topographic aspect is rather that characteristic of the glacial plain to the north.

A minor consequence of the Wisconsin glaciation is seen in the great valley-train along Whitewater valley. This valley is deeply filled with outwash from the Wisconsin ice-sheet, and the partial removal of the valley-train has given rise to terraces flanking the present flood plain.[143] A terrace some 60 feet above low water mark also occurs in the Ohio River valley. This terrace is rarely more than a half-mile in width. Perhaps the greatest valley expanse in the Dearborn upland area occurs below the mouth of the Big Miami River, stretching southwestward to the vicinity of Aurora in southeastern Dearborn County. It comprises some 10 to 12 square miles of flat valley floor. This area is known locally as the "Big Bottoms."[144]

Development of the Muscatatuck Regional Slope.

Unity and Subdivisions.

The Muscatatuck regional slope has been described on a previous page as an upland area which slopes gently to the southwest from the western margin

[141] Ward, 32nd Ann. Rept., Ind. Geol. Surv., 1907, p. 199.

[142] Bigney, Geology of Dearborn County: 40th Ann. Rept., Ind. Geol. Surv., 1915, p. 220.

[143] Leverett, U. S. Geol. Surv., Monogr. 41, 1902, pp. 184-185, 324, 339.

[144] Ward, Soil Survey of Dearborn and Ohio Counties: 32nd Ann. Rept., Ind. Geol. Surv., 1907, pp. 229-230.

of the Dearborn upland, and merges into the Scottsburg lowland lying on the west. The drainage is down this slope and is typically represented by the various branches of the Muscatatuck River in Jefferson and Jennings counties. The northern part of this regional slope is covered by Wisconsin drift, yet the slope of the region has been preserved. Here it has the surface characteristics of a glacial plain, since the underlying bed-rock gives little expression to the surface. Thus the regional slope may be divided into two sections, one lying north of the Wisconsin drift boundary line and the other and more typical portion lying south of this line. This latter section was entirely covered by the Illinoian ice-sheet, and much of its flattish aspect is that of a glacial plain, though the slope is manifestly dependent upon the slope of the underlying bed-rock surface. Bed-rock in the southern section of the slope is usually close to the surface, as the drift covering is thin. The northern section of the regional slope is covered more deeply with glacial drift, and the underlying bed-rock slope is as a whole obscured. The unity of the regional slope is expressed in the fact that the slope of the underlying bed-rock surface is the controlling feature in both sections, for the surface slope of the entire area corresponds to the slope of the bed-rock surface. The later Wisconsin drift of the northern section has been more uniformly and deeply spread over the underlying surface than the thin coating of earlier Illinoian drift over the bed-rock surface of the southern section of the same regional slope. This condition must be understood in order to appreciate the unity of this physiographic region. It may be said then that just one more physiographic event has occurred in the northern section. This event did not destroy the fundamental slope, though the minor surface characteristics were considerably altered.

Influence of Structure and Lithology.

The pre-glacial (or rather pre-Illinoian) development of the area now designated as the Muscatatuck regional slope, well illustrates the control of bed-rock conditions over physiographic regions.[18] The strata dip to the west-southwest at the rate of ten to twenty-five feet per mile. This homoclinal dip on the western limb of the Cincinnati arch in connection with former peneplanation is responsible for the exposure of strata of different ages and degrees of resistance to weathering and erosional agencies. (Fig. 17.) The upper Ordovician limestones and shales dip beneath the relatively resistant Silurian limestones at the eastern edge of the Muscatatuck regional slope. West of the eastern edge of the outcropping Silurian limestones these strata are exposed only in the bottoms of the entrenched valleys. The Silurian and Devonian (chiefly the Jeffersonville) limestones make up some 150 feet of strata which constitute the main bed-rock surface of the Muscatatuck regional slope. These limestones form a resistant unit as compared to the shales which succeed them stratigraphically. The New Albany shale weathers readily, and is present over the limestone on the broad interstream tracts where it is protected from disintegration by a deep soil or by a glacial covering. On the Muscatatuck

[18] Newsom has shown specifically that the drainage direction of this unit, the topographic forms present, and the physiographic unit itself, are due clearly to the dip of the strata and the lithological succession of resistant and non-resistant rock units as they have controlled erosion: Geologic Section across Southern Indiana, Journal Geol., Vol. VI, pp. 251-252; 26th Ann. Rept., Ind. Geol. Surv., pp. 238-239, 243, 287-299, 301-302.

regional slope, it is chiefly confined to the western third. A great thickness of clays, sandy shales, and impure sandstones composing the Borden series (Knobstone: see Dr. Cumings' report on nomenclature and stratigraphy) succeed the New Albany shale. The sandstones of the series come mainly at the top. The lower portion of the series is even less resistant to weathering and erosion than the New Albany shale. None of the strata of the Borden series are present on the Muscatatuck regional slope as it now exists, but probably formed a part of the overlying strata before the present stage of development. See Fig. 17.

The relative resistance of the above mentioned formations may be judged by the topographic forms present on the area of outcrop of each unit and by the relative altitudes of the strata which together form a resistant unit as

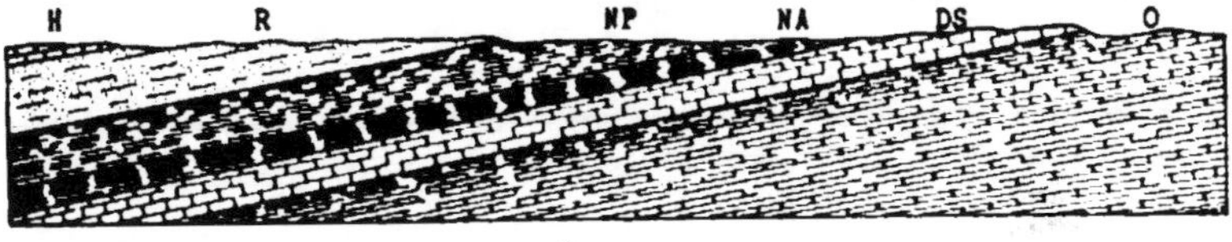

A.

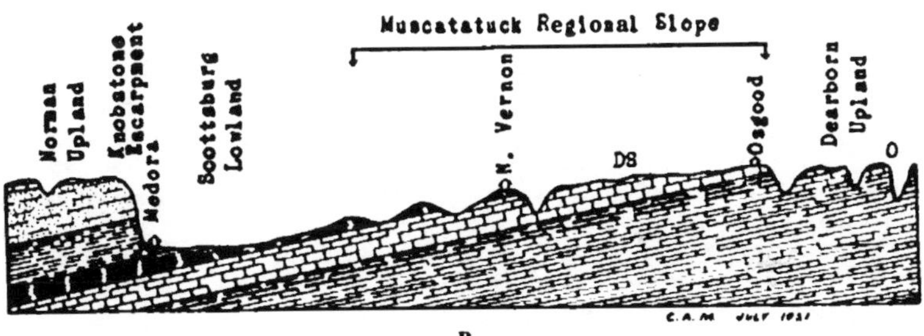

B.

Fig. 17. A. Reconstructed Tertiary peneplain (Highland Rim). B. Diagrammatic section across the Muscatatuck regional slope and the Scottsburg lowland areas, showing the profile of the bed-rock surface and the structural and lithological relationships. H = Harrodsburg Limestone; R = Riverside sandstone (Knobstone); NP = New Providence shale (Knobstone); NA = New Albany shale; DS = Devonian and Silurian limestones; O = Ordovician limestones and shales. Section 70 miles long. Section at base 200 feet above sea level and uppermost surface of section 1,000 feet.

compared with those which form a non-resistant unit. It is largely on the basis of these features that the Muscatatuck regional slope and the Scottsburg lowland, as well as other physiographic units of southern Indiana, have been differentiated.

It is presumed that at the close of the early Tertiary peneplanation there was little difference in the general level of the entire Highland Rim peneplain as developed in southern Indiana. The reconstruction of the peneplain attempted in the accompanying illustration, Fig. 17-A, is not so much for the purpose of showing the peneplain surface as for the representation of the strata on which it was developed. After uplift of the peneplain, weathering and erosion of the different strata proceeded in a very unequal manner, but rather strictly in accordance with their relative resistance. The New Albany shale and the New Providence shale were rapidly stripped off from the underlying Devonian and Silurian limestones. This removal of the non-resistant strata proceeded progressively from east to west. At the same time that the removal of the non-resistant strata was taking place, laying more and more of the limestone surface bare, the slow destruction of the attenuated eastern margin of the limestone mass gave rise to an increase in the area of the Dearborn upland; but the contrasting resistance of the limestone unit and the underlying Ordovician strata gave rise to an escarpment at the eastern edge of the Muscatatuck regional slope, which on a previous page has been described as the Laughery escarpment. The New Albany shale has not yet been entirely removed from the interstream areas of the regional slope on the west. The presence of the shale on the interstream tracts on the western third of the regional slope is in part due to its position and in part due to a protective covering of glacial till, and probably also in part to rejuvenation. But in any case the western margin of the slope would contain some shale on the interstream spaces, since it merges gradually into the Scottsburg lowland on the west.

The surface of the Muscatatuck regional slope is a close approach to a structural upland plain the inclination of which is very near the inclination or dip of the strata on which it is developed.[144] For instance, along the line of the Baltimore and Ohio Railway from Osgood near the eastern boundary of the regional slope to Hayden at the western boundary, a distance of 26 miles, the strata dip 475 feet or at the rate of a little over 18 feet per mile; the surface of the regional slope along the same line inclines 400 feet, or at the rate of a little over 15 feet to the mile. This slight difference in the dip of the strata and the inclination of the regional slope may be accounted for largely by the relative ages of the eastern and western margins of the slope. The eastern edge was already in a beveled condition at the time of the uplift; in addition the eastern margin was uncovered before the western edge. In fact, the western edge is still being stripped of its burden of overlying non-resistant strata. See Fig. 17.

Glaciation and Present Drainage Development.

Glaciation had little to do with the general development of the Muscatatuck regional slope. Glaciation took place chiefly after the slope was developed. But the present drainage lines of most of the area were developed after glaciation. Apparently most of the pre-Illinoian streams north of the Ohio River were small and had approximately the same arrangement down the slope as at present. These pre-Illinoian valleys were probably largely effaced by the deposition of drift, only a thin coating, however, having been left on the upland interstream tracts. The Illinoian glaciation gave rise to a flat plain substantially without valleys. The present drainage lines have been developed subsequently on the glaciated surface. Great areas of the level till plain in southeastern Decatur, western Ripley, Jennings, middle Jefferson and eastern Clark counties still exist, and over much of them the bed-rock lies from 10 to 25 feet below the surface. Occasionally wells penetrate from 50

[144] Siebenthal, 25th Ann. Rept. Ind. Geol. Surv., 1900, pp. 359-360; Newsom 26th Ann. Rept., Ind. Geol. Surv., 1901; pp. 238-239, 287-299.

to 100 feet of glacial drift where the pre-Illinoian valleys were located. But for the most part, it is probable, that the present drainage lines occupy the old valleys, especially in the western half of the regional slope. The present upper branches of the Muscatatuck River in Jefferson, Ripley, and Jennings counties and the upper part of Sand Creek in Decatur and northern Jennings counties, are to all appearances post-Illinoian. These streams are relatively swift and flow over bed-rock surfaces except locally, and they occupy narrow entrenched valleys with prominent cliffs and bluffs of limestone. The Vernon Fork of the Muscatatuck is characteristic. In western Ripley County it is but slightly entrenched, but in middle Jennings County it is conspicuously sunk below the bordering rather flat upland plain. It reaches its maximum entrenchment in the vicinity of Vernon where it is from 140 to 160 feet below

Fig. 18. View of the entrenched Muscatatuck River at Vernon, in middle Jennings County. The general altitude of the Muscatatuck regional slope here is 750 feet, but the Muscatatuck River is about 600 feet in altitude. It is a swift stream and flows over bed-rock in this locality. It has all the aspects of a youthful stream, though here it has some unusual meander curves.

the general level of the upland surface. Here it has a series of remarkable entrenched meander curves,[147] though meander curves are quite characteristic of most of the streams in the middle and eastern part of the regional slope. At the western margin of the regional slope the Muscatatuck River flows in a wide, open valley with no bluffs of any consequence. The limestone is here about at the level of the valley, and the overlying shales seldom give rise to bluffs. It is at this stratigraphic horizon that the Muscatatuck regional slope merges into the Scottsburg lowland with its extraordinarily wide valleys and low-lying uplands.

Some very striking features of valley erosion occur in Jefferson County. The tributaries of Big Creek (one of the forks of Muscatatuck River) head on a remarkably flat divide extending northeast and north from Hanover.

[147] Dryer, The Meanders of the Muscatatuck at Vernon, Indiana: Proc. Ind. Acad. Sci. for 1898, pp. 270-273.

See Figures 3, 19, and 20. This divide ranges in elevation from 850 to 925 feet and is in close proximity to the Ohio River in the vicinity of Hanover.

Fig. 19. View of the flat Muscatatuck Regional slope on the divide between Indian-Kentuck Creek and Big Creek, about 6 miles north of Madison on the Versailles-Madison road. Here the Muscatatuck regional slope is seen in its typical aspects near its high eastern margin. Its altitude here is slightly more than 900 feet.

Fig. 20. View of Big Creek near its source in northern Jefferson County, on the Versailles-Madison road. Big Creek here is more than 900 feet in altitude, and is barely below the general level of the plain of the Muscatatuck regional slope. It flows over bed-rock for miles, as shown in the view. Farther down in its course it becomes intrenched in the inclined plain, and becomes a veritable gorge with steep limestone walls.

Culbertson[149] has called attention to drainage conditions and the peculiarities

[149] Culbertson, Some Peculiarities in the Valley Erosion of Big Creek and Tributaries: Proc. Ind. Acad. Sci. for 1907, pp. 101-103. Also Geology and Natural Resources of Jefferson County: 40th Ann. Rept., Ind. Geol. Surv., 1915, pp. 223-239.

of valley erosion of the region. Most of the tributaries of all the north or south flowing streams flow westward. Only short insignificant streams come in from the west side. See Fig. 21. When the streams flow in a northerly or southerly direction they meander to a marked degree. Recurving is very striking. "The banks of the stream, as well as the sides of the valley, where the meanders are prominent, are almost perpendicular cliffs on the convex side.

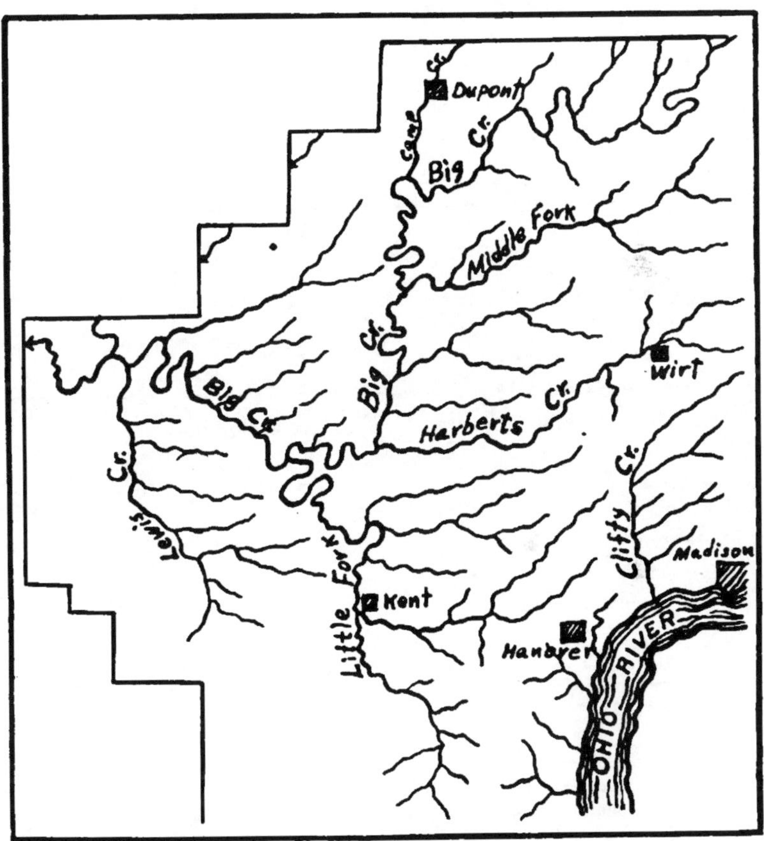

Fig. 21. Sketch map of Big Creek drainage system in Jefferson County (after Culbertson), showing the prominence of the development of westerly flowing streams as contrasted to the notable lack of easterly flowing streams.

These cliffs reach the height of 100 to 150 feet in the lower portions of the stream. The concave sides of the meanders have gentle slopes from one-fourth to one-half a mile in length." The strata in which the valley of Big Creek is developed are largely limestones of Devonian age, and the dip is from 10 to 15 feet per mile in a westerly direction. The underlying Silurian limestones which commonly form the bottoms of the stream beds are much more resistant than the Devonian limestones. "The stream beds, in general, follow the dip of the rocks. In places the bed of the stream is upon the same layer of rock for long distances. Excellent examples of this may be found in the bed of Harberts Creek between Volga and Smyrna church, as well as along parts of Middle Fork and Big Creek. The dip of the rock strata has had much to do with the long, gently sloping streams flowing westward."

"The tributaries that flow in an easterly direction and against the dip of the rock are very short and their gradients are very high. In many of them the water pours into the main valleys over falls located but a few hundred feet from the main stream. . . . The easterly flowing tributaries, eroding their beds largely or entirely in the black shale [New Albany], have cut somewhat longer courses than those in the limestone, but in no case do they even approximate the length of the westward flowing tributaries.

Fig. 22. "Hanging Rock," on the State Highway near Madison, showing the massive Saluda limestone as a cliff former over the weaker strata beneath. Falls 40 to 80 feet are common over this rock on the short, precipitous streams descending from the high upland plain into the deeply intrenched Ohio River in the vicinity of Madison and Hanover. This is the only vicinity along the Ohio River on the southern border of Indiana where water-falls of any significance occur.

"The meanders of these streams are in all probability a consequence of the variable resistance of the rocks followed by maximum erosion of the convex side of streams. They are probably consequent on the slope of the original land surface, although they may have been somewhat modified by the thin mantle left by the glaciers."

The streams which enter the Ohio River from the Muscatatuck regional slope between Indian-Kentuck Creek on the east and Fourteen Mile Creek on the west, are relatively short, and descend over rocky courses to the Ohio River which lies from 250 to 450 feet below the surface of the regional slope. Those in the vicinity of Madison and Hanover plunge over falls ranging from 40 to 100 feet high. In addition to the resistant Silurian limestones, one of the uppermost formations of the Ordovician system of the locality, the Saluda limestone, becomes a very prominent cliff maker, and the chief falls of the

Madison-Hanover region are over this sandy, dolomitic limestone which is from 30 to 50 feet thick. (See Fig. 21.) The region of the falls on Clifty Creek has now become one of the state parks, and is one of the most picturesque localities of the entire state. It is probable that Clifty Creek has grown considerably at the expense of the drainage of Little Fork of Big Creek. (See drainage configuration, Fig. 20.)

The northern section of the Muscatatuck regional slope was, as stated on a previous page, subjected to the Wisconsin glaciation. The bed-rock surface was buried rather deeply, but the main streams descending the regional slope in southern Rush, southeastern Shelby, eastern Bartholomew, and Decatur counties have cut through the drift into the underlying bed-rock. Sand, Clifty, Flatrock creeks, and other streams to the north for the most part occupy valleys made by glacio-fluviatile waters of the Wisconsin glacial stage. These streams are fairly swift, but the valleys are rather broad and entrenchment is slight. The uplands between the main valleys are little modified by the present drainage, and possess characteristics in common with the great glacial plain to the north.

Development of the Scottsburg Lowland.

General Statement.

The Scottsburg lowland has been developed mostly at and near the present base-level of erosion on the New Albany shale and the lower shales of the Borden series. The great extent of valley land and the low-lying uplands are in sharp contrast to the high upland on the west. The altitude of its surface ranges from 425 feet in the broad valley of the Ohio River to somewhat over 700 feet in Johnson County where it loses its indentity as a lowland unit in the adjacent glacial plain. Small areas of greater elevation, such as in the Brownstown Hills in Jackson County, stand as monadnocks or unreduced portions above the general lowland surface. The altitude of the lowland unit will probably average close to 560 feet. The altitude of the sharply contrasting Norman upland surface immediately adjacent on the west will average fully 900 feet, or 340 feet greater.

The Knobstone escarpment facing the lowland area looks from a distance like the side of a great valley. Collett suggested that the lowland was probably the result of a mighty glacial stream which had its origin in the glacial floods from the north, and for a time the name of Collett Glacial River or Valley was applied to the lowland.[149] This stream was supposed to have had its main course down the broad valley of Blue River, heading in northern Henry County, and to have passed through Shelby, Bartholomew, Scott, and Clark counties to the Ohio River. Investigation, however, has shown that no evidence of such a water route exists south of the Muscatatuck River, and the work of Newsom shows definitely that the Scottsburg lowland as a lowland area, is due to normal erosion conditioned by structure and lithology.[150] Newsom unfortunately applied the name of *Eastern Lowland* to the area, a name which would confine the lowland to Indiana and in addition give an erroneous im-

[149] See 11th Ann Rept., Ind. Geol. Surv., 1881, pp. 60-62, 163-166.

[150] Siebenthal, 20th Ann. Rept., Ind. Geol. Surv., 1900, pp. 359-360; Leverett, U. S. Geol. Surv., Monogr. 41, 1902, p. 77; Newsom, 21st Ann. Rept., Ind. Geol. Surv., 1901, pp. 232-302.

pression as to its location. Later Wood[151] suggested the name *Devonian Valley*. This name is unsuitable for the reason that the term Devonian is in common use in geologic sense. Moreover the lowland is perhaps better developed on the lower shales of the Borden series, of Mississippian age. The New Albany shale also is only doubtfully of Devonian age. The name *Scottsburg Lowland*, from its typical development in the vicinity of Scottsburg is subject to none of these objections.

The Scottsburg lowland as it now exists is the product of normal erosion and glaciation. Its designation as a lowland is due wholly to pre-glacial erosion far below the Norman upland on the west, and generally much below the Muscatatuck regional slope on the east. Glaciation by both the Illinoian and Wisconsin ice-sheets greatly modified its northern portion, and completely ef-

Fig. 23. View of the Scottsburg lowland near Scottsburg, Scott County.

faced its pre-glacial continuation to the north-northwest. The history of its development has probably been somewhat as follows:

Pre-glacial Development and Probable Drainage.

At the time of the development of the early Tertiary peneplain in southern Indiana, most of the area now designated the Scottsburg lowland was covered by the upper sandstones of the Borden series. See Fig. 17. If there was much variation in the elevation of the peneplain, it is likely that a lowland slightly depressed below the areas on the east and west existed somewhat east of the present lowland, in conformity with the former position of the non-resistant rocks at the base-level of erosion. The drainage of southern Indiana during this period and until the time of glaciation was considerably different from the present, though the position of the chief drainage lines has not greatly changed. Apparently four main drainage lines flowed westward across

[151] Wood, History of Indiana during the Glacial Period: 40th Ann. Rept., Ind. Geol. Surv., 1915, p. 11.

the area nearly at right angles to its trend. Three of these are represented in the present drainage lines, though with considerable modification. The predecessor of the present Ohio River probably headed near Madison and flowed along the present route of the Ohio to the southwest and crossed into the area now designated the Norman upland and the Mitchell plain. Farther north a drainage system along the approximate line of the present Muscatatuck crossed the area, flowing almost due west, as that stream does today. The Muscatatuck was joined by a stream coming in from the northeast along the line of the present Driftwood or White River in Jackson County. This stream was probably no larger than the old Muscatatuck. The fourth stream system, somewhat more hypothetical, followed the line of the present west fork of White River below Indianapolis. It probably received important tributaries from the east and south, a feature notably lacking in the present drainage. The reason for assuming the existence of such a stream is found in the presence of the wide pre-glacial valley of White River below Indianapolis, and in the known convergence of the pre-glacial lowland and the pre-glacial White River in the vicinity of southern Marion County.

After the uplift of the well developed Tertiary peneplain the drainage became entrenched. The down-cutting of the streams into the resistant strata west of the lowland area gave opportunity for the development of the lowland. The main streams cut wide valleys in the easily eroded shales of the lower portion of the Borden series and in the New Albany shale. Tributary streams, developed during the preceding cycle, conveyed large quantities of the waste material to the main streams which in turn carried it away through the narrow entrenched valleys in the uplands on the west. In time the Scottsburg lowland with its pre-glacial continuation to the north-northwest was eroded far below the surface of the resistant strata on the west and east. The resistant rock which formed the initial Knobstone escarpment was undermined and the escarpment retreated westward to approximately its present position. (See Fig. 17.)

The lowland thus developed was widest and best expressed along the main drainage lines, and protruded somewhat farther west where the main streams entered the upland to the west, as may be seen at present along the Muscatatuck and White rivers in Jackson County. The proximity of the junction of the pre-glacial Muscatatuck and White rivers to the area of non-resistant rock gave rise to the out-lying Brownstown Hills which became isolated from the main upland on the west by the widened and coalescent valleys of these two streams. A correlative of the Brownstown Hills exists in the upland mass of hills between the Ohio River and Floyds Fork of Salt River, Jefferson County, Kentucky. In this case the outlying mass is not so completely set off from the main upland, the area is considerably larger, and the stream valleys setting off the area are not so well balanced in size, though in this latter respect the two areas were probably similar previous to the Pleistocene period.[153]

It is still a question whether or not any portion of the present more elevated interstream surfaces of the lowland occurring between the broad valleys represents the level of the late Tertiary peneplain. It may represent a still later base-level of erosion. The writer favors the belief that a large por-

[153] See Topographic and Geologic Map of Jefferson Co., Ky., accompanying the Report of the Geology of Jefferson Co., by Chas. Butts, Ky. Geol. Surv., Series IV, pt. 2, 1915.

tion of the typical upland flats between the main streams was developed during the earlier part of the glacial period, involving the destruction of the late Tertiary peneplain which had been previously developed. The solution of the question awaits detailed topographic mapping and further careful study.

It is probable that the present topographic condition of the lowland in the region draining into the Ohio River has a rather novel origin. "The lowland plain, stretching north from Louisville consists of a flat to undulating plain varying from 430 feet in the valleys near the Ohio River to something like 600 feet in elevation on the low divide between Silver Creek and the tributaries of the Muscatatuck River. Since there are a large number of hills and flat interstream tracts at an elevation of about 550 feet at the south and coming up to about 600 feet near the above mentioned divide to the north, it has been stated that a local peneplain was formed at that level. The writer concurs in the belief in a base-levelled plain of local area, and that its fur-

Fig. 24. View of the Scottsburg lowland at the foot of the Knobstone escarpment 5 miles north of New Albany. The Scottsburg lowland here has an altitude of about 525 feet. The crest of the Knobstone escarpment reaches an altitude of approximately 1,000 feet, and is representative of the Highland Rim peneplain.

ther development was terminated by rejuvenation. The rejuvenation, however, was not necessarily brought about by uplift. The dissection of the plain was very likely brought about by drainage changes made near the beginning of the Pleistocene. The present Ohio River is a large stream made up of a number of former drainage basins which were more or less destroyed or deranged by combination into a large major stream approximately skirting the outer limits of glacial advance. A very much smaller stream than the present occupied this territory near Louisville. It was able to reduce the area of soft rocks nearly to base-level, but it had a much steeper gradient than the much larger present Ohio. When the present Ohio invaded the basin of the much smaller pre-glacial stream the local peneplain was statically rejuvenated, due to the sinking of the larger stream into the plain on account of its ability to possess a much lower gradient in its grade condition. Such a rejuvenation is here called *static rejuvenation.*

"It may be further stated that the region of the Muscatatuck River to the north still possesses just such a local base-leveled plain as existed in the New Albany locality. It is inferred that the stream which the Ohio dispossessed was somewhat near the size of the Muscatatuck-White River. This stream possesses a gradient in its graded condition slightly less than one foot to the mile, while the Ohio has a gradient below New Albany slightly less than three inches to the mile. It would appear that such a change in gradient would allow a trenching of something like 90 feet, which is approximately the amount of the dissection of the local peneplain in the vicinity of New Albany, using the flood plain as the present local base-level. This figure is derived by taking the difference between the gradients of the Ohio and its assumed predecessor from New Albany to Cannelton, a distance of approximately 120 miles. In the latitude of Cannelton valley filling begins to be rather conspicuous, and this nullifies any difference in the gradients of the former and present streams, assuming that the valley filling of southwestern Indiana and associated regions took place during the Pleistocene. A still further check both on the postulated static rejuvenation and its amount is found in the peculiar gradients of the streams emptying into the Ohio between New Albany and Cannelton. The gradients are approximately as high in their lower reaches as in their middle and upper courses. This is conspicuously true of Blue River and Indian Creek. Other complications, however, enter into the full explanation of these peculiar gradients, making this a problem in itself."[153]

Modifications Due to Glaciation.

The Illinoian ice-sheet covered practically all of the Scottsburg lowland area north of the Ohio River, but over much of the area only a thin coating of glacial drift is present. Evidences of glaciation south of the Muscatatuck River are scanty. Siebenthal[154] in summing up the effects of glaciation in the southern part of the lowland area says: "From these deposits [glacial drift in the vicinity of Charleston, Sellersburg, etc.] and from the glaciated aspects of the country in certain sections, as well as for reasons which will appear later, we think that the whole region, west as far as Silver Creek and south as far as Jeffersonville, was occupied by an ice-sheet which left its impress on the soft Black shale topography without leaving a great amount of drift. No evidence of buried channels has been found, with the exception of that thought to indicate an old channel of the Ohio. If the topography is post-glacial, the pre-glacial topography must have existed in the Black shale and have been carried away entirely by the ice. We find no evidence that this region has been occupied by the great Collett Glacial river as has been urged."

Perhaps one of the most widely noticed effects of glaciation and associated alluviation in the southern part of the region was the change of the Ohio River from its former more deeply eroded bed to the present rocky channel immediately southeast of Jeffersonville. That the Ohio River channel should be immediately underlaid by bed-rock and develop rapids and falls while passing across the Scottsburg lowland is a striking anomaly indeed. Nowhere else

[153] Malott. Static Rejuvenation: Science, N. S. Vol. LII, No. 1338, Aug. 20, 1920, pp. 182-183; Some Special Physiographic Features of the Knobstone Cuesta Region of Southern Indiana; Proc. Ind. Acad. Sci. for 1919, pp. 369-370.

[154] 25th Ann. Rept., Ind. Geol. Surv., 1900, p. 360.

along the southern border of Indiana does this sort of thing occur, though the channel lies in a deep, narrow valley with bordering bluffs of rock for hundreds of miles above and for 120 miles below the Falls of the Ohio at Louisville. In the locality where the falls occur no rock-walled gorge exists, the valley is extraordinarily wide, and the mild upland topography is very low with little rock exposed.

Siebenthal[135] has described a buried pre-glacial channel to the north of Jeffersonville. He infers "that the river must have been dispossessed of this channel by the advancing ice-sheet." But that it should have been forced out of this channel by the ice itself is unlikely. The alluviation, indicated by the terraces which are described by Siebenthal, which was associated with glaciation, was probably the cause of the change to a new course. The entire valley was built up considerably above its present bed by outwash or valley-train material. Later, the river removed much of the valley-fill material, and swinging across the rather broad constructional surface came down immediately over the hard, bed-rock edge of a buried spur of the former upland.[136] It is very likely that the alluviation responsible for this particular change in the position of the channel was associated with the Wisconsin glaciation, because the work done by the river on the buried rock spur seems too insignificant to have occupied all the time since the Illinoian glacial stage.

In Jackson County east of the Brownstown Hills, glacial drift forms a prominent ridge, called Chestnut Ridge, which seems to be of morainic character, though the wells which penetrate the material indicate that it has been mostly deposited by water.[137] Chestnut Ridge is the most prominent mass of ridged drift belonging to the Illinoian glacial stage in Indiana. As a rule the Illinoian drift surface is flat, hummocks and basins being rare indeed. The drift represented in the Chestnut Ridge vicinity was probably massed as a moraine in front of the Brownstown Hills.

The effects of the Illinoian glaciation northward from the vicinity of Columbus cannot be differentiated from those of the Wisconsin glacial stage. The lowland becomes deeply filled with the later drift in the vicinity of Franklin and rather ceases to exist as a lowland, though its pre-glacial continuation must have trended far to the northwest, probably beyond Boone and Clinton counties, near where its direction became more westerly. In short, the northern part of the lowland is well mantled with glacial drift.

The larger effects of the deposition of glacial drift with its increasing thickness toward the north were such as to largely efface the pre-glacial drainage eastward and southward from southern Marion County, and to turn toward the south a large part of the drainage which had formerly gone westward and northward into the old valley of White River southwest from Indianapolis. Thus, the drainage area of East White River was greatly increased.

During the presence of the ice in this portion of the state and immediately to the north, great glacial floods descended the valleys of Flatrock Creek, Blue River, Brandywine, and Sugar creeks. This drainage through Shelby, Johnson, Bartholomew, and Jackson counties constituted a part of the so-called "Collett Glacial River." Instead of the drainage continuing down the low-

---

[135] 25th Ann. Rept., Ind. Geol. Surv., 1900, pp. 363-364.

[136] See Butts, Geology of Jefferson Co., Ky.: Ky. Geol. Surv., Series IV, Vol. III, Pt. IV, 1915, pp. 17, 201-202, 204-205.

[137] See Leverett, U. S. Geol. Surv., Monogr. 41, 1902, pp. 263-265.

land area to the Ohio River, it turned westward through the breach in the upland near the mouth of Muscatatuck River. The great burden of outwash which was deposited along the route of the glacial stream is yet far from having been removed. The gradients of the streams still bear evidence of the existence of the great valley-train which was left along the present line of East White River and its chief tributaries above Columbus. The condition of these streams as compared with the streams outside the influence of the Wisconsin drift is well brought out by Leverett[124] in his discussion of the East White River drainage system. He says:

"Although these headwater tributaries make a great descent in passing down to the basin of Devonian shale, they have carved very insignificant channels. The valleys are usually so shallow that their bridges may be seen for miles back from the borders of the streams. A portion of the Muscatatuck drainage system is, however, characterized by deeper channels, a feature which is probably attributable to the greater age of the system. It lies outside the limits of the newer, or Wisconsin drift, while the principal tributaries of the East White farther north flow through most of their course within the limits of the newer drift sheet.

"In the lower 35 miles of its course, from Shelbyville to Columbus, Blue River has a fall of about 4½ feet per mile, nearly as great as the fall of the headwater portion above Shelbyville. From Columbus to the mouth of the Muscatatuck, a distance of 55 miles, the average fall is very nearly 20 inches per mile. In the remaining 125 miles, where the stream is flowing in a preglacial valley, the fall is about 10 inches per mile.

"The Muscatatuck River in its lower 25 miles has very little fall compared with the neighboring portion of East White River. At the railway crossing south of Seymour the bed of the Muscatatuck is 40 feet lower than at the crossing on the East White immediately north of Seymour. The difference in the gradient is due to a filling of East White valley by deposits of gravel at the Wisconsin invasion. As the Muscatatuck drainage system lies outside the limits of this later ice invasion or the reach of its waters, its valley remains unfilled. The fall on the lower 25 miles of the Muscatatuck is apparently not more than 10 feet, while on East White River in the 25 miles above the mouth of the Muscatatuck, there is a fall of about 50 feet."

It may be added that East White River is in a shallow, constantly changing channel and is over-burdened with sand. The Muscatatuck on the other hand is deeply sunk in a channel that is rarely shifted. The exceedingly low fall of the lower 25 miles of the Muscatatuck may be partly accounted for by its partial ponding owing to the continuation of the White River fill below the mouth of the Muscatatuck.

During and perhaps following the glacial floods along White River, masses of wind-blown sand accumulated in the angle between the junction of Muscatatuck and White rivers, and along the eastern flank of White River valley from Vallonia northward beyond Columbus. The sand varies from a few feet to perhaps 30 feet in thickness and extends occasionally more than a mile back from the valley. It gives rise to a gently hummocky topography.

---

[124] Mono. 41, U. S. Geol. Surv., 1902, pp. 193-194.

Development and Dissection of the Norman Upland.

General Character and Development of the Upland Surface.

The Norman upland represents an upland plain fairly uniformly and deeply dissected by stream action in a thick series of rock strata, which show little abrupt variation in lithology. The upland plain has an elevation which as a whole will average nearly 900 feet, but in some of the interstream areas reaches elevations of 1,000 feet or greater, as in Weed Patch Hill south of Nashville, where a maximum altitude of 1,050 feet is attained in a plateau-like expanse. In other areas the upland surface descends to 700 or 800 feet. Those portions of the upland surface which form the interstream tracts reaching from 900 to 1,000 feet and slightly higher, in all probability represent remnants of the uplifted early Tertiary peneplain or Highland Rim level. Such interstream tracts as the plateau-like expanse of Weed Patch Hill and the wide upland tract in the vicinity of Spearsville in northeastern Brown County, are clearly representatives of an uplifted peneplain surface which is still preserved from dissection, not because of a capping of resistant strata, but because of distance from the main streams. Other areas of less extent and hundreds of interstream ridges approach this same elevation.

The interstream areas which reach the greatest elevations and which represent the Highland Rim level are chiefly between the main streams, such as in Brown County between Camp and Bean Blossom creeks in Brown County, between Bean Blossom and the north fork of Salt Creek, between the north and middle forks of Salt Creek, and between the middle and south forks of Salt Creek. In eastern Lawrence and western Jackson counties this level is reached between Salt Creek and White River drainage, and south of White River mainly in eastern Washington, western Scott, western Clark, and Floyd counties near the crest of the Knobstone escarpment and the great spurs extending westward from the crest between the westerly flowing streams, such as Muscatatuck, north, middle, and south forks of Blue River, and Indian Creek, and between the latter two streams and the anomalous easterly flowing Muddy Fork of Silver Creek.

From these main upland masses minor spurs extend approximately at right angles towards the principal streams. These minor spurs as they approach the main streams do not maintain the altitude of the Highland Rim level, but are commonly from 125 to 250 feet lower at a fairly uniform elevation. The ends of these minor spurs are from 125 to 250 feet above the valleys of the principal streams. This condition is typically represented on the north fork of Salt Creek west of Nashville in Brown County. From a point on the north slope of Kelly Hill on the Bloomington-Nashville road some two miles southwest of Nashville, two distinct topographic levels may be seen above the present valley floor. The upland ridge at Kelly Hill and its correlatives on the north side of Salt Creek reach an altitude of 900 to 950 feet within two miles of Salt Creek valley. Then comes a rather abrupt descent of 125 to 175 feet to a dissected plain which flanks the entrenched valley of Salt Creek. This plain ranges in width from 1 to 2 miles on either side of the valley and has an elevation of approximately 770 feet. It stands about 170 feet above the present valley floor. It extends up the tributary streams for some distance away from Salt Creek valley. This narrow dissected plain is very probably representative of the late Tertiary local peneplanation which in the Norman upland area advanced no farther than a broad gradation plain adjacent to the main streams. Gravels representing the deposits of the streams when flowing at this level are still locally preserved on this gradation plain.

Chief Conditioning Factors in the Dissection of the Upland.

The development of the Norman upland into its present topographic condition has been relatively simple. Its topographic development has consisted largely of normal stream erosion to the point of mature dissection on the uplifted Highland Rim peneplain. The area of its typical development is on a thick series of rock strata which show little abrupt lithological variation; and it is this feature which is chiefly responsible for the uniformity of dissection and similarity of topographic forms over the entire unit. Nevertheless at least three features involving structure and lithology have been important conditioning factors in the development of the Norman upland unit. First, the impure shales and sandstones which make up the upland mass, coupled with the fact that the more resistant strata are towards the top. Second, the westward dip of the rock strata (much modified north of East White River by a major fault). Third, the influence of the rather resistant limestone which now partly covers the upland surface (chiefly south of East White River) or which formerly covered it (chiefly north of East White River).

Influence of the Rock Strata in the Dissection of the Upland.

The strata composing the Norman upland consist of shales or clays, sandy shales, and shaly sandstones of fine-grained texture belonging to the Borden series. (See Dr. Cumings' report on nomenclature and stratigraphy.) The total thickness is greater than 500 feet. The clays and shales predominate at the bottom of the series and the fine-grained rather massive impure sandstones prevail at the top. Thin ledges of relatively pure sandstone rarely more than a foot in thickness are present through the middle and upper part of the series. These stand out in the ravine bottoms giving rise to small waterfalls, but they have little or no influence on the uniform hillside slopes which characterize the area. Bennett[139] in discussing the stratigraphy of the series says: "This formation is made up of sandstones and shales. . . . In the south the shale predominates; in the north, the sandstone. . . . None of the sandstone is pure. It is mixed with considerable quantities of muddy shale and also contains small quantities of iron, which is shown in the weathered rock. The shale is muddy, easily eroded, and contains large quantities of iron nodules. In most places there is a gradual transition from the shale to the sandstone; just where one leaves off and the other begins it is difficult to tell. This is especially true in the northern part of the state. Farther north there is alternating beds of shale and sandstone, but the beds of shale are much the thinnest."

The influence of the rocks upon the land forms present in the Norman upland area is more readily appreciated when consideration is given to the details of weathering and erosion of these rocks. Water is readily absorbed by the impure sandstones and shales, and is tenaciously held by them. Alter-

[139] Four Comparative Cross Sections of the Knobstone Group of Indiana: Proc. Ind. Acad. Sci. for 1897, pp. 258-262.

nate freezing and thawing cause disruption of the exposed rocks. Fragments are broken off parallel with the exposed surfaces, as in exfoliation. The fragments given up by freezing and thawing are readily carried away by the streams. These rocks are also rather easily corraded by running water. Streams even of small size have reached a graded condition close up to their sources. (This is more specifically true south of East White River.) As a result of this, streams heading in a region where these rocks are relatively high above the local base-level have a very steep descent at their very headwaters, but take on a considerably flattened gradient within a short distance of their over-steepened heads. This feature is perhaps largely the result of the more resistant rocks being above the drainage bases. The upper and more resistant impure sandstones do not resist weathering sufficiently to form cliffs; but the slopes are everywhere quite steep, often attaining 30 degrees or more from the horizontal.

Newsom[100] brings out very clearly some important phases of weathering and erosion in the strata of the Borden series. "It is a noticeable fact," says Newsom, "that through the whole Knobstone area where unaffected by glacial material, and where the valley systems are well developed, the south hillsides have gentler slopes than those facing northward, i. e., that erosion is farther advanced on the south-sloping hills than on those sloping northward.

"This feature is most noticeable along the east-west valleys. In north-south valleys the gentler slope, when one is gentler than the other, is usually on the east side of the valley, i. e., on the westward sloping hillside. The difference in the angle of slope between east and west hillsides is not so noticeable as that between north and south slopes.

"This differential weathering of the slopes is attributed to the effect of temperature changes, especially of freezing and thawing, upon the rocks. Changes in temperature probably have a more potent effect in breaking up these rocks than has any other agent.

"The Knobstone strata, being soft, absorb water easily, although they do not permit the free passage of water through them. Owing to their property of absorbing much moisture, they are easily disintegrated by frost action.

"The south hillsides are exposed to many more changes of temperature in the course of a year, and especially in the winter time, than are the hillsides facing northward. During the winter months in this region the nights are cold, often for weeks at a time, freezing a crust over the ground, which next day is thawed out on the south-sloping hillsides. The north slopes, however, being sheltered from the sun's rays, are frozen or covered with snow, and remain so almost the whole winter long. Thus while the rocks of the south slopes will be successively frozen and thawed out many times during the winter, the corresponding north slopes may be frozen and thawed only two or three times.

"One has only to cross this country on a warm winter's day after a cold freezing night, and to see the muddy streams flowing down from the south hill slopes while the north slopes remain solidly frozen, to realize the importance of this process in the wearing away of these rocks. Both the climate and the structure of the rocks are peculiarly favorable for this class of erosion.

[100] A Geologic and Topographic Section Across Southern Indiana: 26th Ann. Rept., Ind. Geol. Surv., 1901, pp. 270-272.

"There is considerable difference also in the quantity of heat that reaches the east and west slopes, the westward slopes receiving more heat than those facing eastward, owing to the fact that the afternoon is the warmest part of the day. Because of this the westward slopes are sometimes more gentle than those facing eastward. These slopes are not so noticeably different, however, as are the north and south hillsides."

Influence of Regional Structure.

The dip of the strata is an important factor controlling the width of the belt of outcrop of a given thickness of strata, upon which unified topographic forms may be developed. In the case of the Norman upland, however, the dip is only partly responsible for the width of the belt. The Norman upland is extremely narrow at the south, considering the great thickness of the strata in which it is developed. North of East White River it is of great width. The extreme narrowness at the south may be ascribed in part to the abundance of soft shale of which the formation is largely composed, in part to the greater amount of erosion that has taken place adjacent the Ohio and Muscatatuck rivers, and in part and perhaps chiefly to the presence of superjacent rather resistant limestones coming out to the very crest of the escarpment which marks the eastern margin of the unit. Its great width to the north of East White River is probably due for the most part to the presence of a major fault which runs parallel to the strike of the strata not far from the western margin of the unit. This fault has an up-throw on the east of probably 200 to 300 feet.[141] This up-throw on the east side of the fault-line has greatly reduced the westward dip of the strata east of it. Such a structural condition raised a broad area of gently dipping limestone strata which capped the Borden series above the base-level of erosion, and as a consequence the limestone was removed from a broad belt during the early Tertiary erosion cycle. This accounts for the wide belt of strata of the Borden series east of the fault-line. Moreover the greater resistance of the upper strata of the higher part of the series has been responsible for retarding the migration of the easterly-facing escarpment down the dip of the strata, as compared to the more rapid migration south of East White River. It should be stated here also that the dip of the strata is much greater in the southern part of the area, and this results in an appreciable narrowing of the southern portion as compared with the northern.

Influence of Presence and Absence of Limestone Capping.

The presence of an unbroken limestone mass covering the strata of the Borden series out to the very crest of the Knobstone escarpment south of Knob Creek, southwest of New Albany, and from there northward to the westward turn of the Ohio River at the southeastern part of Harrison County, practically limits the Norman upland to the escarpment face. Short streams have rarely cut back into it more than 2 miles. The greatest immediate relief of the entire

[141] See Newsom, A Geologic and Topographic Section Across Southern Indiana: 26th Ann. Rept., Ind. Geol. Surv., 1901, pp. 264, 274-275; Ashley, Geology of the Lower Carboniferous Area of Southern Indiana: 27th Ann. Rept., Ind. Geol. Surv., 1902, pp. 90-92; Logan, Petroleum and Natural Gas in Indiana: Dept. of Conservation, State of Indiana, Publication No. 8, 1920, pp. 58-62.

state occurs here along a 15-mile stretch of the escarpment where its deeply dissected edge faces the Ohio River. The maximum elevation at the crest of the escarpment is 1,000 feet, whereas the elevation of the Ohio River is 370 feet. Locust Point Hill in Harrison County, just south of the southeastern corner of Floyd County and some 10 miles southwest of New Albany, reaches an elevation of 980 feet within less than one-half mile of the Ohio River channel.[142]

North of the latitude of a point some 2 miles north of the south line of Floyd County, as far as the valley of Muddy Fork of Silver Creek in middle western Clark County, limestone only locally covers the strata of the Borden series clear out to the crest of the escarpment, and the westwardly flowing streams have cut deeply below the thin limestone strata for several miles back from the escarpment. The valleys here have characteristics in common with those farther north, though the valley sides are less steep. Short streams flowing down the Knobstone escarpment have given the eastern face a tremendous relief. Some of the spurs between the short deeply intrenched valleys retain the elevation of the uplifted Highland Rim surface, but the majority of the short spurs are unevenly reduced much below this level. These spur-ends form the "Knobs" of the region. Viewed from a point some distance out on the Scottsburg lowland, the "Knobs" are not discernible against the wall-like background of the elevated upland.

Between the great valley trench made by the easterly flowing Muddy Fork of Silver Creek and the Muscatatuck River valley at the north line of Washington County, practically all of the limestone capping has been removed for several miles back of the ragged and dissected scarp. Stream dissection is very severe, and the valley sides are exceedingly steep. The interstream ridges are relatively even and have a uniform elevation. The main valleys flowing east and north are cut from 250 to 450 feet below the preserved portions of the uplifted peneplain. The headwater forks of Blue River are much less deeply incised. These southwesterly flowing streams have relatively broad valleys some 200 feet above valleys of similar size which flow into the Scottsburg lowland area. "Knob" development is at its maximum on the scarp facing the Scottsburg lowland, where many of the upland spurs between the easterly flowing streams have been differentially reduced below the level of the uplifted peneplain. The larger valleys which enter the Muscatatuck and White rivers, such as Elk, Delaney, Buffalo, Rush, and other creeks are wide, and bear evidence of having been filled to a considerable depth in their lower courses. Newsom[143] ascribes the silted condition to regional depression, but it is very likely that it may be satisfactorily accounted for by the valley-train which descended White River, and ponded its tributaries.

North of East White River as far as the region of glacial drift, no limestone caps the surface of the Norman upland for several miles west of the escarpment, and in the latitude of Brown County no limestone is present over a strip approximately 30 miles wide. No limestone capping has been present here since the development of the Highland Rim peneplain, and it is here also that the greatest dissection of the upland surface has taken place, regardless of the fact that much more sandstone is present in the upper members

[142] See Kosmosdale Topographic Sheet, U. S. Geol. Surv.

[143] A Geologic and Topographic Section Across Southern Indiana: 26th Ann. Rept., Ind. Geol. Surv., 1901, p. 268.

of the Borden series of the region. Dissection is perfect over much of this great area. The amount of the material removed from the upland block is approximately equal to the amount left in the ridges, or, the amount to be removed before peneplanation is complete in the present erosion cycle. Many of the ridges have been somewhat reduced near the main streams, as for example along the forks of Salt Creek and Bean Blossom. Instead of the very highest part of the upland surface being near the crest of the escarpment on the east, it is here in the heart of the upland. Weedpatch Hill some 3 miles southeast of Nashville is the highest part of the upland, as well as the highest point in the driftless area of Indiana. The escarpment at the eastern margin of the upland is very ragged, and eastward flowing streams penetrate the upland for several miles. Only the ends of the largest spurs rise to the height of the old peneplain surface. At a distance when viewed from the end some of these appear as great conical hills. No "knobs" exist east of the escarpment, such as occur south of Muscatatuck River. The relief of this most typical portion of the dissected Norman upland area may be judged from the depth of the main valleys. Bean Blossom valley at Helmsburg is 660 feet in elevation. Salt Creek valley at Nashville 600 feet, and at the junction of the north and middle forks 525 feet. East White River valley at Fort Ritner is 510 feet.

Influence of Glaciation at the North and East.

Northward from Bean Blossom valley in northern Brown and Monroe counties glacial drift becomes quite noticeable locally. Little or none of southwestern Johnson County exhibits the topographic characteristics of the Norman upland, for glacial drift covers the bed-rock very completely. The area covered by the Illinoian ice-sheet in northern Monroe and southern Morgan counties is quite typical of the Norman upland, though some evidences of glaciation exist. Northward from the Shelbyville moraine of the Wisconsin glacial stage, the bed-rock forms developed in the strata of the Borden series are almost wholly obscured and the surface features are quite characteristic of the great glacial plain which covers so much of Indiana and adjacent states.

Little evidence of glaciation exists along the Knobstone escarpment south of the latitude of Columbus. The ice of the Illinoian stage apparently advanced against the escarpment, but did not over-ride it, though the glacio-fluviatile waters broke over into the south fork of Salt Creek in western Jackson County. From the evidence of glacial till and outwash material at the foot of the escarpment the glacial ice in southwestern Bartholomew County lay against the escarpment to a height of approximately 800 to 850 feet. The foot of the escarpment here is approximately 750 feet A. T. It is evident that the short streams flowing eastward from the Norman upland into the Scottsburg lowland were ponded by the glacial ice and the material carried by it. No details concerning the effects of such ponding are at present available.

Bed-rock Terraces and Their Development.

Bed-rock terraces or benches are quite common along Bean Blossom and Salt Creek valleys. They are quite prominent, especially, along the north fork of Salt Creek which passes from the northeastern part of Brown County southwest past Nashville and thence into Monroe County. These bed-rock terraces or benches are from 20 to 50 feet above the present valley level. They almost always contain a coating of gravels and silt which contain foreign materials of glacial origin. Associated with the bed-rock terraces or benches are less perfectly preserved alluvial terraces which are approximately of the same height above the present valley. Nashville, in Brown County, is built on a terrace which is partly alluvial and partly bed-rock. An excellent example of the bed-rock terrace may be seen one and one-half miles southeast of Nashville on the south side of Salt Creek along the road to Weedpatch Hill. This bed-rock bench rises 40 feet above the valley and is nearly one-fourth mile wide. It is covered with a mantle of coarse gravel and silt. The gravel is much water-worn and contains pebbles of glacial origin.

The origin of these bed-rock benches or terraces constitutes an interesting problem. They cannot be explained as due to harder strata at the horizon of their occurrence. Very frequently they are entirely composed of soft shale, as at Nashville and Belmont. They evidently mark a gradation plain of limited width. It is possible that they represent a gradation plain made during a short period of crustal stability constituting a pause in the general uplift which has been responsible for stream trenching over much of the Mississippi Valley. But it is more likely that they are definitely related to glaciation.

In the opinion of the writer the bed-rock terraces owe their origin to a gradation plain of narrow extent developed by the glacio-fluviatile waters which came down the valley during the Illinoian glacial stage. White River valley was filled some 40 to 60 feet above its present level by the outwash from the glacial floods, and the tributary streams were filled to a similar depth chiefly by indigenous waste. Salt Creek had to adjust its gradient to such a filling, but Salt Creek valley also was one of the chief routes of glacio-fluviatile waters and it is probable that its pre-Illinoian valley was filled to a depth of approximately 50 feet above the present valley floor by outwash material. The glacial floods continued down the valley for a long period of time and much widening of the valley resulted at the level of the top of the valley-fill. After the withdrawal of the ice, the glacial floods ceased to pour down the widened valley. In time the stream cut down into its old pre-Illinoian valley, but left irregular benches on either side which mark the gradation plain made by the glacio-fluviatile waters. The present valley floor below the bed-rock benches and associated alluvial terraces has been excavated out of and below this gradation plain.

The terraces along Bean Blossom valley are more complicated than those along Salt Creek. Great quantities of glacial material were washed down the valley during the Illinoian stage and it is likely that the upper portion of the valley was entered directly by the glacial ice. Moreover the lower part of the valley was obstructed by the ice-sheet extending across it. This valley was also an outlet across the unglaciated portion of Indiana for the glacio-fluviatile waters of the Wisconsin glacial stage.

Unbalanced Drainage Conditions.

The drainage of the Norman upland area is very largely to the west and southwest. The chief streams which arise east of the area flow across the Norman upland in deeply cut valleys of restricted width. West and east forks of White River and the Ohio River are the only streams that pass directly across or through the upland. All other westerly flowing streams such as Camp Creek, Bean Blossom, the three forks of Salt Creek, Guthrie Creek, and the headwater portions of Blue River, head near the crest of the Knobstone escarpment at the eastern margin of the upland, and flow in a southwesterly direction in conformity with the dip of the rock strata. These streams and their tributaries have deeply trenched the upland plain. With one exception, that of Muddy Fork of Silver Creek in western Clark County, all of the streams which flow in an easterly direction, or against the dip of the strata, are short and mostly less than 5 miles in length.

The streams which flow east have a steep gradient from near the crest of the escarpment to the valley floors of the Scottsburg lowland. Their descent in the few miles of their courses is measured in hundreds of feet. The streams flowing in a westerly direction descend an approximately equal vertical distance in from two to ten times the horizontal distance of the easterly flowing streams. Their average gradients are therefore much less than those of the short easterly flowing streams. As a consequence of this unbalanced drainage situation in the Norman upland area, a much greater amount of work is being accomplished in the vicinity of the escarpment by the streams which flow east than by the streams which flow west. This results in shifting the divide, or crest of the escarpment, down the dip of the rock in a westerly direction. In the past such unequal work has been largely responsible for the development of the Scottsburg lowland and in the making of the Knobstone escarpment.

A Notable Case of Successive Stream Piracy.

Piratical Development of Muddy Fork of Silver Creek. The work of the various streams on the eastern margin of the Norman upland has not been equal, and as a result the Knobstone escarpment is somewhat ragged and indented. One stream has notably surpassed all others in eating its way westward. Since this drainage adjustment offers many interesting details, it will be given here in full, closely following the detailed description already published by the writer.[164]

Since the short, steep streams coming down from the Knobstone escarpment have a decided advantage over the back-slope streams, they have a tendency to develop their drainage area by headward erosion into the territory drained by the back-slope streams. The headwaters of the back-slope streams may be expected to be captured by the eastward and northward flowing streams. Search along the escarpment shows that this drainage adjustment as a whole has not taken place, but in a number of places appears imminent. There is one notable case, however, of such piracy. This is along the line of Muddy Fork of Silver Creek, from near Pekin in southeastern Washington County to near Broom Hill in Clark County. Here much reversal of drainage has already taken place, and a great break occurs in the escarpment along the line of this stream. Newsom[165] repeatedly calls attention to this rather unusual opening in the Knobstone escarpment. His maps show a beautiful example of barbed drainage pattern and the broad col at Pekin where the former west-

---

[164] Some Special Physiographic Features of the Knobstone Cuesta Region of Southern Indiana: Proc. Ind. Acad. Sci. for 1919, pp. 373-383.

[165] J. F. Newsom, Geologic and Topographic Section Across Southern Indiana: 26th Ann. Rept., Ind. Geol. Surv., 1901.

wardly flowing stream entered Muddy Fork of Blue River. But it does not appear that Newsom realized the significance of these tell-tale features. Ashley[188] calls attention to the area and the causes of the peculiarity in the following words: " . . . the soft and easily eroded nature of the Knobstone has allowed the erosion to proceed more rapidly so that the gorge has in many cases sunk its bottom down to drainage level, and the point of rapid descent has advanced from the mouth to the headwaters on account of the shortness of the stream. Indeed, in many cases it is evident that, due to their shortness, these northward and eastward flowing streams are cutting down the divides at the expense of the streams flowing the other way. A good illustration of this 'river stealing,' as it is called, is seen about Borden. The valley in which Borden lies originally drained to the northwest, the divide being nearly as far east as Broom Hill. But the Muddy Fork of Silver Creek, having cut down its side of the divide faster than the stream draining to the northwest, has captured all the drainage about Borden and it is only a question of time when it will extend up so far as to tap the Mutton [Muddy] Fork of Blue River at Pekin and divert all the drainage above that point to Silver Creek."

The topographic map accompanying this paper (Fig. 25) shows the topography of a small area in the region of Pekin and Borden. This somewhat restricted region exhibits details of much interest in the drainage adjustment of the region. It lies between 5 and 10 miles back of the general scarp.

Muddy Fork of Blue River flows west-southwest past Pekin, and as a graded stream is entirely in the Knobstone rocks. Muddy Fork of Silver Creek flows southeast, and is characterized by a barbed drainage pattern. This drainage is almost wholly in the Knobstone rocks. Only the long, tongue-like, interstream tracts 800 feet or more in elevation are capped by the Harrodsburg. The slopes are steep and wooded, and are quite characteristic of Knobstone topography where it is in a much dissected condition owing to minor stream development. The uplands between Pekin and Martinsburg have a surface expression typical of the overlying Harrodsburg. The interstream tract east of Martinsburg reaches an elevation of about 950 feet, and is a remnant of the uplifted Tertiary peneplain, being capped by Tertiary gravels and sand.

The valley of Muddy Fork of Blue River at Pekin has an elevation of 700 feet, and seems to be in a graded condition. In the next fifteen miles the valley descends 100 feet, being approximately at an elevation of 600 feet at Fredericksburg. Drainage from approximately 40 square miles flows past Pekin. Muddy Fork of Silver Creek heads in a number of steep ravines a short distance southeast of Pekin. These ravines are sharply trenched below the general level of the upland. Starting from an elevation of 730 feet in a broad, valley-like sag, a mere gravel and silt terrace above Blue River valley at Old Pekin, marking the lowest part of the divide between the two stream systems, one may make a rapid descent into Muddy Fork of Silver Creek. A descent of 100 feet is attained in the first mile, and within one and one-half miles the elevation is down to an elevation of 600 feet. This is the elevation of Blue River 15 miles below Pekin. At Borden the valley of Muddy Fork of Silver Creek is down to an elevation of 560 feet. The stream here has developed a fairly wide, flat valley and is in a graded condition.

[188] Geology of the Lower Carboniferous Area of Southern Indiana: 27th Ann. Rept., Ind. Geol. Surv., 1902, p. 61.

The barbed drainage pattern of Muddy Fork of Silver Creek is a result of stream piracy. Very probably the parent stream of the present Muddy Fork of Silver Creek was a small stream flowing down the eastern face of the Knobstone scarp in a manner very similar to numerous others of the present time. Back-slope streams of the cuesta flowed westward from the crest of the escarpment. The position of the parent stream of Muddy Fork of Silver Creek does not appear to have been more favorable for the development of headwater erosion than that of many streams of the present along the escarpment. But for some reason it has succeeded in capturing practically the entire stream system of a large tributary that formerly flowed northwest and emptied into Blue River at Pekin. It would appear that after having once broken through the divide near the crest of the escarpment further capture of the lower tributaries followed in relatively quick succession.

Fig. 26. View across the valley of Muddy Fork of Silver Creek just west of Borden, Clark County. Here Muddy Fork of Silver Creek is entrenched 180 feet below an old gradation plain (shown by even sky line) of a formerly northwesterly flowing stream which entered Blue River near Pekin. Through stream piracy the drainage has been reversed and deeply entrenched.

The first stream which still shows direct evidence of having once drained into the Blue River system is Dry Fork Branch. This stream now empties into Muddy Fork of Silver Creek, about a mile below Borden. It is the first of the series of barbed tributaries. In succession the tributaries of the old northwest drainage line were annexed to the Silver Creek system. A number of these, especially those coming in from the north, are decidedly barbed. The latest ones to be taken in were those in sections 29 and 32, between Pekin and Borden. Evidence of this successive capture of the tributaries of the northwest flowing stream is not found in the barbed drainage pattern alone. The gradation plain formed by the northwesterly flowing stream has not been entirely destroyed. To the northwest of Borden, just above the town, is preserved the oldest recognized portion of the old gradation plain. Quite a large remnant is preserved here, and it still retains the silts and gravels of the old stream bed. This remnant is shown beautifully on the topographic map (See Fig. 25; also Fig. 26). The elevation of this ancient valley remnant is about 755 feet, whereas the present reversed valley floor is 575 feet. This means that the drainage change permitted the old gradation plain to be trenched at this place something like 180 feet. At the mouth of Dry Fork Branch the entrenchment is not less than 200 feet. Remnants of the gradation plain are perfectly preserved on both sides of the present intrenched valley to the north-

Fig. 27. View of the Old Pekin col, showing the lower end of a valley which has been pirated of its waters by the under-cutting of the rapidly developing Muddy Fork of Silver Creek. Old Pekin in the distance. View taken from the hill in the northern half of Sec. 31 (see Fig. 25).

Fig. 28. Looking eastward up the old pirated valley southeast of Pekin. The trees in the distance mark a deeply intrenched drainage line belonging to Muddy Fork of Silver Creek. The diverted drainage is rapidly destroying the old, now perched, gradation plain of the formerly northwestly flowing stream. View taken from hill in northeast quarter Sec. 31 (See Fig. 25).

west of Borden. The remnants are more extensive farther up the stream where the piracy occurred at successively later periods. Finally the whole of the old valley is seen for a stretch of about three-fourths mile in section 30 stretching southeast from old Pekin. Old Pekin is built on the Blue River margin of it. (See Fig. 27 and Fig. 28.) This portion of the old valley is more than one-half mile wide, and is as wide as the present valley of the Muddy Fork of Blue River. It was made by a stream comparable in size to this Fork of Blue River. It drained an area of approximately 35 square miles while that of Muddy Fork of Blue River drains approximately 40 square miles.

The few tributaries of the old drainage course yet remaining are shallow streams. The largest one comes in from the south. It is not discernibly below the old valley flat in the northeast quarter of section 31. On approaching Blue River it is trenched broadly into the old alluvial deposits, and enters Blue River accordantly. The small tributaries from the north have scarcely been able to transport their load across the old valley flat, and have the appearance of having slightly aggraded the old valley flat where they debouch upon it.

The divide between the present streams on the old valley flat southeast of Old Pekin is only 30 feet above the valley of Blue River. The old valley-flat

Fig. 29. View showing gravel and silt overlying the bed-rock floor of the old abandoned valley southeast of Pekin. Monon Railway cut about 2 miles southeast of Pekin.

projects above the valley of Blue River as a terrace. The appearance is as though Blue River had cut its valley down something like 30 feet since the stream adjustment took place, but such is probably not the case. It is evident that the old valley-flat is composed of alluvium to a considerable depth, probably as much as or even more than the entire 30 feet of its projection above Blue River valley. (See Fig. 29.) This alluvium composed of gravels and silts over the old bed-rock floor is much deeper than a normal stream of its size should have had. With the beheading of this ancient drainage in the earlier stages of successive piracy the drainage remaining would not be able to maintain as low a gradient as the previous larger, more vigorous stream. The result would be aggradation of the valley. If this is a correct interpretation, the fact that the present valley of Blue River is 30 feet below the old valley flat is not altogether a result of erosion downward of its bed since the stream adjustments have been made. Again this aggradation of the lower part of the valley will aid in explaining the exceptionally low gradient of the old northwestwardly flowing stream as determined by the relative elevations of the remnants of the gradation plain. At Borden it is 755 feet, and in the preserved lower portions it is 730 feet. This would make a gradient of less than 5 feet to the mile.

The Potential Future of Muddy Fork of Silver Creek. When one realizes that the larger part of Muddy Fork of Silver Creek has been obtained at the expense of the Blue River drainage system, and that the last annexation was relatively recent, one must inquire whether these drainage adjustments are yet complete. It needs little more than casual observation to see that the piracy is far from complete in the Pekin-Borden region. The tiny system shown on the topographic map in the southeast corner of section 30 was really the latest accession. This was a mere wash leading to the northwest before the Monon Railroad was built through the old col. The cut necessary for a more gradual descent into the Silver Creek system caused the wash to send its waters into the Silver Creek system. The small wash from the north now sends its water into both systems. The rapid headward erosion of the new system will soon cause all of it to be deflected to the southeast. Likewise, the remaining tributaries of the old system must be taken over into the new. Muddy Fork of Blue River would normally remain at its present elevation for a long period. In the meantime the new system will invade farther and farther to the northwest, and in a short time, geologically speaking, Muddy Fork of Blue River itself will be taken over into the Silver Creek system. The headward erosion of the invading system will be relatively rapid, since it has mainly alluvium to work upon in order to capture Blue River. One might go still farther in anticipation of this successive piracy. The invading system will extend itself in the direction of the present flow of Muddy Fork of Blue River and capture tributary after tributary of the present stream, just as it has done in the past after capturing Dry Fork Branch of the old system. By following the line of a graded stream in this manner, the successive stream piracy must be relatively rapid. Such successive stream piracy will continue as long as the stream gradient in the reversed direction is more favorable for headward erosion than the normal direction. Whenever these stream gradients reach a balanced condition the adjustment is complete and the drainage systems have arrived at the beginning of the old age condition.

Development of Special Bed-Rock Terraces. Another result of the above described stream adjustments must be mentioned here. The barbed tributaries of Muddy Fork of Silver Creek have been adjusting themselves to a direction of flow in accord with that of the main stream where they enter it. They have a tendency to adjust themselves in such a manner that the junction of the main stream and the tributary form an acute angle pointing in the downstream direction. Practically all of the barbed tributaries have been and are making this adjustment. Those on the north of the main stream have much more perceptibility orientated themselves in the down stream direction than those on the south. This is because the dip of the rock favors a migration of the main stream against the south bluff, especially in the non-graded portion of the valley. This has resulted in a shortening of the tributaries at the mouth and the consequent nullification of their orientation in the downstream direction of the main stream. While this direction of adjustment of the barbed tributaries has been taking place, the valleys have also been deepened by down cutting. The combination of this direction adjustment at the mouth of the tributaries and the down cutting has caused bed-rock benches or terraces to come into existence on the upstream side near the mouths of the tributaries. (See Fig. 30.) Some of these tributaries have more than the one set of terraces. These terraces range in height from 10 to 25 feet above the present valley flat or above one another. It is probable that new accessions of drainage due to capture above has had something to do with the development of these bed-rock terraces, since the resulting more vigorous stream would permit a lowering of the gradient where previous to drainage accession the valley had become somewhat broadened in a comparatively graded condition. But in any case the adjustment of the barbed tributaries in the downstream direction of the main stream is the responsible factor determining the position of the terraces. These terraces have an origin unlike any others that have come

Fig. 30. View of bed-rock terrace northwest of the mouth of the barbed stream one-half mile northwest of Borden, Clark County. This terrace is an example of those being formed where the barbed tributaries come into the main stream. They are due to an adjustment of the barbed tributaries to the present reversed main stream as down-cutting takes place.

under the observation of the writer. So far as he is aware such terraces have never been described before.

It may be mentioned here that the case of piracy described is one of the same type as that of the famous Kaaterskill Creek in the eastern scarp of the Catskill mountains, New York.[161] The topographic conditions are essentially the same except for magnitude. Kaaterskill Creek has taken over about 12 square miles of the headwater drainage of Schoharie Creek, the back-slope stream of the Catskill mountains cuesta. On the same cuesta an adjacent scarp-stream, Plaaterskill Creek, has added some 5 square miles to its drainage by successive piracy. Farther south, Sawkill Creek has stolen some 10 square miles from a westward flowing stream, but this case is not a case of successive piracy. It was perhaps largely brought about by glacial action. It will be re-called that Muddy Fork of Silver Creek has added something like 35 square miles to its drainage by successive piracy.

[161] For a brief description of this piracy see: N. H. Darton, Bull. Geol. Soc. Amer., Vol. VII, 1896, pp. 505-507. Also, R. D. Salisbury and W. W. Atwood, U. S. Geol. Surv., Prof. Paper 60, 1908, pp. 45-50.

The Mitchell Plain and Some Local Features of Its Development.

General Characteristics and Complexity of the Mitchell Plain.

The Mitchell plain as defined on a previous page is an upland limestone plain characterized chiefly by its sinkhole topography. It is essentially a regional slope with its higher edge at the eastern side adjacent to the Norman upland into which it usually merges as an upland surface. Locally it is sharply set off from the usually higher upland on the east. It inclines westward on the average somewhat less than the regional dip of the limestone strata on and in which it is developed. Local areas of considerable extent, however, incline almost exactly with the westward dip of the limestone strata.

The eastern margin of the Mitchell plain on the interstream areas, where it may be readily discerned, characteristically attains an altitude of 850 to 950 feet, with an average of perhaps 875 feet. Its irregular western margin at the foot of the ragged and very imperfect escarpment formed by the much dissected eastern edge of the clastic Chester strata, ranges in altitude from 525 to 800 feet, and perhaps averages 650 feet, or some 225 feet lower than the average altitude of the eastern margin.

North of East White River the plain is much dissected and locally quite destroyed by major and minor streams which trench it from 80 to 200 feet below the level of the former more extensive upland surface. (See Fig. 9.) South from East White River the plain is still widely preserved and untrenched by minor streams. Areas of considerable extent are very flat and show little sinkhole topography, but for the most part the plain is from 100 to 200 feet above the drainage bases of the main streams, and is characterized by sinkholes and short intermittent streams, the storm waters of which disappear in swallow-holes.

The Mitchell plain as it exists today is the result of strange and complex conditions involving both lithologic control and physiographic development in varying degrees. Portions of it reach nearly if not quite to the level of the Highland Rim peneplain where little subterranean drainage exists. Other portions developed at a lower level, now high above the present local base-level, have all the characteristics of local peneplains, and possess for the most part surface streams little below the general level of the plain. These streams farther down in their courses usually become subterranean. Such areas occur in the headwater portions of streams where entrenchment has not occurred, and where the streams are flowing practically at the level of the late Tertiary peneplain. The headwater portion of Lost River south of Leipsic, Saltillo, and Campbellsburg and north of Livonia and Stampers Creek, is the largest area representing such conditions. Even here it is a strange fact that the rather flat plain as developed almost exactly corresponds in inclination to the dip of the underlying strata.

Other areas occur where the main streams are entrenched, but on the interstream space the level of the late Tertiary peneplain is preserved in a much pitted sinkhole plain, and here again the plain very frequently conforms very closely to the dip of the strata, and is developed far below the top of the Mitchell limestone lithologic unit. Such areas are the most common and most characteristic of the Mitchell plain, and are found in excellent development through the major part of Harrison County, in the interstream spaces between the Ohio River and Buck Creek, between Buck and Indian creeks, and between Indian Creek and Blue River, and also northwestward from Fredericksburg to Orleans and Orangeville, and thence to East White River. The westermost part of the Mitchell plain, however, seems to be developed considerably below the level of the late Tertiary peneplain. The peneplain level is characteristically between 700 and 800 feet, while the westermost part of the Mitchell plain is frequently well developed at a considerably lower elevation.

Still other portions of the Mitchell plain are very much dissected and surface drainage predominates, as for example northward from East White River in middle northern Lawrence, middle Monroe, and eastern Owen counties, where White Rivers, Salt Creek, Bean Blossom Creek, and the tributaries of these streams have strongly dissected the plain. The interstream ridges and occasional rather broad areas represent the level of the nearly destroyed sinkhole plain. The stream dissection that has taken place here belongs to the present cycle of erosion and represents the entrenchment that has taken place since the uplift of the late Tertiary peneplain.

Locally the Mitchell plain consists of broad valleys and long slopes descending from an uneven upland. Such an area is found east and northeast of Salem in Washington County, where the valleys are but little below the level of the late Tertiary peneplain. The undulating upland adjacent to these preserved late Tertiary valleys reaches an elevation of 950 feet or slightly more above sea level, but for the most part it ranges from 800 to 900 feet in altitude, and seems to be composed of the long slopes stretching from the early Tertiary peneplain down to the level of the later Tertiary peneplain level which was barely developed in the headwater portion of Blue River. This region does not yet seem to be affected by the entrenchment which characterizes the present erosion cycle. Little subterranean drainage occurs in this area. Here the Mitchell plain is developed mainly on the Harrodsburg and Salem limestones, with the higher areas capped by the St. Louis limestone.

The complexity of the Mitchell plain is further exemplified along the western margin where it is still in the process of formation. Here it is being enlarged slowly at the expense of the adjacent upland on the west. The shales and sandstones which constitute the chief rock of the nearby upland, and which occur as outliers and ridges upon the western portion of the Mitchell plain are being removed by surface erosion. The surface streams on the clastic rocks frequently become subterranean when they reach the limestone beneath. These streams are usually short and enter swallow-holes on the Mitchell plain. Below the present swallow-holes occur expanses of sinkhole topography which are really local sinkhole plains developed on the limestone after the removal of the clastic rock by normal surface erosion. They are broad, streamless valleys with many surface tributaries reaching to their very margins. The development of these streamless valleys has been progressive. As the tributaries have grown and cut downward, they have progressively reached the limestone. The water then has developed a sink near the margin of the uncovered limestone. Later, when more of the limestone became exposed by the removal of the elastic material, a new sink would appear farther upstream, and the old one would be abandoned. The clastic material removed from the hills and the uplands above the sinkhole plain is taken away by the normal processes of erosion, but it is a strange fact that this clastic material which is usually carried by storm waters only (since the small streams do not flow during dry weather) must pass through subterranean channels before it can leave the area. The sinkhole plain has come into existence not only because it was able to drink up the waters which fell upon it, but also because it has been able to swallow great hills as well. Thus, local sinkhole plains have been developed near the top of the Mitchell limestone unit. These local plains merge with the greater expanse of the Mitchell plain itself. It is very likely that a considerable portion of the western half of the Mitchell plain owes its existence to development in this manner, though such development could not take place unless the main streams of the region were entrenched so as to give an unbalanced static head on the smaller tributary streams.

Fundamental Features of Subterranean Drainage Development.

As stated above, the Mitchell plain is a limestone plain. Its features are largely dependent upon this fact. Sinkholes, swallow-holes, caverns and numerous large springs are characteristic features of limestone regions and are merely phenomena of subterranean drainage.

Underground drainage is developed in limestones through the action of waters which descend through them, usually concentrated along the joints and bedding planes. Limestones are frequently said to be soluble, and sinkholes and caverns are attributed to this fact. Strictly speaking limestones are not soluble in the ordinary sense. Sinkholes and caverns in beds of rock salt and gypsum are due directly to solution. But such is not the case in the development of underground drainage in limestone areas. The action is somewhat more complex and involves a chemical change. The former is entirely physical. Water charged with carbonic acid (H_2CO_3) permeates or comes in contact with the limestone strata, and the calcium carbonate ($CaCO_3$) of which limestone is chiefly composed unites with the carbonic acid. This forms calcium bicarbonate, $H_2Ca(CO_3)_2$, which is a solution. This solution is then drained away in response to the gravitational circulation of the water. This chemical process and drainage is continued at the expense of the limestone strata. Concentration of this process and of drainage along the joints and bedding planes results in cavities and channels being formed. The direct downward descent of the waters gives rise to sinkholes, while the movement of the waters along the bedding planes (which may be nearly horizontal) gives rise to channels and caverns. Once opened by such action, the sinkholes may become swallow-holes and the channels and caverns may become water routes of considerable importance. When open to the free flow of waters descending from the surface, the subterranean water routes are subjected to corrasive erosion which may greatly enlarge the channels and caverns. Caverns may become so enlarged by solution and erosion that they may locally collapse, thus giving rise to another class of sinkholes which may be called collapse sinks.

All limestones are subject to the development of underground drainage, but in varying degrees dependent in part on the structural and textural conditions of the strata and in part upon drainage relationships. Porous limestones are less apt to have well defined channels, and caves are usually small. Dense, compact limestones are usually well jointed and thin bedded, and it is in such limestones that underground drainage is best developed. The waters are concentrated along the joints and bedding planes, and here the channels and cavities occur. In porous limestones the waters permeate the entire stone and this results in dissolution of the whole rather than in the formation of distinct cavities and channels. General drainage relationships are important controls also in the development of underground drainage. It is necessary that the water descend deeply in the limestone, and that it be drained out, as by springs, at lower elevations. A limestone plain or plateau which stands high above the main drainage lines is essential to any considerable development of subterranean drainage. Limestone areas which have been uplifted after peneplanation and which have become entrenched by the major streams offer the best conditions for the development of subterranean drainage.[108]

The Limestone Units of the Mitchell Plain.

The above structural conditions and drainage relationships occur unevenly on the Mitchell plain, and it is due largely to this unevenness that it abounds in complexities. The Mitchell plain is developed in and on the outcropping belt of Mississippian limestones. These limestone strata dip to the west-southwest at a rate of approximately 25 feet to the mile, and have a total thickness of approximately 450 feet in the southern half of the area. The structural conditions of the limestones making up the belt deserve to be discussed here, since the full appreciation of the Mitchell plain is dependent upon a specific knowledge of them.

The Harrodsburg limestone has a total thickness of approximately 80 feet. It is rather hard and crystalline, and consists of thin to massive layers characteristically unevenly bedded. It is rather impure, containing chert and geodes in abundance. The area of its outcrop is generally undulating, and surface drainage dominates. Sinkholes are shallow and confined to the interstream spaces. They make up only a small portion of the area of the outcrop of the underlying limestone. Much of the outcrop of the stone preserves only a small portion of the total thickness of the limestone unit, frequently being only a capping over the clastic and rather impervous Borden series below. It contains no caves of importance. The springs, indicating subterranean drainage are usually quite small and frequently cease to flow during the dry season of the year. Springs, however, are quite common.

The Salem limestone overlying the Harrodsburg, ranges in thickness from a few feet to probably 100 feet. Where typically developed it is a calcareous freestone. It is very massive and unbedded and is not a well-jointed limestone. These structural characteristics prevent it from having many sinkholes formed in it directly. Its enlarged joints which may be seen in the open quarries, however, testify that subterranean drainage is developed to some degree, though such drainage is negligible. Topographically it is characterized by long gentle slopes and fairly broad valleys. It is considerably less resistant to denudational agents than the underlying Harrodsburg, but frequently the topography of one merges rather indistinctly into that of the other.

[108] For further details in the formation of sinkholes and caverns, see the following references: Blatchley, Indiana Caves and Their Fauna: 21st Ann. Rept., Ind. Geol. Surv., 1896, pp. 121-124; Cumings, On the Weathering of the Subcarboniferous Limestones of Southern Indiana: Proc. Ind. Acad. Sci. for 1905, pp. 85-100; Purdue, Origin of Limestone Sinks: Science, N. S., Vol. XXVI, 1907, pp. 120-121; Greene, Caves and Cave Formations of the Mitchell Limestone: Proc. Ind. Acad. Sci. for 1908, pp. 175-183; Beede, The Cycle of Subterranean Drainage as Illustrated on the Bloomington, Indiana, Quadrangle: Proc. Ind. Acad. Sci. for 1910, pp. 91-111; Scott, The Fauna of a Solution Pond: Proc. Ind. Acad. Sci. for 1910, pp. 395-403.

"The Mitchell limestone is a group of limestones toltalling some 350 feet in thickness in the region [south of East White River]. It consists of about 220 feet of St. Louis limestone at the bottom, about 90 feet of the Fredonia Oölite [representing the St. Genevieve], and about 40 feet of Gasper Oölite [distinctly of Chester age] at the top. The Mitchell limestone, though composed of several geologic units, is really a great lithologic unit of compact, thin-bedded, highly jointed limestone layers with occasional thin bands of shale and impure limestone horizons. The limestone in places contains considerable chert. Near the top of the St. Louis limestone, chert is quite conspicuous and bears numerous colonies of the coral *Lithostrotion canadense.*

"The Mitchell limestone is fairly resistant to mechanical denudational agents. It is structurally characterized by its great number of thin beds of very close and compact nature and by its highly jointed condition. These structural characteristics combined with its position above the base level of the region of its outcrop have been responsible for a wide area of subterranean drainage whose perfection of development is probably not excelled anywhere. It is pitted with numerous sinkholes of all sizes and combinations. Only the larger streams in this limestone belt are surface streams. The outcrop of this limestone belt almost everywhere possesses a typical karst topography. Its presence as a fairly resistant stone mechanically and its disposition to drink up the waters which fall upon it by subterranean drainage have caused to come into existence a wide structural plain which has a [total] westward dip somewhat less than the dip of the strata which make up the lithologic unit."[169]

As a sinkhole plain *par excellans*, the Mitchell plain is developed in and on the so-called Mitchell limestone which responds to the forces of weathering and erosion as a single unit, though it is made up of at least three geologic units. The most typical part of the Mitchell plain, however, is rather confined to the St. Louis limestone, or the lower 180 to 240 feet of the Mitchell lithologic unit as it occurs south of East White River. Above this geologic unit occur some 100 to 150 feet of limestone representing the Fredonia Oölite and the lower portion of the Gasper (Paoli) Oölite, in which sinkhole topography is common. The latter two geologic units composing the upper part of the Mitchell limestone unit generally occur as hills on the western third of the Mitchell plain, and rarely is a plain-like expanse developed upon them.

This feature of the Mitchell plain was first noticed by Elrod[170] who discusses and proposes divisions for the Mitchell limestone. He says in part:

"The rocks exposed in place where sinkholes are common in Lawrence, Orange, Washington, and Harrison counties are always members of the upper or middle portion of the Mitchell limestone [referring definitely to the St. Louis geologic unit], and the angular chert masses and fragments scattered over the surface and mixed with the red residual clay come from the same strata or from the Lost River chert stratum [chert horizon near the top of the St. Louis limestone for which Elrod proposes the name Lost River chert]."

In his discussion of Lost River which follows, Elrod notes that the Lost River subterranean channel is not far below the Lost River chert and that it closely follows the dip of the strata to the west. "Comparatively speaking,"

---

[169] Malott, Some Special Physiographic Features of the Knobstone Cuesta Region of Southern Indiana: Proc. Ind. Acad. Sci. for 1919, pp. 365 and 368.

[170] Elrod, The Geologic Relations of Some St. Louis Group Caves and Sinkholes: Proc. Ind. Acad. Sci. for 1898, pp. 258-267.

says Elrod, "sinkholes are rarely seen in the upper Paoli limestone [limestone of the Mitchell series above the St. Louis, or the Fredonia and lower Gasper Oölites], and when they do occur are rough angular openings in the limestone, of limited area. They are not an important feature in the surface drainage of the country, except in the valleys when located near the level of the Lost River chert."

Thus, credit should be given to Elrod for discerning a distinct topographic difference in the Mitchell limestone strata which is related closely to the geologic units of the Mitchell limestone. Elrod discusses the chemical composition of the various Mississippian limestones with a view of giving some hint as to the excellence of the development of sinkholes and caverns in the St. Louis limestone, but arrives at no definite and satisfactory conclusion as to why a sinkhole plain of such excellence should be developed on this limestone as compared with other limestones of equal or greater purity and similar structural characteristics. This problem still awaits a satisfactory solution.

To What Extent Is the Mitchell Plain Structural?

South of East White River the Mitchell plain as developed near the top of the St. Louis limestone is from 5 to 15 miles wide and is strictly a structural plain. Thus along the Monon Railway on an east and west line between Saltillo and Orleans the strata dip to the west at the rate of 20 feet to the mile. The rather flat Mitchell plain at Saltillo has an elevation of 800 feet, and at Orleans 8½ miles due west it is 635 feet in elevation, therefore possessing an inclination of 20 feet to the mile, or exactly the same as the dip of the strata on which it is developed. An east and west line 3 miles farther south extending between a point near Smedleys in Washington County westward to the Dixie Highway south of Orleans also shows an even inclination of the plain to the west at a rate of 20 feet to the mile. Near Smedleys the Mitchell plain is developed at the top of the St. Louis limestone at an elevation of 900 feet. In the 13½ miles west to the Dixie Highway the plain descends 270 feet, or to an elevation of 630 feet. This uniform inclination of the plain is indicated by the chunks of angular, porous, fossiliferous chert strewn over the surface, which occur near the top of the St. Louis limestone. Similarly further south, the same condition exists over a great portion of the sinkhole plain. At the Harrison County line on the Dixie Highway east of Palmyra the plain is 900 feet in altitude, and at Hancock in northwestern Harrison County, 8½ miles due west, the elevation of the plain is 730 feet. Here again the inclination is 20 feet to the mile and exactly equivalent to the westward dip of the top of the St. Louis limestone on which the plain is developed. Large quantities of iron stained, angular chert fragments strew the surface and are occasionally seen in place. North of Lanesville in western Floyd County the Mitchell plain at the top of the St. Louis limestone is 880 feet in altitude, and 11½ miles due west at a point east of Moberly in western Harrison County the plain is 650 feet in altitude. Here again the plain has an inclination to the west of 20 feet to the mile, and is coincident with the chert horizon near the top of the St. Louis limestone. The chert at the surface is present in great abundance, and large angular pieces are piled up along the roads and in the fields where it has been gathered together. Frequently it may be seen in place. Still farther south in Harrison County the Mitchell

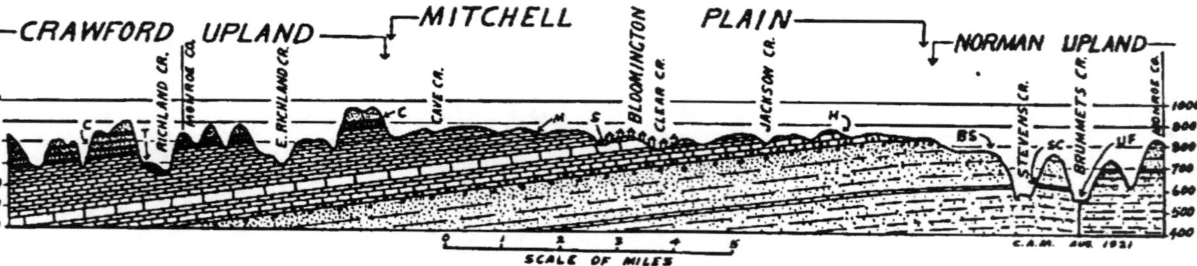

Fig. 31. Section across the Mitchell plain and adjacent portions of the Norman and Crawford [illegible]s from the [illegible]tern line of Monroe County westward through Bloomington to beyond Richland Creek near Whitehall. Note the elevation of the Mitchell plain above the stream base of Richland Creek, thus giving a favorable condition for subterranean drainage. Cave Creek, a former surface tributary of Indian Creek, drains more than 2,000 acres in area into a swallow-hole near Truitt cave, some 6 miles west of Bloomington. The extension of the [illegible]ll plain east of Bloomington is developed on the Harrodsburg limestone which possesses a rolling topography [illegible]th open surface streams and only minor development of rather [illegible]low, interstream sinkholes. The section extends east and west 20 miles. C=Clastic Chester [illegible] [illegible]s, sandstones and thin limestone, 200 [illegible]t; T=Pleistocene terrace; M=Mitchell limestone lithologic unit, embracing thin to massive [illegible]d, well-jointed, compact limestones of the St. Louis, St. Genevieve and lower Chester geologic units, 220 feet; S=Salem limestone: [illegible] [illegible]e and non-bedded, rather porous, 50 feet; H=Harrodsburg li[illegible]stone: hard, semi-crystalline, cherty, uneven-bedded limestone, 90 feet; BS=Borden Series (Knobstone): muddy sandstones, usually massive, and [illegible]dy shales, 600 ± feet; SC=Stevens Creek limestone lentil in the Borden series; UF=Unionville Fault.

plain extends eastward to the very crest of the Knobstone escarpment, and east of Elizabeth and Buena Vista it has an elevation of about 860 feet."[171] The plain descends to the westward at the rate of approximately 20 feet to the mile for a distance of 10 miles or more. In the vicinity of Central the plain near the top of the St. Louis limestone is but slightly more than 600 feet in altitude."[172]

Westward from the above mentioned sections, i. e., westward from Orleans, Dixie Highway some 3 miles south of Orleans, Hancock, Moberly and Central, the Mitchell plain is only partly developed, and its inclination is somewhat less than 20 feet to the mile, except where the sinkhole plain is developed near the top of the St. Louis limestone down a streamless valley to the westward, as down Beaver Creek westward from Mitchell along the Baltimore and Ohio Railway in Lawrence County, and along a number of streamless valleys which reach westward towards Blue River in western Harrison County. Such a condition is shown in the cross section, Fig. 33.

Eastward from the outer edge of the top of the St. Louis limestone the Mitchell plain rises but little, and surface streams are usually fairly well developed and broadly trench the plain. Broad valleys with long, gentle slopes are more characteristic of the area than sharply intrenched valleys with intervening subterranean drainage. The condition east of Bradford in northeastern Harrison County, as shown in Fig. 33, and east of Mill Creek in Washington County, as shown in Fig. 32, and also farther north, east of Smedleys, is that of broadly developed valleys of surface drainage rather than subterranean drainage.

Apparently the development of subterranean drainage to a high degree of perfection near the top of the St. Louis limestone is in itself largely responsible for the structural plain at this horizon. The strata above this horizon are somewhat purer calcium carbonate than the St. Louis strata. Sinkholes develop quite readily, but apparently the strata waste away by chemical denudation very much more readily than the St. Louis strata. Probably the large amount of chert near the top of the St. Louis arrests to a considerable degree the chemical denudation of the stone, and thus aids in giving rise to the structural plain.

Certainly the eastern edge of the structural plain, as at Smedleys, near Bradford, and at the crest of the Knobstone escarpment in southeastern Harrison County, where elevations of 850 to 900 feet are attained, is higher than the late Tertiary peneplain, though this line seems to be a drainage divide. Probably in places the eastern edge of the structural plain is at the level of the early Tertiary plain. The presence of the gravels on the eastern edge in southeastern Harrison County, if they are properly interpreted, indicate that the structural plain reaches quite to the altitude of the old peneplain level. Other areas to the west developed at from 625 to 800 feet have all the aspects of a peneplain, but the lower figure seems rather low for the altitude of the late Tertiary peneplain. It would appear that areas which are less than 700 feet, and locally areas that are considerably higher, depending on drainage relationships, must be due to stripping and wasting away of the upper strata,

[171] See Kosmosdale topographic sheet, U. S. Geol. Surv.

[172] See Ashley's east-west section through New Amsterdam, geological map of the Lower Carboniferous area of southern Indiana (Corydon sheet), 27th Ann. Rept., Ind. Geol. Surv., 1902, p. 71.

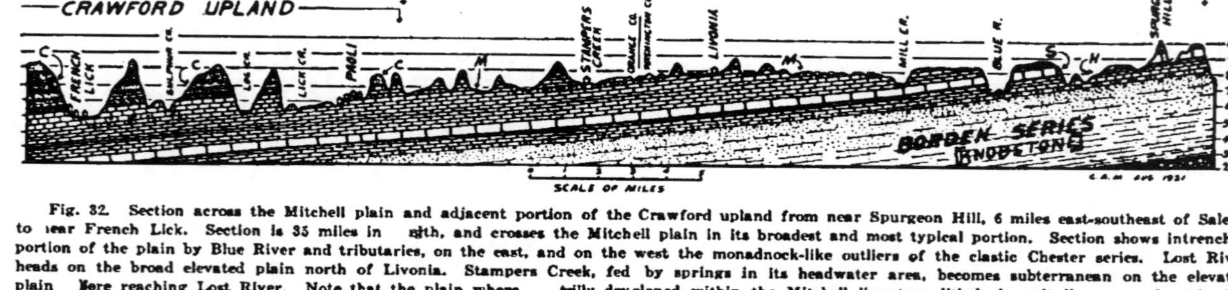

Fig. 32. Section across the Mitchell plain and adjacent portion of the Crawford upland from near Spurgeon Hill, 6 miles east-southeast of Salem, to near French Lick. Section is 35 miles in length, and crosses the Mitchell plain in its broadest and most typical portion. Section shows intrenched portion of the plain by Blue River and tributaries, on the east, and on the west the monadnock-like outliers of the clastic Chester series. Lost River heads on the broad elevated plain north of Livonia. Stampers Creek, fed by springs in its headwater area, becomes subterranean on the elevated plain before reaching Lost River. Note that the plain where fully developed within the Mitchell limestone lithologic unit lies approximately 130 feet below the top of the unit, as in the vicinity of Livonia and Stampers Creek. C=Clastic Chester series; M=Mitchell limestone lithologic unit, 300 feet; S=Salem limestone, 40 feet; H=Harrodsburg limestone, 85 feet.

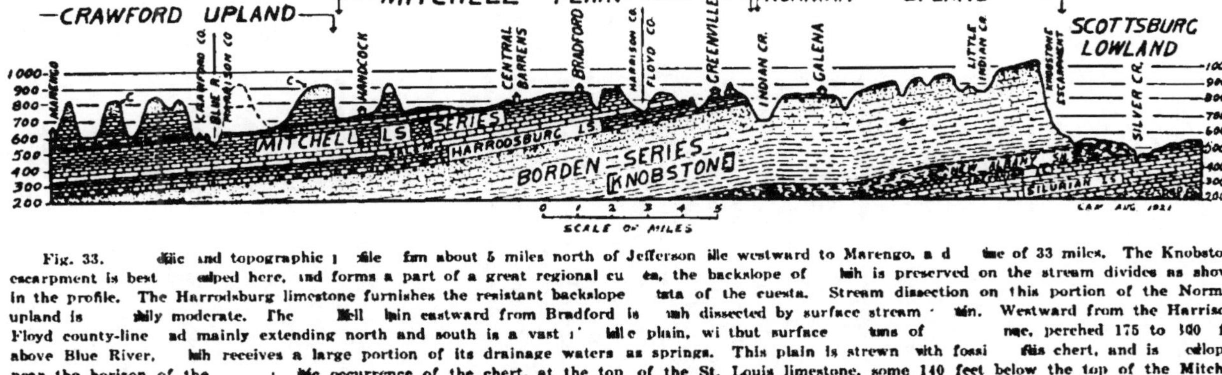

Fig. 33. ...ic and topographic p...file fro... about 5 miles north of Jeffersonville westward to Marengo, a d...ance of 33 miles. The Knobstone escarpment is best ...eloped here, and forms a part of a great regional cu...ta, the backslope of ...ich is preserved on the stream divides as shown in the profile. The Harrodsburg limestone furnishes the resistant backslope ...ata of the cuesta. Stream dissection on this portion of the Norman upland is ...ily moderate. The ...ell ...ain eastward from Bradford is ...ch dissected by surface stream ...ain. Westward from the Harrison-Floyd county-line ...nd mainly extending north and south is a vast s...hole plain, wi...thout surface ...tems of ...age, perched 175 to 300 feet above Blue River, ...ich receives a large portion of its drainage waters as springs. This plain is strewn with fossi...ferous chert, and is ...eloped near the horizon of the ...ile occurrence of the chert, at the top of the St. Louis limestone, some 140 feet below the top of the Mitchell limestone lithologic unit. Local areas within the Crawford Upland on the west possess inliers of Mitchell plain sin...hole topography, though at a much lower altitude than the plain in its ...ide e...anse on the east. The profile ...ws such a l...cal sinkhole plain just east of Blue River. This local sinkhole plain within the Crawford upland area is developed at exactly the same stratigraphic horizon as the great sinkhole plain on the east.

leaving the inclined sink-hole plain as a structural plain. Wide areas of the Mitchell plain in Lawrence, Orange, Washington, and Harrison counties bear little or no evidence of the former existence of surface streams. Apparently after the former surface streams became subterranean, the divides between them wasted away. Clastic material when present must have passed through the sink-holes or swallow-holes, and the limestone divides disappeared by chemical denudation down to the very level of the former surface valleys near the top of the St. Louis limestone. It is certain that such a result could be obtained only by the superior resistance of the horizon near the top of the St. Louis limestone, as this structural level is very apparent.

It must be conceded that much of the Mitchell plain consists of slopes which stretch between the early and later Tertiary peneplain levels 200 feet apart, as exemplified in the vicinity of Salem and elsewhere along the headwaters of the westward and southward drainage on the eastern portion of the Mitchell plain. But portions of the Mitchell plain undoubtedly correspond to the late Tertiary level. Still other portions of it are developed much lower than the later Tertiary level may be conceded to be, and these portions constitute the most perfect areas of subterranean drainage through hundreds of sink-holes. Apparently much of the Mitchell plain as it exists is largely structural, though the rejuvenation at the close of the Tertiary period was of great importance in that it caused the main streams to become entrenched, thus permitting wide expanses of interstream spaces to develop subterranean drainage. Also the rejuvenation of the streams and their consequent intrenchment has permitted subterranean drainage to develop on the limestone as the clastic strata are removed by surface wash and erosion from the western margin of the Mitchell plain, thus giving a greater expanse of sink-hole plain than would otherwise exist. But the relative importance of the structural characteristics and of peneplanation in giving rise to the Mitchell plain cannot at present be stated definitely. Certainly both have been of great importance. Perhaps the more detailed treatment of local areas which follows will show more clearly the part each has played in the development of this important and enigmatical physiographic unit in southern Indiana.

Subterranean Stream Piracy in Western Monroe County.

The Bloomington Quadrangle, covering portions of Monroe, Greene, and Owen counties, shows in detail the topographic features of an area some 233 square miles in extent. Some of the typical topographic and drainage features of the Mitchell plain and the adjacent portion of the Crawford upland are presented in full. The details of the dissected portion of the Mitchell plain from Harrodsburg to Bloomington along Clear Creek are shown with great accuracy. Southwest of Bloomington a considerable portion of the Mitchell plain is well preserved near the divide between East and West White rivers. This region in particular presents a very interesting and complex condition of drainage, and illustrates a very unusual series of captures by subterranean stream piracy. Indian Creek has had some 15 square miles of its headwater portion diverted to the more deeply entrenched streams on either side. Beede[173] in his classic paper on the cycle of subterranean drainage has presented in detail the

[173] The Cycle of Subterranean Drainage as Illustrated in the Bloomington, Indiana, Quadrangle: Proc. Ind. Acad. Sci. for 1910, pp. 81-111.

conditions of drainage in this locality, and the following discussion is either taken directly from his article or follows it rather closely. Relative to the conditions under which subterranean piracy develops Beede says:

"At the time when subterranean drainage is at the maximum it is subjeet to the same accidents as surface drainage, except that the *modus operandi* is different. Subterranean piracy falls under two distinct heads, the capture of one surface stream by another through subterranean drainage, the easiest form to observe, and the capture of one subterranean stream by another. In each case there are minor varieties of capture such as one tributary by another, and self capture. Indeed, these are probably much more common than the capture of one surface stream by another.

"If a surface stream flows a long distance over a rather gentle grade to reach a certain level while a competitor flows a short distance to reach a similar level, it (the latter) may capture the headwaters of the former through subterranean drainage, leaving the divide between the two valleys intact. This tendency is accentuated when the pirate is favored by the dip of the rocks, but frequently occurs in spite of the dip where the dip is gentle. It is probably true that the only essential of such capture is that two streams lie one higher than the other in a region of soluble rocks sufficiently close to each other to permit the final entrance of some of the water of the one to the other. Examples of such piracy are by no means wanting in the Bloomington region."

It appears that the subterranean drainage southwest of Bloomington bears a definite relationship to the late Tertiary peneplain, though drainage relationships are directly responsible. In the Bloomington region monadnocks and ridges occur which reach 950 to 1,000 feet in altitude. These reach nearly to the level of the destroyed early Tertiary peneplain, and some of them may actually be representative of that level. Southwest of Bloomington these are capped with Chester sandstones, and occur as outliers on the Mitchell plain, or interlace with it at its irregular western margin. (See the Bloomington Quadrangle geologic map, 39th Ann. Rept., Ind. Geol. Surv., 1914, p. 190.) The Mitchell plain here on the divide between the two White Rivers reaches an altitude of 900 feet or slightly more. It slopes gently from the higher elevations both to the north and to the south. It is either pitted with sinks or is drained by streams which disappear in swallow-holes in the area under discussion southwest of Bloomington. Beede regards this area as representative of the late Teretiary peneplain, though in the text he has mistakenly called it "the Pleistocene peneplain," making a correction in a footnote at the close of the paper. It is probable that much of the Bloomington Quadrangle which reaches from 750 to 825 feet is representative of the localized late Tertiary peneplain. Towards the divide between the White rivers the peneplain is developed at a considerably higher elevation, though it is likely that the long gentle slopes in the vicinity of Hunter switch do not represent either the early or the late Tertiary peneplains, but are somewhat above one and below the other. It would not be possible for an old peneplain surface to be retained at its original level for a very long period of time on a limestone surface, because it would be reduced by chemical denudation, though untouched by erosive wash. Slopes from outliers to the west descend in rather gentle fashion from near the level of the early Tertiary peneplain and in the course of a mile or two reach the level of the late Tertiary peneplain as preserved in the locality at an elevation somewhat greater than 800 feet. (See Fig. 36.) The site of the city of Bloomington and its immediate vicinity well exemplifies an area which is somewhat below the early Tertiary peneplain with the drainage only

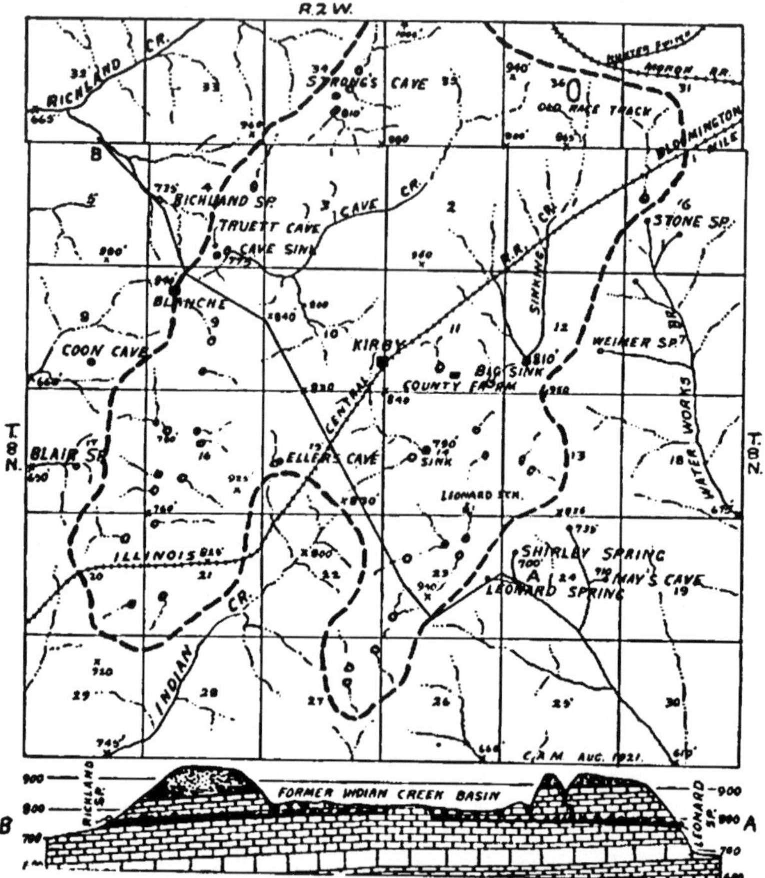

Fig. 34. Drainage conditions near the headwaters of Indian Creek southwest of Bloomington, Monroe County. Some 15 square miles in area of the headwater portion of Indian Creek basin have been diverted by a series of subterranean piracies. The diagrammatic section, BA, reveals the bed-rock conditions and shows that the former basin of Indian Creek is 100 to 150 feet higher than the tributary streams of Richland Creek on the west and Clear Creek on the east. Elevations taken from the Bloomington Quadrangle. Only the main drainage sinks or swallow-holes shown. Heavy dashed line encloses the area which formerly belonged to the surface drainage of Indian Creek.

slightly below the late Tertiary peneplain level. The topographic stage is that of early old age.

Along Richland and Clear creeks to the west and east, respectively, of the broad plain southwest of Bloomington, the late Tertiary peneplain is largely destroyed by the stream dissection that followed the uplift of the late Tertiary peneplain. These streams are from 100 to 250 feet below the peneplain. Such a condition has been of primary importance in the development of the subterranean drainage and the piracy southwest of Bloomington. This is well stated by Beede:

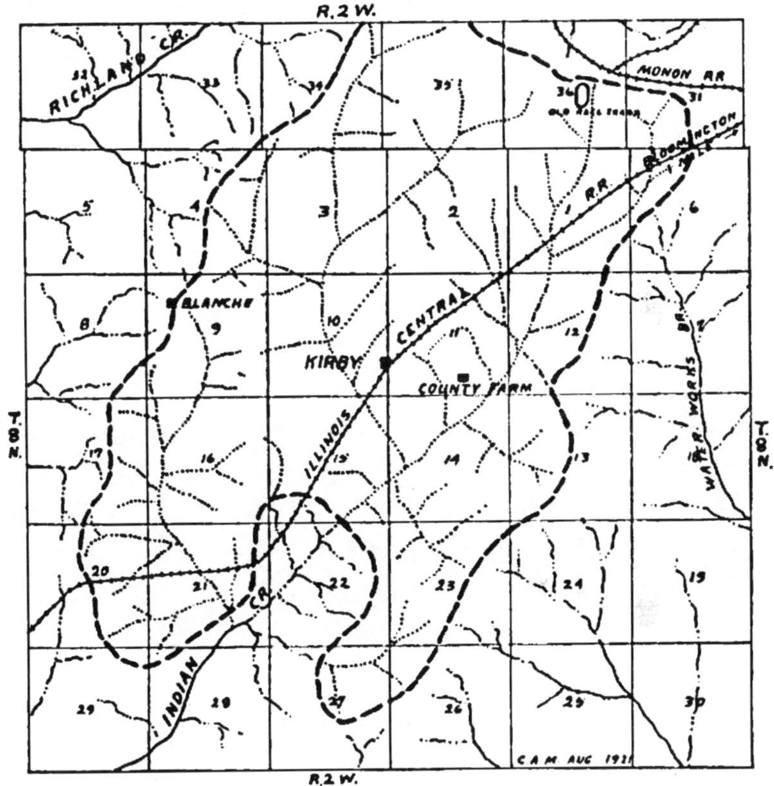

Fig. 35. Surface drainage of the headwater portion of Indian Creek restored. The drainage at the close of the Tertiary period was probably as shown here, though it is likely that some subterranean drainage was present. Heavy dashed line encloses the headwater portion of Indian Creek basin, the drainage of which is now diverted into the tributaries of Richland and Clear creeks through subterranean outlets.

"After the first elevation [at the close of the Tertiary] took place, rapids passed up the main streams cutting gorges in the valleys. As these rapids passed the mouths of the tributaries the latter were left out of adjustment with the master streams and reached them by rushing over high rapids and falls. Some of the larger tributaries reduced the lower parts of their courses with sufficient rapidity to prevent the development of extensive subterranean drainage beneath them, but this was not true of the smaller ones lying on the limestone plain. When the larger streams left the smaller ones hanging in the air, subterranean drainage began in earnest. The rocks were saturated with ground-water and near the mouths of these streams was an unbalanced static head of about a hundred feet. This water gradually flowed into the deeper valleys and was in turn replenished by more from above, and active underground drainage began and continued in the manner already indicated."

"It will be noted," continues Beede, "that the headwaters of the western branches of Clear Creek southwest of Bloomington and the eastern tributaries of Richland Creek nearly west of Bloomington and north of Stanford Station frequently lie in deep valleys with steep heads. On the plain between these two creeks is a region which is drained by great sinks opposite the heads of these streams. A little farther south Indian Creek heads on this plain and continues a little west of south with gentle grade in its headwaters compared with the ones before mentioned. By following the valley at the head of Indian

Fig. 36. View showing a flat portion of the Mitchell plain, without sinkhole development, along Sinking Creek in sections 11 and 12, three miles southwest of Bloomington. View taken from Illinois Central Railway, looking southwest.

Creek northward it will be discovered that the valley extends as far north as the race track west of the northern part of Bloomington, and that the water entering the large sinks just mentioned is really the water of the head of Indian Creek. [See Figures 34 and 35.] The same will be noted of the great sinks northeast and south of Blanche. The water, after entering these sinks, appears in the deeply incised heads of Clear Creek and Richland Creek instead of continuing down Indian Creek. In other words, Richland Creek and Clear Creek have captured the [head] waters of Indian Creek by subterranean piracy.

"This diversion of water was brought about by the location of the streams in question with respect to the rock structure of the region. The strike of the rocks is nearly north and south. The lower rocks in the northeast and southeast part of the region are the soft easily eroded 'Knobstones.' Salt Creek, on account of its very large size, readily etched its lower course to grade and when the soft Knobstone underneath the Mississippian limestone was reached it probably formed falls which rapidly retreated headward and permitted proportionally early deepening of the many tributaries. Throughout the central part of the region the heavy, resistant, Mississippi limestones form the country rock, dipping westward, through which no drainage channels completely penetrated. The headwaters of Indian Creek lie upon these rocks and nowhere do they cut through them. In a large part of its course the soft shales, sandstones and thin limestones of the Mississipian formations form the upland rocks. The result is that Indian Creek with long and gentle grade could not compete with Clear Creek, a branch of Salt Creek, in deepening the channels of its headwaters. In the west part of the region the soft formations of the Upper Mississippian and the basal soft sandstone and soft shales of the Coal Measures or Pennsylvanian rocks form the upland. The Mitchell limestone forms the beds and basal part of the bluffs of the streams in this part of the quadrangle. Richland Creek for the most part lies in these soft formations and flows a short distance to the west fork of White River at Bloomfield, reaching the same elevation as Indian Creek flowing twice the distance to the east fork of White River north of Shoals, in Martin County. Richland Creek being thus favored soon reduced the valleys of its headwaters below the level of Indian Creek. This left the head of Indian Creek 100 to 150 feet above the creeks on either side and its bed resting on soluble rocks. That is, Indian Creek lay upon a table land of soluble rocks with lower streams on either side of it. The divide between Indian Creek and Clear Creek has been cut through and removed much of the way southwest of Bloomington. Thus the headwaters of Richland Creek northeast of Stanford Station are at a level of 680 to 700 feet above tide and were cut into the top of the Mitchell limestone which dips west from the Indian Creek plain into Richland Creek valley, while a west branch of Indian Creek lay at an elevation of 800 feet but a half mile or little more to the eastward. The divide between the two is formed of the shales and sandstones of the Upper Mississippian. The result of this condition was that the water of the western branch of Indian Creek, a mile or more south of Blanche, sank and reappeared in a great spring at the head of Blair Hollow a half mile farther west. A similar thing occurred less than a mile northeast of Blanche and again about a mile and a half farther northeast. These sinks are the largest, or most extensive, on the quadrangle. As we approach the heart of the plain farther east the sinks become smaller and less conspicuous, the smaller ones not being shown upon the map.

"On the eastern side of Indian Creek valley we have large sinks. One of these is just north of the Water Works Pond. Here the drainage entering the sink flows into the pond through Stone spring a few hundred yards farther south, entering Clear Creek valley, being diverted from Indian Creek into which drainage it once flowed. Southwest of this there is a large sink east of the County Farm which receives the drainage of a large region to the north which normally belongs to Indian Creek drainage [see Fig. 36], but appears at the surface as a large spring in the north side of a branch of Clear Creek valley in the N. W. ¼ of Sec. 24, nearly two miles south of the sink [Shirley Spring]. The large sinks south and west of Leonards Schoolhouse have their outlet at Leonards Mill [now torn away and its site made into a part of the lake of the Bloomington water supply] by the house in the head of the deep valley a half mile south of the schoolhouse. Rags put in the upper sink are said to reappear at Leonards Mill.

"From the foregoing it will be seen that the headwaters of Indian Creek have been diverted into Richland Creek and Clear Creek by subterranean piracy. On the west this piracy is favored by the dip of the limestone and on the east it has taken place against the dip, which is very gentle. The sinks near the outlets of the underground streams are large, while those more remote are much smaller. The smallest are not represented on the map. There is another case of piracy near Kirksville which is of the same type as that just described."

Lost River and Its Subterranean Drainage.

Lost River, heading in western Washington County, and passing through northern Orange County and entering White River in southern Martin County, possesses some unusual drainage features on the Mitchell plain. The drainage basin of this stream is approximately 325 square miles in area, and nearly one-half of it is characterized by subterranean drainage. The upper portion of Lost River basin on the Mitchell plain has surface streams, though most of the interstream spaces are pitted with solution sinks characteristic of subterranean drainage. The subterranean drainage in the upper part of the basin is due in part to the entrenchment of Lost River and its tributaries into the gently inclined plain, and in part to the development of subterranean channels below the bed of the upper portion of Lost River. Some 50 or 60 square miles of surface drainage are concentrated into the channel of Lost River above the point where the waters of this stream disappear into developing solution channels or enter great swallow-holes. The waters which enter the subterranean channels appear again as a surface stream about 8 miles west of the place of entrance of the first disappearing waters, but the old bed, occupied by flood-waters during and for a short time after continued heavy rains, may be traced over a very meandering route for a distance of approximately 19 miles. (See Fig. 37.)

Elrod,[174] writing in 1875, gives the following description of Lost River and its drainage:

"The waters of Carters Creek and Lost River increase in volume in their course across the outcrop of the St. Louis limestone until, after uniting and forming Lost River, it strikes the eastern exposure of the concretionary limestone in section 4, township 2 north, range 1 east, where is formed the first sink; the second is in section 8, the third in section 13, township 2 north, range 1 west, and the fourth in section 11.

"During the summer and in dry weather, the first sink takes in all the water, leaving the balance of the channel dry from this place to Orangeville. Light rains will cause this sink to overflow, and very heavy continuous rains for twenty-four hours will carry the water over the whole length of the dry bed. The dry bed or channel extends from the second sink to Orangeville, and is the means by which the excess of rainfall that cannot find passage under ground is carried off, thus preventing an overflow of the surrounding country. The subterranean channel is not a simple, straight, cavernous opening through which the water rushes, but a complex system of mains and leads, a counterpart of the surface drainage. Nor do these underground channels follow the course of the dry bed as might be supposed from there being fre-

[174] Geology of Orange County: 7th Ann. Rept., Ind. Geol. Surv., 1875, pp. 224-225.

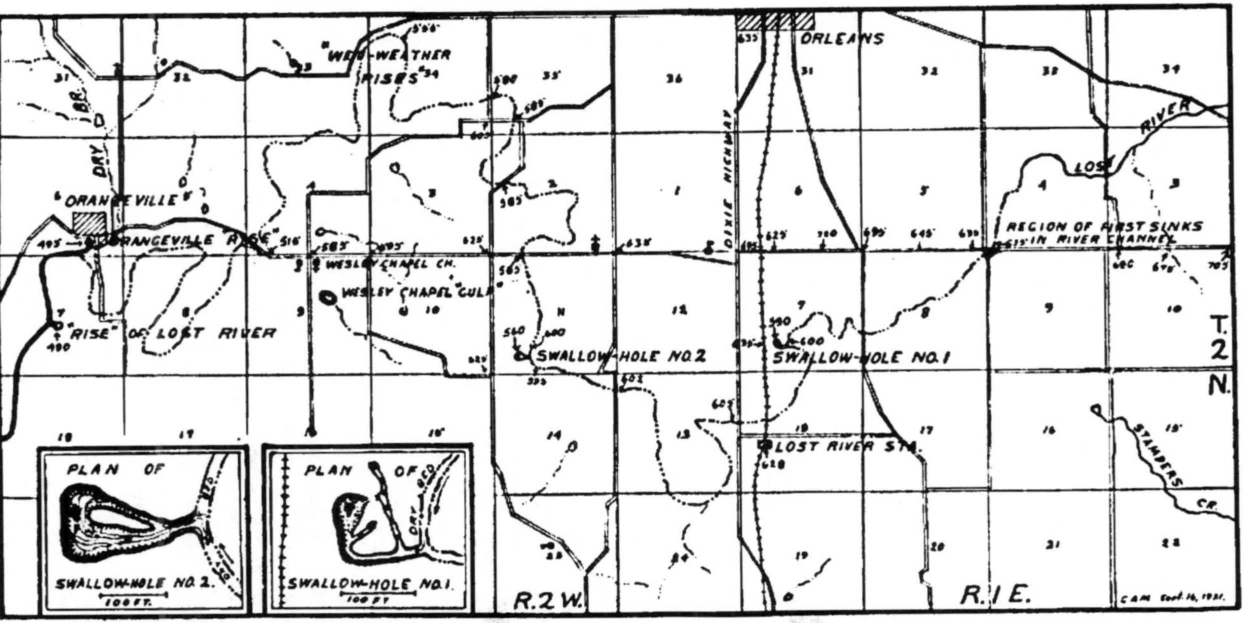

Fig. 37. Sketch map showing the dry-bed channel of Lost River in northern Orange County and ground plans of two important swallow-holes. Lost River has its source some 12 miles east of the beginning of the dry-bed channel. It sinks in its channel and passes through a subterranean channel westward for a distance of 8 miles where it appears in increased volume as a great artesian spring. The old dry-bed channel is very crooked and is 19 miles in length. It contains water during a short time after very heavy and long continued rains, but for the most part great swallow-holes take the waters into the subterranean route, which is nearly directly westward.

quent openings along its banks that connect with them. In sections 23 and 34, township 3 north, range 1 west, are three openings that we may designate as wet weather rises. Whenever the water is running into the fourth sink in force it bursts out at the rises, so that we have water running through both the upper and lower parts of the dry bed and none in the middle channel.

"The dry bed is not an open channel, and has not the vegetal and timber growth common to the margins of streams, but is studded with majestic forest trees, and presents a wild and picturesque appearance when filled with the rushing waters after heavy rains, which seem to be lost in the depth of the forest.

"The underground stream may be reached at the fourth sink, where the cavernous opening is something like eight feet wide and four feet high. The descent is gradual and 590 feet long. The river comes to the surface at Wesley Chapel gulf, in section 9, township 3 north, range 1 west, where the superincumbent rocks have fallen in and forced the stream to the surface. The subterranean stream may also be reached at this point through a cave in the side of the hill. Some years ago a boat was taken in and the channel explored for some distance to a fall, beyond which it was impossible to pass.

"A few yards to the northwest, in the same section, is a dry cave of considerable size, that has quite a local reputation for its numerous large and beautiful stalactites and stalagmites.

"North and near Orangeville is another gulf or rise, where the water runs on the surface for some few yards and again sinks.

"Orangeville is usually spoken of as the "rise" of Lost River, yet it is thought, and doubtless correctly, that the true rise of Lost River is on the farm of Robert Higgins, a mile or more further down the stream. Rains on the headwaters of Carters Creek and Lost River do not affect the rise at Orangeville, but rains on Fultons Branch, that sinks in section 16, township 3 north, range 1 west, do, so that the water at Orangeville is rendered muddy and increased in volume. Yet we must think that Fulton Branch alone is insufficient to account for the whole of this rise, and it is probably fed by other underground streams.

"Stampers Creek, in a small way, is a counterpart of Lost River, lacking the dry bed. It is thought that it again rises at the Spring mills and forms the source of French Lick [Lick Creek]. Sawdust and other refuse from the sawmills situated on the banks of the creek have been worked out at the Spring mills."

Again Elrod, [173] writing in 1898, contributes the following with reference to Lost River and its subterranean drainage conditions:

"The sinkhole area, as a rule, has no surface creeks and branches, and such as reach its limits from without soon find an opening and disappear wholly or in part, except Blue River and Buck Creek. Occasionally the creek or branch is replaced by a dry-bed channel. The dry beds only come into use after heavy rains or when the subterranean passages are burdened beyond their capacity. Lost River through a part of its surface course is a typical dry bed. When it reaches the eastern edge of the sinkhole region it finds a number of underground channels that take in all the water of the perennial stream east of the Orleans and Paoli road. If the first openings are over-

[173] The Geologic Relations of Some St. Louis Group Caves and Sinkholes: Proc. Ind. Acad. Sci. for 1898, pp. 262-263.

taxed, the overplus of water passes through a dry-bed channel farther west into other sinks, but after an excessive rainfall all the sinks fail, and water runs on the surface through the whole extent of the dry-bed system and again becomes a part of the perennial stream a short distance below the Orangeville 'rise.' Indian Creek for a part of the year runs underground, but, unlike Lost River, the greater part of its water passes over a surface channel and a dry bed is only exposed during the summer months. It sinks two miles southwest of Corydon and 'rises' again five below on an air line, and twice that distance following the meanderings of the creek bed. There is ample evidence that Lost River, like Indian Creek, at some period in the past was wholly, or for the greater part of the year, a surface stream over its dry-bed channel.

"Contrary to what might be expected, the subterranean channels do not greatly increase in capacity as they unite and pass under the Kaskaskia [Chester] hills. This is shown four miles west of Orleans at what is called the 'wet-weather rise' of the dry bed. Here water flows out as it is flowing into the upper sinks; hence water may be flowing through two miles of the upper and lower course of the dry bed and not through the middle channel. As soon as the flood-water begins to recede at the 'wet-weather rise' the direction of the flow changes, and, instead of running out, flows back into the opening from which it came. At times the whole underground system of channels is overtaxed and the water finds an outlet at many places, and occasionally through artificial openings, such as the well at Brooktown and another east of Orleans.

"The underground channel of Lost River can be reached at three places through cavernous openings. At the first of these near the first sinks, the superincumbent limestone is about 40 feet thick; at the second opening the channel is not less than 60 feet below the Lost River chert; at Wesley Chapel Gulf it is 30 feet below the chert stratum, and the same at Orangeville. This indicates that the subterranean channel closely follows the dip of the strata to the west."

Lost River rises on the Mitchell plain in western Washington County southeast of Campbellsburg at an elevation of nearly 900 feet. The north fork at Claysville has descended to an elevation of 710 feet. The upland adjacent to Claysville is 775 to 800 feet in altitude. At the junction of the north and south forks of Lost River the elevation of the bed of the stream is 690 feet, and at the entrance of Carters Creek the elevation is 660 feet. Here the adjacent upland is at an altitude of 700 to 750 feet. Bluffs 30 feet high appear along the stream, though the valley is rather wide. Three miles below the entrance of Carters Creek the water in the channel of Lost River begins to disappear and as a perennial stream ceases to exist below the middle of Sec. 8, T. 2 N., R. 1 E. Here the elevation of the stream bed is 625 feet, and the adjacent upland is a distinct sinkhole plain characterized by monotonous shallow sinks of similar shapes with undulations between them. The general altitude of the sinkhole plain here is 650 feet.

The dry-bed channel of Lost River possesses a number of peculiar characteristics. Perhaps the most striking one is the extremely meandering condition of the channel. (See Fig. 37.) From the place where the water first begins to find a subterranean course in Sec. 4, to the "rise" south of Orangeville the air-line distance is about 8 miles. The meandering distance is approximately 19 miles. Apparently the main underground channels descend to the west in a more or less direct line. The total underground descent is approximately 135 feet, and this is probably attained in a distance which exceeds 8 miles but little. The average rate of fall of the underground stream is probably not less than 15 feet to the mile, though the evidence of the swallow-holes and the "gulfs" indicate that the fall of the upper part is much greater than the lower part. The average fall of the dry-bed channel is only about 7 feet per mile, though this fall is uneven, being much greater in the lower portion. The lower 8 miles have a fall of 95 feet, while the upper 10 miles of the dry bed have a descent of only 40 feet. The upper 10 miles of the dry-bed channel is but little entrenched below the sinkhole plain, being only about 25 or 30 feet below it, but in the lower 8 miles the entrenchment is quite distinct, amounting to as much as 75 feet, and even more when the outliers and ridges of sandstone which rise above the Mitchell plain are considered.

Below the "rise" of Lost River the stream is sluggish and occupies a rather deep channel with mud banks lined with trees. The valley is filled quite deeply with alluvium, though the flood plain is rather narrow. The valley itself meanders extensively through the dissected upland. The air-line distance from the "rise" of Lost River to the mouth of the stream is approximately 15 miles, while the channel distance is approximately 36 miles. The total fall of the stream below the "rise" is about 65 feet, or less than 2 feet to the mile.

The sinks, or swallow-holes, and the so-called "gulfs," or localities of collapse of the superincumbent rock above the subterranean channel, deserve some attention. The upper sinks, or the sinks which are in an early stage of development and capable of taking care of the low water flow of Lost River only, are scarcely perceptible openings in the channel of Lost River. It is difficult to tell where the water first begins to sink. Less water runs out of each succeeding pool until finally no water at all flows over the surface of the ripples. These sinks first occur in the stream bed in section 4, T. 2 N., R. 1 E., and continue along the stream to the middle of section 8.

The first large swallow-hole occurs in the southwest quarter of Sec. 7, just east of the Monon Railway and about ½ mile north of Lost River Station. Here the dry bed of the channel of Lost River is at an altitude of about 600 feet, but the descent into the swallow-hole is something like 10 feet. The channel to the swallow-hole turns abruptly westward from the dry-bed channel and after a northward turn along a rocky wall enters the subterranean channel. Branch channels with numerous small holes occur into which the high water pours. The main swallow-hole is much clogged with accumulated logs and brush. Indications of a former swallow-hole just to the north of the present one occur. It was probably filled with debris, and as another more favorable opening was found nearby, the older one was silted up. This swallow-hole receives practically all ordinary showers and even quite heavy rains. Extraordinarily heavy or long continued rains cause considerable ponding and an overflow of the dry-bed channel finally takes place.

The second large swallow-hole is in the southwest quarter of Sec. 11, T. 2 N., R. 1 W. This is the fourth sink of Elrod. It is approximately 2 miles west of the first large swallow-hole, but the course of the dry-bed channel between the two swallow-holes is nearly 5 miles in length. The elevation of the dry-bed channel just below the swallow-hole is 600 feet. For some distance above the swallow-hole the channel is cut below this level. The channel to the swallow-hole, like the one above, turns abruptly to the west from the dry-bed channel, and descends some 25 feet in about 250 feet to the vertical hole which leads to the subterranean channel. This entrance is described above in Elrod's words. At present the swallow-hole is much obstructed by logs and brush, and entrance into the subterranean channel cannot be readily gained. (See Fig. 38.) The lowest level reached by the writer in the bottom of the swallow-hole was 560 feet, or 40 feet below the dry-bed channel. The sinkhole plain adjacent is approximately 620 feet in altitude. Immediately to the west, hills capped with sandstone rise 150 feet above the sinkhole plain. Apparently the subterranean channel passes under these hills.

The Wesley Chapel "Gulf," or the great pit formed by the collapse of the roof of the subterranean channel a short distance southeast of Wesley Chapel Church, in the northeast quarter of section 9, indicates the chief route of the

Fig. 38. Detailed view of the bottom of a swallow-hole into which the waters of Lost River pass. The hole itself is not easily seen because of the logs and brush which have lodged over it. Swallow-hole near the southwest corner of section 11. 4 miles east of the "rise" of Lost River. The bottom of the hole as shown in the view is fully 35 feet below the dry-bed channel of Lost River.

subterranean channel. The collapse which formed this "gulf" was very probably largely brought about by solution of the superincumbent rock at this place. The floor of the "gulf" covers about 4 acres, and it is rimmed by a limestone wall which rises 30 or 40 feet above to the level of the sinkhole floor. Within the rather level floor of this crater-like collapse feature is a circular pit about 40 feet deep within which boils up the water of Lost River. In low water condition of the stream the water immediately re-enters the subterranean channel through openings at one side of the pit. During high water condition the water rises out of the pit and floods the floor of the "gulf," escaping through numerous ill-defined swallow-holes bordering the western edge of floor of the gulf at the foot of the limestone bluff. This "gulf" is the most spectacular feature of the Lost River drainage system.

The "rise" of Lost River, as suggested by Elrod, is south of the reputed "rise" at Orangeville. The volume of water emerging a short distance south of the center of section 7, some half mile below Orangeville, is much greater than the volume emerging from beneath the rock wall at Orangeville. The reputed "rise" of Lost River at Orangeville very probably consists largely of the accumulated waters chiefly from Dry Branch north and northeast of Orangeville. Very likely the waters which disappear in the swallow-holes and sinks west of Orleans and that which goes into the holes in the dry-bed channel of Lost River in its great northward bend come out at the well-known "rise" at Orangeville. The true "rise" of Lost River is altogether unspectacular. The water simply rises from below as a great artesian spring confined by slippery mud banks except on the west where it is open to the main stream coming from the north. The Orangeville "rise" is more striking. (See Fig. 39.) It is shaped like a great half cauldron with a semi-circular wall of lime-

Fig. 39. "Rise" of a subterranean stream at Orangeville reputed to be the "rise" of Lost River. Here a large stream of water comes from beneath the limestone strata from a semi-cauldron-like opening and flows away at one side. In normal and low-water condition some 1,000,000 or 2,000,000 gallons of water per day comes out of this great artesian spring.

stone rising some 10 or 15 feet above the water. The water rises out of the subterranean channel and flows quietly away southward. The artesian condition of the "rise" of Lost River may be ascribed in part to the development of the subterranean channel along definite beds or bedding planes in the St. Louis limestone somewhat corresponding with the dip of the strata, and in part to the filled condition of the valley below the "rise."

The sinkholes in the inclined Mitchell plain in the headwater portion of Lost River are not numerous except in the immediate vicinity of the stream, and in the headwater streams are practically absent. Portions of the upland plain are without sinks and the drainage is normal surface drainage. Apparently the sinks which do occur in the headwater portion of Lost River are surface openings to underground channels which enter Lost River, the water issuing as springs. When the vicinity of the first sinks of Lost River is approached from the east the sinks become more numerous and are present everywhere over the plain. It appears that these sinks are surface feeders to underground channels which lead below the level of the bed of Lost River, Lost River itself finally entering the same system of subterranean channels. Apparently this system of deep subterranean channels has resulted from rejuvenation of the plain upon which Lost River formerly flowed as a surface stream. But it is not granted that the development of the subterranean drainage of Lost River may be ascribed wholly to the favorable condition created by rejuvenation, though this was fundamental and very likely necessary. The subterranean condition has been attained in Lost River partly because of the extremely meandering course of the stream upon the rejuvenated limestone plain. If Lost River had had a more direct route across the plain, it is quite likely that little, if any, subterranean drainage would have developed in the main channel. In other words, the "lost" condition of the stream is largely due to its indirect and meandering route on the rejuvenated limestone plain. Lick Creek, an important tributary on the south somewhat paralleling the course of Lost River has no such subterranean route, though many of its upper tributaries are perched above it and their waters enter it as springs through subterranean channels. It is in this way that the Lost River subterranean drainage differs from most of the other subterranean drainage systems on the Mitchell plain. It may be said to be the only main stream which is actually in a "lost" condition, though Indian Creek in its extremely meandering lower course attains a similar condition during low water stages.

Glacial Modification of the Mitchell Plain.

The Mitchell plain north of Ellettsville in northwestern Monroe County has been modified by glaciation. The chief effect of glaciation has been to mantle the rather dissected and locally nearly destroyed plain with glacial material, and eventually farther north to cover over and entirely obscure the plain. Locally, where the plain has been deeply dissected by surface streams, it has been restored to its original level by glacial material chiefly in the form of lake deposits. Two areas of this kind are of more than ordinary interest. One of these, known as Flatwoods, lies chiefly along the axis of McCormicks Creek between Ellettsville and Spencer in Monroe and Owen counties. The development of this region is given in some detail below.

The other laked area developed in part on the Mitchell plain is quite extensive. It lies along Mill Creek, or Eel River, in northwestern Owen, southeastern Putnam, northwestern Morgan, and southwestern Hendricks counties. It is terminated at its western edge by the double fall of Mill Creek, or Eel River. The lake plain above the fall of Eel River is approximately on the level of the Mitchell plain, which was largely removed in this locality by stream dissection before the region was glaciated. The area of the former lake, or the area affected by ponded drainage conditions, was 50 square miles or more. The altitude of the present lake flat, or lacustrine plain, is 750 to 775 feet. Portions of the Mitchell limestone plain are seen about the margin of the silt-covered area.

The lacustrine plain above the falls of Eel River is due principally to the obstruction of the pre-Illinoian Mill Creek, or Eel River, in the vicinity of Cataract in northeastern Owen County. The Illinoian glacier completely filled the entrenched valley in this locality to or somewhat above the level of the upland Mitchell plain surface. Following the retreat of the ice the waters in the drainage basin above, which also was probably somewhat modified, found outlet at the level of the bed-rock surface high above the old entrenched valley. The waters, however, re-entered the old valley somewhat lower down. Falls were formed where the accumulated waters of the basin above entered the old entrenched valley. In time the basin of the ponded drainage became filled to the level of the bed-rock surface of the stream above the falls, thus giving rise to a lacustrine plain. Inwash in the upper portions of this lacustrine plain have built it up still higher. Portions of it are very flat and poorly drained, as for example along Mud Creek in northwestern Morgan County. This locality west of Monrovia and north of Eminence is known as "The Lakes," or the "Lake Country."

The falls of Eel River near Cataract are very picturesque. The descent of the stream into the entrenched valley takes place over a stretch of a mile or more, and in the series of rapids and two falls of approximately 25 feet each, a total descent of nearly 100 feet occurs. Picturesque gorges of several hundred yards in length occur below each of the falls.

It is not known what influence the Wisconsin glaciation had on the lacustrine plain above the Falls of Eel River. Possibly part of the plain itself is due to the Wisconsin glaciation. Few details concerning the region are at present available. A possible connection of the pre-Illinoian Eel River with Rattlesnake Creek has been suggested,[114] but it is not likely that any such connection existed.

The Flatwoods Region of Monroe and Owen Counties.

The Flatwoods area west of Ellettsville chiefly along the axis of McCormicks Creek in Monroe and Owen counties, offers a very interesting example of the glacial modification of the Mitchell plain near the limits of the Illinoian glacial margin. The area consists of a flat plain some 8 square miles in extent at an altitude of approximately 740 feet. It is partially surrounded by higher land and exhibits knolls of bed-rock material standing above the flat plain. Some of the higher hills partially enclosing the plain reach an altitude of 900 feet or more, and rise 150 feet above the plain. (See Fig. 40.) The area originally consisted essentially of a sinkhole plain considerably dissected by a surface stream which reached westward to West White River near Spencer. This valley was obstructed by glacial material to the level of the sinkhole plain, and in the adjustments which followed, the drainage of the locality sought another route to White River, chiefly through McCormicks Creek, which in its lower course descends 160 feet through a deep gorge, now the site of one of the well known state parks. About the margin of the area sinkholes have developed, and much of the marginal portion, excepting the western side, has a subterranean drainage, thus still retaining the characteristics of the more typical portion of the Mitchell plain. The development of the Flatwoods area and vicinity has been treated in some detail by the writer, and the following is taken with little change from an abstract published in the 39th Ann. Rept., Ind. Geol. Surv., 1914, pp. 217-222.

Just west of Ellettsville is a wide level tract of land, only a portion of which comes within the limits of the Bloomington Quadrangle, which has attracted the attention and study of geologists for some time. It is a low level

[114] Collett, The Geology of Owen County: 7th Ann. Rept., Ind. Geol. Surv., 1875, p. 307.

basin, averaging about 740 feet above sea level, about 2 miles wide and reaching some 6 miles northwest from Ellettsville towards White River near Spencer. It consists of low marshy tracts of land out of which arise occasional island-like monadnocks; a surrounding periphery of hills or higher land; and a striking ash-colored, silty soil, usually containing shot-like concretions of ferrous and ferric oxide, which are locally known as "turkey gravel." In the lower portions, especially near the middle of the basin the ash-colored silt is covered by a black soil containing much vegetable matter. This black soil is very fertile. This peculiar basin, locally known as Flatwoods, is drained mainly by McCormicks Creek, only the very headwaters of which are within the area of the Bloomington Quadrangle.

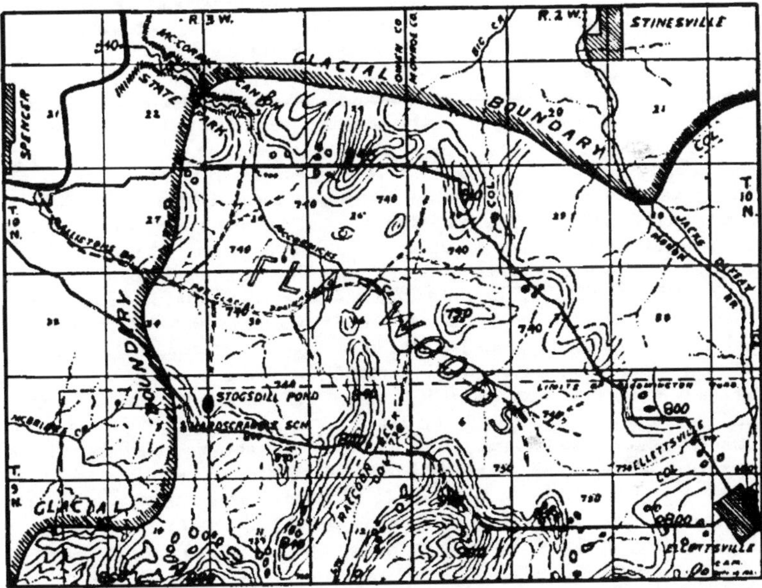

Fig. 40. Sketch map of the Flatwoods region of Monroe and Owen counties, showing the glacial boundary and pre-glacial drainage lines. Flatwoods is a well preserved lacustrine plain developed by the glacial obstruction and ponding of a stream which intrenched the Mitchell plain at the margin of the driftless area. Its surface at an average altitude of 740 feet is at the level of the Mitchell plain, and portions near the margins of the silt-covered plain possess subterranean drainage. It is very likely that the outlet of McCormicks Creek through the deeply intrenched gorge of McCormicks Canyon, State Park, was initiated through subterranean drainage.

About 3½ miles west of Ellettsville in the S. E. quarter Sec. 1, T. 9 N., R. 3 W., is a narrow opening leading from the Flatwoods basin into Raccoon Creek valley. (See Fig. 40.) South of this opening on either side of Raccoon Creek occur remnants of the same flat as observed in the above described Flatwoods basin. In places the remnants are quite extensive, as in Secs. 24, 23, 14, and 13, T. 9 N., R. 3. W. Since the drainage of Raccoon Creek is much lower than the flat-surfaced remnants of the former continuous flat, many V-shaped ravines are cut into it, making the topography rather rough. Nevertheless, the flat tracts remaining form a striking feature.

Evidence of this Raccoon Creek portion of Flatwoods having been a continuous flat, like the present Flatwoods proper, is found in the structure revealed in the steep-sided ravines and in the wells of the region. Stratified sand is very often present not only revealing the fact that there were distinct water currents, but that the valleys were filled up to the level of the present flat-surfaced remnants. In some places the region has been filled as much as a hundred feet. How this took place will be briefly outlined below.

As a matter of history it might be mentioned that Collet in his report on the Geology of Owen County (Seventh Annual Report, Indiana Geological Survey, 1875), in attempting to explain the narrowness of White River valley above Spencer near the northwest end of the Flatwoods basin, asserted that previous to the Illinoian glacial period, White River passed up the narrow gorge of McCormicks Creek, through the Flatwoods basin, through the opening leading into Raccoon Creek valley, and thence down that valley to the present White River valley below Freedom. Siebenthal, commenting on Collett's idea, says: "The Pleistocene terraces of Bean Blossom Creek clearly prove the pre-glacial valley of that creek to have been practically as it is at present. It is impossible to imagine how it could be cut down to its present depth, while White River, into which it emptied, was running at a level 150 feet higher than now, as it is alleged to have done. Moreover, the gorge of McCormicks Creek is clearly post-glacial. And further, it empties into White River at least a mile below the upper end of the 'Narrows,' whose existence it was brought forward to explain." (Twenty-first Annual Report, Ind. Geol. Surv., 1896, p. 302.) Thus Siebenthal makes it clear that Collett's idea is untenable. Siebenthal suggests that the Flatwoods basin was the site of a shallow lake during the Illinoian glacial period and for some time following.

There is no doubt that the phenomena of Flatwoods are to be explained by the ponding of glacial waters in front of the Illinoian ice sheet. It seems that the ice advanced slightly into the area represented by the Bloomington Quadrangle. At its farthest advance it quite probably occupied the extreme northwest sections, very likely entering the quadrangle near the middle of Sec. 33, T. 9 N., R. 3 W., and passing northward past Freeman P. O., and leaving the quadrangle near the Hardscrabble School. Near the residence of Thomas Coble in the N. W. quarter Sec. 3, T. 9 N., R. 3 W., are remnants of an old moraine, the only direct evidence of any continued stand of the ice front upon the quadrangle. From the vicinity of the Hardscrabble School the ice front probably extended irregularly northward for over 2 miles and then swung eastward, crossing the headwaters of Big Creek, Jack's Defeat south of Stinesville, and then over the divide and across Bean Blossom valley. Thus, the stream draining the pre-glacial Flatwoods basin was ice-dammed, as well as the valleys of Jack's Defeat and Bean Blossom. The waters undoubtedly accumulated in Bean Blossom valley until they flowed over the divide at the old col on the farm of Jack Litten about 2 miles southeast of Stinesville. After coming into the valley of Jack's Defeat in this manner, they continued into the next basin west, which is the one now occupied by the Flatwoods. The waters reached this basin through a low opening just west of Ellettsville. Quite an extensive lake was formed in this basin. From this basin the waters found an outlet to Raccoon Creek valley through the opening described above (hereafter this opening will be called the Raccoon Creek col), and passed down this valley, finding an outlet along the edge of the ice sheet or under it in the vicinity of Freeman P. O. Ignoring the work of the post-glacial streams on the old lake-flat (which was the result of the long continued ponded waters), it can be easily seen by the contours of the Bloomington topographic sheet that the general slope is about 10 feet to the mile from the entrance into the Flatwoods basin just west of Elletsville to the vicinity of Freeman P. O. This slope of the old lake-flat is in itself proof of the direction of flow of the glacial waters.

On the withdrawal of the ice, the portion south of the Raccoon Creek col was drained; but the region of Flatwoods proper remained a lake for a long time, with an outlet through the Raccoon Creek col, or through the col just west of Ellettsville leading into Jack's Defeat Creek. Perhaps both of these openings or cols were outlets for a time; but the Raccoon Creek col is slightly lower, and undoubtedly persisted longer. The outlets of lake Flatwoods were at these places because the old buried valley was dammed so effectively at the lower end that it was much higher here than at the outlets above. Very probably a morainal dam was present, for evidences of such are to be seen near the headwaters of Allistons Branch, which was the pre-glacial outlet of the old pre-glacial stream. The steep-sided deep ravines and the wells of this region show that the filling here was considerably over a hundred feet. Moreover, the glacial material itself has all of the appearances of outwash from a moraine. The slope also is in the direction of the Raccoon Creek col, where the waters undoubtedly found an outlet.

The region immediately south of the Hardscrabble School (N. W. corner of the quadrangle) slopes rather rapidly toward the south, to a narrow opening in the line of hills in the middle of the southern half of section eleven. The glacial waters coming from the ice front in the upper portion of the pre-glacial McBrides Creek basin, no doubt flowed through this opening and escaped southward into the Raccoon Creek region. After the withdrawal of the ice the waters continued in this direction, because McBrides Creek valley was filled by a moraine near the western edge of the quadrangle. Remnants of such a moraine have already been mentioned. Post-glacial stream action has been rapidly clearing the morainal and outwash material from the pre-glacial McBrides Creek; but there is much work yet to be done. The present drainage in the region just south of the Hardscrabble School is mainly underground, a condition which was perhaps, already well developed in pre-Illinoian times. Much of the glacial material itself has been carried away through the underground channels.

After the withdrawal of the glacial ice front from the vicinity of Freeman P. O., the streams went to work to clear out the material which had filled their valleys. Gradually Raccoon Creek has cut its way back into the glacial material, or old lake-flat, and at present a tributary is slowly making its way into the Raccoon Creek col, which connects with Flatwoods proper. Beautiful terraces were thus formed in the old lake-flat. These are prominent features on either side of Raccoon Creek and its tributaries, and have a distinct influence upon the position of the contours. But long ago the waters of old lake Flatwoods ceased to find release through the Raccoon Creek col. Underground drainage was well developed in pre-Illinoian times, and soon the old subterranean channels were cleared of glacial debris and the waters went through the subterranean pre-glacial routes. The main stream, however, was through Allistons Branch, and this stream was so thoroughly clogged that it has never been able to be much of a factor in the drainage of the Flatwoods region. The present drainage of Flatwoods is through the rocky gorge of McCormicks Creek, which is distinctly a post-glacial stream. It was initiated through underground drainage, and, having gained an early ascendancy through the peculiar suitability of its rock structure and the great fall, it soon had practically the entire drainage to itself. Later, through weathering and erosion, the drainage became surface drainage, practically as it is today. The lake-flat, however, has always been imperfectly drained; even yet small depressions are either marshy or remain as small lakes; as, for instance, Stogsdill Pond just east of the Hardscrabble School.

McCormicks Creek, which drains the larger part of the intact portion of Flatwoods lacustrine plain, will be given somewhat further consideration. This stream arises on the silt-covered plain west of Ellettsville at an altitude of about 750 feet and flows northwest for a distance of about 5 miles, descending some 50 feet, to the Bloomington-Spencer road. Here it has reached the limestone of the Mitchell plain, and is somewhat trenched below the general level of the lacustrine plain. Within the next 2 miles, or the remainder of its distance to White River, it descends 160 feet, through a rocky gorge with many sharp-angled turns. This gorge runs through the well known McCormicks Canyon State Park. As stated above, it gained this route through the development of subterranean drainage; but this need not be considered as the only means by which this route to White River was attained. It is possible that the old lake overflowed in this vicinity, and that the outlet here deepened more rapidly than the several other outlets, and thus became the permanent outlet. Again, this outlet may have been obtained through stream piracy. A preglacial, deeply-cut valley existed northwestward from near the present falls in the canyon of McCormicks Creek. It is possible that the stream which occupied it developed by headward erosion to the southeast sufficiently to tap the old lake, and so drained it to the northwest. But the writer still holds to the view set forth above, that the canyon was more probably begun by underground drainage. The dip of the rock and the structural conditions of the St. Louis limestone in which the gorge is developed were favorable and such an initial drainage would not be unnatural under such favorable circumstances.

Some Features of Development in the Crawford Upland.

General Features of Erosional Development.

The Crawford upland as described in Chapter III of this paper constitutes the most rugged portion of Indiana. It is an upland plain composed of diverse bed-rock strata in a highly dissected state. Erosive processes have prevailed unequally on the strata of different degrees of resistance, and diversity of topographic form is common throughout the area, though angular forms are more persistent. Little of the upland surface of the supposed uplifted early Tertiary peneplain, or Highland Rim level, exists, though certain areas, usually quite restricted, may be doubtfully referred to it. Dissection has been too severe for the preservation of any considerable areas of this old uplifted surface. Uniformity in the elevation of the higher ridges and hills is far from characteristic of the Crawford upland. Where certain resistant strata are present on some of the major stream divides fairly even upland ridges and surface areas exist, but these extend in a north and south direction in harmony with the strike of the strata on which they are developed. Such surfaces cannot be referred to the Highland Rim level, though some of the higher ridges probably attain this level. Few such upland tracts extend in an east and west direction, because successive strata of different degrees of resistance continually come in as more westerly areas are approached. The east and west ridges are commonly uneven. Where a particular resistant stratum, such as the Hardinsburg or the Cypress sandstones, rests on an east and west ridge, the inclination in the westerly direction is exactly equal to the dip of the rock, which is usually 30 or 35 feet per mile to the west. The topographic forms above the valley flats are usually definitely related to the differential erosion of the rock strata and an appreciation of form development necessarily demands an intimate knowledge of the strata which make up the upland mass. Since the Crawford upland is a rather wide unit and the strata dip 30 or 35 feet per mile to the west, the strata near the eastern margin are different from those on the west, and this again emphasizes the diversity of topographic forms within the unit.

Viewed as a whole the Crawford upland presents a bewildering maze of topographic forms, all indicating the power of stream erosion. Few areas have escaped stream dissection to the stage of maturity or early old age. Relief diversity is one of the chief characteristics and occurs in common with the great variety of topographic forms.

Every stream which passes through the Crawford upland area occupies a meandering valley, and some have exceedingly tortuous and indirect routes. Great horseshoe-shaped curves occur in the deeply entrenched valleys, and frequently a larger regional turn is composed of many smaller curves of the valley, as for example in Lost, Big Blue, and Ohio Rivers. West White River between Spencer and Worthington traverses a river distance of 32 miles, while the direct distance is but 20 miles; East White River in a direct line across the Crawford upland is 30 miles, but the stream course is twice that figure. Indian Creek from its source in western Monroe County to its entrance into East White River above Shoals covers a route 50 miles in length, though the direct distance is only 25 miles. Lost River after leaving the "rise" near Orangeville enters White River 15 miles below, but its waters pass through an exceedingly tortuous course 36 miles in length. Patoka River, heading near the eastern margin of the Crawford upland and passing out near Jasper, traverses a course 60 miles in length, though the air-line distance is but half that. Little Blue River, entirely within the Crawford upland area has a course fully 35 miles in length, though the direct distance is scarcely 19 miles; and Big Blue River between Harrison and Crawford counties, by direct line from the northern edge of Harrison County to the mouth of the stream, is 17 miles, though the meandered course of the stream is not less than 45 miles. Finally, the course of the Ohio River between Mauckport and Troy is said to be 85 miles, while the direct distance is less than 35 miles. All of these meandering valleys are narrow and deeply-set within the hilly upland. Their valley sides are nearly everywhere steep, and sheer cliffs are common features. The great bends have much steeper bluffs on the outside than on the inside of the curves, and the streams usually flow at the foot of these steep bluffs, thus testifying to the continued growth of the meander curves. Usually the steep, rocky valley sides with one or more sets of cliffs are heavily timbered, and in summer the dense foliage somewhat obscures the angular forms which prevail.

As noted above, few local areas may be assigned unequivocally to the Highland Rim peneplain level. It is probable that some of the higher hills or knobs in Perry County, on the dividing ridge between Oil Creek and the Middle Fork of Anderson Creek north of Leopold, reach this level; and it is approximated again in Crawford County in Pilot Knob on the structural ridge between Marengo and Leavenworth, and also on the dividing ridge northwest of English, where elevations of 850 feet or more are attained. In eastern Orange County and the extreme western part of Washington County few places on the sandstone ridges reach an altitude above 900 feet. These ridges tower 150 feet above the Mitchell plain on the east, and probably represent the Highland Rim level. In Martin and western Lawrence counties the higher hills and ridges range from 800 to 900 feet, but no areas exist at these higher elevations which in themselves suggest peneplain affinities. Only by inference from their altitude may one assign them to the Highland Rim level. In western Monroe County the Kirksville Ridge, shown on the Bloomington Quadrangle, reaches an altitude of 900 to 980 feet, and has considerable areas that resemble the remnants of a former peneplain. This ridge is largely structural, but its elevation may permit of its being assigned to the Highland Rim level. The Cincinnati Ridge, shown in part on the Bloomington Quadrangle, may represent the same level, though it is somewhat lower than the Kirksville Ridge. Farther north in Owen County the tops of the highest hills possibly reach the approximate altitude of the uplifted and destroyed Highland Rim peneplain, but none of these areas are in themselves of sufficient breadth to show the characteristics of peneplain remnants.

It is usually difficult to assign any particular area in the Crawford upland to the late Tertiary peneplain stage, for the present cycle of erosion has apparently pretty largely destroyed whatever local peneplains may have been made at that time. Commonly the descent along the minor streams is abrupt and extends from the uneven upland ridges to the present stream levels. This descent frequently amounts to 200 or 300 feet. The ridge surface may not be assignable to any peneplain stage, frequently being too low to warrant its assignment to the Highland Rim level and too high for its assignment to the late Tertiary stage. When extending in a north and south direction these ridges may be fairly even surfaced, corresponding to one of the many sandstone horizons which seem to be chiefly responsible for the retention of the height of the ridges between the deeply intrenched streams.

But local areas do exist which may be definitely assigned to the late tertiary peneplain stage as it has been defined in Chapter IV of this memoir. Frequently along East White River and other main streams the polished chert gravels and smoothed geodes are present on portions of the ridges at elevations approximating 700 feet, though these gravels also occur down to the present stream level. Apparently when they occur at a lower elevation than 700 feet they have been redeposited, having come from the original deposit at the higher level. There are some indications, however, that a second lower horizon of occurrence of the gravels exists. A typical area of the late Tertiary peneplain with its covering of gravels is found in the vicinity of Shoals in Martin County. This area is discussed in some detail below.

It must be emphasized that the Crawford upland is an area which has undergone stream dissection to a mature or post-mature stage. No even upland surfaces of any considerable extent exist. Frequently the ridges are deeply notched, and the higher parts may usually be related to resistant sandstone masses. Occasional broad sags occur much below the uneven surfaced ridges; and these are apparently developed upon resistant sandstones at lower levels. Rock benches or rock terraces are common along the streams, though these benches are never of any considerable width. The more resistant strata commonly give rise to angular forms, and even-sloped valley sides are infrequent. Interstream ridges vary in elevation above the valleys from a few tens of feet to hundreds of feet, but practically all interstream ridges are alike in being uneven on their upper surfaces, having many sags and eminences. Differential erosion is characteristic throughout the Crawford upland. Everywhere one is impressed with the idea of great erosion and dissection, and only infrequently is one reminded of stages of peneplanation where a regional halt has occurred in the erosional activities which have so unevenly dissected the upland mass.

A general decrease in altitude of the interstream ridges from east to west occurs, though it may frequently happen that a ridge on the west is somewhat higher locally than a more easterly ridge. Such a condition may be assigned to the presence of higher resistant sandstone on the western ridge which has caused it to stand high, whereas the more easterly ridge has lost this particular sandstone through erosion, and its surface is now held up by a lower sandstone which may be equally resistant, but lower in elevation as well as lower in geologic occurrence. Outliers of the more massive sandstones frequently cap the prominences on the easterly ridges, miles in advance of the general occurrence of the sandstones. Thus, Pilot Knob, a prominence on the broad divide between Marengo and Leavenworth, is capped by the Mansfield sandstone. This outlier is at least 10 miles east of the more general occurrence of the sandstone.

Thus a study of the development of the Crawford upland reveals two outstanding features: first, surface erosion to a pre-eminent degree, and second, lithologic variety due to alternating resistant and non-resisitant stratigraphic units of variable thicknesses. The former has caused the pronounced relief throughout the upland area and the latter has given diversity and angularity to the forms produced through erosion. Erosive halts depending upon peneplanation are not evident, but structural levels are common and range from the present base level of erosion to the highest parts of the uneven interstream ridges and divides. Few direct evidences of more than the present erosion cycle or more than a single erosion cycle exist. For the study of diversity of reliefs and diversity of topographic forms in a plains region the Crawford upland in its highly dissected state and its diversity of rock strata offers an ideal region. Some of its details will be given below as they occur in selected areas.

General Characteristics of the Rock Strata.

A brief discussion of the rocks which make up the rugged Crawford upland will aid greatly in making clear the chief cause of the diversity and angularity of its land forms. Stream erosion has practically everywhere cut deeply into or through the many geologic units, and exposures are very common, especially in case of the massive sandstones. As noted above, the dip of the strata to the west carries the eastern-most strata and lower strata geologically, below the stream levels in the western part of the area and higher strata of a somewhat different nature make up the ridges of the western part of the upland. Clastic rocks, however, prevail throughout the area.

The lowest formation geologically is the so-called Mitchell limestone, which has been discussed in connection with the Mitchell plain. In the Crawford upland area at least 150 feet of it has to be reckoned with. Practically all of the streams in the eastern half of the area have cut down into it or reach it. But usually outcrops of this limestone do not occur in the interstream areas. The influence of the Mitchell limestone is rather confined to the valley sides and is in close association with erosion near the valleys. Its influence is seen in the local development of sinkholes and subterranean drainage. Inclosed streamless, or semi-streamless valleys, occur near the eastern border of the Crawford upland area. The intrenchment of the streams is partly and principally in this limestone at the east. Indian Creek and the lower part of Big Blue River are intrenched 150 to 200 feet in the Mitchell limestone, and along their courses occur great bluffs which are almost vertical on the outsides of the many meander bends of these valleys. It is in this region of entrenched streams in the Mitchell limestone and overlying clastics of the Chester series that the great caverns occur. The great caves like Wyandotte along Blue River in Crawford County and Mammoth Cave of Kentucky along the deeply entrenched Green River, contain many tiers of abandoned water routes, giving rise to the galleries for which these caves are famous. The series of galleries were very likely formed in succession, as the adjacent streams ate more and more deeply into the thick limestone. Today the cave stream is nearly at the level of the entrenched outer drainage, the upper galleries having been largely abandoned.

The Chester series comprises the greater part of the strata of the Crawford upland area. This series consists of alternating sandstones or shales and limestones, with a total thickness of more than 400 feet. The total thickness is found only in Perry County. Farther north the upper part of the series is absent beneath the overlying Mansfield sandstone of the Pennsylvanian system, because of the erosion which preceded the deposition of the Mansfield sandstone. Eastward the upper formations have been removed by erosion, and the middle and lower formations of the series form the interstream masses of rugged land. Some 12 or 14 recognized divisions occur in the Chester series of southern Indiana, ranging in thickness from 2 feet to 150 feet. Some of these constitute important controls of erosion and the development of land forms, while others are of little importance in this respect.

The sandstones of the Chester series, where they attain a thickness of 20 feet or more usually give rise to benches and bluffs and frequently cap hills and divides. The most important sandstones in this respect are the Cypress and Hardinsburg sandstones, coming in at approximately 110 and 195 feet, respectively, above the Mitchell limestone. The Cypress sandstone is remarkably persistent in occurrence and attains a thickness of 30 or 35 feet throughout southern Indiana at the horizon of its occurrence. It is usually a massive sandstone, and in conjunction with the Beech Creek limestone upon which it rests, gives rise to benches and massive bluffs along both the major and minor streams in the area of its outcrop. (See Figures 11 and 46.) The Hardinsburg sandstone is especially well developed in Crawford County where it gives rise to structural benches along the streams and forms broad, flat divides in the eastern part of the county. Farther north its horizon is occupied by shale and it consequently gives rise to few positive forms. Other sandstones both above and below the Cypress and Hardinsburg sandstones are important locally, and give rise to massive bluffs along the streams. The Sample sandstone along the streams in eastern Crawford County and on the divides in western Harrison County is responsible for much topographic angularity.

The limestones of the Chester series are never of great thickness and are not of first rate importance from a topographic point of view, with the exception of the Beech Creek limestone. This limestone lies immediately beneath the massive Cypress sandstone. It is from 10 to 25 feet thick, and is composed of one or more rather massive ledges, and is broken into blocks by two systems of joints. It is a spring horizon of some importance and locally contains rather extensive caves. Other limestones range from a few feet up to 40 feet in thickness, and locally have given rise to sinkholes in the sinuous areas of their outcrops.

The shales of the Chester series do not give rise to positive forms but rather act as foils to the resistant sandstones between which they frequently occur. The horizon of their outcrops on the steep valley sides is usually covered with rock-slide or talus from the higher ledges of sandstone, and as a consequence the shales are little in evidence. Along the minor streams, however, the shales beneath the sandstone ledges are partly responsible for the small waterfalls.

The Mansfield sandstone of the Pennsylvanian system by virtue of its massiveness and its position capping the Chester series is important physiographically. This massive and frequently non-bedded sandstone ranges from a few feet to more than 200 feet in thickness. It gives rise to massive ridges and great unscalable bluffs which along the streams protrude wall-like through the summer foliage. It occurs as the cap stone on the ridges in much of the outcrop area of the Chester series. A feature which is of some importance physiographically occurs in connection with the unconformity at its base. Locally it is found far down in the Chester series, and yet a short distance away it may be 100 feet or more higher stratigraphically. Apparently it was deposited on an eroded and uneven surface. Further, as a formation it may be locally of little importance as a sandstone, and largely composed of sandy shale. It is best developed as a massive sandstone, here and there containing pebbles, in western Perry, eastern Dubois, western Lawrence, and Martin counties. It is the highest unit in the Crawford upland area, and the surface of the upland descends to the westward in harmony with the westward dip of this massive sandstone.

The Hardinsburg Structural Plain in Eastern Crawford County.

One of the most interesting individual features of the Crawford upland area is the occurrence of broad, flat divides in eastern Crawford County. Flat areas of any considerable extent are so uncommon in the Crawford upland that their occurrence in any locality forms a rather striking anomaly, and requires explanation. The area of particular interest in this respect is shown in the accompanying sketch map, Fig. 41. The chief interstream divide between the Big and Little Blue River systems south of Marengo, and the continuation of the divide between Little Blue and Ohio rivers forms a flat upland surface of considerable breadth upon which are occasional undulations and monadnock-like masses. The divide between Turkey Run fork of Little Blue

River and Bogards Creek and Little Blue River may also be considered in connection with the main north and south divide. The road from Marengo to Leavenworth or to Fredonia follows this flat divide after reaching it some 2

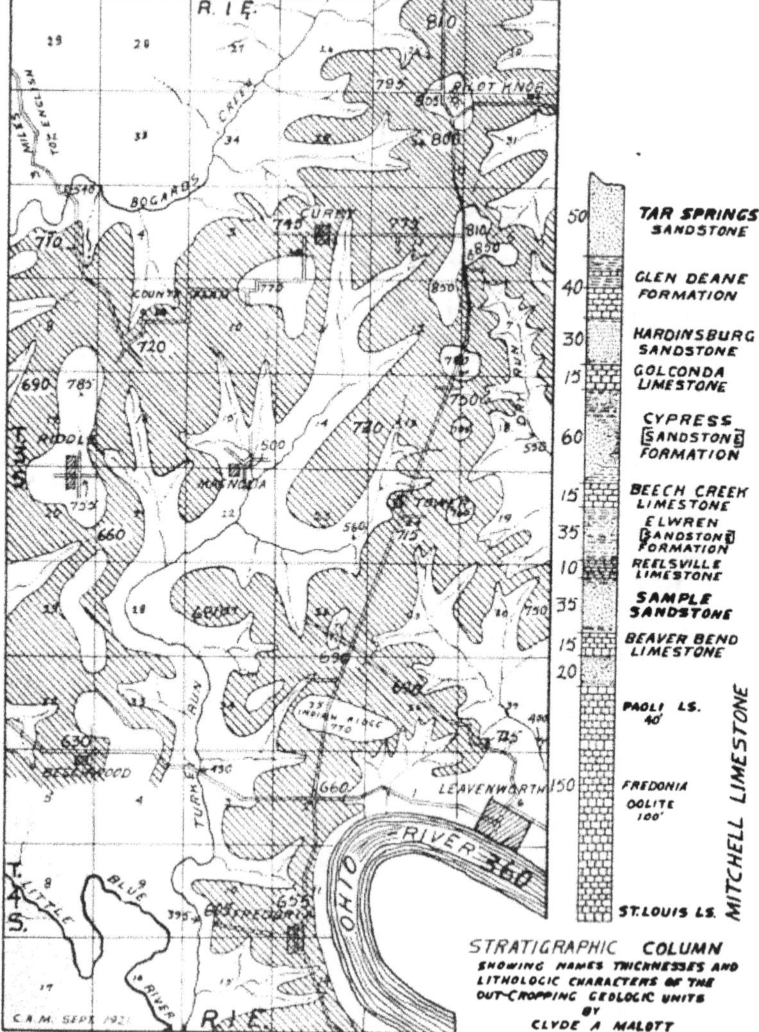

Fig. 41. Sketch map showing the broad, flat divides between the main streams in southeastern Crawford County, developed on the Hardinsburg sandstone. These divides furnish the most extensive flat areas in Crawford County. The Hardinsburg sandstone consists of thin flag-stones and is only 30 feet thick, but is very resistant to erosion, and frequently extensive areas corresponding with its horizon occur, as shown in the above sketch map.

miles south of Marengo. The road to English from Marengo by way of Pilot Knob is an easy but circuitous route running along a considerable portion of these divides. Other roads in Crawford County are far more arduous and difficult routes, as is commonly the case throughout the Crawford upland.

The topographic condition of these broad interstream areas is very similar to that of an uplifted peneplain where extensive remnants have not yet undergone dissection. While much of the area constituting the broad divides is remarkably flat to the eye, undulations occur, as shown on the sketch map, and Pilot Knob, capped by a massive outlier of sandstone, rises 100 feet above its surroundings, attaining an altitude of 905 feet. Indian Ridge, one and one-half miles west of Leavenworth, also rises nearly 100 feet above the flat plain upon which it is situated, attaining an elevation of 770 feet. Other undulations are rather gentle in character and rarely rise more than 50 feet above the flat plain upon which they lie.

The broad divides here are from 200 to 400 feet or more above the streams of the region. At the edges of the flat divide occur precipitous slopes and steep descents. The main streams of the region are cut deeply below the perched plain and the area composing the valley sides with radiating ravines is exceedingly rugged, in harmony with the great relief of the area. The plain itself is inclined and broadly uneven, but the inclinations are usually too slight to be detected with the eye. The elevations given, as shown in Fig. 41, indicate the directions of inclination of this flat upland divide. At the north in the vicinity of Pilot Knob, the altitude is about 800 feet. At Fredonia the altitude is 655 feet. The inclination of the axis of the main ridge is at the rate of about 16 feet to the mile, chiefly southward. To the southwest the inclination on the average is greater, amounting to somewhat more than 20 feet to the mile.

Investigation shows that these broad, flat divides, constituting a veritable perched plain, and rather anomalous in the Crawford upland area, are not necessarily extensive remnants of an uplifted peneplain, but that they represent a rather remarkable structural level developed on the Hardinsburg sandstone. The Hardinsburg sandstone is usually only 30 feet thick and is composed of thin flaggy sandstone layers frequently separated by thin bands of shale. As a unit it is remarkably resistant to erosion. Minor streams running over it rarely cut through in the normal gradual manner, but rather come to its edge, so to speak, and plunge over it in a series of cascades and falls. Its great resistence has considerably retarded erosion and removal of material at the horizon of its occurrence in the interstream areas, and as a result the divides are abnormally broad. The limestone and shales of the Glen Dean formation which overlie it have been largely removed from its surface through water wash, minor stream action, and solution. Pilot Knob with its capping of massive resistant Mansfield sandstone remains as a monument testifying as to the former elevation of the divide. Likewise Indian Ridge, composed of the Glen Dean formation and a capping of Tar Springs sandstone, here rather thicker than common, is an erosion remnant above the structural plain.

This structural plain preserved on the divides is remarkably sensitive to the dip of the Hardinsburg sandstone on which it is developed. The dip of the strata in this section of Crawford County is far from uniform, and the rather flat surface of the plain corresponds almost exactly to the dip of the rock. Local abnormal dips are shown on the plain surface as well as in the rock strata. Where the dip is slight the inclination of the plain is also slight. For instance, west of Fredonia the strata dip 50 feet in the first mile. A very perceptible inclination of the plain occurs here, as it also inclines to the westward, exactly conforming with the dip. Again near Leavenworth the conformity of the plain surface to the dip of the rocks is shown along the road on the steep western limb of a small anticline in sections 31 and 36. Northwestward from the middle of section 36 to near the middle of section 27, northwest of Leavenworth, the dip of the strata is very slight, being less than 5 feet to the mile. The altitude of the plain along this line decreases very little, and is exactly coincident with the surface of the Hardinsburg sandstone.

The structural plain forming the broad, flat divides is in remarkable topographic contrast with its environs. Its edges are formed by precipitous bluffs and steep descents, and the streams which have reached back into the upland plain have cut its edges into deep ravines extremely difficult to cross. Roads are largely confined to the valleys and the divides, and passage from one to the other is far from easy. On the road running south from Marengo one is scarcely aware of the declivities at the margins of the plain. After passing over the monadnock-like Indian Ridge the flat plain seems to be continuous, but on approaching Fredonia from the north one comes suddenly to the edge of the plain. 300 feet below flows the majestic Ohio River, around a magnificent loop conforming to the great meander in the valley trench itself. This view, from nearly 1 mile north of Fredonia, is one of the most impressive in the entire state, and, like the vista from Bear Wallow Hill on the Norman upland of Brown County, is panoramic. A rather striking feature is the preservation of the high plain to the very brink of the bluff. It swings in unbroken sweep around the sharp bend of the deep-set valley. Drainage is directly away from the brink of the bluff. Probably this condition is the result of the unusually steep dip of the strata away from the river at this place. In fact, the exceptional steepness of the dip suggests a partial explanation of the unusually well developed meander loop, since it is likely that in the deepening of the valley the stream migrated somewhat in the direction of the dip.

It may be mentioned here also that the Hardinsburg sandstone commonly gives rise to structural levels in the area west of that represented by the sketch map. In Kentucky on the south, large areas conform to the level of this resistant sandstone. At Hardinsburg, in Breckinridge County, where the sandstone unit received its geologic name, an area of several square miles exactly conform to this horizon. Since this structural plain so closely coincides with the horizon of the Hardinsburg sandstone, it may well be called the *Hardinsburg structural plain.*

Some Local Features Along Indian Creek in Lawrence County.

In the eastern half of the Crawford upland the streams are characteristically deeply intrenched into the Mitchell limestone, though the interstream spaces are composed largely of the sandstones, shales and thin limestones of the Chester series. Local areas bordering the streams frequently possess sinkhole topography where erosion has removed the overlying clastic rocks.

Some of the details of such an area within the Crawford upland are presented in the accompanying topographic map, Fig. 42. This small area along Indian Creek in western Lawrence County is cited chiefly for the purpose of showing a developing subterranean cut-off, but it also shows the topographic condition of a small portion of the Crawford upland where a main stream has cut deeply into the Mitchell limestone. The area lies from 1 to 3 miles north of Silverville, and from 1 to 3 miles east of the Lawrence-Martin County line. It is some 8 or 9 miles west of Bedford.

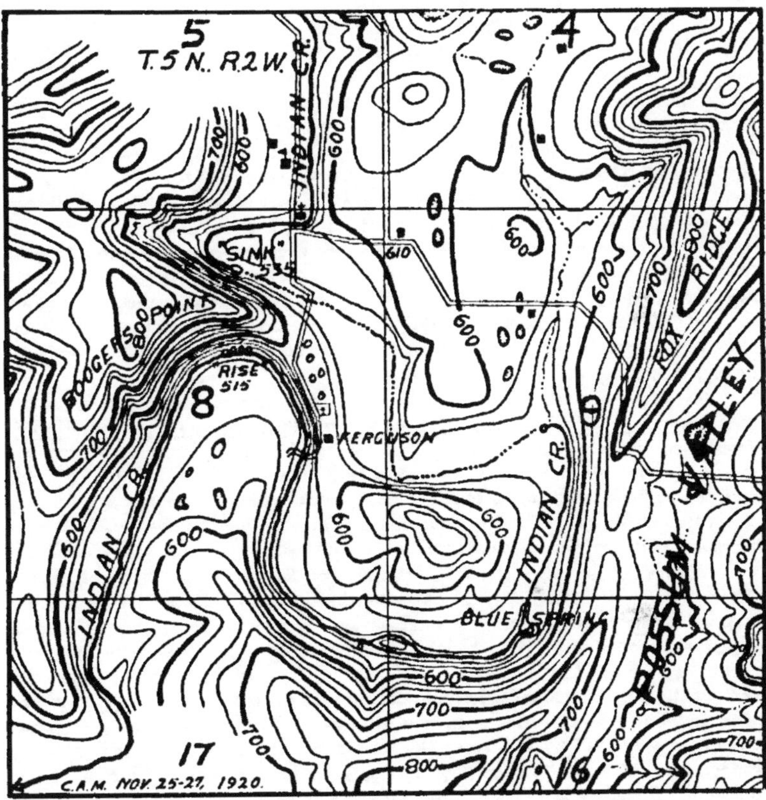

Fig. 42. Topographic sketch of a small area in western Lawrence County, showing a developing subterranean cut-off in one of the great meanders of Indian Creek valley. Indian Creek in low water condition passes beneath "Boogers Point" through a subterranean passage. The fall through the passage is approximately 20 feet. Ferguson Bend, an intrenched meander 3.1 miles around, is being abandoned. The subterranean route is approximately one-fourth mile in length.

Indian Creek in this locality lies in a very sinuous valley deeply sunk below the higher divides. The length of the stream in the 2 miles of the mapped area is 5¾ miles, or nearly three times the direct distance. The stream both above and below the mapped area is but slightly less sinuous. It occupies a narrow valley rarely more than 200 yards in width. It is usually at one side or the other of the narrow valley and at the foot of a bluff or steep rocky slope, varying from 30 to 275 feet in height. The most striking feature of the stream and its valley here is the complex eastward meander. This meander is more than 3 miles in circuit and returns to within one-fourth mile of the place where it begins.

The topographic condition of the area is illustrative of the diversity of form and condition common throughout the Crawford upland. Ridges exist, but the crests are uneven. Sags, saddles, and eminences occur. The ravines are sharp and largely rock-bound. In their upper parts they possess very steep descents. Local isolated sinkhole plains are present, associated chiefly with the entrance of tributary streams into the main valley. They occur far below the sharp and rugged sandstone ridges, and are in great topographic contrast with them. The altitude within the 4 square miles of the topographic sketch ranges from 510 to 875 feet. The maximum relief is 365 feet. Immediate relief is as much as 275 or 300 feet. The chief relief forms are the great bluffs on the outside of the meander turns, sharp, uneven sandstone ridges, and an isolated hill of some prominence within the big meander loop. The curved bluffs on the outsides of the meanders are discontinuous and alternate from side to side of the valley. These relief forms are in great contrast to the local sinkhole plain developed north of the big meander curve. This plain is some 60 or 80 feet above the valley of Indian Creek. A miniature sinkhole plain is also present on the inside of the meander loop in section 8, some 35 or 40 feet above the valley.

The ridges of the area are chiefly composed of massive sandstones, though their lower and more gentle slopes consist of the upper part of the Mitchell limestone. The sinkhole plain itself is developed at a level approximately 100 feet below the top of the Mitchell limestone, or near the top of the St. Louis limestone. The small sinkhole plain within the meander loop in section 8 is somewhat lower than the more extensive plain to the northeast. This is in harmony with the dip of the strata to the southwest.

Possum Valley is a streamless valley which lies to the east of Indian Creek. A small portion of it is shown on the topographic sketch. This valley in itself offers an interesting physiographic feature. As a valley basin it is some 3 or 4 miles in length. It is rimmed by sandstone ridges with the exception of the opening to the south. Its floor is occupied by numerous sinkholes and swallow-holes. Small streams descend from the sandstone ridges and hills and enter swallow-holes in the valley. Some of the ravines or small streams are headed by springs which commonly issue from the foot of steep sandstone bluffs near the top of the ridges. Two such springs are shown on the map. South of the area of the topographic sketch the valley is open, and is occupied by a stream which enters Indian Creek a mile or so below. Little or none of the waters entering the swallow-holes enter the stream below in the open valley. These waters apparently pass to Indian Creek through an underground system which has its terminus at Blue Spring. Blue Spring is a spring of great volume which rises out of a cavernous opening in the edge of the bluff within a few yards of Indian Creek channel. Little of the cavernous opening is visible, because the spring is artesian. After rains the muddy waters rise vigorously and in greatly increased volume. During dry weather the pool at the opening is a deep blue color, and the water rises quietly and flows away at one side practically at the level of Indian Creek.

Possum Valley is characteristic of many valleys of its kind developed in the Mitchell limestone within the eastern part of the Crawford upland. Such valleys are invariably tributaries to a larger and more deeply entrenched main stream. They originated as valley basins through normal surface erosion in the clastic rocks of the Chester series. As the main streams were entrenched through downward erosion the tributary valleys were also cut down, but less rapidly than the main streams. When the tributary streams reached the Mitchell limestone, the main streams were already entrenched well within it, and the tributary streams were somewhat perched in a valley with a limestone floor. Subterranean drainage gradually developed in the tributary valleys, especially some distance from the mouth. In many cases the waters which enter the subterranean channels in the middle and upper portions of the streamless valley, enter the surface stream in its lower course. But more frequently the waters have been diverted through subterranean channels directly to the main stream, passing beneath the divide between the main stream and the tributary. Such valleys are usually east of the main stream. Beaver Valley west of Mitchell in southern Lawrence County is an example of a streamless valley in its upper portion, the subterranean waters of which in part come to the surface lower down in the valley. Possum Valley illustrates the type in which the water has been diverted by subterranean piracy.

Subterranean drainage of this sort takes place as a matter of economy of distance. The subterranean routes are always shorter and more direct than the abandoned surface routes. In the case of Possum Valley the economy of distance is obvious. The subterranean route under the dividing ridge is very short as compared to the old surface route. Streamless valleys of this sort may have one or more underground systems, but the old surface stream system is broken up into a large number of smaller systems. Each tributary of the former surface stream may become a small surface system all to itself, having its own particular swallow-hole into which its waters enter.

It may be added here that such valleys mark the beginning of a sinkhole plain. As the now independent surface streams, formerly tributaries to a larger surface stream occupying the valley, remove the clastic rock from over the limestone, the sinkhole plain becomes larger, and, as previously stated with reference to the origin of the Mitchell plain on the east, considerable areas of sinkhole plain on the growing limestone floor come into existence.

The topographic sketch, Fig. 42, has been prepared especially to show the conditions attending the development of a subterranean cut-off, wherein a great meander loop is being abandoned on account of the development of a subsurface route across the neck of the meander. The waters of Indian Creek in normal and low water condition disappear at the foot of the steep bluff forming the north side of "Boggers Point" spur. The waters reappear one-fourth mile south in a series of springs at the side of the stream where the channel returns from its complex meander loop. Here again is illustrated the economy of distance in subterranean drainage development. The subterranean route beneath the ridge across the meander neck is approximately one-fourth mile in length, while the surface route around the meander loop is more than 3 miles. The fall is approximately 20 feet, and is sufficient to give rise to considerable mechanical erosion along the subterranean route. Such erosion, however, is greatly lessened owing to the lack of concentration in the subterranean route, since it appears that the route is a diffuse one. The waters disappear chiefly in one pool, though other pools below show some indications of water loss. The waters re-enter the channel as broad streams through the talus at the foot of the meander bluff, though holes in the solid rock are also visible. The debouching waters come out along the stream practically at stream level, through a distance of 100 yards or more. There is nothing spectacular about either the "sink" or "rise." In high water the surplus passes through the channel around the meander loop.

One may speculate on the probable future drainage condition here. It does not appear that the subterranean route is likely to become clogged and shut off. The St. Louis limestone is so prone to develop subterranean passages that if for some reason, either through debris carried in or through fallen rock, the route should become obstructed, a new subterranean route would develop. Witness for example the 8-mile journey of Lost River. It is possible and even probable that the passageway will be much enlarged in the future. One may consider it as developing to the stage of an open tunnel with the formation of a natural bridge. The rock of the ridge over the subterranean route is at least 200 feet thick and is competent. The lower 150 feet of it is limestone and the remainder is sandstone. If it should ever reach the open tunnel stage, it is only a step further to the open drainage stage through the destruction of the natural bridge. Such a future is highly probable. Both ends of the subterranean route are situated on the outside curve. These curves may be expected to continue to develop, and the route will thus become shorter. This will only hasten the development of the subterranean route to the open tunnel stage and eventually to the final stage, that of open surface drainage. When it has advanced to either one of these stages the present circuitous meander may be abandoned, and be no longer considered a part of Indian Creek or Indian Creek valley. But the circuitous meander route may be retained through the continued action of flood waters, for it is to be kept in mind that erosion is chiefly accomplished during high water stages in areas of topographic youth and maturity.

This drainage adjustment which is taking place through the development of an underground channel across the neck of a meander loop is designated as a subterranean cut-off. When once effected the result is the same as in a surface cut-off of a meander loop, whether it be a meandering stream or a meandering valley. This drainage adjustment does not come under stream piracy; nor should it be referred to as "self-capture." The term "self-capture" may be inferred to have a definite meaning, but is in itself a rather impossible term. The term "subterranean cut-off" is expressive of the condition of drainage and makes direct reference to the process; and is therefore preferable.

Physiographic Features in the Vicinity of Shoals, Martin County.

A topographic map of several square miles in the vicinity of Shoals and Dover Hill, Martin County (Fig. 43), has been prepared which illustrates the topographic condition of the western portion of the Crawford upland area where the Mansfield sandstone is the controlling rock. As elsewhere the relief forms are those which have been produced through the dissection of an upland area to a state somewhat beyond maturity. The interstream ridges or masses of upland are quite uneven and range in altitude frem 600 to 700 feet or more. In the region mapped a number of the crests reach 700 feet, indicating that the upland of this region before dissection must have been a plain at about this altitude. The presence of polished chert gravels and geodes at this level leads one to interpret it as a broad gradation plain or peneplain.

Near White River the gravels are present at the 700-foot level, where they appear to have been originally deposited. They occur as float on the

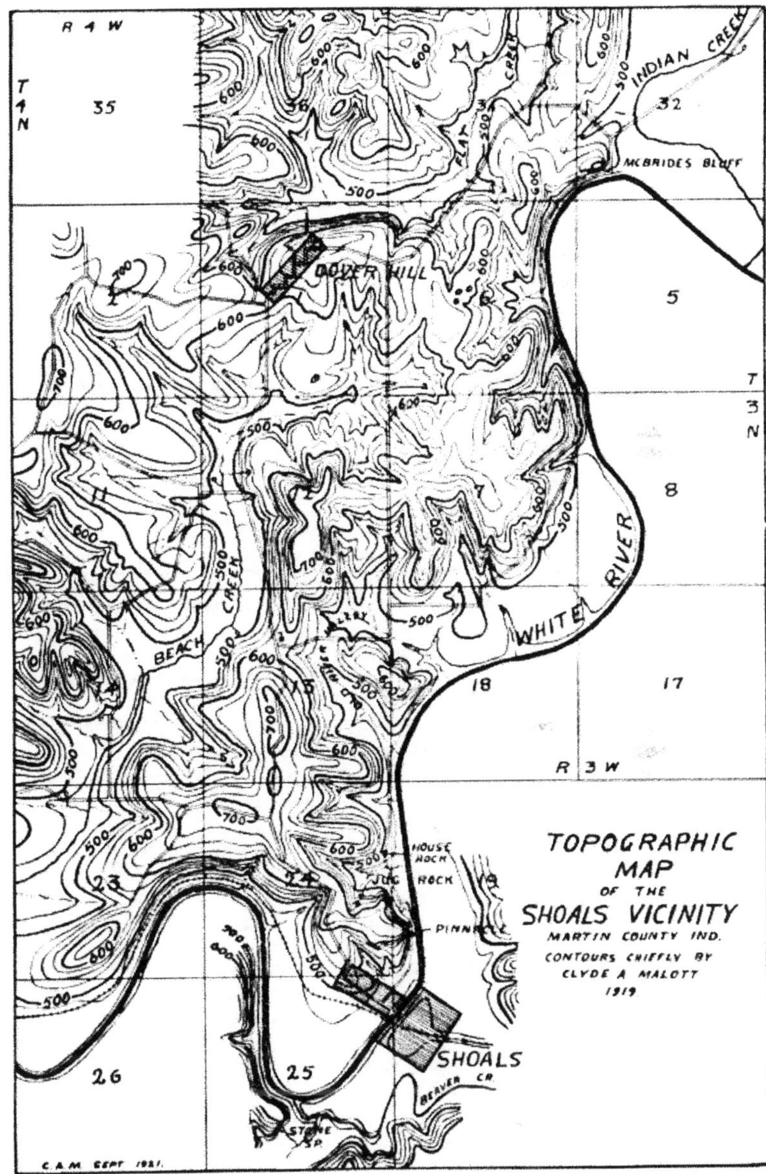

Fig. 43. Topographic map of the Shoals and Dover Hill vicinities, showing the chief erosional forms present in an area where the massive Mansfield sandstone prevails. Note that many of the ridge crests attain an altitude of 700 feet. Late Tertiary gravels are found at this elevation in the area.

flanks of the ridges and as lodged gravels on lower levels where they have accumulated during the erosion of the former extensive plain upon which they were deposited. Between Indian Creek and White River, northeast of the mapped area, these gravels occur on the highest parts of the ridge at 675 to 700 feet. Gravels are present on McBrides Bluff and on the divide to the north at elevations slightly above 600 feet, and below, occurring down to the present valley levels. Here apparently they represent re-deposited gravels or float coming from the original deposit which is now quite completely destroyed in this vicinity. On the divide immediately south of White River, in sections 27 and 28, some 3 miles west of Shoals, the gravels are in place at 700 feet, and locally occur as masses of conglomerate cemented with iron oxide. Geodes 6 inches or more in diameter are conspicuous.

These gravels are undoubtedly equivalent to the cherty gravels and geodes present at similar elevations farther east, as along Salt Creek at Judah's Hill. As suggested in Chapter IV of this paper, they are regarded as having been deposited upon the late Tertiary peneplain. The interstream ridges which reach approximately 700 feet in altitude, and slightly above in the vicinity of Shoals, may be regarded as the representatives of the late Tertiary peneplain level. Some distance south of White River the interstream ridges reach irregular elevations above this level, probably as much as 100 feet or more in a few cases, and such prominences may be regarded as unreduced portions of the late Tertiary peneplain. Probably none of the monadnock residuals reach as high as the destroyed early Tertiary peneplain which is presumed to have formerly been present in the region. Eastward from the Shoals region much of the ridge space rises above the 700-foot level. Westward from Shoals the surface of the upland masses descends, and little of the 700-foot level is preserved. This is very probably because the strata above the Mansfield sandstone are composed largely of non-resistant shales, and have been removed by the present erosion cycle below the level of the older cycle.

White River valley is a meandering valley entrenched some 250 feet below the late Tertiary peneplain. The larger tributary streams have all reached grade in their lower courses and flow in flat valleys. The minor valleys or ravines are quite sharp, and descend from the irregular ridge-crests as steeply inclined V-shaped trenches in which bed-rock is prominent. A general roughness prevails throughout the region, and the angularity of the erosion forms is quite typical of a well dissected upland composed of resistant sandstones which respond readily enough to stream erosion.

The flat valleys present a great topographic contrast to their rock-bound sides where vertical bluffs of massive sandstone are common. White River has a narrow enough valley, but the valleys of the tributaries are rather broad for the size of the streams. The lower part of the valley of Indian Creek, and the valleys of Flat and Beach Creeks are exceedingly flat, and possess all the characteristics of aggraded valleys. Their gradients are very low, the valley flats meet the rocky valley sides with their steep ascents abruptly, and the channels of the streams everywhere have mud banks.

The topographic forms of the region are almost wholly such as are dependent upon the massive sandstones. The Mansfield sandstone in the vicinity of Shoals is nearly 200 feet thick, and its base is close to the valley level of White River. In the Dover Hill vicinity the base of the Mansfield sandstone is considerably higher but is irregular on account of the marked unconformity below the Mansfield, the greatest in the entire state. Locally the Mansfield sandstone is broken by sandy shales and thin sandstones quite unlike the massive, cross-bedded, gritty sandstone which characteristically makes up the formation. In the vicinity of Dover Hill the lower portions of the ridges and the sags are composed of the strata of the Chester series. The Glen Dean limestone here reaches a thickness of 45 feet, and sinkholes and small caves are present. Excellent springs come from this horizon, the water having seeped into the limestone from the overlying sandstone. The sandstones of the Chester series are bluff formers, the Cypress sandstone taking first place in this respect. The sandstone bluffs adjacent to the valley levels east of Dover Hill are composed of this sandstone. In section 31 along Flat Creek the Cypress sandstone is responsible for the beautiful rock benches.

There is scarcely a ravine or valley in the entire area in which massive outcrops of rock are not present. Bluffs from 20 to 50 feet are common. Vertical bluffs 100 feet or more in height occur. On the outsides of the meander curves of White River valley the bluffs are prominent and often precipitous. A slope usually exists, broken by vertical rock-faces of 10 to 40 feet on the steeper slopes. Rarely is the cliff more than 50 feet high. McBrides Bluff east of Dover Hill and "The Pinnacle" near Shoals are notable exceptions. Here sheer descents of 100 feet or more occur. Local benches on sandstone ledges are common. Two areas of considerable extent illustrate such benches. On either side of Beach Creek, in sections 1 and 12 south of Dover Hill, a rock bench is developed on one of the lower ledges of the Mansfield sandstone, and at one or two places on this bench sinkholes occur, due to the presence of the underlying Glen Dean limestone. A second and more prominent sandstone bench occurs on the Cypress sandstone along Flat Creek in section 31 northeast of Dover Hill. This bench resembles a gradation plain in an excellent state of preservation, and the presence of lodged Tertiary gravels from the upper levels lends color to such an erroneous interpretation.

Local topographic phenomena in the vicinity of Shoals have attracted considerable attention. Among the most notable are "The Pinnacle," "Jug Rock," and "House Rock." "The Pinnacle" is a prominent spur which maintains an elevation of 620 feet to the very margin of White River. The upper edge of this eastward-facing spur rises 180 feet above low water of White River. From it one may gain magnificent views of White River valley and the distant ragged uplands. Great angular blocks of sandstone have tumbled down to the foot of the bluff, leaving great scars and joint-faces on the perpendicular wall above.

"Jug Rock" is a denudational remnant of Mansfield sandstone in close proximity to the rock mass from which it has become isolated through the removal of the sandstone immediately surrounding it. It is an upright, roughly cylindrical mass of sandstone rising 40 or 50 feet above its uneven base. (See Fig. 44.) It is from 6 to 12 feet in diameter, being somewhat larger in the middle than towards either the base or top, and is called "Jug Rock" because of its fancied resemblance to an old-fashioned tall jug. It is capped by a large piece of flattened sandstone of triangular outline, which seems to be precariously balanced on a rather small base. "Jug Rock" is situated on the side of a steep hill or bluff. A small ravine descends to the northward immediately on the west side of it. Originally what later became "Jug Rock" was no more than an angular block of sandstone with a northern and western face. Through the presence of joints reaching downward into the massive sandstone, the angular mass of sandstone at the edge of the bluff and the ravine became separated from the main mass, as erosion and weathering proceeded along the joint planes. In its early history as an isolated mass of

Fig. 44. View of "Jug Rock" near Shoals, Martin County, as seen from the upper or south side. "Jug Rock" is a small, but spectacular erosional remnant carved from the massive Mansfield sandstone by natural agencies.

sandstone, it probably little resembled its present shape. Attacked from all sides its size diminished, and it took on a roughly cylindrical shape. It diminished little in height on account of the resistant layer of sandstone which forms its cap. "Jug Rock" stands much higher as seen from the side of the ravine, because the ravine has been deepened more rapidly than the material on the up-hill side has been removed.

"House Rock," across the valley from "The Pinnacle" and "Jug Rock," is simply a grotto in the perpendicular sandstone bluff, where the two sets of joints commonly present in the massive sandstones have been opened widely, but in such a manner that a tilting of the outer massive blocks of sandstone has occurred, nearly closing the joint openings at the top. This condition has been brought about through the creep of the sandstone blocks away from the main mass, chiefly on account of the undermining of the sandstone by solution of the underlying Golconda limestone. (The Mansfield sandstone here rests on and in the Golconda limestone of the Chester series.)

Another feature of physiographic interest illustrated in the Shoals region is the presence of isolated bed-rock masses within abandoned intrenched meander loops of the main valleys. One of these occurs in sections 13 and 18, about 2 miles north of Shoals. Here a rough, angular sandstone hill rises 175 feet above White River. The side facing the river is steep and in places quite perpendicular, forming a prominent bluff. The side facing the abandoned river route is a long slope, in keeping with the gentler slope on the inside of intrenched meander loops in general. Farther up White River other abandoned meander loops of the intrenched valley occur, which are quite similar to the one north of shoals. Down White River from Shoals, in the vicinity of Hindostan Falls, occurs a very unusual meander cut-off. Apparently this cut-off has been completed only recently, for the stream has not had time to cut down the rock barrier along the new route of the cut-off. A total fall of nearly 7 feet is attained in a distance of less than one-fourth mile. The old valley about the abandoned meander is much longer than usual, being between 5 and 6 miles. The course of the river was therefore shortened over 5 miles when the cut-off was completed. The old valley at its greatest width is nearly a mile across. A ragged mass of upland hills is surrounded by valley land within the old meander loop and the present course of the stream over the ungraded bed-rock channel.

A rather unusual cut-off occurs at Shoals. Usually a cut-off is effected at the neck of the meander loop where the two limbs of the meander come together. In the case at Shoals only the end of the sloping point of the upland has been detached, and the prominent meander loop still remains, though less marked than formerly. The main part of Shoals is built on the point of what was formerly the lower end of a sloping spur. The river passes over a bed-rock channel in a somewhat shorter course than formerly. The bed-rock hill which has been isolated from the main spur is low and of small size. During high water when the flood plains are occupied the water passes around the old route, which is continuous with the present flood plain of the river. The configuration of the valley is such that overflow waters would naturally go along the old route, and their passage about it in conjunction with the inflow of Beaver Creek has kept the route open. The grade of the Baltimore and Ohio Railway, however, connects the island on which Shoals is built with the upland during time of high water, except when unusual floods wash the grade away.[184]

[184] Bybee and Malott, The Flood of 1913: Ind. Univ. Studies No. 22, 1914, pp. 167-171.

Apparently valley filling during the Pleistocene was largely responsible for the abandonment of most of the old meanders. The Pleistocene waters which came in unusual amounts may have been able to cause a union of the two limbs of many meanders through erosion of the meander bluffs. But valley filling has caused a union of the valley lands across the necks of many of the meander spurs, which at first probably stood out as bed-rock islands within the flooded river. In the adjustment which followed the glacial period the route maintained by the stream was the short one across the meander neck. Some 30 or 40 feet of the Pleistocene fill deposited by the debris laden waters from the ice-sheets has since been removed and the stream in places is cutting in the bed-rock of the buried meander neck. Some of the meander necks are still deeply buried and the valley is as wide as the old abandoned route. Such a condition exists at Williams above Shoals.

The American Bottoms Region of Eastern Greene County.

Northward from near the southern boundary of Greene County the greater part of the Crawford upland area has been glaciated, and locally is considerably modified by accumulations of glacial drift. General roughness and diversity of relief forms, however, prevail. Local accumulations of glacial drift are prominent, though bed-rock forms produced through erosion and weathering stand out in great relief as far north as the Shelbyville moraine which marks the northern limits of the Crawford upland. The glacial boundary line passes mainly northward through middle eastern Greene County, and it is along this line that the most interesting glacial modifications have occurred. In middle eastern Greene County where thorough stream dissection had taken place, considerable drainage modifications were made through the advance of the Illinoian glacier into the region. Subsequent stream adjustments of an unusual nature have taken place in this region, which from a physiographic point of view makes it one of the most interesting in the state. The region has been described in detail by the writer and the physiographic relationships discussed at some length.[155] The region is known as "The American Bottoms" region. The following presentation is taken in part directly from writer's former publication:

The American Bottoms region lies adjacent to the glacial boundary line in eastern Greene County, chiefly within the driftless area. It receives its name as a region from a broad, flat lacustrine plain bounded by sandstone hills, which is known as the "American Bottoms." The altitude of the region varies from about 525 to 900 feet, and immediate reliefs of 250 feet occur. The region consists chiefly of interstream masses of dissected upland and intervening valleys, which extend chiefly in an east and west direction. The interstream masses of upland are dissected into numerous spurs by minor streams which extend either north or south from the east and west extending main ridges. The interstream ridges are uneven, with gentle eminences and broad sags. The main upland surface varies from 800 to 860 feet in altitude, though where sags occur the altitude descends commonly below 750 feet. The ravines reaching out from the main ridges are very sharp, and bluffs frequently occur. The valleys of Beech, Clifty, and Plummers creeks are flat and in places are rather broad.

[155] Malott, The American Bottoms Region of Eastern Greene County, Indiana—A Type Unit in Southern Indiana Physiography: Ind. Univ. Studies No. 40, 1919, pp. 1-64.

The interstream spaces reaching an altitude of 800 to 850 feet have the appearance of a rolling peneplain surface, and have been interpreted by the writer as being representative of the Highland Rim peneplain level. Such an interpretation, however, is doubtful, as it may be seen that the surface is nearly everywhere developed on sandstones. The more massive the sandstone, usually the higher the eminences upon the uneven ridges. Towards the east the sandstones forming the surface belong to the Chester series. Westward the Mansfield sandstone forms the surface, acting as a capping to the ridge masses.

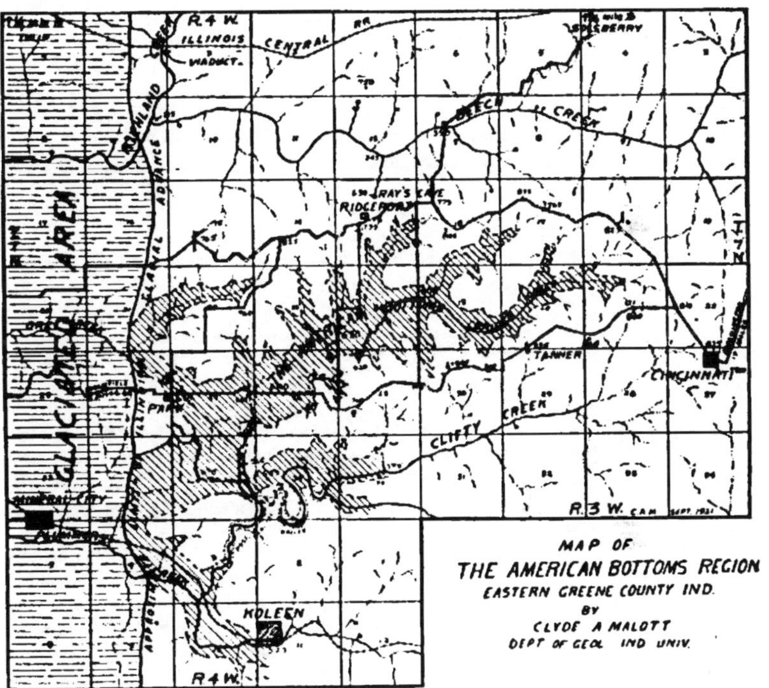

Fig. 45. Sketch map of the American Bottoms region of eastern Greene County, showing the drainage lines, the portion covered by the Illinoian ice sheet, and the obstructed and filled "American Bottoms" valley. Note the position of the post-glacial gorge of Clifty Creek as compared to its preglacial valley. The "American Bottoms" valley is perched approximately 100 feet above the valleys of similar size on either side, and possess a unique subterranean drainage through a stratum of limestone which occurs near the level of the perched valley.

Rather than being a peneplain surface, it is more likely that the surface is reduced considerably below the old peneplain level. No broad areas at elevations consistent with the elevation of the Highland Rim peneplain exist in the region.

"The American Bottoms" is a local lacustrine plain lying between two of the main ridges of the region, and is perched approximately 100 feet above Beech Creek valley on the north and Clifty Creek valley on the south. The flat plain is simply a pre-glacial valley-trench filled some 100 feet with gravel, sand, and silt, the surface of which has been preserved on account of peculiar drainage conditions which have come into existence since the filling of the valley by the outwash from the Illinoian glacier. The plain is narrow at the head of the valley on the east where little filling took place. It gradually becomes wider as the valley becomes larger to the west. Nowhere is the plain more than a mile in width. It passes up the tributary valleys of the main valley in harmony with their size and depth. (See Fig. 45.) The main plain where uneroded by developing post-glacial streams contains several square miles in area. The surface of the plain is in great topographic contrast with its rough environs. The drainage of the plain is chiefly through Bridge Creek which flows over the surface of the plain to its terminus at a large hole in the side of the valley. This hole in the valley-side occurs in the massive Cypress sandstone, and is a rather spectacular opening, differing greatly from the ordinary swallow-hole. Smaller independent streams west of Bridge Creek pass over the plain and into holes in the sandstone wall on the south side of the plain in a manner similar to Bridge Creek. The western portion of the perched plain is considerably eroded and dissected by surface streams reaching into it from Clifty and Plummers Creek.

The larger relief forms of the American Bottoms region, as elsewhere in the Crawford upland area, are the result of unequal dissection. Individual forms are dependent largely upon the more resistant strata which have been differentially cut by stream erosion. The shales of the region have given rise to gentle slopes, and where they occur between resistant sandstone horizons have indirectly given rise to benches. The Beech Creek limestone and the Cypress sandstone of the Chester series and the Mansfield sandstone, unconformably lying upon the Chester series, are of chief importance in giving rise to the positive forms of the region. The strata below the Beech Creek limestone are of importance locally in the eastern part of the region. Along the deeply cut valleys the lower strata give rise to few positive forms, but rather form the lower valley slopes adjacent to the valley flats.

The Beech Creek limestone is nowhere more than 30 feet thick, and over most of the region is not more than 15 feet thick. The Beech Creek limestone outcrops along Beech Creek and its tributaries high above the valley floor and along Clifty Creek to its junction with Plummers Creek valley. In the latter case it comes down to the valley level, and in places is hidden by the valley alluvium. It outcrops only in the easternmost ravines of Bridge Creek, and is below the level of the flat "American Bottoms" valley.

The Beech Creek limestone is of great importance from both the topographic and physiographic standpoints. Along its outcrop it frequently stands out as a wall-like bench, partly on account of the shale which characteristically underlies it, and partly because of its recession *en masse* on weathering. Its highly jointed condition allows it to collect waters from the overlying sandstone into concentrated streams, the outflow of which in steep-headed ravines and gorges makes it perhaps the most extensive and persistent spring-bearing horizon in the entire state. It possesses this character because of its relation to the massive overlying sandstone which has a high porosity. These springs often yield a considerable volume of water, and the characteristic steep-headed gorges with their high walls of solid rock are common throughout the region of the outcrop of this formation, especially where the limestone is rather high above the drainage level. The Ray's Cave gorge is typical. As may be inferred from the presence of these gorges with such large volumes of water coming from their steep-walled heads, the solution of the limestone along the joints has frequently enlarged them to caves of considerable size and length, considering the limited thickness of the formation. Ray's Cave, near Ridgeport, is a cave of very uniform width and height, following the joints strictly, and turning frequently at sharp angles. This cave may be easily followed for a distance of about 1,000 feet, to a point where farther progress is arrested by a mass of great sandstone blocks fallen from above. This distance, however, must represent only a small part of the total length of the cave, since a large volume of water comes from under the fallen debris.

Just how important this limestone is physiographically will be clear when it is realized that the peculiar drainage conditions in the "American Bottoms" basin are due to its presence at critical levels. Bridge Creek and its smaller associates do not empty their waters into the sandstone bluff for any other reason than that the Beech Creek limestone is immediately below it. This limestone is, in fact, only 10 feet below the point where the waters of Bridge Creek enter the sandstone, though the limestone itself is nowhere visible about the margin of the "American Bottoms" flat. Had it not been for the presence of this 20-odd feet of limestone at this particular level, there could have been no subterranean drainage, such as occurs, nor could there have been preserved the unusually broad, filled valley, which, for the most part, is wholly intact. The physiographic effect of this limestone at a critical level is seen again in the southwest quarter of section 35, along Clifty Creek, where the limestone has been carried by the dip to slightly below the level of the stream, permitting the local development of subterranean drainage by the waters of Clifty Creek. Only the flood waters pass around the great double meander at this point. The water going through this subterranean passage, or rather passages, along the enlarged joints in the limestone is lost to view for a distance of about one-fourth of a mile, and nearly 150 feet beneath the crest of the ridge above. This underground passageway of Clifty Creek is in its initial stage, but we can see that it must finally cause the complete abandonment of the surface channel, leaving the great double meander of a dry valley sunk deeply into the strata, the product of a stream which has entrenched itself since the invasion of the Illinois Glacial Lobe into the region.

The Cypress sandstone is probably the most prominent wide-spread sandstone of the Chester series. It is 30 or 40 feet thick and quite massive. Joints are well developed, and its outcrop on the hillsides often exhibits great broad joint-faces and rectangular blocks. The Cypress sandstone is well developed high above Beech Creek valley along its middle and lower course. It is found everywhere along Clifty Creek, reaching the valley level along the lower course of the creek, and in conjunction with the Beech Creek limestone causing characteristic bluffs and cliffs. It everywhere marks the rim of the "American Bottoms" basin, usually causing an abrupt rise from the monotonous valley flat.

The Cypress sandstone, as indicated above, is everywhere an important topographic and physiographic factor, ranking next to the Beech Creek limestone in this respect in the "American Bottoms" region. It has a wall-like appearance along the streams, whether it be next to the valleys or high up the valley slopes. Since it is overlain by shale it gives rise to rather sharp local benches, often of considerable breadth where erosion has removed the overlying material down to its resistant top. These benches have been mistakenly interpreted as local peneplains developed in the region where the sandstone is to be found. They are seen prominently above the abrupt slope produced by the sandstone where it rises above the "American Bottoms." To the west the benches become considerably lower on account of the dip of the strata. Such benches are conspicuously developed on the north side of the "American Bottoms" in section 23. The streams of the "American Bottoms" pass into cavern-like openings developed in this sandstone, owing to the undermining of the sandstone by the weathering and solution of the underlying limestone, and corrasion by the inflowing waters. The cave-like openings of these stream inlets are among the most striking phenomena of the entire region. On the ridges in the eastern part of the area, a number of broad, fairly extensive sags are developed on the Cypress sandstone between the gentle summits of the ridges. Examples of these are to be seen at Cincinnati and Tanner.

The Mansfield sandstone formation locally consists of sandy shales, but typically it is a massive sandstone capping the higher parts of the easterly portions of the ridges and making up an important part of the ridge strata on the western portions of the ridges. Near Ridgeport it is fully 50 feet thick. Its surface is uneven, and bold outcrops of its massive phases are common on the ridge tops within the western portion of the region.

The Illinoian ice-sheet pushed in from the west and greatly modified the thoroughly dissected upland in the American Bottoms region. Where the ice itself was present the ravines and minor valleys were largely obliterated. Outwash spread beyond the ice advance and locally filled and obstructed valleys which were present adjacent to the ice-sheet. Outwash material, composed of sand and gravel, is common along the glacial boundary within the driftless area. Such deposits occur in the vicinity of the viaduct of the Illinois Central Railway across Richland Creek. A partially preserved outwash slope occurs east and south of Park. It would seem that broad sheets of water must have made this outwash deposit of ice-contributed material, carrying it eastward 2 or more miles. The sharp ravines and cuts next to the glacial margin frequently expose layers of coarse gravel inter-bedded with coarse sand. Farther east the gravel disappears, and only the sand, very much cross-bedded and rudely stratified, is seen in the sharp ravines and cuts. Much of the original surface of this outwash fan is excellently preserved in sections 26, 27, 34, and 35. Its gently sloping surface shows very clearly on the topographic map, though deeply trenched by post-glacial erosion.

This outwash material in places almost completely obliterated the pre-glacial topography. A small monadnock-like hill one-half mile southeast is completely surrounded by stratified sand. It would appear that this hill of Mansfield sandstone stood island-like in the floods which swept the sands about and beyond it. The pre-glacial valley of Clifty Creek in the eastern part of section 33, and through the middle of section 34 was completely filled, and, farther east in section 35, it was filled from bluff to bluff to a depth of 100 feet or more. It is easily seen that Clifty Creek never recovered this drainage, but sought an outlet to the southwest across the ridge through section 3. The stream that came in from the pre-glacial "American Bottoms" valley was likewise filled, and, while its lower course has been partially resurrected, much of the old valley is still a sandy flat into which the rain waters readily sink instead of flowing over the surface. The southwest edge of the "American Bottoms" in section 26 is a flat surface of a portion of this outwash plain. The sand of which it is largely made up is exposed in the south flowing stream and ravine-like tributaries near the line between sections 26 and 27.

Just north of Park there is a broad glacial col. It seems that the water from the ice came from the west over this col and was discharged to the southeast through section 27, probably finding its way out toward Plummers Creek along the route of the present Clifty Creek. Proof of the eastward discharge of water through this col is found not only in the direction of flow indicated by the cross-bedding of the water-laid material, but also in the presence of silt terraces due to the ponded condition of the northeast tributary of Ore Creek, in the north half of section 22. These silt terraces are at exactly the same elevation as the broad glacial col. This small pre-glacial valley leading southwest was ponded by the ice and glacial debris to the depth of the water flowing out at the col. The glacial sediment which eddied into this flooded preglacial valley combined with the indigenous material carried into it by the regional streams was sufficient to fill the valley with silt, the remains of which still reveal its glacial history. It is quite probable that the strong discharge over this col resulted in lowering the head of the outwash apron in the vicinity of Park.

The outwash material so effectually filled the lower part of the valley now occupied by the "American Bottoms" that the waters which gathered into it have never transgressed the barrier. The middle and upper part of the valley became a lake with the lowest margin at an elevation of 675 feet above sea level. The southwest rim was the broad outwash plain gently rising to the west. This outwash plain abutted against the abrupt ridge to the south. There are no indications that the water of the lake ever overflowed the barrier at any point.

The lower course of Clifty Creek, as has already been explained in some detail, was also evicted from its pre-glacial channel. The lake produced in its middle and upper reaches, extended to within a short distance of the village of Cincinnati, and the waters rose to an elevation of about 650 feet. At this elevation the water evidently found an outlet across a sag in the ridge to the south, and the draining stream debouched, as it does today, nearly one and one-half miles farther up Plummer's Creek than in pre-Illinoian time.

Both Plummer's and Beech creeks were also dammed by the ice. These streams still discharge through their pre-glacial valleys. The presence of silt terraces in the valley of the former, some 40 feet above the present valley flat, indicates that it must have been effectively dammed for a time, but that the barrier was not sufficiently massive to permanently upset the drainage. In the lower part of Beech Creek valley the terraces rise about 50 feet above the present valley floor, and their presence far up the valley at a consistent elevation of 600 feet shows that the valley was effectively dammed to that height; but this again was not enough to permanently derange the drainage.

The close of the Illinois glacial invasion in the region of the "American Bottoms" found Plummers and Beech creeks ponded some 40 to 50 feet deep, with their waters rising to a level of a barrier that remained long enough to permit the lakes to be filled up by the incoming debris from the drainage basins above. Clifty Creek valley was so effectually obstructed that the water in the glacial lake found an outlet approximately 100 feet above the valley of

Plummers Creek. The pre-glacial valley of the "American Bottoms" was also blocked to such a height that the waters confined to its middle and upper portions never overtopped the barrier. The surface of this lake must have had an elevation of not less than 650 feet and not more than 670 feet above sea level.

Clifty Creek in its lower course exhibits some rather striking features. In section 35, some unusual meanders exist. Here the valley passes in and out of its filled pre-glacial valley. In places the valley is exceedingly narrow and has steep rocky sides, and in other places it is rather wide and possesses valley sides of sand. (See Fig. 45.) Near the southwest corner of section 35 its waters disappear during normal and low water condition. The valley in its southward stretch to Plummers Creek is clearly post-glacial. Here it has steep rocky sides which rise abruptly above the narrow floor of the valley.

Clifty Creek in its initial post-glacial course was perhaps everywhere working in the unconsolidated material of section 35, but on cutting down it came to the underlying bed-rock; except here and there. Wherever it has become entrenched in bed-rock, it is now gorge-like. The wide pre-glacial valley, flanked with terraces, comes to an abrupt end at a gorge just below the mouth of Little Clifty Creek. The present stream has no valley flat here at all, but flows between rock walls composed of Beech Creek limestone below and Cypress sandstone above. This gorge is not a pronounced one, ends rather abruptly, and for a short distance the valley widens out to considerable dimensions. The constriction between the rock walls is repeated where the stream turns north at the beginning of the great meander loop about the middle of section 35. The temporary increase in width just mentioned is due to a small pre-glacial valley which emptied northward into the pre-glacial valley of Clifty Creek. The northward loop, as already described, has a wide valley because here the stream worked in unconsolidated outwash material which has filled the pre-glacial valley. The stream in this very striking meander loop enters into a rock-walled gorge again just south of the center of section 35. The water here is flowing with considerable velocity over a bed-rock floor. After swinging around this northward loop the stream returns to within 125 yards of itself; but the short distance between the two limbs of the meander is separated by a high sandstone ridge. In making the broad southward loop the stream encounters a small fairly broad pre-glacial valley trending northward, and follows the latter to the northward. On rounding this meander the stream hugs a steep cliff of Cypress sandstone. This turn to the northward is due not only to the presence of the pre-glacial valley, but also to the height of the ridge to the west. The initial stream flowed northward until the height of the ridge lessened sufficiently to permit it to cross. It did not come quite far enough at this place to reach the pre-glacial valley of Clifty Creek. Initial flow to the northward was unfavorable because the outwash surface had a southward slope. Clifty Creek follows a double meander through a rock-walled gorge in the southwest corner of section 35. The lower part of this gorge is in the Beech Creek limestone, and the upper part in Cypress sandstone. The gorge here has walls about 50 feet high and a much higher steep ascent on the south. The bed of the stream is in highly jointed limestone, and the stream continues to run over or very close to bed-rock through the remainder of its course to Plummers Creek valley.

One of the remarkable conditions to which this adjustment of drainage has given rise is the present tendency of the stream to seek a subterranean channel. This is an adjustment that belongs to the present. (See Fig. 46.) Only the flood waters of Clifty Creek go around the broad double loop to the north. The regular flow all passes beneath the ridge along the enlarged joints of the Beech Creek limestone, coming out on the other side in a small cave-like opening. This underground passageway is one-fourth of a mile in length, and the ridge reaches a height of nearly 150 feet above it. When the flood waters give sufficient pressure, water comes out on the lower side through numerous openings, and through some of the smaller ones spurts out fountain-like from the limestone wall.

The filled valley leading from the "American Bottoms" has a consequent drainage developed since the fill was made. One of the ravines heads at the col near Park. It descends very steeply into the outwash sand, and to the

Fig. 46. View at the "sink" of Clifty Creek of the American Bottoms region, eastern Greene County, near the southwest corner of section 35, T. 7 N., R. 4 W. Clifty Creek is seeking a subterranean route here through the Beech Creek limestone which lies immediately below stream level beneath the massive Cypress sandstone. The subterranean waters pass beneath a sandstone ridge 150 feet high.

southeast joins a larger stream which comes from the north. The east side of this larger stream is formed by the exposed wall of Cypress sandstone which was also the east side of the pre-glacial valley leading out of the "American Bottoms." The west side of this valley is composed of sand and gravel from which water continually seeps, locally making boggy places at the foot of the steep slope. This stream crosses the bre-glacial Clifty Creek valley at right angles and joins the present Clifty Creek just beyond the south side of the old valley. The stream turns southeast upon striking the rock wall of Cypress sandstone just before it enters the present Clifty Creek. The valley has considerable width where it crosses the old filled valley. In the extreme southeast corner of section 27 the main re-excavating stream from the north in its downward descent came upon a "nose" of the bed-rock ridge, but kept its position, and cut a gorge at that point in the Cypress sandstone.

We have already seen that the "American Bottoms" is a filled valley, due to the spreading of an outwash apron which completely obstructed the lower part of the valley. The middle and upper part of this valley became the site of a lake into which the waters came from all sides, bringing in silt and filling the valley back of the outwash dam to a height of approximately 650 feet above sea level. The conclusion has already been stated that the valley has never been overflowed by the confined lake waters. It would seem that the waters may easily have filtered through the sand deposits into the ravines to the west, in the lower part of the old valley. These ravines now contain

Fig. 47. Near view of cave-inlet or swallow-hole of Bridge Creek. Drainage from approximately 6 square miles in area enters this opening in the Cypress sandstone on the south side of the "American Bottoms" basin.

streams of considerable volume due to the constant seepage of water from the sand at the foot of these steep slopes.

Ashley[148] in his discussion of the stratigraphy of Greene County has the following to say in reference to the Pleistocene: ". . . . In the lowlands and prairies the deposits are found to be of considerable depth, often over 100 feet, these places evidently being old valleys filled up. Some interesting deposits occur along the glacial border in eastern Greene County. As the ice pushed its way southeast across the country it overran the lower

[148] 23rd Ann. Rept., Ind. Geol. Surv., 1898, pp. 768-769.

course of many of the streams flowing west into White River, thus effectively damming them up. Small lakes were thus formed. In time these filled up. Then the ice retreated and the streams resumed their old channels. In most cases they immediately began clearing this lake-deposited material out. As this was in the upper part of their courses where the current had some power, most of the streams have about rid their channels of all vestiges of these deposits. Along Richland Creek in Beech Creek Township however, much material yet remains in the form of gravel terraces mantling the bluffs of the banks of the streams. In places these terraces are over one-quarter of a mile broad. In the ease of a branch of Clifty Creek in the southeastern part of Center Township, instead of clearing out the deposit laid down in 'the ice-bound lake, the water finds its way down through the mass at several places and flows away underground to appear in the old channel farther down. In

Fig. 48. View of the cave-inlet or swallow-hole one-fourth mile west into which the waters of Bridge Creek enter. Drainage waters from approximately 1,110 acres enter this opening.

this case it would appear quite possible that the water found a passage under the ice before the glacier retreated. The result is a flat filling in a valley to which the name of 'American Bottoms' has been given."

Leverett,[18] following Siebenthal's notes has the following to say about the "American Bottoms": "About four miles south from the point where Richland Creek turned westward into the glaciated district, the glacial boundary comes to the west end of another glacial lake whose site is now known as the 'American Bottom.' It extends eastward about five miles from the glacial boundary and has an average width of nearly one mile. This old lake bottom now has subterranean drainage through sand deposits to a tributary of Clifty Creek, where it appears in the form of springs. Because of subterranean drainage the plain is preserved in nearly the condition left by the lake."

From the above it seems quite evident that both Siebenthal and Ashley thought the water capable of filtering through the coarse outwash sands and entering the headward-cutting streams below. Such is not the case at present.

[18] Leverett, The Illinois Glacial Lobe: U. S. Geol. Surv., Monogr. 38, 1898-1899, p. 163.

Bridge Creek and its smaller replicas, as already explained, enter into openings in the massive sandstone bluffs at the south side of the valley. These openings in the sandstone bluffs are very picturesque, and are in themselves remarkable phenomena. (See Figures 47 and 48.) It would appear that the water in the former lake of the "American Bottoms" filtered through the sand, or else entered the already considerably enlarged joints in the Beech Creek limestone, which was well up the side of the pre-glacial valley and under pressure found its way along the magnificent system of joints. It would not appear that there was much opportunity for the discharge of water through the present subterranean passages for a considerable time following the withdrawal of the ice from the region. The water at present enters the openings at an elevation of 620 feet, or slightly lower, and emerges as two or three springs in the sandstone wall at the present valley level at the northeast quarter of section 34, at an elevation of 560 feet above sea level, and at a point slightly less than 2 miles southwest measured from the Bridge Creek inlet. While this is an easy line of discharge at present, it is difficult to believe that it was a line of discharge immediately following the ice withdrawal. The waters could have filtered through the sands after having passed through the limestone passages; but it is more probable that they filtered through the sands all along the western barrier, with sufficient volume to prevent the overtopping of the sand barrier. Certainly, if the waters traversed the limestone passages at all immediately following the ice withdrawal, they went through very slowly, as the passages must have undergone most of their enlargement subsequent to the ice period. Moreover, the valley into which the present outlets debouched was a filled valley, and has only in recent times been excavated to the level where the present springs emerge.

The present "American Bottoms" area consist of a flat, silty soil, everywhere wet and well leached of its lime content. For the most part it is still the bottom of the former shallow lake, the lake having been filled by the material carried in from the surrounding hills. The lake flat slopes westward at a very slight angle from the east, to the two main openings. The slope toward these openings from the west is somewhat greater, but is still at a low angle. The latter slope is the surface of the outwash plain which consists mostly of sand. The most fertile land of the region is in this sandy eastward slope. The present stream of Bridge Creek has a poorly marked valley some 12 feet below the general level of the lake flat where the stream flows into the opening in the sandstone bluff. The second largest opening, a quarter of a mile to the west, is entered by a stream flowing in a slightly depressed swail. The two small openings in the southern half of the southeast quarter of section 26 have independent streams both of which are depressed more sharply below the general level of the lake flat.

In the northeast quarter of section 26 are several small sinks which have their surface expression in the joints of the massive Cypress sandstone. There are several similar sinks in section 24. Some of these are of more than passing interest on account of being on the "nose" of Cypress sandstone in the southeast quarter of the section. Several small ones here reach down some 40 feet, through the whole thickness of the Cypress sandstone before coming to the Beech Creek limestone. These sinks show the results of the passage of water through the joints in the sandstone to the limestone below, which has been partially removed by solution. These sinks are, of course, due to the

Beech Creek below, but superficially they give the appearance of being sinks in the sandstone itself, a thing scarcely possible.

It will be noted by reference to the map (Fig. 45) that the entrances to all the inlets into the subterranean passageways lie against the sandstone bluffs at the south side of the valley in a belt running in a general southwesterly direction. A broad, shallow sinkhole is the last of the series. In line with this belt and two miles to the southwest are the outlets to these passageways in the Cypress sandstone bluff on the east side of the partially re-excavated valley from the north. The chief one of these cutlets is the middle one, where most of the normal flow comes out in a broad, ill-defined opening from under the sandstone at the very level of the valley floor. (See Fig. 49.) A much smaller amount of water comes from the outlet at the south, near the road. The opening to the north, nearly one-quarter of a mile north of the

Fig. 49. View of the main outlet of the subterranean waters of Bridge Creek of the "American Bottoms" basin, nearly 2 miles southwest of where the waters enter the subterranean passageways. The water comes out from beneath the Cypress sandstone. The Beech Creek limestone in which the passageway exists is some 10 feet below the surface at this point.

road, is the most interesting. (Fig. 50.) It is an opening some 2½ feet high and about 10 feet in width, and reaches back under the sandstone to the east. This opening is slightly higher than the openings to the south, but perhaps less than 5 feet higher. The flood waters come out of this opening with great velocity. The opening has a perimeter composed entirely of sandstone, a rather interesting occurrence, showing that it must have been developed along a combined joint and bedding plane, and under pressure from the water within. It is not known how far back one would be able to crawl into the opening, but it is likely that one would come to the passageway, or a series of passageways, along the joints of the Beech Creek limestone within a few yards or a few hundred feet of the opening, after a descent of not more than 10 or 12 feet.

As indicated by the topographic map, the "American Bottoms" for the most part is a basin, and furthermore a basin with holes in it through which the water escapes. The 675-foot contour line is the highest depression contour, and encloses approximately 1,475 acres, or 2 5/6 square miles. The area enclosed by this contour line has a total perimeter of approximately 18 miles. The lake waters were probably much nearer the 650 foot contour line. It may be noted here that some beautiful beach lines are preserved at an elevation of about 645 feet along the road leading east from the church, in the northwest portion of section 34. These gravel beach lines have no relation to the water level of the "American Bottoms" lake. No beaches are preserved in the "American Bottoms" basin. The 625-foot contour line is the third depression contour in the basin, but it is restricted to small areas near where the streams enter the opening. The total depth of the basin from the lowest point in the rims to the hole where the water escapes from the basin, is about 65 feet. Some further figures reveal interesting features. Bridge Creek drains approx-

Fig. 59. View of the flood-water outlet of the waters from the "American Bottoms" basin. This opening has its perimeter entirely in the Cypress sandstone. It is about one-eighth mile north of the chief outlet and is some 3 or 5 feet higher.

imately 3,900 acres or about 6 square miles. The water enters the largest inlet cave near the northwest corner of section 25. Sink-inlet number 2, some one-fourth mile to the west, receives the drainage waters from approximately 1,100 acres. Sink-inlet number 3, or the east one in the southeast portion of the southeast quarter of section 26, receives the drainage waters from about 170 acres. The smallest sink-inlet, number 4, just to the west of number 3, receives the drainage waters from about 60 acres. The sink south of the road in the southern half of section 26 receives the drainage waters from approximately 60 acres. Thus, a total of approximately 5,200 acres or 8 square miles drains into the holes in the "American Bottoms."

It may be noted that the ravines which are re-excavating the pre-glacial valley to the west and southwest of the old lake basin are taking away the material at a very rapid rate. The ravine in the southwest portion of the northwest quarter of section 26 is a great gully or series of gullies, near its abrupt head, eating directly back up the line of the pre-glacial valley. It would appear that this ravine, which at present is so rapidly reaching into the outwash sand by headward erosion, may in time tap the streams leading into the subterranean openings, and thus divert the drainage approximately through the old channel to the present Clifty Creek. While this is a suggestive possibility, it will probably never happen. The present streams of the basin at their subterranean inlets are approximately at an elevation of 620 feet, while the elevation of the ravine where it crosses the road considerably more than a mile to the southwest, is only 20 feet lower. This difference is not enough to give the advantage necessary for piracy, since the small stream would probably require much more than 20 feet fall in the distance it would have to go. It would appear that the subterranean passages must persist unless they should become thoroughly choked, which is unlikely.

When the inlet-sinks of the "American Bottoms" basin were first examined it was thought that the Beech Creek limestone would certainly be visible, and the fact that nowhere is there the least vestige of it exposed at the surface at any of the openings gave cause for considerable surprise. These openings can be entered for only a short distance. It was found however, that the water has to pass over logs, sticks and other trash that has been lodged in the openings near the entrances, and it was found further that the water descends through the trash very sharply, and probably within a short distance from the entrance reaches the Beech Creek limestone.

Just why the water that enters the opening has not cut down to the level of the limestone passageways is not altogether clear. Only two suggestions adequate to explain this condition occur to the writer from observations made in the field. At the openings much foreign material is carried in which continually keeps the throat of the sink in a clogged condition. Thus the water is passing over the logs, trash and other material densely packed with silt at the entrance and has no opportunity to cut down the opening to the level of the limestone passageways below. The second suggestion arises from the nature of the top of the limestone as compared with its deeper portions. The upper few feet of the Beech Creek limestone in the "American Bottoms" region, it will be recalled, consist of coarse yellow limestone, containing considerable sand and clay. This upper portion seems to dissolve much more readily because of these properties. It may be noted that the subterranean drainage which is developed along Clifty Creek does not begin until the top of the limestone is entirely below the level of the stream. The water on the lower side comes out about the middle of the formation. Everywhere that the yellow upper edges are present, the solution joints are much larger than in the purer limestone below. So far as Clifty Creek is concerned, the yellow upper portion is much more favorable for the development of subterranean drainage than the lower portion. Certainly if the middle and lower portions were as favorable, there should be considerable subterranean drainage beneath the high narrow ridge in the constriction southeast of the center of section 35, where the water at present strikes directly against the limestone. The yellow upper portion is apparently not present at this place.

The "American Bottoms" basin is one of the very unusual physiographic phenomena of southern Indiana, and perhaps has no parallel anywhere in the Mississippi Valley. It presents a series of special conditions all of which are specially adjusted. The position of the "American Bottoms" as a filled valley in front of the Illinois Glacial Lobe is not uncommon, but the case becomes special when one recalls that the valley was filled by a particular outwash apron from the ice-sheet to the west. Valleys with the same combination of strata are common enough in the Chester series of southern Indiana, and several of them are filled valleys; but no other has the filling at just the critical height with respect to the Beech Creek limestone, to favor the development of subterranean drainage. The fact that the impounded water which was raised to this critical level has a porous barrier which did not permit it to overflow, but rather encouraged it to filter through in sufficient quantity to take care of the inflow, thus allowing time for the development of subterranean passageways with favorable outlets, is another very special factor entering into the problem. For, had this basin been able to overflow at the lower side, it is quite probable that the surface drainage would have persisted, and the "American Bottoms" basin would not have been preserved. Thus, the "American Bottoms" basin is an unusual physiographic feature, because of the exceptional set of conditions critically adjusted to one another. Only by understanding these conditions can we appreciate how and why the "American Bottoms" has been preserved as a youthful area topographically, which must remain for a long time almost exactly as it is today, standing scores of feet above the neighboring valleys.

Some Features of Development in the Wabash Lowland.

The Wabash lowland area has been described in Chapter III as a region of relatively low relief in which occur great areas of aggraded valley land. The topographic condition of more than half of the Wabash lowland as it exists in southwestern Indiana is shown in detail on the topographic sheets of the U. S. Geological Survey. The area south of White River is almost entirely mapped. North of White River the Vincennes and Clay City topographic sheets clearly depict the regional forms present within their limits. Since the topographic sheets are readily available, detailed description of the area is unnecessary.

Few places in the Wabash lowland area reach an altitude as great as 700 feet. Much of the interstream upland surface is below 600 feet. Great areas of valley land exist below 500 feet, and south of White River much of the valley expanses are at or below 400 feet in altitude. It is very probable that the average altitude of the entire Wabash lowland in Indiana is not more than 525 feet.

The development of the Wabash lowland to its present condition calls into play three physiographic factors. Denudation through weathering and erosion has always been a factor. Glaciation directly gave rise to many of the regional forms in the greater part of the Wabash lowland, and indirectly controlled many of the erosion forms. Valley filling or valley aggradation has been important throughout the entire area. Everywhere valleys have been deeply filled, and masses of upland have been buried as deposition in the lowlands encroached upon the valley sides.

Erosion and weathering of the lands above base level are the most important agencies which give rise to land forms. Erosion in the Wabash lowland is now going on in variable degrees. Some areas are rugged and have considerable relief. Erosion here is at a maximum. Such areas are found largely towards the eastern part of the Wabash lowland, and are illustrated in southeastern Daviess, western Jasper, eastern Pike, and parts of Spencer and Warrick counties. A youthful condition characterizes much of the glacial area, but the relief is not great. In such areas the uplands are rarely more than 100 feet above the adjacent flat valley lands. Valleys are frequently sharp, and the streams are making their way into the flat or rolling glacial plains with considerable rapidity. Locally in such areas dissection has progressed to the stage of maturity. Upland plains largely uneroded exist in Knox, Sullivan, western Greene, Clay and Vigo counties. There are large areas of low uplands with long smooth slopes and rounded forms between the broad valley lands. Local relief is small. Such regions are in the old age stage of the erosion cycle. Erosion is greatly slowed down, and few extravagant examples of it are found. Large areas in such an old erosion stage exist, chiefly below 500 feet in altitude, in southwestern Indiana south of White River. Such areas are shown in particular on the Petersburg, Princeton, New Harmony, Uniontown, Newburg, and Owensboro topographic sheets.

Glaciation has tremendously modified the topographic condition of the greater part of the Wabash lowland. Much of the rolling uplands and the flat youthful plains of Knox, Daviess, Sullivan, western Greene, Clay and Vigo counties owe their present condition to glacial construction. Prior to glaciation this region was apparently a dissected, though low, upland without flat areas between the valleys. The deposit of glacial till and outwash built up and re-shaped the low uplands, and gave rise to a glacial plain from 75 to 125 feet above the present well developed valleys. The area south of Brazil and Staunton in Clay County, as shown on the Clay City topographic sheet, is typical. The developing branches of Otter, Honey, and Birch creeks are at work destroying this glacial plain, but much of it still remains as it was constructed by the Illinoian glacier. Similar conditions exist over large areas north of White River. South of White River there are few areas of the original glacial plain of any considerable size because stream action has largely destroyed it. Local remnants of the glacial plain exist. Examples are shown on the Princeton topographic sheet near Hazleton and Owensville. The Mumford Hills with their coating of loess shown on the New Harmony topographic sheet, and portions north of the Patoka River shown on the Petersburg topographic sheet, are other examples.

Valley filling, or valley aggradation, in southwestern Indiana and adjacent regions has given rise to great expanses of valley land at or near the present base level of erosion. The factors involved in aggradation of the valleys have been discussed in Chapter IV, and will not be repeated here. The flat valleys are in great contrast to the erosion forms of the interstream uplands. All valleys above the ravine order have been aggraded or partly filled. Even small streams on entering a major valley have extensive valley floors which join the major valley floor as a wide extension of the valley of the main stream. The upland spurs between the minor streams adjacent to the major valleys are half buried in the accumulated alluvium, and in places are quite buried, allowing the broad valleys of the tributary streams to merge as one valley expanse. This has given rise to bed-rock hills in the midst of valley alluvium. Such are common along the main streams throughout the Wabash lowland. Typical examples of the topographic effects of valley aggradation are shown on the topographic sheets. Above and below the entrance of Sixmile Creek, south of Bowling Green, and again in the region of Old Hill near the mouth of Splunge

Creek, as shown along Eel River on the Clay City topographic sheet, occur typical examples north of White River illustrating the effects of valley aggradation. The extensive lacustrine plains shown on the Owensboro and Tell City sheets are typical of the effects of aggradation along the Ohio River. Apparently the lake silt, bars, and beaches present in this latter area were the result of ponded waters during the Wisconsin glacial stage. As explained by Shaw[168] the valley-train along the Mississippi River valley was built sufficiently high to pond the waters in the tributary streams, and thus give rise to a peculiar system of lakes. Beaches and bars built during this period are quite discernible on the Owensboro topographic sheet south of the village of Lake, at an elevation of 400 feet. A beach composed chiefly of sand is shown at the same elevation near the foot of the bluff on the valley flat some 3 miles west of Mt. Vernon on the Uniontown topographic sheet.

The Ditney[169] and Patoka[170] areas consisting of about 1,875 square miles south of White River and comprising the area covered by eight topographic sheets, have been described in folio form by the U. S. Geological Survey. The classification and description of the topographic forms are presented here in detail.

As described in the Ditney and Patoka folios the topographic forms exhibit "four rather distinct types of topography: (1) Rugged uplands, (2) rolling uplands, (3) upland plains, and (4) river flats. The last two resulted from the accumulation of unconsolidated material in relatively recent geologic time, while the first two, which embrace by far the greater part of the area, have resulted from the action of stream erosion upon the hard rocks. The resistance of these rocks to erosion has been very nearly the same throughout the quadrangle, the resulting relief depending, therefore, upon the relations of the surface to the drainage lines."

In the Ditney quadrangle the topographic type designated as rugged uplands "is best developed in the eastern half of the area especially in the region between Flat Creek on the north and the valley of Pigeon Creek on the south, but is represented in the western half of the area by a number of more or less isolated peaks rising a hundred feet or so above the level of the surrounding regions. The hills are characterized by relatively sharp summits and the ridges by long, even crests sometimes extending for distances of 2 to 7 miles with change of elevation of only 20 to 40 feet, a feature that is more noticeable because of the fact that the ridges, as a rule, are sharp and narrow and are characterized by steep slopes, which are cultivable only with difficulty. The minor channels, which are exceedingly numerous, are usually more or less V-shaped and are separated from one another by equally sharp divides. They exhibit steep descents in their upper courses.

"The elevation to which the higher points of the upland rise is nearly uniform throughout the area of the quadrangle, and appears to indicate that they are but remnants of an old surface, almost a plain in character, which once extended over the whole of the Ditney area. Within the limits of this area the highest portion of the upland level is in the region a little to the east and northeast of the center, near the point from which the drainage diverges, and where a considerable number of crests at elevations of from 600 to 640 feet

[168] Shaw, Newly Discovered Beds of Extinct Lakes in Southern and Western Illinois and Adjacent States: Ill. State Geol. Surv., Bull. 20, 1915.

[169] Fuller and Ashley, Ditney Folio No. 84, U. S. Geol. Surv., 1902.

[170] Fuller and Clapp, Patoka Folio No. 105, U. S. Geol. Surv., 1904.

above sea level. Isolated hills of similar elevations, however, are found at various points throughout the quadrangle. Among these may be mentioned McGregor and other hills about 3 miles west and 1 mile north of Somerville (elevation, 600 feet); Kennedy Knob, 1 mile northwest of Somerville (600 feet); the hill 1½ miles southeast of Somerville (620 feet); Snake Knob and several other hills to the northeast, north and northwest of Lynnville (620 to 640 feet); Big Ditney Hill, 3 miles north and 1½ miles east of Millersburg (660 feet); and Little Ditney Hill, about 3 miles northwest of the same village (600 feet)."

"In addition to the high upland level just described there appear to be other remnants in the shape of long, even crests or of land surfaces at lower level, for there are a number of rather extensive crests or flats shown by rock hills at an elevation of about 500 feet, while ridges and hills of intermediate elevations are common. Though the evidence is not conclusive, it seems probable that subsequent to the formation of the first a second peneplain was developed at an elevation of from 100 to 150 feet below the former. It was probably much less perfectly developed, however, and it seems likely that in this region it was generally confined to the areas bordering the main drainage lines."

The same type of rugged upland is present to a much less degree in the Patoka quadrangle to the west being represented by an occasional peak rising to an altitude of 600 to 640 feet. "The hills on which the princeton standpipe is built rise to 610 feet, those on the Petersburg road, 2 miles north of the same city, to 645 feet, those north of Maxams Station, southeast of Princeton to 625 feet, and that northeast of St. Joseph to 605 feet."

The second topographic type which is designated as the rolling uplands is described as including upland surfaces of less relief and angularity. "The hills are generally much smaller than in the previous group. Their altitude seldom exceeds 550 feet, and they usually exhibit smooth, gently rounded forms. The valleys are broad, relatively shallow, and are characterized by gently curving cross sections, by the low pitch of their streams, and by broad, flat divides. The rolling uplands are best developed in the vicinity of the older drainage lines, especially in the southern and western portions of the quadrangle, the time since the ice invasion being far too short for the development of a rounded topography by erosion in regions bordering streams that were forced into new channels at that time.

"A rolling upland surface appears to exist between White and Patoka rivers, in the northwestern portion of the quadrangle, but it is largely buried by deposits laid down during the ice invasion, and is now represented mainly by low, rounded hills projecting here and there through the deposits mentioned."

In the Patoka quadrangle the rolling uplands occur in the Claypole, Gordon, Mumford, Foots Pond, and other hills projecting above the Wabash flats, "although the flatter portions of their tops belong to the next group to be described. The sand hills along the eastern border of the Wabash flats, the rock hills southeast of Hazleton, around Owensville, and along Big Creek, and in the morainal ridges between Princeton and Fort Branch, southeast of Owensville, and near Poseyville and Cynthiana belong in the main to the rolling uplands, though the steeper portions approach the previous class in ruggedness."

The third type, classed under the term upland plains, "consist of broad, flat, or gently undulating surfaces standing at an elevation of about 500 feet and composed of deposits which accumulated during the period of the ice invasion. These deposits are of two distinct types. Those of the first type, including those forming the broad, flat uplands in the vicinity of Flat Creek, in the northeast portion of the quadrangle, were laid down as stratified clays, sands, and gravels by streams issuing from the ice sheet into a broad lake, known as glacial Lake Patoka, which then existed in this region. The deposits thus laid down constitute in places an almost featureless plain, above which the bordering uplands rise like bluffs or islands from the sea. Deposits of the second type, known as till, are composed of a heterogeneous mass of clay and sand with some pebbles, which was formed beneath the ice-sheet during its occupancy of the region. These are best developed along the south side of the White River flats, in the northwestern portion of the area. The plain extends southward for several miles, but is more or less broken by rock hills which project above its surface and by streams which have eroded deep channels in its mass."

The fourth topographic type, designated as river flats, constitutes the filled or aggraded valleys which are prevalent throughout the Wabash lowland area. "All of the rivers and large streams and also many of the minor streams, flow through broad, flat plains of silt or very fine sand, which are generally overflowed each spring. Wells sunk for water show that the silts are often of considerable thickness, varying from a few feet in the minor valleys up to 100 feet or more in some of the larger ones. The river flats are widest in those streams which still occupy their original valleys, and are narrowest in those which were forced into new channels during the ice invasion. The flats bordering the principal streams vary but little in elevation throughout the quadrangle, being in general between 380 and 400 feet above sea level. Between the elevations of the flats of Pigeon Creek at the southern border of the quadrangle (390 feet), distant 10 miles or less from the Ohio and the elevation of the Patoka flats (400 feet) north of Oakland City and 75 miles or more from the Ohio, there is a difference of only 10 feet. The meanders of the stream are exceedingly pronounced, and by their resistance to the free flow of the water give rise to annual overflows which cover the adjacent flats to depths of several feet. These conditions are very favorable to changes in the course of streams, and bayous and abandoned channels are common."

In the description of the river flats in the Patoka area some further features are mentioned. Wells show a filling in places as much as 150 feet or more. "No deep wells are known in the portion of the Wabash or White River flats lying within the quadrangle, but the thickness of the deposits is probably 200 feet or more. In the process of upbuilding of this considerable thickness of sediments the minor hills have been entirely obliterated, only the higher prominences rising as 'islands' above the flats. The general level of these flats is very uniform, being a little over 400 feet above the sea in the higher portions of the Wabash flats at the northern edge of the quadrangle, and about the same in the White and Patoka river bottoms. There is, however, a gentle slope southward to the 370-foot level at the southeast corner of the quadrangle."

The tops of the highest hills and the ridges of the rugged uplands which reach 600 or 650 feet in altitude in the Ditney and Patoka areas are thought by Fuller and Clapp to represent remnants of a peneplain, the probable age of which is early Tertiary and which may be correlated with the Highland Rim. Gravels of supposed Tertiary age are described as occurring on the hills north of Princeton in the Patoka area. None were found in the Ditney area. The writer has never examined these gravels, but is inclined to interpret them as belonging to the cherty gravels which he has elsewhere tentatively assigned to the late Tertiary. If such an interpretation should prove correct, the flat topped crests and isolated hills which reach an altitude of 600 or 650 feet in the Ditney and Patoka areas would represent the late Tertiary peneplain rather than the earlier or Highland Rim peneplain, as suggested by Fuller and Clapp.

A second peneplain of more local character, but well represented by the "divides and flat crest at an altitude not far from 500 feet" is thought to have been developed in the Ditney and Patoka areas. The 500-foot level is prominent throughout both quadrangle areas. It is interesting to note that the glacial plain is near this same elevation where its surface has been preserved. But in the unglaciated area the prominences of the level topped ridges at and near 500 feet in altitude certainly suggest the existence of an erosive halt and the development of a local peneplain at this level in the soft shales and loose sandstones of the Coal Measures strata. The writer is inclined towards regarding it as such, and to assign it to the Pleistocene period.

In the Ditney area Big and Little Ditney hills are rather striking prominences and stand out as sharp relief features above the rolling plain at their bases. Big Ditney Hill reaches an altitude of 660 feet and the plain at its base is 500 feet. Little Ditney Hill reaches an altitude of 600 feet and the plain at its base is 450 feet. Thus Big Ditney Hill rises 160 feet above its base and Little Ditney Hill 150 feet. Are both of these hills representatives of the old peneplain which is postulated as having once extended throughout the area? The 500-foot plain at the base of Big Ditney Hill resembles a peneplain, but likewise the 450-foot plain at the base of Little Ditney resembles a peneplain. Is the latter spurious and the other real? The geological map shows that the two hills and the two plains at their bases are geologically equivalent. The structure sheet shows that the strata dip 50 feet in passing from Big Ditney Hill to Little Ditney Hill. One may definitely state that the two prominences and the plains at their bases are geological equivalents, and are developed at levels in accordance with the dip of the strata, but one may not speak so surely with reference to their peneplain affinities.

The Illinoian glaciation produced drastic changes in the drainage lines within the areas covered by the Ditney and Patoka folios. These drainage changes took place both within the area covered by the ice sheet and the adjoining driftless area. These have already been touched upon in Chapter IV. In the accompanying Fig. 51, the present and pre-glacial drainage lines are given, and the extent to which the drainage lines were deranged is depicted. This map is taken chiefly from Fig. 5 of the Patoka Folio.

Changes were most profound in the Patoka River system, Pigeon Creek, Black River, and Big Creek. Associated with the drainage derangement were ponded conditions, or lakes of a temporary nature, in which were accumulated clays, silts, and sands. These accumulations gave rise to extensive lacustrine plains. Some of the lacustrine plains are associated with the maximum advance of the ice, as shown on the glacial map, Plate III, of this report. Other extensive lacustrine deposits are within the glaciated area, and mark subsequent halts and resulting lakes of the retreatal stages of the ice-sheet. These are shown in particular on the geologic sheet of the Patoka Folio.

The genesis of the present Patoka River system was outlined in Chapter IV, and will not be repeated here. In the Ditney Folio a portion of the deposits formed in glacial Lake Patoka are shown, having formed an extensive lacustrine plain a few miles east of Petersburg. The old lake plain was developed at an elevation of 480 to 500 feet. It is now partly destroyed by Flat Creek and other streams. The villages of Cato and Otwell are situated upon typical portions of this old lake flat.

"Next to the Patoka River, Pigeon Creek has suffered the most important change of course. The branches of this stream heading in the regions now

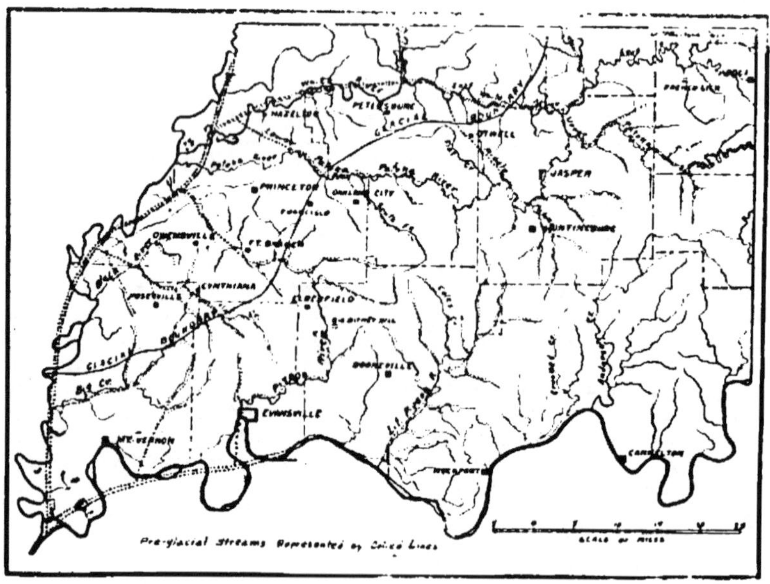

Fig. 51. Map of present and pre-glacial drainage lines in southwestern Indiana. (Taken chiefly from the Patoka Folio, U. S. Geological Survey.)

drained by Snake Run, Sand Creek, Smith Fork, Big Creek, etc., originally united a short distance west of the quadrangle and flowed westward to the Wabash, but during the Illinoian invasion they were deflected by the ice and the drift ridges built up near its margin, and were forced to seek a southward outlet to the Ohio. This outlet was found just east of Elberfeld, the divide over which the waters flowed having an elevation of less than 435 feet, the elevation of the rock rim at the head of the valley of Blue Grass Creek, which otherwise would have served as the outlet. Minor changes in drainage, due to deposition of silt, etc., in the sluggish streams or slack water which formerly existed in the region, have been observed near Tennyson, in the southeastern portion of the area, and northeast of Oakland City. In the former region Coles Creek and Barren Fork turn and flow for several miles parallel with but in the direction opposite that of Little Pigeon Creek, into which they eventually flow. In the latter region South Fork of Patoka River has been deflected from its old channel, which formerly entered the main river in the vicinity of Hurricane Creek north of Oakland City, into the much narrower channel roughly parallel with the first, leading into the Patoka just east of Dongola."

During the stage of maximum ice advance the headwater portion of the present Flat Creek east of Cynthiana in the Patoka area was obstructed, and a lacustrine plain formed. On the retreat of the ice the deposits were continued along this valley to near Cynthiana where a morainic mass accumulated across the valley. The waters were turned to the south near this point into the present Big Creek. The old lake which was formerly present here may be called glacial lake Cynthiana from the village which rests on the morainal dam to the west of it.

At the same stage of the formation of the larger Lake Cynthiana, were several other ice or moraine dammed lakes within the glacial boundary line. A small one was formed just south of Lake Cynthiana along Barr Creek. It may be called Lake St. Wendells from St. Wendells Station, which is located upon it. A small lake was also formed just south of Hazleton, and its site is still marked by a lacustrine plain. The name glacial lake Hazleton is suggested for this former, temporary lake. South of glacial Lake Hazleton a "much larger lakelet accumulated between the ice along the morainal ridge that extended from near Mounts to the highlands southwest of Princeton and the moraine that was formed during the first halt of the ice and extended from the vicinity of Princeton southward to beyond Fort Branch, probably damming the valley south of the latter village. In this lakelet was deposited a considerable thickness of silts, sands, and fine gravels, with an occasional boulder or erratic geode. They reach an altitude of about 430 feet, the probably level of the ponded waters. The outlet was over the morainal barrier south of Fort Branch—which was thereby greatly reduced—and through the valley east of Elberfeld." The name glacial Lake McGary is proposed for this old lake from the village of McGary which is situated on the lake deposits. The name McGary lacustrine plain may well apply to the former floor of this old extinct glacial lake.

On still farther retreat of the ice and an attendant temporary halt, a temporary glacial lake was formed in the lowlands in the vicinity of Poseyville. The lacustrine plain formed here has been largely destroyed by postglacial stream action, in which the wide valley of the upper part of Black Creek occurs and also the valley of Caney Creek which empties into Big Creek. The lacustrine plain itself is but little above the level of the present valleys developed in it. Apparently this lake did not endure a great length of time. The name glacial Lake Poseyville is proposed for this old lake. The village of Poseyville is partly built upon its site.

Many problems of drainage and drainage routes are associated with the aggradation and broadening of valleys in the Wabash lowland area. Not all of these are in the area south of White River. Near Switz City in Greene County, the valley of Lattas Creek on the north coalesces with that of White River on the south through a low opening in which the town is built. Did a stream once pass through this opening, or is it rather due to the joining of the valleys through valley aggradation over a low place in the divide? What relation does the low land to the south of Switz City bear to White River?

Has this low land once been occupied by White River? The "Goose Pond," formerly a lake-like swamp southeast of Linton, Greene County, probably owes its existence to three conditions. First, it is a filled valley; second, White River appears to have migrated miles to the east of its former route, and as a result the tributary stream has been unable to aggrade its valley in harmony with the lengthened stream and has become ponded; and third, White River has been built higher because of valley train material from the Wisconsin glacier, and the tributary streams receiving no such material have become somewhat ponded. Eel River at its entrance into White River presents still another problem in Greene County. This stream enters White River through a narrow opening between the main upland (Point Commerce) and a high bed-rock "island" east of Worthington. Why has it abandoned its broad valley to the west of Worthington? Evidently it formerly entered White River far to the south near the present entrance of Lattas Creek.

The lower course of Eel River is anomalous. The direction is normal, being to the southwest, until near the mouth of Splunge Creek at Old Hill in southwestern Clay County (see Clay City topographic sheet). Here the valley turns rather abruptly to the southeast, and before entering White River at Worthington turns back in its route about 12 miles. In turning back east it again enters the resistant Mansfield sandstone, through which it had just passed. It has been noted that in its normal course to the southwest above Old Hill it is in direct line with Busseron Creek which enters the Wabash River in southwestern Sullivan County. The divide between the two streams is close to Eel River, being near Blackhawk about 4 miles west of the abrupt turn in the river. Splunge Creek comes in at the turn and is very broad, and curiously enough has a valley considerably lower than the valley of Eel River. The divide at Blackhawk is low and broad, and not more than 50 or 60 feet higher than Eel River valley. The natural inference is that for some reason, probably related to glaciation, Eel River has been turned from its former course into its unique southeast course. Investigation shows, however, that it has probably always had its present course. Bed-rock in the low divide in Blackhawk vicinity is at least 40 feet higher than the valley of Eel River. Moreover the broad valley of Eel River is filled with gravels, sands, and silts to a depth of 100 feet or more. The filled condition continues throughout the unique southeasterly route. Its course to the southeast may be due to structural conditions. This awaits investigation.

The route of Raccoon Creek in southwestern Parke County is also unusual. Instead of continuing to the southwest through the broad, open valley coalescing with the great valley of the Wabash, Raccoon Creek turns abruptly to the northwest and passes for about 10 miles through a narrow rock-walled valley and enters the Wabash near Montezuma as a barbed tributary. Undoubtedly it once occupied the broad valley to the southwest, and entered the Wabash River a short distance north of Terre Haute. Why it abandoned the old valley and took its present unusual course is not clear.

General Character and Classification of Problems of Development in Glaciated Indiana.

The development of the Tipton till plain and the Northern Moraine and Lake region of Indiana has been considered in a general way in connection with the description of the topographic conditions. From the standpoint of general development the writer has little to add in this chapter. The simplicity of the glacial plain gives little opportunity for either description or detailed explanation. A till plain is a regional feature that possesses little in the way of *minutiae*. The intimate variety of the moraine offers a great field for minute description, but its development must be considered from the regional and larger point of view, and simplicity again prevails.

Drainage conditions attendant upon glaciation offer a fertile field for investigation involving both description and explanation. Glacio-fluviatile phenomena are largely constructional. Numerous problems exist in middle and northern Indiana which depend upon the determination of the glacio-fluviatile drainage conditions for their interpretation. The relationships of the present drainage are largely dependent upon the conditions imposed by the drainage of glacial waters. Nearly all of the main valleys as they exist today were routes fashioned by glacial waters, and largely formed as the ice retreated northward. Many small streams are misfits in great valleys made by streams of water that disappeared with the ice-sheet. Individual problems may be illustrated in such streams as Cedar Creek, Allen County. It is not clear why this stream abruptly leaves the broad, southwesterly extending valley, which it follows for miles to the southwest, and joins the valley of the St. Joseph River through a veritable gorge. It is not clear whether this gorge was made by this stream or by a glacio-fluviatile stream. In the absence of topographic maps one cannot explain with full confidence the course of Wildcat Creek in Howard and Tipton counties. The drainage pattern of Tippecanoe River is unusual, but a full appreciation of the glacio-fluviatile drainage conditions and the topographic phenomena within and about its basin would probably prove sufficient for a satisfactory explanation. Many of the phenomena of the Wabash River have been presented and explained through the detailed studies of Dryer, yet the absence of topographic maps has prevented a complete presentation of the history and development of this stream.

Another class of problems relating to development in the glaciated portion of Indiana is connected with the underlying bed-rock topography and pre-glacial drainage. Rarely are these features shown in the surface topography in the more deeply drift-covered areas. Yet it is apparent that the larger streams have sought out and partly uncovered some of the pre-glacial drainage lines. Topographic maps in connection with available deep well data would probably prove of great value in the interpretation of the positions of former valleys. Little has been done in Indiana with respect to the determination of the pre-glacial drainage lines, though certainly much deep well data is available.

One of the larger problems associated with the development of the glaciated portion of Indiana is that in connection with the correlation of the moraines. The work of Leverett in this respect probably cannot be improved upon until the publication of detailed topographic maps. Further work along this line must follow the foundations already laid down by Leverett, and consist in detailed refinement of his work. The region north of the Wabash River as a whole offers many problems in this respect. Some of these problems have been discussed in Chapters III and IV of this treatise. The tangle of moraines in Benton, Warren, Fountain, Montgomery and Tippecanoe counties, and again in Wayne and Randolph counties, calls for very precise and detailed work in order that proper correlations may be made.

PART III.

Hydrology of Indiana

By

W. M. Tucker.

17—20642

TABLE OF CONTENTS

LIST OF ILLUSTRATIONS

Introduction

The field work which has been done in the preparation of this paper was carried on during the summers of 1909-10-11, and from September, 1915, to September, 1917. During the summer of 1909 the writer accompanied by J. A. Smith made a survey of the southern part of the state. The object of the investigation was to determine the possibility of water power development. The main streams were traversed by boat from source to mouth. During this summer the following streams were investigated, Whitewater River from Connersville to the state line (no work was done on the East Fork); Blue River from Milltown to the mouth; East Fork of White River from Columbus to the mouth; Muscatatuck River from North Vernon to the mouth; Main White River from the junction of the forks to the mouth; Eel River from Cataract to the mouth. Since this work was accomplished in three months it was not done very thoroughly and was considered only preliminary to a more thorough survey to be carried on in the future.

During the summer of 1910 the author spent one month in establishing gages, most of which were installed in the northern part of the state. At the end of that month gages were in operation at New Trenton on Whitewater, White Cloud on the Blue, Shoals and Tannehill Bridge on the East Fork of White River, Washington, Indianapolis and Anderson on the West Fork of White River, Cataract on the Eel (south), Terre Haute and Logansport on the Wabash, Peoria on the Mississinewa, Logansport on the Eel (north), and South Bend on the St. Joseph.

During the summer of 1911, the writer and J. C. Clark studied the rivers of the northern part of the state in the same manner as has been done in the south. During this time the following streams were investigated, Mississinewa from Marion to the mouth; Sugar Creek from Shades of Death to the mouth; Wabash from Markle to the state line; Eel River from North Manchester to the mouth; Tippecanoe from DeLong to the mouth; Elkhart from Benton to the mouth; Pidgeon River from Howe to the mouth, neglecting that part which is in Michigan; and St. Joseph River from state line to state line.

While engaged in this work only the problem of water power was considered and the amount of work was very great. For this reason few data concerning other important divisions of Hydrology were collected, except as the data taken in the work on water power shed light upon them. The results of the work of these three summers are published in the 35th and 36th Annual Reports of the Indiana Department of Geology and Natural Resources. Much of the material in the present report will be drawn from these sources.

In the summer of 1915, the writer returned to Indiana University to finish the work for the Ph.D. degree, and decided to expand the former work on water power and consider the entire subject of Hydrology as a thesis for this degree. The school work was arranged so that considerable time could be spent in the field. The river gages which had been destroyed in the flood of 1913 were again put into operation. Many discharge measurements were taken and the rating tables revised. After the Commencement of 1916 the writer was given a graduate research fellowship by Indiana University and granted the privilege of working a full year on the thesis before its publication. The year's work was carried on from camps and had to do with the smaller streams of the state, which had not been investigated during previous years.*

During the progress of this work many persons assisted by suggestions, criticism, actual work, defraying expenses and in numerous other ways. Those who have assisted most are Professors E. R. Cumings, J. W. Beede and W. N. Logan, State Geologist, of the Department of Geology of Indiana University, Mr. W. S. Blatchley and Mr. Edward Barrett, former State Geologists, Mr. Robert Reeves, Mr. J. A. Smith, Mr. J. C. Clark, Mr. R. T. Cooke and the gage readers who will be mentioned in connection with the various gages. To these persons the writer wishes to express especial gratitude.

* This manuscript has not been materially revised since its completion in 1917.

Hydrology of Indiana

Hydrology is that branch of Geology which treats of water in all its mechanical relations to the earth. It is very closely related to Physiography in many of its aspects. When rain falls upon the earth, it has three destinations. It either evaporates and becomes water vapor, penetrates the soil and becomes ground water, or runs along the surface and becomes immediate run-off. It either remains upon the surface upon which it falls, or disappears above or beneath it. When the water which penetrates the earth or remains upon its surface is considered merely as water, its discussion belongs to the subject of Hydrology, but if the erosive action of the water is considered, it belongs to the subject of Physiography. Since water is an important factor in the growth of plants and animals and in chemical reactions, Hydrology is more or less closely related to Botany, Zoölogy, and Chemistry. Hydrology treats of the occurrence, movements and uses of water.

Hydrology may be subdivided into several minor subjects, each of which has had many books and articles devoted to it. The principal minor subjects are drainage, irrigation, city water supply, sewage disposal, water power, navigation and flood control. The subtopics under these main divisions are numerous, but many of them will come up in connection with the treatment of the main topics. In order to get clearly before us the topic under consideration, a discussion of these minor subjects will be in order. These discussions will be more or less general with frequent reference to conditions in Indiana, but also going beyond the state for illustrative material.

Drainage.

Drainage is that division of Hydrology which treats of the freeing of the land of its water by movement of the water along or under the surface. The subject of drainage admits of further classification. The following classification will be followed in this paper:

Drainage.

- I. Defined as above.
- II. Classes on basis of manner of development.
  - A. Artificial.
    - 1. Artificial drainage is drainage which is developed by man.
    - 2. Classes on basis of relation to earth's surface.
      - a. Open ditch.
        - (I) Open ditch drainage is artificial drainage which is developed upon the surface of the land.
      - b. Tile.
        - (I) Tile drainage is artificial drainage which is developed beneath the surface of the land through lines of manufactured conduit.
  - B. Natural.
    - 1. Natural drainage is drainage which is developed by the forces of nature.
    - 2. Classes on basis of relation to earth's surface.
      - a. Surface.

(I) Surface drainage is natural drainage which is developed upon the surface of the land.

(II) Classes on basis of manner of escape of water from the land.

(A) Stream.

(1) Stream drainage is natural surface drainage in which the water escapes from the land in continuous depressions (valleys, ravines, gullies).

(B) Sheet.

(1) Sheet drainage is natural surface drainage in which the water escapes from the land in more or less continuous sheets, over perfectly level land or smooth hill slopes.

b. Underground.

(I) Underground drainage is natural drainage which is developed beneath the surface of the land.

(II) Classes on basis of manner in which water moves through the earth.

(A) Ground water.

(1) Ground water drainage is underground drainage in which the water moves through the pores of the rock by the very slow process of seepage.

(B) Cave.

(1) Cave drainage is underground drainage in which the water moves in distinct channels which have been naturally dissolved from the rock and may connect directly with the surface by sinkholes so that the water may never have really permeated the rock at all in reaching the underground channel.

RUNOFF.

All classes of drainage contribute to the runoff of the stream basin. The runoff is the water which passes the mouth or any given point on the stream. It is usually stated in cubic feet per second (or second-feet) which means the number of cubic feet of water which passes a given point in one second. The collection of data to determine the runoff is one of the chief difficulties in the problems of Hydrology.

Means of Determining Runoff. Runoff is determined by several means. The weir method may be used where the water flows over a dam with a perfectly level crest, between vertical abutments. A standard weir is one in which the inner face is a vertical plane and the edge sharp. Weirs are usually made rectangular and will be so considered in this paper. The crest of the weir may extend entirely from one side of the stream, canal or flume to the other, or it may be a notch in a weir plate with edges beveled similar to the crest. In this case the weir is said to have end contraction, while in the former case the contractions are said to be suppressed. If the contractions are not suppressed, the weir plate should extend as a plane on each side of the weir a distance at least three or four times the depth of the water over the crest, and for an equal distance below the crest whether there is contraction or not. It is difficult to measure the depth of water over the crest and instead the height of the surface of the water, before it begins to curve toward the weir, above the crest of the weir is taken. This is done by placing a standard gage above the weir far enough to be in the smooth water and with the zero level with the crest of the weir. The formulae worked out by Francis at Lowell, Mass., in 1854, have been more universally used than any of the many others devised since. The two commonly used formulae are for standard weirs with or without contraction:

With end contractions suppressed:

$$Q = 3.33\ LH^{3/2}.$$

And with two end contractions:

$$Q = 3.33\ (L - 0.2H)H^{3/2}.$$

in which Q = amount of water in second-feet.

L = length of weir crest.

H = height of water on the gage.

This method of determining discharge is very accurate. The length of crest used by Francis in his experiment was ten feet and a head of four to six feet. The formulae are most accurate when used under these conditions.

The gage method is the most satisfactory for large stream calculations. Gages are of several types. The vertical, sloping, and chain and weight gages are most commonly used. The vertical gage consists of a solid piece of wood, upon which are painted or otherwise registered feet and tenths of feet. It is placed in an upright position on the water so that the zero mark will never be exposed above the water. It may be fastened to a bridge pier or other solid anchorage, or it may be painted directly on the bridge pier at very low water, thereby dispensing with the piece of wood. The gage reader at stated intervals, usually once each day, determines the position of the water surface upon the gage. As the stream rises and falls, the amount of fluctuation is shown by the gage reading. The number expressed has nothing to do with the depth of the stream unless the zero of the gage happens to be exactly on the bottom of the stream. The essentials in installing such a gage are, to have the zero mark on the gage below the extreme low water mark, to have the spacings represent vertical distances accurately, and to have the gage securely fastened to its support so that it will not be moved by high water or by other agencies.

The sloping gage is very similar to the vertical gage except that it is placed in a sloping position and the graduations calculated in vertical distances. Sometimes the nature of the stream bank at the point where it is desirable to locate a gage is such that a sloping gage can be installed in a much less exposed position than a vertical gage could be.

The chain and weight gage, where it can be fastened to the hand rail of a bridge or other permanent structure which extends over the stream, is more desirable than the vertical or sloping gages. It consists of a box about six feet long by five inches in width and height. One side of this box is hinged and constitutes the lid. In the bottom of the box is placed a scale five feet long, divided into feet and tenths of feet. One end is marked zero feet and the other five feet with the intermediate feet marked accordingly. At the zero end of the gage a small pulley is placed so that a chain hanging vertically over it will lie horizontally on the scale. A hole one inch in diameter is made under the edge of the pulley on the side opposite the zero of the scale.

The chain is of steel, about the weight of ordinary halter chain, with interlocked flat links. This chain is thoroughly stretched before being used.

Upon it are placed brass markers five feet apart. The markers are made of strips of brass not over one-tenth of an inch in width and long enough to reach entirely around the link. The markers are accurately spaced five feet apart along the chain and soldered securely in place. Upon one end of this chain is placed a weight, which is made from three-quarter-inch iron rod about fourteen inches long. The chain is now ready for use.

The box is securely fastened to the hand rail of a bridge or other firm support which projects over the water. The vertical line from the one-inch hole under the pulley edge to the water must be free from obstruction. The weight is dropped through the hole until its lower end is below the level of the water surface so far that the water surface will not reach it at lowest stage. Then the chain is allowed to run on through the hole until the next marker lies directly over the zero of the gage. The chain is cut off a few inches above this marker and a ring attached to the end. The ring should be more than one inch in diameter so that if the chain becomes free the ring will keep it from running entirely through the hole and dropping into the stream. A hasp and lock are placed upon the lid so that it can be securely fastened when not in use.

The zero of this gage is at the lower end of the weight when the first marker, nearest the ring is on the zero of the scale and the weight is hanging vertically through the hole and over the pulley. If the weight strikes bottom before the first marker lies over the zero of the scale, the zero of the gage is where the lower end of the weight would be if the weight hung free. If the chain is now drawn upward and kept lying horizontally on the scale, until the lower end of the weight just touches the surface of the water, the first marker may lie, say, at four and two-tenths feet on the scale. The four and two-tenths feet are merely used for illustration. In that case the reading at that time would be four and two-tenths feet. If the stream should rise six-tenths of a foot by the next reading the first marker would lie at four feet and eight-tenths on the scale and that would be the reading. If the stream again rises six-tenths of a foot the first marker would lie beyond the five foot mark on the scale and hence could not be used. However, when the first marker passes the five foot mark on the scale the second marker passes the zero mark on the scale, and when the first marker is lying four-tenths of a foot beyond the five foot mark of the scale the second marker would be lying four-tenths of a foot above the zero of the scale. In this case the reading would be taken from the second marker, five feet being added to the direct reading. Thus the reading would be five and four-tenths feet. In like manner when the third marker is used, ten feet is added to the direct reading, etc. After the reading is taken the chain and weight are drawn up and locked in the gage box. In case of long continued use the chain lengthens slightly, due to wearing of the link contacts. The error due to this cause is slight and can be corrected from time to time, by resetting the markers. Care must be taken to keep the initial marker a given distance from the end of the weight.

The Mott type of tape and weight gage which is used to a considerable extent by the U. S. Weather Bureau is very similar to the chain and weight gage. In it a steel tape is used. The scale is one foot long and stands vertically. The tape is wound up on a pulley at the upper end of the scale. The zero end of the scale is the lower end and the zero end of the tape is the inner end when it is wound up. When the weight is lowered until the zero of the tape lies over the zero of the scale the lower end of the weight is at zero of the gage provided that it is hanging free. As the tape is wound up the ascending foot marks on the tape pass the zero of the scale. If the four foot mark on the tape has passed the zero of the scale by two-tenths of a foot when the weight just touches the water, the reading of the gage is four and two-tenths feet. By having hundredths of feet marked on the tape it is easy to read the stage to the hundredth of a foot by this type of gage. There is little to be gained, however, by reading the gage more accurately than to the half tenth.

The city of South Bend, Ind. has a self-recording gage on the St. Joseph River. It consists of a vertical tube set in the river. The tube is perforated so that the water can pass in freely and thus stands within the tube at the same level that it does on the outside. A float is suspended in the tube by a flexible wire which is connected through a system of pulleys and levers to a lever with a recording pen which rests against a circular paper disc. The disc is rotated by clock work and the pen records the rising and falling of the float in the tube. This type of gage is of particular value on the St. Joseph where the control of water by the Oliver dam makes the fluctuations frequent and pronounced. (See Fig. 1.)

When a gage is established, it should be carefully defined and referenced to at least two bench marks. One of these should be near at hand, on an abutment of the bridge or other usually stable position. The other should be located above flood stage of the stream and on a permanent point. With the zero of the gage defined from these points, it is easily possible to replace the gage in case of accident. Even if the bridge is entirely swept away, the gage can be re-established when the bridge is rebuilt.

The reading of the gage must be left to a regularly employed gage reader who visits the gage once a day and records the reading. The self-recording gage records are one week long and must be replaced at some regular time during the week.

Gage readings are of little value unless discharge measurements are made and referred to the gage. After several discharge measurements are made, a rating table can be constructed. After establishing a rating table the gage readings become an index of the discharge as well as of the surface fluctuations.

Manner of Making Discharge Measurements. Discharge measurements are made in a plane at right angles to the stream flow. If referred to a gage the plane should cut through or near the gage. If the gage is on the hand rail of a bridge the discharge measurements can be made from the hand rail of the bridge. An initial point for the work is selected usually flush with one abutment of the bridge. From this point, equal spaces are measured along the hand rail and permanently marked with paint. The spacing varies with the size of the stream. On the largest streams in Indiana twenty feet are used while on the smaller streams ten or even five feet are used. From these points depth readings are taken. At the same time the gage also is carefully read. From the depth readings a cross-section of the stream can be constructed. After the depth readings have been taken, the velocity of the stream is determined at each point. This is done by floats or with a current meter. These will be discussed in a later paragraph. The rate is determined in feet per second at six-tenths the depth. It has been shown that the average rate of flow in general occurs at six-tenths the depth. When these rates

THE DEPARTMENT OF CONSERVATION
DIVISION OF GEOLOGY
STATE OF INDIANA
W. N. LOGAN, STATE GEOLOGIST

DIAGRAM SHOWING EFFECT OF OLIVER DAM UPON STAGE OF ST. JOSEPH RIVER, SOUTH BEND

Fig. 1.

have been determined the discharge is calculated by working out the discharge of each section of the cross-section separately and adding the results. Thus if the depth readings at two consecutive points are 4.6 feet and 5.2 feet, and the distance between points ten feet, the average depth would be 4.9 feet and the cross-sectional area 49 square feet. If the velocities at the two points are 3.7 feet per sec. and 4.1 feet per sec., the average velocity would be 3.9 feet per sec. Thus, this section of the stream with a cross-sectional area of 49 square feet moving forward at an average rate of 3.9 feet per sec. would discharge 3.9 x 49 cubic feet per sec. = 191.1 cubic feet per sec. When the other sections are worked out in similar manner and the results added the discharge of the stream at the existing gage height is given. As long as the stream cross-section is not altered the discharge will always be the same for that gage height provided all other stream conditions remain similar.

A rating table is calculated from the discharge measurements. If at one gage on a stream several discharge measurements are taken during a year at as many gage heights as possible and the gage heights plotted as abscissas and the discharges as ordinates, a curve can be constructed cutting the entire range of gage heights and discharges, whereby the discharge at any gage height can be approximately estimated. From this a table giving the discharges at all gage heights recorded for the year is constructed. By the use of such a rating table the discharge for any length of time may be calculated. In these calculations there are several sources of error. The stream should be straight for at least three hundred feet above and below the point at which the discharge measurements are taken. A bend in the stream causes eddies, reverse currents, and dead water. The stream should be open, free from islands, drift, piers, etc. These obstructions have the same effects as bends. A shifting bed of the stream near the gage makes the discharge at any given gage height irregular. If the gage can be located on a bed-rock stream floor it gives better results. A discharge measurement taken during rising stream at a certain gage height will differ from one taken during a falling of the stream at the same gage height. If the gage is located centrally over the stream the rising discharge will exceed the falling. If the gage is located at the edge of the stream the falling discharge will exceed the rising. This is due to the fact that the surface of a stream is convex during recession and concave during rising. If the gage is located about midway between center and edge of the stream this error is not very important.

The Current Meter. There are many kinds of current meters, all of which are designed to serve the same purpose, i. e., to give the rate of flow of the water. The meter used in this work is a W. and L. E. Gurley meter, pattern No. 617. It consists of a wheel, carrying upon its periphery cone-shaped cups and revolving upon an axis. This wheel is held in a steel frame beneath which a heavy lead weight is fastened to give it stability in the water. The frame is suspended by a cable which also contains a two-wire electric connection with the wheel. At each turn of the wheel a connection is made between the wires. At the other end of the cable is a telephone receiver and a battery. During operations, the receiver is held to the ear of the operator by the ordinary headpiece, and the battery may be carried in the pocket. The number of revolutions of the wheel is counted for one hundred seconds, recorded and then immediately a check of fifty or a hundred seconds is taken. The reading is taken for a hundred seconds for convenience in calculation. The number of revolutions per second can he found by simply moving the decimal point two places to the left. Then by reference to a table accompanying the meter, the rate of flow per second is found. The rating table of the meter is found by drawing the meter through still water by some mechanical device and observing the rate of revolution of the wheel of the meter at various rates of movement through the water. A meter should be rerated occasionally. Care should be taken to keep the bearings of the meter in good condition.

A float is made of a large ball of wood, weighted until it sinks slowly in water, to which is attached by a fine copper wire, a small cigar-shaped float of pine or cork. The small float should be just large enough to carry the weight of the ball. The wire is attached centrally on the side of the float. Opposite to the connection is placed a small pin of white or some other color to suit the observer. To use the floats, two wires are stretched across the stream a known distance apart. The upstream wire is marked into convenient spaces. One observer measures the depth at the marks along this reference wire, adjusts the wire of the float so that the ball will be carried at six-tenths the depth of the stream and starts the float a few feet above the reference wire so that it will have adjusted its movement to the movement of the water by the time it passes under the first wire. At the moment when it passes the upstream wire he signals the second observer, who is at the lower wire with a stop watch. The second observer starts the watch at the signal and by noting the white peg follows the course of the float. When the float passes under the second wire he stops the watch and catches the float. It is convenient to have several floats when this method is used, although a copper wire long enough to reach from reference wire to the other with a float attached and fastened to the upper wire, is convenient for returning the float to the upstream observer. The reference wires are usually placed one hundred feet apart, but in swift streams the distance should be increased. The distance in feet divided by the number of seconds in passing gives the rate in feet per second. This method is impracticable on large streams where the work must be done from boats.

Runoff Records. The runoff records in Indiana are meager and interrupted. The first hydrologic work which was done in this state was carried on by the United States Geological Survey. Gages were established on some of the larger streams as early as 1901-03, and some of these have persisted with interruptions until the present time. The largest amount of available data from any one stream is from Shoals on the East Fork of White River. The interruptions there have been few and short and enough discharge measurements have been made to keep a good up-to-date rating table. The gage is located over a solid rock river bed, so that very little change takes place in the cross-section. The writer established gages on several of the main streams in 1909 and 1910. These were in operation until the flood of March, 1913, which not only swept most of the gages away but in several cases swept away unreported records. At that time no use was being made of the records, reports were not made except on request, and over a year's accumulation of records in some cases had not been reported and were lost. Soon after the flood, the United States Weather Bureau began to install gages on some of the larger streams and these records have been very helpful.

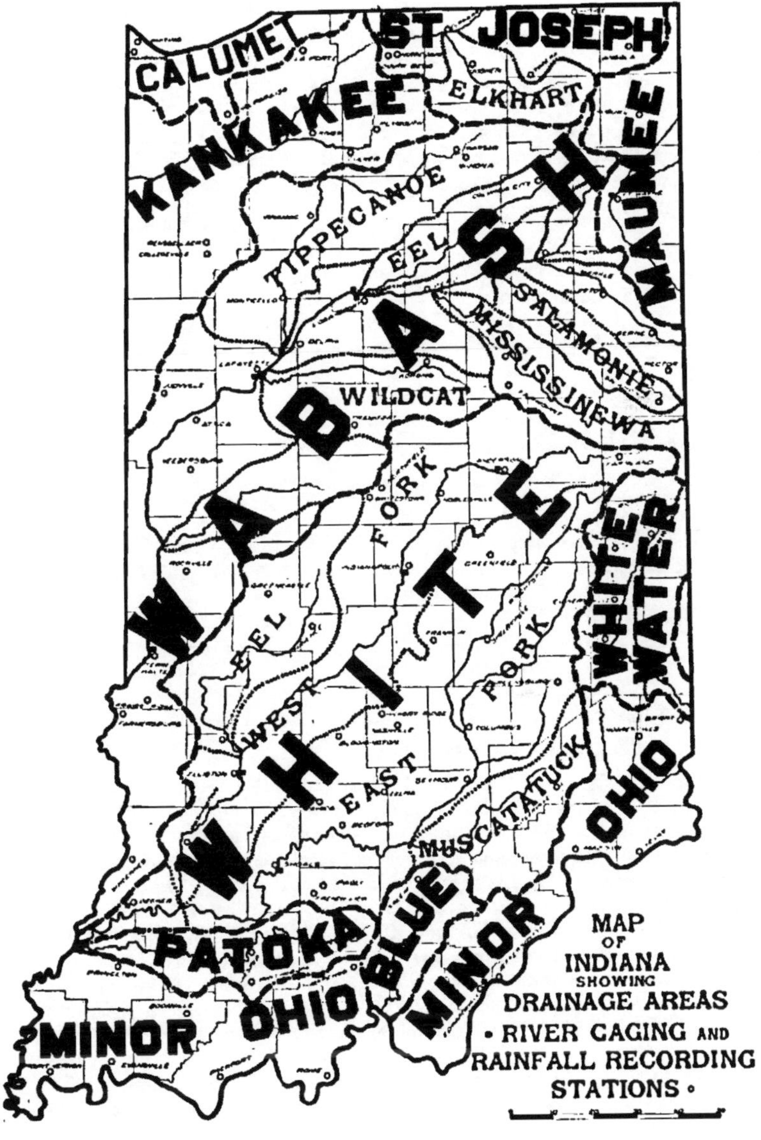
CALUMET
ST. JOSEPH
KANKAKEE
ELKHART
MAUMEE
TIPPECANOE
EEL
WABASH
SALAMONIE
MISSISSINEWA
WILDCAT
WHITE
FORK
WEST
EAST
EEL
WHITE WATER
MUSCATATUCK
OHIO
PATOKA
BLUE
MINOR
OHIO
MINOR
MAP
OF
INDIANA
SHOWING
DRAINAGE AREAS
• RIVER GAGING AND
RAINFALL RECORDING
STATIONS ◦
W. M. TUCKER HYDROLOGIST — W. N. LOGAN STATE GEOLOGIST — J. R. REEVES DRAUGHTSMAN

Fig. 2.

The Natural Drainage of Indiana.

The natural drainage of Indiana falls into the two great systems, the Mississippi and the St. Lawrence. About 89.5% of the state drains to the Mississippi and 10.5% to the St. Lawrence. The entire area of the state including that portion of Lake Michigan which lies within it is 36,550 square miles. The number of square miles draining into the Mississippi is 32,692, and into the St. Lawrence 3,858.

The following table shows the principal sub-systems, their areas in square miles, and the per cent of the area of the state.

Subsystem.	Area Square Miles.	Per Cent of State.
Blue of Washington and Harrison Counties	471	1.29
Minor Ohio Tributaries	4,279	11.71
Whitewater	1,319	3.61
Patoka	813	2.23
White	11,155	30.52
Minor Miami Drainage	73	.20
Wabash	11,481	31.41
Kankakee	3,101	8.48
Total Ohio	32,692	89.45
St. Joseph of Lake Michigan	1,668	4.56
Maumee	1,277	3.49
Calumet and Minor Michigan Tributaries	665	1.82
Surface Lake Michigan	248	.68
Total St. Lawrence	3,858	10.55
Total Indiana	36,550	100.00

Fig. 2 illustrates these river basins.

The drainage of Indiana, in the main, radiates from two hydrographic centers. One of these in Randolph County gives rise to the White, Wabash and Whitewater. The other in Noble County gives rise to the tributaries of St. Joseph, Maumee, and Tippecanoe. Because of the location of these centers the principal drainage is toward the southwest. (See drainage map.) Approximately 64+% of the drainage of the state passes through the mouth of the Wabash, 17% reaches the Ohio by other lines, 8.5% reaches the Mississippi through the Illinois River, while slightly over 6% flows into Lake Michigan, and somewhat less than 4% enters Lake Erie. These are per cents of area. (See Fig. 2.) The rainfall is somewhat heavier in the south and the per cent of runoff would be increased southward, and decreased northward.

The hydrographic center in Randolph County is the highest point in the state (1,285 ft.), and the lowest point is at the mouth of the Wabash (313 ft.). The hydrographic center in Noble County is about 1,000 feet in elevation, Lake Michigan is 581 feet and Lake Erie 573 feet above sea level.

Factors Affecing Drainage. The chief factors which affect the natural drainage of Indiana are the underlying rock, topography, storage, artificial drainage, rainfall and evaporation. These are interdependent. The topography is determined to a great extent by the underlying rock, especially in the southern part of the state. The storage is due to the topography in some parts and to the underlying rock in others. Evaporation is dependent upon the rainfall, topography, mean annual temperature, etc.

The underlying rocks of Indiana belong to the Paleozoic era, of which the representatives of the youngest and oldest periods are not exposed within the state. The following outline shows the systems and series within the state. Systems in parenthesis are not represented in Indiana.

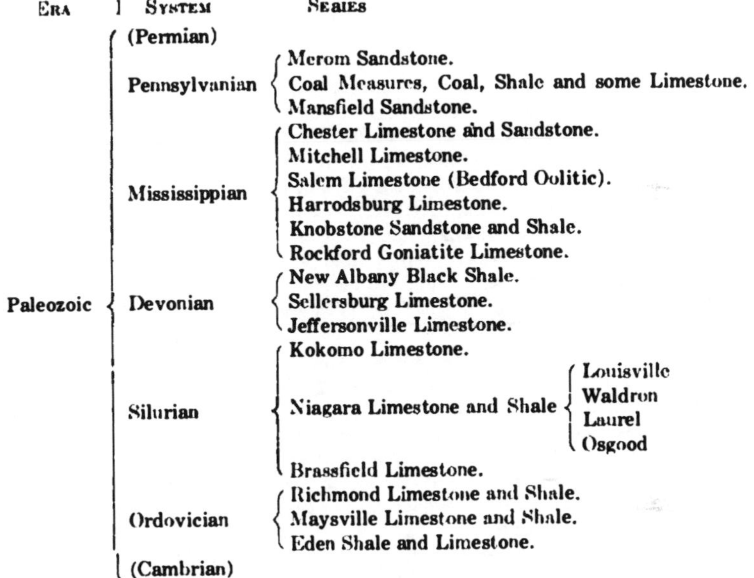

Era	System	Series	
Paleozoic	(Permian)		
	Pennsylvanian	Merom Sandstone.	
		Coal Measures, Coal, Shale and some Limestone.	
		Mansfield Sandstone.	
	Mississippian	Chester Limestone and Sandstone.	
		Mitchell Limestone.	
		Salem Limestone (Bedford Oolitic).	
		Harrodsburg Limestone.	
		Knobstone Sandstone and Shale.	
		Rockford Goniatite Limestone.	
	Devonian	New Albany Black Shale.	
		Sellersburg Limestone.	
		Jeffersonville Limestone.	
	Silurian	Kokomo Limestone.	
		Niagara Limestone and Shale	Louisville
			Waldron
			Laurel
			Osgood
		Brassfield Limestone.	
	Ordovician	Richmond Limestone and Shale.	
		Maysville Limestone and Shale.	
		Eden Shale and Limestone.	
	(Cambrian)		

The entire succession of rocks in Indiana is sedimentary, consisting of limestone, shale, sandstone and coal. In general the strata are horizontal, but there is a noticeable dip toward the southwest which increases in that direction. Thus there is a continual change of formations from east to west across the state. However, each formation may be traced from the Ohio River northward for many miles until it disappears beneath the glacial drift. This arrangement has a peculiar effect upon the drainage of the southeastern part of the state. The Niagara and Brassfield limestones are very hard and form a long high divide almost on a line from Madison to Cambridge City. Whitewater River and some smaller streams skirt the eastern edge of these formations and flow south. West of this divide are the long low grade tributaries of White River. Thus the Whitewater and smaller streams drain the Ordovician formation of the state exclusively. On the other hand the White and Wabash river systems flow directly across the rock formations of the state, and as each formation appears the previous formation disappears beneath it. This has a remarkable influence upon these streams in certain cases. An

18—20642

example of this is the Muscatatuck. For about ten miles below Vernon this stream flows on Jeffersonville limestone. The limestone is hard and forms abrupt bluffs and a rocky bed for the stream. There is no underflow and the stream is of fair size. Near the Euler bridge the limestone disappears beneath the surface and the soft New Albany shale forms the bed of the stream. The valley broadens and is filled with a deep deposit of alluvium. Much of the water disappears as underflow. The diminished stream becomes filled with drift and could scarcely be recognized as the same stream.

The softer formations weather more rapidly and the streams in these formations have broad valleys filled with alluvium. The general level of the country is also greatly reduced by the erosion of these formations. Other formations are harder. In these formations the stream valleys are restricted and the general level of the country is much higher. Thus the state has a series of plateaus extending in a north-south direction across the state and representing the harder formations of rock. There are three of these plateaus which are very distinct. A line from Madison, Jefferson County, to Cambridge City, Wayne County, approximately represents the crest of the plateau formed by the Niagara limestone. A line from Jeffersonville, Clark County, to Danville, Hendricks County, is near the crest of the Knobstone plateau. The other plateau is formed by the Mitchell and Huron limestones and the Mansfield sandstone, and is approximately represented by a line from eastern Perry County to Greencastle, Putnam County. These plateaus are partially or wholly obliterated by the deep glacial deposits in the central and northern parts of the state.

Wherever the underlying rock is limestone the ground water tends to form caves. The texture and composition of the limestone, however, greatly affect the result. The degree of elevation of the country above the level at which the ground water appears at the surface also has a great deal to do with the size of the caves. In the Silurian limestone there are small caves, only a few large enough for exploration. The surface of the country near these caves is pitted with sinkholes which are formed by the ground water causing openings through the rock of sufficient size so that the mantle rock can be carried through into the cave. The continued loss of mantle rock through an opening of this sort soon produces a conical depression above it and this in turn directs more and more water into the sink, which is made larger by the increased flow of water, and so on indefinitely. The water which flows through the sinkholes, directly into the cave, and out at the mouth of the cave can hardly be considered as ground water for it does not penetrate the pores of the earth and carries on erosion by mechanical rather than chemical means.

The caves of the Silurian limestone are small because the strata of limestone are thin and because in much of the region the difference of level of the general surface and the bottom of the stream valleys where the ground water reappears is very small. The latter is not true, however, of the hills of Jefferson County where this difference is about five hundred feet, but even there the caves are small on account of the lack of massive limestone formations. The influence of this drainage is worthy of especial mention because of its effect upon the topography of the country, upon evaporation and upon the water table.

The effect upon the topography has already been mentioned. The sinkholes which are formed are not detrimental to farming, for the slopes are usually gentle and the lack of runoff coupled with the filtering of the water at the bottom of the sinkhole tend to keep the fertility higher than that of similar soil on stream valley slopes. Evaporation is reduced by the underground stream drainage for the stream surface is not exposed to the free moving air currents. When the water reaches the cave it has air over its surface. This air evaporates water from the surface, but since it cannot escape freely from the cave it becomes saturated and retards further evaporation until it is replaced by outside air. The circulation of air in a single mouthed cave is due principally to the breathing of the cave. During the approach of a high pressure area the density of air is increased, and in order to increase the density of the air in the cave more air is forced into it. At such times there is a distinct current of air flowing into the cave. When the pressure passes a maximum and begins to decrease the air of the cave begins to expand and the current at the mouth is reversed. Since the passage of maximum and minimum barometric pressures usually occurs at intervals of three or four days, the air of the cave is partially replaced on the average of once per week and most effectively near the mouth. This process is called cave breathing.

The water table is lowered somewhat by the presence of caves. The amount of this lowering is not definitely known. If we suppose that two regions are alike in all conditions except that one is drained by a cave system whose floor is fifty feet below the general level and the other is drained by a stream system whose stream beds are fifty feet below the general level we have a basis for comparison. It might appear that the water table should be very nearly the same in the two cases and that the only differences would be such as occur in the land forms of the two areas. Since, however, the surface which is drained by streams is depressed in the valleys it is there brought near the water table, while in the cave drained region the land surface bridges the streams and hence lies far above the water table. The cave drained area would therefore be drier than the stream drained area. This would cause the cave drained area to have an advantage in wet seasons, and the stream drained area in dry seasons.

The cave drainage of the state, which occurs wherever the underlying stone is limestone, is best developed in the massive Mitchell limestone. A belt of land seven or eight miles wide, stretching from Greencastle to Corydon and extending on into Kentucky is honey-combed with caves, which are small in the north but which gradually increase in size until the maximum in Indiana is reached in Wyandotte Cave in Crawford County. The increase in size is due to the increase in the depth of the stream valleys. In Putnam County the stream bluffs are much less than 100 feet high in general, while in Crawford County they are 300 feet or more. The caves in the north are single storied wet caves, while Wyandotte is three or more storied with only the lower story wet. The famous Lost River of Orange County has abandoned its surface channel for seventeen miles and flows through an underground passage. The airline distance from the entrance to the exit of the water is about eight miles. The combined length of passages in Wyandotte Cave is stated as twenty-three and one-half miles, although the guide adds that these are cave miles. Just what is meant by a cave mile is something never yet determined, but it is somewhat shorter than the U. S. Standard mile. However, while the real length of the explored passages in this great cave is not accurately known, it is great enough so that the guides are never questioned when they say "twenty-three and one-half miles," except by the unfeeling scientist.

Much of Indiana has been glaciated. Two distinct periods of glaciation are usually recognized, known as the Illinois and Wisconsin glaciations.

The Illinois glacier was the older of these and reached the more southerly limit. Much of the deposit of this glacier has been carried away by the streams. In many places the streams have cut through the drift which was deposited and have their beds in the solid rock beneath. While this glacier in its time modified and probably entirely changed the drainage of the area which it covered, at the present time the remnants of the once extensive drift are scarcely to be considered as a factor which greatly affects the drainage of the state.

The Wisconsin glacier which is more recent, also obliterated to a great extent the previous drainage of the part of the state which it covered. During and since the disappearance of this glacier, a new drainage has developed which has not yet carved its way through the heavy drift to bed-rock. This condition has a marked effect upon the streams. The drift acts as a great storage basin for ground water. The continual appearance of this ground water causes the stream flow to be more uniform and permanent. The presence of many lakes in the Wisconsin glaciated area also tends to regulate the flow of the streams. Many streams flow south off the edge of the Wisconsin glacier and have long valley-trains. A valley-train is a deposit of glacial debris which is found beyond the glacial margin in the valley of the stream which flowed from the edge of the glacier. The deposit was made during the presence of the glacier by the stream which was then overladen with sediment. A valley-train is usually composed of sand and gravel. These valley-trains and the sand and gravel deposits in the glacial area retain a great deal of ground water with which the streams are replenished during dry weather. Glacial till is not as useful in retaining ground water as the sand and gravel. It is composed largely of clay and rock flour which is somewhat impervious to water.

The topography of Indiana bears a close relation to its glacial history. It may be divided roughly into three divisions, to which reference has already been made, i. e., the Wisconsin glacial area, the Illinois glacial area, and the nonglaciated region. The Wisconsin glacial area forms the major part of the state. It is a topographically young region with an undulating surface due to glacial forms. The soil is deep and composed largely of glacial till, sand and gravel. Little rock is exposed. Occasionally the streams have cut through the drift and exposed the underlying rock.

The Illinois glacial area is much older than the Wisconsin and the streams have cut through the drift to the bed-rock beneath. The larger streams have practically reached base level and have begun to widen their valleys. The soil is not so deep as that of the Wisconsin area. It contains little sand and gravel deposit except in the valley-trains from the Wisconsin area.

The nonglaciated area is a typical mature region. Little level land occurs and the drainage is perfect. The streams of this region are flooded during rainy seasons, and dry or very much diminished during dry seasons. The Mitchell limestone belt which extends from Mauckport, Harrison County, to Waveland, Montgomery County, is an exception to the foregoing statement. In this belt the drainage is to a great extent subterranean on account of the extensive development of caves. This condition causes the runoff of this belt to be much more uniform than the runoff of the rest of the nonglaciated region. Little Blue River is wholly in this belt. The East Fork of White

River crosses the nonglaciated region from Seymour, Jackson County, to the west line of Martin County.

The regularity of stream flow depends upon the regularity of the rainfall and upon storage facilities. The natural storage facilities are glaciers, lakes, swamps, deposits of sand and gravel, vegetation, and any mantle rock especially when it is continually cultivated at the surface. All of these are found in Indiana except glaciers. The Rhine, Rhone and other large rivers of Europe are fed by the melting ice of the glaciers of the Alps throughout the drier months of the year, and even if the drought becomes extended in the lowlands these rivers continued to flow without notable diminution. The Mackenzie and Yukon rivers of North America are fed by many glaciers. The glaciers of the United States are small and the largest are near the sea so that they have little effect upon the main streams. The water stored in glaciers and snowfields is estimated to be sufficient to raise the level of the ocean by fifteen feet if it were liberated.

If the ordinary observer were asked what the greatest storage reservoir for land water is, he would probably say the lakes. In lakes and swamps the water as water is visible, and therefore appeals to the ordinary observer. However, the total lake storage of the world is about one-seventh of the total land ice storage.

Indiana has a great many small lakes in the northern part. The storage facilities of these lakes is estimated by the writer to be approximately 6,000,000,000 cubic feet, and is sufficient to make a marked difference between the regularity of flow of the rivers which rise among them and those which do not. English Lake was originally the largest lake in the state and covered about twelve square miles. It was a shallow channel lake along the Kankakee River and has been entirely drained by the dredging of the stream channel. Lake Wawasee, in Kosciusko County, formerly known as Turkey Lake, is now the largest lake in the state. It has an area of 5.5 square miles. It drains into the St. Joseph River through Turkey Creek and the Elkhart River. Lake Maxinkuckee of Marshall County is second in size with an area of 2.5 square miles. It is tributary to the Tippecanoe River. Lake Winona, formerly known as Eagle Lake, while small in size is one of the best known of our lakes. It is a well known summer resort and the site of several educational and religious institutions.

The lakes of Indiana are located principally in two groups or belts within the morainic deposits of the Erie, Saginaw, and Michigan lobes of the Wisconsin glacier. The joint terminal moraine of the Erie and Saginaw lobes extends in a northeast-southwest direction from the northern part of Miami and Wabash counties through Kosciusko, Noble, Steuben and parts of Lagrange, Dekalb and Whitley counties. This area contains the majority of the lakes. The joint terminal moraine of the Michigan and Saginaw lobes extends in a north-south direction from northern Fulton and Carroll counties through Stark, Marshall, St. Joseph and Laporte counties. This area has a great number of lakes. There are few lakes outside of these two areas in Indiana. Along the White, Wabash and Ohio rivers in their lower courses are to be found a few horseshoe lakes and bayous.

When the Wisconsin glacier finally retreated from Indiana, it left a very irregular surface with few stream channels. At that time the lakes of the state were much more numerous and larger than now. Since that time stream erosion has developed new valleys and deepened the ones then existing, thereby draining many of the lakes partially or wholly. Sedimentation on the other hand has been busy filling the lakes and thus destroying them. Wherever the water is reduced to a depth of five feet or less vegetation lays claim to the bed as a habitat and thereafter the refuse matter of the vegetation is added to the deposits of sediment. When a declining lake reaches the stage in which it is largely or entirely filled with growing vegetation it is known as a swamp. Swamps may originate directly without being formed from lakes, but most of the swamps of northern Indiana are the remnants of former lakes. The swamp lands were very extensive when the first white settlers appeared. Not only in the northern part of the state but along the principal rivers of the southern part as well was to be found much swamp land. The land which now constitutes the fine bottom farms of the Wabash and the White was a malaria infected swamp region one hundred years ago. Artificial drainage has destroyed much of the swamp land and changed it into the finest farm land in the world. However, the remaining swamp land of the state greatly exceeds the lake area and because of the large extent of the swamp land the storage within it is correspondingly large.

Another great reservoir of water storage is the great beds of glacial sand and gravel which are found in many parts of the state. In the headwaters of the East Fork of White River, in Shelby and Rush Counties, are unlimited deposits of this material. The streams in this section are more regular in their flow than is the Muscatatuck fork, which rises on the clay soils of Jennings and Ripley Counties. Both of the lake regions are composed largely of glacial gravels. Wherever these deposits are found they are filled with water, which is not only good for all human uses but also feeds the springs and streams and gives them permanency, when streams of equal or greater drainage area in regions with a clay subsoil are left dry half the year.

When the settlement of Indiana began about eighty-five per cent of its area was wooded. This great forest of twenty-seven thousand square miles was principally deciduous. A few evergreen trees flourished on the hilly southland but the giants among the trees were the oak, maple, tulip, sycamore, hickory, ash, elm, linden, walnut, beech and sassafras. The leaf mould from these trees, often with a thickness to be measured in feet, lying in the cool, somber protection of the trees themselves, formed the greatest water storage which this state ever had. This is particularly true of the southern part of the state, in which there is practically no storage at present. The settler destroyed much of this forest simply to clear the land for the plow and he was followed in more recent years by the timber dealer who has further reduced the timberland by part use, part waste to about fifteen per cent of its original extent. This fifteen per cent is represented by farmers' woodlots. There is practically no virgin forest in Indiana where fifty years ago stood twenty thousand square miles of splendid timber. Most of the middle aged people who have spent part of their lives along a stream of water have noticed a great change in the character of the stream within their memory. A stream which once turned a water wheel continually would not now be available for power at all. It would be dry for half the year and flooded part of the other half. Careless observers sometimes attribute this change to change in weather conditions but more careful observers have noticed the disappearance of the wet mosses and leaf moulds in the woodlands, along with the disappearance of the woodlands themselves. Artificial drainage has also contributed to this condition. Indiana can never be restored to its former forested condition as long as civilized races inhabit it, but when the present race becomes more highly civilized much of the waste land which is now of no value whatever will be reset with forest, thereby preserving the land as well as making it produce a fast disappearing commodity. The reforestation of waste land is a subject which should be thoroughly investigated, and some legislative action promoted to require some responsibility in this regard of the land owners.

Since the forests have been destroyed and their water storage obliterated the only storage facility in parts of the state which have no lakes or swamps is the soil, and underlying rock. When the water is stored in the soil or underlying rock, it constantly escapes by two means. It gradually creeps or seeps out toward the lower points of the land surface. Thus from the hilltops there is a slow movement of the ground water toward the valleys. The rate of this movement depends upon the steepness of the slope and the character of the material through which it is moving. It moves faster on steep slopes than on those less steep, and in material composed of large particles and large interspaces than in one of smaller grains and smaller interspaces. If a well is sunk near a stream in alluvial soil the water will stand in the well at about the same level as in the river. The layman thinks that the water in such a well comes from the river but in fact it comes from the other direction entirely. The ground water from the hills is moving toward the river and passes through the well en route. If the well is continually pumped to its limit it forms a depression in the water table at that point and the ground water will flow toward this depression from all directions. If the pumping were great enough to lower the water table as far as the river's edge then some of the water from the river would tend to flow toward the well. This would probably happen in dry weather when the water table had been greatly lowered.

The other means by which the ground water escapes from the soil is by capillarity and evaporation. It has been shown that the soil is filled with capillary tubes which conduct the water from the water table to the surface where the air evaporates it. The nearer the water table is to the surface, the faster the water will be conducted upward by capillarity. The loss of water from the soil by this means is very great and in dry seasons this loss marks the difference between financial gain or loss on the crop. Dry farming methods are based on the idea of preventing as much as possible this loss. The main idea in dry farming is to keep the surface of the ground tilled constantly. The tilling of the first four inches of the surface breaks the capillary system and prevents the water from being conducted entirely to the surface. Instead it is conducted to a point four inches beneath the surface. Whatever evaporation then goes on must be carried on by the air which penetrates the layer of loosened earth and not by the free wind at the surface. The penetration of air to this depth is easily accomplished but is a very slow process when compared with the free movement of air at the surface. Thus the water accumulates beneath the cultivated layer much faster than the air can remove it and so finds itself at the very roots of the small grain plants. Thus cultivation is one means of storing the ground water. In semi-arid regions such as Southern California it is almost criminal for a farmer to allow his soil to crust over, for he is thereby not only lowering the water table under his own land but tending to draw down the water table on all his neighbor's land. In Indiana, however, the idea of keeping as much of the land cultivated as possible has not become current and in dry seasons the hard, uncultivated fields act as veritable aqueducts which conduct the precious ground water to the thirsty atmosphere where it is liberated, while the crops wither.

An interesting illustration of the change in condition of the streams of Indiana is well shown by a map of Carroll County which was drawn and published by a firm known as Skinner and Bennett at Delphi, Indiana, in 1863. This old map was found by Mr. F. J. Breeze among a store of records

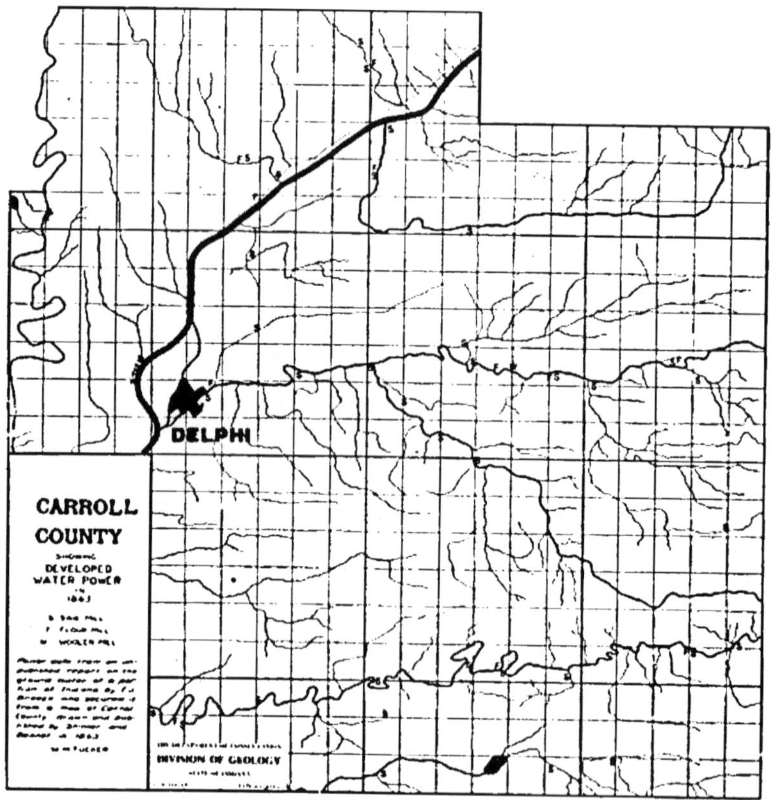

Fig. 3.

in the court house at Delphi and has been used by him in an unpublished report on the Ground Water of a Portion of Indiana. The reproduction of this map (Fig. 3) shows fifty-three points where water power was used at that time. At the present time there are two mills operated by water power in Carroll County. These two mills are located where the flour mills in Sec. 33, R. 1 E., T. 24 N., and Sec. 34, R. 1 W., T. 24 N., were then located. It is true that the sawmills have disappeared because the timber of this section has all disappeared, but it is also true that the power which turned the mills has disappeared as a result of the disappearance of the forests and of artificial drainage. Thus the water power of Carroll County was used for its own self-destruction. While this condition is well shown in the case of Carroll County by this old map, it is no less true of all that part of Indiana which was originally covered with forest.

Rainfall.

The rainfall of Indiana is nonseasonal and very irregular. It varies from year to year and the monthly and geographical distribution also vary greatly. The U. S. Weather Bureau maintains observation stations at seventy-one points in the state where the amount of rainfall is taken daily. Map, Fig. 2 shows the location of the stations where observations have been made during all or a part of the time since 1898. The following table gives the points where stations are now in operation and the mean monthly and mean annual rainfall for each point since the station was started, provided the station has been in operation for at least five years:

Table of points where the U. S. Weather Bureau maintains rain gaging stations in Indiana, and the mean annual and mean monthly rainfall for each station, provided the station has been in operation five years or longer. The table is given in inches of rainfall. (*) denotes a record of less than five years. The means of the northern, central and southern divisions are given. The mean rainfall at a point for five years or longer is considered the normal rainfall for that point.

Stations Northern Division	Jan.	Feb.	Mar.	Apr.	May	June	July	Aug.	Sept.	Oct.	Nov.	Dec.	Annual
Albion	*	....	....	....	....	....	....	....	....	....	....	....	....
Auburn	2 28	2 32	2 54	2 53	3 71	3.62	3.63	3.22	2.17	2.30	2.37	2.13	33.30
Berne	*	....	....	....	....	....	....	....	....	....	....	....	....
Bluffton	2 41	2 28	3.70	3 08	3 79	4 10	3.54	3 20	2 66	2.10	3 02	2.56	36.44
Collegeville	2 39	2.49	3 31	3 50	4 28	3.65	4 33	4.02	3.02	1.97	2.50	2 38	37.83
Crown Point	*	....	....	....	....	....	....	....	....	....	....	....	....
Delphi	2 63	2 44	3.03	3 28	4 63	4 32	3.79	3 07	2.98	....	....	....	....
Ft. Wayne	2 37	2.80	3.60	2 94	3.68	3 84	3 66	3 29	3.21	2 42	2 94	2.45	37.20
Goshen	*	....	....	....	....	....	....	....	....	....	....	....	....
Hammond	2.24	2 29	2.54	2.83	4 04	3 10	2.77	2 25	2.84	1.81	2.31	2.20	31.22
Howe	2.13	2.00	2.13	3 23	4 52	3 47	3.78	3.28	2 79	2.41	2.49	2.20	34.43
Huntington	2 66	2.50	3.39	2 88	4 00	3 99	3 64	3.43	3.05	2.48	3.24	2.79	38 05
Knox	2.98	2.26	2.85	3.80	5.33	3.22	3.88	4.65	3 22	2.45	2 89	2.21	37 90
Kokomo	2 62	2.08	3.09	3.22	3.94	3.25	3.45	3.40	2 71	2.18	2.96	2 36	35.86
Lafayette	2.47	2 79	3.09	3.37	4 41	4.47	3.79	3 32	2.87	2 37	3 06	2.58	38.59
Laporte	2.25	3 81	3.13	2 61	4.11	3 30	3.40	3 22	3 17	2.36	2.78	2.48	35.62
Logansport	2.15	2.67	2.72	3 27	4.24	4.04	3.29	2.83	3.07	2.59	3.15	4.22	36.60
Marion	2.55	2.48	2.99	3.31	3.92	4.08	3.05	3.06	3 09	2.09	3.13	3.14	46.23
Monticello	*	....	....	....	....	....	....	....	....	....	....	....	....
Notre Dame	*	....	....	....	....	....	....	....	....	....	....	....	....
Plymouth	2	1.92	2.75	3 18	4 27	3.06	3 38	3.59	3.27	2.30	2.77	2.22	34.89
Rochester	2.	2 44	3.02	3 41	4.26	3 76	3.70	3.70	3.66	2.38	3.21	....	....
South Bend	2.[illegible]	2 33	2 97	2 68	3 80	3 18	3 47	3 26	3.03	2 42	2.82	2.98	35.67
Valparaiso	1 [illegible]	1.77	2 34	2 38	4 30	3 68	3 26	3.03	2.84	2.00	2.03	1.98	31.43
Wheatfield	*	....	....	....	....	....	....	....	....	....	....	....	....
Whiting	*	....	....	....	....	....	....	....	....	....	....	....	....
Winona Lake	*	....	....	....	....	....	....	....	....	....	....	....	....
Means	2 41	2 39	3 03	3.11	4 14	3 75	3.41	3 54	3.21	2.29	2 67	2 48	36.02

RAIN-GAGING STATIONS—Continued

Central Division	Jan.	Feb.	Mar.	Apr.	May	June	July	Aug.	Sept.	Oct.	Nov.	Dec.	Annual
Anderson	2.62	2.20	3.79	3.23	4.14	3.79	3.55	3.89	3.04	2.53	3.00	2.68	38.46
Bloomington	3.72	3.65	5.45	3.23	4.12	3.78	3.86	3.62	2.69	3.53	3.31	3.79	44.75
Cambridge City	2.89	2.76	4.05	3.38	4.07	4.31	3.60	2.92	3.36	2.60	2.90	2.86	39.70
Connersville	3.06	3.08	3.77	3.22	4.20	4.16	3.32	2.90	2.82	2.61	3.34	2.78	39.26
Crawfordsville	2.97	1.85	5.18	....	....	....	....	....	....	....	....	....	....
Farmersburg	1.86	2.69	3.75	2.79	4.69	4.04	4.25	2.85	2.50	2.96	2.24	2.86	37.38
Farmland	2.73	2.90	3.35	3.30	4.30	4.09	3.41	3.66	3.47	2.37	3.09	2.67	39.34
Frankfort	•	....	....	....	....	....	....	....	....	....	....	....	....
Greencastle	•	....	....	....	....	....	....	....	....	....	....	....	....
Greenfield	3.18	2.59	4.64	3.16	4.06	3.29	3.94	3.25	3.16	3.12	2.69	2.26	39.34
Hickory Hill	•	....	....	....	....	....	....	....	....	....	....	....	....
Indianapolis	2.81	3.20	4.01	3.47	3.94	4.31	4.13	3.33	3.05	2.79	3.52	3.04	41.60
Judyville	•	....	....	....	....	....	....	....	....	....	....	....	....
Mauzy	3.25	3.57	3.84	3.19	4.43	4.29	3.24	2.98	2.89	2.87	3.51	3.10	41.16
Muncie	•	....	....	....	....	....	....	....	....	....	....	....	....
Nashville	•	....	....	....	....	....	....	....	....	....	....	....	....
Noblesville	•	....	....	....	....	....	....	....	....	....	....	....	....
Richmond	2.81	2.63	3.63	2.93	4.01	4.12	3.67	3.74	2.70	2.77	3.09	2.67	38.82
Rockville	2.46	2.49	3.41	3.49	4.21	4.20	3.73	2.88	3.01	2.42	3.44	2.50	38.24
Salamonia	4.22	2.37	3.80	3.33	3.21	3.40	4.55	3.67	3.16	2.50	2.25	2.28	38.74
Shelbyville	3.46	2.59	3.86	3.98	4.15	3.71	1.40	3.84	3.23	3.59	2.65	2.69	42.45
Terre Haute	2.65	2.59	3.49	3.68	4.11	4.16	3.36	3.55	2.78	2.26	3.21	2.63	37.91
Veedersburg	2.61	2.22	3.20	2.89	3.85	4.06	3.54	3.18	2.76	2.69	2.32	2.04	35.36
Whitestown	•	....	....	....	....	....	....	....	....	....	....	....	....
Means	2.82	2.76	3.74	3.01	4.20	4.29	3.83	3.02	2.85	2.69	2.97	2.56	38.69

Southern Division	Jan.	Feb.	Mar.	Apr.	May	June	July	Aug.	Sept.	Oct.	Nov.	Dec.	Annual
Butlerville	4.02	3.62	4.77	3.55	4.61	4.40	3.58	3.51	2.74	2.64	3.70	3.15	44.27
Columbus	3.16	3.15	3.56	3.04	3.35	3.45	3.07	3.30	3.02	2.67	3.28	2.81	37.96
Dam No. 39	•	....	....	....	....	....	....	....	....	....	....	....	....
Decker	•	....	....	....	....	....	....	....	....	....	....	....	....
Elliston	•	....	....	....	....	....	....	....	....	....	....	....	....
Evansville	3.69	3.06	4.60	3.46	3.43	4.17	3.81	3.24	2.66	3.10	4.11	3.83	43.18
Forest Reserve	•	....	....	....	....	....	....	....	....	....	....	....	....
Greensburg	3.18	2.22	4.66	3.23	....	3.62	4.36	2.23	3.01	3.16	3.03	2.93	....
Huntingburg	•	....	....	....	....	....	....	....	....	....	....	....	....
Jeffersonville	3.77	3.90	4.32	3.84	3.96	3.72	3.94	3.28	2.56	2.39	3.91	3.40	42.99
Madison	3.79	3.13	4.81	3.23	4.46	3.99	3.59	2.73	2.57	2.52	3.19	3.35	41.36
Marengo	....	....	....	5.04	5.09	4.93	4.27	3.95	4.01	3.15	4.70	4.22	....
Moores Hill	3.89	3.22	4.79	4.04	4.23	3.02	2.99	3.51	3.02	3.11	2.32	2.83	40.97
Mt. Vernon	3.46	3.09	4.58	4.05	3.73	3.73	4.31	3.11	2.75	3.07	3.93	3.23	43.08
Paoli	3.70	3.21	5.07	4.20	3.93	4.04	3.12	3.89	3.78	3.34	3.03	3.37	44.12
Princeton	3.34	3.44	4.20	3.31	3.63	4.05	3.40	3.17	3.19	2.45	3.51	3.12	40.81
Rome	4.48	4.15	4.78	4.83	3.45	3.56	4.37	3.97	3.38	3.09	3.09	3.66	46.81
Salem	3.63	3.68	4.63	3.32	3.75	3.94	3.37	2.99	2.63	2.95	3.34	3.49	41.72
Scottsburg	3.56	3.13	4.68	2.96	3.91	4.23	3.75	2.79	2.45	2.55	3.05	3.32	40.38
Seymour	3.57	3.63	4.36	3.43	4.17	3.98	3.64	3.43	2.90	2.98	3.48	3.09	42.66
Shoals	•	....	....	....	....	....	....	....	....	....	....	....	....
Vevay	3.98	4.18	4.26	3.30	4.64	4.66	4.11	3.01	2.95	2.30	3.37	3.19	43.84
Vincennes	3.22	3.01	4.95	2.88	3.52	4.72	4.35	3.02	3.06	2.26	3.52	3.26	43.77
Washington	3.41	3.06	4.02	3.53	3.42	4.45	3.92	3.89	3.26	3.04	3.14	3.19	43.33
Worthington	3.01	3.61	4.03	3.52	4.06	4.39	4.04	3.31	2.96	2.77	3.70	3.20	42.60
Means	3.97	3.39	4.66	3.53	3.81	4.16	3.73	3.44	2.82	2.85	3.41	3.32	43.17
State Means	3.05	2.83	3.79	3.23	4.05	4.31	3.63	3.34	2.97	2.63	3.05	2.80	39.21

It will be noted that the foregoing table divides the state into three divisions and that a mean for each division is given. This shows the variation in the average amount of rainfall from north to south. It is found to vary from 36.30 inches in the north to 43.12 inches in the south. The following table shows the mean annual and mean monthly rainfall for the years 1898 to 1915 inclusive. This is found by averaging all the records from observation points for this time. This shows May to be the wettest month, with March as a close second. February is found to be the month of least rainfall, with November slightly greater.

Table showing mean monthly and mean annual rainfall in Indiana for the years 1898 to 1916, inclusive. This table is given in inches of rainfall.

YEAR	Jan.	Feb.	Mar.	Apr.	May	June	July	Aug.	Sept.	Oct.	Nov.	Dec.	Mean An.
1898	5.27	1.87	8.11	2.06	4.49	3.81	3.02	3.36	4.06	4.56	2.87	2.23	45.71
1899	3.23	2.27	4.46	1.60	3.96	2.86	3.28	3.03	1.75	2.91	2.54	3.16	35.05
1900	1.71	3.77	2.06	1.64	4.96	5.54	4.66	3.41	2.06	2.56	4.26	1.20	37.83
1901	1.44	1.68	3.40	2.67	2.54	4.35	1.30	3.10	1.54	3.35	1.30	3.92	30.57
1902	1.41	1.00	3.12	2.05	4.32	7.45	3.38	2.26	4.76	2.58	3.68	4.07	40.08
1903	2.28	4.40	2.95	4.43	3.16	3.72	3.51	3.91	1.85	2.67	1.82	2.16	36.96
1904	4.18	2.54	8.10	3.32	3.33	3.04	2.95	2.48	3.44	1.06	0.36	3.48	38.64
1905	2.16	2.05	2.52	3.74	5.96	3.61	4.50	5.03	3.48	4.89	2.68	2.43	43.70
1906	3.09	1.33	5.16	2.13	2.30	3.44	3.18	4.67	4.07	1.95	4.09	4.20	39.82
1907	6.95	0.48	4.90	2.80	3.71	4.69	4.95	3.83	2.90	2.73	2.79	4.09	44.98
1908	1.63	5.79	4.40	4.40	6.28	2.00	2.94	1.93	0.97	0.34	2.03	1.59	34.70
1909	3.67	5.82	2.88	5.16	4.71	5.16	5.26	3.00	2.66	3.70	3.21	3.09	47.75
1910	2.92	3.07	0.22	3.43	4.07	2.02	5.77	2.14	4.53	5.48	1.87	2.03	37.55
1911	2.91	1.82	2.22	5.43	2.34	3.82	1.71	3.19	5.97	4.07	3.44	2.93	40.06
1912	2.27	2.59	4.46	4.99	4.45	2.81	5.11	5.34	3.33	2.13	1.83	1.41	40.72
1913	7.74	1.56	8.97	3.36	2.79	2.10	3.26	3.10	2.85	3.42	3.92	1.13	44.20
1914	2.28	3.07	2.58	3.16	2.64	2.38	1.85	4.73	2.34	2.37	1.09	3.05	31.54
1915	3.39	1.58	1.15	1.58	5.93	3.33	6.23	6.37	3.69	1.43	2.85	4.24	41.77
1916	7.32	1.33	2.63	2.49	4.47	6.64	2.44	3.15	2.96	2.13	1.85	3.28	40.35
Mean	3.46	2.52	3.91	2.94	4.02	3.83	3.64	3.58	3.11	2.86	2.55	2.83	39.80

Irrigation.

The subject of irrigation in humid lands has never been considered to any great extent in the United States. A great deal has been done recently in the development of irrigation projects in the arid and semi-arid regions of the west. The problem to be considered in this connection is always the cost of development and the increase in production due to the development. In arid regions where the original fertility of the land has not been depleted and where the entire yield of a crop depends upon irrigation, a much greater expense may be incurred in the development than is permissible in even a semi-arid region where dry farming methods will usually produce a crop. In semi-humid regions such as we find in Indiana, where the farming is carried on without regard for even dry farming methods and where a failure due to drought seldom occurs, the cost of any irrigation project must be low. In China where the climate is much the same as ours there are many and widespread irrigation projects. These developments are very old, having been in use for hundreds if not thousands of years. These systems have been developed with great effort by the Chinese and the returns are high. The land is made to yield even at their low prices as much as $200 to $400 per acre per annum. This careful work, both in providing water against any drought and also in caring for all fertilizing material, is necessary to the Chinese, where the rural population in some provinces is as dense as is the population of some Indiana cities. Any general failure to the crop means death by starvation. The farmer of Indiana is never worried for fear he will be in danger of starving if he misses a crop, but he knows a shortage will reduce his bank account or prevent him from procuring some of the luxuries to which he has been accustomed. It is probable that irrigation could be profitably used in large farm projects as it is by gardeners and florists. It has never been tried in Indiana.

The most favorable locations for irrigation projects in Indiana are in the flood plains of the river valleys. The most favorable source for the water supply is from small tributary streams instead of from the river itself. The waterworks pond of Indiana University has a capacity of 40,000,000 gallons and is scarcely ever seriously lowered by the pumping of 90,000 gallons per day. It is designed to supply 200,000 gallons per day even in the driest year. The catchment area for this pond is 200 acres and the gain or loss due to underground flow is considered as negligible. The small run upon which this pond is built is a tributary to Griffey Creek, which has a valley floor in this locality of a thousand feet, more or less, in width and stretching down the stream to its mouth some five miles distant. Suppose this fertile bottom land could be irrigated from this pond during dry seasons. The university pond would furnish, when filled, six inches of available water to 250 acres of land. The cost of the university dam was $40,000, 6 per cent of which would be $2,400. Such a proposition would necessitate the charge of $10 per acre per annum water tax, which is too great. However, a plan is possible by which this cost can be greatly reduced. In the first place a dam can be built at a greatly reduced cost and also one that will impound very much more water. A competent engineer estimated the cost of a dam across Griffey Creek near the North Pike, running north from Bloomington, at about $250,000. The pond would have a capacity of 1,000,000,000 gallons. This amount of water would irrigate 6,130 acres to a depth of six inches, which would be sufficient for ninety days continual drought. At $4 per acre the income would be sufficient to pay 6 per cent on investment and create a sinking fund of $9,500 per year, which would pay for the investment in twenty-six years. If intensive farming methods were practiced on the land in Griffey Creek valley to its mouth and in Bean Blossom valley to the limit of 6,130 acres, attention being given to raising of vegetables which could be canned, the water tax of $4 per year per acre would net several times that amount in increased yield. Such a project would eliminate the possibility of a crop failure for want of water and the storage capacity of the pond would tend to lessen the danger of destructive floods. However, Griffey Creek valley is narrow and therefore offers small storage facilities. There are many places in Indiana where better storage facilities can be found above land which could be profitably irrigated.

The Tippecanoe River offers an opportunity for irrigation on a large scale by the erection of a diversion dam in the vicinity of Oakdale and a feeder canal along the west bluff leading into the Wabash valley. By reference to the data on the Tippecance River, pages 372 to 382, it is seen that the fall from Oakdale to the mouth of the Tippecanoe River is thirty-nine feet, the distance seventeen miles and the minimum discharge at Springboro 269 second feet. If a diversion dam were built and the water carried the seventeen miles with a fall of six inches per mile in the canal, it would reach the valley of the Wabash 30½ feet above the level of the water at the mouth of the Tippecanoe. This does not make any allowance for the height of the dam, which would raise it correspondingly. This elevation would throw the water above much of the bottom land of the Wabash valley and if the canal descended the Wabash valley the elevation above the river would gradually increase. If the average discharge of the Tippecanoe for any three months at Oakdale is considered as 300 second feet, which is considerably underestimating it, the entire discharge for the ninety days would be 2,332,800,000 cubic feet. This discharge would be delivered in a continuous, fairly uniform flow in time of drought. It would be sufficient to furnish six inches of water to 107,107 acres—over 167 square miles. It would thus be more than sufficient to irrigate all the bottom land to which it could be directed along the Wabash River. The agricultural bottom and terrace land along the Wabash from the mouth of the Tippecanoe to the state line is estimated at about 100 square miles. If fifty square miles of this land lies on the west side of the stream and is thus available for irrigation from the proposed canal, slightly over twenty inches of water would be available for any ninety days of continuous drought. This is more than three times as much as needed but since no reduction has been made for leakage and evaporation those losses would come out of the surplus. The entire length of a canal from Oakdale to the state line would be approximately 150 miles. If the average cost of construction be considered at $4,000 per mile, the entire cost would be $600,000, which would be less than $20 per acre for the land to be irrigated. A water tax of one dollar per acre per year would yield over five per cent on the investment. The cost of upkeep and operation has not been considered in these estimates. The upkeep should be small if the canal were well constructed in the beginning, for that part of the canal which would lie within the Wabash valley would be above the usual high water mark. The greatest danger from floods would occur in the Tippecanoe valley. The cost of operation and upkeep should not exceed $50,000 per year. During the remainder of the year when no irrigation was needed the canal would furnish three hundred second feet of water at the state line with a head of about fifty feet. This would develop over 1,300 horse power. With Terre Haute as a convenient point for the sale of this power, the cost of upkeep on the entire proposition would be met by the sale of power. Such a proposition would necessitate the installation of auxiliary steam power to the extent of the power produced by water. The gage at Terre Haute shows the stage of the river during the flood of March, 1913, to have been 30.8 feet at the highest, and the reading practically represents the stage above low water mark, for the lowest record is less than a foot below zero of the gage. Thus the proposed power plant at the state line would be located about twenty feet above the high water mark reached in 1913, which is the highest known record.

A farmer, near the mouth of the Tippecanoe River, in conversation with the writer in the autumn of 1916 concerning a field of corn which would fall within the proposed irrigation project, said: "Last year that field produced sixty bushels per acre, but this year it is running about thirty." When asked for the cause of the reduction, he said that it had been too dry during 1916. He had lost thirty bushels of corn per acre and his cost of planting and tending had not been any less than during the previous year. His cost of harvesting and marketing would be less to the extent of probably $2.50 per acre, but the thirty bushels of corn at the current price (1916) would bring more than $40, which would pay a water rental of $2 per acre per year for twenty years. There is little doubt that the yield of sixty bushels which he produced in 1915 could have been increased by the application of water sometime during the growing and maturing season. There are very few seasons when corn would not be benefited by the application of water at critical times.

Water Power, Navigation and Flood Control.

Artificial light and heat are of equal importance with food, clothing and shelter to the human race in this latitude. The common sources of our artificial light and heat are wood, coal, oil, and gas. Wood has been abandoned as a means of heating except for family use and in very small manufacturing plants. The disappearance of our forests and the slow growth of forest trees make any attempt to produce fuel on a large scale from this source impractical. Authorities on coal have stated that the available supply of coal will last but a short time, at most a few centuries with the present rate of production and increase of production. Gas and oil fields have been found to be even shorter lived than coal fields. The weight of authority seems to indicate that at the present rate of consumption the next two centuries will practically exhaust these four common fuels. A proper conservation of the present supply will greatly increase the life of these fuels, but with the present increasing demand for power the final exhaustion is but a matter of time. In the face of this situation the question naturally arises as to the means of supplying this deficiency. Among the means suggested, the most plausible ones are direct sunlight, wind power and water power. At the present time little has been done along the line of the direct sunlight engine. However, it is possible and probable that an engine may be invented which will run for practical purposes by direct sunlight. It is known that the sunlight that falls on the roof of an ordinary factory is sufficient to produce more power than is used in the factory. If an engine could be invented which would successfully concentrate and utilize this heat, it would still be necessary to store the power for use during the time when the sun is not visible. This could probably be done by a more highly perfected type of storage battery. Wind power has been used for an indefinite time as a means of propelling pumps and other machinery that require but little power. Attempts to use wind power on a large scale have always proved unsuccessful. It is even more inconstant than sunlight. Because of the inconstancy of both wind and sunlight it is probable that neither will ever be used for large scale power purposes.

Water power has long been used for practical purposes. Before the use of steam it was the propelling power of small mills and many of these mills are still in use. Water power is inexpensive, perpetual, and requires less attention than other power when it is once properly installed. While the water power of Indiana must be used on a low head, it is a resource from which thousands of dollars could be realized if it were properly installed and utilized. The New York Water Power Commission estimates a saving over steam in the State of New York by the development of additional water power through reservoirs of twelve dollars per horse power per annum. It requires at least ten tons of coal to produce a horse power for a year. If Indiana could substitute 50,000 horse power of water power for as much now produced by steam, which in all probability could be done, it would mean a saving of 500,000 tons of coal per annum in addition to the $12.00 per horse power saved by the substitution. This would be of great economic importance in increasing the life of coal. The amount of developed water power in Indiana at the minimum stage of the streams upon which the power is developed is between seven and eight thousand horse power. At present there is much interest in water power. The valuable farm lands in the valleys of the Wabash and White rivers are a great hindrance to the full development of the water power of the state. If in the future the fuels are exhausted and the use of direct sunlight is not found to be feasible, the lowlands along the rivers will be condemned and used for storage basins for water for power purposes. Until the demand for power becomes imperative the entire water power of the state will not be developed.

A proper development of the water power of Indiana would bring about several important results. The navigation facilities would be greatly increased, the increased storage would tend to purify the water, and the reservation of water in the storage basins would tend to lessen the damage wrought by floods. The three problems—water power, navigation and protection from floods, are very closely related. The great problem in each case is to bring about a regular stream flow. The following statement of Van Hise bears directly on this point:

"The greatest difficulty of navigation is the unequal stream flow. At one time the stream is in flood, overflowing its banks, rolling down with great velocity toward the sea; at another time it is comparatively small, indeed often being divided into several small streams trickling over its bed. The conditions in either case are not favorable to navigation; in the first, because of the velocity of the stream, and in the second, insufficient depth to carry a vessel. In the projected improvements, according to Leighton; the first and most important step is to so control the streams as to get a nearly uniform flow.

"The holding of flood waters, and therefore securing greater regularity, may be accomplished to a considerable extent by levees on each side of the river bank at some distance from the low water river channel, so as to make a basin. At time of flood the water rises above the banks, and so makes between the levees a long, narrow, temporary lake which may require several days to fill and empty. Such intermittent levee reservoirs prevent damage from floods and to a reasonable extent regularize the flow of the stream.

"In many cases, in addition to the system of levees such as indicated, it will be necessary to construct at the headwaters of the great navigable stream adequate systems of reservoirs. We have seen that the development of reservoirs is of immense importance with reference to water power. Also it is of equal importance with reference to navigation."

It is probable that the rivers of Indiana will never be made use of for navigation on any large scale. The only navigation now carried on is that of motor boats and house boats. The lower Wabash, lower White and St. Joseph are the only streams on which these are seen. However, if the streams were of more uniform volume the use of these craft for pleasure and fishing purposes would be greatly encouraged. The greatest need for storage in Indiana is because of the flood conditions. This is shown by a comparison of the St. Joseph River with the White or lower Wabash. The St. Joseph rises in the lakes of Michigan and northern Indiana and never has destructive floods as do the streams of southern Indiana. The Niagara is an extreme example of the effect of storage. It fluctuates but a few inches in the wettest and dryest weather.

The most reasonable project for controlling floods seems to be the self-controlling dam scheme. At a favorable point on the river, discharge measurements are taken at times of reasonable flood, that is a flood which is outside the river banks but not with damaging effect. At this point a dam is built which has vents to allow the amount of measured discharge to escape but to retain the discharge in excess of this amount. The main vent should be over the stream channel, and others for roads, etc., if convenient or necessary. The height of the dam should be governed by the amount of storage necessary and by the conditions above the dam. The essential points in selecting a site for such a project would be that there should be a constriction in the valley and a wide valley above in which is situated no town and no homes that could not be placed above dam level. In ordinary weather the stream would flow through the dam just as it had flowed before the dam was there. So it would flow until the discharge becomes greater than the capacity of the vent. When the precipitation ceased the stream would continue to discharge as much as the vent would accommodate until the pond was reduced to ordinary flood stage. This arrangement would take the crest off the flood below the dam and if several such reservoirs were located in the headwaters of a large stream it would do much to take the crest off the floods on the whole stream. The project would necessarily have to be carried out by the government. The occupants of the land within the pond area would both lose and gain by the presence of the pond. The loss of crop would not be much greater than if the dam were not present, for the dam is not calculated to restrict the water until after the river has reached ordinary flood stage; and crops are destroyed in any case when streams go above ordinary flood stage. The water would rise higher on the river bluffs behind the dam and thus the occupant would need to remove his possessions above the level of the dam. The restriction of the water would tend to keep it from washing except at the points where the vents are placed. These places could be protected by concrete aprons. The water would deposit sediment within the pond area which would increase the fertility of the land.

Another proposal which would do much to allay the destructiveness of floods is that of reforesting waste lands. No survey has been made in the state to determine how much land could be reforested. The principal part of the waste land of the state lies in the southern part where the prevention of floods is most needed. In general the reforestation of waste land with timber which is quick growing and useful at an early age would be a good financial investment even without considering its effect upon floods. The rate at which the timber is being cut and the price to which timber products have advanced offers profitable returns on capital which could be invested in the reforestation of waste land. The black locust tree is probably best adapted to this purpose. It is quick growing, beneficial to the soil in which it grows, and is one of the best of timbers for posts and railroad ties.

City Water Supply.

Water for domestic purposes should be free from organic impurities and objectionable mineral matter. Nature sometimes furnishes water which leaves nothing to be desired in these respects. Such water is usually derived from the underground supply or from mountain streams. Since there are no mountain streams in Indiana the only source of pure water without artificial treatment is from ground water. In many parts of the state the supply of underground water is limited in amount and therefore insufficient for the use of large cities. For this reason many cities are forced to draw their supply from streams and lakes. This surface supply, especially from the streams, is liable to be polluted with sewage and as the population increases the danger of pollution grows. There are circumstances under which it is necessary to use bad water or none at all, and in such cases some means of purification must be devised. This responsibility falls upon the individual or corporation which furnishes the water, although the responsibility is not always met. Thus the purpose of a city water corporation is to furnish plenty of pure water to individuals desiring it.

In isolated dwellings and in villages and small towns not yet provided with a public water supply, drinking water must, as a rule, be obtained either by collecting the rain water from the roofs or by sinking wells. Because of the superior qualities of the ground water, namely, clearness and coolness, the well water is usually preferred. Too often the location of the well is determined merely by convenience and without any thought of sanitary conditions. Thus it is often located near a privy, cesspool, stable or barn. Because of the draught of water from the well, the ground water table is slightly lowered about the well and the ground water on all sides tends to flow in that direction. Thus it is merely a question of time when the well will receive water from the neighboring source of pollution. On the other hand a well may remain pure when surrounded by sources of pollution for great lengths of time, when it seems well nigh impossible for it to do so. Chemical and bacteriological examination is the only means of determining when a well is polluted, unless the pollution is very pronounced.

The solid content of water exists either in suspension or solution. The suspended matter may be organic or inorganic. The organic matter may be either animal or vegetable. The method of determining the presence of these is of value in connection with the possible means of sedimentation and filtration. To determine the amount of suspended matter, a filter paper is dried at 100°C and weighed. A carefully measured amount of water is passed through the paper, which is again dried at 100°C and weighed. The difference in weight is the weight of the suspended matter. The paper and residue are then ashed and the weight of the combined ash minus the weight of the ash of the paper, which is known, is the weight of the inorganic portion of the suspended matter. Various other tests are made to determine the settling qualities of the suspended matter and whether the organic part is of animal or vegetable origin. The amount of suspended matter is expressed as parts per million which is also the number of milligrams per liter of water.

Most of the chemical tests are conducted by titrating methods. The details of these tests may be found in any good textbook on water analysis. The tests are usually made for total solids, free ammonia, albumenoid ammonia, chlorine, nitrogen as nitrites and nitrogen as nitrates. The presence of chlorine is always looked upon with suspicion. While common salt is in itself entirely harmless in quantities commensurate with the chlorine usually found in water, yet its presence indicates that it may have come from household sewage. It may be that the salt came from a source that is entirely unobjectionable but its source should be known. The occurrence of any of the constituents mentioned in great quantities is always questionable. However, no fixed standard has ever been agreed upon by chemists, because an analysis which would condemn a water from one source would not indicate objectionable impurity in water from another source.

Perhaps the most trustworthy means of determining the purity of a given sample of water is by a bacteriological examination. The most critical chemical analysis will seldom reveal anything more than the sources and amount of food supply for the pathogenic organisms which may exist in very large numbers. It is then the bacteriologist who must make the final examination and give the final verdict. The test most commonly used is the Koch Culture Media Test. Koch early observed that, if a potato were sliced and kept in a warm, even temperature for a few days, the microscope would reveal an abundance of micro-organisms which had developed thereon. This and similar experiments led to the use of the gelatin peptone solid medium as a means of cultivation of bacteria. The water is collected in a sterilized bottle and kept plugged with sterilized cotton wool. Definite amounts of this sample are then introduced into sterilized flasks or test tubes and kept in an oven of constant temperature for about five days. The culture media are then removed and examined under a microscope to determine the colonies of bacteria produced. In actual practice the slides are either prepared permanently or photographic records are kept.

Previous to Koch's experiments it was supposed that any treatment which rendered the water chemically pure also rendered it practically pure organically. Since Koch's experiments this has been found to be not strictly the case. In general, however, it may be said that a water which tests satisfactorily from the bacteriologist's standpoint is safe.

Filtration has long been practiced as a means of purifying water for domestic use. Sand has always been used for filtering purposes because it is usually cheap and convenient. In fact sand is a very poor filter material. The following analysis of water before and after sand filtration will illustrate the point:

Table showing the effect of sand filtration upon river water—expressed in parts per million:

	Before	After
Total solid matter	284.00	262.00
Organic carbon	1.23	1.19
Organic nitrogen	0.25	0.22
Ammonia	0.00	0.00
Nitrogen as nitrites-nitrates	0.87	0.89
Total combined nitrogen	1.02	1.11
Chlorine	16.00	16.00

The slight change in the composition of the water here indicated led to the belief that sand filtration would probably not be effective in taking out disease germs. Direct tests proved this to be the case. While the filtration of the above sample shows little effect, nevertheless on other samples it shows much more marked effects. The above sample was a comparatively clear water, most of the solid matter being in solution. The filter produces a much greater change when the solid matter is largely in suspension. However the change in such cases is almost entirely in the inorganic constituents of the water.

In localities where there is a public water supply, it is without doubt the duty of the water board or company to deliver the water to the consumers in a condition fit for household use. If filtration is absolutely necessary, it should be done on a large scale by the authority controlling the work. Practically, however, in the case of most existing water supplies, the water as supplied to the consumer may be appreciably improved by filtration. Household filtration is often necessary in country residences and in the smaller towns, where there is no public supply, and where it is necessary to use rain water which has been stored in tanks and cisterns.

For filtration on the household scale, numerous devices have been devised and patented and a great variety of materials has been proposed. Some of the materials are: various sorts of porous stone, sand, powdered glass, brick, iron turnings and iron in other forms, vegetable and animal charcoal, sponge, wool, flannel, cotton, straw, sawdust, excelsior, and wire gauze. A material suitable for a household filter must not communicate any injurious or offensive quality to the water which passes through it, it must remove from the water all suspended matter and leave the water bright and clear, and it must be of a material which can either be readily cleaned or speedily renewed. It is quite generally conceded that on the whole there is nothing better suited for filtering purposes than well burned animal charcoal. This material possesses great power of removing organic matter from solution and is used in the arts to decolorize solutions. On many organic matters it acts not simply by adhesion but apparently by bringing them into contact with oxygen and thus entirely destroying them. Its power does not last indefinitely and a bone coal filter, like any other filter, requires cleansing or renewal at proper intervals.

Of the numerous devices for retaining the filtering substance we will first consider the smallest type, that which is to be attached to the faucet, where the water is brought in pipes either from the service mains of a general supply or from a tank in the building. Considering the amount of water that must flow through a very limited amount of material, no filter capable of being screwed upon an ordinary water tap can act in any other way than as a strainer, and all that can be required of such a filter is that it shall remove suspended particles, and be readily cleansed or renewed at trifling expense. The older forms which could be cleansed only by unscrewing from the tap and reversing either the whole apparatus or some inside receptacle of the filtering medium, were all open to objection and no one of them is to be recommended as superior to the primitive and unpatentable device of attaching to the faucet a bag of cotton flannel to be frequently washed and renewed.

The so-called portable filters are intended to occupy a permanent position in the room, or in some cases to be placed in the tank from which the supply is drawn. Sand has been more generally used than other materials in these filters although certain kinds of very porous sandstone, pumice stone, and unglazed earthenware are much used and have the advantage of permanency of shape. Slabs of this material are used as partitions in a box or tank. The action is in the main mechanical. The material collects on the upper or further side of the block of stone and is occasionally removed by washing. A material which is used to a considerable extent in London is the so-called silicate of carbon. It is the residue of the distillation of certain varieties of bituminous shale. In the common form of household filter using this material, it is cemented as a partition in an earthen jar and is not readily cleaned, but there is no doubt that until the filter becomes clogged it is very efficient in purifying water even from the dissolved organic matter.

In another type an earthenware box is filled with animal charcoal in the form of charred bones broken into small bits and freed from dust. There is no chamber for storing the filtered water, and the water is filtered at the time it is drawn off for use. The filter is readily cleansed and the charcoal renewed. The water filters upward through the filtering material in order that the matter spontaneously settling down may not be deposited upon the filtering material and may not therefore help to clog its pores, and further in order that the suspended matter strained from the water may tend to fall away from the surface and be deposited elsewhere. Thus the filtering material requires less frequent cleansing.

Another material which has been used with very favorable results is what is known as spongy iron. The water is supplied from an inverted bottle, which may be kept filled by being connected with a service pipe regulated by means of a ball cock attachment. In this filter it is claimed that a considerable portion of the dissolved organic matter is removed and that bacteria are completely removed.

Wood-charcoal is sometimes used in household filters, but it has little or no chemical action. It does little more than remove suspended matter and its usefulness for that purpose is of short duration.

The collection of water from the roofs of houses involves the collection of dust and dirt more or less objectionable in character, especially where coal is burned. Although it is possible by mechanical contrivances to avoid collecting the first portion of water coming from the roof, yet these do not perfectly accomplish their object and are not at all commonly employed. Moreover, the construction of ordinary cisterns is such that after the water is once collected it is liable to deterioration and to contamination by various foreign matters which may fall into it, so that filtration if not absolutely necessary is certainly very desirable. Underground cisterns are sometimes constructed so that the water is not drawn directly from the cistern itself but from a pump well to enter which the water must pass through a wall of porous bricks. When the wall is new it is without doubt of much service but it soon becomes clogged and covered on the outside with a coating of organic matter, so that after a time the water which passes through the brick wall must first have an opportunity to leach what it can from this mass of decaying matter. As a rule the interiors of cisterns are not very accessible and when the cistern is relied upon as the sole or principal supply for the household it is impossible to renew frequently the filtering wall or even to thoroughly clean the surface. If the body of the cistern be divided by a partition wall into two compartments which can be made to communicate or not at will, the two may be cleaned at different times and thus the danger of water famine be averted.

Water on a large scale is usually filtered in one of three ways, or by some combination of the three. The three ways are filtration by sedimentation, by chemical means or by the sand process.

When a stream or other body of surface water is used as the source of supply, it is best, from a sanitary point of view, to pump directly from the source into the distribution without the intervention of settling basins and reservoirs. This is, however, not generally practicable. In some cases the generally or occasionally turbid character of the water renders sedimentation necessary and in other cases storage basins are necessary in order that the stream may furnish a sufficient quantity of water during a dry season. The particles of suspended matter which render a stream muddy are of greater specific gravity than the water and settle out more readily and completely if the water be allowed to remain quiet for a time. Lakes and ponds are natural settling basins and have the advantage over running streams, that they are less liable to become turbid in times of freshets. They are not, however, entirely free from floating matter, but this can only be removed by filtration. It is desirable to remove the floating matter, not simply on aesthetic grounds, but because particles of gritty mineral matter, or even of clay, often cause diarrhea, especially in the case of persons not accustomed to the daily use of the water.

The chemical agent most used in the filtration of water is aluminum sulphate. The natural alkalinity of the water will cause a precipitate of aluminum hydrate which in its coagulated mass carries much of the suspended matter with it. This is followed by filtration in a sand filter.

Sand filter beds, as usually constructed, are water-tight basins some ten feet or more in depth, the sides built of masonry, and the bottom puddled or made of concrete or brick. The area may vary from 20,000 to 150,000 square feet. In building up the filtering bed provision is first made for the ready collection of the water by constructing upon the floor of the basin drains or channelways of stone, brick or tiling. Then follows a layer of broken stone and layers of gravel of gradually diminishing size. Upon the gravel rests a deeper layer of carefully graded sand. The thickness of the various layers is subject to considerable variation.

The water stands several feet deep over the sand and is allowed to flow down through the filter at such a rate as experience shows to be most advantageous. Naturally when the sand is clean a greater quantity of water can be passed in a given time than when the sand becomes clogged. Practice differs as to the maximum rate but it is seldom over six inches vertically per hour and often less. At the rate mentioned each square foot of surface would deliver 12 cubic feet or 90 gallons per day. When the beds become clogged so that the rate of filtration is too slow, the water is drawn out from the beds and the uuper layer of sand for a couple of inches is removed and washed.

Another type of sand filter known as the rapid sand filter is used in some cases to remove suspended matter. It is constructed very much on the order of the type previously discussed except that the sand used is of uniform size throughout, usually that which will pass through a sixteen mesh and lie on a thirty mesh sieve. The depth of sand is usually about three feet. The rate of water passage is governed by the head of water above the filter. A head of five feet will filter approximately 1.5 gallons per square foot per minute. This would give the filter a capacity of 2,160 gallons per day per square foot of surface. Since the pumpage of city plants usually runs about 100 gallons per capita per day, a filter of this type should have approximately one square foot to each 20 inhabitants. It is necessary to wash this type of filter frequently. This is done by forcing water upward through the sand.

The action which takes place as the water passes through an ordinary filter is threefold. The most obvious action is the arresting of suspended particles which are too large to pass through the pores of the filter. The second action partakes somewhat of the character of sedimentation. The interstices between the particles of the filter serve as so many settling chambers, and the particles of clay which are small enough to enter these interstices, and therefore could pass entirely through the filter, are deposited, not only on the floors of these chambers but also on the sides and roofs. This action is probably due to adhesion on contact. The third sort of action which takes place in the porous material of a filter is the removal of substances which are actually in solution. As far as mineral matter is concerned this action is small although not inappreciable with certain filtering media. Dissolved organic substances are removed in appreciable amounts by well conducted sand filtration. Most porous substances possess the power of removing certain kinds of organic matter by presenting surface upon which bacteria which feed upon the organic substances can work. This point will be more fully discussed under sewage disposal. Direct oxidation of organic matter will take place in a filter faster than in standing water, especially if the filter is frequently emptied and refilled.

The factors which influence the efficiency of a filter are the thickness of the medium, the rate of water passage, and the frequency of renewing the medium. It is obvious that the thicker the medium the more complete will be the removal of solid substances and the less frequent the necessity of renewing or cleaning the medium. In estimating the thickness of the medium, the fine sand only should be considered, because that is the part which is effective in removing solids. The removal of micro-organisms is less perfect when the rate of filtration is increased. This has been shown by exhaustive experiment. A filter should be frequently washed and the medium should be changed at longer intervals. Washing should be more frequent in summer than in winter because of the presence of algae in the water which tend to clog the filter. If settling basins precede the filters the addition of copper sulphate to them will prevent the growth of algae and will in no way injure the water.

The covered filter bed is the one now most commonly in use, for the exposure of the surface of the water to the open air has some disadvantages. In summer the water becomes unduly warm and the growth of the lower orders of plant life is thus favored. In winter the freezing of the water causes inconvenience.

The most valuable accessible data upon the expense of filtration as drawn from actual experience, are found in the report of the Poughkeepsie, N. Y. Water Works. These data indicate that the expense varies between $2.50 and $3.00 per million gallons, which does not include the cost of pumping nor the interest on the money invested in the plant. The original cost of the beds was $54,000.00, the interest on which would exceed the cost of maintenance.

The most important filtering plant in Indiana is the Indianapolis plant. The water is drawn from the west branch of White River at Broad Ripple, about seven miles north of the center of Indianapolis. It is conducted to the city by canal. Part of the water is used for water power and the power is used for pumping water into the city mains. Sufficient power to pump 14,000,000 gallons per day is developed. The filtration plant has a capacity of 30,000,000 gallons per day. The water is treated by four processes: First, with aluminum sulphate; second, in settling basins; third, in slow sand filters, and fourth, with chlorine gas. The first three processes are for the purpose of removing suspended matter and the fourth for its germicidal effect.

The sedimentation basins have a capacity of 45,000,000 gallons. The daily pumpage of the city averages 24,000,000 gallons. Thus the water remains in the sedimentation basins for an average period of forty-five hours. The filter beds cover seven and a half acres, which at the present rate of pumpage affords a capacity of approximately three gallons per square foot per hour. The filtration plant is located on Twenty-first Street and comprises one hundred and twenty-five acres. Analysis of the water from this plant shows it to be equal to the bottled water which is sold on the market.

In the previous paragraphs the entire pumpage of 24,000,000 gallons per day was considered as being filtered. This is not the case however, for part of the water is drawn from deep wells and filtration is unnecessary. There are three pumping stations operated by the Indianapolis Water Company, and each station pumps water from wells. The Broad Ripple station has several wells with a total capacity of 1,000,000 gallons. One of these wells is now operated. The Fall Creek station has sixteen wells, of which twelve are operated. The capacity of these wells is about 12,000,000 gallons per day and the present pumpage about 3,000,000 gallons per day. The Riverside station has thirty-five wells with ten under construction. The capacity of these wells is 18,000,000 gallons per day and the pumpage varies. The wells are approximately three hundred and fifty feet deep, through gravel into bed-rock.

The Indianapolis Water Company operates four hundred and fifty miles of pipe line which varies in diameter from four to forty inches. It has 40,000 customers and serves 71% of the population of Indianapolis. The rest have private wells. The company operates on both flat and meter rates. The flat rate for a five or six room modern house, with sprinkling privileges, for the year is $14.85. The meter rates vary from 16c to 4c per thousand cubic feet. The average for the first million cubic feet per month is 5.6c per thousand and above the first million 4c per thousand.

The Indianapolis Water Company is a private corporation. The vice-president and general manager is Mr. C. L. Kirk.

The finest water in Indiana for all purposes is found in the glacial gravels and sands. These gravels and sands are largely insoluble in water and for this reason the water coming from them contains little mineral matter. If sufficient depth is reached in these materials to make filtration perfect and the temperature low, the water is ideal. Many cities in the eastern and northern portions of the state find an inexhaustible supply of such water wherever it is convenient to sink a well. Such cities are very fortunate and little complaint is heard concerning the water. A few cities in Indiana are able to get water from the bed-rock, especially the sandstone. This water is usually very good, but is sometimes highly charged with mineral matter. The largest cities of the state with the exception of South Bend and Fort Wayne take their water from streams. Many other cities also depend upon streams for their supply. Several cities in the lake districts depend upon the lakes for their supply. A few procure their supply from springs. A few more depend upon artificial ponds. A list of the county seats and a few other cities follow with the source of water, pumpage in gallons per day, manner of treatment, number of patrons, number of gallons (nearest ten) per patron, population, number of gallons (nearest ten) per capita population, and the ratio of patronage to population.

CITY	Source of Water, (Depth of wells) given in feet.	Pumpage, gallons per day	Treatment of Water	Patrons	Gallons per patron	Population	Gallons per capita	Patronage of population, %
Albion.........	Wells 98'........	200,000	None........	260	770	1,200	170	22
Anderson.......	White River.....	2,500,000	Coagulation.. Sedimentation Filtration.... Chlorinization	6,000	420	25,000	100	24

CITY	Source of Water, (Depth of wells given in feet.	Pumpage, gallons per day	Treatment of Water	Patrons	Gallons per patron	Population	Gallons per capita	Patronage of population, %
Angola	Wells 114'	700,000	None	650	1,080	2,600	270	25
Attica	Wells 110'	750,000	None	950	790	3,500	210	27
Auburn	Wells 230'	750,000	Gravel Filter	935	800	4,500	170	21
Aurora	Ohio River	500,000	Filtration	700	720	5,100	100	14
Bedford	White River	1,300,000	Coagulation, Sedimentation	1,500	870	11,000	120	14
Bicknell	Wells 50'	1,000,000	None	200	5,000	2,900	340	7
Bloomfield	Wells 300'	350,000	None	375	930	2,500	140	15
Bloomington	Springs, Artificial Reservoirs	1,500,000	None	2,200	680	11,000	140	20
Bluffton	Wells 325'	250,000	None	825	300	5,000	50	17
Boonville	Artificial lakes	225,000	None	700	320	4,000	60	18
Brazil	....	....	....	....	....	....	....	....
Brookville	Wells 150'	250,000	None	400	625	2,200	110	18
Brownstown	Wells 30'	190,000	None	205	720	1,400	140	19
Cannelton	Wells 110'	100,000	None	225	440	2,000	50	11
Clinton	Wells 70'	640,000	None	1,400	460	9,000	70	16
Columbia City	Wells 300'	500,000	None	1,047	480	4,000	125	26
Columbus	White River	1,900,000	Coagulation, Sedimentation, Filtration, Chlorinization	1,600	1,190	9,400	200	17
Connersville	Wells 180'	1,500,000	None	1,600	940	10,000	150	16
Corydon	Spring	90,000	None	350	260	1,700	50	21
Covington	*	....	....	....	....	....	....	....
Crawfordsville	Springs and well*	750,000	None	2,200	340	11,000	70	20
Crown Point	Wells 100'	250,000	None	600	420	2,800	90	21
Danville	Wells 110'	300,000	None	640	470	1,700	640	38
Decatur	Wells 250'	500,000	None	783	640	5,000	100	16
Delphi	Springs	500,000	None	650	770	2,200	230	30
Elkhart	Wells 40'	1,500,000	None	4,500	330	22,000	70	20
Elwood	Wells*	1,000,000	None	1,400	710	11,500	90	12
English	Springs	70,000	None	135	520	550	130	25
Evansville	Ohio River	9,000,000	Filtration	12,000	750	80,000	110	15
Ft. Wayne	Wells 250'	4,500,000	None	16,000	280	83,000	50	19
Fowler	Wells 850'	200,000	None	400	500	1,500	130	27
Frankfort	Wells 85'	1,000,000	None	2,000	500	9,800	100	20
Franklin	Wells 100'	375,000	None	1,080	350	4,800	80	23
Gary	Lake Michigan	4,876,500	Chlorinization	4,600	1,060	60,000	80	8
Goshen	Wells 150'	1,500,000	None	2,100	710	9,500	160	22
Greencastle	Walnut Creek	450,000	50 feet of sand	1,150	390	3,800	120	30
Greenfield	Wells 150'	2,500,000	None	1,200	2,080	4,400	570	27
Greensburg	Wells 200'	500,000	None	1,200	415	5,250	90	23
Hammond	Lake Michigan	10,000,000	Chlorinization	7,850	1,270	30,000	330	26
Hartford City	Wells 300'	250,000	None	1,200	210	6,300	40	19
Huntington	Wells 200'	1,200,000	None	2,300	520	11,000	110	21
Indianapolis	White River and wells 350'	24,000,000	Coagulation, Sedimentation, Filtration, Chlorinization	40,000	600	320,000	75	12
Jasper	*	....	....	....	....	....	....	....
Jeffersonville	Wells 39'	1,200,000	None	1,400	860	10,000	120	14
Kentland	Wells 85'	27,500	None	240	110	1,200	20	20
Knox	Wells 250'	50,000	None	153	330	1,800	30	9
Kokomo	Wells 130'	2,000,000	None	5,000	400	17,250	120	29
Lafayette	Wells*	4,900,000	None	6,200	790	21,000	230	30
Lagrange	Wells 120'	500,000	None	375	130	1,800	280	21
Laporte	Wells*	2,250,000	None	2,500	900	13,000	170	19
Lawrenceburg	Wells 100'	*	None	185	*	3,500	*	5
Lebanon	Wells 150'	370,000	None	750	490	5,700	60	13
Liberty	Springs	45,000	None	340	130	1,300	30	26
Logansport	Eel River	6,000,000	Coagulation, Sedimentation, Filtration, Chlorinization	4,200	1,430	21,000	290	20
Madison	Wells 72'	1,000,000	None	1,400	710	6,500	150	22
Marion	Wells 250'	1,900,000	None	4,100	460	20,000	95	21
Martinsville	....	....	....	....	....	....	....	....
Michigan City	Lake Michigan	8,000,000	None	4,000	2,000	23,500	340	17
Monticello	Wells 50'	250,000	None	550	450	2,100	120	26
Mt. Vernon	Ohio River	777,000	Sedimentation, Filtration	1,238	630	5,600	140	22
Muncie	White River and wells*	3,000,000	Filtration	6,000	500	27,000	110	22
Nashville	a	....	....	....	....	340	....	....
New Albany	Ohio River	2,000,000	Sedimentation, Filtration	4,100	490	22,000	90	19
Newcastle	Wells 200'	1,350,000	None	3,000	450	12,000	110	25
Newport	a	....	....	....	....	800	....	....

CITY	Source of Water, (Depth of wells) given in feet .	Pumpage, gallons per day	Treatment of Water	Patrons	Gallons per patron	Population	Gallons per capita	Patronage of population, %
Noblesville	Wells*	600,000	None	750	800	5,000	120	15
North Vernon	Muscatatuck River	450,000	None	485	930	2,900	160	17
Paoli	Lick Creek	100,000	Filtration	200	500	1,400	70	14
Peru	Wells 300′	1,000,000	None	3,600	230	13,000	80	28
Petersburg	White River	100,000	None	390	260	2,500	40	16
Plymouth	Wells 200′	400,000	None	850	470	4,000	100	21
Portland	Wells 205′	600,000	None	1,100	550	5,400	110	20
Princeton	Patoka River	1,000,000	Coagulation, Filtration	1,100	910	6,900	140	16
Rensselaer	Wells 1,200′	240,000	None	600	400	2,500	100	24
Richmond	Infiltration Galleries	2,750,000	None	5,500	500	26,000	110	21
Rising Sun	Wells 100′	150,000	None	150	1,000	1,500	100	10
Rochester	Lake Manitou	500,000	None	850	590	3,000	150	26
Rockport	Wells 95′	200,000	None	500	400	2,500	80	20
Rockville	Wells 40′	60,000	None	450	130	1,900	30	24
Rushville	Wells 40′	480,000	None	1,300	370	5,300	90	25
Salem	*	....	....	....	....		....	....
Scottsburg	a	....	....	....	....	2,000	....	....
Seymour	White River	750,000	Filtration	1,250	600	6,300	120	20
Shelbyville	Wells 30′	1,200,000	None	1,750	690	10,000	120	18
Shoals	....	....	....	....	....	1,200	....	....
South Bend	Wells 110′	10,000,000	None	12,000	830	68,000	150	18
Spencer	Wells 85′	60,000	None	325	180	2,200	30	15
Sullivan	Infiltration Galleries	150,000	None	400	375	4,500	30	9
Tell City	Wells 80′	300,000	None	420	710	4,000	75	11
Terre Haute	Wabash River	5,150,000	Coagulation, Sedimentation, Filtration, Chlorinization	7,574	680	65,000	80	12
Tipton	Wells 400′	675,000	None	1,100	610	4,600	150	24
Valparaiso	Flint Lake	850,000	Filtration	1,900	450	7,500	110	25
Veedersburg	Wells 32′	30,000	None	250	120	1,900	20	13
Vernon	*	....	....	....	....	425	....	....
Versailles	a	....	....	....	....	475	....	....
Vevay	Ohio River	125,000	None	123	1,020	1,100	110	11
Vincennes	Wabash River	2,500,000	Filtration	3,500	710	18,000	140	19
Wabash	Wells 75′ and 300′	970,000	None	2,008	480	8,700	110	23
Warsaw	Wells* and Center Lake	750,000	None	1,300	580	4,800	160	27
Washington	West Branch of White River	120,000	Coagulation, Sedimentation, Chlorinization	1,350	90	7,500	20	18
Williamsport	Wells 190′	35,000	None	375	90	1,250	30	30
Winamac b	Tippecanoe River	200,000	None	300	870	1,600	125	19
Winchester	Wells 180′	250,000	None	600	420	4,800	50	13

*No data.
(a) No water works.
(b) Commercial purposes only.

It will be noticed in the foregoing table that many of the waterworks are furnishing between 75 and 150 gallons per capita per day. This is especially true of those plants which are well established in cities where manufacturing is not abnormally developed and where the rates are based on meters. The flat rate system is always conducive to extravagance on the part of the consumer. This is particularly well shown in the case of Evansville and Ft. Wayne which are of about the same population, have about the same manufacturing conditions and appear in the table consecutively. Ft. Wayne with 16,000 patrons on meter rates pumps half as much water as Evansville with 12,000 patrons, part of whom have meter and part flat rate. Auburn and Tipton are smaller cities of the same approximate population. Auburn has only flat rate and Tipton only meter service. The pumpage per patron in Tipton is 190 gallons per day less, and per capita, 20 gallons per day less than in Auburn. Consumers usually prefer the flat rate system but defeat their own case by wasting great quantities of water when the flat rate is given them. The only fair system is the meter system by which the consumer pays for just the amount of water he uses.

Sewage Disposal.

One of the necessities of a community is that it dispose of its waste material. The accumulation of waste material in a community is as dangerous to the community as the accumulation of waste in an organism is dangerous to the organism. Sickness and death are the penalties imposed upon both the organism and the community for failure to observe the fundamental law of waste disposal.

The Health of Towns Commission of Great Britain in 1844 published a report which brought about the first attempt at systematic sewage disposal. It is true that there were great drainage sewers in some of the old Roman cities but it is probable that these sewers were used only for the purpose of carrying away storm water. Such sewers were used in London before the report of the Health of Towns Commission, but there was also a law prohibiting waste material from being introduced into them. The report of this committee showed such a vast accumulation of putrid matter within the city limits that it aroused the people to great interest and effort. The committee advanced the idea that putrid matter gave rise to disease germs, not simply by harboring them and furnishing a perfect medium for their development, but by actually creating them. We know the latter idea to be erroneous today, but nevertheless it served its purpose in that time, and within three years after the report it was a crime punishable by law if all waste material were not dumped into the sewers.

Germany early followed the lead of England, and France and the Latin countries have since taken up the work. The first great sewage system in United States was that of Chicago, which was built in 1855. Several other cities had storm sewers and partial sewage disposal prior to this date. At the present time few cities of more than 5,000 inhabitants are without a well developed system of sewage drainage.

With the development of the sewerage system another problem presents itself. The sewage which is accumulated in the drain must be disposed of in some manner. It may be carried without the city and there dumped into a stream or cesspool. In time the point at which the sewage is dumped becomes excessively foul, and if the sewage is dumped into a stream the stream carries this filth to the countryside and to cities in its lower course. As long as the cities are small and far apart nature tends to purify the water between them and little inconvenience is experienced, but when the cities become large and near together it becomes necessary to require that some sort of deodorization of sewage be carried on by each city.

Sewage as it appears at the mouth of any large sewer is somewhat milky in color, shows the presence of much grease, and is attended with very disagreeable odors. Chemical analyses show that the sewage from many cities varies but little in its chemical composition and therefore an approximate analysis can be used as a basis for further considerations. If we consider that the sewage discharge from a city is 100 gallons per capita, which is very close to the average in Indiana, as shown by the pumpage of the city waterworks, its analysis will show that it is approximately 998 parts water, one part solid mineral matter and one part animal and vegetable matter. Sixty or seventy per cent of the solid matter is in solution and the rest in suspension. The animal and vegetable matter is volatile on ignition except for a small amount of ash which remains with the mineral matter. The sewage problem resolves itself into disposing of the one per cent of organic matter. The following table taken from Water Supply and Irrigation Paper No. 185, U. S. Geological Survey, page 15, shows the analysis of an ideal sewage:

PARTS PER MILLION

	Total	In Solution	In Suspension
Residue on evaporation	800	500	300
Mineral and ash	400	300	100
Organic and volatile	400	200	200
Nitrogenous	150		
Nitrogen	15		
Carbon	75		
H, O, S, P, etc.	60		
Non-nitrogenous	250		
Fats etc.	50		
Carbon	35		
H, O	15		
Carbohydrates	200		
Carbon	90		
H, O, etc.	110		
Total carbon	200		
Total nitrogen	15		
Total H, O, S, P, etc.	185		

In order to change these numbers to grams per capita per day multiply by 0.38, a daily flow of 100 gallons per capita being assumed.

The amount of oxygen necessary to change the organic substances to mineral form is considerable. Dibden estimates it to be from one to three times the weight of the organic matter to be acted upon. The following table taken from the same source as the previous table and credited to Dibden shows the parts of oxygen necessary to oxidize one part of the various organic substances:

OXYGEN REQUIRED

SUBSTANCE	By the nitrogen	By the hydrogen	By the carbon	Total	Oxygen already present
Gelatin	0.523	0.528	1.333	2.384	0.251
Chondrin	0.411	0.586	1.310	2.289	0.294
Albumen	0.457	0.568	1.414	2.439	0.220
Cellulose, woody fibre		0.506	1.184	1.680	0.494
Starch		0.496	1.184	1.680	0.494
Fat, stearic acid		1.016	2.025	3.041	0.133

The problem of sewage disposal is to supply this required oxygen and to supply it under such conditions that it will unite with the organic matter to be eliminated.

The following outline shows the methods in common use for the disposal of household and other wastes:

Methods of Disposal.

A. Midden system.
B. Pail system.
C. Water carriage system.
  1. Combined system.
  2. Separate system.
    a. Discharge into convenient stream.
    b. Chemical treatment.
    c. Bacteriological treatment, septic tank.
    d. Combination of b and c.
    e. Electrical treatment.

The midden system is the old method of having a cesspool into which the sewage is introduced, after which the liquid parts filter away through the ground, oft-times to appear at some undesirable point such as the well or spring. The solid parts are disintegrated by anaerobic bacteria and for the most part changed into liquid.

The pail system is used in some small cities. The sewage is accumulated in pails, which are removed by the city about once per week usually in the night. Another pail is left in the place of the one removed. The city of Mitchell, Indiana, uses this system. The sewage is taken to some point well without the city and there burned or buried. The Goux pail is sometimes used for transporting the sewage. The Goux pail is a wooden pail into which a mixture of factory waste and sulphate of lime are thrown. A mould is then inserted and left until the bucket is ready for use. When the mould is removed the mixture has taken the shape of the mould and lines the bucket. This mixture is said to be very effective in absorbing water and also in preventing decomposition. When the bucket is dumped the mixture is dumped with it and the presence of the sulphate of lime aids in the final destruction of the sewage. The Goux pail is shown in Fig. 4.

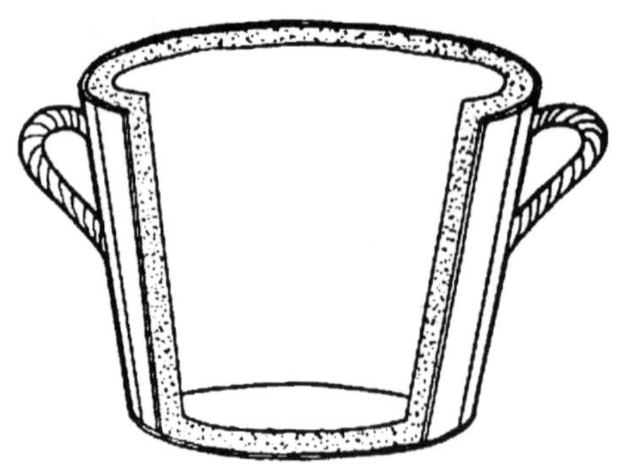

COUX PAIL

Fig. 4.

The water carriage system is used in most large cities. The combined system is one in which both the storm water and sewage are carried through the same sewers. The system is poor because the storm water is ofttimes as much as fifty times the usual sewage drainage and the sewers must be large enough to accommodate both. In seasons of drought the household sewage flows through these large, dark, damp sewers and has ideal conditions for the development of disease germs and foul gases. The trapping of these large sewers is hard to keep in perfect condition, and the result is that the foul gases often find their way into houses, especially of the poorer class. If any system of purification of sewage is used it requires the handling of all the storm water, which is very difficult. New York City uses this system and discharges her sewage into the Atlantic Ocean.

The separate system is one in which the sewage is handled through a set of sewers separate from the storm sewers. In this case the storm water can be discharged into a convenient stream and the sewage disposed of in any other suitable manner. Baltimore uses the separate system because it is necessary to purify its sewage in order that the oyster beds in Chesapeake Bay may not be contaminated.

When sewage is discharged into a stream it contaminates the stream in proportion to the relative amounts of sewage and stream water. A small amount of sewage in a large stream is negligible, but when a large city discharges its sewage into a relatively small stream it produces a very serious nuisance. From the point of discharge of sewage the stream gradually becomes purer down stream on account of self-purification. Self-purification of streams is a slow but sure process. The organic matter is gradually oxidized as the stream absorbs the necessary oxygen. This oxidation is largely due to the action of bacteria. Another factor in self-purification is the devouring of large particles of organic matter by animal forms other than bacteria. It has been shown that a measure of the albuminoid ammonia in a stream is a measure of its contamination and also that the dissolved oxygen in a stream

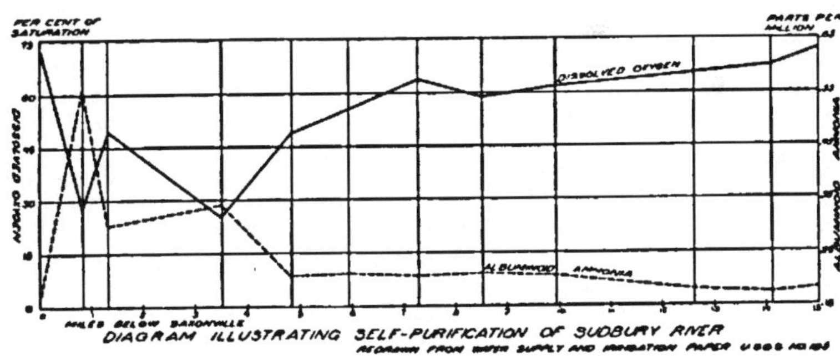

Fig. 5.

is inversely proportional to the albuminoid ammonia present. Fig. 5 is reproduced from Water Supply and Irrigation Paper No. 185, U. S. Geological Survey, page 17, and shows the relation between the dissolved oxygen and the albuminoid ammonia in the Sudbury River at several points within fifteen miles below the point where Saxonville, Mass., discharges its sewage.

The disposal of the sewage of Indianapolis presents the greatest problem along this line in Indiana. Until the present time most of the sewage of Indianapolis has been discharged into the West Branch of White River without treatment of any kind. This produced a very serious condition of contamination, especially in periods of drought. In 1909, Mr. J. A. Smith and the writer descended White River from Indianapolis and found the condition such that it produced extreme nausea. Night camp was pitched twenty miles by river below Indianapolis, and one-fourth of a mile from the river on a tributary stream, but the effects of sewage were still very disagreeable. The decaying carcasses of several hogs which had been thrown into the river by the packing houses of Indianapolis greatly aggravated the situation. The sewage of Indianapolis at this time formed practically half the volume of the stream. The bed of the stream was covered with a coating of dark, greasy sludge, largely organic matter, to a depth of one inch or more. This condition had practically disappeared at Martinsville, about fifty miles by river below Indianapolis.

The chemical treatment of sewage is usually expensive and is for the purpose of disinfection. There are various methods of treatment. The principal methods are by heat, alum, lime, acids, chlorine, chlorine compounds, and copper compounds. Lime is used more extensively than the others because of its lower cost. Its action as a germicide is very slight and is probably due entirely to its action as a precipitant by which means it settles out the organic matter. Alum was extensively used until recently when the advance in price became prohibitive. Its action is also that of a precipitant. The acids are much more effective than alkalis as germicides. Chlorine and copper are used very little in Indiana.

Bacteriological treatment of sewage by means of the septic tank seems to be the ideal method of rendering the sewage nonputrescent, which is the first desideratum in sewage treatment. The next is to render it sterile. This the septic tank does not do although it greatly lessens the number of disease germs. The principle of the septic tank is simply to offer conditions under which the bacteria which change organic matter to mineral matter can work. Before this method is discussed it is well to understand two things, first, the small amount of organic matter in sewage to be disposed of, and second, the great number of bacteria which develop under favorable conditions to destroy this organic matter. From the previous table showing the composition of an ideal sewage, four hundred parts per million are shown as organic and volatile. This is one part of organic matter in 2,500 parts of sewage, the other 2,499 parts being either water or harmless mineral matter. Thus it is seen that the amount of organic matter in sewage is very small indeed. On the other hand, it has been shown and proved that an ounce of sewage-soaked soil may contain more than three billion bacteria of the type which destroy organic matter. Their rate of propagation is almost beyond conception. If there were no forces to interfere one bacterium would have 16,500,000 descendants in twenty-four hours. If unhindered, in five days the descendants of this one bacterium would cover the surface of the earth to a depth of over two feet.

There are two kinds of bacteria, aerobic and anaerobic. Aerobic bacteria thrive in the presence of light and oxygen, and the anaerobic cannot live in the presence of oxygen and thrive best in dark places. Sewage which is thrown upon the earth and passes through the soil is acted upon by both classes of bacteria. The oldest of all forms of bacterial sewage treatment is known as land treatment, broad irrigation or sewage farming. This was done before bacterial action was understood. There was the double purpose of purification and fertilization. In some cases very successful results have been obtained by this method. In Pasadena, Cal., a four hundred and fifty acre farm is planted in English walnuts and yields from $7,000.00 to $10,000.00 annual profit. All the sewage of Pasadena is filtered through this farm. The project is a recent one and what the ultimate result will be upon the trees remains to be seen. The land is liable to receive so much more liquid than the ordinary rainfall that it will become sewage sick and useless for either purification or for raising crops. The amount of land necessary to purify a small amount of sewage is enormous and most places which have tried it have found it to result in a financial loss. It is in operation in many parts of England and is successful in Berlin, Germany, where 19,000 acres, an area larger than the city itself in size, are employed. Land treatment is on the whole haphazard, uncertain, and expensive.

In 1895, Donald Cameron, of Exeter, England, brought the septic tank into prominence. It consists of a large tank in which sewage is allowed to remain for some time to be acted on by the anaerobic bacteria. By the action of these bacteria part of the solid matter becomes liquefied, and goes into solution; part rises to the top as scum and part settles to the bottom as sludge. The inlet and outlet of the tank are placed below the surface so that the liquid may pass through with as little commotion as possible. The scum which rises to the top becomes oxidized after a time and passes off into the air as harmless gas. A certain amount of decomposition takes place in the sludge. In large tanks the sludge accumulation is very slow. At Mansfield, Ohio, only a few inches were drawn off after the tank had been in use for a year and a half. However, the use of the septic tank alone is not successful. Some provision must be made for the work of the aerobic bacteria. This is best accomplished by the contact beds and filter beds. These can best be explained by concrete example. Since one of the best types of plants is in operation in Bloomington, Ind., it will be used as an illustration.

Bloomington is located within the basin of the east branch of White River, but almost on the divide between the east and west branches. The divide lies near the north limits of the city. The city site slopes gently toward the south and is drained naturally by two small streams which are tributary to Clear Creek. The mantle rock upon which the city is located is of residual clay and very thin. The usual depth to bed-rock is probably on the average ten feet. The underlying rock is mostly porous Indiana oölite of the Salem formation. Springs are very abundant in this stone.

The water supply before 1890 was from wells and springs. In this year the city waterworks was established, but the sewage was still disposed of by the midden system. By 1900 the city had become saturated with the filth until all wells were dangerous, although many were still in use. Some of the progressive citizens at about this time began a campaign of agitation for some system of sewage disposal. This agitation continued and finally resulted in the establishment of the present system in 1907-8.

It was found that the conditions were ideal for the establishment of a disposal plant because of the slope which makes the system entirely under the control of gravity. It was decided to put in a separate system. The storm sewers will not be considered here. The entire system for the disposal of sewage consists of 56,550 feet (10.7 miles) of sewer and the disposal plant.

The sewer is distributed as follows: 38,217 feet of eight-inch sewer pipe which form the upper extreme of the system, 2,243 feet of ten-inch sewer pipe which accommodate the water when two of the eight-inch sewers unite, 3,079 feet of twelve-inch sewer pipe, 1,010 feet of fifteen-inch sewer pipe, 4,576 feet of eighteen-inch sewer pipe, 4,912 feet of twenty-four-inch sewer pipe and 2,550 feet of twelve-inch cast-iron pipe. The cast-iron pipe leads to the disposal tank which is located one and one-half miles south of the city near the Monon Railroad. This connects with the twenty-four-inch sewer pipe just outside the city limits on the south. At this point there is an overflow which allows the water which the twelve-inch castiron pipe will not carry to escape into Clear Creek. The twelve-inch castiron pipe forms the pressure head for the disposal plant. The twenty-four-inch sewer pipe subdivides upon reaching the city into the smaller sizes until the smallest are found in the suburban parts of the city. At the upper ends of these small sewers are flush tanks which fill and empty automatically and as often as desired. A constant stream of water enters the tank which upon filling to a certain point automatically dumps the charge into the sewer thus flushing out any material which might have lodged in the sewer. By regulating the size of the stream entering the tank the rate of flushing is regulated. There are twenty-three of the flush tanks in the system. There are 321 manholes entering the sewers and flush tanks. Most of the sewers are laid on or near bed-rock and the main sewer is embedded in the rock with cement in much of its course.

The capacity of the system and other points are shown by discharge measurements made on five different dates and under different conditions. On November 19, 1913, during the time when the weather was dryest and the city water was shut off for lack of water, the discharge was 9,000 gallons in one hour between 3:00 and 4:00 p. m. There was very little water flowing from house connections and a slight rain during the previous week was insufficient to cause any surface flow. This is taken as the normal ground water flow in dry weather. On July 9, 1913, between 2:00 and 3:00 p. m. the flow was 20,000 gallons. At this time the conditions were very similar to the conditions in November, except that the city water was on and the house connections were running. On July 10 at the same hour the flow was 20,000 gallons. This would indicate a flowage from house connections of 11,000 gallons per hour assuming it to be constant for that hour. On April 26, 1913, the flowage between 2:00 and 3:00 p.m. was 30,000 gallons at the disposal plant and an overflow of 10,000 gallons at the head of the pressure pipe. This would indicate a leakage of 29,000 gallons per hour, assuming that the house flow was 11,000 gallons. This was in medium wet weather. Thus it is assumed that the ground water leakage varies between 9,000 and 29,000 gallons per hour except in flood times. On January 8, 1914, when the conditions were similar to those on April 26, 1913, the flowage at the disposal plant was 34,000 gallons and the overflow at the head of the pressure pipe 6,000 gallons. This shows the pressure pipe carrying 4,000 gallons per hour more than at the previous date and is due to the fact that the pressure line had been raised two feet at the upper end. It was found that the original head was not sufficient for the best action of the disposal plant. There has been no trouble with the pressure line since it was raised.

Figures 6 and 7 are drawn to represent the ground plan and cross-section of the disposal plant. It will be seen from the ground plan that the entire plant consists of a square which is surrounded by a moat. The dimensions to the outside of the moat are 150 feet each way. The moat is in general 12.5 feet wide. Within the moat are the filter beds which are composed of crushed stone which reach to the depth of the moat. A wall of stone separates the filter beds from the moat. This can be seen in the cross-section. Upon the filter beds are four systems of castiron pipes which connect with siphons in four wells in the corners of the contact beds. These pipe systems have trickling filters at irregular intervals which spray the liquid from the pipes upon the filter beds.

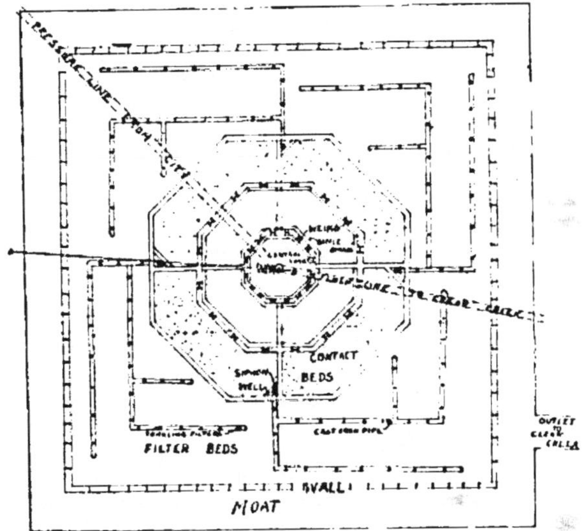

Fig. 6. Ground Plan Bloomington Sewage Disposal Plant.

Fig. 7. Section.

Rising some ten feet above the filter beds is an octagonal structure of concrete which contains the contact beds and septic tank. It will be noticed from the drawing, Fig. 6, that the inner division of this structure is undivided and that the middle and outer divisions are each divided into four divisions. The inner and middle divisions compose the septic tank. The capacity of the inner division is 13,600 cubic feet (101,728 gallons), of the pressure line 2,000 cubic feet (14,960 gallons), and of the twenty-four-inch sewer pipe 31,056 cubic feet (232,300 gallons). This totals approximately 350,000 gallons while the hourly flow as shown in the previous paragraph is 34,000 gallons. Thus when the main pipe lines and the central tank are full they hold ten hours' discharge. The pressure line enters the side of the central tank two or three feet above the bottom as shown in Fig. 7. The liquid rises in this tank and flows over sharp edged weirs into the middle divisions. There are baffle boards on each side of the weir so that the water must rise from the inner tank and sink into the middle division. It leaves the middle division through another set of weirs which lead into the contact beds. The contact beds are built of stone which gradually becomes finer toward the top. The bottom stones are very large and are placed far apart, leaving much room between. Upon these are placed two-man stones which are followed by smaller and smaller stones until the upper layer should be about the size of walnuts. The bottoms of these beds are connected with the wells at the corners. These have no stone in them and are occupied by the goose-neck siphon alone. When the liquid flows from the middle division upon the contact beds it runs down, gradually filling the bed and displacing the air. When the bed fills above the top of the siphon it immediately discharges and empties the contact bed entirely allowing it to fill with air again. When this discharge is made the liquid is sprayed upon the filter bed through which it finds its way into the moat and thence into Clear Creek.

The action of the anaerobic bacteria takes place in the main sewer lines and the inner division of the septic tank. In the inner division where the culmination of this work occurs is where the scum and sludge occur. The scum accumulates on the top of the liquid between the baffle boards and forms a covering which shuts out the light and air, thereby creating the proper condition for the work of the anaerobic bacteria. This scum has never been removed and does not accumulate beyond a certain limit. Sometimes it is thick enough to bear the weight of a man but again it is thinner. The sludge which accumulates in the bottom at a very slow rate is occasionally thrown into Clear Creek by a direct outlet for this purpose. This outlet is manipulated at will by a valve at its inner end and is only opened when Clear Creek is in flood. Some sludge also occurs in the middle division and there are vents between the middle and inner divisions on the floor to allow this sludge to be drained off by the outlet from the inner division. These vents are kept closed except when the sludge is being drained.

When the liquid flows over the weirs from the inner division it falls in a thin sheet for ten or twelve inches. The purpose of this arrangement is to aerate the liquid so that the aerobic bacteria will be able to complete the work which has been begun by the anaerobic. They carry on their work in the middle division, through the contact beds, through the filter beds and even into the moat. When the system is well attended the effluent which leaves the moat is clear and nonputrescent. Its bacterial content has never been determined. The care of the plant is very simple. The four divisions are made in the middle compartment and the contact beds for the purpose of resting the contact beds. Two parts will easily take care of the sewage while the other two parts rest. In two or three days these should be reversed. This is done by adjusting the weirs. Once a month the top layer of rock in the contact beds should be raked over with a long-toothed rake and stirred to a depth of six or eight inches. The plant is simple and very effective when properly managed.

The entire original cost of constructing the Bloomington sewerage plant was $86,000, distributed as follows: disposal tank, $19,500, and pipe lines, $66,500. Within five years the additional costs were as follows: raising pressure line, $5,500; three or four miles of additional sewer, $7,000; maintenance for first five years, $200. Thus the total cost at the end of five years was $98,700. When the present city administration began its work $3,000 was spent to remove the top layers of rock from the contact beds and replace it with new stone. This was entirely unnecessary from the standpoint of the sewage disposal plant.

One of the most important problems connected with sewage disposal is the conservation of certain mineral substances which are continually being wasted thereby. The principal one of these is phosphorus. The percentage of phosphorus in sewage is very small and therefore the problem of its economical recovery is extremely difficult. The supply of phosphorus is so limited in the world that the escape of even a small amount becomes a grave matter. This is a problem for applied chemistry to solve.

A list of the principal cities of Indiana and their manner of sewage disposal follows:

City	System	Disposal Plant	Discharge
Albion*			
Anderson	Sewer	None	White River
Angola	Sewer	Small septic tank(a)	Pigeon River
Attica	Sewer	None	Wabash River
Auburn	Sewer	None	Cedar Creek
Aurora	Sewer	None	Ohio River
Bedford	Sewer	Anaerobic septic tank	White River
Bloomfield	Sewer	None	White River
Bloomington	Sewer	Complete disposal plant	Clear Creek
Bluffton	Sewer	None	Wabash River
Booneville	Sewer	None	Cypress Creek
Brazil*			
Brookville	Sewer	None	Whitewater River
Brownstown	Cesspool	None	
Cannelton	Sewer	None	Ohio River
Clay City	Cesspool	None	
Clinton	Sewer	None	Wabash River
Columbia City	Sewer	None	Blue River
Columbus	Sewer	None	White River
Connersville	Sewer	None	Whitewater River
Corydon	Cesspool and pail	None	
Covington*			
Crawfordsville	Sewer	None	Sugar Creek
Crown Point	Sewer	Septic tank	A small creek
Danville	Sewer	Septic tank(b)	A small creek
Decatur	Sewer	None	St. Mary's River
Delphi	Sewer	None	Deer Creek
Elkhart	Sewer	None	St. Joseph River
English	Pail	None	
Evansville	Sewer	None	Ohio River
Ft. Wayne	Sewer	None	Maumee River
Fowler	Sewer	None	A dredge ditch
Frankfort	Sewer	None	Prairie Creek
Franklin	Sewer	None	Hurricane Creek
Gary	Sewer	None	Little Calumet River
Goshen	Sewer	None	Elkhart River
Greencastle	Sewer	Septic tank	A small creek
Greenfield	Sewer	None	Brandywine Creek
Greensburg	Sewer	Complete disposal plant	A small creek
Hammond	Sewer and pump	None	Grand Calumet River
Hartford City	Sewer	None	Lick Creek
Huntington	Sewer	None	Little Wabash River
Indianapolis	Sewer	None(c)	White River
Jasper*			
Jeffersonville	Sewer	None	Ohio River
Kentland	Sewer	None	Kent State Ditch
Knox	Sewer	None	A small creek
Kokomo	Sewer	None	Wild Cat Creek
Lafayette	Sewer	None	Wabash River
Lagrange	Sewer	None	Fly Creek
Laporte	Sewer	None	Kankakee River
Lawrenceburg	Cesspool and a few sewers	None	Ohio River
Lebanon	Sewer	None	A small creek
Liberty	Cesspool	None	
Logansport	Sewer	None	Wabash River
Madison	Sewer	None	Ohio River
Marion	Sewer	None	Mississinewa River
Martinsville*			
Michigan City	Sewer	None	Lake Michigan

City	System	Disposal Plant	Discharge
Monticello	Sewer	None	Tippecanoe River
Mount Vernon*			
Muncie	Sewer	None	White River
Nashville	Cesspool	None	
New Albany	Sewer	None	Ohio River
Newcastle	Sewer	None	Blue River
Newport	Cesspool	None	
Noblesville	Sewer	None	White River
North Vernon	Sewer	None	Muscatatuck River
Paoli	Cesspool(d)	None	Lick Creek
Peru	Cesspool and pail (e)	None	Country
Petersburg	Partial sewer	None	Prides Creek
Plymouth	Sewer	*	*
Portland	Sewer	None	Salamonie River
Princeton	Partial sewer	*	*
Rensselaer	Sewer	None	Iroquois River
Richmond	Sewer	Septic tanks (a)	Whitewater River
Rising Sun	Cesspool and gutter	None	Ohio River
Rochester	Sewer	None	Ditch to Tippecanoe River
Rockport*			
Rockville	Cesspool	None	
Rushville	Sewer	None	Flat Rock River
Salem*			
Scottsburg	Cesspool	None	
Seymour	Sewer	None	White River
Shelbyville	Sewer	None	Blue River
Shoals*			
South Bend	Sewer	None	St. Joseph River
Spencer	Cesspool and partial sewer	None	White River
Sullivan	Sewer	None	Buck Creek
Tell City	Sewer	None	Ohio River
Terre Haute	Sewer	None	Wabash River
Tipton	Sewer	None	Cicero Creek
Valparaiso	Sewer	None	Salt Creek
Vernon*			
Versailles	Cesspool(f)	None	
Vevay	Cesspool and gutter	None	Ohio River
Vincennes	Sewer	Complete disposal plant	Wabash River
Wabash	Sewer	None	Wabash River
Warsaw	Sewer	None	Center Lake
Washington	Sewer	None	White River
Williamsport	Sewer	None	Wabash River
Winamac	Sewer	None	Tippecanoe River
Winchester	Sewer	None	White River

*No data.

(a) Septic tank disposes of part of sewage.

(b) Septic tank designed for city of 6,000 population but sewers near edge of city do not enter it and flow directly into stream.

(c) Plans under way for complete disposal system.

(d) Sewer from court house and jail.

(e) Garbage hauled by contract at city's expense.

(f) Sewer from court house.

Section II.

RIVER SYSTEMS OF INDIANA.

The Whitewater System.

Whitewater River is located in southeastern Indiana. It rises by two main branches in southern Randolph and Wayne counties. The west branch flows in a general southerly direction past Cambridge City and Connersville. Between Laurel and Metamora in Franklin County it bends toward the east and flows in that direction for eleven miles to Brookville, where it is joined by the east branch. The main stream bends immediately toward the southeast and flows in that direction to its mouth at Valley Junction, Ohio, where it empties into the Big Miami River. The east branch flows in a general southerly direction through Richmond to Brookville, where it joins the west branch. It is parallel to the west branch and about ten miles east of it.

Whitewater valley is situated in the rocks of the Cincinnati series. The west bluff of the west branch, throughout its course above Metamora, is capped by a considerable thickness of limestone of Silurian age (Brassfield and Niagara). The Niagara forms a distinct divide parallel to and just west of the west branch, along its upper course. The crest of this divide forms the western edge of the Whitewater basin. This condition causes the western tributaries of the west branch to be very short and very swift streams.

The Whitewater basin lies entirely within the area covered by the Illinois glacier. The Wisconsin glacial boundary makes a great bend northward in this vicinity. It crosses the west branch near Alpine and the east branch near Fairfield in Franklin County. All the larger tributaries have their sources in the Wisconsin glacial area. The main parts of the trunk streams, however, lie outside of this area. A great valley train, which fills the valley to a depth of approximately one hundred feet, extends throughout its course south of the Wisconsin glacial line. The valley-train is composed of sand and gravel. The headwaters of both branches are within the deep glacial deposits of the Wisconsin glacial area. These conditions make the discharge of the stream fairly constant.

The drainage basin of the Whitewater River in Indiana occupies practically four counties, Wayne, Fayette, Union and Franklin. Small portions of these counties drain to other streams and small portions of other counties drain into Whitewater. The drainage area of Whitewater River in Indiana is 1,319 square miles, of which the basin of the East branch comprises 314, the West branch 847, and the main basin below the junction of the branches 158 square miles. This forms 3.61% of the area of the state. The East branch receives the runoff from about fifty square miles of land in Ohio.

The U. S. Weather Bureau has five observation stations in or near this basin, and the record of these stations will be used in computing the amount of water furnished to the basin. The stations are Richmond and Cambridge City, Wayne County; Connersville, Fayette County; Mauzy, Rush County; and Greensburg, Decatur County. The following table shows the mean annual precipitation in inches for these stations for the years 1900 to 1915, inclusive.

	1900	1901	1902	1903	1904	1905	1906	1907
Richmond	40.59	26.97	37.56	34.08	35.65	41.72	31.71	48.78
Cambridge City	39.65	31.55	40.25	41.69	40.39	*	34.21	46.54
Connersville	40.26	26.13	38.75	36.09	37.61	46.96	40.93	43.80
Mauzy	39.17	31.11	43.23	40.11	40.26	48.94	41.20	47.08
Greensburg	*	*	44.59	42.29	41.66	43.88	37.88	43.09
Mean	39.92	28.94	40.88	38.85	39.11	45.37	37.19	45.86
	1908	**1909**	**1910**	**1911**	**1912**	**1913**	**1914**	**1915**
Richmond	33.64	48.38	36.28	37.53	32.22	49.21	34.05	38.23
Cambridge City	32.57	48.99	39.59	41.13	39.56	57.23	27.04	44.16
Connersville	37.32	42.73	36.77	29.57	31.99	48.16	29.03	43.35
Mauzy	33.96	45.70	42.31	37.49	35.86	51.83	30.86	42.22
Greensburg	32.61	*	38.38	*	38.12	*	*	45.31
Mean	34.02	46.45	38.67	36.43	35.55	51.61	30.24	42.65
Average mean annual for sixteen years								39.48

*No data.

The rainfall of this basin as shown by this record is slightly higher than the average rainfall of the state for the same period. If all the water were carried away it would represent a discharge of 4,537 cubic feet per second for the ten years. The runoff from any region in Indiana is on the average about 35% of the rainfall. This would give an average runoff of about 1,500 cubic feet per second for the ten years. The rest is lost by evaporation, etc.

By evaporation is meant both the direct evaporation into the air and that taken up by plants and animals. Another source of loss is through seepage in the underlying strata. The conditions for such loss are good in this valley in one respect, i. e., the dip of the underlying strata is about thirty-five feet per mile toward the west and the basin skirts the edge of a thickness of three to four hundred feet of these strata. On the other hand, the underlying strata are composed of limestone and shale which are almost impervious to water. Hence it is probable that there is little loss from this cause. A greater source of loss in this valley is from the underflow which penetrates the valley-train, to which reference has already been made. The loss from this cause is great because the sand and gravel are very clean and hence very porous. This loss could be overcome by building a subsurface dam to the solid rock beneath the valley-train. This would entail heavy expense and would not be profitable for any hydraulic project at the present time.

Gaging Station on Whitewater River at Cedar Grove, Indiana.—This station is located at a wagon bridge which crosses Whitewater River at Cedar Grove. The gage is of the chain and weight type and is attached to the west hand rail of the bridge fifteen feet north of the south pier. The length of the chain from the initial marker to the end of the weight is 44.96 feet. The zero of the gage is 39.14 feet below a bench mark, which is three inches each way from the east corner of the concrete east wing of the south abutment of the bridge. The channel of the stream is straight for three hundred feet above and five hundred feet below the bridge. The bed of the stream is of sand and gravel and is very unstable. This causes considerable changing in the cross-section of the stream. The original gage was located on the wagon bridge at New Trenton, but the gage and records were swept away by the flood of 1913. The present gage was installed on November 5, 1915. It has been read daily since its installation by Alfred and Bess Brown.

Discharge measurements are made from the west hand rail of the bridge, every ten feet from the initial point, which is flush with the south abutment of the bridge.

Daily Gage Heights on Whitewater River at Cedar Grove, Indiana, for 1915.

Day	November	December	Day	November	December
1	....	3.4	17	3.0	9.8
2	....	3.3	18	3.0	7.3
3	....	3.3	19	6.3	7.1
4	*	3.3	20	5.4	6.1
5	3.0	3.3	21	5.4	6.0
6	3.0	3.3	22	4.8	5.8
7	3.0	3.2	23	4.8	5.5
8	3.0	3.1	24	3.9	5.5
9	3.0	3.1	25	3.6	5.3
10	3.0	3.0	26	3.5	5.3
11	3.0	3.1	27	3.6	5.2
12	3.1	3.2	28	3.6	5.5
13	3.1	3.2	29	3.5	8.5
14	3.1	3.2	30	3.4	7.4
15	3.0	3.8	31	....	6.7
16	3.0	4.0			

*Gage installed Nov. 5.

Daily Gage Heights on Whitewater River at Cedar Grove, Indiana, for 1916

Day	Jan.	Feb.	Mar.	Apr.	May	June	July	Aug.	Sept.	Oct.	Nov.	Dec.
1	10.3	10.7	4.1	4.5	3.9	3.4	2.6	1.9	1.9	1.8	1.6	1.8
2	13.3	6.7	3.9	4.3	3.6	3.3	2.5	1.8	1.8	1.8	1.6	1.7
3	7.2	5.1	3.6	4.2	3.8	6.1	2.4	1.8	1.8	1.8	1.6	1.7
4	6.2	5.0	3.4	4.0	4.7	6.8	2.4	2.6	1.8	1.7	1.6	1.7
5	6.0	4.7	3.2	3.8	4.2	6.6	2.3	2.6	1.8	1.7	1.6	2.0
6	5.8	4.3	3.5	3.6	3.7	6.5	2.3	2.4	1.8	1.7	1.6	2.0
7	5.5	4.2	7.0	3.5	6.2	6.3	2.2	2.4	1.9	1.7	1.6	1.9
8	5.2	4.0	5.1	3.5	4.8	5.6	2.2	6.6	1.8	1.6	1.6	2.0
9	5.0	4.9	4.3	3.7	4.2	5.1	2.4	5.8	1.8	2.0	1.6	2.4
10	5.0	4.9	4.3	3.7	3.9	4.8	2.4	3.2	1.8	1.9	1.6	2.4
11	7.0	4.8	3.9	3.7	3.5	4.7	2.4	2.8	1.8	1.7	1.6	2.3
12	13.5	8.0	3.6	3.5	3.5	4.2	2.2	2.6	1.8	1.7	1.6	2.2
13	10.7	6.1	3.7	3.4	3.3	3.8	2.2	2.5	1.7	1.7	1.6	2.1
14	6.3	5.7	3.7	3.3	3.1	3.5	2.2	2.5	1.7	1.6	1.6	2.0
15	5.8	4.3	3.8	3.2	3.0	3.5	2.2	2.4	1.7	1.6	1.6	1.9
16	*	4.4	3.7	3.0	2.9	3.3	2.2	2.2	1.6	1.6	1.6	1.9
17	*	4.6	3.6	2.9	2.8	3.0	2.2	2.1	1.6	1.6	1.6	1.9
18	*	5.0	3.5	2.8	2.7	3.0	2.2	2.0	1.6	1.6	1.6	1.9
19	5.8	4.2	3.5	2.8	2.7	3.2	2.1	2.0	1.6	1.8	1.7	1.9
20	6.0	4.1	3.5	2.9	2.6	3.0	2.1	2.7	1.6	3.0	1.7	1.9
21	6.5	4.2	3.5	10.9	2.5	2.9	2.1	2.5	1.6	2.6	1.7	1.9
22	8.1	3.9	3.7	6.3	2.4	2.9	2.1	2.3	1.6	2.2	1.7	1.9
23	6.4	3.8	4.0	5.4	2.4	2.8	2.1	2.0	1.6	1.9	1.7	1.8
24	6.1	4.4	4.2	4.7	2.6	2.7	2.0	1.9	1.6	1.8	1.7	1.8
25	5.7	4.2	4.6	4.5	2.8	2.8	2.0	1.9	1.6	1.8	1.9	1.8
26	*	4.0	4.8	4.7	2.9	2.9	2.0	1.8	1.6	1.8	1.8	1.9
27	*	4.4	8.7	4.8	3.0	2.9	2.0	1.9	1.6	1.7	1.8	6.7
28	*	4.6	7.2	4.7	3.4	2.8	2.0	1.9	1.9	1.7	1.8	5.0
29	*	4.5	5.8	4.5	4.0	2.7	2.0	1.9	2.0	1.7	1.8	4.3
30	13.2	....	4.9	4.2	3.7	2.6	1.9	1.9	1.8	1.7	1.8	3.2
31	17.2	....	4.8	....	3.5	.....	1.9	1.9	....	1.7	....	3.0

*No record.

Discharge Measurements on Whitewater River at Cedar Grove, Indiana, during 1915-1916.

Date	Hydrographer	Gage Height, Feet	Discharge, Sec.-Ft.
November 5, 1915	W. M. Tucker	3.00	991
December 15, 1915	W. M. Tucker	3.90	1,703
December 28, 1915	W. M. Tucker	7.20	6,212
February 5, 1916	W. M. Tucker	4.45	2,247
April 22, 1916	W. M. Tucker	5.90	4,064
April 23, 1916	W. M. Tucker	4.90	2,651
July 4, 1916	W. M. Tucker	2.60	603
July 7, 1916	W. M. Tucker	2.40	469
November 12, 1916	W. M. Tucker	1.60	88

Rating Table for Whitewater River at Cedar Grove, Indiana, for 1915-1917

Gage Height, Feet.	Discharge, Sec.-Ft.	Gage Height, Feet.	Discharge, Sec.-Ft.	Gage Height, Feet.	Discharge, Sec.-Ft.
1.6	90	3.1	986	4.6	2,358
1.7	135	3.2	1,071	4.7	2,454
1.8	183	3.3	1,158	4.8	2,553
1.9	230	3.4	1,248	4.9	2,656
2.0	278	3.5	1,341	5.0	2,763
2.1	325	3.6	1,434	5.1	2,876
2.2	374	3.7	1,526	5.2	2,994
2.3	421	3.8	1,619	5.3	3,123
2.4	469	3.9	1,711	5.4	3,260
2.5	518	4.0	1,803	5.5	3,406
2.6	585	4.1	1,895	5.6	3,560
2.7	662	4.2	1,986	5.7	3,716
2.8	738	4.3	2,078	5.8	3,877
2.9	818	4.4	2,170	5.9	4,041
3.0	905	4.5	2,264	6.0	4,208

Rating Table for Whitewater River at Cedar Grove, Indiana, for 1915-1917 —Continued.

Gage Height, Feet.	Discharge, Sec.-Ft.	Gage Height, Feet.	Discharge, Sec.-Ft.	Gage Height, Feet.	Discharge, Sec.-Ft.
6.1	4,374	7.6	6,872	9.1	9,369
6.2	4,541	7.7	7,038	9.2	9,536
6.3	4,707	7.8	7,205	9.3	9,702
6.4	4,874	7.9	7,371	9.4	9,869
6.5	5,040	8.0	7,538	9.5	10,035
6.6	5,207	8.1	7,704	9.6	10,202
6.7	5,373	8.2	7,871	9.7	10,368
6.8	5,540	8.3	8,037	9.8	10,535
6.9	5,706	8.4	8,204	9.9	10,701
7.0	5,873	8.5	8,370	10.0	10,868
7.1	6,039	8.6	8,537	....	....
7.2	6,206	8.7	8,703	....	....
7.3	6,372	8.8	8,870	....	....
7.4	6,539	8.9	9,036	....	....
7.5	6,705	9.0	9,203	....	....

NOTE—This table is applicable only for open channel conditions. It is tangent above ten feet with a variation of 167 sec. ft. per tenth.

1916	1 Average Rainfall, Inches	2 Rainfall per acre Cu. Ft.	3 Discharge Cu. Ft. per Sec.	4 Discharge Cu. Ft per Acre	5 Per Cent 4 of 2
January	7.01	25,446	7,289	23,658	92.97
February	1.17	4,247	2,972	9,023	12.46
March	2.98	10,817	2,342	7,601	70.26
April	3.18	11,543	2,135	6,705	58.09
May	3.90	14,157	1,327	4,306	30.41
June	5.27	19,131	1,912	6,004	31.38
July	1.99	7,224	368	1,194	16.53
August	3.94	14,302	665	2,159	15.09
September	3.00	10,890	149	466	4.28
October	1.81	6,570	195	634	9.65
November	1.71	6,207	119	374	6.02
December	3.08	11,180	607	1,969	17.61
Annual	....	141,714	....	64,093	45.23

Whitewater River is one of the most picturesque rivers in Indiana. The presence of the vast amount of sand and gravel of the valley-train makes it clean and pleasant to the eye. The wide, level valley floor is very fertile and with the bluffs usually wooded the view presented to one riding through the valley is of surpassing beauty. The river itself is a succession of reaches and rapids, which make it an excellent stream for boating and fishing, thereby attracting the summer camper. Many picturesque fishing camps are located between Connersville and the state line.

While this stream is quiet and safe in times of ordinary discharge, it becomes very dangerous during flood periods. The small, swift tributaries on the west discharge very quickly into the West branch and the discharge from the East branch having a shorter distance to travel they arrive simultaneously at the junction, causing severe floods in the main valley. During the flood of 1913 in March the water was constricted by bridges at Brookville on the East branch. These bridges finally gave way and the resulting rise in the main valley was almost like the rush of a tidal wave. Houses were submerged and carried away before the occupants were able to escape. At Cedar Grove the entire farm of a Mr. Wilhelm was swept clear of buildings which had never been in danger before. The loss to this man alone was approximately $10,000. He has rebuilt his buildings on ground about thirty feet higher than the previous site.

Near the railroad station at New Trenton was an old covered bridge which had seen service for many years. At the end of the bridge was an elevator, coal yard, lumber yard and saloon belonging to Mr. Brown. In previous floods the water had reached the floors in some of these buildings. During the flood of 1913 the bridge, station, and entire business of Mr. Brown were swept away almost instantly. It was stated by an eyewitness that at one moment everything was in place and in two minutes not a stick was left. Not only were the buildings swept away but the ground on which they had stood had disappeared after the flood receded. One man was drowned, another swept away to be rescued over twenty-four hours later from an island at Harrison, Ohio, where he had managed to cling to some drift. Mr. Brown's loss was about $30,000. These incidents are merely a few of many such, and illustrate something of the danger and hardship which must be endured in time of flood on streams of this character.

The bottom land along Whitewater River lies in terraces called first and second bottoms. The first bottoms are low and are overflowed during any ordinary flood. The second bottom lies about thirty feet higher and is only covered during the most extreme floods. These bottoms are very fertile and are used almost exclusively for corn, although some wheat, oats, rye and vegetables are also raised. The tributary valleys have not been studied with a view to using them as storage reservoirs for irrigation water, but a general knowledge of their character seems to lead to the conclusion that such reservoirs could be constructed to irrigate the entire valley with little trouble. This problem will be investigated more fully.

Three cities draw their water supply from Whitewater River. Connersville takes its supply from the West branch by a canal seven miles long and uses the water for power to pump the city supply. The cities of Hagerstown and Cambridge City are located about twenty and fifteen miles, respectively, above Connersville, and as a safeguard against the contamination of the water by these cities the Connersville water supply is carefully filtered before using. Richmond takes its supply from the East branch. No cities are located on this fork above Richmond, but the water is filtered to remove solid matter and as a safeguard against possible contamination. Brookville takes its supply from the East branch one-half mile above the junction of the two branches. The water of Whitewater except in flood is very clear and apparently pure. The presence of the great quantity of sand through which it flows tends to make it an exceptionally clear stream.

The water power now used on the Whitewater River is a very small per cent of the available power. The East branch and the main stream have no developed power. The West branch has two developed systems. One is at Connersville and the other at Metamora and Brookville. Both are of the feeder dam type. In the early part of last century, a commercial canal was built by the government along the main stream and the West branch. It extended from the Ohio River up the Whitewater River and northward. In the latter part of the last century this canal was abandoned for transportation purposes. Hydraulic companies have taken advantage of this abandoned canal for the construction of power systems. Seven miles north of Connersville a dam has been constructed across the West branch and the water turned into this canal. The canal conducts the water to Connersville, where it is used for power. The total fall on this canal from the crest of the dam to the tail race of the Connersville Hydro-electric Plant is seventy-seven feet. Of this fall about fifty-four feet are used for producing power. All of this power is employed by the Connersville Hydro-electric Company, except 8.1 feet which is used by the McCann Milling Company. The Connersville Hydro-electric Company has two plants, one of which is above the McCann Milling Company's plant and the other below. The following table gives the elevations of several points along the canal and the fall between them. The elevations are based on the elevation of the L. E. & W. railroad station, which is given in the U. S. Geological Survey Dictionary of Altitudes as 828 feet above sea level.

	Elevation feet	Fall feet
Tail race lower hydro-electric plant	784.48	
Head race lower hydro-electric plant	812.28	27.80
Tail race McCann Milling Company plant	813.80	1.52
Head race McCann Milling Company plant	821.90	8.10
Tail race upper hydro-electric plant	823.34	1.44
Head race upper hydro-electric plant	841.22	17.88
Crest of dam, seven miles from city	861.43	20.21
Base of dam	852.19	
Height of dam	9.24	

The power produced on this system is approximately 400 h. p., of which the McCann Milling Company uses about 50 h. p. and the Connersville Hydro-electric Company the balance.

One mile below Laurel, Franklin County, is another feeder dam which turns the water from the river into the abandoned commercial canal. The canal conducts the water sixteen miles to Brookville, where it empties into the river. The total fall from the crest of the dam to the tail race at Brookville is eighty-five feet. Of this fall twenty-eight feet are used for power. At Metamora, five miles below the dam, the Metamora Flour Mill utilizes eight feet and produces thirty horse power. At Brookville, sixteen miles below the dam, the Thompson and Norris Paper Mill utilizes twenty feet and produces two hundred seventy-five horse power. The usual flow of the canal is one hundred and fifty second feet, although the amount of water in times of drought is not over one-third of this amount. The low water discharge at the gage at Cedar Grove is eighty-eight second feet, which represents the combined flow of the east and west branches of the river.

There is no developed power on the East branch nor upon the main stream. Because of the shifting nature of the bed of this stream, the frequency of floods, and the small discharge in times of drought, it is impractical to develop more power.

White River System.

White River drains the south central part of Indiana. Its basin comprises somewhat less than one-third of the area of the state, 30.5%. It rises in central and southeastern Indiana by numerous branches, which unite to form two main branches. The general direction of drainage is toward the southwest. The east and west branches unite at the southwest corner of Daviess County. The main stream flows from this point to the Wabash River at Mt. Carmel, Illinois. The drainage area of the whole system is approximately 11,155 square miles. The main stream drains 182 square miles, the east branch 5,689 square miles and the west branch 5,284 square miles. Each branch has one large tributary which joins it in its middle course. The Muscatatuck is tributary to the east branch, and Eel River to the west branch. Muscatatuck has a drainage area of 1,112 square miles nd Eel River of 1,161 square miles.

The U. S. Weather Bureau maintains gaging stations at Anderson, Indianapolis and Elliston on the west branch, at Shoals on the east branch and at Decker on the main stream. The State Department of Geology and Natural Resources maintains a gaging station at Anderson on the west branch. The location of these gages is shown in fig. 2. A private gage was kept for a short time in 1909 and 1910 on the east branch at Tannehill Bridge, which is one mile west of Taylorville in Bartholomew County. The data from this gage is given in the 35th Annual Report of the Indiana Department of Geology and Natural Resources, pages 46 and 47. The readings from the gages mentioned are taken daily. Previous to 1916 the gage readings at Elliston and Decker were kept for seven months of the year only, December to June, inclusive. This arrangement was carried on by the Weather Bureau in order to procure data on floods, most of which occur in these months. In 1916 an arrangement was made between the Weather Bureau and the Department of Geology and Natural Resources of Indiana by which the gages are to be read daily throughout the year. The Department of Geology and Natural Resources established a gage at Anderson in 1910 before the establishment of the one by the Weather Bureau. When the Weather Bureau did establish a gage at this point the readings were only taken for seven months, so the State Department continued to maintain its gage, which fact accounts for the duplication for these gages.

The gage with the best record is that at Shoals. The records at this gage have been kept since 1903 with a few short interruptions. The area which discharges through this gage is 4,914 square miles. The Anderson gage has almost a continuous record since 1911. The area drained through this gage is 499 square miles. The continuous records at Elliston and Decker began in December, 1915, and the records taken by the Weather Bureau for 1916 are not yet available, so that the available data is as yet useless. The area drained through the Elliston gage is 4,066 square miles and through the Decker gage, 10,717 square miles. The gage at Indianapolis is read every day and has a long record, but for the purposes of this paper the records are of little use. The gage is located below the dam of the Indianapolis Water Company at Broad Ripple, where most of the water is drawn off in dry seasons. Thus the gage gives no record of the minimum discharge of the river at this point. The area of White River basin above this gage is 1,610 square miles.

East Branch of White River. The east branch of White River rises along the crest of the Niagara escarpment in Henry, Fayette, Rush, Decatur, Ripley and Jefferson counties. The tributaries from this escarpment are long streams with slight fall. The largest tributary is known as Blue River, in Henry, Rush and Shelby counties. It rises in the Wisconsin glacial area, which it leaves in Bartholomew County. It then flows for a short distance in the Illinois glacial area and enters the unglaciated region in Jackson County. It flows directly across the unglaciated region and re-enters the Illinois glacial area in Daviess County, in which it continues to its mouth. The Wisconsin glacial deposits at the source of this stream tend to regulate the flow so that it never ceases, even in northern Rush County, where the stream is very small. A long valley-train of glacial material occurs in Bartholomew and Jackson counties, diminishing in Lawrence and Martin counties. This valley-train covers the underlying strata and leaves few bed rock dam sites. This stream flows across every rock formation in the state except the Ordovician, but only occasionally is bed rock exposed in the river bed. These exposures occur where the stream meanders against one of its bluffs.

The valley of the east branch of White River is everywhere broad and level. It is an excellent farming region. Frequent floods occur which cover the lowland for great distances. During these floods the river often makes radical changes in its course. The loose sand and gravel of which the bed of the stream is composed is easily shifted by the flood water. Gradual changes are constantly going on whereby the stream in time entirely changes its course. These conditions hinder the installation of water power stations.

The rainfall in the White River basin is very near the average of the state. Several observation stations are located in the basin and the average of these for the last eighteen years is slightly more than forty inches. What part of this is discharged is not definitely known. Records of the east branch at Shoals, is furnished by the U. S. Geological Survey gage, and by the rainfall records of the U. S. Weather Bureau, indicate that the runoff at this point averaged 35.04% of the rainfall for the eight years during which continuous records were kept. These data are given in the tables which are computed entirely from government reports. They show the average depth of monthly rainfall and the number cubic feet per acre for the basin of the east branch of White River above the gage at Shoals for eighteen years, 1898 to 1915, inclusive. These averages were taken from twenty U. S. Weather Bureau recording stations. These stations are listed on page 329.

Table showing average depth of monthly rainfall in inches; and cubic feet per acre of rainfall for east branch of White River above the gage at Shoals, Ind., 1898-1915 inclusive.

Drainage Area 4,913.5 sq. mi.—3,144,640 Acres.

	1898		1899		1900		1901		1902		1903	
	In.	Cu. ft.	In.	Cu. ft.	In.	Cu. ft.	In.	Cu. ft.	In.	Cu. ft.	In.	Cu. ft.
January	7.08	25,700.4	3.81	13,830.3	2.57	9,329.1	1.50	5,445.0	1.92	6,969.6	2.86	10,381.8
February	1.85	6,715.5	2.67	9,692.1	3.78	13,721.4	1.82	6,606.6	1.00	3,630.0	5.89	21,380.7
March	9.05	32,851.5	4.59	16,661.7	1.97	7,151.1	3.35	12,160.5	2.99	18,053.7	4.31	15,645.3
April	1.49	5,408.7	2.25	8,167.5	1.82	6,606.6	3.39	12,305.7	2.45	8,895.5	3.88	14,084.4
May	3.53	12,813.9	3.77	13,685.1	4.85	17,605.5	2.29	8,312.7	5.24	19,021.2	3.16	11,470.8
June	3.42	12,414.6	3.36	12,196.8	4.17	15,137.1	4.20	15,246.0	6.68	24,248.4	5.12	18,585.6
July	3.44	12,487.2	3.03	10,998.9	4.87	17,678.1	1.33	4,827.9	2.91	10,563.3	3.29	11,942.7
August	3.49	12,268.7	3.28	11,906.4	2.72	9,873.6	3.35	12,160.5	2.21	8,022.3	3.47	12,596.1
September	5.01	18,186.3	1.14	4,138.2	1.49	5,408.7	1.57	5,699.1	4.76	17,278.8	1.33	4,827.9
October	4.36	15,826.8	2.73	9,909.9	2.21	8,022.3	2.14	7,768.2	3.04	11,035.2	2.73	9,909.9
November	2.76	10,018.8	2.59	9,401.7	3.66	13,285.8	1.20	4,356.0	4.39	15,935.7	2.22	8,058.6
December	2.87	10,055.1	3.44	12,487.2	1.51	5,481.3	4.24	15,391.2	4.80	17,424.0	2.05	7,441.5

	1904		1905		1906		1907		1908	
	In.	Cu. ft.	In.	Cu. ft.	In.	Cu. ft.	In.	Cu. ft.	In.	Cu. ft.
January	5 01	18,186.3	2.23	8,091.9	2.76	10,018.8	8.19	29,729.7	1.79	6,497.7
February	2.76	10,018.8	2.27	8,240.1	1 38	5,009 4	0.51	1,851.3	6.16	22,360.8
March	9.08	32,960.4	2 83	10,272.9	6 66	24,175.8	5.58	20,255 4	4.57	16,589.1
April	2 38	6,639.4	3.64	13,213.2	1.90	6,897.0	2.80	10,164.0	4.76	17,278 8
May	3 35	12,160.5	5.75	20,872.5	2.03	7,368.9	3.47	12,596.1	7.02	25,482.6
June	3.68	13,358.4	3.08	11,180.4	3.43	12,450.9	4.74	17,206 2	2.15	7,804 5
July	2 00	7,260.0	3 79	13,757.7	2.39	8,675.7	5.06	18,367.8	1.99	7,223.7
August	1.44	5,227.2	5.92	21,489.6	4.43	16,080.9	3.09	11,216.7	2.06	7,477.8
September	3.34	12,124.2	2.44	8,857.2	5.23	18,984.9	2.58	9,365 4	0.78	2,831.4
October	1.26	4,573.8	6 43	13,176.9	1.45	5,263.5	3.67	13,322.1	0.29	1,052.7
November	0.59	2,141 7	2 29	8,312.7	4.52	16,480.2	3.17	11,507.1	1.55	5,626.5
December	4.99	18,113.7	2.66	9,655.8	4.37	15,863.1	4.11	14,919.3	1.47	5,336.1

	1909		1910		1911		1912		1913	
	In.	Cu. ft.	In.	Cu. ft.	In.	Cu. ft.	In.	Cu. ft.	In.	Cu. ft.
January	2 57	9,329.1	3 76	13,648.8	3.78	13,721.4	2 76	10,018 8	9 53	34,593.9
February	6.51	23,631.3	4.21	15,282.3	2.00	7,260.0	3.15	11,434.5	[illegible] 82	6,696.6
March	3.14	11,398.2	0.21	762.3	3.17	11,507.1	5 50	19,965 0	1 .86	39,385.5
April	4 97	18,041.1	3.43	12,450.9	6.40	23,232.0	5 87	21,308.1	4 23	15,354.9
May	5.67	20,582.1	4.42	16,044.6	3.19	11,579.7	5 12	18,585.1	1.86	6,751.8
June	4.07	14,774.1	2.46	8,929.8	3.79	13,757.7	2 48	9,002 4	2.20	7,986.0
July	6.41	23,268.3	7.80	28,314.0	1.56	5,662 8	5 63	20,436.9	2 76	10,018.8
August	2.30	8,349.0	1.72	6,243.6	2.55	9,256.5	6 33	22,977 9	3.11	11,289.3
September	2.29	8,312.7	3.54	12,850.2	7.08	25,700.4	3.62	13,140 6	2.75	9,982.5
October	4 72	17,133.6	8 10	29,403.0	4.74	17,206.2	1.83	6,642 9	3.26	11,833.8
November	2.61	9,474.3	1.83	6,642.9	3.56	12,922.8	1.02	3,702 6	4.49	16,289.7
December	3.46	12,559.8	2.21	8,022.3	3.55	1 ,886.5	2.01	7,296 3	1.28	4,646.4

	1914		1915	
	In.	Cu. ft.	In.	Cu. ft.
January	2.20	7,986.0	3.30	11,979.0
February	3.87	14,048.1	1.59	5,771.7
March	2.85	10,345.5	1.51	5,481.7
April	3.72	13,503.6	1.25	4,537.5
May	1.53	5,553.9	5.10	18,513.0
June	2.34	8,494.2	2.72	9,873.6
July	2.53	9,183.9	6.19	22,469.7
August	5.58	10,255.4	8.58	31,000.2
September	2.32	8,421.6	3.48	12,632.4
October	2.25	8,167.5	1.82	6,606.6
November	1.00	3,630.0	3.10	11,253.0
December	3.05	11,071.5	5.00	18,150.0

White River (East Branch) at Shoals, Ind. This station was established June 25, 1903, by A. C. Lootz. It is located at the highway bridge, in the village of Shoals, Ind., four hundred feet above the Baltimore and Ohio Southwestern Railroad bridge. There are rapids just below this station and also about 5½ miles below. The gage is read once each day by Mr. O. H. Greist. The standard chain gage is fastened to the railing and metal posts of the down-stream side of the first span on the left end of the highway bridge. The length of the chain from the end of the weight to the marker is 46.41 feet. This gage is established to take the place of the original vertical gage, which was fastened to one of the piers. Discharge measurements are made from the three-span highway bridge, to which the gage is attached. The initial point for sounding is the face of the left abutment. The channel is straight above and below the station and the current is swift. The right bank is a high, rocky road-embankment, and never overflows; the left bank is a steep, rocky bluff and does not overflow. The bed of the stream is rocky and the channel is divided into three parts by the bridge piers. Bench mark No. 1 is the stone cap on the down-stream end of the first pier from the left bank. Its elevation is 100 feet above gage datum.

*Mean Daily Gage Height, in Feet, of White River (East Branch) at Shoals, Indiana, for 1903.**

Day	June	July	Aug.	Sept.	Oct.	Nov.	Dec.
1	....	64.5	63.8	63.5	63.4	63.5	63.7
2	....	64.4	64.7	63.5	63.4	63.5	63.7
3	....	64.3	64.8	63.5	63.4	63.5	63.7
4	....	64.1	65.3	63.5	63.5	63.5	63.6
5	....	64.0	66.1	63.5	63.5	63.5	63.6
6	....	64.2	66.9	63.5	63.5	63.5	63.5
7	....	64.2	66.8	63.5	63.6	63.5	63.5
8	....	64.1	66.0	63.5	63.7	63.5	63.5
9	....	64.1	65.5	63.5	63.8	63.5	63.6
10	....	64.1	65.1	63.5	63.9	63.5	63.6
11	....	64.0	64.9	63.5	63.9	63.5	63.5
12	....	64.0	64.6	63.5	63.9	63.5	63.5
13	....	63.9	64.4	63.5	64.0	63.5	63.5
14	....	63.9	64.1	63.5	64.0	63.5	63.5
15	....	63.8	64.1	63.5	63.9	63.5	63.5
16	....	63.8	64.0	63.4	63.9	63.5	63.5
17	....	63.8	64.0	63.4	63.8	63.5	63.5
18	....	63.9	64.0	63.4	63.7	63.7	63.5
19	....	63.8	63.9	63.4	63.7	63.7	63.5
20	....	63.7	63.9	63.4	63.6	63.7	83.6
21	....	63.7	63.9	63.4	63.6	63.8	64.1
22	....	63.5	63.8	63.4	63.5	63.9	64.2
23	....	63.5	63.8	63.4	63.5	61.0	64.2
24	....	63.5	63.7	63.4	63.5	63.9	64.5
25	....	63.5	63.7	63.4	63.5	63.8	64.6
26	....	63.5	63.7	63.4	63.5	63.7	64.9
27	61.3	63.5	63.6	63.4	63.5	63.7	65.2
28	61.3	63.6	63.6	63.4	63.5	63.6	65.0
29	64.4	63.6	63.6	63.4	63.5	63.6	65.0
30	64.5	63.6	63.5	63.4	63.5	63.7	64.8
31	....	63.6	63.5	....	63.5	....	64.6

*Water Supply and Irrigation Paper, U. S. Geological Survey No 98, page 217.

The observations of this station during 1903 have been made under the direction of E. Johnson, Jr., district hydrographer.

*Discharge Measurements of White River (East Branch) at Shoals, Indiana, in 1903.***

Date	Hydrographer	Gage Height Feet	Discharge Second-feet
June 22	A. C. Loots	....	*2,000
August 4	L. R. Stockman	65.07	3,392
September 24	L. R. Stockman	63.40	511

*Float Measurements.

**Water Supply and Irrigation Paper, U. S. Geological Survey No. 98, page 216.

*Rating Table for White River (East Branch) at Shoals, Indiana, from June 22 to December 31, 1903.**

Gage Height Feet	Discharge Sec.-Ft.	Gage Height Feet	Discharge Sec.-Ft.
63.4	510	66.8	6,590
63.5	640	67.0	6,970
63.6	770	67.2	7,350
63.7	910	67.4	7,730
63.8	1,050	67.6	8,110
63.9	1,200	67.8	8,490
64.0	1,350	68.0	8,870
64.1	1,510	68.2	9,250
64.2	1,680	68.4	9,630
64.3	1,860	68.6	10,010
64.4	2,045	68.8	10,390
64.5	2,230	69.0	10,770
64.6	2,415	69.2	11,150
64.7	2,600	69.4	11,530
64.8	2,790	69.6	11,910
64.9	2,980	69.8	12,290
65.0	3,170	70.0	12,670
65.2	3,550	70.5	13,620
65.4	3,930	71.0	14,570
65.6	4,310	71.5	15,520
65.8	4,690	72.0	16,470
66.0	5,070	72.5	17,420
66.2	5,450	73.0	18,370
66.4	5,730	73.5	19,320
66.6	6,210	74.0	20,270

*Water Supply and Irrigation Paper, U. S. Geological Survey No. 98, page 218.

Table made from measurements of August 4 and September 24, 1903, and January 21, 1904. Table should be accurate to limiting height in 1903.

*Mean Daily Gage Height, in feet, of White River (East Branch) at Shoals, Indiana, for 1904.**

Day	Jan.a	Feb.a	Mar.a	Apr.	May	June	July	Aug.	Sept.	Oct.	Nov.	Dec.b
1	64.6	66.2	67.6	91.0	67.0	64.8	65.3	63.8	63.3	63.5	63.2	63.2
2	64.5	65.5	68.1	88.8	66.5	65.6	65.1	63.7	63.3	63.5	63.2	63.2
3	64.3	65.2	68.2	87.2	65.9	66.0	65.0	63.7	63.3	63.5	63.2	63.2
4	64.2	64.9	68.7	85.6	65.6	66.0	64.8	63.7	63.3	63.4	63.2	63.2
5	64.1	64.8	68.6	84.4	65.4	65.6	64.7	63.7	63.3	63.4	63.2	63.2
6	64.1	68.5	69.1	83.4	65.3	65.4	64.5	63.7	63.2	63.3	63.2	63.2
7	64.1	71.5	72.3	80.2	65.1	65.0	64.4	63.7	63.2	63.3	63.2	63.2
8	64.0	72.8	72.9	73.2	65.0	64.9	64.3	63.7	63.2	63.3	63.2	63.2
9	64.0	74.3	71.5	68.4	64.9	64.7	64.3	63.7	63.2	63.3	63.2	63.2
10	64.0	76.1	70.3	67.7	64.8	64.6	64.3	63.6	63.3	63.3	63.2	63.2
11	64.0	77.0	70.5	67.5	64.7	64.5	64.4	63.6	63.3	63.3	63.2	63.2
12	64.0	76.0	69.8	67.3	64.7	64.4	64.5	63.6	63.3	63.3	63.2	63.2
13	64.0	72.5	69.1	67.0	64.7	64.4	64.5	63.5	63.3	63.3	63.2	63.2
14	64.0	68.5	68.9	66.8	64.6	64.3	64.5	63.5	63.3	63.3	63.2	63.2
15	64.0	66.5	68.7	66.5	64.6	64.4	64.4	63.5	63.3	63.3	63.2	63.0
16	64.0	66.0	68.0	66.2	64.5	64.3	64.5	63.5	63.2	63.3	63.2	63.0
17	64.0	65.9	67.6	66.0	64.5	64.5	64.3	63.5	63.2	63.3	63.2	63.1
18	64.0	65.6	67.5	65.8	66.5	64.6	64.2	63.5	63.4	63.3	63.2	63.1
19	64.1	65.4	67.6	65.6	66.5	64.7	64.2	63.5	63.7	63.3	63.2	63.2
20	64.1	65.2	67.7	65.4	64.5	64.9	64.2	63.5	63.6	63.3	63.2	63.2
21	64.6	65.1	67.7	65.3	64.5	64.9	64.2	63.6	63.5	63.3	63.2	63.2
22	69.4	67.3	68.0	65.2	64.5	64.9	64.2	63.6	63.5	63.2	63.2	63.2
23	73.3	68.3	74.4	65.1	64.6	64.8	64.2	63.6	63.4	63.2	63.2	63.2
24	73.5	69.6	75.2	65.1	64.7	64.8	64.1	63.5	63.4	63.2	63.2	63.3
25	74.2	70.9	78.4	65.2	64.6	64.8	64.1	63.4	63.4	63.2	63.2	63.4
26	74.5	70.6	87.1	66.5	64.6	64.8	64.0	63.4	63.7	63.2	63.2	63.9
27	74.8	69.4	87.7	67.5	64.6	64.8	64.0	63.3	63.8	63.2	63.2	64.5
28	75.0	68.0	92.8	68.6	64.6	65.2	63.9	63.3	63.8	63.2	63.2	65.2
29	72.2	67.3	95.0	68.4	64.6	64.8	63.9	63.3	63.7	63.2	63.2	66.0
30	67.8	......	94.9	67.8	64.6	65.4	63.8	63.3	63.5	63.2	63.2	66.3
31	66.5	......	93.4	......	64.7	......	63.8	63.3	......	63.2	......	66.0

*Water Supply and Irrigation Paper, No. 128, page 94.
aIce conditions January, February and March; uncertain.
bFrozen December 15 to 31.

*Discharge Measurements of White River (East Branch) at Shoals, Indiana, in 1904.**

Date	Hydrographer	Width Feet	Area of Section Sq. Ft.	Mean Vel. Ft. per sec.	Gage Height Ft.	Discharge Sec. Ft.
January 24	F. W. Hanna	....	4,105	4.61	73.47	19,010
March 5	F. W. Hanna	375	2,321	4.99	68.64	11,590
March 30	F. W. Hanna	427	13,410	6.00	95.20	79,820
May 5	F. W. Hanna and Johnson	356	1,124	3.72	65.43	4,180
June 16	F. W. Hanna	349	789	2.30	64.53	1,812
July 28	F. W. Hanna	307	515	1.60	63.88	823
August 24	F. W. Hanna	295	379	1.28	63.32	484
September 15	F. W. Hanna	295	373	1.06	63.24	397
October 20	F. W. Hanna	295	371	1.07	63.23	396
November 3	F. W. Hanna	288	324	.99	63.17	320

*Water Supply and Irrigation Paper, No. 128, page 93.

*Rating Table for White River (East Branch) at Shoals, Indiana, from January 1 to December 31, 1904.**

Gage Height Feet	Discharge Sec.-Ft.	Gage Height Feet	Discharge Sec.-Ft.
63.0	215	67.2	7,610
63.1	296	67.4	8,110
63.2	360	67.6	8,610
63.3	440	67.8	9,110
63.4	520	68.0	9,610
63.5	605	68.5	10,860
63.6	695	69.0	12,110
63.7	790	69.5	13,360
63.8	890	70.0	14,610
63.9	990	70.5	15,860
64.0	1,100	71.0	17,110
64.1	1,210	71.5	18,360
64.2	1,330	72.0	19,630
64.3	1,450	72.5	20,930
64.4	1,580	73.0	22,230
64.5	1,710	73.5	23,530
64.6	1,850	74.0	24,830
64.7	2,000	75.0	27,430
64.8	2,150	76.0	30,030
64.9	2,310	77.0	32,630
65.0	2,470	80.0	40,430
65.1	2,640	83.0	48,230
65.2	2,820	84.0	50,830
65.3	3,010	85.0	53,430
65.4	3,210	87.0	58,630
65.6	3,640	88.0	61,230
65.8	4,110	90.0	66,430
66.0	4,610	91.0	69,030
66.2	5,110	92.0	71,630
66.4	5,610	93.0	74,230
66.6	6,110	94.0	76,830
66.8	6,610	95.0	79,430
67.0	7,110		

*Water Supply and Irrigation Paper, No. 128, page 95.

The above table is applicable only for oben-channel conditions. It is based upon thirteen discharge measurements made during 1903 and 1904. It is well defined between gage heights 63.2 and 65.4 feet. Above gage height 72 feet the rating curve is a tangent, the difference being 260 per tenth.

Two flood measurements above 65.4 feet gage height define the tangent. The table has been extended beyond these limits.

*Daily Gage Height in Feet, of East Branch of White River at Shoals, Indiana, for 1905.**

Day	Jan.	Feb.	Mar.	Apr.	May	June
1	65.5	63.9	66.4	66.0	67.0	66.4
2	65.0	64.0	68.4	66.0	67.2	66.8
3	64.7	63.8	67.9	65.9	67.5	66.4
4	64.5	63.8	68.0	65.6	67.0	65.8
5	64.3	63.7	68.1	65.3	65.8	65.4
6	64.3	63.7	67.1	65.0	66.1	65.2
7	64.1	63.7	66.7	64.9	67.0	65.9
8	63.9	63.7	67.1	64.8	67.3	64.9
9	63.7	63.7	70.6	64.7	67.5	64.9
10	63.6	63.8	71.5	64.5	68.0	65.0
11	63.6	63.9	71.7	65.1	68.8	64.9
12	65.5	64.0	70.0	65.3	69.6	64.8
13	65.5	64.2	68.9	65.4	71.6	64.6
14	66.4	64.4	67.2	65.4	72.8	63.5
15	66.0	64.5	66.4	65.2	74.0	64.4
16	65.7	64.6	66.0	65.0	75.2	64.3
17	66.8	64.4	65.8	64.8	76.2	64.3
18	67.2	64.2	65.5	64.6	75.3	64.3
19	65.8	64.0	65.4	64.5	72.3	64.3
20	65.2	63.9	65.2	64.4	69.7	64.8
21	64.7	64.2	65.2	64.6	67.4	65.0
22	64.5	65.0	65.1	66.6	66.4	65.4
23	64.4	65.2	65.0	67.7	66.0	65.6
24	63.8	66.0	65.0	68.7	65.6	65.8
25	63.8	67.1	65.0	67.5	65.4	65.4
26	63.8	70.3	65.0	67.0	65.2	65.1
27	63.9	71.6	65.0	66.9	65.1	64.9
28	64.0	68.2	65.7	66.4	65.0	64.7
29	64.0	....	65.2	66.2	64.9	64.6
30	64.0	....	65.6	66.4	65.0	64.4
31	63.9	....	60.0	....	65.0	....

Day	July	Aug.	Sept.	Oct.	Nov.	Dec.
1	64.3	63.8	65.1	63.9	65.6	66.3
2	64.2	63.8	65.6	64.0	65.4	69.0
3	64.1	63.8	65.0	64.0	65.3	70.0
4	64.1	63.7	64.7	64.1	65.2	71.0
5	64.0	63.6	64.3	64.1	66.0	70.8
6	64.0	63.6	64.2	64.3	66.5	69.3
7	64.0	63.5	64.2	64.5	66.5	68.4
8	64.0	63.6	64.0	64.6	66.8	67.0
9	63.9	63.7	64.0	64.6	67.0	66.5
10	63.9	63.7	64.0	64.4	66.7	65.6
11	64.0	63.8	64.2	64.2	66.2	65.4
12	64.2	64.0	64.2	64.1	65.8	65.2
13	64.3	64.1	64.1	64.0	65.5	65.2
14	64.3	64.8	64.1	63.9	65.2	65.1
15	64.2	66.2	64.2	63.8	65.2	65.0
16	64.2	65.7	64.2	63.8	65.0	64.9
17	64.2	66.1	64.2	64.0	64.9	64.8
18	64.1	66.1	64.1	65.3	64.8	64.7
19	64.0	66.0	64.0	69.1	64.8	64.8
20	64.0	66.7	64.0	71.1	64.8	65.0
21	63.9	66.1	64.2	69.4	64.9	65.8
22	63.9	65.9	64.3	69.8	65.0	67.0
23	64.0	65.8	64.3	68.7	64.9	68.2
24	64.4	66.0	64.1	67.3	64.8	68.4
25	64.4	66.0	64.0	67.0	64.7	68.5
26	64.3	65.9	63.9	68.0	64.7	67.9
27	64.3	65.7	63.8	67.9	64.7	66.9
28	64.2	65.4	63.8	68.0	64.9	66.5
29	64.0	65.3	63.8	67.3	65.1	66.0
30	63.9	65.2	63.8	66.6	65.8	65.9
31	63.8	64.9	....	66.0	....	65.8

*Water Supply and Irrigation Paper, No. 169, page 87.

*Discharge Measurements of East Branch of White River at Shoals, Indiana, 1905.**

Date	Hydrographer	Width Feet	Area of Section Sq. Ft.	Mean Velocity Feet per Second	Gage Height Feet	Discharge, Second-Feet
March 16	S. K. Clapp	355	1,421	4.28	66.00	6,090
May 15	M. S. Brennan	406	4,248	4.26	73.58	18,120
June 15	S. K. Clapp	330	744	2.47	64.40	1,838
October 16	M. S. Brennan	313	564	1.74	63.80	982

*Water Supply and Irrigation Paper, No. 169, page 87.

*Station Rating Table for East Branch of White River at Shoals, Indiana, from January 1 to December 31, 1905.**

Gage Height Feet	Discharge Sec.-ft.	Gage Height Feet	Discharge Sec.-ft.	Gage Height Feet	Discharge Sec.-ft.	Gage Height Feet	Discharge Sec.-ft.
63.50	570	65.40	4,041	67.30	8,390	70.40	15,520
63.60	670	65.50	4,266	67.40	8,620	70.60	15,980
63.70	790	65.60	4,492	67.50	8,850	70.80	16,440
63.80	918	65.70	4,718	67.60	9,080	71.00	16,900
63.90	1,050	65.80	4,945	67.70	9,310	71.20	17,360
64.00	1,190	65.90	5,172	67.80	9,540	71.40	17,820
64.10	1,338	66.00	5,400	67.90	9,770	71.60	18,280
64.20	1,500	66.10	5,630	68.00	10,000	71.80	18,740
64.30	1,673	66.20	5,860	68.20	10,460	72.00	19,200
64.40	1,856	66.30	6,090	68.40	10,920	72.50	20,350
64.50	2,051	66.40	6,320	68.60	11,380	73.00	21,500
64.60	2,259	66.50	6,550	68.80	11,840	73.50	22,700
64.70	2,479	66.60	6,780	69.00	12,300	74.00	23,900
64.80	2,700	66.70	7,010	69.20	12,760	74.50	25,100
64.90	2,922	66.80	7,240	69.40	13,220	75.00	26,300
65.00	3,145	66.90	7,470	69.60	13,680	75.50	27,500
65.10	3,365	67.00	7,700	69.80	13,140	76.00	28,700
65.20	3,592	67.10	7,930	70.00	14,600	76.50	29,900
65.30	3,816	67.20	8,160	70.20	15,060	....	....

NOTE—The above table is applicable only for open channel conditions. It is based on 14 discharge measurements made during 1903-1905. It is fairly well defined between gage heights 63.2 feet and 69.0 feet. The table has been extended beyond these limits being based on one measurement at 95.2 feet. This measurement may be considerably in error owing to backwater.

*Water Supply and Irrigation Paper, No. 169, page 88.

*Discharge Measurements of East Branch of White River at Shoals, Indiana, in 1906.***

Date	Hydrographer	Width Feet	Area of Section Sq. ft.	Gage Height, Feet	Discharge, Sec. ft.
February 15*	Brennan & Kriegsman	341	943	64.90	2,550
March 1	E. F. Kriegsman	331	967	65.08	3,200
March 29	E. F. Kriegsman	406	4,390	74.01	20,000
April 2	E. F. Kriegsman	430	9,400	85.62	37,800
April 15	E. F. K iegsman	353	2,510	69.30	12,400

*Thin ice running.

**Water Supply and Irrigation Paper, No. 205, page 69.

*Daily Gage Height, in Feet, of East Branch of White River at Shoals, Indiana, for 1906.**

Day	Jan.	Feb.	Mar.	Apr.	May	June
1	65.8	65.5	65.2	84.5	65.0	64.3
2	66.2	65.4	65.1	86.0	65.0	64.3
3	67.8	65.3	67.1	87.4	65.0	64.2
4	76.4	65.2	69.0	88.0	64.9	64.2
5	75.0	65.0	70.8	87.5	64.9	65.1
6	73.5	64.8	70.2	85.7	65.0	64.8
7	73.0	64.6	68.8	82.8	65.1	64.7
8	72.5	64.3	67.4	77.0	65.0	64.5
9	71.3	64.3	66.3	73.2	64.8	64.5
10	69.5	64.4	66.6	72.0	64.7	64.7
11	67.8	64.5	66.6	71.0	64.6	64.6
12	66.6	64.7	66.6	70.5	64.4	64.4
13	66.4	64.7	66.6	70.0	64.5	64.3
14	66.2	64.7	66.7	69.8	64.5	64.2
15	67.5	64.7	66.9	69.5	64.5	64.1
16	69.1	64.8	67.0	69.2	64.4	64.0
17	70.0	64.7	67.1	69.3	64.4	64.0
18	69.8	64.7	66.7	68.8	64.4	63.9
19	69.5	64.7	66.7	67.8	64.3	63.9
20	69.2	64.6	67.0	67.0	64.3	63.9
21	68.0	64.7	68.1	66.7	64.3	63.8
22	67.6	64.9	69.7	66.4	64.3	63.8
23	67.9	65.5	69.8	66.0	64.3	63.9
24	67.7	65.9	69.7	65.8	64.3	63.8
25	67.5	65.8	69.7	65.7	64.2	63.8
26	67.1	65.5	70.0	65.6	64.2	63.8
27	66.7	65.3	70.7	65.4	64.2	63.8
28	66.3	65.2	73.7	65.3	64.2	63.8
29	66.0	....	74.4	65.2	64.1	63.8
30	65.8	....	78.2	65.1	64.1	64.8
31	65.6	....	82.5	....	64.1	....

NOTE—Slight ice conditions during part of February, but flow was probably not much affected thereby
*Water Supply and Irrigation Paper, No. 205, page 69.

*Rating Table for East Branch of White River at Shoals, Indiana, for 1905 and 1906.**

Gage Height, Feet	Discharge Sec.-Ft.	Gage Height, Feet	Discharge Sec.-Ft.
63.8	880	66.2	6,360
63.9	1,000	66.3	6,580
64.0	1,130	66.4	6,800
64.1	1,270	66.5	7,020
64.2	1,410	66.6	7,240
64.3	1,560	66.7	7,460
64.4	1,720	66.8	7,680
64.5	1,890	66.9	7,900
64.6	2,070	67.0	8,100
64.7	2,260	67.2	8,500
64.8	2,460	67.4	8,900
64.9	2,680	67.6	9,300
65.0	2,920	67.8	9,700
65.1	3,180	68.0	10,080
65.2	3,460	68.2	10,400
65.3	3,750	68.4	10,840
65.4	4,050	68.6	11,220
65.5	4,360	68.8	11,590
65.6	4,670	69.0	11,950
65.7	4,980	70.0	13,750
65.8	5,280	71.0	15,400
65.9	5,580	72.0	17,000
66.0	5,860	73.0	18,500
66.1	6,120	74.0	20,000

NOTE—The above table is applicable only for open channel conditions. It is based on discharge measurements made during 1903 to 1906. It is well defined between gage heights 63.2 feet and 65.4 feet. Above gage height 72.0 feet the rating curve is tangent, the difference being 150 per tenth.

*Water Supply and Irrigation Paper, No. 205, page 69.

The following measurement was made October 12, 1908; Width, 275 feet; area, 331 sq. ft.; gage height, 63.2 feet; discharge, 345 second-feet.

*Daily Gage Height, in Feet, of East Branch of White River at Shoals, Indiana, for 1908.**

Day	May	June	Aug.	Oct.	Nov.	Dec.
1	....	65.5	....	....	63.2	63.4
2	67.4	65.5	....	....	63.2	63.4
3	67.8	65.4	....	....	63.2	63.4
4	68.0	65.3	....	....	63.2	63.4
5	77.5	65.2	....	....	63.2	63.3
6	81.8	65.0	....	....	63.2	63.3
7	83.8	64.9	....	....	63.2	63.3
8	85.6	64.8	....	....	63.2	63.3
9	87.1	64.9	....	....	63.2	63.3
10	87.9	64.8	....	....	63.2	63.3
11	88.2	64.7	....	....	63.2	63.3
12	88.2	64.7	....	63.2	63.2	63.3
13	87.5	64.7	64.8	63.2	63.2	63.3
14	85.9	64.6	....	63.2	63.2	63.3
15	82.6	64.6	....	63.2	63.2	63.3
16	76.6	64.5	....	63.2	63.2	63.3
17	69.7	64.5	....	63.2	63.2	63.3
18	67.4	64.4	....	63.2	63.2	63.3
19	66.7	64.4	....	63.2	63.2	63.3
20	66.5	64.3	....	63.2	63.2	63.3
21	66.3	64.3	....	63.2	63.2	63.3
22	66.2	64.2	....	63.2	63.2	63.3
23	66.0	64.2	....	63.2	63.2	63.3
24	66.0	64.2	....	63.2	63.3	63.3
25	65.7	64.6	....	63.2	63.2	63.3
26	65.5	64.6	....	63.2	63.3	63.3
27	65.4	64.4	....	63.2	63.4	63.3
28	65.3	64.3	....	63.2	63.5	63.3
29	65.4	64.2	....	63.2	63.4	63.3
30	65.4	64.1	....	63.2	63.4	63.3
31	65.5	....	....	63.2	....	63.3

*Water Supply and Irrigation Paper, No. 243, page 102.

*Rating Table for East Branch of White River at Shoals, Indiana, for 1906 to 1908.**

Gage Height Feet	Discharge Sec.-ft.	Gage Height Feet	Discharge Sec.-ft.	Gage Height Feet	Discharge Sec.-ft.	Gage Height Feet	Discharge Sec.-ft.
63.20	340	64.60	2,070	66.00	5,860	67.80	9,700
63.30	410	64.70	2,260	66.10	6,120	68.00	10,080
63.40	490	64.80	2,460	66.20	6,360	68.20	10,460
63.50	580	64.90	2,680	66.30	6,580	68.40	10,840
63.60	670	65.00	2,920	66.40	6,800	68.60	11,220
63.70	770	65.10	3,180	66.50	7,020	68.80	11,590
63.80	880	65.20	3,460	66.60	7,240	69.00	11,950
63.90	1,000	65.30	3,750	66.70	7,460	69.20	12,310
64.00	1,130	65.40	4,050	66.80	7,680	69.40	12,670
64.10	1,270	65.50	4,360	66.90	7,900	69.60	13,030
64.20	1,410	65.60	4,670	67.00	8,100	69.80	13,390
64.30	1,560	65.70	4,980	67.20	8,500	70.00	13,750
64.40	1,720	65.80	5,280	67.40	8,900	71.00	15,400
64.50	1,890	65.90	5,580	67.60	9,300	72.00	17,000

NOTE—The above table is not applicable for ice or obstructed-channel conditions. It is based on 22 discharge measurements made during 1903 to 1906 and 1908. It is well defined between gage heights 63.2 feet and 65.4 feet. Above gage height 72.0 feet the rating curve is a tangent, the difference being 150 per tenth.

*Water Supply and Irrigation Paper, No. 243, page 102.

*Daily Gage Height, in Feet, of East Branch of White River at Shoals, Indiana, for 1909.**

Day	Jan.	Feb.	Mar.	Apr.	May	June	July	Aug.	Sept.	Oct.	Nov.	Dec.
1	2.3	2.2	14.7	3.8	9.1	11.4	4.45	3.65	2.85	2.4	3.4	4.1
2	2.3	2.2	13.6	3.7	8.0	10.1	4.15	3.6	2.8	2.4	3.3	3.95
3	2.3	2.3	11.1	3.7	7.4	7.4	3.95	3.85	2.8	2.3	3.15	3.8
4	2.3	2.4	8.4	3.5	6.6	6.0	3.65	4.3	2.7	2.3	3.05	3.75
5	2.3	2.6	6.4	3.5	6.1	5.4	3.5	4.1	2.7	2.3	3.0	3.6
6	2.3	2.7	5.5	3.6	6.1	5.3	4.5	3.75	2.6	2.3	3.0	3.5
7	2.3	2.8	5.0	5.3	6.0	5.7	7.2	3.45	2.55	2.3	2.9	3.6
8	2.3	2.7	4.8	5.6	5.6	6.0	6.25	3.35	2.5	2.3	2.9	3.9
9	2.3	2.7	6.8	6.7	4.9	5.9	5.45	3.2	2.5	2.3	2.85	3.65
10	2.3	2.6	14.4	7.9	5.0	5.9	5.15	3.1	2.5	2.3	2.8	3.3
11	2.3	2.7	15.2	7.0	6.4	6.0	4.95	3.05	2.75	2.3	2.8	3.65
12	2.3	2.7	14.8	6.1	8.0	6.0	5.45	3.1	3.0	2.3	2.75	5.05
13	2.3	2.8	16.2	5.4	8.8	7.1	8.8	3.0	3.25	2.3	2.7	7.35
14	2.2	3.1	17.2	5.4	8.7	6.7	10.9	3.0	3.05	2.3	2.7	8.7
15	2.3	4.7	15.6	5.3	8.0	5.4	8.85	3.0	2.85	2.2	2.7	8.75
16	2.3	5.1	12.2	5.2	6.8	4.8	7.9	3.0	2.75	2.2	2.8	9.2
17	2.4	5.2	8.8	5.0	6.7	4.4	8.0	3.1	2.65	2.3	3.05	8.65
18	2.4	5.6	6.4	4.8	7.8	4.2	7.9	3.1	2.5	2.4	3.2	8.05
19	2.4	5.6	5.3	4.5	7.7	4.0	7.1	3.15	2.5	2.4	3.2	6.8
20	2.4	5.7	4.8	4.7	6.5	3.9	5.85	3.25	2.4	2.4	3.1	5.4
21	2.4	5.7	4.6	5.3	5.6	3.8	4.8	3.3	2.4	2.85	3.25	4.4
22	2.3	5.8	4.5	8.5	5.0	3.8	4.2	3.3	2.4	3.55	3.45	4.4
23	2.3	6.3	4.3	8.4	4.7	3.7	3.95	3.2	2.4	3.55	3.8	4.3
24	2.3	10.4	4.2	8.4	4.6	3.6	3.75	3.2	2.4	3.65	4.0	4.0
25	2.3	12.2	4.0	7.2	4.4	3.6	3.6	3.25	2.4	3.8	4.2	3.7
26	2.4	12.2	4.0	6.2	4.4	3.6	3.6	3.35	2.4	3.95	4.6	4.05
27	2.4	13.2	4.1	5.6	5.3	3.7	3.6	3.4	2.4	4.25	5.15	4.3
28	2.4	14.3	4.0	5.0	7.6	4.1	3.55	3.4	2.4	4.25	5.15	3.7
29	2.4	....	4.0	5.4	8.6	4.7	3.35	3.3	2.4	4.05	4.8	4.0
30	2.4	....	3.9	7.4	9.0	4.6	3.5	3.1	2.4	3.75	4.4	4.0
31	2.3	....	3.9	....	10.2	....	3.65	2.95	....	3.55	....	4.1

*Water Supply and Irrigation Paper, No. 263, page 111.

To correlate the above table with previous gage readings add 61 feet. The gage was re-established Jan. 1, 1909 and the datum raised 61 feet.

*Daily Gage Height, in Feet, of East Branch of White River at Shoals, Indiana, for 1910.**

Day	Jan.	Feb.	Mar.	Apr.	May	June	July	Aug.	Sept.	Oct.	Nov.	Dec.
1	3.1	5.2	14.5	3.5	3.8	3.6	3.45	4.3	3.05	2.7	3.6	5.4
2	3.2	4.9	15.2	3.4	3.8	3.5	3.45	3.95	3.15	2.6	3.5	4.9
3	3.5	4.8	17.9	3.4	3.9	3.4	3.35	3.85	3.05	2.6	3.5	4.4
4	4.6	4.6	20.5	3.4	4.2	3.3	3.15	3.7	3.2	2.9	3.4	4.05
5	4.0	4.6	22.2	3.4	4.7	3.3	3.1	3.5	3.3	5.5	3.4	3.85
6	4.8	4.6	22.3	3.4	4.7	3.2	3.3	3.4	4.5	15.5	3.4	3.7
7	5.4	4.6	20.6	3.3	4.7	3.2	4.05	3.3	4.45	18.65	3.4	3.6
8	5.6	4.5	16.3	3.5	4.9	3.2	4.5	3.2	4.0	17.55	3.4	3.5
9	5.2	4.3	11.2	3.4	5.3	3.1	5.0	3.1	3.95	17.2	3.4	3.4
10	4.8	4.2	7.7	3.4	5.9	3.1	5.1	3.1	3.65	19.55	3.4	3.4
11	4.6	4.1	6.1	3.3	6.0	3.1	5.1	3.0	3.5	21.95	3.3	3.35
12	3.9	4.1	5.5	3.4	5.8	3.1	5.0	3.0	3.35	23.55	3.3	3.3
13	4.4	3.9	5.2	3.5	5.6	3.1	5.2	2.9	3.15	23.65	3.2	3.2
14	8.9	3.8	4.9	3.4	5.2	3.2	4.75	2.9	3.1	22.15	3.2	3.2
15	9.9	3.8	4.7	3.6	4.9	3.2	4.55	2.9	3.0	18.55	3.2	3.15
16	8.4	3.8	4.5	3.7	4.6	3.1	5.05	2.8	2.9	11.8	3.2	3.2
17	11.2	4.0	4.4	4.5	4.3	3.1	6.7	2.8	2.8	7.5	3.2	3.2
18	12.5	4.8	4.3	4.4	4.1	3.1	6.9	2.9	2.7	6.0	3.1	3.1
19	16.1	5.6	4.2	5.0	4.1	2.9	8.15	2.8	2.7	5.15	3.1	3.1
20	17.3	5.6	4.1	5.7	4.0	2.9	8.65	2.8	2.7	4.9	3.1	3.0
21	17.7	5.7	4.0	5.7	4.0	2.8	9.2	2.8	2.6	4.6	3.1	3.0
22	18.0	6.8	4.0	5.6	4.3	2.8	9.45	2.75	2.6	4.5	3.1	2.85
23	17.6	8.1	3.9	5.2	5.0	2.7	7.85	2.95	2.6	4.35	3.1	3.0
24	16.0	9.1	3.8	4.8	5.4	2.7	6.25	2.8	2.6	4.3	3.1	3.0
25	12.8	8.6	3.7	4.5	5.3	2.7	5.2	2.7	2.7	4.2	3.05	2.9
26	9.4	8.3	3.7	4.2	5.1	2.7	4.45	2.7	2.8	4.1	3.0	2.8
27	7.3	8.3	3.7	4.1	4.8	2.8	4.3	2.8	2.8	4.0	3.0	2.8
28	6.3	12.3	3.6	4.0	4.6	3.2	4.25	2.9	2.7	3.9	3.75	3.25
29	5.8	....	3.6	3.9	4.3	3.1	4.6	3.0	2.7	3.8	4.85	4.25
30	5.8	....	3.6	3.9	4.0	3.2	5.3	2.9	2.7	3.7	5.45	6.0
31	5.7	....	3.5	....	3.8	....	5.0	2.9	....	3.7	....	7.7

*Water Supply and Irrigation Paper, U. S. G. S. No. 283, page 106.

*Daily Gage Height, in Feet, of East Branch of White River at Shoals, Indiana, for 1911.**

Day	Jan.	Feb.	Mar.	Apr.	May	June	July	Aug.	Sept.	Oct.	Nov.	Dec.
1	9.0	12.8	7.0	7.2	5.7	3.4	3.05	2.45	2.2	5.4	3.6	5.85
2	11.3	12.9	6.3	6.8	8.8	3.5	3.15	2.45	2.2	8.6	3.6	5.8
3	11.8	11.8	5.7	6.1	10.2	3.4	3.15	2.4	2.3	12.4	3.45	5.35
4	10.5	10.2	5.2	9.6	10.4	3.5	3.05	2.4	2.2	12.1	3.5	5.05
5	9.1	8.5	4.9	12.2	10.2	3.7	2.95	2.3	2.75	11.3	3.5	4.8
6	7.6	7.0	4.8	14.7	9.9	3.4	2.85	2.4	3.4	9.8	3.75	4.45
7	6.4	6.3	5.1	15.6	9.0	3.5	2.85	2.3	3.1	8.75	5.15	4.35
8	5.4	6.7	8.6	16.4	8.7	3.8	2.85	2.4	3.15	7.6	6.2	4.2
9	5.0	7.0	10.6	17.1	6.1	4.4	2.85	2.5	3.25	7.1	6.2	4.05
10	4.8	7.0	11.6	16.3	5.2	4.3	2.85	2.4	3.55	6.1	6.2	4.05
11	4.7	6.4	11.7	14.1	4.2	4.0	2.95	2.5	3.3	5.25	5.75	4.7
12	4.6	5.7	10.6	12.1	4.4	3.7	2.95	2.4	3.2	4.7	(a)	6.6
13	4.9	5.3	9.4	9.1	4.4	3.4	2.85	2.4	3.4	4.4	(a)	7.15
14	7.5	5.0	8.1	13.4	4.2	3.3	2.75	2.5	3.65	4.3	(a)	8.8
15	9.9	4.9	6.6	16.7	4.2	3.2	2.75	2.4	3.75	(a)	(a)	10.9
16	12.3	4.8	5.6	16.6	4.1	3.0	2.75	2.4	3.7	(a)	(a)	11.0
17	12.6	4.8	5.1	17.5	3.9	3.05	2.75	2.4	4.45	(a)	(a)	10.6
18	11.0	5.0	4.8	18.8	3.9	3.05	2.75	2.3	6.5	5.35	(a)	10.15
19	10.2	5.3	4.5	19.0	3.7	3.95	2.75	2.15	8.45	6.05	(a)	9.55
20	8.7	5.9	4.5	18.3	3.7	3.65	2.75	2.3	8.35	6.5	(a)	8.85
21	7.2	7.0	4.5	16.2	4.1	3.15	2.65	2.2	6.2	6.3	5.4	8.0
22	6.5	7.9	4.5	13.0	4.3	3.05	2.55	2.2	5.3	6.05	5.55	7.0
23	7.0	7.6	4.4	10.1	4.8	3.05	2.45	2.15	4.8	5.5	5.3	6.25
24	7.9	6.8	4.2	8.5	4.9	3.15	2.45	2.3	4.4	4.85	(a)	5.95
25	7.7	6.4	4.2	7.5	4.7	3.25	2.45	2.2	3.95	4.3	5.35	5.7
26	7.0	6.4	4.2	6.6	4.6	3.15	2.45	2.3	3.6	4.1	5.3	5.5
27	6.3	6.8	4.2	6.0	4.2	3.05	2.45	2.2	3.45	4.0	6.0	5.5
28	6.7	7.2	4.7	5.5	4.0	2.95	2.45	2.2	3.45	3.85	5.7	5.75
29	8.0	......	5.6	5.2	3.7	2.95	2.45	2.3	5.3	3.7	5.45	6.2
30	10.7	......	6.5	5.1	3.6	2.95	2.45	2.2	5.65	3.65	5.5	6.45
31	12.1	......	6.8	......	3.6	......	2.45	2.3	......	3.6	......	6.5

(a) No record.
*Water Supply and Irrigation Paper, U. S. G. S. No. 303, page 73.

*Daily Gage Height, in Feet, of East Branch of White River at Shoals, Indiana, for 1912.**

Day	Jan.	Feb.	Mar.	Apr.	May	June	July	Aug.	Sept.	Oct.	Nov.	Dec.
1	6.8	8.9	21.5	18.8	17.3	5.0	3.4	2.8	4.2	3.4	3.0	2.7
2	7.7	8.0	22.8	19.4	16.9	4.6	3.8	2.9	4.1	3.4	3.0	2.4
3	8.2	7.4	23.7	20.4	15.7	4.3	3.5	2.8	3.7	3.0	2.9	3.1
4	7.6	6.6	22.5	21.0	13.9	4.1	3.8	2.7	3.5	3.1	2.9	2.9
5	6.6	5.3	17.4	18.4	11.1	3.9	4.1	2.8	3.3	2.9	3.0	2.5
6	5.3	4.6	9.5	17.2	10.7	3.8	4.2	2.6	3.2	2.9	2.9	2.5
7	4.4	4.5	6.5	17.6	10.1	3.8	4.1	2.7	3.1	2.8	3.1	2.7
8	4.1	4.0	5.7	15.5	10.5	3.7	5.0	2.8	3.1	3.0	3.1	2.9
9	4.4	4.1	6.7	12.8	9.9	3.6	5.6	2.5	2.8	2.9	3.2	2.6
10	4.5	4.3	7.9	10.4	8.5	3.5	5.3	2.6	3.0	2.9	3.1	3.2
11	4.5	5.1	8.8	8.8	7.9	3.5	5.0	2.7	3.2	2.6	2.8	3.3
12	4.5	5.0	8.6	7.6	12.3	3.4	4.7	2.7	3.0	2.7	3.2	3.1
13	4.5	4.2	8.7	6.7	15.1	3.4	4.6	3.2	2.8	2.8	3.0	2.9
14	4.5	4.8	8.9	6.2	14.0	3.4	4.5	3.7	2.8	2.5	3.0	2.7
15	4.0	3.8	10.6	6.2	13.7	3.7	4.5	3.7	2.8	2.8	3.0	2.5
16	4.0	3.8	13.2	6.1	12.8	3.9	4.7	4.2	2.4	2.6	2.9	2.5
17	4.4	3.8	13.2	6.2	13.0	3.8	4.7	5.8	2.8	2.8	3.0	3.1
18	4.0	4.0	14.0	7.6	12.8	3.6	4.4	5.7	2.8	2.8	3.0	2.5
19	8.4	4.3	14.5	9.2	11.4	3.7	4.4	5.2	2.9	2.8	3.1	2.5
20	12.2	5.2	14.2	10.6	10.9	3.7	4.5	3.9	3.0	2.8	3.1	2.8
21	11.6	6.1	13.9	10.1	9.3	3.7	4.5	4.7	3.0	2.4	3.1	2.7
22	12.1	7.2	14.8	9.3	7.6	3.9	5.2	5.2	2.8	2.0	3.0	2.8
23	16.9	8.0	15.0	8.3	6.2	3.7	5.0	6.5	3.0	2.8	3.0	2.5
24	12.3	7.8	15.8	7.5	5.5	3.5	4.9	6.2	3.0	2.9	2.8	2.5
25	14.0	7.2	17.6	7.0	5.1	3.5	4.8	5.5	3.0	3.0	2.5	2.6
26	12.9	11.2	18.4	6.4	4.8	3.5	4.6	5.0	3.5	3.0	3.0	2.6
27	9.2	18.6	18.9	13.3	4.6	3.4	4.2	4.5	3.9	2.7	3.0	2.8
28	7.9	19.1	19.0	17.0	4.4	3.4	3.7	4.4	4.0	2.6	3.0	2.8
29	8.0	19.0	18.9	17.0	4.7	3.4	3.5	4.0	3.7	3.0	2.5	2.7
30	8.6	......	19.7	17.2	4.7	3.4	3.4	4.7	3.5	2.8	2.7	2.6
31	8.8	......	19.0	......	4.8	......	2.7	4.5	......	3.0	......	2.6

*Daily River Stages, Principal Rivers of the United States, U. S. Weather Bureau, Part XI, page 288.

*Daily Gage Height, in Feet, of East Branch of White River at Shoals, Indiana, for 1913, except for October and November.**

Day	Jan.	Feb.	Mar.	Apr.	May	June	July	Aug.	Sept.	Oct.	Nov.	Dec.
1	2.8	9.8	6.9	33.6	4.2	3.2	3.05	2.7	2.4	......	......	3.8
2	3.1	6.9	9.0	30.3	4.1	3.2	2.95	2.6	2.35	......	......	4.6
3	3.2	6.4	9.7	27.8	4.0	3.2	2.95	2.6	2.25	......	......	5.3
4	3.5	5.9	9.2	26.6	3.9	3.1	2.7	2.45	2.65	......	......	5.5
5	3.4	5.0	8.8	25.6	4.0	3.0	2.5	2.15	2.5	......	......	5.5
6	4.1	5.2	7.7	22.3	3.8	2.8	2.85	2.05	2.05	......	......	5.3
7	5.7	5.0	6.7	19.9	3.7	2.9	2.9	2.6	2.4	......	......	4.8
8	9.8	4.7	6.0	20.7	3.6	3.0	3.0	2.65	2.55	......	......	4.6
9	11.9	4.5	5.8	19.5	3.5	2.8	2.55	2.85	2.1	......	......	4.2
10	11.6	4.1	5.5	18.8	3.5	3.0	2.55	2.65	1.85	......	......	4.0
11	13.4	4.4	6.2	17.8	3.5	3.1	2.8	2.55	2.3	......	......	3.9
12	17.8	4.2	5.1	17.2	3.5	3.1	2.4	2.35	2.5	......	......	3.7
13	18.7	4.5	5.2	17.7	3.3	3.1	2.65	2.2	2.5	......	......	3.5
14	18.7	4.1	7.1	19.1	3.3	2.7	2.3	2.2	2.6	......	......	3.7
15	18.9	4.1	9.7	19.7	3.2	2.9	2.8	2.25	2.6	......	......	3.3
16	19.7	3.9	10.0	19.4	3.2	2.6	2.7	2.2	2.4	......	......	3.4
17	21.8	4.1	10.0	18.1	3.2	3.1	2.65	2.75	2.2	......	......	3.4
18	23.4	4.0	9.4	15.2	3.3	3.0	2.7	2.9	2.6	......	......	3.2
19	23.0	4.1	6.3	11.9	3.6	2.8	2.65	3.1	2.45	......	......	3.2
20	22.2	4.1	6.1	8.8	3.5	2.7	2.65	2.85	2.15	......	......	3.2
21	23.3	4.4	6.0	6.9	3.4	2.7	2.75	2.9	2.4	......	......	3.2
22	24.4	4.4	7.3	6.0	3.3	2.9	3.05	4.4	2.5	......	......	3.2
23	25.1	4.6	8.0	5.5	3.3	2.7	2.9	4.5	2.25	......	......	3.3
24	26.3	4.6	8.7	5.2	3.1	2.7	2.85	3.45	2.45	......	......	3.4
25	26.3	4.5	21.6	4.9	3.1	2.6	2.8	3.8	2.2	......	......	3.4
26	25.8	4.2	29.4	4.8	3.3	2.6	2.85	3.5	2.2	......	......	3.8
27	25.3	4.2	37.0	4.7	3.2	2.9	2.65	3.15	2.1	......	......	3.9
28	25.0	4.7	42.0	4.6	3.0	3.1	2.55	2.8	2.6	......	......	3.7
29	23.8	......	41.5	4.6	3.2	3.0	2.35	2.7	2.4	......	......	3.8
30	21.0	......	39.4	4.2	3.2	2.5	2.35	2.6	2.15	......	......	3.7
31	15.1	......	36.6	......	3.2	......	2.3	2.5	......	......	......	3.5

*Water Supply and Irrigation Paper, U. S. G. S. No. 353, page 82, and Daily River Stages on Principal Rivers of U. S., Part XII, p. 304.

*Daily Gage Height, in Feet, of East Branch of White River at Shoals, Indiana, for 1914.**

Day	Jan.	Feb.	Mar.	Apr.	May	June	July	Aug.	Sept.	Oct.	Nov.	Dec.
1	3.4	4.5	6.2	15.4	4.7	3.2	3.0	2.6	3.7	2.2	2.5	2.7
2	3.4	5.3	6.5	16.5	4.4	3.2	2.8	2.6	3.7	2.6	2.0	2.7
3	3.4	6.9	6.2	16.2	4.3	3.3	3.0	2.0	3.2	2.4	2.0	2.7
4	3.3	6.5	6.7	15.4	4.1	3.2	2.9	2.0	3.3	2.2	1.9	2.5
5	3.4	6.0	6.9	14.5	4.1	3.1	2.8	2.6	3.2	2.9	2.6	2.6
6	3.5	5.8	7.2	12.5	4.3	3.2	2.5	2.6	2.9	2.2	2.7	2.2
7	3.5	5.6	8.0	10.5	4.8	3.3	2.8	2.8	2.6	2.0	2.5	1.7
8	3.7	5.9	8.9	8.9	5.6	3.1	2.0	2.4	3.0	2.2	2.5	2.3
9	3.8	5.8	9.2	8.6	5.6	3.6	2.6	2.3	3.2	2.7	2.4	2.4
10	3.9	5.5	8.9	9.3	5.2	3.3	2.6	2.0	3.0	2.6	2.5	2.4
11	4.0	5.0	8.5	10.2	5.2	3.2	2.8	2.4	3.1	3.2	2.4	2.4
12	4.3	4.5	8.1	10.2	5.0	3.3	2.2	2.8	3.1	2.5	2.5	2.3
13	4.2	4.0	8.0	9.5	4.8	3.0	1.9	3.5	3.0	2.5	2.5	2.3
14	4.0	4.0	8.0	9.4	4.4	3.2	2.5	3.5	2.9	2.8	2.5	1.7
15	3.8	4.1	7.8	8.5	4.1	2.8	3.0	3.5	2.6	2.6	2.5	2.3
16	3.7	3.5	8.0	8.4	3.9	3.0	2.9	3.6	2.3	2.8	2.0	2.2
17	3.5	4.1	8.0	9.8	3.8	3.2	2.0	2.9	2.2	2.7	2.5	2.2
18	3.5	3.6	8.0	9.6	3.9	2.9	3.3	2.8	2.2	2.6	2.5	2.2
19	3.5	5.0	7.0	8.4	3.7	3.0	2.6	2.1	2.6	2.0	2.6	2.3
20	3.4	10.0	6.2	6.8	3.6	3.2	2.1	2.6	3.0	2.5	2.7	2.4
21	3.3	9.5	5.8	5.9	3.5	3.0	3.2	2.5	2.8	2.6	2.3	2.2
22	3.3	9.7	5.3	5.6	3.5	2.0	3.3	2.8	2.2	2.6	2.5	2.3
23	3.3	9.8	5.0	5.2	3.5	2.9	3.3	3.2	2.0	2.7	2.0	2.2
24	3.2	8.3	4.8	5.0	3.5	2.9	2.7	2.1	2.5	2.7	2.5	2.4
25	3.3	7.5	4.5	4.8	3.3	2.8	2.5	2.2	2.9	2.5	2.4	2.4
26	3.2	7.2	4.4	4.9	3.7	2.8	2.6	2.7	2.9	2.2	2.5	2.3
27	3.2	6.7	4.4	5.0	3.5	2.8	2.7	2.7	2.7	2.6	2.4	2.4
28	3.2	6.2	8.3	5.1	3.4	2.8	2.7	2.6	2.4	2.6	2.6	2.3
29	3.1	......	11.0	5.1	3.3	2.0	2.6	2.6	2.3	2.7	2.7	2.2
30	3.2	......	13.6	4.9	3.3	1.9	2.7	3.4	2.2	2.5	2.1	2.8
31	3.3	......	14.8	......	2.9	......	2.7	3.3	......	2.6	......	2.9

*Daily River Stages on Principal Rivers of U. S., U. S. Weather Bureau Bulletin 549, Part XII, page 304.

*Daily Gage Height, in Feet, of East Branch of White River at Shoals, Indiana, for 1915.**

Day	Jan.	Feb.	Mar.	Apr.	May	June	July	Aug.	Sept.	Oct.	Nov.	Dec.
1	2.5	8.0	3.6	2.7	2.8	6.0	2.7	3.1	4.5	2.8	2.2	3.9
2	2.3	13.5	3.6	2.9	3.2	5.0	4.0	2.8	4.4	3.5	3.0	3.5
3	2.8	13.5	3.4	3.2	2.8	4.5	4.0	3.2	4.0	3.2	2.8	3.5
4	2.6	13.0	3.3	3.0	3.9	4.9	4.0	3.6	3.8	3.5	2.8	3.5
5	2.6	13.9	3.3	2.6	4.2	5.0	4.4	3.6	3.5	3.7	2.9	3.5
6	2.5	17.2	3.7	3.2	4.5	5.1	4.7	3.4	3.7	2.9	2.9	3.3
7	2.6	18.0	4.7	2.8	4.3	4.9	4.6	3.4	3.5	3.3	2.7	3.1
8	3.0	17.2	5.3	2.8	4.4	4.4	8.3	3.4	3.3	2.8	2.5	3.2
9	2.8	15.2	5.4	3.0	5.0	4.0	13.0	3.8	3.5	2.8	2.8	3.1
10	3.4	12.8	5.0	2.9	5.5	3.8	13.0	3.5	3.4	3.2	2.8	3.2
11	3.7	10.0	4.4	3.1	4.8	3.7	13.2	3.6	4.9	2.8	2.9	3.0
12	3.0	7.0	4.1	2.7	4.0	3.6	14.0	4.1	4.5	2.7	2.7	3.6
13	3.2	5.5	3.8	2.8	3.5	3.3	13.0	4.0	4.4	2.8	2.8	4.1
14	3.0	5.0	3.7	3.2	3.4	3.4	11.0	5.0	4.0	2.9	2.6	4.2
15	2.8	5.0	3.9	2.8	3.0	3.6	9.2	5.8	3.8	2.7	2.8	4.2
16	2.8	4.9	3.5	3.2	3.0	4.0	7.6	6.2	3.5	2.7	2.7	4.1
17	2.9	4.8	3.4	3.0	2.9	4.5	6.0	6.5	3.5	2.6	2.8	6.1
18	2.8	4.7	3.5	3.2	3.0	4.8	4.7	6.3	3.6	2.9	2.9	12.5
19	2.8	4.5	3.4	2.6	2.7	4.7	4.5	5.3	3.9	3.0	3.1	13.0
20	2.9	4.3	2.8	2.8	2.7	4.5	4.1	4.8	4.6	2.9	4.1	12.7
21	3.1	4.2	3.2	2.8	2.9	5.1	4.0	7.1	4.4	2.9	5.4	14.2
22	2.9	4.3	3.4	2.7	3.9	4.7	4.7	9.4	4.0	3.5	6.2	14.7
23	2.9	4.2	3.4	2.9	4.5	4.3	4.6	9.0	3.8	4.0	6.2	13.7
24	3.6	4.0	3.0	3.3	5.3	4.0	3.6	9.3	3.8	2.6	5.5	11.8
25	2.8	3.7	2.9	2.8	5.1	3.7	3.5	8.3	3.6	3.6	4.8	10.1
26	3.4	3.7	3.2	3.1	4.8	3.6	3.6	7.3	3.4	3.0	4.5	9.2
27	3.4	3.7	3.0	3.3	6.2	3.4	3.3	6.3	3.1	3.3	4.1	8.4
28	3.2	3.6	3.1	3.3	7.2	2.9	2.9	6.5	3.0	3.0	3.8	8.0
29	3.5	....	3.2	3.2	8.0	3.4	3.5	5.0	3.7	3.1	3.8	9.1
30	3.0	....	3.2	3.2	7.9	3.5	3.0	5.3	2.7	3.0	3.7	15.0
31	3.0	....	3.2	....	7.2	....	3.4	4.7	....	3.5	....	15.6

*Daily River Stages on Principal Rivers of U. S., U. S. Weather Bureau Bulletin No. 582, page 126.

Discharge Rating Table for East Branch of White River at Shoals, Indiana, 1906-16.

Gage Height Feet	Discharge Sec.-Ft.	Gage Height Feet	Discharge Sec.-Ft.	Gage Height Feet	Discharge Sec.-Ft.	Gage Height Feet	Discharge, Sec.-Ft.	Gage Height Feet	Discharge Sec.-Ft.
2.2	340	5.2	6,360	8.2	12,310	11.2	17,300	14.2	21,800
2.3	410	5.3	6,580	8.3	12,490	11.3	17,450	14.3	21,950
2.4	490	5.4	6,800	8.4	12,670	11.4	17,600	14.4	22,100
2.5	580	5.5	7,020	8.5	12,850	11.5	17,750	14.5	22,250
2.6	670	5.6	7,240	8.6	13,130	11.6	17,900	14.6	22,400
2.7	770	5.7	7,460	8.7	13,210	11.7	18,050	14.7	22,550
2.8	879	5.8	7,680	8.8	13,390	11.8	18,200	14.8	22,700
2.9	996	5.9	7,900	8.9	13,570	11.9	18,350	14.9	22,850
3.0	1,122	6.0	8,100	9.0	13,750	12.0	18,500	15.0	23,000
3.1	1,258	6.1	8,300	9.1	13,915	12.1	18,650	15.1	23,150
3.2	1,403	6.2	8,500	9.2	14,080	12.2	18,800	15.2	23,300
3.3	1,557	6.3	8,700	9.3	14,245	12.3	18,950	15.3	23,450
3.4	1,720	6.4	8,900	9.4	14,410	12.4	19,100	15.4	23,600
3.5	1,890	6.5	9,100	9.5	14,575	12.5	19,250	15.5	23,750
3.6	2,070	6.6	9,300	9.6	14,740	12.6	19,400	15.6	23,900
3.7	2,260	6.7	9,500	9.7	14,905	12.7	19,550	15.7	24,050
3.8	2,460	6.8	9,700	9.8	15,070	12.8	19,700	15.8	24,200
3.9	2,680	6.9	9,800	9.9	15,235	12.9	19,850	15.9	24,350
4.0	2,920	7.0	10,080	10.0	15,400	13.0	20,000	16.0	24,500
4.1	3,180	7.1	10,240	10.1	15,560	13.1	20,150	16.1	24,650
4.2	3,460	7.2	10,400	10.2	15,720	13.2	20,300	16.2	24,800
4.3	3,750	7.3	10,620	10.3	15,880	13.3	20,450	16.3	24,950
4.4	4,050	7.4	10,840	10.4	16,040	13.4	20,600	16.4	25,100
4.5	4,360	7.5	11,030	10.5	16,200	13.5	20,750	16.5	25,250
4.6	4,670	7.6	11,220	10.6	16,360	13.6	20,900	16.6	25,400
4.7	4,980	7.7	11,450	10.7	16,520	13.7	21,050	16.7	25,550
4.8	5,280	7.8	11,590	10.8	16,680	13.8	21,200	16.8	25,700
4.9	5,580	7.9	11,770	10.9	16,840	13.9	21,350	16.9	25,850
5.0	5,860	8.0	11,950	11.0	17,000	14.0	21,500	17.0	26,000
5.1	6,120	8.1	12,130	11.1	17,150	14.1	21,650	17.1	26,150

Discharge Rating Table for East Branch of White River at Shoals, Indiana, 1906-16—Continued.

Gage Height Feet	Discharge Sec.-Ft.	Gage Height Feet	Discharge Sec.-Ft.	Gage Height Feet	Discharge Sec.-Ft.	Gage Height Feet	Discharge Sec.-Ft.	Gage Height Feet	Discharge Sec.-Ft.
17.2	26,300	18.6	28,400	20.0	30,500	23.0	35,000	30.0	45,500
17.3	26,450	18.7	28,550	20.1	30,650	23.5	35,750	31.0	47,000
17.4	26,600	18.8	28,700	20.2	30,800	24.0	36,500	32.0	48,500
17.5	26,750	18.9	28,850	20.3	30,950	24.5	37,250	33.0	50,000
17.6	26,900	19.0	29,000	20.4	31,100	25.0	38,000	34.0	51,500
17.7	27,050	19.1	29,150	20.5	31,250	25.5	38,750	35.0	53,000
17.8	27,200	19.2	29,300	20.6	31,400	26.0	39,500	36.0	54,500
17.9	27,350	19.3	29,450	20.7	31,550	26.5	40,250	37.0	56,000
18.0	27,500	19.4	29,600	20.8	31,700	27.0	41,000	38.0	57,500
18.1	27,650	19.5	29,750	20.9	31,850	27.5	41,750	39.0	59,000
18.2	27,800	19.6	29,900	21.0	32,000	28.0	42,500	40.0	60,500
18.3	27,950	19.7	30,050	21.5	32,750	28.5	43,250	41.0	62,000
18.4	28,100	19.8	30,200	22.0	33,500	29.0	44,000	42.0	63,500
18.5	28,250	19.9	30,350	22.5	34,250	29.5	44,750	43.0	65,000

The following tables are calculated to show the relation between the rainfall and the runoff in the basin of White River (East Branch) above the gage at Shoals, Indiana. Column 1 shows the average monthly rainfall in inches in the basin. The data was derived by taking the average of the monthly reports of the U. S. Weather Bureau from the following observation stations: Franklin, Greenfield, Knightstown, Nashville, Shelbyville, Avoca, Bedford, Butlerville, Columbus, Greensburg, Scottsburg, Seymour, Zelma, Bloomington, Hickory Hill, Mauzy, Forest Reserve, Madison, Salem, and Shoals, which are in or near the basin. The last seven, which are on the margin of the basin, were given in the average half the weight of the rest, which are well within the basin. Column 2 gives the average rainfall per acre in cubic feet within the basin. It is derived directly from Column 1. Column 3 gives the average discharge in cubic feet per second at the gage at Shoals, and is derived from the daily gage readings and the rating tables. Column 4 gives the total average discharge per acre from the basin and is derived from Column 3. Column 5 gives the relation between the rainfall and runoff and is derived by dividing Column 2 into Column 4, and expressing it as per cent. The drainage area of the basin is considered as 4,913.5 square miles. This area was derived by carefully checked planimeter readings from the drainage map accompanying this report.

1904	1 Average Rainfall, Inches	2 Rainfall per Acre, Cu. Ft.	3 Discharge Cu. Ft. per Sec.	4 Discharge Cu. Ft. per Acre	5 Per Cent. 4 of 2
January	5.01	18,186	7,353	6,268	34.5
February	2.76	10,019	11,714	9,342	93.2
March	9.08	32,960	25,284	21,555	65.4
April	2.38	8,639	18,208	15,022	173.9
May	3.35	12,160	2,441	2,081	17.1
June	3.68	13,358	2,376	1,960	14.7
July	2.00	7,260	1,540	1,313	18.1
August	1.44	5,227	647	512	10.6
September	3.34	12,124	525	433	3.6
Octbber	1.26	4,574	435	371	8.1
November	.59	2,142	360	297	13.9
December	4.99	18,114	932	795	4.4
Annual	....	12,064	....	4,996	41.4

1905	1 Average Rainfall, Inches	2 Rainfall per Acre, Cu. Ft.	3 Discharge Cu. Ft. per Sec.	4 Discharge Cu. Ft. per Acre	5 Per Cent. 4 of 2
January	2.23	8,095	2,694	2,297	28 4
February	2.27	8,240	3,216	2,476	30.1
March	2.83	10,273	7,320	6,240	60.7
April	3.64	13,213	4,697	3,875	29.3
May	5.75	20,872	10,392	8,859	42.4
June	3.08	11,180	3,378	2,787	24.9
July	3.79	13,758	1,351	1,154	8.4
August	5.92	21,490	3,190	2,720	12.7
September	2.44	8,857	1,590	1,312	14.8
October	6.43	13,177	5,264	4,488	34.1
November	2.29	8,313	4,106	3,387	40.7
December	2.66	9,656	7,264	6,193	64.1
Annual	....	12,260	....	3,816	31.1

1909	1 Average Rainfall, Inches	2 Rainfall per Acre, Cu. Ft.	3 Discharge Cu. Ft. per Sec.	4 Discharge Cu. Ft. per Acre	5 Per Cent. 4 of 2
January	2.57	9,329	431	367	3 93
February	6 51	23,631	6,019	4,630	19 59
March	3.14	11,398	11,050	9,411	82.56
April	4.97	18,041	6,924	5,707	31.63
May	5.67	20,582	9,399	8,005	38.89
June	4.07	14,774	6,215	5,122	34.66
July	6.41	23,268	6,024	5,131	22 05
August	2.30	8,349	7,584	6,459	77.36
September	2.29	8,312	681	561	6.75
October	4.72	17,133	1,107	943	5.50
November	2.61	9,474	1,995	1,644	17.35
December	3.46	12,559	5,112	4,354	34.66
Annual	....	176,852	....	52,334	29.59

1910	1 Average Rainfall, Inches	2 Rainfall per Acre, Cu. Ft.	5 Discharge Cu. Ft. per Sec.	4 Discharge Cu. Ft. per Acre	5 Per Cent. 4 to 2
January	3.76	13,648	11,801	10,051	73.64
February	4.21	15,282	6,673	5,134	33.59
March	.21	762	10,771	9,174	1,203.93
April	3.43	12,450	3,198	2,636	21.17
May	4.42	16,044	4,833	4,116	25.65
June	2.46	8,929	1,255	1,034	11.58
July	7.80	28,314	6,124	5,216	18.42
August	1.72	6,243	1,291	1,100	17.62
September	3.54	12,850	1,378	1,136	8.84
October	8.10	29,403	12,799	10,901	37.07
November	1.83	6,642	1,818	1,498	22.55
December	2.21	8,022	2,470	2,104	2.62
Annual	....	158,625	....	54,100	34.11

1911	1 Average Rainfall, Inches	2 Rainfall per Acre, Cu. Ft.	3 Discharge Cu. Ft. per Sec.	4 Discharge Cu. Ft. per Acre	5 Per Cent. 4 of 2
January	3.78	13,721	11,985	10,208	74.39
February	2.00	7,260	10,025	7,712	106.22
March	3.17	11,507	8,136	6,929	60.22
April	6.40	23,232	18,130	14,943	64.32
May	3.19	11,579	6.353	5,411	46.73
June	3.79	13,757	1,765	1,455	10.58
July	1.56	5,662	777	662	11.69
August	2.55	9,256	437	372	4.02
September	7.08	25,700	3,524	2,904	11.30
October	4.74	17,206	7,399	6,302	36.63
November	3.56	12,922	5,657	4,662	36.08
December	3.55	12,886	8,894	7,577	58.80
Annual	....	164,693	....	69,137	41.98

1912	1 Average Rainfall, Inches	2 Rainfall per Acre, Cu. Ft.	3 Discharge Cu. Ft. per Sec.	4 Discharge Cu. Ft. per Acre	5 Per Cent. 4 to 2
January	2.76	10,018	10,482	8,928	89.12
February	3.15	11,434	9,267	7,384	64.58
March	5.50	19,965	22,083	18,808	94.20
April	5.87	21,308	18,094	14,913	69.99
May	5.12	18,585	14,768	12,578	67.68
June	2.48	9,002	2,402	1,980	22.00
July	5.63	20,436	4,093	3,486	17.06
August	6.33	22,977	3,405	2,900	12.62
September	3.62	13,140	1,497	1,234	9.39
October	1.83	6,642	942	802	12.07
November	1.02	3,702	1,087	896	24.20
December	2.01	7,296	825	703	9.64
Annual	....	164,505	....	74,612	45.36

1913	1 Average Rainfall, Inches	2 Rainfall per Acre, Cu. Ft.	3 Discharge Cu. Ft. per Sec.	4 Discharge Cu. Ft. per Acre	5 Per Cent. 4 of 2
January	9.53	34,593	24,778	21,103	61.00
February	1.82	6,606	4,960	3,831	57.99
March	10.86	39,385	20,540	17,494	44.42
April	4.23	15,354	22,818	18,807	122.49
May	1.86	6,751	1,871	1,594	23.61
June	2.20	7,986	1,014	836	10.47
July	2.76	10,018	578	492	4.91
August	3.11	11,289	1,030	877	7.77
September	2.75	9,982	439	362	3.63
October	3.26	11,833	....	....	....
November	4.49	16,289	....	....	....
December	1.28	4,646	2,928	2,494	53.68
Annual	....	174,739	....	....	....

1914	1 Average Rainfall, Inches	2 Rainfall per Acre, Cu. Ft.	3 Discharge Cu. Ft. per Sec.	4 Discharge Cu. Ft. per Acre	5 Per Cent. 4 of 2
January	2.20	7,986	1,962	1,671	20.92
February	3.87	14,048	7,845	6,027	42.90
March	2.85	10,345	10,804	9,202	88.95
April	3.72	13,504	13,197	10,877	80.55
May	1.53	5,554	3,387	2,885	51.94
June	2.34	8,494	1,154	951	11.20
July	2.53	9,184	815	694	7.56
August	5.58	20,255	864	736	3.63
September	2.32	8,422	948	781	9.27
October	2.25	8,167	633	539	6.60
November	1.00	3,630	522	430	11.85
December	3.05	11,071	480	409	3.69
Annual	....	120,660	....	35,202	29.17

1915	1 Average Rainfall, Inches	2 Rainfall per Acre, Cu. Ft.	3 Discharge Cu. Ft. per Sec.	4 Discharge Cu. Ft. per Acre	5 Per Cent. 4 of 2
January	3.30	11,979	1,122	956	7.98
February	1.59	5,772	11,008	8,468	146.68
March	1.51	5,482	2,359	2,009	36.65
April	1.25	4,537	1,106	912	20.10
May	5.10	18,513	4,229	3,602	19.46
June	2.72	9,874	3,606	3,046	30.85
July	6.19	22,470	7,214	6,144	27.34
August	8.54	31,000	6,066	5,166	16.67
September	3.48	12,632	2,622	2,161	17.11
October	1.82	6,607	1,274	1,085	16.42
November	3.10	11,253	2,347	1,934	17.19
December	5.00	18,150	9,657	8,225	45.32
Annual	....	158,269	....	43,708	27.62

West Branch of White River. The west branch of White River rises by many tributaries on the Niagara escarpment in Randolph, Delaware, Madison and Tipton counties. Its source is within the Wisconsin glacial area. It flows in a general southwesterly direction throughout its course, and unites with the east branch at the southwest corner of Daviess County. Its entire course lies within the glaciated areas. It leaves the Wisconsin glacial area near Martinsville and flows in the Illinois glacial area from there to its junction with the east branch. Below Martinsville is a great valley-train from the Wisconsin glacier. Because of the glacial drift above and the valley-train below, rock exposures are rare. These occur where the river meanders against one of the bluffs. The river crosses all the rock formations of the state except the Ordovician, and in the Knobstone, Mitchell and Mansfield plateaus, bluffs of solid rock are seen continually from the river, yet the river seldom cuts into one of them.

The valley is broad and flat. The banks of the river range from ten to fifty feet in height. These banks are usually composed of clay, sand and gravel. The bottom land is very valuable farm land, although much of it is subject to overflow.

The mean annual rainfall above the gage at Indianapolis on west branch of White River as shown by the station records from this section for eighteen years (1898-1915, inclusive) is 38.76 inches. The following tables, which are computed entirely from government reports, show the average depth of monthly rainfall and the cubic feet per acre rainfall of the basin of the west branch of White River for the eighteen years above noted. These averages were taken from six U. S. Weather Bureau recording stations. The following is a list of the stations used: Anderson, Fairmount, Farmland, Indianapolis, Northfield and Whitestown. Anderson was given twice the weight of the other stations in the average, because it is near the center of the region while the others are near the margin.

Table showing average depth of monthly rainfall in inches; and cubic feet per acre of rainfall for the West Branch of White River above the gage at Indianapolis, Indiana, 1898-1915 inclusive.

Drainage Area, 1,609.1 sq. mi.—1,029,824 Acres.

	1898		1899		1900		1901	
	In.	Cu. ft.	In.	Cu. ft.	In.	Cu. ft.	In.	Cu. ft.
January	4.62	13,770.6	3.39	12,305.7	1.81	6,570.3	1.45	5,263.5
February	2.09	7,586.7	2.14	7,767.2	3.49	12,668.7	1.47	5,336.1
March	.41	30,528.3	4.25	15,427.5	2.37	8,603.1	3.44	12,487.2
April	56	9,292.8	1.15	4,174.5	2.09	7,586.7	2.79	10,127.7
May	.79	17,387.7	3.24	11,761.2	5.71	20,727.3	3.00	10,890.0
June	.37	8,603.1	1.90	6,897.0	4.39	15,935.7	3.95	14,338.5
July	42	16,044.6	4.00	14,520.0	4.13	14,991.9	1.13	4,101.9
August	65	13,249.5	3.19	11,579.7	3.73	13,539.9	3.52	12,813.9
September	.29	11,942.7	3.64	13,213.2	2.24	8,131.2	.96	3,484.8
October	58	16,625.4	2.47	8,966.1	2.31	8,385.3	3.05	11,071.5
November	89	10,490.7	3.28	11,906.4	3.92	14,229.6	1.04	3,775.2
December	.10	7,623.0	2.80	10,164.0	1.20	4,356.0	4.29	15,572.7

	1902		1903		1904		1905	
	In.	Cu. ft.	In.	Cu. ft.	In.	Cu. ft.	In.	Cu. ft.
January	2.86	10,381.8	2.17	7,877.1	4.45	16,153.5	1.09	3,956.7
February	5.89	21,380.7	3.55	12,886.5	2.61	9,474.3	1.89	6,860.7
March	4.31	15,645.3	2.02	7,332.6	9.15	33,214.5	2.88	10,454.4
April	3.88	14,084.4	3.36	12,196.8	4.72	17,133.6	4.04	14,665.2
May	3.16	11,470.8	5.49	19,928.7	3.61	13,104.3	5.00	18,150.0
June	5.12	18,585.6	3.63	13,176.9	2.73	9,909.9	2.74	9,946.2
July	3.29	11,942.7	2.33	8,457.9	3.42	12,414.6	2.82	10,236.6
August	3.47	12,596.1	3.88	14,084.4	2.36	8,566.8	5.23	18,984.9
September	1.33	4,827.9	1.59	5,771.7	2.94	10,672.2	4.61	16,734.3
October	2.73	9,909.9	3.50	12,705.0	1.14	4,138.2	4.11	14,919.3
November	2.22	8,058.6	1.62	5,880.6	.15	544.5	1.72	6,243.6
December	2.05	7,441.5	1.61	5,844.3	3.70	13,431.0	2.10	7,823.0

	1906		1907		1908		1909	
	In.	Cu. ft.	In.	Cu. ft.	In.	Cu. ft.	In.	Cu. ft.
January	2.30	8,349.0	6.72	24,393.6	1.90	6,897.0	2.63	9,546.9
February	1.12	4,065.6	.40	1,452.0	4.15	15,064.5	[illegible].04	18,295.2
March	5.27	19,130.1	4.42	16,044.6	5.07	18,404.1	.76	10,018.8
April	2.21	8,022.3	2.36	8,566.8	3.39	12,305.7	.04	18,295.2
May	2.73	9,909.9	3.26	11,833.8	5.27	19,130.1	70	20,691.0
June	2.45	8,893.5	4.77	17,315.1	2.59	9,401.7	.84	24,829.2
July	2.54	9,220.2	5.65	20,509.5	8.02	29,112.6	.57	20,219.1
August	5.52	20,037.6	3.61	13,104.3	1.48	5,372.4	.79	13,757.7
September	3.29	11,942.7	2.35	8,530.5	.90	3,267.0	.46	8,929.8
October	2.30	8,349.0	2.35	8,530.5	51	1,851.3	.65	13,249.5
November	3.11	11,289.3	2.17	7,877.1	3.68	13,358.[illegible]	.53	12,813.9
December	3.55	12,886.5	3.52	12,777.6	1.61	,844.	.72	8,973.6

	1910		1911		1912	
	In.	Cu. ft.	In.	Cu. ft.	In.	Cu. ft.
January	3.05	11,071.5	3.09	11,216.7	1.86	6,751.8
February	3.26	11,833.8	1.34	4,864.2	2.22	8,058.6
March	.07	254.1	2.41	8,748.3	4.73	17,169.9
April	3.33	12,087.9	4.93	17,895.9	4.48	16,262.4
May	3.58	12,995.4	1.74	6,316.2	4.81	17,460.3
June	2.76	10,018.8	2.80	10,164.0	2.05	7,441.5
July	4.48	16,262.4	1.13	4,101.9	6.12	22,215.6
August	1.54	5,590.2	2.78	10,091.4	4.96	18,004.8
September	6.28	22,796.4	4.68	16,988.4	3.78	13,721.4
October	4.88	17,714.4	4.44	16,117.2	1.84	6,679.2
November	1.98	7,187.4	3.41	12,378.3	1.35	4,900.5
December	1.76	6,388.8	2.53	9,183.9	1.23	4,464.9

	1913		1914		1915	
	In.	Cu. ft.	In.	Cu. ft.	In.	Cu. ft.
January	7.36	24,716.8	2.31	7,731.9	2.94	10,672.2
February	1.59	5,771.[illegible]	2.72	9,873.6	1.35	4,900.5
March	9.08	32,960	1.76	6,388.8	1.41	5,118.3
April	3.31	12,015.	3.12	11,325.6	1.03	3,738.9
May	2.55	9,256	2.29	8,312.7	5.07	18,404.1
June	3.41	12,378.	2.22	8,058.6	3.18	11,543.4
July	4.58	16,625.	1.37	4,973.1	7.45	27,043.5
August	4.49	16,298.	5.02	18,222.6	5.95	21,598.5
September	2.09	7,686.	2.30	8,349.0	3.74	13,576.2
October	2.88	10,454	1.93	7,005.9	2.05	7,441.5
November	3.63	13,176.	1.45	5,263.5	3.00	10,890.0
December	.67	2,432.	3.34	12,124.2	4.14	15,028.2

West Branch of White River at Anderson, Indiana. This station was established August 15, 1911, by W. M. Tucker. It is located on the Eighth Street bridge in Anderson. The gage, which is of the weight and chain variety, is located on the upstream handrail of the bridge about twenty feet from the west bank of the stream. The datum of the gage is 821.3 feet above sea level as taken from the elevation of the C., C., C. & St. L. Ry. station in Anderson; 42.95 feet below the rail of the P., C., C. & St. L. track in front of the station; and 22.2 feet below a bench mark (843.5 feet) which is located on the west side of the viaduct by which the P., C., C., & St. L. Ry. passes over Eighth Street. The length of the chain from end of weight to first marker is 27.78 feet.

The river at this point is straight for 300 feet above and below the bridge. The bed of the stream is sandy but only subject to slight changes. The current is steady. There are no piers to break the channel, except when the river overflows the west bank.

Not enough current readings have been taken from this gage to make up a good rating table. The current readings have been taken at low gage readings and are sufficient for a rating table between gage heights 5.00 and 7.00 feet.

The gage has been read daily since 1911 by Mr. Lowell Dilts and Mr. Horace Malone of Anderson, Ind.

The following data have been taken from this station:

Daily Gage Heights on White River (West Branch) at Anderson, Indiana, for 1911.

Day	Aug.	Sept.	Oct.	Nov.	Dec.	Day	Aug.	Sept.	Oct.	Nov.	Dec.
1	......	5.4	5.7	5.7	5.7	17	5.4	7.0	6.4	6.0	6.2
2	......	5.3	5.9	5.6	5.6	18	5.5	6.2	7.9	6.0	6.3
3	......	5.3	5.8	5.6	5.6	19	5.6	6.0	7.4	8.3	6.3
4	......	5.3	5.7	5.6	5.7	20	5.5	5.9	7.1	8.3	6.2
5	......	5.4	5.7	5.6	5.9	21	5.5	5.7	6.8	8.2	6.0
6	......	5.5	5.7	5.6	5.8	22	5.3	5.7	6.9	8.1	6.3
7	......	5.4	9.4	5.7	5.8	23	5.3	5.6	6.7	7.5	6.0
8	......	5.4	10.7	5.7	5.8	24	5.4	5.5	5.8	7.0	6.4
9	......	5.5	8.5	5.6	5.9	25	5.5	5.5	5.7	6.6	6.2
10	......	5.9	7.0	5.6	5.9	26	5.4	5.5	5.7	6.4	6.2
11	......	5.7	6.7	5.6	6.0	27	5.4	5.5	5.7	6.2	6.1
12	......	5.6	6.4	5.6	6.0	28	5.5	5.6	5.6	5.8	6.0
13	......	5.5	6.4	5.6	6.0	29	5.4	5.7	5.6	5.8	5.9
14	*	5.6	6.3	5.6	6.0	30	5.4	5.7	5.6	5.7	5.9
15	5.5	7.2	6.1	6.1	6.1	31	5.4	......	5.6	.....	5.8
16	*	7.1	6.3	6.1	6.2						

*No record.

Daily Gage Heights on White River (West Branch) at Anderson, Indiana, for 1912.

Day	Jan.	Feb.	Mar.	Apr.	May	June	July	Aug.	Sept.	Oct.	Nov.	Dec.
1	5.8	.....	7.5	8.9	7.0	6.0	5.8	5.7	5.7	5.6	5.7	5.5
2	5.8	.....	6.8	9.4	6.8	6.0	5.9	5.7	5.7	5.6	5.7	5.6
3	5.8	.....	6.5	12.0	6.5	6.0	6.0	5.7	5.6	5.6	5.6	5.6
4	5.8	.....	6.3	10.7	6.4	5.9	6.1	5.6	5.6	5.6	5.6	5.6
5	5.7	.....	6.3	9.1	6.3	5.9	6.0	5.6	5.6	5.6	5.6	5.7
6	5.7	.....	6.0	8.3	6.3	5.9	5.9	5.6	5.5	5.6	5.6	5.7
7	5.7	.....	6.1	7.5	6.3	5.9	5.7	5.6	5.6	5.6	5.6	5.7
8	5.7	.....	6.2	8.1	6.2	5.9	5.7	5.8	5.5	5.6	5.6	5.6
9	5.7	.....	7.7	7.4	6.1	5.9	6.0	5.8	5.5	5.6	5.6	5.6
10	5.7	.....	6.5	7.0	6.1	5.9	6.8	6.7	5.5	5.8	5.7	5.7
11	5.7	.....	6.8	6.8	6.2	5.9	6.8	6.5	5.6	5.8	5.7	5.7
12	5.7	.....	6.7	6.7	6.4	5.9	7.8	6.7	5.5	5.8	5.6	5.6
13	5.8	.....	6.5	6.7	6.6	5.8	7.0	7.0	5.6	8.6	5.7	5.6
14	5.8	.....	8.5	6.7	7.4	5.8	7.5	6.7	5.7	7.5	5.7	5.5
15	5.8	.....	11.6	6.6	8.8	5.9	7.5	6.2	5.6	6.2	5.8	5.5
16	5.8	.....	12.0	6.6	8.9	6.0	8.0	6.0	5.7	5.8	5.7	5.4
17	*	.....	8.5	6.7	9.0	6.4	7.2	5.6	5.6	5.8	5.6	5.3
18	.....	.....	8.5	8.2	8.4	6.2	6.0	5.5	5.8	5.6	5.7	5.5
19	.....	.....	8.1	6.7	7.8	6.0	6.5	5.5	5.8	5.6	5.7	5.3
20	.....	.....	7.7	6.7	7.4	5.9	7.8	5.6	5.7	5.7	5.7	5.5
21	.....	.....	7.7	6.7	7.2	5.9	7.4	7.0	5.6	5.8	5.6	5.5
22	.....	.....	9.4	6.5	7.0	6.3	7.0	7.8	5.8	5.6	5.7	5.5
23	.....	.....	8.9	6.4	6.9	6.1	6.4	6.2	5.7	5.6	5.6	5.5
24	.....	.....	9.2	6.4	6.8	5.9	6.0	6.0	5.7	5.7	5.6	5.3
25	.....	.....	8.5	6.3	6.1	5.9	5.9	5.9	5.8	5.6	5.6	5.3
26	.....	.....	7.6	6.3	6.6	5.8	5.8	5.7	5.7	5.6	5.6	5.6
27	.....	.....	8.6	6.3	6.5	5.8	5.8	5.9	5.8	5.6	5.6	5.6
28	.....	11.6	10.4	6.2	6.4	5.8	5.7	6.2	5.7	5.6	5.5	5.5
29	.....	8.7	12.5	6.7	6.3	5.8	5.7	5.5	5.6	5.7	5.5	5.5
30	.....	.....	16.5	7.3	6.2	5.8	5.6	5.6	5.6	5.8	5.5	5.5
31	.....	.....	10.3	.....	6.1	.....	5.8	5.5	.....	5.8	.....	5.6

*No record from January 17 to February 27.

Daily Gage Heights on White River (West Branch) at Anderson, Indiana, for 1913.

Day	Jan.	Feb.	Mar.	Apr.	May	June	July	Aug.	Sept.	Oct.	Nov.	Dec.
1	5.5	6.5	7.9	9.4	6.3	6.2	6.5	5.6	5.4	5.4	5.6	6.0
2	5.5	6.0	7.6	9.2	6.3	6.2	6.3	5.7	5.4	5.4	5.6	5.6
3	5.6	*	7.0	9.1	6.3	6.2	5.9	5.7	5.4	5.4	5.5	5.9
4	5.6	*	6.9	10.2	6.2	6.2	5.9	5.7	5.4	5.4	5.5	5.9
5	5.7	*	6.5	10.9	6.2	6.2	5.9	5.6	5.4	5.4	5.4	5.9
6	7.7	*	6.8	11.8	6.1	6.1	5.9	5.6	5.4	5.4	5.4	5.8
7	10.6	*	7.4	10.3	6.1	6.6	5.8	5.6	5.4	5.4	5.4	5.8
8	11.4	*	6.8	8.5	6.0	7.2	5.8	5.6	5.4	5.4	5.5	5.8
9	8.0	*	6.8	7.6	6.0	6.7	5.8	5.6	5.4	5.4	5.5	5.8
10	12.0	*	6.6	9.9	6.0	6.2	5.8	5.6	5.4	5.4	5.5	5.7
11	10.1	*	6.8	11.8	6.0	6.0	5.8	5.6	5.4	5.4	5.6	5.7
12	8.5	*	7.2	9.4	6.0	5.9	5.8	5.7	5.5	5.4	5.7	5.6
13	7.6	*	7.6	8.6	6.2	5.9	5.8	5.7	5.5	5.4	5.7	5.6
14	7.3	*	8.0	8.3	7.4	5.9	5.8	5.7	5.5	5.4	5.6	5.6
15	10.0	*	7.5	9.4	7.2	5.9	5.8	5.7	5.5	5.5	5.5	5.6
16	10.7	*	7.0	9.6	7.2	5.8	5.8	5.7	5.4	5.5	5.9	5.6
17	10.3	*	6.6	10.1	7.6	5.8	6.3	5.7	5.4	5.5	5.9	5.6
18	8.5	6.0	6.4	10.3	6.9	5.8	6.4	5.7	5.4	5.5	5.9	5.6
19	11.4	6.0	6.3	10.2	6.2	5.8	7.5	5.7	5.4	5.5	5.8	5.5
20	12.5	6.1	6.3	7.3	6.2	5.9	9.4	5.6	5.5	5.6	5.9	5.5
21	10.5	6.0	7.6	6.9	6.1	5.9	8.2	5.7	5.5	5.7	5.6	5.4
22	11.9	6.1	8.3	6.6	6.0	5.9	7.3	5.6	5.5	5.7	5.6	5.4
23	9.7	6.0	9.9	6.5	6.1	6.1	6.6	5.5	5.5	5.7	5.6	5.4
24	8.3	6.0	15.1	6.5	6.2	7.2	6.0	5.5	5.5	5.7	5.6	5.4
25	7.7	5.9	25.1	6.5	6.2	6.7	5.9	5.5	5.5	5.7	5.6	5.4
26	8.7	5.9	19.1	6.5	6.2	6.4	5.8	5.5	5.5	5.6	5.7	5.4
27	7.8	6.1	14.3	6.5	6.6	6.1	5.8	5.5	5.5	5.6	5.7	5.4
28	7.0	7.4	13.5	6.5	6.3	6.0	5.7	5.5	5.4	5.5	5.9	5.5
29	6.7	.....	11.8	6.5	6.2	6.0	5.8	5.5	5.4	5.5	5.9	5.5
30	6.5	.....	9.9	6.4	6.2	5.9	5.8	5.5	5.4	5.5	5.9	5.5
31	6.7	.....	9.5	.....	6.2	.....	5.8	5.4	.....	5.5	.....	5.5

*No record.

Daily Gage Heights on White River (West Branch) at Anderson, Indiana, for 1914.

Day	Jan.	Feb.	Mar.	Apr.	May	June	July	Aug.	Sept.	Oct.	Nov.	Dec.
1	5.6	6.2	5.9	7.6	5.9	5.6	5.5	5.3	5.7	5.4	5.4	5.5
2	5.6	6.9	6.0	9.0	5.9	5.6	5.5	5.3	5.9	5.4	5.4	5.5
3	5.6	7.5	6.2	7.8	5.9	5.6	5.5	5.3	6.3	5.4	5.4	5.5
4	5.6	7.1	6.4	7.2	5.9	5.6	5.5	5.3	6.0	5.4	5.4	5.6
5	5.6	6.8	6.9	6.7	6.0	5.6	5.5	5.3	5.9	5.4	5.4	5.6
6	5.6	6.5	7.5	7.5	6.0	5.8	5.5	5.3	5.7	5.4	5.4	5.6
7	5.6	6.3	8.6	9.9	6.0	5.8	5.4	5.3	5.6	5.4	5.4	5.6
8	5.7	6.2	8.0	12.4	6.1	5.8	5.4	5.3	5.5	5.3	3.4	5.6
9	5.8	6.1	8.4	10.7	6.2	5.7	5.4	5.6	5.5	5.3	5.4	5.6
10	5.9	6.0	8.7	9.4	6.1	5.7	5.4	6.2	5.5	5.3	5.6	5.6
11	5.9	5.9	8.5	8.6	6.1	5.7	5.4	6.0	5.5	5.3	5.5	5.6
12	5.9	5.8	8.3	7.8	6.1	5.6	5.8	5.8	5.4	5.3	5.5	5.6
13	5.9	5.8	8.0	7.4	6.0	5.6	6.2	5.6	5.4	5.3	5.5	5.6
14	5.9	5.8	7.8	7.0	6.0	5.6	6.4	5.6	5.4	5.3	5.4	5.7
15	5.9	5.8	7.5	6.7	6.0	5.6	6.9	5.4	5.4	5.3	5.4	5.7
16	5.8	5.7	7.1	6.6	6.0	5.6	6.7	5.4	5.4	5.3	5.3	5.6
17	5.8	5.7	6.4	6.4	5.9	5.5	6.1	5.4	5.4	5.3	5.3	5.6
18	5.8	5.7	6.4	6.3	5.9	5.5	5.9	5.4	5.4	5.5	5.3	5.6
19	5.9	5.7	6.3	6.2	5.9	5.5	5.5	5.5	5.4	5.5	5.3	5.6
20	5.9	5.7	6.3	6.2	5.9	5.5	5.5	5.6	5.4	5.7	5.3	5.5
21	5.9	5.6	6.2	6.1	5.9	5.5	5.5	5.6	5.4	5.7	5.3	5.5
22	5.9	5.6	6.2	6.1	5.8	5.4	5.5	5.5	5.4	5.6	5.3	5.5
23	6.0	5.6	6.1	6.1	5.8	5.4	5.4	5.4	5.4	5.6	5.3	5.5
24	6.0	5.7	6.0	6.0	5.8	5.4	5.4	5.4	5.4	5.6	5.4	5.6
25	6.0	5.7	5.9	6.0	5.7	5.4	5.4	5.6	5.4	5.5	5.4	5.6
26	6.0	5.7	6.0	6.0	5.7	5.6	5.4	5.6	5.4	5.5	5.3	5.5
27	6.0	5.9	6.2	6.0	5.7	5.6	5.4	5.5	5.4	5.5	5.4	5.5
28	5.9	5.9	8.3	6.0	5.7	5.6	5.4	6.0	5.4	5.5	5.4	5.5
29	5.9	.....	8.6	5.9	5.7	5.5	5.4	6.6	5.4	5.4	5.4	5.5
30	5.8	.....	8.3	5.9	5.6	5.5	5.4	6.3	5.4	5.4	5.4	5.6
31	5.8	.....	8.0	.....	5.6	.....	5.4	5.9	.....	5.4	.....	5.6

Daily Gage Heights on White River (West Branch) at Anderson, Indiana, for 1915.

Day	Jan.	Feb.	Mar.	Apr.	May	June	July	Aug.	Sept.	Oct.	Nov.	Dec.
1	5.6	6.3	5.8	5.5	5.4	7.1	6.0	6.5	5.7	5.5	5.6	5.5
2	5.6	12.7	5.8	5.5	5.4	7.8	6.0	7.1	5.7	5.5	5.6	5.5
3	5.5	11.4	5.8	5.5	5.5	7.7	5.9	7.8	5.7	5.5	5.5	5.5
4	5.5	9.6	5.8	5.6	5.6	6.8	5.9	10.7	5.7	5.6	5.5	5.7
5	5.5	8.9	5.8	5.6	5.6	6.4	5.9	8.5	5.6	5.6	5.5	5.9
6	5.5	8.4	5.7	5.6	5.6	6.1	5.8	7.8	5.6	5.6	5.8	6.1
7	5.5	7.6	5.7	5.6	5.7	6.4	5.8	7.3	5.6	5.6	5.9	6.3
8	5.6	7.1	5.7	5.6	5.7	6.6	10.5	7.0	5.6	5.6	6.2	6.6
9	5.6	6.9	5.7	5.6	5.6	6.8	11.8	6.5	5.6	5.6	6.6	6.9
10	5.6	6.7	5.7	5.6	5.6	6.5	9.6	6.4	5.6	5.6	7.1	7.2
11	5.8	6.5	5.7	5.6	5.6	6.2	8.5	6.3	5.6	5.6	7.3	7.0
12	6.1	6.8	5.7	5.9	5.6	6.0	8.5	9.5	5.5	5.6	7.0	6.8
13	6.6	7.0	5.7	5.8	5.5	5.8	7.4	8.6	5.5	5.6	6.9	6.7
14	7.3	6.9	5.6	5.8	5.5	6.3	6.9	7.3	5.5	5.6	6.6	6.7
15	7.7	6.8	5.6	5.7	5.5	6.8	7.2	7.0	5.5	5.6	6.1	6.6
16	6.7	6.5	5.6	5.7	5.5	6.1	9.8	6.8	5.6	5.6	5.8	6.5
17	6.4	6.3	5.6	5.7	5.4	5.8	10.0	6.8	5.6	5.6	5.6	6.3
18	6.1	6.1	5.6	5.7	5.4	5.9	8.3	7.0	5.7	5.6	5.6	6.2
19	5.9	6.0	5.6	5.7	5.4	6.0	8.1	7.3	5.6	5.6	5.8	6.2
20	5.7	6.0	5.6	5.6	5.4	6.1	7.8	7.6	5.6	5.6	6.0	6.0
21	5.6	6.0	5.6	5.6	5.5	6.2	7.8	7.8	5.6	5.6	6.2	6.1
22	5.5	6.0	5.6	5.6	5.5	6.3	7.5	8.1	5.6	5.6	6.4	6.1
23	5.5	5.9	5.6	5.6	5.6	6.1	7.2	8.3	5.6	5.6	6.3	6.3
24	5.5	5.9	5.6	5.5	5.6	6.0	6.9	7.0	5.6	5.8	6.2	6.0
25	5.5	5.9	5.6	5.5	5.7	5.9	6.4	6.6	5.6	5.5	6.0	5.9
26	5.6	5.9	5.5	5.5	5.8	5.8	6.1	6.3	5.6	5.5	5.9	6.2
27	5.6	5.8	5.5	5.5	5.9	5.8	5.9	6.2	5.5	5.5	5.7	6.4
28	5.6	5.8	5.5	5.5	6.2	5.8	5.8	6.1	5.5	5.8	5.6	6.2
29	5.6	.....	5.5	5.4	6.4	5.9	5.6	6.0	5.5	5.8	5.5	6.3
30	5.6	.....	5.5	5.4	6.7	5.9	5.6	5.8	5.5	5.7	5.5	6.1
31	5.9	.....	5.5	.....	6.9	.....	5.8	5.8	.....	5.6	.....	6.0

Daily Gage Heights on White River (West Branch) at Anderson, Indiana, for 1916.

Day	Jan.	Feb.	Mar.	Apr.	May	June	July	Aug.	Sept.	Oct.	Nov.	Dec.
1	9.3	17.4	6.1	7.2	6.0	6.0	5.6	5.1	5.2	5.1	5.0	5.0
2	16.6	11.9	6.1	6.9	6.0	5.9	5.6	5.1	5.1	5.1	5.0	5.0
3	18.0	8.4	6.0	6.6	6.1	6.1	5.6	5.1	5.1	5.1	5.0	5.1
4	10.2	7.8	5.8	6.4	6.7	6.6	5.5	5.2	5.0	5.0	5.0	5.2
5	9.5	7.3	5.9	6.3	6.5	6.2	5.4	5.2	5.0	5.0	5.0	5.5
6	8.6	7.0	5.9	6.1	6.2	6.0	5.4	5.2	5.0	5.0	5.1	5.7
7	8.2	6.8	7.0	6.1	9.1	7.9	5.4	5.2	5.0	5.0	5.1	5.4
8	7.5	6.5	7.4	6.1	11.6	8.1	5.4	5.3	5.0	5.0	5.1	5.3
9	6.9	6.6	6.8	6.0	8.6	7.3	5.3	5.8	5.3	5.0	5.1	5.4
10	6.6	6.5	6.5	6.0	7.3	6.9	5.3	5.5	5.1	5.0	5.1	5.7
11	7.1	6.5	6.2	6.0	6.7	6.6	5.3	5.9	5.0	5.0	5.1	5.5
12	9.3	6.6	6.1	6.0	6.3	6.3	5.3	5.5	5.0	5.0	5.0	5.4
13	12.3	7.5	6.1	5.9	6.2	6.1	5.3	5.4	5.0	5.0	5.0	5.3
14	11.8	6.8	6.1	5.9	6.1	6.1	5.3	5.4	5.0	5.0	5.0	5.0
15	9.3	6.6	6.1	5.9	6.0	6.1	5.3	5.3	5.0	5.0	5.0	5.1
16	7.8	6.4	6.0	5.8	5.9	6.1	5.2	5.3	5.0	5.0	5.0	5.1
17	7.1	6.4	6.0	5.7	5.9	6.2	5.2	5.2	5.0	5.1	5.0	5.1
18	6.6	6.7	6.0	5.8	5.8	6.2	5.2	5.2	5.0	5.1	5.0	5.1
19	6.2	6.7	5.9	5.8	5.7	6.1	5.3	5.1	5.0	5.1	5.0	5.1
20	6.0	6.4	5.9	5.8	5.7	6.2	5.3	5.2	5.0	5.1	5.0	5.1
21	6.2	6.3	6.0	5.8	5.7	7.2	6.5	5.1	5.0	5.1	5.0	5.1
22	6.2	6.4	6.3	5.8	5.7	8.9	6.6	5.1	5.0	5.1	5.0	5.1
23	6.4	6.4	7.5	6.4	5.8	7.2	5.6	5.6	5.0	5.1	5.1	5.0
24	6.7	7.0	7.0	6.1	5.8	6.2	5.3	5.3	5.0	5.0	5.2	5.1
25	6.8	6.9	6.7	6.0	5.8	6.1	5.2	5.2	5.0	5.0	5.1	5.1
26	7.1	6.7	7.0	6.1	5.8	6.0	5.2	5.4	5.0	5.0	5.1	5.2
27	7.4	6.3	10.0	6.4	5.8	5.9	5.1	5.3	5.0	5.0	5.0	5.9
28	6.8	6.2	11.1	6.4	6.1	5.8	5.1	5.2	5.1	5.0	5.0	7.6
29	7.1	6.2	9.8	6.1	6.3	5.7	5.1	5.2	5.1	5.0	5.1	6.9
30	14.9	.....	8.3	6.1	6.6	5.7	5.1	5.2	5.1	5.0	5.0	6.3
31	19.1	.....	7.6	....	6.2	....	5.1	5.2	.....	5.0	...	6.0

Discharge Measurements on West Branch of White River at Anderson, Indiana, 1911-1916, Inclusive.

Date	Hydrographer	Gage Height, Feet	Discharge, Sec.-Ft.
September 19, 1911	W. M. Tucker	6.0	281.8
April 10, 1912	W. M. Tucker	6.9	757.4
October 15, 1915	W. M. Tucker	5.6	175.3
October 30, 1915	W. M. Tucker	5.6	171.8
February 23, 1916	W. M. Tucker	6.5	533.9
July 25, 1916	W. M. Tucker	5.2	63.6
August 7, 1916	W. M. Tucker	5.2	68.0

Partial Rating Table for West Branch of White River at Anderson, Indiana, 1911-1916, Inclusive.

Gage Height, Feet	Discharge, Sec.-Ft.	Gage Height, Feet	Discharge, Sec.-Ft.	Gage Height, Feet	Discharge, Sec.-Ft.
5.0	40	5.7	200	6.4	479
5.1	51	5.8	227	6.5	532
5.2	65	5.9	254	6.6	586
5.3	92	6.0	281	6.7	641
5.4	119	6.1	328	6.8	698
5.5	146	6.2	377	6.9	757
5.6	173	6.3	427	7.0	819

This table is applicable only for channel conditions. It is well defined within the limits given.

Data from the gages at Elliston on the West Branch of White River and at Decker on the Main Branch have only been collected for one full year (1916) and but one current reading has been taken at each of these points. For these reasons the data are of no value in determining the amount of water which the streams discharge, and will not be included in the present paper. Information concerning the water power of White River may be found on pages 48 to 54 and 66 to 75 of the 35th Annual Report of the Indiana Department of Geology and Natural Resources.

Blue River System.

The Blue River system drains a portion of three counties, Washington, Crawford and Harrison, and flows into the Ohio River near Leavenworth, Crawford County. It has a drainage area of 471 square miles, 1.29% of the area of the state. The reason for giving it special consideration is because its tributary drainage is largely underground. The basin lies almost wholly within the cave district of the Mitchell limestone. The effect of this drainage upon evaporation and regularity of stream flow has been considered to some extent. The lowest recorded discharge on Blue River, which is registered at gage height 1.4 feet, is 50 second-feet. The highest discharge recorded on this stream was on October 7, 1910, when the gage reading was 14.7 feet and the discharge 10,914 second-feet. The ratio of the low to the high discharge is 1:218. On the West branch of White River at Shoals during the same period the lowest recorded reading, registered at gage height 2.0 feet, is 215 second-feet. The highest discharge during the flood of October, 1910, occurred at Shoals on October 13, with the gage reading 23.65 feet and a discharge of 36,000 second-feet. The ratio of the low to the high discharge is 1:167. These figures would indicate that Blue River is more erratic than White River but the comparison is not well founded. The greater drainage area of the White River tends to flatten the crests of the floods. The rainfall during the flood of October, 1910, was not on the average as heavy over White River basin as over Blue River basin. It seems that no comparison can be made with the data in hand which will show whether or not the underground drainage has an effect on the regularity of stream flow.

The effect of underground drainage on evaporation cannot be shown by the data in hand. The only two full years of discharge data which have been kept on White River at Shoals and on Blue River at White Cloud are 1910 and 1911. The records for 1915 are complete for the two gages except for seventeen days in February when the gage at White Cloud was out of order. The following comparisons of the percentage relations of discharge to rainfall on the two streams for these three years are interesting but prove nothing.

Per cent of rainfall which was discharged:

	1910.	1911.	1915.
White River	34.11	41.98	27.62
Blue River	33.5	41.9	33.2*

*Except February.

February was a month of high discharge at Shoals and was probably also at White Cloud, which would tend to increase the percentage for Blue River and therefore make the discrepancy greater for this year. If complete data could be gathered for several years it would probably show some interesting comparisons in this respect.

A gaging station was established one mile above White Cloud near the home of Mr. Julius Rothrock on August 18, 1909. The gage is constructed of heavy oak planks securely spiked to a large oak stump and to the roots of a large sycamore tree. The plank is placed with the slant of the river bank, which is about thirty degrees. The center is securely supported by heavy oak posts set in the bank. The scale is made of brass-headed tacks on the upstream side of the gage. The base of this gage is three feet below a nail in the root of the sycamore tree to which it is attached, and 23.7 feet below a nail in the corner of a barn which stands fifty feet south of the gage. The gage has been read since its installation up to August, 1917, by Russell Rothrock. Discharge measurements have been made by the writer from time to time and a fairly accurate rating table constructed.

Further information concerning this stream may be found on page 30 of the 35th Annual Report of the Indiana Department of Geology and Natural Resources.

Daily Gage Heights on Blue River at White Cloud, Indiana, for 1909.

Day	Jan.	Feb.	Mar.	Apr.	May	June	July	Aug.	Sept.	Oct.	Nov.	Dec.
1	....	....	....	....	....	....	....	....	1.6	1.6	1.7	1.8
2	....	....	....	....	....	....	....	....	1.6	1.5	1.7	1.9
3	....	....	....	....	....	....	....	....	1.6	1.5	1.7	2.2
4	....	....	....	....	....	....	....	....	1.6	1.5	1.7	2.2
5	....	....	....	....	....	....	....	....	1.6	1.5	1.6	2.2
6	....	....	....	....	....	....	....	....	1.6	1.5	1.6	2.1
7	....	....	....	....	....	....	....	....	1.6	1.5	1.6	2.1
8	....	....	....	....	....	....	....	....	1.6	1.5	1.7	2.3
9	....	....	....	....	....	....	....	....	1.6	1.6	1.7	2.4
10	....	....	....	....	....	....	....	....	1.6	1.6	1.8	2.2
11	....	....	....	....	....	....	....	....	2.4	1.7	1.8	2.2
12	....	....	....	....	....	....	....	....	2.0	1.7	1.7	3.5
13	....	....	....	....	....	....	....	....	2.0	1.8	1.7	4.3
14	....	....	....	....	....	....	....	....	1.8	1.7	1.7	4.8
15	....	....	....	....	....	....	....	....	1.7	1.7	1.7	3.7
16	....	....	....	....	....	....	....	....	1.7	1.7	1.7	3.3
17	....	....	....	....	....	....	....	*	1.7	1.7	1.8	3.0
18	....	....	....	....	....	....	....	2.0	1.7	2.0	1.8	2.8
19	....	....	....	....	....	....	....	1.9	1.7	2.0	1.8	2.6
20	....	....	....	....	....	....	....	1.9	1.7	2.1	1.8	2.3
21	....	....	....	....	....	....	....	1.6	1.6	2.4	1.9	2.2
22	....	....	....	....	....	....	....	1.7	1.6	2.2	1.8	2.2
23	....	....	....	....	....	....	....	1.7	1.7	2.2	2.2	2.2
24	....	....	....	....	....	....	....	1.7	1.6	2.5	2.2	2.1
25	....	....	....	....	....	....	....	1.7	1.6	2.5	2.3	2.1
26	....	....	....	....	....	....	....	1.6	1.6	2.2	2.1	2.0
27	....	....	....	....	....	....	....	1.6	1.6	2.0	2.0	1.9
28	....	....	....	....	....	....	....	1.6	1.5	1.9	2.0	1.9
29	....	....	....	....	....	....	....	1.6	1.6	1.9	1.9	2.0
30	....	....	....	....	....	....	....	1.6	1.5	1.8	1.9	1.9
31	....	....	....	....	....	....	....	1.7	....	1.7	....	1.8

*Gage installed August 18.

Daily Gage Heights on Blue River at White Cloud, Indiana, for 1910.

DAY	Jan.	Feb.	Mar.	Apr.	May	June	July	Aug.	Sept.	Oct.	Nov.	Dec.
1	1.8	2.7	7.0	2.0	2.7	2.6	2.3	3.1	1.7	1.6	2.0	2.5
2	2.2	2.6	6.2	2.0	2.6	2.6	2.0	2.8	1.7	1.6	2.0	2.4
3	3.0	2.6	4.4	2.0	2.5	2.5	2.1	2.7	1.7	1.6	2.0	2.3
4	2.7	2.6	4.0	2.0	4.0	2.4	2.0	2.4	1.6	1.6	2.0	2.2
5	2.5	2.6	3.7	2.0	3.1	2.7	2.2	2.4	2.9	1.9	2.0	2.2
6	3.2	2.6	3.5	2.0	2.8	2.5	2.9	2.3	2.3	14.2	2.0	2.2
7	2.7	2.5	3.3	2.0	2.8	2.5	7.0	2.3	2.3	14.7	1.9	2.2
8	2.3	2.4	3.1	2.0	5.7	2.3	4.2	2.2	2.2	6.9	2.0	2.1
9	2.3	2.4	3.0	2.0	5.2	2.3	3.4	2.2	2.1	5.1	1.9	2.1
10	2.2	2.4	2.9	2.0	5.0	2.1	3.6	2.1	2.0	4.2	2.0	2.1
11	2.2	2.4	2.8	1.9	3.5	2.3	3.5	2.1	1.9	3.8	1.9	2.1
12	2.3	2.3	2.7	2.0	3.6	2.2	3.7	2.0	1.8	3.5	1.9	2.0
13	2.5	3.3	2.7	3.1	3.9	2.8	4.3	2.0	1.8	3.3	1.9	2.0
14	9.6	2.2	2.6	3.2	3.4	2.6	4.7	2.0	1.7	3.1	1.8	2.0
15	5.2	2.2	2.5	2.8	3.1	2.5	3.6	1.9	1.7	3.0	1.9	2.0
16	4.0	2.3	2.5	2.8	2.9	2.4	3.8	1.9	1.7	2.8	1.9	2.0
17	3.5	3.0	2.4	4.9	2.8	3.3	9.0	1.9	1.7	2.8	1.9	2.0
18	4.8	3.1	2.4	4.9	2.9	2.3	5.3	1.9	1.7	2.7	1.8	2.0
19	7.1	2.9	2.3	4.5	2.9	2.2	4.3	1.8	1.7	2.6	1.8	2.0
20	4.5	3.0	2.3	4.8	2.8	2.2	3.6	1.8	1.7	2.5	1.8	2.0
21	4.0	3.3	2.3	4.4	5.1	2.0	3.3	1.8	1.7	2.4	1.8	1.9
22	3.8	5.2	2.3	3.8	3.8	2.0	3.0	2.1	1.7	2.4	1.8	2.0
23	3.6	5.4	2.3	3.5	3.5	2.0	2.8	2.1	1.6	2.4	1.8	2.0
24	3.4	4.6	2.2	3.3	3.3	2.1	2.7	2.0	1.6	2.3	1.8	2.0
25	3.3	3.8	2.2	3.1	4.4	2.0	2.7	2.0	1.6	2.2	1.8	1.8
26	3.1	3.4	2.1	2.9	3.8	2.0	2.8	1.9	1.7	2.2	1.8	1.9
27	3.1	7.1	2.1	2.9	3.4	2.0	2.5	1.9	1.7	2.2	4.3	1.9
28	3.0	12.1	2.1	2.9	3.1	3.2	2.4	1.8	1.7	2.1	3.4	2.2
29	3.0	......	2.1	2.9	2.9	3.7	3.2	1.7	1.6	2.1	2.8	3.9
30	2.9	......	2.0	2.8	2.8	2.5	5.1	1.7	1.6	2.1	2.8	6.8
31	2.8	......	2.0	......	2.8	......	3.6	1.7	......	2.0	......	5.0

Daily Gage Heights on Blue River at White Cloud, Indiana, for 1911.

DAY	Jan.	Feb.	Mar.	Apr.	May	June	July	Aug.	Sept.	Oct.	Nov.	Dec.
1	4.0	3.5	3.2	2.8	4.7	2.1	1.8	1.5	1.7	1.7	1.8	2.8
2	4.5	3.3	3.1	2.7	9.7	2.0	1.8	1.5	1.7	3.6	1.8	2.6
3	4.0	3.1	3.0	2.8	4.9	2.0	1.7	1.6	1.6	3.6	1.7	2.6
4	3.9	3.0	2.9	3.7	4.1	2.0	1.7	1.6	1.6	2.8	1.7	2.6
5	3.5	2.9	2.8	6.3	3.6	1.9	1.7	1.6	1.6	2.5	1.7	2.5
6	3.2	3.3	2.7	7.6	3.3	2.0	1.7	2.0	1.6	2.3	2.3	2.3
7	3.1	4.8	3.0	4.8	3.1	2.2	1.6	1.6	1.6	2.1	3.8	2.2
8	3.0	3.9	5.8	4.1	3.0	2.0	1.6	1.6	1.5	2.0	3.2	2.2
9	2.9	3.6	4.3	3.7	2.9	2.0	1.6	1.6	1.6	1.9	2.8	2.2
10	2.9	3.3	3.8	3.4	2.8	1.8	1.6	1.5	2.0	1.9	2.4	2.1
11	2.8	3.1	3.5	3.2	2.9	1.9	1.6	1.6	2.3	1.8	2.4	2.2
12	2.8	3.0	3.3	3.3	2.7	1.8	1.6	1.7	2.0	1.8	2.4	2.5
13	2.7	2.9	3.1	3.9	2.5	1.8	2.0	1.6	2.1	1.8	4.0	4.1
14	3.0	2.8	3.0	10.5	2.5	1.7	1.7	1.6	2.8	1.8	3.1	5.0
15	3.2	2.8	2.9	8.4	2.5	1.7	1.6	2.0	2.7	1.8	2.8	5.1
16	3.2	2.8	2.8	5.4	2.4	1.7	1.7	1.8	2.6	1.7	2.7	4.7
17	3.0	2.7	2.7	4.5	2.4	1.7	1.6	1.7	2.7	2.2	2.7	4.1
18	2.9	2.6	2.7	4.0	2.3	1.8	1.6	1.7	2.4	2.6	2.5	3.6
19	2.8	2.8	2.6	3.7	2.3	2.4	1.6	1.5	2.3	2.8	2.7	3.2
20	2.8	5.0	2.6	3.6	2.3	2.0	1.6	1.5	2.9	2.5	2.6	3.0
21	2.8	5.5	2.5	3.4	2.3	1.8	1.6	1.5	2.3	2.3	2.6	2.9
22	2.9	3.9	2.5	3.2	2.6	1.8	1.6	1.5	2.1	2.3	2.4	2.9
23	3.3	3.6	2.4	3.1	3.2	1.7	1.8	1.5	2.0	2.2	2.3	2.8
24	3.1	3.4	2.4	3.0	3.2	2.0	1.8	1.5	1.8	2.1	2.7	2.8
25	3.0	3.4	2.3	2.9	2.7	3.2	1.7	1.6	1.9	2.0	3.1	2.7
26	3.0	3.5	2.3	2.8	2.4	2.5	1.6	1.5	1.7	2.0	2.9	2.7
27	3.0	3.5	2.9	2.8	2.3	2.2	1.6	1.5	1.7	2.0	2.7	3.2
28	3.4	3.3	3.9	2.7	2.2	2.0	1.6	1.7	1.7	1.9	2.7	3.7
29	3.7	......	2.8	2.7	2.2	1.9	1.6	2.2	1.6	1.9	2.9	3.1
30	3.9	......	2.8	3.4	2.1	1.9	1.5	1.9	2.0	1.9	2.9	3.0
31	3.8	......	3.0	......	2.2	......	1.6	1.8	......	1.8	......	3.6

Daily Gage Heights on Blue River at White Cloud, Indiana, for 1912.

Day	Jan.	Feb.	Mar.	Apr.	May	June	July	Aug.	Sept.	Oct.	Nov.	Dec.
1	4.4	3.8	4.0	4.0	4.9	2.5	2.0	1.8	.....	.....	.....	.....
2	3.7	3.4	3.3	4.1	4.2	2.4	2.0	1.8	.....	.....	.....	.....
3	3.0	3.2	3.1	5.2	3.8	2.4	2.1	1.8	.....	.....	.....	.....
4	3.0	3.0	3.0	4.7	3.5	2.2	2.1	1.8	.....	.....	.....	.....
5	2.7	2.9	3.0	4.1	3.2	2.2	2.2	1.8	.....	.....	.....	.....
6	2.5	2.9	2.9	3.8	2.9	2.2	2.1	1.7	.....	.....	.....	.....
7	2.5	2.9	2.9	3.6	4.3	2.1	2.1	1.7	.....	.....	.....	.....
8	2.4	2.9	3.1	3.7	4.4	2.1	2.0	1.7	.....	.....	.....	.....
9	2.4	2.5	5.8	3.5	3.8	2.0	1.9	2.6	.....	.....	.....	.....
10	2.4	2.2	4.5	3.4	3.4	2.0	1.9	3.4	.....	.....	.....	.....
11	2.3	2.2	4.0	3.2	3.3	2.0	1.9	3.0	.....	.....	.....	.....
12	2.2	2.2	4.5	3.0	7.7	1.9	1.9	2.7	.....	.....	.....	.....
13	2.0	2.2	5.0	3.2	6.3	2.0	1.9	2.6	.....	.....	.....	.....
14	2.0	2.2	4.1	3.4	4.3	2.0	1.9	2.6	.....	.....	.....	.....
15	1.7	2.2	5.9	3.2	4.0	2.0	1.9	2.6	.....	.....	.....	.....
16	1.7	2.4	6.5	2.9	3.7	2.0	2.0	2.5	.....	.....	.....	.....
17	1.6	2.5	4.8	2.8	3.8	2.7	2.2	2.2	.....	.....	.....	.....
18	1.8	2.5	4.0	4.7	2.9	2.5	2.2	2.2	.....	.....	.....	.....
19	8.3	2.6	3.9	4.1	3.3	2.2	2.3	2.3	.....	.....	.....	.....
20	5.5	2.7	3.9	3.6	3.2	2.1	2.2	*...	.....	.....	.....	.....
21	3.9	6.9	4.1	3.7	3.0	2.1	2.0	.....	.....	.....	.....	.....
22	3.0	9.2	6.0	4.7	2.9	2.1	3.2	.....	.....	.....	.....	.....
23	3.5	5.9	4.8	4.5	2.8	2.0	3.0	.....	.....	.....	.....	.....
24	3.9	3.9	5.3	3.6	2.7	2.0	2.4	.....	.....	.....	.....	.....
25	3.8	3.9	7.0	3.3	2.7	2.0	2.1	.....	.....	.....	.....	.....
26	3.6	5.0	5.1	3.4	2.6	2.0	3.1	.....	.....	.....	.....	.....
27	3.0	7.1	4.9	9.8	2.5	2.0	2.1	.....	.....	.....	.....	.....
28	3.0	5.1	4.0	6.5	2.4	1.9	2.0	.....	.....	.....	.....	.....
29	4.9	4.1	5.1	5.7	2.8	2.0	1.9	.....	.....	.....	.....	.....
30	6.5	.....	4.7	6.2	3.3	2.1	1.9	.....	.....	.....	.....	.....
31	4.2	.....	4.6	...	2.7	.....	1.8	.....	.....	.....	.....	.....

*No record from August 19, 1912 to August 18, 1914.

Daily Gage Heights on Blue River at White Cloud, Indiana, for 1914.

Day	Jan.	Feb.	Mar.	Apr	May	June	July	Aug.	Sept.	Oct.	Nov.	Dec.
1	.....	.....	.....	.....	.....	.....	.....	.....	2.3	1.4	1.5	1.4
2	.....	.....	.....	.....	.....	.....	.....	.....	2.2	1.4	1.4	1.5
3	.....	.....	.....	.....	.....	.....	.....	.....	2.0	1.4	1.4	1.5
4	.....	.....	.....	.....	.....	.....	.....	.....	1.9	1.5	1.4	1.5
5	.....	.....	.....	.....	.....	.....	.....	.....	1.8	1.5	1.4	1.6
6	.....	.....	.....	.....	.....	.....	.....	.....	1.8	1.5	1.5	1.6
7	.....	.....	.....	.....	.....	.....	.....	.....	1.8	1.4	1.5	1.6
8	.....	.....	.....	.....	.....	.....	.....	.....	1.7	1.5	1.4	1.5
9	.....	.....	.....	.....	.....	.....	.....	.....	1.7	2.3	1.4	1.6
10	.....	.....	.....	.....	.....	.....	.....	.....	1.6	1.7	1.5	1.6
11	.....	.....	.....	.....	.....	.....	.....	.....	1.9	1.8	1.5	1.6
12	.....	.....	.....	.....	.....	.....	.....	.....	1.7	1.7	1.5	1.5
13	.....	.....	.....	.....	.....	.....	.....	.....	1.7	1.6	1.4	1.6
14	.....	.....	.....	.....	.....	.....	.....	.....	1.6	1.7	1.5	1.5
15	.....	.....	.....	.....	.....	.....	.....	.....	1.6	1.6	1.4	1.5
16	.....	.....	.....	.....	.....	.....	.....	.....	1.5	1.6	1.4	1.4
17	.....	.....	.....	.....	.....	.....	.....	.....	1.5	1.6	1.4	1.4
18	.....	.....	.....	.....	.....	.....	.....	1.7	1.5	1.6	1.4	1.4
19	.....	.....	.....	.....	.....	.....	.....	1.6	1.5	1.6	1.4	1.5
20	.....	.....	.....	.....	.....	.....	.....	1.6	1.4	1.6	1.4	2.0
21	.....	.....	.....	.....	.....	.....	.....	1.8	1.4	1.6	1.4	2.2
22	.....	.....	.....	.....	.....	.....	.....	1.7	1.5	1.6	1.4	2.5
23	.....	.....	.....	.....	.....	.....	.....	1.7	1.5	1.5	1.4	2.4
24	.....	.....	.....	.....	.....	.....	.....	1.9	1.5	1.5	1.4	2.2
25	.....	.....	.....	.....	.....	.....	.....	2.4	1.5	1.5	1.4	2.1
26	.....	.....	.....	.....	.....	.....	.....	4.6	1.6	1.5	1.4	1.7
27	.....	.....	.....	.....	.....	.....	.....	3.1	1.6	1.5	1.4	1.8
28	.....	.....	.....	.....	.....	.....	.....	3.6	1.5	1.5	1.4	1.6
29	.....	.....	.....	.....	.....	.....	.....	3.4	1.5	1.5	1.5	1.7
30	.....	.....	.....	.....	.....	.....	.....	3.1	1.5	1.5	1.5	3.3
31	.....	.....	.....	.....	.....	.....	.....	2.6	.....	1.5	.....	3.2

Daily Gage Heights on Blue River at White Cloud, Indiana, for 1915.

Day	Jan.	Feb.	Mar.	Apr.	May	June	July	Aug.	Sept.	Oct.	Nov.	Dec.
1	2.7	11.0	2.1	1.8	1.7	3.1	3.3	2.0	1.8	2.2	1.5	2.0
2	2.4	11.3	2.0	1.7	1.7	3.1	4.1	2.0	1.8	2.0	1.5	2.0
3	2.2	[a]...	2.0	1.8	1.8	3.3	3.4	2.2	1.8	1.9	1.4	2.0
4	2.2	.....	2.0	1.8	3.4	3.4	3.5	2.2	1.7	1.7	1.4	2.0
5	2.2	.....	2.2	1.8	2.7	2.9	3.9	2.1	1.7	1.7	1.5	2.1
6	2.1	.....	2.8	1.8	2.3	2.9	3.8	2.0	1.7	1.6	1.4	2.0
7	2.9	.....	3.1	1.7	4.4	3.1	3.0	1.8	1.7	1.6	1.4	1.9
8	3.3	.....	2.8	1.7	7.4	2.8	5.6	1.7	1.7	1.5	1.4	1.9
9	2.9	.....	2.6	1.7	4.1	2.6	7.1	1.6	1.7	1.5	1.5	1.9
10	2.7	.....	2.6	1.7	3.4	2.5	3.8	1.6	1.6	1.5	1.5	1.9
11	2.5	.....	2.5	1.8	2.9	2.4	5.0	1.7	1.7	1.5	1.5	1.9
12	2.4	.....	2.4	1.8	2.7	2.3	5.1	3.2	1.7	1.5	1.6	2.6
13	2.4	.....	2.3	1.8	2.6	2.2	6.8	2.7	1.6	1.5	1.7	3.1
14	2.4	.....	2.3	1.7	2.4	2.3	4.4	2.3	1.6	1.6	1.8	2.9
15	2.3	.....	2.3	1.7	2.4	2.3	3.7	2.0	1.6	1.6	2.0	2.6
16	2.4	.....	2.2	1.7	2.2	2.4	3.5	1.8	1.5	1.5	4.0	2.7
17	2.3	.....	2.2	1.6	2.0	2.3	4.0	2.5	1.5	1.5	1.9	12.2
18	2.3	.....	2.1	1.6	2.9	2.2	3.0	2.9	1.5	1.6	1.9	15.1
19	2.5	.....	2.1	1.6	2.0	2.3	2.9	2.2	1.5	1.6	4.2	6.8
20	2.4	2.5	2.1	1.6	2.0	4.9	2.7	2.0	1.5	1.5	4.3	5.1
21	2.4	2.4	2.1	1.6	4.8	3.9	2.5	3.4	1.5	1.5	3.2	4.3
22	2.3	2.3	2.1	1.6	3.5	2.9	2.4	3.9	1.5	1.5	2.8	3.8
23	3.6	2.3	2.0	1.7	4.2	2.6	2.3	3.4	1.4	1.6	2.3	3.6
24	4.0	2.3	2.0	1.8	4.5	2.4	2.2	2.7	1.4	1.6	2.1	3.4
25	3.2	2.2	2.0	1.8	3.9	2.2	2.1	2.5	1.4	1.5	2.0	3.2
26	3.0	2.2	1.9	1.8	4.1	2.1	2.1	2.3	1.4	1.5	1.9	3.2
27	2.8	2.2	1.9	1.8	4.1	2.1	2.0	2.2	1.4	1.5	1.9	3.1
28	2.7	2.1	1.9	1.8	4.2	2.1	1.9	2.1	1.4	1.5	1.9	3.3
29	2.6	.....	1.8	1.8	4.0	2.0	1.8	2.0	1.4	1.5	2.0	4.1
30	2.5	.....	1.8	1.7	3.6	2.6	1.8	1.9	1.4	1.5	2.0	7.1
31	3.3	.....	1.8	.....	3.4	.....	1.8	1.8	.....	1.5	.....	5.7

[a] Gage out of order from February 2 to 20.

Daily Gage Heights on Blue River at White Cloud, Indiana, for 1916.

Day	Jan.	Feb.	Mar.	Apr.	May	June	July	Aug.	Sept.	Oct.	Nov.	Dec.
1	5.1	8.2	2.9	2.7	3.1	2.5	2.2	1.9	2.0	1.7	1.4	1.4
2	7.8	5.8	3.0	2.6	3.0	2.3	2.3	1.8	3.7	1.6	1.4	1.4
3	5.6	4.6	3.1	2.8	3.0	2.2	3.6	1.9	3.2	1.6	1.4	1.5
4	4.5	3.8	3.0	3.3	6.0	2.3	4.3	1.9	2.6	1.6	1.4	1.4
5	4.0	3.3	3.1	3.1	4.4	2.2	2.9	1.8	2.3	1.5	1.4	1.9
6	3.7	3.6	3.1	3.0	3.8	3.5	2.5	1.7	2.2	1.5	1.4	1.8
7	3.5	4.2	5.0	2.9	3.4	7.2	2.3	1.7	2.0	1.5	1.4	1.9
8	3.2	3.7	5.6	2.9	3.1	4.3	2.2	1.7	2.0	1.4	1.4	1.9
9	3.1	3.4	4.3	3.0	3.0	3.4	2.1	1.7	2.0	1.4	1.5	2.3
10	3.0	3.2	3.7	3.2	2.8	3.0	2.0	1.8	1.8	1.4	1.5	2.3
11	4.6	3.0	3.4	3.2	2.7	3.3	1.9	1.8	1.8	1.4	1.5	2.3
12	11.1	3.0	3.1	3.1	2.6	3.0	2.0	1.9	1.8	1.4	1.4	2.1
13	11.0	6.6	3.1	3.0	2.5	2.8	2.9	1.8	1.7	1.4	1.5	2.0
14	6.9	5.9	3.0	3.0	2.4	2.5	6.1	1.7	1.7	1.4	1.5	2.0
15	5.0	4.5	4.4	2.9	2.4	2.5	3.8	2.0	1.7	1.4	1.4	1.8
16	4.3	4.1	3.9	2.7	2.5	2.5	3.8	6.5	1.7	1.5	1.5	1.7
17	3.9	4.0	3.7	2.7	2.5	3.1	2.9	2.9	1.6	1.5	1.4	1.8
18	3.3	3.8	3.5	2.6	2.4	2.6	2.7	2.3	1.6	1.4	1.4	1.7
19	3.2	3.6	3.3	2.6	2.3	6.1	2.9	2.1	1.6	1.7	1.7	1.7
20	3.0	3.3	3.1	2.5	2.2	5.0	2.9	2.0	1.6	2.0	1.5	1.7
21	3.0	3.3	3.0	6.3	2.1	3.9	3.1	1.9	1.5	2.1	1.5	2.0
22	4.1	3.1	3.0	5.8	2.2	5.9	4.4	1.9	1.5	2.2	1.4	1.9
23	4.9	3.1	2.9	4.2	2.2	3.5	3.1	1.8	1.5	2.1	1.4	1.8
24	4.1	3.8	2.9	3.6	2.1	3.1	2.9	1.7	1.5	1.8	1.4	2.0
25	3.7	4.0	2.8	3.2	2.1	2.8	2.7	1.8	1.5	1.7	1.4	2.0
26	3.4	3.7	2.9	3.2	2.0	2.6	2.4	1.8	1.5	1.6	1.4	2.0
27	3.2	3.4	3.1	3.7	2.0	2.5	2.3	1.8	1.5	1.6	1.5	4.3
28	3.3	3.1	3.0	3.9	2.0	2.4	2.1	2.2	1.5	1.6	1.5	5.9
29	4.0	3.1	3.1	3.8	2.1	2.4	2.1	4.0	2.0	1.5	1.4	4.7
30	11.6	.....	3.0	3.2	2.6	2.3	2.0	2.8	1.8	1.5	1.5	3.5
31	10.1	.....	2.9	.....	3.0	.....	1.9	2.4	.....	1.5	.....	3.0

Discharge Measurements on Blue River at White Cloud, Indiana, during 1909-1916.

Date	Hydrographer	Gage Height, Feet	Discharge, Sec.-Ft.
August 18, 1909	Tucker and Smith	2.05	183
July 27, 1910	W. M. Tucker	2.50	350
July 28, 1910	W. M. Tucker	3.55	1,109
July 28, 1910	W. M. Tucker	4.00	1,498
January 2, 1911	W. M. Tucker	4.60	2,021
April 15, 1911	W. M. Tucker	8.00	5,030
May 2, 1911	W. M. Tucker	5.50	2,828
August 1, 1911	W. M. Tucker	1.50	62
February 2, 1912	W. M. Tucker	8.80	5,710
June 23, 1916	W. M. Tucker	3.30	904

Discharge Rating Table on Blue River at White Cloud, Indiana, during 1909-1916.

Gage Height, Feet	Discharge, Sec.-Ft.	Gage Height, Feet	Discharge, Sec.-Ft.	Gage Height, Feet	Discharge, Sec.-Ft.	Gage Height, Feet	Discharge, Sec.-Ft.
1.4	50	3.4	986	5.4	2,730	7.4	4,490
1.5	63	3.5	1,069	5.5	2,818	7.5	4,578
1.6	79	3.6	1,153	5.6	2,906	7.6	4,666
1.7	95	3.7	1,238	5.7	2,994	7.7	4,754
1.8	113	3.8	1,324	5.8	3,082	7.8	4,842
1.9	134	3.9	1,411	5.9	3,170	7.9	4,930
2.0	162	4.0	1,498	6.0	3,258	8.0	5,018
2.1	197	4.1	1,586	6.1	3,346	8.1	5,106
2.2	227	4.2	1,674	6.2	3,434	8.2	5,194
2.3	263	4.3	1,762	6.3	3,522	8.3	5,282
2.4	304	4.4	1,850	6.4	3,610	8.4	5,370
2.5	350	4.5	1,938	6.5	3,698	8.5	5,458
2.6	401	4.6	2,026	6.6	3,786	8.6	5,546
2.7	458	4.7	2,114	6.7	3,874	8.7	5,634
2.8	522	4.8	2,202	6.8	3,962	8.8	5,722
2.9	592	4.9	2,290	6.9	4,050	8.9	5,810
3.0	667	5.0	2,378	7.0	4,138	9.0	5,898
3.1	744	5.1	2,466	7.1	4,226	....	....
3.2	823	5.2	2,554	7.2	4,314	....	....
3.3	904	5.3	2,642	7.3	4,402	....	....

NOTE—This table is applicable only for open channel conditions. It is tangent above 4.0 feet with 88 Sec. ft. per 0.1 ft. variation.

The following tables are calculated to show the relation between the rainfall and the runoff in the basin of Blue River above the gage at White Cloud, Indiana. Column 1 shows the average monthly rainfall in inches in the basin. The data was derived by taking the average of the monthly reports of the U. S. Weather Bureau from the following observation stations: Edwardsville, Forest Reserve, French Lick, Jeffersonville, Paoli, Scottsburg, Marengo and Salem, which are in or near the basin. In making this average, Salem was given a weight of three, Marengo of two and the other stations of one, because of their relative relations to the basin. Column 2 gives the average rainfall per acre in cubic feet within the basin. It is derived directly from column 1. Column 3 gives the average discharge in cubic feet per second at the gage at White Cloud and is derived from the daily gage readings and the rating table. Column 4 gives the total average discharge per acre from the basin and is derived from column 3. Column 5 gives the relation between the rainfall and runoff and is derived by dividing column 2 into column 4, and expressing it as per cent. The drainage area of the basin is considered as 434.4 square miles. This area was derived by carefully checked planimeter readings from the drainage map accompanying this report.

1910	1 Average Rainfall, Inches	2 Rainfall per Acre, Cu. Ft.	3 Discharge Cu. Ft. per Sec.	4 Discharge Cu. Ft. per Acre	5 Per Cent. 4 of 2
January	3.65	13,249	1,119	10,787	81.4
February	4.45	16,153	1,134	9,880	61.1
March	0.31	1,125	723	6,976	520.0
April	5.02	18,222	737	6,877	37.7
May	5.52	20,037	1,071	10,328	51.5
June	3.59	13,031	371	3,468	26.6
July	9.05	32,851	1,247	12,027	36.6
August	1.08	3,920	213	2,054	52.4
September	3.48	12,632	134	1,257	10.0
October	7.86	28,531	1,286	12,407	43.5
November	2.18	7,913	245	2,289	28.9
December	2.70	9,801	426	4,111	42.0
Annual	....	158,594	....	53,095	33.5

1911	1 Average Rainfall, Inches	2 Rainfall per Acre, Cu. Ft.	3 Discharge Cu. Ft. per Sec.	4 Discharge Cu. Ft. per Acre	5 Per Cent. 4 of 2
January	2.82	10,236	861	8,305	81.1
February	2.52	9,147	1,014	8,837	96.6
March	2.56	9,292	729	7,030	75.6
April	6.14	22,288	1,598	14,912	66.9
May	3.59	13,031	801	7,730	59.3
June	3.96	14,374	178	1,665	11.6
July	1.73	6,279	90	871	13.9
August	2.75	9,982	92	890	8.9
September	5.62	20,400	200	1,873	9.2
October	4.07	14,774	267	2,578	17.5
November	4.01	14,556	473	4,417	30.3
December	4.09	14,846	788	7,601	51.2
Annual	....	150,211	....	66,715	41.9

1915	1 Average Rainfall, Inches	2 Rainfall per Acre Cu. Ft.	3 Discharge Cu. Ft. Per Sec.	4 Discharge Cu. Ft. per Acre	5 Per Cent. 4 of 2
January	4.84	17,569	472	4,554	25.9
February	1.05	3,811	....	....	....
March	1.17	4,247	251	2,422	57.0
April	1.07	3,884	104	978	25.2
May	7.84	28,459	1,005	9,697	34.1
June	3.99	14,483	535	4,992	34.5
July	5.62	20,400	1,123	10,829	53.1
August	6.28	22,796	325	3,142	13.8
September	2.26	8,203	78	733	8.9
October	1.40	5,082	82	798	15.7
November	4.06	14,737	302	2,820	19.1
December	7.58	27,515	1,521	14,667	53.3
Annual	....	167,378*	....	55,635*	33.2*

*Except February.

1916	1 Average Rainfall, Inches	2 Rainfall per Acre Cu. Ft.	3 Discharge Cu. Ft. per Sec.	4 Discharge Cu. Ft. per Acre	5 Per Cent. 4 of 2
January	8.48	30,782	2,363	22,765	73.95
February	2.54	9,220	1,518	13,681	148.38
March	3.11	11,289	963	9,278	82.19
April	4.00	14,520	927	8,643	59.52
May	4.20	15,246	566	5,453	35.77
June	6.94	25,192	956	8,913	35.38
July	5.21	18,912	644	6,204	32.80
August	4.76	17,279	323	3,112	18.01
September	2.88	10,454	180	1,678	16.05
October	1.89	6,861	84	809	11.79
November	1.35	4,901	56	522	10.65
December	3.91	14,193	397	3,825	26.95
Annual	....	178,849	....	84,883	47.46

The Wabash River System.

The Wabash River drains parts of Ohio, Indiana and Illinois. This report has to do with that part which lies within Indiana. The state line of that part of the state which is bounded by the Wabash is defined as the center of the river. For this reason no work has been done upon that part of the river. Exclusive of White River and other tributaries entering the Wabash farther south, the Wabash system drains 11,480 square miles of Indiana. Its main stream rises in the Grand Reservoir, an artificial lake, at Celina, Mercer County, Ohio. It flows northwest across the state line to Huntington, west to Logansport, southwest to Covington and south to its junction with the state line, 14.6 miles south of Terre Haute. The main tributaries on the north and the points at which they enter the Wabash are: Little Wabash at Huntington, Eel at Logansport, and Tippecanoe near Battle Ground. On the south are: Salamonie at Lagro, Mississinewa near Peru, Wild Cat near Lafayette and Rock near Newport. Rock River is also known as Sugar Creek. The following table gives the areas drained by each of the tributaries:

Stream.	Area of Basin, Sq. Mi.
Little Wabash	313
Eel	852
Tippecanoe	2,026
Salamonie	508
Mississinewa	861
Wild Cat	834
Rock	803

The basin of the Wabash lies entirely within the area of the Wisconsin glacier except that part which is south of the north line of Vigo County. The drift is much heavier in the northern part of the basin than in the southern part. The Tippecanoe and Eel Rivers rise among the lakes of Kosciusco and Noble counties, while the main stream and southern tributaries have no lakes within their basins.

As a whole the Wabash system lies near bed rock. Many exposures occur along the stream beds and in the bluffs. The source of the system is in the Silurian, and it crosses the Devonian, Mississippian and part of the Pennsylvanian before reaching the state line on the west. The rock exposures form good dam sites for power development. A full discussion of the power conditions along the Wabash and its tributaries will be found on pages 492 to 538 of the 36th Annual Report of the Indiana Department of Geology and Natural Rèsources.

The mean annual rainfall above the gage at Terre Haute on the Wabash River, as shown by the station records from this section for eighteen years (1898-1915, inclusive), is 37.78 inches. The following tables, which are computed entirely from government reports, show the average depth of monthly rainfall and the cubic feet per acre rainfall for the basin of the Wabash River for the eighteen years above mentioned. These averages are taken from twenty-six U. S. Weather Bureau recording stations. The following is a list of the stations used: Berne, Bluffton, Columbia City, Culver, Delphi, Fairmount, Huntington, Kokomo, Lafayette, Logansport, Marion, Markle, Monticello, Peru, Rochester, Warsaw, Winamac, Attica, Crawfordsville, Farm-

land, Frankfort, Hector, Judyville, Rockville, Salamonie, Terre Haute, and Veedersburg.

Table showing average depth of monthly rainfall in inches and cubic feet per acre of rainfall for the Wabash River basin above the gage at Terre Haute, Indiana, 1898-1915 inclusive:

	1898		1899		1900		1901	
	In.	Cu. Ft.	In.	Cu. Ft.	In.	Cu. Ft.	In.	Cu. Ft.
January	3.88	14,084.4	2.82	10,236 6	0.76	2,758.8	1.33	4,827.9
February	1.84	6,679.2	2.00	7,260.0	3.44	12,487.2	1.43	5,190 9
March	6.67	24,212.1	4.42	16,044.6	2.30	8,349.0	3.47	12,596.1
April	2.24	8,131.2	1.00	3,630.0	1.60	5,808 0	2.42	8,784.6
May	6.20	22,506.0	3.87	14,048.1	4.53	16,443.9	3.51	12,741.3
June	4.21	15,282.3	2.41	8,748.3	6.53	23,703.9	4.68	16,988.4
July	3.55	12,886.5	3.49	12,668.7	4.04	14,665.2	0.72	2,613.6
August	2.92	10,599.6	1.94	7,042.2	3.72	13,503 6	3.50	12,705.0
September	3.55	12,886 5	2.12	7,695 6	2.28	8,276.4	1.14	4,138.2
October	4.44	16,117.2	2.84	10,309.2	2.69	9,764.7	4.02	14,592.6
November	3.16	11,470.8	2.52	9,147.6	4.37	15,863.1	1.27	4,610.1
December	1.90	6,897.0	3.11	11,289.3	0.86	3,121 8	3.56	12,922.8

	1902		1903		1904		1905	
	In.	Cu. Ft.	In.	Cu. Ft.	In.	Cu. Ft.	In.	Cu. Ft.
January	1.00	3,630.0	1.88	6,824.4	4 68	16,988.4	1.86	6,751.8
February	1.00	3,630 0	3.59	18,031.7	2.13	7,731.9	1.68	6,098.4
March	3.14	11,398.2	1.82	.606.6	7.61	27,624.3	2.09	7,586.7
April	1.74	6,316.2	4.97	1 .041.1	4.17	15,137.1	3.80	13,794.0
May	3.36	12,196.8	3.49	1.,668.7	3.40	12,342 0	5.39	19,565.7
June	8.82	32,016.6	3.57	12,959.1	2.61	9,474.3	3.47	12,596.1
July	6.13	22,251.9	3.19	11,579.7	3.64	13,213.2	3 43	12,450.9
August	1.98	7,187.4	3.84	13,939.2	2.76	10,018.8	4.27	15,500.1
September	5.10	18,513.0	2.02	7,332.6	3.42	12,414.6	3.84	13,939.2
October	2.64	9,583.2	2.95	10,708.5	1.17	4,247.1	3.44	12,487.2
November	3.17	11,507.1	1.89	6,860.7	0.22	798.6	2.96	10,744.8
December	3.42	12,414.6	1.83	6,642.9	2.70	9,801.0	1.90	6,897.0

	1906		1907		1908		1909	
	In.	Cu. Ft.	In.	Cu. Ft.	In.	Cu. Ft.	In.	Cu. Ft.
January	2.92	10,599.6	6.21	22,532.3	1.45	5,263.5	2.65	9,619.5
February	1.03	3,738.9	0 25	907.5	4.68	16,988.4	5.04	18,295.2
March	4.21	15,282 3	4.29	15,572.7	4.31	15,645.3	2.58	9,365.4
April	2.36	8,566.8	2.32	8'421 6	3.71	13,467.3	5.47	19,856.1
May	2.65	9,619.5	3.39	12 305,7	5.55	20,146.5	4 50	16,335.0
June	3.13	11,361.9	5.63	20 436 9	1.99	7,223.7	6.17	22,397.1
July	3.14	11,398.2	5.29	19 202 7	3.58	12,995.4	4.36	15,826.8
August	5.74	20,836.2	3.52	12 777 6	1.74	6,316.2	3.80	13,794.0
September	3.39	12,305 7	2.82	10 236 6	0.90	3,267.0	3.32	12,051.6
October	2.28	8,276.4	2.08	7 550 4	0.35	1,270.5	3.26	11,833.8
November	3.43	12,450.9	2.20	7 986 0	2.23	8,094.9	3.52	12,777.6
December	3.89	14,120.7	4.01	14 556 3	1.69	6,134.7	2.81	10,200.3

	1910		1911		1912		1913	
	In.	Cu. Ft.	In.	Cu. Ft.	In.	Cu. Ft.	In.	Cu. Ft.
January	2.71	9,837.3	3.33	12,087.9	1.94	7,042.2	6.88	24,974.4
February	2.33	8,457.9	1.59	5,771.7	2.28	8'276.4	[illegible].35	4,900.[illegible]
March	0.21	762.3	1.83	6,642.9	3.88	14 084.4	.45	30,673.
April	3.13	11,361.9	4.53	16,443 9	4.45	16 153.5	.92	14,229.
May	3.54	12,850.2	1 80	6,534.0	3.72	13 503 6	.14	11,398.
June	1.52	5,517.6	3.46	12,559.8	3.20	11 616.0	.05	7,441.
July	3.68	13,358.4	2.12	7,695.6	4.99	18 113.7	.01	14,556.
August	2.46	8,929.8	3.79	13,757.7	4.66	16 915.8	3.33	12,087.
September	5.40	19,602.0	5.61	20,364.3	2.83	10 272 9	[illegible].64	9,583.
October	3.21	11,652.3	3.61	13,104.3	2.26	8 203.8	.56	12,922.
November	2.11	7,659.3	3.34	12,124.2	2.32	8 421.6	.55	12,886.5
December	3.51	12,741.3	2.34	8,494.2	0.98	3 557.4	.90	3,267.0

	1914		1915	
	In.	Cu. Ft.	In.	Cu. Ft.
January	2.51	9,111.3	2.85	10,345.5
February	2.50	9,075.0	1.78	6,461.4
March	2.25	8,167.5	1.05	3,811.5
April	3.12	11,325.6	1.87	6,788.1
May	3.37	12,233.1	4.81	17,460.3
June	1.93	7,005.9	2.55	9,256.5
July	1.62	5,880.6	8.16	29,620.8
August	4.01	14,556.3	5.44	19,747.2
September	2.12	7,695.6	3.99	14,483.7
October	2.17	7,877.1	1.39	5,045.7
November	1.20	4,356.0	2.64	9,583.2
December	2.89	10,490.7	3.07	11,144.1

The United States Geological Survey maintained a gage at Logansport, from April 27, 1903, to July 8, 1916. The following data is taken from the U. S. Water Supply Papers:

Wabash River at Logansport, Indiana. This station was established April 27, 1903, by Mr. Geo. E. Waescbe. It is located at the Cicott Street bridge, about 1 mile from the center of the city of Logansport, 1¼ miles from the Pennsylvania station, 1¾ miles from the Wabash Railroad station, four blocks from the street car line, and 1,000 feet below the mouth of Eel River. The standard chain gage is placed on the second span of the bridge, at the third panel from the second pier, and is supported by the bridge pins, and is between the lower chord bars. It is reached through a trapdoor in the floor planks of the bridge. The distance from the end of the weight to the marker is 20.78 feet. The gage is read once each day by Mr. C. B. Woodruff. Discharge measurements are made from the downstream side of the bridge to which the gage is attached. The initial point for soundings is the back wall of the bridge seat for the north abutment. The channel is nearly straight for 1,000 feet above and for 1,500 feet below the station. The distance between abutments is 550 feet, and the channel is broken by three bridge piers. The right bank is high and is not subject to overflow at the bridge. The left bank is submerged only at extreme high water. The bed of the stream is covered with small bowlders and is rough. The stream is shallow and the current is never sluggish. Benchmark No. 1 is the top of the north abutment, under the fourth board of the downstream sidewalk. Its elevation is 18.814 feet above gage datum. From Pennsylvania Railroad levels its elevation above sea level has been found to be 591 feet. Benchmark No. 2 is the top of the third course of masonry from the top of the north abutment. Its elevation above the zero of the gage is 15.31 feet.

The observations at this station during 1903 have been made under the direction of Mr. E. Johnson, Jr., district hydrographer.

*Discharge Measurements of Wabash River at Logansport, Indiana, in 1903.**

DATE	Hydrographer	Gage Height, Feet	Discharge, Sec.-Ft.
April 27	Geo. E. Waesche	2.50	2,367
June 8	Geo. E. Waesche	4.55	7,180
June 16	Geo. E. Waesche	....	1,444
July 8	Geo. E. Waesche	2.30	1,358
July 16	E. C. Murphy	1.54	719
August 15	E. Johnson, Jr.	1.35	418
September 30	L. R. Stockman	1.30	349
November 10	L. R. Stockman	1.38	452
December 28	E. Johnson, Jr.	2.75	1,285

*Water supply paper U. S. Geol. Survey, No. 98, page 225.

*Mean Daily Gage Height, in Feet, of Wabash River at Logansport, Indiana, for 1903.**

Day	Apr.	May	June	July	Aug.	Sept.	Oct.	Nov.	Dec.
1		2.15	2.30	1.55	1.70	1.70	†	†	1.30
2		2.10	2.25	10.15	1.60	1.70			1.30
3			2.70	7.25	1.50	1.70			1.40
4		2.00	2.50	4.80	1.60	1.70			1.50
5		2.00	3.25	3.65	1.60	1.70			1.60
6		1.95	4.15	3.00	1.55	1.92			1.70
7		1.90	5.50	2.55	1.55	1.72			1.50
8		1.90		2.30	1.50	1.72			1.45
9		1.90	4.30	2.15	1.45	1.70			1.40
10			3.65	2.00	1.40	1.40		1.40	1.40
11		1.80	3.10	1.90	1.40	1.40		1.40	1.85
12		1.80	2.65	1.95	1.35	1.35		1.40	1.85
13		1.80	2.40	1.90	1.35	1.35		1.40	2.40
14		1.75	2.25	1.25	1.35	1.40		1.40	2.90
15		1.75	2.10	1.70	1.35	1.40		1.40	2.90
16		1.75	1.95	1.60	1.45	1.40		1.50	2.90
17			1.85	1.60	1.40	1.40		1.50	2.70
18		1.70	1.80	1.75	1.40	1.40		1.80	2.40
19		1.65	1.80	1.85	1.30	1.40		1.60	2.40
20		1.65	1.75	1.85	1.35	†		1.40	2.40
21		1.60	1.70	1.80	1.35			1.50	2.70
22		1.65	1.80	1.70	1.30			1.50	2.90
23		1.70	2.05	1.80	1.30			1.40	2.80
24		2.15	2.20	1.75	1.30			1.40	2.70
25		2.80	2.00	1.70	1.30			1.40	2.70
26		2.70	1.85	1.70	1.30			1.40	2.70
27	2.50	3.00	1.75	1.70	1.35			1.40	2.80
28	2.40	3.60	1.70	1.60	1.90			1.40	2.75
29	2.30	3.30	1.60	1.60	2.10			1.40	2.68
30	2.20	2.85	1.60	1.65	1.90			1.30	2.60
31		2.60		1.70	1.70				2.53

*Water supply paper U. S. Geol. Survey, No. 98, page 226. †Observer absent.

*Rating Table for Wabash River at Logansport, Indiana, from April 27 to December 31, 1903.**

Gage Height, Feet	Discharge, Sec.-Ft.	Gage Height, Feet	Discharge, Sec.-Ft.	Gage Height, Feet	Discharge, Sec.-Ft.	Gage Height, Feet	Discharge, Sec.-Ft.
1.3	380	2.6	1,990	4.8	8,100	8.0	22,780
1.4	460	2.7	2,170	5.0	8,890	8.5	25,130
1.5	550	2.8	2,350	5.2	8,720	9.0	27,480
1.6	650	2.9	2,560	5.4	10,590	9.5	29,830
1.7	750	3.0	2,770	5.6	11,500	10.0	32,180
1.8	860	3.2	3,200	5.8	12,440	10.5	34,530
1.9	980	3.4	3,670	6.0	13,380	11.0	36,880
2.0	1,100	3.6	4,180	6.2	14,320	11.5	39,320
2.1	1,230	3.8	4,740	6.4	15,260	12.0	41,580
2.2	1,370	4.0	5,340	6.6	16,200	12.5	43,930
2.3	1,510	4.2	5,970	6.8	17,140	13.0	46,280
2.4	1,660	4.4	6,640	7.0	18,080		
2.5	1,820	4.6	7,350	7.5	20,430		

*Water supply paper U. S. Geol. Survey, No. 98, page 226.

*Discharge Measurements of Wabash River at Logansport, Indiana, in 1904.**

Date	Hydrographer	Gage Height, Feet	Discharge, Sec.-Ft.
January 22b	F. W. Hanna	13.11	46,660
March 2c	F. W. Hanna	8.20	23,660
March 29c	F. W. Hanna	10.50	32,480
May 3	F. W. Hanna and Johnson	2.98	3,744
June 17	F. W. Hanna	1.66	878
July 21	F. W. Hanna	1.76	920
August 23	F. W. Hanna	1.50	723
September 14	F. W. Hanna	1.48	542
October 21	F. W. Hanna	1.30	401
November 4	F. W. Hanna	1.27	379

*Water supply paper U. S. Geol. Survey, No. 128, page 84.

*Mean Daily Gage Height, in Feet, of Wabash River at Logansport, Indiana, for 1904.**

Day	Jan.†	Feb.†	Mar.†	Apr.	May	June	July	Aug.	Sept.	Oct.	Nov.	Dec.†
1	2.53	3.10	9.90	12.22	3.75	2.90	1.80	1.30	1.30	1.35	......	1.30
2	2.50	3.00	7.88	13.00	3.25	3.00	1.73	1.30	1.33	1.35	1.22	1.35
3	2.33	2.30	7.65	13.02	2.95	3.00	1.75	1.35	1.27	1.30	1.25	1.45
4	2.23	2.50	7.88	11.06	2.75	2.90	1.68	1.33	1.21	1.29	1.27	1.22
5	2.23	2.50	7.30	7.65	2.58	2.70	1.79	1.30	1.25	1.26	1.23	1.28
6	2.20	3.00	6.50	5.73	2.50	2.40	1.90	1.35	1.30	1.25	1.25	1.22
7	2.20	10.20	8.19	4.88	2.40	2.30	3.10	1.29	1.24	1.28	1.27	1.22
8	2.15	9.65	7.35	4.45	2.30	2.20	5.12	1.29	1.20	1.35	1.27	1.28
9	2.15	7.90	5.83	4.70	2.30	2.00	4.95	1.32	1.25	1.45	1.22	1.25
10	2.15	6.35	5.00	4.95	2.35	1.90	4.71	1.25	1.25	1.46	1.30	1.18
11	2.15	5.60	5.00	4.70	2.25	1.90	4.43	1.25	1.25	1.46	1.30	1.31
12	2.10	5.10	5.15	4.68	2.13	1.82	4.14	1.24	1.25	1.45	1.30	1.20
13	2.10	4.20	.....	4.30	2.10	1.78	3.53	1.20	1.26	1.45	1.30	1.40
14	2.20	3.60	4.20	3.90	......	1.73	2.99	1.25	1.48	1.40	1.31	......
15	2.20	2.70	3.70	3.60	......	1.70	2.69	1.21	1.41	1.37	1.31	......
16	2.15	2.50	3.42	3.33	2.00	1.70	2.50	1.25	1.25	1.35	1.25	......
17	2.15	2.30	3.45	3.15	1.95	1.65	2.00	1.25	1.30	1.35	1.31	......
18	2.20	2.34	5.40	2.90	2.00	1.73	1.99	1.25	1.41	1.29	1.30	......
19	2.25	2.38	6.75	2.90	2.15	1.75	1.85	1.25	1.49	1.28	1.30	......
20	2.60	2.10	7.00	2.63	2.40	1.90	1.81	1.50	1.49	1.25	1.33	......
21	10.85	2.25	6.20	2.52	2.40	2.56	1.78	1.61	1.51	1.30	1.30	......
22	13.19	2.21	8.30	2.48	2.35	3.30	1.65	1.76	1.39	1.30	1.35	......
23	12.55	2.35	8.10	2.85	2.43	3.15	1.65	1.50	1.37	1.30	1.30	1.31
24	10.00	3.50	6.68	2.15	2.40	2.97	1.61	1.45	1.37	1.30	1.28	1.35
25	7.45	4.80	6.85	2.60	2.40	2.49	1.55	1.40	1.40	1.30	1.31	1.41
26	6.00	4.50	13.45	7.05	2.75	2.20	1.53	1.29	1.56	1.30	1.30	1.50
27	4.75	3.70	14.84	7.12	2.70	2.00	1.53	1.30	1.50	1.30	1.30	1.73
28	4.38	3.88	13.05	5.90	2.53	1.93	1.50	1.27	1.56	1.24	1.34	......
29	4.00	9.22	10.32	5.25	2.40	1.87	1.50	1.30	1.46	1.22	1.30	4.12
30	3.70	......	7.50	4.50	2.43	1.80	1.41	1.33	1.40	1.22	1.28	3.95
31	3.50	......	9.10	......	2.55	......	1.39	1.34	......	......	......	3.41

†Ice conditions from January 1 to March 3 and December 14 to 31, 1904.
*Water supply paper U. S. Geol. Survey, No. 128, page 85.

*Rating Table for Wabash River at Logansport, Indiana, from January 1 to December 31, 1904.**

Gage Height, Feet	Discharge, Sec.-Ft.	Gage Height, Feet	Discharge, Sec.-Ft.	Gage Height, Feet	Discharge, Sec.-Ft.	Gage Height, Feet	Discharge, Sec.-Ft.
1.2	260	2.5	2,540	4.2	7,380	8.0	22,770
1.3	360	2.6	2,770	4.4	8,050	8.5	25,060
1.4	480	2.7	3,010	4.6	8,730	9.0	27,410
1.5	620	2.8	3,260	4.8	9,430	9.5	29,790
1.6	770	2.9	3,510	5.0	10,140	10.0	32,200
1.7	930	3.0	3,770	5.2	10,880	10.5	34,650
1.8	1,100	3.1	4,030	5.4	11,640	11.0	37,130
1.9	1,280	3.2	4,300	5.6	12,420	12.0	42,130
2.0	1,470	3.3	4,580	5.8	13,220	13.0	47,130
2.1	1,670	3.4	4,870	6.0	14,040	14.0	52,130
2.2	1,880	3.6	5,460	6.5	16,170	15.0	57,130
2.3	2,090	3.8	6,080	7.0	18,320	....	....
2.4	2,310	4.0	6,720	7.5	20,520	....	....

*Water supply paper U. S. Geol. Survey, No. 128, page 85.

The preceding table is applicable only for open-channel conditions. It is based upon 19 discharge measurements made during 1903 and 1904. It is well defined between gage heights 1.3 feet and 8.2 feet. The table has been extended beyond these limits. Above gage height 10.7 feet the rating curve is a tangent the difference being 500 per tenth.*

*Water supply paper U. S. Geol. Survey, No. 128, page 86.

*Discharge Measurements of Wabash River at Logansport, Indiana, in 1905.**

Date	Hydrographer	Gage Height, Feet	Discharge, Sec.-Ft.
March 21	S. K. Clapp	3.75	5,450
May 27	M. S. Brennan	2.22	1,700
June 14	S. K. Clapp	3.79	5,480
July 14	S. K. Clapp	2.75	2,436
August 23	M. S. Brennan	1.95	1,100
September 4	M. S. Brennan	1.74	748

*Water supply paper U. S. Geol. Survey, No. 169, page 75.

*Daily Gage Height, in Feet, of Wabash River at Logansport, Indiana, for 1905.**

Day	Jan.	Feb.	Mar.	Apr.	May	June	July	Aug.	Sept.	Oct.	Nov.	Dec.
1	2.95	.....	6.12	4.82	4.25	3.14	1.67	1.40	1.63	1.48	1.60	4.50
2	2.42	.....	5.88	4.25	3.70	2.88	1.70	1.40	1.72	1.55	1.63	4.05
3	2.29	2.15	5.55	3.45	3.20	2.70	1.70	1.36	1.75	1.60	1.58	3.68
4	2.00	.....	5.20	2.93	2.68	2.52	1.62	1.25	1.75	1.65	1.55	3.60
5	1.90	.....	4.69	2.60	2.55	2.40	1.62	1.35	1.60	1.89	1.60	3.50
6	1.85	.....	4.35	2.40	2.80	5.10	1.63	1.58	1.55	1.87	2.00	3.10
7	1.90	.....	3.95	2.28	3.49	4.00	1.63	2.45	1.50	1.80	2.35	2.84
8	.....	.....	3.70	2.20	3.39	3.55	1.63	2.17	1.45	1.75	2.50	2.58
9	.....	2.10	8.85	2.15	2.95	3.22	1.63	1.80	1.42	1.65	2.39	2.48
10	.....	.....	4.00	2.10	2.73	3.23	1.70	1.70	1.42	1.60	2.33	2.40
11	.....	.....	3.86	2.05	3.80	3.78	2.25	1.53	1.50	1.60	2.20	2.30
12	2.70	.....	3.60	2.05	10.50	3.78	3.17	1.45	1.49	1.54	2.05	2.25
13	.....	.....	3.10	2.05	10.05	3.80	2.95	1.40	3.33	1.50	1.85	2.22
14	.....	.....	2.85	2.00	8.45	3.72	2.63	1.49	3.48	1.50	1.81	2.15
15	.....	.....	2.70	2.00	7.10	3.33	2.40	1.77	2.96	1.45	1.80	2.10
16	.....	2.20	2.50	1.95	6.12	3.00	2.39	1.90	2.83	1.42	1.78	1.99
17	.....	.....	2.85	1.90	5.10	2.85	2.10	1.70	3.58	1.40	1.75	2.10
18	3.00	.....	3.40	1.86	4.50	2.68	1.92	2.30	4.42	2.13	1.70	2.05
19	.....	.....	3.60	1.86	3.88	2.65	1.91	2.56	3.78	2.45	1.70	2.00
20	.....	.....	3.60	1.88	3.50	2.60	1.75	2.70	3.28	2.30	1.65	1.90
21	.....	.....	3.71	5.65	3.23	2.40	1.99	2.43	3.00	2.20	1.58	2.80
22	.....	.....	3.47	6.45	2.85	2.72	1.73	2.21	2.80	2.05	1.55	4.65
23	.....	2.23	3.10	5.98	2.60	2.69	1.68	1.99	2.00	2.00	1.63	4.58
24	.....	.....	3.00	5.15	2.45	2.30	1.50	2.00	2.20	1.85	1.62	4.10
25	2.25	.....	3.58	4.48	2.30	2.19	1.50	2.57	2.00	1.80	1.59	3.60
26	.....	.....	3.45	3.75	2.25	2.19	1.43	2.60	1.80	1.78	1.60	3.20
27	.....	7.90	2.95	4.65	2.20	2.15	1.40	2.30	1.80	1.70	1.58	2.90
28	.....	6.45	2.65	4.12	2.15	2.00	1.37	1.98	1.70	1.65	1.93	2.63
29	.....	.....	2.45	5.55	2.10	1.88	1.40	1.82	1.55	1.60	5.80	2.60
30	.....	.....	2.50	4.95	2.25	1.78	1.40	1.72	1.50	1.53	6.10	2.75
31	.....	.....	5.32	.....	2.61	.....	1.43	1.55	.....	1.60	.....	3.00

*Water supply paper U. S. Geol. Survey, No. 169, page 75.

Note—Gage heights interpolated March 2, 9, November16, and December 7. Ice conditions January 1 to February 27. From January 12 to February 27 river was entirely frozen across. Gage was read to the surface of the water in a hole in the ice.

*Station Rating Table for Wabash River at Logansport, Indiana, from January 1 to December 31, 1906.**

Gage Height, Feet	Discharge, Sec.-Ft.	Gage Height, Feet	Discharge, Sec.-Ft.	Gage Height, Feet	Discharge, Sec.-Ft.	Gage Height, Feet	Discharge, Sec.-Ft.
1.20	330	2.30	1,805	3.40	4,440	4.50	7,700
1.30	360	2.40	2,010	3.50	4,715	4.60	8,020
1.40	420	2.50	2,225	3.60	4,995	4.70	8,340
1.50	510	2.60	2,445	3.70	5,280	4.80	8,670
1.60	630	2.70	2,675	3.80	5,570	4.90	9,000
1.70	770	2.80	2,910	3.90	5,865	5.00	9,340
1.80	920	2.90	3,150	4.00	6,160	5.20	10,020
1.90	1,080	3.00	3,395	4.10	6,460	5.40	10,720
2.00	1,250	3.10	3,645	4.20	6,760	5.60	11,430
2.10	1,425	3.20	3,905	4.30	7,070	5.80	12,160
2.20	1,610	3.30	4,170	4.40	7,380	6.00	12,910

*Water supply paper U. S. Geol. Survey, No. 169, page 76.

Note—The above table is applicable only for open channel conditions. It is based on discharge measurements made during 1903-1905. It is well defined between gage heights 1.2 and 4.00 feet. The table has been extended beyond these limits. Above 6 feet the discharge is estimated, being based on three high-water measurements of 1904.

*Discharge Measurements of Wabash River at Logansport, Indiana, in 1906.**

Date	Hydrographer	Gage Height, Feet	Discharge, Sec.-Ft.
February 9	Brennan and Kriegsman	2.02	1,270
March 10	E. L. Kriegsman	2.78	3,010
April 3	E. L. Kriegsman	5.42	11,800
May 10	E. F. Kriegsman	1.72	1,070

*Water supply paper U. S. Geol. Survey, No. 205, page 60.

*Daily Gage Height, in Feet, of Wabash River at Logansport, Indiana, for 1906.**

Day	Jan.	Feb.	Mar.	Apr.	May	June	July
1	2.73	2.58	2.08	7.90	1.90	1.72	1.52
2	2.57	1.40	1.88	6.85	2.40	1.66	1.55
3	2.48	1.18	2.48	5.60	1.90	1.66	1.60
4	2.38	1.58	3.83	4.85	1.90	1.70	1.60
5	4.88	1.88	3.30	3.90	1.80	1.70	1.75
6	4.43	2.03	2.93	3.80	1.80	1.72	1.87
7	3.88	2.03	2.78	3.60	1.76	1.70	1.75
8	3.20	2.06	2.70	4.40	1.72	1.50	1.75
9	2.63	1.97	......	6.70	1.76	2.70	1.60
10	2.00	1.88	2.78	7.15	1.73	2.40	1.60
11	2.18	1.83	2.68	5.90	1.66	2.06	1.86
12	2.38	1.78	2.58	4.85	1.65	1.80	1.89
13	2.18	1.83	2.53	4.80	1.60	1.86	1.83
14	2.15	1.88	2.41	4.70	1.72	1.80	1.82
15	2.08	2.08	2.28	5.90	1.72	1.76	1.80
16	2.68	2.18	2.23	5.25	1.80	1.72	1.80
17	2.88	2.23	2.18	3.80	1.68	1.70	1.80
18	3.18	1.88	2.08	3.40	1.60	1.72	1.80
19	3.08	1.73	1.93	3.30	1.60	1.66	1.80
20	2.88	1.78	1.98	2.86	1.40	1.60	1.80
21	3.88	1.78	1.98	......	1.43	1.60	1.80
22	8.21	1.83	2.03	2.60	1.53	1.72	1.80
23	6.88	1.78	2.08	2.40	1.52	1.36	1.78
24	5.88	1.88	1.88	2.30	1.56	......	......
25	4.88	2.08	1.88	2.20	1.76	1.19	......
26	4.13	2.38	1.98	2.20	1.50	1.19	......
27	3.88	2.71	8.48	2.20	1.50	1.30	......
28	3.68	2.58	9.03	2.10	1.60	1.37	......
29	3.03	......	8.38	2.15	1.44	1.40	......
30	2.88	......	8.18	2.10	1.47	1.57	......
31	......	......	8.03	......	1.84	......	......

*Water supply paper U. S. Geol. Survey, No. 205, page 61.

Rating Table for Wabash River at Logansport, Indiana, for 1906.

Gage Height, Feet	Discharge, Sec.-Ft.	Gage Height, Feet	Discharge, Sec.-Ft.	Gage Height, Feet	Discharge, Sec.-Ft.	Gage Height, Feet	Discharge, Sec.-Ft.
1.20	310	2.30	1,900	3.40	4,690	5.00	9,770
1.30	400	2.40	2,110	3.50	4,980	5.20	10,480
1.40	500	2.50	2,330	3.60	5,270	5.40	11,210
1.50	610	2.60	2,560	3.70	5,570	5.60	11,950
1.60	730	2.70	2,800	3.80	5,870	5.80	12,710
1.70	860	2.80	3,050	3.90	6,170	6.00	13,500
1.80	1,000	2.90	3,310	4.00	6,480	7.00	17,600
1.90	1,160	3.00	3,580	4.20	7,110	8.00	22,100
2.00	1,330	3.10	3,850	4.40	7,750	9.00	26,800
2.10	1,510	3.20	4,130	4.60	8,410	9.10	27,270
2.20	1,700	3.30	4,410	4.80	9,080	....	....

NOTE—The above table is applicable only for open channel conditions. It is based on discharge measurements made during 1903 to 1906. It is well defined.

On July 16, 17, 18, 1910, a gaging station was established by the writer on the Cicott Street bridge in Logansport. A discussion of this gage is given on page 77 of the Thirty-fifth Annual Report of the Indiana Department of Geology and Natural Resources. The following data have been collected at this station:

Discharge Measurements on Wabash River at Logansport, Indiana, for 1910-11.

Date	Hydrographer	Gage Height, Feet	Discharge, Sec.-Ft.
July 18, 1910	W. M. Tucker	5.4	442.72
Apri 15, 1911	W. M. Tucker	9.0	8,865.96
July 6, 1911	Tucker and Clark	5.6	545.23

These three readings are not sufficient data to formulate a rating table, but taken with the United States Geological Survey rating tables show that the base of this gage is approximately 4.1 feet lower than the base of the United States Geological Survey gage was. Thus the gage readings can be used with the United States Geological Survey rating tables if 4.1 feet be subtracted from the gage reading in each case and the discharge for that gage height be taken from the rating table.

Daily Gage Readings on Wabash River at Logansport, Indiana, from July 18, 1910, to July 17, 1911, Inclusive.

Day	July	Aug.	Sept.	Oct.	Nov.	Dec.	Jan.	Feb.	Mar.	Apr.	May	June	July
1	.....	5.2	5.3	5.9	5.4	8.3	9.0	9.9	6.5	6.0	7.7	6.1	5.9
2	.....	5.2	5.2	5.8	5.4	8.1	9.4	8.0	6.4	6.0	8.1	6.1	5.8
3	.....	5.2	5.2	5.7	5.5	7.5	†	7.9	6.4	6.0	7.2	6.1	5.7
4	.....	5.2	5.3	5.8	5.6	6.0	†	7.9	6.3	6.3	6.8	6.1	5.6
5	.....	5.2	5.4	5.9	5.8	5.8	†	7.8	6.2	10.2	6.6	6.1	5.6
6	.....	5.2	6.0	6.0	5.8	5.7	†	7.7	6.1	10.6	6.3	6.2	5.6
7	.....	5.1	6.4	6.2	5.8	5.7	†	7.6	6.1	9.3	6.2	6.2	5.5
8	.....	5.1	6.2	10.3	5.7	5.7	†	7.5	6.1	8.7	6.2	6.3	5.4
9	.....	5.1	5.9	10.6	5.7	5.8	†	7.3	6.0	7.9	6.2	6.0	5.4
10	.....	5.1	5.9	9.5	5.7	5.9	†	6.4	6.0	7.9	6.2	6.9	5.4
11	.....	5.1	5.8	9.4	5.6	5.9	†	6.6	6.2	6.2	6.2	5.8	5.5
12	.....	5.1	5.8	8.5	5.6	5.8	†	6.4	6.3	6.7	6.2	5.6	5.4
13	.....	5.1	6.3	7.9	5.6	5.8	†	6.6	6.9	6.8	6.2	5.6	5.4
14	.....	5.1	7.2	7.5	5.5	5.6	†	6.7	6.5	8.0	6.1	5.6	5.4
15	.....	5.1	6.6	7.3	5.5	5.6	†	9.2	6.3	8.9	6.1	5.6	5.4
16	.....	5.1	6.5	6.0	5.5	5.5	†	10.7	6.2	8.6	5.9	5.7	5.4
17	.....	5.1	6.2	5.9	5.5	5.5	†	9.0	6.1	8.1	5.7	6.1	5.6
18	5.4	5.2	6.1	5.9	5.4	5.5	†	8.5	6.2	7.3	5.8	6.1	.....
19	5.3	5.2	6.2	5.7	5.4	5.5	†	8.5	6.1	7.2	5.8	6.1	.....
20	5.3	5.5	6.1	5.7	5.3	5.5	†	8.2	6.0	10.0	5.8	6.1	.....
21	5.3	5.6	6.1	5.7	5.3	5.6	†	8.1	5.9	10.1	5.9	5.7	.....
22	5.2	5.5	6.0	5.6	5.3	5.7	†	8.0	5.9	9.3	5.9	5.8	.....
23	5.2	5.5	6.0	5.7	5.3	5.8	7.0	7.2	5.9	8.6	5.9	5.9	.....

Daily Gage Readings on Wabash River at Logansport, Indiana, from July 18, 1910, to July 17, 1911, Inclusive—Continued.

DAY	July	Aug.	Sept.	Oct.	Nov.	Dec.	Jan.	Feb.	Mar.	Apr.	May	June	July
24	5.2	5.5	6.0	5.7	5.4	6.0	6.8	7.0	5.9	8.8	5.8	6.0	......
25	5.2	5.5	6.0	5.5	5.4	5.8	6.7	6.8	5.9	7.2	5.7	6.1	......
26	5.2	5.5	6.2	5.6	5.4	5.8	6.6	6.7	5.9	7.2	5.6	7.6	......
27	5.2	5.5	6.2	5.6	5.5	5.8	7.4	6.6	6.0	7.2	5.6	6.9	......
28	5.2	5.4	6.1	5.5	5.5	7.3	11.9	6.5	6.0	7.2	5.6	6.0	......
29	5.4	5.3	6.0	5.5	8.4	8.1	11.7	......	6.0	7.1	5.6	6.0	......
30	5.4	5.3	5.9	5.5	8.8	9.0	11.5	......	6.0	7.0	5.5	5.9	......
31	5.2	5.3	......	5.4	......	9.0	11.3	......	6.0	......	5.5	......	......

† No records taken.

The Cicott Street bridge was carried away during the flood of 1913. The records from the gage which were kept by Mr. William Sehrt were also destroyed when the flood swept through his dwelling. The bridge was replaced by a handsome concrete structure. Upon this bridge the U. S. Weather Bureau established a gage of the Mott type. It was read but seven months of the year until the summer of 1916, when arrangements were made whereby the Department of Geology and Natural Resources of Indiana was to take records from the gage for the remaining five months. The gage is read each day by Mr. F. L. C. Boerger. The following data has been collected at this station:

Daily Gage Heights on Wabash River at Logansport, Indiana, for 1915.

DAY	Jan.	Feb.	Mar.	Apr.	May	June	July	Aug.	Sept.	Oct.	Nov.	Dec.
1*	......	......	......	......	......	......	......	......	......	......	0.7	1.2
2	......	......	......	......	......	......	......	......	......	......	0.6	1.1
3	......	......	......	......	......	......	......	......	......	......	0.6	1.1
4	......	......	......	......	......	......	......	......	......	......	0.6	1.0
5	......	......	......	......	......	......	......	......	......	......	0.6	1.0
6	......	......	......	......	......	......	......	......	......	......	0.6	0.9
7	......	......	......	......	......	......	......	......	......	......	0.6	0.8
8	......	......	......	......	......	......	......	......	......	......	0.6	0.8
9	......	......	......	......	......	......	......	......	......	......	0.6	0.8
10	......	......	......	......	......	......	......	......	......	......	0.6	0.8
11	......	......	......	......	......	......	......	......	......	......	0.6	0.8
12	......	......	......	......	......	......	......	......	......	......	0.9	0.8
13	......	......	......	......	......	......	......	......	......	......	0.9	0.7
14	......	......	......	......	......	......	......	......	......	......	0.9	0.4
15	......	......	......	......	......	......	......	......	......	......	0.7	0.3
16	......	......	......	......	......	......	......	......	......	......	0.7	0.2
17	......	......	......	......	......	......	......	......	......	......	0.7	0.1
18	......	......	......	......	......	......	......	......	......	......	0.7	0.9
19	......	......	......	......	......	......	......	......	......	......	1.5	3.5
20	......	......	......	......	......	......	......	......	......	......	4.1	2.4
21	......	......	......	......	......	......	......	......	......	......	4.3	1.8
22	......	......	......	......	......	......	......	......	......	*	3.3	1.6
23	......	......	......	......	......	......	......	......	......	1.4	2.7	1.5
24	......	......	......	......	......	......	......	......	......	1.2	2.1	1.2
25	......	......	......	......	......	......	......	......	......	1.1	1.8	1.1
26	......	......	......	......	......	......	......	......	......	1.0	1.5	1.0
27	......	......	......	......	......	......	......	......	......	0.9	1.4	1.5
28	......	......	......	......	......	......	......	......	......	0.8	1.3	2.0
29	......	......	......	......	......	......	......	......	......	0.8	1.3	1.8
30	......	......	......	......	......	......	......	......	......	0.7	1.2	1.7
31	......	......	......	......	......	......	......	......	......	0.7	......	1.7

*Daily gage readings began on October 23.

Daily Gage Heights on Wabash River at Logansport, Indiana, for 1916.

Day	Jan.	Feb.	Mar.	Apr.	May	June	July	Aug.	Sept.	Oct.	Nov.	Dec.
1	1.7	14.6	1.6	4.0	2.3	1.5	0.9	0.0	—.5	0.0	—.5	—.2
2	9.7	12.8	1.4	3.6	1.9	1.4	0.8	—.1	—.5	—.3	—.5	—.2
3	12.3	9.5	1.3	3.2	1.9	1.2	0.8	—.1	0.0	—.3	—.5	—.2
4	12.6	5.7	1.2	2.8	3.1	1.1	0.6	0.0	—.3	—.3	—.5	—.2
5	10.8	4.4	1.3	2.4	3.3	1.3	0.5	0.0	—.3	—.3	—.5	0.0
6	10.3	3.4	1.2	2.1	2.8	1.1	0.4	0.2	—.3	—.3	—.5	0.4
7	8.3	2.8	1.6	1.9	3.6	1.5	0.4	0.1	0.0	—.3	—.5	0.7
8	5.5	2.1	2.4	1.7	5.6	3.5	0.3	0.0	0.8	—.4	—.5	0.7
9	4.2	1.8	2.7	1.6	5.1	3.7	0.3	0.0	0.0	—.4	—.3	0.5
10	3.3	2.1	2.3	1.5	4.1	3.0	0.3	0.0	0.0	—.4	—.3	0.5
11	2.9	1.9	1.9	1.4	3.3	2.4	0.2	—.1	0.0	—.4	—.3	0.5
12	3.2	1.9	1.6	1.4	2.4	2.1	0.3	1.0	0.0	—.4	—.3	0.5
13	8.8	1.7	1.4	1.4	1.8	1.3	0.3	0.6	—.2	—.4	—.3	0.3
14	8.0	1.6	1.4	1.5	1.9	1.1	1.3	0.5	—.2	—.4	—.3	0.0
15	6.1	1.6	1.5	1.5	3.7	1.1	1.3	0.2	—.2	—.4	—.3	0.0
16	4.2	1.6	1.5	1.4	3.3	1.0	0.8	0.1	—.2	—.5	—.3	0.0
17	3.9	1.6	1.4	1.4	2.6	1.1	0.5	0.0	—.2	—.5	—.3	0.0
18	2.9	1.8	1.3	1.3	2.1	1.1	0.6	—.1	—.2	—.5	—.3	0.0
19	2.5	1.9	1.2	1.2	1.7	1.2	0.5	—.1	—.2	—.5	—.3	0.0
20	1.9	1.8	1.2	1.1	1.5	1.7	0.4	—.1	—.3	—.3	—.5	0.0
21	4.3	1.8	1.2	1.0	1.3	1.6	0.3	—.1	—.4	0.0	—.5	0.0
22	7.3	1.6	1.2	1.1	1.3	5.0	0.7	—.1	—.5	0.0	—.5	0.0
23	6.5	1.6	6.2	1.1	1.3	4.4	0.6	—.1	—.4	0.0	—.5	0.0
24	5.3	3.1	5.7	1.3	1.5	3.5	0.5	—.2	—.4	—.1	—.2	0.0
25	4.2	2.9	5.2	1.3	1.4	2.5	0.4	—.2	—.4	—.1	0.0	0.0
26	3.6	2.6	4.7	1.2	1.2	1.8	0.2	0.0	—.4	—.1	0.0	0.0
27	3.4	2.1	6.2	3.6	1.5	1.6	0.1	—.2	0.0	—.1	0.0	7.0
28	4.8	1.8	7.3	4.0	1.9	1.3	0.1	—.2	0.0	—.1	0.0	7.0
29	4.8	1.7	7.7	3.5	3.1	1.0	0.0	—.3	0.0	—.1	0.0	6.0
30	6.3	.....	6.7	2.8	2.3	0.8	0.0	—.3	0.0	—.3	0.0	5.0
31	12.9	.....	5.2	.....	1.8	.....	0.0	—.5	.....	—.3	.....	4.5

Discharge Measurements on Wabash River at Logansport, Indiana, during 1915-1916.

Date	Hydrographer	Gage Height, Feet	Discharge, Sec.-Ft.
October 23, 1915	W. M. Tucker	1.37	3,592
November 27, 1915	W. M. Tucker	1.40	3,681
December 19, 1915	W. M. Tucker	3.20	9,261
January 6, 1916	W. M. Tucker	9.40	35,996
January 7, 1916	W. M. Tucker	7.20	26,328
March 10, 1916	W. M. Tucker	2.20	6,002
May 9, 1916	W. M. Tucker	5.00	16,148
August 12, 1916	W. M. Tucker	1.00	2,716
October 12, 1916	W. M. Tucker	—0.40	377

Rating Table for Wabash River at Logansport, Indiana, for 1915-1917.

Gage Height, Feet	Discharge, Sec.-Ft.	Gage Height, Feet	Discharge, Sec.-Ft.	Gage Height, Feet	Discharge, Sec.-Ft.	Gage Height, Feet	Discharge, Sec.-Ft.
—0.5	290	1.4	3,690	3.3	9,690	5.2	16,820
—0.4	380	1.5	3,950	3.4	10,080	5.3	17,220
—0.3	480	1.6	4,220	3.5	10,480	5.4	17,630
—0.2	590	1.7	4,490	3.6	10,870	5.5	18,040
—0.1	710	1.8	4,770	3.7	11,260	5.6	18,460
0.0	840	1.9	5,050	3.8	11,640	5.7	18,880
0.1	980	2.0	5,340	3.9	12,010	5.8	19,310
0.2	1,130	2.1	5,630	4.0	12,370	5.9	19,740
0.3	1,290	2.2	5,930	4.1	12,720	6.0	20,180
0.4	1,460	2.3	6,230	4.2	13,070	6.2	21,070
0.5	1,640	2.4	6,540	4.3	13,420	6.4	21,990
0.6	1,830	2.5	6,850	4.4	13,780	6.6	22,930
0.7	2,030	2.6	7,170	4.5	14,140	6.8	23,890
0.8	2,240	2.7	7,500	4.6	14,510	7.0	24,870
0.9	2,460	2.8	7,840	4.7	14,880	8.0	29,870
1.0	2,690	2.9	8,190	4.8	15,260	9.0	34,870
1.1	2,930	3.0	8,550	4.9	15,640	10.0	39,870
1.2	3,180	3.1	8,920	5.0	16,030	....	....
1.3	3,430	3.2	9,300	5.1	16,420	....	....

NOTE—This table is applicable only for open channel conditions.

From the foregoing data, a fair idea of the discharge of the stream can be gained. The lowest discharge in 1903 was 342 second-feet on July 14, and the highest about 33,000 second-feet on July 2. The lowest discharge in 1904 was 260 second-feet on August 13 and September 8, and the highest was 56,140 second-feet on March 27. In 1905, the lowest discharge was 345 second-feet on August 4, and the highest estimated at 33,750 second-feet on May 12. In 1906, the lowest was 294 second-feet, on February 3, and the highest 26,900 second-feet on March 28. In 1910-11, the lowest was less than 310 second-feet, from August 7 to August 17, 1910, and the highest 21,200 second-feet on January 28, 1911. This shows the high discharge to be practically one hundred times the low discharge. Since this station is situated one mile below the mouth of Eel River, the discharge from it must be subtracted in each case to determine the discharge of the Wabash above the junction of the two streams. The discharge of Eel River is not known for the years prior to 1910 and 1911. Between August 7 and 17, 1910, it is found from the data of the Third Street Gaging Station to have varied from 125 second-feet to 210 second-feet, which indicates that the discharge on the Wabash was less than 100 second-feet during part of this time. On January 28, 1911, the discharge of Eel River was 1,895 second feet, which indicates that the discharge of the Wabash was 19,305 second-feet. Thus the maximum discharge of the Wabash at the mouth of Eel River is approximately two hundred times the minimum. The United States Geological Survey maintained a gage on the Wabash River at Terre Haute, Indiana, from February 25, 1905, to July 20, 1906. The following data is taken from the United States Water Supply Papers:

*Wabash River at Terre Haute, Indiana.** This station was established February 25, 1905. It is located at the Vandalia Line railway bridge near the city waterworks. There are no tributaries nor any islands, falls or dams in the river near the station.

The channel is practically straight for 700 feet above and below the station. There is a considerable angle of approach on the right bank, but practically none at the left bank. The right bank is comparatively low and alluvial, but is protected by a levee that does not overflow. The left bank is high, covered with buildings, and does not overflow. All of the water passes between the abutments of the bridge. The bed of the stream is composed of hard, permanent material, is clean of vegetation, and always consists of but one channel. The current has a rather swift velocity at low stages.

Discharge measurements are made from the downstream lower chord of the bridge. The initial point for soundings is the center of the truss pin through the left end of the downstream lower chord of the left span of the bridge.

A standard chain gage is fastened to the downstream side of the bridge on the first span from the left bank; length of chain, 42.97 feet. During 1905 the gage was read by Mr. Albert Shewmaker. The gage is referred to bench marks as follows: (1) The Terre Haute city bench mark at the northeast corner of First and Chestnut streets; elevation 0.33 feet. (2) The southeast corner of the left abutment of the Vandalia Line railway bridge; elevation 36.40 feet. (3) The base of the railroad rail immediately opposite the pulley at the gage box; elevation 42.51 feet. Elevations refer to the datum of the gage. This is 0.33 foot below the city datum, which is 438.72 feet above mean sea level.

*Discharge Measurements of Wabash River at Terre Haute, Indiana, in 1905.**

Date	Hydrographer	Gage Height, Feet	Discharge, Sec.-Ft.
May 12	M. S. Brennan	6.90	15,750
June 25	M. S. Brennan	3.74	7,367
July 29	M. S. Brennan	1.42	2,648
August 25	M. S. Brennan	2.00	3,700
September 11	M. S. Brennan	1.81	3,311
October 15	M. S. Brennan	1.00	2,066

*Water supply paper U. S. Geol. Survey, No. 169, page 77.

*Daily Gage Height, in Feet, of Wabash River at Terre Haute, Indiana, for 1905.***

Day	Feb.	Mar.	Apr.	May	June	July	Aug.	Sept.	Oct.	Nov.	Dec.
1	.....	13.7	5.1	9.6	6.4	2.2	1.72	1.5	1.42	1.9	9.55
2	.....	14.3	7.0	8.0	6.2	2.02	1.58	1.75	1.4	1.82	9.85
3	.....	14.4	7.2	6.7	6.6	2.88	1.38	1.75	1.3	1.85	8.62
4	.....	14.7	6.2	5.8	5.2	3.8	1.28	1.6	1.22	1.8	7.05
5	.....	15.0	5.2	5.3	4.6	2.78	1.22	1.85	1.18	1.88	6.02
6	.....	13.3	4.4	4.0	4.05	2.48	1.15	1.95	1.18	2.85	5.83
7	.....	10.6	3.9	4.7	4.7	2.28	1.3	1.88	1.18	3.4	5.3
8	.....	9.0	3.6	4.5	6.0	1.98	1.22	1.7	1.32	3.9	4.88
9	.....	8.2	3.3	4.9	5.6	2.35	1.48	1.5	1.3	3.98	4.4
10	.....	7.3	3.1	5.6	4.9	2.58	1.82	1.4	1.28	3.88	4.06
11	.....	6.6	3.0	5.3	4.6	4.0	1.68	1.92	1.18	3.62	3.8
12	.....	6.5	3.05	6.8	4.7	4.75	1.45	3.0	1.1	3.32	3.62
13	.....	6.2	3.1	12.7	5.2	4.25	1.28	3.1	1.02	3.05	3.42
14	.....	5.6	2.95	14.1	5.4	4.4	1.52	2.72	1.0	3.75	3.18
15	.....	5.0	2.72	15.1	5.1	4.4	2.62	2.4	1.0	2.52	2.85
16	.....	4.6	2.55	16.3	5.0	3.85	2.8	3.32	.95	2.38	2.65
17	.....	4.3	2.4	17.0	4.6	3.4	2.62	3.95	.9	2.22	2.48
18	.....	4.0	2.32	16.8	4.2	3.0	2.55	4.6	2.0	2.12	2.52
19	.....	4.35	3.15	15.7	4.8	2.68	2.25	5.0	3.4	2.02	2.55
20	.....	5.1	2.02	12.8	5.4	2.75	2.1	5.1	4.15	1.92	2.48
21	.....	5.6	2.82	9.5	7.3	4.6	2.38	4.8	4.92	1.82	2.62
22	.....	5.6	5.9	7.8	4.8	3.4	2.58	4.2	4.68	1.75	3.2
23	.....	5.5	9.2	6.6	4.1	2.65	†2.15	3.6	4.12	1.65	4.9
24	.....	5.3	10.9	5.7	3.9	2.28	1.72	3.2	3.68	1.65	7.08
25	.....	4.8	10.6	5.0	3.7	1.98	1.68	2.72	3.3	1.6	7.4
26	9.7	4.8	9.4	4.6	3.3	1.78	1.78	2.32	2.95	1.65	6.7
27	12.3	5.0	8.4	4.3	2.9	1.6	2.1	2.05	2.7	1.58	6.0
28	11.3	5.0	7.5	4.1	2.65	1.5	2.38	1.8	2.45	1.42	5.3
29	.....	4.5	8.0	3.9	2.5	1.42	2.3	1.68	2.2	2.1	4.9
30	.....	4.1	8.2	6.7	2.4	1.78	2.02	1.5	1.12	8.0	4.5
31	.....	4.5	.....	7.1	.....	1.68	1.75	.....	1.95	.....	4.48

†Gage height interpolated. **Water supply paper U. S. Geol. Survey, No. 169, page 78.

Station Rating Table for Wabash River at Terre Haute, Indiana, from February 25 to December 31, 1905.

Gage Height, Feet	Discharge, Sec.-Ft.	Gage Height, Feet	Discharge, Sec.-Ft.	Gage Height, Feet	Discharge, Sec.-Ft.	Gage Height, Feet	Discharge, Sec.-Ft.
0.90	1,945	2.50	4,670	4.20	8,430	7.40	17,200
1.00	2,065	2.60	4,870	4.40	8,910	7.60	17,800
1.10	2,195	2.70	5,080	4.60	9,390	7.80	18,400
1.20	2,335	2.80	5,290	4.80	9,890	8.00	19,000
1.30	2,485	2.90	5,505	5.00	10,400	8.20	19,600
1.40	2,643	3.00	5,720	5.20	10,920	8.40	20,300
1.50	2,808	3.10	5,940	5.40	11,450	8.60	21,000
1.60	2,978	3.20	6,160	5.60	11,990	8.80	21,700
1.70	3,152	3.30	6,380	5.80	12,540	9.00	22,400
1.80	3,330	3.40	6,605	6.00	13,100	9.50	24,150
1.90	3,513	3.50	6,830	6.20	13,660	10.00	25,900
2.00	3,700	3.60	7,055	6.40	14,240	10.50	27,800
2.10	3,890	3.70	7,280	6.60	14,820	11.00	29,800
2.20	4,080	3.80	7,510	6.80	15,400	11.50	31,800
2.30	4,275	3.90	7,740	7.00	16,000	12.00	33,800
2.40	4,470	4.00	7,970	7.20	16,600	....	....

NOTE—The above table is applicable only for open channel conditions. It is based on six discharge measurements made during 1905. It is well defined between gage heights 1 foot and 4 feet. The table has been extended beyond these limits, being based on one measurement at 6.9 feet and also on well-defined area and mean-velocity curves to 12 feet. Above this point the curve is more uncertain as the river overflows.

Discharge Measurements of Wabash River at Terre Haute, Indiana, in 1906.

Date	Hydrographer	Width, Feet	Area of Section, Sq. Ft.	Gage Height, Feet	Discharge, Sec.-Ft.
February 16†	Brennan and Kriegsman	482	3,790	4.40	6,710
March 28	E. F. Kriegsman	606	10,700	15.84	40,600
March 31	E. F. Kriegsman	714	13,000	19.20	62,800
Apri 18	E. F. Kriegsman	580	8,520	12.02	25,500
April 19	E. F. Kriegsman	564	7,740	10.80	22,200
April 20	E. F. Kriegsman	557	6,790	9.30	19,300
April 20	E. F. Kriegsman	557	6,650	8.92	18,200
April 21	E. F. Kriegsman	552	6,210	7.95	16,800
April 21	E. F. Kriegsman	549	5,950	7.65	16,200
April 23	E. F. Kriegsman	541	5,190	6.35	13,200
June 9	E. F. Kriegsman	535	4,500	4.98	11,400

†Partial ice conditions.

Daily Gage Heights, in Feet, of Wabash River at Terre Haute, Indiana, for 1906.

Day	Jan.	Feb.	Mar.	Apr.	May	June	July
1	4.68	8.22	5.02	19.8	3.88	5.72	1.08
2	4.72	7.28	4.88	19.22	3.8	4.22	1.22
3	4.85	6.0	7.38	18.45	3.78	3.5	1.35
4	5.88	5.88	9.9	17.7	3.75	2.88	1.32
5	6.5	5.15	10.78	16.8	3.78	2.88	1.38
6	7.35	4.55	10.48	15.22	3.52	3.2	1.7
7	7.85	2.45	9.28	12.9	3.3	3.12	1.72
8	7.18	2.35	8.48	10.75	3.1	2.52	1.62
9	5.82	2.7	8.3	14.22	2.95	4.22	1.6
10	4.78	2.78	8.18	15.02	2.9	5.55	1.52
11	3.85	2.92	7.82	15.1	2.82	4.52	1.42
12	4.02	3.02	7.28	15.3	2.72	4.08	1.35
13	3.62	3.2	6.75	15.42	2.62	3.38	1.32
14	3.58	3.72	6.38	14.95	2.52	2.92	1.32
15	3.55	4.42	5.98	12.85	2.45	2.6	1.25
16	2.75	4.15	5.58	12.52	2.35	2.32	1.22
17	3.82	3.4	5.28	12.75	2.28	2.12	1.4
18	3.88	3.12	4.98	12.25	2.15	1.98	1.75
19	3.95	3.6	4.82	10.7	2.12	1.82	1.7
20	4.65	4.02	4.55	9.1	2.08	1.75	1.45
21	5.72	4.2	4.35	7.9	2.02	1.75	......
22	13.68	3.78	4.42	6.95	1.92	1.7	......
23	15.95	3.82	4.6	6.28	1.9	1.68	......
24	16.82	3.78	4.52	5.72	1.85	1.6	......
25	17.12	4.18	1.48	5.38	1.9	1.5	......
26	17.38	5.0	6.35	5.05	1.98	1.42	......
27	17.22	5.22	14.28	4.78	1.9	1.3	......
28	16.25	5.3	15.88	4.48	2.12	1.28	......
29	14.12	......	16.42	4.25	2.08	1.2	......
30	11.18	......	17.02	4.05	1.88	1.15	......
31	9.3	......	18.95	......	4.32	......	......

NOTE—Ice gorge at railroad bridge, a short distance above the gaging section, February 7 to 18. The flow at the gaging section was probably affected by ice on only February 15 and 16.

Rating Table for Wabash River at Terre Haute, Indiana, for 1905 and 1906.

Gage Height, Feet	Discharge, Sec.-Ft.	Gage Height, Feet	Discharge, Sec.-Ft.	Gage Height, Feet	Discharge, Sec.-Ft.	Gage Height, Feet	Discharge, Sec.-Ft.
1.00	2,050	2.50	4,510	4.00	7,620	7.00	14,700
1.10	2,180	2.60	4,700	4.20	8,060	8.00	17,200
1.20	2,320	2.70	4,900	4.40	8,500	9.00	19,800
1.30	2,460	2.80	5,100	4.60	8,940	10.00	22,500
1.40	2,610	2.90	5,300	4.80	9,380	11.00	25,300
1.50	2,770	3.00	5,500	5.00	9,830	12.00	28,400
1.60	2,940	3.10	5,700	5.20	10,300	13.00	31,800
1.70	3,110	3.20	5,900	5.40	10,780	14.00	35,500
1.80	3,280	3.30	6,100	5.60	11,260	15.00	39,500
1.90	3,450	3.40	6,310	5.80	11,740	16.00	43,700
2.00	3,620	3.50	6,520	6.00	12,220	17.00	48,100
2.10	3,790	3.60	6,740	6.20	12,700	18.00	52,700
2.20	3,970	3.70	6,960	6.40	13,200	19.00	57,400
2.30	4,150	3.80	7,180	6.60	13,700	20.00	62,100
2.40	4,330	3.90	7,400	6.80	14,200	....	....

NOTE—The above table is applicable only for open-channel conditions. It is based on discharge measurements made during 1905 and 1906. It is well defined.

On July 21 and 22, 1910, the writer visited the Terre Haute waterworks station and found that daily gage readings had been kept at that point since 1901. These gage readings were secured through the kindness of Mr. Taylor, chief engineer of the Terre Haute Water Works Company, and will be given in this report. Two current readings were taken from the Wabash Avenue bridge in Terre Haute, and these will be used with the readings of the United States Geological Survey to determine a rating table to accompany the gage readings:

Discharge Measurements on Wabash River at Terre Haute, Indiana, from 1905 to 1911.

Date	Hydrographer	Gage Height, Feet	Discharge, Sec.-Ft.
May 12, 1905	M. S. Brennan	6 00	15,750
June 25, 1905	M. S. Brennan	2.84	7,367
July 29, 1905	M. S. Brennan	.52	2,648
August 25, 1905	M. S. Brennan	1.10	3,700
September 11, 1905	M. S. Brennan	.10	2,066
February 16, 1906†	Brennan and Kriegsman	3.50	6,710
March 28, 1906	E. F. Kriegsman	14.94	40,600
March 31, 1906	E. F. Kriegsman	18.30	62,800
April 18, 1906	E. F. Kriegsman	11.12	25,500
April 19, 1906	E. F. Kriegsman	9.90	22,200
April 20, 1906	E. F. Kriegsman	8.40	19,300
April 20, 1906	E. F. Kriegsman	8.02	18,200
April 21, 1906	E. F. Kriegsman	7.05	16,800
April 21, 1906	E. F. Kriegsman	6.75	16,200
April 23, 1906	E. F. Kriegsman	5.45	13,200
June 9, 1906	E. F. Kriegsman	4.08	11,400
July 23, 1910	W. M. Tucker	.67	2,962
August 28, 1911	Tucker and Clark	.25	2,249

† Partial ice conditions.

NOTE—The readings taken by Messrs. Brennan and Kriegsman were based on the government gage, which was placed .9 feet lower than the waterworks gage. Thus the reduction of each gage reading by .9 ft. gives data for a rating table for the waterworks gage. The reduction has been made in this table.

Gage Readings on Wabash River at Terre Haute Waterworks Station for 1901.

Day	June	July	Aug.	Sept.	Oct.	Nov.	Dec.
1	7.0	3.7	1.4	1.3	1.5	1.0	1.2
2	6 0	3.0	1.4	1.3	1.5	1.0	1.2
3	5.0	2.8	1.5	1.5	1.5	1.2	1.2
4	4.0	2.5	1.5	1.5	1.5	1.2	1.2
5	3.5	3.0	1.5	1.5	1.5	1.2	1.2
6	4.0	2.0	1.5	1.5	1.5	1.2	1.2
7	6.0	2.0	1.5	1.5	1.5	1.3	1.2
8	4.5	1.8	1.5	1.5	1.5	1.3	1.2
9	4.0	1.5	1.5	1.5	1.5	1.3	1.2
10	3.5	1.0	1.5	1.5	1.5	1.3	1.0
11	3.0	1.0	1.5	1.5	1.5	1.5	1.0
12	2.5	0.5	1.5	1.5	1.5	1.3	1.0*
13	2.0	0.0	1.5	1.5	1.0	1.3	.9
14	1.8	0.0	1.5	1.5	1.0	1.3	3.5
15	1.5	0.0	1.5	1.5	1.2	1.3	6.7
16	1.5	0.5	1.5	1.5	2.0	1.3	6.0
17	1.5	0.8	1.5	1.5	2.5	1.3	*
18	1.5	0.8	1.5	1.5	2.3	1.3	*
19	1.5	1.0	1.5	1.5	1.8	1.3	*
20	1.0	1.0	1.3	1.5	1.3	1.3	*
21	1.5	1.0	1.3	1.5	1.0	1.3	5.0
22	4.0	1.0	1.3	1.5	0.5	1.3	5.3
23	5.5	1.0	1.3	1.5	0.5	1.2	5.5
24	10 0	1.3	1.3	1.5	0.0	1.2	6.0
25	7.0	1.3	1.3	1.5	0.0	1.2	6.0
26	5.5	1.3	1.3	1.5	0.5	1.2	6.3
27	8.0	1.3	1.3	1.5	0 5	1.2	6.3
28	6.0	1.3	1.3	1.5	0.7	1.2	6.3
29	5.0	1.3	1.3	1.5	0.8	1.2	6.5
30	4.5	1.4	1.3	1.5	1.0	1.2	6.5
31	......	1.4	1.3	......	1.0	......	6.3

*Ice conditions.

Gage Readings on Wabash River at Terre Haute Waterworks Station for 1902.

Day	Jan.	Feb.	Mar.	Apr.	May	June	July	Aug.	Sept.	Oct.	Nov.	Dec.
1	6.3	1.5	10.0	8.7	2.7	1.7	15.5	3.8	0.6	5.0	1.0	4.0
2	6.0	1.5	12.0	9.5	2.5	2.0	17.0	3.5	0.6	5.0	1.0	4.4
3	6.0	1.5	12.5	9.0	2.3	5.5	18.7	5.3	0.4	5.0	1.0	6.0
4	6.0	1.5	10.0	8.0	1.8	4.7	19.1	5.0	0.4	5.5	1.0	7.0
5	6.0	1.5	8.0	6.5	1.5	5.5	18.9	4.0	0.4	6.7	1.2	7.8
6	5.8	1.5	5.5	5.5	2.0	7.0	18.2	5.0	0.8	7.3	1.3	8.4
7	5.5	1.5	5.0	5.3	3.0	6.3	17.4	4.3	0.8	7.8	1.7	8.6
8	5.0	1.5	4.5	5.0	3.0	6.7	16.2	4.0	0.8	8.5	2.3	7.3
9	4.5	1.5	5.7	4.4	3.3	6.7	15.0	3.6	0.6	8.3	3.2	6.0
10	3.5	1.5	5.9	4.0	3.0	8.0	11.2	3.2	0.4	7.0	3.2	5.0
11	3.0	1.5	5.5	3.8	2.7	8.0	8.8	3.0	0.4	5.8	3.0	4.5
12	2.5	1.5	5.7	3.5	2.3	6.3	7.6	2.6	0.2	4.5	3.9	5.3
13	2.0	1.5	7.0	3.3	1.8	6.3	6.4	2.3	1.0	4.0	4.0	5.8
14	1.5	1.5	8.2	3.0	1.5	5.5	5.4	2.3	0.5	5.8	3.8	6.5
15	1.0	1.5	8.5	2.8	1.2	5.3	4.6	2.6	0.4	4.5	3.7	5.7
16	1.0	1.5	8.2	2.5	1.0	5.2	4.0	2.8	0.4	4.0	3.4	7.8
17	0.5	1.5	8.7	2.0	0.8	5.8	3.7	2.7	0.4	3.8	3.4	8.9
18	0.5	1.5	8.3	2.0	0.7	5.8	3.3	2.8	0.4	3.0	3.7	9.0
19	0.0	1.5	8.0	1.7	0.5	5.5	3.2	2.3	0.4	2.8	4.2	9.8
20	0.5	1.5	7.0	1.5	0.3	4.8	4.2	2.6	0.3	2.5	5.0	10.2
21	0.8	1.5	6.0	1.5	1.0	3.6	4.8	2.8	0.3	2.3	5.7	12.3
22	1.0	1.5	5.0	1.5	2.3	2.5	5.8	2.8	0.3	2.7	5.3	14.5
23	1.5	1.5	4.5	1.3	3.2	2.0	5.7	2.6	0.3	2.7	4.8	15.2
24	2.0	1.5	4.2	1.2	2.7	1.7	5.5	2.3	0.3	2.6	4.3	15.5
25	2.0	1.5	3.7	1.2	2.5	1.5	4.6	2.0	2.0	2.2	3.8	15.8
26	2.0	1.2	3.5	1.2	4.0	1.5	4.4	1.6	3.8	1.8	3.4	16.0
27	2.0	0.0	3.3	1.0	4.2	1.7	3.6	1.2	2.8	1.5	3.3	15.1
28	1.5	5.0	3.0	1.0	4.0	1.8	3.2	1.2	4.0	1.5	3.4	14.0
29	1.5	.....	4.3	1.0	3.3	5.0	3.2	1.0	5.0	1.4	2.8	10.0
30	1.5	.....	6.0	1.3	2.7	13.5	3.8	0.8	3.5	1.3	3.5	8.0
31	1.5	.....	6.3	.....	1.9	.....	3.8	0.8	.....	1.2	.....	7.0

Gage Readings on Wabash River at Terre Haute Waterworks Station for 1903.

Day	Jan.	Feb.	Mar.	Apr.	May	June	July	Aug.	Sept.	Oct.	Nov.	Dec.
1	6.0	14.5	15.0	3.8	5.2	5.5	1.8	1.2	1.5	-.1	.4	.3
2	5.5	15.2	15.8	3.4	4.7	5.0	1.5	1.0	1.7	-.1	.5	.3
3	6.3	15.5	16.0	3.4	4.7	4.7	1.5	1.0	1.7	-.1	.5	.3
4	7.7	15.7	17.0	6.5	4.2	5.7	5.5	1.0	1.5	-.1	.6	.4
5	8.0	15.8	18.0	9.0	3.9	5.5	9.8	5.0	1.0	.1	.6	.4
6	8.5	15.8	18.7	10.8	3.8	5.8	9.7	4.3	.8	.2	.5	.4
7	9.0	15.8	18.2	12.5	3.3	7.5	7.8	3.0	.8	.1	.5	.4
8	8.5	15.9	17.5	12.7	3.1	8.8	6.2	2.0	.7	.4	.5	.5
9	8.0	15.7	17.2	12.0	2.8	9.3	5.0	1.7	.7	.5	.6	.5
10	6.5	15.0	16.6	10.3	2.8	9.0	4.0	1.6	.4	1.0	.6	.5
11	3.5	13.5	16.7	8.8	2.7	7.8	3.7	1.4	.3	1.5	.6	.5
12	3.0	14.0	16.7	11.7	2.6	6.5	3.7	1.3	.1	1.7	.6	.6
13	2.7	14.3	16.2	13.5	2.4	5.5	3.8	1.3	.0	1.7	.5	.5
14	2.5	15.9	15.4	15.7	2.2	4.0	3.3	1.1	.0	1.5	.5	.5
15	2.5	16.7	15.0	18.0	2.3	3.5	2.0	.8	.0	1.0	.4	.5
16	2.2	17.3	13.5	19.5	2.3	3.0	2.0	.8	-.5	.7	.4	.2
17	2.7	17.3	13.3	20.0	2.3	2.5	1.8	.7	-.2	.5	.4	1.0
18	4.3	16.9	12.4	19.8	2.1	2.3	1.8	.5	.0	.5	.2	1.3
19	4.6	15.2	10.4	19.2	1.9	2.0	1.7	.4	.0	.3	.1	1.6
20	4.7	11.0	9.4	18.2	1.8	2.0	1.6	.4	-.3	.2	.2	1.6
21	4.0	8.2	8.7	16.6	1.7	2.0	1.6	.3	-.3	.1	.0	2.7
22	4.2	7.1	8.2	14.0	5.5	2.2	1.8	.2	-.4	.0	.1	3.0
23	3.5	6.6	7.6	11.3	7.7	2.1	1.8	.2	-.4	.0	.0	3.6
24	3.4	6.8	7.2	9.0	4.5	2.3	1.8	0	-.3	.1	.0	3.5
25	3.3	7.0	6.3	7.9	4.0	2.3	1.7	.0	-.3	.1	.1	3.6
26	3.1	6.5	5.6	7.3	4.3	2.3	1.6	-.1	-.3	.2	.1	4.0
27	3.2	7.1	5.3	7.2	4.5	2.3	1.4	-.2	-.2	.2	.2	4.0
28	5.3	12.8	5.2	6.8	5.3	2.2	1.3	-.3	-.1	.3	.3	3.8
29	8.0	.....	4.8	6.3	5.6	2.0	1.3	2.0	.0	.3	.3	3.5
30	14.0	.....	4.4	5.8	5.7	1.8	1.3	1.8	.0	.3	.3	3.5
31	14.3	.....	4.2	.....	6.0	.....	1.3	1.5	.0	.4	.....	3.5

Gage Readings on Wabash River at Terre Haute Waterworks Station for 1904.

DAY	Jan.	Feb.	Mar.	Apr.	May	June	July	Aug.	Sept.	Oct.	Nov.	Dec.
1	3.3	9.0	12.8	23.1	14.3	3.9	2.0	.3	.6	1.8	—.3	—.5
2	2.3	7.5	14.7	23.7	12.7	4.0	2.3	.0	.5	1.5	—.3	—.5
3	2.5	6.5	15.8	23.1	10.3	4.0	2.0	.0	.0	.8	—.4	—.5
4	2.0	6.0	17.2	22.6	8.3	4.0	2.0	—.1	.0	.8	—.4	—.6
5	2.0	5.0	17.8	22.0	7.1	3.9	1.9	—.1	—.2	.7	—.3	—.6
6	1.7	4.7	17.5	21.3	6.1	3.8	2.3	—.2	—.3	.5	—.3	—.6
7	1.5	12.0	17.3	20.3	5.9	3.8	3.6	—.3	—.4	.4	—.3	—.6
8	1.5	15.6	17.0	19.2	5.7	3.5	9.7	—.3	—.5	.3	—.3	—.7
9	1.3	16.7	16.9	17.8	5.5	3.3	10.4	—.3	—.6	.3	—.3	—.7
10	1.3	17.1	16.7	16.4	5.5	3.0	9.8	—.3	.2	.2	—.3	—.7
11	1.2	17.3	16.6	15.3	5.4	3.0	8.8	—.3	.3	.2	—.3	—.7
12	1.0	17.4	16.3	14.8	5.3	2.5	8.0	—.4	.3	.3	—.3	—.7
13	1.0	16.6	15.3	14.2	4.8	2.2	7.8	—.4	.3	.4	—.3	—.8
14	.7	14.0	13.5	13.5	4.0	2.2	6.8	—.5	.3	.7	—.3	—.8
15	.7	10.4	12.0	12.0	4.0	2.0	5.7	—.5	.2	.6	—.3	—.9
16	.6	7.2	10.4	10.8	3.9	1.8	4.5	—.6	.2	.4	—.3	—.9
17	.7	6.3	9.0	9.5	3.9	1.9	4.0	—.6	.1	.3	—.3	—.9
18	.7	5.1	12.3	8.7	3.8	1.9	3.7	—.7	.5	.3	—.3	—.9
19	.6	4.0	13.5	7.8	3.7	1.9	3.0	—.6	.4	.2	—.3	—.9
20	1.0	3.3	13.8	7.1	3.6	2.0	3.0	—.4	.4	.2	—.3	—1.0
21	4.1	3.0	14.3	6.5	3.7	1.9	2.3	3.8	.3	—.1	—.3	—1.0
22	15.5	4.5	14.8	6.1	4.0	2.1	1.8	2.8	.3	—.2	—.3	—1.0
23	17.3	4.8	16.3	6.0	4.0	2.2	1.7	2.2	.3	—.2	—.4	—1.0
24	19.0	6.5	17.3	5.7	4.4	3.0	1.5	1.8	.3	—.3	—.4	—.7
25	20.2	6.8	17.8	5.8	4.3	3.7	1.3	1.8	.3	—.3	—.4	.0
26	20.6	6.8	19.9	9.6	4.1	3.7	1.0	1.5	.6	—.3	—.4	.2
27	20.5	6.9	25.1	13.0	4.1	3.0	1.0	1.3	1.5	—.3	—.4	.5
28	20.0	6.9	24.9	13.9	4.1	2.6	.7	1.0	1.7	—.3	—.5	.2
29	18.3	9.5	24.0	14.3	4.0	2.2	.7	1.0	2.3	—.3	—.5	.2
30	16.0	......	23.3	14.5	4.0	1.9	.7	.7	2.4	—.3	—.5	.3
31	12.5	......	22.8	......	4.0	......	.5	.6	......	—.3	......	.3

Gage Readings on Wabash River at Terre Haute Waterworks Station for 1905.

DAY	Jan.	Feb.	Mar.	Apr.	May	June	July	Aug.	Sept.	Oct.	Nov.	Dec.
1	.5	.7	13.5	3.8	8.2	5.6	1.3	.9	.7	.7	.8	8.5
2	.7	.3	14.0	5.8	7.8	5.6	1.3	.8	.7	.5	.8	9.2
3	.7	.3	14.5	6.0	6.7	5.8	1.4	.8	1.0	.5	.7	8.2
4	.0	.2	14.6	5.2	5.5	5.0	3.5	.8	.8	.4	.6	6.5
5	.0	.2	15.2	4.8	4.8	4.2	1.5	.7	.8	.3	.8	5.0
6	.0	.2	14.0	4.2	4.4	5.2	1.3	.6	.9	.3	2.0	5.0
7	.2	.2	11.2	3.2	4.2	5.4	1.2	.6	1.0	.3	2.5	4.7
8	.3	.3	9.1	2.7	3.8	4.9	1.0	.4	.8	.3	3.0	4.0
9	.4	.3	8.2	2.2	4.0	4.8	1.2	.4	.7	.3	3.0	3.8
10	.5	.3	7.5	1.8	4.8	4.0	1.5	1.0	.7	.3	2.9	3.3
11	.5	.5	6.7	1.8	4.8	5.7	1.9	.8	1.0	.0	2.8	3.0
12	1.0	.5	6.3	1.7	4.6	5.6	4.0	.8	2.0	.0	2.5	2.8
13	1.1	.7	6.0	1.9	10.8	4.1	3.3	.6	2.5	.0	2.3	2.8
14	2.0	.7	5.5	1.7	13.6	4.4	5.4	.5	2.0	.0	1.9	2.6
15	1.8	.7	4.8	1.5	14.7	4.3	4.3	1.2	1.6	.0	1.8	2.3
16	1.7	.7	4.0	1.3	15.9	4.1	3.0	2.0	2.3	.0	1.7	2.0
17	1.6	.8	3.5	1.2	16.9	3.8	2.8	1.8	2.8	—.2	1.6	1.8
18	1.6	.8	3.0	1.0	17.0	3.3	2.3	1.7	3.7	—.2	1.4	1.7
19	1.8	.8	3.0	.6	16.2	3.4	2.0	1.5	4.2	2.7	1.3	1.7
20	1.8	.8	3.0	.8	13.6	4.3	1.7	1.2	4.2	3.0	1.0	1.6
21	1.3	.8	5.0	1.1	10.0	7.0	3.7	1.5	4.0	4.0	.8	1.6
22	2.0	.8	4.8	3.8	7.8	4.3	3.0	1.5	3.5	4.0	.8	2.0
23	2.0	.5	4.8	6.9	6.1	3.2	2.0	1.8	2.8	3.5	.8	3.5
24	1.7	3.7	4.7	10.2	5.1	2.8	1.6	1.5	2.3	3.0	.8	6.0
25	2.0	5.0	4.4	10.3	4.3	2.8	1.3	1.2	2.0	2.5	.7	6.8
26	2.0	7.0	4.2	9.3	3.8	2.4	1.1	.8	1.7	2.2	.7	6.0
27	1.7	11.0	4.2	8.3	3.5	2.0	.8	.8	1.2	2.0	.7	5.3
28	.9	11.5	4.3	7.2	3.3	1.8	.7	1.3	1.0	1.7	.7	4.7
29	.9	......	3.8	7.3	3.1	1.6	.5	1.5	.7	1.5	.6	4.0
30	.8	......	3.3	7.7	3.9	1.5	.4	1.3	.7	1.2	6.7	3.8
31	.7	......	3.3	......	7.7	......	.3	.8	......	1.0	......	3.8

Gage Readings on Wabash River at Terre Haute Waterworks Station for 1906.

Day	Jan.	Feb.	Mar.	Apr.	May	June	July	Aug.	Sept.	Oct.	Nov.	Dec.
1	3.7	7.5	4.2	19.0	3.0	5.5	.3	—.3	.8	.0	—.5	3.6
2	3.9	6.7	4.0	18.6	3.0	3.7	.2	—.3	.5	.0	—.5	3.1
3	4.0	5.7	5.8	17.8	3.0	2.8	.4	—.3	.3	.7	—.3	2.8
4	5.0	5.0	8.8	17.1	3.0	2.0	.3	—.3	.2	.7	—.3	2.6
5	5.5	4.7	10.0	16.2	3.0	2.0	.3	—.3	.....	.7	—.3	2.4
6	6.3	3.8	9.9	15.0	2.8	1.8	.7	—.3	—.2	.8	—.3	4.1
7	7.0	2.7	8.8	12.5	2.8	2.5	.8	—.3	—.3	.6	—.3	9.3
8	6.8	1.5	7.8	10.8	2.8	1.8	.8	—.2	—.3	.3	—.3	10.3
9	5.7	1.5	7.5	13.0	2.8	1.9	.7	.3	—.3	.1	—.2	11.3
10	4.5	1.3	7.3	14.0	2.3	5.0	.6	.3	—.3	—.1	—.2	11.0
11	3.7	1.2	7.0	14.0	2.0	4.0	.6	.3	—.3	—.1	—.2	9.8
12	3.7	1.2	6.5	14.2	1.8	3.5	.5	.7	—.4	—.3	—.2	8.7
13	3.5	1.3	6.0	14.5	1.8	2.8	.5	.5	—.3	—.3	—.2	7.8
14	2.7	2.0	5.5	14.4	1.7	2.0	.5	.4	—.3	—.3	—.2	6.3
15	2.7	3.7	5.3	13.0	1.7	1.8	.4	.2	—.4	—.5	—.2	8.0
16	3.0	3.2	4.8	11.5	1.5	1.5	.4	.0	—.4	—.5	—.2	10.2
17	3.0	2.7	4.5	11.8	1.3	1.3	.4	—.2	—.4	—.5	—.2	10.6
18	2.8	1.8	4.0	11.8	1.3	1.2	.6	—.2	—.5	—.5	—.2	10.8
19	3.0	2.3	3.8	10.2	1.3	1.1	.8	.2	—.5	—.4	.0	10.0
20	3.7	3.0	3.8	8.7	1.2	.8	.7	.0	—.5	—.4	.0	8.7
21	4.0	3.7	3.6	7.5	1.0	.8	.7	.5	—.5	—.4	1.5	7.0
22	11.5	3.0	3.4	6.6	1.0	.8	1.5	.4	—.5	—.4	.5	5.8
23	14.9	3.0	3.7	5.7	1.0	.8	.9	.4	—.3	—.5	8.8	5.0
24	16.0	3.0	3.7	5.0	1.0	.7	.7	.6	—.2	—.5	10.0	4.0
25	16.3	3.0	3.7	4.7	1.0	.6	.2	1.0	—.3	—.5	9.3	3.0
26	16.7	4.0	4.3	4.3	1.0	.5	.0	1.0	—.3	—.5	8.3	3.0
27	16.7	4.3	12.5	3.8	1.2	.5	.0	1.8	—.3	—.5	7.0	2.9
28	15.0	4.5	15.0	3.7	1.3	.4	—.1	1.8	—.4	—.5	5.5	3.0
29	13.8	.....	15.7	3.5	1.3	.3	—.2	1.6	—.4	—.5	4.6	3.0
30	11.0	.....	16.0	3.3	1.0	.3	—.2	1.0	—.2	—.5	3.8	3.0
31	8.8	.....	17.4	.....	2.0	.....	—.2	.8	.....	—.5	.....	5.5

Gage Readings on Wabash River at Terre Haute Waterworks Station for 1907.

Day	Jan.	Feb.	Mar.	Apr.	May	June	July	Aug.	Sept.	Oct.	Nov.	Dec.
1	7.9	7.3	2.8	12.8	3.4	7.5	4.5	2.4	1.2	.8	.3	1.3
2	9.8	6.5	2.8	12.8	4.0	11.3	3.7	2.3	2.0	.8	.7	1.3
3	12.2	6.5	3.0	11.6	5.0	13.0	3.0	2.0	2.9	1.0	1.1	.9
4	15.0	5.5	3.2	9.3	5.0	13.3	2.8	2.0	2.8	2.9	1.1	.8
5	16.0	4.8	3.7	7.7	4.9	13.8	2.4	2.0	2.3	3.6	2.2	.6
6	16.3	3.7	3.8	6.7	4.3	14.0	2.3	2.2	1.9	3.5	2.8	.6
7	16.2	3.5	3.7	5.8	3.0	14.0	2.0	2.3	1.3	2.9	2.8	.5
8	16.2	3.5	3.5	5.7	3.7	13.7	2.0	4.0	1.3	2.8	2.5	.4
9	17.3	3.5	3.3	5.5	3.9	12.7	2.2	3.9	1.3	2.7	2.0	.5
10	17.5	3.5	3.2	5.2	3.8	11.0	2.2	3.0	1.7	2.1	1.9	1.1
11	17.4	3.8	3.2	4.9	3.8	9.5	2.9	2.8	1.5	1.9	1.7	1.2
12	17.2	3.8	4.0	4.8	3.5	8.9	4.2	2.3	1.2	1.8	1.3	2.3
13	16.3	3.8	10.5	4.5	3.4	7.9	5.4	2.1	1.0	1.3	1.0	3.5
14	16.0	3.8	14.3	4.3	3.2	6.9	6.9	2.0	.9	1.2	.9	3.9
15	15.0	4.0	15.2	4.0	3.0	6.0	6.9	1.8	.7	1.0	.8	3.8
16	14.6	4.0	16.3	4.0	3.0	6.0	6.7	1.7	.7	.8	.8	3.3
17	15.0	4.0	16.7	3.9	2.9	5.8	6.0	1.7	.7	.8	.8	2.8
18	15.7	4.0	17.1	3.9	2.8	5.3	10.5	2.7	.5	.8	.8	2.8
19	16.4	4.4	17.3	3.8	2.8	4.8	12.2	2.8	.4	.7	.7	2.8
20	18.7	4.7	16.8	3.7	2.8	4.8	12.2	2.3	.3	.7	.7	2.7
21	21.2	4.7	15.7	3.3	2.8	4.5	10.0	2.0	.3	.7	1.0	2.3
22	24.0	4.5	13.8	3.1	2.1	4.3	7.9	1.9	1.8	.7	1.2	2.0
23	24.6	4.0	11.7	3.0	2.0	4.8	6.4	1.9	2.0	.6	1.3	2.3
24	23.3	3.5	9.6	3.0	1.9	4.9	5.6	1.7	1.8	.6	1.7	4.5
25	21.2	3.2	8.1	3.0	4.0	5.4	4.9	1.3	1.5	.6	2.0	5.3
26	20.0	3.0	7.2	2.9	4.8	6.3	4.0	1.2	1.2	.6	2.0	8.5
27	19.8	3.0	6.3	2.9	6.7	7.0	3.6	1.0	.8	.6	1.8	10.0
28	15.0	3.0	6.1	2.9	7.3	6.7	3.3	.9	.7	.6	1.7	11.5
29	11.5	.....	7.0	2.9	7.0	6.0	3.3	.8	.7	.7	1.5	13.8
30	8.6	.....	9.9	3.0	6.7	5.2	2.9	.8	.6	.4	1.5	14.9
31	7.3	.....	11.8	.....	5.8	.....	2.6	.8	.....	.3	.....	15.5

Gage Readings on Wabash River at Terre Haute Waterworks Station for 1908.

DAY	Jan.	Feb.	Mar.	April	May	June	July	Aug.	Sept.	Oct.	Nov.	Dec.
1	16.5	2.7	12.8	9.4	7.0	5.8	1.7	1.0	.9	—.1	—.3	.3
2	17.3	2.0	15.0	11.2	6.6	5.5	1.6	.8	.0	—.2	—.3	.3
3	17.3	1.7	16.4	10.5	5.8	5.5	1.5	.8	.0	—.2	—.3	.3
4	16.8	1.4	17.3	8.9	7.0	5.0	1.6	.2	—.1	—.2	—.3	.3
5	15.8	2.0	17.4	7.8	11.2	4.7	1.5	.2	—.2	—.2	—.3	.3
6	14.1	7.5	18.6	7.8	18.5	3.5	1.5	.2	—.2	—.2	—.3	.3
7	11.6	8.8	19.0	7.0	15.5	3.5	1.7	.2	—.2	—.2	—.3	.3
8	9.3	8.3	19.3	7.5	18.7	2.8	2.0	.3	—.2	—.2	—.3	.3
9	7.8	8.3	20.9	12.0	19.5	2.8	1.9	.3	—.2	—.2	—.3	.2
10	6.8	7.0	22.8	13.0	20.0	2.8	1.7	.4	—.2	—.2	—.3	.2
11	6.0	7.0	22.8	13.9	19.2	2.7	1.3	.3	—.2	—.2	—.3	.2
12	6.3	8.0	21.8	14.5	18.3	2.5	1.3	.3	—.2	—.2	—.3	.2
13	9.2	12.3	20.5	14.6	18.1	2.8	1.1	.3	—.2	—.3	—.3	.2
14	10.4	12.8	19.3	14.6	18.2	2.7	1.0	.3	—.2	—.3	—.3	.2
15	10.7	14.5	18.2	13.8	18.0	2.5	1.0	1.7	—.2	—.3	—.3	.2
16	10.7	15.9	16.8	11.8	17.9	2.5	1.0	2.8	—.2	—.3	—.3	.2
17	9.7	17.0	15.2	9.5	17.8	2.3	1.0	2.0	—.2	—.3	—.3	.2
18	8.1	17.7	13.5	8.4	17.6	2.3	1.0	1.5	—.2	—.3	—.3	.2
19	6.8	18.3	11.9	7.8	17.0	2.0	.9	1.0	—.2	—.3	—.3	.2
20	6.0	18.8	13.7	17.3	16.8	2.3	.8	.8	—.2	—.3	—.3	.2
21	5.3	18.0	14.5	6.8	16.3	2.5	1.5	.7	—.2	—.3	—.3	.2
22	5.2	16.8	15.1	6.0	15.8	2.0	1.0	.5	—.2	—.3	—.3	.2
23	5.1	14.5	15.8	5.7	15.0	1.9	1.0	.4	—.2	—.3	—.3	.2
24	4.8	11.3	16.5	6.0	13.0	1.9	1.0	.4	—.2	—.3	—.3	.2
25	4.7	9.7	16.7	7.9	10.5	2.0	1.3	.3	—.2	—.3	—.3	.2
26	4.2	13.6	15.9	9.1	8.8	2.2	2.3	.6	—.2	—.2	.3	.2
27	4.5	14.3	14.0	8.8	7.9	2.0	2.0	.6	—.2	—.3	.3	.2
28	4.3	14.0	11.0	7.9	7.5	1.8	1.8	.7	—.1	—.3	.3	.2
29	3.9	13.5	9.7	7.7	7.0	1.8	1.8	.7	—.1	—.3	.3	.2
30	3.5	......	9.0	7.0	6.4	1.7	1.5	.8	—.2	—.3	.3	.2
31	2.5	......	9.4	.....	6.0	......	1.3	.8	......	—.3	......	.2

Gage Readings on Wabash River at Terre Haute Waterworks Station for 1909.

DAY	Jan.	Feb.	Mar.	Apr.	May	June	July	Aug.	Sept.	Oct.	Nov.	Dec.
1	.2	1.1	17.7	5.3	11.0	7.9	6.0	2.0	1.0	2.0	1.8	4.6
2	.2	1.1	17.0	4.7	13.0	6.7	5.0	2.0	1.0	1.0	1.8	4.1
3	.2	.9	16.4	4.1	13.5	5.9	4.2	1.9	.9	.8	1.5	3.9
4	.2	.9	14.2	3.8	14.2	7.8	3.6	1.8	.8	.8	1.4	3.5
5	.2	.9	10.8	3.4	15.0	10.2	3.0	2.2	.8	.7	1.3	3.1
6	.2	2.3	8.3	3.3	14.9	11.2	3.0	2.0	.7	.7	1.1	3.5
7	.0	2.8	6.8	10.2	13.7	11.3	6.5	1.8	.7	.7	1.1	3.7
8	.0	3.0	5.0	14.0	10.8	10.0	5.9	1.5	.7	.7	1.1	3.5
9	.0	3.7	7.2	14.4	10.0	8.3	5.0	1.3	.9	.5	1.0	ice
10	.0	4.0	9.8	14.8	11.0	9.2	4.7	.9	.7	.5	1.0	ice
11	.0	4.0	8.8	15.0	10.8	11.2	3.8	.9	.5	.3	1.0	1.0
12	.0	4.2	9.0	15.1	10.8	12.4	4.1	.9	.5	.2	1.0	2.0
13	.0	4.7	10.3	14.4	11.2	13.8	8.2	2.2	.9	.2	1.0	5.0
14	.0	4.3	10.2	13.8	11.2	15.0	11.0	4.0	.9	.2	1.3	8.9
15	0	6.4	8.8	14.8	10.3	15.0	10.7	3.9	.7	.0	1.3	10.8
16	.0	7.7	7.2	13.6	9.0	13.7	8.8	3.5	.8	.0	1.3	12.3
17	.0	7.0	6.0	12.3	7.8	11.0	7.1	3.0	.6	.0	1.5	13.0
18	.0	6.0	5.0	9.8	7.3	10.9	5.8	2.5	.6	.3	2.3	12.8
19	.0	6.5	4.7	8.2	6.7	8.3	6.3	2.3	.6	.2	3.4	11.0
20	.0	11.2	4.1	7.2	5.8	6.7	7.0	2.0	.5	.2	3.5	ice
21	.0	13.0	4.0	6.7	4.8	5.7	6.3	1.8	.5	.3	3.5	ice
22	.1	13.7	4.0	8.0	4.3	4.8	5.0	1.7	.5	.2	3.3	ion
23	.8	14.3	4.0	9.6	3.8	4.2	4.3	1.5	.7	.7	5.4	2.0
24	.9	15.1	3.9	9.1	3.5	4.0	3.5	1.2	2.2	1.0	7.2	2.0
25	.8	16.3	3.9	7.9	3.3	2.8	2.9	.8	4.3	1.3	9.8	2.0
26	.8	16.9	3.9	6.8	3.3	11.0	2.5	1.2	4.7	2.0	11.0	1.5
27	.9	17.5	4.9	5.8	4.7	12.4	2.3	2.0	4.1	3.0	10.8	2.0
28	.9	17.7	6.6	5.8	8.5	11.0	2.0	1.5	3.3	3.3	9.0	2.0
29	1.1	....	6.9	4.8	10.0	10.0	1.9	1.2	2.7	2.9	7.2	*
30	1.1	....	6.9	4.8	9.3	7.8	1.8	.9	2.0	2.3	6.5	2.0
31	1.1	....	6.0	.....	8.8	....	1.7	.8	....	2.0	.....	2.3

*No record.

Gage Readings on Wabash River at Terre Haute Waterworks Station for 1910.

DAY	Jan.	Feb.	Mar.	Apr.	May	June	July	Aug.	Sept.	Oct.	Nov.	Dec.
1	2.3	7.0	14.8	1.9	4.0	2.7	.7	—.2	—.2	.8	.3	1.0
2	2.3	6.0	15.5	1.8	3.7	2.5	.5	—.3	—.3	.8	.3	1.0
3	2.3	5.5	16.0	1.8	4.5	2.3	.3	—.3	—.3	.7	.2	1.0
4	2.5	5.3	16.3	1.8	5.2	2.0	1.0	—.3	—.3	.8	.2	1.0
5	2.5	6.0	16.5	1.8	4.7	1.8	.9	—.3	—.3	1.9	.2	1.5
6	2.7	7.2	16.3	1.8	4.6	1.8	.8	—.3	.2	4.8	.2	.8
7	2.7	6.5	15.5	1.8	4.3	1.8	.5	—.3	4.3	3.3	.2	.8
8	2.7	5.5	14.5	1.8	4.3	1.7	.5	—.3	2.8	3.0	.2	.7
9	2.7	4.8	12.5	1.8	4.1	1.5	.4	—.3	4.7	2.8	.2	.3
10	4.9	4.3	10.7	1.8	3.8	1.3	.4	—.3	4.7	4.2	.2	.3
11	4.3	4.0	8.8	1.8	3.3	1.3	.3	—.3	3.0	6.3	.2	.3
12	3.8	3.8	7.5	1.7	3.7	1.0	.5	—.3	2.0	6.0	.2	.3
13	6.0	3.3	6.5	1.7	3.8	.9	1.0	—.3	1.6	4.8	.2	.3
14	12.0	2.8	5.8	1.6	3.4	.8	.6	—.4	1.3	3.5	.0	.3
15	12.5	2.7	5.3	1.5	2.8	.8	.4	—.4	1.0	2.8	.0	.3
16	13.2	2.7	4.9	1.7	2.5	.7	2.0	—.4	1.0	2.0	.0	.3
17	13.5	3.0	4.5	1.8	2.3	.6	5.7	—.4	1.7	1.8	.0	.3
18	14.7	3.0	4.0	1.8	2.3	.5	4.5	—.3	1.5	1.2	.0	.3
19	17.5	2.7	3.8	1.9	2.0	.4	2.8	—.3	1.2	1.0	.0	.3
20	18.5	2.7	3.7	2.0	1.8	.4	2.0	—.3	1.0	.9	.0	.3
21	17.5	5.0	3.4	2.5	1.5	.3	1.5	—.3	1.2	.8	.0	.0
22	19.0	4.3	3.1	3.7	3.0	.3	1.0	—.3	1.1	.8	.0	.0
23	19.3	4.9	3.0	4.8	5.8	.3	.7	—.3	1.0	.7	0	.0
24	18.8	5.0	2.9	4.6	4.5	.3	.5	—.3	.9	.7	.0	0
25	17.8	4.0	2.8	4.2	5.7	.2	.4	.2	.8	.6	.0	.0
26	16.5	4.3	2.8	4.2	5.6	.1	.3	.2	.8	.5	.0	.0
27	14.5	8.0	2.6	4.0	5.0	.1	.2	.1	.8	.5	.0	.0
28	11.0	14.5	2.5	3.8	4.0	.5	.2	.1	.8	.4	1.7	.0
29	9.7	......	2.3	3.8	3.3	1.0	.0	.0	.8	.4	1.8	4.2
30	9.7	......	2.2	4.0	3.0	.8	.0	—.1	.8	.3	1.2	8.2
31	7.8	......	2.0	.....	2.8	......	.0	—.1	.....	.3	......	6.0

*Gage Readings on Wabash River at Terre Haute Waterworks Station for 1911.**

DAY	Jan.	Feb.	Mar.	Apr.	May	June	July	Aug.	Sept.	Oct.	Nov.	Dec.
1	6.0	13.7	4.0	2.6	4.3	0.7	1.2	—.3	—.1	2.6	1.6	5.3
2	8.8	13.7	3.9	2.2	4.3	0.7	1.2	—.3	—.1	2.8	1.6	5.7
3	9.0	12.3	3.4	4.0	4.9	0.9	1.2	—.3	—.1	3.0	1.5	5.5
4	7.4	9.0	3.4	4.2	5.2	1.0	0.9	—.3	—.1	3.0	1.4	4.4
5	6.8	7.8	2.9	6.8	5.0	1.0	0.8	—.3	—.1	3.5	1.4	4.2
6	4.6	6.8	2.9	7.4	4.3	1.2	0.6	—.3	.0	......	1.4	3.7
7	4.0	6.6	2.9	8.8	3.9	1.0	0.5	—.3	.0	3.4	1.3	3.3
8	3.0	5.9	5.5	10.3	3.0	1.0	0.5	—.4	—.3	3.4	1.3	2.3
9	7.0	5.1	4.0	9.9	2.9	1.0	0.4	—.4	—.2	2.7	1.3	2.3
10	7.0	4.7	4.0	8.0	2.8	0.9	0.4	—.5	—.2	2.5	1.4	3.1
11	6.6	4.0	4.0	6.6	2.6	0.9	0.4	—.5	0.4	......	1.4	4.0
12	6.4	3.4	4.1	5.6	2.4	0.9	0.3	—.5	0.6	......	1.9	5.7
13	3.4	3.4	3.9	5.0	2.0	0.8	0.3	....	0.6	2.0	3.0	7.3
14	5.0	3.6	3.9	10.0	1.9	0.6	0.2	—.6	0.4	2.0	4.3	8.8
15	10.6	4.0	3.9	12.5	1.8	0.4	......	—.6	0.1	1.8	5.4	9.9
16	11.6	5.4	3.9	11.5	1.7	0.2	......	—.6	0.4	1.7	5.9	9.5
17	12.0	9.6	3.9	10.8	1.5	0.2	0.1	—.1	1.1	1.7	5.2	8.7
18	11.4	10.0	3.9	10.0	1.3	0.2	—.2	—.3	2.8	1.9	8.3	7.7
19	8.9	10.0	3.2	8.6	1.2	0.2	—.1	—.4	2.6	2.5	8.8	7.6
20	6.6	9.8	3.0	9.6	1.4	0.2	.0	—.4	2.7	2.6	11.0	7.6
21	5.4	9.9	2.9	8.8	1.2	0.4	0.2	—.4	2.7	2.8	11.9	7.4
22	5.4	9.2	2.6	9.0	1.0	0.6	.0	—.4	2.4	2.7	11.6	6.8
23	5.0	8.0	2.2	10.6	1.4	0.6	—.2	—.4	1.9	3.2	10.0	6.2
24	4.3	6.6	2.0	9.3	1.4	0.5	—.2	—.4	1.4	4.3	8.2	5.8
25	3.9	5.9	1.8	8.0	1.4	0.5	—.3	—.4	1.3	4.3	6.7	5.8
26	3.9	5.0	1.8	7.8	1.2	0.5	—.3	—.1	1.8	3.7	5.8	5.8
27	4.6	4.9	1.6	5.8	1.2	0.5	—.3	—.3	2.3	3.0	5.2	5.4
28	5.6	4.4	2.0	5.0	1.0	0.7	—.3	—.3	3.9	2.6	5.3	4.9
29	12.0	......	2.0	4.8	0.9	2.0	—.3	—.3	4.6	2.4	5.4	4.6
30	13.0	......	2.4	4.4	0.9	2.2	—.3	—.1	3.5	1.9	5.1	4.1
31	13.6	......	2.6	......	0.8	......	—.3	—.3	......	1.9	......	4.2

*U. S. Department of Agriculture, Weather Bureau. Daily River Stages at River Gage Stations on the Principal Rivers of the United States. Part XI. for the years 1911-1912. W. B. No. 507, page 291.

*Gage Readings on Wabash River at Terre Haute Waterworks Station for 1912.**

DAY	Jan.	Feb.	Mar.	Apr.	May	June	July	Aug.	Sept.	Oct.	Nov.	Dec.
1	4.3	5.0	14.5	18.1	13.0	3.1	1.4	1.9	1.9	0.6	0.7	0.8
2	4.1	4.4	14.8	18.8	13.6	3.1	1.6	1.8	1.8	0.6	2.3	0.9
3	5.3	3.6	15.0	19.7	13.8	2.9	1.5	1.5	1.5	0.5	2.0	0.9
4	5.6	3.6	15.0	19.7	13.6	2.8	4.5	1.2	1.5	0.4	2.5	0.9
5	4.6	3.6	14.0	18.8	13.0	2.8	3.4	1.2	1.4	0.4	2.6	0.9
6	†	3.3	11.7	17.8	11.5	2.5	2.6	1.1	1.3	0.4	2.3	1.0
7	†	3.0	8.6	16.9	10.6	2.2	2.2	0.9	1.1	0.3	2.3	0.9
8	†	2.9	6.6	15.9	8.8	2.0	1.8	1.0	1.1	0.3	2.8	0.9
9	†	2.9	5.4	14.8	7.4	1.9	2.8	0.9	1.0	0.3	3.0	0.9
10	†	2.9	4.8	12.9	6.4	1.8	2.4	1.0	0.9	0.3	3.7	0.9
11	†	2.6	4.5	10.8	5.8	1.8	2.9	1.1	0.9	0.4	3.7	0.8
12	†	2.0	4.3	9.0	6.6	1.7	2.8	1.4	0.8	1.0	3.3	0.8
13	†	2.0	5.3	8.0	7.0	1.6	3.5	1.8	0.9	0.7	3.0	.0
14	†	1.9	7.0	8.6	8.6	1.6	9.5	4.3	0.9	0.6	2.7	.0
15	†	1.9	11.4	9.7	10.6	1.7	13.8	3.9	0.7	0.5	2.5	0.2
16	†	1.6	14.4	9.7	11.7	2.0	10.7	4.1	0.7	0.5	2.3	0.6
17	3.2	1.6	14.2	8.8	11.1	2.2	6.5	3.7	0.6	0.5	2.2	0.8
18	3.2	1.9	14.6	7.9	10.1	3.3	6.0	3.0	0.6	0.5	2.3	0.9
19	8.7	3.0	15.5	8.0	9.1	5.6	5.8	2.8	0.8	0.6	1.9	0.9
20	10.9	5.4	16.8	8.5	9.1	5.5	5.8	2.5	0.8	0.6	1.8	0.8
21	11.3	5.4	18.8	8.4	9.5	5.4	5.6	7.9	0.7	0.5	1.6	0.7
22	11.3	6.6	19.2	8.4	8.1	4.9	6.2	8.2	0.6	0.5	1.4	0.7
23	11.3	5.8	19.0	7.7	7.4	4.1	6.0	8.1	0.6	0.5	1.3	0.4
24	12.0	5.6	19.0	7.2	5.4	3.5	4.5	7.8	0.6	0.5	1.3	0.2
25	12.2	5.0	18.3	6.9	4.8	2.8	3.9	5.9	0.6	0.5	1.0	0.2
26	12.2	9.4	17.4	6.4	3.7	2.3	4.8	4.6	0.6	0.4	1.0	0.4
27	11.3	13.6	16.3	6.3	3.7	2.0	4.5	3.8	0.6	0.5	0.9	0.5
28	9.0	14.4	15.9	5.8	3.2	1.9	3.8	3.3	0.6	0.4	1.0	0.2
29	7.6	14.6	15.3	9.6	3.9	1.7	3.1	3.0	0.5	0.4	0.8	0.2
30	6.6	......	16.6	11.6	3.4	1.5	2.6	2.5	0.6	0.4	0.7	0.4
31	5.6	......	17.4	......	3.2	......	2.2	2.1	......	0.3	......	0.4

† Frozen.

*U. S. Department of Agriculture, Weather Bureau. Daily River Stages at River Gage Stations on the Principal Rivers of the United States. *Part XI.* for the years 1911-1912. W. B. No. 507, page 291.

*Gage Readings on Wabash River at Terre Haute Waterworks Station for 1913.**

DAY	Jan.	Feb.	Mar.	Apr.	May	June	July	Aug.	Sept.	Oct.	Nov.	Dec.
1	0.4	14.8	4.6	22.0	4.9	5.6	1.6	1.0	0.8	0.5	1.2	2.9
2	0.3	12.1	4.5	20.7	4.7	4.6	1.6	1.0	0.8	0.5	1.0	4.9
3	0.3	10.8	4.3	19.4	4.4	4.2	1.8	1.1	0.7	0.5	0.9	4.9
4	0.1	†	5.1	18.6	4.4	3.8	1.7	1.0	0.7	0.5	0.9	5.4
5	0.0	†	5.1	17.6	4.2	3.6	1.4	0.9	0.6	0.5	0.9	5.2
6	0.3	†	5.8	16.9	4.0	3.2	1.5	0.9	0.6	0.5	0.8	4.6
7	1.3	†	6.4	16.0	3.9	3.0	1.3	0.9	0.6	0.5	0.7	4.1
8	5.1	†	6.0	15.5	3.8	2.9	1.2	0.8	0.5	0.5	0.8	3.7
9	7.2	†	6.0	14.7	3.6	2.7	1.1	0.9	0.5	0.5	0.8	3.2
10	9.0	†	5.6	14.9	3.5	2.6	1.0	0.8	0.5	0.5	0.6	2.9
11	10.0	3.9	5.8	15.4	3.2	2.5	1.0	0.7	0.4	0.5	0.7	2.7
12	12.7	4.0	7.3	15.8	3.1	2.4	1.0	0.7	0.4	0.5	0.8	2.5
13	13.2	3.8	8.3	15.9	3.0	2.4	1.4	1.5	0.4	0.5	0.8	2.3
14	13.3	3.5	9.1	16.0	3.0	2.3	1.2	2.0	0.4	0.6	0.9	2.2
15	12.8	......	9.7	16.2	3.0	2.3	1.2	2.6	0.3	0.6	1.0	2.1
16	12.2	3.9	10.0	16.4	5.0	2.1	1.2	2.3	0.5	0.6	1.5	2.0
17	12.8	4.0	10.4	15.8	5.5	2.4	1.2	1.9	0.7	0.6	1.7	1.9
18	14.3	3.9	9.9	14.6	4.9	2.1	2.4	1.9	0.7	0.6	1.8	1.8
19	14.9	3.8	8.1	12.2	4.8	2.0	4.9	1.6	0.7	0.6	1.8	1.8
20	15.1	4.0	6.8	9.7	4.4	1.9	6.0	1.5	0.7	0.6	1.9	1.8
21	17.0	4.8	6.0	8.4	4.0	1.8	5.2	1.4	0.6	0.6	1.9	1.7
22	19.0	4.5	7.1	7.5	3.8	1.7	4.0	1.6	0.6	0.6	1.8	2.0
23	19.5	4.6	7.0	6.7	3.7	2.0	3.2	2.0	0.6	0.7	2.9	1.8
24	20.5	4.7	14.5	6.2	4.3	2.0	2.7	1.8	0.6	0.9	1.7	1.7
25	21.2	4.3	19.5	5.9	4.1	1.9	2.4	1.6	0.6	0.9	1.6	1.7
26	21.0	4.0	27.0	5.9	3.7	2.2	2.0	1.5	0.6	0.9	1.6	1.7
27	20.4	4.2	31.2	5.7	3.5	2.3	1.7	1.5	0.6	1.3	1.2	1.6
28	19.6	4.7	30.8	5.5	3.3	2.1	1.4	1.6	0.5	1.5	1.5	1.6
29	18.8	......	29.2	5.3	4.2	2.0	1.2	1.5	0.5	1.5	2.2	1.6
30	17.8	......	26.8	5.1	6.4	1.7	1.1	1.3	0.5	1.3	2.3	1.6
31	16.5	......	24.0	......	6.1	......	1.0	1.0	......	1.2	......	1.5

† Frozen.

*U. S. Department of Agriculture, Weather Bureau. Daily River Stages at River Gage Stations on the Principal Rivers of the United States. *Part XII.* for the years 1913-1914. W. B. No. 549, page 309.

*Daily Gage Heights, in Feet, of Wabash River at Terre Haute, Indiana, for 1914.**

Day	Jan.	Feb.	Mar.	Apr.	May	June	July	Aug.	Sept.	Oct.	Nov.	Dec.
1	1.5	4.0	†	15.4	4.0	3.0	0.9	0.2	0.3	—.2	0.0	0.0
2	1.4	3.7	†	15.6	3.9	2.9	1.0	0.2	0.3	—.2	—.1	0.0
3	1.4	3.6	†	15.6	3.8	2.7	1.0	0.1	0.3	—.2	—.1	0.0
4	1.4	4.7	†	15.6	3.6	2.6	1.0	0.1	0.3	—.2	—.1	0.0
5	1.4	5.3	†	15.5	3.5	2.4	0.9	0.1	0.8	—.2	—.2	0.0
6	1.4	6.4	†	14.9	3.4	2.4	0.8	0.0	0.8	—.2	—.2	0.0
7	1.4	7.8	4.7	13.9	3.4	2.3	0.8	0.0	0.9	—.2	—.2	0.1
8	1.4	7.5	5.8	15.2	3.7	2.4	0.8	0.0	1.1	—.2	—.1	0.1
9	1.4	‡5.5	6.8	16.1	5.9	2.2	0.8	0.0	0.9	—.2	—.1	0.2
10	1.4	‡2.5	7.8	16.7	6.1	2.2	0.7	0.0	0.7	0.0	—.1	0.2
11	1.4	‡2.5	9.0	16.9	7.6	2.0	0.7	—.1	0.6	0.1	—.1	0.3
12	1.4	‡2.5	9.7	17.1	8.7	1.9	0.6	0.0	0.5	0.2	—.1	0.4
13	1.4	†	9.4	16.9	9.1	1.7	0.6	0.0	0.4	0.2	—.1	0.4
14	1.4	†	11.0	16.6	11.9	1.7	0.5	0.0	0.2	0.3	—.1	0.4
15	1.0	†	14.2	15.2	13.1	1.5	0.5	0.0	0.2	0.7	—.1	0.4
16	1.4	†	14.7	11.9	13.8	1.4	0.5	0.0	0.1	0.9	—.1	0.2
17	1.7	†	15.2	9.0	14.0	1.4	0.5	—.1	0.1	0.7	—.1	0.2
18	1.7	†	15.5	7.5	14.0	1.2	0.4	—.1	0.0	0.6	—.1	0.2
19	1.8	†	15.7	6.6	12.8	1.3	0.4	—.2	0.0	0.4	—.1	0.2
20	1.8	†	15.6	5.9	9.7	1.3	0.4	—.1	0.0	0.3	—.1	0.2
21	1.8	†	15.1	5.1	7.6	1.0	0.7	—.1	—.1	0.2	—.2	0.2
22	1.8	†	12.8	4.8	6.3	1.0	0.6	—.1	—.1	0.2	—.2	0.2
23	1.7	†	9.5	4.5	5.6	1.0	0.5	0.0	0.0	0.1	—.4	0.2
24	1.7	†	7.5	4.2	4.8	1.0	0.4	—.1	—.1	0.1	—.3	0.2
25	1.7	†	6.4	3.9	4.2	1.0	0.3	0.4	—.2	0.1	0.0	0.2
26	3.4	†	5.8	3.7	3.9	1.0	0.3	0.4	—.2	0.1	0.0	0.2
27	4.9	†	5.6	4.0	3.6	1.5	0.3	0.3	—.2	0.1	—.1	0.2
28	5.2	†	10.0	4.3	3.3	1.2	0.3	0.1	—.2	0.1	—.1	0.2
29	4.8	......	12.0	4.4	3.0	1.1	0.3	0.4	—.2	0.0	0.0	0.2
30	4.4	......	14.2	4.3	2.9	1.0	0.3	0.6	—.2	0.0	0.0	0.2
31	4.0	......	14.9	......	2.9	......	1.2	0.3	......	0.0	......	0.2

† Frozen.
‡Estimated account ice around gage.
*U. S. Department of Agriculture, Weather Bureau. Daily River Stages at River Gage Stations on the Principal Rivers of the United States. Part XII. for the years 1913-1914. W. B. No. 549, page 309.

Daily Gage Heights, in Feet, of Wabash River at Terre Haute, Indiana, for 1915.

Day	Jan.	Feb.	Mar.	Apr.	May	June	July	Aug.	Sept.	Oct.	Nov.	Dec.
1	0.2	3.2	4.1	1.5	1.3	5.4	1.0	7.7	4.0	5.6	1.4	2.8
2	0.2	5.0	3.8	1.4	1.0	8.0	2.0	9.2	3.4	5.2	1.4	2.6
3	0.2	7.0	3.5	1.4	1.0	8.0	2.0	10.8	3.2	4.4	1.3	2.5
4	0.2	7.0	3.5	1.4	1.0	7.0	1.6	11.5	2.9	4.4	1.2	2.4
5	0.2	8.0	2.9	1.3	0.9	6.1	1.6	10.1	2.7	4.9	1.1	2.3
6	0.2	11.2	3.0	1.2	0.9	5.5	2.0	9.1	2.6	4.4	1.1	2.2
7	0.7	12.8	3.1	1.1	0.9	4.8	1.9	9.2	2.5	3.7	1.0	2.0
8	0.7	12.8	3.1	1.0	0.9	4.0	11.3	9.1	2.5	3.2	1.0	1.9
9	0.8	13.4	3.1	1.0	1.0	3.6	14.8	9.1	2.4	2.8	0.9	1.9
10	0.8	13.7	3.0	1.0	1.1	2.8	12.6	7.7	2.6	2.5	0.9	1.7
11	1.0	13.2	2.9	1.0	1.1	2.8	7.8	6.8	2.5	2.2	0.9	1.7
12	1.3	10.3	2.9	1.2	1.0	2.8	7.6	10.7	2.7	2.0	1.4	2.5
13	2.2	8.3	2.8	1.3	0.9	2.7	9.9	9.0	2.8	1.9	1.9	2.5
14	2.2	7.4	2.6	1.3	0.9	2.7	11.1	8.3	2.7	1.9	1.8	2.4
15	2.1	8.5	2.5	1.2	0.8	2.5	10.6	10.4	2.2	1.9	1.6	2.2
16	1.8	9.2	2.3	1.3	0.7	2.3	9.1	10.4	2.2	1.8	1.5	1.8
17	1.9	8.8	2.2	1.3	0.7	2.6	9.4	9.4	3.2	1.7	1.5	2.4
18	1.9	7.9	2.1	1.3	1.0	4.0	10.8	8.0	5.5	2.9	1.3	3.3
19	1.9	6.7	2.1	1.2	1.2	4.9	11.8	6.7	4.3	3.7	3.1	3.2
20	1.9	5.6	2.0	1.2	1.3	5.2	12.4	5.6	5.0	3.6	3.4	2.7
21	1.9	4.8	2.0	1.2	1.3	4.0	12.1	9.2	5.0	3.5	3.5	2.1
22	1.9	4.2	1.9	1.2	1.2	3.1	11.1	10.8	6.6	3.8	5.2	3.2
23	1.9	4.2	1.9	1.3	1.1	2.8	9.7	10.8	5.8	3.8	6.6	3.3
24	1.9	4.8	1.9	1.3	0.9	2.3	7.7	11.9	4.7	3.4	6.0	3.0
25	1.9	5.8	1.8	1.3	0.9	1.9	6.3	12.2	3.8	2.9	5.1	3.2
26	1.9	6.0	1.8	1.3	1.4	1.8	5.5	11.5	3.4	2.4	4.1	3.1
27	1.7	5.5	1.8	1.4	1.9	1.6	6.0	9.6	3.0	2.1	3.7	2.6
28	1.7	4.7	1.8	1.5	2.5	1.4	4.7	7.8	2.7	1.9	3.1	1.9
29	1.5	......	1.7	1.5	2.7	1.2	4.0	6.2	2.8	1.7	3.0	1.7
30	1.5	......	1.7	1.4	2.9	1.1	3.6	5.2	4.4	1.6	2.9	2.3
31	1.5	......	1.6	......	3.4	......	4.6	4.5	......	1.5	......	3.1

Daily Gage Heights, in Feet, of Wabash River at Terre Haute, Indiana, for 1916.

Day	Jan.	Feb.	Mar.	Apr.	May	June	July	Aug.	Sept.	Oct.	Nov.	Dec.
1	6.2	22.0	6.0	13.8	6.4	7.8	4.0	0.8	0.3	0.0	0.2	0.5
2	12.5	23.0	5.5	13.8	5.5	6.4	4.8	0.7	0.3	0.1	0.2	0.4
3	15.0	22.8	5.0	12.8	4.8	5.7	4.3	0.9	0.2	0.1	0.1	0.4
4	16.0	22.2	4.6	10.6	4.6	5.1	3.9	0.9	0.2	0.1	0.1	0.4
5	16.4	21.0	4.1	9.1	4.8	4.4	3.9	0.9	0.2	0.0	0.1	0.5
6	17.0	19.8	4.0	8.0	5.6	3.9	3.6	0.9	0.3	0.0	0.1	0.7
7	17.8	18.4	4.7	6.9	5.9	6.4	3.2	0.9	0.1	—0.1	0.2	0.8
8	18.2	17.9	5.7	6.2	6.1	7.9	3.0	0.8	0.2	—0.2	0.2	1.2
9	18.5	15.0	5.8	5.6	7.0	8.4	2.6	0.8	0.4	—0.2	0.2	1.5
10	17.8	12.0	5.9	5.1	8.9	9.3	2.3	0.9	0.5	0.0	0.3	1.6
11	17.4	8.1	6.0	4.7	9.0	10.4	2.1	0.9	0.8	0.0	0.3	1.8
12	16.4	7.5	5.8	4.5	8.8	9.9	2.0	0.9	0.6	—0.1	0.3	1.7
13	16.8	7.5	5.0	4.2	6.5	8.5	1.9	1.1	0.4	—0.1	0.3	1.8
14	17.3	7.1	4.7	4.0	5.6	6.9	1.9	1.3	0.2	0.0	0.2	1.2
15	17.6	6.0	4.5	3.8	4.7	6.4	2.0	1.8	0.2	0.0	0.1	0.6
16	17.6	5.8	4.3	3.8	7.7	6.4	2.4	1.5	0.1	—0.1	0.1	0.2
17	17.7	5.8	4.1	3.8	11.4	6.5	3.0	1.3	0.0	—0.1	0.1	0.4
18	17.3	6.1	4.1	3.9	12.8	6.0	3.0	1.0	0.0	—0.1	0.1	0.6
19	15.6	6.6	4.1	3.8	12.5	5.4	2.6	0.8	0.0	0.0	0.1	0.5
20	10.0	6.7	4.0	3.7	12.0	5.1	2.8	0.8	—0.1	—0.1	0.0	0.5
21	8.2	6.7	3.8	3.5	11.6	5.7	2.5	0.7	—0.1	0.2	0.0	0.5
22	12.3	6.4	3.7	3.5	8.3	7.4	4.1	0.6	—0.1	0.3	0.0	0.5
23	14.5	6.2	3.7	3.5	6.9	7.5	2.7	0.6	—0.2	0.3	0.1	0.5
24	14.9	7.2	4.2	3.1	6.0	9.8	2.0	1.0	—0.2	0.3	0.2	0.5
25	15.4	8.1	8.9	2.9	5.6	10.1	2.0	0.6	—0.2	0.5	0.2	0.6
26	15.7	8.9	10.2	2.9	5.4	9.0	1.7	0.3	—0.2	0.5	0.2	0.6
27	15.8	8.7	10.7	3.0	5.0	8.4	1.5	0.3	—0.2	0.5	0.2	0.7
28	15.2	7.6	10.7	3.3	5.4	6.9	1.4	0.2	0.0	0.4	0.4	1.8
29	14.6	6.6	11.8	4.8	5.9	6.5	1.2	0.2	0.0	0.3	0.6	2.6
30	15.4	......	12.7	6.5	7.7	6.0	1.0	0.2	0.0	0.2	0.6	3.4
31	18.7	......	13.4	......	8.4	......	0.9	0.1	......	0.2	......	3.6

Daily Gage Heights, in Feet, of Wabash River at Terre Haute, Indiana, for 1917.

Day	Jan.	Feb.	Mar.	Apr.	May	June
1	4.1	2.3	4.1	6.3	5.1	11.4
2	3.9	2.3	3.2	6.7	5.2	11.8
3	3.7	2.5	2.9	7.4	5.3	9.8
4	3.4	2.5	2.6	9.3	5.2	8.0
5	3.5	*	2.2	11.1	5.9	11.3
6	3.8	*	1.9	12.4	6.7	14.4
7	5.2	*	1.6	12.9	7.2	13.9
8	5.1	*	1.5	13.4	8.4	13.3
9	7.4	*	1.5	13.8	8.6	12.9
10	9.3	*	1.4	14.1	8.5	14.4
11	9.7	*	1.4	14.3	6.4	15.2
12	7.0	*	1.4	13.0	5.3	15.1
13	6.2	*	2.3	9.7	4.6	14.5
14	2.0	*	12.7	7.4	4.0	13.8
15	*	*	14.5	6.1	3.6	11.7
16	*	*	14.9	5.3	3.1	9.0
17	*	*	15.0	4.7	2.9	7.6
18	*	*	15.3	4.0	2.7	6.4
19	*	*	15.4	4.0	2.6	5.8
20	*	2.2	14.9	4.0	2.3	4.9
21	*	2.3	12.9	3.9	2.1	4.3
22	*	2.4	9.6	3.9	4.0	3.7
23	*	2.1	7.7	3.8	8.0	3.4
24	*	2.6	8.2	3.6	8.0	3.0
25	*	3.3	8.3	3.5	10.1	3.0
26	*	3.9	8.9	3.3	10.1	3.7
27	*	4.8	9.4	3.3	11.8	3.9
28	*	5.0	8.9	3.6	8.3	4.5
29	*	......	7.8	4.3	6.7	9.3
30	2.3	......	7.7	4.9	6.4	9.8
31	2.4	......	7.3	......	8.4	......

*Frozen.

Rating Table for Wabash River at Terre Haute, Indiana, for 1901-11, Inclusive.

Gage Height, Feet	Discharge, Sec.-Ft.	Gage Height, Feet	Discharge, Sec.-Ft.	Gage Height, Feet	Discharge, Sec.-Ft.	Gage Height, Feet	Discharge, Sec.-Ft.
0.0	1,930	1.5	4,330	3.0	7,400	5.9	14,200
0.1	2,050	1.6	4,510	3.1	7,620	6.1	14,700
0.2	2,180	1.7	4,700	3.3	8,060	7.1	17,200
0.3	2,320	1.8	4,900	3.5	8,500	8.1	19,800
0.4	2,460	1.9	5,100	3.7	8,940	9.1	22,500
0.5	2,610	2.0	5,300	3.9	9,380	10.1	25,300
0.6	2,770	2.1	5,500	4.1	9,830	11.1	28,400
0.7	2,940	2.2	5,700	4.3	10,300	12.1	31,800
0.8	3,110	2.3	5,900	4.5	10,780	13.1	35,500
0.9	3,280	2.4	6,100	4.7	11,260	14.1	39,500
1.0	3,450	2.5	6,310	4.9	11,740	15.1	43,700
1.1	3,620	2.6	6,520	5.1	12,220	16.1	48,100
1.2	3,790	2.7	6,740	5.3	12,700	17.1	52,700
1.3	3,970	2.8	6,960	5.5	13,200	18.1	57,400
1.4	4,150	2.9	7,180	5.7	13,700	19.1	62,100

The data contained in these tables give a very good idea of the nature of the river at Terre Haute. The data are not entirely accurate. The readings were recorded in the Terre Haute Waterworks record as feet and inches. The inches were reduced to tenths by letting each equal the nearest tenth. However, the error due to this cause is small. An examination of these data shows that the Wabash is a very erratic river. The following table shows some striking features:

Table Showing Annual Fluctuations of Wabash River at Terre Haute, and Ratio between Maximum and Minimum Discharge.

YEAR	Date	Minimum Discharge, Sec.-Ft.	Date	Maximum Discharge, Sec.-Ft.	Times Minimum
1901	October 24	1,930	June 24	25,000	13
1902	February 27	1,930	July 4	62,100	32
1903	November 4	1,200	April 17	66,330	55
1904	December 20	900	March 27	90,300	100
1905	January 4	1,930	May 18	52,000	26
*	October 17	1,920	May 17	50,000	....
1906	October 15	1,300	April 1	61,630	47
*	July 1	2,160	April 2	61,200	....
1907	September 20	2,320	January 23	87,950	37
1908	October 13	1,530	March 10	79,490	52
1909	January 7	1,930	February 28	55,520	29
1910	August 14	1,410	January 23	63,040	45
1911	August 13	1,300	February 1	37,500	21

*U. S. Geol. Survey data.

This table shows the maximum annual discharge to have occurred twice in each of the first four months and once in each of the next three during the past eleven years. The minimum discharge is also distributed among seven months with October three, August and January two each, and September, November, December and February one each. The lowest ratio between the minimum and maximum discharge of any year in which complete data is recorded is 1 to 26 in 1905 and in the previous year the ratio was 1 to 100. Both the absolute minimum and maximum discharges occurred in 1904. The profile of the Wabash shows a much more uniform gradient than any of the smaller streams of the state. The fall is slight throughout its course. The profile is taken from Water Supply and Irrigation Paper No. 169, U. S. Geological Survey, page 74. Several of the altitudes have been checked by the writer:

Tables of Altitudes and Distances Along Wabash River.

LOCATION	Estimated Distance, Miles	Altitude, Feet	Fall Per Mile, Inches
Source	0.0	1,000.0	0.0
Huntington	100.0	699.0	36.0
Mouth of Salamonie River	15.0	667.0	25.6
Mouth of Mississinewa River	20.0	633.0	20.4
Logansport	20.0	583.0	30.0
Lafayette	50.0	506.0	18.5
Attica	25.0	487.0	9.1
Covington	20.0	470.0	10.2
Terre Haute	55.0	447.7	4.9
State Line	14.6	440.6	5.8
Hutsonville, Ill.	29.0	424.6	6.6
Vincennes	46.4	398.8	6.7
Mouth of White River	32.5	376.5	8.2
Grayville, Ill.	28.0	365.0	4.9
Mouth of Little Wabash River	46.0	323.0	11.0
Mouth of Wabash River	16.0	311.0	9.0

Eel River.

Eel River rises by several branches in the lake district of southern Noble, northern Whitley, and northwestern Allen counties. It flows in a general southwest direction, across Whitley, Wabash, Miami and Cass counties, and debouches into the Wabash River at Logansport, Indiana. Its total length is approximately 110 miles.

The valley is young and has little bottom land. The bluffs are from 30 to 75 feet in height. The valley lies entirely within glacial drift and does not touch bed rock until within nine miles of the mouth. At Adamsboro, the stream first touches bed rock and a rapid is formed. This rapid is wide and shallow. It is about one-fourth mile in length, with a total fall of six feet. From this point to the mouth, the stream flows on bed rock much of the way.

The drainage basin of Eel River is long and narrow. The tributaries are short and those that do not rise in lakes are intermittent streams. Three small tributaries which debouch in Wabash County rise in small lakes of southern Kosciusko. The tributaries near the source of the stream rise in the lakes of southern Noble County. The storage offered by these lakes makes the discharge regular. The drainage area of Eel River is 777 square miles.

On July 14 and 15, 1910, a gaging station was established on the bridge which crosses Eel River on Third Street, Logansport. It is a chain gage, and is located on the down stream hand-rail of the south span of the bridge. The length of the chain from the end of the weight to the first marker is 18 feet 6.7 inches. The gage is read once each day by Mr. H. J. Kruck of the Logansport Water Works Company. The stream is straight for 500 feet above this station, and a similar distance below. The channel is about 360 feet wide between the bridge abutments. Two piers divide the channel into three parts. The depth is fairly regular. The bed of the stream is of sand, which is deposited at the head of the backwater from Uhl's mill dam. The backwater reaches a short distance above the station, but the current is regular and steadily down stream. The bed has not changed perceptibly during the years since the gage was established. All the water must pass between the bridge abutments except during extremely high water, when the north bank overflows. The initial point for soundings is three and one-half feet from the north end of the down stream hand-rail. The discharge and gage readings from this station appear in the following table:

Discharge Measurements on Eel River at Logansport, Indiana, for 1910-11.

Date	Hydrographer	Gage Height, Feet	Discharge, Sec.-Ft.
July 15, 1910	W. M. Tucker	2.8	207.16
August 12, 1910	W. M. Tucker	2.4	125.33
February 21, 1911	W. M. Tucker	3.6	1,390.2
April 15, 1911	W. M. Tucker	3.9	2,147.89
July 5, 1911	Tucker and Clark	2.9	263.28

Gage Readings on Eel River at Logansport, Indiana, from July 16, 1910, to July 15, 1911.

Day	July	Aug.	Sept.	Oct.	Nov.	Dec.	Jan.	Feb.	Mar.	Apr.	May	June	July
1	.....	2.5	2.7	2.8	2.8	*2.8	3.2	3.4	3.0	3.0	3.4	3.2	2.9
2	.....	2.3	2.7	2.9	2.8	2.8	3.3	3.3	3.0	3.0	3.4	3.1	2.9
3	.....	2.6	2.7	2.8	2.6	2.8	4.0	3.2	3.0	3.0	3.2	2.9	2.9
4	.....	2.7	2.8	2.8	2.6	2.8	†	3.2	3.0	3.1	3.1	3.0	2.9
5	.....	2.5	3.0	3.1	2.8	2.8	†	3.1	3.0	3.7	3.1	3.2	2.9
6	.....	2.6	3.5	3.1	2.8	2.8	†	3.1	2.9	3.9	3.1	3.0	2.9
7	.....	2.8	3.3	3.1	2.7	2.8	†	3.1	2.9	3.6	3.0	3.0	2.9
8	.....	2.7	3.1	3.1	2.7	2.8	†	3.0	2.9	3.3	2.9	2.9	2.9
9	.....	2.6	2.9	3.0	2.6	2.8	†	2.9	2.9	3.2	2.9	2.9	2.9
10	.....	2.6	3.0	2.9	2.6	2.8	†	2.9	2.9	3.1	2.9	2.8	2.8
11	.....	2.5	2.9	2.8	2.7	2.8	†	2.8	2.9	3.0	2.9	2.8	2.7
12	.....	2.4	2.8	2.8	2.6	2.8	†	3.0	3.0	3.0	2.9	2.8	2.7
13	.....	2.7	3.2	2.8	2.9	2.8	†	3.0	3.0	3.1	2.9	2.8	2.7
14	.....	2.8	3.6	2.8	2.9	2.8	†	3.3	3.0	3.9	2.9	2.8	2.6
15	.....	2.5	3.2	2.8	2.7	2.8	†	4.7	2.9	3.9	2.8	2.8	2.7
16	2.8	2.6	3.0	2.9	2.7	2.8	†	4.3	2.9	3.5	2.8	2.8	.....
17	2.9	2.7	2.9	2.8	2.6	2.8	†	4.1	3.0	3.4	2.8	2.9	.....
18	2.8	2.7	3.0	2.8	2.7	2.8	†	4.0	3.0	3.2	2.8	2.9	.....
19	2.8	2.6	2.9	2.8	2.7	2.8	†	4.0	3.0	3.9	2.8	2.9	.....
20	2.7	2.7	3.0	2.8	2.9	2.8	†	3.9	3.0	4.4	2.8	2.8	.....
21	2.7	2.8	2.9	2.8	2.8	2.8	†	3.6	3.0	4.2	2.8	2.8	.....
22	2.7	2.8	2.8	2.8	2.8	2.8	†	3.3	2.9	3.9	2.8	2.7	.....
23	2.7	2.8	2.8	2.9	2.8	2.8	†	3.3	2.9	3.9	3.0	2.7	.....
24	2.8	2.9	2.9	2.9	2.8	2.8	†	3.1	2.9	3.7	3.0	2.7	.....
25	2.5	2.8	3.3	2.9	2.8	2.8	2.9	3.1	2.9	3.3	2.9	2.9	.....
26	2.5	2.8	3.2	2.5	2.8	2.8	2.9	3.1	3.0	3.2	2.8	4.3	.....
27	2.7	2.8	3.1	2.6	2.9	2.8	3.1	3.1	3.0	3.1	2.8	3.8	.....
28	2.6	2.8	2.9	2.7	2.8	2.9	3.8	3.1	3.0	3.1	2.8	3.5	.....
29	2.7	2.7	2.7	2.7	2.8	3.2	4.0	.....	3.0	3.2	2.8	3.2	.....
30	2.8	2.6	2.8	2.9	2.9	3.1	4.0	.....	3.0	3.2	2.8	2.9	.....
31	2.7	2.7	.....	2.6	.....	3.1	3.6	.....	2.9	.....	3.0	.....	.....

*Ice conditions during all of December. Ice ½ to 6 inches thick. Water over ice on Dec. 29, 30 and 31.

†No record.

Rating Table for Eel River at Logansport, Indiana, for 1910-11.

Gage Height, Feet	Discharge, Sec.-Ft.	Gage Height, Feet	Discharge, Sec.-Ft.	Gage Height, Feet	Discharge, Sec.-Ft.	Gage Height, Feet	Discharge, Sec.-Ft.
2.3	115	3.0	332	3.7	1,645	4.4	3,395
2.4	125	3.1	425	3.8	1,895	4.5	3,645
2.5	137	3.2	550	3.9	2,145	4.6	3,895
2.6	155	3.3	720	4.0	2,395	4.7	4,145
2.7	178	3.4	925	4.1	2,645	....	....
2.8	210	3.5	1,155	4.2	2,895	....	....
2.9	260	3.6	1,395	4.3	3,145	....	....

The readings from this gage were discontinued in July, 1911, and the gage was damaged during the flood of 1913. It was repaired and re-established on October 23, 1915. The datum of the gage was raised 0.98 feet. The initial point for soundings was changed to a point on the down stream hand-rail directly over the inside face of the south abutment of the bridge. Readings are taken at intervals of ten feet from this point. The width of the stream at this point is 360 feet. The gage has been read during 1915, 1916, and 1917 by Mr. F. L. C. Boerger. The following data have been accumulated during 1915 and 1916.

Discharge Measurements on Eel River at Logansport, Indiana, for 1915 and 1916.

Date	Hydrographer	Gage Height, Feet	Discharge, Sec.-Ft.
October 23, 1915	W. M. Tucker	3.9	271.8
November 27, 1915	W. M. Tucker	4.1	456.8
January 30, 1916	W. M. Tucker	7.1	7,478.9
May 17, 1916	W. M. Tucker	5.0	2,463.2
August 12, 1916	W. M. Tucker	3.85	251.6

Daily Gage Heights on Eel River at Logansport, Indiana, for 1915.

Day	Oct.	Nov.	Dec.	Day	Oct.	Nov.	Dec.
1	.....	3.9	4.1	17	.....	3.9	3.9
2	.....	3.9	4.0	18	.....	3.9	3.9
3	.....	3.9	4.0	19	.....	4.0	3.9
4	.....	3.9	4.0	20	.....	4.5	3.8
5	.....	3.9	4.0	21	.....	4.5	3.8
6	.....	3.9	4.0	22	*	4.2	3.9
7	.....	3.9	4.0	23	3.9	4.1	3.9
8	.....	3.9	3.9	24	3.9	4.1	3.9
9	.....	3.9	3.9	25	3.9	4.0	3.9
10	.....	3.9	3.9	26	3.9	4.0	3.9
11	.....	3.9	3.9	27	3.9	4.1	3.9
12	.....	4.0	3.9	28	3.9	4.0	3.9
13	.....	3.9	3.9	29	3.9	4.1	3.9
14	.....	3.9	3.9	30	3.9	4.1	3.9
15	.....	3.9	3.9	31	3.9	.....	3.9
16	.....	3.9	3.9				

*Gage repaired and reinstalled October 23.

Daily Gage Height on Eel River at Logansport, Indiana, for 1916.

Day	Jan.	Feb.	Mar.	Apr.	May	June	July	Aug.	Sept.	Oct.	Nov.	Dec.
1	3.9	8.8	4.2	4.9	4.3	4.1	4.3	3.8	3.8	3.8	3.9	4.0
2	6.7	6.6	4.2	4.9	4.2	4.1	4.3	3.8	3.8	3.8	3.9	3.9
3	6.8	5.7	4.1	4.7	4.6	4.1	4.2	3.8	3.8	3.8	3.9	3.9
4	5.8	5.3	4.0	4.5	5.1	4.0	4.2	3.8	3.8	3.8	3.8	3.9
5	5.4	5.1	4.0	4.3	4.7	4.0	4.2	3.8	3.8	3.8	3.8	4.2
6	5.7	4.9	4.0	4.3	4.4	4.0	3.9	3.8	4.0	3.7	3.8	4.2
7	5.2	4.6	4.3	4.2	5.1	4.4	3.9	3.8	3.9	3.7	3.8	4.0
8	5.0	4.8	4.5	4.2	5.1	4.7	3.9	3.8	3.8	3.7	3.8	4.0
9	4.8	6.6	4.3	4.2	4.6	4.7	3.9	3.8	4.0	3.7	3.8	4.0
10	4.6	6.0	4.2	4.2	4.4	4.5	3.9	3.8	3.9	3.7	3.9	4.0
11	4.5	5.4	4.1	4.2	4.3	4.3	3.9	3.8	4.0	3.7	3.9	4.0
12	5.0	5.0	4.0	4.1	4.1	4.2	3.9	3.8	3.7	3.8	3.9	4.0
13	5.8	4.8	4.1	4.1	4.1	4.1	3.9	3.8	3.8	3.8	3.9	4.0
14	5.2	4.7	4.1	4.2	4.7	4.1	4.3	3.8	3.8	3.8	3.9	4.0
15	4.8	4.3	4.1	4.2	6.5	4.1	4.7	3.7	3.8	3.8	3.9	4.0
16	4.6	4.3	4.0	4.2	5.8	4.1	4.1	3.8	3.8	3.8	3.9	4.0
17	4.2	4.2	4.0	4.2	5.1	4.1	4.1	3.8	3.8	3.8	3.9	4.0
18	4.3	4.3	4.0	4.1	4.8	4.1	4.0	3.8	3.8	3.8	3.9	4.0
19	4.2	4.2	4.0	4.0	4.5	4.2	3.9	3.7	3.8	3.8	3.9	4.0
20	4.2	4.2	3.9	4.0	4.3	4.2	3.9	3.7	3.6	3.8	3.8	4.0
21	7.6	4.2	4.0	4.0	4.2	4.3	3.9	3.8	3.8	4.0	3.8	4.0
22	6.7	4.2	5.3	4.0	4.2	5.6	3.9	3.5	3.8	4.0	3.8	4.0
23	5.9	4.2	5.9	4.0	4.2	4.9	3.9	3.7	3.8	3.9	3.8	4.0
24	5.4	4.3	5.2	4.0	4.3	4.6	3.8	3.7	3.8	3.9	3.9	4.0
25	5.1	4.2	5.0	4.0	4.2	4.3	3.7	3.5	3.8	3.9	3.9	4.0
26	5.0	4.2	5.0	4.1	4.2	4.2	3.8	3.5	3.8	3.9	3.8	4.0
27	5.0	4.2	5.8	5.4	4.9	4.3	3.8	3.8	3.9	3.9	3.8	4.5
28	5.5	4.2	5.7	4.9	4.8	4.2	3.8	3.8	3.8	3.9	3.8	4.5
29	5.1	4.2	5.5	4.6	4.5	4.1	3.8	3.8	3.8	3.9	3.9	4.3
30	6.6	......	5.2	4.4	4.3	4.1	3.8	3.8	3.8	3.9	4.0	4.3
31	9.0	......	5.0	.....	4.2	....	3.8	3.8	....	3.9	......	4.3

Rating Table for Eel River at Logansport, Indiana, for 1915-16.

Gage Height, Feet	Discharge, Sec.-Ft.	Gage Height, Feet	Discharge, Sec.-Ft.	Gage Height, Feet	Discharge, Sec.-Ft.	Gage Height, Feet	Discharge, Sec.-Ft.
3.5	140	4.3	761	5.1	2,695	5.9	4,695
3.6	160	4.4	971	5.2	2,945	6.0	4,945
3.7	184	4.5	1,203	5.3	3,195	6.5	6,195
3.8	220	4.6	1,445	5.4	3,445	7.0	7,445
3.9	274	4.7	1,695	5.5	3,695	7.5	8,695
4.0	350	4.8	1,945	5.6	3,945	8.0	9,945
4.1	450	4.9	2,195	5.7	4,195	8.5	11,195
4.2	584	5.0	2,445	5.8	4,445	9.0	12,445

This table is based on ten discharge measurements made in 1911, 1912, 1915 and 1916. It is approximately accurate throughout and well established below 5 feet.

Since the gaging station from which this data is taken is in the backwater from the Uhl dam, the manipulation of the water at the mill has considerable effect upon the data. When the gage stands at 2.7 feet the water is just at the crest of the dam. Whenever the gage registers lower than that, the head is being pulled down at the mill. The gage is read about 7:00 a. m. each day and it is probable that the head has been reduced on some days at the time of reading. From the data it seems that the minimum continuous discharge is about 150 second-feet and the maximum is over 4,000 second-feet. A discharge of from 200 to 300 second-feet occurred most of the year.

The profile of Eel River is very irregular and shows the greatest fall near its mouth. After the stream bed crosses the edge of the hard Niagara limestone at Adamsboro, it has heavy fall to its mouth. The following table shows the profile of the river.

PROFILE OF EEL RIVER.

Location	Estimated Distance, in Miles	Altitude, in Feet	Fall Per Mile, Inches
Source		900	
North Manchester	50	732	40.32
Laketon	9	715	22.66
Stockdale	9	701	18.66
Chili	9	682	25.33
Denver	4	671	33.00
Mexico	5	653	43.20
Hoover	7	629	41.14
Adamsboro	6	617	28.00
Mouth	9	574	57.33

Information concerning water power on Eel River may be found on pages 494 to 501 of the 36th Annual Report of the Indiana Department of Geology and Natural Resources. Since that report was made two other power plants have been investigated above North Manchester.

Liberty Mills.—A flour mill which is operated by water power is located at Liberty Mills in the northeastern part of Wabash County on Eel River. It is of the feeder dam and race type. The dam is situated in Sec. 22, R. 7 E., T. 30 N., and a race one-half mile in length leads to the mill in Sec. 27 of the same township and range. The dam is 5.2 feet high, with stone abutments and wooden structure. The head at the mill is 9.3 feet. One Dolan wheel is used, producing a power of 75 h. p. The power is used to operate a flour and feed mill. The proprietor of this mill is Mr. E. S. Writtenhouse.

Collimer.—A small water power plant is located at Collimer on Eel River in the Southwestern part of Whitley County. The plant is located in Sec. 7, R. 8 E., T. 30 N. It is situated at the end of the dam. The dam is 6.5 feet in height. It is built of concrete throughout. The head at the plant is 6.5 feet. Three American wheels are in use, producing a power of 90 h. p. The power is used for grinding flour and feed. The plant is owned and operated by Mr. H. L. Hauptmeyer under the name of the Collimer Milling Company.

Mexico.—The power plant at Mexico is discussed on page 497 of the previously mentioned report. Since the report of 1911 a woolen mill has been located on the west end of the dam at this point and is employing about 150 h. p. of the water power.

Dennis Uhl & Co.—The flour mill of Dennis Uhl & Co. is discussed on page 500 of the previously mentioned report. Since the report of 1911 this mill has burned and has not been rebuilt.

Tippecanoe River.

The Tippecanoe River rises among the lakes of Noble, Whitley and Kosciusko counties. It is a very crooked stream and flows in a general westward direction to the northeastern corner of Pulaski County and then in a general southward direction to its mouth in Tippecanoe County. It is about 166 miles in length.

The valley contains much swamp land in its upper parts, but below DeLong little swamp land occurs. The valley is narrow and the bluffs low in the middle course. The level upland at DeLong is about 30 feet above the river, at Monterey 16 feet, at Ora 23 feet, and at Winamac 32 feet. At Monticello the bluffs are 67 feet high, and at Oakdale 100 feet. This is approximately the maximum depth of the valley. The upper and middle courses of the valley lie entirely in glacial drift. It first touches bed rock at Norway, three miles above Monticello. In this part of its course the stream crosses the Cincinnati anticline, whose surface formation in this locality is the hard Niagara limestone. After touching bed rock at Norway, it is near the rock throughout the rest of its course and rock exposures are frequent. The banks below Norway are usually high and there is little overflow.

The basin of Tippecanoe River embraces almost all of White, Pulaski and Fulton counties, half of Kosciusko, and small portions of several other counties. The entire drainage area is about 1,900 square miles. It is overlaid entirely with glacial drift and many small glacial lakes occur in the northern and eastern parts of the basin. There are no large tributaries, but most of the tributaries drain either lakes or swamps and for this reason have a more or less continuous discharge.

The mean annual rainfall above the gage at Springboro on the Tippecanoe River, as shown by the station records from this section for eighteen years (1898-1915, inclusive), is 37.15 inches. The following tables which are computed entirely from government reports show the average depth of monthly rainfall and the cubic feet per acre rainfall for the basin of the Tippecanoe River for the eighteen years above mentioned. These averages are taken from six U. S. Weather Bureau recording stations. The following is a list of the stations used: Culver, Monticello, Rochester, Warsaw, Winamac and Winona.

Table Showing Average Depth of Monthly Rainfall in Inches, and in Cubic Feet per Acre, for the Tippecanoe River Basin Above the Gage at Springboro, Indiana, 1898-1915, Inclusive.

	1898		1899		1900		1901	
	In.	Cu. Ft.	In.	Cu. Ft.	In.	Cu. Ft.	In.	Cu. Ft.
January	2.67	9,692.1	2.47	8,966.1	0.44	1,597.2	1.39	5,045.7
February	1.76	6,388.8	1.59	5,771.7	3.10	11,253.0	1.60	5,808.0
March	2.50	9,075.0	3.98	17,387.4	2.26	8,203.8	3.07	11,144.1
April	1.62	5,880.6	0.87	3,158.1	0.47	5,336.1	2.47	8,966.1
May	5.07	18,404.1	4.63	16,806.9	3.57	12,959.1	2.47	8,966.1
June	3.49	12,668.7	2.70	9,801.0	4.79	17,387.7	3.21	11,652.3
July	1.72	6,243.6	3.64	13,213.2	5.20	18,876.0	0.69	2,504.7
August	7.00	25,410.0	1.74	6,316.2	3.57	12,959.1	3.16	11,470.8
September	3.59	13,031.7	1.89	6,860.7	2.05	7,441.5	0.91	3,303.3
October	6.17	22,397.1	3.25	11,797.5	2.95	10,708.5	5.19	18,839.7
November	4.13	14,991.9	2.23	8,094.9	4.63	16,806.9	1.20	4,356.0
December	0.82	2,976.6	2.19	7,949.7	0.36	1,306.8	4.02	14,592.6

	1902		1903		1904		1905	
	In.	Cu. Ft.	In.	Cu. Ft.	In.	Cu. Ft.	In.	Cu. Ft.
January	1.86	6,751.8	1.65	5,989.5	5.03	18,258.9	1.90	6,897.0
February	1.10	3,093.0	3.55	12,886.5	2.02	7,332.6	1.72	6,243.6
March	2.75	9,982.5	1.26	4,573.8	7.87	28,568.1	1.62	5,880.6
April	1.51	5,481.3	5.23	18,984.9	4.73	17,169.9	3.95	14,338.5
May	4.33	15,717.9	3.06	11,107.8	3.39	12,305.7	6.16	22,360.8
June	11.21	40,692.3	4.36	15,826.8	2.54	9,220.2	3.99	14,483.7
July	5.22	18,948.6	3.68	13,358.4	3.67	13,322.1	3.48	12,632.4
August	1.54	5,590.2	3.89	14,120.7	2.59	9,401.7	3.40	12,342.0
September	3.01	10,926.3	3.07	11,144.1	4.24	15,391.2	3.97	14,411.1
October	2.75	9,982.5	3.56	12,922.8	1.14	4,138.2	3.13	11,361.9
November	2.94	10,672.2	1.36	4,936.8	0.23	864.9	3.22	11,688.6
December	2.16	7,840.8	2.80	10,164.0	2.59	9,401.7	1.74	6,316.2

	1906		1907		1908		1909	
	In.	Cu. Ft.	In.	Cu. Ft.	In.	Cu. Ft.	In.	Cu. Ft.
January	3.10	11,253.0	4.74	17,206.2	1.31	4,755.3	3.08	11,180.4
February	0.82	2,976.6	0.22	798.6	5.32	19,311.6	5.40	19,602.0
March	2.97	10,781.1	4.46	16,189.8	4.11	14,919.3	2.78	10,091.4
April	2.34	8,494.2	2.58	9,365.4	3.77	13,685.1	5.24	19,021.2
May	2.19	7,948.7	3.38	12,269.4	6.71	24,357.3	4.12	14,955.6
June	4.06	14,737.8	5.39	9,565.7	1.60	5,808.0	6.37	23,123.1
July	4.07	14,774.1	5.28	9,166.4	3.43	12,450.9	3.97	14,411.1
August	3.87	14,048.1	4.16	5,100.8	2.55	9,256.5	4.00	14,520.0
September	3.49	12,668.7	3.58	2,995.4	1.00	3,630.0	3.29	11,942.7
October	2.37	8,603.1	1.52	5,517.6	0.32	1,161.6	2.31	8,385.3
November	4.05	14,701.5	2.37	8,693.1	2.26	8,203.8	3.80	13,794.0
December	3.95	14,338.5	4.87	17,678.1	1.55	5,626.5	2.99	10,853.7

	1910		1911		1912		1913	
	In.	Cu. Ft.	In.	Cu. Ft.	In.	Cu. Ft.	In.	Cu. Ft.
January	2.63	9,346.9	2.90	10,527.0	1.40	5,082.0	5.77	20,945.1
February	1.84	6,679.2	1.95	7,078.5	1.83	6,642.9	1.11	4,029.3
March	0.34	1,234.2	1.82	6,606.6	2.62	9,510.6	6.91	25,083.3
April	3.86	14,011.8	4.46	16,189.8	3.56	12,922.8	2.58	9,365.4
May	3.86	14,011.8	2.86	10,381.8	4.27	15,500.1	5.49	19,928.7
June	0.90	3,267.0	4.21	15,282.3	3.70	13,431.0	1.31	4,755.3
July	2.63	9,510.6	2.31	8,385.3	5.32	19,311.6	4.17	15,137.1
August	3.23	11,724.9	3.11	11,289.3	4.97	18,041.1	3.76	13,648.8
September	5.39	19,565.7	5.32	19,311.6	2.62	9,510.6	2.42	8,784.6
October	2.69	9,764.7	4.43	16,080.9	2.18	7,913.4	3.69	13,394.7
November	1.52	5,517.6	3.81	13,830.3	3.39	12,305.7	3.38	12,269.4
December	1.89	6,860.7	1.89	6,860.7	0.94	3,412.2	0.96	3,484.8

	1914		1915			1914		1915	
	In.	Cu. Ft.	In.	Cu. Ft.		In.	Cu. Ft.	In.	Cu. Ft.
Jan	2.79	10,127.7	2.45	8,893.5	July	1.14	4,138.2	7.29	26,462.7
Feb	1.38	5,009.4	2.19	7,949.7	Aug	3.94	14,302.2	4.71	17,097.3
Mar	2.81	10,200.3	0.66	2,395.8	Sept	2.10	7,623.0	3.81	13,830.3
Apr	2.71	9,837.3	2.40	8,712.0	Oct	1.80	6,534.0	0.98	3,557.4
May	4.85	17,605.5	5.73	20,799.9	Nov	0.72	2,613.6	2.30	8,349.0
June	2.57	9,329.1	1.82	6,606.6	Dec	2.68	9,728.4	1.90	6,697.0

A gage was maintained by the United States Geological Survey at Springboro, five miles west of Delphi, during parts of the years 1903-06, inclusive. The data from this gaging station give a good idea of the discharge of the stream. The following data are taken directly from the U. S. G. S. Water Supply and Irrigation papers:

Tippecanoe River at Springboro.—"This station was established March 14, 1903, by George E. Waesche. The station is located at the highway bridge at Springboro, Ind. The nearest railroad station is Delphi, five miles east of Springboro. The standard chain gage is located on the second span from the east bank, one panel length beyond the center of the span. The length of the chain from the end of the weight to the marker is 25.66 feet. The gage is read once each day by Lois Imler. Discharge measurements are made from the downstream side of the bridge to which the gage is attached. The initial point for soundings is the face of the east abutment. The channel is straight for about 1,600 feet above and about 2,000 feet below the station. Its width at ordinary stages is 350 feet, broken by two piers, and at high water is 510 feet, broken by three piers. Both banks are high and can not overflow to any considerable extent. The bed of the stream is rocky and rough; the current is swift. The bench mark is the head of an anchor bolt in the east abutment; it is the outside anchor of the downstream truss. Its elevation above the zero of the gage is 22.25 feet.

The observations at this station during 1903 have been made under the direction of E. Johnson, Jr., district hydrographer.

Discharge Measurements of Tippecanoe River at Springboro, Indiana, in 1903.

Date	Hydrographer	Gage Height, in Feet	Discharge, Sec.-Ft.
March 16	G. F. Waesche	4.73	3,448
May 18	G. F. Waesche	3.00	747
June 22	G. E. Waesche	2.87	662
July 17	E. C. Murphy	3.03	892
August 14	E. Johnson, Jr.	2.75	584
September 30	L. R. Stockman	3.05	705
November 11	L. R. Stockman	3.03	653
December 29	E. Johnson, Jr.	4.00	*1,083

*Partly frozen.

Mean Daily Gage Height, in Feet, of Tippecanoe River at Springboro, Indiana, for 1903.

Day	Jan.	Feb.	Mar.	Apr.	May	June	July	Aug.	Sept.	Oct.	Nov.	Dec.
1	......	......	......	3.50	3.82	3.42	2.78	2.99	2.94	3.10	3.08	3.10
2	......	......	......	3.50	3.80	3.42	4.20	3.00	3.53	3.20	3.06	3.09
3	......	......	......	3.72	3.73	3.28	7.03	3.03	3.40	3.40	3.09	3.08
4	......	......	......	4.94	3.65	3.16	7.07	3.11	3.32	3.50	3.10	3.10
5	......	......	......	5.54	3.57	3.50	7.18	2.99	3.28	3.47	3.09	3.12
6	......	......	......	5.37	3.52	4.15	6.40	3.16	3.02	3.54	3.08	3.18
7	......	......	......	5.00	3.49	4.59	5.54	3.07	2.90	3.70	3.06	3.22
8	......	......	......	4.64	3.43	4.63	5.05	......	2.88	3.90	3.05	3.28
9	......	......	......	4.18	3.36	4.12	4.50	2.96	2.70	4.17	3.04	3.26
10	......	......	......	4.21	3.32	3.77	4.10	2.98	2.81	4.07	3.02	3.24
11	......	......	......	5.25	3.24	3.67	3.88	2.99	2.87	4.00	3.03	3.21
12	......	......	......	6.69	3.20	3.26	3.76	3.00	2.94	3.93	3.10	3.20
13	......	......	......	8.55	3.18	3.28	3.56	2.80	2.80	3.87	3.09	3.29
14	......	......	5.18	9.65	3.16	3.16	3.40	2.77	2.90	3.81	3.05	3.95
15	......	......	4.90	9.30	3.13	3.09	3.27	2.73	3.20	3.74	*	4.15
16	......	......	4.75	8.53	3.10	3.04	3.15	2.84	3.60	3.67	*	†
17	......	......	4.58	7.86	3.09	3.02	3.07	2.85	3.70	3.69	*	†
18	......	......	4.58	7.10	3.04	3.00	3.10	2.82	3.79	3.58	*	†
19	......	......	4.52	6.02	2.99	2.93	3.66	2.81	3.83	3.40	*	†
20	......	......	4.35	4.89	2.95	2.96	4.00	2.84	3.60	3.39	*	4.20
21	......	......	4.54	3.30	2.93	2.98	3.87	2.86	3.50	3.38	*	4.18
22	......	......	4.40	4.85	2.99	2.87	3.45	2.85	‡	3.36	3.02	4.16
23	......	......	4.22	4.62	3.10	2.93	3.30	2.83	‡	3.35	3.05	4.14
24	......	......	3.97	4.47	3.28	2.96	3.14	2.81	‡	3.34	3.08	4.15
25	......	......	3.87	4.33	3.72	2.92	3.10	3.01	‡	3.31	3.07	4.13
26	......	......	3.88	4.47	4.05	2.87	3.05	2.94	‡	3.25	3.05	4.17
27	......	......	3.82	4.38	4.08	2.93	2.97	2.96	‡	3.24	3.04	4.16
28	......	......	3.78	4.14	3.80	2.86	2.93	2.97	‡	3.20	3.07	4.15
29	......	......	3.70	4.05	3.70	2.83	2.91	2.98	‡	3.19	3.09	4.14
30	......	......	3.60	3.90	3.54	2.82	2.94	3.01	3.05	3.14	3.11	4.13
31	......	......	3.50	......	3.38	......	2.96	2.97	......	3.09	......	4.13

*Gage reader absent. †Frozen. ‡Gage stolen.

Rating Table for Tippecanoe River at Springboro, Indiana, from March 14 to December 31, 1903.

Gage Height, Feet	Discharge, Sec.-Ft.	Gage Height, Feet	Discharge, Sec.-Ft.	Gage Height, Feet	Discharge, Sec.-Ft.	Gage Height, Feet	Discharge, Sec.-Ft.
2.7	551	4.0	1,987	5.3	4,230	7.2	7,650
2.8	611	4.1	2,151	5.4	4,410	7.4	8,010
2.9	680	4.2	2,320	5.5	4,590	7.6	8,370
3.0	758	4.3	2,490	5.6	4,770	7.8	8,730
3.1	845*	4.4	2,660	5.7	4,950	8.0	9,090
3.2	940	4.5	2,830	5.8	5,130	8.5	9,990
3.3	1,043	4.6	3,000	5.9	5,310	9.0	10,890
3.4	1,154	4.7	3,170	6.0	5,490	9.5	11,790
3.5	1,273	4.8	3,345	6.2	5,850	10.0	12,690
3.6	1,400	4.9	3,520	6.4	6,210	10.5	13,590
3.7	1,535	5.0	3,695	6.6	6,570	11.0	14,490
3.8	1,678	5.1	3,870	6.8	6,930		
3.9	1,829	5.2	4,050	7.0	7,290	....	....

Discharge Measurements of Tippecanoe River near Delphi, Indiana, in 1904.

Date	Hydrographer	Width, Ft.	Area of Section, Sq. Ft.	Mean Velocity, Ft. per Sec.	Gage Height, Feet	Discharge, Sec.-Ft.
January 23	F. W. Hanna	448	2,404	4.10	8.48	9,863
March 2*	F. W. Hanna	466	3,920	2.55	13.00	10,010
March 28	F. W. Hanna	449	2,560	4.78	8.80	12,240
May 2	Hanna and Johnson	343	917	4.04	4.95	3,708
June 18	F. W. Hanna	253	328	2.12	2.98	694
July 23	Hanna and Johnson	238	292	1.83	2.92	534
August 22	F. W. Hanna	255	370	2.38	3.20	882
September 13	F. W. Hanna	241	278	1.62	2.85	451
October 22	F. W. Hanna	249	273	1.77	2.90	484
November 5	F. W. Hanna	240	266	1.49	2.86	396

*Ice jam.

Mean Daily Gage Height, in Feet, of Tippecanoe River near Delphi, Indiana, for 1904.

Day	Jan.*	Feb.*	Mar.*	Apr.	May	June	July	Aug.	Sept.	Oct.	Nov.	Dec.*
1	......	......	14.60	9.55	6.59	4.25	3.49	2.85	2.93	3.10	2.83	2.83
2	......	......	13.44	9.43	5.07	4.21	3.37	2.82	2.90	3.07	2.82	2.82
3	......	......	13.21	8.49	4.80	3.87	3.21	2.80	2.87	3.05	2.84	2.80
4	......	......	12.90	8.34	5.60	3.74	3.09	2.81	2.86	3.00	2.84	2.80
5	......	......	12.37	8.21	5.12	3.65	3.17	2.81	2.84	2.95	2.85	2.91
6	4.10	8.25	12.25	8.03	4.96	3.61	3.22	2.77	2.85	2.91	2.85	2.96
7	......	......	12.80	7.95	4.72	3.40	3.75	2.75	2.83	2.90	2.84	2.97
8	3.80	......	13.20	8.20	4.57	3.45	4.03	2.73	2.83	2.93	2.80	2.98
9	......	......	12.47	8.75	4.35	3.49	4.40	2.73	2.84	3.00	2.85	3.00
10	......	......	7.05	8.15	4.05	3.52	4.37	2.72	2.85	3.07	2.87	3.02
11	......	......	7.00	7.75	4.00	3.43	4.26	2.71	2.85	3.05	2.86	3.00
12	......	......	6.95	7.08	3.93	3.18	4.20	2.69	2.84	3.00	2.85	2.99
13	......	7.85	6.91	5.53	3.84	3.17	4.17	2.74	2.81	3.00	2.86	2.98
14	......	......	7.00	4.98	3.89	3.13	3.98	2.73	3.00	2.98	2.84	......
15	......	......	6.87	5.02	3.91	3.13	3.96	2.72	2.98	2.97	2.87	......
16	3.85	......	6.77	5.17	3.89	3.12	3.71	2.71	3.01	3.03	2.89	......
17	......	......	7.00	5.17	3.85	3.54	3.55	2.78	3.05	3.00	2.88	3.04
18	......	......	7.20	5.14	4.01	3.27	3.47	2.80	3.21	2.97	2.86	......
19	......	......	7.23	5.12	3.99	3.20	3.38	2.81	3.10	2.95	2.86	......
20	4.75	8.20	7.10	5.09	3.95	3.16	3.30	2.90	3.05	2.92	2.85	......
21	11.05	......	7.15	4.98	3.90	3.10	3.21	3.05	3.00	2.90	2.83	......
22	11.25	......	7.77	4.91	3.81	3.09	3.00	3.35	2.98	2.98	2.80	......
23	10.95	......	8.10	4.67	3.70	3.07	2.95	3.42	2.97	2.95	2.85	3.05
24	......	......	8.54	4.47	3.64	3.06	2.91	3.41	2.95	2.94	......	3.09
25	......	......	8.92	4.01	3.73	3.00	2.90	3.35	2.93	2.97	......	3.12
26	......	......	11.20	5.80	3.72	2.96	2.90	3.28	3.10	2.95	......	......
27	15.00	8.20	9.41	7.03	3.71	3.00	2.99	3.18	3.15	2.92	......	......
28	......	......	7.94	6.70	3.70	3.83	2.98	2.98	3.23	2.81	......	......
29	......	15.20	7.02	6.35	3.60	3.71	2.93	3.00	3.20	2.85	2.84	......
30	7.25	......	6.76	6.10	3.74	3.79	2.91	2.93	3.16	2.85	2.83	3.05
31	......	......	8.90	......	4.20	......	2.89	2.90	......	2.84	......	......

*Frozen January 1 to March 9 and December 14 to 31.

Rating Table for Tippecanoe River near Delphi, Indiana, from January 1 to December 31, 1904.

Gage Height, Feet	Discharge, Sec.-Ft.	Gage Height, Feet	Discharge, Sec.-Ft.	Gage Height, Feet	Discharge, Sec.-Ft.	Gage Height, Feet	Discharge, Sec.-Ft.
2.7	280	3.8	1,770	4.9	3,640	7.0	8,110
2.8	390	3.9	1,930	5.0	3,830	7.5	9,220
2.9	510	4.0	2,090	5.2	4,230	8.0	10,370
3.0	630	4.1	2,250	5.4	4,650	8.5	11,520
3.1	760	4.2	2,410	5.6	5,070	9.0	12,670
3.2	890	4.3	2,580	5.8	5,490	9.5	13,820
3.3	1,030	4.4	2,750	6.0	5,910	10.0	14,970
3.4	1,170	4.5	2,920	6.2	6,350	10.5	16,120
3.5	1,320	4.6	3,090	6.4	6,790	11.0	17,270
3.6	1,470	4.7	3,270	6.6	7,230	12.0	19,570
3.7	1,620	4.8	3,450	6.8	7,670	13.0	21,870

The above table is applicable only for open-channel conditions. It is based upon discharge measurements made during 1903 and 1904. It is well defined between gage heights 2.8 feet and 8.8 feet. The table has been extended beyond these limits. Above gage height 7.4 feet the rating curve is a tangent, the difference being 230 per tenth.

Discharge Measurements of Tippecanoe River near Delphi, Indiana, in 1905.

Date	Hydrographer	Width, Ft.	Area of Section, Sq. Ft.	Mean Velocity, Ft. per Sec.	Gage Height, Feet	Discharge, Sec.-Ft.
March 21	S. K. Clapp	319	713	3.9	4.25	2,782
May 27	M. S. Brennan	325	682	3.23	4.29	2,203
June 13	S. K. Clapp	285	672	3.66	4.30	2,459
July 14	S. K. Clapp	272	617	3.44	4.10	2,120
August 24	M. S. Brennan	256	294	1.89	2.98	556
October 5	M. S. Brennan	251	276	1.76	2.98	486

Daily Gage Heights, in Feet, of Tippecanoe River near Delphi, Indiana, for 1905.

Day	Jan.	Feb.	Mar.	Apr.	May	June	July	Aug.	Sept.	Oct.	Nov.	Dec.
1	3.04	......	7.8	4.1	7.9	4.4	3.4	3.35	3.9	3.02	3.25	4.4
2	3.08	......	6.9	4.0	7.8	4.2	3.35	3.3	3.5	3.09	3.25	4.4
3	3.09	......	6.1	3.95	7.8	4.0	3.3	3.3	3.6	3.08	3.2	4.25
4	3.08	3.06	5.9	3.7	7.6	3.95	3.25	2.25	3.75	3.04	3.2	4.2
5	3.06	......	5.7	3.65	7.4	3.9	3.15	3.15	3.7	3.02	3.15	4.0
6	3.03	......	5.6	3.6	7.3	3.85	3.15	3.06	3.55	3.0	3.8	3.85
7	3.0	......	5.5	3.5	7.2	3.75	3.25	3.0	3.45	2.98	4.1	3.8
8	3.0	......	5.4	3.55	6.8	3.75	3.25	3.0	3.35	2.98	4.0	3.7
9	......	......	5.1	3.55	6.8	3.9	3.2	2.98	3.15	2.97	3.95	3.65
10	......	......	4.7	3.55	6.5	4.0	3.3	2.98	3.1	2.96	3.95	3.55
11	......	3.1	4.1	3.5	5.6	4.5	3.7	2.97	3.09	2.96	3.9	3.5
12	......	......	4.1	3.5	10.1	4.3	4.3	2.99	3.06	2.95	3.9	3.45
13	......	......	4.05	3.7	9.4	4.4	4.1	3.0	3.1	2.94	3.8	3.45
14	3.0	......	4.05	3.9	10.2	4.4	4.0	3.09	3.3	2.93	3.7	3.65
15	......	......	4.2	3.85	10.0	4.4	3.9	3.2	3.25	2.91	3.6	3.6
16	......	......	4.2	3.85	9.8	4.3	3.7	3.45	3.15	2.9	3.5	3.5
17	......	......	4.2	3.8	8.9	5.0	3.6	3.3	3.15	2.9	3.35	3.45
18	......	3.12	4.2	3.75	7.6	5.6	3.45	3.15	3.15	3.15	3.35	3.4
19	......	......	4.1	5.2	6.5	5.0	3.45	3.15	3.1	4.4	3.3	3.4
20	......	3.1	4.1	7.0	4.8	4.3	3.35	3.45	3.1	4.4	3.3	3.25
21	3.03	......	4.05	6.1	4.8	4.2	3.25	3.4	3.09	4.3	3.3	3.3
22	......	......	4.1	6.0	4.7	4.3	3.2	3.35	3.08	4.1	3.3	3.4
23	......	......	4.3	6.0	4.7	4.05	3.2	3.3	3.1	4.0	3.25	4.8
24	......	......	4.3	6.59	4.6	3.85	3.15	3.25	3.1	3.95	3.25	4.65

Daily Gage Heights, in Feet, of Tippecanoe River near Delphi, Indiana, for 1905—Continued.

DAY	Jan.	Feb.	Mar.	Apr.	May	June	July	Aug.	Sept.	Oct.	Nov.	Dec.
25	......	3.15	4.2	5.8	4.6	3.7	3.1	3.25	3.09	3.85	3.2	4.2
26	......	......	4.2	5.5	4.4	3.6	3.07	3.2	3.08	3.8	3.15	3.9
27	......	......	4.2	5.4	4.4	3.65	3.05	3.15	3.06	3.6	3.1	3.9
28	3.1	......	4.1	7.8	4.4	3.7	3.02	3.08	3.05	3.35	3.1	3.85
29	......	......	4.1	8.0	4.5	3.5	3.35	3.01	3.04	3.3	4.3	3.35
30	......	......	4.2	7.9	4.5	3.45	3.65	2.95	3.03	3.3	4.4	3.8
31	......	4.90	4.1	......	4.7	......	3.4	2.98	......	3.25	......	3.75

NOTE—There were ice conditions during January and February. From January 9 to February 11 th river was frozen over except for a narrow channel near the west bank. February 12-28 river frozen entirely across. Gage heights are to surface of water in hole in ice. The following comparative readings were also made:

DATE	Water Surface, Feet	Top of Ice, Feet	Thickness of Ice, Ft.
January 14	3.0	2.9	0.3
January 21	3.03	3.03	.35
January 28	3.1	3.12	.5
February 4	3.06	....	.5
February 11	3.1	3.12	1.0
February 18	3.12	3.12	1.0
February 25	3.15	3.19	1.0

Station Rating Table for Tippecanoe River near Delphi, Indiana, from January 1, 1904, to December 31, 1905.

Gage Height, Feet	Discharge, Sec.-Ft.	Gage Height, Feet	Discharge, Sec.-Ft.	Gage Height, Feet	Discharge, Sec.-Ft.	Gage Height, Feet	Discharge, Sec.-Ft.
2.70	280	4.20	2,410	5.70	5,280	8.40	11,290
2.80	390	4.30	2,580	5.80	5,490	8.60	11,750
2.90	510	4.40	2,750	5.90	5,700	8.80	12,210
3.00	630	4.50	2,920	6.00	5,910	9.00	12,670
3.10	760	4.60	3,090	6.20	6,350	9.20	13,130
3.20	890	4.70	3,270	6.40	6,790	9.40	13,590
3.30	1,030	4.80	3,450	6.60	7,230	9.60	14,050
3.40	1,170	4.90	3,640	6.80	7,670	9.80	14,510
3.50	1,320	5.00	3,830	7.00	8,110	10.00	14,970
3.60	1,470	5.10	4,030	7.20	8,550	10.50	16,120
3.70	1,620	5.20	4,230	7.40	8,990	11.00	17,270
3.80	1,770	5.30	4,440	7.60	9,450	11.50	18,420
3.90	1,930	5.40	4,650	7.80	9,910	12.00	19,570
4.00	2,090	5.50	4,840	8.00	10,370	12.50	20,720
4.10	2,250	5.60	5,070	8.20	10,830	13.00	21,870

NOTE—The above table is applicable only for open-channel conditions. It is based on 23 discharge measurements made during 1903-1905. It is not very well defined.

Discharge Measurements for Tippecanoe River at Delphi, Indiana, in 1906.

DATE	Hydrographer	Width, Feet	Area of Section, Sq. Ft.	Gage Height, Feet	Discharge, Sec.-Ft.
February 10*	Brennan and Kriegsman	272	522	3.86	1,450
March 10	E. F. Kriegsman	335	790	4.74	3,320
April 3	E. F. Kriegsman	325	779	4.62	3,090
May 9	E. F. Kriegsman	257	361	3.27	962

*Slush and cake ice running.

Daily Gage Height, in Feet, of Tippecanoe River near Delphi, Indiana, for 1906.

DAY	Jan.	Feb.	Mar.	Apr.	May	June	July
1	3.75	5.17	4.18	5.47	3.49	3.1	3.0
2	3.73	4.8	4.76	5.25	3.45	3.04	3.02
3	3.7	4.2	5.34	4.97	3.57	3.0	3.0
4	3.72	4.16	5.27	4.63	3.51	2.99	2.99
5	3.78	4.09	5.21	4.37	3.44	2.97	2.97
6	3.74	4.02	4.99	4.6	3.43	3.04	2.96
7	3.69	3.96	4.96	4.5	3.4	3.0	2.94
8	3.66	3.93	4.89	4.42	3.33	2.95	2.9
9	3.61	3.9	4.83	5.7	3.27	3.1	2.88
10	3.58	3.87	4.82	5.27	3.25	3.08	2.85
11	3.55	3.84	4.79	5.06	3.21	3.11	2.84
12	3.52	3.81	4.76	4.89	3.19	3.32	2.87
13	3.47	3.78	4.72	4.56	3.22	3.26	2.9
14	3.43	3.76	4.68	4.63	3.18	3.17	2.81
15	3.57	3.68	4.66	5.17	3.16	3.13	2.85
16	3.51	3.64	4.59	4.95	3.15	3.1	2.98
17	3.48	3.61	4.37	4.8	3.14	3.04	2.97
18	3.46	3.66	4.19	4.67	3.11	3.0	2.9
19	3.42	3.64	3.94	4.43	3.09	2.99	2.84
20	4.73	3.6	3.88	4.37	3.08	3.0	2.82
21	5.8	3.57	3.75	4.19	3.05	2.95	......
22	6.12	3.52	3.73	4.0	3.03	2.96	......
23	6.36	3.48	3.7	3.96	3.1	2.94	......
24	6.39	3.45	3.75	3.91	3.15	2.93	......
25	6.3	4.4	3.78	3.84	3.13	2.93	......
26	6.15	4.33	4.12	3.76	3.11	2.9	......
27	6.1	4.28	6.18	3.65	3.07	2.87	......
28	6.0	4.21	6.0	3.6	3.05	2.85	......
29	5.85	......	5.94	3.57	3.03	2.95	......
30	5.57	......	5.86	3.53	3.1	3.02	......
31	5.25	......	5.71	......	3.15	......	......

NOTE—Flow slightly affected by ice conditions February 5 to 10.

Rating Table for Tippecanoe River near Delphi, Indiana, for 1904 to 1906.

Gage Height, Feet	Discharge, Sec.-Ft.	Gage Height, Feet	Discharge, Sec.-Ft.	Gage Height, Feet	Discharge, Sec.-Ft.	Gage Height, Feet	Discharge, Sec.-Ft.
2.80	390	3.70	1,620	4.60	3,090	5.50	4,860
2.90	510	3.80	1,770	4.70	3,270	5.60	5,070
3.00	630	3.90	1,930	4.80	3,450	5.70	5,280
3.10	760	4.00	2,090	4.90	3,640	5.80	5,490
3.20	890	4.10	2,250	5.00	3,830	5.90	5,700
3.30	1,030	4.20	2,410	5.10	4,030	6.00	5,910
3.40	1,170	4.30	2,580	5.20	4,230	6.20	6,350
3.50	1,320	4.40	2,750	5.30	4,440	6.40	6,790
3.60	1,470	4.50	2,920	5.40	4,650	....	....

NOTE—The above table is applicable only for open-channel conditions. It is based on 25 discharge measurements made during 1903 to 1906. It is well defined.

Hog Point.—This station was established by the writer on February 12, 1916. The gage is of the chain and weight type and is located on the upstream truss rods of the first span from the east end of the Hog Point bridge, Sec. 9, R. 3 W., T. 24 N. The zero of the gage is 17.62 feet below the top surface of the northwest corner of the east abutment. The gage is read daily by Paul Booth. The following data have been collected from this station.

Discharge Measurements at Hog Point Bridge during 1916.

DATE	Hydrographer	Gage Height, Feet	Discharge, Sec.-Ft.
February 12	W. M. Tucker	7.6	10,992
February 13	W. M. Tucker	7.7	11,227
May 11	W. M. Tucker	3.8	3,038
May 14	W. M. Tucker	10.7	20,702
July 26	W. M. Tucker	3.1	1,807
August 19	W. M. Tucker	3.3	2,079
September 5	W. M. Tucker	2.0	409
September 6	W. M. Tucker	1.9	305
September 25	W. M. Tucker	1.9	299

Daily Gage Heights on Tippecanoe River at Hog Point, near Battle Ground, Indiana, for 1916.

DAY	Feb.	Mar.	Apr.	May	June	July	Aug.	Sept.
1	......	3.1	5.4	3.5	5.5	4.5	3.1	3.1
2	......	3.1	5.5	3.6	5.7	4.5	3.1	2.8
3	......	3.1	5.2	3.6	6.2	4.5	3.1	2.6
4	......	3.5	4.8	3.6	5.7	4.4	3.1	2.3
5	......	3.4	4.6	3.6	5.5	4.3	3.1	2.0
6	......	3.3	4.3	3.6	5.2	4.3	3.2	2.0
7	......	3.2	4.1	3.6	5.3	4.3	3.2	1.9
8	......	3.2	4.0	3.6	5.4	4.2	3.1	1.9
9	......	3.1	3.7	4.8	5.4	4.2	3.1	1.9
10	......	3.1	3.6	3.9	5.6	4.0	3.1	1.9
11	*	3.1	3.5	3.8	5.5	3.6	3.1	1.9
12	7.6	3.1	3.4	3.7	5.5	3.4	3.2	1.9
13	7.7	3.1	3.5	6.4	5.4	3.3	3.1	1.9
14	7.7	3.1	3.5	10.6	5.3	3.1	3.1	1.9
15	7.4	3.1	3.5	11.4	5.3	3.1	3.1	1.9
16	7.1	3.1	3.4	11.0	5.4	3.2	3.1	1.9
17	7.3	3.0	3.3	10.0	6.4	3.2	3.1	1.9
18	6.5	3.0	3.2	9.5	6.5	3.1	3.2	1.9
19	6.8	3.0	3.1	9.4	6.0	3.1	3.3	1.9
20	6.2	3.0	3.2	8.6	5.9	3.1	3.1	1.9
21	5.9	2.9	3.2	7.1	5.7	3.2	3.1	1.9
22	5.6	3.0	3.2	6.7	5.4	3.1	3.1	1.9
23	5.5	6.2	3.2	6.2	5.2	3.0	3.1	1.9
24	5.1	6.6	3.2	5.9	5.1	3.1	3.1	1.9
25	4.6	6.3	3.2	5.6	4.7	3.2	3.1	1.9
26	4.0	5.8	3.2	5.2	4.6	3.1	3.1	1.9
27	3.6	6.2	4.4	5.1	4.6	3.1	3.1	1.9
28	3.4	7.3	4.2	5.1	4.6	3.1	3.2	2.0
29	3.2	8.5	4.0	5.2	4.6	3.1	3.3	2.2
30	......	8.2	3.7	5.1	4.5	3.1	3.4	2.2
31	......	6.5	......	5.0	......	3.1	3.2	†

*Gage installed February 12.

†No record during rest of year.

Rating Table for Tippecanoe River at Hog Point Bridge for 1916.

Gage Height, Feet	Discharge, Sec.-Ft.	Gage Height, Feet	Discharge, Sec.-Ft.	Gage Height, Feet	Discharge, Sec.-Ft.
1.9	302	3.6	2,600	5.3	6,110
2.0	414	3.7	2,870	5.4	6,340
2.1	526	3.8	3,050	5.5	6,580
2.2	641	3.9	3,230	5.6	6,820
2.3	756	4.0	3,410	5.7	7,060
2.4	878	4.1	3,600	5.8	7,300
2.5	1,000	4.2	3,790	5.9	7,550
2.6	1,125	4.3	3,990	6.0	7,800
2.7	1,250	4.4	4,190	6.5	9,100
2.8	1,380	4.5	4,390	7.0	10,450
2.9	1,510	4.6	4,590	7.5	11,750
3.0	1,650	4.7	4,800	8.0	13,125
3.1	1,790	4.8	5,010	8.5	14,500
3.2	1,940	4.9	5,220	9.0	15,900
3.3	2,100	5.0	5,440	9.5	17,300
3.4	2,260	5.1	5,660	10.0	18,700
3.5	2,430	5.2	5,880	10.5	20,100
				11.0	21,500

This table is applicable only for open-channel conditions.

From the preceding data the minimum discharge is found to be 269 second-feet, and to have occurred on August 12, 1904. The lowest in 1903 was 600 second-feet on July 1; in 1905, 510 second-feet on October 16 and 17; and in 1906, 400 second-feet on July 14. This data leads to the conclusion that a discharge of about 300 second-feet could be taken as a safe minimum.

The United States Weather Bureau shows the slightest rainfall for the time during which this gage record was kept to have been in November, 1904, when the mean monthly precipitation for northern Indiana was but .23 inch. During this month, however, the discharge at Springboro did not fall below 390 second-feet. The mean monthly rainfall for July, 1904, was 3. 46 inches in northern Indiana, and in August was 3.17 inches, which is a normal rainfall in each case. A problem presents itself as to the reason for the discharge being so low on August 12 and remaining so high throughout November. The answer is unknown and cannot be derived from the data. The natural storage produced by the lakes would keep the discharge up during drought, but why the discharge should fall so low in August, when 30 per cent of the precipitation in the Tippecanoe basin would have given a continuous discharge of over 1,200 second-feet; while the maximum discharge, which was on the 23d, was only 1,200 second-feet, according to the gage readings, is a point which cannot be accounted for except by the conclusion that the data in one or the other case is incorrect. It seems very probable that the gage readings are low, because the rainfall data are collected from twenty-one stations and the error at any one station would be divided by twenty-one in striking the mean, and it is unlikely that twenty-one people reading rain gages should make an error which would be very large in the average. For these reasons the minimum discharge is estimated to be somewhat higher than the data show on August 21, 1904.

The profile of the Tippecanoe River is peculiar in that the fall in the upper course is much less than in the lower course. The greatest fall occurs in the vicinity of Monticello. This is due to the fact that the stream flows over the Cincinnati anticline in northern White County and has been unable to cut through the hard rock at Norway. The Wabash River has been deepening its valley at the mouth of the Tippecanoe faster than the Tippecanoe could deepen its valley in the vicinity of Norway. The result of this condition is a gradually increasing fall in the lower course of the stream. The following table shows the profile of the river:

PROFILE OF TIPPECANOE RIVER.

LOCATION	Estimated Distance, Miles	Altitude, Feet	Fall per Mile, Inches
Source	0	950	......
De Long	80	711	20.85
Monterey	7	703	13.71
Ora	5	698	12.00
Winamac	15	678	16.00
Monticello	31	599	30.58
Oakdale	11	563	37.09
Springboro	8	544	28.50
Mouth	9	524	26.66

A discussion of the water power conditions along the Tippecanoe River is given on pages 508 to 512 of the 36th Annual Report of the Indiana Department of Geology and Natural Resources.

Mississinewa River.

The Mississinewa River rises in Darke County, Ohio, enters Indiana in Randolph County and flows in a northwest direction to its mouth near Peru. It is very crooked, has no large tributaries and is 105 miles in length. The valley lies between two glacial moraines which were deposited by the Erie Lobe of the Wisconsin Glacier. It is carved into bed rock at several points. The rock is Silurian. No natural storage occurs along this stream and the discharge is very irregular. The drainage area is 861 square miles. A gage was in operation at Peoria from July 8, 1910, until March 26, 1913.

Peoria.—This station was established by the writer on July 8, 1910. It is a direct reading gage made of heavy oak bridge planks and securely spiked to the root of a small tree and to two oak posts. It is set with the slant of the river bank, which is about thirty degrees. The scale is made of brass-headed tacks on the upstream side of the gage. Zero of this gage is 18.69 feet below a wire nail in the north side of a small hackberry tree which stands 37 feet southwest of the upper end of the gage. The tree is beside a large bowlder and has a woven wire fence attached. The gage was read each day by H. F. Whisler. The data from this gage follow:

Discharge Measurements on Mississnewa River at Peoria, Indiana, during 1910-1913.

DATE	Hydrographer	Gage Height, Feet	Discharge, Sec.-Ft.
July 8, 1910	W. M. Tucker	1.0	65
April 5, 1911	W. M. Tucker	4.6	3,191
June 23, 1911	Tucker and Clark	1.8	183
June 26, 1911	Tucker and Clark	1.9	203
November 19, 1911	W. M. Tucker	6.6	7,153
December 11, 1911	W. M. Tucker	2.5	399
March 13, 1912	W. M. Tucker	3.3	1,016
March 16, 1912	W. M. Tucker	7.8	9,738
May 4, 1912	W. M. Tucker	3.5	1,279

Daily Gage Heights on Mississinewa River at Peoria, Indiana, for 1910.

Day	Jan.	Feb.	Mar.	Apr.	May	June	July	Aug.	Sept.	Oct.	Nov.	Dec.
1	....	....	....	....	....	....	*....	1.1	1.0	1.7	1.3	3.0
2	....	....	....	....	....	....	....	1.1	1.0	1.5	1.3	2.8
3	....	....	....	....	....	....	....	1.0	1.0	1.4	1.2	2.4
4	....	....	....	....	....	....	....	1.0	1.0	1.4	1.2	2.2
5	....	....	....	....	....	....	....	1.0	1.3	1.3	1.2	2.0
6	....	....	....	....	....	....	....	1.0	1.7	1.3	1.2	2.9
7	....	....	....	....	....	....	*	1.0	1.7	6.6	1.2	2.8
8	....	....	....	....	....	....	1.0	1.0	1.5	8.4	1.2	1.7
9	....	....	....	....	....	....	1.0	1.0	1.6	7.8	1.1	1.7
10	....	....	....	....	....	....	1.0	1.0	1.6	7.8	1.1	1.7
11	....	....	....	....	....	....	1.1	1.0	1.6	7.4	1.1	1.7
12	....	....	....	....	....	....	1.0	1.0	1.5	6.4	1.1	1.8
13	....	....	....	....	....	....	1.0	1.1	2.1	4.0	1.1	1.8
14	....	....	....	....	....	....	1.0	1.1	2.0	3.2	1.1	1.7
15	....	....	....	....	....	....	1.0	1.1	2.0	2.1	1.1	1.7
16	....	....	....	....	....	....	1.0	1.1	2.1	1.8	1.1	1.6
17	....	....	....	....	....	....	1.0	1.1	2.2	1.8	1.1	1.6
18	....	....	....	....	....	....	1.1	1.0	2.3	1.7	1.1	1.6
19	....	....	....	....	....	....	1.0	1.0	2.3	1.7	1.1	1.5
20	....	....	....	....	....	....	1.0	1.0	2.3	1.6	1.1	1.5
21	....	....	....	....	....	....	1.0	1.0	2.3	1.6	1.1	1.5
22	....	....	....	....	....	....	1.1	1.0	2.3	1.6	1.1	1.5
23	....	....	....	....	....	....	1.0	1.0	2.2	1.5	1.1	1.6
24	....	....	....	....	....	....	1.0	1.0	2.2	1.5	1.1	1.6
25	....	....	....	....	....	....	1.1	1.0	2.2	1.5	1.1	1.6
26	....	....	....	....	....	....	1.0	1.1	2.1	1.5	1.1	1.6
27	....	....	....	....	....	....	1.0	1.2	2.1	1.5	1.1	1.6
28	....	....	....	....	....	....	1.0	1.1	2.0	1.5	1.5	1.9
29	....	....	....	....	....	....	1.0	1.1	2.0	1.4	2.9	2.8
30	....	....	....	....	....	....	1.0	1.0	1.8	1.4	3.3	5.0
31	....	....	....	....	....	....	1.1	1.0	....	1.4	....	4.8

*Gage installed July 7.

Daily Gage Heights on Mississinewa River at Peoria, Indiana, for 1911.

Day	Jan.	Feb.	Mar.	Apr.	May	June	July	Aug.	Sept.	Oct.	Nov.	Dec.
1	4.7	4.2	2.3	2.0	3.7	1.2	1.5	1.2	2.1	2.7	1.5	2.7
2	4.5	3.6	2.2	2.0	2.9	1.2	1.5	1.2	1.9	3.1	1.5	2.7
3	4.5	3.4	2.1	2.0	2.8	1.2	1.9	1.2	1.9	2.7	1.5	2.7
4	4.3	3.1	2.0	2.1	2.1	1.2	1.3	1.0	1.5	2.1	1.5	2.7
5	4.0	3.0	2.0	2.1	2.2	1.2	1.3	1.0	1.7	1.8	1.5	2.7
6	3.4	2.8	1.9	5.3	2.2	1.6	1.3	1.0	1.7	2.3	1.6	2.7
7	3.0	2.8	1.8	5.2	2.3	2.5	1.3	1.1	1.7	2.7	1.4	2.4
8	2.9	2.8	1.9	4.8	2.0	2.5	1.3*	1.1	1.7	3.1	1.4	2.2
9	2.8	2.7	1.9	4.0	2.4	2.1	1.2	1.4	1.7	3.5	1.4	2.2
10	2.8	2.7	2.2	3.6	2.4	2.0	1.1	1.0	1.3	3.1	1.7	2.3
11	2.7	2.7	2.8	3.1	2.2	1.7	1.2	1.0	1.3	2.6	1.9	2.5
12	2.7	2.5	3.2	2.6	2.2	1.7	1.1	1.3	1.3	2.6	3.1	2.5
13	2.6	2.4	2.8	2.4	2.0	1.7	1.0	1.5	1.5	2.1	2.1	2.9
14	2.7	2.8	2.5	3.2	2.0	1.4	1.1	1.5	3.7	2.5	2.1	3.7
15	2.9	3.8	2.3	4.6	2.0	1.4	1.1	1.4	5.0	2.0	1.9	4.5
16	5.0	3.7	2.1	4.4	2.0	1.3	1.1	1.2	4.4	1.9	2.3	4.1
17	6.4	3.3	2.1	2.7	1.7	1.3	1.1	1.2	3.4	3.2	4.1	4.0
18	5.6	3.2	2.1	2.3	1.7	1.1	1.0	1.2	3.1	3.0	6.7	3.9
19	4.9	3.0	2.1	2.0	1.8	1.1	1.0	1.2	2.7	2.6	6.7	3.3
20	4.4	2.9	2.2	3.8	1.8	1.4	1.2	1.1	2.1	2.6	6.1	3.5
21	4.1	2.8	2.2	3.8	1.6	1.4	1.0	1.1	2.3	2.6	5.3	2.6
22	3.8	2.6	2.1	3.5	1.6	1.5	1.1	1.1	2.7	2.7	3.2	2.6
23	3.4	2.5	2.0	3.7	1.6	1.8	1.2	1.2	3.1	2.8	3.2	2.6
24	3.6	2.4	2.0	3.2	1.5	1.6	1.3	2.1	3.0	2.3	3.2	2.5
25	3.0	2.4	1.9	2.3	1.4	2.2	1.3	2.3	2.8	2.3	3.0	2.1
26	2.8	2.4	1.9	2.1	1.4	1.9	1.3	2.3	2.8	2.3	2.7	2.1
27	3.4	2.3	2.1	1.8	1.4	1.7	1.2	2.0	2.9	1.8	2.9	2.1
28	7.1	2.3	2.0	1.5	1.4	1.6	1.2	2.1	2.3	1.8	2.9	2.0
29	7.0	....	1.9	1.5	1.4	1.6	1.2	2.1	1.9	1.8	2.8	2.0
30	6.6	....	2.0	2.2	1.2	1.5	1.2	2.1	2.1	1.5	2.8	2.1
31	5.0	....	2.1	....	1.2	....	1.2	2.1	....	1.5	....	2.7*

*No record, reading interposed.

Daily Gage Heights on Mississinewa River at Peoria, Indiana, for 1912.

Day	Jan.	Feb.	Mar.	Apr.	May	June	July	Aug.	Sept.	Oct.	Nov.	Dec.
1	3.2	2.1	5.0	4.7	3.3	1.9	1.3	1.1	2.0	1.3	1.9	1.4
2	3.0	2.1	4.7	4.3	3.1	1.9	1.3	1.3	2.0	1.3	1.5	1.3
3	2.7	2.1	4.7	4.0	3.2	1.9	1.3	1.3	2.0	1.3	1.5	1.3
4	2.7	2.1	3.1	3.9	3.3	1.9	1.3	1.4	2.0	1.3	1.5	1.7
5	2.3	1.7	3.1	3.9	4.2	1.7	1.4	1.9	2.1	1.4	1.3	1.8
6	2.0	1.7	3.1	3.7	3.7	1.9	1.4	3.5	1.7	1.3	1.0	2.0
7	2.0	1.6	2.0	3.3	3.1	1.4	1.7	4.7	1.7	1.3	1.0	2.3
8	2.0	1.6	2.0	3.3	3.1	1.4	3.4	4.8	1.5	1.3	1.0	2.3
9	1.9	1.6	2.0	3.1	3.1	1.4	3.7	4.8	1.5	1.3	1.0	2.3
10	1.6	1.6	2.7	2.7	2.3	1.4	3.7	4.8	1.5	1.1	1.0	2.3
11	1.6	1.6	3.1	2.7	2.3	1.4	3.7	4.7	1.2	1.1	1.0	2.1
12	1.6	1.6	3.0	2.9	2.3	1.1	3.1	4.7	1.0	1.1	1.0	2.5
13	1.6	1.6	3.0	2.9	2.3	1.1	3.0	5.1	1.0	1.1	1.0	2.4
14	1.6	1.6	3.9	2.6	2.7	1.1	2.4	5.7	1.0	1.1	1.0	2.4
15	1.6	1.6	8.1	2.6	3.1	1.8	2.1	6.6	1.0	1.1	1.2	2.6
16	1.6	1.6	7.3	2.6	3.1	1.9	1.3	6.6	1.0	1.0	1.2	2.3
17	1.6	1.9	7.1	2.6	4.3	1.5	1.3	6.6	1.0	1.0	1.1	2.3
18	1.9	2.9	7.0	2.8	4.3	1.2	1.3	6.3	1.0	1.2	1.0	2.1
19	6.7	3.1	6.0	3.5	4.1	2.7	1.3	6.0	1.0	1.2	1.1	2.1
20	5.1	4.3	4.3	2.7	3.6	2.9	1.3	5.3	2.6	1.2	1.1	2.1
21	4.7	4.7	3.3	2.7	3.3	2.2	1.0	5.2	2.5	3.0	1.1	2.1
22	4.7	4.2	3.1	2.7	3.0	2.2	1.0	5.0	2.3	3.0	1.1	2.1
23	4.5	4.2	3.1	2.6	2.7	2.2	1.0	4.7	2.3	2.6	1.1	2.1
24	4.1	7.1	3.0	2.6	2.7	1.9	1.0	4.7	2.3	2.4	1.1	2.3
25	4.1	7.9	3.7	2.7	2.3	1.7	1.0	3.6	2.3	2.5	1.1	2.4
26	3.6	8.3	3.7	2.3	2.3	1.7	1.1	3.6	2.0	2.2	1.1	2.7
27	3.1	8.1	3.9	2.3	1.9	1.7	1.1	3.1	1.7	2.2	1.1	2.7
28	2.9	7.4	4.1	2.3	2.0	1.7	1.1	2.5	1.7	2.2	1.1	2.9
29	2.9	6.2	8.2	2.3	2.1	1.7	1.1	2.5	1.3	1.9	1.2	3.1
30	2.2	......	8.7	4.1	2.0	1.3	1.1	2.5	1.3	1.9	1.2	3.1
31	2.1	......	6.3	......	2.0	......	1.1	2.2	......	1.9*	......	3.1

*No record, reading interposed.

Daily Gage Heights on Mississinewa River at Peoria, Indiana, for 1913.

Day	Jan.	Feb.	Mar.	Apr.	May	June	July	Aug.	Sept.	Oct.	Nov.	Dec.
1	3.0	2.2	1.8	......	......	......	......	......	......	......	......	......
2	3.0	2.2	2.3	......	......	......	......	......	......	......	......	......
3	3.0	2.3	2.3	......	......	......	......	......	......	......	......	......
4	3.0	2.3	3.4	......	......	......	......	......	......	......	......	......
5	2.7	2.3	3.9	......	......	......	......	......	......	......	......	......
6	2.3	2.0	5.1	......	......	......	......	......	......	......	......	......
7	2.3	3.0	5.1	......	......	......	......	......	......	......	......	......
8	2.3	3.1	5.3	......	......	......	......	......	......	......	......	......
9	2.3	3.4	4.4	......	......	......	......	......	......	......	......	......
10	2.5	3.7	4.5	......	......	......	......	......	......	......	......	......
11	2.9	4.1	4.1	......	......	......	......	......	......	......	......	......
12	2.9	4.1	4.0	......	......	......	......	......	......	......	......	......
13	2.9	4.3	3.7	......	......	......	......	......	......	......	......	......
14	2.9	4.3	3.3	......	......	......	......	......	......	......	......	......
15	3.3	4.3	3.1	......	......	......	......	......	......	......	......	......
16	3.1	4.0	3.1	......	......	......	......	......	......	......	......	......
17	3.1	3.7	4.6	......	......	......	......	......	......	......	......	......
18	3.5	2.3	4.6	......	......	......	......	......	......	......	......	......
19	3.5	2.3	3.9	......	......	......	......	......	......	......	......	......
20	3.6	2.1	3.7	......	......	......	......	......	......	......	......	......
21	3.7	2.1	3.3	......	......	......	......	......	......	......	......	......
22	3.5	1.7	3.3	......	......	......	......	......	......	......	......	......
23	3.5	2.3	4.1	......	......	......	......	......	......	......	......	......
24	3.5	2.3	4.8	......	......	......	......	......	......	......	......	......
25	3.1	2.5	7.8	......	......	......	......	......	......	......	......	......
26	3.1	2.2	9.9	......	......	......	......	......	......	......	......	......
27	3.1	2.1	*	......	......	......	......	......	......	......	......	......
28	2.7	1.8	......	......	......	......	......	......	......	......	......	......
29	2.6	......	......	......	......	......	......	......	......	......	......	......
30	2.9	......	......	......	......	......	......	......	......	......	......	......
31	2.9	......	......	......	......	......	......	......	......	......	......	......

*Gage out, flood of 1913. Gage never reinstalled.

Rating Table for Mississinewa River at Peoria, Indiana, for 1910 to 1913, Inclusive.

Gage Height, Feet	Discharge, Sec.-Ft.	Gage Height, Feet	Discharge, Sec.-Ft.	Gage Height, Feet	Discharge, Sec.-Ft.	Gage Height, Feet	Discharge, Sec.-Ft.
1.0	66	3.3	1,023	5.6	5,126	7.9	9,955
1.1	77	3.4	1,149	5.7	5,325	8.0	10,175
1.2	89	3.5	1,285	5.8	5,525	8.1	10,395
1.3	102	3.6	1,430	5.9	5,726	8.2	10,615
1.4	116	3.7	1,583	6.0	5,928	8.3	10,835
1.5	131	3.8	1,743	6.1	6,131	8.4	11,055
1.6	147	3.9	1,910	6.2	6,335	8.5	11,275
1.7	164	4.0	2,083	6.3	6,540	8.6	11,495
1.8	183	4.1	2,261	6.4	6,746	8.7	11,715
1.9	204	4.2	2,443	6.5	6,953	8.8	11,935
2.0	227	4.3	2,628	6.6	7,161	8.9	12,155
2.1	252	4.4	2,815	6.7	7,370	9.0	12,375
2.2	280	4.5	3,003	6.8	7,580	9.1	12,595
2.3	312	4.6	3,191	6.9	7,791	9.2	12,815
2.4	349	4.7	3,380	7.0	8,003	9.3	13,035
2.5	391	4.8	3,570	7.1	8,216	9.4	13,255
2.6	439	4.9	3,761	7.2	8,430	9.5	13,475
2.7	494	5.0	3,953	7.3	8,645	9.6	13,695
2.8	557	5.1	4,146	7.4	8,861	9.7	13,915
2.9	629	5.2	4,340	7.5	9,078	9.8	14,135
3.0	711	5.3	4,535	7.6	9,296	9.9	14,355
3.1	804	5.4	4,731	7.7	9,515	....	....
3.2	908	5.5	4,928	7.8	9,735	....	....

NOTE—This table is applicable only for open-channel conditions. It is a tangent above 7.8 feet with 220 sec.-ft. per 0.1 ft. variation. It is not very accurate above 7.8 feet.

A discussion of the power conditions along the Mississinewa is given on pages 516 to 518 of the 36th Annual Report of the Department of Geology and Natural Resources. In the same article will be found a discussion of other tributaries of the Wabash from which no gage or discharge data have been taken.

St. Joseph River.

The St. Joseph River rises in the lake district of south central Michigan and flows southwest into Indiana, where it makes a great bend to the west and then to the north, and finally re-enters Michigan, where it flows into Lake Michigan at St. Joseph. The total length in Indiana is about forty-two miles.

The valley is narrow and is from forty to sixty feet deep. There is no swamp land within the valley and no storage is developed in the stream.

The drainage basin lies entirely within glacial drift and no bed rock is exposed. About four hundred lakes varying in size from five square miles to one-eighth of a square mile or less form a great amount of natural storage. The entire drainage area is approximately 4,600 square miles, of which 1,668 are in Indiana.

The mean annual rainfall above the gage at South Bend on the St. Joseph River as shown by the station records from this section for eighteen years (1898-1915, inclusive) is 35.78 inches. The following tables which are computed entirely from government reports show the average depth of monthly rainfall and the cubic feet per acre rainfall for the basin of the St. Joseph River for the eighteen years above mentioned. These averages are taken from nine U. S. Weather Bureau recording stations. The following is a list of the stations used: Angola, Elkhart, Goshen, Howe, Lima, Notre Dame, South Bend, Syracuse and Topeka. Lima and Howe are names for the same station at different times.

Table Showing Average Depth of Monthly Rainfall, in Inches; and Cubic Feet per Acre of Rainfall, for the St. Joseph River Basin Above the Gage at South Bend, Indiana, 1898-1915, Inclusive.

	1898		1899		1900		1901	
	In.	Cu. Ft.	In.	Cu. Ft.	In.	Cu. Ft.	In.	Cu. Ft.
January	3.29	11,942.7	2.17	7,877.1	0.95	3,448.5	2.55	9,256.5
February	2.46	8,929.8	2.62	9,510.6	4.18	15,173.4	1.72	6,243.6
March	5.77	20,945.1	3.54	12,850.2	2.45	8,893.5	3.27	11,870.1
April	1.03	3,738.9	0.90	3,267.0	1.72	6,243.6	2.08	7,550.4
May	3.67	13,322.1	4.76	17,278.8	2.22	8,058.6	2.07	7,514.1
June	2.61	9,474.3	1.59	5,771.7	3.50	12,705.0	2.72	9,873.6
July	3.30	11,979.0	3.80	13,794.0	5.86	21,271.8	2.95	10,708.5
August	4.04	14,665.2	1.83	6,642.9	5.34	19,384.2	3.23	11,724.9
September	3.24	11,761.2	3.05	11,071.5	1.69	6,134.7	0.93	3,375.9
October	6.09	22,106.7	3.87	14,048.1	3.24	11,761.2	6.01	21,816.3
November	3.90	14,157.0	1.77	6,425.1	5.51	20,001.3	1.80	6,534.0
December	1.72	6,243.6	3.46	12,559.8	0.63	2,286.9	3.59	18,031.7

	1902		1903		1904		1905	
	In.	Cu. Ft.	In.	Cu. Ft.	In.	Cu. Ft.	In.	Cu. Ft.
January	0.93	3,375.9	1.92	6,969.6	4.00	14,520.0	2.19	7,949.7
February	1.55	5,626.5	2.88	10,454.4	3.14	11,398.2	1.85	6,715.5
March	3.44	12,487.2	1.69	6,134.7	4.88	17,714.4	2.40	8,712.0
April	1.92	6,969.6	4.42	16,044.6	2.57	9,329.1	3.70	13,431.0
May	5.62	20,400.6	1.46	5,299.8	3.17	11,507.1	5.65	20,509.5
June	7.74	28,096.2	3.36	12,196.8	1.79	6,497.5	3.70	13,431.0
July	4.63	16,806.9	5.68	20,618.4	2.25	8,167	4.67	16,592.1
August	1.89	6,860.7	5.16	18,730.8	4.26	15,463.8	4.30	15,609.0
September	5.00	18,150.0	3.65	13,249.5	2.49	9,038.7	3.72	13,503.6
October	1.39	5,045.7	1.85	6,715.5	1.88	6,824.4	3.30	11,979.0
November	2.72	9,873.6	1.88	6,824.4	0.24	871.2	2.31	8,385.3
December	2.90	10,527.0	2.74	9,946.2	1.49	5,408.7	1.66	6,025.8

	1906		1907		1908		1909	
	In.	Cu. Ft.	In	Cu. Ft.	In.	Cu. Ft.	In.	Cu. Ft.
January	2.85	10,345.5	3.58	12,995.4	1.39	5,045.7	2.93	10,708.5
February	0.99	3,593.7	0.29	1,052.7	3.94	14,302.2	3.86	14,011.8
March	2.71	9,837.3	4.21	15,282.3	3.36	12,196.8	1.79	6,497.7
April	2.56	9,292.8	3.21	11,652.3	3.41	12,378.3	4.38	15,899.4
May	2.31	8,385.3	4.60	16,698.0	6.82	24,756.6	3.85	13,975.5
June	4.83	17,605.9	4.29	15,572.7	2.11	7,659.3	5.62	20,400.6
July	3.40	12,342.0	5.71	20,727.3	1.92	6,969.6	3.44	12,487.2
August	3.84	13,939.2	3.06	11,107.8	2.43	8,820.9	3.84	13,939.2
September	2.19	7,949.7	4.11	14,919.3	1.03	3,738.9	2.94	10,672.2
October	2.74	9,946.2	2.59	9,401.7	0.34	1,234.2	2.20	7,986.0
November	3.48	12,632.4	1.95	7,078.5	2.29	8,312.7	4.48	16,262.4
December	3.86	14,011.8	4.53	16,443.9	1.48	5,372.4	3.26	11,833.8

	1910		1911		1912		1913	
	In.	Cu. Ft.	In.	Cu. Ft.	In.	Cu. Ft.	In.	Cu. Ft.
January	2.26	8,203.8	1.72	6,243.6	1.23	4,462.9	3.10	11,253.0
February	1.63	5,916.9	1.40	5,082.0	2.10	7,623.0	1.09	3,956.7
March	0.53	1,923.9	2.01	7,296.3	1.38	5,009.4	4.39	15,935.7
April	3.33	12,087.9	3.95	14,338.5	3.54	12,850.2	1.96	7,114.8
May	4.08	14,810.4	1.95	7,078.5	5.52	20,037.6	4.87	17,678.1
June	1.52	5,517.6	5.25	19,057.5	3.41	12,378.3	0.86	3,121.8
July	2.99	10,853.7	3.02	10,962.6	5.87	21,308.1	3.37	12,233.1
August	2.59	9,401.7	3.24	11,761.2	4.69	17,024.7	2.95	10,708.5
September	5.02	18,222.6	3.11	11,289.3	3.08	11,180.4	2.49	9,038.7
October	3.04	11,035.2	4.12	14,955.6	2.32	8,421.6	3.85	13,975.5
November	2.18	7,913.4	3.60	13,068.0	2.92	10,599.6	2.65	9,619.5
December	1.96	7,114.8	1.93	7,005.9	1.22	4,428.6	0.90	3,267.0

	1914		1915			1914		1915	
	In.	Cu. Ft.	In.	Cu. Ft.		In.	Cu. Ft.	In.	Cu. Ft.
Jan.	2.88	10,454.4	2.35	8,530.5	July	1.19	4,319.7	5.23	18,984.9
Feb.	1.47	5,336.1	1.72	6,243.6	Aug.	3.99	14,483.7	4.11	14,919.3
March	3.35	12,160.5	0.52	1,887.6	Sept.	1.51	5,481.3	5.24	19,021.2
April	2.57	9,329.1	2.11	7,659.3	Oct.	1.86	6,751.8	1.54	5,590.2
May	4.80	17,424.0	4.26	15,463.8	Nov.	1.44	5,227.2	2.07	7,514.1
June	3.45	12,523.5	2.54	9,220.2	Dec.	2.63	9,546.9	1.25	4,537.5

The discharge of the St. Joseph River is determined from the gaging station which was established on the Leaper bridge at the North Michigan Street crossing in South Bend, on July 11 and 12, 1910. The gage is a chain gage and is located on the up-stream hand-rail of the south span of the bridge. The initial point for soundings and current readings is 16 inches along the hand-rail from the inner edge of the iron post on the up-stream side of the south end of the Leaper bridge. The bridge is on a broad bend in the river, which throws the main current near the north bank. There are no islands or vegetation to interfere with the readings. The current is swift and very steady. During high water, the discharge is between the two bridge abutments except at extremely high stages, when it flows over the levee upon which Michigan Street is built south of the bridge. This has not occurred since the gaging station has been established. The gage is read daily by M. J. W. Fisher, chief engineer of the South Bend Water Works Company.

Discharge Measurements on St. Joseph River at South Bend, Indiana, during 1910-1911.

Date	Hydrographer	Gage Height, Feet	Discharge, Sec.-Ft.
July 13, 1910	W. M. Tucker	2.20	2,470
February 20, 1911	W. M. Tucker	4.05	8,820
April 4, 1911	W. M. Tucker	3.05	4,230
July 7, 1911	Tucker & Clark	2.25	2,546
July 29, 1911	Tucker & Clark	1.85	1,857
July 30, 1911	Tucker & Clark	1.15	912

Gage Readings on St. Joseph River at South Bend, Indiana, for 1910.

Day	July	Aug.	Sept.	Oct.	Day	July	Aug.	Sept.	Oct.
1	.....	1.4	1.3	1.6	17	1.4	1.7	1.9	1.5
2	.....	2.1	1.5	1.1	18	1.9	1.8	1.6	1.7
3	.....	1.8	1.8	1.4	19	2.0	2.0	1.6	1.4
4	.....	1.7	1.4	1.5	20	2.1	1.4	2.0	1.5
5	.....	1.8	1.7	1.4	21	1.7	1.4	1.7	1.8
6	.....	1.8	1.4	1.8	22	2.2	1.4	1.7	1.8
7	.....	1.7	2.0	2.0	23	1.9	1.5	1.8	1.2
8	.....	1.5	2.0	1.9	24	1.4	1.2	1.9	*
9	.....	2.1	1.9	1.3	25	1.3	1.5	1.1	.....
10	.....	2.0	1.9	1.6	26	1.7	2.0	1.9	.....
11	.....	1.9	1.5	1.9	27	2.0	1.5	1.9	.....
12	.....	2.0	1.8	1.7	28	1.8	1.3	1.7	.....
13	3.0	1.5	1.9	1.7	29	1.9	1.1	1.4	.....
14	1.9	1.4	1.9	1.6	30	1.6	1.9	1.7	.....
15	2.0	1.0	1.8	1.6	31	1.5	1.7	.....	.....
16	1.9	1.7	2.0	1.1					

*Chain stolen October 24 and replaced February 4, 1911.

Gage Readings on St. Joseph River at South Bend, Indiana, for 1911.

Day	Jan.	Feb.	Mar.	Apr.	May	June	July	Aug.	Sept.	Oct.	Nov.	Dec.
1	.....	.....	2.8	1.9	2.8	1.2	2.4	1.4	1.8	1.8	1.7	3.4
2	.....	.....	2.9	1.4	2.6	1.2	2.1	1.4	1.3	1.8	1.9	3.1
3	.....	.....	2.4	1.8	2.1	1.6	2.2	1.7	1.3	2.0	1.8	3.0
4	.....	2.6	2.4	2.1	2.5	1.3	2.1	1.7	1.6	1.9	1.8	3.1
5	.....	2.0	2.2	2.7	2.4	1.3	1.9	1.7	1.6	2.2	1.7	2.5
6	.....	2.5	2.2	3.6	2.3	1.6	1.7	1.1	1.4	2.0	1.9	2.6
7	.....	2.4	2.2	3.2	1.3	1.4	1.8	1.4	1.8	2.3	1.8	2.6
8	.....	2.4	2.1	2.9	2.0	2.2	1.5	1.5	1.9	2.0	1.9	2.6
9	.....	2.3	2.0	2.8	1.4	2.6	1.5	1.6	1.5	2.1	1.7	2.5
10	.....	1.7	2.0	2.6	2.0	2.1	1.7	1.3	1.2	2.0	1.9	2.5
11	.....	2.1	2.0	2.4	1.6	1.3	1.6	1.4	1.3	2.0	1.8	3.0
12	.....	2.3	1.9	2.6	1.3	*	1.7	1.3	1.6	2.1	1.9	3.4
13	.....	2.7	2.1	2.5	1.1	2.2	1.8	1.4	1.8	2.0	2.2	3.1
14	.....	2.0	2.8	2.6	1.2	1.6	1.2	1.8	1.7	2.1	2.5	2.9
15	.....	4.0	2.7	2.6	1.6	1.7	1.4	1.9	2.3	1.7	2.4	3.0
16	.....	3.8	2.5	2.5	1.7	1.6	1.4	2.0	1.5	1.5	2.3	3.1
17	.....	3.7	2.5	2.8	1.5	1.7	1.5	1.8	1.8	1.9	2.5	2.5
18	.....	4.0	2.3	2.7	1.7	1.4	1.5	1.7	2.0	1.8	3.5	2.8
19	.....	4.6	2.2	2.7	1.5	1.2	1.4	1.8	2.0	1.9	4.2	2.6
20	.....	4.5	2.2	3.5	1.5	1.3	1.4	1.3	1.9	1.8	4.0	2.6
21	.....	3.9	2.1	4.7	1.2	1.4	1.3	1.6	1.9	2.0	3.9	2.5
22	.....	3.7	2.0	4.2	1.4	1.4	1.4	1.9	1.9	1.9	3.7	2.4
23	.....	3.4	1.9	4.0	1.2	1.6	1.7	2.0	1.3	1.9	3.8	2.3
24	.....	3.2	1.8	3.9	1.6	1.7	1.5	1.9	1.8	1.8	3.8	2.4
25	.....	3.1	1.8	3.6	1.3	1.4	1.6	2.0	1.8	2.0	3.7	2.8
26	.....	3.1	1.7	3.3	1.1	2.7	1.6	2.0	1.9	2.0	3.8	2.7
27	.....	3.1	1.9	3.1	1.2	3.2	1.6	1.4	2.0	1.9	3.4	2.7
28	.....	2.9	2.0	2.9	1.0	2.9	1.5	1.9	1.8	2.0	2.9	2.5
29	.....	.....	2.0	2.8	1.0	2.9	1.3	1.9	2.0	1.3	3.5	1.9
30	.....	.....	1.9	2.5	1.4	2.7	1.7	1.9	1.7	1.6	3.5	2.1
31	.....	.....	1.9	.....	1.6	.....	1.6	1.9	.....	1.7	.....	2.2

*No report.

Gage Readings on St. Joseph River at South Bend, Indiana, for 1912.

Day	Jan.	Feb.	Mar.	Apr.	May	June	July	Aug.	Sept.	Oct.	Nov.	Dec.
1	2.6	1.7	2.1	7.5	3.6	3.3	2.6	2.4	1.9	2.2	2.0	1.8
2	2.7	1.8	2.4	7.3	3.3	2.5	2.5	2.3	2.4	1.2	2.2	1.6
3	2.5	1.7	1.6	7.7	3.1	3.1	2.2	2.2	2.0	2.2	1.7	1.7
4	2.2	1.6	2.6	7.6	3.7	3.2	2.2	2.1	2.1	2.2	2.2	1.4
5	2.1	1.5	2.3	7.6	3.6	3.0	1.9	2.2	3.1	2.2	2.1	1.6
6	2.2	1.7	2.4	7.4	3.5	2.9	1.8	2.1	2.1	2.0	2.2	2.0
7	2.2	1.7	1.8	7.1	3.5	2.6	1.2	1.7	2.2	1.8	2.4	1.7
8	1.8	1.7	1.8	7.2	3.3	2.4	1.3	2.4	2.0	2.0	2.3	1.3
9	1.7	1.8	1.9	6.7	3.3	2.4	1.9	1.8	1.6	2.1	2.5	1.5
10	2.2	1.8	1.3	6.4	3.2	1.5	2.5	2.2	1.9	2.2	2.6	1.9
11	2.0	1.1	1.8	6.0	3.2	1.3	2.5	2.8	1.8	2.3	3.0	1.5
12	2.3	1.4	1.7	4.7	3.4	1.9	2.1	1.7	2.2	2.2	2.5	1.7
13	2.2	1.7	1.6	4.5	5.9	2.0	2.2	2.1	2.2	1.8	2.6	1.6
14	1.3	1.4	1.7	5.0	5.9	2.3	3.2	2.1	2.2	1.9	2.7	1.6
15	1.9	1.5	1.8	4.3	5.2	1.5	3.5	2.4	1.7	2.2	2.6	1.4
16	1.6	1.5	1.8	4.3	4.9	2.0	3.4	2.3	2.4	2.2	3.0	1.6
17	2.1	1.5	1.5	3.7	4.7	2.8	2.9	2.0	2.2	2.0	2.2	1.6
18	1.8	1.2	2.1	3.8	4.5	2.7	3.3	3.7	1.8	1.6	2.4	1.7
19	1.5	1.6	4.4	3.7	4.0	1.4	2.0	7.9	2.3	1.8	2.6	1.6
20	1.9	1.8	6.2	3.5	3.7	1.9	1.7	7.0	2.2	1.5	2.5	1.5
21	1.8	2.3	4.7	3.4	3.8	1.5	2.1	4.9	2.3	1.7	2.6	1.7
22	1.9	2.2	4.1	3.9	3.7	1.6	2.4	3.3	1.5	1.8	2.4	1.0
23	1.8	1.1	4.3	3.3	3.2	1.5	2.7	4.1	2.1	1.7	2.0	1.8
24	1.5	1.4	4.2	3.2	3.4	2.3	2.6	3.2	2.3	1.7	1.8	1.6
25	1.6	1.7	4.5	3.2	3.1	2.4	2.3	3.2	2.4	1.6	2.0	1.3
26	1.5	2.4	4.7	2.9	2.8	2.4	2.2	3.2	2.4	1.8	1.8	1.6
27	1.7	2.0	5.9	2.9	2.9	2.1	2.2	2.0	2.0	1.7	1.8	1.8
28	1.7	2.4	6.5	2.6	2.9	2.4	2.7	2.3	2.2	2.0	2.0	1.7
29	1.7	2.2	6.5	3.2	3.2	2.1	2.3	2.3	1.6	2.2	1.7	1.7
30	1.8	.....	6.2	3.4	3.5	2.1	2.1	2.2	1.9	2.1	1.8	1.8
31	1.8	.....	6.5	.....	3.3	.....	2.5	2.3	.....	2.3	.....	1.7

Gage Readings on St. Joseph River at South Bend, Indiana, for 1913.

Day	Jan.	Feb.	Mar.	Apr.	May	June	July	Aug.	Sept.	Oct.	Nov.	Dec.
1	1.7	4.1	2.3	8.3	2.7	3.3	2.1	2.2	1.8	1.8	1.9	2.9
2	1.8	3.8	2.2	7.8	3.0	3.1	1.8	2.1	1.9	1.6	2.3	2.2
3	1.7	2.5	1.6	7.3	2.5	3.4	1.9	1.9	2.0	1.8	2.1	2.0
4	1.6	2.8	1.9	7.3	2.5	3.2	1.7	1.8	1.9	2.0	1.8	2.0
5	1.5	3.1	2.0	7.3	2.1	3.0	2.0	1.9	2.0	1.8	1.9	2.1
6	1.7	2.9	2.0	7.2	2.3	2.5	1.8	2.0	1.8	2.1	2.1	1.9
7	2.3	2.0	1.9	7.3	2.4	2.6	1.4	1.7	1.8	2.0	1.8	1.7
8	2.2	2.8	1.8	7.1	2.1	2.4	1.6	2.1	1.6	1.6	2.0	1.9
9	1.9	3.1	1.9	6.8	2.2	2.1	1.6	2.3	2.0	1.9	1.9	1.7
10	2.2	3.5	3.8	6.5	2.1	1.9	1.7	2.0	1.9	1.7	1.7	1.7
11	1.8	3.3	4.6	6.2	2.0	1.9	2.0	2.1	1.9	1.9	1.9	1.6
12	2.0	3.4	4.4	6.3	1.8	1.8	1.7	1.8	2.0	1.7	1.7	1.7
13	2.1	2.9	4.3	5.9	1.6	1.8	1.5	1.7	1.9	2.0	1.7	1.6
14	2.0	2.4	4.4	5.7	1.7	1.8	1.4	1.7	1.8	1.9	1.8	1.6
15	2.2	2.3	4.4	5.2	1.7	1.7	1.8	1.9	2.0	1.8	1.7	1.9
16	2.4	2.4	4.0	4.9	1.8	2.1	2.2	1.8	1.8	1.8	1.8	1.6
17	4.1	2.0	4.6	4.7	2.1	2.0	1.9	2.1	1.9	1.7	1.9	1.8
18	4.6	2.2	4.2	4.6	2.7	1.8	2.1	1.9	1.9	1.7	1.7	1.6
19	4.9	2.1	3.9	4.3	1.8	1.8	2.3	2.2	1.8	1.5	1.9	1.9
20	4.8	2.1	3.8	4.1	1.9	1.7	1.8	2.4	1.8	1.7	1.8	1.7
21	6.4	2.4	3.9	4.8	2.4	1.8	2.4	2.7	1.6	1.9	2.0	1.6
22	6.3	2.4	4.1	4.4	2.1	2.2	1.9	2.0	1.5	1.8	1.8	1.8
23	5.8	2.2	3.9	3.1	2.3	2.1	2.1	2.1	1.9	2.0	1.9	1.8
24	6.1	2.4	5.3	3.1	2.4	1.9	2.0	2.2	2.0	1.8	1.9	1.4
25	6.0	2.5	8.1	3.4	2.2	2.0	2.1	2.0	1.8	1.9	1.9	1.5
26	6.0	2.5	10.3	2.6	1.9	2.1	2.0	2.1	1.8	*	1.8	1.4
27	5.9	2.2	9.5	2.7	4.1	2.1	2.2	1.9	1.6	1.8	1.6	1.4
28	5.8	2.4	9.0	3.1	5.6	1.7	2.4	1.9	1.7	1.9	1.5	1.3
29	4.8	.....	8.4	2.8	4.0	1.7	2.2	2.0	1.9	2.1	1.7	1.5
30	4.6	.....	8.2	2.8	3.8	1.9	2.2	1.9	1.7	1.9	1.4	1.4
31	4.4	.....	8.4	.....	3.2	.....	2.0	1.7	.....	2.0	.....	1.5

*No record.

Gage Readings on St. Joseph River at South Bend, Indiana, for 1914.

DAY	Jan.	Feb.	Mar.	Apr.	May	June	July	Aug.	Sept.	Oct.	Nov.	Dec.
1	1.4	2.3	1.8	5.5	2.1	3.0	1.7	*	*	6.6	6.9	7.7
2	1.5	2.6	1.8	5.6	2.0	2.7	1.8	*	*	6.5	6.9	7.7
3	1.4	2.6	1.4	5.5	2.0	2.5	1.6	*	*	6.5	7.0	7.2
4	1.6	3.0	1.7	5.2	2.5	2.6	1.3	*	*	7.0	7.1	7.1
5	1.6	2.5	1.6	4.6	2.1	2.5	1.8	*	*	6.9	7.0	7.1
6	1.6	2.3	1.6	5.0	2.0	2.7	1.6	*	*	6.7	7.0	7.0
7	1.5	2.1	2.2	4.6	2.2	2.6	2.1	*	*	6.9	6.8	7.2
8	1.6	1.9	1.4	4.2	2.5	2.7	1.9	*	*	6.9	6.9	7.2
9	1.9	1.9	2.1	4.0	2.5	2.6	2.0	*	*	6.8	7.0	7.3
10	1.6	2.0	1.7	4.0	2.3	2.5	1.8	*	*	6.5	7.0	7.3
11	1.4	2.1	1.7	3.8	3.2	2.3	1.5	*	*	6.8	7.0	7.4
12	1.6	1.9	1.9	3.6	3.8	2.3	1.8	*	*	6.7	6.8	7.2
13	1.5	2.1	1.8	3.6	5.7	2.2	1.9	*	*	6.8	6.8	7.2
14	1.6	2.0	2.4	3.3	6.2	2.1	1.6	*	*	6.8	6.8	7.3
15	1.6	1.9	2.8	3.2	6.6	2.3	1.7	*	*	6.8	6.8	7.3
16	1.5	1.8	3.3	3.0	6.8	2.2	1.8	*	*	6.7	6.8	7.4
17	1.7	1.8	3.5	3.0	6.7	2.1	1.7	*	*	6.7	6.7	7.3
18	1.5	1.8	2.6	2.6	6.4	2.2	1.5	*	*	6.7	6.6	7.1
19	1.6	1.8	3.5	2.5	5.2	2.0	1.4	*	†7.4	6.7	6.8	7.0
20	1.6	1.8	3.0	2.6	5.4	2.2	1.8	*	7.7	6.9	6.6	7.4
21	2.0	1.9	2.3	2.4	4.7	1.8	1.6	*	7.7	6.8	6.6	7.3
22	1.8	1.4	2.9	2.1	4.6	1.8	*	*	7.7	6.8	6.6	7.4
23	1.8	2.0	2.9	2.0	4.2	1.8	*	*	7.7	6.7	7.3	7.4
24	1.8	1.8	2.2	2.2	*	1.9	*	*	7.7	6.7	7.3	7.2
25	2.9	1.8	2.0	2.3	3.5	1.9	*	*	7.8	7.0	7.3	7.2
26	3.6	1.7	2.1	2.1	3.1	1.9	*	*	7.4	7.0	7.3	7.2
27	2.7	1.8	3.0	2.3	2.8	1.8	*	*	7.6	7.1	7.2	7.2
28	2.6	2.0	4.9	2.2	2.9	1.6	*	*	7.6	7.1	7.2	7.3
29	2.6	......	5.3	2.3	3.0	1.7	*	*	7.7	7.0	7.9	7.5
30	2.4	......	5.5	2.1	2.6	1.9	*	*	7.7	7.0	7.4	7.5
31	2.5	......	6.0	......	2.8	......	*	*	......	6.8	......	7.2

*No record. Gage destroyed by removal of bridge, July 22.

†Record taken from City Engineer's records which were recorded by an automatic gage two city blocks below Leeper bridge.

Gage Readings on St. Joseph River at South Bend, Indiana, for 1915.

DAY	Jan.	Feb.	Mar.	Apr.	May	June	July	Aug.	Sept.	Oct.	Nov.	Dec.
1	7.2	7.7	9.0	8.5	7.5	8.3	8.4	9.0	9.5	9.2	*	*
2	7.5	8.0	7.8	8.0	8.0	8.1	8.3	9.4	9.0	9.2	*	*
3	7.1	8.0	8.5	7.6	8.1	8.0	*	9.3	8.7	8.9	*	*
4	7.5	7.7	7.7	8.3	8.3	7.8	*	*	8.4	8.2	*	6.6
5	8.2	8.4	8.5	8.2	8.0	7.7	*	*	8.6	9.0	*	7.4
6	7.5	8.7	8.4	8.1	8.3	7.8	8.5	9.1	8.5	9.0	*	7.8
7	7.3	8.3	8.0	8.1	7.8	8.0	8.2	9.1	9.0	8.9	*	7.1
8	7.4	8.5	8.5	8.2	7.5	7.5	8.1	8.2	9.1	8.6	*	7.6
9	7.5	8.7	8.2	7.9	8.5	7.5	8.4	9.1	9.1	8.5	*	7.9
10	7.3	8.8	8.2	7.6	8.0	*	8.5	9.0	8.8	7.5	*	8.0
11	7.7	9.0	8.1	8.7	8.2	*	8.6	9.0	8.5	8.7	*	7.4
12	7.6	9.3	8.3	8.3	8.5	7.5	9.1	9.0	9.2	8.5	*	8.2
13	7.8	9.5	8.6	8.3	8.0	8.0	9.4	8.8	9.3	*	*	8.7
14	7.8	10.5	8.0	8.3	8.0	*	9.5	8.9	8.5	8.6	*	7.5
15	7.7	9.7	8.2	8.3	7.5	7.7	*	8.3	9.1	8.6	*	7.5
16	*	9.9	8.5	8.0	8.4	7.8	9.3	9.2	9.2	8.5	*	7.2
17	*	9.8	8.5	7.7	8.0	8.0	9.3	9.0	9.4	7.5	*	7.6
18	8.3	9.8	8.4	8.5	8.0	8.0	8.6	8.9	9.1	8.5	*	7.2
19	8.0	10.1	8.0	8.5	8.3	7.5	9.1	8.9	8.0	8.6	*	6.8
20	7.9	10.2	8.5	8.0	8.1	8.3	8.7	8.9	9.4	8.6	*	7.3
21	8.0	9.6	8.0	8.2	8.1	8.8	9.0	8.5	9.0	9.3	*	7.3
22	7.7	10.0	8.4	8.0	7.5	8.5	9.0	8.1	9.0	8.0	*	7.4
23	7.7	10.2	8.5	8.0	8.1	8.7	8.6	9.3	9.0	7.6	*	7.5
24	7.4	10.2	8.0	7.6	8.0	8.5	8.2	9.4	9.0	8.5	*	7.4
25	7.7	9.7	8.3	8.3	8.0	8.5	8.8	8.6	9.0	8.6	*	7.2
26	7.5	9.2	7.9	8.4	8.0	7.4	8.8	8.5	8.0	7.7	*	7.1
27	7.5	9.3	8.5	8.2	8.0	8.4	*	9.0	9.8	7.7	*	7.5
28	7.8	8.8	7.7	8.3	7.7	8.5	8.6	8.4	9.1	7.8	*	7.5
29	7.5	......	8.1	8.2	7.6	8.5	9.0	8.0	9.2	7.8	*	7.7
30	7.4	......	8.0	7.9	8.0	8.0	9.0	9.0	9.0	7.8	*	7.5
31	*	......	8.1	......	8.3	......	8.1	8.1	......	7.0	......	*

*No record. These records were lost from the City Engineer's office.

Rating Table for St. Joseph River at South Bend, Indiana, for 1910-14.

Gage Height, Feet	Discharge, Sec.-Ft.	Gage Height, Feet	Discharge, Sec.-Ft.	Gage Height, Feet	Discharge, Sec.-Ft.	Gage Height, Feet	Discharge, Sec.-Ft.
1.0	760	2.0	2,110	3.0	4,140	4.0	8,490
1.1	873	2.1	2,280	3.1	4,430	4.1	9,200
1.2	988	2.2	2,460	3.2	4,760	4.2	9,965
1.3	1,105	2.3	2,640	3.3	5,110	4.3	10,780
1.4	1,227	2.4	2,825	3.4	5,475	4.4	11.650
1.5	1,365	2.5	3,020	3.5	5,870	4.5	12,570
1.6	1,495	2.6	3,225	3.6	6,300	4.6	13,540
1.7	1,640	2.7	3,440	3.7	6,770	4.7	14,560
1.8	1,790	2.8	3.670	3.8	7,280	....	....
1.9	1,950	2.9	3,900	3.9	7,850	....	....

From the above data, which shows the lowest gage height to have been one foot, the smallest discharge was 767 second-feet. The gage was read each day at 6:00 o'clock a. m. with the idea that at that time the river would have filled all the storage above the several dams which are located between South Bend and Elkhart, and would have resumed the natural flow by overflow over the dams. Mr. Fisher expressed the idea, after a year's observation, that this was not the case, but that the discharge at that time of the day was less than normal. To ascertain the real condition, gage readings were taken every hour for 24 hours, from 6 a. m., July 28, to 6 a. m., July 29. The results are shown in the following tables:

A. M. July 28.
6:00........................ 1.5
7:00........................ 1.75
8:00........................ 2.25
9:00........................ 2.3
10:00........................ 2.3
11:00........................ 2.3
12:00........................ 2.3

P. M. July 28.
1:00........................ 1.6
2:00........................ 1.9
3:00........................ 2.0
4:00........................ 2.0
5:00........................ 2.6
6:00........................ 2.9
7:00........................ 2.8
8:00........................ 2.0
9:00........................ 1.6
10:00........................ 1.6
11:00........................ 1.55
12:00........................ 1.5

A. M. July 29.
1:00........................ 1.4
2:00........................ 1.3
3:00........................ 1.2
4:00........................ 1.25
5:00........................ 1.35
6:00........................ 1.3

When the discharges for these twenty-five gage heights are taken from the rating table and averaged, the average is found to be 1991 second-feet. Such a discharge is represented in the rating table by a gage height of 1.925 feet, which is more than .5 foot higher than the average (1.4 feet) of the two 6:00 o'clock a. m. readings. This demonstration corresponds to Mr. Fisher's idea, but is insufficient to prove the exact error. An average result of several such demonstrations could be taken as a correcting factor. See Figure 1. It is safe to conclude, however, that the discharge as shown by the gage height is somewhat low. During the time which this station has been in operation, there has probably been no day in which the average discharge has not been below 1,000 second-feet. The estimated discharge at Elkhart, above the mouth of the Elkhart River, is .6 as great as that at South Bend. The profile of the St. Joseph in Indiana shows a fairly regular fall, which varies from 1.99 below South Bend to 2.93 between Elkhart and Mishawaka. The following table shows the profile of the St. Joseph River in Indiana:

Location	Estimated Distance, Miles	Altitude, Feet	Fall Per Mile, Feet
Upper state line........................	0	750.51	2.03
Elkhart River mouth..................	15	720.01	2.03
Mishawaka (below dam)...............	12	684.82	2.93
South Bend (below dam)...............	6	671.98	2.14
Lower state line........................	9	654.00	1.99

A discussion of the water power conditions on the St. Joseph and its tributaries is given on pages 470-490 of the 36th Annual Report of the Indiana Department of Geology and Natural Resources.

The Maumee System.

Maumee River is located in the northeastern part of the state. It rises by two main branches, the St. Joseph and the St. Marys rivers. The St. Joseph River rises in Steuben County, flows south along the state line into Dekalb County, and joins the St. Marys River at Ft. Wayne, in Allen County. It drains 636 square miles of Indiana in Steuben, Dekalb and Allen counties and 690 square miles of Ohio and Michigan. The St. Marys River rises in Ohio and enters Indiana in Adams County and flows in a general northwest direction into Allen County, where it joins the St. Joseph at Ft. Wayne. It drains 326 square miles of Indiana in Adams, Wells and Allen counties, and 450 square miles of Ohio. From the junction of the St. Joseph and St. Marys rivers the Maumee flows east by northeast into Ohio and finally into Lake Erie at its western extremity.

This river system is situated in the glacial deposits and so far as known has not cut through this deposit at any point. No investigation of this stream has been made by the writer and the data herein given was received from the U. S. G. S. The area of the two main branches as given by the U. S. G. S. is 2,104 square miles. This area drains through the gage which is established on the Maumee about three hundred feet below the junction. Discharge measurements were taken at this gage in 1912-13-14 and the results are given in the following table. The gage record is being kept daily by the local weather observer and the data will be available when a new rating table is established.

TABLE G-16.

Maumee River, Fort Wayne, Indiana, Drainage Area, 2,104 Square Miles.

1912

Month	Discharge in Sec.-Ft.			Run-off		Rain in Inches
	Maximum	Minimum	Mean	Sec. Ft. Per Sq. Mile	Depth in Inches	
January	4,885	375	1,552	.737	.860	1.49
February	3,390	290	1,677	.797	.829	2.22
March	18,625	875	6,502	3.091	3.560	2.88
April	19,800	1,035	5,303	2.520	2.810	....
May	5,005	550	2,290	1.088	1.255	3.25
June	1,840	250	624	.296	.330	2.83
July	3,230	210	1,167	.554	.639	4.27
August	4,575	210	1,182	.561	.647	4.10
September	505	210	316	.151	.168	2.66
October	760	135	460	.218	.251	2.21
November	2,725	290	873	.398	.444	1.88
December	420	25	152	.072	.083	1.39

1913

January	16,175	170	7,274	3.453	3.983	5.81
February	3,390	710	1,261	.599	.623	1.36
March	33,600	1,160	10,061	4.760	5.489	7.90
April	14,345	1,160	5,372	2.551	2.846	3.24
May	4,500	465	1,341	.637	.734	3.47
June	3,150	250	728	.346	.386	1.01
July	1,625	210	598	.284	.327	3.52
August	2,310	135	687	.326	.376	2.90
September	420	95	230	.109	.122	1.99
October	1,035	170	483	.229	.264	3.17
November	760	250	352	.167	.186	2.48
December	2,940	95	887	.423	.488	.72

1914

January	2,240	210	553	.263	.303	2.77
February	2,170	170	675	.322	.335	1.85
March	8,145	605	3,615	1.718	1.982	2.18
April	9,730	250	4,236	2.012	2.242	3.54
May	15,500	760	5,033	2.380	2.745	5.27
June	1,460	210	733	.348	.388	3.09
July	......	......	......	......	......	2.24
August	505	50	206	.098	.113	4.52
September	920	135	379	.180	.201	2.17
October	1,035	95	432	.205	.236	2.83
November	170	135	143	.068	.076	1.09
December	1,400	135	492	.234	.270	3.15

SECTION III.

LAKES OF INDIANA.

General Discussion.

A lake is a body of standing water which is not connected with the sea at sea level. Salt lakes are not connected with the sea and fresh lakes are connected with the sea by streams. Small lakes are called ponds. The essential conditions for the existence of a lake are: first, a depression of such shape that water will stand therein; and, second, water sufficient to partially or entirely fill the depression. On the basis of permanency there are two classes, permanent and temporary or intermittent lakes. A permanent lake is a lake which retains water permanently. An intermittent or temporary lake is a lake which disappears in dry seasons. On the basis of salinity there are two classes of lakes, fresh and salt. A fresh lake is a lake which discharges water through a stream or underground passage. A salt lake is a lake which discharges water by evaporation only. On the basis of cause of the basin there are many classes, each class corresponding to the manner in which the depression which the lake occupies was formed. Some of the common causes of lakes are: wind action, glacial action, damming of streams, cutting off of meanders of streams, and diastrophism.

The lakes which have existed in Indiana in recent geological history have been fresh lakes, usually permanent and with basins formed by six or seven agencies. The classes due to cause of basin are marginal, irregular, kettlehole, channel, oxbow, wind scoured, beaver dam and artificial. A marginal lake is a lake which occurs along the margin of a glacier and one side of whose basin is composed of the ice of the glacier. The extinct lakes, Chicago and Maumee, illustrate this class, although there were many other small ones along the edge of the ice sheet when it was present in Indiana. An irregular lake is a lake whose basin is formed by the irregular deposit of glacial drift. Lake Wawasee and other large irregular shaped lakes of the state are examples of this class. A kettlehole lake is a lake whose basin is formed by the melting of a large detached piece of ice which has been buried in glacial drift. These lakes are usually small and their basins are the shape of an inverted cone. Many small lakes of Indiana are of this type. A channel lake is a lake whose basin is formed by the obstruction of a river channel by glacial drift or other means. Crooked Lake in Steuben County is probably an example of this type of lake. An oxbow lake is a lake whose basin is formed by a river cutting off one of its meanders and silting up the ends of the abandoned channel. Many such lakes are found along the lower courses of White and Wabash rivers. A wind scoured lake is a lake whose basin is formed by the wind which scours out the depression in a dune region. Such lakes are often of the intermittent type. Small lakes of this type are found among the sand dunes of northern Indiana. A beaver dam lake is a lake whose basin is formed by the work of beavers. Such lakes were common in the Kankakee basin fifty years ago. An artificial lake is a lake whose basin is formed by the work of man. The lake at Rome City in Noble County is of this type.

A lake is one of the most short lived of topographic features. Practically all natural agents tend to destroy lake basins. The inflowing water carries silt into the basin where it is settled out of the standing water. This tends to fill up the basin. Vegetation which grows in the shallow water of the lake and on the water surface decays and falls to the bottom, thus filling up the basin. Wind blown sand and dust which falls upon the surface of the lake settles to the bottom. The salt in salt lakes is precipitated and fills the basin. Any stream flowing from a lake tends to deepen its channel, a condition which tends to diminish the basin. All these and other agencies are active always in the destruction of the lake basins.

The result of the above action is to reduce the lake to a swamp, which later becomes a morass and finally comes to be solid earth. When this earth becomes dry enough for the burrowing animals and earth worms to occupy they bring up clay and other mineral substances from below and mix with the vegetable matter at the top and in time the land becomes the finest of farm land. It is level, fine grained, easily tended and very fertile. Some of the finest farm land of Indiana is of this type.

It is estimated that about one-fifth of Indiana is either occupied by lakes or has been in recent geological time. This time dates from the Illinois glaciation. The lakes which were formed during this ice invasion are practically all extinct. The region which they occupied is the region lying south of the Wisconsin glacial limit and west of the great re-entrant between the lobes of the Illinois glacier. No doubt lakes were formed by the Illinois glaciation in the region now covered by Wisconsin drift, but these were obliterated by the Wisconsin invasion. During the Wisconsin ice invasions many marginal lakes were formed and many of other types were left upon its retreat. All the large lakes of northern Indiana were formed by this agency.

Great changes have taken place in the lake conditions of the state since its settlement. The last fifty years have witnessed the reduction of actual lake surface by at least half. This is due to natural causes coupled with artificial drainage. The systematic drainage of the land has not only been carried on to the extent of draining many small, shallow lakes but the effect of the drainage has been to lower the water table in many places to such an extent that the surface of the lakes has been lowered. Large areas in the northern part of the state which were swamps when this region was first occupied are now fine farming regions. The original drainage was attempted by individuals, later by communities, and still later by county, state and nation. Considerable sums of money have been expended by the National Government in deepening the channel of the Kankakee River and this coupled with the work carried on by the state and counties has drained English Lake, the largest lake shown on the early state maps. The draining of this lake has also drained the adjacent swamps and rendered hitherto worthless land highly valuable.

The existing lakes of Indiana are now practically confined to three locations. First, the small river lakes which are along the lower courses of the Ohio, Wabash and White rivers. Second, a belt of glacial lakes extending from northern Wabash and Miami counties to Steuben County and occupying Wabash, Miami, Fulton, Kosciusko, Whitley, Noble, Dekalb, Lagrange and Steuben counties. Third, a belt of glacial lakes extending from Fulton to Laporte counties and occupying Fulton, Marshall, Starke, St. Joseph and Laporte counties. The second and third occupy the marginal moraines formed by the Saginaw lobe of the Wisconsin glacier in conjunction with the moraines of the Erie lobe on the east and the Michigan lobe on the west.

The survey of the lakes which has been undertaken by the writer is as yet only begun. Accurate plane table maps of five lakes or groups of lakes have been made and soundings of two are complete. The accompanying maps show the extent of the work. The soundings on Lake Manitou are taken from the work of Prof. Will Scott of Indiana University. A few checks were made. The soundings on Lake Wawasee have been compared with the work of Chancey Juday which is published in the Proceedings of the Indiana Academy of Science and are found to correspond in general.

Lake Wawasee. Lake Wawasee is located in T. 34 N., R. 7 E., the extreme northeast corner of Kosciusko County. It consists of a main lake and three northern extensions, Johnson's Bay, Boner Lake and Syracuse Lake. The distance from Buttermilk Point at the extreme southeast point to Syracuse on Lake Syracuse on the extreme northwest point is 5⅝ miles. The extreme width is slightly less than 1¼ miles. The length of the main lake is on a curve from Buttermilk Point to the cement company's marl slip, about 4⅝ miles. The extreme depth of the lake is slightly more than 60 feet. Three points reach below 60 feet, near Buttermilk point, near Vawter Park and ½ mile east of Lake View Hotel. The entire area of the lake is 5.95 square miles. The volume of the lake, excluding Boner Lake, which was not sounded, is approximately 1,860 million cubic feet.

Lake Wawasee is one of the principal pleasure resorts of Indiana. It is reached by the B. & O. Ry. by detraining at Syracuse, Wawasee Station or Cromwell, according to the part of the lake to be visited. Cromwell is two miles east of the east end of the lake. Wawasee Station is on the north bank and Syracuse at the extreme west end of the lake. There were three hundred and ten cottages and ten hotels about the lake in 1917. The cottages may be rented furnished for $15 to $30 per week. Many of the cottages are occupied by the owners during the summer months. The rates at the hotels vary according to the accommodations and range between $2 and $3 per day. Bathing, boating and fishing are the advertised attractions. The number of visitors increases yearly. The relief from business cares which these visitors enjoy here is without doubt of great value to them when they return to their work. The tendency of the people of Indiana to spend a week or two during the year in some one of the pleasant resorts of the state will no doubt make a saner and healthier population which in time will manifest itself in better business methods, cleaner legislation and in general more wholesome state conditions.

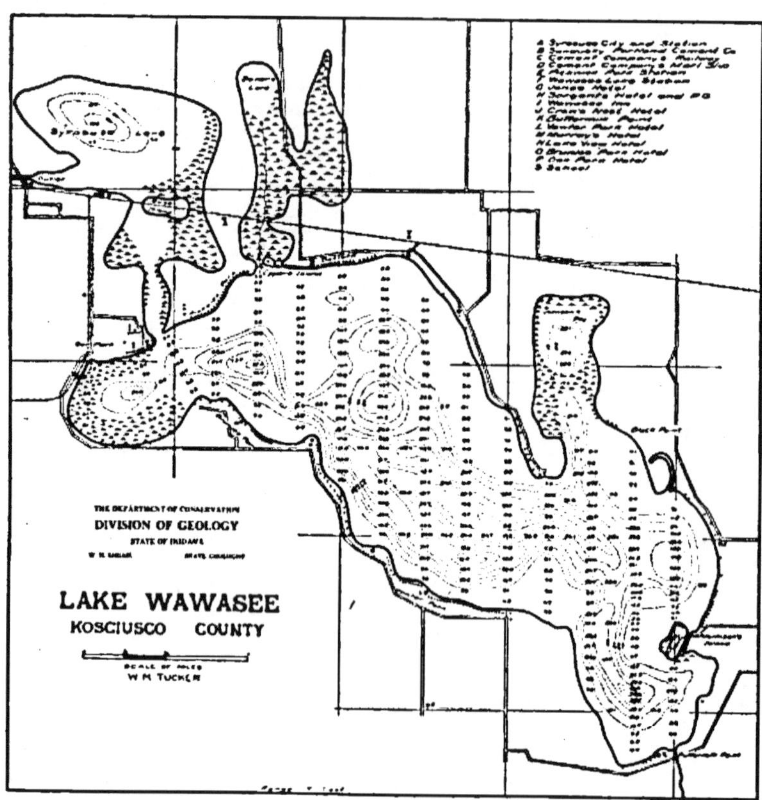

Fig. 8

This lake is of considerable economic importance. In 1917 the Sandusky Portland Cement Plant had a slip in Conklings Bay for the handling of marl which was dredged from the bottom of the lake and used in the manufacture of cement. The company had its own railroad, which touched the slip on its way to Lake Wabee toward the west, where marl was also collected. This plant has since been dismantled and the marl rights relinquished. The water which is used by the town of Syracuse is taken from Syracuse

Lake. Large quantities of ice are harvested from the lakes for summer use in the hotels. A small amount of water power is developed from the discharge of Lake Wawasee in Syracuse. It is used to produce electricity for the city. In dry seasons the power is not used because of a lack of water. The discharge from Lake Wawasee drains into the Elkhart River.

Lake Manitou. Lake Manitou is located in T. 30 N., R. 3 E., near the city of Rochester in Fulton County. It is 2¼ miles long and 1 1/3 miles wide in its greatest dimensions. A long, curved extension stretches south and east

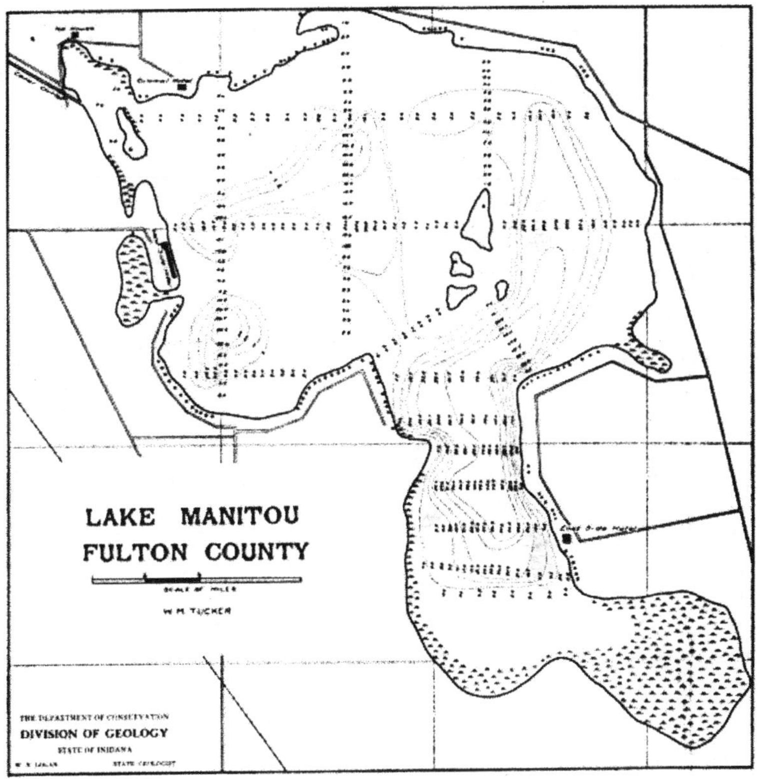

Fig. 9

from the main body of the lake. The deepest point in the lake is somewhat less than 50 feet and occurs near the entrance to this extension. The entire area of the lake, excluding the islands is 1.2 square miles. Its volume is approximately 303 million cubic feet.

Lake Manitou, like Lake Wawasee, is a favorite pleasure resort. It is reached from Rochester, through which the Chicago and Erie and the Lake Erie and Western railroads pass. A splendid boulevard leads to the lake from the city, which is 1½ miles away. There were about one hundred forty cottages and three large hotels along the shores of the lake in 1917. The rates are the same as at Lake Wawasee. This lake is reputed to be one of the best if not the best fishing lake in the state. Many water fowl also frequent it

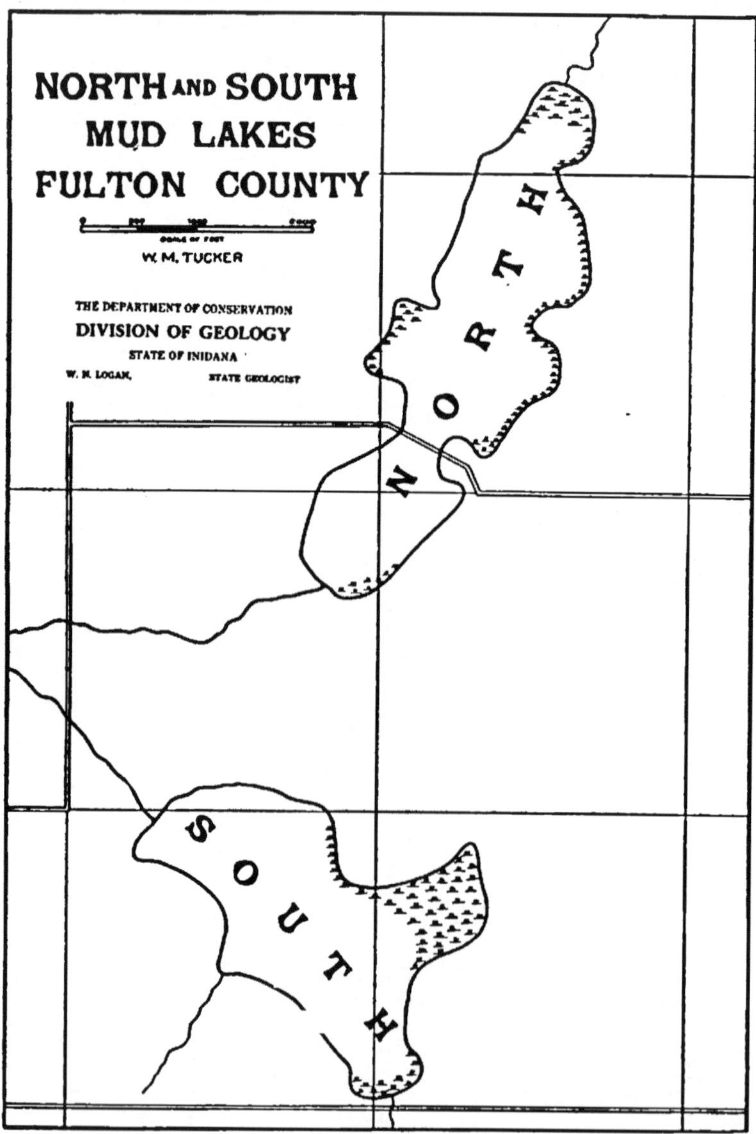

Fig. 10

during the migration season. The abundance of fish and fowl is attributed to the water rice which grows in the shallow water. It furnishes food for the water fowls and protection for the small fish from the larger ones.

Rochester receives its water supply from Lake Manitou. Large quantities of ice are harvested from this lake for summer use about the hotels and cottages. A small amount of water power was formerly employed from the water of this lake. It was conducted into the city through the canal which now carries water to the city water plant. The drainage from Lake Manitou is discharged into Tippecanoe River.

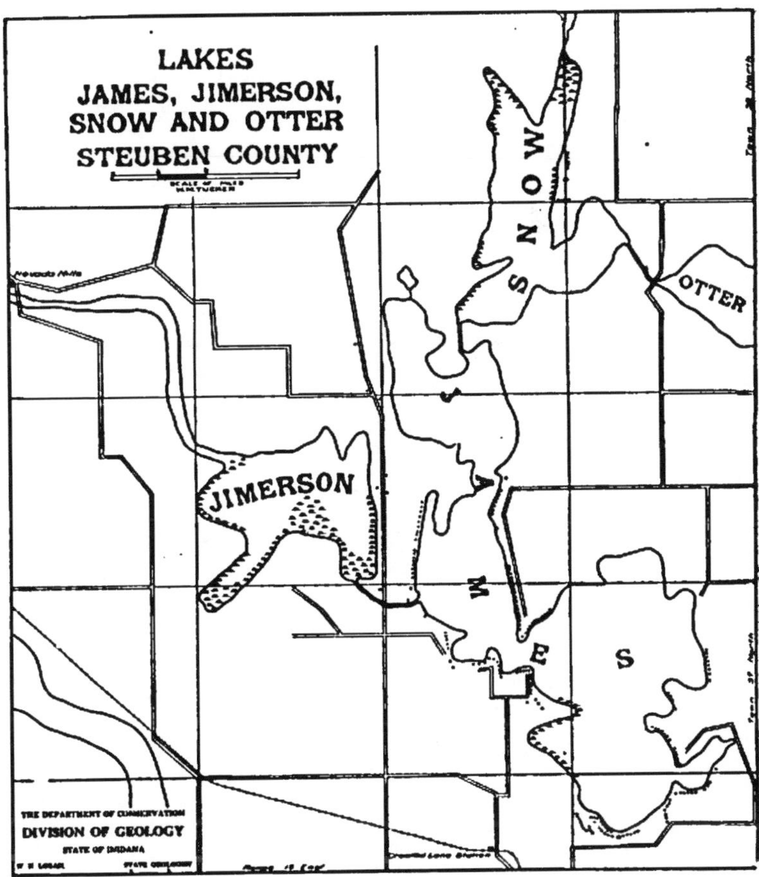

Fig. 11

North and South Mud Lakes. North and South Mud lakes are small lakes in T. 29 N., R. 3 E., near the southeast corner of Fulton County. North Mud Lake is .9 mile long and an average of about ¼ mile wide. The area of this lake is .158 square mile. It is a shallow lake with muddy banks.

It has attraction as a summer resort. Three cottages are located near the south end. South Mud Lake is even smaller than North. It is .6 mile long and ¼ mile wide. Its area is .144 square mile. It is also very shallow and muddy. Six cottages are grouped near the east shore. These lakes are three miles west of Macy, Miami County, through which the Lake Erie and Western Railroad passes.

No soundings were taken on these lakes. Both lakes have been reduced considerably by artificial drainage. They drain by Mud Creek into Tippecanoe River.

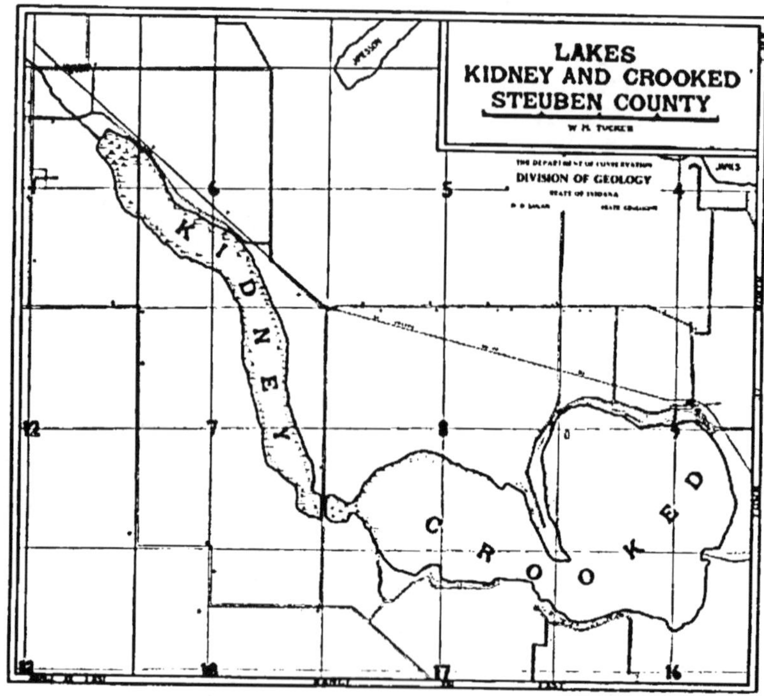

Fig. 12

The two groups of lakes shown in Figures 11 and 12 are located in Steuben County. Crooked and Kidney Lakes drain into the St. Joseph River through Pigeon River. James, Snow, Otter and Jimerson lakes drain into the St. Joseph River through Fawn River. While James and Crooked lakes are but ½ mile apart, the surface of Crooked Lake is 23.5 feet higher than James Lake. The water power of the Crooked River System is treated in the 36th Annual Report of the Department of Geology and Natural Resources. Fawn River has considerable developed water power, much of which is in Michigan. Further work on the lakes of the State is now being carried on by the writer.

Bibliography.

1. H. E. Barnard: The Water Supply of Indiana. Eighth Annual Report of the Indiana State Board of Health, 1913, pages 142-359.
2. Chas. Brossman: Sewage Disposal. Proceedings of the Indiana Academy of Science, 1914, pages 365-372.
3. H. P. Bybee and C. A. Malott: The Flood of 1913 in the Lower White River Region of Indiana. Indiana University Bulletin, Vol. XII, No. 11, Indiana University Press, Bloomington, Indiana, 1914.
4. Climatological Data, Indiana Section, 1898-1915. U. S. Weather Bureau, Weather Bureau Office, Indianapolis, Indiana.
5. E. R. Cumings: The Geological Conditions of Municipal Water Supply in the Driftless Area of Southern Indiana. Proceedings of the Indiana Academy of Science, 1911, pages 112-146.
6. Drainage Laws of Indiana. (Typewritten) Indiana Bureau of Legislative Information, State House, Indianapolis, Ind.
7. C. G. Elliot: A Report upon the Drainage of Agricultural Lands in the Kankakee Valley, Indiana. Office of Experiment Stations, Circular 80, Government Printing Office, Washington, D. C., 1909.
8. C. G. Elliot: Engineering and Land Drainage. John Wyley and Sons, New York, 1903.
9. Prescott Folwell: Water-supply Engineering, John Wyley and Sons, New York, 1905.
10. Joseph P. Frizell: Water Power. John Wyley and Sons, New York, 1905.
11. John R. Haswell: Land Drainage in Maryland. The Maryland Agricultural Experiment Station, College Park, Md., 1914.
12. F. H. King: Farmers of Forty Centuries, or Permanent Agriculture in China, Korea and Japan. Madison, Wis., 1911.
13. M. O. Leighton: National Aspect of Swamp Drainage. Sixty-second Congress, Second Session, Senate Document No. 877, 1912.
14. S. W. McCallie: Drainage Reclamation in Georgia. Foote and Davies Co., Atlanta, Ga., 1911.
15. Walter McCulloh: Water Conservation. Addresses delivered in the Chester S. Lyman Lecture Series, 1912, before the Senior Class of the Sheffield Scientific School, Yale University. Yale University Press, New Haven, Conn., 1913.
16. R. D. Marsden: Work of Drainage Investigations, 1909-1910. Reprint from the Annual Report of the Office of Experiment Stations for the year ending June 30, 1910. Government Printing Office, Washington, D. C., 1911.
17. Daniel W. Mead: Waterpower Engineering, the Theory, Investigation and Development of Water Powers. McGraw-Hill Book Co., New York, 1915.
18. John H. Nolen: Missouri's Swamp and Overflowed Lands and their reclamation. Hugh Stephens Printing Co., Jefferson City, Mo., 1913.
19. Edmund T. Perkins: Address to Second Annual National Drainage Congress. American Printing Co., Ltd., New Orleans, La., 1912.
20. George W. Rafter: Hydrology of the State of New York. Publication of the New York State Education Department, Bulletin 85, 1905.

21. Leonard S. Smith: The Water Powers of Wisconsin. Publication of the Wisconsin Geological and Natural History Survey, Bulletin XX, Economic Series No. 13. Published by the State, Madison, Wis., 1908.
22. Arthur A. Stiles, State Levee and Drainage Commissioner of Texas: Reclamation of Overflowed Lands. Von Boeckmann-Jones Co., Austin, Texas, 1913.
23. George Fillmore Swain: Conservation of Water by Storage. Addresses delivered in the Chester S. Lyman Lecture Series, 1914, before the Senior Class of the Sheffield Scientific School, Yale University. Yale University Press, New Haven, Conn., 1915.
24. W. M. Tucker: Water Power of Indiana. Thirty-fifth Annual Report of the Indiana Department of Geology and Natural Resources, pages 11-77. Indianapolis, Ind., 1910.
25. W. M. Tucker: Water Power of Indiana. Thirty-sixth Annual Report of the Indiana Department of Geology and Natural Resources, pages 469-538. Indianapolis, Ind., 1911.
26. Chas. R. Van Hise: Conservation of Natural Resources. The MacMillan Co., New York, 1910.
27. C. C. Vermeule: Report on Water-supply, Water-power, the Flow of Streams and Attendant Phenomena. Publication of the Geological Survey of New Jersey, Vol. 3, 1894.
28. Water Supply and Irrigation Papers, Nos. 97, 98, 128, 129, 169, 170, 200, 238, 243, 254, 264, 283, 284, 303, 304, 313, 334, 337, 353, 354, 371. Stream Measurements on Mississippi and St. Lawrence River Systems. U. S. Geol. Surv., Government Printing Office, Washington, D. C., 1903-1915.
29. H. M. Wilson: Irrigation Engineering, Fifth Edition, Revised and Enlarged. John Wyley and Sons, New York, 1905.
30. James Wilson, former Secretary of Agriculture: Swamp Lands of the United States. Sixtieth Congress, First Session, Senate Document No. 443. Government Printing Office, Washington, D. C., 1908.
31. James Wilson, former Secretary of Agriculture: Expenditures for Drainage Investigations. Sixty-second Congress, Second Session, Senate Document No. 466. Government Printing Office, Washington, D. C., 1911.
32. J. O. Wright: Address before the Southern Commercial Congress, Nashville, Tenn. Sixty-second Congress, Second Session, Senate Document No. 743. Government Printing Office, Washington, D. C., 1905.

PART IV

Nomenclature and Description of the Geological Formations of Indiana

By

E. R. Cumings

TABLE OF CONTENTS

TABLE OF CONTENTS—Continued

LIST OF ILLUSTRATIONS

* Figs. 1, 2 and 3 are on folded plate at beginning of Part IV.

REVISED GEOLOGICAL TIME-SCALE OF INDIANA

Pennsylvanian	Post-Alleghenian	Merom	Merom ss.	
		Shelburn	Somerville ls. Coal VIII	
	Alleghenian	Petersburg	Coal VII, Millersburg Coal VI Coal Va Coal V Coal IVa	
		Staunton	Coal IV, Linton Coal IIIa Coal III, Seelyville	
	Pottsvillian	Brazil	Coal II, Up. Minshall Lower Minshall Upper Block Coal Lower Block Coal	
		Mansfield	Coal I	
Mississippian	Chesterian	Upper	Buffalo Wallow fmn. Siberia ls. Tar Springs ss.	
		Middle	Glen Dean ls. Hardinsburg ss. Golconda ls. Indian Springs sh. Cypress ss.	
		Lower	Beech Creek ls. Elwren ss. Reelsville ls. Sample ss Beaver Bend ls. Mooretown ss.	
			Paoli ls.	Mitchell
	Meramecian	St. Genevieve	Fredonia oölite	Mitchell
		St. Louis	St. Louis ls.	Mitchell
		Salem	Salem ls.	
	Osagian	Warsaw	Harrodsburg ls.	
		Keokuk	Riverside ss. Rosewood sh. Kenwood ss.	Borden (or Knobstone)
		Burlington	New Providence sh.	Borden (or Knobstone)
	Kinderhookian	Kinderhook	Rockford ls.	
Devonian	Senecan	Genesee and Portage	New Albany sh.	
	Erian	Hamilton	Beechwood ls. Silver Creek ls.	Sellersburg
		Onondaga	Jeffersonville ls.	
		Schoharie	Pendleton ss. (Geneva?)	
Silurian	Cayugan	Bertie	Kokomo ls.	
	Niagaran	Guelph	Huntington ls.	
		Lockport	Noblesville dol. Louisville ls. Waldron sh. Laurel dol.	
		Rochester	Osgood fmn.	
	Medinan	Cataract	Brassfield ls.	
Ordovician	Cincinnatian	Richmond	Elkhorn Whitewater Saluda Liberty Waynesville Arnheim	
		Maysville	Mt. Auburn Corryville Bellevue Fairmount Mt. Hope	
		Eden	McMicken Southgate Economy	
	Mohawkian	Trenton	Cynthiana	

INTRODUCTION.

The history of Indiana geological surveys has been ably summarized by Blatchley (1917, 89-177); and brief reviews of the stratigraphy are given in a number of the annual reports. In the 28th report there appears a rather full summary of our knowledge of the Indiana geological formations, written by Hopkins (1904, 15-77) and Foerste to accompany the large-scale geological map of Indiana published in 1904. The geological time-scale for Indiana published by Hopkins (p. 17) represents substantially the terminology of Indiana stratigraphy as worked out by the earlier surveys and continued down to the present organization. It is presented herewith for comparison with the revised terminology of the present report:

Cenozoic	Quaternary-Pleistocene	Recent	
		Glacial	
	Tertiary		
Paleozoic	Permo-Carboniferous	Merom sandstone	Upper Barren
		Productive Coal Measures	Upper Productive
			Lower Barren
			Lower Productive
		Mansfield	Pottsville
	Mississippian or Lower Carboniferous	Huron	Chester-Kaskaskia
		Mitchell	St. Louis
		Bedford Oölitic	Mauch Chunk-Warsaw
		Harrodsburg	Burlington*
		Knobstone	Keokuk
		Goniatite	Chouteau
	Devonian	New Albany	Genesee
		Sellersburg	Hamilton
		Silver Creek	Hamilton
		Jeffersonville	Corniferous
		Pendleton	Schoharie
	Silurian (Upper Silurian)	Waterlime	Lower Helderberg (?) Salina
		Louisville	Niagara
		Waldron	Niagara
		Laurel	Niagara
		Osgood	Niagara
		Clinton	
	Ordovician	Richmond	Hudson River (Cincinnati)
		Lorraine	Hudson River (Cincinnati)
		Utica	
		Trenton	
		St. Peter (?)	Chazy
		Lower Magnesian (?)	Calciferous
	Cambrian	Potsdam (?)	

*The inversion of the Burlington and Keokuk in this table by Hopkins was probably not intentional.

The paleogeography of Indiana has received incidental treatment in the memoirs of Schuchert (1910, 427-606), Ulrich (1911, 281-680; 1919, 76-162), Stauffer (1909, pls. xiv, xv), and Savage and Van Tuyl (1919, 354, 369, 371,

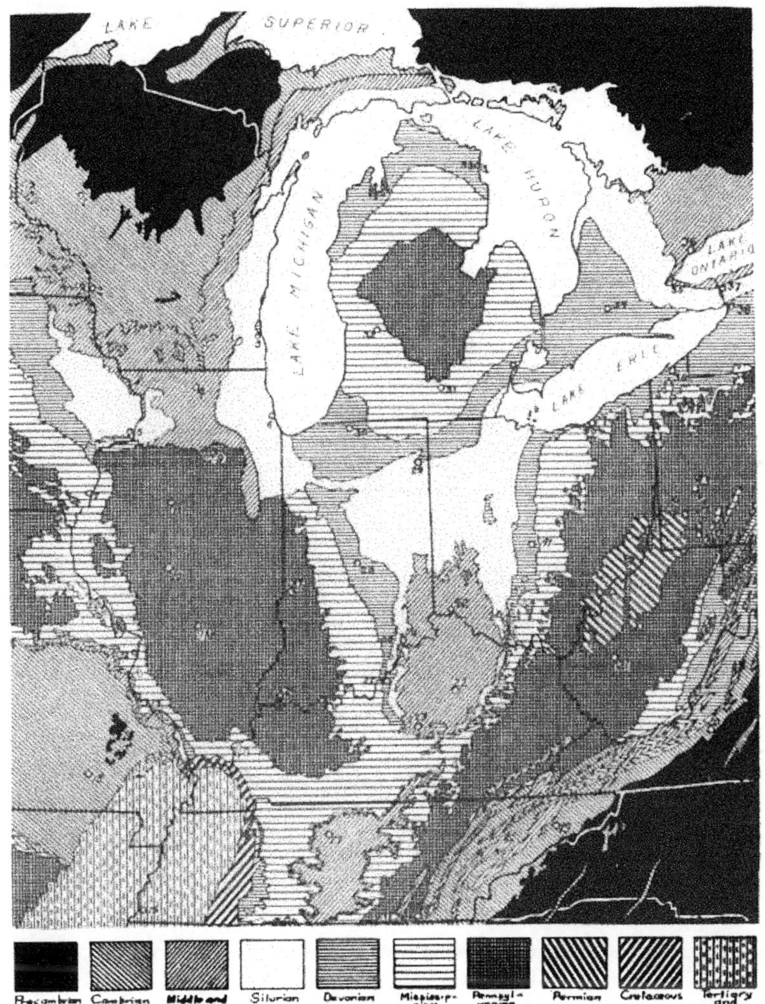

Fig. 4. Geological map of the Central States.

376); and in the text books of Schuchert (1915), Chamberlin and Salisbury (1906), and Grabau (1921), and the Outlines of Geologic History by Bailey

Willis and others (1910). Only one attempt has been made to deal specifically with Indiana paleogeography, namely, that by Blatchley in his "Indiana of Nature" (1904, 39-59). The maps published by Blatchley are in reality merely outcrop maps, and his discussion of the relations of land and water during the successive geological periods betrays a misconception of both the data and principles of paleogeographic interpretation.

Fig. 6. Index map showing type localities of geological formations in the Mississippi-Ohio valley region.

The maps presented herewith are based in part on those of Schuchert, Ulrich, and Stauffer, and in part on the writer's interpretation of the faunistic and diastrophic data. Like all such maps they are merely suggestive, and lay no claim to finality. The general geological map of the central United States will aid in the interpretation of the paleogeography and especially of the chief structural features of this region, such as the Nashville and Cincinnati arches

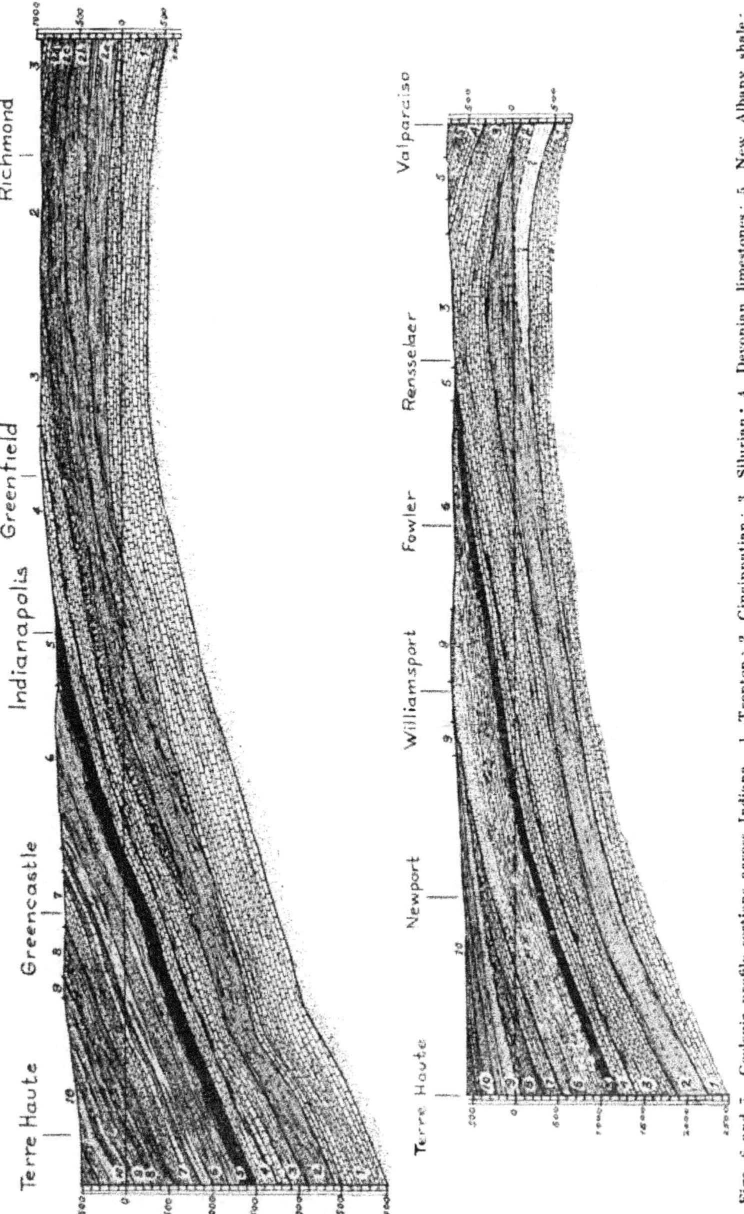

Figs. 6 and 7. Geologic profile sections across Indiana. 1, Trenton; 2, Cincinnatian; 3, Silurian; 4, Devonian limestones; 5, New Albany shale; 6, Borden (Knobstone); 7, Harrodsburg, Salem and Mitchell; 8, Chester; 9, Mansfield; 10, Coal Measures.

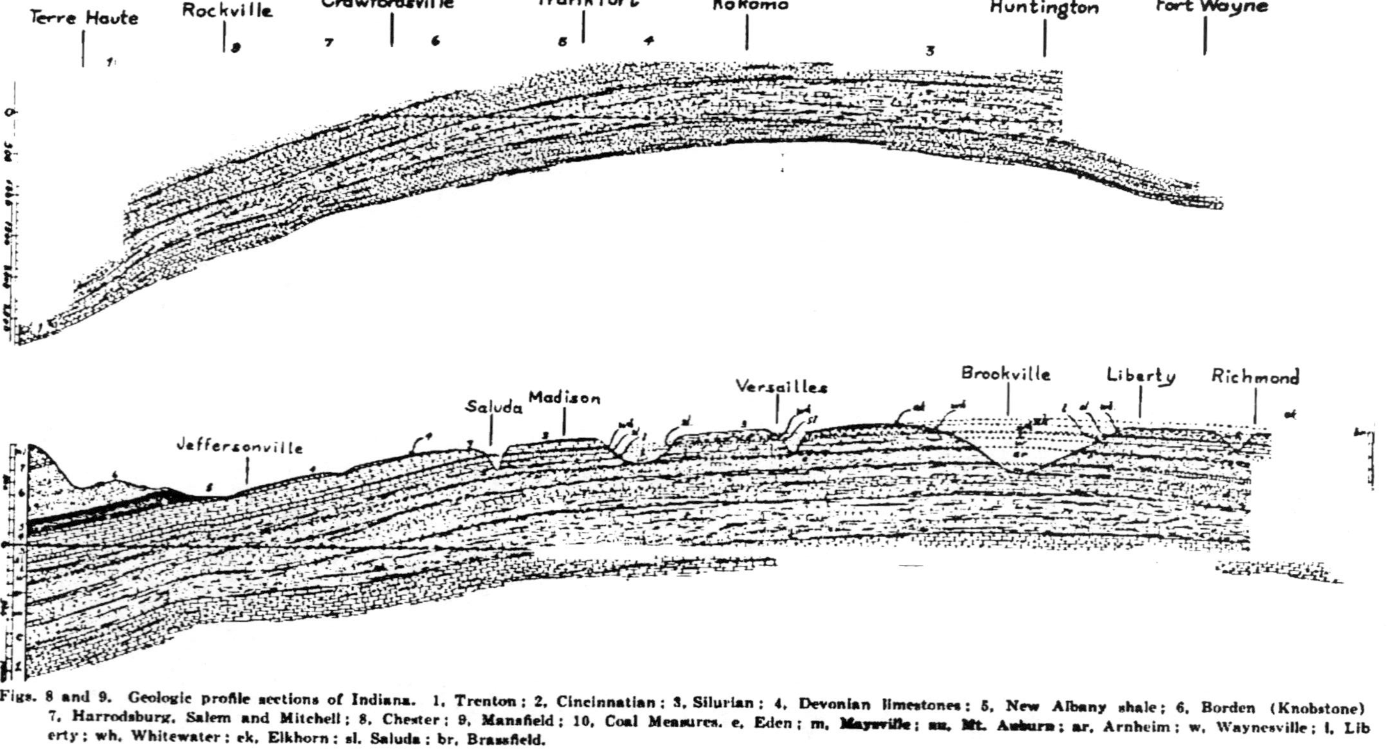

Figs. 8 and 9. Geologic profile sections of Indiana. 1, Trenton; 2, Cincinnatian; 3, Silurian; 4, Devonian limestones; 5, New Albany shale; 6, Borden (Knobstone) 7, Harrodsburg, Salem and Mitchell; 8, Chester; 9, Mansfield; 10, Coal Measures. e, Eden; m, Maysville; au, Mt. Auburn; ar, Arnheim; w, Waynesville; l, Liberty; wh, Whitewater; ek, Elkhorn; sl, Saluda; br, Brassfield.

and the great coal basins. It is drawn with special reference to the delineation of these features and is based on the geologic map of North America by Bailey

Fig. 10. Paleogeography of the Utica. After Ulrich. Broken lines, probably sea; solid lines, sea.

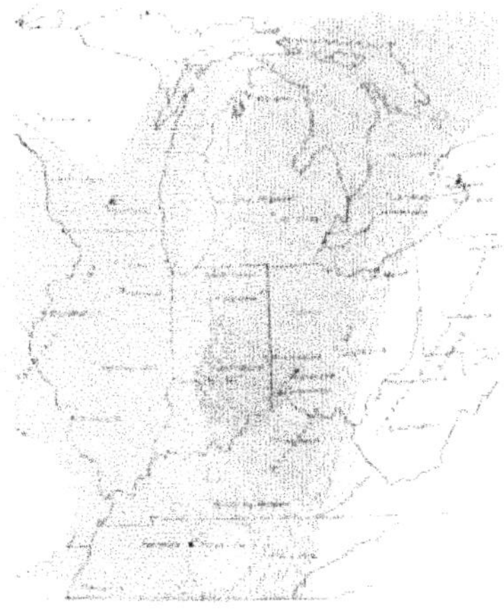

Fig. 11. Paleogeography of the Waynesville. Dotted area, non-marine sediments. Vertical lines, eastern or Waynesville waters. Horizontal lines, western (northern) or Fernvale waters. Dashed lines, probably land. Broken lines, probably sea.

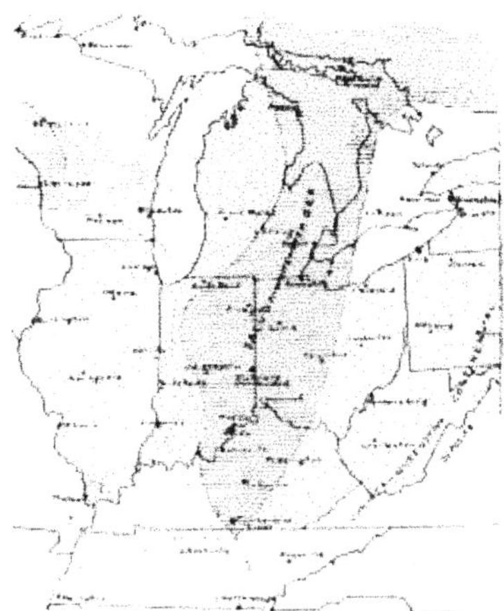

Fig. 12. Paleogeography of the Saluda-Whitewater. Dotted area, non-marine sediments. Dashed lines, probably land. Broken lines, probably sea; solid lines, sea.

Fig. 13. Paleogeography of the Brassfield. Modified from Schuchert. Broken lines, probably sea; solid lines, sea.

Willis and George W. Stose, 1911. As an aid to the understanding of geologic terminology an index map of type localities is also included.

The general stratigraphic chart or time-scale, includes a graph drawn to scale, exhibiting the main lithologic characters and thicknesses of the various formations.

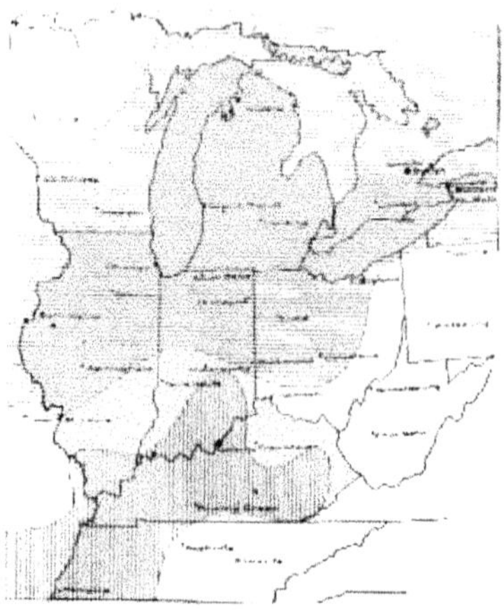

Fig. 14. Paleogeography of the Louisville-Guelph. After Schuchert. Horizontal lines, northern fauna. Vertical lines, southern fauna.

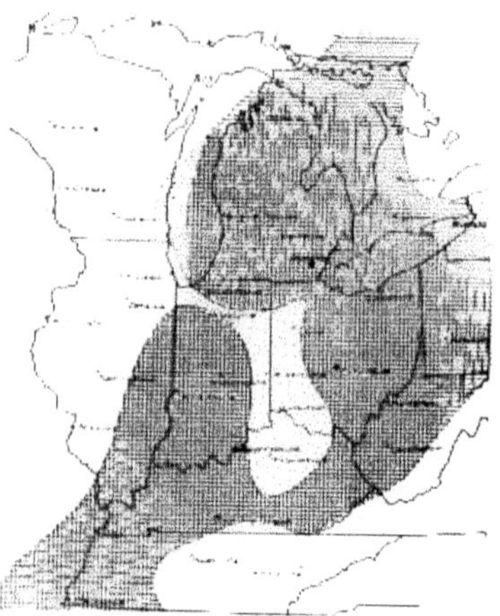

Fig. 15. Paleogeography of the Middle Devonian. Slightly modified from Stauffer. Vertical lines, southern faunas. Horizontal lines, northern faunas.

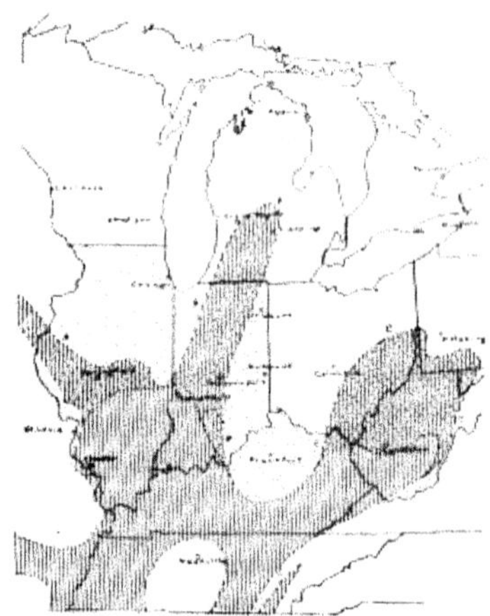

Fig. 16. Paleogeography of the St. Genevieve and Chester. After Ulrich. Dotted lines (A), St. Genevieve. Dotted spaces (B), (C), early Chester. Solid lines, late Chester.

Fig. 17. Paleogeography of the Mansfield. Modified from Schuchert. Dotted area, mostly non-marine. Solid lines, Mansfield sea.

Nomenclature and Description of the Geological Formations of Indiana

CHAPTER I.

ORDOVICIAN.

Lapworth (1879, 14) proposed the name *Ordovician*, from the Roman name of a tribe of Wales, in the following words: "It should be called the ORDOVICIAN SYSTEM, after the name of this old British tribe [the Ordovices]". To these rocks Murchison (1835, 46-52) had previously applied the name Lower Silurian, but owing to the controversy between Murchison and Sedgwick as to the delimitation of the names Cambrian and Silurian it has seemed best to most geologists, especially in America, to adopt Lapworth's suggestion.

Main subdivisions.—The geologists of the first geological survey of New York—Emmons, Mather, Hall and Vanuxem—subdivided their "New York System" into formations to which they gave geographic names, such as Potsdam, Chazy, Trenton, Niagara, etc.; and later grouped these formations into divisions known as the Champlain, Ontario, Helderberg and Erie divisions. Unfortunately, owing to the prior publication of the names proposed by Murchison and Sedgwick, (Cambrian, Silurian, Devonian) these New York division names fell into disuse until Clarke and Schuchert attempted to revive them in 1899 (1899, 874-878). It has not been possible, however, to counteract the force of priority and long usage back of the British names.

On the other hand there has been some success in the attempt to use the names of the New York divisions in a somewhat subordinate sense, and in this sense the name Champlain, for example, has been suggested and used by Schuchert (1915) and Grabau (1921) for the middle Ordovician rocks. In their 1899 paper Clarke and Schuchert proposed to call these rocks *Mohawkian*, a name that has been in rather common use in American literature ever since. This name is preferred in the present report, and is made to include the rocks from the Chazy to the top of the Trenton.

The Mohawkian is in turn divided into the Chazy, Lowville, Black River and Trenton series, and of these only the extreme upper part of the Trenton is exposed in Indiana (the Cynthiana).

The name Cincinnatian was approved by Clarke and Schuchert (1899) for the upper division of the Ordovician, and has been in general use ever since. The term "Cincinnati group" had long previously been in use for the upper Ordovician of the Ohio valley. The Cincinnatian is again subdivided into the Utica, Eden, Maysville and Richmond. All of these with the exception of the Utica are present in Indiana.

The principal source of information on the Ordovician rocks of Indiana is in the papers by Cumings (1901, 1901b, 1908, 1913) and Foerste (1897, 1903b, 1904a, 1905, 1909c, 1909d, 1910, 1912, 1912a).

MOHAWKIAN.

TRENTON.—(Point Pleasant, Cynthiana). While the Trenton formation has long been recognized as the chief oil and gas rock of Indiana, it was not

R
M
E
T
Geologic map
of
INDIANA
ORDOVICIAN
T Trenton (Cynthiana)
E - Eden
M - Maysville
R - Richmond
Department of Conservation
Division of Geology
State of Indiana
W N Logan, State Geologist
MILES

Fig. 18.

till the publication of the 1904 map of Indiana, and the accompanying summary of Indiana geological formations that Foerste (1904b, 22) announced the presence of Trenton outcrops in Switzerland County. The oil-bearing rock is undoubtedly considerably lower in the Trenton formation than the beds outcropping in Switzerland County.

The name Trenton was first used by Vanuxem (1837)*.

As early as 1829 Vanuxem (1829, 256) called attention to the resemblance between certain rocks of the Ohio valley and the rocks of Trenton Falls, New York; and the blue limestones of Cincinnati and vicinity were thought by him and by Conrad (1841, 27), to be the western continuation of the "black limestones of Trenton Falls". The first really definite correlation of an Ohio valley formation with the Trenton was by James Hall (1842, 61), who so correlated the rocks seen at low water of the Ohio at Newport, Kentucky, and now known as the Point Pleasant beds.

Conrad says (1842, 228-235) the Trenton forms "the bed of the Ohio from Cincinnati to Louisville". He was mistaken; but it does form the bed of the river at Cincinnati, and again between North's Landing and Florence, Indiana.

Point Pleasant Beds.—In 1873 Edward Orton, Sr. (1873, 370), proposed to call the rocks along the Ohio river 15 to 20 miles above Cincinnati the Point Pleasant beds from the town of that name on the river.** These beds were supposed to be below the "River Quarry" beds (Orton, 1873, 371) at Newport. Linney (1882) says the stone quarried at Point Pleasant is the equivalent of the Gray limestone in the upper part of the Trenton of New York. In 1888 (1888, 5) and again in 1889 and 1890 (1889, 546; 1890, 12) Orton recognized the Trenton age of these Point Pleasant beds. Ulrich (1888, 307, and footnote) says the beds "XI-a" of his section "are nearly equivalent to the River Quarry beds of Prof. Orton", and that the whole thickness of these beds may be seen "near Point Pleasant". He is not, however, certain as to the exact position of the Point Pleasant beds. In the report of the 1879 Committee of the Cincinnati Society of Natural History on correlation, it is stated that the Trenton group is not exposed at Cincinnati, but is probably "represented in the banks of the Ohio river a few miles east of the city" (Miller, 1879, 193-194).

J. F. James in 1891 (1891, 104) concludes that "as far as present information goes, there seems to be no more reason for assigning the Point Pleasant beds to the Trenton than there would be in making a similar disposition of the lowest beds at Cincinnati". He concludes that the Point Pleasant beds and those at Ludlow, Kentucky, (opposite Cincinnati) are of the same age. "There is no good reason to say that the Trenton outcrops at the surface at any locality within her [Ohio's] borders".

Nevertheless, Orton in volume VII of the Ohio survey (1893, 5) again refers the beds to the Trenton, and this correlation is adopted by Winchell and Ulrich in the Geology of Minnesota (1897, xcviii) and by Nickles in the Geology of Cincinnati (1902, 52-60), as well as by Prosser (1905, 3, 35), Bassler (1906, 8), and others. Nickles (1905, 19) in a later paper, and A. M. Miller (1919, 30) refer the Point Pleasant beds to the Cincinnatian.

* It is credited by Weeks (1902) to Vanuxem, 1810.

** Weeks mistakenly credits the names to James, 1891. Proc. Am. Assoc. Adv. Sci., vol. 40, p. 283; and Jour. Cin. Soc. Nat. Hist., vol. 14, p. 93.

Cynthiana.—The first to correlate the limestone beds at river level near Patriot, Indiana, with the Point Pleasant beds of Ohio, was Foerste (1905, 151). In 1906 Foerste (1906, 10, 13) proposed the name *Cynthiana* for certain beds capping the Trenton rocks of Kentucky, and included under this designation the Point Pleasant beds. He places the Cynthiana in the Cincinnatian. A. M. Miller (1919, 30) maps the outcrop near Warsaw, Kentucky, (opposite Patriot) as Cynthiana. Following Foerste, he subdivides the formation into the Greendale and Point Pleasant beds. Foerste (1906, 10) correlates the Cynthiana with the lower Winchester (Campbell, 1898). Nickles (1905, 15) also correlated the Point Pleasant with the Winchester. Foerste later (1912, 23; 1912a, 430) uses the name Catheys (Campbell, 1898) in place of Cynthiana, while Miller (1915) says the Cynthiana is in part equivalent to the Catheys. According to Bassler (1915) the Catheys is all older than any part of the Cynthiana. He also regards the latter as a member of the Trenton, rather than of the Cincinnatian. A. M. Miller (1913, 1915, 1917, 1919) places the Cynthiana in the Cincinnatian. Fenneman (1916, 60, 61) places it in the Trenton and says the upper 50 feet or more in the Cincinnati region contains much shale. Beneath the shaly part are 60 feet of limestone (Point Pleasant limestone). A fragmental limestone caps the Cynthiana. Fenneman thinks this fragmental limestone represents a redeposited portion of the Cynthiana caused by an invading sea, and that the Cynthiana was elevated and eroded slightly before the deposition of the Eden. Ninety per cent of the fossils of the Cynthiana, he says, fail to reappear in the Eden. Raymond (1916, 258) regards the Cynthiana fauna as "very closely allied to that of the Eden and Maysville". In his correlation diagram on page 257, and again on the chart, plate 8, he places the Cynthiana opposite the Stewartville dolomite of Minnesota, Upper Picton of Ontario, and Utica of central New York. Linney as far back as 1882, referred the formation to the Hudson. I have followed Ulrich, Bassler and Fenneman in placing the Cynthiana in the Mohawkian.

In 1906 Bassler (1906, 8, 9) used the name Bromley for what appears to be the lower part of the Cynthiana; and in 1919 Foerste (1909c, 210) proposed the name Nicholas for the upper part of the rocks at Point Pleasant, formerly called Point Pleasant. Bassler (1915, 1919) divides the Cynthiana into Greendale, Bromley, Gratz and Rogers Gap (in the ascending order). He regards it as higher than the Catheys and correlates it with the Collingwood of Ontario and upper part of the Schenectady beds of New York. This places it considerably higher than the Stewartville and Picton.

Description.—The Cynthiana outcrops in a narrow belt along the Ohio river in Switzerland County between the towns of Florence and North's Landing. It is about 50 feet thick and consists of dark blue to gray layers of limestone from a few inches to a foot or more thick. It is quite fossiliferous. The following species are characteristic of the formation:

Agelacrinites vetustus Foerste, *Bucania frankfortensis* Ulrich, *Calymene abbreviata* Foerste, *Columnaria alveolata interventa* Foerste, *C. alveolata minima* Foerste, *Constellaria emaciata* Ulrich and Bassler, *C fisheri* Ulrich, *Crepipora spatiosa* Ulr., *Cyphotrypa frankfortensis* Ulrich and Bassler, *Dekayella* (=*Heterotrypa*) *foliacea* U. and B., *Hebertella maria parkensis* Foerste, *Peronopora milleri* Nickles, *Platystrophia colbiensis* Foerste, *P. precursor* Foerste, P. *precursor latiformis* McEwan, P. *precursor angustata* McEwan, *Plectambonites curdsvillensis* Foerste, *Rafinesquina declivis* (James), *R. winchesterensis* Foerste, *Saccospongia rudis* Ulrich, and *Zygospira obsoleta* (Foerste).

Trenton (?) of the Kentland dome.—In 1883 Collett (1883, 58, 59) mentioned fossils from the inclined strata in the "Niagara dome" east of Kentland, Indiana, which on the authority of Mr. G. K. Greene of New Albany, he identified as of Ordovician age, and he, therefore, referred the rocks in question to the Ordovician. This reference was repeated in the 15th Indiana report by Gorby (1886, 236). Mr. Greene always insisted that the fossils in question were Ordovician forms, a view which was not favored by most of the geologists who saw the locality. It appears, however, that Greene was correct, and I understand that Foerste now regards the rock containing the fossils as of Trenton age. This is interesting as the only instance in all northern Indiana of an outcrop of Ordovician rocks, and sheds new light on the question of the origin of the "Niagara domes".*

CINCINNATIAN.

History and correlation.—The name in this form was first used by Clarke and Schuchert (1899); but the name Cincinnati group or formation had been in use for many years, having been formally proposed by Meek and Worthen in 1865 (1865, 155). Weeks (1902) credits the term "Cincinnati limestone" to Mather 1859, but this cannot be considered as a formal proposal of the name.

The rocks outcropping at Cincinnati were described as far back as 1815 by Daniel Drake and again by the same author in 1825 in a "Geological account of the Ohio valley" (1825, 124-139). In 1829 Vanuxem correlated these rocks with the Trenton of New York (1829, 256). The "blue limestone" of southwestern Ohio was again described by Locke in 1838 (1838, 207) and by Clapp in 1843, and correlated with the Caradoc of Wales. It was correlated by Conrad (1841, 27) with the limestone of Trenton Falls.

In 1842-43 Hall recognized the Trenton at river level near Cincinnati and therefore correlated the main mass of the "blue limestone and marlite" with the Hudson formation of New York (1842, 61; 1843, 267-293). Vanuxem (1842, 14) says that the "lowest mass observed in the states of Ohio and Indiana, for example, belongs to the upper part of the Champlain division of the New York system and extends from the Trenton limestone to the top of the sandstone and shale of Pulaski. It is concluded to be one mass; contains the fossils of these two periods the lower part called the blue limestone, the other the blue marlite". Owen (1844) correlated the "blue limestone" with the Trenton, Utica and Hudson river groups, and substantially the same correlation has prevailed down to recent years.

The name Cincinnati group, proposed by Meek and Worthen, was objected to by certain geologists on the ground that the rocks at Cincinnati and vicinity had been shown to be exactly equivalent to the so-called Utica and Hudson river formations of New York, and in 1879 a committee of the Cincinnati Society of Natural History, with Mr. S. A. Miller as chairman, proposed to drop the name Cincinnati and use the names Utica and Hudson. This report was adopted by the Society, though not without the protest of Mr. U. P. James, who continued to use the name Cincinnati group (1879, 27-28). Mr. Miller, himself, had earlier used the latter name (1874, 97-115). Writing in 1881, however, he calls Cincinnati a synonym of Hudson river group (1881, 268). Nevertheless, though most geologists came to speak of the Cincinnati

* In a paper published after this memoir was in the hands of the printer, Foerste describes several cephalopods, and lists other fossils from Kentland. He correlates the rocks with the Black River.

Fig. 19. Upper left, Upper Eden (McMicken) near Guilford, Indiana. Upper right, Liberty formation, type section at Liberty. Lower left, Saluda resting on Liberty, in the cut at Madison, Indiana. Lower right, Upper Eden (McMicken), and base of Maysville in cut near Guilford, Indiana.

rocks as the Hudson formation, the name Cincinnati group or formation was often used, as, for example, by Foerste (1885), Hall and Sardeson (1892), Winchell and Ulrich (1897), Kindle (1898), J. F. James (1891-1896), Newsom (1898), Clarke and Schuchert (1899), Orton (1888, 1889). Walcott (1890, 335-356) urged the use of Hudson, and Winchell and Ulrich (1897) used both names.

In the meantime the shales of the Hudson valley had been subjected to a careful restudy under the auspices of the New York geological survey, and in 1901, Ruedemann announced (1901, 564-568) that the typical Hudson shales are older than the Lorraine, Pulaski, etc., of western New York, with which they had been correlated, and that in fact nearly the whole of the Ordovician is represented in the great shale series of eastern New York. The name Hudson was therefore dropped and since 1901 practically all geologists have used the term Cincinnatian. (See for example, Nickles 1902, 1903, 1905; Ulrich, 1904, etc.; Foerste, 1905, 1906, etc.; Cumings, 1908, 1913; Bassler, 1906, etc.; Schucbert, Ulrich, etc.)

Ruedemann's results have been signally verified by his later exact studies in the Hudson and Mohawk valleys (1912). It is likely that nothing later than Trenton is present in the shales of the Hudson and Mohawk valleys. Attention should here be called to the fact that Worthen (1866, 137) in explaining why he proposed the name Cincinnati, said "As it is now acknowledged that the rocks along the Hudson river valley to which the name 'Hudson River Group' has been applied belong, as long ago maintained by Emmons, to a different horizon from the so-called Hudson river rocks of western New York and the states farther westward, it seems to be an awkward misnomer to continue to apply the name 'Hudson River Group' to these western deposits". And again (1866a, xv-xix) he suggests that the rocks of the Hudson valley, claimed as true Hudson by Hall, *may be Trenton*. The description of the Cincinnatian will be given under the discussion of its subdivisions.

Subdivisions.—Prior to 1873 the Cincinnati rocks were not subdivided into separate formations, though in 1842 Hall recognized Utica fossils in the shales of the old First Ward at Cincinnati; and the Cincinnati collectors thereafter were disposed to apply the name Utica as well as the name Hudson to the Cincinnati rocks. In 1873 Orton (1873, 370-371) subdivided the Cincinnati rocks into the *Point Pleasant beds (q. v.), River Quarry beds, Eden shales, Hill Quarry beds* and *Lebanon beds*. These are about equivalent to the Cynthiana, Eden, Maysville and Richmond of the present classification. The River Quarry beds are included in the present paper under the Cynthiana formation *(q. v.)*.

Eden and Utica.—Orton's Eden shales included at the base the beds called Utica by Hall and extended up through the 250 feet of shales to the conspicuous limestones capping the hills at Cincinnati. In 1888, Ulrich published in the American Geologist an extensive paper on the stratigraphy and correlation of the Ordovician of Kentucky and Ohio (1888) and included the Eden beds in his division XI-b, correlating the beds with the black shale of the Findlay oil wells (see Orton 1888, 1889; Phinney 1891). He says (*loc. cit.* 315) "These shales [in the Findlay well] because they are black or brownish and contain *Leptobolus insignis*, have been correlated by Professor Orton with the Utica shale of New York. . . . judging from the position held by them and from a series of drillings kindly sent me by Professor Orton, I fully concur with him in thus placing them". In volume VI of the Ohio survey (1888, 8), Orton notes that the Utica shale of the wells thins to the southward, and expresses doubt as to the presence of the Utica in the Cincinnati section. Nevertheless, the name Utica was applied to the Eden shales in the Geology of Minnesota (1897, lxxxix, cii) by Winchell and Ulrich, and thereafter came into common use. (See Blatchley and Ashley 1898a; Cumings 1901, 1908; Nickles 1902, 1903, 1905; Foerste 1904a; Ulrich and Schuchert 1902; Ulrich 1904; Bassler 1903; Ellis 1906; Ward 1908; Bigney 1916).

In 1905 Foerste proposed a return to Orton's name Eden (1905, 150), and the latter name has been in common use ever since. (See later papers by Nickles, Bassler, Cumings Prosser, Foerste, Ulrich, Schuchert, etc.). In a later paper Foerste (1909c, 210) says "The Eden was described by Orton, but formerly included also the Mt. Hope beds now referred to the Maysville."

For the green shales seen immediately above the Trenton (Cynthiana) in the old First Ward of Cincinnati, and containing several Utica species *(Triarthrus becki, Leptobolus insignis)* Foerste (1905, 151) proposed the name Fulton, from the present name of the First Ward. This name has been in common use ever since that date. The Fulton bed is not present in Indiana.

Still a different arrangement of names is given by Fenneman (1916, 63) in his Geology of Cincinnati, where he uses the name *Latonia* shale for the 200 or more feet of shale above the Fulton and below the Maysville, and says this name will be used in the forthcoming Cincinnati Folio of the U. S. Geological Survey. He uses Eden for the combined Fulton and Latonia shales.

In his paper on the Geology of Cincinnati, Nickles (1902) divided the Eden into three faunal zones, called in ascending order, the *Aspidopora newberryi* (80 ft.), *Batostoma jamesi* (120 ft.) and *Dekayella ulrichi* (60 ft.) beds. Later, Bassler (1906, 8-10) proposed the names *Economy, Southgate* and *McMicken*, respectively, for these three divisions, and these names have come into common use. (See later papers by Bassler, Ulrich, Foerste, Cumings, etc.)

The Eden has commonly been correlated with the Frankfort of New York (So, Ulrich and Schuchert 1902, 643; Ulrich 1904, 90; 1911, etc.; Bassler 1915; Raymond 1916). Foerste, on the other hand, (1916, 10-13) says the shales in the "Gulf" at Lorraine, New York, "up to within a short distance of the level of Lorraine village correspond approximately to the Eden division of the Cincinnatian series of rocks. . . . This suggests the division of the Lorraine into at least two divisions, an upper and a lower one, the lower including the *Eden* part. . . . [The] Eden division cannot be identified with the Frankfort, now that the fossil content of the Frankfort has been worked out by Dr. Ruedemann [1912, 34-37]. . . . The Frankfort may be regarded as above the Utica but below the typical Eden." (See also Foerste (1914) on the Lorraine of New York and Quebec.) In his latest correlation table, however, Bassler (1919, 51) still correlates the Eden with the Frankfort. Grabau (1921, II, 283) includes the Frankfort and Pulaski in the Lorraine. Until the final results of Ruedemann's work are known, this question must remain undecided.*

* There is still some doubt as to the age of the typical Utica shale of New York. Within recent years a certain tendency has developed to consider it as a shaly facies of the upper Trenton. See, for example, Grabau 1919a, 232-233; Ruedemann 1908, 44. Raymond (1916, 248-251) appears to correlate it with upper Picton (Cobourg) and Cynthiana.

Description.—The Eden formation consists of about 200 to 240 feet of soft blue to greenish, marly shales, which weather very readily to marly clays of a yellowish to greenish color, interstratified with occasional thin layers of very fossiliferous limestone or (occasionally) of sandstone. The shales contain few fossils other than graptolites and trilobites, which are sometimes abundant. The fauna of the limestone layers is on the other hand very prolifie, consisting of Bryozoa, Brachiopoda, crinoids, etc., in profusion. In the upper or McMicken division there is an increase in the amount of limestone, and several species of fossils *(Dalmanella multisecta* and *Heterotrypa ulrichi)* occur in enormous numbers. The best localities in Indiana for the study of the Eden are the cuts along Tanner's Creek in Dearborn County, near Guilford, and the exposures near Vevay and Patriot.

Fauna.—The following species are especially characteristic of the Eden of Indiana:— *Climacograptus typicalis* Hall (S).* *Iocrinus subcrassus* Meek and Worthen (S). *Nereidavus varians* Grinnel (S). *Amplexopora persimilis* Nickles (E. S.). *A. Petasiformis* (Nich.) (E. S. M.). *A. Petasiformis welchi* (James) (E. S. M.). *A. Septosa* (Ulrich) (M). *Arthropora cleavelandi* (James) (E. S. M.). *Arthrostylus tenuis* (James) (E. S. M.). *Aspidopora areolata* Ulrich (S). *A. newberryi* (Nich.) (E. S. M.). *Atactopora angularis* Ulrich and Bassler (E). *A. Hirsuta* Ulrich (S). *A. intermedia* Cumings and Galloway (M). *Atactoporella newportensis* (Ulrich) (E). *A. typicalis* Ulr. (E). *Batostoma implicatum* (Nich.) (E. S. M.) *B. jamesi* (Nich.) (E. S. M.). *Berenicea vesiculosa* Ulr. (S). *Bythopora arctipora* (Nich.) (E. S. M.). *B. parvula* (James) (S. M.) *Ceramoporella triloba* Cumings and Galloway (M). *C. tubulosa* C. & G. (M). *Coeloclema alternatum* (James) (S. M.). *C. commune* Ulrich (E. S. M.). *Crepipora simulans* Ulr. (S). *C. venusta* (Ulr.) (E). *Dekayia obscura* (Ulr.) (M). *D. maculata* (James) (M). *Dicranopora meeki* (James) (S. M.). *Hallopora onealli* (James) (E. S. M.). *H. onealli communis* (James) (E. S. M.). *H. onealli sigillarioides* (Nicholson) (S. M.). *Heterotrypa ulrichi* Nich. (M). *Homotrypa curvata praecipta* Bassler (S). *H. glabra* Cumings and Galloway (M). *Peronopora vera* Nickles (S. M.). *Chasmatopora variolata* (Ulr.) (S. M.). *Proboscina confusa* (Nich.) (S). *Rhynidictya paralella* (James) (S). *Stictoporella flexuosa* (James) (S). *Stigmatella nicklesi* Ulrich and Bassler (S). *Crania albersi* Miller (E. S.). *Dalmanella multisecta* (Meek) (E. S. M.). *Plectambonites rugosus* (Meek) (E. S. M.). *Rafinesquina squamula* (James) (S. M.). *Hormotoma gracilis* (Hall) (S). *Orthoceras junceum* Hall (E). *Acidaspis ceralepta* (Anthony) (E). *Proetus spurlocki* Meek (E. S.). *Bollia persulcata* Ulr. (S). *Bythocypris cylindrica* (Hall) (S). *Ceratopsis oculifera* (Hall) (E. M.). *Primitia centralis* Ulr. (S).

MAYSVILLE.—(Hill Quarry, Lorraine). In the upper part of the Eden, layers of limestone become more numerous, and finally predominate to such an extent as to give rise to a conspicuous bed of limestone to which the name Hill Quarry beds was applied by Orton (1873, 371). He says (376) that in these beds "the solid rock [limestone] has risen again to as high a proportion as one foot in five or six of ascent", whereas in the Eden, the proportion is seldom more than one foot of limestone to ten feet of ascent. Orton evidently, as noted by Foerste, began the Hill Quarry division with the rocks now called Fairmount. He gives the thickness as 150 feet, and evidently includes in it everything up to and comprising the highest strata of the Cincinnati hills. The "Lynx bed" of Cincinnati collectors containing *Platystrophia ponderosa auburnensis* Foerste ("Orthis lynx") and exposed on the Auburn hills, is regarded by Orton (p. 395) as the top of the Cincinnati section or base of the Lebanon division (Richmond). This bed is now known as the Mt. Auburn formation.

The first advance over Orton's classification was the proposal of Winchell and Ulrich (1897, cii) to call the Hill Quarry beds Lorraine (Emmons 1842, 119, etc.) correlating them with the well-known formation of that name in

* S=Southgate; E=Economy; M=McMicken.

western New York. While this correlation of the Cincinnati formation had been implied in the former use of the name Hudson, the more specific name Lorraine had not previously been applied to these rocks. In 1900-1901 Cumings proposed the faunal designations *Rafinesquina* and *Platystrophia* zones for these beds. The name Lorraine at once met with favor and appears in papers by Nickles (1902, 1903, 1905), Bassler (1903), Foerste (1902, 1903, 1903b, 1904a, b), Cumings (1908, as an alternative of Maysville). It occurs even in recent papers by Ellis (1906), Ward (1908), Tucker (1911), Bigney (1916) and Culbertson (1916). In 1904 Foerste also suggested the faunal names *Plectorthis*, *Platystrophia* and *Rafinesquina* zones (1904b, 23-24) for the Lorraine of Indiana.

In his 1905 paper Foerste (1905, 150) proposed to drop the New York name and to adopt the name *Maysville*, from Maysville, Kentucky, for these rocks, on the grounds that the Cincinnati rocks were deposited in a distinct province. The latter name has, therefore, been in common use since that date. (See Nickles 1905; Cumings 1908, 1913; Foerste 1909c, 1909d, 1910, 1910a, 1912, 1912a, etc.; Bailey Willis 1912; Schuchert 1910; Ulrich 1911, 1914, 1919; Bassler 1915, 1919, etc.)

The term *Covington* group was proposed by Bassler (1906, 8, 9) on the suggestion of Ulrich, for the combined Utica, Eden and Maysville divisions, but seems later to have been discarded even by these authors. In 1913 Cumings suggested (1913, 354) that the upper part of the Maysville, including the Arnheim formation, commonly included under the Richmond, might appropriately be called the *Harmon* formation.

The first suggestion of a subdivision of the Maysville series into named formations was by Nickles (1902) in a paper on the "Geology of Cincinnati", though Ulrich (1888) and Cumings (1901) had previously suggested subdivision on faunal grounds, and the latter author had named three faunal zones corresponding roughly to the lower middle and upper divisions of the Maysville. Nickles (1902) recognized six subdivisions which he named in the ascending order the *Mt. Hope* (50 ft.), *Fairmount* (80 ft.), *Bellevue* (20 ft.), *Corryville* (60 ft.), *Mr. Auburn* (20 ft.) *and Warren* (80 ft.). To these he also attached faunal names, which have not, however, met with much favor. These formation names have come into general use.

In 1906 Bassler (1906, 8, 10), acting on a suggestion of Ulrich, proposed to divide the Maysville into two formations: the *Fairview*, including the Mt. Hope and Fairmount, and the *McMillan*, including the Bellevue, Corryville and Mt. Auburn. The Warren beds, renamed *Arnheim* by Foerste (1905, 150), he places in the Richmond, a return to the usage of Orton. These two names have been used by these authors in later papers and by Foerste, Cumings, Fenneman and others, with Nickles' names as subdivisions. Still a different grouping, especially applicable to Indiana, has been used by Cumings and Galloway (1913, 354). These authors follow Bassler in grouping the Mt. Hope and Fairmount together as Fairview, but place the Corryville, Mt. Auburn and Arnheim together as the *Harmon* formation, and separate the Bellevue as a distinct formation. The reason for this change is that in Indiana the Corryville and Mt. Auburn lose their distinctive characters and the Arnheim, especially in its lower half, is faunistically very much more closely related to the Maysville than to the Richmond. Cumings and Galloway also propose to call their three divisions (from below upward) the *Plectorthis plicatalla, Rafinesquina ponderosa* and *Rafinesquina fracta* zones. (See also Foerste, 1904b, 24.)

Mt. Hope-Fairmount (Fairview).—In Ohio and the region of Maysville, Kentucky, the Mt. Hope and Fairmount are fairly distinct divisions and the base of the latter is characterized by a conspicuous bed containing great numbers of *Strophomena planoconvexa*. The Fairmount is predominantly blue limestone layers wth especially heavy beds near the top, while the Mt. Hope is mostly soft, marly shale similar to the Eden. In Indiana there is less distinction both lithologically and faunally between the two, though the Fairmount still largely consists of limestones, some layers being somewhat sandy (dolomitic ?), and there is no sharp boundary between the two formations. The combined Mt. Hope-Fairmount in Indiana is about 75 feet thick. There is, furthermore, no sharp lithologic boundary between the Mt. Hope and the McMicken shales, beneath it. The fauna also shows a transition from one to the other, such very common species as *Amplexopora septosa*, *Peronopora vera*, *Hallopora communis*, *Dalmanella multisecta* and *Plectorthis plicatella* being common to both. *Constellaria constellata* and its varieties rather constantly characterize the lower part of the division and mark the transition from the Eden to the Maysville. (For detailed sections the reader is referred to the author's 1908 paper on the Ordovician of Indiana.) The faunas of this and the other divisions of the Maysville will be given at the close of the dicussion of the Maysville. The best localities for collecting fossils and for the study of the thickness and lithology of these beds are the railroad cuts along Tanner's creek in Dearborn County and the outcrops in the vicinity of Lawrenceburg, Aurora and Vevay. The top of the Fairmount is well exposed in the quarries in Madison, Indiana.

Bellevue.—This formation was included in the *Platystrophia* zone of the writer's earlier papers (1901) and the 1904 paper of Foerste (1904b, 23), and constituted the *Rafinesquina ponderosa* zone of Cumings and Galloway's 1913 paper. The name Bellevue, as already indicated, was propoŝed by Nickles (1902, 82). The formation consists of irregular rubbly blue limestones with shaly partings and immense numbers of fossils which freely weather out, so that beautiful specimens may be collected in abundance. In fact, many layers are a veritable coquina rock. It is in striking contrast to the rather heavy, and much less fossiliferous, layers of the upper Fairmount, a feature especially well shown at Maysville, Cincinnati and Madison. In fact the Bellevue is one of the easiest of the Maysville formations to recognize in the field. In Indiana it is also conspicuously set off by its fauna which is characterized by a very constant association of the brachiopods, *Platystrophia ponderosa*, *P. laticosta*, *Rafinesquina ponderosa* and *Hebertella sinuata* with a great abundance of the Bryozoa *Heterotrypa frondosa*, *Hallopora ramosa* and (usually) *Monticulipora mammulata*. In Ohio and Kentucky these species especially the *Platystrophias*, go up into the Corryville, but in Indiana they do not occur in the same association above the Bellevue. According to Mrs. McEwan (1919, 402-403) the typical *Platystrophia ponderosa* Foerste is confined to the Bellevue, though several varieties are found both below and above this horizon. The best locality in Indiana for the study of the Bellevue is at Madison and vicinity (Clifty Creek). Many good exposures may also be seen near Vevay, Lawrenceburg, Aurora and Manchester Station (on Tanner's creek). The formation is very attractive to collectors on account of the abundance and beauty of the large brachiopods and Bryozoa. The Bellevue is about 25 feet thick in Indiana.

Corryville.—The *Corryville* formation in the Cincinnati and Maysville region is rather shaly but somewhat resembles the Bellevue in appearance, though less fossiliferous. It is also less rubbly and there are a good many well defined limestone layers. Its fauna also strongly resembles the Bellevue, containing abundantly such common species as *Rafinesquina fracta, Platystrophia laticosta, P. sublaticosta, P. unicostata, Heterotrypa frondosa, Hallopora ramosa, H. rugosa, Bythopora gracilis, Homotrypa curvata,* etc. *Chiloporella flabellata* is regarded as the most characteristic bryozoan and occurs at certain levels in great abundance. *Bythopora gracilis* is also much more abundant in the Corryville than elsewhere. *Rafinesquina fracta, Amplexopora filiasa, Hallopora andrewsi, Dekayia appressa* and *Heterotrypa paupera* are common and rather characteristic fossils. *Cyclora minuta* is often very abundant in the hard limestone layers of the Corryville. In Indiana the *Platystrophias* are much less abundant and in some places entirely lacking in the Corryville, and its resemblance to the Bellevue is far less striking than in Ohio and Kentucky. The Bryozoa mentioned above nevertheless occur rather persistently and serve to identify the formation. It is probably not over 40 or 50 feet thick in Indiana. Foerste says (1904a, 90) the "Corryville bed at Madison, therefore, cannot exceed 54 feet and possibly may not exceed 44 feet." In fact Butts (1915, 69) does not regard the Corryville and Mt. Auburn as present at all in the Madison section. With this opinion, Mrs. McEwan (1920), who has made a very careful study of the section, does not agree.

Mt. Auburn.—The tops of the hills at Cincinnati are capped by a bed of irregular rubble limestone containing great numbers of *Platystrophia ponderosa auburnensis,* the form spoken of by Cumings (1903, 26-30) as the gerontic *Platystrophia lynx,* and commonly called *Orthis Lynx,* ("double headed Dutchman") by collectors. These rocks were accordingly called the "lynx beds". As already noted, Orton took the lynx beds as the top of his "Cincinnati rocks proper", or Eden and Hill Quarry divisions; because they were the highest beds exposed at Cincinnati, and not because he regarded this horizon as marking any stratigraphic or faunal break. Nickles (1902, 85) called the "lynx bed" the *Mt. Auburn,* from the Auburn hills in Cincinnati, where it is about 20 feet thick. In Indiana there is no such bed of abundant *Platystrophia auburnensis,* and only occasional specimens of this variety are found in the rocks capping the Corryville. Other common Mt. Auburn species, however, such as *Coeloclema oweni* and *Homotrypa pulchra* are found in abundance through several feet of rubble limestone at this level on Tanner's Creek, and this horizon undoubtedly marks the level of the Mt. Auburn. Whether the Mt. Auburn is present at Madison and vicinity is somewhat doubtful, though according to Mrs. McEwan (1920) *Coeloclema oweni* and other members of the fauna occur there at the appropriate level. Foerste says (1904a, 90) "At Madison only three feet can be definitely assigned to this [Auburn] bed". Cumings (1908, 636) stated, on the basis of the absence of Platystrophia, that the Mt. Auburn is absent from the Madison section.

Arnheim.—Though this division is closely related to the Corryville-Mt. Auburn in Indiana, it is considered separately, on account of its distinctive characters elsewhere, and because of the doubt as to whether it should be placed in the Maysville or Richmond series.

As stated above, the formation immediately succeeding the Mt. Auburn beds was called *Warren* by Nickles (1902, 86) and placed in the Lorraine (Maysville). This disposition of the formation was followed by Bassler (1903, 567), Nickles (1903, 1905) and Foerste (1903b, 1904b). In 1905 Foerste finding that the name Warren was preoccupied, proposed the name *Arnheim*, from Arnheim near Georgetown, Ohio, for these beds (1905, 150). Bassler and Foerste subsequently placed the Arnheim formation in the Richmond, a usage now generally adopted (see later papers by Foerste, Bassler, Ulrich, Schuchert, Miller, Cumings (1908), McEwan (1920). In 1913 Cumings and Galloway (1913, 359-360) argued that the fauna of the Arnheim in Indiana is dominantly Maysville and so again placed the formation in that series.

Foerste (1910, 1912a) has elaborately investigated the Arnheim in Ohio, Indiana and Kentucky and divides it into two divisions, a lower or *Sunset* division (1910, 18, 19), named from Sunset, Kentucky, and an upper or *Oregonia* division (1910, 18), named from Oregonia, Ohio. In southern Ohio east of the Cincinnati axis and in Kentucky the lower division is calcareous, consisting of rather strong beds of limestone, often cross-bedded and sometimes wave-marked and (in Kentucky) rather barren of fossils. At Arnheim it abounds in the valves of *Rafinesquina alternata* often standing edgewise in the layers of rock. Foerste says (1910, 19) "It is suspected that this lower part has closer affinities with the Mt. Auburn fauna than with the upper Arnheim"; and again (1912a, 429) he says that the lower part of the Arnheim in southeastern Ohio and Kentucky is comparatively unfossiliferous and the transition to the upper part is rather abrupt. This level may be traced with greater exactness than any other horizon in the Richmond on the east side of the Cincinnati axis.

The Oregonia division contains *Dinorthis carleyi* (*D. retrorsa* of authors), *Rhynchotrema dentata* and *Leptaena richmondensis* associated with *Platystrophia ponderosa* vars. and certain Richmond Bryozoa such as *Batostoma varians*, *Hallopora subnodosa*, etc. East of the Cincinnati axis and in Kentucky it is usually a limestone rubble with much shale and quite different in appearance from the Sunset beds. The combined thickness of lower and upper Arnheim in Indiana is at least 80 feet. West of the axis and in Ohio and Indiana there is no such lithologic contrast between the lower and upper Arnheim, the lower as well as the upper part containing a large amount of shale and large numbers of fossils. Nevertheless, the faunal contrast between the two parts is well maintained, even in Indiana.

The Richmond fauna comes in with the invasion of *Dinorthis carleyi*, *Leptaena richmondensis* and *Batostoma varians*, particularly the latter. The fauna of the lower division, especially in Indiana and Ohio west of the Cincinnati arch, is overwhelmingly Maysville, and is especially noteworthy for the abundance of *Rafinesquina fracta* and such species of Bryozoa as *Dekayia appressa*, *D. multispinosa*, etc., and *Heterotrypa frondosa*. None of these pass into the Waynesville. The *Dekayias* with the exception of *D. obscura* and *D. maculata* are pre-eminently Maysville forms. The abundance of *Rafinesquina fracta* in the lower Arnheim of Indiana and Ohio is also significant since it also is a characteristic middle and upper Maysville form. On the other hand, the Richmond species mentioned above are lacking in the lower Arnheim. As pointed out by Cumings and Galloway (1913) the great Richmond invasion comes in the Waynesville, which presents an almost complete faunal contrast to the Maysville. While in Kentucky there is evidence of local diastrophic movements preceding and during Arnheim time, the widespread readjustment that admitted the Richmond species into the Ohio basin certainly did not occur earlier than the upper Arnheim. It may be that the dividing line between the Maysville and Richmond should be placed at the base of the Oregonia beds where *Batostoma varians* first appears. It should not be placed at the base of the Arnheim. The fact that the lowest Richmond fauna of the Ontario-Quebec province, as shown by Foerste (1916, 96), is of Waynesville age, points to the same conclusion. (See also on the Arnheim, Schuchert 1910, 530; Ulrich 1911; 1914, 612; Butts 1915, 39-45; A. M. Miller, 1919; Fenneman 1916.)

Fauna of the Maysville.—The characteristic species of the Maysville are as follows:—*Agelacrinites cincinnatiensis* (Roemer) (F. B. C.).* *Allonychia jamesi* (Meek) (C). *Allonychia subrotunda* Ulrich (C). *Amplexopora ampla* U & B (F.). *A. cingulata* Ulrich (F. S.). *A. filiasa* (D'Orbigny) (B. C. A.). *A. robusta* Ulrich (Maysville). *A. septosa* and varieties (F. H.). *Anomalodonta plicata* Ulrich (C). *Aparchites minutissimus* (Hall) (Maysville). *Arabellites acirulatus* James (F). *Arthropora cincinnatiensis* (James) (H). *A. shafferi* (Meek) (Maysville). *A. hirsuta* Ulrich (F.). *Atactopora maculata* (Ulrich) (F.). *Atactoporella multigranosa* (Ulrich) (F. B. C. S.). *A. ortoni* (Nicholson) (B. C. A. S.). *Bàtostoma maysvillense* Nickles (F. C.). *B. gracilis* (Maysville). B. *striata* Ulrich (B. C. S.). *Ceramoporella distincta* Ulrich (F.). *Byssonychia acutirostris* Ulrich (F.). *V. alreolata* Ulrich (C). B. *imbricata* Ulrich (C.). B. *radiata* Hall (Maysville, etc.). *B. praecursa* Ulrich (F.). *Bythropora dendrina* (James) (F. C.). B. *gracilis* (Maysville). B. *striata* Ulrich (B. C. S.). *Ceramporella distincta* Ulrich (Maysville). C. *granulosa* Ulrich (Maysville). C. *whitei* (James) (F. B. C. S.). *Ceratopsis* (F. C.). *B. gracilis* (Maysville). *B. striata* Ulrich (B. C. S.). *Ceramoporella distincta* Ulrich (C. S.). *Clathrospira conica* (U. & S.) (F.). *Coeloclema oweni* (James) (A.). *Constellaria florida* Ulrich (H. F.) [=*C. constellata*]. *C. florida prominens* Ulrich (H. F.). *Crepipora impressa* Ulrich (F.). *C. simulans* Ulrich (F.). *Ctenodonta obliqua* (Hall) (Maysville). *Cuneamya elliptica* Miller (C). *Cyclonema pravile* Ulrich (H.). *C. gravile striatulum* Ulrich (H.). *C. humerosum* Ulrich (B. C. A.). *C. fluctuatum* Ulrich (S.). *C. inflatum* Ulrich (F.). *C. mediale* Ulrich (F.). *C. simulans* Ulrich (C.). *C. sublaere* Ulrich (H.). *C. transversum* Ulrich (F.). *Cymatonota recta* Ulrich (B. C.). *Cyrtoceras conoidale* Wetherby (F.). *Cyrtolites ornatus* Con., (Maysville). *Dalmanella fairmountensis* Foerste (B.). *Dekayia appressa* Ulrich (F. B. C. S.). *D. aspera* (M-E & H.) (H. F.). *D. magna* Gumings (B. S.). *D. multispinosa* Ulrich (C. S.). *Dendrocrinus cincinnatiensis* (Meek) (F.). *Dicranopora emacerata* (Nicholson) (H. F.). *D. meeki* (James) (H.). *Discotrypa elegans* (Ulrich (F.). *Eridonychia apicalis* Ulrich (F.). *E. paucicostata* Ulrich (F.). *Escharopora falciformis* (Nicholson) (H. F.). *E. maculata* (Ulrich) (F.). *E. pavonia* (D'Orbigny) (H. F.). *Eunicites simplex* Hinde (H.). *Glyptocrinus decadactylus* Hall (F). G. *dyeri* Meek (C). G. *subglobosus* (Meek) (C.). *Gomphoceras cincinnatiense* Miller (C.). *G. faberi* Miller (C.). *Hallopora andrewsi* (Nicholson) (F. B. C.). *H. dalei* (M-E. & H.) (H. F.). *H. ramosa* (D'Orbigny) (B. C. A. S.). *H. rugosa* (M-E. & H.) (B. C. A. S.). *H. subplana* (Ulrich) (H. F.). *Hebertella sinuata* (F. B. C.). *Hemicystites stellatus* Hall (F.). *Heterocrinus juvenis* Hall (C.). *Heterotrypa frondosa* (D'Orbigny) (F. B. C. A. S.). *H. inflecta* (Ulrich) (C.). *H. lobata* (Cumings) (H. F.). *H. Paupera* (Ulrich) (C.). *H. solitaria* Ulrich (F. B.). *H. subfrondosa* (Cumings) (H. F.). *H. subpulchella* (Nicholson) (H. F.). *Homotrypa alta* C. & G. (H. F.). *H. cincinnatiensis* Bassler (H. F.). *H. curvata* Ulrich (F. B. C.). *H. dumosa* Bassler (H. F.). *H. flabellaris spinifera* Bassler (F.). *H. grandis* Bassler (Upper Maysville). *H. obliqua* Ulrich (Maysville). *H. pulchra* Bassler (C. A. S.). *H. spinea* Cumings & Galloway (H. F.). *Homotrypella dubia* (Cumings & Galloway) (B. C. S.). *Iocrinus subcrassus* (M. & W.) (Maysville). *Isotelus pigas* Dekay (Maysville). *Isotelus maximus* Locke (Maysville). *Lepidocoleus jamesi* (H. & W.) (Maysville). *Leptotrypa calceola* (M. & D.) (C.). *L. clavacoidea* (James) (C.). *L. minima* Ulrich (B.). *L. ornata* Ulrich (C.). *Lingula cincinnatiensis* (H. & W.) (F.). *Lophospira bicincta* Hall (Maysville). *Modiolodon obtusus* Ulrich (B.). M. *truncatus* (Hall) (F. B. C.). *Modiolopsis faberi* Miller (F.). *M. Modiolaris* (Conrad) (F.). M. *subparallela* Ulrich (F.). *Monticulipora cincinnatiensis* (James) (C.). *M. mammulata* (D'Orbigny) (F. B.). *Nicholsonella vaupeli* (Ulrich) (Maysville). *Opisthoptera ampla* Ulrich (F.). *Orthoceras byrnesi* Miller (F.).

* F=Fairmount, H=Mt. Hope, B=Bellevue, C=Corryville, A=Mt. Auburn, S=Lower Arnheim.

O. meeki Miller (F.). *Orthodesma parvum* Ulrich (C.). *Orthonotella faberi* Miller (F.). *Orthorhynchula linneyi* (James) (F.). *Peronopora vera* (H. F.). *P. decipiens* (Rominger) (B. C. A. S.). *Petigopora asperula* Ulrich (B.). *P. gregaria* Ulrich (F. B. C.). *P. petechialis* (Nicholson) (Maysville). *Platystrophia jurensis* McEwan (H. F.). *P. pauciplicata* Cumings (H. F.). *P. strigosa* McEwan (F.). *P. nitida* McEwan (F.). *P. morrowensis* (James) (C.). *P. corryvillensis* McEwan (C.). *P. sublaticosta* McEwan (C.). *P. ponderosa* Foerste (B.). *P. ponderosa auburnensis* Foerste (A.). *P. ponderosa* vars (Maysville). *P. profundosulcata* (Meek) (H. F.). *P. profundosulcata hopensis* Foerste (H.). *P. crassa* (James) (F. B. C.). *P. laticosta* Meek (B. C.). *P. unicostata crassiformis* McEwan (B. C.). *Plectorthis aequivalvis* Hall (F.). *P. aequivalvis latior* Foerste (F.). *P. aequivalvis pervagata* Foerste (F.). *P. fissicosta* (Hall) (F.). *P. fissicosta triplicatella* (Meek) (H. F.). *P. neglecta* (H.). *P. plicatella* (H. F.). *P. sectostriata* (Ulrich) (F.). *Proetus parviusculus* Hall (C.). *Protowarthia cancellata* Hall (H. F. B. C.). *Psiloconcha sinuata* Ulrich (B.). *Pterinea cincinnatiensis* (M. & F.) (F.). *Pterygometopus carleyi* (Meek) (F.). *Rafinesquina alternata* (Conrad) (Maysville). *R. alternata alternistriata* (Hall) (Maysville). *R. alternata fracta* (B. C. S.). *R. alternata ponderosa* (B.). *R. squamula* (James) (F.). *Raphistoma halli* (Miller) (F.). *Rhytimya compressa* Ulrich (F.). *R. convexa* Ulrich (F.). *R. munda* (M. & F.) (C.). *R. producta* Ulrich (F.). *Spatiopora aspera* Ulrich (B.). S. *lineata* Ulrich (B.). *S. maculosa* Ulrich (F. B. C.). *S. tuberculata* (M-E. & H.) (F. C.). *Stigmatella alcicornis* Cumings & Galloway (F.). S. *dychei* (James) (A.). *S. irregularis* (Ulrich) (Lower Maysville). *S. nicklesi* U. & B., (F.). S. *sessilis* Cumings & Galloway (F.). S. *catenulata* Cumings & Galloway (S.). *Strophomena maysvillensis* Foerste (F. H.). S. *planoconvexa* Hall (F.). S. *sinuata* James (F.). *Technophorus faberi* Miller (F.). *T. punctostriatus* Ulrich (F.). *Trematis dyeri* (F.). *Whiteavesia pholadiformis* (Hall) (Maysville?) *Zygospira cincinnatiensis* Meek (H. F.). *Z. modesta* Hall (Maysville, etc.).

RICHMOND.—(Lebanon, Madison.) The name Richmond was proposed by Winchell and Ulrich (1897, ciii) in 1897 for the Lebanon beds of Orton (1873, 371) for the reason that the name Lebanon had already been used for an older Ordovician formation of Tennessee.* The same name has also been used by R. T. Hill (1899, 53-6) for a Tertiary formation in Jamaica. Winchell and Ulrich *(loc. cit.)* state that the formation is "350 feet thick in southwestern Ohio and southeastern Indiana [and] almost the entire series is excellently exposed at Richmond, Indiana." This is hardly true, in as much as only the Liberty and Whitewater beds are seen at Richmond. The Elkhorn is however exposed within a few miles of the city and part of the Waynesville may be seen some nine miles south. It is unfortunate that the name Lebanon could not have been retained. The name Richmond has been in general use since 1897 (for example, Nickles 1902, 1903, 1905; Foerste 1904, a, b, 1905, 1905a, 1906, 1909, c, d, e, 1910, 1910a, 1912, a, b, 1916; Cumings 1908, 1913; Bassler 1903, 1906, 1911, 1915, etc.; Miller 1919; Fenneman 1916; Butts 1915; Schuchert 1910, etc.; Ulrich 1911, 1914, etc.; M. Y. Williams 1919; Savage 1919, etc., etc.). The first detailed work on the Richmond of Indiana was done by Foerste (1897, 1898, 1900, 1900a) and Cumings (1901). Foerste's earlier work concerns only the upper part of the formation and recognizes the so-called Madison bed, now known as Saluda. Cumings gave sections of the whole thickness of the formation except the Elkhorn beds and part of the Whitewater. He suggested subdivision into faunal zones as follows (in ascending order): *Dalmanella meeki* zone, *Strophomena* zone, *Rhynchotrema* zone. The first two have been adopted into the literature as faunal designations. In Nickles' 1902 paper, the Richmond is divided into lower, middle and upper divisions, corresponding in a general way to the *Dalmenella meeki, Strophomena* and *Rhynchotrema* beds, and the Madison beds (as he supposed). In 1903 he published a second paper dealing exclusively with the Richmond stratigraphy and named his divisions in ascending order: *Waynesville* (50 feet)

* Weeks (1902) mistakenly gives the credit to Campbell (1898); but the latter credits the name to Ulrich.

Fig. 20. Upper left. Waynesville formation, near Weisburg, Indiana. Upper right Whitewater formation at Richmond. Lower left. Liberty formation at Richmond. Lower right. Saluda formation, Butler Falls, near Hanover, Indiana.

(1903, 205), *Liberty* (35 feet) (*Ibid.* 207), and *Whitewater* (45-50 feet) (*Ibid.* 208). He called the Waynesville, which corresponds exactly to the *Dalmanella meeki* zone of Cumings, the *Bythopora meeki* zone; the *Liberty*, the *Strophomena planumbona* zone (*Strophomena* zone of Cumings) and the Whitewater, the *Homotrypa wortheni* zone. The higher portion of the Whitewater (not exposed at Richmond) and the overlying beds on Elkhorn creek were evidently overlooked or regarded as equivalent to the Madison beds of Foerste. The base of the Whitewater is not accurately defined by Nickles. He says the Liberty passes rather gradually into the overlying beds but that on the east side of the Cincinnati axis a layer with large streptelasmas marks the top (of the Liberty). In 1903 Nickles (1903, 210-214) definitely refers the upper beds on Elkhorn creek to the Madison formation of Foerste. Nickles' names appear in subsequent publications of himself and of Foerste, Bassler, Ulrich, Cumings, etc., and are in general use.

The exact relations of the upper beds near Richmond were not disclosed until the publication of Cuming's paper on the Ordovician rocks of Indiana (1908). It is necessary, however, first to discuss the horizon of the Madison bed (Saluda) of Poerste.

In 1897 Foerste published a report on the middle and upper Silurian rocks of Indiana (1897, 213-288) and in connection with the discussion of the Ordovician-Silurian boundary, gave considerable detailed information on the upper beds of the Richmond in the vicinity of Madison, Indiana. The massive banded sandy magnesian limestone seen in the great railroad cut at Madison he proposed to call the Madison bed (*loc. cit.*, p. 218). This name had been loosely applied to the Cincinnatian rocks of Indiana by Borden (1874, 139). Even as far back as 1837 Owen (1837, 28) had used the name "Madison water limestone" for a bed that seems to occupy the position of the lower part of the Madison or Saluda of Foerste.* Inasmuch as neither Owen nor Borden defined their names in such a way as to make it possible to exactly identify the beds intended, Foerste's use of the name must be regarded as prior. In 1875, however, Irving (1875, 442) had applied the name Madison sandstone to a Cambrian formation of Wisconsin. Foerste therefore later (1902a, 369) substituted the name *Saluda* for Madison bed, and this name has been in common use ever since. He does not make it clear in the earlier paper whether he included the coral (*Favistella* or *Columnaria*) beds in the Madison; but in the 1902 paper he states (1902a, 369) that the Saluda "includes all the material overlying the coral beds, the latter forming the base." While this definition still leaves something to be desired, we may infer that he meant to include the coral beds in the Saluda. With the Madison or Saluda he correlated the Cumberland sandstone (Shaler, 1877, 159-160) of Tennessee and Kentucky (1900, 57-63); but later concluded that the Cumberland includes more than the Saluda (1901, 436; 1902a, 368). Foerste met his chief difficulty in attempting to trace the Saluda northward and correlate it with the rocks of the Richmond and Ohio areas, and in this he was not very successful. In 1897, in proposing the name Richmond, Winchell and Ulrich (1897, ciii) had stated that "At Richmond the upper part shows an increase of arenaceous matter while the uppermost layers of shale have become harder and include one or two heavy beds of impure limestone. South-

* Weeks (1902) credits the name Madison to Borden (1874). The name has also been used by Peale (1893, 33) for a carboniferous limestone of the west.

ward from this locality in Ripley and Jefferson counties (Indiana) the heavy layers are increased." This evidently was an attempt to correlate the upper rocks on Elkhorn creek, Richmond, with the Madison bed, a correlation again made by Nickles in 1903. Foerste, who also says the Madison is about 30-60 feet thick on the east side of the Cincinnati arch, evidently had some such correlation in mind for he says (1900, 63) "In Ohio the Madison beds are replaced by clays and clayey shales which are at times mottled with purple and reddish purple and are usually devoid of fossils." Evidently the beds referred to in Ohio constitute the Belfast of Foerste (1896, 164).

In 1904 Foerste (1904, 328-334) returned to the Saluda problem giving many details in regard to the faunas and stratigraphy of the upper members of the Richmond. He here definitely takes the lower *Favistella* reef as the base of the Saluda. In tracing the formation northward to Versailles, he notes that *Strophomena vetusta*, *Streptelasma divaricans* and other Whitewater species occur in a fossiliferous bed *above* the coral reefs and Saluda proper, and says (p. 332) "there is evidently a recurrence of species usually found more abundantly in the Whitewater bed." This confusion is reflected in Foerste's later suggestion (1905, 150) that the Liberty and Whitewater formations be grouped together as the *Versailles* formation, because their distinctness could not be recognized in the more southern localities. He was still under the impression that all the rocks in the Whitewater river gorge at Richmond were stratigraphically lower than the Saluda bed. He says further (p. 334) "The Silurian, therefore, instead of resting upon lower beds of Richmond age in the region in which the Clinton is absent, acutally rests on beds representing the latest deposits of Madison age". That is, he referred these upper fossiliferous beds to the Madison or Saluda. The persistence of this error is shown in the fact that as late as 1915 Butts (1915, 59-68) fails to recognize the fact that the upper fossil bed of the Saluda (Hitz bed) is Whitewater.

Now the first to notice this upper faunal horizon was W. T. S. Cornett, (1874) who gave an account of it in the Indianapolis Journal in 1874. Cornett's discovery is mentioned later by Cox (1879, 18-19) who gives a list of fossils said to occur "resting upon the banded rock" (Saluda), among which are *Orthis acutilirata*, *O. retrorsa*, *O. insculpta*, *O. subquadrata*, *Strophomena planumbona*, *S. sulcata*, *Rhynchonella capax*, *R. dentata*, etc. Most of these identifications are undoubtedly in error. Attention was again called to it by G. C. Hubbard in 1892 (pp. 68-70) who stated that he had independently come to the conclusion that the beds immediately below the "Salmon-colored" stone (Brassfield) are Ordovician, (the Saluda having been originally referred by Owen and others to the Silurian.)

This upper fossiliferous horizon was later named the *Hitz bed* by Foerste (1903b, 347) and has also been known as the *Lophospira, (or Murchisonia) hammeli* bed (Foerste 1897, 222; Nickles 1903, 213). It forms the upper fossil bed and part of the upper "mottled beds" of Cumings (1908). Butts (1915, 65) has recently revived the superfluous name Hitz bed.

We may now return to the discussion of the higher formations at Richmond. In his detailed report on the Cincinnatian rocks of Indiana (1908, 607-1189) Cumings traced the various members of the Saluda (of Foerste) northward and proved by both stratigraphic and faunal evidence that (1) this Hitz bed northward becomes an important and extensive fossiliferous horizon, especially from Versailles northward, containing a typical Whitewater fauna; (2) the banded rock forming the main overhanging mass in the Madison cut thins northward, becoming more calcareous and shaly and finally enters the Richmond section as a six foot bed of massive limestone and shale capping the Liberty formation; (3) the *uppermost* of the two coral beds seen at the base of the Saluda in the Madison section can easily be traced northward to the vicinity of Richmond where it runs beneath the Whitewater, *Tetradium* everywhere characterizing the bed except at Madison, where *Favistella (Columnaria)* also occurs. From this evidence it was concluded (*op. cit.*, p. 679) that the "mottled bed" (containing the Hitz bed as its upper member) is a thinned representative of the Whitewater formation of Richmond; and Foerste's conclusion that there is a disconformity at the top of the Richmond was corroborated. These results clearly showed that the forty feet of rocks lying just beneath the Silurian on Elkhorn creek, near Richmond, could not be correlated with the Saluda as Nickles supposed, and Cumings therefore named them the Elkhorn beds (*op. cit.*, p. 678). This name has been generally adopted. (See Foerste 1909, b, c, d, 1910, 1912, etc., Ulrich 1911, 1914, etc.; Schuchert 1910, etc.; Cumings 1913; Bailey Willis 1912; Butts 1915, etc.) The upper part of the Elkhorn contains a bed of clay, which had previously been correlated by Foerste with the Belfast bed of Ohio (1896, 193). The latter he had in the previous year referred to the Medina (1895, 181). He inclines in later papers to place it in the Ordovician, and says (1900, 68) his Belfast is *not* the typical "Medina" rocks of Orton but the so-called "Medina" is the clay bed *below* the Belfast bed. Nickles (1903, 32) correlated the Belfast bed with the Saluda. Orton (1878, 384, 385) referred the sandy (Belfast) bed to the Clinton. Prosser (1916, 336) says Cumings and Foerste refer the Belfast to the Richmond, while Shideler thinks it may be related to the Brassfield. (See also Orton 1871, 1874; Foerste 1888, 1900; Prosser 1905.)

Recent studies of the Bryozoa from the Hitz bed at Madison fully confirm the correlation of this horizon with the Whitewater farther north. The Elkhorn is totally absent from the Madison section. Coryell (1915, 389-393) has shown that the Whitewater is present in typical development and thickness (75 feet) in the region of Weisburg and Spades in Dearborn County. From this point it thins rapidly toward Madison and in Kentucky (Butts 1915) disappears entirely, the arenaceous Saluda being there capped by the Brassfield without any intervening fossiliferous bed. It is obvious from the above discussion that the Saluda (exclusive of the Hitz bed) and Whitewater consist of two oppositely directed wedges, the Saluda thinning northward and the Whitewater thinning southward and overriding the Saluda. Further discussion of this point will be given in the sections on the Liberty and Whitewater.

Waynesville.—The *Waynesville* (Nickles 1903, 205) formation or *Dalmanella meeki* zone of Cumings (1901, 215) and Foerste (1904, 25, 26) consists largely of soft blue shale with thin layers of highly fossiliferous blue limestone. The weathered shales and limestones are, as thruout the Cincinnatian, of a buff-yellow or yellowish-brown color. It is about 50 feet thick at Madison and 80 to 100 feet in Dearborn and Franklin counties. The formation has been divided by Foerste (1909f) into the *Blanchester*, *Clarksville* and *Ft. Ancient* divisions (in descending order) and these divisions have been generally recognized. The Blanchester division as defined by Foerste is largely limestone and includes all between the upper and lower *Hebertella insculpta* horizons. It has a very rich fauna in which *Strophomena nutans*, *S. neglecta*, *S. vetusta*, etc., are common, together with numerous specimens of *Hebertella insculpta*, *Leptaena rhomboidalis*, *Rhynchotrema capax*, *Platystrophia clarksvillensis*, *P. cumingsi*, etc. The upper part of the division (upper *Hebertella insculpta* zone) was originally placed by Foerste in the Liberty as the base of that formation; but later (1909f, 290) he placed it in the Waynesville with the following statement: "There are two *Hebertella insculpta* horizons in Ohio. Only the upper one of these horizons was known at the time the name Waynesville bed was introduced, and this upper *Hebertella insculpta* horizon was chosen as the base of the Liberty bed. In reality, there is a greater stratigraphic break immediately above the upper *Hebertella insculpta* horizon, so that the latter should form the top of the Waynesville bed." He might have added that this is the horizon taken by Cumings as the base of the Liberty in tracing the Waynesville-Liberty boundary through southern Indiana (1908, 640-641, and chart opposite, p. 637) though in deference to Foerste and Nickles he allowed their definition of the Liberty to stand, in summing up his discussion of the formations (1908, 671). Cumings later (1913) definitely adopted the top of the upper *insculpta* bed as the boundary.

The second division of the Waynesville, the *Clarksville*, as defined by Foerste, extends from the *Orthoceras fosteri* horizon to the lower *Hebertella insculpta* layer. The Clarksville according to Foerste represents the introduction of the fauna regarded as typically Richmond and not found in the Lower Richmond or Ft. Ancient. Common species are *Strophomena planumbona*, *Streptelasma rugans*, *Rhynchotrema* sp. and *Leptaena richmondensis* together with *Platystrophia clarksvillensis*, etc. It would appear that the Blanchester and Clarksville beds are about equivalent to the *Leptaena rhomboidalis (=L. richmondensis)* zone of Cumings (1908, 671, 672). The Clarksville is somewhat less calcareous than the Blanchester but still very fossiliferous, containing, besides the fossils mentioned above, large numbers of large Bryozoa *(Heterotrypa prolifica, etc.).*

The Fort Ancient division is still more shaly and is characterized by immense numbers of *Dalmanella meeki (D. jugosa* of some authors) though this species is found more or less commonly to the top of the Waynesville, and also in the Arnheim. In the Fort Ancient it occurs almost to the exclusion of other Brachiopods except *Rafinesquina loxorhytis*.

A large number of Richmond Bryozoa are introduced in this division, for example, *Bythopora meeki*, *Hallopora subnodosa*, *Homotrypella hospitalis*, *H. rustica*, *Helopora harrisi*, *Stigmatella crenulata*, *S. interporosa*, *S. spinosa*, *Heterotrypa prolifica*, and varieties, etc. The presence of this group of *Heterotrypas* of Black River (Decorah) and Trenton affinities in the Waynesville is particularly significant. It has been a common error to associate them with *H. frondosa* of the Maysville, but they are in reality not at all closely related to the *frondosa group*. (On this point see Cumings & Galloway 1913, pp. 414-419.) A large number of characteristic pelecypods *(Anomalodonta gigantea*, *Modiolopsis concentrica*, *M. pholadiformis*, *Pterinea demissa*, etc.) occur in this division. Several of the Waynesville species are also found in the upper Arnheim (Oregonia). For example, *Batostoma varians*, *Eridotrypa simulatrix*, *Hallopora subnodosa*, *var.*, *Leptaena richmondensis* and a number of Pelecypods, etc., and it may be that the upper Arnheim and Lower Waynesville should be associated together in a single division, as suggested by Foerste (1909f, 293). This would agree with the suggestion made elsewhere in this paper that the boundary between the Maysville and Richmond might be drawn at the base of the Oregonia bed of the Arnheim.

The Waynesville formation has wide distribution, being known in the Quebec-Ontario region (Foerste 1914, 1916) and the Manitoulin Islands (Foerste 1912b) as well as in Ohio, Indiana and Kentucky. It is probable also that the western Richmond (Fernvale-Maquoketa) is of about the age of this formation and the lower Liberty. Many of the species from Wilmington, Illinois, are identical with Waynesville and lower Liberty forms. In Kentucky the formation becomes unfossiliferous, argillaceous, suncracked and shows obvious evidence of in-shore or playa deposition. In fact it presents much the appearance of the Saluda.

Because of this relationship of the Waynesville and Liberty faunas and the Fernvale-Maquoketa faunas Foerste (1912, 22) has suggested grouping the two formations under the name *Laughery*, from Laughery creek in Indiana.

Liberty.—The Liberty formation (Nickles 1903, 207) or *Strophomena-Rhynchotrema* bed of Cumings (1901) (*S. planumbona* bed of Nickles 1903), consists more largely of limestone than any other member of the Richmond in Indiana. The layers are often several inches or even a foot thick, and shale is much less in evidence than in the other formations. It is from 25 to possibly 50 feet thick, usually from 35 to 40 feet, and very constant in its thickness, as well as in its lithologic characters. It is extraordinarily fossiliferous throughout, especially abounding in Brachiopods and small branching Bryozoa. No other Cincinnatian formation rivals it in the beauty and abundance of the specimens. The most characteristic fossils are *Strophomena planumbona*, *S. vetusta*, *Plectambonites sericeus* (in the lower part), *Rhynchotrema capax*, *Dinorthis subquadrata*, *Platystrophia cumingsi*, *Amplexopora pumila*, *A. granulosa*, *Constellaria polystomella*, *Homotrypa austini*, *H. cylindrica* and *Rhombotrypa quadrata*. The Liberty formation extends throughout the range of the Richmond in Ohio and Indiana and far into Kentucky and because of its persistent fauna is an easy bed to identify.

In the fine section along Whitewater river in Richmond, Indiana, the formation is capped by a bed of six feet of heavy limestone layers and intercalated shale representing the banded Saluda of the Madison section, and in sections near Richmond, containing the coral *Tetradium*. In Ohio this bed fails, and some dispute has consequently arisen as to the proper upper limit of the Liberty. Nickles, as noted above, left this point in doubt for the region west of the Cincinnati axis, and the first, therefore, to definitely fix the upper limit of the Liberty was Cumings (1908), who selected the Saluda wedge as the dividing stratum between the Liberty and the Whitewater. A glance at the chart opposite page 637 of Cuming's paper will show the compiete justification for this procedure. If the base of the Saluda be taken just below the lower *Columnaria* reef at Madison, and this horizon be traced into the Richmond section, it will be noted that it comes in with the above mentioned *heavy* limestone layer ("massive bed") capping the Liberty. This is due to the fact that, going northward, the interval between the two lower Saluda coral beds, of the southern sections, becomes constantly less and the lower reef finally disappears completely north of Ballston; the upper reef entering Richmond region as a part of the heavy limestone bed or "massive bed" of Cumings (1908). The same relationships are still better displayed on Tanner's creek, in the railroad cut (No. 18) just above Weisburg Station. Here the "massive bed" still has its southern associate, the "shale bed" or thin slabby, suncracked limestone layers, so characteristic of the sections between Versailles and Madison, and there are 75 feet of Whitewater above it (Coryell 1915, 389-93). Shideler's contention (1914, 229-235), therefore, that the Saluda is above the Whitewater, is not borne out by the facts and I understand he has now receded from it; while Ulrich's view that the Saluda and Whitewater are coterminous can be given a status only by dropping the Liberty-Whitewater boundary to an horizon somewhere below the perfectly natural one marked by the base of the "massive bed" or Saluda wedge.*

Saluda.—The Saluda formation (Foerste 1902a, 369) has been described in part in the foregoing discussion of the delimitation of the upper Richmond. It fails entirely in Ohio, but in Jefferson and Ripley counties, Indiana, and especially in Kentucky it is an important formation. Lithologically the Saluda as defined by Foerste consists in its type region of three divisions: (1) a lower bed of ash-colored argillaceous rock containing two conspicuous coral layers *(Favistella stellata = Columnaria alveolata,* and *Tetradium minus);* (2) above this bed comes a great mass of brownish to greenish banded sandy dolomite practically unfossiliferous. The layers of this bed are usually beautifully ripple marked and sun-cracked. (3) At the top and between this massive dolomite and the Brassfield formation of the Silurian is a less resistant mass of somewhat rubbly mottled impure limestone containing a conspicuous fossil bed ("Hitz" or "*Lophospira hammeli* bed") at the top. The middle and lower divisions thin notably to the northward running in as a thin bed between the Liberty and Whitewater, as explained elsewhere. The upper division expands to the northward becoming the Whitewater formation. Aside from the "Hitz bed" which carries a Whitewater fauna, the fauna of the Saluda is meagre and is especially characterized by the columnar corals mentioned above, and several species of Ostracods, namely, *Leperditia caecigina, Leperditella glabra* and *Eurychilina stiatomarginata.* Besides these a few Bryozoa and the Brachiopods *Hebertella occidentalis* var. and *Strophomena sulcata* occur. The Hitz fauna besides the characteristic *Lophospira hammeli* contains *Holopea hubbardi, Hebertella occidentalis* var., a large number of Ostracods, especially *Leperditia caecigina* and numerous Bryozoa, among which *Homotrypa constellariformis* and its relatives and *Amplexopora* n. sp. (a characteristic lower Whitewater form) are especially abundant.

The Saluda becomes more arenaceous to the southeastward and probably is represented by part of the Cumberland sandstone of southern Kentucky and Tennessee. In the Appalachian region and in western New York the Saluda and Whitewater are represented by part of the Juniata-Queenston red continental deposits, which lie unconformably below the Medina (Albion-Tuscarora) sandstone.

Whitewater.—The Whitewater formation (Nickles 1903, 208) or *Homotrypa wortheni* bed constitutes, according to Nickles, the upper half of his

* Ulrich contends that the Whitewater fauna comes in some 10 or 15 feet below the Saluda wedge, and, therefore, regards the Saluda as a bed or wedge *within* the Whitewater. As a matter of fact there is so little difference between the Liberty and Whitewater faunas that a real faunal boundary between the two can hardly be said to exist; while on the other hand the Saluda wedge forms a *conspicuous* boundary in Indiana.

middle Richmond of his 1902 paper, and consists of rough concretionary nodular blue limestone and shale. On the weathered outcrop the rock appears yellowish or brown. The thickness as stated by Nickles is 45 to 50 feet; but as a matter of fact the formation is about 80 feet thick in the vicinity of Richmond. It is extremely fossiliferous. Nickles defines the formation as comprising the outcrops exposed in the banks of the Whitewater river "at the north end of Richmond, Indiana, and between the bridges; numerous exposures [he says] are also found along Short creek three miles south of Richmond." All of the rock exposed at Richmond, as a matter of fact, above the Liberty formation belongs to the Whitewater division. The highest bed can best be studied on Elkhorn creek, three and one-half miles southeast of Richmond. At the latter place the superjacent Elkhorn formation is also exposed. Nickles does not accurately define either the base or the top of his formation. These were subsequently fixed, as already described, by Cumings (1908). Comparatively few species of fossils are restricted to the Whitewater, most of the fauna being the same as that of the Liberty. *Strophomena sulcata* though not confined to this formation, is one of the most characteristic Brachiopods; but from Ballston northward *Rhynchotrema dentata* becomes abundant and very characteristic, especially of the upper half of the formation. Varieties of this species also occur sparingly in the Arnheim and Liberty. *Platystrophia acutilirata* (typical form) with its relatives *P. acutilirata prolongata* and *P. senex*, and *Hebertella occidentalis* are very characteristic of the Whitewater. *Rhynchotrema capax* is very abundant and attains its maximum size in the Whitewater. Among the Pelecypods *Byssonychia obesa*, *Ischyrondonta decipiens*, *I. truncata* and *Ortonella hainesi*, are characteristic. *Bucania crassa*, *B. simulatrix*, *Lophospira tropiodophora*, *L. hammeli*, and *Salpingostoma richmondensis*, are characteristic Gastropods. Numerous Bryozoa occur in the Whitewater, but only a few are restricted to it, namely *Batostoma variabile*, *Homotrypa constellariformis*, *H. nitida*, *H. nicklesi* and a new species of *Amplexopora*. Other common Bryozoa are *Bythopora delicatula*, *Homotrypa cylindrica*, *H. ramulosa*, *H. wortheni*, *Homotrypella rustica* and *Monticulipora epidermata*. The coral *Streptelasma divaricans* is abundant in the Whitewater, and in the more southern localities constitutes in company with *Strophomena sulcata* the best zone marker.

The Whitewater formation, as already intimated, barely reaches the Madison section where it is represented by the "Hitz bed" and probably some of the underlying impure, mottled limestone. In Kentucky it fails entirely. In Ohio it is present on both sides of the Cincinnati arch.

Elkhorn.—The Elkhorn formation (Cumings 1908, 678) comprises the highest beds of Richmond age known in the Ohio valley region. At the type locality, Elkhorn creek, three and one-half miles southeast of Richmond, Indiana, the formation consists of about 50 feet of blue shale and limestone, as follows: at the top is a bed of soft clay-shale four feet thick, underlain by six feet of hard brownish limestone layers abounding in the brachiopods *Platystrophia moritura* and *Hebertella sinuata*. Below this come 25 feet of blocky, arenaceous limestone, underlain in turn by 15 feet of very soft blue, barren shale. Beneath this occur the lumpy Whitewater limestones abounding in *Rhynchotrema dentata*. The fauna of this formation has been little studied and may be expected to furnish some very interesting species especially of Bryozoa most of which will be new. *Homotrypa wortheni prominens* seems to be a common form. It has been intimated by Foerste, Bassler and others that the Elkhorn, or some part of it, may be the equivalent of the Belfast bed of the eastern side of the Cincinnati arch. Until further faunal and stratigraphic evidence is forthcoming, the relationships of these two interesting formations must remain in doubt.

The Elkhorn bed can be traced southward to the region of Ballston, Indiana, but probably pinches out before Versailles is reached. There is no vestige of it in the Madison section.

The facts noted above in regard to the marked thinning and disappearance of the Elkhorn and Whitewater beds to the southward make it evident that there is a pronounced disconformity at the top of the Richmond. The faunistic and stratigraphic relationships of the Brassfield to the Richmond confirm this conclusion. Nickles (1903, 212-13) did not recognize this disconformity, but it was noted by Foerste (1904, 334) and afterwards amply verified by him and by Cumings (1908). It would appear, however, that the first to recognize the disconformable relations of the Silurian and Ordovician in Indiana was G. C. Hubbard in 1892 (1892, 68-70) who noted that there is an "abrupt paleontological break between the two strata" [i. e., the Brassfield and Hitz bed], and concluded that the two beds are unconformable. As a matter of fact, this horizon below the Brassfield and its equivalents, the Medina, Sexton creek, Cataract, etc., appears to be marked by nonconformities everywhere in America, and furnishes an admirable systemic boundary.

(*Marble Hill Bed.*—In 1859, D. D. Owen (1859, 128) spoke of a bed of crystalline limestone, quarried at Marble Hill in southern Jefferson county, Indiana, as the "Marble Hill Bed". This name is also used by Borden (1874, 139-141) and other Indiana geologists. Foerste (1897, 129) places the bed at 140 feet below the Silurian-Ordovician contact. The Marble Hill bed is characterized by large numbers of gastropods and a pseudocrystalline texture and pleasing color. It was also known as the *Murchisonia* bed.)

Fauna of the Richmond.—The following species are characteristic of the Richmond of the Ohio valley region: *Agelacrinites austini* (Foerste) (W.)*, *Agelacrinites faberi* (Miller) (W.), *Amplexopora granulosa* Cumings & Galloway (L.), *A. pumila* C. & G. (L.), *A. pustulosa* Ulrich (W.), *Anomalodonta casei* (Meek & Worthen) (Wh.), *A. costata* (Meek) (W.), *A. gigantea* Miller (W.), *Aparchites minutissimus* (Hall) (Richmond), *Arabellites procursus* Foerste (E.), *Archinacella richmondensis* Ulrich (Wh.), *Atactopora angularis* Ulrich & Bassler (W.), *Atactoporella schucherti* Ulrich (W. Wh.), *Batostoma prosseri* C. & G. (W. L.), *B. variabile* Ulrich (Wh.), *B. varians* (James) (Oregonia, W. L. Wh.), *Beatricea nodulifera* Foerste (L.), B. *nodulifera intermedia* (Foerste) (L.), B. *nodulosa* (Billings) (S.), *B. undulata* (Billings) (S.), *B. undulata cylindrica* Foerste (L. E.), *Bellerophon mohri* Miller (Wh.), *B. subangularis* Ulr. (Wh.), *Bernicea primitiva* Ulrich (Richmond), *Beyrichia parallela* (Ulrich) (Wh.), *B. tumida* (Ulrich) (W.), *Bollia persulcata* (Ulrich) (Richmond), *B.* pumila Ulrich (W. L.), *Bucania crassa* Ulrich (W.), B. *simulatrix* Ulrich (W.), *Bythotrephis gracilis* Hall (Richmond), *Byssonychia grandis* Ulrich (W. Wh.). *B. obesa* Ulrich (Wh.), B. *radiata* (Hall) (Cincinnatian), *B. richmondensis* Ulrich (W. L. Wh.), B. *robusta* (Miller) (Wh.), B. *subrecta* Ulrich (W. L.), B. *tenuistriata* Ulrich (Wh.), *Bythocypris cylindrica* (Hall) (Richmond), *Bythopora delicatula* (Nicholson) (W. L. Wh.), B. *meeki* (James) (W. L. Wh.), B. *striata* Ulrich (W.), *Calapoecia cribriformis* (Nicholson) (L. Wh.), *Calymene meeki* Foerste (Richmond), C. *meeki retrorsa* Foerste (W.), *Calazyga headi schuchertana* (Ulrich), (W.), *Ceramoporella granulosa* Ulrich (W.), *C. ohioensis* (Richmond, etc.), *C. whitei* James (W.), *Ceratopsis robusta* (Ulrich) (Richmond), *Ceraurus miseneri* Foerste (Wh.), *Chasmops breviceps* (Hall) (W. L.), *Clidophorus faberi* Hall (Richmond), C. *fabula* (Hall) (Richmond), Clionychia *excavata* Ulrich (Wh.), *Columnaria alveolata* (S.) [=*Favistella stellata* of authors], *Constellaria* limitaris (Ulrich) (W.), C. *polystomella* Nicholson (W. L.), *Conularia formosa*

* W=Waynesville; L=Liberty; S=Saluda; Wh=Whitewater; E=Elkhorn.

Miller & Dyer (W. Wh.). *Cornulites richmondensis* (Miller) (Richmond). *Corynotrypa delicatula* (James) (Richmond). C. *inflata* (Hall) (Richmond). *Crania laelia* Hall (Richmond). C. *scabiosa* Hall (Richmond). *Ctenobolbina hammelli* (M. & F.) (W.). *Ctenodonta cingulata* Ulrich (W.), *Ctenodonta hilli* (Miller) (Wh.). *Cuneamya curta* Whitfield (W.). C. *neglecta* (Meek) (W.). *Cycloconcha milleri* (Meek) (W.). *Cyclonema bilix* (Wh.). C. *bilix fluctuatum* (W.). *Cyclora depressa* Ulrich (Richmond). C. *minuta* Hall (Richmond). C. *parvula* Hall (Richmond). *Cymatonota typicalis* Ulrich (W. Wh.). C. *attenuata* Ulrich (W.). *C. Constricta* Ulrich (W.). C. *semistriata* Ulrich (W.). *Cyphotrypa stidhami* (Ulrich) (Wh.). *Cyrtoceras amoenum* Miller (Wh.). C. *faberi* J. F. James (W.). *C. hitzi* Foerste (Wh. S.). C. *tenuisceptum* Faber (W.). *Cyrtodonta cuneata* (Miller) (Wh.). *Cyrtolites ornatus* Conrad (W.). *Dalmanella meeki* (Miller) (W. L.). *Dendrocrinus caduceus* (Hall) (W. L.). *D. casei* Meek (Wh.). *D erraticus* Miller (W.). *Dermatostroma canaliculatum* Parks (W.). *D. corrugatum* (Foerste) (Wh.). *D. glyptum* (Foerste) (Wh.). *Dicranopora emacerata* (Nich.) (Richmond). *Dinorthis carleyi* (Hall [=*D. retrorsa* of authors] (Upper Arnheim). *D. Carleyi insolens* Foerste (W.). *D. subquadrata* (Hall) (L. Wh.). *Dystactospongia madisonensis* Foerste (Wh. S.). *Eridotrypa simulatrix* (Ulrich) (W.). *Eurychilina striatomarginata* (Miller) (Wh. S.). *Fenestella granulosa* Whitfield (W. L. Wh.). *Girvanella richmondensis* (Miller) (Wh.). *Gomphoceras indianense* M. & F. (W.). *Graptodictya perelegans* (Ulrich) (W. L.). *Gyroceras baeri* (M. & W. (L.). *Hallopora frondosa* Cumings (Wh.). H. *subnodosa* Ulrich (Richmond). *Hebertella insculpta* (Hall) (W.). H. *occidentalis* Hall (L. Wh.). H. *sinuata* Hall (E.). *Helicotoma marginata* Ulrich (Wh.). *Helopora elegans* Ulrich (L.). H. *harrisi* James (W.). *Heterotrypa microstigma* C. & G. (W.). H. *subramosa* Ulrich (W.). H. *prolifica* Ulrich (W.). H. *singularis* Ulrich (W.). *Holopea hubbardi* Miller (Wh. S.). *H. oxfordensis* Ulrich (W.). *Homotrypa austini* Bassler (W. L. Wh.). H. *communis* Bassler (W. L.). H. *cylindrica* Bassler (L. Wh.). H. *dawsoni* (Nich.) (W. Wh.). H. *flabellaris* Ulrich (W. L. Wh.). *H. frondosa* Cumings (W.). *H. flabellaris spinifera* Bassler (W. Wh.). H. *nicklesi* Bassler (Wh.). H. *nodulosa* Bassler (W. L.). *H. ramulosa* Bassler (L. Wh.). *H. nitida* Bassler (Wh.). *H. constellariformis* Cumings (Wh.). *H. richmondensis* Bassler (W. L. Wh.). *H. wortheni* (James) (L. Wh.). *H. wortheni prominens* Bassler (E.). *Homotrypella hospitalis* (Nicholson) (W. L. Wh.). H. *rustica* Ulrich (W. L. Wh.). Hyolithes *versaillesensis* M. & F. (Richmond). *Iocrinus subcrassus* M. & W. (W.). *Ischyrodonta decipiens* Ulrich (E.). I. *elongata* Ulrich (Wh.). I. *misneri* Ulrich (Wh.). I. *modioliformis* Ulrich (Wh.). *I. ovalis* Ulrich (Wh.). I. *truncata* Ulrich (Wh.). *Isotelus gigas* Dekay (Richmond?). *I.* Maximus Locke (Richmond). *Labechia montifera* [=*Stromatocerium montiferum* (Ulrich)] (Wh.). *Lepadocystis moorei* (Meek) (Wh.). *Leperditia caecigena* (Miller) (Wh. S.). *Leptaena richmondensis* Foerste (W.) [=*L. rhomboidalis* of authors]. *L. richmondensis praecursor* Foerste (upper Arnheim). *Lichenocrinus affinis* Miller (W.). L. *tuberculatus* Miller (Wh.). *Lophospira acuminata* Ulrich (Wh.). L. *bicincta* (Hall) (Richmond). *L. bowdeni* Safford (Richmond). *L. hammeli* (Wh. S.). L. *perlamellosa* Ulrich (W.). *L. tropidophora* (Meek) (Richmond). *Mesotrypa orbiculata* C. & G. (Arnheim). *M. patella* (Ulrich) (Wh.). *Microceras inornatum* Hall (Richmond). *Modiolodon declivis* Ulrich (Wh.). *M. subovalis* Ulrich (W.). *M. subrectus* Ulrich (Wh.). *Modiolopsis concentrica* H. & W. (W.). *M. versaillesensis* Miller (W.). *Monticulipora cleavelandi* James (Wh.). M. *epidermata* U. & B. (L. S. Wh.). *M. laevis* Ulrich (Wh.). *M. parasitica* Ulrich (W. L. Wh.). *Nereidavus varians* Grinnel (W. L. Wh.). *Nicholsonella peculiaris* C. & G. (Arnheim). *Opisthoptera casei* (M. & W.) (Richmond). *O. fissicosta* (Meek) (W.). *O. obliqua* Ulrich (Richmond). *Orthoceras carleyi* H. & W. (W.). *O. duseri* H. & W. (W.). *O. hitzi* Foerste (Wh. S.). *Orthodesma curvatum* (H. & W.) (W.). *O. rectum* (H. & W.) (W.). *O. subangulatum* Ulrich (Wh.). *Ortonella hainesi* (Miller) (Wh.). *Pachydictya fenestelliformis* (Nicholson) (W. L.). *Peronopora decipiens* (Rominger) Richmond). *Platystrophia alternata* McEwan (W.). P. *foerstei ampla* McEwan (W. L.). P. *clarksvillensis* Foerste (W.). *P. cumingsi* McEwan (Richmond). *P. annicana* Foerste (W.). *P. moritura* Cumings (E.). P. *acutilirata* (Conrad) (L. Wh.). P. *acutilirata prolongata* Foerste (Wh.). P. *acutilirata senex* Cumings (Wh.). P. *elkhornensis* McEwan (E.). P. *cypha bellatula* McEwan (W.). *Plectambonites sericeus* (Sowerby) (W. L.). *Proboscina auloporides* (Nicholson) (Richmond). *Protaraea richmondensis* Foerste [=*P. Vetusta* of authors] (Richmond). *P. richmondensis papillata* Foerste (Wh.). *Psiloconcha elliptica* Ulrich (W.). P. *grandis* Ulrich (W.). P. *subrecta* Ulrich (W.). *Pterinea corrugata* (James) (W.). *P. demissa* (Conrad) (Richmond). *Ptilodictya flagellum* Nich. (L.). P. *magnifica* Miller (Wh.). P. *plumaria* James (Wh.). *Rafinesquina alternata* (Conrad) (Richmond, etc.). *R. alternata loxorhytis* (Meek) (W.). *Raphistoma richmondense* Ulrich (Wh.). *Rhynidictya lata* Ulrich (W.). *Rhombotrypa quadrata* (Rominger) (W. L. Wh.). *R. subquadrata* (Ulrich) (W. L.). *Rhynchotrema capax* (Conrad) (W. L. Wh.). *R. dentatum* (Hall) (W. L. Wh.). *Rhytimya* byrnesi Miller (W. Wh.). *R. faberi* (Miller) (Richmond). *Salpingostoma rich-*

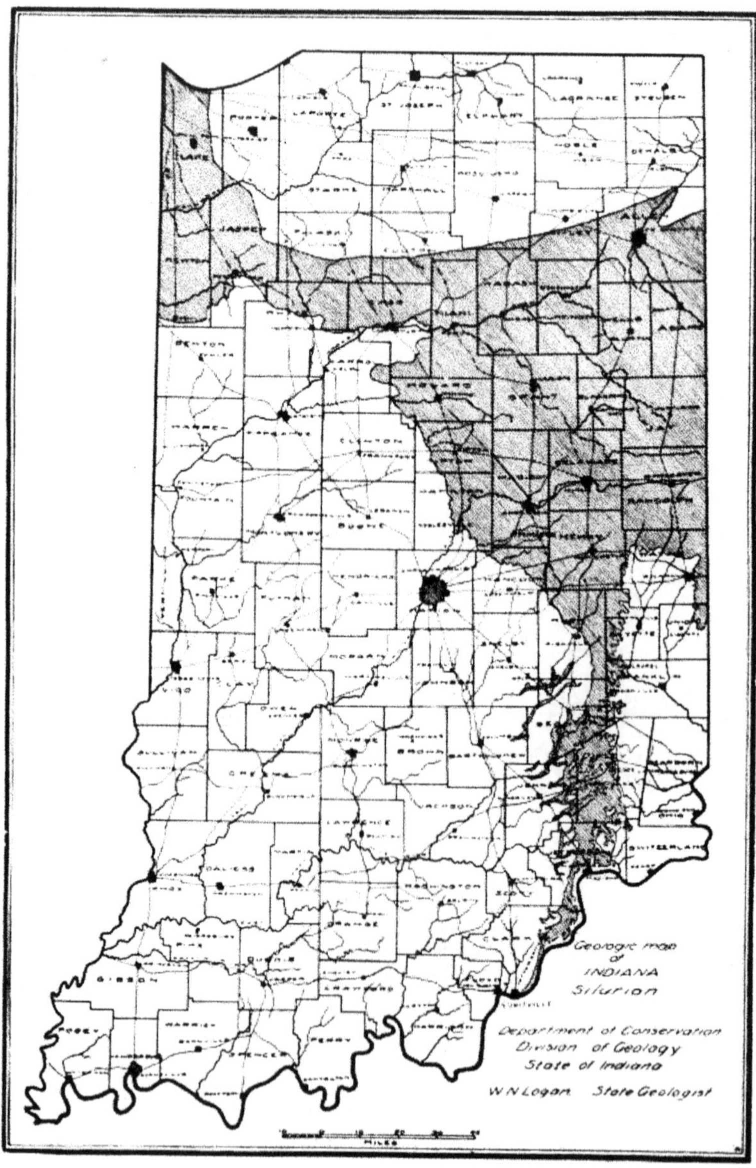
Geologic map
of
INDIANA
Silurian
Department of Conservation
Division of Geology
State of Indiana
W. N. Logan State Geologist

Fig. 21.

mondensc Ulrich (Wh.). *Schizolopha moorei* Ulrich (Wh.). *Spatiopora corticans* (Nicholson) (W.). S. *montifera* Ulrich (W.). S. *tuberculata* (M-E. & H.) (W.). *Stigmatella crenulata* U. & B. (W.). S. *personata* U. & B. (W.). S. *spinosa* U. & B. (W.). *Streptelasma dispandum* Foerste (W.). S. *divaricans* (Nicholson) (W. L. Wh.). S. *divaricans augustatum* Foerste (Wh.). *S. insolitum* Foerste (Wh.). *S. rusticum* Billings (Richmond). *Stromatocerium montiferum* (Ulrich) (Wh. S.). *Strophomena concordensis* Foerste (W.). S. *neglecta* (James) (W.). S. *nutans* (Meck) (W.). *S. planumbona* (Hall) (W. L.). S. *planumbona elongata* (James) (W.). *S. planumbona gerontica* Foerste (W.). S. *planumbona subtenta* (Hall) (W. L.). S. *sulcata* (Verneuil (L. Wh.). S. *vetusta* (James) (L. Wh.). S. *vetusta praecursor* Foerste (W.). *Tetradella lunatifera* (Ulrich) (Richmond). T. *quadrilirata* (H. & W.) (Richmond). *T. simplex* (Ulrich) (Richmond). *Tetradium approximatum* Ulrich (S. W. E.). *Trematis millipunctata* Hall (Richmond). *Trochonema madisonense* Ulrich (S. Wh.). *Whiteavesia pholadiformis* (Hall) (W.). *Whitella obliqua* Ulrich (W.). *W. ohioensis* Ulrich (W.). *W. quadrangularis* (Whitefield) (W.). *W. suborata* Ulrich (W.). *W. umbonata* Ulrich (W.). *Xenocrinus baeri* (Meek). (L.). *X. penicillus* Miller (L.). *Zygospira modesta* Hall (Richmond, etc.).

CHAPTER II.

SILURIAN.

Murchison proposed the name Silurian in 1835 (1835, 47) for a group of rocks in Wales lying below the Old Red Sandstone. It has been in practically universal use since that date for the younger portion of the Silurian rocks of Murchison, who also included under this designation the Lower Silurian or Ordovician. The lower and middle portions of the Silurian *(sensu stricto)* were comprised in the Ontario division of the first geological survey of New York (final reports 1842-43) but the upper portion fell in the Helderberg division. Clarke and Schuchert (1899) suggested the revival of the name Ontario, in the form Ontaric and this latter name has been used to some slight extent in the United States.

The New York geologists, Hall and Vanuxem, referred to their Ontario division the following named formations (in the ascending order): *Medina, Oneida, Clinton* and *Niagara.* Hall (1843b) regarded the Oneida as below the Medina, while Vanuxem (1842) placed the Oneida above the Medina, where, as shown by Hartnagle (1907, 29-38), it belongs; and included the gray sandstone of Oswego in the Ontario division. In the Helderberg division (Mather, 1840, 236-46) were placed the following formations (in ascending order): The Onondaga salt group (=Salina), the Waterlime group (=Roundout, Cobleskill and Manlius), the Pentamerus limestone (=Coeymans), the Catskill (or Delthyris) shaly limestone (=New Scotland), the Oriskany sandstone, the Cauda-galli grit (=Esopus shale), the Schoharie grit, the Onondaga limestone and the Corniferous limestone. Of these, the Onondaga salt group, now known as the Salina (Dana, 1863, 246) and the Waterlime group, belong in the Silurian as at present constituted.

In 1899 Clarke and Schuchert proposed to group the Silurian formations of New York into three series as follows: (in ascending order), *Oswegan,* including the Medina and Oneida and Shawangunk; *Niagaran,* including the Clinton, Rochester, Lockport and Guelph; and *Cayugan,* including the Salina, Roundout and Manlius formations. As will appear presently several of these formations have within recent years been more minutely subdivided.

The Silurian rocks of Indiana fall largely in the Niagaran division, though both the Oswegan and Cayugan have meagre representation. Recently Ulrich and Bassler (Bassler 1915, chart) have proposed to call the lower or Oswegan division of the Silurian, in which they include the Richmond, the *Medinan.*

The chief and indeed almost the only source of accurate information on the Silurian rocks of Indiana is in the various papers by Foerste, (1895, 1896, 1897, 1898, 1900, 1904, 1904b, 1909, 1909b, 1917a) and Kindle (1904, 1913).

MEDINAN.—(Oswegan, Alexandrian, Anticostian (in part)). When the name Oswegan (Clarke and Schuchert 1899, 876-77) was proposed for the lower division of the Silurian, the "Lower Medina Shales" now known as Queenston* (Grabau 1908, 622-23) in New York and Juniata shales (Darton, 1896) in Pennsylvania were included in the Silurian, and even the Oswego sandstone was commonly regarded as belonging to this system. Grabau (1908, 1909, 1913, etc.) has shown that the Queenston is of Richmond age and disconformable with the Medina sandstane; and this conclusion has been amply confirmed by Foerste (1914, 1916, 1917c), Schuchert (1914), Kindle (1913b, 1914), M. Y. Williams (1919) and others. Grabau (1908, 622) also restricted the term Medina to the Medina *sandstone* or Upper Medina of Hall, but Kindle proposed to call this formation the *Albion*, a name which is adopted in the Niagara Falls folio of the U. S. Geological Survey (1913b, 6) by Taylor and Kindle. The latter, however, (1914, 915) did not himself approve of this procedure, which was in reality suggested by the Committee on geologic names, of the U. S. Geological Survey (*Ibid.*). Schuchert (1914, 286) concurs in restricting the name Medina to the upper Medina (sandstone). Since this division of the Silurian is represented by diverse types of sedimentation in various parts of the continent and is capable of subdivision into several formations and since the name Oswegan is now inappropriate, with the exclusion of the Oswego sandstone from the Silurian, the writer suggests that the term *Medinan* replace the name Oswegan as a series name for that portion of the Silurian lying beneath the New York Clinton; and that the name Albion be restricted to that portion of the Medina sandstone lying between the Whirlpool and Thorold sandstones or between the Queenston and Thorold where the Whirlpool is absent. This use of the term Modinan differs from that of Ulrich (1911, 1914) and Bassler (1915, 1919) in excluding the Richmond from the Silurian. This old and appropriate name should have the right of way over such innovations as Alexandrian and Anticostian. The dominantly calcareous phase containing the Cataract fauna may appropriately be called the Cataract phase (Schuchert 1913, 107), while the Brassfield, Sexton creek, and associated Medinan rocks containing a dominantly southern fauna may be conveniently grouped under Savage's very appropriate name Alexandrian (Savage 1908, 434).**

Brassfield.—(Clinton, Ohio Clinton, Niagaran (in part), Medina of Indiana, Kentucky and Ohio reports.) In Kentucky the oölitic iron ore bed and associated limestones capping the Cincinnatian were recognized as a distinct formation as long ago as 1857 (Owen, 1857) and doubtfully referred to the Clinton; and in Ohio the formation has been referred to the Clinton from 1869 (Orton) nearly to the present time. (See Orton 1871, 1873, 1874, 1878, 1884, 1888; Foerste 1885, 1887, 1888a, 1890, 1891, 1893, 1897, 1898, 1900,

* Chadwick later in the same year (1908, Sci. vol. 28, 347) proposed the name *Lewiston* for the same formation.

** Crabau (1921) includes the Medinan along with the Niagaran of Clarke & Schuchert in his Niagaran or "Lower Silurian" division.

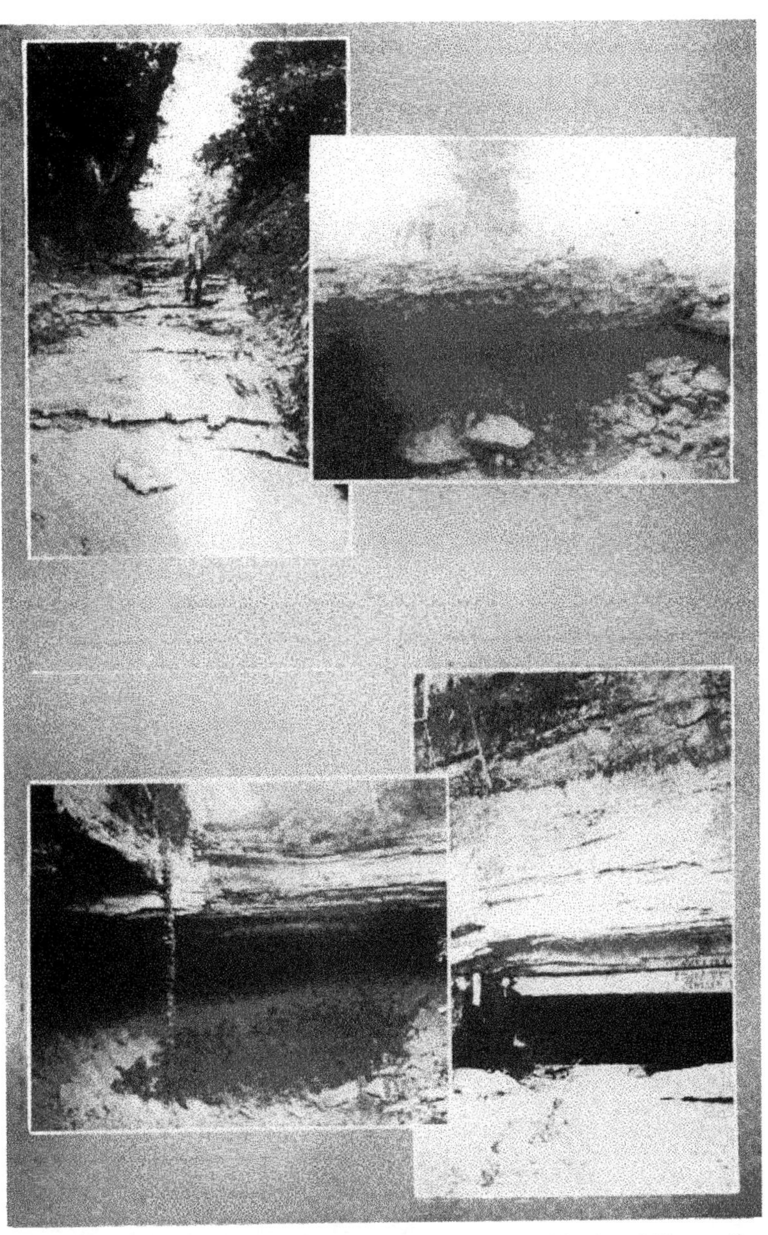

Fig. 22. Upper left, Cynthiana, near Patriot, Indiana. Upper right, Brassfield, near Versailles. Lower left, Saluda, near Hanover. Lower right, Laurel (foreground), Waldron (in tunnel), Louisville (projecting ledges above heads of students), Geneva (upper ledge of picture); Tunnel Mill, Vernon, Indiana.

1904, 1904b, etc.) In Indiana, on the contrary, the formation was not generally recognized as distinct from the overlying or underlying rocks till 1889 when Prof. A. H. Young sent a collection of fossils from the salmon-colored limestone layer near Hanover, Indiana, to Dr. Foerste and suggested that the fauna was of Clinton character. In 1896 Foerste published an extensive paper on the "Clinton" of Ohio and Indiana (1896, 161-200) and traced the formation southward from Richmond to Osgood. This was the first careful work on the "Clinton" (Brassfield) of Indiana. Previous Indiana geologists had included the formation sometimes with the overlying Niagara rocks and sometimes with the underlying Ordovician. D. D. Owen (1837-1859, 26-27) evidently included the horizon in his "Magnesian limestones." Richard Owen (1862, 40) says he was informed by Dr. Plummer of Richmond that the latter "found Upper Silurian fossils not many miles off from Richmond on Elkhorn creek." Since the Brassfield is the chief Silurian rock exposed on Elkhorn creek he probably had this formation in mind. Again (*ibid.* pp. 45-46) he says he found a "reddish silico-calcareous rock, apparently of upper Silurian age" in western Fayette County about four miles from the Rush County line. This may also have been Brassfield. Borden (1875, 158-169) gives many sections in Jefferson County. He says "the Clinton here, as in New York, is in the main a sandstone, and of variable texture; it is very soft in the lower part where it is composed largely of sand and clay" (*Ibid.* 159). "The prevailing color of the rock is light yellow, with salmon and pink shades" (*Ibid.*, 160). Judging from his sections his "Clinton" is mostly Saluda. In the 7th Indiana Report, Borden (1876, 154) referred the lower part of the Niagara along with the Saluda to the Clinton. Cox in 1879 (20-21) expresses the belief that the so-called "Clinton" (of Borden) should not be separated from the Niagara, and (p. 178) sees no reason for calling the rock at Elkhorn falls (near Richmond) Clinton. He consequently refers it to the Niagara*. In the 12th Indiana report Elrod (1883, 108) referred to the "Clinton group" a bed of "calcareous sandrock, or shale, ranging in thickness from six inches to three feet" and lying below the "Lower Niagara shale". Since this "Lower Niagara shale" corresponds approximately to the Osgood formation, Elrod seems, judging from his sections, rather by accident, to have included the present Brassfield in his "Clinton". Again in the 14th report (1884a, 47) Elrod refers to the "Clinton" 20 feet of siliceous bastard limestone in Fayette County, much discolored with iron ore. This includes rocks both below and above the Brassfield. Elrod's error was repeated by Maurice Thompson in the 15th Report (1886). Finally it may be well to state that the "Clinton" reported in Indiana well records from 1888 to the present time is quite certainly neither Clinton nor Medina. In 1892 S. A. Miller (1892, 611) states that he is unable to find any evidence of the existence of the Clinton or Medina in Indiana. In 1893 Joseph Moore of Richmond (1893, 28-29) mentioned outliers (probably very large glacial boulders) of "Clinton" limestone "2 miles southwest of the central part of Richmond."

With the publication of Foerste's papers (1895, 1896, 1897, 1898, 1900, 1901a, 1902, 1904, etc.) the so-called "Clinton" of Indiana received definite status as a distinct formation, correlated with the "Clinton" of Ohio and Kentucky and carrying a characteristic fauna. Foerste worked out in detail

* This statement implies that the bed at Elkhorn Falls *had* been referred to the Clinton, which would make this the first reference of an Indiana formation to the Clinton, rather than the Hanover rock cited by Foerste.

the lithologic and faunistic characters, distribution and thickness of the rock, and its disconformable relations to the underlying and overlying formations. In 1906 Foerste (1906, 18) proposed to rename the formation *Brassfield*, from Brassfield, Madison County, Kentucky; and this name has been generally adopted (see later papers by Foerste; also by Savage 1908, 1909, 1910, 1912, 1913, 1916, 1919, etc.; Schuchert 1910, 1913, 1914, 1915; Prosser 1916; Bassler 1915, etc.; Ulrich 1911, 1914, etc.; Parks 1913; M. Y. Williams 1914, 1919; etc., etc.). In Schuchert's Paleogeography of North America (1910, 489) and in a paper by Schuchert and Twenhofel (1910, 704-710) the formation is designated "Ohio Clinton" and stated to be below the typical Clinton of New York. That this formation is not really equivalent to the New York Clinton had been recognized by Foerste as long ago as 1896, when he said (1896, 189) "Although I have been accustomed to call the Ohio formation the Clinton, yet I should be willing to recognize the fact that the identity is not very marked, by giving it a name of its own, for instance, the *Montgomery formation*, on account of its typical development in Montgomery County in Ohio."* This suggestion was not followed up and the name Montgomery lapsed. In 1904 Foerste (1904, 340), stated that "the Clinton of Ohio, Indiana and Kentucky appears to have attained the stage of development equivalent to that of the Clinton of New York, below the lenses, but does not contain such species as *Pentamerus oblongus*, *Atrypa reticularis*, *Spirifer radiatus*, *S. niagarensis*, which in the west begin their existence in the Osgood bed." In proposing the name Brassfield Foerste after reviewing the fauna (1906, 35) commented again as follows on the relations of the formation to the New York Clinton: "The identification of the Brassfield limestone of Kentucky, and its northern extension in Ohio and Indiana, in former years, with the Clinton limestone of New York, rests rather upon a somewhat similar facies of the two faunas, and upon the general absence of the more typical species of the Rochester shale fauna of New York in these limestones at the base of the Silurian in Ohio, Indiana and Kentucky, than upon the presence of any considerable number of species common to both areas.** On closer inspection, the fauna of the Brassfield limestone of Ohio, Indiana and Kentucky appears to differ sufficiently from the fauna of the Clinton limestone of New York to warrant the assumption of the presence of some sort of barrier between the two areas."

In 1913 Schuchert (1913, 107) proposed the name Cataract formation for the fossiliferous limestones, shales and dolomites of Ontario, 48 miles northwest of Toronto, formerly called Clinton. He then supposed that the new formation lay below the Medina. This name was accordingly adopted in the Guidebook of the International Congress of Geologists, by Parks (1913, 128), with the statement that the Cataract "represents an invasion from the south and west at the commencement of Silurian time. The upper limestones and shales of this formation are highly fossiliferous and present a fauna comparable with that of the Brassfield formation of Ohio and Kentucky." In 1914 M. Y. Williams (1914a, 40) stated that the "Medina sandstones of Niagara Gorge (125 feet thick) are represented farther north by dolomite and

* This name, it appears, had previously been used by Rogers in 1837 and 1884, for a grit and coal formation of Virginia and by Diller in 1892 for a Silurian limestone of California.

** Bassler, however, has shown that the Rochester fauna is present in the Osgood formation of Indiana and Kentucky. Bull. U. S. Geol. Survey. No. 292. p. 5. 1906.

shales (Cataract formation)"; and in the same year Kindle (1914, 918) reported that the "Examination by the writer of a number of sections holding this [Cataract formation] fauna, in connection with a review of the Niagara section, has convinced him that all of the terranes associated with the Cataract fauna are represented in the Medina of the Niagara section". Schuchert later (1914, 294) admitted that "the typical Medina formation shades through lateral alteration into the typical Cataract." He also *(Ibid.)* regarded "the Medina, Cataract and Brassfield as correlates of one another," but did not mean to imply that "each one is wholly equivalent of any other." "Each formation [fauna] invades eastern North America from a different direction and each one has its own peculiar faunal assemblage." These facts are brought out on the paleogeographic map presented herewith, based on Schuchert's map (*loc. cit.*, p. 295). In spite of this definite abandonment of the name Clinton for the Indiana-Kentucky-Ohio formation in 1906, the name recurs in Indiana reports as late as 1916. (See Kindle and Barnett 1909, 396; Tucker 1911, 15; Culbertson 1916, 226-227.)

Description.—The Brassfield formation in Indiana consists of rather coarsely crystalline, pink, salmon-colored, brownish red or sometimes bluish or greenish gray mottled limestone, appearing in most outcrops to consist of a single bed from less than a foot to 10 or 12 feet thick. Where weathering has brought out the bedding planes it appears to be made up of layers several inches (5-8) thick. In some sections in Ripley County and at Elkhorn Falls in Wayne County, the Brassfield is immediately underlain by a very soft clay-shale. It is an easy stratum to identify. The thickness is very variable in Ripley County where the formation is occasionally totally absent. In some of the Ripley and Jefferson county localities the rock is brecciated, containing angular fragments of a light blue, dense limestone, derived from the underlying Richmond. There is unmistakable evidence of disconformity both below and above the Brassfield. The rock does not appear to be very fossiliferous, but nearly every outcrop will afford a few specimens, and some localities, especially in the vicinity of Laurel, Indiana, are quite fossiliferous. The fauna, which is characteristically Silurian and totally distinct from the Richmond fauna, is best known, however, from the Ohio localities about Dayton, where Dr. Foerste has collected about 100 species.

Fauna.—The more common and characteristic species as determined by Foerste, are as follows: *Illaenus daytonensis* H. & W., *I. ambiguus* Foerste, *I. madisonianus-elongatus* Foerste, I. *madisonianus depressus* Foerste, *Proetus determinatus* Foerste, *Cyphaspis clintoni* Foerste, *Odontopleura ortoni* Foerste, *Encrinurus thresheri* Foerste, *Calymene niagarensis* (Whitfield), *C. vogdesi* Foerste, *Deiphon foerstei* (Barrande), *Phacops pulchellus* Foerste, *Dalmanites wortheni* Foerste, *Plectambonites prolongatus* (Foerste), *P. transversalis* (Wahlenberg), *Leptaena rhomboidalis* (Wilckens), *Strophomena daytonensis* Foerste, *S. striata* Hall, *S. hanoverensis* Foerste, *Orthis flabellites* Foerste (and varieties), *Hebertella fausta* (Foerste), *H. daytonensis* Foerste, *Platystrophia reversata* (Foerste), *P. daytonensis* Foerste, *P. brachynota* (Hall), *Dalmanella elegantula* (Dalman), *D. elegantula parva* (Foerste), *Rhipidomella hybrida* (Sowerby), *Triplecia ortoni* (Meek), *Atrypa marginalis* (Dalman), *Camarotoechia convexa* (Foerste), *Stricklandinia triplesiana* Foerste, *Homotrypa confluens* (Foerste), *Aspidopora parvula* (Foerste), *Lioclemella ohioensis* (Foerste), *Hallopora magnopora* (Foerste), *Chasmatopora angulata* (Hall), *Hemitrypa ulrichi* (Foerste), *Ptilodictya Whitfieldi* (Foerste), *Ptilodictya expansa* Hall, *Clathropora frondosa clintonensis* (H. & W.), *Phaenopora expansa* H. & W., *P. fimbriata* (James), *P. magna* (H. & W.), *P. multifida* Hall, *Pachydictya bifurcata* (Hall), *P. bifurcata instabilis* (Foerste), *P. crassa* (Hall), *P. emaciata* (Foerste), *Trigonodictya eatonensis* Ulrich, *Rhinopora verrucosa* Hall, *Dictyonema pertenue* Foerste, *D. scalariforme* Foerste, *Corynotrypa elongata* (Vine).

NIAGARAN.—(Cliff rock, Catenipora bed (Lyon), Magnesian limestone.) In all the earlier Indiana reports, the succession of formations between the Ordovician or blue limestones and the coral rock of the Ohio Falls was referred to the "cliff rock" or Niagara group of the Silurian, without subdivision. The name "cliff rock" was first applied to these limestones, more particularly to the division now known as the Laurel formation, by Locke (1838, 211) in Ohio, and was used for many years (Clapp 1843, Hall 1851, Owen 1840, etc.) as a general designation of the Niagara and other limestones of the Ohio-Mississippi valley region. Owen (1844, 116-117) gives as synonyms of Cliff rock: galeniferous limestone, cornetiferous limestone, magnesian limestone, mountain limestone. Clapp (1843) correlates the "Cliff rock" with the Niagara, Gypeous shale, Waterlime and Onondaga of New York; while Hall (1851) says the "Cliff" equals the Niagara, Clinton and Corniferous, and, again, (1843, 279) "In the Cliff limestone we have the Helderberg series of New York or at least two persistent members, the Onondaga and Corniferous, with the Waterlime and perhaps a meagre representation of the salt group, together with the Niagara limestone." Locally the rock was known as the Catenipora [Halysites] bed (Lyon), Chain coral beds and Magnesian limestones (Owen 1839, 1856, etc.) and was described at the Falls of the Ohio by Lapham as early as 1828.

The name Niagaran was proposed by Clarke and Schuchert (1899, 876-877); but the older name Niagara has been in general use since it was proposed by Hall (1842, 58) and Vanuxem (1842, 90-94). The earliest connection of the name Niagara with a geological formation, however, was Conrad's definite use of the name (1837) for the Medina sandstone of western New York. By a curious turn of fortune this use of the name was entirely lost sight of and the sandstone still bears the name with which it was rechristened by Vanuxem.

The first member of the Niagaran series of rocks of Indiana to receive a distinctive geographic name was the conspicuous band of shale seen near the top of the formation in Shelby and Bartholomew counties, which is 1883 was named the *Waldron* shale by Moses Elrod (1883, 116). The name "Utica limestone", from Utica in Clark County, had previously been loosely applied by Borden (1874, 172) to the Niagara rocks of Clark and Floyd counties. No other subdivisions were named till 1896 when Foerste (1896, 190) proposed the name *Laurel* limestone (Franklin County, Indiana) for the main cliff rock member of the Niagara, and called the underlying shales and limestones the "Osgood phase" of the Laurel. In the 21st Indiana Report (1897, 218) he added the name *Louisville* for the Niagaran limestone immediately above the Waldron shale and used the term *Osgood* as a distinct formation name.

The greater part of the limestones and dolomites of northern Indiana in the White and Wabash valleys has been referred to the Niagara since the days of the Owens (see R. Owen 1862, 57), though Richard noticed that "some of the fossils found seem assignable to yet higher groups, the Onondaga Salt Group of the New York geologists, and their Lower Helderberg Limestones". He assigns a thickness of 1500 feet *(Ibid.)* to the Silurian of Indiana. This tendency to place certain rocks of Indiana in the Lower Helderberg division has persisted down to recent years, and dates from the 1842 New York Reports when all of the Silurian above the Niagara was relegated to that division. The correlation of these higher rocks (Kokomo limestone, etc.) will be discussed in the proper place.

Formation names were not assigned to the Niagaran of northern Indiana till Kindle (1904) gave the first adequate account of these rocks in 1904. He then proposed the name *Noblesville dolomite (Ibid.,* p. 407) for the lower part and *Huntington limestone (Ibid.,* p. 408) for the upper part. Undoubtedly further study will make a minuter subdivision possible. The name Huntington stone was in common use for the rock quarried at Huntington (Cox 1879, 59-60). Phinney in 1884, (1884, 142, 147, 148) correlated the lower strata of the northern Indiana Niagaran with the Springfield beds of Ohio and the upper portion with the Cedarville dolomite or Guelph.

The unusual dips of the Niagaran rocks of northern Indiana, indicating "cone" or "dome" (quaquaversal) structure, were noticed as far back as 1862 by R. Owen (1862, 98) and have been often mentioned in the Indiana state reports and elsewhere (see Collett 1872, 307; Gorby 1886, 228-241; 1889, 180-183; Thompson 1889, 41-53; 1892, 177-186; Phinney 1891, 653; Elrod and Benedict 1892, 200-238; Ashley and Siebenthal 1899, 190; Elrod 1902, 205-215; Kindle 1902, 221-224; 1904, 402-413; Ward 1906, 216-217). They have been explained as mud-lump structures (Kindle), giant cross-bedding (Collett), coral-reef structure (Phinney), a narrow anticline ("Wabash arch") running across the state (Gorby), "irregularities of the sea bottom", a variety of cleavage (Elrod), volcanic agencies (Ashley), etc. Recent unpublished studies of numerous well records by Dr. W. N. Logan, indicate that the Niagara and underlying Ordovician rocks are characterized by numerous domes and irregular anticlinal and synclinal structures; and the presence of Ordovician strata in the Kentland dome also indicates that the structures are genuinely diastrophic or vulcanic and involve not only the Niagaran rocks but the underlying formations. The possibility of cryptovolcanic action, such as described by Bucher (1920, unpublished) in an area near Cincinnati, Ohio, should not be overlooked in any further study of these domes. Cross-bedding, cleavage, coral-reef structure, mud-lump structure and the "Wabash arch" may, with the data at present available, be ruled out of account. Kindle's view that the structures are post-Niagaran and pre-Devonian is undoubtedly correct.

Osgood.—Foerste first used the name Osgood in 1896, but definitely delimited the formation in his paper on the middle Silurian rocks of Indiana the next year (1897, 218 and 227-230). The details of its distribution were given in this and a succeeding paper (1898, 197, 208-211, 217-253); and the history of this and other Silurian formations in 1900 (1900, 41-47, 78-80). Dr. Elrod, whose important place in Indiana geology has not been sufficiently appreciated, was the first to apply a separate designation to the shaly beds at the base of the Niagaran section in southern Indiana. In the 12th Indiana Report (1883, 108-109) he designated as the "Lower Niagara shale" the "thin strata of shale, flag, and thicker beds of marl" below the building [Laurel] stone, to distinguish them from the "Upper or Waldron shale." The Osgood has been identified in Kentucky (Foerste 1906, 18, 58, etc.), Tennessee (Foerste, 1903a, 554-715; Pate and Bassler 1908, 409-410, etc.) and Ohio (Prosser 1916, 334-365; Foerste 1919, 373-374). It is correlated with the Rochester shale of New York by Bassler (1906a, 5) who says "The Osgood is chiefly if not exactly comparable with the Rochester shale." Foerste two years previously (1904, 340) had reached about the same conclusion. He correlates the Osgood with "the lenses at the top of the Clinton and . . . the lower half of the Rochester shale of New York." This correlation has been generally approved. It should be noted in this connection that a certain degree of nomenclatorial confusion has been introduced here by the suggestion of Ulrich and Bassler (1915) that the Rochester shale be placed in the Clinton division instead of Niagaran, as in all previous classifications. Chadwick (1918a) regards the Rochester shale and Irondiquoit limestone as constituting a distinct division between the Clinton and Niagara, while M. Y. Williams (1919, 52, footnote) prefers to retain the Rochester in the Niagaran series and says the fauna differs from that of the typical Clinton below.

Description.—The Osgood formation of Indiana consists typically of two beds of shale and two beds of limestone which have been called by Foerste the lower and upper Osgood shales and lower and upper Osgood limestones. The lower Osgood limestone is sometimes absent or at least very thin. It has been called the basal Osgood limestone (Foerste 1897, 226-227) and correlated with the Dayton limestone of Ohio (see below). This basal limestone, where present, is from 8 to 15 inches thick. The lower Osgood shale or clay immediately above it is a barren soft blue clay 11 to 16 feet thick. In places (Ripley and Jennings counties) the lower part has changed into a rubbly limestone. In the southern counties (Ripley, Jefferson, etc.) the lower shale is overlain by from 3 to 5 feet of brownish to whitish, sometimes conoidal limestone. Overlying this second limestone is the upper shale or clay from 3 to 5 feet thick and similar in appearance to the lower clay. These four divisions cannot be recognized in the more northerly localities (Franklin, Fayette, etc.).

The *Dayton* limestone.—In Ohio the Brassfield is overlain by several feet of hard, fine grained limestone, which was quarried for many years under the name of "Dayton stone." This designation was adopted by Orton (1871, 301, etc.) as the geological name of the formation, though the name had been used in a commercial sense before that date. (See Ohio Survey Reports of Progress for 1869 and 70 and vols. I to VII, 1873-1893.) The name first occurs in Indiana Reports in 1869, when Haymond (1869, 1st Ind. Rpt., 193) stated that "the most valuable building stone in the county [Franklin] and probably in the state is found in Laurel and Posey townships. It is of the same character and belongs to the same formation as the Dayton stone so extensively used in Cincinnati and other places, and the same as that found at Greensburg and St. Paul." He evidently had in mind the Laurel limestone. In 1896 Foerste referred three or four feet of limestone resting on the "Clinton" (Brassfield) at Elkhorn Falls to the Dayton limestone (1896, 193), and in the same paper called the lower part of the Laurel ("Osgood phase") the Dayton limestone. The name was not repeated in the 1897 report (21st Indiana), but the lower limestone of the Osgood, as already indicated, was called basal Niagara or lower Osgood limestone. In the 24th Report (1900, 80) he says "At the base of the Niagara of Ohio are several feet—two to five—of hard, fine grained, white limestone, which have been called the Dayton stone. . . . Above the Dayton is a series of rocks . . . which evidently correspond stratigraphically to the Osgood beds of Indiana." In 1904 (1904b, 29) he definitely correlated the Lower Osgood with the Dayton limestone. This correlation has been repeated by Butts (1915, 81), who correlates the lower Osgood limestone with the Dayton, and by Prosser (1916, 363) who, apparently following Foerste, makes a similar correlation. Foerste (1906, 17, 50; 1909, 1, 4) also correlates the Dayton with the Oldham limestone of Kentucky.

Fauna of the Osgood.—Fossils are not very abundant in the Osgood formation, cystids being the most common. Bassler (1915, 1487-1488) lists the following species, not all of which are found in Indiana. Some of those listed were probably obtained from the Laurel. *Acanthoclema asperum* Bassler, *Allocystites hammelli* Miller, *Atrypa reticularis* Linnaeus, *Batostomella granulifera* (Hall), *Bernicea consimilis* Lonsdale, *Calymene niagarense* (Hall), *Calymenella nasuta* (Ulrich), *Camarotoechia indianensis* (Hall), *C. neglecta* (Hall), *Caryocrinites ellipticus* (Miller and Gurley), *C. hammelli* (M. & G.), *C. indianensis* (Miller), *C. ornatus* Say, *Ceramopora imbricata* Hall, *C. niagarensis* Bassler, *Chasmatopora asperato-striata* (Hall), *Chilotrypa ostiolata* (Hall), *Coeloclema cavernosa* Bassler, *Corynotrypa dissimilis* (Vine), *Cyrtia exporrecta* (Wahlenberg), *C. myrtia* (Billings), *Cyrtoceras cinctulum* Foerste, *Dalmanella elegantula* Dalman, *Dalmanites limulurus* (Green), *Diaphorostoma cliftonense* Foerste, *D. niagarense* Hall, *Duncanella borealis* Nicholson, *Enterolasma caliculum* (Hall), *Eridotrypa striata* (Hall), *Favosites cristatus* E. & H. *F. forbesi* E. & H., *Gomphocystites indianensis* Miller, *Hallopora clausa* (Bassler), *H. elegantula* (Hall), *Holocystites adipatus* Miller, *H. affinis* Miller & Faber, H. *amplus* Miller, *H. asper* M. & G., *H. baculus* Miller, H. *benedicti* Miller, H. *brauni* Miller, H. *canneus* Miller, *H. colletti* Miller, *H. commodus* Miller, H. *dyeri* Miller, *H. elegans* Miller, H. *faberi* Miller, H. *globosus* Miller, H. *gorbyi* Miller, *H. gyrinus* M. & G., H. *indianensis* Miller, H. *madisonensis* Miller, *H. ornatissimus* Miller, *H. ornatus* Miller, H. *papulosus* Miller, H. *parvulus* Miller, *H. perlongus* Miller, *H. plenus* Miller, H. *rotundus* Miller, H. *scitulus* Miller, *H. spangleri* Miller, *H. sphaeroidalis* M. & G., *H. splendens* M. & G., H. *subovatus* Miller, H. *subrotundus* Miller, H. *tumidus* Miller, H. *turbinatus* Miller, *H. ventricosus* Miller, H. *wetherbyi* Miller, *H. wykoffi* Miller, *Hyolithes cliftonensis* Foerste, *Idiotrypa punctata* Hall, *Labechia delicatula* Parks, *Leptaena rhomboidalis* (Wilckens), *Lichenalia concentrica* Hall, *Lioclema asperum* (Hall), L. *explanatum* Bassler, L. *multiporum* Bassler, *Loculipora ambigua precursor* Bassler, *Mesotrypa nummiformis* (Hall), *Monotrypa benjamini* Bassler, *Nicholsonella florida* (Hall), *Nucleospira pisiformis* Hall, *Orthis flabellites* Foerste, *O. interplicata* Foerste, *Orthostrophia fasciata* (Hall), *O. fissistriata* (Foerste), *Pachydictya crassa* (Hall), *Phaenopora ensiformis* Hall, *Pisocrinus gemmiformis* Miller, *Platyceras pronum* Foerste, *Ptiloporella nervata* (Nicholson), *Rhipidomella hybrida* (Sowerby), *Rhombotrypa spinulifera* Bassler, *Rhynchotreta cuneata americana* (Hall), *Schuchertella subplana* (Conrad), *Semicoscinium tenuiceps* (Hall), *Spirifer eudora* (Hall), *S. niagarensis* (Conrad), S. *radiatus* (Sowerby), *Stephanocrinus cornetti* Miller, *Stephanocrinus deformis* Rowley, S. *elongatus* Miller, *S. gemmiformis* Hall, *S. hammelli* Miller, S. *obpyramidalis* Miller, *S. osgoodensis* Miller, *S. quinquepartitus* Rowley, *Striatopora flexuosa* Hall, *Thecia major* Rominger, *Trematocystis hammelli* (Miller), *T. subglobosus* (Miller), *Trematopora spiculata* Miller, *T. tuberculosa* Hall, *Triplecia (Cliftonia) tenax* Foerste, *Whitfieldella quadrangularis* Foerste.

Laurel.—The important quarry rock of Franklin, Shelby, Decatur and Ripley counties, constituting the main "cliff" rock of the Ohio bluffs below Madison, was named Laurel limestone, from the town of Laurel in Franklin County, Indiana, by Foerste in 1896 (1896, 18, 190), and this name has been used consistently since that date. (Foerste 1897, 1898, 1900, 1901, 1901a, 1903, 1904b, 1906, 1909b, etc.; Pate and Bassler 1908; Kindle and Barnett 1909; Newsom 1903; Elrod 1902; Taylor 1906; Culbertson 1916; Blatchley, Bassler, Ulrich, Miller, Prosser, Butts, etc., etc.). Elrod (1902, 209-215) has suggested that there is a disconformity between the Laurel and the Waldron shale, but this has not been generally recognized.

The Laurel limestone, or better, dolomite, (Butts, 1915, 82) is an evenly bedded almost white, hard rock from 40 to 50 feet thick in Indiana. The layers vary from 4 to 16 inches thick and in the quarries furnish dimension stone of the proper thickness for various purposes such as lintels, curbing, sills, sidewalks, etc. In Jefferson County the formation is yellower in color, drusy, irregularly knotty and unfit for quarrying. In Clark County it is

Fig. 23. Upper left, Geneva limestone, Muskatatuck creek, near North Vernon, Indiana. Upper right, Brassfield limestone, on the Madison-Hanover road. Lower left, Laurel limestone in quarry near Osgood. Lower right, Silver creek and Beechwood, with New Albany shale overlying.

softer and more argillaceous and weathers rapidly. Little is known of the fauna, though numbers of large coiled cephalopods have been found and numerous crinoids and cystids. The rock has been extensively quarried at Osgood, St. Paul, Laurel, etc. The laurel formation should probably be correlated with the Lockport division of New York.

Fauna of the Laurel Limestone.—The Laurel is very sparingly fossiliferous. Bassler (1915, 1491-92) lists the following species, not all of which are from Indiana:—*Acacocrinus americana* Wachsmuth & Springer, *Aethocystites sculptus* Miller, *Allocrinus benedicti* Miller, *Amplexus cinctus* Miller, *Anastrophia internascens* Hall, *Atrypa reticularis* Linnaeus, *A. nodostriata* Hall, *Barrandeocrinus canaliculatus* (Miller), *B. (?) indianensis* (Miller & Gurley), *Callicrinus brachleri* W. & S., *Calymene niagarensis* Hall, *Conocardium elrodi* Miller, *Corydocephalus byrnesanus* (M. & G.), *C. phlyctainoides* (Green), *Cyphocrinus gorbyi* Miller, *Cyrtoceras howardi* Miller, *C. indianense* Miller, *C. nashvillense* Miller, *Dalmanella elegantula* (Dalman), *Dawsonoceras annulatum* (Sowerby), *Deltacrinus indianensis* (Miller), *Emperocrinus indianensis* M. & G., *Favosites spinigerus* Hall, *Gazacrinus inornatus* Miller, *G. ventricosus* W. & S., *Gyroceras elrodi* White, *Habrocrinus howardi* (Miller), *Holocystites pustulosus* Miller, *Indianocrinus punctatus* M. & G., *Macrostylocrinus indianensis* M. & G., *Mariacrinus aureatus* Miller, *M. (?) granulosus* Miller, *Melocrinus aequalis* Miller, *M. oblongus* W. & S., *Periechocrinus ornatus* (H. & W.), *P. umbrosus* (M. & G.), *Petalocrinus longus* Bather, *Pisocrinus baccula* M. & G., *P. gemmiformis* Miller, *Plasmopora follis* E. & H., *Spirifer crispus simplex* (Hall), *S. radiatus* (Sowerby), *Streptelasma spongaxis* Rominger, *S. gorbyi* Miller, *Stribalocystites gorbyi* Miller, *S. sphaeroidalis* M. & G., *S. tumidus* Miller, *Subulites benedicti* Miller, *Uncinulus stricklandi* (Sowerby), *Zophocrinus howardi* Miller.

Waldron.—Elrod proposed the name Waldron shale from the famous Hall quarry locality on Conn's creek in Shelby County, in 1883 (1883, 106, 111). The names "Waldron bed" and "Waldron fossil bed" had been used locally for many years previously and appear in the 11th Indiana Report (1882). It seems doubtful therefore whether Elrod should receive credit for the name.* The marvellously preserved and abundant fossils of the formation originally described by Hall (1862, 1879, 1882), and made the basis of Beecher and Clarke's classic memoir on the development of Paleozoic Brachiopoda, have made Waldron one of the most famous fossil localities in America; and many splendid collections have been made there and at Hartsville, Tarr Hole, St. Paul, Vernon and other outcrops of the Waldron bed.** The literature of the Waldron is extensive; but the principal papers are those of Foerste (1888, 1896, 1897, 1898, 1900, 1901, 1901a, 1903a, 1909b); Cumings (1900); Price (1900); Kindle and Barnett (1909); Hall (1879b, 1882); Pate and Bassler (1908); Elrod (1883, 1884, 1902); Bassler (1906a); and Butts (1915). The formation extends with characters substantially intact into Kentucky and Tennessee. It cannot be identified, however, north of the southern boundary of Rush County in Indiana.

The Waldron shale varies from a very soft light greenish gray to bluish fragile clay-shale to a calcareous knotty shale. Toward the north the clay gives way to limestone. Its thickess varies from a foot to 8 to 10 feet. In Kentucky it is still thicker. The famous fauna is found only in occasional "colonies" of limited extent. For example a few yards away from the celebrated "Hall quarry" the shale is absolutely barren. At Tarr Hole on Clifty Creek, half way between Hartsville and Newbern in Bartholomew County, fossils are exceedingly abundant, while a mile farther up the creek, near Hartsville, a beautiful outcrop of the shale, by the side of the road, is practically barren. In Kentucky (Butts, 1915) the shale is barren, while

* Weeks mistakenly credits the name to Elrod 1884, 13th Indiana Report, pp. 93, 95.

** The name "Waldron sandstone" was later used by Gallaher, Missouri Geological Survey, Biennial report, 1898, p. 52, for an upper Carboniferous formation of Missouri.

in Tennessee Foerste (1901) has collected a very extensive fauna. In the latter state the formation is known as the Newsom shale.

The Waldron is disconformably overlain by the Louisville limestone.

Fauna of the Waldron.--This formation has the most abundant and well preserved fossils of any Silurian formation of Indiana. The following list of species, includes the forms found in Indiana: *Acacocrinus elrodi* Wachsmuth and Springer, *Allonema waldronense* Ulrich and Bassler, *Amphieristocrinus ? calyx* (Hall), *A. typus* Hall, *Amphicoelia leidyi* Hall, *Anastrophia internascens* Hall, *Arctinurus occidentalis* (Hall), *Atrypa reticularis* (Linnaeus), *A. reticularis niagarensis* Nettleroth, *Ascodictyon siluriense* Vine, *Astylospongia praemorsa pusilla* Rauff, *Atrypina disparilis* (Hall), *Aulopora precius* Hall, *Barrphyllum fungulus* White, *Batostomella granulifera* (Hall), *Bellerophon tuber* Hall, *Bernicea consimilis* (Lonsdale), *Beyrichia granulosa* Hall, *B. waldronensis* Ulrich and Bassler, *Bilobites* bilobus (Linnaeus), *Botryocrinus nucleus* (Hall), B. polyx (Hall), *Bumastus armatus* (Hall), *B. ioxus* Hall, *Buthotrephis gracilis crassa* Hall, *Calymene niagarensis* Hall, *Camarotoechia (?) acinus* (Hall), *C. (Stegerhynchus) whitei* (Hall), *Caryospongia juglans nuxmoschata* (Hall), *Ceramopora ? confluens* Hall, *C. explanata* Hall, C. *notha* Hall, C. *raripora* Hall, *Ceratocephala fimbriata* (Hall), *Chaunograptus novellus* (Hall), *Cheirurus niagarensis* (Hall), *Chilotrypa varia* (Hall), *C. variolata* (Hall), *Chonetes undulatus* Hall, *Cladopora sarmentosa* Hall, *Clorinda fornicata* (Hall), *Coleolus spinulus* Hall, *Conularia infrequens* Hall, C. *proprius* Hall, *Corynotrypa dissimilis* (Vine), C. *elongata* (Vine), *Crania setifera* Hall, C. *siluriana* Hall, C. *springera* Hall, *Cyathocrinus ? aemulus* Hall, C. *benedicti* Miller, *Cycloceras amycus* (Hall), *Cyphaspis christyi* Hall, *Cypricardinia arata* Hall, C. *subovata* M. & D., *Cyrtia myrtia* (Billings), *Cyrtolites sinuosus* Hall, *Dalmanella elegantula* (Dalman), *Dalmanites bicornis* (Hall), *D. halli* Weller, *D. verrucosus* (Hall), *Dawsonoceras annulatum* (Sowerby), *Deltacrinus stigmatus* (Hall), *D. tunicatus* Hall, *Diamesopora infrequens* (Hall), *D. osculum* (Hall), *D. subimbricata* (Hall), *Diaphorostoma niagarense* (Hall), *D. plebium* (Hall), *Dictyonella reticulata* Hall, *Dimerocrinus inornatus* (Hall), *D. occidentalis* (Hall), *D. waldronensis* (M. & D.), *Duncanella borealis* Nicholson, *Entomis waldronensis* Ulrich, *Eridotrypa echinata* (Hall), *Eucalyptocrinus crassus* Hall, *E. ellipticus* Miller, *E. elrodi* Miller, *E. ovalis* Hall, *E. tuberculatus* Miller & Dyer, *Euomphalopterus alatus obsoletus* Ulrich, *Favosites forbesi occidentalis* (Hall), *F. spinigerus* Hall, *Fenestella bellistriata* Hall, *F. parvulipora* Hall, F. *pertenuis* Hall, *F.* prolixa Hall, Fistulipora *halli* Rominger, *F. neglecta* Rominger, *F. neglecta maculata* (Hall), *Goniophora speciosa* Hall, *Gyroceras abruptum* Hall, *Hallopora elegantula* (Hall), *Homalonotus delphinocephalus ?* Hall, *Homoeospira evax* Hall, *Homoeospira sobrina* (Beecher and Clarke), *Homocrinus ancilla* (Hall), *Ichthyocrinus subangularis* Hall, *Inocaulis divaricatus* Hall, *Ischadites subturbinatus* Hall, *Kionoceras cancellatum* (Hall), *Lampterocrinus parvus* Hall, *Lecanocrinus pusillus* (Hall), *Leperditia faba* Hall, *Leptaena rhomboidalis* (Wilckens), *Leptotrypa ? Sphaerion* (Hall), *Lingula gibbosa* Hall, *Lioclema ? exsul* (Hall), *Loculipora ambigua* (Hall), *Lyriocrinus melissa* (Hall), *Macrostylocrinus fasciatus* (Hall), *M. granulosus* (Hall), M. *striatus* Hall, *Mariacrinus carleyi* (Hall), *Melocrinus obconicus* Hall, *Meristina maria* (Hall), *M. rectirostris* Hall, *Metopolichas breviceps* (Hall), *Mimulus waldronensis* (M. & D.), *Modiolopsis perlatus* Hall, M. *subalatus* Hall, *Monotrypella ? consimilis* (Hall), *Myelodactylus gorbyi* Miller, *Mytilarca acutirostra* (Hall), *M. sigilla* Hall, *Nautilus ?? oceanus* Hall, *Nematopora macropora* (Hall), *N. minuta* (Hall), *Nucleospira pisiformis* Hall, *Orthis benedicti* Miller, *Orthis (?) subnodosa* Hall, *Orthoceras simulator* Hall, *Palaeomanon bursa* (Hall), *Paleschara ? incrassata* Hall, *P. offula* Hall, *Periechocrinus whitfieldi* (Hall), *Pholidops ovalis* Hall, *Platycrinus siluricus* Hall, *Plectambonites transversalis* (Wahlenberg), *Polypora conferta* (Hall), *P. punctostriata* (Hall), *P. tantula* (Hall), *Protokionoceras medulare* (Hall), *Pseudohornera niagarensis* (Hall), *Pterinea brisa* Hall, *Ptilodictya angusta* (Hall), *Receptaculites sacculus* Hall, *Reticularia bicostata petila* (Hall), *Rhipidomella hybrida* (Sowerby), *Rhynchotreta cuneata americana* (Hall), *Saccocrinus americanus* Springer, *Schuchertella subplana* (Hall), *S. tenuis* (Hall), *Semicoscinium acmeum* (Hall), *Spatiopora maculata* (Hall), *Spirifer crispus* (Hisinger), S. *crispus simplex* (Hall), *S. eudora* (Hall), S. *radiatus* (Sowerby), *Spirorbis ? flexuosus* Hall, S. *inornatus* Hall, *Stephanocrinus gemmiformis* Hall, *S. pentalobus* (Hall), *Stictotrypa orbipora* (Hall), *S. similis* (Hall), *Streptelasma radicans* Hall, *Stropheodonta profunda* Hall, *S. semifasciata* Hall, *S. striata* (Hall), *Strophostylus cyclostomus* Hall, *S. cyclostomus disjunctus* Hall, *Trematopora halli* Ulrich, *T. ? singularis* (Hall), T. *spiculata* Miller, *T. whitfieldi* Ulrich, *Trochoceras waldronense* Hall, *Uncinulus stricklandi* (Sowerby), *Vinella radiciformis* (Vine), *V. radiciformis conferta* Ulrich, *Whitfieldella nitida* (Hall), *Zaphrentis celator* Hall, *Zygospira (?) minima* Hall.

Louisville.—The name Louisville limestone was introduced by Foerste in 1897 (1897, 218) from the fine exposures of this rock east of the city of Louisville, Kentucky. The same name had been used by Hall (1879a, 139-147) for a Devonian formation of New York, but was dropped, so that Foerste's name has been generally adopted (see Foerste, 1898, 1900, 1901, 1903a, 1904b, 1906, 1909b, 1919, etc.; Pate and Bassler 1908; Bassler 1906a, 1915, etc.; Kindle 1901, 1904; Kindle and Barnett 1909; Newsom 1903; Hopkins 1904; Price 1900). Newsom (1903, 235) gives "Louisville limestone and Utica lime-rock." In 1900 J. A. Price, in a paper on the Waldron horizon (1900, 84, 85) proposed the name "Hartsville bed" for a layer of limestone overlying the Waldron shale, which he says is the "stratigraphical equivalent of the Louisville limestone." It was a quite unnecessary name, and has not come into general use. (See Kindle 1901, 537; Hopkins 1904, 39.) Hopkins, however, *(loc. cit.)* makes the Hartsville bed a subdivision of the Jeffersonville limestone.

In western Tennessee, according to Pate and Bassler (1908), there intervene between the Waldron and Louisville (Lobelville of Tennessee) 270 feet of shale and limestone comprised in the Lego, Dixon, Beech River and Bob formations (in ascending order). This sequence is totally unrepresented in Indiana and Kentucky where the Louisville rests directly, and obviously disconformably, on the Waldron. The Louisville formation is 63 feet and five inches thick at the quarries in the eastern part of Louisville (Butts, 1915), and varies from 42 to 100 feet in Jefferson County, Kentucky. In Indiana it is, according to Foerste (1904b, 32) 55 feet at Louisville, 57 at Charlestown Landing, 30-35 at Hanover, 25 at Paris Crossing and 10 at Vernon. East of Greensburg it appears to be absent and it cannot be recognized as a distinct formation north of Rush County. It is disconformably overlain by the Devonian limestone (see below).

At its type locality the formation is, according to Butts (1915, 88-89) a gray, finegrained, thick-bedded, low magnesian limestone. The different layers vary somewhat in composition. Magnesian carbonate runs from 2.16% to 29.76%; Calcium carbonate from 48.85% to 91.8%; Silica 3.40% to 26.56%; Alumina 3% and less. The extreme percentages of Magnesia and Silica and low percentage of lime, are not representative. The Silica in the limestone is largely in the form of chert. In Indiana the Louisville is a light gray to yellowish or buff, massive, magnesian limestone, practically barren of fossils north of Jefferson County, Indiana. In the Louisville region it is very fossiliferous especially abounding in corals.

There is some difference of opinion about the correlation of the Louisville. Foerste (1904b, 34, 35) says "The evidence of the presence of the Louisville and overlying beds [in Northern Indiana], however, is conclusive." The highest horizons exposed in Howard, Carroll, Cass, Miami and Hamilton counties are "characterized by the presence of various species of *Conchidium*, in addition to numerous other species frequently found in the upper part of the Louisville bed at Louisville, or at still higher horizons in Tennessee." These *Conchidium* beds apparently fall in Kindle's *Lower* or Noblesville division (Kindle 1904, 406-408), above which he places the Huntington limestone, correlated with the Guelph of Ontario. Foerste *(loc. cit.)* seems to have regarded the *Conchidium* beds as higher than the *Pentamerus* (Huntington) beds. Butts (1915, 99) cites Foerste and Kindle as above, and says

"Foerste expresses the opinion that the limestone which he correlates with the Louisville lies at a slightly lower horizon than the limestone at Huntington, Indiana, in which Kindle reports several species of the Guelph fauna. If such a relation of the Louisville and Guelph exists the equivalence of the Louisville and part of the upper half of the Lockport dolomite of New York is indicated, for wedges of Guelph dolomite carrying its peculiar fauna are included in the upper half of the Lockport dolomite. Moreover, the Louisville fauna partakes largely of Rochester (normal) forms, and many of these forms pass into the Lockport, where they occur in portions of the Lockport intercalated between the wedges carrying the Guelph fauna. Furthermore, in as much as neither the Lockport nor Louisville contains any considerable representation of a Cayugan fauna, but almost exclusively a Niagara one there seems no reason for assigning the Louisville a younger age than upper Lockport. . . . It would follow that the Laurel, Waldron and Louisville are about equivalent to the whole of the Lockport."

Nevertheless, Ulrich (1911, 560-561) places the Louisville *above* the Guelph, which he regards as of identical age with the Lockport; and this correlation is repeated by Bassler (1915).

The Louisville is believed by the writer to be of upper Lockport age; while the Guelph faunas which began to invade in Lockport time continued to live on in a sea of gradually increasing salinity, long after Lockport time and conditions had passed. (See M. Y. Williams, 1919, 75-76.)

Fauna of the Louisville.- The species listed below represent the Louisville fauna of Kentucky and Indiana (after Bassler, 1915). Not all of them have as yet been identified in Indiana, where less collecting has been done than across the Ohio. Since most of them are from near Louisville, they probably occur in the Clark county localities: *Alveolites fibrosus* Davis, *A. louisvillensis* Davis, *A. thoroldensis* Parks, *A. undosus* Miller, *Amplexus shumardi* (Edwards and Haime), *Anastrophia internascens* Hall, *A. interplicata* Hall, *Anisocrinus greeni* (Miller and Gurley), *Anisophyllum ? bilamellatum* Hall, *A. trifurcatum* Hall, *A. unilargum* Hall, *Astraeospongia meniscus* (Roemer), *Atrypa calvini* Nettleroth, *A. marginalis* Dalman, *A. nodostriata* Hall, *A. reticularis niagarensis* Nettleroth, *A. rugosa* Hall, *Aulopora precius* Hall, *A. pyyamaea* Davis, *A. vanclevi* Hall, *Blothrophyllum cinctosum* Greene, *B. niagarense* Davis, *Bumastus ioxus* Hall, *Calceola (Rhizophyllum) attenuatus* (Lyon), *C. corniculum* (Lyon), *Camarotorchia acinus* (Hall), *C. indianensis* (Hall), *C. pisa* (Hall and Whitfield), *Caryocrinites kentuckiensis* (Miller and Gurley), *Chonophyllum ? capax* Hall, *C. vadum* Hall, *Cladopora aculeata* Davis, *C. complanata* Davis, *C. equisetalis* Davis, *C. laqueata* Rom., *C. menis* Davis, *C. ordinata* Davis, *C. proboscidalis* Davis, *C. reticulata* Davis, *C. striata* Davis, *Clathrodictyon drummondense* Parks, *C. rectum* Parks, *C. vesiculosum* Nicholson and Murie, *Clorinda ventricosa* Hall, *Coelocaulus petila* (H. & W.), *Cornites crassus* (Rominger), *C. laminatus* Hall, *C. verticillatus* (Winchell and Marcy), *Conchidium crassiplicata* Hall & Clarke, *C. exponens* H. & C., *C. knappi* (Hall and Whitfield), *C. littoni* (Hall), *C. nettlerothi* H. & C., *C. nysius* (H. & W.), *C. tenuicostatum* (H. & W.), *C. unguiforme* (Ulrich), *Corynotrypa dissimilis* (Vine), *Cyathophyllum flos* Davis, *C. intertrium* Hall, *C. radiatum* Rominger, *Cyathospongia excrescens* Hall, *Cyclonema cancellata* (Hall), *Cyrtia exporrecta* (Wahlenberg), *Cyrtia myrtia* (Billings), *Cystiphyllum gemmulum* Greene, *C. granilineatum* Hall, *C. incurvum* Davis, *C. lineatum* Davis, *C. louisvillense* Greene, *C. niagarense* Hall, *Dalmanella elegantula* (Dalman), *Desmograptus pergracilis* (H. & W.), *Diaphorostoma niagarense* (Hall), *Dictyostroma undulatum* Nicholson, *Dimerocrinus halli* (Lyon), *Diorychopora tenuis* Davis, *Dipyphyllum billingsi* Greene, *D. huronicum* Rominger, *Discoceras marshi* (Hall), *Eridophyllum cruciforme* Davis, *E. dividuum* Davis, *E. louisvillense* Greene, *E. rugosum* E. & H., *E. sentum* Davis, *Favosites cristatus* E. & H., *F. cristatus major* Davis, *F. discus* Davis, *F. favosus* Goldfuss, *F. forbesi* E. & H., *F. hisingeri* E. & H., *F. louisvillensis* Davis, *F. niagarensis* Hall, *F. spinigerus* Hall, *Gypidula globulosa* (Nettleroth), *G. knotti* (Nettleroth), *G. (Sieberella) nucleus* (H. & W.), *G. (Sieberella) uniplicata* (Nettleroth), *Hallia divisa* Hall, *H. scitula* Hall, *Hallopora elegantula* (Hall), *Halysites catenularia* Linnaeus, *H. labyrinthicus* Goldfuss, *Heliolites elegans* Hall, *H. interstinctus* (Linnaeus), *H. megastoma* McCoy, *H. pyriformis* Guettard, *H. subtubulatus* McCoy, *Heliophyllum denticulatum* Hall, *H. flos* Greene,

H. gemmiformum Hall, *H. mitellum* Hall, *H. parvum* Hall, *H. pulcatum* Hall, *Illaenus cornigerus* H. & W., *Leptaena rhomboidalis* (Wilckens), *Lindströmia ? herzeri* (Hall), *Lophiostroma spindicandum* (Parks), *Lophospira casii* (M. & W.), *Lyellia americana* E. & H., *L. discoidea* Davis, *L. glabra* Owen, *L. papillata* Rominger, *L. parvituba* Rominger, *L. puella* Davis, *Macrostylocrinus meeki* (Lyon), *Melocrinus oblongus* Wachsmuth & Springer, *Meristina maria* (Hall), *Michelinia louisvillensis* Greene, *M. niagarensis* Davis, *M. prima* Davis, *Nucleospira elegans ?* Hall, *N. pisiformis* Hall, *Omphyma verrucosa* Rafinesque and Clifford, *Orthis flabellites* Foerste, *O. nettlerothi* Foerste, *O. (?) rugiplicata* H. & W., *O. subnodosa* Hall, *Orthostrophia (Schizoramma) nisis* (H. & W.), *Pachydictya crassa* (Hall), *Pentamerus cylindricus* H. & W., *P. oblongus* Sowerby, *P. pergibbosus* H. & W., *Plasmopora follis* E. & H., *Platyceras unguiforme* Hall, *Pleurotomaria casii* Meek & Worthen, *Poleumita rugaelineata* (H. & W.), *Ptychophyllum benedicti* Greene, *P. fulcratum* Hall, *P. invaginatum* Davis, *P. ipomaea* Davis, *P. stokesi* E. & H., *Reticularia dubia* (Nettleroth), *Rhipidomella hybrida* (Sowerby), *Rhynchonella (?) bellaforma* Nettleroth, *R. rugicosta* Nettleroth, *Rhynchospira (?) helena* Nettleroth, *Rhynchotreta cuneata americana* (Hall), *Romingeria uva* Davis, *R. vannula* Davis, *Schuchertella subplana* (Conrad), *S. tenuis* (Hall), *Spirifer crispus simplex* Hall, *S. eudora* (Hall), *S. foggi* (Hall), *S. radiatus* (Sowerby), *S. rostellum* (H. & W.), *Streptelasma conulus* Rominger, *S. patula* Rominger, *S. spongaxis* Rominger, *Striatopora huronensis* Rominger, *Stricklandinia (?) louisvillensis* Nettleroth, *Strombodes incertus* Davis, *S. mammillaris* (Owen), *S. pentagonus* Goldfuss, *S. pygmaeus* Rominger, *S. quadrangularis* Davis, *S. separatus* Ulrich, *S. sinemurus* Davis, *S. striatus* (D'Orbigny), *S. unicus* Davis, *Stropheodonta profunda* (Hall), *Strophonella costatula* H. & C., *S. striata* (Hall), *Strophostylus cancellatus* (Hall), *Syringopora (Drymopora) fascicularis* (Davis), *S. fibrata* Rominger, *Thecia major* Rominger, *T. minor* Rominger, *Trochonema fatuum* Hall, *Troostocrinus reinwardti* (Troost), *Uncinulus stricklandi* (Sowerby), *U. tennesseensis* (Roemer), *Whitfieldella nitida* (Hall), *Wilsonia saffordi* (Hall), *W. saffordi depressa* (Nettleroth), *W. wilsoni* (Sowerby), *Zaphrentis obliqua* Davis, *Z. scutella* Davis, *Z. socialis* Davis, *Z. subvesicularis* Hall, *Z. umbonata* Rominger.

Noblesville and Huntington.—The Niagaran formations of southern Indiana thin to the northward and run under the drift in Rush County, so that there is no visible physical connection between them and the great mass of Niagaran dolomites and limestones of northern indiana, and it is possible that they were not deposited over this intervening area. The extent and thickness of these latter deposits are known largely from subsurface data supplied by the numerous oil and gas well records of the region, since outside of the Wabash and White River valleys, outcrops are very few in northern Indiana. The Niagara age of the rocks was early recognized (R. Owen, 1862); but no detailed work of importance was done prior to that of Dr. E. M. Kindle (1904, 397-486, pls. I-XXV) although frequent mention of the outcrops occurs in the Annual Reports. Kindle's report is, therefore, the only important source of information about these rocks and their fauna. The literature of the curious Niagara "domes" or quaquaversal structures of the Wabash valley has already been cited under the general discussion of the Niagaran. Foerste (1904b, 33-36) gives a brief summary account of the Niagaran of the northern Indiana area and recognizes two divisions (p. 35) an "Upper horizon in which *Conchidium* is common and a lower horizon, in which *Pentamerus oblongus* is common." He also states (p. 34) that "none of the lists of fossils so far published from any part of northern Indiana give any evidence of the presence of the Clinton, Osgood, Laurel or Waldron beds." On the other hand he says "the evidence of the presence of the Louisville and overlying beds . . . is conclusive". (See above.)

Kindle (*loc. cit.*, 406-408) also divides the Niagaran of northern Indiana into two divisions which he names the *Noblesville* dolomite (below) and the *Huntington* limestone (above). The earlier beds are best exposed at Connor's Mill in Hamilton County, an outcrop mentioned by R. Owen in 1862 (1862, 102). The rock is, according to Kindle, a "hard, thin-bedded buff-colored dolomite lying in strata three to ten inches thick, of which there are 25 feet exposed at the type locality. . . . It is probable . . . that this formation includes not less than 100 feet of strata. Faunally the beds are characterized by such well-known Niagara (Lockport limestone) fossils as *Spirifer nobilis*, *S. radiatus*, *S. crispa var. simplex*, *Conchidium multicostatum*, *Sphaerexochus romingeri*, etc. No trace of the Guelph* fauna appears in it."

"In the quarries at Huntington is found a fauna of later age, and very different from that of the Noblesville dolomite. . . . The [characteristic] cephalopod and gastropod element of the fauna appears to be of distinctly Guelphic affinities". The Huntington limestone is a "light gray, or cream colored, granular dolomitic limestone of saccharoidal texture". About 80 feet are exposed at Huntington, and the formation is probably from 150 to 200 feet thick. It is the same rock as that called Huntington stone by Cox (1879, 59-60), though the latter did not use the name as a geologic term.

Phinney had previously (1884, 142-148) correlated the upper part of the northern Indiana Niagaran with the lower part of the Guelph of Canada and the Cedarville beds (Orton, 1871, 271-277) of Ohio; and the lower strata (of Grant and Delaware counties) with the Springfield (Orton 1871, 271-274) beds of Ohio. S. A. Miller (1892, 612) could see no evidence of the Guelph in Indiana. Kindle and Barnett (1909, 396) conclude that the Guelph is absent in *southern* Indiana.

Fauna of the Niagaran of northern Indiana.- Bassler (1915, 1502-1503) gives the following list of species of the Niagaran of northern Indiana: *Amphicoelia neglecta* (McChesney), *Anastrophia internascens* Hall, *Anodontopsis wabashensis* Kindle & Breger, *Ascoceras indianensis* Newell, *A. newberryi* Billings, *Atrypa calvini* Nettleroth, *A. reticularis* (Linnaeus), *Bumastus armatus* (Hall), *B. insignis* (Hall), *B. ioxus* Hall, *Calymene vogdesi* Foerste, *Camarotoechia ? acinus* (Hall), *Ceratocephala goniata* Warder, *Cheirurus niagarensis* (Hall), *Conchidium laqueatum* (Conrad), *C. littoni* (Hall), *C. multicostatum* (Hall), *C. trilobatum* Kindle and Breger, *C. unguiformis* (Ulrich), *C. multistriatum* K. & B., *C. oweni* K. & B., *Cyclonema cancellatum* (Hall), *Cyrtia exporrecta myrtia* (Billings), *Dalmanella elegantula* (Dalman), *Dalmanites vigilans* Hall, *Dawsonoceras annulatum* (Sowerby), *Diaphorostoma niagarense* (Hall), *Dinobolus conradi* Hall, *Discoceras marshii* (Hall), *Eatonia goodlandensis* K. & B., *Encrinurus indianensis* K. & B., *Eotomaria laphami* (Whitfield), *Euomphalopterus alatus americanus* K. & B., *E. alatus limatoideus* K. & B., *Gomphoceras lineare* Newell, *G. projectum* Newell, *G. scrinium* Hall, *G. wabashense* Newell, *Gypidula (Sieberella) paleata* (Dalman), *G. (S.) nucleus* Hall and Whitfield, *Hexameroceras cacabiforme* Newell, *H. delphicolum* Newell, *Kionoceras cancellatum* (Hall), *K. delphiense* (K. & B.), *K. kentlandense* (K. & B.), *Leptaena rhomboidalis* (Wilckens), *Meristina maria* (Hall), *M. rectirostris* Hall, *Nucleospira pisiformis* Hall, *Odontopleura ortoni* (Foerste), *Orthis flabellites* Foerste, *Orthis (?) subnodosa* Hall, *Orthoceras niagarense* Hall, *O. abstructum* Newell, *O. rigidum* Hall, *O. unionense* Worthen, *Pentameroceras mirum* (Barrande), *Pentamerus oblongus compressa* K. & B., *P. oblongus cylindricus* K. & B., *Phragmoceras angustum* (Newell), *P. ellipticum* Hall & Whitfield, *P. parvum* H. & W., *Pholidostrophia niagarensis* K. & B., *Pisocrinus benedicti* Miller, *Plectoceras bickmoreanum* (Whitfield), *Plethomytelus cuneatus* K. & B., *Pleurotomaria (?) axion* Hall, *P. elevoidea* K. & B., *P. hoyi* Hall, *P. ? idia* Hall, *Poleumita huntingtonensis* (K. & B.), *P. huntingtonensis alternata* (K. & B.), *P. plana* (K. & B.), *Protokionoceras medullare* (Hall), *Protophragmoceras hercules carrollense* (K. & B.), *Reticularia proxima* K. & B., *Rhipidomella circulus* Hall, *R. hybrida* (Sowerby), *Rhynchonella coletti* Miller, *Rhynchorthoceras dubium* Hyatt, *Schizolopha (?) prosseri* K. & B., *Schuchertella subplana* (Conrad), *Sphaerexochus romingeri* Hall, *Spirifer crispus* Hisinger, *S. crispus simplex* Hall, *S. foggi* (Nettleroth), *S. nobilis* (Barrande), *S. radiatus* (Sowerby), *Streptomytilus wabashensis* K. & B., *Strophonella striata* Hall, *S. williamsi* K. & B., *Strophostylus cancellatus* (Hall), *S. elevatus* Hall, *Trimeroceras gilberti* K. & B., *Trochoceras desplainense* McChesney, *Whitella (?) siluriana* K. & B., *Whitfieldella nitida* Hall, *Wilsonia saffordi* (Hall).

* The name Guelph was proposed by Sir William Logan in 1863 (Can. Geol. Surv. 1843-63, pp. 298-344).

CAYUGAN.—(Monroan). The upper division of the Silurian was named Cayugan by Clarke and Schuchert (1899, 876-877), and this name has met with general favor. It includes the Salina or Onondaga Salt Group and Waterlime beds of the older New York reports; and comprises the so-called Waterlime, Lower Helderberg, etc., of the Indiana Reports. The name "Cayuga dolomite" was also used by Chapman (1864, 190) in his "Popular and practical exposition of the minerals and geology of Canada." The Cayugan rocks of Indiana have been named the Kokomo limestone by Foerste (1904, 33) from the excellent exposures of this so-called Waterlime formation or Eurypterid bed at Kokomo in Howard County.

Kokomo.—The use of the term "Lower Helderberg" in Indiana geology has been unfortunate on account of the changing connotations of that term, and has resulted in more or less confusion. For example, Blatchley (1898a, 19) says the waterlime *and* Lower Helderberg "are closely related limestones, whose known thickness in the State varies from 15 to 150 feet and 25 to 250 feet, respectively. The Lower Helderberg . . . outcrops along the Wabash River near Logansport and farther northwest near Monon and Rensselaer. The Waterlime is chiefly represented in northern Indiana, outcropping near Kokomo, where it is extensively quarried for building material, and also near Logansport. Farther north it so merges into the Lower Helderberg that the two are difficult to distinguish." Again (1904, 42-43) he says in Indiana the Waterlime and Lower Helderberg "so merge as to be difficult to distinguish." The latter "represents the final epoch of the Upper Silurian time. . . . It directly overlies the Niagara limestone where the Waterlime is absent. Outcrops occur at Logansport," etc.. On the other hand Elrod (1902, 209) says "If the Waterlime is excluded from the Lower Helderberg, it is probable no true representative of that period is to be found in Indiana."

The first recognition of the so-called "Lower Helderberg" strata in rocks west of the New York area was by Hall (1843, 512) who referred certain rocks in the vicinity of Macanac (Mich.) to the Onondaga Salt Group or Waterlime. Chapman (1864, 190) recognized rocks of this age in the Lake Erie region and gave to them the name of "Bertie dolomite." The name Lower Helderberg appears to have been introduced into northern Indiana geology by Orton (1889, 563), who stated that "In Indiana it [Lower Helderberg] has not heretofore been recognized as distinct from the Niagara limestone, but, though it lacks here the importance that it has in Ohio, it certainly exists in considerable force and covers an extended but at present undefined belt of country." He explains that the rock in question is the formation called Waterlime [part of the present Monroan] in the Put-in-Bay region of Ohio. Further on (p. 566) he says "It enters Indiana from Ohio in Adams and Allen counties, with a probable breadth of outcrop of eighteen miles. . . . At Decatur it is worked in several quite large quarries . . . the junction of the Niagara and Lower Helderberg limestones occurs at this point, the upper quarries of the town being in Niagara." He mentions, however, the occurrence of the Niagara fossils *Schuchertella subplana* and *Halysites labyrinthica* in his "Lower Helderberg." Kindle (1904, 422-423) refers these rocks to the Niagaran. Orton furthermore (p. 567) refers to the Lower Helderberg part of the rocks at Huntington, the "well known Wabash flagging"; 66 feet of rock at Logansport; and traces the formation as far west as Delphi. The Waterlime beds at Kokomo (Kokomo limestone) are referred to as "thoroughly and unmistakably" characteristic of the formation. Substantially the same correlations are repeated by Phinney (1891, 633-634), who gives a geological map of Indiana (pl. LXIII) on which the Lower Helderberg is represented as a narrow belt extending westward through Adams, Wells, Huntington, Wabash, Miami and Cass counties and southward in Miami through Howard and Tipton into Hamilton. Another small area is mapped in Pulaski, and a larger area in Lake, Newton and Jasper. The greater part of the Silurian north and west of the Adams-Cass-Hamilton "Lower Helderberg" is mapped as "Upper Helderberg" (Corniferous and Hamilton). Grabau and Sherzer (1910, 21-22) state, following Orton and Phinney that "From Ohio all the main divisions of the Monroe [Lane, 1895, 66] may be traced into Indiana but only limited exposures of the strata occur." But they also note (p. 22) that "In the geological map of the state compiled 1901-1903, the Waterlime and the Niagara are represented by a single color and the 'Kokomo limestone' is referred by Foerste to the Bertie or Lower Waterlime." They make no mention of Foerste's further observation (1904b, 33) that "the Cayugan of New York corresponds in a general way with the rocks formerly known as the Lower Helderberg . . . [and is] represented by the Waterlime at Kokomo. In Indiana, the Waterlime division . . . has been definitely identified at only one locality. While it may be present also elsewhere along the flanks of the Indiana extension of the Cincinnati geanticline, it has been identified by means of its fossils only at Kokomo." Kindle (1904, 413) also, though he makes no mention of the "Lower Helderberg" implies its general absence in his statement that "The evidence at hand points to a general elevation of the sea bottom at the close of the Niagara in the area around the northern end of the Cincinnati geanticline." This area as he elsewhere points out (*Ibid.* 411-413) was resubmerged in middle Devonian time. A brief resubmergence in Bertie time must be assumed, however, to account for the Kokomo limestone (*q. v.*). Schuchert (1910, plate 70) has represented this Bertie sea, but evidently took his outcrop for Indiana from the Phinney map. Lane, Prosser, Sherzer and Grabau (1907, 551) state that the Salina embayment extended from Maryland over Ohio, Michigan, Wisconsin and probably into Indiana. Foerste (1909, 6-7) was also aware of these necessary relationships, for he says "It is impossible, at present, to determine what are the relationships between the Greenfield [Lower Monroe] limestone of Ohio and the Salina horizons at Kokomo. Both areas undoubtedly were connected during early [late] Cayugan times with lower Michigan, northern Ohio and western New York." Foerste regards the brachiopod horizon overlying the Eurypterid beds at Kokomo as a distinct formation, but also of Salina age. He does not suggest a name for it. Kindle, in 1913 (1913, 282-288) published an important paper on the age of the Kokomo limestone in which he disagrees with Clarke and Ruedeman (1912, 320, 351, 215) who regard the Eurypterid fauna as of Lockport age, and concludes that the "Kokomo Eurypterid fauna represents a horizon of Salina age which is as yet faunally unknown in New York." He cites the following papers on the Kokomo limestone: Claypole, 1890, 259-260; Phinney, 1891, 632-633; S. A. Miller, 1894, 312; Hall and Clarke, 1894, pl. 66; David White 1901, 269-270; Kindle 1904, 445; Foerste 1909, 1-39; Clarke and Ruedemann 1912, 1-628.

Terminology has been still further complicated by the proposal of the name *Monroan* by Grabau (1909, 351-356) for the Monroe group of Michigan (Lane, 1893) and Ohio, and the suggestion that this term he used in place of Cayugan, for the upper part of the Silurian, exclusive of the "non-marine" Salina (Shawangunk and Longwood), but including the water-limes of Ohio, or "marine" Salina. He held that the Cayugan of New York exclusive of the Salina included only the uppermost portions of the "Monroan". As thus defined the Monroan included, in the ascending órder, the following formations of Northwestern Ohio and Southeastern Michigan:—Greenfield dolomite [possible equivalent of the Kokomo limestone] Tymochtee shales, Put-in-bay dolomite, Raisin River dolomite, Sylvania sandstone, Flat Rock dolomite, Anderdon limestone, Amherstburg dolomite, Lucas dolomite. These formations are therefore so grouped by Grabau and Scherzer (1910), Schuchert (1910), Scherzer (1911), Ulrich (doubtfully) 1911, and Grabau (1913). Stauffer (1915) has raised a doubt about the Silurian age of the upper half of this group, and M. Y. Williams (1919) has definitely placed the Sylvania-to-Lucas inclusive in the Devonian. Bassler (1915) places all except the Greenfield dolomite in the Devonian and correlates the latter about with the Manlius of New York, or somewhat above the Kokomo which he correlates with the Bertie of Ontario. He thus excludes the Kokomo completely from the "Monroan" and apparently does not regard any part of the latter as present in Indiana. Chadwick (1918, 173), who regards the Put-in-Bay as above the Raisin River dolomite, suggests that the Bertie either be made to include, as originally, the entire series of beds between the Camillus and Akron, inclusive of the latter, or else be restricted to the Buffalo cement bed; and, as quoted by M. Y. Williams (1919, 93) correlates the Greenfield dolomite with the lower part of the Bertie. Williams, however, inclines to the opinion that the Bertie may be correlated with the Put-in-Bay dolomite and perhaps in part with the Tymochtee and Greenfield, if the Put-in-Bay be regarded as below the Raisin River. Dr. O'Connell (1916, 118-120) discusses the Kokomo Eurypterid beds, and gives a section of the Defenbaugh quarry at Kokomo. She argues that the Kokomo sediments came from a land to the west and were river-borne, and gives a paleogeographic map (after Grabau) showing the Bertie-Kokomo sea which entered Indiana from the north. In his recently published Textbook of Historical Geology, Grabau (1921) retains the name "Monroan" series; and evidently includes in the upper Monroan the same formations as in his 1909 paper. He regards the Devonian-like corals as an invasion from the Arctic realm.

Fauna of the Kokomo.—The Kokomo limestone has a small but distinctive fauna in which certain Eurypterids hold a prominent place. It is as follows: *Amplexus septatus* Foerste, *Buthotrephis divaricata* David White, B. *newlini* D. White, *Chonetes colliculus* Foerste, *Coelospira congregata* Kindle and Breger, *Conchidium coletti* Miller, *Dalmanella elegantula* (Dalman), *Eurypterus (Onychopterus) kokomoensis* Miller and Gurley, *Eurypterus ranilarva* Clarke and Ruedemann, *Eusarcus newlini* (Claypole), *Favosites pyriformis kokomoensis* Foerste, *Isochilina musculosa* Foerste, *Kloedenia kokomoensis* Foerste, *Leptaena rhomboidalis* Wilckens, *Pentamerus divergens* Foerste, *Spirifer corallinensis* Grabau, *Spirifer exiguus* Foerste, *Stylonurus (Drepanopterus) longicaudatus* Clarke & Ruedemann, *Whitfieldella erecta* Foerste, *Wilsonia kokomoensis* Foerste. The Eurypterids and Brachiopods are said by Dr. O'Connell (1916, 119) not to occur in association.

Department of Conservation
Division of Geology
State of Indiana
W. N. Logan State Geologist

Fig. 21.

CHAPTER III.

DEVONIAN.

The name Devonian was proposed by Sedgwick and Murchison in 1839 in a paper "On the classification of the older rocks of Devonshire and Cornwall" (1839, 121-123) and includes also the rocks commonly known as the "Old Red Sandstone" in England and Scotland. In the final reports of the first Geological Survey of New York, the rocks now called Devonian were included in the middle and upper portions of the Helderberg division, and in the Erie division of the New York System (Catskill division of Mather 1840).

As already explained, the Onondaga salt group and Waterlime group of the Helderberg are now placed in the Silurian. The balance of the Helderberg, i. e., the Pentamerus (Coeymans), Delthyris shaly (New Scotland), Upper Pentamerus (Becraft, Kingston), Oriskany, Caudagalli grit (Esopus), Schoharie grit, Onondaga and Corniferous (Onondaga) belong to the lower and middle Devonian. The names Coeymans, New Scotland, Becraft and Esopus were proposed by Clarke and Schuchert (1899, 876-877).

The Erie (or Catskill) division included the Marcellus, Hamilton, Tully, Genesee, Portage or Nunda and Chemung. The Catskill ("Old Red Sandstone") was sometimes placed as a distinct formation at the top of this divisioh; or else in a separate division. It was later recognized that the Catskill is in the main a non-marine phase of the Chemung. These constitute the upper Devonian.

In 1899-1900 Clarke and Schuchert arranged the Devonian formations as follows:

Chautauquan	Chemung beds—Catskill sandstone facies	
Senecan	Portage	Naples
		Ithaca
		Oneonta
	Genesee shale	
	Tully limestone	
Erian	Hamilton beds	
	Marcellus shale	
Ulsterian	Onondaga limestone	
	Schoharie grit	
	Esopus grit	
Oriskanian	Oriskany beds	
Helderbergian	Kingston beds	
	Becraft limestone	
	New Scotland beds	
	Coeymans limestone	

The Kingston (Port Ewan) bed was subsequently transferred to the Oriskanian division by Chadwick (1908), and the Esopus by Hartnagel (1912). Schuchert (1910, 541; 1915) omits the Ulsterian division, and places the

Esopus in the Oriskanian,* and the Schoharie and Onondaga in the Erian division, though Dunbar, (1919, 27), evidently with Schuchert's approval, uses Ulsterian, in which he places the rocks of Onondaga age. Schuchert's usage is adopted in the present report.

The upper coral reef of the Falls of the Ohio and the overlying black shale were recognized as of Devonian age by David Dale Owen in the 1859 reprint of the 1837 Reconnaisance of Indiana *(op. cit.* p. 2*)*, and have ever since been referred to this system.

The principal source of information on the Devonian rocks and fossils of Indiana is the well known papers by Kindle (1899, 1901).

Erian. As noted above, the Schoharie and Onondaga formations are, following Schuchert, placed in the Erian division. The Schoharie (Hall 1843b, 18, 151) is represented by a single outcrop of sandstone at Pendleton, Indiana; but the Onondaga (Hall, 1839, 290-293, Corniferous of Indiana authors) and Hamilton (Vanuxem, 1840, 380) are present over extensive areas in both the northern and southern parts of the State.

With the exception of the *Pendleton* sandstone and *Geneva* limestone, no consistent use of local geographic designations for these formations occurs in the Indiana Reports prior to the papers of Kindle in 1899, 1901, who proposed the name *Jeffersonville* limestone (1899, 8) for the so-called Corniferous or Onondaga rocks, and *Sellersburg* beds (1899, 8, 110) for the overlying Hamilton or Cement beds and Encrinital limestone of authors. In the year 1901 Siebenthal (1901, 341) proposed to call the cement beds of Clark County, included in Kindle's Sellersburg, the *Silver Creek* hydraulic limestone; while very recently Butts (1915, 120) has proposed the name *Beechwood* for the "Encrinital" bed at the top of the Sellersburg. The Geneva limestone of Collett (1882, 63, 78, 81, 82) may represent a facies of some part of the Jeffersonville; while the Shelby bed of Foerste (1898, 234-235) is according to Kindle (1901, 536) the same as the Geneva. The name "North Vernon limestone" or "North Vernon bluestone" though in use as a trade name seems to have been definitely proposed by Borden (1876, 148, 160) for the limestone immediately below the Black Shale in the North Vernon region, supposed to be of the age of the Cement beds in Clark County.

Pendleton.—The name "Pendleton sandstone" was evidently a local and trade name before it was adopted as a geologic designation. Cox, for example, (1869, 7) says "I have been informed that several barrels of the Pendleton sandstone were subsequently shipped to Pittsburg." He again uses the name in 1879 (1879, 60-62).** The name occurs frequently in the Indiana reports (Brown, 1884a, 26, 34; Thompson 1892, 39; Siebenthal 1901, 348; Kindle 1901, 559-561, etc.). It should probably not be credited to any particular geologist since it came into use merely as a trade name. The first adequate description and list of fossils is given by Kindle (1901, 560-561) who approves Hall's correlation (1879c, 60) of the rock with the Schoharie grit of New York. Siebenthal (1901, 348) mistakenly correlates a sandstone immediately above the Waldron shale in Jennings County and a similar sandstone on Graham Creek, with the Pendleton. Hopkins (1904, 39-40) somewhat hesitatingly correlates the Pendleton with the Schoharie.

* The name Oriskany was first used by Hall in 1839; 3d Annual Report of the fourth district, pp. 308, 309.

** Weeks (1902, 317) gives this as the first use of the name. See also Kindle (1901, 560).

Kindle (1901, 560) gives the following section at the quarry near the falls just below the Big Four railroad bridge at Pendleton:—

	Ft.	In.
"1. Hard gray limestone	3	6
2. Massive white sandstone with 10 to 12 inch strata	6	8
3. Bluish drab calcareous fine grained sandstone		10"

He also regards a conglomerate seen at a number of places in Pendleton as a "bed of local development in the upper part of the Pendleton sandstone." This conglomerate consists of rounded pebbles of chert in a matrix of coarse sand. Kindle points out that Cox's section (1879, 60-62) is inverted from the true order of succession of the strata. No. 1 of Kindle's section outcrops "at a few points along the creek between Pendleton and Huntsville and is frequently a sandy dirty buff rock containing casts of a large Gastropod" *(Pleurotomaria)*.

From No. 1 of the above section Kindle reports the following fauna:— *Reticularia fimbriata, Eunella sp., Martinia subumbona, Pleurotomaria sp.* The fauna of No. 2 is as follows: *Reticularia fimbriata* var., *Martinia subumbona, Eunella sp., Pentamerella arata, Atrypa reticularis, Tentaculites dexitheca, Bellerophon curvilineatus, Callonema bellatula, Schizodus contractus (?), Conocardium trigonale, Conocardium cuneus, Proetus curvimarginatus, P. latimarginatus, Cyrtoceras eugenium, Zaphrentis giganteum, Favosites limitaris.*

From bed No. 3 he lists the following: *Dalmanites verrucosus, Leptaena rhomboidalis, Cornulites proprius ?, Murchisonia sp., Schizotreta tenuilaminata ?, Euomphalus sp., Orthoceras sp., Streptelasma (?) sp.*

Jeffersonville.—(Upper Coralline limestone (in part), Coralline falls limestone, Devonian limestone (in part), Corniferous limestone, Upper Helderberg limestone, Geneva (?) limestone, Shelby bed (?)). The famous coral beds at the Falls of the Ohio river attracted attention in the early part of the 19th century and along with the other rocks exposed at the Falls are mentioned by Lapham (1828), Owen (1837, 1843, 1856, 1859, etc.), Clapp (1843), Hall (1842, 1848-49, etc.), Lyon (1857, 1859-60), Lyon and Casseday (1859), Yandell and Shumard (1847), etc. The name most commonly applied to the Devonian limestones of Indiana is "Corniferous limestone" (Eaton 1823, 1839), a name still used by the oil driller. (See Cox, 1878, 138; Borden 1875, 115, 128; Collet 1882, 69-71; Elrod 1882, 174, 186-195; Elrod 1883, 106, 112-113; Elrod 1884, 92; Thompson 1886, 78; Gorby 1889, 179; Scovell 1897, 513; Blatchley & Ashley 1898, 19; Blatchley 1904, 43-44; Ward 1908, 198; Culbertson 1916, 227-228). In this common usage the true Corniferous (Jeffersonville) is not always distinguished from the overlying Sellersburg limestone. No local geographic name was applied to the rocks at the Falls until 1899, when Kindle proposed the name *Jeffersonville* limestone for the "lowest formation of the Devonian as developed at the Falls of the Ohio lying between the Sellersburg beds and the Catenipora [Halysites] beds of the Niagara"; and the name *Sellersburg* limestone for the beds "from the New Albany [black] shale down to the lowest beds worked at the cement quarries." The fossil coral reef at the Falls occurs in the lower part of the Jeffersonville.

Farther north in Indiana various names had already been applied to portions of these formations. Borden (1876, 148, 160) used the name "*North*

Vernon Bluestone" or "North Vernon limestone" for beds which he correlated with the hydraulic limestone (Silver Creek) of Clark County; but the name as commonly used evidently included the Jeffersonville, or some part of it, also.

In Jennings, Decatur, Bartholomew and Shelby counties the "Corniferous" consists of a chocolate colored massive sandy dolomite nearly devoid of fossils, to which Collett (1882, 63-82 gave the name "*Geneva limestone*." This limestone or dolomite has been traced as a southwardly thinning stratum as far as Charlestown, Clark County, where it is three feet thick and lies between the Jeffersonville limestone and the Louisville limestone. In its typical development to the north, the Geneva is 20 to 30 feet thick and appears to entirely supplant the typical Jeffersonville and Sellersburg formations, which thin out to the northward. What the actual relationships of the Geneva and Jeffersonville-Sellersburg formations are, has not been satisfactorily determined. The Geneva may be a lithologic facies of one or both of these formations or, more likely, a distinct formation older than the Jeffersonville. Ulrich (1911, pl. 28) apparently correlates it with the Schoharie, though his reasons for the procedure are not stated. Dunbar (1919, 89) thinks that the Camden chert of Tennessee and Clear Creek chert of Illinois may be partially equivalent to the Schoharie and Esopus of New York; and it is probable, therefore, that the Schoharie sea may have had a much wider spread than has heretofore been suspected. The isolated remants at Pendleton may after all have relatives comparatively near at hand.

The fact that more than one of the New York Devonian formations is represented in the limestones in the vicinity of the Falls was not appreciated by most of the earlier geologists who referred the rocks to the Silurian (Clapp 1843, 177-178), Helderburg (Clapp 1843, Owen 1843), Corniferous (Onondaga) (Hall 1843, etc.), etc. Further details of correlation will be discussed under the Silver Creek and New Albany formations.

The Jeffersonville limestone is about 25 feet thick at its type locality and thins both to the north and south. It is a massive, crystalline, light colored and rather coarse-grained, cherty corralline limestone. Butts (1915, 104) gives the following section at the "Whirpool" one mile west of Jeffersonville, Indiana. (Number 5 is the top of the section.)

"5. Limestone, massive, rather coarse-grained, light gray, a little black chert locally. Whitish and shelly on weathering. Exfoliates diagonally to the bedding. Spirifer acuminatus about in top. Stropheodonta hemispherica, abundant and conspicuous, Bryozoa abundant includes bryozoan and Nucleocrinus zones of authors. Sp. acuminatus zone.......................... 9 ft.

4. Limestone, siliceous, cherty, bluish-gray, fine-grained, very hard. Spirifer gregarius, very abundant. Favosites hemisphericus, Turbo shumardi, &c. Spirifera gregarius zone.............. 2 ft.

3. Limestone, 6 inches to 1 foot layers, medium, coarse grained, light pinkish, bluish, or brownish gray. In places largely made up of Stromatopora .... 7 ft.

2. Limestone medium thick-bedded, very coarse crystalline, brownish in part. Crowded with corals. Coral layer................ 7 ft.
Certainly Jeffersonville ....25 ft.

1. Limestone, covered with water, not identified.................. 3 ft."

Fauna of the Jeffersonville limestone. Butts (1915, 106-115) gives the following list of fossils most of which are from the Louisville region of the Jeffersonville formation:- *Acrophyllum clarki* Davis, *A. ellipticum* Davis, *A. oneidaense* Bill., *Alveolites constans* Davis, *A. minimus* Davis, *A. mordax* Davis, *A. squamosus* Davis, *Aulacophyllum conigerum* Davis, *A. insigne* Davis, *A. mutabile* Davis, *A. parvum* Davis, *A. sulcatum* D'Orbigny, *A. unguloideum* Davis, *Aulopora cornuta* Billings, *A. culmula* Davis, *A. edithana*, *A. procumbens* Davis, *A. serpens* Goldfuss, *Blothrophyllum approximatum* Nicholson, B. *cinctum* Davis, *B. corium* Davis, *B. decorticatum* Billings, B. *liratum* Davis, *B. louisvillense* Davis, B. *parvulum* Davis, *B. sessile* Davis, *B. zaphrentiforme* Davis, *Chonophyllum magnificum* Billings, C. *multiplicatum* Davis, *Cladopora acupicata* Davis, C. *alpenensis* Rominger, C. *aspera* Rominger, C. *bifurca* Davis, C. *crassa* Davis, C. *cryptodens* Billings, C. *dentata* Davis, C. *desquamata* Davis, C. *dispansa* Davis, C. *expatiata* Rominger, C. *fibrata* D., C. *francisci* D., C. *gracilis* D., C. *imbricata* Rom., C. *iowaensis* Owen, C. *labiosa* Bill., C. *pinguis* Rom., C. *pulchra* Rom., C. *radula* D., C. *ricta* D., C. *rimosa* Rom., C. *robusta* Rom., C. *tela* D., *Cyathophyllum brevicorne* D., C. *corniculum* Leseur, C. *davidsoni* E. & H., C. *detextum* D., C. *fimbriatum* D., C. *flos* D., C. *greeni* D., C. *ligatum* D., C. *multigemmatum* D., C. *oedipus* D., C. *ovoideum* D., C. *pocillum* D., C. *pumilus* D., C. *robustum* Hall, C. *rugosum* Hall, C. *winchelli* D., C. *cicatriciferum* D., C. *edwinanum* D., C. *grande* Billings, C. *hispidum* D., C. *limbatum* D., C. *lineatum* D., C. *nettlerothi* D., C. *os* D., C. *plicatum* D., C. *squamosum* Nich., C. *sulcatum* Hall, C. *theissi* D., C. *tumidosum* D., C. *vesiculosum* Goldfuss, *Dendropora elegantula* Billings, *D. proboscidalis* Rominger, *Diphyllum coagulatum* Davis, *D. coalescens* Davis, *D. conjunctum* D., *D. gigas* Rom., *D. panicum* Davis, *D. strictum* E. & H., *D. verneuilianum* E. & H., *Drymopora (Syringopora) commensalis* D., *D. (S.) fascicularis*, *D. (S.) intermedia* Nich., *D. (S.) nobilis* Bill., *Eridophyllum arundinaceum* Bill., *E. simcoense* Bill., *Favosites amplissimus* D., F. *arbor* D., F. *baculus* D., F. *canadensis* Bill., *F. cariosus* D., *F. clausus* Rom., *F. clelandi* D., *F. convexus* D., *P. cymosus* D., *F. digitatus* Rom., F. *emmonsi* Rom., F. *epidermatus* Rom., F. *frutex* D., F. *fustiformis* D., F. *hemisphericus* and vars. Troost, F. *impeditus* D., F. *intertextus* Rom., *F. limataria* Rom., *F. mundus* D., *F. ocellatus* D., *F. pirum* D., F. *proximus* D., F. *quercus* D., *F radiatus* Rom., F. *radiciformis* Rom., *F. ramulosus* D., *F. spiculatus* D., F. *tuberosus*, *Hadrophyllum d'Orbignyi* E. & H., *Heliophyllum (Cyathophyllum) collegatum* Bill., *H. (C.) exiguum* Bill., *H. (C.) halli* E. & H., *H. (C.) infoveatum* D., *H. (C.) multirrena* D., *Michelinia corrugata* D., *Platyaxum (Cladopora, Pachypora) canadense* Rom., P. *corioideum* D., *P. fischeri* Bill., *P. foliatum* D., *P. turgidum* Bill., *P. undosum* D., *Procteria michelinoidea* D., *P.* papillosa D., *Ptychophyllum coniferum* D., *P. diaphragma* D., *P. tropeum* D., *P. typicum* D., *Romingeria incrustans* D., *R. umbellifera* Bill., *R. uva* D., *Striatopora alba* D., S. *linneana* Bill., *Syringopora bouchardi* Nich., S. *hisingeri* Bill., *S. perelegans* Bill., *S. straminea* D., S. *tabulata* E. & H., S. *tubiporoides* Yandell & Shumard, *Thecia ramosa* Rom., *Zaphrentis compressa* Edwards, *Z. (Clisiophyllum) conigera* Bill., *Z. exilis* D., *Z. gigantea* Leseur, *Z. greenana* D., *Z. immanis* D., *Z. linneyi* D., *Z. maconathi* D., *Z. nodulosa* Rom., *Z. prolifica* Bill., *Z. rafinesquei* E. & H., *Z. romingeri* D., *Z. torquata* D., *Z. trigemma* D., *Ancyrocrinus spinosus* Hall, *Codaster americanus* Shumard, C. *pyramidatus* Shumard, *Dolatocrinus lacus* Lyon, *D. marshi* Lyon, *Megistocrinus knappi* Lyon & Casseday, *M. spinulosus* Lyon, *Nucleocrinus angularis* Lyon, N. *greeni* Miller & Gurley, *N. venustus* M. & G., *N. verneuilli* Troost, *Poteriocrinus cylindricus* Lyon, *P. simplex* Lyon, *Botryllopora socialis* Nich., *Buskopora bistriata* Hall, *B. dentata* Ulr., *B. pyriformis* Hall, *Clonopora semireducta* Hall, *Coscinium cribriforme* Prout, *Cystopora geniculata* Hall, *Cystodictya gilberti* Meek, C. *ovatipora* Hall, C. *vermicula* Hall, *Dekayia (?) devonica* Ulrich, *Eridopora denticulata* Hall, *Fenestella aequalis* Hall, F. *cultrata* Hall, F. *curvijunctura* Hall, *F. depressa* Hall, F. *perplexa* Hall, *F. proutana* Miller, *F. pulchella* Ulrich, *F. serrata* Hall, F. *singularitas* Hall, F. *stellata* Hall, F. *tenella* Hall, F. *variopora* Hall, *F. verrucosa* Hall, *Fenestrapora infraporosa* Ulrich, *Fistulipora alternata* Hall, F. *conulata* Hall, F. *geometrica* Hall, F. *granifera* Hall, F. *normalis* Ulrich, *F. ovata* Hall, F. *subcava* Hall, *F. substellata* Hall, *Glossotrypa paliformis* Hall, *Hederella adnata* Davis, *H. canadensis* Nicholson, *H. cirrhosa* Hall, *Helicopora ulrichi* Claypole, *Hemitrypa cribrosa* Hall, *Hernodia humifusa* Hall, *Intrapora puteolata* Hall, *Lichenotrypa longispina* Hall, *Lioclema intercellatum* Hall, *Orthopora regularis* Hall, *O. rhombifera* Hall, *Phractopora cristata* Hall, *Phyllopora aspera* Ulrich, *Polypora aculeata* Hall, *P.* blandida Ulrich, *P. celsipora minor* Hall, *P. intermedia* Prout, *P. laevistriata* Hall, *P. levinodata* Hall, *P. quadrangularis* Hall, *P. shumardi* Prout, *P. striatopora* Hall, *P. submutans* Hall, *P. transversa* Ulrich, *Prismopora sparsipora* Hall, *P. triquetra* Hall, *Reteporidra adnata* Hall, *Rhombopora lineinoidea* Ulrich, *R. lineinoidea-humilis* Ulrich, *Scalaripora scalariformis* Hall, *S. subconcava* Hall, *Selenopora circincta* Hall, *S. complexa* Hall, *Semicoscinium biimbricatum* Hall, *S. biserrulatum* Hall, S. *interruptum* Hall, *S. latijuncturum* Hall, S. *lunulatum* Hall, *S. permarginatum* · Hall, S.

planodorsatum Ulrich. *S. rhomboideum* Prout. *S. semirotundum* Hall. *S. tortum* Hall. *S. tuberculatum* Prout. *Stratopora perminuta* Ulrich. *Thamniscus nanus* Hall. *Trematella annulata* Hall. *T. arborea* Hall. *Unitrypa acaulis* Hall. *U. anonyma* Hall. U. *fastigata* Hall. *tegulata* Hall. *Arthyris fultonensis* Swallow [=*A. vittata*]. *Atrypa ellipsoidea* Nettleroth. *A. reticularis* Linnaeus. *Camarotoechia carolina* Hall. C. *tethys* Billings. *Chonetes mucronatas* Hall. *Cranaena (Terebratula) romingeri* Hall. *Cyrtina crassa* Hall. *Eunella (Terebratula) harmonia* Hall. *E. (T.) lincklaeni* Hall. *Meristella nasuta* Conrad. *Nucleospira concina* Hall. *Parazyga (Trematospira) hirsuta* Hall. *Pentamerella arata* (Conrad). *P. pavilionensis* Hall. *P. thusnelda* Nettleroth. *Pholidostrophia iowaensis* Owen [=*Stropheodonta nacrea*]. *Productella semiglobosa* Nettleroth. *Rhynchonella louisvillensis* Nettleroth. *R. tenuistriata* Nettleroth. *Schuchertella (Streptorhynchus) chemungensis arctistriata* Hall. *Spirifer acuminatus* (Conrad). S. *arctisegmentus* Hall. *S. audaculus* (Conrad). S. *davisi* Nettleroth. S. *divaricatus* Hall. *S. duodenarius* Hall. *S. fornaculus* Hall. S. *gregarius* Clapp. *S. grieri* Hall. *S. varicostus* Hall. *S. segmentus* Hall. *S. varicosus* Hall. *Stropheodonta demissa* (Conrad). S. *hemispherica* Hall. *S. inequistriata* (Conrad). S. *perplana* (Conrad). S. *plicata* Hall. *Terebratula jucunda* Hall. *Actinopteria boydi* Conrad. *Aviculopecten fasciculatus* Hall. *A. pecteniformis* (Conrad). *A. princeps* (Conrad). *Conocardium cuneus* Hall. *Cypricardinia cataracta* Conrad. *Glyptodesma cancellatum* Nettleroth. *G. occindentale* Hall. *Goniophora truncata* Hall. *Modiomorpha affinis* Hall. *M. mytiloides* Conrad. *Paracyclas elliptica* Hall. Bucania *devonica* Hall & Whitfield. *Callonema bellatulum* Hall. C. *clarkei*. C. *imitator* H. & W., *Cyclonema multilirata* Hall. *Murchisonia desiderata* Hall. *Platyceras bucculentum* Hall. *P. compressum* Nettleroth. *P. conicum* Hall. *P. dumosum* Conrad. *P. dumosum rarispinum* Hall. *P. erectum* Hall. *P. milleri* Nettleroth. *P. multispinosum* Meek. *P. rictum* Hall. *P. symmetricum* Hall. *P. thetis* Hall. *P. ventricosum* Conrad. *Platyostoma turbinatum* Hall. *Pleuronotus (Euomphalus) decewi* Billings. *Pleurotomaria arabella* Nettleroth. *P. lucina* Hall. *P. procteri* Nettleroth. *P. sulcomarginata* Conrad. *Strophostylus varians* Hall. *Trochonema rectilatera* Hall. T. *yandellana* H. & W., Turbo *shumardi* DeVerneuil. *Gomphoceras sp.*, *Goniatites discoideus* Hall. *Gyroceras inelegans ?* Meek. *Leperditia subrotunda* Ulrich. *Isochilina rectangularis* Ulrich. *Aparchites inornatum* Ulrich. *Beyrichia lyoni* Ulrich. *B. kolmondini* Jones. *Ctenobolbina spinulosa* Ulrich. C. *armata* Ulrich. C. *carimarginata* Ulrich. C. *insolens* Ulrich. C. *papillosa* Ulrich. C. *informis* Ulrich. *C. antespinosa* Ulrich. *Kirkbya subquadrata* Ulrich. *K. paralella* Ulrich. *K. semimuralis* Ulrich. *K. cymbula* Ulrich. *K. germana* Ulrich. *Bollia ungula* Jones. *B. obesa* Ulrich. *Halliella retifera* Ulrich. *Octonaria stigmata* Ulrich. *O. stigmata loculosa* Ulrich. *O. ovata* Ulrich. *O. clavigera* Ulrich. *Bythocypris devonica* Ulrich. B. *indianensis* Ulrich. *Pachydomella tumida* Ulrich. *Barychilina punctostriata* Ulrich. *B. punctostriata curta* Ulrich. B. *pulchella* Ulrich. *Calymene platys* Green. *Dalmanites (Odontocephalus) aegeria* Hall. *D. anchiops* Green. *D. aspectans* Conrad. *D. helena* Hall. *D. pleuroptyx* Green. *D. selenurus* Hall & Clarke. *Phacops cristata* Hall. *P. cristata* pipa Hall & Clarke. *P. rana* Green. *Proetus canaliculatus* Hall & Clarke. *Proetus clarus* Hall. *P. crassimarginatus* Hall. *P. microgemma* Hall & Clarke.

To the above list should be added the following from Kindle's paper (1901, 579-758):- - *Crania crenistria* Hall. *Crania granosa* Hall & Clarke. *Craniella hamiltoniae* Hall. *Cyclorhina nobilis* Hall. *Camarotoechia sappho* Hall. C. *congregata* (Conrad). C. *nitida* Kindle. *Rhynchonella gainesi* Nettleroth. *R. gainesi cassensis* Kindle. *R. depressa* Kindle. *Ambocoelia umbonata* (Conrad). *Cyrtina hamiltonensis* Hall. C. *hamiltonensis recta* Hall. *Leptaena rhomboidalis* (Wilckens). *Chonetes arcuatus* Hall. *Chonetes coronatus* (Conrad). C. *acutiradiatus* Hall. C. *subquadratus* Nettleroth. C. *vicinus* (Castelnau). *Pentagonia unisulcata* (Conrad). *Stropheodonta concava* Hall. *Rhipidomella vanuxemi* Hall. *Schizophoria Striatula* (Schlotheim). *Martinia williamsi* Kindle [Geneva]. *Productella spinulicosta* Hall. *Spirifer gregarius greeni* Kindle. *Spirifer manni* Hall. *Spirifer byrnesi* Nettleroth. *Reticularia fimbriata* (Conrad). *R. wabashensis* Kindle. *Gypidula romingeri indianensis* Kindle. *Meristella nasuta* (Conrad). *M. barrisi* Hall. *Eunella sullivanti* Hall. *Cryptonella lens* Hall. *C. ovalis* Miller. *Camarospira eucharis* Hall. *Aviculopecten exacutus* Hall. *A. terminalis* Hall. *Limoptera cancellata* Hall. *Pterinopecten reflexus* Hall. *P. undosus* Hall. *P. nodosus* Hall. *Glyptodesma erectum* Hall. *Leptodesma rogersi* Hall. *Schizodus contractus* Hall. *Nucula hanoverensis* Kindle. *Nucula lamellata* Hall: *Sanguinolites ? sanduskyensis* Meek. *Goniophora hamiltonensis* Hall. *Cypricardinia indenta* Conrad. *Clinopistha subnasuta* H. & W., *Bellerophon leda* Hall. *B. pelops* Hall. *B. patulus* Hall. *B. curvilineatus* Conrad. *Callonema conus* Kindle [Geneva]. C. *lichas* Hall. *Loxonema rectistriatum* Hall. *Macrocheilina hebe* Hall. *M. carinatus* Nettleroth. *Naticopsis levis* Meek. *Polyphemopsis louisvillae* H. & W., *Pleurotomaria lucina perfasciata* Hall. *P. arbella* Nettleroth. *Trochonema meekanum* (Meek). *Cyclonema crenulata* Meek. *Aclisina barnetti* Kindle. *A. barnetti elongata* Kindle. *Euomphalus planodiscus* Hall. *E. (Straparollus) exiguus* Kindle. *Capulus cassensis* Kindle. *Platyceras carinatum* Hall. *P. dumosum pileum* Kindle. *P. fornicatum* Hall. *P. indian-*

ense Miller and Gurley, *P. rictum spinosum* Kindle, *P. echinatum* Hall, *P. arctiostoma* Ulrich, *P. (Orthonychia) fluctuosum* Ulrich, *P. blatchleyi* Kindle, *P. lineare* Kindle, *P. subcirculare* Kindle, *Coleolus tenuicinctum* Hall, *Tentaculites scalariformis* Hall, *Gyroceras jason* Hall, *G. indianense* Kindle, *Orthoceras thoas* Hall, *Cyrtoceras expansum* Kindle, *Proetus folliceps* Hall & Clarke, *P. macrocephalus* Hall, *Dalmanites (Cryphaeus) boothi calliteles* Green, *D. (Odontocephalus) selenurus* (Hall & Clarke), *D. (Hausmannia) pleuroptyx* Green (Hall), *Lichas sp.*

Sellersburg, Silver Creek and Beechwood.—(Encrinital, Crinoidal, Waterlime, Upper Helderberg, Hydraulic limestone, Cement beds, Hamilton, North Vernon). In 1899 Kindle (1899, 8, 110) proposed to call the limestones of Clark County from the base of the black shale down to the lowest bed quarried at the cement quarries, the *Sellersburg* formation. In 1901 Siebenthal (1901, 331-389) prepared a very careful paper on the cement beds, under the title "The Silver creek hydraulic limestone of southeastern Indiana." Although aware of the sense in which Kindle had proposed the name Sellersburg (338-339), he nevertheless restricts Kindle's name (341) to the white and gray crystalline limestone [crinoidal layers] *overlying* the cement beds; and applies the name *Silver Creek* (345) to the Cement beds. He also discards Borden's name North Vernon, on the ground that more than one formation had been marketed under that name. It must be admitted, however, that Borden applied the name in good faith as a geologic term, though as first used it more nearly coincides with the entire Sellersburg. Were it not for the fact that there is still doubt about the correlation of the various beds at North Vernon, Borden's name would certainly have the right of way over all others which have been applied to the beds under discussion. As the matter stands, Kindle's name must be recognized in its original meaning with the Silver Creek and Beechwood (proposed by Butts 1915, 120, for the crinoidal bed) as subdivisions. This procedure has been adopted, very wisely, by Butts (1915, 118). The "water lime" beds and the overlying crinoidal limestone of the Falls of the Ohio region, were described as early as 1828 by Lapham (1828, 65-69). Owen (1837-1859, 25) recognized the "hydraulic limestone" interposed between the crinoidal and the coralline (Jeffersonville) limestones. Hall (1842, 58) regarded the cement beds and underlying coral beds as the western continuation of the Upper Helderberg group (Corniferous) of New York. Clapp (1843) refers the rock to the middle or upper part of the Upper Helderberg. Owen (1843, 151-152) correlated it with the Upper Helderberg. Yandell & Shumard (1847) correlated the limestones above the Catenipora beds (Silurian) and below the Black Shale with the Onondaga-Corniferous-Hamilton. Lyon mentions the cement beds in 1857 (1857, 484) and in 1859 Lyon and Casseday (1859, 244) correlated the crinoidal [Beechwood] limestone with the Hamilton. This appears to be the first definite reference of a formation at the Falls to the Hamilton. Borden (1874, 172; 1875, 115, 126), however, should be given chief credit for correlating these formations with the Hamilton, for in 1874 he referred the hydraulic limestones and crinoidal (and New Albany shale doubtfully) to that division, a correlation that has been exactly verified. He also expressly restricts the term Corniferous to the beds below the cement rock. The fact should not be lost sight of, furthermore, that E. DeVerneuil in 1847 recognized that the Devonian limestones of the Ohio valley carry the Onondaga-Corniferous-Hamilton faunas (see Hall, 1848-1849, 370). Hall (1879a, 139-154) also refers the hydraulic and crinoidal to the Hamilton, and states (p. 142) that "The beds from *r* to *w* [Encrinital, hydraulic, spirifer and coral beds, etc., down to the Silurian] have been regarded, I believe, by all geologists as the equivalent of the Upper Helderberg limestones of New York. . . . I have heretofore accepted this determination, and aided in the dissemination of this opinion." After making a study of the Falls section in the summer of 1877 be concluded that the hydraulic and encrinital limestones "are the equivalent of the Hamilton group of New York". On page 154 of his paper, however, Hall quotes Borden's table of formations published in 1874 (which, by the way, he attributes to Cox) and says "Prof. Cox doubtfully refers the Crinoidal and Hydraulic limestones to the Hamilton group." In a footnote he says "Since this reference does not appear in succeeding reports, the view then entertained may have subsequently been modified," and acknowledges that he had at the first overlooked Cox's [Borden's] report. As a matter of fact, Borden repeats his correlation in the 1875 report, and supports it by the citation of Hamilton fossils from the Crinoidal limestone. Siebenthal (1901, 343) has already called attention to these facts, and the further misplacing of a question mark in Hall's quotation of Borden's 1874 table, so as to make the latter appear to be in doubt about correlating the Hydraulic and Crinoidal with the Hamilton, whereas Borden's doubt applied only to the New Albany shale, which he subsequently correlated with the Genesee of New York. The fact is, Hall here for the first time corrected his own mistaken correlation of all the Devonian limestones at the Falls with the Corniferous, a correlation for which he and not "all geologists" was chiefly responsible. With the great weight of Hall's opinion in its favor, this revised correlation has been generally accepted down to the present day, though some of the Indiana geologists continued to speak of the whole mass of limestones as Corniferous. Kindle's splendid work (1901, 529-758) has completely verified this correlation.*

The following are additional references on the Silver Creek and Sellersburg:—R. Owen 1862, 107 "celebrated waterlime"; Borden 1876, 148, uses name North Vernon for the equivalent of the hydraulic cement rock of Clarke County; Cox 1879, 88, uses North Vernon; Elrod 1882, 174, uses North Vernon for "Upper Corniferous," i. e., the rock immediately below the Black Shale; Elrod 1883, 106, North Vernon equals "Upper Corniferous"; Foerste 1898, 213-288, describes these limestones in the vicinity of Charlestown, Clarke County; Grabau 1898, 80, Hamilton represented in Indiana, Iowa, etc., by limestone; Newsom 1903, 247-249, describes the formations; Schuchert 1903, 144-152, the corals of the Hamilton of New York, Western Ontario and Louisville belong to one province. The Hamilton of this area received many of its species from the Onondaga below; communication was also established between the Mississippian sea and Brazil; Hopkins 1904, 39, 41, described the formations; Ellis 1906, 761, table of formations; Blatchley 1907, 60, hydraulic limestone of Silver Creek formation; Tucker 1911, 15, table of formations; Butts 1915, 118-129, correlation and careful description: Names the crinoidal limestone, Beechwood: finds a stratigraphic hiatus between Jeffersonville and

* The writer takes this opportunity to pay a deserved tribute to Professor W. W. Borden, who founded the Borden Institute and museum at the town of New Providence, now Borden, and served for many years with distinction on the Indiana Geological Survey. He brought together a valuable collection, still housed at the Borden museum, and contributed much to the advancement of science and education in Indiana. It would be eminently fitting if his name and that of the town of Borden could be henceforth attached, in place of the obsolete name "Knobstone", to the great group of shales and sandstones of the Burlington and Keokuk, so well exposed about Borden, and this division of the Mississippian in Indiana be called the *Borden group*.

Sellersburg, Marcellus shale being absent; Stauffer 1915, 225-227 discusses origin of the Hamilton fauna and cites literature. See also papers by Schuchert 1910, Ulrich 1911, Stauffer 1909, Dunbar 1919, A. M. Miller 1919, etc.

The Silver Creek member of the Sellersburg formation as described by Siebenthal (1901, 345) ranges from 15 to 16 feet thick in the type region, and 8 to 10 at Charlestown, thinning rapidly to the north and disappearing as a formation in the northern part of Scott County. The present writer would, however, question its supposed absence from Jennings County where it was identified by Borden. It is a homogeneous, finegrained, bluish to drab or dark gray argillaceous magnesian limestone which when calcined forms a water or hydraulic lime. According to Butts (1915, 119) it ranges from 50% to 60% $CaCO_3$, 16% to 35% $MgCO_3$, 10% to 15% Silica and 2% to 5% Alumina. A large part of the silica is in the form of chert.

The Beechwood (Crinoidal) is from 3 to 8 feet thick, thick bedded, light gray, coarsely crystalline, and very fossiliferous, especially abounding in crinoid plates and segments. It contains black phosphatic nodules, especially in the lower layer.

Fauna of the Sellersburg.—Butts (1915, 124) lists the following species from the Silver Creek limestone: *Zaphrentis* sp., *Fenestella* sp., *Lichenalia* sp., *Athyris fultonensis* (Swallow), *Atrypa reticularis* (Linnaeus), *A. spinosa* Hall, *A. subquadrata* Nettleroth, *Chonetes yandellanus* Hall, *Cyrtina hamiltonensis* Hall, *C. hamiltonensis recta* Hall, *Eunella lincklaeni* Hall, *Glossina triangulata* Nettleroth, *Meristella haskinsi* Hall, *Spirifer byrnesi* Nettleroth, *S. fornacula* Hall, *S. iowaensis* (Owen), *S. oweni* Hall, *S. varicosus* Hall, *Stropheodonta concava* Hall, *S. perplana* (Conrad), *Tropidoleptus carinatus* (Conrad), *Aviculopecten crassicostatus*, H. & W., *Paracyclas elliptica* Hall, *P. lirata* Conrad, *Polyphemopsis louisvillae* H. & W., *Nautilus maximus* Conrad.

From the Beechwood, Butts (1915, 126-128) lists the following: *Alveolites goldfussi* Billings, *A. scandularis* Davis, *Antholites speciosus*, *Aulacophyllum conigerum* D., *Aulopora cornuta* Bill., *Chonophyllum nanum* D., *Cladopora alcicornis* D., *C. gulielmi* D., *C. pinguis* Rominger, *Cyathophyllum ethelanum* D., *C. insigne* D., *C. pustulosum* D., *C. scyphus* Rom., *C. tornatum* D., *C. trauthanum* D., *Cystiphyllum americanum* E. & H., *C. ohioense* Nicholson, *Dendropora alternans* Rom., *D. neglecta* Rom., *D. ornata* Rom., *D. osculata* D., *Diphyllum (Crepidophyllum archiaci* Bill., *Drymopora auloporoidea* Davis, *D. fruticosa* D., *Favosites cavernosus*, *F. digitatus* Rom., *F. goodwini* D., *F. eximus* D., *F. placenta* Rom., *F. rotundituba* D., *Heliophyllum juvene* Rom., *H. infoveatum* D., *Michelinia insignis* Rom., *M. plana* D., *Zaphrentis cornalba* D., *Z. explanata* D., *Z. gallicalcar* D., *Z. nettlerothi* D., *Z. nodulosa* Rom., *Z. reynoldsi* D., *Z. trigemma* D., *Z. ungula* Rom., *Ancyrocrinus bulbosus* Hall, *Dolatocrinus bulbaceus* Miller & Gurley, *D. greeni* M. & G., *D. tuberculatus* Wachsmuth & Springer, *Gennaeocrinus kentuckiensis* Shumard, *Megistocrinus depressus* Hall, *M. rugosus* Lyon and Casseday, *Ambocoelia umbonata* (Conrad), *Athyris fultonensis* (Swallow), *Atrypa spinosa* Hall, *Camarotoechia sappho* Hall, *Centronella glans-fagea* Hall, *Chonetes acutiradiatus* Hall, *Crania sheldoni* White [=*C. bordeni* H. & W.], *Cyrtina hamiltonensis* Hall, *C. hamiltonensis recta* Hall, *Delthyris sculptilis* Hall, *Orbiculoidea doria* Hall, *Pentagonia unisulcata* Conrad, *Pholidostrophia iowaensis* (Owen), *P. spinulicosta* Hall, *Rhipidomella goodwini* Nettleroth, *R. liria* Billings, *R. vanuxemi* Hall, *Schizophoria striatula* (Schlotheim), *Schuchertella chemungensis arctistriatus* (Hall), *Spirifer audaculus* (Conrad), *S. kobbsi* (Nettleroth), *S. iowaensis* Owen, *S. macconathi* Nettleroth, *S. oweni* Hall, *S. segmentum* Hall, *S. varicosus* Hall, *Stropheodonta concava* Hall, *S. demissa* (Conrad), *S. perplana* (Conrad), *Aviculopecten princeps*, *Clinopistha antiqua* Meek, *C. striata* Nettleroth, *C. subnasuta* H. & W., *Grammysia gibbosa* H. & W., *Limoptera cancellata* Hall, *Modiomorpha affinis* Hall, *M. alta* Conrad, *M. charlestownensis* Nettleroth, M. *concentrica* Conrad, *M. mytiloides* Conrad, *Nucula herzeri* Nettleroth, *N. neda* H. & W., *N. niotica* H. & W., *Paracyclas elongata* Nettleroth, *P. ohioensis* Meek, *Ptychodesma knappiana* Hall, *Yoldia valvulus* H. & W., *Bellerophon leda* Hall, *Euomphalus samsoni* Nettleroth, *Loxonema hamiltoniae* Hall, *L. hydraulicum* H. & W., *L. leviusculum* Hall, *L. rectistriatum* Hall, *Platyceras conicum* Hall, *P. dumosum* Conrad, *P. echinatum* Hall, *R. varispinum* Hall, *Platyostoma lineatum* Conrad, *Tentaculites scalariformis* Hall, *Gomphoceras oviforme* Hall, *G. turbiniforme* Meek and Worthen, *Kirkbya* sp., *Dalmanites calypso* Hall and Clarke, *Phacops rana* Green, *Proetus macrocephalus* Hall.

To these lists the following species described by Kindle (1901) from the Sellersburg should be added:—*Crania doria* Hall. *Crania* sp., *Camarotoechia tethys* (Billings). *Rhynchonella louisvillensis* Nettleroth. *R. painei* Nettleroth. *R. tenuistriata* Nettleroth. *Leptaena rhomboidalis* (Wilckens). *Parazyga hirsuta* Hall. *Athyris spiriferoides* (Eaton). *Chonetes manitobensis* Whiteaves. *C. coronatus* (Conrad). *C. mucronatus* Hall. *Delthyris raricosta* Conrad. *Pentamerella pavilionensis* Hall. *Stropheodonta inequistriata* (Conrad). *Rhipidomella leucosia* Hall. *Martinia subumbona* (Hall). *Nucleospira concinna* Hall. *Productella spinulicosta* Hall. *Vitulina pustulosa* Hall. *Spirifer divaricatus* Hall. *S. gregarius* Clapp. *S. granulosus* (Conrad). *S. arctisegmentum* Hall. *S. duodenarius* (Hall). *S. mucronatus* Hall. *S. macrus* Hall. *Reticularia fimbriata* (Conrad). *Pterinea flabella* (Conrad). *Glyptodesma occidentale* Hall. *Paracyclas elliptica* Hall. *Cardiopsis crassicostata* H. & W. *Nucula corbuliformis* Hall. *Modiomorpha recta* Hall. *Grammysia subarcuata* Hall. *Cypricardinia ? cylindrica* H. & W. *Solemya vetusta*, *Bellerophon lyra* Hall. *Bellerophon* sp., *Platyostoma pleurotoma* Hall. *Platyostoma* sp., *P. lineata callosum* Hall. *Isonema humilis* Meek. *Macrocheilina carinatus* Nettleroth. *Pleurotomaria sulcomarginata* Conrad. *P. lucina* Hall. *Trochonema emacerata* H. & W. *T. rectilatera* H. & W. *Platyceras carinatum* Hall. *P. milleri* Nettleroth. *P. thetis* Hall. *P. bucculentum* Hall. *P. symmetricum* Hall. *P. indianensis* Miller & Gurley. *P. attenuatum* Hall. *P. erectum* Hall. *P. rictum spinosum* Kindle. *P. compressum* var., *Coleolus tenuicinctum* Hall. *Conularia* sp., *Gomphoceras minum* Hall. *G. raphanus* Hall. *Goniatites discoideus ohioensis* Hall. *Proetus crassimarginatus* Hall. *Dalmanites (Cryphaeus) pleione* Hall & Clarke.

SENECAN. *New Albany.*—(Devonian Black Shale, Black Slate, Black Lingula Shale, Louisville-Delphi black shale, Genesee shale or slate, Marcellus shale, Black bituminous shale and slate of authors.) The name New Albany shale was proposed by Borden (1874, 158) in 1874 for the formation previously known as the "Black slate" or "Black Lingula slate" in Indiana and Kentucky. Two years previously Collett had used the name "Louisville-Delphi black slate"; and he repeats the designation as late as 1883. This is an altogether unusual form of geologic name and although Collett appears to have employed it definitely as a geologic term, it can scarcely receive serious consideration. The New Albany fell in the Subcarboniferous of Owen (1837) as he at first defined the term; but he subsequently (1844) restricted the term to the lower calcareous division of the Carboniferous, thus excluding the New Albany and lower formations; and in 1859 he defined the Subcarboniferous as consisting of the rocks *between* the Devonian black slate and the coal measures. The formation in Indiana has subsequently been commonly regarded as of upper Devonian age, though its exact correlation has been and remains a matter of uncertainty.

Clapp (1843, 18-19) correlated it with the Marcellus of New York. Hall (1842, 57, 62; 1843, 280; 1843b, 519) at first correlated it with the Marcellus shale of New York, stating that it "is the only representative [in the Indiana region] of that rock, the Hamilton and the Genesee slate", and repeated this correlation with more or less hesitation down to 1862. This correlation was followed by D. D. Owen (1843, 152; 1844), H. D. Rogers (1843, 161-162; 1844, 137-161) and others. Owen, however, in 1847 changed his mind, and correlated it with the Genesee of New York (1847, 72). In 1847 de Verneuil (Hall 1848-1849, 370) also stated that the "Black shists" of Kentucky and Indiana represent the Genesee slate. This led Hall (1851, 285-318) to express some doubt as to whether the black shale is Marcellus or Genesee. Yandell & Shumard (1847, 16) also favor the Genesee. Nevertheless in 1860 we find Hall (1860, 96) still correlating the Indiana formation with the Marcellus. Safford (1856, 158) and Lyon (1860, 620) were of the opinion that the Black shale belongs in the Subcarboniferous, though Safford had in mind the Chattanooga shale of Tennessee. This correlation has recently been revived by Ulrich (1911; 1915, etc.) for the Chattanooga and the upper part of the

Fig. 25. Left, New Albany shale. Right, Rockford limestone, with New Albany shale below, and New Providence shale above.

New Albany. Lyon's view, however, was based on an erroneous determination of the horizon of the Goniatite limestone. Meek and Worthen (1861, 167-177) definitely correlated the New Albany with the Genesee and this view has prevailed generally down to recent years.

(See Meek 1867, 65, Genesee; Cox 1869, 29-30, refers it to the Marcellus; Borden 1874, 158 is in doubt; Borden 1875, 115, refers it to Genesee; so also 1876, 169; Whitfield 1875, 181, Genessee, Hamilton and Marcellus; Hall 1879a, 152, "The Black slate of the west is the equivalent, and even the absolute continuation of the black shales succeeding the Hamilton group of New York (the Genesee slate)"; Elrod 1882, 195, correlates Genesee = Louisville black shale = Delphi black shale = Huron shale of Ohio = Devonian black shale of the West; R. T. Brown 1883, 84, Genesee; McCaslin 1884, 131-132, Genesee; Thompson 1886, 16-17, Genesee; H. S. Williams 1888, 232-233, the New Albany shale is upper Devonian, "The black shale and its fauna may be regarded as terminating the Devonian"; H. S. Williams 1891, numerous references and literature; Cubberly 1893, 219 gives thickness; H. S. Williams 1897, 393-403, New Albany may vary with reference to the time scale in different regions of the south and contains at least locally eocarboniferous faunas; Hans Duden 1897, 108-119, Description, Chemical composition and Bitumen content. Plants described by White; Scovell 1897, 513 Genesee; Blatchley and Ashley 1898, 19-20, Genesee; Blatchley 1898, 27-28 Genesee, gives thickness at several places; Girty 1898, 384-394, careful review of the literature, and concludes that at least the lower part of the New Albany may be correlated with the Genesee but by implication rather favors the idea of Williams and Clarke that not all portions of the formation will bear this correlation, may extend into *Portage*, etc.; Kindle 1899, 7-8, 10-33, 102-111, gives many sections, fauna, stratigraphy, history and correlation, concludes that "the weight of the evidence seems therefore in favor of the equivalence of the fauna of the Genesee and the New Albany shale"; Kindle 1901, 569, thinks it may be equivalent to both the Genesee and the *Portage*; Siebenthal 1901, 339-341, description; Newsom 1903, 247, 251-255 historical sketch and description; Hopkins, 1904, 39, 41, "Genesee, Marcellus"; Blatchley 1906, distribution, thickness and uses as road metal; Ellis 1906, 761 table of formations; Ward 1908, 198, mention; Tucker 1911, 15, table of formations.)

Since the publication of Schuchert's monumental treatise on Paleogeography in 1910, the discussion of the correlation of the New Albany shale has entered a new phase, with a tendency to regard at least the upper portion as belonging to the Mississippian. Schucbert (1910, 548) refers the upper New Albany to the Mississippian and correlates it with the Hannibal (Kinderhookian) of Missouri. Ulrich (1911, pl. 29 between 608-609) correlates the upper New Albany with the Chattanooga, Cleveland-to-Sunbury, etc., as lower Mississippian (Waverlyn) and repeats this correlation in 1912 (1912, 157). The lower New Albany he (1911, pl. 29) correlates with the Genesee-Portage. Kindle (1912, 187-213) takes issue with Ulrich as to the age of the Ohio formations, but does not discuss the age of the New Albany. Butts (1915, 130-137) is "strongly inclined" to maintain the upper Devonian age of the New Albany shale; and repeats this opinion in a later paper (1918, 10). Dunbar (1918, 754-755; 1919, 25; 1920, 114) while expressing some doubt, inclines to place the Chattanooga shale in the Chautauquan division of the Devonian, but says (p. 93) that the Tennessee Survey had decided that the upper part is early Mississippian. The present writer inclines to the opinion that the New Albany shale in Indiana is upper Devonian (Senecan?) throughout; but that to the southward in Tennessee, etc., black shale deposition extended on into the Mississippian. (On the black shale problem see also Grabau, 1921, 440-447.)

The New Albany black shale averages about 100 feet thick in Indiana and does not ordinarily depart widely from that thickness. Blatchley (1898, 27-28) gives 104 feet at New Albany, 103 at Salem, 115 at Seymour, 124 at Bridgeport, 124 at Goshen, 95 at South Bend, 65 at Valparaiso and 112 at Crown Point. These figures are based on well records. Mr. Reeves of the

Indiana Survey has measured sections as great as 140 feet thick and as low as 80 feet.

This rock is an evenly laminated, conspicuously and regularly jointed, deep brown to black, fissile, carbonaceous shale, containing, according to recent analyses of dried samples by Mr. Ray of the Survey, 50% to 60% SiO_2, about 11% Fe_2O_3, 10% Al_2O_3, 5% MgO, P_2O_5, SO_3, TiO_2, Na_2O and K_2O, and 24% fixed and volatile carbon, etc. It weathers into thin plates with a brownish to gray color. It is an oil shale, yielding as high as 25 gallons of oil to the ton by dry distillation. Disseminated specks of pyrite are common. About 15 feet above the base in a number of localities in southern Indiana (North Vernon, etc.) are several thin layers of fine-grained calcareous sandstone, showing extraordinarily perfect jointing. Several thin bands of greenish or blue shale occur in the lower 30 feet. In the upper part of the formation fossil tree trunks have been found. The Marine fossils listed from the New Albany are all from the lower ten feet. The formation is best exhibited in Clark, Scott and Jennings counties.

Fossils of the New Albany.—Butts (1915, 133) lists the following species: *Paleophycus newalbanense* Duden, *P. lineare* Duden, *Parenchymophycus asphalticum* Duden, *Pseudobornia ? inornata ?* Dawson, *Sporangites huronensis* Dawson, *Sporangites radiatus* Duden, *Chonetes lepidus* Hall, *Leiorhynchus quadricostatus* (Vanuxem), *Lingula melie* Hall, *Schizobolus concentricus* (Vanuxem), *Stiliola fissurella* Hall.

To this list the following should be added from Kindle's (1901, 571-579) list: *Leiorhynchus limitare* (Vanuxem), *Lingula spatulata* Vanuxem, *Barroisella subspatulata* Meek & Worthen, *Orbiculoidea lodiensis* (Vanuxem) ?, *Stropheodonta* sp., *Paleoneilo* sp., *Panenka radians* (Hall), *Panenka* sp., *Plethospira socialis* Girty ?, *Pleurotomaria* sp., *Macrocheilina* sp., *Straparollus* sp., *Goniatites wabashensis* Kindle, *G. delphiensis* Kindle, *Orthoceras* sp., *Spathiocaris emersoni* Clarke.

CHAPTER IV.

MISSISSIPPIAN.

The rocks now known as *Mississippian*, *Pennsylvanian* and *Permian* in this country were originally included more or less loosely in the Carboniferous group or system of Conybeare (1822, 333). While the name Carboniferous is still in use as a general and popular designation of the upper Paleozoic, it is no longer in good standing as a systemic term, and should not be used in that sense.

The first suggestion in America of a limitation of the term Carboniferous to the Coal bearing rocks proper or Coal measures, was Owen's proposal of the name Subcarboniferous (1837-39, 12), which he at first used as a designation of all the rocks below the coal measures, as follows:

Sub-Carboniferous Group:

1. Oölitic limestone.
2. Silico-Calcareous series with occasional beds of clay.
3. Black bituminous aluminous slate.
4. Fossiliferous and inferior strata of the Cubcarboniferous group consisting (1) of fossiliferous beds of the Falls of the Ohio, (2) Waterlime and variegated strata, (3) Sand and burrstone, (4) bluish or brownish limestone.

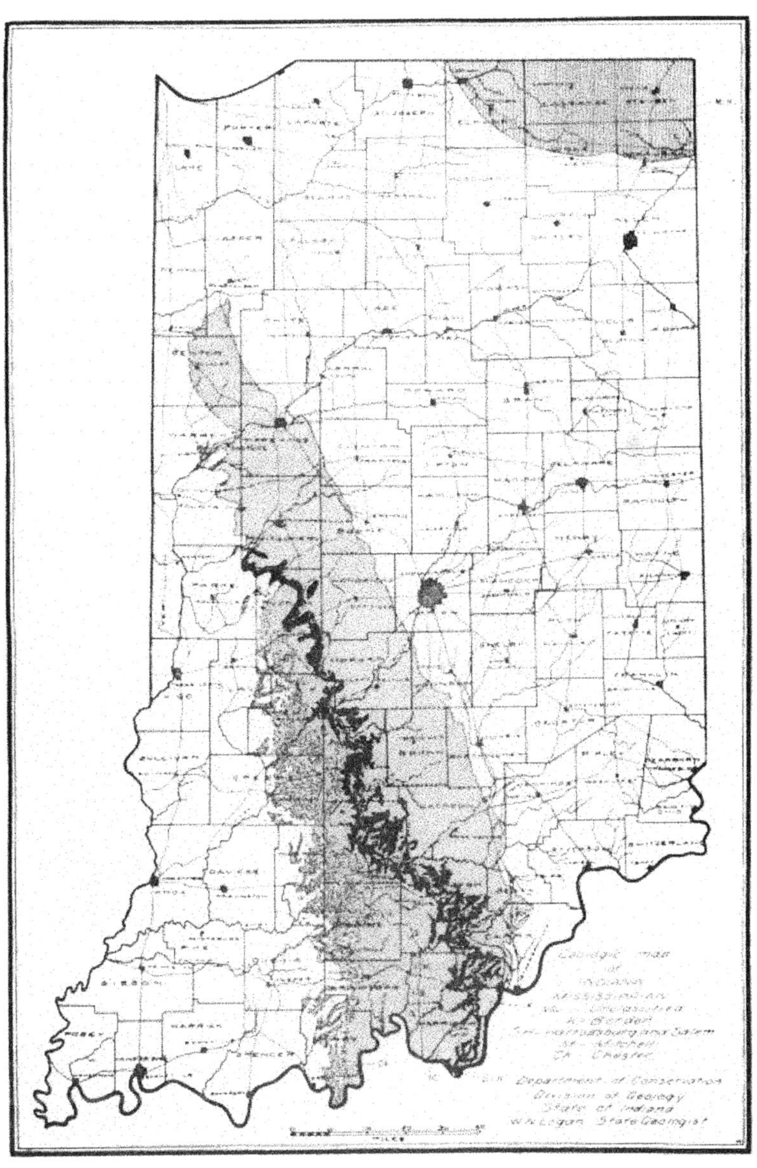
Department of Conservation
Division of Geology
State of Indiana
W.N. Logan State Geologist

Fig. 26.

In 1844 he restricted the term to the lower calcareous division of the Carboniferous and in 1859 defined it as follows: the Subcarboniferous consists of "series of limestones with subordinate fine-grained sandstones and shales" lying between the Devonian Black Slate and the Coal measures, or (1856, 16) the strata extending down to the "Black lingula shale" on which the Coal measures repose. To these rocks Owen at first, and the earlier geologists generally in this country and England, had applied the name "Mountain limestone". Hall (1843, 267-293) has independently applied the name subcarboniferous to certain beds of limestone intercalated in the Knobstone rock. Perhaps the earliest recognition of the common character of these rocks in the Mississippi valley was by Thomas Nuttall (1821, 14-52) in his paper on the "Geological structure of the valley of the Mississippi." In 1838 Caleb Briggs (1838, 79) gave the name "Waverly sandstone" or "Waverly series" to the rocks in Ohio between the black shale and the Carboniferous conglomerate.* De Verneuil (1847) recognized the "Carboniferous" age of the rocks thus referred to the Subcarboniferous and Waverly. Lyon (1860, 612-621) used Subcarboniferous in the sense in which Owen first employed it in 1837, and Hall (1857-59, etc.) used it for his so-called Chemung group of the west and employed the term "Carboniferous limestone" for most of the rocks commonly referred to the Subcarboniferous. In the 1850's and 60's the geologists Owen, Hall, Swallow, Norwood, Meek, Worthen, Shumard, C. A. White, Engelmann and A. Winchell made great advances in deciphering the stratigraphy and correlation of the rocks of the Mississippi valley, and in these years proposed a set of names which constitutes the framework of our present terminology. Among these names was one suggested by Alexander Winchell (1871, 79) for the entire mass of Subcarboniferous rocks. This was the term "Mississippi group". This important proposal attracted little attention until 1891 when in his important correlation paper on the Devonian and Carboniferous (1891, 135), Professor H. S. Williams proposed to revive this appropriate name in the form "Mississippian" and to discard the non-geographic term Subcarboniferous. Winchell's name has therefore come by common consent to stand for this system of rocks in America.

The main subdivisions of the Mississippian were named in rapid succession by these workers between the years 1846 and 1862. Apparently the first to receive a geographic name was the thin band of limestone immediately above the Black shale in Indiana which was known to collectors as the Goniatite or *Rockford* Goniatite limestone. The name Rockford seems to have been first published by Owen and Norwood in 1846. The term was, however, like Waldron and Spergen, a popular designation and appears never to have been explicitly proposed as a geologic formation name. The shale and sandstone rocks next above the Rockford received from D. D. Owen (1857, 351; 1859, 21) the name "*Knobstone*" or "Knob sandstone" rocks from the "Knob" region of Kentucky. The Kinderhook or lower division of the Mississippian was named by Meek and Worthen (1861, 288) from the town of that name in western Illinois. Hall (1856a, 53-56; 1857, 187-203) named

* Weeks (1902) credits Waverly to Mather 1838, American Journal of Science, Ser. 1, vol. 34, p. 356.

the *Burlington* formation from Burlington, Iowa, in 1856; and in the same paper called the rocks next higher, the *Keokuk*, crediting the name to Owen.* The formation next above the Keokuk he called *Warsaw (Ibid.)*. The name *St. Louis* is another example of those designations which first came into use as local and trade names and were subsequently adopted into geologic nomenclature. H. S. Williams says (1891, 147) "St. Louis limestone had been used as a general term, and was technically applied by Dr. Owen as a discovery of Dr. Shumard in 1849." The reference to Owen is to a letter to de Verneuil in the Bulletin of the Geological Society of France. I have not seen this paper, but assume that Williams is correct in his statement. It is credited by Weeks (1902, 358) to Engelmann (1847, 119-120)** and by H. S. Williams (1891, 144) to H. King (1851, 182-199), though, on another page, he appears to credit it, as cited above, to Owen. He says (p. 150) "The name [St. Louis] was definitely proposed and defined by Dr. H. King of St. Louis in 1851." Hall (1859, 55) credits the name to Swallow (1st Missouri Rept.). It is evident that the Mississippi valley geologists were all more or less in the habit of employing an already locally familiar name, and that no one in particular should be credited with the authorship of the term.

The name *Chester* presents a problem of another sort. It was first used in manuscript by Worthen in 1853. Worthen says (1866, 42) that he proposed the name in his field notes of 1853 and made this known to Norwood and to Hall. In spite of this, he says, Hall in his 1856 paper, proposed the name Kaskaskia. "His reasons for substituting the name Kaskaskia", says Worthen, "for Chester limestone do not appear." To those familiar with Hall's methods, further inquiry as to his "reasons" is scarcely profitable. Weeks (1902) credits the name to Swallow (1857, 5). It is here credited to Worthen 1853. Kaskaskia, as intimated, was proposed by Hall (1856, 56). This term has been superseded by Chester (but see under Chester). St. Genevieve was named by Shumard (1857 [1860], 406).

David Dale Owen, James Hall and A. H. Worthen were largely responsible for the subdivision of the Mississippian into the Kinderhook, Burlington, Warsaw, St. Louis and Chester (Kaskaskia) divisions (in ascending order). The rudiments of this classification are seen in Owen's section of

* D. D. Owen (1852, 92) in a section of subcarboniferous limestones of Iowa., has "Encrinital group of Burlington", "Reddish brown encrinital group of Hannibal", and "Cherty limestone of Keokuk," or "Keokuk cherty limestone." It can scarcely be argued that he meant these as formation names, though Hall (1856, 1857) credits Keokuk to him. In the text, however, (p. 91) he speaks of all these localities in precisely the same manner, i. e., "Encrinital group of Burlington"; "The mural escarpment below the town of Hannibal" is composed of these beds, etc.; "The cherty limestones . . . below the Keokuk landing." If one of these named be credited to Owen, all must. Hall, on the other hand, defines the terms explicitly and employs them definitely as geologic names. He should receive the credit for them.

** Engelmann says *(loc. cit.)* "Remarks on the St. Louis Limestone; by G. Engelmann, M. D. . . . The St. Louis limestone underlies the western edge of the great Illinois coal field. . . . The St. Louis limestone forms the uppermost bed of the carboniferous or mountain limestone on the Mississippi . . . our St. Louis limestone is its [Carboniferous limestone] upper part," etc.

the Subcarboniferous limestones of Iowa (1852, 92).* In 1857 Hall published a paper (1857, 187-203), afterward republished with additional material in the first report of the Iowa Survey (1858), in which he gives the following subdivisions of the Subcarboniferous or "Carboniferous limestone":—

VII. Coal Measures.
VI. Kaskaskia limestone or upper Archimedes limestone.
V. Gray, brown, or ferruginous sandstone, overlying the limestone of Alton and St. Louis.
IV. "St. Louis limestone" or "Concretionary limestone."
III. "Arenaceous bed", Warsaw or second Archimedes limestone. "Magnesian limestone," Spergen Hill, Bloomington, Iowa.
II. Keokuk limestone, or lower Archimedes limestone. Beds of passage, cherty beds 60 to 100 feet.
I. Burlington limestone.

*Owen's section is as follows:

COAL MEASURES		Shale Coal Shale	Bituminous shales Bed of coal 6 to 18 inches thick Gray argillaceous shales
UPPER SERIES	f¹	Upper concretionary limestone	White, brittle, close textured limestone usually irregularly bedded and concretionary.
	e¹	Gritstones	Brown and white gritstones containing locally remains of lepidodendron, calamites and other carboniferous plants.
	d¹	Lower concretionary limestone	Lower, compact, white limestone, usually concretionary; sometimes magnesian with earthy matter and marlite in the interstices; containing two or more species of *Lithostrotion* (*Stylina* of Lesueur); including the more evenly bedded limestones of St. Louis, containing *Melonites multipora*, several species of *Productus*, *Spirifer* (*bisulcatus ?*) and reticulated corals of the genera *Retepora* and *Fenestella*.
	c¹	Gritstones	Sandstones sometimes with small pebbles embedded.
	b¹	Magnesian limestones	Argillaceous, cherty, and marly partings. Locally hydraulic limestone. Cellular magnesian building stone; locally with vermicular ramifications and green particles disseminated. Containing reticulated corals and *Terebratula Roysii*.
	a¹	Geodiferous beds.	Impure limestones containing cavities filled with spars and geodes.
LOWER SERIES OF SUBCARBONIFEROUS LIMESTONES	f	Archimedes limestones	Thin-bedded limestones with marly partings, containing archimedes; also brown calcareous beds with cells lined with spars and impure limestones charged with minute *Spirifers*, together with light gray limestone, containing *Terebratula Roysii*, *Orthis umbraculum*, etc.
	e	Shell beds	Gray crystalline limestones, containing *Spirifer striatus*, *S. cuspidatus*, *S. rotundatus*, *Productus punctatus*, *P. semi-reticulatus*, *etc.*
	d	Keokuk cherty limestone	Cherty limestones of Keokuk.
	c	Reddish-brown encrinital, group of Hannibal	Brown encrinital limestone alternating with bands of chert, as near Hannibal.
	b	Encrinital group of Burlington	White crystalline and semioölitic limestones containing *Productus cora* and *Spirifer cuspidatus* Brown and flesh-colored encrinital limestones containing several species of *Pentremites*, *Platycrinus*, *Actinocrinus* and *Poteriocrinus*. Brown earthy crinoidal limestones with crystalline specks. Band of cellular, buff, magnesian limestone. Oolitic limestone containing *Gyroceras burlingtonensis*. Dark gray argillaceous limestones (locally hydraulic ?)
	a	Argillo-Calcareous group, Evans Falls	Buff, finegrained siliceous rock, containing casts of *Chonetes*, *Posidonomya*, *Allorisma*, *Spirifer*, *Phillipsia*. Ash-colored earthy marlites. Slope. Rocks hidden by talus and alluvium of the Mississippi River.

Next below the Burlington limestone (No. I) he reported: "Oölitic limestone and argillaceous sandstone of the age of the Chemung group of New York", and outcropping at Burlington, Iowa; Evans Falls and Hannibal, Missouri. This mistaken lithologic correlation of these lower Burlington rocks with Chemung, though never accepted by Worthen and other Mississippi valley geologists, plagued western geology for many years. It caused Hall to restrict the name "subcarboniferous" to the so-called Chemung rocks, *below* the "Carboniferous limestone". To him the Chemung was, of course, Devonian. This subject will be discussed further under the Burlington.

This scheme of formation is repeated in abbreviated form in the third volume of the Paleontology of New York (1859, 53) and the "great Carboniferous limestones of the Mississippi valley" are correlated with the "Red shales, conglomerate and Catskill Mountain group" of New York and Pennsylvania. Below the Burlington he recognized the "Chemung and Portage groups" and the "Hamilton".

Sidney Lyon's classification, published the succeeding year (1860, 614), follows the earlier practice of Owen in placing everything below the Coal Measures in the "Sub-Carboniferous Series". He subdivides the series as follows (modern equivalents in brackets):—

Series			
Sub-Carboniferous Series	*n*.	Cavernous Limestone	[Mitchell]
	o.	Middle Limestone	[Harrodsburg, Salem]
	p.	Sandstones and Shales	[Knobstone]
	q.	Black Shale	[New Albany]
	r.	Encrinital Limestone	[Beechwood]
	s.	Hydraulic Limestone	[Silver Creek]
	t.	Spirifer bed	[Jeffersonville]
	u.	Nucleocrinus bed	[Jeffersonville]
	v.	Turbo bed	[Jeffersonville]
	w.	Coral beds	[Jeffersonville]
	x.	Catenipora beds	[Louisville]

The balance of the Mississippian (Chester) was placed in his "Millstone Grit Series" as below:

Series			
Millstone Grit Series	*a*.	See Rep. Geol. Surv. of Kentucky	[Coal]
	b.	5th Sandstone	[Mansfield]
	c.	4th Limestone	[Chester]
	d.	Beds of Colored Clays	[Chester]
	e.	4th Sandstone	[Chester]
	f.	3rd Limestone	[Chester]
	g.	Aluminous Shale	[Chester]
	h.	3d Sandstone	[Chester]
	i.	2d Limestone	[Chester]
	k.	2d Sandstone	[Chester]
	l.	1st Limestone	[Chester]
	m.	1st Sandstone	[Chester]

Worthen's Reports of the Illinois Geological Survey began to appear in 1866 (volume I) and thereafter in 1866 (II), 1868 (III), 1870 (IV), 1873 (V), 1875 (VI), 1883 (VII), 1890 (VIII). He was assisted at various times by F. B. Meek, J. S. Newberry, Leo Lesquereux, James Shaw, G. C. Broadhead, E. T. Cox, Orestes St. John, Charles Wachsmuth, W. H. Barris, S. A. Miller, H. Engelmann, H. C. Freeman, F. H. Bradley, H. A. Green, Frank Springer, E. O. Ulrich and Oliver Everett,—a distinguished group. This series of volumes constitutes a grand contribution to our knowledge of the Mississippian formations, and especially of their faunas.

Worthen (1866, 40) adopted the Hall classification somewhat modified; recognizing the Kinderhook of Meek and Worthen below the Burlington, rather than Hall's "Chemung" and combining the Warsaw with the St. Louis (p. 43) because of the intimate relationships of the Warsaw and St. Louis fossils. In the 3d report (1868) the term "Lower Carboniferous" was substituted for "Subcarboniferous" and appears in all later reports. Chester is used in all these reports instead of Hall's Kaskaskia and is made to include the Ferruginous sandstone. The "geode" and "cherty" beds are united to the Keokuk.

Englemann (1868, 188-190) retains Warsaw and uses Ferruginous sandstone but says farther south the "Kaskaskia or Chester limestone" comes in between the Ferruginous sandstone and the Coal measures. Safford (1869) has "Siliceous group" and "Mountain limestone." C. A. White (1870) adopts Worthen's scheme of names, omitting Chester and does not recognize Hall's "Chemung". A. Winchell, in 1871, published an elaborate and exhaustive review of the literature of the Subcarboniferous, proposing to call it the "Mississippi group"; and completely set aside Hall's correlation of the lower Burlington rocks with the Chemung. He proposed also to call the lower member of the subcarboniferous the Marshall group, from the town of Marshall in southern Michigan. His Marshall is, therefore, a synonym of Kinderhook. (Winchell 1871, 1871a). This classification is adopted by N. H. Winchell in the first report of the Minnesota Survey (1873).

A Report of the Geological Survey of Missouri was also published in 1873 by Raphael Pumpelley. Opposite page 292 is the following section by B. F. Shumard, who reported on St. Genevieve County (1873, 292-293):—

Carboniferous System	*e*. Coal Measures	Hard siliceous limestone Blue, purple and drab shale Micaceous sandstone
	h. Archimedes Limestone or Kaskaskia Limestone	Archimedes group
	f. Sandstone	
	h'. Archimedes limestone	
	g. St. Louis Limestone	
	h''. Archimedes Limestone or Warsaw Limestone	
	i. Encrinital Limestone	
	j. Chouteau Limestone	Chemung group
	k. Vermicular Sandstone and Shale	

This is the origin of the terms 1st, 2d and 3d Archimedes limestone, the 3d (h''') being the second, or Warsaw, of Hall. The second Archimedes limestone is called *St. Genevieve*, a name first used by Shumard in 1857. *Chouteau*, used here for part of the "Chemung" of Hall was proposed by Swallow in 1855 in the 1st and 2d Missouri Reports (pp. 101, 176, 194). It appears to be another synonym of Kinderhook.

No original contribution to the general terminology of the Mississippian is contained in the Indiana Reports from the days of David Dale Owen down to the present. In all the earlier reports down to 1897, the terminology as worked out by the geologists of Illinois, Iowa and Missouri was adopted in whole or in part, the lower division (with the exception of the thin Rockford limestone) being known as the Knobstone formation, and usually correlated with the Kinderhook, or occasionally more correctly with the Burlington and Keokuk. In 1897 and 1903 a new and not altogether happy set of names was proposed for the Warsaw-to-Chester formations, two of the proposals being, unfortunately, already preoccupied by well-known names of long standing. The details of this part of the discussion will appear under the heads of the several Indiana formations. A recrudescence of the ancient name Subcarboniferous appears as late as 1916 in VanGorder's report of Greene County. The name *Salem* was proposed by Cumings (1901a, 232-233) for the formation in Indiana variously known as Warsaw, Bedford, Indiana oölite, Spergen hill bed, etc.

It remains to speak of the series names used in the present revision.

In 1891 H. S. Williams exhaustively reviewed the literature of the Devonian and Carboniferous (1891, 1-279) and proposed the following grouping of the Mississippian formations:—

Mississippian Series	St. Genevieve group	Chester St. Louis Warsaw (in part)
	Osage group	Keokuk Burlington
	Chouteau group	Chouteau and the "Vermicular" and "Lithographic" of Broadhead (1874)

The term *Osage* from the Osage river in Missouri, used here as a series name, was first proposed by Branner (1888, XII) and revived in the present sense by Williams. Williams regards the Mississippian as a series under the Carboniferou system.

In 1892 C. R. Keyes (1892, 283-300) reviewed the "principal Mississippian sections" and grouped the formations as follows:

Mississippian Series	Kaskaskia group	Chester shales Kaskaskia Limestone Aux Vases sandstone
	St. Louis group	St. Genevieve limestone St. Louis limestone Warsaw limestone (in part, not typical
	Osage group	Warsaw shales and limestones (typical) Geode bed Keokuk limestone Upper Burlington limestone Lower Burlington limestone
	Kinderhook group	Chouteau limestone Hannibal shales Louisiana limestone

Commenting on these classifications in 1898, Dr. Weller, who has since become the leading authority on the Mississippian, says (1898, 305) Keyes' classification "is a stratigraphic classification which is nothing more than a further elaboration of Hall's earlier one, uniting some of his divisions and dividing others". He defends Williams' subdivision into three groups, as above.

In 1904 Dr. Ulrich published a chapter (1904, 90-113) on the correlation of the geologic formations of the lead and zinc region of Northern Arkansas. In his correlation chart he groups the Mississippian rocks as follows:

Chester	Kaskaskia	Birdsville Tribune
	Cypress	
	St. Genevieve	
Meramec	St. Louis Spergen [Salem] Warsaw	
Osage	Keokuk Burlington	
Kinderhook		

The subdivisions of the Chester and Meramec will be discussed below. The name *Meramec* is here proposed by Ulrich for the Warsaw-to-St. Louis divisions. This classification is repeated and discussed in some detail by Ulrich in a later paper (1905, 15-70).

In 1906 there appeared volumes II and III of Chamberlin and Salisbury's great treatise on geology. Volume II deals with the pre-Mesozoic systems, and on page 498 the authors state that "There seem to be sufficient reasons . . . for regarding the Mississippian or Early Carboniferous as a system coördinate with the Silurian, Devonian, etc., and this classification is here adopted." This practice has since been generally adopted in America. On page 500 they subdivide the Mississippian as follows:

4. Kaskaskia or Chester.
3. St. Louis.
2. Osage or Augusta (including the Burlington, Keokuk and Warsaw).
1. Kinderhook or Chouteau.

Weller (1908, 90) in 1908 discusses the Salem, Warsaw and St. Louis divisions in Illinois, and objects to Ulrich's Meramec division which he regards as unnecessary and especially to the inclusion of the Warsaw in the Meramec. The Warsaw, he maintains, "should from both physical and faunal reasons be more properly joined with the subjacent Keokuk formation, and there is no more reason for associating the Salem and St. Louis in one larger division, than in bringing the St. Louis and the superjacent St. Genevieve formations together."

Schuchert (1910, 547-555) discusses the paleogeography of the Mississippian and divides it into two diastrophic periods, the Mississippic (emended) including the Kinderhook and Osage divisions, and the Tennesseic including the Meramec and Chester. In the Meramec he places the Spergen [Salem] or upper beds of Hall's Warsaw Section, and the St. Louis. He transfers the

St. Genevieve to the Chester division, and uses Kaskaskia for the Chester proper, adopting Ulrich's 1905 subdivisions.

Ulrich (1911, 582 and pl. 29) also splits the Mississippian into two systems which he designates "Waverlyan" and "Tennessean". In the former he places the Chattanoogan, Kinderhookian and Osagian, and in the latter the Meramecian and Chesterian. The Chattanoogan is a new group in which he places the Cleveland, Bedford, Berea, and Sunbury of Ohio, Chattanooga shale, upper New Albany shale, Bradford of Pennsylvania, etc. The Burlington and Keokuk fall in the Osage, and the Warsaw, Spergen [Salem], St. Louis and Moorefield shale of Arkansas in the Meramec. The St. Genevieve is placed in the Chesterian. Ulrich's reason for placing the Warsaw in the Meramec seems to be on account of "diastrophism and the introduction of new types." (p. 517).

The rather revolutionary procedure of dividing the Mississippian into two systems has met with very little favor.

Weller in 1914 published a beautiful monograph of the Mississippian Brachiopoda, the result of many years of painstaking work. This constitutes another of the great landmarks in Mississippian studies. In his introductory chapter on the stratigraphy, he arranges the Mississippian formations as follows:

V. Chester group	Clore formation Palestine formation Menard formation Okaw formation Ruma formation Paint Creek formation Yankeetown formation Renault formation Brewerville formation
IV. St. Genevieve Limestone	
III. Meramec group	St. Louis Salem
II. Osage group	Warsaw Keokuk Burlington
I. Kinderhook	Containing many formations more or less local in their geographical distribution

Of the Meramec group he says (1914, I, 19) "In the older reports of the Illinois Survey by Worthen, this series of limestones was commonly called the St. Louis group, but this name, although proposed many years before the one here used, is inadvisable because the use of the same name for a subordinate formation and for the group leads to much confusion and therefore is not good practice." He excludes the Warsaw from the Meramec and places it in the Osage group on the grounds that its relationships with the latter are closer and that there is a "line of unconformity" between the Warsaw and the Salem.

In Butts' report on Jefferson County, Kentucky, (1915) the Warsaw (Harrodsburg) is placed in the Meramec division (Chart op. p. 31).

Ulrich returned to the discussion of the Mississippian in 1918, contributing a section on the correlation of the Chester formations to the Report by Butts on the Mississippian formations of western Kentucky (1918, pt. 2, 1-272, pls. 1-11). In this paper and in Butts' section of the same volume, the Mississippian is divided into the Kinderhook, Osage, Meramec and Chester, the Warsaw being placed in the Meramec, and the St. Genevieve in the Chester. Neither author gives his reasons for this disposition of the Warsaw. Butts, however, (pp. 35-36) recognizes an unconformity (disconformity) at the base of the St. Louis between it and the Warsaw, the Salem being absent in parts of the area.

The latest contribution to this discussion is by Weller (1920, 282, 408-416 and 1920a, 91-244) who suggests (1920) dividing the Mississippian into a lower and an upper division, calling the lower the "Iowa series," (p. 282) and the upper the "Chester series." This grouping might therefore be presented as follows:

Mississippian System	Chester series: A large group of formations	
	Iowa series	St. Genevieve
		St. Louis
		Salem
		Warsaw
		Keokuk
		Burlington
		Kinderhook

He objects strongly to Schuchert and Ulrich's two-system arrangement stating (p. 415) that "In his 'Revision of the Paleozoic Systems' Ulrich has split the Mississippian into two so-called systems, the Waverlyn below and the Tennessean above, the line of cleavage between the two being placed between the Warsaw and Keokuk formations. From the evidence afforded by the Mississippi Valley section of the Mississippian which is the type section of these strata, there is less reason for placing a major dividing line at this horizon than at almost any other position in the entire succession of formations, and there is no basis whatsoever for the recognition of the so-called Waverlyn and Tennesseean as systems." He also says that Ulrich "has made a grave error in including the Ste. Genevieve limestone in the Chesterian." A far greater faunal break occurs between the St. Genevieve and the Chester than below the St. Genevieve.

Objection may be made to the use of the name Iowa Series for the lower Mississippian series on the ground that *Iowan* has long (since 1895) been used for one of the glacial stages; and, although there is some doubt about the reality of the "Iowan glacial stage", it would appear to the writer unwise to confuse geological terminology by employing the name in this new sense.

In the Geology of Hardin County, Illinois, (1920a, 96-97) Butts and Weller place the Warsaw in the Meramec division, in which they also include the St. Genevieve. Weller does not explain this inclusion of the Warsaw in the Meramec division in opposition to the contentions of his 1914 paper, and to his recognition (1920, 410) of the disconformable relations of the Warsaw and Salem,—a relation seen also in Indiana.

The classification adopted in the present report follows the usual fourfold grouping into *Kinderhook*, *Osage*, *Meramec* and *Chester*, placing the Warsaw in the Osage and the St. Genevieve in the Meramec.

The principal sources of detailed information on the Mississippian formations of Indiana are: Hopkins and Siebenthal 1897; Hopkins 1896; Kindle 1896, 1899; Ashley 1903; Newsom 1903; Cumings, Beede, etc., 1906; F. C. Greene 1911; Beede 1915; Malott 1919. For general reviews of Mississippian literature see A. Winchell 1871, 1871a and H. S. Williams 1891.

KINDERHOOKIAN. *Rockford.*—(Goniatite limestone, Rockford Goniatite limestone, Chouteau). This formation, though thin and local in its distribution, was recognized very early because of its unusual fauna of well-preserved Goniatites; and came to be known locally at the "Goniatite limestone of Rockford." The name Rockford was presently adopted into geologic nomenclature, Owen and Norwood (1847, 12 pp.) apparently being the first to publish it.*

Hall (1851, 309) placed the formation in the Marcellus division. Christy in 1851 wrote a paper on the "Goniatite limestone of Rockford Indiana" in which he mistakenly located the limestone centrally in the Black Shale, a mistake which for a time created considerable confusion of correlation. Sidney Lyon in 1860 correlated the rock on paleontological evidence with the Subcarboniferous; and Meek and Worthen (1860, 447) regarded it as "probably of upper Devonian age but containing Carboniferous Goniatites." Meek and Worthen the next year (1861, 167) in a paper on the "Goniatite limestone of Rockford" correlated it with the Chouteau of Missouri, and showed that it is *not* central in the Black Shale but *above* it. Hall (1860, 95) misled by Christy's error, correlated the Rockford with a band of limestone in the Marcellus shale of New York. White and Whitfield (1862, 289-306) however, objected to correlating it with the Carboniferous and referred the Burlington and associated beds to the Chemung, largely on the authority of Hall. In 1863 Winchell (1863, 61-62) correlated the Goniatite limestone with the Chemung=Marshall=Waverly=Burlington=(Kinderhook) and Chouteau. In 1866 Worthen (1866, 109-110) placed it in the Kinderhook division correlating it about with the Chouteau. Winchell later (1871) correlates it with the Marshall group of Michigan.

The formation is frequently mentioned in the Indiana Reports, but little further is given on the correlation until Kindle's important paper in 1899. He here reviews the literature and concludes that the Rockford should be regarded "as the sole representative of the Kinderhook in Indiana" (1899, 99). He says further "The Rockford evidently disappears by thinning toward the south and is represented in the Kentucky sections by the greenish-blue argillaceous New Providence shale. . . . We have therefore conclusive evidence that the Rockford fauna and the New Providence fauna were contemporaneous and existed side-by-side over a portion of this area at the end of the Black Shale epoch." He does not agree with Meek and Worthen's correlation of the Rockford with the Chouteau but thinks it more nearly of the age of the Louisiana limestone of Missouri. He denominates it the "*Muensteroceras oweni* bed". Newsom (1903, 255-260) gives a list of fossils and agrees with Kindle's

* Weeks (1902) credits "Rockford" to Webster, who used the name "Rockford shales" for a Devonian formation in Iowa in 1889, Proc. Davenport Academy of Science, Vol. V, pp. 100-109; and "Rockford Goniatite limestone" to Hopkins and Siebenthal 1897, 21st Annual Report of the Indiana Department of Geology and Natural Resources, p. 296. The name Rockford had been in common use as a formation name in Indiana publications and elsewhere long before either of these dates.

correlation. Hopkins (1904, 44) correlates it with the Chouteau. Schuchert (1910, 548) correlates the Rockford about with the Fern Glen (Upper Kinderhookian), while Ulrich (1911, pl. 29) correlates it with the Glen Park-Hannibal, or somewhat lower than Schuchert places it. Weller (1914) places it in the Kinderhook.

The Rockford formation is a ferruginous limestone of brownish or greenish gray color, usually from one to three feet thick. It is typically exposed in the bed of White River at Rockford, Jackson County, but may be seen whenever the contact of the Knobstone and New Albany formations comes to view, between Rockford and the Ohio river. It does not extend into Kentucky.

Its fauna is interesting, containing the Goniatites *Prodromites praematurus*, *Prolecanites greeni*, *P. lyoni*, *Aganides rotatorius*, *Muensteroceras oweni* and *M. parallelum*. Weller (1914) lists the Brachiopods: *Brachythyris semiplicata* (Hall) and *Ambocoelia unionensis* Weller. From Kindle's list (1898, 408-488) the following are added: *Amplexus (?) rockfordensis* Miller & Gurley, *Synbathocrinus oweni* Hall, *Shumardella missouriensis* (Shumard), *Shumardella obsolens* (Hall), *Anatina leda* Hall, *Cardiopsis radiata* Meek and Worthen, *Cypricardia ventricosa* Hall, *Megambonia lyoni*, *Bellerophon cyrtolites* Hall, *B. lineolatus* Hall, *Murchisonia (Pleurotomaria) limitaris* Hall, *Platyceras haliotoides* M. & W., *Pleurotomaria mitigata* Hall, *P. vadosa* Hall, *Straparollus lens* Hall, *S. spirorbis* Hall, *Hyolithes aculeatus* Hall, *Gyroceras gracile* Hall, *Solenochcilus rockfordensis* (M. & W.), *Oncodoceras rockfordensis* (M. & W.), *Stroboceras trisulcatus* (M. & W.), *Orthoceras heterocinctum* Winchell, *O. indianense* Hall, *O. icarus* Hall, *O. marcellensis* Vanuxem [?], *O. Whitii* Winchell, *Trematoceras discoidalis* (Winchell), *Tribloceras digonum* (M. & W.), *Proetus doris* Hall, *Phillipsia rockfordensis* Winchell.

OSAGIAN.—*Knobstone or Borden.*—(Asage, Kinderhook, Waverly, Chemung, Calcareo-siliceous (in part), Marshall, Keokuk sandstones, Chouteau, etc.) The name "Knobstone" or "Knob sandstone" was applied to this mass of rocks by D. D. Owen (1857, 351; 1859, 21), from its outcrops in the conical hills or "Knobs" of Kentucky. It presents a curious example of the persistence of a non-geographic term—one of very few that have survived the nomenclatorial vicissitudes of the past half century. It would perhaps seem unnecessarily ruthless to uproot this sturdy survivor of an ancient race; but if a local name is to be continued for this interesting series of rocks, which bids fair to be ultimately minutely subdivided into named formations, the selection of a geographic term should not longer be postponed. This series does indeed demand a distinctive name, since it is not exactly coterminous with either the Osage or the Waverly; and unless an appropriate geographic designation is employed, geologists will persist in calling it Knobstone. I, therefore, propose to call the series of formations, comprised between the top of the Rockford, where present, or top of the New Albany where the Rockford is absent; and the base of the Harrodsburg or Warsaw limestones and shales, the *Borden* series. The name is taken from the town of Borden, in Clark County, and from the name of W. W. Borden, geologist, teacher and philanthropist, after whom the old town of New Providence was renamed.

Owen at first (1837) included this mass in his "Calcareo-siliceous" or "encrinital" group which also took in the Harrodsburg of the present classification. In the 1859 reprint and the early Kentucky reports (1857, etc.) he recognized the distinctness of the "Knobstone" from the overlying and underlying rocks. It was placed in his subcarboniferous group, and formed the base of the group as redefined in 1859. It was also referred to the Carboniferous by Owen and Norwood (1847). Hall, however, (1843, 267-293) had correlated it with the Portage and Chemung of New York, regarding the green shales of the lower part (New Providence) as Portage and the overlying sandstones as Chemung. That this correlation was based solely on lithologic resemblance and the assumed westward persistence of the New York formations, is shown by the fact that Hall mentions the discovery of shells of Carboniferous type in the "Knobstone", and was led to question the inference as to absolute identity. Nevertheless, he repeated this correlation down to 1860. In the Foster and Whitney report (1850, II, 292-309) he expresses some doubt of his correlation, stating that "the green shales and sandstones of Ohio and Indiana which succeed this Black Shale, have been recognized as Carboniferous by their fossils, though there is still some doubt whether the lower part may not represent the Chemung group of New York." Hall's opinion influenced Owen (1844) to adopt this erroneous correlation. De Verneuil (1847) correctly associated the Knobstone rocks with the Carboniferous, being influenced by the fossils and not by lithology or by stratigraphic prepossessions. The decided Carboniferous affinities of the Rockford goniatites, had, after the correction of Christy's mistake of placing the Rockford in the Black Shale, a decided influence in support of the Mississippian affinities of the Borden group and since 1860 there has been little disposition to question the correctness of de Verneuil's correlation. (See Worthen 1861; 1866, 116-117; Meek and Worthen 1861, 167-177; White and Whitfield 1862, 289; Winchell 1869-70 [1871] etc.)

There has been much less unanimity as to the exact position of the beds in the Mississippian.

White and Whitfield (1862, 289), judging from the fossils, regard the upper part as Keokuk, and Worthen (1866, 116-117) agreed with this correlation. On the other hand Winchell (1871, 57-82) says "It appears from observations made by others and by myself that the Knobstone formation of Indiana and Kentucky with the associated shales and limestones is substantially restricted to the horizon of the Keokuk division of the Mississippi limestone series or 'Mississippi' group." This, by the way, is the first use of the name "Mississippi group." Beachler (1887, 1106-1109) listed the crinoids from the famous fossil bed at Crawfordsville, Indiana, and referred the beds to the Keokuk. The next year (1888, 407-412) he gave a list of species, and sections of the beds, and evidently regarded the whole Knobstone of that region as Keokuk. It should be noted here that not all geologists seem to have been aware of the fact that these famous crinoid beds are in the upper Knobstone. The Indiana Survey geologists, however, from 1874 on, usually correlated the formation with the Kinderhook of Illinois, and Waverly of Ohio.

(So, Collett 1874, 198; Borden 1875, 115; Cox 1875, 44; Collett 1883, 68 Waverly; Gorby 1886, 124, correlates upper part with Waverly and lower part with Kinderhook; Thompson 1892, 33; S. A. Miller 1892, Waverly=Kinderhook=Chouteau=Marshall; Blatchley and Ashley 1898a, 20, Waverly; Newsom 1903, 277; "lower layers" Kinderhook; Blatchley 1905, 47-52 Waverly). Collett (1875, 85; 1876, 318; 1879, 303), Elrod (1882, 174, 198-201), Brown (1884, 75-78), M'Caslin (1884, 128-131), and Scovell (1897, 513) do not definitely correlate the Knobstone, but all place it below the "Keokuk group", by which they evidently mean the Harrodsburg and the Crawfordsville beds.

H. S. Williams in 1888 also correlated the Knobstone with the Kinderhook (1888, 232).

On the other hand, Kindle (1899, 100-102) and Siebenthal (1901, 338) correlate it with the Osage (Burlington and Keokuk), and Hopkins (1904, 44-49) with the Keokuk; which is a return to the views of Worthen, Winchell and Beachler. The Kinderhook age of the Knobstone has, nevertheless, up to very recent years, been rather generally accepted, and especially the Kinderhook age of the lower (New Providence) portion (see, for example, Weller 1914). Schuchert (1910) and Ulrich (1911), however, place it in the Osage, the former as Burlington and Keokuk and the latter as Burlington only. It would appear from the recent detailed work of Butts in Kentucky (1915, 137-163; 1918, 10) that the older correlation with the Burlington and Keokuk (Osage) is correct, and that the overlying (Harrodsburg) limestone is not Keokuk but Warsaw. A. M. Miller (1919, 97-101) regards the Borden or Knobstone as the equivalent of the Cuyahoga and Logan formations of Ohio which constitute all but the lower part of the Waverly, and are correlated by him with the Burlington and Keokuk. Butts and Weller (1920a, 92-96) correlate the Borden with the Osage.

(On the Borden (Knobstone) see the following additional papers: Newsom 1898, 253-256, map and discussion; Bennett 1898, 258-262, sections; Price 1898, 262-266, Mt. Carmel fault, etc.; Jones 1898, 257-258, upper limit; Bennett 1899, 283-287, eastern escarpment; Newsom and Price 1899, 289-291, distribution and correlation; Blatchley and Ashley, 1898, 20, thickness and correlation; Ashley 1903, 87-89, thickness; Reagan 1904, 205-223, description and map in Monroe County; Newsom 1903, 260-280, extended description and map; Blatchley 1905, 47-52, thickness, economic uses, correlation; Blatchley 1904, 47-48, thickness, distribution, etc.; Blatchley 1907, 38-39, economic; Shannon 1908, 99-105, lithology, distribution and soils in Monroe County; Siebenthal 1908, 305, thickness; Tucker 1911, 15, table of formations; Cumings 1912, 121-123 good account of physical properties; Beede 1915, 191 brief description.)

The Borden has been divided into the *New Providence, Kenwood, Rosewood* and *Holtsclaw* formations. The first of these names was proposed in 1874 by W. W. Borden (1874, 161) from the old name of the town of Borden in Clark County, and has been used until recently for the lower and middle shaly portion of the Borden. Butts, however, limits the term to the lower 120 to 160 feet of green shale below the lower sandstones, which is evidently what Borden had in mind when he proposed the name. This lower sandstone capping the New Providence shale in the Louisville region is called *Kenwood* sandstone by Butts (1915, 148), from Kenwood hill at Louisville; and the second or "Knob" shale is named *Rosewood* shale from Rosewood, Harrison County, Indiana (1915, 150). The Kenwood sandstone does not appear to be present north of the Louisville area and is probably a very local bed. For the upper sandstone capping the "knob shale" in the Ohio River counties, Butts proposed the name *Holtsclaw* sandstone, from Holtsclaw Hill, Jefferson County, Kentucky (1915, 151).

In 1896 Hopkins, in his report on Carboniferous sandstones of Indiana, (1896, 196, 287) used the name *Riverside sandstone* for a finegrained sandstone quarried at Riverside, near Williamsport, in Warren County, Indiana. Later (1904, 45) he recognized that this sandstone is of Knobstone age and suggested that the name Riverside be dropped. It has been retained, nevertheless, as a designation of the upper sandstone of the Borden, and is so used by Kindle (1899, 100, 101), Newsom (1903, 262), Reagan (1904, 206), Ulrich (1911, pl. 29), Cumings (1912), Beede (1915, 191-194) and others. Kindle says *(loc. cit.)* "The sandstone in northern Indiana designated as the Riverside sandstone by Mr. Hopkins, has been correlated with the upper sandstone and sandshales of the Knob region in southern Indiana on the evidence of fossils collected at and near the typical locality." He gives a list of the species. The present writer has verified Kindle's faunal evidence and also notes that the thick upper sandstone of the Bloomington area which is correlated with the Riverside contains the same fauna as the Holtsclaw. It seems very probable, therefore, that Holtsclaw is a synonym of Riverside, and the writer has, therefore, given the latter name preference. Still another name has been applied to this formation. In his 1870 paper on the Marshall group, A. Winchell (1871, 385-418) speaks of the Williamsport (Riverside) locality and correlates the sandstone there with the Rockford Goniatite limestone of southern Indiana, calling it on his chart (p. 415) the "*Williamsport gritstone.*" He seems to think this sandstone an especially close correlate of the Marshall of Michigan. There is no certain indication that Winchell intended Williamsport as a formation name, and it would serve no useful purpose to revive it.

New Providence.—This formation comprises the lower green shales about 120 to 150 feet thick at the base of the Borden series. The history of its correlation has been covered under the general discussion above. Butts (1915, 144-147), on the authority of Drs. Girty and Springer correlates it with the Burlington, though Weller (1914, etc.) has been inclined to correlate it with the Kinderhook. Girty, however, admits that the lower part might prove to be Kinderhook. The stratigraphic relations and the preponderance of faunistic evidence seem to the writer to favor the Burlington age of this formation.

Description.—The New Providence consists of soft greenish, rapidly weathering, clay-shale. Thin limestones and bands of iron carbonate occur distributed through the formation. When dry the slacked shale pulverizes readily. It retains its lamination, even after it has weathered to a plastic clay. Fossils are common in the calcareous and iron carbonate layers, but rare in the rest of the rock. The iron carbonate often takes the form of concretions, arranged in bands parallel with the bedding. The formation is best developed in Clark, Scott and Washington counties.

Fauna.—There is no systematic list of the fossils of the New Providence as here restricted, other than the one given by Butts from the Louisville region (1915, 140-142) and the crinoids listed by Springer (1911), the majority of collectors not having distinguished between the fauna of this rock and the balance of the Borden series. Butts' list follows:—*Amplexus fragilis* White and St. John, *A.* sp., *Cladochonus* near *gracilis* Keyes, *C.* near *longi* Rowley, *Cyathaxonia arcuata* Weller, *C. cynodon* Rafinesque & Clifford, *C.* sp., *Favosites valmeyerensis* ? Weller, *Monilopora amplexa* Rowley, *Palaeacis cavernosa* Miller, *P.* sp. *Striatopora* near *carbonaria* White, *Trochophyllum verneuilianum* M-E. & H., *Triplophyllum centralis* (M-E. & H.), *T. cliffordana* (M-E. & H.), *T. (?) declivis* Miller, *T.* sp., species of *Actinocrinus*, *Agaricocrinus*, *Amphoracrinus*, *Barycrinus*, *Cactocrinus*, *Cyathocrinus*, *Euryocrinus*, *Halysiocrinus*, *Megistocrinus*, *Mespilocrinus*, *Platycrinus*, *Scaphiocrinus*, *Taxocrinus* and *Wachsmuthicrinus*; and *Catillocrinus tennesseeae* Shumard, *Eretmocrinus yandelli*, *Forbesiocrinus nobilis*, *Gilbertsocrinus* cf. *tenuiradiatus* M. & W., *Metichthyocrinus tiaraeformis* Hall, *M. clarkensis* M. & G., *Orphocrinus* cf. *stelliformis* Owen & Shumard, *Platycrinus americanus* O. & S., *P. burlingtonensis* ? O. & S., *P.* cf. *granosus*, *P. planus* O. & S., *P. sculptus* Hall, *P. verrucosus* White, *P.* cf. *yandelli* O. & S., *Schizoblastus decussata*, *Stemmatocrinus trautscholdi* W. & S., *Synbathocrinus angularis* M. & G., *S. robustus* Shumard, *Chainodictyon* ? sp., *Cystodictya americana* Ulrich, *C. lineata* Ulr., *C. pustulosa* Ulr., *Fenestella compressa* Ulr., *F. regalis* Ulr., *F. triserialis* Ulr., *F.* sp., *Fistulipora* sp., *Lioclema punctatum* ? Hall, *Meekopora* ? *aperta* Ulr., *Polypora* sp., *Ptilopora cylindracea* Ulr., *Rhombopora angustata* Ulr., *R. elegantula* Ulr., *R. incrassata* Ulr., *R.* sp., *Stenopora* sp., *Streblotrypa major* Ulr., *Thamniscus divaricans* Ulr., *T. sculptilis* Ulr.,

Brachythyris suborbicularis (Hall), *Chonetes planumbonus* M. & W., *C. shumardanus* de Koninck, *Cliothyridina* near *glenparkensis* Weller, *Productus semireticulatus* Martin, *Ptychospira sexplicata* White & Whitfield, *Pustula* n. sp., *P. punctata* Martin, *Rhipodomella oweni* Hall and Clarke, *Schuchertella* near *lens* White, *Spirifer* near *floydensis* Weller, *S.* near *logani* Hall, *S.* near *Moorefieldanus* Girty, *S.* near *veronensis*, Swallow, *Spiriferina subelleptica* McChesney, *Paleoneilo* sp., *Posidoniella ?* sp., *Platyceras* sp., *Goniatites* sp., *Brachymetopus* sp., *Griffithides* sp., sp. of *Ostracods*.

The following Brachiopod is added from Weller's list (1914): *Strophalosia cymbula* H. & C.

Kenwood.—According to Butts (1915, 148-150) the Kenwood sandstone has a constant thickness of about 40 feet throughout Jefferson County, Kentucky, and extends southward in Kentucky and northwestward into the river counties of Indiana. It does not, however, according to our observations, persist very far into Indiana. Butts' description is as follows: "At most localities the Kenwood sandstone is composed of thin sandstone layers alternating with shale, but at a few points the sandstone becomes rather massive and includes a larger proportion of the whole. . . . The prevailing character of the formation is indicated by the following section:

	Ft.	In.
Rosewood shale		
17. Shale		
Kenwood sandstone	Ft.	In.
16. Sandstone, chocolate, hard, finegrained		3
15. Shale, blue	2	
14. Sandstone, blue, finegrained	2	
13. Shale, blue		8
12. Sandstone, reddish calcareous?		3
11. Shale, blue, calcareous nodules	2	6
10. Sandstone, bluish finegrained		4
9. Shale, blue	1	3
8. Sandstone, rusty		2
7. Shale, blue	1	3
6. Sandstone, rusty		2
5. Shale, blue	5	
4. Sandstone, bluish, finegrained with iron carbonate nodules up to 1 ft. in diameter (turtle stones)		8
3. Shale, blue	1	6
2. Sandstone, iron carbonate nodules		2
1. Shale, blue	2	6
	23	5

The sandstone layers seem to be almost wholly composed of very small quartz grains bound by a ferruginous cement. Occasionally small flakes, apparently of mica, are present. The sandstone is greenish where fresh, but rusty when weathered."

Productus wortheni Hall, is the only fossil listed from the Kenwood.

Rosewood.—Butts (1915, 150-151) defines the Rosewood shale as including in a general way the "Knob shale" of authors, nearly 200 feet thick. It makes the chief slopes of the Knob region about Louisville. It is bluish-gray, unevenly fissile, and siliceous, having about 68% silica and 14% alumina with some lime, iron oxide, and K_2O. Limestone lenses and iron nodules occur about 70 feet below the top. A characteristic feature of the upper half of the formation is fucoid-like, whitish markings, regarded as worm-trails by Borden. The upper part of the Rosewood is somewhat fossiliferous. The formation has not been certainly identified north of Clark County.

Fauna.—Butts (1915, 153-154) lists the following species from the Rosewood in the Louisville area: Fucoids ?, *Amplexus fragilis* White & St. John, *Cladochonus* near *longi* Rowley, *Triplophyllum* sp., *Cystodictya* sp., *Fenestella* several sp., *Hemitrypa* sp., *Pinnatopora* sp., *Rhombopora* several sp., *Brachythyris suborbicularis* (Hall), *Chonetes shumardanus* de Koninck, *C.* sp., *Cliothyridina parvirostris* (M. & W.), *Cyrtina* near *burlingtonensis* Rowley, *Dielasma* sp., *Productus* near *arcuatus* Hall, *P. crawfordsvillensis* ? Weller, *P. wortheni* ? Hall, *P. ovatus* Hall, *Reticularia pseudolineata* (Hall), *Rhynchopora beecheri* ? Greger, *Schuchertella* near *chemungensis* (Conrad), *Spirifer floydensis* Weller, *Spirifer rostellatus* Hall, *Spiriferina subelleptica* ? (McChesney), *Conocardium*, 2 sp., *Crenipecten* sp., *Cypricardinia* near *scitula* Herrick, *Leda* sp., *Paleoneilo* near *bedfordensis* Meek, *Bembexia* n. sp., *Bucanopsis* ? sp., *Oxydiscus* sp., *Platyceras* sp., *Orthoceras* sp., *Phillipsia* sp., *Brachymetopus elegans* ?

Riverside (or Holtsclaw).—As intimated in the general discussion above, the formation called Holtsclaw by Butts is probably the equivalent of the Riverside sandstone farther north, and the older name is, therefore, preferred. The sandstones quarried at Riverside, Warren County, Indiana, and named "Riverside sandstone" by Hopkins (1896, 196, 287) attracted attention many years ago and were correlated with the Kinderhook and Waverly by Worthen (1866, 117), and by Winchell (1871), the latter speaking of them as Williamsport gritstones. Gorby (1886, 86) also used the name Williamsport of the sandstone quarried near there, but not as a formation name. Kindle (1899) showed the probable equivalence of these sandstones to the upper Knob sandstone farther south; and Newsom, Beede and others have applied the name consistently to this upper Knobstone formation. It would be both unwise and unfair to adopt Butts' name unless and until it can be shown that his Holtsclaw is a distinct formation.

Hopkins states (*loc. cit.* 287) that the Riverside "corresponds lithologically to the Knobstone group near the base of the lower Carboniferous, and possibly belongs to that group. . . . . It is a very fine grained sandstone, blue on a fresh surface, weathering buff to dark gray on long exposure. It is quite evenly stratified, in many places on natural exposures the stratification planes becoming quite abundant, even grading into shale." It contains about 93% silica.

Farther south, in Monroe County, the Riverside is a less resistant formation and not adapted to use as a commercial stone. It is a very fine grained tough sandstone with clay binder, and suffers severely from frost action. Its color is light blue, weathering dirty gray or brown. In dry seasons an abundant efflorescence of magnesium sulphate comes out on clean exposures. Long continued weathering gives a shaly appearance to the outcrops unless they are kept clean by stream action. The rock produces a lean sandy soil, with a scrubby growth of small oaks, Judas tree, dogwood, etc. In Monroe County the formation is at least 100 feet thick.

Butts (1915, 151) defines the Holtsclaw as comprising the "Knob sandstone" of authors, with a thickness in the Louisville area of 15 to 25 feet. It is described as "a bluish-gray or buffish, rather loosely cemented, soft and easily disintegrated, very fine-grained, thick to massive bedded stratum." It will be noted how perfectly this description fits the Riverside farther north.

The Borden has usually been assigned a thickness of from 400 to 600 feet in Indiana. Unpublished data indicate that in the Monroe-Brown county area the group is as much as 800 feet thick.

Fauna.—Butts lists the following species from his "Holtsclaw sandstone":—Fucoids, *Cladochonus* near *gracilis* Keyes, C. sp., *Triplophyllum* sp., *Batostomella* ? sp., *Cystodictya lineata* ? Ulrich, C. sp., *Dichotrypa* sp., *Fenestella* several sp., *Hemitrypa* sp., *Lioclema* sp., *Polypora* several sp., *Rhombopora* sp., *Stenopora* sp., *Athyris lamellosa* l'Eveille, *Brachythyris suborbicularis* Hall, *Chonetes shumardanus* de Koninck, C. sp., *Crania* sp., *Dielasma* near *fernglenensis* Weller, *Eumetria verneuiliana* Hall, *Lingulodiscina* sp., *Orthothetes Keokuk* Hall, *Productus arcuatus* Hall, *Productus crawfordsvillensis* Weller, P. n. sp., P. near *ovatus* Hall, P. near *parvus* M. & W., P. sp., P. *wortheni* Hall, *Pustula alternata* ? Norwood & Pratten, *Reticularia pseudolineata* (Hall), R. sp., *Rhynchopora beecheri* Greger, *Spirifer crawfordsvillensis* Weller, *Spirifer floydensis* Weller, S. *keokuk* Hall, S. *montgomeryensis* Weller, S. *rostellatus* Hall, S. *tenuicostatus* Hall, *Spiriferina subelliptica* ? McChesney, S. *sp. Syringothyris texta* Hall, *Aviculopecten* sp., *Conlu'aria* sp., *Cypricardinia* near *scitula* Herrick, *Myalina Keokuk* ? Worthen, *Platyceras* sp., *Phillipsia* sp.

Syringothyris texta and *Orthothetes keokuk* which Butts says are limited to the Holtsclaw are the commonest fossils of the Riverside in Central Indiana and at Riverside. At Borden *Syringothyris texta* occurs more than 150 feet below the Harrodsburg contact.

The following species from the entire Borden (Knobstone), some of which are very doubtful, are listed by Kindle (1898, 408-488)*: *Lithostrotion proliferum* Hall (??), *Zaphrentis cornucopia* (???), Z. *dalei* E. & H. (?), *Catillocrinus bradleyi* M. & W., *Sabatocrinus swallowi* Hall (?), *Chonetes illinoiensis* Worthen [listed as C. *logani* Hall], *Leiorhynchus quadricostatus* Vanuxem (?), *Rhipidomella oweni* Hall & Clark [listed by Kindle as *Orthis michilini* Hall], *Rhipidomella penelope* Hall [??, a Hamilton species], *Productus reticulatus* Gabb [??, a South American species], *Liorhynchus greenianum* (Ulrich), *Shumardella missouriensis* (Shumard), *Spirifer asper* Hall, [??, a Hamilton species], *Spirifer biplicatus* Hall (?), *Spirifer carteri* Hall [see *Syringothyris texta* Hall], *Spirifer marionensis* Shumard (?), *Brachythyris peculiaris* (Shumard) (?), *Brachythyris semiplicata* (Hall) [probably from the Rockford], *Spiriferina cristata* (Schlotheim) [S. *kentuckyensis* as listed by Kindle. An Upper Carboniferous species], *Orthothetes crenistria* (Phillips) [??, probably does not occur in America], *Streptorhynchus pectinaceum* Hall [evidently *Strophomena pectinacea* Hall = *Orthotetes chemungensis*: not a Mississippian species.], *Streptorhynchus umbraculum* Von Buch [not an American species, see *Orthothetes umbraculum* Hall, probably does not occur in the Borden. Forms referred to this species are probably O. *keokuk*.], *Terebratula calvini* [this is *Dielasma calvini* (H. & W.), a Chemung species], *Cardiomorpha subglobosa* Meek (?), *Grammysia rhomboidalis* M. & W. [=G. *rhomboides* Meek], G. *ventricosa* Meek. [Waverly sp.], *Paleoneilo bedfordensis* Meek [Bedford shale of Ohio], *Schizodus medinaensis* Meek [Waverly of Ohio], *Cyclonema pulchellum* Miller & Gurley, *Holopea grandis* Miller & Gurley, *Murchisonia indianensis* M. & G., *Igoceras pabulocrinus* Owen [*Platyceras infundibulum* M. & W., as listed. Crawfordsville], *Orthonychia acutirostra* Hall, [*Platyceras uncum* M. & W. as listed. Crawfordsville?], *Pleurotomaria mississippiensis* White & Whitfield [P. *textiligera* Meek, as listed. Burlington?], *Conularia micronema* Meek [Keokuk], C. *newberryi* Winchell [Waverly, Marshall], *Goniatites browmensis* Miller, G. *greenii* Miller, G. *indianensis* Miller, *Remeliceras clarkense* M. & G., *Solenocheilus henryvillensis* M. & G., *Caulerpites marginatus* Lesquereux.

To these lists should be added the species listed by Wachsmuth & Springer, Springer, Beachler and others from the Crawfordsville crinoid beds. (See species marked "Crawfordsville" in faunal list under Warsaw).

Warsaw.—(Harrodsburg, Encrinital (part), Keokuk, Burlington, Geode beds, Knob limestone, Middle limestone (part), etc.). The geode-bearing crystalline limestones capping the "Knob sandstone" in Indiana were commonly correlated with the Keokuk of the Mississippi valley section by the older Indiana geologists, apparently largely because of the abundance of geodes; and this correlation has persisted in Indiana literature down to recent years.

(See for example, Collett 1874, 265; Collett, 1875, 85, refers to "Keokuk beds" 36 feet of reddish crinoidal beds on Weed Patch Hill in Brown county. [there is no Harrodsburg in Brown County]; Borden 1874, 172, Knob lime-

* The generic name now in use is given, in most cases without further explanation.

Fig. 27. Upper left, St. Genevieve limestone near Trinity Springs, Indiana. Upper right, St. Louis limestone, near Bloomington. Lower left, weathered joint in the Salem, Hunter Valley, near Bloomington, Indiana. Lower right, Riverside sandstone, near Harrodsburg, Indiana.

stone=Keokuk group; Collett 1876, 317-318; Collett 1879a, 303, etc.; Collett 1880, 417; Greene 1880, 432-438; Collett 1883, 68; R. T. Brown 1884, 78; Thompson 1886, 18, includes crinoid beds at Crawfordsville; Gorby 1886, 124, 133-136, 205-206; Thompson 1889, 63-64; Beachler 1889, 65-70, Crawfordsville beds; Kindle 1895, 52-53; Siebenthal 1897, 296-297, Burlington and Keokuk; Scovell 1897, 513, Crawfordsville beds; Blatchley and Ashley 1898, 20-21; Ashley 1899, 79; Ashley 1903, 66, 86-87, Burlington and Keokuk; Newsom 1903, 280-281 Keokuk; Crawfordsville crinoid beds are in the Knobstone; Hopkins 1904, 49-52, Burlington.)

In 1897, recognizing the uncertainties of correlation, Hopkins and Siebenthal named this formation the Harrodsburg limestone from Harrodsburg, Monroe County, Indiana.* This name has been in general use since that date. (See Blatchley 1898, 20-21; Jones 1898, 257; Ashley 1899, 74, 79; Ashley 1903, 86-87; Newsom 1903, 280-281; Blatchley 1904, 48; Reagan 1904, 206-216, sections; Hopkins 1904, 49-52; Blatchley 1906, 139-140, etc.; Ellis 1906, 761; Cumings 1906, 85, weathering; Shannon 1906, 957-959; Shannon 1908, 106-108, lithology, thickness, etc.; Siebenthal 1908, 305-306, thickness and description; Tucker 1911, 15; Beede 1915, 194-203, good description, with section: redescribes type section.)

Consistently with the correlation of the Harrodsburg with the Keokuk, the next overlying formation, now known as the Salem, was called *Warsaw*. In fact the early correlation of the Spergen bed with the Warsaw by Hall and others had much to do with placing the Harrodsburg in the Keokuk.

Schuchert (1910, 548) in his paleogeography adheres to this older correlation; but Ulrich (1911, pl. 29) places the Harrodsburg opposite both Keokuk and Warsaw, Weller (1908, 87) having proven in the meantime that the Salem and Warsaw are distinct in the Warsaw type section and elsewhere in Illinois. Nevertheless Weller in 1914 (1914, I, 353-368, etc.) still places the Harrodsburg as equivalent to the Keokuk, though he also refers the Crawfordsville beds to the Keokuk (*op. cit.*, 145, etc.).

Butts (1915, 157-163; 1918, 26-32) appears to be the first to definitely break away from the older correlation and to correlate the Harrodsburg with the Warsaw. He says (1915, 163) "On the basis of lithic similarity and stratigraphic position, the formation here under discussion [Harrodsburg] is correlated with the Warsaw limestone typically exposed at Warsaw, Illinois. The Warsaw at that place, including the lower 40 feet of magnesian limestone with geodes, formerly included in the Keokuk, but now by the best authorities placed in the Warsaw, succeeds the Keokuk as in Jefferson County [Kentucky]." This transfer of the "geode beds" to the Warsaw is further explained by Weller (1920a, 97) as follows: "As originally used, the name [Warsaw] was applied only to beds which overlie the so-called 'geode beds' of the section at Keokuk and Warsaw, these lower beds being included in the Keokuk formation. More recent studies have shown that the true line of demarkation between the Keokuk and Warsaw is more properly at the base of the 'geode bed' instead of at the summit of it, and such an emended definition of the Warsaw formation has been accepted by the United States Geological Survey." On page 100 *(Ibid.)* he correlates the Harrodsburg with the Warsaw as thus emended.

* The type section is where the Dixie highway crosses Judah's creek, one mile due south of Harrodsburg village.

A. M. Miller (1919, 102) includes both the Harrodsburg and Salem in the Warsaw.

Description.—The Warsaw or Harrodsburg limestone is a rather coarse, crinoidal, crystalline, fossiliferous, limestone from 60 to 90 feet thick in central Indiana, thinning to the northward and disappearing in Montgomery County, though beds in Fountain and Newton have sometimes been correlated with it. It contains some shale especially in the lower part. The Crawfordsville crinoid beds belong in the Borden. Southward the formation becomes more shaly, especially in the lower part, but retains about the normal thickness (60 to 70 feet). In the Louisville region, according to Butts (1915, 158) it is 65 to 82 feet thick. Siebenthal (1897, 297) places 65 feet of rock in the type section, which is as follows:

Harrodsburg type section (descending order)	Ft.	In.
Massive fossiliferous limestone	6	0
Gray heavy-bedded limestone	16	0
Blue argillaceous shale	2	0
Limestone	0	4
Chert	0	3
Heavy-bedded blue to gray crystalline limestone	6	0
Yellow calcareous shale with geodes	1	3
Fine, heavy-bedded blue crystalline limestone	11	0
Flaggy limestone	1	0
Gray argillaceous limestone	0	10
Calcareo-argillaceous shale with bands of limestone and some geodes	18	0
Heavy limestone, weathering shaly	3	0
Calcareous shale in bed of creek	?	?
	67*	8

Beede (1915, 199) gives a much more detailed section north of Harrodsburg station, in which he finds 95 feet 3 inches of Harrodsburg. This is the best section so far published.

Geodes are characteristic of the formation, but not confined to it, occurring abundantly in the upper part of the Riverside.

Fauna.—In places, as at Harrodsburg and north of Bloomington, the formation is very fossiliferous, the upper bed sometimes amounting to a veritable coquina of Bryozoan fronds. The older lists of fossils from the "Keokuk" and "Burlington and Keokuk" include species not obtained from the Harrodsburg, and are in other respects too inaccurate to be of use. A critical review of Kindle's (1898, 408-488) list from the "Burlington and Keokuk" is appended:—***Amplexus fragilis* White and St. John [Burlington, Warsaw, Keokuk], *Aulopora gigas* Rominger [??], *Lithostrotion proliferum* Hall [=*L. canadense* Castlenau. (?) Probably wrongly identified], *Palæacis obtusus* M. & W. [Keokuk, Illinois, Missouri], *Zaphrentis cornucopia* [??], *Z. dalei* M-E. & H. [Warsaw, Keokuk], *Z. spinulifera* Hall [=*Z. dalei*], *Abrotocrinus cymosus* M. & G., *Actinocrinus agassizi* Troost [*Forbesiocrinus agassizi* Hall?], *Actinocrinus brontes* Hall [=*A. lowei* Hall; Keokuk Iowa, Missouri, Illinois], *Actinocrinus coreyi* Lyon and Casseday [=*Agaricocrinus coreyi* (L. & C.)]. *Actinocrinus gibsoni* M. & G., [Crawfordsville]***

* 65 feet 8 inches. Siebenthal added wrong. See Beede 1915, p. 198.

** In this revised list the name used by Kindle is given first, with the modern equivalent in brackets.

*** The species marked [Crawfordsville] are from the crinoid beds in the Borden or Knobstone rocks, and are of Keokuk age.

A. grandis M. & G., *A. magnificus* W. & S. [Crawfordsville]. *A. nashvillae* Troost [=*Lobocrinus nashvillae* (Hall); Keokuk, Illinois, Iowa, Kentucky]. *A. pernodosus* Hall [Keokuk, Warsaw, Illinois, etc.]. *Agaricocrinus gorbyi* Miller [=*A. splendens* M. & G., Crawfordsville]. *A. indianensis* Miller [=*A. splendens* M. & G.]. *A. nodosus* M. & W. [Burlington]. *A. pentagonus* Hall [=*A. bullatus* Hall, Upper Burlington]. *A. springeri* White [=*A. coreyi* L. & C.]. *A. tuberosus* Troost [=*A. americanus var tuberosus*, Keokuk, Illinois, Iowa, Tennessee]. *A. wortheni* Hall [Keokuk, Iowa, Illinois]. *Agelacrinus squamosus* M. & W. [Crawfordsville]. *Amphoracrinus viminalis ?* Hall [=*A. viminalis* Hall, Waverly, Ohio]. *Archaeocidaris agassizi* Hall [Burlington, Iowa, Missouri]. *A. keokuk* Hall [Keokuk, Illinois, Iowa, Missouri]. *A. norwoodi* Hall [Chester]. *A. wortheni* Hall [St. Louis]. *Barycrinus formosus* M. & G., B. *herculeus* M. & W., [Crawfordsville]. *Barycrinus hoveyi* Hall [Crawfordsville]. B. *magister* Hall [Keokuk, Iowa]. B. *pentagonus* Worthen [Keokuk, Illinois]. B. *princeps* Miller & Gurley [Crawfordsville]. *B. spectabilis* M. & W., [St. Louis]. *B. stellatus* Troost [*B. stellatus* Hall ?]. B. *stellifer* Miller, B. *tumidus* Hall [Keokuk, Illinois]. B. *washingtonensis* M. & G., *Batocrinus aequalis* Hall [Burlington, Iowa, Missouri]. *B. agnatus* Miller [=*Macrocrinus jucundus* (M. & G.), Crawfordsville]. B. *biturbinatus* (Hall) [=*Dizygocrinus biturbinatus* (Hall), Keokuk, Illinois, Iowa, Missouri, Kentucky]. *B. cantonensis* Miller & Gurley, B. *crawfordsvillensis* Miller [=*Dizygocrinus crawfordsvillensis* (Miller), Crawfordsville]. *B. facetus* Miller [=*Dizygocrinus facetus* (M. & G.)]. B. *indianensis* Lyon and Casseday [=*Dizygocrinus indianensis* (L. & C.), Crawfordsville]. B. *jucundus* M. & G. [See *Macrocrinus jucundus* (M. & G.)]. *B. marinus* M. & G. [Crawfordsville]. *B. montgomeryensis* Worthen [=*Dizygocrinus montgomeryensis* (Worthen), Crawfordsville]. *B. mundulus* Hall [Keokuk, Ill.]. *B. pistillus* M. & W. [Burlington, Iowa]. B. *pyriformis* Shumard [=*Lobocrinus pyriformis* (Shumard), Kinderhook, Burlington]. *B. wachsmuthi* (White) [=*B. grandis* W. & S., Crawfordsville]. B. *whitei* W. & S. [=*Dizygocrinus whitei* W. & S., Keokuk and Warsaw (Salem) Indiana]. *Catillocrinus bradleyi* M. & W. [Crawfordsville]. *C. wachsmuthi* M. & W. [Burlington]. *Cyathocrinus arboreus* M. & W. [Crawfordsville]. C. *crawfordsvillensis* Miller [Crawfordsville]. C. *decadactylus* L. & C. [=*Scytalocrinus gorbyi* Miller]. C. *gurleyi* Miller, C. *harrisi* Miller [Crawfordsville]. C. *hexadactylus* L. & C. [=*Vasocrinus hexadactylus* (L. & C.), Crawfordsville]. C. *insperatus* Lyon [Crawfordsville]. C. *multibrachiatus* L. & C. [Crawfordsville]. C. *opimus* Miller & Gurley [Crawfordsville]. C. *poterium* M. & W. [Crawfordsville]. C. *signatus* M. & G., C. *tumidulus* M. & G. [Crawfordsville]. *Dichocrinus expansus* M. & W. [=*D. polydactylus* L. & C., Crawfordsville]. *D. ficus* L. & C. [Crawfordsville]. *Dichocrinus lineatus* M. & W. [Burlington]. *D. sculptus* L. & C. [=*D. ornatus* W. & S., St. Louis]. *D. simplex* Shumard [=*Talarocrinus simplex* (Shumard), Warsaw=Salem]. *D. striatus* O. & S. [Burlington]. *D. ulrichi* M. & G., *Dolatocrinus bradleyi* [=*Halysiocrinus bradleyi* M. & W., Crawfordsville]. *Eretmocrinus adultus* W. & S. [=*Dizygocrinus originarius var. adultus* W. & S.]. *E. intermedius* W. & S., *E. ramulosus* Hall [Keokuk, Illinois, Tennessee]. *E. praegravis* Miller, *Evactinopora* [This is a genus of Bryozoa]. *Forbesiocrinus speciosus* [=*F. multibrachiatus* L. & C., Crawfordsville and elsewhere probably in the Borden]. *F. washingtonensis* M. & G., *F. wortheni* Hall [Crawfordsville]. *Goniasteroidocrinus lyonanus* M. & G. [=*Gilbertsocrinus lyonanus* (M. & G.) Crawfordsville]. G. *tuberosus* L. & C. [=*Gilbertsocrinus tuberosus* (L. & C.), Crawfordsville]. *Granatocrinus ficus* [??] *G. granulosus* M. & W. [=*Schizoblastus granulosus* (M. & W.), Keokuk, Illinois]. G. *norwoodi* O. & S. [Burlington]. *Graphiocrinus wachsmuthi* M. & W. [Burlington]. *Ichthyocrinus clarkensis* M. & G., *Lepidesthes colletti* White, *L. coreyi* M. & W. [Crawfordsville]. *Melonites multiporus* O. & N. [St. Louis]. *Oligoporus danae* M. & W. [Keokuk, Illinois, Iowa, Missouri]. *O. nobilis* M. & W. [Burlington and Keokuk, Illinois]. *Ollacrinus tuberosus* L. & C. [=*Gilbertsocrinus tuberosus* (L. & C.), *q. v.*]. *Onychaster flexilis* M. & W. [Crawfordsville]. *Onychocrinus cantonensis* Miller & Gurley [=*O. ulrichi* M. & G.]. *O. exculptus* L. & C. [Crawfordsville]. *O. parvus* M. & G. [Chester]. *O. ramulosus* L. & C. [Crawfordsville]. *O. ulrichi* M. & G. [Crawfordsville]. *Palaeaster crawfordsvillensis* Miller [Crawfordsville]. *Pentremites burlingtonensis* M. & W. [Burlington]. *P. Conoideus* Hall [Salem]. *P. pyramidatus* Hall [*Metablastus bipyramidates ?*]. *P. wortheni* Hall [=*Metablastus wortheni* (Hall), Crawfordsville]. *Platycrinus bonoensis* White, *P. discoideus* O. & S. [Burlington]. *P. halli* Shumard [Burlington]. *P. hemisphericus* M. & W. [Crawfordsville]. *P. lodensis* M. & W. [*P. lodensis* H. & W., Cuyahoga of Ohio]. *P. planus* O. & S. [Burlington and Keokuk]. *P. yandelli* O. & S. [Burlington]. *Poteriocrinus amoenus* Miller [Crawfordsville]. *P. arcanus* Miller & Gurley, *P. bisselli* Worthen [=*Scytalocrinus bisselli* (Worthen) Chester]. *P. cantonensis* Miller & Curley, *P. circumtextus* M. & G. [Crawfordsville]. *P. concinnus* M. & W. [=*Woodocrinus concinnus*, Crawfordsville]. *P. coreyi* Worthen [=*Scytalocrinus grandis* W. & S., Crawfordsville]. *P. coryphaeus* Miller [Crawfordsville]. *P. crawfordsvillensis* M. & G. [Crawfordsville]. *P. depressus* M. & W. [=*Decadocrinus depressus* M. & W., Crawfordsville]. *P. divaricatus* Hall [Warsaw]. *P. gibsoni* White [Crawfordsville]. *P. grandis* W. & S. [=*Scytalocrinus grandis* W. & S. Crawfordsville]. *P. granilineus* M. & G. [Crawfordsville]. *P. gurleyi* White [=*Scaphiocrinus gurleyi* White, Crawfordsville]. *P. hoveyi* [=*Barycrinus hoveyi* q. v.]. *P. indianensis* M. & W. [=*Scytalocrinus indianensis* (M. & W.). Crawfordsville]. *P. nodobrachiatus* (Hall) [=*Scaphiocrinus nodobrachiatus* Hall, Crawfordsville]. *P. penicilliformis* Worthen [=*Decadocrinus penicilliformis* (Worthen). Keokuk, Illinois]. *P. robustus* (Hall) [=*Scytalocrinus robustus* (Hall). Crawfordsville]. *P. subaequalis* W. & S. [*Scaphiocrinus aequalis* ?]. *P. subramosus* M. & G. [Crawfordsville]. *P. unicus* (Hall) [=*Scaphiocrinus unicus* Hall, Crawfordsville]. *P. verus* M. & G. [Crawfordsville]. *Protaster gregarius* M. & W. [=*Aganaster gregarius* (M. & W.) Crawfordsville]. *Rhodocrinus benedicti* Miller, *Scaphiocrinus aequalis* Hall, [Crawfordsville]. *S. arrosus* M. & G., *S. bellus* M. & G., *S. bonoensis* M. & G., *S. coreyi* M. & W. [Crawfordsville]. *S. decadactylus* [=*Scytalocrinus decadactylus* (M. & W.). Keokuk, Illinois]. *S. depressus* M. & W. [=*Decadocrinus depressus* (M. & W.). Crawfordsville]. *S. disparilis* M. & G. [Crawfordsville]. *S. gibsoni* White [Crawfordsville]. *S. gurleyi* White [Crawfordsville]. *S. granuliferus* M. & G. [Crawfordsville]. *S. graphicus* M. & G. [Crawfordsville]. *S. lacunosus* M. & G., *S. lyoni* Miller [Crawfordsville]. *S. maniformis* Miller, *S. manus* M. & G. [Crawfordsville]. *S. nodobrachiatus* Hall [Crawfordsville]. *S. praemorsus* M. & G., *S. repertus* M. & G., *S. robustus* Hall [=*Scytalocrinus robustus* (Hall). Crawfordsville]. *S. unicus* Hall [Crawfordsville]. *Steganocrinus benedicti* Miller. *Synbathocrinus robustus* Shumard, *S. swallovi* Hall [St. Louis, Keokuk]. *S. wortheni* Hall [Burlington]. *Taxocrinus colletti* M. & G. [Crawfordsville]. *Taxocrinus crawfordsvillensis* M. & G. [=*Parichthyocrinus crawfordsvillensis* (M. & G.)]. *T. meeki* Hall [=*Parichthyocrinus meeki* (Hall). Keokuk, Iowa, Illinois]. *T. multibrachiatus* (L. & C.) [=*Forbesiocrinus multibrachiatus* L. & C. Crawfordsville]. *T. ramulosus* (Hall) [Burlington]. *T. shumardianus* Hall [St. Louis]. *T. ungula* M. & G. [Crawfordsville]. *T. suboratus* M. & G. [=*Parichthyocrinus suboratus* (M. & G.)]. *T. splendens* M. & G. [=*T. colletti* M. & G., Crawfordsville. The most abundant species at Crawfordsville]. *Troostocrinus lineatus* (Shumard) [=*Metablastus lineatus* (Shumard). Burlington, Iowa, Illinois, Missouri]. *T. nitidulus* M. & G. [=*Metablastus nitidulus* (M. & G.) St. Louis]. *Vasocrinus lyoni* Hall [=*V. hexadactylus* (L. & C.). Crawfordsville]. *Zeacrinus doverensis* M. & G. [Chester]. *Z. dubius* M. & G., *Z. ramosus* Hall [=*Woodocrinus ramosus* (Hall). Burlington, Iowa]. *Z. salemensis* M. & G., *Z. troostianus* M. & W. [=*Woodocrinus troostanus* (M. & W.). Burlington, Illinois, Iowa]. *Archimedes owenanus* (Hall). *A. reversus* Hall [=*A. wortheni* (Hall). Warsaw]. *Cleodictya gloriosa* Hall [not a Bryozoan]. *C. mohri* Hall [not a Bryozoan]. *Coscinium asterium* Prout [=*Fistulipora asteria* (Prout). Keokuk, Illinois]. *C. elegans* Prout [=*Glyptopora elegans* (Prout). Warsaw, Illinois Missouri]. *C. escharoides* [*C. escharense* ?, not recognizable]. *C. michelinia* Prout [=*Glyptopora michelinia* (Prout). Warsaw]. *C. wortheni* Prout [Scarcely recognizable]. *Dictyophyton cylindricum* Whitf. [Not a Bryozoan]. *Ectenodictya eccentrica* Hall [not a Bryozoan]. *Evactinopora grandis* M. & W. [Burlington, Iowa]. *E. sexradiata* M. & W. [Burlington, Iowa]. *Fenestella delicata* Meek [Waverly, Ohio]. *F. shumardi* Prout [Pennsylvanian of New Mexico]. *Lyrodictya romingeri* Hall [not a Bryozoan]. *Phragmodictya catilliformis* (Whitfield) [Not a Bryozoan]. *P. patelliformis* Hall [not a Bryozoan]. *Physospongia dawsoni* Whitf. [not a Bryozoan]. [The above five species of sponges are all from Crawfordvsille]. *Prismopora serrata* (Meek) [Chester]. *Ptilopora prouti* Hall [Warsaw]. *Stictopora carbonaria* Meek [=*Cystodictya carbonaria* (Meek). Pennsylvanian]. *Athyris lamellosa* (l'Eveille) [Kinderhook, Burlington, Keokuk]. *A. royssi* l'Eveille [not an American species. See *Cliothyridina prouti* Swallow]. *A. sublamellosa* Hall [=*Cliothyridina subamellosa* (Hall) St. Genevieve and Chester]. *Camarophoria subtrigona* (M. & W.) [=*Tetracamera subtrigona* (M. & W.) Keokuk]. *Chonetes logani* N. & P. [=*C. illinoisensis* Worthen. Burlington, Keokuk]. *C. planumbonus* M. & W. [Upper Keokuk]. *Derbya ruginosa* Hall & Clarke [=*Streptorhynchus ruginosum* (H. & C.). Salem, St. Louis]. *Lingula crawfordsvillensis* [=*Dielasma sinuata* Weller, Keokuk, Salem]. *L. indianensis* M. & G. [Crawfordsville]. *Orthis dubia* Hall [=*Rhipidomella dubia* (Hall). Keokuk, Warsaw, Salem]. *Orthis keokuk* Hall [=*Orthothetes keokuk* (Hall). Keokuk]. *O. michilini* l'Eveille [not an American species. See *Rhipidomella oweni* H. & C., New Providence shale]. *Productus alternatus* N. & P. [=*Echinoconchus alternatus* (N. & P.). Burlington to Chester]. *P. burlingtonensis* Hall [Burlington]. *P. cora* d'Orbigny [not an American species. The forms referred by various American authors to *P. cora* comprise a considerable section of the genus *Productus* as at present constituted.]. *P. flemingi* Sowerby [not an American species. Several American species have been referred to this form.]. *P. keokuk* Hall [=*P. viminalis* White. Burlington and Keokuk]. *P. magnus* M. & W. [upper Keokuk]. *P. mesialis* Hall [Keokuk]. *P. punctatus* Martin [not an American species]. *P. semireticulatus* *Martin* [not an American species. See especially *P. viminalis* White, *P. setigerus* Hall, and *P. curtirostris* Winchell]. *P. setigerus* Hall [Crawfordsville]. *P. wortheni* Hall [Keokuk]. *Rhynchonella mutata* Hall [=*Camarotoechia mutata*

(Hall), Salem]. *R. subcuneata* Hall [=*Tetracamera subcuneata* (Hall), Salem]. *Spirifer carteri* Meek [=*Syringothyris textus* Hall, Borden]. *S. cuspidatus* Meek [=*Syringothyris typus* ?]. *Spirifer grimesi* Hall [Burlington]. *S. keokuk* Hall [Keokuk]. *S. lateralis* Hall [Warsaw and Salem]. *S. lineata* Martin [not an American species]. *S. logani* Hall [Keokuk]. *S. mortonanus* Miller [Crawfordsville]. *S. neglectus* Hall [=*Spiriferella neglecta* (Hall), Keokuk]. *Spirifer plenus* Hall [=*Spiriferella plena* (Hall), Burlington]. *S. pseudolineatus* Hall [=*Reticularia pseudolineata* (Hall), Keokuk]. *S. suborbicularis* Hall [=*Brachythyris suborbicularis* Hall, Burlington and Keokuk]. *S. striatus* Martin [not an American species. Various species have been referred to this form.]. *Streptorhynchus crenistria* Phillips [Several forms have been referred to this European species. See *Orthothetes keokuk* (Hall), *Streptorhynchus minutum* (Cumings), etc.]. *Streptorhynchus keokuk* Hall [=*Orthothetes keokuk* (Hall), Keokuk]. *S. pectinaceum* Hall [=*Orthothetes chemungensis* (Hall), a Devonian species]. *Strophomena rhomboidalis* of authors [=*Leptaena analoga* (Phillips) and *L. convexa* Weller, Kinderhook and Burlington]. *Terebratula gorbyi* Miller [Commonly referred to *Dielasma*. See *D. sinuata* Weller. Miller's type is a Cretaceous species]. *Terebratula hastata* Sowerby [=*Dielasma bovidens* of American authors. Pennsylvanian]. *T. inornata* McChesney [Pennsylvanian]. *T. sacculus* Martin [=*Dielasma sacculus* (Martin), Carboniferous of Nova Scotia]. *Aviculopecten amplus* M. & W. [Keokuk, Illinois]. *A. colletti* Worthen [Crawfordsville]. *A. indianensis* M. & W. [Crawfordsville]. *A. oblongus* (M. & W.) [Warsaw, Ill.]. *A. winchelli* Meek [=*Crenipecten winchelli* (Meek), Waverly]. *Conocardium indianense* Miller [Crawfordsville]. *Cypricardella nucleata* Hall [=*Microdon nucleata* (Hall), Salem]. *Lithophaga lingualis* M. & W. [=*L. illinoisensis* Worthen, Crawfordsville]. *Monotis gregaria* M. & W. [Pennsylvanian]. *Myalina keokuk* Worthen [Keokuk, Iowa, Illinois, Missouri]. *Pinna subspatulata* Worthen [Warsaw]. *Sanguinolites multistriatus* Worthen [Crawfordsville]. *Dentalium primarium* Hall [Warsaw]. *D. primevum* [??]. *Bellerophon sublaevis* Hall [Warsaw, Salem, St. Louis]. *Capulus equilateralis* (Hall) [Crawfordsville]. *C. infundibulum* M. & W. [=*Igoceras pabulocrinus* (Owen), Crawfordsville]. *C. sulcatinus* Keyes [Crawfordsville]. *Platyceras equilateralis* (Hall) [=*Capulus equilateralis* (Hall) Keokuk]. *P. fissurellum* Hall [=*Igoceras fissurellum* (Hall), Burlington and Keokuk]. *P. infundibulum* [=*Igoceras pabulocrinus*]. *P. uncum* M. & W. [=*Orthonychia acutirostre* (Hall), Keokuk to St. Louis]. *Pleurotomaria shumardi* M. & W. [Warsaw]. *Conularia crawfordsvillensis* R. Owen [Crawfordsville]. *C. intertexta* Miller. *C. subcarbonaria* M. & W. [Keokuk, Illinois, Missouri]. *Orthoceras expansum* M. & W. [St. Louis?]. *Streptodiscus indianensis* Miller [=*Nautilus indianensis* (Miller)]. *Phillipsia bufo* (M. & W.) [=*Griffithides bufo* M. & W., Crawfordsville]. *P. meramecensis* Shumard [Keokuk to Chester], *P. portlocki* M. & W., [Keokuk, Illinois, Iowa, Missouri]. *Macrocaris gorbyi* Miller.

MERAMECIAN. *Salem*.—(Oölitic limestone, Warsaw, St. Louis, Middle limestone (part), Spergen bed, Spergen, Bedford, Bedford oölitic limestone, Indiana oölite, White River limestone, Salem stone, etc.) An entirely unnecessary amount of confusion has arisen in regard to the geologic name of this famous building stone of Indiana. Being an important quarry rock, many trade names, such as Bedford stone, White River stone, Salem stone, Bloomington stone, etc., have been applied to it. On the other hand the famous fossil bed at Spergen's (or Spurgeon's) hill, near Salem, early became known to collectors as the "Spergen Hill Bed." (So, for example, Lyon 1860, 619; Whitfield 1882, 42-49; H. S. Williams 1900, 348; Weller 1898, 313; Meek 1873, 383; Collett 1879, 272, etc.) Hall, however, who described the fauna (1856, 1-36; 1883, 319-375) from the "Spergen fossil bed" correlated the formation with the Warsaw of Illinois, and it came thereafter to be known as the Warsaw. Worthen (1866, *et. seq.*) discarded the name Warsaw, placing the formation in the St. Louis, and Indiana geologists influenced by this arrangement of the Subcarboniferous rocks, thereafter came to speak of the Spergen fossil bed as St. Louis. During no part of this period down to 1897 was there any definite proposal of Spergen or any other local name as a formation name nor any usage of the name in that sense. In 1897 Hopkins and Siebenthal (1897, 289-427) published an elaborate account of the Indiana oölitic limestone and quarry industry, and in accordance with their custom of suggesting local names for the Indiana formations, proposed to call this rock the *Bedford* limestone or

Fig. 28. Upper left, St. Louis, McCormick's Gorge, Spencer, Indiana. Upper right, Salem limestone, near Bloomington. Lower left, Salem limestone, near Victor. Lower right, Harrodsburg limestone, near Harrodsburg, Indiana.

Bedford oölitic limestones from Bedford, Lawrence County, Indiana, where the most extensive quarries are located. This name was used in subsequent Indiana reports during Mr. Blatchley's incumbency as State Geologist, as well as in various reports and Bulletins of the U. S. Geological Survey. This was the first local name definitely proposed and used as a geologic formation name for this rock. It had been used long before as a trade name; and even as a geologic name by Elrod (1876, 209). Unfortunately this same name Bedford, from Bedford, Ohio, near Cleveland, had been definitely proposed by Newberry in 1871 for the well known shale just below the Berea sandstone of Ohio (1871, 22) and has been so used by the Ohio survey ever since.

In 1900 Professor C. S. Presser, who was studying the Mississippian and Devonian rocks of Ohio, wrote to the author of the present report that the Indiana name Bedford would have to yield to the Ohio name, and suggested that the writer propose a new name for the Indiana formation. The writer thereupon (1901a, 232-233) proposed, and carefully defined, the name *Salem limestone*, taking the name from the well known town of Salem in Washington County, Indiana, a place famous in the early days for its quarries in the oölitic limestone, and the nearest town of importance to the Spergen fossil bed. The name Spergen was considered and dismissed by the writer on account of the small thickness and non-typical lithologic character of the Spergen Hill outcrop, and the fact that the name Spergen Hill is entirely forgotten even by the people of that locality, and does not occur on any map of the region. In the same number of the Journal of Geology with the writer's proposal, statements were published by Prosser, Siebenthal and Dr. T. C. Chamberiain on the name of this formation and the general principles of geologic nomenclature.

In 1905 (1905, 19-20) Prosser again discussed the name of the Bedford shale of Ohio and quoted a decision of the Committee on Geologic Names of the United States Geological Survey as follows: "(1) That Bedford rock was used by Owen in 1862 in a *Report of Geological Reconnaisance of Indiana* 1859-60, p. 137, but the usage is so indefinite as not to constitute a preemption of the term for stratigraphic purposes. (2) Bedford shale is a term first employed by Newberry in *Ohio Geological Survey Report of Progress*, 1869, p. 21 and this usage should stand. Furthermore it is understood here that Mr. Cumings has recently proposed to drop the name Bedford limestone of Indiana, and substitute for it Salem limestone."

In spite of this clear and explicit proposal and definition of the name Salem in 1901, and the implied acceptance of the name by the Committee on Geologic names of the United States Geological Survey, as above quoted, and in spite of the fact that there was no prior definite proposal or usage of any other name for this formation except Bedford, Mr. Ulrich in 1904 (1904, 90, chart) substituted the name "Spergen Hill" as the name of this formation without comment or explanation. The following year he again used "Spergen", stating: "As this name [Bedford] has been constantly used for another formation in Ohio since 1871, Cumings rejects 'Bedford' and proposes 'Salem' instead. However, both of these names are objectionable as formation names, for they are widely employed as trade names of quarried stone." It may be submitted that if geologic formation names are to be rejected on this basis a general overhauling of geologic nomenclature will be necessary, and such names as Potsdam, Medina, Niagara, Laurel, Silver

Creek, St. Louis, Columbus, Berea, Sharon, Dakota, etc., etc., to mention a few only, will have to be discarded. It may also be remarked further that Salem is *not* and never has been widely used as a trade name.

In 1906 Cumings, Beede, *et al.* called attention to Mr. Ulrich's unwarranted use of the name Spergen (1906a, 1189); and Weller (1908, 82) characterized Ulrich's substitution of the name Spergen as "wholly unwarranted."

Nevertheless the U. S. Survey, reversing its 1901 decision, has apparently adopted this name: and in recent publications (Butts 1915, 164; 1918, 32-37), evidently realizing the flimsy pretext on which the name Spergen was proposed, has attempted to justify the name by quoting prior usage.

It may not be out of place, therefore, to review this part of the case also. The usages quoted are Lyon, 1860, Siebenthal, 1897, Weller 1898, and H. S. Williams 1900. Sidney Lyon (1860, 612-621) gives a section and description of the "Subcarboniferous" rocks of Kentucky in which occurs the following statement (p. 619) in a paragraph headed "*o. Middle of the Carboniferous Limestone.*—": "About one hundred and eighty feet above the base of o lies the equivalent of the 'Spurgen's Hill' beds of Washington County, Indiana, which Professor James Hall considers the equivalent of the beds at Warsaw, the beds above Alton, Illinois, and those near Bloomington, Indiana." If this is to be regarded as a proposal of "Spergen" as a geologic formation name, a second general overhauling of geologic nomenclature will be necessary with this precedent in mind. The Siebenthal usage of 1897 is as follows (1897, 298): "By other writers it has previously been called Bedford stone . . . Spergen Hill limestone," etc. Since Siebenthal himself proposed the name Bedford, he evidently did not regard Spergen as a prior name, and certainly cannot be quoted as authority for the use of a name which he specifically refused to use. Weller in 1898 (1898, 313) says "One of the best known of the St. Louis faunas is that of the Spergen Hill beds of Indiana." In 1908, as above quoted, he characterized Ulrich's use of Spergen as "wholly unwarranted," and himself (as also in 1914) used Salem. H. S. Williams 1900, 348) says "As has already been announced,* the age of the Spring Creek limestone [of Arkansas] is about equivalent to the Warsaw, St. Louis or Spergen Hill formations." This is merely a repetition of the familiar usage initiated by James Hall, as is indicated further by the fact that Weller, who collaborated with Professor Williams on the Arkansas report, does not, as indicated by his statements quoted above, regard Williams' usage as a formal proposal of the name.

In view of the above, and of Weller's emphatic statements of 1908 and 1914,** one is led to wonder for what reasons the latter eminent authority on the Mississippian has, in his 1920 papers, abandoned the name Salem. If a geologic name cannot be said to have been "proposed" until it has been twice passed on by the Committee on Geologic names of the U. S. Geological Survey, it is high time American geologists were apprised of this rule. The name *Salem limestone* is the name officially adopted by the Indiana Geological Survey for this formation.

The Salem limestone was correlated with the Warsaw by Hall (1856, 4-34) and called Warsaw by Collett (1876, 315-317; 1879, 272); Gorby (1886, 124, 136-146); Elrod (1899, 258-260); Hopkins (1904, 17), and other geologists; but by Worthen (1866, etc.) was placed in the St. Louis division. The

* Am. Jour. Sci. Ser. 3, Vol. XLIX, pp. 94-97, 1895.

** Weller (1914, 20), says "Cuming's name clearly has priority."

term oölitic limestone, customarily applied to this rock (though strictly speaking a misnomer), was at first used by Owen (1837) for all the limestones of the Mississippian, but afterwards restricted (1859) to the quarry rock. It is not in reality an oölite, though the abundance of small ostracods and foraminifera gives it the appearance of an oölite. Occasionally true oölite grains are present and the name "oölitic limestone" or "Indiana Oölite" (Blatchley 1908, 307) is an appropriate enough trade name, but cannot have any standing as a geologic name. Weller's work in Illinois (1908, 81-102) has shown conclusively that this formation is distinct from the typical Warsaw and constitutes a division of the Mississippian of cognate rank with the Keokuk, Warsaw, St. Louis, etc. In fact it had already been given such distinct status in the Meramec division by Ulrich (1904, 90; 1905, 30). As already pointed out, the Harrodsburg limestone is in reality the true representative of the Warsaw of Illinois.

The name Bedford still clings to the formation as a trade name (see, for example, Beede 1915, 204-206), in spite of Blatchley's attempt to substitute the name "Indiana oölitic limestone" (1908, 307); and much of the literature of the past twenty-five years having to do with the quarry rock and industry will be found under that name, or the older term "oölitic limestone."

The chief source of information on the stratigraphy and paleontology of the Salem limestone of Indiana is the memoir on the Salem by Cumings, Beede, Branson and Smith in 1906 (1906a, pp. 1189-1486). Other useful references are as follows:—Hali 1856, 2-36, first description of the fauna; Owen 1859, oölitic limestone; R. Owen 1862, 137, uses name "Bedford rock"; Lyon 1860, 619; "Spergen's Hill bed"; Worthen 1866; Meek 1873, 383, Spergen fossils; Collett 1874, 276, "famous Bedford stone"; Elrod and McIntire 1876, 209 "Bedford Limestone", apparently used as a formation name; Collett 1879, 272, equals Warsaw; Whitfield 1882, 39-97, redescribes the fauna with illustrations; Hall 1883, 319-375, Fauna, with Whitfield's plates; Gorby 1886, "Salem stone", describes the section at Salem; Thompson 1886, 28-32; Thompson 1889, 61, describes Spergen fossil bed; Thompson 1892, 43-53, describes building stone, conditions of deposition, and section at Salem; Hopkins and Siebenthal 1897, 289-427, detailed and exhaustive description of the formation and its uses as a commercial stone: Propose name Bedford; Elrod 1899, 258-260, regards it as the lowest (Warsaw) member of the St. Louis; Blatchley 1900, 22-23, calls it "Indiana oölitic stone"; Blatchley 1901, 325, area of outcrop and lithology; Siebenthal 1901a, 390-393, quarry development, statistics, etc.; Cumings 1901a, 232-233, proposes name Salem; Siebenthal 1901b, 234-235 defends name Bedford; Prosser 1901, 270-272 states case for Bedford shale of Ohio; Chamberlin 1901, 267-270 discusses general principles of geologic nomenclature; Newsom 1903, 281, puts it in St. Louis; Ashley 1903, 83-86, 103-119, sections; Ulrich 1904, 90 (chart), proposes name "Spergen"; Reagan 1904, 206-216, analyses; Blatchley 1904, 48, mention; Hopkins 1904, 52-57, synonyms, etc.; Blatchley 1904a, use for lime; Prosser 1905, 19-20 on use of name "Bedford"; Ulrich 1905, 30, again uses Spergen; Blatchley 1906, 884, road material; Shannon 1906, 955-957 "Salem or Bedford oölitic stone", for road metal; Cumings 1906, 85-100, weathering; Shannon 1907, 56, mention; Blatchley 1907, 28-31, calls it "Indiana oölitic limestone"; Blatchley, 1908, 307, says "Indiana oölitic limestone" is the official name for future Indiana reports; Hopkins 1908, 310-355, texture and economic value; R. S. Blatchley 1908, 356-459, detailed description of quarry industry; Siebenthal 1908, 307, discusses name, still prefers "Bedford"; Shannon 1908, 108-110, describes topography, weathering, soils, etc.; Weller 1908, 82, name Spergen "unwarranted", uses Salem; Tucker, 1911, 15, mention; Van Tuyl, 1912, 167-168, uses Salem; Weller 1914, 13, 19-21, Salem "clearly has priority"; Beede 1915, 204-206, description and stratigraphy; Mance 1915, 230-312, detailed discussion of the quarry industry from the standpoint of utilization of waste stone; Butts 1915, 164, uses "Spergen"; Mance 1917, 1-204, elaborate discussion of quarry economics; Butts 1918, 32, uses "Spergen"; Barton, 1918, 362, etc., uses Salem; Weller 1920, 408, uses Spergen; Weller 1920a, 80, 96, "Spergen (Salem)"; Illinois Geological Survey, Press Bulletin, 1921, uses Salem.

Description.—The Salem limestone is, in its typical development, what Grabau would call a "calcarenyte" or calcareous freestone. It is a very massive bedded, granular pure carbonate of lime rock of a very light bluish gray color in deep unweathered material, but changing to light buff near the surface. Most fresh quarry faces show very little evidence of bedding other than the occasional stylolite seams. When the strata are cross-bedded, as is commonly the case, and especially in weathered exposures, the thick beds stand out conspicuously. The rock often weathers with a pitted or "honey-combed" surface which gives the natural outcrops a most interesting and picturesque appearance. Not uncommonly, however, the weathered surfaces are quite smooth and clear cut. On hill sides the smooth rounded weathered outcrops may have the appearance of huge glacial boulders. The above description applies to the quarry rock. Beede points out, however, (1915, 204-206) that associated with this phase of the stone are bituminous marls of a dark buff or brown to almost black color and fetid odor, bedded like shale, and occurring either below or above the quarry stone, or sometimes both below and above. This rock goes by the name of bastard stone among the quarrymen. Beede consequently regards the formation as consisting of great lenses of calcareous sandrock (the quarry rock) imbedded in a matrix of finer, more or less calcareous, material. The finer material occupies the place of the "oölitic" where the latter is missing. He believes that the fine grained rock consists of material washed out of the calcareous sand by wave and current action and deposited in the quieter pools. The writer would suggest that a modern parallel for the deposition of the Salem may be found in the calcareous sands and oozes of the Florida Keys and banks.

True oölite is not as common in the Salem as the well known name "oölitic limestone" would lead one to suppose. Most of the oölitic appearance is due to small round fossils belonging to the Foraminifera and Ostracoda. Some true oölitic or concentric grains do occur, however.

The thickness varies from practically nothing up to 90 or 100 feet. The best outcrops may be seen in the vicinity of Salem, Bedford, Harrodsburg, Victor, Sanders, Bloomington, Ellettsville and Stinesville.

Fauna.—The Salem is nearly always fossiliferous, but not many localities exhibit the fauna in its perfection. Such localities are the Spergen Hill cut at Harristown near Salem, Lanesville, the old Cleveland quarry near Harrodsburg, Dark Hollow near Bedford, the old quarries just north of Elletsville, and the Hunter valley quarries near Bloomington. With very few exceptions, the specimens are very small and represent a dwarfed fauna. In some cases, as at the Cleveland quarry, they seem to be worn, and the shells, etc., probably had been rolled about on the sea beach or possibly blown by the wind into calcareous sand dunes.

The subjoined list of species is taken from the paper by Cumings, Beede, Branson, and Smith (1906, 1373-1375): *Endothyra baileyi* Hall, *Amplexus blairi* Miller, *Bordenia zaphrenti-*

formis Greene, *Ceratopora agglomerata* Grabau, *Cyathaxonia venustum* Greene, *Cystelasma lanesvillensis* Miller, *C. rugosum* Ulrich, *C. septatum* Greene, *C. tabulatum* Beede, *Enallophyllum grabaui* Greene, *Michelinia indianensis* Beede, *Monilopora beecheri* Grabau, *Palaeacis cuneiformis* M-E. & H., *Syringopora monroensis* Beede, *Zaphrentis casseday* M-E. & H., *Z. clinatus* Greene, *Z. compressus* M-E. & H., *Batocrinus calyculus* (Hall), *B. crassitestus* Rowley, *B. davisi* Rowley, *B. davisi lanesvillensis* Rowley, *B. davisi sculptilis* Rowley, *B. icosidactylus* Casseday, *B. irregularis* Casseday, *B. magnirostris* Rowley, *B. sacculus* M. & G., *B. salemensis* M. & G., *Dichocrinus blatchleyi* Beede, *D. oblongus* W. & S., *D. striatus* O. & S., *D. sp.*, *Dizygocrinus decoris* (Miller), *D. cuconus* (M. & W.), *D. unionensis* (Worthen), *D. whitei* (W. & S.), *D. sp.*, *Forbesiocrinus sp.*, *Icthyocrinus clarkensis* M. & G., *Platycrinus bonoensis* White, *P. boonevillensis* Miller, *P. sp.*, *Poteriocrinus coryphaeus* Miller, *Synbathocrinus swallowi* Hall, *Talarocrinus simplex* (Shumard), *T.* cf. *trijugus* M. & G., *Pentremites conoideus* Hall, *P. conoideus amplus* Rowley, *P conoideus perlongus* Rowley, *Tricoelocrinus meekianus* Etheridge and Carpenter, *Metablastus wortheni* (Hall), *Tricoelocrinus woodmani* (M. & W.), *Archaeocidaris norwoodi* Hall, *Ortonia blatchleyi* Beede, *Spirorbis annulatus* Hall, *S. imbricatus* Ulrich, *S. nodulosus* Hall, *Cystodictya lineata* Ulrich, *C. ocellata* Ulrich, *Dichotrypa flabellum* (Rominger), *D. sp.*, *Fenestella compressa elongata* Cumings, *F. exigua* Ulrich, *F. multispinosa bedfordensis* Cumings, *F. rudis* Ulrich, *F. serratula* Ulrich, *F. serratula quadrata* Cumings, *Fenestella tenax multinodosa* Cumings, *Fenestella nodosa* Prout, *Fenestella tenuissima* Cumings, *Fenestralia sancti-ludovici* Prout, *F. sancti-ludovici compacta* Ulrich, *Fistulipora spergenensis* Rominger, *F. spergenensis minor* Cumings, *Glyptopora michelinia* (Prout), *Hemitrypa beedei* Cumings, *H. plumosa* (Prout), *H. proutana* Ulrich, *Hemitrypa proutana nododorsalis* Cumings, *Pinnatopora sp.*, *Polypora biseriata* Ulrich, *P. internodata* Cumings, *P. maccoyana* Ulrich, *P. simulatrix* Ulrich, *P. spininodata* Ulrich, *P. striata* Cumings, *Rhombopora bedfordensis* Cumings, *R. sp.*, *Stenopora rudis* Ulrich, *S. tuberculata* (Prout), *S. sp.*, *Worthenopora spinosa* Ulrich, *W. spatulata* (Prout), *Athyris densa* Hall, *Tetracamera subcuneata* (Hall), *Centronella crassicardinalis* Whitfield [=*Athyris densa* Hall], *Cleiothyris hirsuta* (Hall) [=*Cliothyridina hirsuta* (Hall)], *Dielasma formosum* (Hall), *Dielasma gorbyi* Beede non Miller [=*D. sinuata* Weller, not *Terebratula gorbyi* Miller. The latter is a Cretaceous species], *Dielasma turgidum* Beede non Hall [=*D. sinuata* Weller, and *Girtyella turgida* (Hall)], *Eumetria marcyi* Girty, Beede, Morse, etc., non Shumard [=*E. verneuiliana* (Hall)], *Orthothetes minutus* Cumings, [=*Streptorhynchus minutum* (Cumings), Beede's figure 7, pl. 20, *op. cit. supra*, is *Streptorhynchus ruginosum* (Hall & Clarke).] *Productus biseriatus* Whitfield, [=*Echinoconchus biseriatus* (Hall)], *P. burlingtonensis* var. Beede [=*P. altonensis* Norwood and Pratten], *P. gallatinensis* Beede non Girty [=*P. altonensis* N. & P.)], *P. indianensis* Hall, *Pugnax grosvenori* (Hall), [=*Camarotoechia grosvenori* (Hall)], *Pugnax quadrirostris* Beede, *Reticularia pseudolineata* Beede non Hall [=*R. setigera* (Hall)], *R. setigera* (Hall), *Rhipidomella dubia* (Hall), *Rhynchonella ricinula* Hall [=*Allorhynchus macra* (Hall)], *R. mutata* Hall [=*Camarotoechia mutata* (Hall)], *Seminula trinuclea* (Hall), [=*Composita trinuclea* (Hall)], *Spirifer bifurcatus* Hall, *S. horizontalis* Rowley (?), *S. lateralis delicatus* Rowley [=*S. lateralis* Hall] *S. subaequalis* Beede non Hall [=*Spirifer tenuicostatus* Hall], *S. subcardiiformis* Hall [=*Brachythyris subcardiiformis* (Hall)], *S. suborbicularis* Hall [=*Brachythyris suborbicularis* (Hall)], *Spiriferina norwoodana* (Hall), *Conocardium carinatum* Hall, *C. castastomum* Hall, *C. prattenanum* Hall, *Cypricardinia indianensis* Hall, *Edmondia subplana* (Hall), *Goniophora plicata* (Hall), *Macrodon sp.*, *Microdon ellipticus* Whitfield, *M. oblongus* (Hall), *M. subellipticus* (Hall), *Nucula shumardana* Hall, *Nuculana nasuta* (Hall), *Pteronites spergenensis* Whitfield, *Acmaea sp.*, *Anomphalus rotuliformis* Cumings, *Bellerophon gibsoni* White, *B. sublaevis* Hall, *B. sp.*, *Bembexia elegantula* (Hall), *Bucanopsis textilis* (Hall), *Bulimorpha bulimiformis* (Hall), *B. canaliculata* (Hall), *B. elongata* (Hall), *Coelocconus sp.*, *Conularia greeni* M. & G., *C. missouriensis* Swallow, *Cyclonema leavenworthanum* (Hall), *Eotrochus concavus* (Hall), *Gryphochiton* (?) *parvus* Stephens, *Holopea proutana* Hall, *Loxonema yandellanum* Hall, *Macrocheilus littonanus* (Hall) [=*Sphaerodoma littonana* (Hall)], *Macrochelius stinesvillensis* Cumings, *Murchisonia terebriformis* Hall, *M. vincta* (Hall), *M. sp.*, *Orthonychia acutirostre* (Hall), *Pleurotomaria conula* Hall, *P. meekana* Hall, *P. nodulostriata* Hall, *P. piasensis* Hall, *P. subglobosa* Hall, *P. swallowana* Hall, *P. trilineata* Hall, *P. wortheni* Hall. [Probably none of these species in strictness belong to the genus *Pleurotomaria*]. *Polytremaria* (?) *solitaria* Cumings, *Soleniscus glaber* Cumings, *Solenospira attenuata* (Hall), *S. turritella* (Hall), *S. vermicula* (Hall), *Strophostylus carleyana* (Hall), *Straparollus spergenensis* (Hall), *S. planorbiformis* (Hall), *S. planispira* (Hall), *S. quadrivolvis* (Hall), *Subulites harrodsburgensis* Cumings, *Nautilus clarkanus* Hall, *Orthoceras epigrus* Hall, *O. sp.*, *Temnocheilus sp.*, *Griffithides bufo* (M. & W.), *Leperditia carbonaria* (Hall), *L. sp.*, *Cytherellina glandella* Whitfield. (The fish remains described by Branson are not included.)

St. Louis and St. Genevieve.—(Mitchell, St. Louis, Barren, Lithostrotion limestone, Cavernous limestone, Lawrence-Crawford (part), Concretionary limestone, Middle limestone (part), Paoli (in part), Mammoth Cave formation of A. M. Miller). This great limestone formation was long ago recognized as a distinct unit in the Indiana Mississippian and was consistently correlated on both faunal and lithologic grounds with the St. Louis of the Mississippi river section. It was at first called the "Barren" limestone by Owen (1859, 22-23) who also mentions the common occurrence of chert and the lithographic character of the rock. In 1856 he speaks of it as the "Lithostrotion or Barren" limestone, and Lyon (1860, 612-621) gave it the name Cavernous limestone. R. Owen (1862, 124-132) apparently includes these rocks in his "Middle or Lawrence-Crawford" limestone, though the precise limits of his subdivisions are not clear. He mentions *Lithostrotion* as a common fossil and notes the abundance of chert and the caves in this formation. Worthen (1866, 83-89) includes in the St. Louis group the "oölitic limestone which outcrops at the river's edge, three miles above Alton, and the equivalent beds at Bloomington, and Spergen Hill, Indiana, and the blue calcareo-argillaceous shale and magnesian and arenaceous limestones at Warsaw". He gives *Lithostrotion canadense* and *L. proliferum* as the characteristic fossils. The inclusion of the Salem limestone and Warsaw in the St. Louis group has already been commented on. From 1872 (Cox in 3d and 4th Indiana Reports) to 1897, when Siebenthal proposed the name Mitchell, this group of limestones was invariably referred to as the St. Louis formation. (So, Cope 1872, 157; Borden 1874, 168; Collett 1874, 195, etc.; Cox 1875, 14; Elrod and McIntire 1876, 207, etc.; Collett 1876, 315-317; Collett 1879, 303, etc.; Brown 1884, 78; Gorby, 1886, 136-147, 205; Thompson 1886, 78, etc.; Thompson 1892, 512; Scovell 1897, 513.)

In 1897 Siebenthal (1897, 298-299), evidently believing that the mass of limestones between the well known quarry stone (Salem) and the Chester sandstones, could not be so exactly correlated with the St. Louis (See Hopkins 1904, 57), though as a matter of fact he gives no explanation of his procedure, named the formation *Mitchell* limestone, from the town of Mitchell in Lawrence County, and this name has been used consistently in Indiana geology ever since. No further advance in deciphering the precise stratigraphy of the Mitchell was made by the Indiana Survey till 1915, when Beede (1915, 206-212) pointed out the fact that both the typical St. Louis and the St. Genevieve are represented in this formation. Dr. Elrod, however, (1899, 258-267) in a paper generally overlooked by other geologists, had soon after the name Mitchell was proposed, called attention to the conspicuous chert bed midway in the Mitchell limestone, which he called "Lost River chert," and had named the upper Mitchell limestone above the chert, the "*Paoli* limestone" (*loc. cit.*, 259). He defined the Paoli limestone as including "all the rocks found below the first Kaskaskia sandstone and above the Lost River Chert." He proposed, since Siebenthal did not define the upper limit of the Mitchell, to restrict the name Mitchell to the limestone *below* the chert. As so redefined the Mitchell would be almost exactly equivalent to the St. Louis, and the Paoli limestone to the St. Genevieve and Gasper oölite. It should be borne in mind, therefore, that if local names are to be used for these recognizable subdivisions of the "Mitchell" in Indiana, Elrod's names must receive consideration.* His name "Paoli" would have prior claims over Ulrich's "Fredonia." It is proposed here, however, to restrict the term Paoli to the representative of the lower Gasper in Indiana. Elrod's paper is the first clear recognition of the dual character of the Mitchell,—sixteen years before Beede again pointed it out. Between the two dates, however, Ulrich also (1911, pl. 29) had correlated the Mitchell with both St. Louis and St. Genevieve.

Matters have not been bettered so far as concerns the terminology of this limestone succession, by Professor A. M. Miller's recent proposal (1917; 1919, 103) of the name Mammoth Cave formation for the exact equivalent in Kentucky of the Mitchell limestone. It would seem best to follow Butts' example (1918) discarding both Mitchell and Mammoth Cave, and returning to the long established names St. Louis and St. Genevieve; and the latter procedure is adopted in the present revision. Mitchell may still be used as a convenient handle for the whole mass of cavernous limestones between the Salem and the clastic Chester. The extreme top of the Mitchell appears to be a member of the Gasper oölite and is here called Paoli limestone. The Fredonia of Ulrich (1905, 39-53) appears to be about equivalent to the St. Genevieve (or Paoli limestone of Elrod) as represented in Indiana.

The St. Genevieve is here included, with the St. Louis in the Meramec series, an arrangement in accordance with that recently approved by Weller (1920, 283-284; 1920a, 96).

(On the Mitchell see also Blatchley and Ashley 1898, 21; Ashley 1898, 74-79; Newsom 1903, 282, St. Louis; Ashley 1903, 66 and 77-83; Hopkins 1904, 57-64 corresponds "in part" to the St. Louis; Blatchley 1904a, 217-218, mentions the oölite near top of Mitchell; Reagan 1904, 206, 216; Blatchley 1906, 884, 894, 915-918 Road metal; Cumings 1906, 85-100 weathering; Ellis 1906, 761; Shannon 1906, 960-962; Taylor 1906, 1005; Blatchley 1907, 32; Shannon 1908, 111-113; Siebenthal 1908, 308-309, detailed sections; Shannon 1909, 277-278; Greene 1909, 175-184 caves; Tucker 1911, 15; Beede 1915, 206-212 description, sections, correlations; Logan 1919, 26, 45, etc.).

Description.—The lower or St. Louis division of the Mitchell limestone is a close-grained, hard, compact, rather thin-bedded, light dove-colored, semilithographic limestone breaking with conchoidal fracture, with occasional beds of soft blue shale and shale partings. It is conspicuously jointed and cavernous, the whole area of its outcrop abounding in caves and sinkholes. The bedding planes are usually marked by stylolites. Chert occurs commonly in stringers and layers. The shale beds are especially characteristic of the lower 40 feet of the Mitchell. This division is capped by a conspicuous and persistent bed of chert or cherty limestone,—the "Lost River Chert" of Elrod. Above this chert horizon the limestone is commonly oölitic and more heavily bedded, though of the same light blue or drab color. *Lithostrotion canadense* which characterizes the beds below the Lost River chert, is totally lacking in the beds above, which seem to correlate with the Fredonia member of the St. Genevieve. *Diaphragmus elegans* is a common fossil.

The total thickness of the Mitchell formation is about 200 feet in Monroe County and 300 feet or more in the Ohio River counties.

Fauna.—The various older lists of fossils from the St. Louis of Indiana given in the Indiana reports do not discriminate between the fauna of the St.

* Dr. Moses N. Elrod, like W. W. Borden, is another Indiana geologist to whom belated justice should be done. His reports on Shelby, Bartholomew and Orange counties, etc., show a grasp of stratigraphic relationships beyond that evinced in most of the Indiana county reports.

Louis, *sensu strictu*, and the Salem ("Warsaw") faunas. No adequate lists from the St. Louis and St. Genevieve of Indiana are as yet available. Butts' lists of species from the Kentucky area immediately adjoining the river counties of Indiana on the south, are given below in the belief that most of these species will also be found in Indiana.

Fauna of the St. Louis limestone of Meade, Hardin and Breckenridge counties, Kentucky, according to Butts (1918, 41-42):—*Syringopora* sp., *Cladochonus* sp., *Lithostrotion proliferum* Hall, L. *basaltiforme* Owen, [*L. canadense* (Castlenau)], *Zaphrentis carinata* Worthen, *Meniscophyllum* sp., *Hapsiphyllum calcariforme* (Hall) [*Zaphrentis calciformis* Hall ?], *Melonites multiporus* Norwood and Owen, *Platycrinus huntsvillae* W. & S., *Pentremites cavus* Ulrich, P. *conoideus* Hall, *Stenopora* sp., *Batostomella* sp., *Fistulipora* sp., *Fenestella* several sp., *Polypora* several sp., *Cystodictya lineata* Ulrich, C. *pustulosa* Ulrich, *Dichotrypa* sp., *Worthenopora spinosa* Ulrich, [From the writer's observations, it is likely that most of the Bryozoa listed under the Salem also occur in the St. Louis]. *Rhipidomella* dubia Hall, *Schuchertella* sp., *Orthotetes kaskaskiensis* (McChesney), Productus *scitulus* M. & W., *P. altonensis* N. & P., P. near *viminalis* White, P. *ovatus* Hall, *Pustula indianensis ?* (Hall), P. *biseriata* (Hall), *Camarotoechia mutata* Hall, *Girtyella indianensis* (Girty), *Tetracamera acutirostra* (Swallow), *Spiriferella* n. sp. [Butts], *Spiriferina salemensis* Weller, *Spirifer pellaensis* Weller, S. *bifurcatus* Hall, S. *tenuicostatus* Hall, *Brachythyris subcardiiformis* (Hall), *Reticularia setigera* (Hall), *R.* n. sp. [Butts], *Composita trinuclea* (Hall), *Cliothyridina sublamellosa* (Hall), *Eumetria verneuiliana* (Hall), *Allorisma* sp., *Streblopteria* near *herzeri*, *Aviculopecten* sp., *Schizodus* sp., *Bellerophon sublaevis ?* Hall, *Bucanopsis textilis* (Hall), *Euomphalus* sp., Platyceras sp., *Nautilus* sp., *Griffithides* sp.

From the Fredonia oölite member of the St. Genevieve Butts (*loc. cit.*, pp. 56-57) lists the following: *Chaetetes* sp., *Syringopora* sp., *Lithostrotion harmodites* E. & H., *Triplophyllum spinulosum* E. & H, *Campophyllum* n. sp. [Butts], *Pentremites* princetonensis Ulrich, *Mesoblastus glaber* M. & W., *Platycrinus huntsvillae* W. & S., *Batocrinus ?* sp., Fenestella several sp., Polypora several sp., *Cystodictya* sp., *Rhipidomella dubia* (Hall), *Orthothetes kaskaskiensis* (McChesney), *Productus ovatus* Hall, *P.* parvus M. & W., Diaphragmus *elegans* Girty, Pustula (*Echinoconchus*) *genevievensis* (Weller), *Pugnoides ottumwa* (White), *Girtyella indianensis* (Girty), Spirifer *pellaensis* Weller, S. bi*furcatus ?* Hall, Composita *trinuclea* (Hall), *Cliothyridina sublamellosa* (Hall), Eumetria *verneuiliana* (Hall), *Allorisma* near *marvillense* Whitfield, Sphenotus *monroensis*, *Deltopecten monroensis?* (Worthen), *Conocardium* near *meekanum* Hall, *Schizodus depressus* Worthen, *Astartella* n. sp. [Butts], *Pleurotomaria*, several sp., *Solenospira* sp., *Bellerophon sublaevis* Hall, *Euomphalus similis* (M. & W.), *E. similis var planus* (M. & W.), *Straparollus spergenensis* (Hall), *Cyclonema subangulatum* Hall, *Naticopsis* n. sp. [Butts], *Platyceras* sp., *Cystelasma rugosum* Ulrich, C. *quinqueseptatum* Ulrich, *Michelinia princetonensis* Ulrich, M. *subramosa* Ulrich.

CHESTERIAN.—(Kaskaskia, Ferruginous sandstone (part), Archimedes limestones of Owen (part), Upper or Orange-Martin limestone, Pentremital limestone, Huron of Hopkins). As already indicated, the name Chester was proposed by Worthen in manuscript in 1853. Weeks (1902) credits the name to Swallow (1858, 5), who was the first to use the name in print. Worthen himself published the name in 1860 (1860, 697), and in 1866 in the first volume of the Illinois reports, Hall having in the meantime used Kaskaskia (1856a, 26; 1857, 187-203) for substantially the same rocks. In the 1866 report Worthen used the term "Chester Group," including in it the ferruginous sandstone as well as the higher beds. By general consent Worthen's name has been adopted as the standard name of this series of formations. Ulrich (1904) however, uses Kaskaskia for the upper division of the Chester, above the Cypress sandstone. Kaskaskia has been rather commonly employed for the Chester in whole or in part by Indiana geologists, and especially for the upper part of the series. (So, Collett 1874, 272-273, upper Chester; Collett 1876, 314, upper Chester limestone; Thompson 1886, 19, Upper Chester; Kindle, 1896, 331, etc., Chester; Siebenthal 1897, 300, Chester or Kaskaskia; Blatchley and Ashley 1898, 21, Chester or Kaskaskia.)

Of the Kaskaskia, Weller says (1920, 282-283): "Hall gave the name Kaskaskia limestone to the whole of the succession above a conspicuous sandstone formation in the Mississippi River bluffs of Randolph County [Illinois] which was commonly called the 'ferruginous sandstone'. Worthen used the name Chester limestone for the same beds which Hall called Kaskaskia, but included this Chester limestone with the underlying sandstone in what he called the 'Chester Group'." This "ferruginous sandstone" of the older workers is the Aux Vases (Keyes 1892, 296) or basal Chester sandstone of the present terminology; and consequently the use of Kaskaskia for the upper portion

Fig. 29. Chester formations at Ray's cave, Greene county.

only of the Chester is not in accordance with Hall's usage. Much of the uncertainty in regard to the relation of the so-called Kaskaskia to the Chester has arisen from the confusion of the Aux Vases or ferruginous sandstone with the Cypress sandstone, considerably higher up in the section. It would appear that no useful purpose can be subserved by the retention of the name Kaskaskia, without warping it entirely out of its original meaning; and the wiser procedure seems to be to abandon the term altogether, as a synonym of Chester.

Still further confusion was introduced into Indiana geological literature in 1903 when Ashley (1903, 71-77) on the suggestion of Hopkins, proposed the name "Huron", from the little town of Huron in Lawrence County, Indiana, for the rocks formerly called Chester in this State. Hopkins (1904, 17, 42, 64-67) gives Chester, Kaskaskia, Archimedes and Pentremital as synonyms of Huron. This new name was proposed in the belief that the "Illinois terms as originally used were more comprehensive and did not correspond with this remaining division of the lower carboniferous rocks of Indiana" *(loc. cit.)*. The name Huron was pecularily unfortunate, first because it tended to cause geologists to lose sight of the intimate connection of these rocks with the well-known Illinois series, and more particularly because the name Huron was already in common use for a Devonian formation of Ohio,—the Huron shale of Newberry (1871, 19).* Attention was very soon called to this duplication, and Prosser (1905, 22-24) gave a careful review of the use of the name Huron in Ohio and Michigan. It was clear that the Indiana name Huron could not be accepted, and in work undertaken under the direction of the Indiana University Department of Geology by Mr. F. C. Greene, it was suggested to him that if the rocks could not be shown to be approximately equivalent to the Illinois Chester, an alternative name be proposed. Mr. Greene's work (1911, 269-288), however, quite conclusively proved the equivalence, and he, therefore, reintroduced the older name Chester. R. S. Blatchley (1911, 88) in his quarry report uses both names; but Beede (1915, 212-216) says the name Huron must be abandoned. It occurs in the Indiana Reports and other literature from 1903 to 1914. (See Ashley 1903, 66, 71-77; Hopkins 1904, 17, 42, 64-67; Blatchley 1904, 50; Blatchley 1906, 144-146, 889, 894-896, 904-906, 918, 926; Taylor 1906, 1001; Blatchley 1907, 32; Shannon 1908, 113-114; Shannon 1909, 279-280; Tucker 1911, 15; Barrett 1914, 10.)

Up to 1905 the Chester series was not subdivided into named formations, though one of the conspicuous sandstones of southern Illinois and adjacent parts of Kentucky had long been known as the Cypress sandstone (Engelmann 1868, 189-190) from Cypress creek in Southern Illinois. This was mistakenly correlated by Engelmann with the "Ferruginous sandstone" of authors. The possibility of subdivision was, however, recognized, and lithologic names were commonly applied to the succession of sandstones and limestones. One of the first of these lithologic arrangements is that of Sidney Lyon in 1860 (1860, 614, etc.) who divided the succession into the 1st., 2d., 3d., and 4th. limestones (reading from the base upward) and the 1st., 2d., 3d., and 4th. sandstones. Lyons 5th. sandstone appears to be the Mansfield or Coal measures conglomerate. It is not possible to certainly identify Lyon's limestones and sandstones in the multiplicity of named formations of the present scale. Engelmann (in 1863, pub. 1868, 189) gave a section in southern Illinois in which the Chester was divided into ten lithologic members numbered from 1 to 10 in descending order, the limestones receiving the odd numbers, 1, 3, 5, etc., and the sandstones the even numbers 2, 4, 6, etc. No. 8 of his section he called the Cypress sandstone.

In 1876 Elrod and McIntire (1876, 207) gave a good account of the Chester rocks of Orange County, Indiana, and divided the succession into the

* See also Winchell. 1st Biennial report of Progress of the Geological Survey of Michigan, 1861, pp. 71-139, where he uses "Huron group" for all the rocks from the top of the Hamilton to a conglomerate overlying the Point Aux Barques gritstones.

Lower (No. 1), Middle (No. 2) and Upper (No. 3) limestones, and Lower (No. 1), Middle (No. 2) and Upper (No. 3) sandstones. This scheme was repeated by Kindle in 1896 (1896, 329-368), who gives the following section in Orange County (modern equivalents in brackets):—

6. Coarse sandstone and shale, 100 feet [Mansfield and Tar Springs].
5. Upper Kaskaskia limestone, 13 feet [Golconda].
4. Upper Kaskaskia sandstone, 40 feet [Cypress].
3. Middle Kaskaskia limestone, 15 feet [Beech Creek].
2. Lower Kaskaskia sandstone, 35 feet [Brandy Run, Reelsville, Elwren].
1. Lower Kaskaskia limestone, 18 feet [Beaver Bend, Sample, Upper Mitchell].

Siebenthal (1897, 300) also uses Lower, Middle and Upper limestone and Lower, Middle and Upper Sandstone; while Blatchley (1898, 21), Newsom (1903, 283-284), Greene (1911, 269) and Beede (1915, 212-216) use a similar terminology. Inasmuch as different authors have not always selected the same layers for their 1st., 2d., and 3d. or Lower, Middle and Upper formations, the scheme has not worked well. The fact is, as now ascertained, there are far more than three or four limestones and three or four sandstones in the Indiana Chester, and this fact is at the bottom of the discrepancies of the older work.

In most of the Indiana reports, however, no attempt is made to subdivide the Chester.

In 1905 Dr. Ulrich published a discussion of the stratigraphy of the Chester formations of western Kentucky (1905, 36-66) and made the first serious attempt to disentangle the Chester succession and to name the various subdivisions. His scheme is exhibited in the following table of formations (*loc. cit.* 24):

<table>
<tr><td rowspan="3">Chester Group</td><td>Kaskaskia Limestone</td><td>Birdsville formation
Tribune Limestone</td><td>Kaskaskia limestones of Hall.
Chester limestone of authors.
Huron formation of recent Indiana reports is same as Birdsville formation.</td></tr>
<tr><td colspan="2">Cypress Sandstone</td><td>Aux Vases sandstone of Keyes.
Probably also Big Clifty sandstone of Norwood.</td></tr>
<tr><td>St. Genevieve Limestone</td><td>O'Hara Limestone
Rosiclare Sandstone
Fredonia Oolitic Limestone</td><td>Two upper members referred to as lower Chester by Worthen and Engelmann. Entire formation referred to as St. Louis by Norwood.</td></tr>
</table>

Greene (1911, 269) tried to fit the Indiana succession to this scheme as follows:

Chester Group	Upper (3d) Limestone Upper Sandstone	Birdsville formation
	Middle (2d) Limestone	Tribune
	Middle Sandstone	Cypress
	Lower (1st) Limestone Lower Sandstone Oölitic upper part of the Mitchell Limestone	O'Hara Rosiclare } St. Genevieve Fredonia
	Remainder of the Mitchell	St. Louis

Dr. Weller began his important studies of the Chester in 1906, and has continued them down to the present. In 1913 (1913, 118-129) he published the following names of the Chester formations of Randolph and Monroe counties in Illinois (in descending order):

9. Clore limestone.
8. Palestine sandstone.
7. Menard limestone.
6. Okaw limestone.
5. Ruma formation.
4. Paint Creek formation.
3. Yankeetown formation.
2. Renault formation.
1. Brewerville sandstone (later referred to Aux Vases sandstone).

The St. Genevieve was not included by Weller in the Chester. The same list of formations was repeated in the Monograph of Mississippian Brachiopoda (1914, 12-13).

Weller found it impossible to fit his scheme of formations to that of Ulrich, and after several seasons of field work, succeeded in showing that the latter had made several serious mistakes. In the first place the limestones referred to the Tribune by Ulrich include three different horizons. The typical Tribune, at Tribune, Kentucky, is a limestone high up in the "Birdsville" and the equivalent of the Menard. "The Tribune east of Princeton, Kentucky, lies above the sandstone that had been mistakenly identified as Cypress by Ulrich. . . . The third limestone referred to the Tribune . . . proved to have a position beneath the sandstone which Ulrich had called the Cypress" and to belong to the O'Hara member of the St. Genevieve. Ulrich had also been "mistaken in his interpretation of the Cypress sandstone of Engelmann, the true Cypress being at the base of the Birdsville formation instead of beneath the so-called Tribune" (See Weller 1920a, 129). These errors lead to the additional error of including the St. Genevieve in the Chester, since Ulrich included two distinct formations in his O'Hara or upper St.

Genevieve limestone; his upper O'Hara being the equivalent of the Renault limestone, while the lower O'Hara is genuine St. Genevieve. The two formations are disconformable. Butts in conjunction with Weller also proved that a shale formation (the Paint Creek shale) lies between the true Cypress sandstone and the so-called Cypress of Ulrich. Southward in Kentucky the Paint Creek formation is represented by a part of the "Tribune" of Ulrich, renamed *Gasper oölite* by Butts (1918, 64).

In 1918 the Kentucky survey published a volume on the Mississippian of western Kentucky, to which Ulrich contributed a section on the Chester formations. In this publication Ulrich corrected some of his errors of 1905, but fell into other errors, especially that of correlating the Cypress sandstone with the lower Okaw of Weller, which is in reality the equivalent of the Golconda. Ulrich's "Standardized time scale" (1918, plate E) is as follows (reading down):

Chester	Clore formation	
	Palestine formation	
	Menard limestone	
	Tar Springs sandstone	
	Glen Dean limestone	
	Hardinsburg sandstone	
	Golconda formation	
	Cypress sandstone	
	Gasper limestone	
	Aux Vases sandstone	O'Hara limestone
	St. Genevieve Limestone	Rosiclare sandstone
		Fredonia oölite

Weller's complete general time-scale of the Chester (1920a, 80) is inserted below for comparison:

Chester	Upper Chester	Kinkaid limestone
		Degonia sandstone
		Clore limestone
		Palestine sandstone
		Menard limestone
		Waltersburg sandstone
		Vienna limestone
		Tar Springs sandstone
	Middle Chester	Glen Dean limestone
		Hardinsburg sandstone
		Golconda limestone
		Cypress sandstone
	Lower Chester	Paint Creek formation
		Yankeetown formation
		Renault formation
		Aux Vases sandstone

The Aux Vases of this classification is the Brewersville of Weller's 1913 and 1914 papers. The Gasper of Butts (part of Ulrich's Tribune) falls in the Paint Creek formation of Weller. The chief differences between Weller's and Ulrich's interpretations are in the lower part of the section.

In Meade and Breckenridge counties, Kentucky, Butts (1918, pl. 2, etc.) recognizes the *Sample* sandstone, which he places between his lower and upper Gasper formations, and the *Buffalo Wallow* formation which he correlates about with the Palestine and Clore of Illinois (pp. 112-113).

In 1919-1920 Dr. C. A. Malott of the Indiana Survey, published a revised section of the Chester of Indiana, based on several years' field work in Indiana and a careful study of the typical outcrops in Illinois and Kentucky. He adopted the Illinois and Kentucky names so far as reasonably exact correlations could be made, and added a number of new names for formations not thus far recognizable outside of Indiana. His scheme of formations is as follows: (1919, 8-18; 1920, 521-522):—

Chester Series of Indiana	Buffalo Wallow formation
	Siberia limestone
	Tar Springs sandstone
	Glen Dean limestone
	Hardinsburg sandstone
	Golconda limestone = [3d Limestone of Indiana authors]
	Indian Springs shale
	Cypress sandstone
	Beech Creek limestone = [2d Limestone of Indiana authors]
	Elwren sandstone and shale
	Reelsville limestone = [1st Limestone of Indiana authors]
	Brandy Run sandstone
	Beaver Bend limestone = [Upper Mitchell Ledge of authors]
	Sample sandstone
	Gasper oölite

The names *Siberia, Indian Springs, Beech Creek, Elwren, Reelsville, Brandy Run* and *Beaver Bend* are proposed by Malott.

This arrangement of formations is adopted with certain modifications suggested by Dr. Malott in the present revision.

Professor Allan D. Hole, of Earlham College, also published a paper in 1919 (1919, 183-185) in which on the basis of the faunas he identifies the lowest of the "three" Chester limestones in Orange County, Indiana, with the Renault of Illinois. The middle limestone he correlates with the Paint Creek of Illinois and the upper with the Okaw.

Dr. Malott and the writer grouped the Chester formations in three divisions as indicated on the Chart (p. 408) on the basis of the principal disconformities within the series; and have made use of this grouping in their teaching and in the work of the Indiana survey. It coincides exactly with the grouping recently published by Drs. Weller and Butts (1920a, 80).*

The new formation names are used by Dr. Logan, the State Geologist, in his recent reports on Kaolin deposits (1919) and Petroleum and Natural Gas (1920) and are officially adopted by the Indiana Survey.

The Chester of the Mississippi valley has been correlated with the Mauch

*The author wrote to Dr. Weller early in 1920 suggesting that these three subdivisions of the Chester be given geographic names, and proposed the names *West Baden* and *Stephensport* for the lower and middle divisions, respectively, suggesting that Dr. Weller propose a name for the upper division. The latter, however, expressed a desire to postpone the naming of these divilisons; and in deference to his wishes they are referred to merely as lower, middle and upper.

Chunk of Pennsylvania by Fontaine and White (1880), Lesquereux (1880), Lesley (1886, 618, 656) and others, and with the Maxville limestone of Ohio by Andrews (1871), Newberry (1874) and Herrick (1888, 21-23). Schucbert (1910, 553) correlates it with the Maxville of Ohio and with the Greenbrier and Mauch Chunk of Pennsylvania. Ulrich (1911, pl. 29) correlates the St. Genevieve with the Maxville and the upper Chester with the Mauch Chunk. These correlations cannot be regarded as possessing any claims to finality.

(Other references to the Chester of Indiana are as follows: Cox 1872, 81; Collett 1874, 195, 199-200, 272-274, 279; Cox 1875, 14; Collett 1876, 314-315, Kaskaskia limestone=upper member of Chester group; Collett 1879, 303-306, 425, 430-432; Gorby 1886, 147-149; Thompson 1886, 78, 84-85; Beede and Shannon 1907, 383-422, many detailed well sections.)

Paoli.—The lowest member of the Chester series in Indiana is a compact oölitic limestone, forming the top of the so-called Mitchell limestone. It is regarded as a member of the Gasper of Butts (1918, 64), the name Gasper taking the place of Ulrich's name "Tribune". The name Gasper comes from the Gasper river in Warren County, Kentucky. The name *Paoli* limestone is proposed for this member in Indiana. The rock is a typical oölite containing an aboundance of shot-like grains with concentric structure. Its color is dark gray to nearly white.

The Paoli is very fossiliferous in Places, and Butts and Ulrich (1918) list an extensive fauna from the Gasper formation of Kentucky. No adequate list is as yet available from Indiana.

Mooretown.—The first clastic unit of the Chester series of southern Indiana consists of shale or sandstone, or shale and sandstone, ranging from 10 to 60 feet in thickness. The typical thickness is about 20 feet. This horizon was at first thought to be equivalent to the Sample sandstone of Butts, and was so designated by Malott (1919) in his paper on the American Bottoms region. Later investigations, as yet unpublished, by Butts and Malott in southern Indiana and in Meade and Breckenridge counties, Kentucky, have shown that Malott's Brandy Run sandstone is equivalent to the Sample sandstone, and that the first clastic member of the Chester in Indiana is an as yet unnamed formation below the Beaver Bend limestone. It is present locally in Meade and Breckenridge counties, Kentucky, and is an important horizon throughout southern Indiana. It is used to mark the top of the Mitchell lithologic unit. Locally, as in the sandstone ridge south of New Amsterdam, overlooking the Ohio river, it is a very massive sandstone. Here it attains a thickness of 60 feet. Malott suggests that this important clastic horizon be named the *Mooretown* after the village of Mooretown, near which excellent exposures occur, which may be differentiated from the Beaver Bend limestone above and the Paoli limestone below.

Beaver Bend.—The Beaver Bend limestone of Malott (1919, 9) is another member of the lower Gasper of Butts. It is highly oölotic, often massive, and conspicuously jointed, forming along its outcrop an important spring line. The name comes from the big bend of Beaver creek, near Huron, Lawrence County, Indiana.

Sample.—In Kentucky Butts recognizes a sandstone formation in the midst of the "Gasper", which he calls the *Sample* sandstone (1918, 71). The name is from Sample station, Breckenridge County, Kentucky. At that point the formation is 40 feet thick and is a massive medium coarse-grained siliceous sandstone: but it varies notably in lithology and thickness from point to point, at times becoming quite shaly. It is a persistent horizon in Indiana, partly shale and partly sandstone, 30 to 50 feet thick.

Reelsville.—The Reelsville of Malott (1919, 10) is a thin limestone from 2 to 10 feet thick of a compact oölitic to sub-oölitic texture. It weathers to a reddish color. It is a very persistent stratum extending from the type locality in Putnam County, Indiana, into Meade and Breckenridge counties in Kentucky, where it constitutes a member of the Gasper formation.

Elwren.—The interval between the Reelsville and the next higher limestone (Beech Creek) is occupied by 40 to 50 feet of sandstone and shale, named Elwren by Malott (1919, 11), from Elwren station on the Illinois Central railroad in Monroe County, Indiana. It is exposed in the cuts near the station. Usually it consists of two masses of sandstone separated by a shale horizon. The sandstone is usually distinctly bedded; but is occasionally massive. It is of a rusty brown color on the weathered outcrop.

Beech Creek.—Malott gives the name Beech Creek (1919, 11) to a persistent limestone formation from 8 to 24 feet thick, to which the name "middle" or "second" limestone has commonly been applied in the Indiana reports. Beech Creek is a small stream in eastern Greene County, Indiana. The Beech Creek formation consists of two or more thin bedded strata, attaining a maximum thickness of 24 feet, but averaging about 12 feet. On the weathered outcrop it presents a ragged face made up of cubical chunks of limestone. "It is a gray, compact to sub-oölitic, and often semi-crystalline limestone, frequently locally quite completely oölitic, and contains large numbers of brachiopods, especially of the genus *Productus*. Of the succession of Chester limestones, none contains such a large number of large well-preserved crinoid stems standing out prominently on the weathered faces as the Beech Creek" (Malott 1919, 13). In the Greene County area the upper ledge of the formation contains considerable sand and clay and weathers yellow. In this ledge are also found several species of *Archimedes*. The outcrop of the Beech Creek limestone forms the most important spring bearing horizon of any of the Chester formations. This limestone forms the upper member of the "Gasper" of Butts.

Cypress.—This is one of the most conspicuous subdivisions of the Chester of Illinois, Indiana and Kentucky, and the first to receive a geographic name. The name Cypress was proposed by Engelmann in 1862 for "No. 8" of his section (1868, 189-190).* The name was used in the early Illinois reports by Engelmann and Worthen (1866, 356). In Kentucky Norwood (1876, 369) calls it the "Big Clifty" sandstone. It was often confused with the "Ferruginous" (Aux Vases) sandstone at the base of the Chester and was wrongly correlated with the Bethel (Yankeetown) by Ulrich. Ulrich revived the name in 1905 and it has been in common use since (see Greene 1911; Weller 1914; Ulrich & Butts 1918; Malott 1919, 1920; Logan 1919, 1920; Weller 1920, 1920a, etc.)

In Indiana it is usually a massive, non-bedded or indistinctly bedded, medium to coarse grained, yellowish to whitish sandstone, about 30 feet thick. It weathers reddish brown, especially along the joints. Being massive and strong, it is a cliff-making rock, and has often been mistaken for the Mans-

* Englemann's paper was read in 1862. The final publication of the volume of transactions seems to have been in 1868. Weeks (1902) gives 1868, and Weller 1863.

field, from which it differs in always lacking the quartz pebbles of the latter. In some sections the interval of the Cypress is occupied by shales. Toward the top the formation becomes shaly and grades into the Indian Springs shales. The Cypress is separated by a disconformity from the underlying formations.

Indian Springs.—Malott (1920, 521-522) has proposed the name Indian Springs shale for the blue-gray to olive shales about 20 feet thick commonly seen between the Cypress sandstone and the overlying Golconda limestone.

Golconda.—This name was proposed by Butts (1918, 91)* for the persistent limestone bed overlying the Cypress sandstone and Indian Spring shales. Butts includes in the Golconda a considerable thickness of limestone and shale, the lower shale evidently corresponding to our Indian Springs. Above this shale he has 10 feet of coarsely crystalline fossiliferous limestone which he calls the *Pterotocrinus capitalis* zone. Then come 80 feet of shale, and limestones weathering to a lemon yellow color, capped by 30 to 40 feet of solid, fairly pure limestone.

In Indiana Malott recognizes some 30 feet of limestone and shaly partings in this formation. The limestone is coarse, semi-crystalline, often oölitic, and contains large numbers of Crinoids and Blastoids and several species of *Archimedes*. It is the upper or 3d limestone of Indiana authors. It is the only Chester limestone containing chert, which in this rock is quite abundant.

Hardinsburg.—The conspicuous sandstone seen above the Golconda limestone in Southern Indiana and Kentucky has been named Hardinsburg by Butts (1918, 96) from Hardinsburg in Breckenridge County, Kentucky. At the type locality it is at least 30 feet thick. In Indiana it is a somewhat shaly, flaggy sandstone from 25 to 55 feet thick. The flagstones are ripple-marked and very resistent to erosion, forming small benches on the outcrop.

Glen Dean.—This limestone is named by Butts (1918, 97) from the town of Glen Dean in southern Breckenridge County, Kentucky, where it is exposed along the railroad. In its type locality it consists of varying proportions of limestone and shale and some sandstone. The limestone is bluish-gray, medium thick-bedded, and characterized by the fossils *Prismopora serratula* and *Archimedes laxus*. The lower layers are crowded with bases of *Agassizocrinus* and a slender *Archimedes*.

In Indiana the formation is from 10 to 45 feet thick, massive, cream-colored, oölitic, and abounding in specimens of large *Pentremites*. It is more characteristically oölitic than any other Chester limestone. It resembles lithologically the Beaver Bend limestone.

Tar Springs.—Butts, who proposes the name, credits it to Owen (1857). It does not appear that Owen used the name in a very definite way and the writer prefers to credit it to Butts (1918, 103). At Tar Springs, Breckenridge County, Kentucky, the springs issuing from the sandstone yield petroleum and this has given rise to the name. At the type locality the formation, according to Butts, is a thick-bedded, massive, medium to coarse grained, firmly cemented sandstone. Locally it is highly cross-bedded. The sand grains are commonly well rounded and about 0.3 mm. in average diameter. The formation varies from a few feet up to 50 feet thick. Plant remains are common in the rock.

* Butts *(loc. cit.)* says the name was first used by Ulrich in a paper read before the Geological Society of America in December, 1915. I do not find any reference to the name in the report and abstract of papers of that meeting.

In Indiana it is irregular in thickness and character, varying from a massive sandstone to bands of shale sandstone and limestone, in some sections scarcely recognizable. It rests disconformably on the Glen Dean.

Siberia.—Malott (1920, 521-522) proposes the name *Siberia* limestone from the town of Siberia, in northern Perry County, Indiana, for a thin limestone lying between the Tar Springs sandstone and the overlying sandy and variegated shales of the Buffalo Wallow formation. It is a rather persistent coarse-grained siliceous, somewhat crossbedded limestone varying from a few feet to 25 feet in thickness. It is probably of about the age of the Vienna limestone of Illinois.

Buffalo Wallow.—On account of the difficulty or impossibility of recognizing the higher formations of Southern Illinois and Western Kentucky, in Breckenridge and Warren counties, Kentucky, Butts (1918, 112) proposed the name Buffalo Wallow formation for the succession of variegated shales, limestones and sandstones above the Tar Springs formation. Owing to similar difficulties it has seemed best to use this designation for the highest Chester beds in Southern Indiana, and not attempt further subdivision at present. These shales are very variable in color, being yellow, maroon, green, blue, etc., and often very soft and slippery, especially in weathered outcrops.

Faunas.—No adequate lists of fossils are as yet available from these newly defined formations in Indiana. Butts and Ulrich (1918) have given lists from the equivalent Chester limestones in Kentucky; and the student is referred to their memoir. The faunas of the Chester of Illinois have been minutely studied by Dr. Weller, and his papers, particularly on the Mississippian Brachiopoda (1914) and the Geology of Hardin County, Illinois, (1920a) should be consulted. The older lists from the entire Chester would be of little benefit in the present stage of our knowledge, and are, therefore, not included. (See Kindle 1898, 408-488.)

From Butts' (1918) lists a few of the more characteristic species are selected and enumerated below:

Gasper oölite.—*Triplophyllum spinulosum* E. & H., *Talarocrinus simplex* Shumard, *T. cornigerus* (Shumard), *T. symmetricus* L. & C., *T. patei* M. & G., *Agassizocrinus conicus* O. & S., *Pentremites pyriformis* Say, *P. godoni* de France, *P. sulcatus* Roemer, *Archimedes meekanus* Hall, *A. laxus* Hall, *Lyropora quincunxialis* Hall, *Cystodictya nicklesi* Ulrich, *Orthothetes kaskaskiensis* (McChesney), *Productus inflatus* McChesney, *P. ovatus* Hall, *Diaphragmus elegans* (N. & P.), *Girtyella indianensis* (Girty), *G. brevilobata* Swallow, *Dielasma illinoisense* Weller, *Camarophoria explanta* (McChesney), *Spiriferina transversa* McChesney, *S. spinosa* (N. & P.), *Reticularia setigera* (Hall), *Composita subquadrata* (Hall), *C. trinuclea* (Hall), *Cliothyridina sublamellosa* (Hall), *Eumetria vera* (Hall).

Golconda limestone.—*Triplophyllum spinulosum* E. & H., *Pterotocrinus capitalis* Lyon, *P. depressus* L. & C., *Pentremites obesus* Lyon, *P. godoni* de France, *P. pyriformis* Say, *Archimedes communis* Ulrich, *A. meekanus* Hall, *A. swallowanus* Hall, *A. laticolvis* Ulrich, *A. terebriformis* Ulrich, *A. confertus* Ulrich, *Septopora subquadrans* Ulrich, *Lyropora quincunxialis* Hall, *Productus ovatus* Hall, *Diaphragmus elegans* N. & P., *Pustula (Echinoconchus) genevievensis* Weller, *Camarophoria explanata* (McChesney), *Spirifer leideyi* N. & P., *S. pellaensis ?* Weller, *Spiriferina spinosa* N. & P., *S. transversa* (McChesney), *Cliothyridina sublamellosa* (Hall).

Glen Dean limestone.—*Triplophyllum spinulosum* (E. & H.), *Pterotocrinus depressus* L. & C., *P. acutus* Wetherby, *P. acutus var. bifurcatus* Wetherby, *Agassizocrinus conicus* O. & S., *Pentremites pyramidatus* Ulrich, *P. pyriformis* Say, *P. godoni* deFrance, *Meekopora clausa* Ulrich, *Stenopora ramosa* Ulrich, *Anisotrypa solida* Ulrich, *Septopora cestriensis* Prout, *S. subquadrans* Ulrich, *Lyropora ranosculum* Ulrich, *Archimedes meekanus* Hall, *A. swallowanus* Hall, *A. laxus* Hall, *A. sublaxus* Ulrich, *Prismopora serratula* Ulrich, *Streblotrypa nicklesi* Ulrich, *Orthothetes kaskaskiensis* (McChesney), *Productus ovatus* Hall, *P. inflatus* McChesney, *Diaphragmus elegans* (N. & P.), *Camarophoria explanata* (McChesney), *Dielasma illinoisense* Weller, *Girtyella indianensis* (Girty), *Spirifer increbescens* Hall, *S. leidyi* N. & P., *Martinia sulcata* Weller, *Reticularia setigera* (Hall), *Spiriferina spinosa* (N. & P.), *S. transversa* (McChesney), *Composita trinuclea* (Hall), *C. subquadrata* (Hall), *Cliothyridina sublamellosa* (Hall), *Eumetria vera* (Hall), *E. costata* Hall.

Buffalo Wallow formation.—*Triplophyllum spinulosum* E. & H., *Pentremites pyramidatus* Ulrich, *Pterotocrinus depressus* L. & C., *Agassizocrinus conicus* O. & S., *Archimedes laxus* Hall, *A. swallowanus* Hall, *A. meekanus* Hall, *Septopora subquadrans* Ulrich, *Streblotrypa nicklesi* Ulrich, *Orthothetes kaskaskiensis* (McChesney), *Productus ovatus* Hall, *Diaphragmus elegans* (N. & P.), *Camarophoria explanata* (McChesney), *Spirifer leidyi* N. & P., *S. increbescens* Hall, *Spiriferina transversa* (McChesney), *S. spinosa* (N. & P.), *Reticularia setigera* (Hall), *Martinia sulcata* Weller, *Composita subquadrata* (Hall), *Cliothyridina sublamellosa* Hall, *Eumetria vera* (Hall), *Sulcatipinna missouriensis* Swallow, *Myalina* near *swallowi* McChesney, *Bucanopsis textilis* (Hall), *Euomphalus planidorsatus* M. & W.

CHAPTER V.

PENNSYLVANIAN.

The term Pennsylvanian or Pennsylvania series was first used by J. C. Branner (1891, xiii) and H. S. Williams (1891, 83, etc.). It was the proposal of Professor Williams; and supersedes the older term "Coal Measures". H. D. Rogers (1858) divided the Coal Measures or "Seral" series* into the Seral Conglomerate, Lower Productive, Lower Barren, Upper Productive, and Upper Barren Coal Measures. The basal conglomerate had long been known both in this country and in Europe as the Millstone Grit, a name which occurs occasionally in Indiana literature down to 1916 (ex. VanGorder, 1916, 248). The terms Lower Productive, Lower Barren, etc., have also been commonly used in America, especially in the Appalachian coal fields. It should be noted, however, that the barren limestone of Owen and Barren Group of Safford and Killebrew have nothing to do with these coal measures groups, and belong in the Mississippian. The Indiana Coal Measures have been commonly referred to the Lower Productive and Lower Barren divisions.

The Millstone Grit or Seral Conglomerate, was, in report HH of the Pennsylvania survey (Lesley 1877, xxvi) renamed the Pottsville conglomerate by Lesley.** The Pottsville of the Appalachian region is a composite series

* The so-called "transcendental nomenclature" of the Rogers Brothers (H. D. and W. B.) is now a matter of antiquarian interest only. Their names, referring literally to parts of the day, are as follows (modern equivalent in brackets) —:

XII. Seral	[Pottsville and Coal Measures.]
XI. Umbral X. Vespertine	[Mississippian.]
IX. Ponent	[Catskill.]
VIII. Vergent Cadent	[Portage and Chemung, Marcellus, Hamilton, Genesee.]
VII. Post-Meridian Meridian	[Oriskany, Schoharie, Corniferous.]
VI. Premeridian	[Lower Helderberg.]
V. Scalent Surgent	[Clinton, Salina.]
IV. Levant	[Oneida, Medina.]
III. Matinal	[Trenton, Utica, Lorraine.]
II. Auroral	[Beekmantown to Black River.]
I. Primal	[Cambrian.]

** It appears that Lesley suggested the name, although it first occurs in a report by F. and W. G. Platt. The latter credit it to Lesley, as does also Ashburner (1877, 533), to whom Weeks (1902) credits the name.

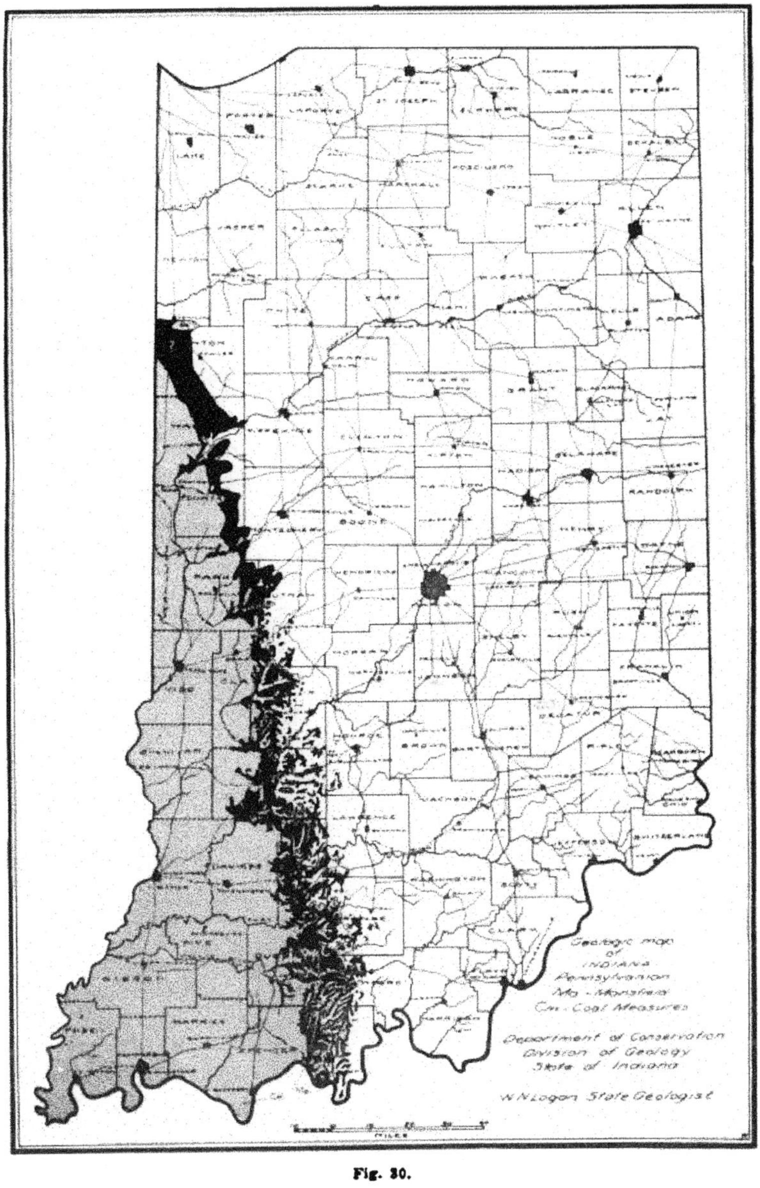

Geologic map
of
INDIANA
Pennsylvanian
Mo - Mansfield
Cm - Coal Measures
Department of Conservation
Division of Geology
State of Indiana
W N Logan State Geologist

Fig. 30.

of sandstones, conglomerates, shales and coals, of great thickness, and has in later years been subdivided into a large number of named formations, as indicated below.

The Lower Productive Coal Measures were also called the Allegheny series by Rogers (1840, 177) and this name used in all the later Pennsylvania reports, is the generally accepted designation of the lower coal measures proper. The Allegheny also contains a very large number of named formations, which will be listed below.

In Pennsylvania these important coal bearing beds are overlain by a series of formations (Lower Barren rocks) in which less commercially important coal is found, and to these beds Platt (1875, 8) applied the name Conemaugh series, without, however, adequately defining the term. It was carefully redefined by O'Hara (1900, 118) of the Maryland Survey. The Upper coal measures of Indiana possibly belong in this division.

The Upper Productive Measures have also been commonly known as the Monongahela series, a term first used by Rogers in 1840 (1840, 150). These rocks are probably not represented in Indiana.

The Upper Barren of authors is the Dunkard or Dunkard Creek series of I. C. White (1891, 20) and is generally conceded to be of Permian age. No recognizable representative of the Permian has been found in Indiana. Certain beds near Danville, Illinois, have been very doubtfully correlated with the Permian.

The typical development of the Pennsylvanian in the Appalachian province may be expressed in detail, as follows:—

Monongahela (Upper Productive)	Waynesburg coal Brownstown sandstone Little Waynesburg Waynesburg limestone Uniontown sandstone Uniontown coal Uniontown limestone Ritchie red beds Tyler red beds Benwood limestone Upper Sewickley coal Sewickley sandstone Lower Sewickley coal Fishpot limestone Redstone coal Redstone limestone Pittsburg sandstone Pittsburg coal

Conemaugh (Lower Barren)	Little Pittsburg limestone Little Pittsburg coal Little Clarksburg coal Morgantown sandstone Elk Lick coal Washington reds Ames limestone Harlem coal Pittsburg reds Barton coal Cowrun sandstone Anderson coal Combridge limestone Buffalo sandstone Brush creek limestone Brush creek coal Upper Mahoning Gallitzin coal Mahoning limestone Lower Mahoning Uffington shale
Allegheny (Lower Productive)	Upper Freeport coal Upper Freeport limestone Butler sandstone Lower Freeport coal Lower Freeport limestone Freeport sandstone Upper Kittanning Johnstown limestone Middle Kittanning Lower Kittanning Van Port limestone Clarion coal Clarion sandstone Putnam Hill limestone Brookville coal

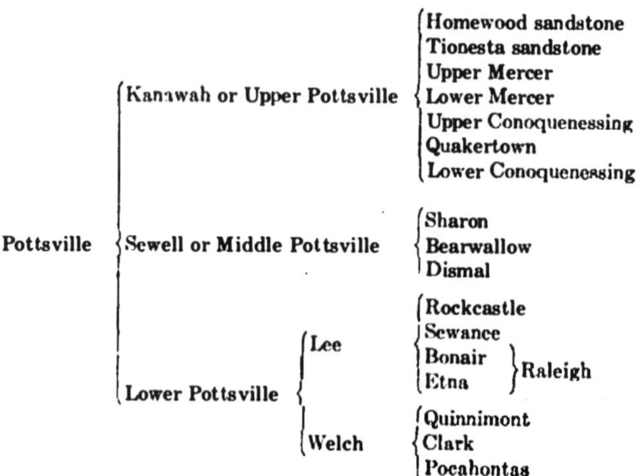

According to Ashley, Colonel Crogan noticed coal on the Wabash River in 1763. Robert Fulton dug coal at Fulton, Perry County, Indiana, in 1812. Owen mentions coal repeatedly in his 1837 Reconnaissance; and Lawrence Byrem indicates it on his geological map of the western states in 1843. Hall's geological map of eastern North America was also published in 1843, with the final report of the 4th New York District, and shows the outlines of the coal basins in a general way. The coal measures of the west are mentioned by Hall in 1843 in his section from Cleveland to the Mississippi River. The Geological Survey of Iowa, Wisconsin and Illinois by D. D. Owen in 1844 gives a chart of the Illinois-Indiana coal field. Richard C. Taylor published a map of the Indiana-Illinois coal field in 1848 in his "Statistics of Coal". In 1848 also David Christy in his "Letters on Geology" gave a section of the lower Carboniferous as exposed between Paoli and French Lick and noted the geosynclinal structure of the coal basin.

David Dale Owen in the 1859 reprint of the 1837 geological Reconnaissance of Indiana gives several sections of the Indiana coal measures (for example p. 32) in which the coals are numbered from above downward. In the third volume of the Kentucky reports, Owen (1857a, 18-24) presents a "Connected section of upper and lower Coal Measures of western Kentucky", in which the coals are numbered from 1 to 18, beginning with the lowest coal and numbering upward. Just above his number 12 coal is a hard sandstone called in Kentucky the "Anvil Rock". The coals of workable thickness are said to be all below this "Anvil Rock". "It would appear," he states, (p. 21) "from the foregoing section of the five hundred and fifty feet above the Anvil Rock, that we have, in this position, the only real Barren Coal Measures in the west".

D. D. Owen's nomenclature was accepted by Lesquereux (1862, 279) in the Richard Owen Reconnaissance Report; and he attempted to apply it in southern Indiana. He identified the 50 foot sandstone at the "Cut-off" of the Wabash near New Harmony as the equivalent of the sandstone between coals 14 and 15 of Owen's section. This sandstone is probably the equivalent of the Merom of the present classification (but see Fuller, Patoka Quadrangle). The Anvil Rock may probably be identified as the Busseron sandstone above Indiana coal VII, and Owen's coal 9 is quite certainly the well-known coal V of Indiana and Illinois. Indiana VI and VII may be correlated with Kentucky 11 and 12. The coals below No. 9 of the Kentucky section cannot be matched with the Indiana coals with any assurance. Lesquereux correlated the Brazil coals with Owen's No. 1.

In the First Annual Report of the Indiana Survey, Cox (1869, 19-46) abandoned Owen's numerical system altogether, and substituted a set of letters from A to N. The letter O was added subsequently by J. T. Scovell (1897, 517). Cox's system was inelastic and was never applied with any degree of consistency in the various counties. It was no improvement on Owen's terminology. Cox's K usually refers to our coal V. He uses the term "Block Coals" in a different sense from its present meaning. This letter scheme occurs in the Annual Reports down to the great coal report of Ashley in 1899.

In the latter year the Indiana Survey, under Dr. W. S. Blatchley, State Geologist, published a gigantic volume of 1741 pages, the 23d. Annual Report, on coal. Mr. G. H. Ashley was chiefly responsible for this report. The coal veins are discussed in detail by counties, with a bibliography of previous work on each county. Ashley here proposed to divide the Indiana Coal Measures into nine divisions numbered from I to IX (Roman), each division containing a prominent coal, except No. I, which stood for the Mansfield formation, with its minor coal veins (Cannelton coal, etc.), and division IX, which is the Merom Sandstone. Other coal seams than the main one of the division were given sub letters; thus coal Va, Vb, etc. These numbers have come into general use.

In 1909 a supplementary coal report was published by the Blatchley Survey (the 33d Annual Report), also over the name of Dr. Ashley, this time in cooperation with the U. S. Geological Survey.* Certain errors in the 1899 report were corrected. It was found that the coals had not always been properly correlated in the several counties. For example, coal VI of Clay County corresponds to coal III of most of the State. Ashley, therefore, proposed to retain the 1899 numbers from III to VII as used in Sullivan County, and to use II for the upper coal at Minshall. Coal V is the bed outcropping at Alum Cave, and VI is coal VI of the 1899 report, 50 to 75 feet above coal V. VII is old coal VII of Sullivan County, outcropping on Busseron creek. Coal IV is old coal IV of Linton, and coal VII of northeastern Vigo County, of the 1899 report. Coal III is the lowest coal worked at Linton. It is correlated with the Rock Creek coal of the Ditney Folio. For the coals below II, local names are used in the 1909 report. The lower coal at Minshall is called the Minshall or Lower Minshall coal; and the two important block coals at Brazil are called Lower and Upper Block coals. The coal at Cannelton (II of the 1899 report) is called the Cannelton coal, or Shoals coal, or frequently, coal I. This scheme is still in use, and is followed in the present revision. Some errors in correlation, however, still remain to be corrected. Where local names have come to have sufficient standing, they are indicated on the chart accompanying the present report, as alternative designations of the coals. Thus, Petersburg

*A beautiful colored map was prepared for this report, showing the several divisions of the coal measures mapped separately, and containing an immense amount of valuable tabulated data on the margins. This map does not appear to have been distributed with the report.

or Alum Cave coal for V, Millersburg coal for VII, etc. The intervening sandstones, shales and limestones have not, with a few exceptions, been given names in Indiana. The exceptions are the Mansfield sandstone, Merom sandstone, Somerville limestone, Inglefield sandstone (probably the same as the Merom), etc.

In his report on coal measures sandstones, Hopkins (1896, 199) proposed the name *Mansfield*, from the town of Mansfield, Parke County, for the rock previously known as Millstone grit, Conglomerate sandstone, etc.

The name *Brazil* has long been applied to the famous block coals mined at that town. In the Ditney Folio, however, Fuller and Ashley (1902, 2) proposed the name for a formation extending from the "top of the Pottsville" (Mansfield) to the bottom of the Petersburg coal (coal V). As so used the name corresponds to nothing significant in the Coal Measures stratigraphy, overlapping from the Pottsville into the Allegheny, and includes beds unknown in the vicinity of Brazil. Either the name should be abandoned, or it should be restricted to limits consonant with stratigraphic requirements, and in keeping with the rocks exhibited at Brazil. The writer prefers the latter procedure, and has accordingly emended the name Brazil to include only the Brazil block coal intervals and the Minshall coals up to the marked disconformity above coal II. The Mansfield and the Brazil, as emended, together constitute the Pottsville of Indiana.

The name *Petersburg* was also proposed by Fuller (1902, 2) for the interval from the bottom of the Petersburg coal (coal V) to the bottom of coal VII, or the top of the limestone immediately beneath it. Here again the interval included has no stratigraphic justification. Coal VII should have been included, since another important disconformity occurs immediately above it. There is no important break just below coal V. The writer, therefore, emends the term Petersburg to include the interval from the disconformity at the top of coal IV to the disconformity at the top of coal VII.

For the interval remaining, after these emendations, between the Petersburg and Brazil formations, the name *Staunton* is proposed, from the town of Staunton, Clay County. This formation, therefore, includes the interval from the disconformity above coal II, to the disconformity above coal IV, and contains coals III, IIIa and IV.

Fuller's name Millersburg is dropped, since the Millersburg coal naturally falls in the Petersburg formation, as emended. The interval between the disconformity above coal VII (Millersburg coal) and the base of the Merom sandstone is here named the *Shelburn* formation, from the town of Shelburn in Sullivan County. This comprises coal VIII. Fuller's Ditney formation (1902, 2), comprising the beds between the top of the Somerville limestone and the base of the Inglefield (Merom) sandstone is included in the Shelburn.

Fuller and Ashley (1902, 3) applied the name Inglefield sandstone to the sandstone capping the Coal Measures of the Ditney Folio, stating that "recent field work has served to throw grave doubts upon the exactness of the correlation [with the Merom], and the formation has, therefore, been given the name Inglefield". Our studies, however, indicate that the Inglefield and Merom are probably the same formation. The name Inglefield is, therefore, dropped.

Merom was proposed by Collett in 1871, (1871, 199) and has been used by Collett (1874, 320, etc.; 1875, 251, etc.; 1883, 8, 48, etc.); Blatchley and

Ashley (1898, 16, 22); Ashley (1899, 74, 909); Newsom (1903, 287); Hopkins (1904, 74, etc.); Ashley (1909); Tucker (1911, 15), etc.

In the Patoka Folio (1904) Fuller and Clapp repeat the 1902 statement in regard to the Inglefield and recognize what they regard as a higher formation than the Inglefield or Merom, under the name *Wabash* formation. This latter formation includes the "Shales and sandstones, with occasional thin limestone or coal, lying above the top of the Inglefield formation within the limits of the Patoka quadrangle." The Parker's coal is chosen as the line of

Fig. 51. Cross-bedded Mansfield sandstone, Shoals, Ind.

demarkation between the Inglefield and the Wabash. The Friendville and Aldrich coals are placed in this formation. This formation is recognized previsionally, pending the final determination of its relations to the Merom. On the graph accompanying the present report, however, the Merom is placed as the terminal member of the Pennsylvanian of Indiana. It may be suggested that the sandstones at the "Cut-off" of the Wabash near New Harmony and at Skelton cliff near Owensville will probably be found to belong to the Merom formation.

The Illinois Geological Survey (Lines 1912, 73-74) uses the name *Carbondale* for the lower or Allegheny [?] coal measures, and *McLeansboro* (DeWolf 1910, 181) for the remainder of the Pennsylvanian. In Bulletin 16 of the Illinois Survey (DeWolf 1910, 180) the Carbondale interval had been subdivided into the *LaSalle* and *Petersburg* formations. The latter name as used by the Illinois Survey correponds nearly to the Petersburg of Fuller. The LaSalle formation does not fit the stratigraphic conditions of Indiana. The Illinois Survey seems subsequently to have abandoned both of these names. As proposed, the LaSalle included the rocks from the base of coal 2 to the base of coal 5; and the strata between coal 5 and the top of coal 6 constituted the Petersburg formation. Coal 6 of Illinois is regarded by the Illinois Survey as the probable equivalent of coal VII of Indiana. It appears to us that if the Staunton and Petersburg can be correlated with the Allegheny with any assurance, the name Carbondale may well be dropped and the interval be spoken of merely as Allegheny. For the remaining Coal Measures (McLeansboro) we prefer the more general term Post-Allegheny. This part of the section possibly represents the Conemaugh of Pennsylvania. If the Wabash formation be recognized, the Post-Allegheny division would consist of the Shelburn, Merom and Wabash formations.

POTTSVILLE. *Mansfield.*—(Millstone Grit, Conglomerate sandstone, Coal Measures Conglomerate, Pottsville, etc.). As already noted, this formation was so named by Hopkins (1896, 199). In all the older literature it is spoken of as the Millstone Grit, Conglomerate or Conglomerate sandstone. It is unquestionably of Pottsville age, and has been correlated by David White (1908, 268-272; see also DeWolf 1910, 179; Ashley 1909, 58-59), and T. F. Jackson (1915a, 395-398; 1917, 405-428) with the upper part of the middle Pottsville (Sharon). White regards the higher Pottsville of Illinois as perhaps as high as the Conoquenessing, and the lowest Pottsville of Southern Illinois as perhaps as low as the upper part of the lower Pottsville. This higher Pottsville of White would fall in the Brazil formation. White places the boundary between the Pottsville and the Allegheny between the two Minshall coals, and the top of the Allegheny just above coal VII (Illinois coal 6). On the basis of the latest available evidence, the writer would place both Minshall coals in the Pottsville. An important disconformity occurs above coal II (Upper Minshall), sometimes cutting the coal out entirely. This horizon is taken as the base of the Allegheny. The interval between the conglomerate sandstone and the base of the Allegheny as thus defined, is here called the Brazil formation (emend).

White is also of the opinion, according to Ashley, that the Mansfield of Fountain County is much younger than the Mansfield of Martin and Orange counties. This would be in line with his determinations in Illinois as indicated above. Evidently the Pottsville sea invaded from the south, as White intimates.

Description.—The Mansfield is a brownish to yellowish, whitish, or gray, often pebbly, rather coarse-grained sandstone, where typically developed. Sometimes it is a typical conglomerate. It is often conspicuously crossbedded and ripple-marked. Where of considerable thickness, as at "Turkey Run" and the "Shades" in Parke and Montgomery counties, or at Shoals in Martin County, it forms conspicuous cliffs, rock-houses, "Devil's Lanes" and other picturesque features. It weathers with pitted or honey-combed surfaces.

In many places, however, its thickness is greatly reduced, or it may consist mostly of shales and thin flaggy sandstones.

The formation rests with marked unconformity and overlap on the subjacent rocks, being found in contact in different places with practically every division of the Chester series, and to the north overlapping across the St. Genevieve and St. Louis, and coming finally to rest on the Riverside sandstone. It is probable that the Mansfield originally extended across the Cincinnati arch into the Michigan basin. At many places in Indiana along the Mansfield-Mississippian unconformity, Kaolin (Indianaite) is found. The best known deposits are in Lawrence County.

Several minor coal seams occur in the Mansfield formation, one of which, the Cannelton coal, is important. The Shoals coal of Martin County may be doubtfully correlated with the Cannelton. It has a maximum thickness of 4 feet and was formerly worked on a small scale near Shoals. The coal at Cannelton on the Ohio River in Perry County, is in places 4 feet thick, varying from that figure to nothing. It is a compact "cannel" coal, breaking with conchoidal fracture and has been mined at Cannelton for nearly a century.

The base of the Lower Block coal at Brazil is taken as the top of the Mansfield formation. The Mansfield varies from a few feet to 280 feet in thickness.

(Additional references on the Mansfield are as follows: Kindle 1896, 329-368; Blatchley and Ashley 1898, 21; Ashley 1899, 95-96; Blatchley 1900, 26; Newsom 1903, 284-285; Ashley 1903, 66; Blatchley 1904, 51; Hopkins 1904, 67-71; Blatchley 1907, 37-38; Shannon 1907, 56, etc.; Shannon 1908, 114-115; Shannon 1909, 280-281; R. S. Blatchley 1911, 88, columnar sections; Tucker 1911, 15; Shannon 1912, 155; Barrett 1914, 41-59, as glass sand; Jackson 1915, 223-229, Description on Bloomington quadrangle; Barrett 1917, 79-89, several beautiful pictures of the Mansfield; Logan 1919, 24 *et. seq.*, many details.)

Brazil.—The Brazil formation, as emended, includes the Lower and Upper Block coals of Brazil, Clay County, Indiana, and the lower and upper Minshall coals. The upper Minshall, or coal II is, however, local, being absent in some of the sections. In such cases it has been cut out along the unconformity at the top of the Pottsville. The "Block" coals are rather hard, jointed, slabby coals, which do not break readily across the laminae and are won in slabs or joint blocks of large size, requiring much splitting and breaking to bring them to lumps of the smaller commercial sizes. The coal is more expensive to mine than most Indiana coals and brings a higher price on the market. Both coals often have sandstone roofs. The upper vein runs from 2 to 4 feet thick, and the lower up to 5 feet, averaging about 3 feet.

The lower Minshall coal (Minshall of Ashley) ranges up to 6 feet thick, and averages about 4. It is overlain by a black shale and a limestone and underlain by a fire clay. The limestone forms the roof in some places. It is a firm shiny black coal of semi-block type.

The upper Minshall coal, or coal II, is usually 10 to 20 feet above the Minshall. It is worked in but few places, and is from 2 to three and one-half feet thick. It usually has partings which cut it into benches and has a black shale roof. The unconformity above sometimes descends, cutting out the coal and the shale beneath it.

ALLEGHENIAN. *Staunton.*—The name Staunton is proposed, as indicated above, for the interval between the disconformity above coal II and the disconformity above coal IV. It includes coal III and IV, the latter being one of the most important Indiana coals. Coal III is characterized by conspicuous partings and by much pyrite. Typically it is split by these partings into three beds. The interval between coals II and III is usually about 75 or 80 feet. Coal IIIa, about 20 feet above coal III, is a thin coal seldom of workable character. Coals III and IIIa are often underlain by limestones and another limestone occurs above IIIa. The Staunton division is terminated by a sandstone, surmounted by coal IV, or the Linton coal. This coal averages 5 feet thick with a maximum of 7. The roof is sandstone or shale; and often gives evidence of disconformity, the coal being entirely cut out in places. It is a bituminous coal of excellent quality running from 45% to 47% fixed carbon, and 5¾% to over 8% ash.

Petersburg.—The name Petersburg of Fuller and Ashley (1902, 2) is here emended to include the interval from the disconformity over coal IV to the disconformity over coal VII. It thus includes coals IVa, V, VI and VII, and several shale, sandstone and limestone formations. Coal IVa is usually about 25 feet above coal IV, but does not occur in every place where coal IV is present. Coal V, about 120 feet above coal IV, is the most important and persistent coal bed of Indiana, and is correlated with coal 5 of Illinois and coal 9 of Kentucky. It is usually a solid bed overlain by a black shale containing marine fossils, and many pyrite concretions. Occasionally the coal is overlain by sandstone. The black roof-shale is nearly always overlain by a limestone from 4 to 6 feet thick. The rider of coal V occurs above this limestone. Coal V will average 6 feet thick in Sullivan County, and is as much as 11 feet thick in some places. It is 8 feet thick at Alum Cave, which may be regarded as the type locality of the coal. It is also known as the Petersburg coal.

Coal VI is uniformly from 6 to 8 feet thick and has very persistent thin partings near the center of the vein. It is normally about 75 feet above coal V. The roof is usually a crumbling shale, with overlying sandstone. The coal seems to be generally absent north of Sullivan County. It may probably be correlated with coal 11 of Kentucky.

Coal VII, forming the top of the Petersburg formation, is from 2 to 6 feet thick, and is overlain disconformably by a sandstone or "rolly roof", and underlain by fire clay, and a limestone from 2 to 10 feet thick. The latter sometimes comes within a few feet of the coal. Coal VII is known also as the Millersburg coal. The Illinois geologists regard Indiana VII as the equivalent of Illinois 6, and place the top of the Allegheny series at this level.

POST-ALLEGHENIAN. *Shelburn.*—The name Shelburn is proposed for the interval from the top of coal VII to the base of the Merom sandstone; and includes coals VIII and IX and the Somerville limestone, with subordinate coal seams, shales and sandstones. The lower member is the Busseron sandtone, a possible equivalent of the "Anvil Rock" of Kentucky. This sandstone and sandy shale rests with marked disconformity on coal VII, which is sometimes cut out completely. Coal VIII comes about 70 feet above coal VII in Sullivan County, and Coal IX about 50 feet above coal VIII.

The *Somerville* limestone, probably the most important limestone formation in the Indiana coal measures, is a double bed of hard, gray limestone, with shale parting, 15 to 20 feet thick. It is usually quite fossiliferous.

Merom.—Collett used the name "Merom sandstone" in 1871 (1871, 199) for the sandstone capping the hills at Merom, Sullivan County, Indiana. His statement "The stone work of the college edifice was quarried from massive ledges of the 'Merom sandstone', north of town", would not indicate that he intended to propose the name in a technical sense. Farther on he speaks of the "upper member of the 'Merom sandstone'," and describes the Merom formation as follows: "This massive sandstone is here, at its northern terminus [at Graysville] well developed. . . . Deep, narrow gorges, with precipitous or overhanging sides, give a romantic boldness to the scenery, and afford good exposures for observation. It may be characterized as a very coarse grained sandstone, varying in color from brown to yellowish red, with occasional strata of snowy whiteness irregularly laminated. False and diagonal bedding and coarseness of materials show that it was deposited by strong currents of water subject to frequent change of direction and cross-currents. Portions are compact quarry rock, which, however, generally tends to disintegrate. The coloring matter is derived from small partings and veins of iron which, being harder than their sandy matrix, fret the sides and overhanging arches of the gorges with irregular tracery of network in relief."

In the Fifth Report Collett (1874, 320-325) again describes the "'Merom' --Fort Knox Sandstone", in Knox County. This is an excellent detailed description, and calls especial attention to the loosely aggregated character of the sandstone, which causes it to weather readily to coarse sand.

In the Seventh Report (1876, 251-254) he definitely speaks of it as the Merom sandstone, saying that he had given it this name in 1870. He suggests that it may be of Triassic age. This suggestion is repeated in the Thirteenth Annual Report (1884, 48-49); and revived by Hopkins in the 28th. Report (1904, 75), though on page 17 of the same report he correlates it with the Dunkard (Permian) of Pennsylvania. Hopkins (1904, 74) calls attention to the basal conglomerate at the bottom of the Merom and to the disconformable relations of the formation with the underlying rocks. The Merom is 40 to 60 feet thick and generally unfossiliferous. Its relations with the so-called Inglefield have already been discussed.

(Additional references on the Merom are as follows:—Blatchley and Ashley 1898, 16, 22; Ashley 1899, 74, 906, 909, 913; Newsom 1903, 287; Ashley 1909; Case 1918, 503, 504.)

APPENDIX A.

Some of the Problems Requiring Further Study.

1. The fauna of the Cynthiana of Indiana is comparatively unknown.
2. The subdivisions of the Eden in Indiana remain to be carefully delimited and mapped.
3. The relation of the Maysville formations of Indiana to those of Cincinnati, and the type section, requires much further study. This applies especially to the Corryville and Mt. Auburn.
4. The stratigraphy and paleontology of the Arnhelm and the proper position of the Maysville-Richmond boundary.
5. The lower boundary of the Waynesville is not yet definitely fixed.
6. The Whitewater fauna of the area between Batesville and Madison should be carefully studied.
7. Fauna of the Elkhorn comparatively unknown.
8. Detailed mapping of all the Cincinnatian formations.
9. Careful collecting and study of the Brassfield fauna.
10. Detailed stratigraphic and paleontologic study of the Osgood and Laurel.
11. Careful determination of the boundary between the Louisville and Devonian in part of the southern Indiana area.
12. Detailed stratigraphy of the Devonian limestones of southern Indiana, with special reference to the Geneva limestone problem.
13. Detailed study of the Silurian and Devonian of northern Indiana with close attention to structural relations (Niagara domes).
14. Consistent attempt to determine the age of the New Albany shale.
15. Detailed stratigraphy and paleontology of the Knobstone. This series requires a vast amount of detailed work.
16. Fauna and stratigraphy of the Harrodsburg (Warsaw). Fauna comparatively unknown.
17. Detailed stratigraphy of the Salem.
18. Much detailed work should be done on the stratigraphy and paleontology of the Mitchell (St. Louis and St. Genevieve), with subdivision into formations. The faunas are little known.
19. Further detailed work on the Chester, and especially on the faunas of its subdivisions.
20. Further study of the Pottsville and coal floras.
21. Much detailed stratigraphic and faunistic study of the coal measures.

APPENDIX B

Analytical Index to Indiana Geological Survey Reports.

David Dale Owen....	1837 (1859)	Geological Reconnaissance.
Richard Owen........	1862	Geological Reconnaissance.
E. T. Cox............	1869	1st Report Geological Survey of Indiana.
E. T. Cox............	1871	2d Report Geological Survey of Indiana.
E. T. Cox............	1872	3d and 4th Report Geological Survey of Indiana.
E. T. Cox............	1874	5th Report Geological Survey of Indiana.
E. T. Cox............	1875	6th Report Geological Survey of Indiana.
E. T. Cox............	1876	7th Report Geological Survey of Indiana,

E. T. Cox	1879	8th, 9th, 10th Report Geological Survey of Indiana.
John Collett	1879	1st Report Indiana Department of Statistics and Geology.
John Collett	1880	2d Report Indiana Department of Statistics and Geology.
John Collett	1882	11th Report Indiana Department of Geology and Natural History.
John Collett	1883	12th Report Indiana Department of Geology and Natural History.
John Collett	1884	13th Report Indiana Department of Geology and Natural History.
John Collett	1884	14th Report Indiana Department of Geology and Natural History.
Maurice Thompson	1886	15th Report Indiana Department of Geology and Natural History.
Maurice Thompson	1889	16th Report Indiana Department of Geology and Natural History.
S. S. Gorby	1892	17th Report Indiana Department of Geology and Natural Resources.
S. S. Gorby	1894	18th Report Indiana Department of Geology and Natural Resources.
S. S. Gorby	1894	19th Report Indiana Department of Geology and Natural Resources.
W. S. Blatchley	1896	20th Report Indiana Department of Geology and Natural Resources.
W. S. Blatchley	1897	21st Report Indiana Department of Geology and Natural Resources.
W. S. Blatchley	1898	22d Report Indiana Department of Geology and Natural Resources.
W. S. Blatchley	1899	23d Report Indiana Department of Geology and Natural Resources.
W. S. Blatchley	1900	24th Report Indiana Department of Geology and Natural Resources.
W. S. Blatchley	1901	25th Report Indiana Department of Geology and Natural Resources.
W. S. Blatchley	1903	26th and 27th Reports Indiana Department of Geology and Natural Resources.
W. S. Blatchley	1904	28th Report Indiana Department of Geology and Natural Resources.
W. S. Blatchley	1905	29th Report Indiana Department of Geology and Natural Resources.
W. S. Blatchley	1906	30th Report indiana Department of Geology and Natural Resources.
W. S. Blatchley	1907	31st Report Indiana Department of Geology and Natural Resources.
W. S. Blatchley	1908	32d Report Indiana Department of Geology and Natural Resources.
W. S. Blatchley	1909	33d Report Indiana Department of Geology and Natural Resources.
W. S. Blatchley	1910	34th Report Indiana Department of Geology and Natural Resources.

W. S. Blatchley......1911 35th Report Indiana Department of Geology and Natural Resources.

Edward Barrett......1912 36th Report Indiana Department of Geology and Natural Resources.

Edward Barrett......1913 37th Report Indiana Department of Geology and Natural Resources.

Edward Barrett......1914 38th Report Indiana Department of Geology and Natural Resources.

Edward Barrett......1915 39th Report Indiana Department of Geology and Natural Resources.

Edward Barrett......1916 40th Report Indiana Department of Geology and Natural Resources.

Edward Barrett......1917 41st Report Indiana Department of Geology and Natural Resources.

Edward Barrett......1918 Administrative Report, Indiana Year Book.

W. N. Logan.........1919 Kaolin Deposits, Department of Conservation, Division of Geology.

W. N. Logan.........1920 Petroleum and Natural Gas, Department of Conservation, Division of Geology.

W. N. Logan.........1920 Administrative Report, Indiana Year Book.

W. N. Logan.........1921 Administrative Report, Indiana Year Book.

W. N. Logan.........1922 Handbook of Indiana Geology [this volume], Department of Conservation, Division of Geology.

Bibliography of Indiana Stratigraphy and Paleontology

Including Papers Bearing on Certain General Problems of Correlation

By

E. R. CUMINGS.

Andrews, E. B., 1871. [Report of labors in the second district]. Geol. Surv. Ohio, Rept. of Progress in 1870, pp. 55-251.

Ashburner, C. A., 1877. Section of the Paleozoic rocks of central Pennsylvania. Proc. Am. Phil. Soc., vol. 16, pp. 519-560.

1887. Natural gas in the United States. Am. Inst. Min. Eng., vol. 15, pp. 505-542.

1888. Coal. U. S. Geol. Surv., Min. Res. for 1887, pp. 168-382.

Ashley, G. H., 1898. Note on an area of compressed structure in western Indiana. Geol. Soc. Amer., Bull., vol. 9, pp. 429-430.

1898a. Note on fault structure in Indiana. Ind. Acad. Sci., Proe. for 1897, pp. 244-250.

1898b. [And Blatchley, W. S.] Geological scale of Indiana. 22d Indiana Rept., pp. 17-23.*

1899. Coal deposits of Indiana. 23d. Indiana report, pp. 1-1741.

1900. Geological results of the Indiana coal survey. Geol. Soc. Amer., Bull., vol. 11, pp. 8-10.

1902. Eastern Interior coal field. U. S. Geol. Surv., 22d. Ann. Rept., pt. 3, pp. 265-305.

1903. The geology of the Lower Carboniferous area of southern Indiana. 27th. Indiana Report, pp. 49-122.

1903a. [And Fuller, M. L.] Recent work in the coal field of Indiana and Illinois. U. S. Geol. Surv., Bull. No. 213, pp. 284-293.

1909. Supplementary report on the coal deposits of Indiana and Illinois. 33d. Indiana Report, pp. 13-150.

1909a. Stratigraphy and coal beds of the Indiana coal field. U. S. Geol. Surv., Bull. No. 381-A, pp. 5-14.

Ashley, G. H., and Siebenthal, C. E., 1899. [See Ashley, 1899].

Barrett, Edward, 1914. Glass sands of Indiana. 38th. Indiana Report, pp. 41-59.

1917. The Beautiful Shades. 41st. Indiana Report, pp. 79-89.

1912-1917. [The 36th. to 41st. Indiana Reports are by Edward Barrett].

Barton, D. C., 1918. Notes on the Mississippian chert of the St. Louis area. Jour. Geol., vol. 26, pp. 361-374.

Bassler, R. S., 1903. The structural features of the Bryozoan genus Homotrypa. U. S. Nat. Mus., Proc., vol. 26, pp. 565-591.

* For the sake of economy of space the Indiana Reports are referred to uniformly by number, as above. The full titles will be found in the Index to Indiana Reports, pp. 531-533.

1906. A study of the James types of Ordovician and Silurian Bryozoa. U. S. Nat. Mus. Proc., vol. 30, pp. 1-66.

1906a. The Bryozoan fauna of the Rochester shale. U. S. Geol. Surv., Bull. No. 292, 136 pp., 31 pls.

1909. The Nettleroth collection of invertebrate fossils. Smithsonian Misc. Coll., vol. 52, pp. 121-152, 3 pls.

1911. The early Paleozoic Bryozoa of the Baltic province. U. S. Nat. Mus., Bull. No. 77, 382 pp., 13pls.

1911a. Corynotrypa, a new genus of tubuliporoid Bryozoa. U. S. Nat. Mus., Proc., vol. 39, pp. 497-527, 27 figs.

1915. Bibliographic Index of North American Ordovician and Silurian fossils. U. S. Nat. Mus., Bull. 92, two vols. [See correlation table at the end of volume II].

1919. Maryland Geological Survey: Cambrian and Ordovician. 424 pp., 58 pls.

Beachler, C. S., 1887. Crinoid beds at Crawfordsville, Indiana. Amer. Nat., Dec. 1887, pp. 1106-1109.

1888. Keokuk group at Crawfordsville, Indiana. Am. Geol., vol. 2, pp. 407-412.

1889. Corrected list of fossils found at Crawfordsville. 16th. Indiana Report, pp. 65-76.

1889a. Notice of some new and remarkable forms of Crinoidea from the Niagara limestone at St. Paul, Decatur Co., Indiana. Am. Geol., vol. 4, pp. 102-103.

1891. The rocks at St. Paul, Indiana, and vicinity. Am. Geol., vol. 7, pp. 178-179.

1892. Keokuk group of the Mississippi valley. Am. Geol., vol. 10, pp. 88-96.

Beede, J. W., 1906. [See Cumings, Beede, *et. al.*]

1907. [And Shannon, C. W.] Iron ores of Martin County, Indiana. 31st. Indiana Report, pp. 383-424.

1911. The cycle of subterranean drainage on the Bloomington quadrangle. Ind. Acad. Sci., Proc. for 1910, pp. 81-103.

1915. Geology of the Bloomington quadrangle. 39th. Indiana Report, pp. 190-312. [With G. C. Mance, T. F. Jackson, etc.]

Bennett, L. F., 1898. Four comparative cross sections of the Knobstone group of Indiana. Ind. Acad. Sci., Proc. for 1897, pp. 258-262.

1899. Notes on the eastern escarpment of the Knobstone formation of Indiana. Ind. Acad. Sci., Proc. for 1898, pp. 283-287.

1900. Headwaters of Salt creek in Porter county. Ind. Acad. Sci., Proc for 1899, pp. 164-166.

Bigney, A. J., 1892. Preliminary note on the geology of Dearborn county. Ind. Acad. Sci., Proc. for 1891, pp. 66-67.

1911. A new bed of trilobites. Ind. Acad. Sci., Proc. for 1910, p. 139.

1915. Geology of Dearborn county. 40th. Indiana Report, pp. 211-222.

Blatchley, R. S., 1908. The Indiana oölitic limestone industry in 1907. 32d. Indiana Report, pp. 299-459.

Blatchley, R. F., 1911. The Oakland City, Indiana, oil field in 1910. 35th. Indiana Report, pp. 81-143.

Blatchley, W. S., 1896. A preliminary report on the clays and clay industries of the coal-bearing counties of Indiana. 20th. Indiana Report, pp. 23-185.

1898. Geology of Lake and Porter counties; the clays and clay industries of northwestern Indiana. 22d. Indiana Report, pp. 25-104, 105-153.

1898a. [And Ashley, 1898b.] Geological scale of Indiana. 22d. Indiana Report, pp. 17-23.

1900. Natural resources of the State of Indiana. 24th. Indiana Report, pp. 3-40.

1901. Oölite and oölitic stone for portland cement manufacture. 25th. Indiana Report, pp. 322-330.

1904. The Indiana of Nature: its Evolution. Ind. Acad. Sci., Proc. for 1903, pp. 33-59.

1904a. The lime industry in Indiana. 28th. Indiana Report, pp. 211-257.

1905. The clays and clay industries of Indiana. 29th. Indiana Report, pp. 13-657.

1906. The geologic distribution of road materials of Indiana. 30th. Indiana Report, pp. 120-160.

1907. The natural resources of Indiana. 31st. Indiana Report, pp. 13-72.

1908. 32d. Indiana Report, p. 307, footnote.

1917. A century of geology in Indiana. Ind. Acad. Sci., Proc. for 1916, pp. 89-177.

1896-1911. The annual reports of the Indiana Geological Survey from 1896-1911, vols. 20-35, were published by W. S. Blatchley.

Blatchley, W. S., and Ashley, G. H., 1898. [See Blatchley, 1898a].

Borden, W. W., 1874. Report of a geological survey of Clark and Floyd counties. 5th. Indiana Report, pp. 134-189.

1875. Scott and Jefferson counties. 6th. Indiana Report, pp. 112-186.

1876. Jennings and Ripley counties. 7th. Indiana Report, pp. 146-202.

Branner, J. C., 1888-1891. Arkansas Geological Survey, vols. I-IV, Report for 1888, etc. [References to Branner 1888 are to this report].

Bradley, F. H., 1869. Geology of Vermilion county. 1st. Indiana Report, pp. 138-174.

1876. Geological chart of the United States east of the Rocky Mountains and of Canada. Am. Jour. Sci., 3d ser., vol. 12, pp. 286-291.

Briggs, Caleb, 1838. Report on work between Scioto and Hockhocking. Geol. Surv. Ohio, 1st. Annual Report by W. W. Mather, pp. 71-98.

Brown, R. T., 1854. Geological Survey of the State of Indiana. Indiana Board of Agriculture, 3d. Report, for 1853, pp. 299-332.

1882. Geology of Fountain County. 11th. Indiana Report, pp. 89-125.

1883. Report of a geological and topographic survey of Marion county. 12th. Indiana Report, pp. 79-99.

1884. Geology of Morgan county. 13th. Indiana Report, pp. 71-85.

1884a. Geological and topographical survey of Hamilton and Madison counties. 14th. Indiana Report, pp. 20-40.

1886. Hancock county. 15th. Indiana Report, pp. 187-197.

Butts, Charles, 1915. Geology of Jefferson county [Kentucky]. Kentucky Geol. Surv., Ser. 4, vol. 3, pt. 2, pp. i-xiv and 1-270.

1918. [1917]. Mississippian formations of Western Kentucky. [With a section on stratigraphy by E. O. Ulrich; see Ulrich, 1918]. Kentucky Geol. Surv., 1918 [1917 on title page], pp. 1-119.

Byrem, Lawrence, 1843. A concise description of the geological formations and mineral localities of the western states, designed as a key to the geological map of the same, pp. 1-48.

Campbell, M. R., 1898. Richmond folio, Kentucky. U. S. Geol. Surv., Geol. Atlas, folio No. 46.

Case, E. C., 1918. Permo-Carboniferous conditions versus Fermo-Carboniferous time. Jour. Geol., vol. 26, pp. 500-506.

Casseday, Samuel A., 1854. Translation of a description of a species of Batocrinus from Spergen Hill. Abdruck a. d. Zeitschrift der Deutch. Geol. Gesellschaft, Jahrg. 1854, p. 237.

Casseday and Lyon, 1859. [See Lyon and Casseday].

Chadwick, G. H., 1908. Revision of the New York Series. Science, N. S., vol. 28, pp. 346-348.

1918. Cayugan waterlimes of western New York. [Abstract]. Geol. Soc. Amer., Bull., vol. 28, pp. 173-174.

1918a. Stratigraphy of the New York Clinton. Geol. Soc. Amer., Bull., vol. 29, pp. 327-368.

Chamberlain, T. C., 1901. Geologic terminology. Jour. Geol., vol. 9, pp. 267-270.

Chamberlain, T. C., and Salisbury, R. S., 1906. Text book of Geology, vol. 2, xxvi and 692 pp.

Chapman, E. J., 1864. A popular exposition of the minerals and geology of Canada. Originally pub. in Can. Jour., N. S., vols. 6 to 8. 1861-1863.

Christy, David, 1848. Letters on geology, being a series of communications originally addressed to Dr. John Locke of Cincinnati, etc. 68 pp.

1851. On the Goniatite limestone of Rockford, Jackson county, Indiana. Amer. Assoc. Adv. Sci., Proc., vol. 5, pp. 76-80.

Clapp, A., 1843. The geological equivalents of the vicinity of New Albany, Indiana, as compared with those described in the Silurian system of Murchison. Phila. Acad. Sci., Proc., vol. 1, pp. 18-19, 177-178.

Clarke, John M., 1897. Cephalopoda of Minnesota. Minn. Geol. and Nat. Hist. Surv., Paleontology, vol. 3, pt. 2, pp. 694-759.

1897a. The Lower Silurian Trilobites of Minnesota. Minn. Geol. Nat. Hist. Surv., Paleontology, vol. 3, pt. 2, pp. 695-759, 82 figures.

Clarke, John M., and Ruedemann, R., 1903. Guelph fauna in the State of New York. N. Y. State Mus., Mem. 5, 195 pp., 21 pls.

1903a. Catalog of type specimens of Paleozoic fossils in New York Museum. N. Y. State Mus., Bull. No. 65, 847 pp.

1912. The Eurypterida of New York. N. Y. State Mus., Mem. 14, vol. 1, Text, 439 pp.; vol. 2, plates, pp. 441-628, 88 pls.

Clarke, John M., and Schuchert, Charles, 1899. The nomenclature of the New York series of geological formations. Science, N. S. Vol. 10, pp. 874-878.

Claypole, E. W., 1890. Paleontological notes from Indianapolis. Am. Geol., vol. 5, pp. 255-260.

1894. A new species of Carcinosoma. Am. Geol., vol. 13, pp. 77-79.

Collett, John, 1871. Geology of Sullivan county, Indiana. 2d. Indiana Report, pp. 191-240.

1872. Geology of Dubois, Pike, Jasper, White, Carroll, Cass, Miami, Wabash and Howard counties. 3d. and 4th. Indiana Reports, pp. 192-287 and 291-337.

1874. Warren, Knox, Lawrence and Gibson counties. 5th. Indiana Report, pp. 191-312, 315-424, 426-428.

1875. Geology of Brown county. 6th Indiana Report, pp. 77-110.

1876. Geological reconnaissance of the coal measures of Putnam county. 7th. Indiana Report, pp. 463-468.

1876a. Vanderburg, Owen and Montgomery counties. 7th. Indiana Report, pp. 240-462. [List of fossils from Crawfordsville].

1879. Harrison and Crawford counties. 8th., 9th., 10th. Indiana Reports, pp. 291-522.

1879a. 1st. Report of the Department of Statistics and Geology of Indiana, 514 pp.

1880. 2d. Report of the Department of Statistics and Geology of Indiana, 544 pp., 11 pls.

1882. Geology of Indiana. Mines and Quarries. 11th. Indiana Report, pp. 16-33; Shelby county, pp. 55-88.

1883. Economic geology; outline of geology of Indiana; geology of Newton and Jasper counties. 12th. Indiana Report, pp. 17-25, 44-78.

1884. Outline of the geology of Indiana; economic geology; stone coals; geology of Posey county. 13th. Indiana Report, pp. 2-70. [Colored geological map of Indiana].

1884a. Indiana Geological Survey. 14th. Indiana Report. [Geological map of Indiana]. 63 pp.

Conrad, Timothy A., 1837. First annual report on the geological survey of the third district of New York. N. Y. Geol. Surv., 1st. Ann. Rept., pp. 155-186.

1841. Fifth report on the paleontology of the State of New York. N. Y. Geol. Surv., 5th. Ann. Rept., pp. 25-27.

1842. Observations on the Silurian and Devonian systems of the United States, with descriptions of New Organic remains. Phila. Acad. Sci., Jour., vol. 8, pp. 228-280.

1855. Description of a new species of Pentamerus. Phila. Acad. Nat. Sci., Proc., vol. (1855), p. 441.

Conybeare, W. D., 1822. Outlines of the geology of England and Wales.

Cope, E. D., 1872. Report on Wyandotte cave and its fauna. 3d. and 4th. Indiana Reports, pp. 157-182.

1875. [Description of Reptiles and Fishes from Vermillion County, Illinois, indicating strata of Permian or Triassic age]. Phila. Acad. Sci., Proc. vol. (1875), pp. 404-411.

1879. The relations of the horizons of extinct vertebrata of Europe and North America. U. S. Geol. and Geogr. Surv. of the Territories, F. V. Hayden in charge, Bull., vol. 5, pp. 33-54.

1884. [And Wortman, J. L.]. Post-Pliocene vertebrata. 14th. Indiana Report, pp. 3-41.

Cornett, W. T. S., 1874. Geology of the Madison Hills. [Announcement of the discovery of Lower Silurian fossils in the Upper Silurian of Borden]. Indianapolis Journal, July 10th, 1874.

1875. List of fossils in Jefferson county. 6th. Indiana Report, pp. 182-186.

1876. Letter to W. W. Borden on the discovery of Lower Silurian fossils in rocks formerly considered as Upper Silurian. 7th. Indiana Report, p. 183.

Coryell, H. N., 1915. Correlation of the outcrop at Spades, Indiana. Ind. Acad. Sci., Proc. for 1914, pp. 389-393.

Cox, E. T., 1869. First annual report on the geological survey of Indiana, made during the year 1869, 240 pp.

1871. Second report of the geological survey of Indiana, made in the year 1870, 303 pp.

1872. Third and Fourth reports of the geological survey of Indiana, for 1871 and 1872.

1872a. Western coal measures and Indiana coal. Amer. Assoc. Adv. Sci., Proc., vol. 20, pp. 236-252.

1874. Fifth report of the geological survey of Indiana, for 1873, 494 pp. Maps.

1875. Sixth report of the geological survey of Indiana, for 1874, 287 pp. Maps.

1876. Seventh report of the geological survey of Indiana, for 1875. 599 pp.

1879. Eighth, Ninth and tenth reports of the geological survey of Indiana, made during the years 1876-1878, 541 pp.

1869-1879. The 1st. to 10th. Indiana Reports are by E. T. Cox.

Cozzens, L, 1846. Discovery of three fossils from the Falls of the Ohio. Ann. N. Y. Lyceum Nat. Hist., vol. 4, pp. 157-159.

Cubberly, E. P., 1894. Indiana's structural features as revealed by the drill. 18th. Indiana Report, pp. 219-255, map and 16 colored geological sections.

Culbertson, Glenn, 1916. The geology and natural resources of Jefferson county, Indiana. 40th. Indiana Report, pp. 223-239.

Cumings, E. R., 1900. The Waldron fauna at Tarr Hole, Indiana. Ind. Acad. Sci., Proc. for 1899, pp. 174-176.

1901. Notes on the Ordovician rocks of southern Indiana. Ind. Acad. Sci., Proc. for 1900, pp. 200-215.

1901a. The use of Bedford as a formational name. Jour. Geol., vol. 9, pp. 232-233.

1901b. A section of the upper Ordovician at Vevay, Indiana. Am. Geol., vol. 28, pp. 361-380.

1902. A revision of the Bryozoan genera Dekayia, Dekayella and Heterotrypa of the Cincinnati group. Am. Geol., vol. 29, pp. 197-218.

1903. The morphogenesis of Platystrophia: A study of the evolution of a Paleozoic Brachiopod. Am. Jour. Sci., 4th. ser., vol. 15, pp. 1-48, 121-136, 27 figures.

1906. Weathering of the Subcarboniferous limestones of southern Indiana. Ind. Acad. Sci., Proc. for 1905, pp. 85-100.

1906a. Fauna of the Salem limestone of Indiana [with J. W. Beede, Essie A. Smith, and E. B. Branson]. 30th. Indiana Report, pp. 1187-1486.

1908. Stratigraphy and Paleontology of the Cincinnati Series of Indiana. 32d. Indiana Report, pp. 607-1189.

1912. The geological conditions of municipal water supply in the driftless area of southern Indiana. Ind. Acad. Sci., Proc. for 1911, pp. 111-146.

1912a. [And J. J. Galloway]. A note on the Batostomas of the Richmond series. Ind. Acad. Sci., Proc. for 1911, pp. 147-167.

1913. [And Galloway, J. J.]. The stratigraphy and paleontology of the Tanner's creek section of the Cincinnati series of Indiana. 37th. Indiana Report, pp. 353-479.

Dana, James D., 1863. Manual of Geology, with special reference to American geological history. 1st. ed. 1863, xiv and 798 pp.

Darton, N. H., 1896. Franklin Folio, Virginia-West Virginia. U. S. Geol. Surv., Atlas, Folio No. 32.

Davis, W. J., 1885. Monograph of the fossil corals of the Silurian and Devonian rocks of Kentucky. Pt. 2, Geol. Surv. of Kentucky, pp. i-xii, pls. 1-139. [Species figured but not described].

Dennis, D. W., 1888. The east-west diameter of the Silurian island about Cincinnati. Amer. Nat., vol. 22, p. 94.

1889. An analytical key to the fossils of Richmond, Indiana. 48 pp.

1899. Two cases of variation of species [of Brachiopoda] with horizon. Ind. Acad. Sci., Proc. for 1898, pp. 288-289.

DeWolf, Frank W., 1910. Administrative report and studies of Illinois coal fields. Illinois Geol. Surv., Bull. No. 16, pp. 10-23 and 177-301.

Diller, J. S., 1892. Geology of the Taylorville region, California. Geol. Soc. Amer., Bull., vol. 3, pp. 369-394.

Drake, Daniel, 1815. Picture of Cincinnati.

1825. Geological account of the valley of the Ohio. Am. Phil. Soc., Trans., N. S., vol. 2, pp. 124-139.

Dryer, C. R., 1889. Geology of DeKalb county. 16th. Indiana Report, pp. 98-104. Geology of Allen county, *ibid.*, pp. 105-130.

1892. Geology of Steuben and Whitley counties. 17th. Indiana Report, pp. 114-134 and 160-170.

1894. Geology of Noble and Lagrange counties. 18th. Indiana Report, pp. 17-32 and 72-82.

1899. Meanders of the Muscatatuck at Vernon, Indiana. Ind. Acad. Sci., Proc. for 1898, pp. 270-273.

Duden, Hans, 1897. Some notes on the black slate or Genesee shale of New Albany, Indiana. 21st. Indiana Report, pp. 108-119. [With description of several fossil plants].

Dunbar, C. O., 1918. Stratigraphy and correlation of the Devonian of western Tennesee. Amer. Jour. Sci., vol. 46, pp. 732-756.

1919. Stratigraphy and correlation of the Devonian of western Tennessee. Geol. Surv. of Tennessee, Bull. No. 21, pp. 1-127.

1920. New species of Devonian fossils from west Tennessee. Connecticut Acad. Arts and Sci., Trans., vol. 23, pp. 109-158.

Eaton, Amos, 1824. A geological and agricultural survey of the district adjoining the Erie canal. 163 pp.

1839. Cherty lime rock or corniferous limerock proposed as the line of reference for state geologists of New York and Pennsylvania. Am. Jour. Sci., vol. 36, pp. 61-71, 198.

Ellis, R. W., 1906. The roads and road materials of southeastern Indiana. 30th. Indiana Report, pp. 757-871.

Elrod, Moses N., 1876. [And McIntire, E. S.]. Geological Survey of Orange county. 7th. Indiana Report, pp. 203-239.

1882. Bartholomew county. 11th. Indiana Report, pp. 150-213.

1883. Decatur county. 12th. Indiana Report, pp. 100-152.

1884. Rush county. 13th. Indiana Report, pp. 86-115.

1884a. Union and Fayette counties. 14th Indiana Report, pp. 41-60, 61-72.

1892. [And Benedict, A. C.]. Geology of Wabash county. 17th Indiana Report, pp. 192-272.

1899. The geologic relations of some St. Louis caves and sinkholes. Ind. Acad. Sci., Proc. for 1898, pp. 258-267.

1902. Niagara group unconformities in Indiana. Ind. Acad. Sci., Proc. for 1901, pp. 205-215.

Elrod and Benedict. [See Elrod, 1892].

Elrod and McIntire. [See Elrod, 1876].

Emmons, Ebenezer, 1842. Geology of New York, Part II, comprising the survey of the Second Geological District. 437 pp.

Engelmann, G., 1847. Remarks on the St. Louis limestone. Am. Jour. Sci., 2d. ser., vol. 3, pp. 119-120.

Engelmann, H, 1868. On the Lower Carboniferous system as developed in southern Illinois. St. Louis Acad. Sci., Trans., vol. 2, pp. 188-190. [Paper read in 1862].

1866. [And Worthen, A. H.] Geology of Hardin county, Illinois. Geol. Surv. Ill, vol. 1, pp. 350-366, 372-375.

Etheridge, R., 1878. Paleontology of the coasts of the Arctic lands visited by the British expedition under Capt. Geo. Nares, etc. Quar. Jour. Geol. Soc., London, vol 34, pp. 568-639.

Featherstonaugh, G. W., 1836. Report of a geological Reconnaissance made in 1835 from the seat of the Government by the way of Green Bay and the Wisconsin territory, etc., 168 pp.

Fenneman, N. M., 1916. Geology of Cincinnati and vicinity. Ohio Geol. Surv., 4th ser., Bull. 19, 207 pp.

Foerste, A. F., 1885. The Clinton group of Ohio with description of new species. Sci. Lab., Denison Univ., Bull., vol. 1, pp. 63-120.

1887. Clinton group, etc., pt. II, Denison Univ. Sci. Lab., Bull., vol. 2, pp. 89-110, 148-176.

1888. Notes on a geological section at Todds Fork, Ohio. Am. Geol., vol. 2, pp. 412-419.

1888a. The Clinton group of Ohio, pt. IV. Denison Univ. Sci. Lab., Bull., vol. 3, pp. 3-12.

1888b. A study of the head of typical forms of Lichas breviceps from the Niagara of Waldron, Indiana. Denison Univ. Sci. Lab., Bull., vol. 3, pl. xiii, fig 21.

1890. Notes on Clinton group fossils with special reference to collections from Indiana, Tennessee and Georgia. Boston Soc. Nat. Hist., Proc., vol. 24, pp. 263-355.

1891. On the Clinton oölitic iron ores. Am. Jour. Sci., 3d. ser., vol. 41, pp. 28-29.

1891a. The age of the Cincinnati anticlinal. Am. Geol., vol. 7, pp. 97-109.

1893. Fossils of the Clinton group in Ohio and Indiana. Ohio Geol. Surv., vol. 7, pp. 516-601.

1893a. Remarks on specific characters of Orthoceras. Am. Geol., vol. 12, pp. 232-236.

1895. On Clinton conglomerates and wave marks in Ohio and Kentucky, etc. Jour. Geol. vol. 3, pp. 50-60, 169-197.

1896. An account of the Middle Silurian rocks of Ohio and Indiana, including the Niagara and Ohio Clinton, and the bed at the top of the Lower Silurian strata, formerly considered Medina. Jour. Cin. Soc. Nat. Hist., vol. 18, pp. 161-199.

1897. A report on the geology of the Middle and Upper Silurian rocks of Clark, Jefferson, Ripley, Jennings and southern Decatur counties, Indiana. 21st. Indiana Report, pp. 213-288.

1898. A report on the Niagara limestone quarries of Decatur, Franklin and Fayette counties, etc. 22d. Indiana Report, pp. 195-255.

1899. Age and development of the Cincinnati anticline. Sci. N. S., vol. 10, p. 488. [Abstract].

1900. General discussion of the Middle Silurian rocks of the Cincinnati anticlinal, with their synonomy. 24th. Indiana Report, pp. 41-80.

1900a. Further studies on the history of the Cincinnati geanticline. Sci. N. S., vol. 11, p. 145.

1901. Silurian and Devonian limestones of Tennessee and Kentucky. Geol. Soc. Amer., Bull., vol. 12, pp. 395-444.

1901a. Niagara group along the western side of the Cincinnati anticline. Sci. N. S., vol. 13, pp. 134-135.

1902. Bearing of the Clinton and Osgood formations on the age of the Cincinnati anticline. Sci. N. S., vol. 15, p. 90.

1902a. The Cincinnati anticline in southern Kentucky. Am. Geol., vol. 30, pp. 359-369.

1903. The Cincinnati group in Western Tennessee between the Tennessee river and the Central Basin. Jour. Geol., vol. 11, pp. 29-45.

1903a. Silurian and Devonian limestones of western Tennessee. Jour. Geol., vol. 11, pp. 679-715.

1903b. The Richmond group on the western side of the Cincinnati anticline in Indiana and Kentucky. Am. Geol., vol. 31, pp. 333-361.

1904. The Ordovician-Silurian contact in the Ripley Island area of southern Indiana, etc. Am. Jour. Sci., 4th ser., vol. 18, pp. 321-342.

1904a. Variation in thickness of the subdivisions of the Ordovician of Indiana. Am. Geol., vol. 34, pp. 87-102.

1904b. Description of the rocks formed in the different geological periods in Indiana. 28th. Indiana Rept., pp. 21-39.

1905. The classification of the Ordovician rocks of Ohio and Indiana. Science, N. S., vol. 22, pp. 149-152.

1905a. Note on the distribution of the Brachiopoda in the Arnheim and Waynesville beds. Am. Geol., vol. 36, pp. 244-250.

1906. The Silurian, Devonian and Irvine formations of east-central Kentucky, with an account of their clays and limestones. Kentucky Geol. Surv., Bull. No. 7, 369 pp.

1909. Silurian fossils from the Kokomo, West Union, and Alger horizons of Indiana, Ohio and Kentucky. Jour. Cin. Soc. Nat. Hist., vol. 21, pp. 1-41.

1909a. The Bedford fauna at Indian Fields and Irvine, Kentucky. Ohio Nat., vol. 9, pp. 515-523.

1909b. Fossils from the Silurian formations of Tennessee, Indiana and Kentucky. Denison Univ. Sci. Lab., Bull., vol. 14, pp. 61-116.

1909c. Preliminary notes on Cincinnatian fossils. Den. Univ. Sci. Lab., Bull., vol. 14, pp. 208-232.

1909d. The Brachiopoda of the Richmond group. Science, N. S., vol. 29, p. 635.

1909e. [See Morse and Foerste, 1909].

1909f. Preliminary notes on Cincinnatian and Lexington fossils. Denison Univ. Sci. Lab., Bull., vol. 14, pp. 289-334.

1910. Preliminary notes on Cincinnatian and Lexington fossils of Ohio, Indiana, Kentucky and Tennessee. Denison Univ. Sci. Lab., Bull., vol. 16, pp. 17-100, 6 pls.

1910a. Brachiopoda of the Richmond group [Abstract]. Geol. Soc. Amer., Bull., vol. 20, p. 699.

1912. Strophomena and other fossils from Cincinnatian and Mohawkian horizons, chiefly in Ohio, Indiana and Kentucky. Denison Univ., Sci. Lab., Bull., vol. 17, pp. 17-172.

1912a. The Arnheim formation within the areas traversed by the Cincinnati geanticline. Ohio Nat., vol. 12, pp. 429-456.

1912b. The Ordovician section in the Manitoulin area of Lake Huron. Ohio Nat., vol. 13, pp. 37-48.

1914. Notes on the Lorraine faunas of New York and the Province of Quebec. Denison Univ. Sci. Lab., Bull., vol. 17, pp. 247-339.

1914a. Notes on Agelacrinidae and Lepadocystinae, with descriptions of Thresherodiscus and Brockocystis. Denison Univ. Sci. Lab., Bull., vol. 17, pp. 399-744.

1916. Upper Ordovician formations in Ontario and Quebec. Canadian Geol. Surv., Mem. No. 83, 279 pp.

1916a. Notes on Cincinnatian fossil types. Den. Univ. Sci. Lab., Bull., vol. 18, pp. 285-355.

1917. Intraformational pebbles in the Richmond group at Winchester, Ohio. Jour. Geol., vol. 25, pp. 289-306.

1917a. Notes on Silurian fossils from Ohio and other central states. Ohio Jour. Sci., vol. 17, pp. 187-204, 233-267.

1917b. Notes on Richmond and related fossils. Jour. Cin. Soc. Nat. Hist., vol. 22, pp. 42-55.

1917c. Richmond faunas of Little Bay de Noquette in Northern Michigan. Ottawa Nat., vol. 31, pp. 97-103.

1919. Silurian fossils from Ohio with notes on related species from other horizons. Ohio Jour. Sci., vol. 19, No. 7, pp. 367-404, 3 pls.

Fontaine, W. M., [and White, I. C.] 1880. The Permian and Upper Carboniferous flora of West Virginia and southwestern Pennsylvania. 2d. Geol. Surv. Pa., Rep. Prog. ix, 143 pp.

Fuller, M. L., 1902. [and Ashley] Ditney folio, Indiana. U. S. Geol. Surv., Geol. Atlas, Folio No. 84.

Fuller, M. L., 1904. [and Clapp] Patoka folio, Illinois-Indiana. U. S. Geol. Surv., Geol. Atlas, Folio No. 105.

Galloway, J. J., 1912. [See Cumings and Galloway, 1912a.]

Galloway, J. J., 1913. [See Cumings and Galloway, 1913].

Girty, G. H., 1895. Development of the corrallum of Favosites forbesi var. occidentalis. Am. Geol., vol. 15, pp. 131-146.

1898. Description of a fauna found in the Devonian black shale of eastern Kentucky. Am. Jour. Sci., 4th. ser., vol. 6, pp. 384-394.

Gorby, S. S., 1886. Geology of Tippecanoe, Washington and Benton counties and description of the Wabash Arch. 15th. Indiana Report, pp. 61-96, 117-153, 198-220, 228-241.

1885. A new Crinoid. Hoosier Mineralogist and Archeologist, vol. 1, No. 10.

1889. Geology of Miami county: Natural gas and petroleum. 16th. Indiana Report, pp. 165-188, 189-301.

1892. Natural Resources of Indiana. Geological Report. 17th. Indiana Report. 705 pp. 20 pls.

1894. Geological Report. 18th. Indiana Report. 356 pp. 12 pls.

1894a. Geological Report. 19th. Indiana Report, 296 pages.

1892-1894. The 17th., 18th., and 19th. Indiana Reports are by S. S. Gorby.

Gordon, C. H., 1888. The well at Keokuk, Iowa. Am. Geol., vol. 2, p. 362.

1889. Notice of a deep boring at Keokuk. Am. Geol., vol. 4, p. 127.

1890. On the Keokuk bed at Keokuk, Iowa. Am. Jour. Sci., 3d. ser., vol. 40, pp. 295-300.

1890a. On the brecciated character of the St. Louis limestone. Am. Nat., vol. 24, pp. 305-313.

Grabau, A. W., 1898. Geology and Paleontology of Eighteen Mile creek and the lake shore sections of Erie county, New York. Buffalo Soc. Nat. Hist., Bull., vol. 6, pp. iii-xxiv and 1-403.

1908. A revised classification of the North American Siluric system. [Abstract]. Science N. S., vol. 27, pp. 622-623.

1909. A revised classification of the North American lower Paleozoic. [Abstract]. Science N. S., vol. 29, pp. 351-356.

1909a. Physical and Faunal evolution of North America during Ordovicic, Siluric and early Devonic time. Jour. Geol., vol. 17, pp. 209-252.

1913. Early Paleozoic delta deposits of North America. Geol. Soc. Amer., Bull., vol. 24, pp. 399-528.

1921. Textbook of Geology: Part II, Historical Geology. 976 pp.

Grabau, A. W., and Sherzer, W. H., 1910. The Monroe formation of southern Michigan and adjoining regions. Mich. Geol. Surv., Geol. Ser. 1, pub. 2, 248 pp.

Greene, F. C., 1909. Caves and cave formations of the Mitchell limestone. Ind. Acad. Sci., Proc. for 1908, pp. 175-184.

1911. The Huron group in Western Monroe County, and eastern Greene county, Indiana. Ind. Acad. Sci., Proc. for 1910, pp. 269-288.

1911a. The fauna of the Brazil limestone. Ind. Acad. Sci., Proc. for 1910, pp. 169-171.

Greene, G. K., 1880. Geology of Monroe county. Indiana Department of Statistics and Geology, 2d. Annual Report, pp. 427-449.

1898-1906. Contributions to Indiana Paleontology. Parts 1 to 20, and vol. 2, pts. 1-3. Ewing and Zeller, New Albany, Indiana: Published privately. [Descriptions and figures of numerous species from the Silurian, Devonian and Carboniferous. Many unrecognizable and many synonyms].

1906. On the age of the rocks near Kentland, Indiana. Contr. to Indiana Paleontology, vol. 2, pt. 1, pp. 11-17.

Gurley, W. F. E., 1888. New Carboniferous fossils, Bulletin No. 1, Danville, Illinois, pp. 1-9.

Haines, Mary P., 1879. List of fossils found in the Lower Silurian rocks in the vicinity of Richmond, Indiana. 8th., 9th., 10th. Indiana Reports, pp. 201-204.

Hall, James, 1839. Third Annual Report of the Fourth Geological District of New York. N. Y. Geol. Surv., 3d. Ann. Rept., pp. 287-339.

1842. Notes on the geology of the western states. Am. Jour. Sci., vol. 42, pp. 51-62.

1843. Notes explanatory of a section from Cleveland, Ohio, to the Mississippi River, in a southwest direction, with remarks upon the identity of the western formations with those of New York. Assoc. Amer. Geol., Trans., pp. 267-293.

1843a. On the geographical distribution of fossils in the older rocks of the United States. Am. Jour. Sci., vol. 45, pp. 157-160, 162-163.

1843b. Geology of New York, Pt. IV, [Final report of the Fourth District]. xxvii and 685 pp.

1844. Geographical distribution of fossils. Am. Jour, Sci., vol. 47, pp. 117-118.

1847. Nature of the strata and geographical distribution of the organic remains in the older formations of the United States. Boston Jour. Nat. Hist., vol. 5, pp. 1-20.

1847a. Paleontology of New York, vol. I, xxiii and 333 pp., pls. 1-87.

1848-1849. [Translation] On the parallelism of the Paleozoic deposits of North America with those of Europe, etc., etc., [by Ed deVerneuil: Bull. Soc. Geol., France, vol. 4, 2d. ser.], Am. Jour. Sci., vol. 5, pp. 176-183, 359-370; vol. 7, 45-51, 218-231.

1851. Parallelism of the Paleozoic deposits of the United States and Europe. Foster and Whitney Report on the Lake Superior Land District, part 2; Senate Exec. Doc. No. 4, 1851, pp. 285-318.

1852. Key to a chart of the successive geological formations, etc., Boston, 1852.

1852a. Paleontology of New York, vol. II, viii and 358 pp., 241 plates.

1852b. Remarks on the westward extension of the New York Silurian formations. Am. Assoc. Adv. Sci., Proc., vol. 6, pp. 255-256.

1856. [Spergen Hill fossils] Trans. Albany Inst., vol. 4, 1856, pp. 14-34. [The date of this paper is variously given].

1856a. On the Carboniferous limestones of the Mississippi Valley. Amer. Assoc. Adv. Sci., Proc., vol. 10, pp. 51-69.

1857. On the Carboniferous limestones of the Mississippi Valley. [Abstract]. Am. Jour. Sci., 2d. ser., vol. 23, pp. 187-203.

1857a. Description of Paleozoic fossils. 10th Report Regents of the State of N. Y., pp. 41-186.

1858. [And Whitney] Report on the Geological Survey of the State of Iowa, etc., vol. 1, 724 pp.

1859. Paleontology of New York, vol. III. 2 vols.: Text xii, and 532 pp.; 120 plates.

1860. On the fossils of the Goniatite limestone in the Marcellus shale, etc. Report Regents of the State of New York, No. 13, pp. 95-112.

1860a. Description of new species of Crinoidea from the Carboniferous rocks of the Mississippi valley. Bos. Jour. Nat. Hist., vol. 7, pp. 261-328.

1861. Supplementary note, etc. 14th Ann. Rept. Regents, State of N. Y., p. 81.

1862. Preliminary note on the Trilobites and other Crustaceans of the Upper Helderberg, Hamilton and Chemung. 14th. Ann. Report of the Regents, State of N. Y., pp. 82-144.

1863. Notice of new species of fossils from a locality of the Niagara group in Indiana, etc. Trans. Alb. Inst., vol. 4, pp. 195-228. [Waldron fauna].

1864. Description of new species of fossils from the Carboniferous limestone of Indiana and Illinois. Trans. Alb. Inst., vol. 4, pp. 1-36.

1867. Paleontology of New York, Vol. IV. 428 pp. 63 pls.

1872. On the relations of the Middle and Upper Silurian rocks of the United States. Geol. Mag., vol. 9, pp. 509-513.

1874. On the relation of the Niagara and Lower Helderberg formations, and their geographical distribution in the United States and Canada. Am. Assoc. Adv. Sci., Proc., vol. 22, pt. 2, pp. 321-335.

1878. Note upon the history and value of the term Hudson River group in American Geological nomenclature. Am. Assoc. Adv. Sci., Proc., vol. 26, pp. 259-265.

1879. The Hydraulic beds and associated limestones of the Falls of the Ohio. Trans. Alb. Inst., vol. 9, pp. 169-180.

1879a. Paleontology of New York, Vol. V., Pts. 1 and 2. Pt. I, 268 pp., 92 pls. Pt. II, Text, 492 pp., 113 pls.

1879b. The fauna of the Niagara group in central Indiana. N. Y. State Mus. Nat. Hist., 28th Ann. Rept., pp. 99-199.

1879c. [Correlation of the Pendleton sandstone] In 8th., 9th., 10th. Indiana Reports, p. 60.

1882. The fauna of the Niagara group in Indiana. 11th. Indiana Report, pp. 217-345.

1883. [Spergen Hill fauna]. 12th. Indiana Report, pp. 319-375. [Published with Whitfield's plates].

1883a. Description of fossil corals from the Niagara and Upper Helderberg groups of Indiana. 12th. Indiana Report, pp. 239-318.

1883b. Contributions to the geological history of the American Continent. Am. Assoc. Adv. Sci., Proc., vol. 31, pp. 29-69.

1884. Description of fossil corals from the Niagara and Upper Helderberg groups. N. Y. State Mus., 35th. Ann. Rept., pp. 707-464.

1884a. Description of species of fossil reticulate sponges constituting the family Dictyospongia. N. Y. State Mus., 35th. Ann. Report, pp. 465-481.

1887. Paleontology of New York, Vol. VI. Corals and Bryozoa. Text and plates; 298 pp., 66 plates.

1888. Paleontology of New York, Vol.. VII. Trilobites and other Crustacea. Text and plates; lxiv and 236 pp., 36 pls.

1888a. [In same volume] Paleontology of New York, Vol. V, pt. II, Supplement. Pteropoda, Cephalopoda and Annelida, pp. 1-42, pls. 114-129.

1892. Paleontology of New York, Vol. VIII, pt. I. Brachiopoda. 367 pp., pls. 1-20. [Hall and Clarke].

1894. Paleontology of New York, Vol. VIII, pt. II, Brachiopoda. 394 pp., pls. 21-84. [Hall and Clarke].

Hall, James, and Clarke, J. M., 1892 and 1894. [See Hall 1892 and Hall 1894].
Hall, James, and Whitfield, R. P., 1872. Description of new species of fossils from the vicinity of Louisville, Kentucky, and the Falls of the Ohio. 24th. Ann. Rept. Regents, Univ. State of N. Y., pp. 223-239.

1875. Descriptions of invertebrate fossils, mainly from the Silurian system. Geol. Surv. Ohio, vol. 2, pt. 2 [Paleontology], pp. 65-161, pls.

1875a. Description of Crinoidea from the Waverly group. Geol. Surv. Ohio, vol. 2, pt. 2 [Paleontology], pp. 162-179, pls.

Hall, C. W., and Sardeson, F. W., 1892. Paleozoic formations of southeastern Minnesota. Geol. Soc. Amer., Bull., vol. 3, pp. 251-268.

Harper, G. W., and Bassler, R. S., 1896. Catalog of the fossils of the Trenton and Cincinnati periods occurring in the vicinity of Cincinnati, Ohio. Cincinnati, O., 34 pp.

Hartnagle, C. A., 1907. Stratigraphic relations of the Oneida Conglomerate. N. Y. State Mus., Bull., No. 107, pp. 29-38.

1912. Classification of the geologic formations of the State of New York. New York State Mus., Handbook 19 [of the state of New York Education department], 96 pp.

Hay, Oliver P., 1912. The Pleistocene age and its vertebrata. 36th. Indiana Report, pp. 539-784, 2 pls. (Maps) 80 figs.

Haymond, Rufus, 1844. Notice of remains of Megatherium, Mastodon and Silurian fossils. Amer. Jour. Sci., vol. 46, pp. 294-296.

1869. Geology of Franklin county. 1st. Indiana Report, pp. 175-202.

Herrick, C. L., 1888. The geology of Licking county, Ohio. [One of a series of papers on the same subject published between 1885 and 1888]. Denison Univ. Sci. Lab., Bull., vol. 3, pp. 13-110. [See also *ibid.*, vols. 2 to 4].

Hill, R. T., 1899. The geology and physical geography of Jamaica; study of a type of Antillean development, based upon surveys made for Alexander Agassiz. Harvard College, Mus. Comp. Zool., Bull., vol. 34, pp. 1-256.

Hobbs, B. C., 1872. Report of a geological survey of Parke county. 3d. and 4th. Indiana Reports, pp. 341-384.

Hole, Allan D., 1919. Notes on the paleontology of certain Chester formations of southern Indiana. Ind. Acad. Sci., Proc. for 1918, pp. 183-185.

Hopkins, T. C., 1896. The Carboniferous sandstones of western Indiana. 20th. Indiana Report, pp. 188-327.

1896a. The sandstones of western Indiana. U. S. Geol. Surv., 17th. Ann. Rept., pt. 3, pp. 780-787.

1896b. The sandstones of western Indiana. Mineral Industry, for 1895, pp. 559-564.

1897. Origin of conglomerates of western Indiana. Geol. Soc. Amer., Bull., vol. 8, pp. 14-15 [abstract].

1897a. Stylolites. Am. Jour. Sci., 4th. ser., vol. 4, pp. 142-144.

1897b. [and Siebenthal, C. E.]. The Bedford Oölitic limestone of Indiana. 21st. Indiana Report, pp. 291-427.

1902. The Lower Carboniferous area in Indiana. Sci. N. S., vol. 15, p. 83 [Abstract].

1903. Lower Carboniferous area in Indiana. Geol. Soc. Amer., Bull., vol. 13, pp. 519-521. [Abstract].

1904. A short description of the topography of Indiana and of the rocks of the different geological periods; to accompany the geological map of the State. 28th. Indiana Report, pp. 15-77. [The part on the Ordovician and Silurian was written by A. F. Foerste].

1904a. The geological map of Indiana. 28th. Indiana Report, pp. 11-14.

1904b. Contents of the published volumes of the Reports of the Indiana Geological Surveys, and Department of Geology and Natural History, and the Department of Geology and Natural Resources. 28th. Indiana Report, pp. 487-495.

1904c. General Index to all the publications of the Indiana Geological Survey, etc. [as above]. 28th. Indiana Report, pp. 497-553.

1908. General structural and economic features of the Indiana oölitic limestone. 32d. Indiana Report, pp. 310-335.

Hopkins and Siebenthal, 1897. [See Hopkins 1897b].

Hubbard, G. C., 1892. The upper limit of the Lower Silurian at Madison, Indiana. Ind. Acad. Sci., Proc. for 1891, pp. 68-70.

1892a. Hudson River fossils of Jefferson county, Indiana. Ind. Acad. Sci., Proc. for 1891, p. 68.

1892b. The Cystidians of Jefferson county, Indiana. Ind. Acad. Sci., Proc. for 1891, p. 67.

Irving, R. D., 1875. Note on some new points in the elementary stratification of the Primordial and Canadian rocks of south central Wisconsin. Am. Jour. Sci., 3d. ser., vol. 9, pp. 440-443.

Jackson, T. F., 1915. [Geology of the Bloomington Quadrangle by Beede and others]. Report on the Pennsylvanian or Coal Measures. 39th. Indiana Report, pp. 224-229.

1915a. The paleobotany of the Bloomington, Indiana, Quadrangle. Ind. Acad. Sci., Proc. for 1914, pp. 395-398.

1917. The description and stratigraphic relationships of fossil plants from the lower Pennsylvanian rocks of Indiana. Ind. Acad. Sci., Proc. for 1916, pp. 405-428, 10 pls.

James, J. F., 1886. Cephalopoda of the Cincinnati group. Jour. Cin. Soc. Nat. Hist., vol. 8, pp. 235-253.

1886a. Protozoa of the Cincinnati group. Jour. Cin. Soc. Nat. Hist., vol. 9, pp. 244-252.

1891. On the age of the Point Pleasant beds. Jour. Cin. Soc. Nat. Hist., vol. 14, pp. 93-104.

1892. Manual of the Paleontology of the Cincinnati group. Jour. Cin. Soc. Nat. Hist., vol. 14, pp. 45-72; 149-163; vol. 15, pp. 88-100, 144-159.

1894. Manual of the Paleontology of the Cincinnati group. Jour. Cin. Soc. Nat. Hist., vol. 16, pp. 178-208.

1895. Manual of the Paleontology of the Cincinnati group. Jour. Cin. Soc. Nat. Hist., vol. 18, pp. 67-88.

1897. Manual of the Paleontology of the Cincinnati group, Part VIII. Jour. Cin. Soc. Nat. Hist., vol. 19, pp. 99-118.

James, U. P., 1874. Description of new species of fossils from the Lower Silurian formations, Cincinnati group. Cin. Quar. Jour. Sci., vol. 1, pp. 239-242.

1879. The Paleontologist, No. 4. [This author published privately a series of papers, called The Paleontologist, in which appear many lists and descriptions of fossils, without illustrations, often very difficult to recognize].

Jones, Lee, 1898. The upper limit of the Knobstone in the region of Borden, Indiana. Ind. Acad. Sci., Proc. for 1897, pp. 257-258.

Keating, W. H., 1824. Narrative of an expedition to the sources of the St. Peters River, etc., etc. 2 vols. 439 and 459 pp.

Keyes, C. R., 1889. The Carboniferous Echinodermata of the Mississippi basin. Am. Jour. Sci. 3d. ser., vol. 3, p. 186.

1890. Synopsis of American Carbonic Calyptraeidae. Phila. Acad. Sci., Proc. for 1890, pp. 150-181.

1892. The principal Mississippian section. Geol. Soc. Amer., Bull., vol. 3, pp. 283-300.

1901. A depositional measure of unconformity. Geol. Soc. Amer., Bull., vol. 12, pp. 173-196.

1901a. A schematic standard for the American Carboniferous. Am. Geol., vol. 28, pp. 299-305.

1901b. The value of provincial Carboniferous terranes. Am. Jour. Sci., 4th. ser., vol., 12, pp. 305-309.

Kindle, E. M., and Marsters, V. F., 1894. [See Marsters and Kindle, 1894].

Kindle, E. M., 1895. [On the dip of certain rocks]. Ind. Acad. Sci., Proc. for 1894, pp. 52-53.

1896. The Whetstone and Grindstone rocks of Indiana. 20th. Indiana Report, pp. 329-368.

1898. A catalog of the fossils of Indiana, accompanied by a bibliography of the literature relating to them. 22d. Indiana Report, pp. 407-514.

1899. The Devonian and Lower Carboniferous faunas of southern Indiana and central Kentucky. Amer. Paleont., Ithaca, N. Y., Bull. No. 12, pp. 1-112.

1901. The Devonian fossils and stratigraphy of Indiana. 25th. Indiana Report, pp. 529-763, pls. 1-31.

1902. The Niagara limestones of Hamilton county, Indiana. Am. Jour. Sci., 4th. ser., vol. 14, pp. 221-224.

1903. The Niagara domes of northern Indiana. Am. Jour. Sci., 4th. ser., vol. 15, pp. 459-468.

1904. The stratigraphy and Paleontology of the Niagara of northern Indiana. 28th. Indiana Report, pp. 397-486, pls. 1-28 [Paleontology with C. L. Breger].

1912. The stratigraphic relations of the Devonian shales of northern Ohio. Am. Jour. Sci., 4th. ser., vol. 34, pp. 187-213.

1913. The age of the Eurypterids of Kokomo, Indiana. Am. Jour. Sci., 4th ser., vol. 36, pp. 282-288.

1913a. The unconformity at the base of the Onondaga limestone in New York and its equivalent west of Buffalo. Jour. Geol., vol. 21, pp. 301-319.

1913b. [And Taylor, Frank B.]. The Niagara Quadrangle. U. S. Geol. Surv., Geol. Atlas, Folio No. 190, 25 pp.

1914. What does the Medina sandstone of the Niagara section include? Sci., N. S., vol. 39, pp. 915-918.

Kindle, E. M., and Barnett. V. H., 1909. The stratigraphic and faunal relations of the Waldron fauna in southern Indiana. 33d. Indiana Report, pp. 393-416.

King, Henry, 1851. Some remarks on the geology of the State of Missouri. Amer. Assoc. Adv. Sci., Proc., vol. 5, pp. 182-199.

Kingsley, J. S., 1888. [Suggestions in regard to the age and History of the Caves of the Indiana, Kentucky and Tennessee region.] Am. Nat., vol. 22, pp. 1104-1106.

Knowlton, F. K., 1889. Description of a problematic organism from the Devonian at the Falls of the Ohio. Am. Jour. Sci., 3d. ser., vol. 37, pp. 202-209.

Lane, A. C., 1893. Michigan Geol. Surv., Report State Board for 1891-92, [p. 66].

Lane, A. C., Prosser, C. S,. Sherzer, W. H., and Grabau, A. W., 1907. The nomenclature and subdivisions of the Upper Siluric strata of Michigan, etc. Geol. Soc. Amer., Bull., vol. 19, pp. 540-556.

Lapham, Increase A., 1828. Notice of the Louisville and Shippingsport canal and of the geology of the vicinity. Am. Jour. Sci., vol. 14, pp. 65-69.

Lapworth, Charles, 1879. [Name of the Ordovician]. Geol. Mag., N. S., Dec. 2, vol. 6, p. 14.

Lesley, J. P., 1856. Manual of coal and its topography, illustrated by original drawings, etc., 224 pp.

1877. [In a report by F. and W. G. Platt on the Cambria and Somerset districts of the bituminous coal fields of western Pennsylvania, the name Pottsville is credited to Lesley]. 2d. Geol. Surv. Pennsylvania, Rept. H H, xxx and 104 pp., pls. and maps.

1886. Annual Report of the Geological Survey of Pennsylvania for 1885. 769 pp., 18 pls. and atlas.

Lesquereux, Leo, 1862. Report on the distribution of the geological strata in the Coal Measures of Indiana. Geol. Reconn. of Indiana, by R. Owen, pp. 273-341.

1870. Description of new species and an enumeration with remarks on species already known. Geol. Surv. Ill., vol. 4, pp. 379-477.

1876. Species of fossil marine plants from the Carboniferous measures. 7th. Indiana Report, pp. 134-145.

1880. Description of the Coal flora of the Carboniferous formation in Pennsylvania and throughout the United States. Vol. 1, Cellular cryptogamous plants, Fungi Thalassophytes. Vol. 2, Vascular cryptogamous plants, Calamaria, Filicacea [Ferns]. 694 pp., 87 pls.

1884. Principles of Paleozoic Botany. 13th. Indiana Report, pt. 2, pp. 1-106.

Levette, G. M., 1874. Geological survey of Dekalb, Steuben, Lagrange, Elkhart, Noble, St. Joseph and Laporte counties, Indiana. 5th. Indiana Report, pp. 430-474.

Lines, E. F., 1912. The stratigraphy of Illinois with reference to Portland-cement materials. Illinois State Geol. Surv., Bull. No. 17, pp. 59-76.

Linney, W. M., 1882 (?). Notes on the rocks of Central Kentucky. 19 pp. [Date uncertain. Darton gives 1889].

Locke, John, 1838. [Report on the geology of southwestern Ohio]. Geol. of Ohio by W. W. Mather, 2d. Ann. Rept., pp. 201-274.

Logan, Sir W. E., 1863-65. Geological Survey of Canada, report of progress from its commencement to 1863, 983 pp. Atlas of maps and sections, 42 pp. 13 pls., 1865.

Logan, Sir W. E., and Hall, James, 1866. Geological map of Canada and part of the United States, etc. [Published on smaller scale in Geology of Canada, 1863].

Logan, W. N., 1919. Kaolin in Indiana. Publication of the Indiana Department of Conservation, Division of Geology, 131 pp.

1920. Petroleum and Natural Gas in Indiana. Publication of the Indiana Department of Conservation, Division of Geology, 227 pp.

Long, S. H., 1823. Account of an expedition from Pittsburg to the Rocky Mountains performed in the years 1819-20; compiled by Edward James, vol. 1, 5 and 503 pp., vol. 2, 442 pp., Atlas of 11 sheets.

Lyell, Sir Charles, 1849. A second visit to the United States, Vol. II, pp. 269-278. [Describes the coral reef at Louisville].

Lyon, Sidney S., 1857. Description of new species of Organic Remains. Kentucky Geol. Surv., vol. 3, pp. 467-498.

1860. Remarks on the stratigraphical arrangement of the rocks of Kentucky, from the Catenipora escharoides horizon of the Upper Silurian Period, in Jefferson county, to the base of the Productive Coal Measures in the eastern edge of Hancock county, Ky. St. Louis Acad. Sci., Trans., vol. 1, pp. 612-621.

1866. Description of new species of Paleozoic fossils from Kentucky and Indiana. Phila. Acad. Nat. Sci., Proc., vol. (1866), pp. 409-414.

Lyon, S. S., and Casseday, S. A., 1859. Description of nine new species of Crinoidea from the Subcarboniferous rocks of Indiana. Am. Jour. Sci., 2d. ser., vol. 28, pp. 233-246.

1859a. Description of nine new species from the Subcarboniferous rocks of Indiana and Kentucky. Am. Jour. Sci., 2d. ser., vol. 29, pp. 68-79.

McCaslin, David S., 1883. Geology of Jay county. 12th. Indiana Report, pp. 153-176.

1884. Johnson county. 13th. Indiana Report, pp. 116-137.

McChesney, I. T., 1859. New Paleozoic fossils. 64 pp., Chicago, 1859.

McEwan, (Mrs.) Eula D., 1919. A study of the Brachiopod genus Platystrophia. U. S. Nat. Mus., Proc., vol. 56, pp. 383-448, pls. 42-52.

1920. The Ordovician at Madison, Indiana. Am. Jour. Sci., 4th. Ser., vol. 50, pp. 154-158.

Macfarlane, James, 1873. Coal regions of America: Their topography, geology and development. xxvi and 676 pp. Second and Third editions, 1874, 1877.

1879. An American geological railway guide, etc. First edition, 219 pp.

1890. Same, second edition, 426 pp.

McClure, William, 1809. Observations on the geology of the United States [explanatory of a geological map, etc.]. Am. Phil. Soc., Trans., vol. 6, pp. 411-428.

1818. Observations on the geology of the United States. [With remarks on the effects of rock decomposition on the nature and fertility of soils, etc.]. Am. Phil. Soc., Trans., vol. 1, N. S., pp. 1-91.

1822. Comparative features of American and European Geology. Am. Jour. Sci., vol. 5, pp. 197-198.

Malott, C. A., 1919. The American Bottoms region of eastern Greene county, Indiana: A type unit in southern Indiana physiography. Indiana University Studies, vol. 6, pp. 1-61.

1920. The stratigraphy of the Chester series of Southern Indiana. Science, N. S., vol. 51, pp. 521-522.

Mance, G. C., 1915. [Geology of the Bloomington Quadrangle]. The utilization of waste stone, etc. 39th. Indiana Report, pp. 230-312.

1917. Power Economy and the Utilization of waste in the quarry industry of southern Indiana. Indiana University Studies, vol. 4, pp. 1-204.

Marcou, Jules, 1853. Geological map of the United States and British provinces of North America. [With explanatory text, etc.]. 92 pp. 8 pls. Boston, 1853.

Marsters, V. F., and Kindle, E. M., 1894. Geological literature of Indiana (Stratigraphic and Economic). Proc. Ind. Acad. Sci. for 1893, pp. 156-191.

Mather, W. W., 1840. Fourth annual report of the geological survey of the first geological district of New York. N. Y. Geol. Surv., 4th. Ann. Rept., pp. 211-258.

1859. Report of the State House artesian well at Columbus, Ohio. 42 pp.

Maximillian (Prince) Alexander Philip, 1838-43. "Reise durch Nord-Amerika," Coblenz, 1838-43. [Spent a winter at New Harmony, Indiana.]

Meek, F. B., 1865. Description of nine species of Crinoidea, etc., from the Paleozoic rocks of Illinois and some of the adjoining states. Phila. Acad. Nat. Sci., Proc., vol. (1865), pp. 143-166.

1867. Remarks on the geology of the valley of the Mackenzie river with descriptions of fossils. Chicago Academy of Science, Trans., vol. 1, pt. 1, pp. 61-114.

1869. Description of new species of Crinoidea and Echinoidea from the Carboniferous rocks of the western states. Phila. Acad. Nat. Sci., Proc., vol. (1869), pp. 67-83.

1871. Description of new western Paleozoic fossils mainly from the Cincinnati group of the Lower Silurian series of Ohio. Phila. Acad. Nat. Sci., vol. (1871), pp. 308-336.

1871a. On some new Crinoids and shells. Am. Jour. Sci., 3d. ser., vol. 2, pp. 295-302.

1872. Description of two new star fishes and a crinoid from the Cincinnati group of Ohio and Indiana. Am. Jour. Sci., 3d. ser., vol. 3, p. 260.

1873. Spergen Hill fossils identified among specimens from Idaho. Am. Jour. Sci., 3d. ser., vol. 5, pp. 383-384.

1873a. Description of invertebrate fossils of the Silurian and Devonian systems. Geol. Surv. Ohio, vol. 1, pt. 2, [Paleontology] pp. 1-243. pls.

Meek, F. B., and Worthen, A. H., 1861. Remarks on the age of the Goniatite limestone at Rockford, Indiana, and its relation to the black slate of the western states and to some of the succeeding rocks above the latter. Am. Jour. Sci., 2d. ser., vol. 32, pp. 167-177, 288.

1860. Description of new Carboniferous fossils from Illinois and the western States. Phila. Acad. Nat. Sci., Proc., vol. (1860), p. 447.

1865. Description of new species of Crinoidea, etc., from the Paleozoic rocks of Illinois and some adjacent states. Phila. Acad. Sci., Proc., vol. (1865), pp. 143-155. [At the end of the paper, p. 155, is a special note entitled "Note in regard to the name CINCINNATI GROUP used in the foregoing paper"].

1865a. Description of new Crinoidea, etc., from the Carboniferous rocks of Illinois and some of the adjacent States. Phila. Acad. Nat. Sci., vol. (1865), pp. 155-166.

1865b. Contributions to the paleontology of Illinois and other western states. Phila. Acad. Nat. Sci., Proc., vol. (1865), pp. 251-275.

1866. Geological Survey of Illinois, Vol. II, Paleontology, Introduction, pp. iii-xix.

1868. Geological Survey of Illinois, Vol. III, Paleontology.

1870. Description of new species and genera of fossils from the Paleozoic rocks of the western States. Phila. Acad. Sci., Proc., vol. (1870), pp. 22-64.

1873. Geological Survey of Illinois, Vol. V., Paleontology.

Merrill, G. P., 1891. Stones for Building and Decoration. 453 pp.

Middleton, W. G., and Moore, J., 1900. Skull of fossil Bison. Ind. Acad. Sci., Proc. for 1899, pp. 178-181.

Miller, A M., 1913. Geology of the Georgetown Quadrangle, Kentucky. Kentucky Geol. Surv., 4th. ser., vol. 1, pp. 317-351.

1915. The Ordovician Cynthiana formation. Am. Jour. Sci., 4th. Ser., vol. 40, pp. 651-657.

1917. Table of geological formations of Kentucky. 7 pp. Lexington, Ky., March, 1917.

1919. Geology of Kentucky. Kentucky, Dept. of Geol. and Forestry, Ser. 5, Bull. 2, 392 pp.

Miller, S. A., 1874. The position of the Cincinnati group in the geological column of fossiliferous rocks of North America. Cin. Quar. Jour., Sci., vol. 1, pp. 97-115.

1874a. Monograph of the Gasteropoda of the Cincinnati group. Cin. Quar. Jour. Sci., vol. 1, pp. 302-321.

1874b. Monograph of the Lamellibranchiata of the Cincinnati group. Cin. Quar. Jour. Sci., vol. 1, pp. 211-236.

1874c. Notice of Modiolopsis pholadiformis (Foster and Whitney). Cin. Quar. Jour. Sci., vol. 1, p. 282.

1875. Crania reticularis. Cin. Quar. Jour. Sci., vol. 2, p. 280.

1875a. Class Cephalopoda as represented in the Cincinnati group. Cin. Quar. Jour. Sci., vol. 2, pp. 121-134.

1875b. Monograph of the class Brachiopoda of the Cincinnati group. Cin. Quar. Jour. Sci., vol. 2, pp. 6-62.

1878. Description of new genera and eleven new species of fossils. Jour. Cin. Soc. Nat. Hist., vol. 1, pp. 100-108.

1878a. Description of eight new species of Holocystites from the Niagara group of Indiana. Jour. Cin. Soc. Nat. Hist., vol. 1, pp. 129-136.

1879. [And others]. Report of a Committee on geological nomenclature [of the Cincinnati group]. Jour. Cin. Soc. Nat. Hist., vol. 1, pp. 193-194. [This report was also published in the 8th., 9th., 10th. Indiana Report, pp. 23-25, 1879].

1879a. Description of twelve new species and remarks upon others. Jour. Cin. Soc. Nat. Hist., vol. 2, pp. 104-118.

1879b. Description of two new species of fossils from the Niagara group and five from the Keokuk group. Jour. Cin. Soc. Nat. Hist., vol. 2, pp. 254-259.

1879c. Remarks upon the Kaskaskia group and description of new species of fossils from Pulaski county, Kentucky. Jour. Cin. Soc. Nat. Hist., vol. 2, pp. 31-42.

1879d. Catalog of fossils found in the Hudson River, Utica slate, and Trenton groups exposed in southeastern Indiana and southwestern Ohio, and the northern part of Kentucky. 8th., 9th., 10th. Indiana Reports, pp. 26-56.

1881. Observations on the unification of geological nomenclature, with special reference to the Silurian formation of North America. Jour. Cin. Soc. Nat. Hist., vol. 4, pp. 267-293.

1881a. Description of new species of fossils. Jour. Cin. Soc. Nat. Hist., vol. 4, pp. 259-262.

1882. Description of eight new species and two new genera of fossils from the Hudson River group, with remarks upon others. Jour. Cin. Soc. Nat. Hist., vol. 5, pp. 34-44.

1882a. Description of ten new species of fossils. Jour. Cin. Soc. Nat. Hist., vol. 5, pp. 79-88.

1882b. Description of three new orders and four new families in the class Echinodermata, and eight new species from the Silurian and Devonian formations. Jour. Cin. Soc. Nat. Hist., vol. 5, pp. 221-231.

1889. The structure, classification and arrangement of American Paleozoic Crinoids into families. 16th. Indiana Report, pp. 302, 326.

1890. North American Geology and Paleontology for the use of amateurs, students and scientists. 664 pp.

1892. Paleontology. 17th. Indiana Report, pp. 611-705, 20 pls. [Silurian, Devonian and Carboniferous fossils].

1892a. Some new species and new structural parts of fossils. Jour. Cin. Soc. Nat. Hist., vol. 15, pp. 79-87.

1894. Paleontology. 18th. Indiana Report, pp. 257-333, 12 pls. [Carboniferous and Silurian fossils].

1894a. Description of some Cincinnati fossils. Jour. Cin. Soc. Nat. Hist., vol. 17, pp. 137-158.

Miller, S. A., and Dyer, C. B., 1878. Contributions to Paleontology. Jour. Cin. Soc. Nat. Hist., vol. 1, pp. 24-39.

Miller, S. A., and Faber, C. L., 1894. New species of fossils from the Hudson River group, and remarks upon others. Jour. Cin. Soc. Nat. Hist., vol. 17, pp. 22-33.

1892. Description of some Çincinnati fossils. Jour. Cin. Soc. Nat. Hist., vol. 15, pl. 1.

Miller, S. A., and Gurley, W. F. E., 1889. Description of some new species and genera of Echinodermata from the Coal Measures and Subcarboniferous rocks of Indiana, Missouri and Iowa. 16th. Indiana Report, pp. 326-373.

1890. Description of some new genera and species of Echinodermata from the Coal Measures and Subcarboniferous rocks of Indiana, Missouri and Iowa. Jour. Cin. Soc. Nat. Hist., 1890, pp. 3-25.

1894. Description of some new species of invertebrates from the Paleozoic rocks of Illinois and the adjacent states. Illinois State Mus. Nat. Hist., Bull., No. 3, pp. 1-81, 8 pls.

1895. New and interesting species of Paleozoic fossils. Illinois State Mus. Nat. Hist., Bull. No. 7, p. 89, 5 pls.

1896. New species of Echinodermata and a new crustacean from the Paleozoic rocks. Illinois State Mus. Nat. Hist., Bull. No. 10, pp. 1-91, 5 pls.

1896a. New species of Paleozoic invertebrates from Illinois and other states. Illinois State Mus. Nat. Hist., Bull. No. 11, pp. 1-50, 5 pls.

1897. New species of crinoids cephalopods and other Paleozoic fossils. Illinois State Mus., Bull. No. 12, 69 pp., 5 pls.

Moore, D. R., 1886. Fossil corals of Franklin county. Brookville Soc. Nat. Hist., Bull. No. 2, pp. 50-51.

1886. Two hours among the fossils of Franklin county. Brookville Soc. Nat. Hist., Bull. No. 1, pp. 44-45.

Moore, Joseph, 1890. Concerning a skeleton of the great fossil beaver Casteroides ohioensis. Jour. Cin. Soc. Nat. Hist., vol. 13, pp. 138-169.

1890a. Concerning some features of Casteroides ohioensis not heretofore known. Amer. Assoc. Adv. Sci., Proc. vol. 39, pp. 265-267.

1893. Glacial and preglacial erosion in the vicinity of Richmond, Indiana. Ind. Acad. Sci., Proc. for 1891, pp. 27-29.

1893a. An inquiry as to the cause of the variety in rock deposits as seen in the Hudson River beds at Richmond, Indiana. Ind. Acad. Sci., Proc. for 1892, pp. 26-27.

1893b. The recently found Casteroides in Randolph county, Indiana. Am. Geol., vol. 12, pp. 67-74.

1897. Account of a morainal stone quarry of Upper Silurian limestone near Richmond, Indiana. Ind. Acad. Sci., Proc. for 1896, pp. 75-76.

1897a. The Randolph Mastodon. Ind. Acad. Sci., Proc. for 1896, pp. 277-278.

1900. A cranium of Casteroides at Greenfield, Indiana. Ind. Acad. Sci., Proc. for 1899, p. 171.

1900a. [And Middleton, W. G.]. Skull of fossil bison. Ind. Acad. Sci., Proc. for 1899, pp. 178-181.

Moore, Joseph, and Hole, Allan D., 1902. Concerning well-defined ripple-marks in the Hudson River limestone, Richmond. Ind. Acad. Sci., Proc. for 1901, pp. 216-220.

Morse, W. C., and Foerste, A. F., 1909. The Waverly formation of east central Kentucky. Jour. Geol., vol. 17, pp. 164-177.

Murchison, Sir R. I., 1835. [Proposes the name Silurian]. Phil. Mag., vol. 7, pp. 46-52.

Nettleroth, Henry, 1889. Kentucky fossil shells. A monograph of the fossil shells of the Silurian and Devonian rocks of Kentucky. Kentucky Geol. Surv., 245 pp., pls.

Newberry, J. S., 1870. Geological Survey of Ohio, Report of Progress for 1869. 60 pp. [Address delivered to the Ohio Legislature].

1871. Geological Survey of Ohio, Pt. 1, Report of Progress for 1869, 167 pp.

1873. [Description of fossil fishes; and other papers]. Geology of Ohio, Vol. I, Pts. I and II, pp. 1-222; 247-355 [pt. II].

1874. The Carboniferous system. Geology of Ohio, Vol. II, pp. 81-180.

1879. List of fossils of Harrison county. 8th., 9th., 10th. Indiana Reports, pp. 341-349.

1890. The Paleozoic fishes of North America. U. S. Geol. Surv., Monogr. No. 16, 340 pp.

Newberry, J. S., and Worthen, A. H., 1866. Description of new species of vertebrates mainly from the Subcarboniferous limestones and Coal Measures of Illinois. Geol. Surv. Ill., vol. 2, pp. 11-141.

Newell, Frederick H., 1888. Niagara cephalopods from northern Indiana. Boston Soc. Nat. Hist., Proc., vol. 23, pp. 466-486.

Newsom, J. F., 1897. A geological section across southern Indiana from Hanover to Vincennes. Jour. Geol., vol. 6, pp. 250-256. [Abstract in Ind. Acad. Sci., Proc. for 1897, pp. 250-253].

1898. The Knobstone group in the region of New Albany. Ind. Acad. Sci., Proc. for 1897, pp. 253-256.

1898a. A geological section across southern Indiana from Hanover to Vincennes. Ind. Acad., Proc. for 1897, pp. 250-253.

1899. [And Price, J. A.]. Notes on the distribution of the Knobstone group in Indiana. Ind. Acad. Sci., Proc. for 1898, pp. 289-290.

1902. Drainage of southern Indiana. Jour. Geol., vol. 10, pp. 166-181.

1903. A geologic and topographic section across southern Indiana, from the Ohio river at Hanover, to the Wabash River at Vincennes, with a discussion of the general distribution and character of the Knobstone group in the State of Indiana. 26th. Indiana Report, pp. 227-302, maps and figures.

Nicholson, H. A., 1875. Description of the corals [and Bryozoa] of the Silurian and Devonian systems. Geol. Surv. Ohio, Vol. II, Pt. II, pp. 183-242, pls.

1875a. Description of Amorphozoa from the Silurian and Devonian formations. Geol. Surv. Ohio, Vol. II, pt. II, pp. 245-255, pls.

Nickles, J. M., 1902. The geology of Cincinnati. Jour. Cin. Soc. Nat. Hist., vol. 20, pp. 49-100, 1 pl., map.

1902a. Description of a new Bryozoan "Homotrypa hassleri" n. sp. from the Warren beds of the Lorraine group. Jour. Cin. Soc. Nat. Hist., vol. 20, pp. 103-105, figure.

1903. The Richmond group in Ohio and Indiana and its subdivisions, with a note on the genus Strophomena and its type. Am. Geol., vol. 32, pp. 202-218.

1905. The upper Ordovician rocks of Kentucky and their Bryozoa. Kentucky Geol. Surv., Bull. No. 5, pp. 1-64, 3 pls.

Nickles, J. M., and Bassler, R. S,. 1900. A synopsis of American fossil Bryozoa, including bibliography and synonomy. U. S. Geol. Surv., Bull. No. 173, 663 pp.

Norwood, C. J., 1876. Report on the geology of the region adjacent to the Louisville, Paducah and Southwestern railway, with a section. Geol. Surv. Ky., Repts. of Progress, vol. 1, N. S., pp. 355-448. [Bottom pagination].

Norwood, J. G., 1848. Letter announcing the discovery of fossil fishes. Boston Soc. Nat. Hist., Proc., vol. 2, p. 102.

Norwood, J. G., and Owen, D. D., 1846. Researches among the Protozoic and Carboniferous rocks of central Kentucky, etc.

1846a. Description of a new fossil fish from the Paleozoic rocks of Indiana. Am. Jour. Sci., 2d. ser., vol. 1, pp. 367-371.

1848. [Devonian fossils near the falls of the Ohio]. Boston Soc. Nat. Hist., Proc., vol. 2, p. 116.

Norwood, J. G., and Pratten, Henry, 1858. Notice of Producti found in the western states, with description of twelve new species. Phila. Acad. Nat. Sci., Proc., vol. 3, 2d. ser., pp. 5-20.

1858a. Notice of the genus Chonetes as found in the western states and territories, with description of eleven new species. Phila. Acad. Nat. Sci., Proc., vol. 3, 2d. ser., pp. 23-31.

Nuttall, Thomas, 1821. Observations on the geological structures of the valley of the Mississippi. Phila. Acad. Nat. Sci., Proc., vol. 2, pt. 1, pp. 14-52.

O'Connell, Marjorie, 1916. The habitat of the Eurypterida. Buffalo Soc. Nat. Hist., Bull., vol. 11, No. 3, pp. 1-277.

O'Harra, C. C., 1900. The geology of Allegany county, Maryland. Maryland Geol. Surv., Allegany county, pp. 57-164, pls. 7-16.

Orton, Edward, 1871. The geology of Highland county. Ohio Geol. Surv., Rep. of Progress for 1870, pp. 255-310.

1873. Report on the third geological district. Geol. Surv. Ohio, Vol. I, pp. 365-480.

1874. Report on third district. Geol. Surv. Ohio, Vol. II, pp. 611-696.

1878. [Geology of Warren, Butler, Preble and Madison counties, Ohio]. Geol. Surv. Ohio, Vol. III, pt. I, pp. 381-428.

1886. Indiana building stones. 10th. Census of the United States. [Building stones of the U. S., and quarry industry]. pp. 215-219. Part of vol. X.

1888. Geology of Ohio [Economic Geology]. Geol. Surv. of Ohio, Vol. VI, 792 pp.

1888a. The Trenton limestone as an oil formation. Am. Geol., Vol. 1, p. 133.

1888b. Geological Survey of Ohio, Preliminary report upon petroleum and inflammable gas, with a supplement. 200 pp., Columbus, 1887.

1889. The Trenton limestone as a source of petroleum and inflammable gas in Ohio and Indiana. U. S. Geol. Surv., 8th. Ann. Report, pp. 475-662.

1889a. The new horizons of oil and gas in the Mississippi valley. Amer. Assoc. Adv. Sci., Proc., vol. 37, pp. 181-182.

1890. Geological survey of Ohio [third organization] first annual report, x and 323 pp., 2 maps.

1890a. The origin of the rock pressure of natural gas of the Trenton limestone of Ohio and Indiana. Am. Jour. Sci., 3d. ser., vol. 39, pp. 225-229.

1890b. [Same]. Geol. Soc. Amer., Bull., vol. 1, pp. 87-94, 96.

1893. Geological scale and structure of Ohio. Geol. Surv. of Ohio, Vol. VII, pp. 3-44.

1873-1893. [Volumes I to VII of the Ohio Geological Survey were published between these dates by Newberry and Orton, and issued in the years 1873, 1874, 1878, 1884, 1888 and 1893].

1894. Geological Surveys of Ohio. [Historical sketch]. Jour. Geol., vol. 2, pp. 502-516.

Owen, D. D., 1837 [1838]. Report of a geological reconnaissance of the State of Indiana, made in the year 1837 in conformity to an order of the legislature. 34 pp., Indianapolis, 1838. Republished, 63 pp., 1859. [This report is commonly referred to as the 1837 or 1837-59 Reconnaissance].

1839. Continuation of report of a geological reconnaissance of the State of Indiana, made in the year 1838, etc., 54 pp. Republished 69 pp., 1859. [Commonly referred to as the 1839 Reconnaissance, or 1839-59 Reconnaissance. These reports are bound together, with some additional matter, in the 1859 volume. The original 1838 and 1839 reports are now extremely difficult to obtain].

1839a, [1844]. Report on a geological exploration of Iowa, Wisconsin and Illinois, in 1839. President's Message to the 26th Congress, 1st Session; House Exec. Doc., No. 239, pp. 9-115; 28th Congress, 1st Session, Senate Exec. Doc., No. 407, pp. 15-145, 1844. [Referred to under both dates].

1840. [See 1839a].

1843. On the geology of the western states. Am. Jour. Sci., vol. 45, pp. 151-152.

1843a. Fossil palm trees in Posey county, Indiana. Am. Jour. Sci., vol. 45, pp. 336-337.

1844. [See 1839a].

1846. On the geology of the western states of North America. Quar. Jour. Geol. Soc., London, vol. 2, pp. 433-447.

1847. Review of the New York geological Reports. Am. Jour. Sci., vol. 3, p. 72, etc. [D. D. O.]

1849. [Letter to de Verneuil]. Bull. Soc. Geol. de France, II, vol. 6, pp. 419-441. [Quoted by H. S. Williams].

1852. Report of a geological survey of Wisconsin, Iowa and Minnesota, and incidentally, of a portion of Nebraska Territory, etc. xxxviii and 638 pp., maps and plates.

1856. Report of the geological survey in Kentucky, made in the years 1854 and 1855, 416 pp. [Known as the 1st. Kentucky Report].

1857. Second report of the geological survey of Kentucky, made in the years 1856 and 1857, 390 pp.

1857a. Third report of the geological survey of Kentucky, made in the years 1856 and 1857, 589 pp.

1859. Republication of the 1837 and 1839 Reconnaissance Reports of Indiana. [This is the report now commonly referred to].

1861. Fourth report of the geological survey of Kentucky, made during the years 1858 and 1859, 617 pp.

Owen, Richard, 1862. Report of a geological reconnaissance of Indiana made during the years 1859 and 1860 under the direction of the late David Dale Owen. xvi and 368 pp. [Includes reports by Peters, Lesquereux and Lesley].

1881. On the unification of geologic nomenclature. Science [Edited by Michels], vol. 2, pp. 438-440.

Owen, D. D., and Norwood, J. G., 1846. Researches among the Protozoic and Carboniferous rocks of central Kentucky, etc., 12 pp.

Parks, W. A., 1913. 12th. Internat. Geol. Congress, Guide book No. 4. [Issued by the Canadian Geological Survey].

Pate, W. F., and Bassler, R. S., 1908. The late Niagaran strata of West Tennessee. U. S. Nat. Mus., Proc., vol. 34, pp. 407-432.

Peale, A. C., 1893. The Paleozoic section in the vicinity of Three Forks, Montana. U. S. Geol. Surv., Bull. No. 110, pp. 9-45.

Phinney, A. J., 1882. Geology of Delaware County, Indiana. 11th. Indiana Report, pp. 126-149.

1883. Randolph county. 12th. Indiana Report, pp. 177-195.

1884. Grant county. 13th. Indiana Report, pp. 138-153.

1886. Henry, Randolph, Wayne and Delaware counties. 15th. Indiana Report, pp. 97-116.

1888. [Natural gas in Indiana]. U. S. Geol. Surv., Mineral Resources, for 1887, pp. 485-489.

1891. The natural gas field of Indiana. U. S. Geol. Surv., 11th. Ann. Rept., pt. 1, pp. 589-742.

Platt, Franklin, 1875. Report of progress in the Clearfield and Jefferson district of the bituminous coal fields. 2d. Geol. Surv. of Pennsylvania, Rept. H, viii and 296 pp., maps and pls.

Plummer, John T., 1843. Suburban geology, or rocks, soil, and water about Richmond, Wayne county, Indiana. Am. Jour. Sci., vol. 44, pp. 281-313.

Price, J. A., 1898. Notes on Indiana geology. Ind. Acad. Sci., Proc. for 1897, pp. 262-266.

1900. A report on the Waldron shale and its horizon in Decatur, Bartholomew, Shelby and Rush counties, Indiana. 24th. Indiana Report, pp. 81-143, map and pls.

Prosser, C. S., 1901. [On the use of the term Bedford limestone]. Jour. Geol., vol. 9, pp. 270-272.

1903. Revised nomenclature of the Ohio geological formations. Jour. Geol., vol. 11, pp. 519-547.

1905. Revised nomenclature of the Ohio geological formations. Geol. Surv. Ohio., 4th. ser., Bull. No. 7, pp. i-xv, 1-36.

1916. The classification of the Niagaran formations of western Ohio. Jour. Geol., vol. 24, pp. 334-365.

1916a. The stratigraphic position of the Hillsboro sandstone, [Ohio]. Am. Jour. Sci., 4th. ser., vol. 41, pp. 435-448.

Prout, H. A., 1858. Description of Bryozoa from the Paleozoic rocks of the western States and Territories. St. Louis Acad. Sci., Trans., vol. 1, pp. 443-452.

Raymond, P. E., 1916. Expedition to the Baltic provinces of Russia and Scandinavia: Pt. I, the correlation of the Ordovician strata of the Baltic basin with those of eastern North America. Harvard College, Mus. Comp. Zool., Bull., vol. 56, (Geol. Ser., vol. 10), No. 3, pp. 179-286, 8 pls.

1921. A contribution to the description of the fauna of the Trenton group. Canada Dept. of Mines, Mus. Bull. No. 31, pp. 1-64.

Reagan, A. B., 1904. Geology of Monroe county, Indiana, north of the latitude of Bloomington. Ind. Acad. Sci., Proc. for 1903, pp. 205-223, map.

Rogers, Henry D., 1838. Some facts in the geology of the central and western portion of North America; Collected from the Statements and unpublished notices of recent travelers. Geol. Soc., London, Proc., vol. 2, pp. 103-106.

1840. Fourth Annual Report on the geological survey of the State of Pennsylvania. 252 pp.

1843. [On Marcellus and Hamilton of the south and west]. Am. Jour. Sci., vol. 45, pp. 161-162.

1844. Address [On American geology and present condition of geologic research in the United States]. Am. Jour. Sci., vol. 47, pp. 137-160, 247-278.

1851. On the coal formations of the United States, and especially as developed in Pennsylvania. Amer. Assoc. Adv. Sci., Proc., vol. 4, pp. 65-70.

1856. Geological map of the United States and British North America. Physical Atlas of natural phenomena, by A. K. Johnston, Edinburgh, plate 8, folio.

1858. The Geology of Pennsylvania; two vols., 586 pp., and 1045 pp., pls. and Atlas.

1858a. Conditions of the physical geography attending the productions of the Paleozoic strata of the United States. Geol. Pennsylvania, vol. 2, pp. 776-815.

Rogers, W. B., 1884. Geology of the Virginias. A reprint of Annual Reports and other papers on the geology of Virginia, published between the years 1836 and 1882. 832 pp., maps in pocket.

Rominger, Carl, 1866. Observations on Chaetetes and some related genera, in regard to their systematic position, with an appended description of some new species. Phila. Acad. Sci., Proc., vol. (1886), pp. 113-123.

1876. Fossil corals. Geol. Survey Michigan, vol. 3, pt. 2, pp. 1-155, pls. 1-55.

1892. On the occurence of typical Chaetetes in the Devonian strata at the Falls of the Ohio, and likewise in the analogous beds of Germany. Am. Geol., vol. 10, pp. 56-62.

Rowley, R. R., 1890. Some observations on the natural casts of Crinoids and Blastoids from the Subcarboniferous limestones of Indiana, Iowa, Illinois, Kentucky and Alabama. Am. Geol., vol. 6, pp. 66-67.

1893. Range of Chouteau fossils. Am. Geol., vol. 12, pp. 49-50.

1906. Descriptions of new fossils. Contributions to Indiana Paleontology, vol. 2, pt. 1, pp. 7-11.

1906a. Descriptions of fossils. Contr. to Indiana Paleontology, vol. 2, pt. 2, pp. 21-31.

Ruedemann, Rudolph, 1901. Hudson River beds near Albany and their taxonomic equivalents. N. Y. State Mus., Bul., No. 42, pp. 489-587, 2 pls.

1904. Graptolites of New York, Pt. I. Graptolites of the lower beds. N. Y. State Mus., Mem., No. 7, pp. 455-803, 17 pls.

1908. Graptolites of New York, Pt. II, Graptolites of the higher beds. N. Y. State Mus., Mem., No. 11, 583 pp., 31 pls.

1909. Types of inliers observed in New York. N. Y. State Mus., Bull. 133, pp. 164-193.

1912. The Lower Siluric Shales of the Mohawk valley. N. Y. State Mus., Bull., No. 162, pp. 5-151.

Ruedemann and Clarke. [See Clarke and Ruedemann].

Safford, J. M., 1856. A geological reconnaissance of Tennessee, first biennial report, 164 pp.

1869. Geology of Tennessee. 550 pp., 7 pls.

Savage, T. E, 1908. On the lower Paleozoic stratigraphy of southwestern Illinois. Am. Jour. Sci., 4th. ser., vol. 25, pp. 431-443.

1908a. Lower Paleozoic stratigraphy of southwestern Illinois. Ill. State Geol. Surv., Bull. No. 8, pp. 103-116.

1909. The Ordovician and Silurian formations in Alexander county, Illinois. Am. Jour. Sci., 4th. ser., vol. 28, pp. 509-519.

1910. The faunal succession and correlation of the pre-Devonian formations of southern Illinois. Ill. State Geol. Surv., Bull. No. 16, pp. 302-341.

1912. The Channahon and Essex limestones in Illinois. Ill. Acad. Sci., Trans., vol. 4, pp. 97-103.

1913. Stratigraphy and paleontology of the Alexandrian series in Illinois and Missouri. Ill. State Geol. Surv., extract from Bull. 23, 124 pp., 7 pls.

1916. Alexandrian rocks of northeastern Illinois and eastern Wisconsin. Geol. Soc. Amer., Bull., vol. 27, pp. 305-324.

Savage, T. E., and Van Tuyl, F. M., 1919. Geology and stratigraphy of the area of Paleozoic rocks in the vicinity of Hudson and James Bays. Geol. Soc. Amer., Bull., vol. 30, pp. 339-421.

Sayler, N., 1865. Geological map of Indiana; scale, 5 miles to one inch. Cincinnati, 1865.

Schuchert, Charles, 1897. A synopsis of American Fossil Brachiopoda, including bibliography and synonomy. U. S. Geol. Surv., Bull., No. 87, 464 pp.

1903. On the faunal provinces of the middle Devonic of America and the Devonic coral sub-provinces of Russia, with two paleogeographic maps. Am. Geol., vol. 32, pp. 137-162.

1904. [Review]. The Stratigraphy and paleontology of the Niagara of northern Indiana. [By E. M. Kindle]. Am. Jour. Sci., 4th. Ser., vol. 18, pp. 465-469.

1905. [Assisted by Dall, Stanton and Bassler]. Catalog of the type specimens of fossil invertebrates in the department of geology, U. S. National Museum. U. S. Nat. Mus., Bull., No. 53, 704 pp.

1910. Paleogeography of North America. Geol. Soc., Amer., Bull., vol. 20, pp. 427-506, 56 pls.

1913. The Cataract; a new formation at the base of the Siluric in Ontario and New York. [Abstract]. Geol. Soc. Amer., Bull., vol. 24, p. 107.

1913a. The delimination of the geologic periods illustrated by the paleogeography of North America. Internat. Cong. Geol. Twelfth, Canada, 34 pp. [Advance copy]. Same, 1914.

1914. Medina and Cataract formations of the Siluric of New York and Ontario. Geol. Soc. Amer., Bull., vol. 25, pp. 277-320.

1915. A textbook of geology. 2 vols. [and 2 vols. bound as one] 1026 pp. [Vol. 1 by L. V. Pirsson, Vol. 2 by C. Schuchert].

1915a. Revision of Paleozoic Stelleroidea with special reference to North American Asteroidea. U. S. Nat. Mus., Bull. 88, 301 pp., 38 pls.

1916. Correlation and chronology in geology on the basis of paleogeography. Geol. Soc. Amer., Bull., vol. 27, pp. 491-514.

1918. A century of geology; the progress of historical geology in North America. Am. Jour. Sci., 4th. ser., vol. 46, pp. 45-103.

Schuchert and Winchell, 1894, 1895. [See Winchell and Schuchert].

Schuchert and Clarke, 1899. [See Clarke and Schuchert, 1899].

Schuchert and Ulrich, 1902. [See Ulrich and Schuchert, 1902].

Schuchert, Charles, and Twenhofel, W. H., 1910. Ordovicic-Siluric section of the Mingan and Anticosti Islands, gulf of St. Lawrence. Geol. Soc. Amer., Bull., vol. 21, pp. 677-716.

Schuchert, Charles, and Barrell, Joseph, 1914. A revised geologic time-table for North America. Am. Jour. Sci., 4th. Ser., vol. 38, pp. 1-27.

Scovell, J. T., 1897. Geology of Vigo county, Indiana. 21st. Indiana Report, pp. 507-576.

Sedgwick, Adam, and Murchison, Sir R. I., 1839. [Name Devonian proposed]. Geol. Soc. London, Proc., vol. 3, pp. 121-123.

Seely, H. M., 1886. The genus Strephochetus; distribution and species. Am. Jour. Sci., 3d. ser., vol. 32, p. 31.

Shaler, N. S., 1877. Geological Survey of Kentucky, Report of progress, vol. 3, new series. 451 pp., 5 maps.

[No date]. On the fossil Brachiopoda of the Ohio valley. Publication of the Kentucky Geol. Surv., pp. 1-44, pls. 1-8.

Shannon, C. W., 1906. The roads and road materials of Monroe county, Indiana. 30th. Indiana Report, pp. 941-967.

1907. Drainage area of the east fork of White River, Indiana. Ind. Acad. Sci., Proc. for 1906, pp. 53-70.

1907a. The iron ore deposits of Indiana. 31st. Indiana Report, pp. 299-428.

1908. [Shannon and others]. A soil survey of seventeen counties of southern Indiana. 32d. Indiana Report, pp. 15-298, maps and plates.

1909. Soil survey of Dubois, Perry and Crawford counties, Indiana. 33d. Indiana Report, pp. 277-342.

1912. Soil survey of Morgan and Owen counties. 36th. Indiana Report, pp. 135-280.

Shannon and Beede, 1907. [See Beede and Shannon, 1907].

Sherzer, W. H., 1911. Geology of Wayne county, Michigan. Mich. Geol. Surv., Geol. Ser. 9, pub. 12, see pp. 208-215.

Shideler, W. H., 1914. The upper Richmond beds of the Cincinnati group. Ohio Nat., vol. 14, pp. 229-235.

Shumard, B. F., 1857. [1860] Observations on geology of county of St. Genevieve [Abstracts]. St. Louis Acad. Sci., trans., vol. 1, pp. 404-415.

1858. Description of new fossil Crinoidea from the Paleozoic rocks of the western and southern portions of the United States. St. Louis Acad. Sci., Trans., vol. 1, pp. 71-80.

1858a. Description of new species of Blastoidea from the Paleozoic rocks of the western states, with some observations on the summit of the genus Pentremites. St. Louis Acad. Sci., Proc., vol. 1, pp. 238-248.

1873. Geology of Missouri, 1855-71, Geology of St. Genevieve county. Geol. Surv. Mo., Repts. for 1855-71, pp. 290-303. [Edited by R. Pumpelley].

Siebenthal, C. E., 1897. [See Hopkins and Siebenthal, 1897].

1898. The Bedford oölitic limestone [Indiana]. U. S. Geol. Surv., 19th. Ann. Rept., pt. 6, [continued], pp. 292-296.

1899. The Bedford oölitic limestone. Mineral Industry, for 1898, vol. 8, pp. 479-482.

1901. The Silver creek hydraulic limestone of southeastern Indiana. 25th. Indiana Report, pp. 331-389.

1901a. The Indiana oölitic limestone industry in 1900. 25th. Indiana Report, pp. 390-393.

1901b. The use of the term Bedford limestone. Jour. Geol., vol. 9, pp. 234-235.

1908. General geographical and stratigraphical features of the Indiana oölitic limestone. 32d. Indiana Report, pp. 303-309.

Smith, S. L, 1871. Notice of a fossil insect from the Carboniferous formations of Indiana. Am. Jour. Sci., 3d. Ser., vol. 1, pp. 44-46.

Smith, Essie A,. 1906. [See Cumings, Beede, etc., 1906].

Springer, Frank, 1911. The Crinoid fauna of the Knobstone formation. U. S. Nat. Mus., Proc., vol. 41, pp. 175-208.

1920. The Crinoidea Flexibilia. Smithsonian Institution Monograph, Publication No. 2501, 2 parts, text 486 pp., and Atlas of 76 plates.

Stauffer, C. R., 1907. The Devonian limestones of central Ohio and southern Indiana. Ohio Nat., vol. 7, No. 8, pp. 184-186.

1909. The middle Devonian of Ohio. Ohio Geol. Surv., 4th. ser., Bull. 10, pp. 1-204, 17 pls.

1915. The Devonian of southwestern Ontario. Canadian Geol. Surv., Mem. No. 34, 341 pp., 20 pls., map.

1916. Relative age of the Detroit river series [with a discussion by A. C. Lane]. Geol. Soc. Amer., Bull., vol. 27, pp. 72-78. [Abstract].

Stevens, R. P., 1858. Description of new Carboniferous fossils from the Appalachian, Illinois and Michigan coal fields. Am. Jour. Sci., 2d. ser., vol. 25, pp. 258-265.

Stevenson, J. J., 1888. [Report of a sub-committee on the Upper Paleozoic.] Am. Geol., vol. 2, pp. 248-256.

Stitson, W. B., 1818. Sketch of the geology and mineralogy of a part of the State of Indiana. Am. Jour. Sci., vol. 1, pp. 131-133.

St. John, O. H., and Worthen, A. H., 1883. Description of fossil fishes. Geol. Surv. Ill., vol. 7, pp. 57-264.

Swallow, G. C., 1855. Geology of Missouri, etc. Geol. Surv. Mo., 1st. and 2d. Ann. Repts., pp. 57-207.

1858. Explanations of the geological map of Missouri, and a section of its rocks. Amer. Assoc. Adv. Sci., Proc., vol. 11, pp. 1-21.

Taylor, A. E., 1906. The roads and road materials of portions of central and eastern Indiana. 30th. Indiana Report, pp. 315-570. The roads and road materials of a portion of southwestern Indiana. *Ibid.*, pp. 969-1006.

Thompson, Maurice, 1886. Compend of the geology of Indiana; building stones; clays; chalk beds; glacial desposits; survey of Clinton county; natural gas; terminal moraine in central Indiana. 15th. Indiana Report, pp. 5-333.

1889. Drift beds of Indiana; the Wabash Arch; gold, silver and precious stones; formation of soils; etc. 16th. Indiana Report, pp. 20-97.

1892. Indiana building stones; geological and natural history report of Carroll county. 17th. Indiana Report, pp. 19-113 and 171-191.

1886-1889. [The 15th. and 16th. Indiana Reports are by Maurice Thompson].

Thompson, W. H., 1886. Marshall county. 15th. Indiana Report, pp. 177-182.

1886a. A geological survey of Starke county. 15th. Indiana Report, pp. 221-227.

Tucker, W. M., 1911. The water power of Indiana. 35th. Indiana Report, pp. 11-77. [Table of geological formations on p. 15].

Ulrich, E. O., 1879. Description of a trilobite from the Niagara group of Indiana. Jour. Cin. Soc., Nat. Hist., vol. 2, pp. 131-134.

1879a. Descriptions of new genera and species of fossils from the Lower Silurian about Cincinnati. Jour. Cin. Soc. Nat. Hist., vol. 2, pp. 8-30.

1880. Catalog of fossils occurring in the Cincinnati group of Ohio. 31 pp., published privately.

1882-1884. American Paleozoic Bryozoa. Jour. Cin. Soc. Nat. Hist., vol. 5, pp. 121-177, 232-259 (1882); vol. 6, pp. 82-92, 148-168, 245-279 (1883); vol. 7, pp. 24-51 (1884). All illustrated by plates and figures.

1886. Description of new Silurian and Devonian fossils. Contributions to American Paleontology, vol. 1, No. 1, pp. 3-35, pls. [The only issue of this publication].

1888. Correlation of the Lower Silurian horizons of Tennessee and part of the Ohio and Mississippi valleys with those of New York and Canada. Am. Geol., vol. 1, pp. 100-110, 179-190, 305-315; vol. 2, pp. 39-44 (to be continued). [No further parts of this paper were ever published].

1890. New and little known American Paleozoic Ostracoda. Jour. Cin. Soc. Nat. Hist., vol. 13, pp. 104-137, 173-211.

1890a. Paleontology of Illinois: Sponges, pp. 211-282, and Bryozoa, pp. 285-688. Plates in separate volume. Geol. Surv. of Illinois, vol. 8, Text and Plates, pp. 211-282, and 285-688.

1890b. New Lamellibranchiata [from the Lower Silurian]. Am. Geol., vol. 5, pp. 270-284; vol. 6, pp. 173-181.

1892. New Lamellibranchiata. Am. Geol., vol. 10, pp. 96-104.

1892a. New Lower Silurian Ostracoda, No. 1. Am. Geol., vol. 10, pp. 263-270.

1892b. New Lower Silurian Lamellibranchiata, chiefly from Minnesota rocks. Minn. Geol. and Nat. Hist. Surv., 19th. Ann. Rept., pp. 211-248.

1893. New and little known Lamellibranchiata from the Lower Silurian rocks of Ohio and adjacent states. Ohio Geol. Surv., vol. 7, pp. 627-693, pls.

1895. On Lower Silurian Bryozoa of Minnesota. Minn. Geol. and Nat. Hist. Surv., final report, vol. 3, pt. 1, [Paleontology], pp. 96-332, 28 pls.

1897. The Lower Silurian Lamellibranchiata of Minnesota (pp. 475-628); The Lower Silurian Ostracoda of Minnesota (pp. 629-693). Minn. Geol. and Nat. Hist. Surv., final report, vol. 3, pt. 2, [Paleontology], pp. 475-628, 629-693, 13 pls.

1897a. [Ulrich and Scofield]. The Lower Silurian Gastropoda of Minnesota. Minn. Geol. and Nat. Hist. Surv., final report, vol. 3, pt. 2, [Paleontology], pp. 813-1081, 11 pls.

1897b. [And Winchell, See Winchell and Ulrich, 1897].

1900. New American Paleozoic Ostracoda. Jour. Cin. Soc. Nat. Hist., vol. 19, pp. 179-186.

1904. Determination and correlation of formations [of northern Arkansas]. U. S. Geol. Surv., Professional Paper, No. 24, pp. 90-113.

1905. Lead, zinc and fluorspar deposits of western Kentucky. Part I. Geology and general relations. U. S. Geol. Surv., Professional Paper, No. 36, pp. 15-105.

1911. Revision of the Paleozoic systems. Geol. Soc. Amer. Bull., vol. 22, pp. 281-680. [Correlation charts between pp. 608 and 609].

1912. The Chattanoogan series with special reference to the Ohio shale problem. Am. Jour. Sci., 4th. ser., vol. 34, pp. 157-183.

1913. The Ordovician-Silurian boundary. Internat. Geol. Congress, Twelfth, Canada, 50 pp. [Advance copy].

1914. The Ordovician-Silurian boundary. Internat. Geol. Congress, XII, Canada, 1913, C. R. pp. 593-667. [Same paper as above, final publication].

1915. The Kinderhookian age of the Chattanoogan series. Geol. Soc. Amer., Bull., vol. 26, pp. 96-99. [Abstract].

1916. Correlation by displacements of the strand-line and the function and proper use of fossils in correlation. Geol. Soc. Amer., Bull., vol. 27, pp. 451-490.

1918. The formations of the Chester series in western Kentucky and their correlates elsewhere. Kentucky Geol. Surv., Mississippian formations of western Kentucky, 272 pp., pls. and tables. [1917 on the title page].

1919. [Paleogeographic maps in the Geology of Maryland, Cambrian and Ordovician]. Geol. of Maryland, Cambrian and Ordovician, by R. S. Bassler. [See Bassler, 1919].

Ulrich, E. O., and Bassler, R. S., 1904. A revision of the Paleozoic Bryozoa, Pt. I. On genera and species of Ctenostomata. Smithsonian Miscell. Coll., vol. 45, pp. 256-294, pls. and figs.

1904a. A revision of the Paleozoic Bryozoa. Pt. II. On genera and species of Trepostomata. Smithsonian Misc. Coll., vol. 47, pp. 15-55, pls.

1906. New American Paleozoic Ostracoda. Notes and descriptions of upper Carboniferous genera and species. U. S. Nat. Mus., Proc., vol. 30, pp. 149-164, pls.

1908. New American Paleozoic Ostracoda. Preliminary revision of the Beyrichidae, with descriptions of new genera. U. S. Nat. Mus., Proc., vol. 35, pp. 277-340, pls.

Ulrich, E. O., and Schuchert, C., 1902. Paleozoic seas and barriers in eastern North America. N. Y. State Mus., Bull. No. 52, pp. 633-663.

Van Gorder, W. B., 1916. Geology of Greene County, Indiana. 40th. Indiana Report, pp. 240-266.

Van Tuyl, F. M., 1912. The Salem limestone and its stratigraphic relations in southeastern Iowa. Iowa Acad. Sci., Proc. vol. 19, pp. 167-168.

Vanuxem, Lardner, 1829. Remarks on the character and classification of certain American rock formations. Am. Jour. Sci., vol. 16, pp. 254-256.

1837. First Annual Report of the geological survey of the fourth district of New York. Geol. Surv. N. Y., 1st. Ann. Rept., pp. 187-212.

1839. Third annual report of the third district. Geol. Surv. N. Y., 3d. Ann. Rept., pp. 241-285.

1840. Fourth annual report of the geology of the third district. Geol. Surv. N. Y., 4th. Ann. Rept., pp. 355-383.

1842. Geology of New York, part 3, comprising the survey of the third district. [Final Report], 307 pp., pls.

Verneuil, Ed. de, 1847. Note sur le parallélisme des roches des dépôts palèozoïques de l'Amérique septentrionale avec ceux de l'Europe, suivie d'un tableau des espèces fossils communes aux deux continents, avec l'indication des étages où elles se rencontrent, et terminée par un examin critique de chacune de ces espèces. Soc. Géol. France, Bull. 2 sér., vol. 4, pp. 646-709. [Translated by James Hall. See Hall, 1849].

Vogdes, Anthony W., 1888. The genera and species of North American Carboniferous Trilobites. New York Acad. Sci., Annals, vol. 4, pp. 69-105, 2 pls.

1893. A classed and annotated bibliography of the Paleozoic Crustacea, 1698-1892. California Acad. Sci., occasional papers, 4, pp. 1-412.

1917. Paleozoic Crustacea; the publications and notes on the genera and species during the past twenty years. 1895-1917. San Diego Soc. Nat. Hist., Trans., vol. 3, No. 1, pp. 1-141, 5 pls., text figures.

Wachsmuth, Charles, and Springer, Frank, 1881. Revision of the Paleocrinoidea, Pt. II, Phila. Acad. Nat. Sci., Proc., vol. (1881), pp. 177-414.

1892. Description of two new genera and eight new species of camerate crinoids from the Niagara group. Am. Geol., vol. 10, pp. 135-144.

1897. The North American Crinoidea camerata. Harvard Coll. Mus. Comp. Zool., Mem. 21, 837 pp, 83 pls.

Walcott, C. D., 1883. The Utica slate and related formations of the same geological horizon. Alb. Inst., Trans., vol. 10, pp. 1-17.

1890. The value of the term "Hudson River group" in geologic nomenclature. Geol. Soc. Amer., Bull., vol. 1, pp. 335-353, 354-355.

1893. Geologic time as indicated by the sedimentary rocks of North America. Jour. Geol., vol. 1, pp. 639-676.

1894. Geologic time as indicated by the sedimentary rocks of North America. Amer. Assoc. Adv. Sci., Proc., vol. 42, pp. 129-169.

Wallace, S. W., 1878. On the geodes of the Keokuk formation and the genus Biopalla, with some species. Am. Jour. Sci., 3d. ser., vol. 15, p. 366. [See also Indiana geological report for 1873, p. 278].

Ward, L. C., 1906. The roads and road materials of the northern third of Indiana. 30th. Indiana Report, pp. 161-274.

1908. A soil survey of Decatur, Jennings, Jefferson, Ripley, Dearborn Ohio and Switzerland counties. 32d. Indiana Report, pp. 197-244.

Warder, Robert B., 1872. Geology of Dearborn, Ohio, and Switzerland counties, 3d. and 4th. Indiana Reports, pp. 387-434.

Weeks, F. B., 1902. North American Geologic Formation Names; bibliography, synonomy and distribution. U. S. Geol. Surv., Bull. No. 191, 448 pp.

Weller, Stuart, 1895. A circum-insular Paleozoic fauna. Jour. Geol., vol. 3, pp. 903-917.

1898. Classification of the Mississippian series. Jour. Geol., vol. 6, pp. 303-314.

1898a. The Silurian fauna interpreted on the epicontinental basis. Jour. Geol., vol. 6, pp. 692-703.

1898b. A bibliographic index of North American Carboniferous invertebrates. U. S. Geol. Surv., Bull. No. 153, 653 pp.

1898c. The Batesville sandstone of Arkansas. N. Y. Acad. Sci., Trans., vol. 16, pp. 251-282, 3 pls.

1899. Kinderhook faunal studies. I. The fauna of the Vermicular sandstone at Northview, Webster County, Missouri. St. Louis Acad. Sci., Trans., vol. 9, No. 2, pp. 9-51, 5 pls.

1900. The Paleontology of the Niagaran limestone in the Chicago area. The Crinoidea. Chicago Acad. Sci., Bull. No. 4, pp. 1-153, 15 pls.

1900a. The succession of fossil faunas in the Kinderhook beds at Burlington, Iowa. Iowa Geol. Surv., vol. 10, pp. 63-79.

1900b. Kinderhook faunal studies. II. The fauna of the Chonospectus sandstone at Burlington, Iowa. St. Louis Acad. Sci., Trans., vol. 10, No. 3, pp. 57-129, 9 pls.

1901. Kinderhook faunal studies. III. The faunas of beds Nos. 3 to 7, at Burlington, Iowa. St. Louis Acad. Sci., Trans., vol. 11, pp. 147-214, 9 pls.

1905. The northern and southern Kinderhook faunas. Jour. Geol., vol. 13, pp. 617-634.

1906. Kinderhook faunal studies. IV. The fauna of the Glen Park limestone. St. Louis Acad. Sci.. Trans., vol. 16, No. 7, pp. 435-471, 2 pls.

1907. The paleontology of the Niagaran limestone in the Chicago area: The Trilobita. Chicago Acad. Sci., Nat. Hist. Surv., Bull. No. 4, pp. 161-281, 10 pls.

1907a. The pre-Richmond unconformity in the Mississippi valley. Jour. Geol., vol. 15, pp. 519-525.

1908. The Salem limestone. Illinois Geol. Surv., Bull. No. 8, pp. 81-102.

1909. Kinderhook faunal studies. V. The fauna of the Fern Glen formation. Geol. Soc. Amer., Bull., vol. 20, pp. 265-332, 6 pls.

1909a. Correlation of the middle and upper Devonian and Mississippian faunas of North America. Jour. Geol., vol. 17, pp. 257-285.

1910. Internal characters of some Mississippian Rhynchonelliform shells. Geol. Soc. Amer., Bull., vol. 21, pp. 497-516, 18 figures.

1911. Genera of Mississippian loop-bearing Brachiopoda. Jour. Geol., vol. 19, pp. 439-448, 7 figures.

1913. Stratigraphy of the Chester group in southwestern Illinois. Ill. Acad. Sci., Trans., vol 6, pp. 118-129.

1914. The Mississippian Brachiopoda of the Mississippi valley basin. Ill. Geol. Surv., Monograph 1, 508 pp. 83 pls. [in separate volume].

1920. The Chester series in Illinois. Jour. Geol., vol. 28, pp. 281-303, 395-416.

1920a. The geology of Hardin county and the adjacent parts of Pope County, Illinois. [In collaboration with L. W. Currier and R. D. Salisbury]. Ill. Geol. Surv., Bull. 41, 416 pp.

Wetherby, A. G., 1878. Description of a new family and genus of Lower Silurian Crustacea. Jour. Cin. Soc. Nat. Hist., vol. 1, pp. 162-166.

White, C. A., 1870. Report on the geological survey of the state of Iowa, etc., containing results of work done in the years 1866, 1867, 1868 and 1869, vol. 1, viii and 391 pp., pls.; vol. 2, viii and 443 pp., pls., map.

1878. Description of new species of invertebrate fossils from the Carboniferous and Upper Silurian Rocks of Illinois and Indiana. Phila. Acad. Nat. Sci., Proc., vol. (1878), pp. 29-37.

1880. Fossils of the Indiana rocks. 2d. Report, Dept. Statistics and Geology, Indiana, pp. 471-522.

1884. Fossils of the Indiana rocks. 13th. Indiana Report, pp. 107-180.

White, C. A., and Whitfield, R. P., 1862. Observations on the rocks of the Mississippi valley which have been referred to the Chemung group of New York, together with descriptions of new species of fossils from the same horizon at Burlington, Iowa. Boston Soc. Nat. Hist., vol. 8, pp. 289-306.

White, David, 1896. Fossil plants of the Hindostan whetstone beds, [Indiana]. 20th. Indiana Report, pp. 354-355.

1901. Two new species of algae from the Upper Silurian rocks of Indiana. U. S. Nat. Mus., Proc., vol. 24, pp. 265-270.

1908. Report on field work done in 1907. Ill. Geol. Surv., Bull. No. 8, pp. 268-272.

White, I. C., 1891. Stratigraphy of the bituminous coal fields of Pennsylvania, Ohio and West Virginia. U. S. Geol. Surv., Bull. No. 65, 212 pp., 12 pls.

White and Fontaine, 1880. [See Fontaine and White, 1880].

Whitfield, R. P., 1875. List of fossils from the Black Shale. 6th. Indiana Report, pp. 179-182.

1875a. [See Hall and Whitfield, 1875].

1881. Remarks on Dictyophyton and description of new species of allied forms from the Keokuk beds at Crawfordsville, Indiana. Amer. Mus. Nat. Hist., Bull., vol. 1, pp. 10-20.

1882. On the fauna of the limestones of Spergen Hill, Indiana, with a revision of the descriptions of fossils, etc. Am. Mus. Nat. Hist., Bull., vol. 1, pp. 39-97, pls. [The plates of this paper were republished by Hall in the 12th. Indiana Report].

1885. On a new cephalopod from the Niagara rocks of Indiana. Amer. Mus. Nat. Hist., Bull., vol. 1, No. 6, p. 192.

1893. Republication of descriptions of Lower Carboniferous Crinoidea from the Hall collection now in the American Museum of Natural History, with illustrations of the original type specimens not heretofore figured. Am. Mus. Nat. Hist., Mem., vol. 1, pt. 1, pp. 1-37, 3 pls.

1893a. Contributions to the paleontology of Ohio. Ohio Geol. Surv., vol. 7, pp. 407-494.

1898. [Assisted by E. O. Hovey] Catalog of the type and figured specimens in the paleontological collections of the geological department, American Museum of Natural History. Am. Mus. Nat. Hist., Bull., vol. 11, pp. i-vii and 1-71.

1899. [Assisted by E. O. Hovey]. Catalog of the type and figured specimens in the paleontological collections of the geological department, American Museum of Natural History. Am. Mus. Nat. Hist., Bull., vol. 11, pp. 74-188.

1900. Description of a new crinoid from Indiana. Am. Mus. Nat. Hist., Bull., vol. 13, pp. 23-24, pl.

1900a. List of fossils, types and figured specimens used in the paleontological work of R. P. Whitfield, showing where they are probably to be found at the present time. N. Y. Acad. Sci., Annals, vol. 12, pp. 139-186.

1900b. [And E. O. Hovey]. Catalog of the types and figured specimens in the paleontological collections of the department of geology, American Museum of Natural History. Am. Mus. Nat. Hist. Bull., vol. 11, pp. 190-356.

1901. [And E. O. Hovey]. Catalog of the types and figured specimens in the paleontological collections of the department of geology, American Museum of Natural History. Am. Mus. Nat. Hist., Bull., vol. 11, pp. 357-500.

1905. Description of new fossil sponges from the Hamilton group of Indiana. Am. Mus. Nat. Hist., Bull., vol. 21, pp. 297-300, 3 pls.

Williams, H. S., 1886. On the classification of the upper Devonian. Amer. Assoc. Adv. Sci., Proc., vol. 34, pp. 222-234.

1888. Report of the subcommittee of the upper Paleozoic [Devonic]. Am. Geol., vol. 2, pp. 225-239.

1888a. On the different types of the Devonian system in North America. Am. Jour. Sci., 3d. Ser., vol. 35, pp. 51-59.

1891. Correlation papers; Devonian and Carboniferous. U. S. Geol. Surv., Bull. No. 80, 279 pp.

1893. The making of the geological time scale. Jour. Geol., vol. 1, pp. 180-197.

1893a. The elements of the geological time scale. Jour. Geol., vol. 1, pp. 283-295.

1895. On the recurrence of Devonian fossils in strata of Carboniferous age. Am. Jour. Sci., 3d. ser., vol. 49, pp. 94-101.

1897. On the southern Devonian formations. Am. Jour. Sci., 4th. ser., vol. 3, pp. 393-403, map.

1900. The Paleozoic faunas of northern Arkansas. Ark. Geol. Surv., Ann. Rept. for 1892, vol. 5, pp. 268-362.

Williams, M. Y., 1914. Stratigraphy of the Niagara escarpment of southwestern Ontario, (pp. 178-188); Thedford and vicinity, Ontario (pp. 282-285); The Silurian of Manitoulin island and western Ontario (pp. 275-281). Canada Geol. Surv., Summary Repts. 1912-13, pub. 1914.

1914a. Sections illustrating the lower part of the Silurian system of southwestern Ontario. [Abstract]. Geol. Soc. Amer., Bull., vol. 25, pp. 40-41.

1919. The Silurian Geology and faunas of Ontario peninsula, and Manitoulin and adjacent islands. Canada Dept. of Mines, Mem. No. 111, 195 pp.

Willis, Bailey, 1910. [And Salisbury]. Outlines of geologic history with especial reference to North America. [A symposium by various authors, presented before section E of the Amer. Assoc. Adv. Sci., in Baltmore, Dec. 1908, and republished in book form by the University of Chicago press]. 306 pp. [Paleogeographic maps by Bailey Willis].

1912. Index to the stratigraphy of North America. U. S. Geol. Survey, Profess. Paper No. 71, 894 pp., 19 figs., 5 inserts; colored geological wall map of North America in four sheets, in separate case.

Winchell, Alexander, 1861. [Geology]. Geol. Surv., Mich., 1st. Biennial report of progress, pp. 19-206.

1863. On the identification of the Catskill red sandstone group with the Chemung. Am. Jour. Sci., 2d. ser., vol. 35, pp. 61-62.

1871. [1869-70]. On the geological age and equivalents of the Marshall group. Am. Phil. Soc., Proc., vol. 11, pp. 57-82, 385-418. [Paper read in 1869, and sometimes referred to under that date, or 1870].

1871a. Notes and descriptions of fossils from the Marshall group of the western States, with notes on fossils from other formations. Am. Phil. Soc., Proc., vol. 12, pp. 245-260.

Winchell, N. H., 1873. [Chart of geological nomenclature; and general sketch of the geology of Minnesota]. Geol. and Nat. Hist. Surv., Minn., 1st. Ann. Rept., pp. 38-118, chart and map.

Winchell, N. H., and Ulrich, E. O., 1897. The Lower Silurian deposits of the Upper Mississippi: A correlation of the strata with those in the Cincinnati, Tennessee, New York and Canadian Provinces, and the stratigraphic and geographic distribution of the fossils. Minn. Geol. and Nat. Hist. Surv., Paleontology, vol. 3, pt. 2, pp. lxxxiii-cxxix.

Winchell, N. H., and Schuchert, Charles, 1895. Sponges Graptolites and corals from the Lower Silurian of Minnesota. Minn. Geol. and Nat. Hist. Surv., Final Rept., vol. 3, pt. 1 [Paleontology], pp. 55-95, 2 pls.

1895a. The Lower Silurian Brachiopoda of Minnesota. Minn. Geol. and Nat. Hist., Surv., Final Rept., vol. 3, pt. 1, [Paleontology], pp. 333-474, 6 pls.

Wood, Elvira, 1909. A critical summary of Troost's unpublished manuscript on the crinoids of Tennessee. U. S. Nat. Mus., Bull. No. 64, 150 pp., 16 pls.

Worthen, A. H., 1860. Review of some points in B. F. Shumard's report on the geology of St. Genevieve county. St. Louis Acad. Sci., Trans., vol. 1, pp. 696-698.

1866. [Various papers], Geol. Surv. Ill., vol. 1, 504 pp.

1866a. [Various papers], Geol. Surv. Ill., vol. 2, 460 pp., 50 pls.

1866-1890. [Volumes 1 to 8 of the geological survey of Illinois were published by Worthen between these dates].

1884. Description of two new species of Crustacea, 51 species of Mollusca and 3 species of crinoids from the Carboniferous of Illinois and adjacent states. Ill. State Mus. Nat. Hist., Bull. No. 2, pp. 1-27.

1889. Catalog of American Paleozoic fossils: the collection of Prof. A. H. Worthen, deceased. 75 pp., published privately at Warsaw, Ill.

Yandell, L. P., and Shumard, B. F., 1847. Contributions to the Geology of Kentucky. 36 pp. Louisville.

PART V.

Economic Geology of Indiana

By

W. N. Logan

CONTENTS

CHAPTER III. Coal—Continued

CHAPTER X. Oil and Gas—Continued. Page

LIST OF ILLUSTRATIONS

List of Illustrations—Continued. Page

List of Illustrations—Continued.

Economic Geology of Indiana

CHAPTER I.

BUILDING STONES.

Building stones usually include those used in the construction of edifices, those used for ornamental purposes in construction, and those used for roofing, flagging and curbing.

The kinds of rock included under building stones are varieties of the three great divisions of rocks, sedimentary, igneous and metamorphic. The only igneous and metamorphic rocks in Indiana are those occurring in the glacial drift. The boulders of igneous and metamorphic rock are frequently used for interior and exterior rubble-work in the glaciated area of the State. The main supply of building stone in Indiana is obtained from the sedimentary division of rocks.

Properties of Building Stones. The value of a stone for building purposes depends upon its chemical and physical properties. Color is one of the important properties of building stones. Permanency and uniformity of color are very desirable qualities. A stone may have a very pleasing color in the quarry and change to a very different one when placed in a structure. The color may be in the constituent grains or in the cementing material which binds the grains together. Sedimentary rocks often contain oxides of iron which produce red, yellow or buff colors.

A good building stone must have two kinds of strength. It must have compressive or crushing strength so that it can withstand the weight placed upon it without crumbling. The stone must also have transverse strength so that when it is supported at each end and loaded in the middle as in window and door sills, it will not break.

A good building stone should have a low porosity so that it will not absorb much water and thus will be better able to resist the action of frost.

A building stone should have good resistance to heat. It should not crumble easily under the temperatures of ordinary fires. It should also be able to withstand sudden changes of temperature so that if its temperature is raised suddenly and then suddenly lowered the stone will not crumble.

The chemical analysis may reveal little concerning the value of a building stone. The presence of minerals which may produce detrimental colors during weathering may be revealed, and whether the stone should be classed as calcareous, silicious, ferruginous or argillaceous.

The life of a building stone depends upon the climatic conditions of the region in which it is used. The same stone will last for a much longer period in a dry climate than in a humid climate. Its length of life will also be determined by its position in the structure. A porous rock in the foundation will have a much shorter life than when placed higher in the wall.

Geology of Indiana Building Stone.

The Ordovician limestones. These are generally thin and irregularly bedded limestones. Serviceable in some outcrops for building purposes and flags but not contributing largely to our supplies of good building stone. The Ordovician limestones outcrop in the counties in the southeastern portion of the State. The position of outcrop is indicated on the geological map.

Silurian limestone (Niagara) is widely distributed in Indiana. Its out-

Plate I. Showing the use of the Indiana oölitic limestone in structural and ornamental work. Observe steps, porch and trimmings.

crop is largely concealed in the north central and eastern Indiana by glacial drift. In the southeastern portion of the State it is exposed in many places.

The unweathered stone is often blue in color but white on weathered surfaces. It is generally thinly bedded and is much used for curbing, guttering and flagging. It has also been used for structural purposes in buildings and bridges. The Silurian limestone has been quarried at the following points: Anderson, Alexandria, Bluffton, Buena Vista, Delphi, Eaton, Greensburg, Harper, Holton, Huntington, Kokomo, Laurel, Longwood, Marion, Markle, Montpelier, Newpoint, Osgood, Peru, Sardina, St. Paul and Westport.

Plate II. Blocks of oölitic being unloaded at the mill after having been taken from the quarry. Note the absence of lines of stratification.

Devonian limestones occur both north and south of the Silurian area. The limestones are thin bedded and often blue in color. They have been used for macadam, rubble, curbing, flagging and bridge piers. They have been quarried at Decatur, Logansport, North Vernon, Speed and other places.

Mississippian limestones. The Mississippian system of rocks in Indiana contain a wealth of good building stone. The divisions which contain limestones are the Harrodsburg (Warsaw), the Salem (oölitic), the Mitchell, and the Chester limestones which include the Beaver Bend, the Reelsville, the Beech Creek, the Golconda and the Glen Dean. The Harrodsburg limestone overlies the Knobstone shales and sandstones and underlies the Salem when that formation is present. It is often thin and irregularly bedded but beds of three or more feet in thickness are not uncommon. It is often very fossiliferous containing crinoid stems and bryozoans. It has been used locally for building purposes but has had a more extended use for rubble and macadam. Its high calcium carbonate content makes it desirable for the manufacture of lime and cement. For its distribution see the accompanying geological map.

Plate III. In the limestone belt. Taking off the overburden preparatory to opening a quarry in the Indiana oölitic limestone.

The most widely used and widely known of the Mississippian limestones of Indiana is the Salem, or Indiana oölitic. This is the most widely used limestone for building purposes in the United States. The demands for it are constantly increasing.

Building Stones.

Indiana Oölitic. The superior qualities of the oölitic limestone of Indiana are now widely recognized. It is a medium fine-grained stone with even texture, composed of minute shells, the fragments of shells and concretionary grains (oölites) cemented by calcite. When first taken from the quarry it is soft and easily carved but under the action of atmospheric agents accompanied by the loss of quarry water it becomes harder.

Composition. The Indiana oölitic consists essentially of calcium carbonate which rarely falls below ninety-eight per cent and ranges even higher than ninety-nine. It also contains on the average more than three-quarters of a per cent of magnesium carbonate, a small amount of iron, probably in the form of a sulphide and a small amount of insoluble matter, largely silica. A detailed quantitative analysis reveals traces or minute quantities of other elements such as aluminium, carbon, potassium, phosphorus, sodium and sulphur. Some compounds formed from these elements would be detrimental if they occurred in sufficient quantities but this rarely ever happens. The

Plate IV. View of the Indiana oölitic limestone district showing mills and quarries.

iron present is partly in the form of minute crystals of pyrite which are widely distributed through the rock. Rarely are they so concentrated as to form blotches by their oxidation.

Color. The color of the limestone varies through shades of buff, gray and blue. The prevailing color in the oxidized zone above the level of ground water is buff. The prevailing color below the level of ground water is blue. Gray shades occur both above and below the level of ground water. The blue color is due to the presence of compounds which are oxidized in the zone of weathering the oxidation producing the buff color which is permanent in so far as any change in the constituent materials of the stone is concerned. The compounds present in the blue stone which are oxidized are probably compounds of iron and of organic matter. The oxygen is carried into the stone by ground waters which were originally meteoric and in that state secured their oxygen. Since the level of ground water is a vacillating irregular line, and the penetration of oxygen carrying waters was greater at some points than others and the amount of oxygen carried was variable the line of contact between buff and blue is a very irregular one. It requires only a short period of time for stones of slightly different colors to assume the same hue after being placed in a building. Light buff and deep blue stones may be used in the same building if care is taken to select the proper blending and avoid the juxtaposition of strong contrasts.

Texture. The oölitic limestone is composed of shells, fragments of shells and true oölites. These constitute the grains of the stone. These grains were deposited under sea water, the size and uniformity of the grain being dependent upon the sorting action of the water. In the quieter waters small grains of fairly uniform size were deposited; where currents prevailed larger particles

Plate V. View of a stone mill in the Indiana oölitic district. Stone in the foreground being unloaded from cars.

were deposited and where currents were shifting fine and coarse grains succeeded each other in rapid succession. The finer particles consist largely of the shells of foraminifera which are spherical, conical or disc shaped. Around some of these shells there are concentric layers of calcium carbonate thus forming a pseudoölite. There are also true oölitics present but the number is not large.

In the coarser varieties of stone larger shells or fragments of shells are present. Bryozoans, brachiopods, gastropods and other fossil forms predominate over the foraminifera. These are generally young forms of small size but occasionally a large form is present both valves being present and closed indicating that it was probably floated to its position due to its hollow condition.

The shells are composed of calcium carbonate and are cemented together with crystals of the same mineral. The value of the stone for building purposes is greatly enhanced by the proper balance between the softer grains and the harder cement.

Porosity. The property of having pores, minute spaces not filled with mineral matter it possessed by most rocks. In the oölitic stone both visible or macroscopic and invisible or microscopic pores exist. The percentage of pore space varies with the size, shape and arrangement of the grains. The pores, both visible and invisible, are of irregular shape and are due to incomplete cementation in part and in part to cavities in the interior of shells. There are bands of stone in which cementation has been carried to a more perfect state resulting in high density and minor porosity. In such instances circulating water carrying calcium carbonate derived from upper layers have been the agents of cementation. Porosity decreases the weight of a stone and excess porosity decreases its strength. It is also a measure of its absorptive power and in a measure its resistance to frost.

Absorption. The porosity of a stone is measured by the quantity of water it will absorb. The ratio of absorption is the ratio between the weight of the stone and the weight of the water absorbed. The ratio of absorption ranges from 1/13 to 1/95 in the Indiana oolitic. The resistance of a stone to frost action depends upon its absorptive power. The more water a stone absorbs the greater the disrupting effect when the water freezes in the pores of the rocks.

Specific Gravity. The specific gravity of the air dried oolitic stone varies between 2.25 and 2.65. The average of a large number of samples determined by the writer gave a specific gravity of 2.45. The specific gravity of the unseasoned stone is higher, being nearer 3. The shipping weight of the stone is estimated at from 175 to 185 pounds per cubic foot. The actual weight per cubic foot may be determined by multiplying the specific gravity of the stone by the weight of a cubic foot of water (62½ pounds).

Crushing Strength. The compression or crushing strength of a stone is measured in terms of load per square inch of surface required to crush the stone. Tests made upon samples of Indiana oölitic limestone indicate that its crushing strength ranges from 4,000 to 10,000 pounds per square inch. The stone of the minimum strength would sustain a wall constructed of oölitic stone to a height of 329 feet and the stone of maximum strength, a wall 822 feet high.

Transverse Strength. The transverse or cross-breaking strength of a building stone is measured in terms of the modulus of rupture which is the weight necessary to break a bar of one inch cross section when resting on supports one inch apart the weight being applied in the middle. The load required to produce rupture in the oölitic stone varies from 81 to 130 pounds. The transverse strength of the stone is sufficient to meet ordinary structural conditions. Cross-breaking is produced by differential settling of structures and where this is excessive few stones are able to resist it.

Resistance to Heat. Building stones are sometimes subjected to high temperature and should be able to withstand not only the heat but the contractional effects of sudden cooling. The failure of a stone may be due to the failure of the individual grains or to the failure of the cement. The grains and the cement of the oölitic stone are composed of calcium carbonate and there is probably little difference in the temperature required to produce failure in these two constituent parts of the stone. At a temperature varying between 800° and 1000° limestone is converted into lime and this is the temperature required to produce complete failure in the oölitic stone. Water suddenly thrown upon the stone when heated to a temperature of approximately 1000°F only caused a slight crumbling of the stone. It may be said that the Indiana oölitic has as great fire resisting properties as any stone of similar composition and higher resistance than many limestones.

Resistance to Frost. The power of a stone to resist frost action depends upon its composition, structure and density. These properties all affect the amount of absorption of the stone. The higher the amount of clay or organic matter contained in a stone the more easily it is affected by frost action. A layered rock resists frost action less easily than one of homogenous structure. The greater the density of the stone the more easily it resists the action

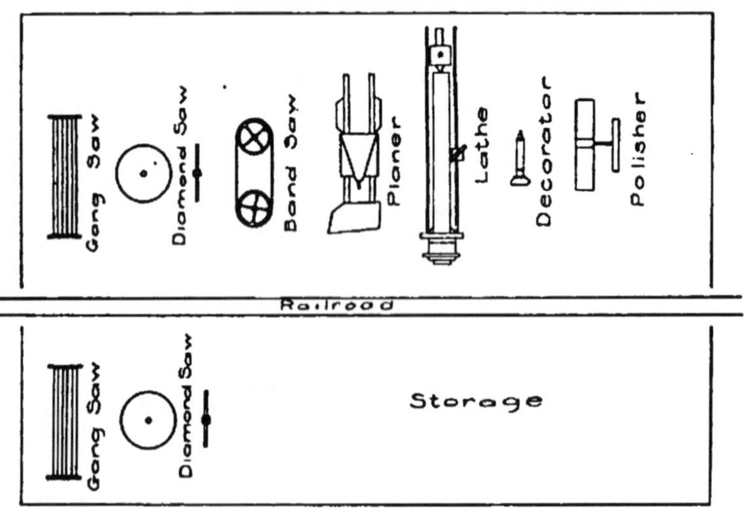

Plate VI. Flow sheet of Indiana oölitic stone mill.

of frost. Being of fairly uniform composition and homogenous in structure the property of resisting frost in the Indiana oölitic is dependent upon its density or porosity which is somewhat variable.

Distribution. The outcrop of the Salem formation which contains the Indiana oölitic limestone extends from Putnam County southward to the Ohio River. The belt extends through Putnam, Owen, Monroe, Lawrence, Washington and Harrison counties. The width of the outcrop varies from a few rods to fifteen miles. Quarries of the oölitic stone occur near Romona, Stinesville, Ellettsville, Hunter Valley, Bloomington, Clear Creek, Sanders, Oolitic, Bedford, Salem, Corydon and Georgetown. The main quarry district lies in Lawrence and Monroe counties extending from Stinesville to Bedford. Seventy-six large quarries are located in these two counties and supply fifty-five mills which mill between ten and twenty million cubic feet per year.

The Salem limestone dips southwestward at the rate of about thirty-five feet to the mile. At a short distance from the outcrop the bed passes under an over-burden of Mitchell and Chester rocks too heavy to be removed. So quarrying is confined at the present time to the outcrop and to places where the over-burden is thin. When these areas are exhausted sub-surface quarrying will doubtless be undertaken.

Mitchell limestone. Overlying the Salem limestone is the bedded Mitchell limestone. The layers of this stone vary in thickness from a few inches to four feet. Perhaps on the average the layers range from one to two feet. The stone is harder than the Salem and in some zones contains much chert. It furnishes a very serviceable building stone and is especially well adapted for basements and foundations as it resists frost action well. It is one of the best limestones in the United States for macadam and has had an ex-

Plate VII. The Pinnacle, Shoals, Indiana. An exposure of Mansfield sandstone.

tensive use in Indiana for this purpose. It has also been used extensively in the manufacture of lime and Portland cement. It has been quarried at many points along its area of outcrop. It has been quarried and used extensively in buildings at Spencer. Other quarries are located at Bedford, Bloomington, Corydon, Abydell, Milltown, Mitchell, Greencastle, Putnamville, Marengo and Salem. Many quarries are located convenient to highways which are to be macadamed, the length of haul determining in most cases the location of the quarry. The distribution of the Mitchell is indicated on the geological map.

Chester Limestones. Overlying the Mitchell in Indiana is a series of interstratified limestones, sandstones and shales of Chester age. The lowermost bed of limestone, the Beaver Bend, is a cream colored oölitic limestone, varying from massive in some outcrops to thin bedded in others. Its thickness is usually about ten to fourteen feet though in places it attains more than twenty feet. It may be used for rubble, macadam and lime.

The Reelsville is usually about four feet thick but attains a thickness of ten feet. It is generally pyritiferous and on that account and because of its thinness, is of limited usefulness.

The Beech Creek limestone is normally about fifteen feet thick but attains in places a thickness of twenty-five feet. It is usually jointed, breaking up into cubical blocks. It is usually massive but sometimes thin bedded. The limestone has been used for rubble, macadam and burned locally for lime.

The Golconda limestone is coarsely crystalline in the lower part and oölitic in the upper portion. It generally occurs in two ledges which are separated by a stratum of shale. It attains a thickness of thirty feet. It has been used locally for building, rubble and macadam. The Glen Dean is the uppermost Chester limestone and occurs only very locally.

The Chester limestones have been used locally in Owen, Greene, Monroe, Lawrence, Martin, Orange, Crawford and Perry counties.

Pennsylvanian limestones. Beds of limestone are interstratified with beds of shale, sandstone and coal in the coal measures of Indiana. These limestones are usually irregularly and thinly bedded. They are serviceable locally for the cruder structural purposes and for concrete and macadam. One of these has been quarried near Farmersburg.

Sandstones for Buildings.

There are a number of geological formations in Indiana which contain sandstones suitable for structural uses. The St. Peters sandstone lies too deeply buried to be utilized in Indiana. The Pendleton sandstone of the Devonian outcrops in a few places and is used for structural purposes and glass sand.

The Mississippian rocks contain a number of sandstones some of which are suitable for structural purposes. The Knobstone contains ledges of sandstone which are thick enough and sufficiently indurated to be serviceable for building purposes. The sandstone from the Riverside division of the Knobstone has been quarried at St. Anthony near New Albany, and at Riverside. It has served a local demand in a number of places within the area of its outcrop.

The Chester division of the Mississippian contains a number of sandstones which are locally useful for structural purposes. The Brandy Run sandstone which lies between the Beaver Bend limestone and the Reelsville limestone is in many places only a sandy shale or a thin bedded, fissile sandstone. But in places ledges of indurated sandstone occur.

The Elwren sandstone which occupies the interval between the Reelsville limestone and the Beech Creek limestone is like the Brandy Run shaley in places and not often sufficiently indurated for good building stone.

The Cypress sandstone lies upon the Beech Creek limestone and is more uniform in thickness and properties than any of the other Chester sandstones. Indurated ledges of good structural qualities are of common occurrence. The Hardinsburg and the Tar Springs are other sandstones of the Chester which exhibit local phases suitable for building stone. The Chester sandstones have been quarried at Cannelton, Fountain, Williamsport and other places.

Pennsylvanian Sandstones. The principal sandstone occurs at the base of the coal measures and is called the Mansfield. In places it forms a basal conglomerate which has an iron oxide cement. In many places it contains ledges of iron stone which are suitable for building purposes. It varies in thickness from a few feet to two hundred feet. It has been quarried at Mansfield, Attica, Pottsville, Shoals and other places.

The coal measures contain other sandstones which are useful for structural purposes and are so used locally. Such sandstones have been quarried at Middleton, Jasper and other places. One such sandstone is the Merom, occurring in the southwestern part of the State.

LOCATION	Calcium Carbonate ($CaCO_3$)	Magnesium Carbonate ($MgCO_3$)	CHEMICAL ANALYSES OF HYDRAULIC LIMESTONE (Devonian)							
			Silica (SiO_2)	Ferric Oxide (Fe_2O_3)	Alumina (Al_2O_3)	Lime (CaO)	Magnesia (MgO)	Organic—Water, Undetermined	Total	AUTHORITY
Silver Creek, Indiana, "Ohio Valley"......	54.31	16.90	18.33	1.67	4.98	0.14	0.33	1.19	97.85	W. A. Noyes, Analyst
Silver Creek, Indiana, "Black Diamond"..	51.98	32.97	9.69	1.95	2.77	0.10	0.11	0.36	99.90	W. A. Noyes, Analyst
Silver Creek, Indiana, "Belknap's Falls City"...............	52.50	35.09	9.80	1.40	2.03	0.04	0.11	0.47	101.44	W. A. Noyes, Analyst
Silver Creek, Indiana, "Speed's"..........	61.70	16.74	13.65	1.45	3.46	0.15	0.25	0.45	97.85	W. A. Noyes, Analyst
Silver Creek, Indiana, "Hansdale".........	60.69	15.90	15.21	1.44	4.07	0.07	0.32	0.86	98.56	W. A. Noyes, Analyst

Niagara Limestone.

Analyses of Cement Rock from Derbyshire Falls, Franklin County, Indiana.

	I	II	III
Lime (CaO)..	27.13	38.11	24.44
Magnesia (MgO)..	15.15	7.71	9.81
Silica (SiO_2)..	11.57	9.25	21.51
Ferric oxide (Fe_2O_3)....................................	1.56	1.85	2.69
Alumina (Al_2O_3)....................................	4.58	2.95	8.32
Potash (K_2O)..	1.03	0.84	1.91
Soda (Na_2O)..	0.10	0.07	0.16
Loss by ignition (CO_2 and H_2O).............. ..	38.81	39.03	31.38
Total..	99.93	99.81	100.22

ANALYSIS OF OÖLITIC LIMESTONE.

Locality	Company	Year	Authority	Lime Carbonate $CaCO_3$	Magnesia Carbonate $MgCO_3$	Insoluble	Iron Oxide Fe_2O_3	Alumina Al_2O_3	Alkalies K_2O, Na_2O	Water H_2O	Total
Harrison County											
Twin Creek	Twin Creek Stone & Land Co.	1896	W. A. Noyes, Rose Polytechnic Inst.	98.16	.97	.76	.15				100.04
Salem ...		1886	Ind. Geol. Rept. 1886, p. 144	96.04	.72	1.13	1.06		.15	.10	
Harrison Co.	Stockslager's Quarry	1878	Ind. Geol. Rept. 1878, p. 96	98.09		.31	.18	.14	.40	.12	
Lawrence County											
Bedford	Ind. Stone Co.	1896	W. A. Noyes, Rose Polytechnic Inst.	98.27	.84	.64	.15				99.90
Bedford	Chi. & Bed. Stone Co.	1878	Ind. Geol. Rept. 1878, p. 95	96.60	.27	.50	.98		.40	.61	100.00
Bedford	Hoosier Quarry, buff		Bedford Quarries Co., Circular	98.20	.39	.63	.39				99.61
Bedford	Hoosier Quarry, buff		Bedford Quarries Co., Circular	97.26	.37	1.69	.49				99.81
Monroe County											
Hunter Valley	Hunter Bros. Quarry	1896	W. A. Noyes, Rose Polytechnic	98.11	.92	.86	.16				100.05
Big Creek	Ind. Steam Stone Works		L. H. Streaker, State University	93.80	4.01	.15	.64			1.00	100.00
Big Creek	Ind. Steam Stone Works		L. H. Streaker, State University	93.07	4.22	.50	.71			1.19	100.00
Bloomington	Dunn & Dunn Quarry—white	1881	Ind. Geol. Rept. 1881, p. 32	95.62	.89	1.74	.23	.06		.59	99.45
Bloomington	Dunn & Dunn Quarry—blue	1881		95.55	.93	1.60	.09	.09		.42	99.37
Bloomington	Dunn & Co.	1878	Ind. Geol. Rept. 1878, p. 95	95.54	.40	.65	1.00		.55	.25	
Stinesville	Monroe Marble Co.	1862	Ind. Geol. Rept. 1862, Owens, p. 137	95.00	.22	.90	3.00		.83	.05	100.00
Owen County											
Romona	Romona Oölitic Stone Co.	1896	W. A. Noyes, Rose Polytechnic	97.90	.65	1.26	.18				99.99
4 Miles East of Spencer	Simpson & Archer	1878	Ind. Geol. Rept. 1878, p. 91	96.79	.23	.70	.91		.32	.41	99.90

Table Showing Chemical Composition of Sandstones.

LOCALITY	Insoluble Residue	Alumina	Iron Oxide	Lime	Total
Mansfield	92.16	....	6.29	.05	*98.59
Judson	93.21	.51	4.91	.12	98.75
Hillsboro	91.65	.56	6.60	.12	98.93
Fountain	91.66	.60	6.41	.05	98.72
Bloomfield	85.29	.19	11.83	.06	97.43
St. Anthony	88.41	.63	8.40	.13	97.57
Riverside	93.16	1.60	2.69	.13	97.58
Williamsport	98.57	.05	.65	.02	99.29
Greenhill	98.73	.28	.86	.03	99.40
Cannelton	96.18	.54	1.56	.15	98.43

*Includes .09 per cent. alkalies.

CHAPTER II.

CEMENT.

Cement is a calcined or cinerated material which has the property of setting when mixed with water and of hardening in the air or under water.

The class of cements which harden only in the air are called simple cements and those which will harden under water are called complex cements.

Simple Cements. Simple cements are calcined materials which may be divided into two classes: 1. Hydrate cements which are manufactured from gypsum by driving off a part of its water crystallization. They include such cements as plaster of Paris, Keene's cement, Parian cement and cement plaster. They differ from each other in the addition of small amounts of sand, limestone and clay and in slight variations in methods of manufacture. 2. Carbonate cements. These consist of quick limes produced by calcination from various varieties of limestone, marble or dolomite, the temperature of decarbonation being reached in the process.

Complex Cements. These are cements, the materials of which have been subjected to temperatures high enough to form new chemical compounds. Four classes have been suggested: (a) Natural cements include such brands as Roman and Rosendale cements which are manufactured by burning a silico-aluminous limestone at a temperature between decarbonation and clinkering. They exhibit no free lime, possess hydraulic properties, and do not slake unless ground very fine. They are of lighter weight, burn at lower temperatures, set quicker, have less ultimate strength, greater variation in composition and usually contain a higher per cent of magnesia than Portland cement.

The composition of the silico-aluminous limestones from which natural cements are derived is variable, the constituents varying between moderately wide limits. The silica from 9 to 25 per cent; alumina from 1.5 to 17; ferric oxide from 1.34 to 6.30; lime from 22 to 36; magnesia from

2 to 18; potash and soda from 1 to 6; sulphur trioxide from 0 to 2; volatile matter (CO_2+ H_2O) 32 to 34 per cent.

The Devonian limestones of southeastern Indiana have been used extensively in the manufacture of natural cement but production has fallen off until at the present time it is being produced at only one plant.

(b) Hydraulic limes. These cements are manufactured by burning a silicious limestone at a temperature a little above decarbonation. They are of a yellowish tint and contain considerable free lime. They set slowly and have little strength in the neat but are of greater strength when mixed with sand. They are of little economic importance in the United States.

(c) Pozzuolan Cements. Cements of this class are made from an uncalcined mixture of slaked lime and a silico-aluminous substance such as volcanic ash or blast-furnace slag. The composition of the mixture may vary between the following limits: Silica, 52 to 60 per cent; alumina, 9 to 21 per cent; ferric oxide, 5 to 22 per cent; lime, 2 to 10 per cent; magnesia

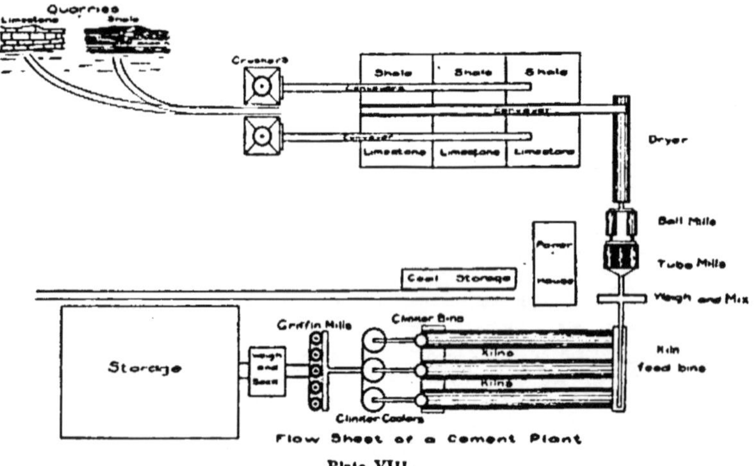

Flow Sheet of a Cement Plant

Plate VIII.

0 to 2 per cent; potash and soda, 3 to 16 per cent; moisture 0 to 12 per cent. Cements of this type are manufactured in Ohio and Alabama.

(d) Portland Cement. Portland cement is the product obtained from burning at a high temperature an artificial mixture of calcareous and silico-aluminous rocks or slags. The mixture consists essentially of lime, silica, alumina and oxide of iron, though small quantities of magnesia and other compounds are usually present. It was first manufactured in England and so named because of its resemblance to the Portland building stone of that country.

Raw Materials. The raw materials used in the manufacture of Portland cement vary widely in physical and chemical characteristics. In the selection of materials the attempt is usually made to select for one constituent a rock high in calcium carbonate content and for the other one highly aluminous in composition. The mixture may be marl and clay; limestone and clay or shale; chalk and clay or shale; pure limestone and argillaceous limestone or limestone and slag. Sometimes it is desirable to mix two limestones if one contains too high a per cent of magnesia but is so situated in the quarry as to make its use essential to economy. It is also desirable in some plants to mix two kinds of shale, say a highly silicious with a highly aluminous one. In Indiana the raw materials used in the manufacture of Portland cement are marl and clay; shale and limestone, and slag and limestone.

Quarrying Raw Materials. When surface clays are used the open pit method is employed for securing the clay which is loaded, by steam shovels on cars for transportation to the plant. The marl is taken from the bottom of the lakes by steam dredges.

The shale is taken from the face of an outcrop by blasting down the shale and loading it on cars by the use of steam shovels. The limestone is obtained from the outcrop of a ledge of considerable thickness. Holes about six inches in diameter are drilled about ten feet back from the face of the quarry at intervals of twelve or fourteen feet to the depth of the quarry which is often sixty feet or more. Each of these holes is charged at top and bottom with from 300 to 350 pounds of blasting powder. The larger blocks of limestone are drilled and broken up with explosives and the smaller masses loaded on cars by the use of steam shovels.

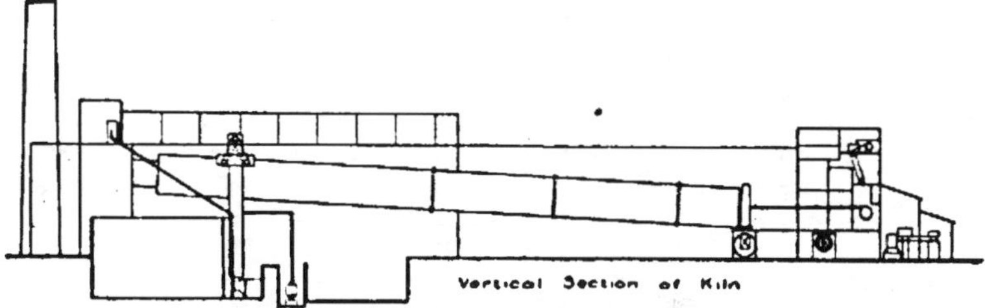

Plate IX. Vertical section of kiln used in wet process.

Crushing and Grinding. The limestone is crushed first in large gyrotary crushers, then in small ones.

The crushed limestone and shale are dried in horizontal cylindrical ovens and then weighed and mixed in a proportion depending upon the chemical composition of the limestone and shale.

The mixture is first ground in a tube mill until it is fine enough to pass through a sieve of 65 meshes to the inch and then in a ball mill until it passes a 95 mesh sieve.

Burning. The pulverized raw materials are burned in horizontal cylindrical revolving kilns which are constructed of a steel shell lined with fire brick. The kiln is placed on a slant and the raw material is admitted at the stack or higher end and works its way downward toward the lower end where the flames of the kiln are fed by crushed coal under an air pressure of 55 pounds per square inch. The temperature attained is from 2,000 to 2,500 degrees F, which is sufficient to convert the raw material into a clinker. The clinker drops through an opening in the lower end of the kiln into a pit from which it is elevated into a vertical cylinder where it is cooled. Gypsum is added and it is ground until it is fine enough to pass a 250 mesh sieve. Being elevated to storage bins it is then weighed and sacked at one operation. This is the dry process used in most of the plants in Indiana. The wet process is described under the discussion of the Indiana Portland Cement Company.

Distribution of Raw Materials.

Raw Materials. The raw materials for the manufacture of cement are widely distributed in Indiana. The formations which contain suitable materials range in age from the Ordovician to the Pleistocene. The outcrop of these materials cover a large area in the State and the materials are easily accessible to transportation facilities and to fuel supplies. Many small streams and a few large ones cross the outcrop of the raw materials and from these a sufficient water supply may be obtained.

Ordovician Limestones and Shales. Limestones of this age attain a thickness of thirty feet or more in a single ledge in the southeastern portion of Indiana. Excellent outcrops occur in Clark and Jefferson counties. The limestone is often argillaceous but in many places is a soft calcareous stone of whitish color. Associated with the Ordovician limestones are beds of shale which in places are highly calcareous, in other places silicious. There is little doubt that in these limestones and shales there are the proper materials for the manufacture of Portland Cement. Chemically some of the limestones contain more than ninety per cent of the carbonates of calcium and magnesium the latter varying from six to ten per cent.

Silurian Limestones and Shales. The Silurian limestones form the bed rock of a large area in the southeastern, eastern and north central parts of Indiana. Much of the limestone of this age in the eastern and north central portions of the State contain too high a magnesia content to be used and care be exercised in selecting a location. Shales of a calcareous nature are to be found associated with these limestones and Ordovician and Mississippian shales are near at hand.

Devonian Limestones. Devonian limestones suitable for the manufacture of both natural and Portland cement outcrop in southeastern Indiana. These Devonian limestones attain a thickness of ninety feet. They outcrop near beds of shale of Mississippian age which may be used to obtain the proper Portland cement mixture. The Devonian limestones are being used in the manufacture of natural and Portland cement at Speed and the New Providence shales of the Mississippian to form the mixture for the latter.

Mississippian Limestones. The Mississippian rocks of Indiana occupy a large area of outcrop in Indiana and their strata contains an inexhaustible supply of excellent materials for the manufacture of Portland cement.

Limestones. The limestones of the Mississippian which are suitable for use in the manufacture of Portland cement may be grouped under the Harrodsburg (Warsaw, the Salem (Oölitic), the Mitchel and the Chester. The Harrodsburg, the oldest and lowermost member of this group of limestones, attains a thickness of 150 feet. It is a highly calcareous limestone suitable for use in the manufacture of Portland cement. Overlying the Harrodsburg is the Salem limestone which is of high calcium carbonate composition and an excellent building stone. Only the grades of this stone unsuited to building purposes and the waste from the quarries should be used for cement. The thickness of the Salem lies usually between forty and one hundred feet. The percentage of calcium carbonate in the Salem frequently runs as high as ninety-eight per cent and one per cent or more of magnesium carbonate and less than one per cent of silica, iron and alumina.

The Mitchell limestone which overlies the Salem where both are present is a compact bedded stone which attains in places, a thickness of two or three hundred feet. This limestone contains a higher per cent of magnesia than the Salem but rarely sufficient to render it unsuited to the manufacture of cement. It is being used at present in the manufacture of Portland cement at Limedale and at Mitchell, a form of utilization for which it is well suited. It is also valuable in the manufacture of lime and for road metal.

Plate X. Shale quarry. Just after a heavy rain, Louisville Cement Co.

The Chester division of the Mississippian contains a number of limestones such as the Beaver Bend, the Beach Creek and the Golconda which could be used in the manufacture of Portland cement. These beds range in thickness from ten to thirty feet. Because of the nearness of these limestones to the thicker beds of the lower horizon the necessity for their use will probably not arise except in very isolated instances. Their high carbonate content, freedom from detrimental impurities and nearness to excellent beds of shale warrant their consideration if greater nearness to fuel supplies or other considerations should suggest their use.

Mississippian Shales. The Mississippian formations contain an abundance of shales suitable for use in the manufacture of Portland cement. They are as widely distributed as the limestones of that period and are within easy reach of their outcrop. These shales may be grouped under the Knobstone and the Chester division.

Knobstone Shales. These shales lie below the Harrodsburg limestone and their outcrop lies parallel with the outcrop of the Harrodsburg and the succeeding limestone and at accessible distances from these limestones. The Knobstone shale contains facies which are highly arenaceous in character and others which are highly aluminous. In securing the proper mixture for Portland cement it is desirable to use a certain portion of each of these shales. The shales are being used in the Lehigh plant at Mitchell and the plant at Speeds in the manufacture of Portland cement.

Chester Shales. Shales occur in the Chester at the horizon of the Brandy Run, the Elwren, the Golconda, the Hardinsburg, the Indian Springs and the

Plate XI. Rock quarry, Devonian limestone, Louisville Cement Co., Speeds, Ind.

Buffalo Wallow. These shales vary in thickness from ten to forty feet and some of them are in places wholly replaced by sandstones. Both aluminous and arenaceous facies of the shales prevail so that is is not a difficult matter to make selections for the proper mixture. These shales are being used by the plant located at Limedale.

Pleistocene and Post Pleistocene Marls. The glacial lake basins of northern Indiana contain large quantities of marl, the calcareous secretion of the plant chara. This marl is essentially calcium carbonate containing small quantities of magnesium carbonate, iron carbonate, silica, alumina and organic matter. An analysis of a sample of the marl is given as follows:

Analysis of Sample of Marl from James Lake.

Calcium carbonate	92.41
Magnesium carbonate	2.38
Calcium sulphate	.15
Ferric oxide	.29
Insoluble (silica, etc.)	1.16
Organic matter	1.97
Total	98.36

Workable deposits of marl occur in lake basins in counties in the northern part of Indiana for the distribution of which see the article on marl in this report.

Pleistocene and Post Pleistocene Clays. Near the beds of marls in the northern part of Indiana there are beds of clay of Pleistocene and Post Pleistocene age which are adapted to use in cement mixtures. The average of the analyses of eight samples of glacial clay which was used by the Wabash Cement Company is given by Oglesbey as follows:

Analysis of Pleistocene Clay.

Silica (SiO_2)	56.74
Alumina (Al_2O_3)	19.43
Ferric oxide (Fe_2O_3)	4.83
Lime (CaO)	7.27
Magnesia (MgO)	3.05
Loss on ignition	10.39
Total	101.71

Furnace Slag. From the iron smelters in Indiana a large supply of slag is obtained which could be used in the manufacture of Pozzuolan cement which is made from granulated blast furnace slag ground with dried quicklime or hydrated lime. The blast furnace slag may also be used in the manufacture of Portland cement in which the slag takes the place of the silico-aluminous material of the shale. Ground limestone, instead of quicklime, is mixed with slag and the mixture cinderated. The Universal cement plant at Buffington uses slag from furnaces of the Illinois Steel Company.

Cement Plants.

Indiana Portland Cement Company. The plant of this company is located south of the Vandalia and Monon station at Limedale, near Greencastle. The plant began operations in 1919. The raw materials used are Mitchell limestone 78%, Chester or Allegheny shale 11% and surface clay 11%. The capacity of the plant is 1200 to 1500 bbls. per day. The process used is the wet process. The limestone and shale are crushed in gyratory crushers. The clay is disintegrated in the wash-mill. The clay slurry, crushed shale and limestone are passed through the kominuter and then through a vertical rotary screen into the tube mill. After being ground in the tube mill the slurry is discharged to correcting basins from which it goes to the mixing basin. From the mixing basin it goes to the storage basin and from that through the

Plate XII. Universal Portland Cement Company's Plant at Buffington, Indiana. The capacity is 100,000 sacks per day of 24 hours.

feed tank into the rotary kiln which has a length of 240 feet and a diameter of ten feet. The clinker produced in the kiln is cooled in a rotary cooler, ground in a kominuter and tube mill and stored in concrete storage bin of the silo type. The following table shows the composition of the raw materials and the product as furnished by the president of the company, Mr. Adam H. Beck:

	Clay	Shale	Limestone	Cement
Loss on ignition	7.23	12.41	43.44	1.65
Silica (SiO_2)	71.77	51.32	1.88	22.32
Alumina (Al_2O_3)	12.10	23.53	.25	6.19
Ferric oxide (Fe_2O_3)	4.62	7.21	.29	3.27
Lime (CaO)	.80	.35	53.99	63.01
Magnesia (MgO)	1.00	1.60	.44	.90
Sulphur trioxide (SO_3)				1.53

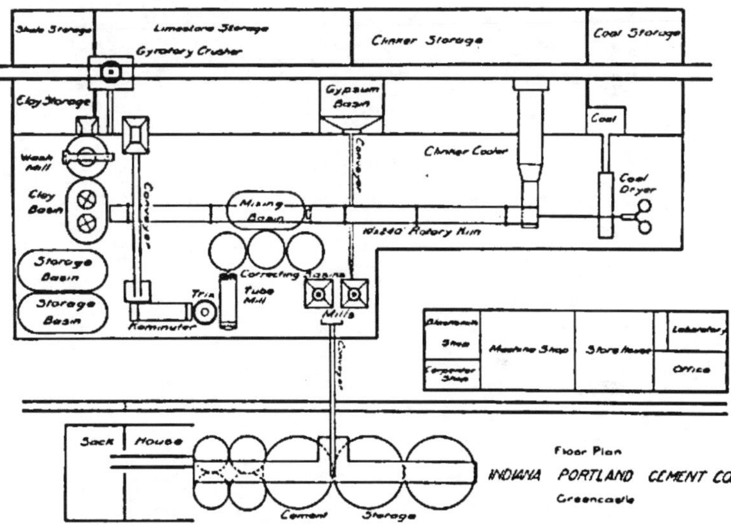

Plate XIII. Floor plan of Indiana Portland Cement Company's plant, Greencastle, Indiana.

Lehigh Portland Cement Company. A large plant of this company is located at Mitchell. The plant consists of two 10 kiln units and has a total capacity of two and a quarter million barrels of cement per year. The raw materials consist of the Mitchell limestone and two varieties of Knobstone shale, one a silicious and the other an aluminous variety. The limestone used runs high in calcium carbonate, rarely falling below 96 per cent. The limestone is first crushed in large gyratory crushers then in smaller ones. The crushed limestone and shale are dried in horizontal rotary ovens, then weighed and mixed. The mixture is ground in a tube mill until it is fine enough to pass through a sieve of 65 meshes to the inch, then in a ball mill until it passes a 95 mesh screen. The pulverized mixture is burned in rotary kilns which are fired with powdered coal under pressure. The clinker formed is cooled in vertical cooling cylinders. Gypsum is added and the clinker ground, stored and sacked.

Plate XIV. View of Indiana Portland Cement Plant, Greencastle, Indiana.

Louisville Cement Company. The plant of this company is located at Speeds near Sellersburg in Clark County. The raw materials used consist

Plate XV. General mill view of Louisville Cement Plant, Speeds, Ind.

of hydraulic limestone of Devonian age and Knobstone (New Providence) shale of Mississippian age. The limestone quarry is located near the plant but the shale is brought from a distance. The company manufactures a considerable amount of natural cement but the larger part of its product is Portland cement. The production of the former for 1919 was 300,000 bbls., and of the latter 800,000 bbls.

The composition of the raw materials and the finished product of the Portland cement is given in the following table:

Constituent	Lime-stone	Shale	Raw Mixture	Portland Cement
Silica (SiO_2)	4.78	65.54	14.94	21.32
Alumina, etc. (Al_2O_3)	1.38	23.52	6.34	8.78
Cal. carb. ($CaCO_3$)	91.25			
Calcium oxide (CaO)		1.67	41.85	63.26
Magnesia (MgO)	2.59	5.00	1.94	3.72
Loss on ignition			34.93	.94
Undetermined		.27		
Insoluble res.				.42
Sulphur trioxide (SO_3)				1.56

The composition of the hydraulic limestone and the natural cement is given as follows:

Constituent	Natural Cement Rock	Natural Cement
Silica (SiO_2)	19.02	21.92
Alumina, etc. (Al_2O_3)	.98	9.42
Lime (CaO)	34.68	47.63
Magnesia (MgO)	8.47	12.47
Loss on ignition	36.85	8.56

Syracuse Plant of the Sandusky Cement Company. This plant was established in 1900 and continued in operation for eighteen years. It was shut down in 1919 but resumed operation in 1920. The materials used during the first period of operation of the plant were marls taken from Lakes Syracuse and Wawasee and clay from near Lapaz, about thirty miles west of the plant at Syracuse. The composition of these raw materials and the product is given in the Twenty-fifth Annual Report of the Indiana Survey as follows, by S. B. Newberry:

	Marl	Clay	Cement
Silica	1.74	55.27	22.06
Alumina	.90	10.20	4.80
Iron oxide	.28	3.40	1.66
Lime	49.84	9.12	65.44
Magnesia	1.75	5.73	3.82
Sulphur trioxide (SO_3)	1.12		0.90
Loss	46.01		

The marl contained organic matter making necessary the use of a large quantity of water to form a slurry. The expense of removing the extra water in burning rendered it difficult for the manufactured cement to compete with that produced by cheaper methods. Owing to labor conditions and other difficulties the plant was run at a loss during 1917 and 1918. Arrangements have been made for using limestone from Logansport and the proper machinery has been installed for grinding this raw material. The capacity of the plant is about 50,000 barrels per month.

Universal Portland Cement Company. The plant of this company is located at Buffington. The plant has three separate producing units, mills 3, 4 and 6. Power is supplied by transmission lines from Gary and South Chicago and from the waste heat power plant at Buffington, 12,000 K W by the former and 8,000 by the latter. Coal is used for fuel, that which is used in the kilns being pulverized.

Plate XVI. Kiln room, showing two kilns, each 10 feet in diameter and 150 feet long. Capacity 1,100 barrels per day each.

Kilns of mills No. 4 and 6 are equipped with Cottrell dust precipitators, using energy transformed and rectified to 65,000 volts direct current. The efficiency of these collectors is more than 90 per cent, and dust collected per kiln per hour is about 1,000 pounds.

Raw materials used in manufacture are granulated slag obtained from the blast furnaces of the Illinois Steel Company and Illinois limestone.

Slag and limestone are dumped into their respective bins at opposite ends of the raw material building. When crushed to about 1½ inches in size, the limestone is dried in rotary driers, and is given a preliminary grinding before delivery to hoppers above automatic, electrically-operated scales. Slag is fed direct to the drier, then pulverized and elevated to hoppers above the automatic scales.

At the scales the slag and limestone are proportioned to assure a uniform mixture, which is then ground in tube mills and elevated to hoppers above the rotary kilns. In these kilns, the mixture attains a temperature of about 2,500 degrees Fahrenheit and burns to a hard "clinker".

After curing for about ten days the clinker is reduced to about ½ inch size, and then is pulverized to pass a slanting screen with ⅛ inch openings. Gypsum is automatically added to regulate and retard the setting. Other tube mills complete the grinding and deliver the cement to an inclined belt which conveys it to the storage bins. On the conveyor the cement is automatically sampled every eight seconds. Every hour the accumulated sample is carried to the laboratories where tests and analyses are made to assure uniformity.

From the storage bins the cement is drawn, weighed, sacked and trucked to the cars. (From the Guide Book of the American Institute of Mining and Metallurgical Engineers.)

The capacity of the plant is 100,000 sacks per twenty-four hours. The analyses of the raw materials and finished product is given below:

Analyses of Raw Materials and Product.

	Slag	Limestone	Gypsum	Cement
Silica (SiO_2)	34.48	2.22	2.06	19.61
Alumina (Al_2O_3)	12.99	1.27	.40	7.71
Ferric oxide (Fe_2O_3).....	.99	1.07	.40	2.22
Iron (Fe)	1.20			
Calcium oxide (CaO)....	42.78	52.58	32.17	64.23
Calcium sulphide (CaS)..	2.90			
Magnesia (MgO)	3.28	.78		2.29
Manganese oxide (MnO_2).	1.38			.71
Carbon dioxide (CO_2)....		42.08		
Sulphur trioxide (SO_3)..			44.25	1.73
Moisture (H_2O)				
Volatile matter			20.72	1.50

Wabash Portland Cement Company. The plant of this company is located at Stroh in Lagrange County. The plant is located between Big and Little Turkey lakes and uses the marl from the lakes and glacial clays for its raw materials. The composition of the raw materials and the products as given by the company is as follows:

	Marl	Clay	Cement
Lime (CaO)	50.20	11.24	62.86
Magnesia (MgO)	.45	3.60	2.34
Alumina (Al_2O_3)			6.58
Ferric oxide (Fe_2O_3)............	1.20	17.12	2.14
Silica (SiO_2)	1.24	52.98	23.01
Loss on ignition..................	46.12	14.16	.40
Sulphur trioxide (SO_3)...........			1.30

The annual production is from 300,000 to 350,000 barrels of cement.

With the exception of the chemical laboratory and office, the whole plant is under one roof. It is equipped with one steel rotary dryer, tube mill for coal grinding, clay pulverizer, four continuous rotary kilns, size 5x60 feet, daily capacity 480 bbls., three pug mills, three tube mills and four ball mills.

The marl is quarried by dredging with a clam shell bucket, loaded directly into narrow gauge cars for haulage to the mill. The clay is loaded with steam shovel into standard gauge cars and unloaded directly into the plant. These raw materials are washed to remove deleterious matter from clay and marl. Mixed in the proper proportions the raw materials are ground in tube mills, burned in the kilns and the clinker ground into the finished product.

CHAPTER III.

COAL.

Coal is one of the most important natural resources of Indiana. The industrial development of the State is dependent in large measure upon its coal mining industry. For this reason the development and conservation of this resource becomes of paramount interest to the citizens of the State.

Definition. Coal is a brown or black mineral substance of organic origin. It is combustible, burning generally with a smoky flame and leaving an unconsumed residue called ash which consists of mineral matter.

Varieties. Many varieties of coal exist but only a few classes are generally recognized. The classification is based upon the chemical and physical properties of the coal but there are many gradations between the classes. The common varieties are peat, lignite, sub-bituminous, bituminous, semi-anthracite and anthracite. Peat is supposed to represent vegetable accumulations which are at the beginning of the coalification process. The peat beds of Indiana are discussed under a separate head. Lignite does not occur in Indiana in quantity, a small amount of lignitized wood is found in the glacial drift in a few places. The coal of Indiana is of the bituminous or soft coal variety. No anthracite or hard coal is found in the State.

Composition. The compositions of the varieties of coal differ greatly. The elementary analyses show that the amount of fixed carbon increases from peat through the other varieties, in the order named above, to anthracite and that the volatile matter decreases. In the following table the percentage of the various elements found in the different varieties of coal is given.

Elementary Analyses of Coal Varieties.

KIND	Carbon	Hydrogen	Oxygen	Nitrogen	Sulphur	Ash	Water
Peat	59.47	6.52	31.51	2.51	....	....	.22
Lignite	52.66	5.22	27.15	.71	2.02	12.24	....
Sub-bituminous	58.41	5.06	28.99	1.09	.63	4.79	....
Bituminous	82.70	4.77	9.39	1.62	.45	1.07	....
Semi-bituminous	83.14	4.58	4.65	1.02	.75	5.86	....
Anthracite	90.45	2.43	2.45		....	4.67	....

Fuel Value. The fuel value of a coal is dependent upon the amounts of fixed carbon and hydrocarbons in it. These compounds are called the fuel constituents of coal. The fuel ratio of a coal is determined by dividing the fixed carbon percentage by the percentage of volatile hydrocarbons which the coal contains. For example if a coal contains sixty per cent of fixed carbon and thirty per cent of volatile hydrocarbons then its fuel ratio is expressed as two.

The volatile hydrocarbons burn readily but produce little heat. However, they aid in the combustion of the coal. The fixed carbon is the heat producing constituent of coal. The heating power of coal increases with the fuel ratio to a certain point beyond which the difficulties of ignition more than counter-

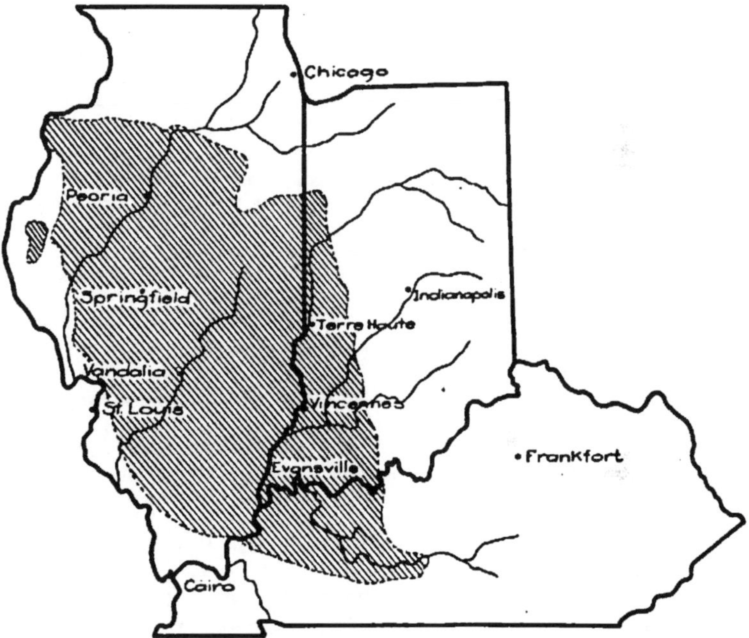

Plate XVII. Diagram to show the extent of the Eastern Interior coal field.

balance the extra heat units derived from the higher percentage of fixed carbon. Such is the condition in certain very graphitic anthracite coals which are found in some Atlantic Coast States.

Origin. The generally accepted theory of the formation of coal is that it is of vegetable origin. The possibility of small accumulations of carbonaceous matter from inorganic sources is not denied but that our thick beds of coal had any other than an organic origin is not accepted. The organic theory of origin is supported by a great many facts, among others the following: 1. There is a gradual gradation from the green fibers of plants to the compact lignite-like brown peat. 2. There is a close relationship between the elementary analysis of a coal and of vegetable fiber. 3. Many impressions of plants and parts of plants, leaves, twigs, stems and seeds are found in coal. 4. Microscopic sections of coal reveal the presence of vegetable cell structure, spores, spore cases and parts of vegetation. 5. The trunks, branches and roots of forms of vegetation which have been converted into coal have been found in coal beds.

Mode of Accumulation of Organic Matter. A difference of opinion exists among scientists as to the mode of accumulation of the vegetable matter which formed the beds of coal. A great many modes of accumulation have been suggested such as flood plains, in deltas, in peat bogs, in marine marshes, in sargassa seas or as fresh or brackish water deposits on coastal plains. It does not seem as though accumulations sufficiently free from impurities could be formed either on the flood plains of rivers or in delta deposits. It would seem as though coals thus formed would contain a great deal of land derived inorganic material. Peat bogs are too limited in extent as we know them today to furnish the area comprised in some of our widely distributed coal beds though it is possible that peat bogs may have been more extensive during the coal forming periods. Some coals may have originated from accumulations of peat in bogs but probably not the more extensive beds. The accumulation of marine algae in sargassa seas does not seem to have been the mode of accumulation for coals are rarely formed of that type of vegetation and land plants of large size are found in coals. The accumulation in low areas on coastal plains in fresh or brackish water swamps seem to conform best to the conditions under which the more extensive coal beds were formed.

The facts which seem to support the coastal plain theory of accumulation are that the rocks with which coal beds are associated were deposited on the borders of land areas. They show alternation between marine and terrestrial conditions. The large areas of some coal fields seem to postulate conditions similar to coastal plains. The absence of much sediment indicates the absences of waves, currents and rivers. The succession of coal beds is more easily explained by the assumption of an oscillating coastal plain where fresh and brackish water conditions of deposition were followed by intervals of marine deposition. In order to account for the thickness of vegetation required to form the coal beds it seems necessary to assume that the accumulation took place upon a gradually sinking area and to account for their extent to assume that an area of large extent was involved in the movement.

Coalification Processes. The changing of vegetable matter into coal involves processes some of which are biological, some chemical and some dynamical. The first stage called putrefaction is probably largely bio-chemical. The vegetable matter is macerated and the fibers broken down into a jelly-like mass by the action of organisms among which bacteria played an important part. The chemical reactions involved the breaking up of cellulose into marsh gas, carbon dioxide, carbon monoxide, water and other compounds. The second stage involves metamorphic changes of a dynamo-chemical origin. These changes are the result of heat and pressure produced by superincumbent beds and of crustal movements. It is assumed that fissuring of the superincumbent rocks was necessary in order to permit thorough oxidation.

Mode of Occurrence. Coal occurs in beds which lie in positions governed by the movements of their enclosing rocks. In regions of little crustal disturbance they lie nearly horizontal or with moderate degrees of dip. In mountain areas they are often in steeply inclined beds, in complex folds and often badly faulted. The individual coal bed may be fairly unifrom in thickness or it may vary greatly in short distances. Lens-like beds which are thin in a marginal area and thicken toward a central area are of common occurrence. In Indiana as a rule the coal beds dip toward the southwest at about thirty feet to the mile but there are local areas where greater and even reverse dips occur.

Associated Rocks. The coal of Indiana occurs in beds in sedimentary rocks. The beds of coal are intercalated with beds of shale, sandstone, limestone and clay. The coal usually rests on beds of white or gray clays called fire clay but not always possessing fire-proof qualities. In some places the coal rests on shale or sandstone. The rocks overlying the coal beds are generally shales. These are often black thinly laminated shales splitting readily into sheets of considerable size. The shales contain considerable carbonaceous matter which yields often as much as twenty gallons of oil to the ton. Some

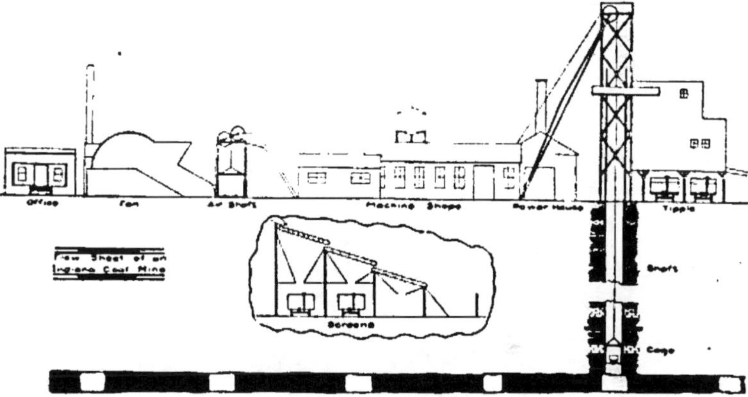

Plate XIX. Flow sheet of an Indiana coal mine.

of the shales contain large spherical concretions of lime carbonate or pyrite. In some places they are interstratified with thin beds of fossiliferous limestones. In places sandstone rests upon the upper surface of the bed of coal and near the outcrop, residual clays or glacial debris may occupy the surface. Coal IV rests in places upon and beneath sandstone. Limestones rarely rest immediately above or beneath the coal but they occur frequently within a few feet of the coal. It is important that the rock immediately underlying the coal should not be subject to creep or it may rise in the entries and cause difficulties in mining. It is important also that the rock immediately above the coal should have proper tensile strength and a certain degree of induration so as to be able to remain unsupported over small areas.

Indiana Coal.

The coal of Indiana belongs to the bituminous division of coals. There are two varieties that have characteristics which distinguish them from the common bituminous type. The first is known as "block" coal. It is cut into block-like masses by vertical joints and splits readily along bedding planes or laminations. The beds are made up of alternate layers of bright and dull coal. The coal splits easily along the dull layer and reveals charcoal-like surfaces which frequently show impressions of plant structure. The coal does not break readily across the laminae and in weathering the blocks split rather than crumble. The coal does not cake in burning and generally leaves a slight gray ash. Another variety is called "Cannel" coal and is thought to have originated from the accumulations of the spores of plants. The appearance of its surface is dull and resinous and physical composition homogeneous. It is slightly laminated and jointed like block coal but breaks with a conchoidal fracture. It contains a high per cent of volatile matter and for this reason is very useful in the manufacture of producer gas.

Plate XX. Tipple of an Indiana coal mine. A mine with modern equipment.

The greater part of the coal of Indiana belongs to the common bituminous type. It is generally bright on unweathered surfaces, breaks in cubical masses and is not strongly laminated. In burning it cakes and forms either a red or a gray ash depending upon the amount of iron which it contains. These bituminous coals contain a high per cent of volatile matter and moisture and a medium amount of ash and sulphur. They do not coke readily when used alone but when mixed with other coking coals have been used successfully. Coal IV is considered the best coking coal. They may be used successfully in the manufacture of producer gas. There are about thirty-five coal beds in Indiana, eight of these are workable over large areas and one over practically all of the coal area it occupies, other beds are workable locally.

Indiana Coal Beds.

The coal beds of Indiana occur in the rocks of Pennsylvania age, a few beds in the Pottsville division and many more in the Allegheny division and perhaps a few to a higher division (Conemaugh). The coals belonging to the Pottsville include the Cannelton coal, the Kirksville, the Shoals, the Lower Block, the Upper Block, the Minshall and Coal II. An unconformity exists in places, at least, between Coal II and Coal III. In places the unconformity occurs above the Minshall and at other places above Coal II.

Cannelton Coal. The type locality of this coal is at Cannelton in Perry County. It occurs near the base of the Pottsville and was deposited in basins and as a result thins from a central point of maximum thickness toward the marginal areas. The maximum thickness of the coal is four feet and the average probably not over three. In places it forms a solid bed, in others it is split by a clay parting. The underlying formation is shale, clay, bone coal or cannel coal. The overlying formation is shale or sandstone.

Plate XXI. A coal mine which produced 128 cars (6,128 tons) of coal in eight hours. The American mine at Bicknell, Knox County, Indiana.

Shoals Coal. This coal occurs near the base of the Pottsville in the vicinity of Shoals in Martin County. West of Shoals it occurs in a shale beneath the Mansfield sandstone. It was deposited at this point in a depression formed by the erosion of the Mississippi rocks. The unconformity between the Pennsylvania and the Mississippian is marked as a few rods east of the coal the Mississippian strata are at a higher elevation than the coal. The coal is from one to three feet thick and often very impure. It may be of the same age as the Cannelton coal or somewhat younger.

Kirksville Coal. The Kirksville coal occurs in the southwestern part of Monroe County near Kirksville and in the eastern part of Greene County. It outcrops on the Koontz, Coleman, Wampler and other places in Indian Creek Township and has been mined in this locality in a limited way. The bed is lenslike and has a thickness varying from two to three feet. It is usually solid but in some places has a clay parting dividing it into two beds of nearly equal thickness. It is a firm hard coal of a semi-block character and not smutty. It lies upon clay or shale and has a sandstone or shale roof. The chemical composition of the coal is as follows:

Fixed carbon	32.96
Volatile matter	42.74
Sulphur	2.76
Ash	4.30
British Thermal Units	14,599.70

A coal of similar stratigraphical position occurs on the State Farm at Putnamville. It is from a foot to eighteen inches thick, has a shale roof and a fire clay bottom. It occurs below the Mansfield sandstone a distance of about forty feet.

Lower Block Coal. This coal is typically developed in the northern part of Clay County and has been mined extensively in the Brazil district where it has an average thickness of three feet. It reaches a maximum thickness of five feet in this district. The upper portion of the bed consists of from six to ten inches, of bituminous coal through which the joint crevices of the block coal below do not extend with any degree of regularity. Below the thickest portion of the coal there is often a layer or two of bone coal or soft coal. These are generally separated from the block coal by clay. The typical Lower Block coal has been recognized from southern Parke County to Greene County. A coal bed in Martin County at Sampson Hill and elsewhere occupies the same stratigraphical horizon. Coals in Perry, Spencer and Dubois counties may be equivalent to the Lower Block coal.

The composition of several samples of the Lower Block coal is given below:

	No. 1	No. 2	No. 3	No. 4	No. 5	No. 6	No. 7
Fixed Carbon	49.16	49.96	50.42	48.23	46.08	46.05	38.87
Volatile matter	36.11	35.16	36.32	36.34	32.56	33.19	26.85
Moisture	11.20	13.82	9.80	11.26	15.39	15.91	16.91
Ash	3.53	1.06	3.46	4.16	5.88	4.85	7.37
Sulphur	.62	1.47	.34	.56	1.95	1.22	1.89
Calories	6,774	6,888	7,050	6,858	6,489	.....	5,291
B. T. U.	......	......	......	......	11,680	.....	9,524

No. 1 from Cardonia near Gart No. 5, Brazil Block Coal Co. *No. 2* Brazil, Mine No. 1, B. B. C. Co. *No. 3* Carbon, near Eureka, B. B. C. Co. No. 1. *No. 4* Asherville, near Crawford, Block Coal Co. No. 3. *Nos. 5, 6, 7* southwest of Perth. Analyst of first four, W. A. Noyes; last three, F. M. Stanton.

Upper Block Coal. This coal is typically developed in Clay and southern Parke counties. Coals of equivalent age may occur north in Fountain County and south in Martin, Daviess, Dubois and Spencer counties. The average thickness of the bed is about three feet but thicknesses of five feet are recorded. The upper Block coal was deposited in basins which range in area from a few acres to several square miles. It is generally continuous between the basins but very thin. The joint crevices between the blocks is more open at the top than at the bottom of the bed which is the reverse in the Lower Block bed. The Upper Block is characterized by the presence of a layer of hard brittle coal about two inches thick a little below the middle of the bed. Clay generally underlies the coal and the overlying rock may be shale or sandstone. The Upper Block is separated from the Lower Block where both are present by about thirty feet of clay, shales and sandstones though the interval is sometimes much less. The analyses of a sample of Upper Block from Woodside, Owen County, is reported by W. A. Noyes as follows:

Analysis of Upper Block Coal.

Fixed carbon	47.40
Volatile matter	36.45
Moisture	12.73
Ash	3.42
Sulphur	.55
Calories	6,636

Minshall Coal. This coal is typically developed around Minshall, Mecca and Sand Creek in Parke County. The coal occurs in basins and the average thickness in the basin is about four feet and the maximum thickness about six feet. Probably the best areas are to be found in Warren, Parke and Fountain counties. The coal usually has fire clay underlying it and shale above it. The overlying shale contains a fossiliferous, thin-bedded limestone which lies only a few feet above the coal or in some places rests on the coal. Usually from twenty to thirty feet of clay, shales and sandstone separate the Minshall from the Upper Block. The coal is a firm coal with a shiny surface, not smutty and a semi-block. The analysis of a sample of the Minshall is reported by W. A. Noyes, from Gifford No. 1, Williamstown, as follows:

Analysis of Minshall Coal.

Fixed carbon	39.48
Volatile matter	37.67
Moisture	13.12
Ash	9.75
Sulphur	2.95
Calories	6,107

Coal II. This coal occurs above the Minshall coal and is called Upper Minshall. The interval between the coals is generally from ten to twenty feet but in places the fire clay under Coal II rests upon the Minshall. Coal II varies greatly in thickness and workable thicknesses are rare in many localities. In Warren County it varies from two to three and one-half feet. In many places it is separated into layers by partings. In some places Coal II has been cut out by erosion preceding the deposition of the Allegheny rocks. Throughout the stratigraphical horizon of Coal II there are local beds of coal which may be the equivalent of the bed above the Minshall.

Coal III. This coal has an extensive development between Coxville in Parke County and Linton in Greene County. It is typically developed in the northeastern part of Vigo County. Over large areas Coal III will average six feet in thickness and in some localities maximum thickness of eleven feet are reported. Pyrite and clay partings are common in the coal. These partings average five or six inches but are in places over a foot in thickness. Horses of pyrite also occur in the coal in places. Ashley has suggested that the basin in which Coal III was laid down in the northern part of Indiana area has an east and west direction. South of the Greene County line the coal which occupies the apparent stratigraphical position of Coal III is much thinner, ranging from two feet to three and one-half feet. The composition of Coal III is given in the following table. W. A. Noyes and F. M. Stanton, analysts.

Analyses of Samples of Coal III.

	No. 1	No. 2	No. 3	No. 4	No. 5	No. 6
Fixed carbon	46.45	39.35	40.80	41.35	41.85	40.49
Volatile matter	41.88	39.49	38.62	38.62	37.76	36.09
Moisture	6.49	11.54	12.26	10.45	9.22	10.45
Ash	5.18	9.62	8.32	9.58	11.17	12.62
Sulphur	2.93	4.41	4.71	4.04	3.94	4.39
Calories	6,897	6,475	..	6,525	.....	6,214
B. T. U.	.....	11,655		11,745	.....	11,185

No. 1 Coxville, Cox No. 3, B. B. C. Co. *Nos. 2* and *3* Rosedale, Parke County Coal Co. *Nos. 4, 5* and *6* Hymera, No. 4, Consolidated Indiana Coal Co. See 33d Ann. Rept. Ind. Geol. Surv. Coal III usually has a red ash due to the presence of iron pyrite in the coal.

The interval between Coal II and Coal III is about seventy-five or eighty feet, usually.

Coal IIIa. This coal lies about twenty feet above Coal III but at Hillsdale only a bed of clay separated the two coals. It also lies close to Coal III in the Brazil districts. Over large areas Coal IIIa is only 18 inches thick. It is workable in only local areas.

Coal IV. This coal is typically developed in the neighborhood of Linton in Greene County. The average thickness of the bed is about five feet with a maximum of seven. It is often split into two beds by a parting and even when solid usually has a smooth parting near the center. It is of workable thickness in Parke, Vermillion, Vigo, Sullivan, Greene, Daviess, Knox, Pike and Warrick counties. The roof over IV is usually shale or sandstone and the floor is sandstone or sandy clay. The interval between Coal IV and Coal III is normally about seventy-five feet, though much less in the northern part of the coal area. The composition of Coal IV is given in the following analyses made by F. M. Stanton and J. W. Graves.

Analyses of Samples of Coal IV.

	No. 1	No. 2	No. 3	No. 4	No. 5	No. 6
Fixed carbon	45.38	46.35	46.20	44.45	47.01	46.23
Volatile matter	35.54	32.57	32.07	35.94	33.04	33.48
Moisture	13.53	13.98	13.58	13.70	14.23	12.15
Ash	7.55	7.10	8.15	5.91	5.72	8.14
Sulphur	.95	.96	.nl	2.66	.89	1.41
Calories	6,521	....	6,344	6,628	6,512	6,534
B. T. U.	11,738	......	11,419	11,930	11,722	11,761

Nos. 1, 2 and 3 from N. W. of Linton, Greene Co. No. 4, Diamond, N. W. of Mine No. 9, B. B. C. Co., Parke County. Nos. 5 and 6, Dugger No. 4 mine, Island Coal Co., Sullivan County.

Coal IV usually has a gray or white ash as the per cent of pyrite is low and there is not enough iron to color the ash.

Coal IVa. A bed of coal occurs typically about twenty feet above Coal IV and is designated Coal IVa. It is not present at all places where Coal IV is present and seems to be present in a few places where Coal IV is absent. It is workable locally over small areas.

Coal V. This is the most important bed of coal in Indiana. Its importance is due to its extensive distribution, it extends almost continuously from the Ohio River into Vermillion County. In only a few places where it should be found normally is it absent. Its uniformly workable thickness is another factor which contributes to its importance. It has an average thickness approaching five feet, and over large areas its thickness will run from six to eight feet while locally it has thicknesses of eleven feet. The quality of the coal varies but taken as a whole it is a good steam coal and in many localities furnishes a good coal for domestic use. In the majority of places Coal V is overlain by a sheety black shale which contains concretions of pyrite which are often spherical and of large size. They sometimes project downward into the coal. The shale also contains the impressions of fish fins and marine shell fish. In some places the shale is replaced by sandstone and in some places a lens of gray shale lies between the coal and the black shale. A bed of fire clay usually underlies the coal but in a few places it rests on shale or sandstone. At a distance of from three to ten feet above there is a bed of limestone varying in thickness from four to six feet. This limestone contains numerous gastropod casts and shells. The coal contains in many places thin partings of pyrite which are irregularly distributed even in local areas. The interval between Coal IV and Coal V is normally 100 to 125 feet.

The composition of Coal V is shown in the analyses of the samples given below. Analyses made by W. A. Noyes, F. M. Stanton and E. E. Somermeir.'

Analyses of Samples of Coal V.

	No. 1	No. 2	No. 3	No. 4	No. 5	No 6
Fixed carbon	49.16	41.64	44.42	46.27	42.17	43.53
Volatile matter	37.99	36.31	32.48	34.71	42.60	40.64
Moisture	6.50	10.30	12.08	12.88	6.49	7.06
Ash	6.35	11.75	11.02	6.14	8.74	8.77
Sulphur	1.85	4.23	3.65	1.70	3.18	3.64
Calories	6,981	6,232	6.117	6,556	7.002	6,811
B. T. U.	.	11,218	11,011	11,801	......	12,260

No. 1 Cabel and Kauffman, Mine No. 9, three miles southwest of Washington, Daviess County. No. 2 West of Linton, Greene County. No. 3 Bicknell, Knox County. No. 4 Ayrshire, Pike County. No. 5 Alum Cave, Phoenix No. 1, New Pittsburg Coal and Coke Co., Sullivan County. No. 6 Boonville, Big Four Mine, Warrick County.

Coal Va. In some places two thin beds of coal occur above Coal V, one about ten feet above and the other at about thirty-five feet. The upper is generally the thicker and is designated as Coal Va. Locally it is of workable thickness. It is almost as persistent as Coal V.

Coal VI. The interval between Coal V and Coal VI is normally about seventy-five feet. Coal VI is typically developed in Sullivan County to a fairly uniform thickness of from six to eight feet. Usually about one foot of the bottom of the coal is bone. The coal is characterized by the presence near the central portion of the vein of two thin shale partings about five inches apart and of about one-half inch in thickness. Coal VI usually rests on clay or shale and has a shale roof. North of Sullivan County the bed of coal becomes variable occasionally splitting or pinching out and probably disappearing altogether north of the central part of Vigo County. South of Sullivan County in Knox it is also irregular. In Gibson County it has a thickness of four feet and is workable at Francisco and elsewhere. In Warrick County it apparently is thinner and comes near enough Coal VII so that the two may be mined together in some places. The composition of Coal VI may be judged from the following analyses made by R. E. Lyons, W. A. Noyes and F. M. Stanton.

Analyses of Samples of Coal VI.

	No. 1	No. 2	No. 3	No. 4	No. 5	No. 6
Fixed carbon	52.90	48.54	46.14	42.29	42.08	40.91
Volatile matter	32.60	35.22	31.65	29.40	34.80	34.80
Moisture	10.04	7.61	13.99	12.17	12.82	13.53
Ash	4.46	8.63	14.32	9.16	10.30	10.76
Sulphur	1.39	1.67	2.31	4.66	3.27	3.15
Calories	6,844	6,489	6,291	5,732	6,117	6,082
B. T. U.	......	......	11,324	10,318	11,119	10,948

No. 1 Fort Branch Coal Co., Fort Branch, Gibson County. No. 2 Bicknell Coal Co., Knox County. Nos. 3 and 4, Star City Mine 29, Consolidated Indiana Coal Co. Nos. 5 and 6 West Terre Haute, Fauvre Coal Co., Vigo County.

Coal VII. This coal bed lies normally about forty-five feet above Coal VI. It has its best development in Sullivan County where it ranges in thickness from three to six feet. The roof of the coal is shale or sandstone. Where it is formed by sandstone it is rolling and irregular causing the coal to vary much in thickness. A bed of fire clay underlies the coal and in places rests on limestone which is nearly always present a short interval below the coal. The coal is soft in places, in others firm and generally free from sulphur. It has been mined extensively west of Terre Haute. It occurs also in Vermillion County. It is present in Knox County south of Bicknell. The coal bed thins toward the south and the interval between it and Coal VI diminishes until the two coals are thought to form a single bed in the southern part of Gibson County.

Coal VIII. About seventy feet above Coal VII in Sullivan County near Merom there is a bed of coal which Ashley designates as Coal VIII and which he says is not known to be workable elsewhere. In the southwestern part of Indiana a coal bed occurs at the same stratigraphical level as Coal VIII.

Coal IX. This coal occurs about fifty feet above Coal VIII in the southwestern part of Indiana. This coal outcrops in the hills about Somerville in Gibson County. In the southwestern part of Indiana there are three thin coal beds lying about Coal IX, all of them have been worked locally. The first one, Parker Coal, lies 150 feet above Coal IX; the second, Friendsville Coal, about forty-five feet higher, and the third, Aldrich Coal, about thirty-six feet higher.

Distribution of Coal in Indiana.

By consulting the map accompanying this chapter it will be seen that coal occurs in twenty-six counties of the State. The following counties lie wholly within the area of coal bearing rocks: Parke, Vermillion, Vigo, Clay, Sullivan, Greene, Knox, Daviess, Martin, Gibson, Pike, Dubois, Posey, Vanderburgh, Warrick and Spencer. The coal bearing rock occupy parts of the following counties: Warren, Fountain, Montgomery, Putnam, Owen, Monroe, Lawrence. Orange, Crawford and Perry.

In Parke County coal occurs at the horizon of Lower Block. Upper Block and Minshall are both well developed, each reaching workable thicknesses. Coal III and Coal IV occur west of the lower part of Raccoon Creek and dip toward the southwest where in the southwestern part of the county they dip below Coal V which occurs only in this part of the county.

Vermillion County contains most of the workable beds of coal. Minshall coal and Coal II outcrop on Big Vermillion River. Coal III extends from Hillsdale toward the southwest passing under Coal IV which extends as far north as Little Vermillion River. Coal V occurs in the south and southwest portions of the county. Coal VI is not represented by workable deposits. Coal VII occurs only in the southwest corner.

Vigo County contains nearly all the workable coal beds. Well records show the presence of several coal beds below Coal III, these probably represent Coal II, Minshall and the Block coals. Lower Block and Minshall are worked near Fontanet. Coal III underlies the county except near Foleyville. Coal IV is absent only in the northeast corner. Coal V occurs west of the river to Durkus Ferry and west of a line drawn from that point to the southeast corner of the county. Coal VI does not occur in workable quantities if at all. Coal VII is found only in the southwestern part of the county extending farthest east along the Sullivan county line.

Clay County contains all the coals including and below Coal V. The Block coals reach the surface in the northern and eastern parts. Minshall coal near Brazil and southwest apparently reaching the surface in some unexpected places. Coal III is mined along the western border of the county. An outlier of it occurs at Middlebury near Clay City on the east side of Eel river valley. The remainder of it lies on the west side of the river. Coal IV is found along the western border of the county in basins. Coal V outcrops in the southwest corner.

Sullivan County probably contains all the coals to Coal IX. Deep well records show the presence of coals which probably correspond to most of the lower coals. Coal IV is the lowest coal outcropping in the county. It outcrops along the eastern edge of the county. Coal VI outcrops about one-fourth to one-half of a mile west of the outcrop of Coal V. Coal VII occurs almost to the eastern border of the county along the south line but lies west of Busseron Creek farther north. Coal VIII has a workable thickness near Merom in the western part of the county.

Greene County contains nearly all of the coals below Coal VIII. The Minshall coal seems to be absent. Coals which correspond in stratigraphical position with the Block coals occur along the western bluffs of White River. Coal II is thought to occur to the west of the river bluffs. Coal III and IV underlie the western tier of townships, and Coal V underlies the west half of the tier. Coal VI and Coal VII occur along the western edge.

Knox County has all of the coal beds of large workable extent outcropping within the county or underlying it. Coal V is absent only in the northeast part of the county. Coal VI outcrops in an irregular line a few miles west of the West Fork of White River and underlies the remainder of the county. Coal VII outcrops a short distance west of Coal VI and dipping below the surface rapidly it occurs at a depth of 382 feet at Vincennes, or fifty-three feet above sea level.

Daviess County contains representatives of nearly all the coals from Upper Block to Coal V, inclusive. A coal at the horizon of Upper Block outcrops along the eastern border of the county. A coal at about the horizon of the Minshall coal outcrops about two miles west of the eastern boundary of the county. A coal that is thought to be Coal III outcrops four or five miles west of the east boundary. Coal IV outcrops a short distance west of the outcrop of Coal III. These two coals underlie the central part of the county at shallow depths. Coal V outcrops in the western third of the county and Coal VII at the western edge.

Martin County, only the lower coals are represented. Coal I (Shoals Coal) and representatives of the Block coals occur at outcrops in the hills along a north and south line drawn through Shoals. Sampson Hill and Dover Hill are mining localities. Coal I is only a short distance above the river west of Shoals.

Gibson County contains accessible beds of the higher coals. Coal V may be reached by shafts in the eastern part of the county. Coal VII outcrops in the hills in that region and Coals V, VI and VII are all above sea level at Princeton. The Parker, Friendsville and Aldrich coals are present in the hills of the western part of the county.

Pike County contains outcrops of Coals III to VII, inclusive. Coal III outcrops in the northeastern and southeastern corners. Coal IV in the eastern part of the county. Coal V outcrops in the southeastern part of the county and underlies the central and western parts. Coal VII occupies the upland in the northwest part of the county.

Dubois County. Coal I (Cannelton Coal) occurs in the tops of the ridges in the eastern part of the county. Two or three beds of coal lie above Coal I and below Coal III which outcrops along the western boundary of the county. Minshall and Coal II are probably represented by beds which occupy those horizons.

Posey County contains Coals V and VII at considerable depth, perhaps throughout the entire county. Coal VIII is probably 600 feet below the surface at Mt. Vernon. The higher coals outcrop in the uplands in the northwestern part of the county.

Vanderburgh County. A coal at about the horizon of Upper Block, Coal V and Coal VII are apparently all the coals of workable thickness in this county. Coal V lies 125 feet above sea level and at a depth of 260 feet below the surface around Evansville.

Warrick County. At Boonville four coals lie below the surface. These are probably Coals III, IV, IVa and V. Coals III and IV outcrop in the eastern part of the county and Coal V near the central part just east of Boonville. Coals VI and VII are very close together in the western part of the county.

Spencer County. Coal I occurs in the eastern part of the county. Coals III and IV outcrop in the southwest part of the county and are found at shallow depths elsewhere. Coal V outcrops in the tops of some of the hills in the southwest part of the county.

In the following counties small areas of the lower coals are found. They occur in workable thickness in many places.

Warren County. Coal beds occur in the western two tiers of townships chiefly west of Pine Creek. The Minshall Coal is present in this area and attains a thickness of four feet. A few feet above the Minshall is Coal II which attains a thickness of about three feet. Other coals may be present under the glacial drift in the western part of the county.

Fountain County. The coals represented in this county are Upper Block, Minshall and Coal II. The coals lie in limited basins and the outcrop is concealed largely by glacial drift. The Upper Block extends farther to the east and probably underlies the largest area though it must be deeply buried in the Western part where some of the higher coals may come in under the glacial drift.

Montgomery County. A thin bed of coal lying under a sandstone is found in the western part of Montgomery County. It is rarely over a foot in thickness.

Putnam County. A thin bed of coal lies under the Mansfield sandstone in this county. An outcrop occurs in some shales on the State Farm. The coal has a thickness of one and one-half to two feet and is probably at about the horizon of the Kirksville Coal. Lower Block Coal occurs in some of the higher hills of the west part of the county.

Owen County. Coal I occurs about the central part of the county and the Block coals in the southwestern portion of the county lie near the surface and the coal is mined from strip pits.

Monroe County. Coal occurs in the southwest part of the county. This coal is probably Coal I though it may be at the horizon of Lower Block. The coal lies in a basin and has a parting at one point which consists of about one foot of clay. It divides the bed into two layers about sixteen inches thick. In other places the coal is solid and about three feet thick.

Lawrence County. Coal I outcrops in the hills in the southwest part of the county.

Orange County. Coal I and possibly a higher coal in the hills of the western part of the county.

Crawford County. In the western part of the county and the southwestern corner the Cannelton Coal occurs.

Perry County. The lower coals are present, the Cannelton coals (Coal I) is the principal one but in the west part of the county there are one or two thin coals above.

CHAPTER IV.

CLAYS.

Definition. Clay is a soft rock composed largely of minerals belonging to the kaolin group. When mixed with the proper proportion of water it may be easily molded into forms which have the property of retaining their shape. This property, called plasticity, is not possessed in a high degree by other rocks, and is therefore one of the determinative characters of clay. Clay is a mechanical mixture of minerals. The proportion of these mineral constituents may vary; hence the composition of clays varies greatly. Aluminous clays are those containing a large proportion of the mineral kaolinite, which is the basis of all clays. Arenaceous clays contain a large quantity of sand. Calcareous clays contain much carbonate of lime or gypsum. Ferruginous clays contain a considerable proportion of some iron compound.

Clay Group of Rocks. The rocks which contain clay in such abundance as to be classed in the clay or argillaceous group of rocks are: kaolin, clay, shale, loam, loess, till, adobe and soil. The term kaolin is applied to beds of pure clays, largely of residual origin. Common clay is a mixture of kaolin and sand and other impurities. Shale is a consolidated clay which after deposition was compressed and indurated. Loam is a mixture of clay and sand. Till is a clay which has been deposited by glaciers. Adobe is a silty clay largely wind-deposited. Soil is usually a mixture of clay and other rock particles with organic matter.

Origin of Clay. The decomposition and the alteration of rocks containing silicates of aluminium is the source of clay. The group of silicates known as feldspars constitutes the most fruitful primary source of clay. Feldspar is one of the principal constituents of granite and other igneous or metamorphic rocks of the granitoid group. For this reason the primary formation of residual deposits of clay is closely associated with the disintegration of granite and the subsequent alteration of its silicate minerals.

The disintegration of granite and the decomposition of its minerals is accomplished by the various mechanical and chemical agents which are actively engaged in rock weathering. The alteration of the silicates is accomplished by the action of mineral and vegetable acids carried through the pores of the rock by circulating waters.

One of the most destructive of these acids is carbonic acid (H_2CO_3). This acid first attacks the potash and soda, hence silicates containing these bases are the first to be broken up. Lime and magnesia compounds are next attacked, then the silicates containing iron, and lastly the aluminium silicates, the most stable of the compounds. These complex compounds having been broken up into their component elements, reactions between the elements occur and new compounds are formed. Aluminium uniting with silicic acid forms new silicates which are free from the other bases which, since they are more readily soluble, are carried away by circulating waters.

The aluminium silicates thus formed are kaolinite (chiefly), cimolite, halloysite and others; also some oxides and hydroxides of aluminium such as bauxite and gibbsite. These aluminous minerals form beds of rock called kaolin. Kaolin is the basis of all clays. The purity of a clay and often its usefulness depends upon the percentage of Kaolin which it contains.

Clays of secondary origin may result from the decomposition of a large number of rocks containing clay, such as shale, slate, argillaceous sandstone, limestone and others.

Chemical Composition of Clay.

The purest form of clay contains the following chemical elements: Aluminium, hydrogen, oxygen and silicon. Clays usually contain impurities and so the chemical elements composing the minerals commonly present in clay are aluminium, calcium, carbon, hydrogen, iron, magnesium, oxygen, potassium, silicon, sodium, sulphur and titanium. Carbon and sulphur may occur as simple elementary substances uncombined. The other elements are combined to form such compounds as water, quartz and others. In the chemical determination of these elements they are represented as combined with oxygen to form oxides.

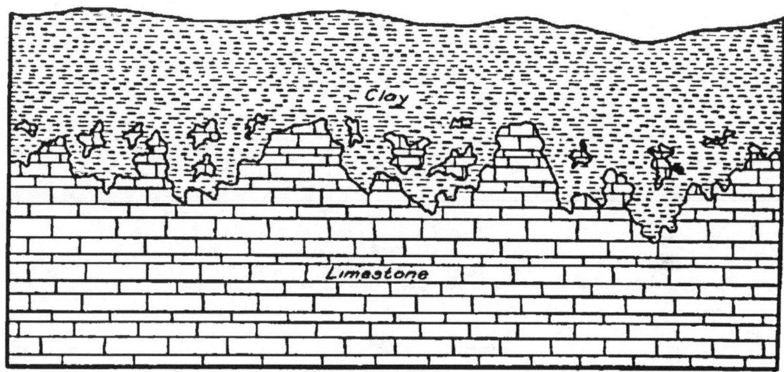

Plate XXII. Diagrammatic section to illustrate the formation of residual clay from limestone by solution.

Chemical Components of Clay.

Name of Component	Chemical Symbol
Silica	SiO_2
Alumina	Al_2O_3
Ferric oxide	Fe_2O_3
Lime	CaO
Magnesia	MgO
Potash	K_2O
Soda	Na_2O
Titanic acid	TiO_2
Sulphur trioxide	SO_3
Carbon dioxide	CO_2
Water	H_2O

Iron oxide, lime, magnesia, potash and soda are classed as fluxing impurities because they lower the fusion point of the clay. The lime is usually combined with carbon dioxide (CO_2) to form calcite or with water and sulphur trioxide to form gypsum. Other combinations also exist so that an ultimate chemical analysis such as the above does not present, for instance, the amount of calcite or gypsum present in the clay, but merely the amount of water, lime and sulphur trioxide that is present in the clay. The determination of the percentage of the different mineral compounds in the clay is called its rational analysis. The rational analysis may be computed from the ultimate analysis and is useful in making clay mixtures.

The chemical analysis of a complex substance, such as ordinary clay, may reveal the presence of a large number of chemical elements but only a limited number of such elements is present in such quantities as to have any

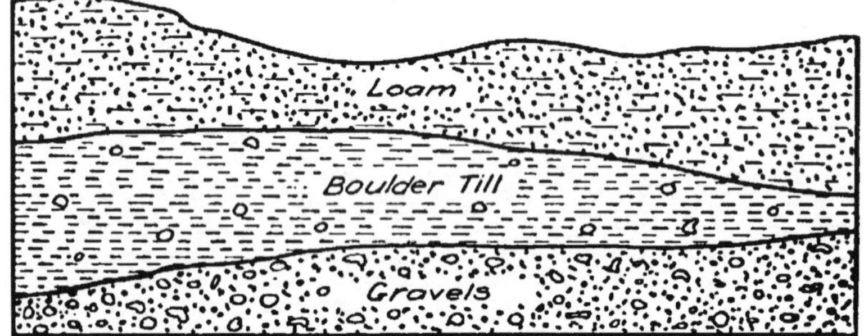

Plate XXIII. Diagram to show the occurrence of glacial clay (till) in Pleistocene deposits.

influence on the physical properties of the clay. The following table presents the ultimate and rational analyses of some of the clays of Indiana. They were selected to show the variation in the composition of the clays.

Analyses of Indiana Clays.

Ultimate Analyses.

Constituent	No. 1	No. 2	No. 3	No. 4	No. 5
Moisture (H_2O)	20	20.04	......	......	1.95
Volatile matter (CO_2)	......	......	8.15	6.62	.....
Silicon dioxide (SiO_2)	40	40 48	56.67	64.41	52.65
Alkalies (Na_2O, K_2O)	......	......	3.12	5.00	3.09
Aluminium oxide (A_2O_3)	40	39.60	21.35	18.23	13.21
Titanic oxide	......	......	1.00	.66	1.13
Iron oxide (Fe_2O_3)	......	.11	7.67	5.88	4.67
Calcium oxide (CaO)	......	......	......	1.77	8.50
Magnesium oxide (MgO)	......	......	.71	1.81	4.95
Carbon dioxide (CO_2)	......	......	......	......	8.19
Total	100	100.23	99.49	100 26	98.34

Rational Analyses.

	No. 1	No. 2	No. 3	No. 4	No. 5
Clay substances	100	100.18	54.01	46.12	33.42
Free silica	......	......	31.57	42.98	37.02
Fluxing impurities	......	.11	12.50	15.12	17.67

No. 1 Allophane. No. 2 Mine run white halloysite. No. 3 Chester shales. No. 4 Knobstone shale. No. 5 Average of four glacial clays.

The clay substance is computed as kaolinite by taking from the ultimate analyses the necessary amount of alumina and silica to form it. The remaining silica is called free silica and is supposed to exist as quartz sand. The same of the iron, lime, magnesia and alkalies forms the fluxing impurities.

Chemical Compounds in Clay.

The chemical compounds which are usually included in the ultimate analysis of a clay are the compounds which exert the greatest influence upon the physical properties of clay. For this reason such compounds are of sufficient importance to merit an examination of their form of occurrence in clays, their properties and the effects which their presence may produce upon clay wares. Following is a discussion of the more important of these compounds.

Silica. The silica, the percentage of which is expressed in the ultimate analysis, may be divided, in respect to its influence upon the clay, into three parts. The first part is that which is combined with the alumina to form

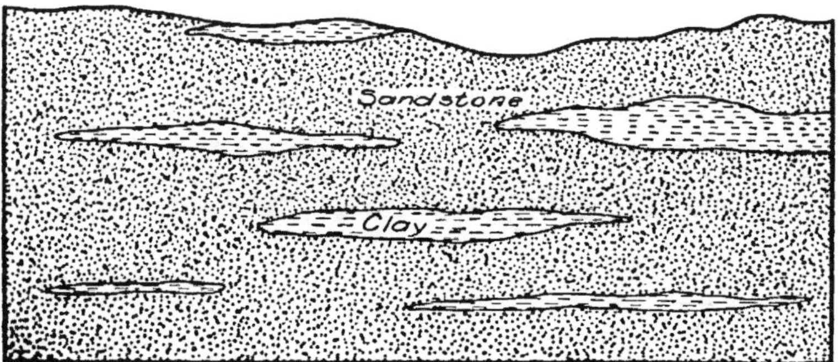

Plate XXIV. Diagram showing the occurrence of lenses of clay in sandstone.

the kaolin group of minerals. The second and smaller part is combined in other silicates such as hornblende, feldspar and mica. The third part is uncombined silica called free silica which may be largely in the form of quartz grains (sand) or to a small extent of colloidal silica. In making a rational analysis of a clay the last two parts are rarely separated. The usual method is to compute the amount of silica combined to form kaolinite. This amount called combined silica is deducted from the total amount of silica as revealed by the ultimate analysis and the remainder is called free silica. This method is not entirely satisfactory from the clayworker's standpoint, since some of the silicates have very different properties from quartz and may exert a very different influence on the clay ware.

The effects produced upon clay by the presence of free silica are to influence its texture, its bonding power, its plasticity, its strength, its fusibility and other physical properties. The collodial silica increases the plasticity of the clay and increases the hardness of dry clay. These effects are discussed under the properties of clays.

Alumina. The alumina revealed by the chemical analysis is derived largely from the kaolin group of minerals in the clay, but a part may be derived from feldspar and other aluminous minerals. The amount of aluminium in Indiana clays reaches its maximum in the halloysite deposits where allophane occurs containing as much as forty per cent of alumina. Alumina is the most refractory substance found in clays and one of the most refractory found in nature. Besides contributing to the refractoriness of the clay it also furnishes the bonding material for holding together the inert particles of quartz and other substances. Without its presence the clay could not be fashioned into the desired form. Since alumina forms the basis of kaolin or clay substance the amount of alumina present has an important influence on the usefulness of a clay. The amount of alumina in a clay is revealed by the ultimate analysis and is partly a guide to the refractoriness and plasticity of the clay. It is not an infallible guide because there are other factors of plasticity and the presence of other substances may cause a low degree of fusibility even with a high per cent of alumina. The percentage of alumina runs lowest in some of the residual surface clays and in some of the sandy shales.

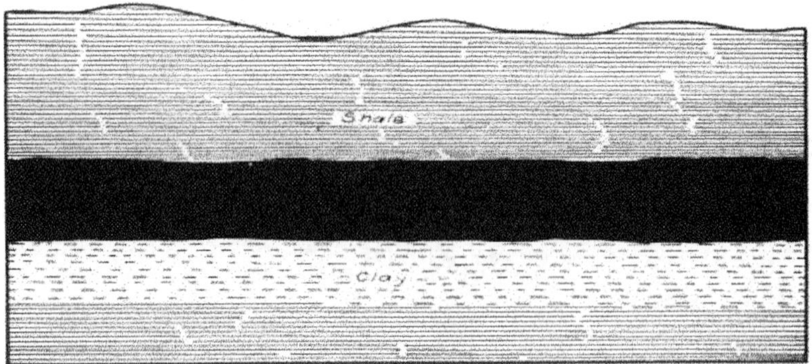

Plate XXV. Diagram to show the occurrence of fire clay (under clay) beneath a bed of coal which has an overburden of shale.

Iron Oxide. The amount of iron oxide varies in different clays. It is generally least in kaolins and highest in common red-burning residual clays. In Indiana clays it is least in the pure white halloysite and highest in some varieties of mahogany clays. The chief source of the iron oxide in clay is from iron ore minerals, but a small amount may be derived from ferro-magnesian minerals such as mica. The iron compounds such as hematite, limonite and siderite may exist either in a finely divided state or as concretions in the clay. Limonite on the application of heat loses its water of crystallization and becomes red oxide of iron. It is to this last compound that the red color of clay ware is due. Siderite, the carbonate of iron, under the influence of heat gives up its carbon dioxide and becomes ferrous oxide. In the presence of oxygen the ferrous iron may be changed to the ferric oxide, the red oxide.

The sulphide of iron which is a common impurity in clays may be reduced to the ferric oxide under the action of heat. Iron is also a fluxing ingredient of clays. When the iron compound is reduced to the ferrous state in the absence of oxygen it will unite with silica forming a ferrous silicate. In the presence of other easily reducible compounds the ferrous silicate may act as a rapid solvent. If there is plenty of oxygen present the ferrous oxide will be further oxidized to the more refractory ferric state. Black or blue colors in clay wares may be caused by ferrous oxide but the red color is due to the presence of ferric oxide.

Calcium Oxide (Lime). The amount of lime in workable clays is generally well under five per cent. The origin of the lime is from limestone (calcium carbonate) and gypsum (calcium sulphate). Small amounts of lime may be derived from lime-bearing silicates some of which are of common occurrence in clays. The effect produced by the presence of lime in clay, will depend on the distribution of the lime and the amount present. Concentration of lime in concretions may produce cracks or flaws in clay wares, by absorbing water and slaking after the wares are burned. In the presence of iron these concretions may fuse and cause cavities or slaggy masses in the wares. The same amount of lime, finely divided and more uniformly distributed through the clay, would have no detrimental effect. However, since lime acts as a flux, its presence in appreciable quantities tends to lower the fusion point of the clay. For this reason vitrifying clays or fire clays should not contain much lime. In the presence of a considerable quantity of iron, the fluxing action of lime may be rapid and effective. With only a small increase of temperature above incipient fusion, the wares may be reduced to a slaggy mass. Lime in considerable quantities in a commonly red-burning clay, may also prevent the development of a red color in the ware. Serious difficulties are sometimes encountered by the clayworkers of Indiana in attempting to use residual clays derived from limestones or glacial clays containing limestone concretions in the manufacture of common brick and drain tile. If the quantity of the lime concretions is not large the difficulties may be overcome by fine grinding and a uniform mix in pugging, but if the quantity is excessive it may be necessary to abandon that part of the clay pit.

Magnesia. The source of magnesia in clay is from magnesian carbonate, from magnesium sulphate and more rarely from silicates containing magnesium. Dolomite or magnesium is the chief source. This mineral is a calcium-magnesium carbonate in which one atom each of calcium and magnesium is united with the carbonate radical. By the decomposition of pyrite in clays, sulphuric acid may be formed. The latter may attack the magnesium carbonate and form magnesium sulphate. The sulphate is soluble in water and if the drainage of the clay bed is perfect, the sulphate will be carried out by circulating waters. If the sulphate is not separated from the clay, it will be brought to the surface of the ware, either in drying or burning and produce efflorescence. The action of magnesia under heat is said to correspond to that of lime, with the exception that at high temperatures the magnesia is not as rapid a fluxing agent as lime. The amount of magnesia in the majority of the clays used by clayworkers of Indiana is small.

Alkalies. The alkalies commonly found in clays are potash (K_2O) and soda (Na_2O). The per cent of alkalies contained in clays is generally small and of little influence in common clays but in clays used for high grade ware and fire clays even small quantities may be very detrimental. Alkalies in clays are commonly derived from a silicate mineral, such as feldspar. The compounds of potassium and sodium formed by the breaking down of these complex compounds are sulphates, carbonates and chlorides. These compounds being soluble, are removed from the clay under perfect drainage conditions. Imperfectly drained clay beds may contain a considerable amount of these soluble alkalies. In well drained clays the alkalies are contained in silicates like feldspars. The alkalies act as powerful fluxes. They fuse at low temperatures; the soluble salts at about red heat; the silicates at higher temperatures. The soda silicates fuse at lower temperatures than the potash silicates. The feldspars are considered an aid to vitrification since they produce a longer period between incipient fusion and complete vitrification. They are detrimental to a high degree of refractoriness.

Titanium Dioxide. The titanium which the ultimate analysis of a clay reveals is derived largely from rutile, TiO_2, and ilmenite, $(TiFe)O_3$. Small quantities of titanium will produce a yellow color in kaolin. In large amounts

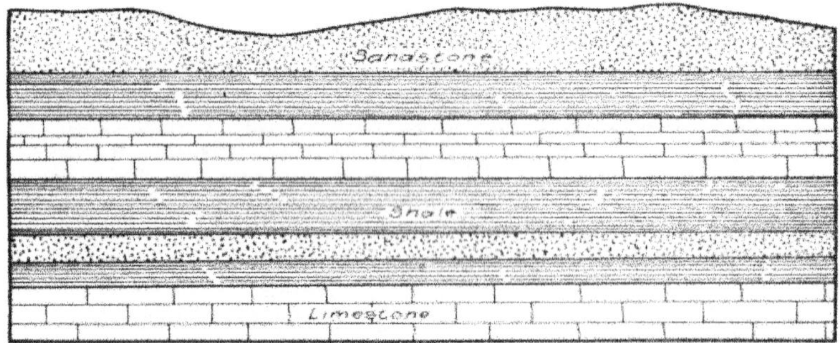

Plate XXVI. A diagram to illustrate the relation of shale to stratified deposits.

it will give a bluish tint to hard bodies. It is thought also, to reduce the refractoriness of a clay. The amount in Indiana clays is generally too small to be detrimental.

Sulphur Trioxide. The sulphur which is present in clay may be derived from a number of sources. It may be from organic matter in the clay, or from gypsum or from sulphide of iron and more rarely from other sources, such as soluble sulphates. When iron pyrites is heated to a red heat the sulphur is set free and uniting with oxygen, it is expelled from the clay in the form of sulphur trioxide. At a high temperature the sulphur trioxide which is combined with water and lime to form gypsum is driven off. The effects of the evolution of this gas is to cause the ware to become porous, if not vitrified, and to cause the ware to swell, if the outer surface of the ware is softened before the gas has all been expelled from the interior of the ware. Care in increase of temperature is required for the burning of clays containing compounds likely to produce sulphur trioxide.

Carbon Dioxide. The source of carbon in clays is from organic matter, calcium and magnesium carbonates. The organic matter is consumed during the burning of the clay and the carbon, being converted into carbon dioxide, is expelled as a gas. The calcium carbonate gives up its carbon dioxide and is converted into quicklime at about 900 degrees C. The effect of the expulsion of this gas is to make the body of the clay more porous, unless the burning is carried to the point of vitrification. Swelling of the ware results if the temperature is not carried above the 900 degree point very gradually.

Water. The water which exists in air dried clays is either mechanically combined or chemically combined. The mechanically combined water is converted into water vapor and expelled when the clay is heated at a temperature of 100°C. This water must be expelled before the clay ware is subjected to high temperatures. The sudden or rapid conversion of the water into water vapor will cause bursting of the clay wares. After expelling the water, the clay wares should be kept at a temperature of 100°C or in a dry atmosphere until burned, to prevent reabsorption of water. The amount of mechanically combined water is given in the ultimate analysis under the head of moisture. A part of the chemically combined water in clay is combined with alumina and silica to form kaolinite. But there are other minerals containing water of crystallization, which are of common occurrence in clays. Such, for example,

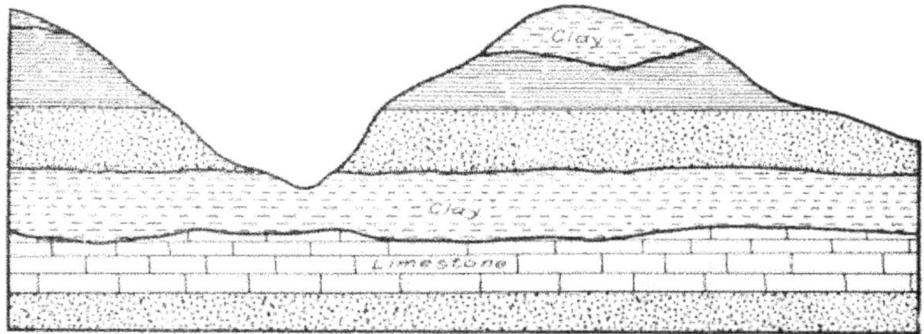

Plate XXVII. A diagram to illustrate the occurrence of clay on a hill and its outcrop in a valley.

as gypsum and limonite. The chemically combined water is usually driven off at a temperature ranging from 400°C to 600°C. The chemically combined water is included under the head of volatile matter in the chemical analysis.

Soluble Salts. Nearly all clays contain some soluble salts and the amount is sometimes great enough to cause serious imperfections in the ware or in the glaze. Sulphates of calcium, magnesium, potassium, sodium and aluminium are often present in clays and may be brought to the surface of the ware in drying. The calcium sulphate is often developed by oxidation of iron pyrites in clay producing sulphuric acid which in turn attacks the calcium carbonate and forms calcium sulphate. Chemical reactions between the fuel gases and compounds clay may also result in the formation of soluble salts.

Physical Properties of Clay.

The physical properties of clay which, from the clayworker's standpoint, are most valuable are plasticity, strength and refractoriness. Plasticity enables the worker to fashion the clay into the desired form. The strength of the clay permits the clay ware to be handled during the drying and burning processes without danger of breakage. The power of the clay to withstand high temperatures permits it to be burned to a compact, hard body of permanent form. While the properties mentioned above are, for the majority of wares, the most important physical properties, there are other properties of very great importance in the manufacture of some wares. The succeeding pages contain brief discussions of the following physical properties of clays: Color, odor, feel, taste, hardness, porosity, structure, shrinkage, slaking, plasticity, fusibility, specific gravity, fineness of grain, bonding power and tensile strength.

Color. The color of clays is an exceedingly variable property. Many shades and tints are represented. The color may be due to presence of organic matter or to the presence of minerals of iron, manganese and others. Yellow, buff, red and brown colors are generally due to the presence of iron; pink and purple to manganese compounds. Blue and slate colors may be due to the presence of carbonate of iron or organic matter and black to the latter. White clays are devoid of perceptible coloring matter, but some white clays have color developed by burning.

By the color of the raw clay it is not possible to predict the color of the burned product unless the nature of the coloring matter and its amount are known. Some white clays contain enough iron to produce a dark shade when burned in an oxidizing flame. Titanium may produce a purple tint when burned at a high temperature. Some black clays are found to be very white after burning. The dark coloring matter in many clays is organic matter which when burned out leaves a white product. Some yellow or red clays containing an excess of iron may burn to an iron black.

The color to which a clay will burn often has an important bearing on its value. A clay may be of high value as a stoneware clay because of the presence of coloring matter which would develop dark shades or splotches during burning. The most satisfactory test to determine the color of the burned product is to subject a sample of the clay to the same conditions of temperature to which the proposed ware is to be subjected. The shades of the burned clay are almost as variable as the natural clay.

The oxidation of iron compounds in the clay produces light reds, cherry reds, dark reds, chocolates and iron blacks, the latter being produced by an excessive amount of iron or of ferrous iron due to lack of oxygen. Clays may contain a slight per cent of iron and still be white or yellow in color. Vitrified wares contain iron silicates which may impart a green, brown or black color to the ware. Spots on white, yellow or red wares are often produced by sprinkling the surface of the wet ware with particles of iron or manganese compounds. The oxidation or reduction of these particles produces black, brown or red specks on the ware.

The Indiana kaolin is white or stained yellow, red, purple or mahogany. The fire clays are usually gray or bluish gray. The New Albany shale black, the Knobstone gray or bluish-gray. The residual clays usually yellow or red.

Feel. Clay containing particles of sand is harsh or gritty to the touch. The grit in some clays is easily detected by rubbing the clay between the fingers. In other clays the grit can be detected only by moistening the clay between the teeth. Clays having a large percentage of clay base are smooth to the touch or feel. Kaolin is somewhat like soapstone to the touch and it is a very common practice for people to refer to an unctuous clay or shale as a soap stone. These clays may be shaved with a knife to a perfectly smooth surface, while a clay containing grit will have minute pits on its surface where the blade of the knife has pulled out the sand grains. The moistened surface of the unctuous clay feels smooth like soap. As a rule the gritty clays are least plastic and are called "lean" or "short" clays, while the more unctuous clays are more plastic and are called "fat" clays. Blue clays are commonly referred to as soapstone by well drillers and white clays as chalk.

Odor. The odor which emanates from the moistened surface of clay is distinct and characteristic. It is the odor which rises from a dusty road at the beginning of a thunder shower, or from a dusty floor when it is being sprinkled. A very similar odor is given from the surface of some minerals when they are rubbed and they are said to have an argillaceous odor. Some clays containing decaying organic matter have a fetid odor. Some very silicious clays contain such a small amount of clay substance that the argillaceous odor is not distinct. Therefore, this property cannot be counted a safe guide to the amount of clay substance.

Hardness. The property, by virtue of which one mineral is able to scratch another mineral, is called hardness. Clays are soft rocks. They usually range in the scale of hardness from one to three. The maximum degree is attained in the chalk-like kaolin. This property refers to the ease with which the rock

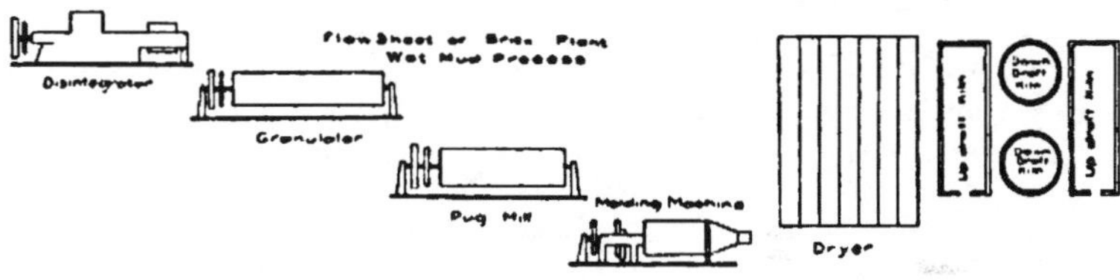

Plate XXVIII.

may be scratched. The individual particles in a clay may be a great deal harder than the rock. For instance, the quartz grains would have a hardness of seven, while the feldspar would have a hardness of six and the kaolinite might vary in hardness from one to three.

Burnt clay has a much higher degree of hardness than raw clay. Vitrified clay products reach a hardness equal to that of quartz, which will scratch glass. Hardness is a property very essential in all clay wares which are to be subjected to abrasion as are paving brick; or to compression as are building brick; or to chemical action and compression as are sewer pipe and silo blocks. Many of the vitrified clay products assume a hardness at which they are untouched by steel.

Taste. The presence of certain soluble salts in clay may be detected by tasting the clay. Common salt, alum, epsom salts and ferrous sulphate are not infrequently detected in this way. Clay prospectors sometimes place clay between the teeth in order to determine its proportions of sandy matter. They may also employ this method to determine the texture and degree of plasticity, a method often referred to as "tasting".

Slaking. The crumbling of a clay under the action of water is termed slaking. When a clay slakes it breaks up into small pigments. Slaking takes place wherever an air-dried clay surface is exposed to the action of water. The size of clay fragments or grains into which the clay mass is separated is fairly uniform for the same clay, but varies greatly in different clays. The shape of the particles is variable. Some are flat, some cubical, others irregular. As the particles of the clay separate they absorb water and increase in size.

The speed of slaking varies in different clays. Clays of marked density, such as shale and flint clays, slake very slowly, while the leaner surface clays slake very rapidly. Wet or puddled clays do not slake as rapidy as air-dried clay to which water is suddenly applied. The speed of slaking may be determined by taking samples of natural clays of equal size, placing them in water and observing the time elapsing until they are completely crumbled.

Clays used for any purpose requiring molding without grinding ought to possess at least a moderate slaking speed. A clay possessing a low slaking speed causes loss of time when tempered either in wet pan or pug mill. Such clays must be pulverized in the disintegrator and the granulator before they can be tempered and molded. Clays that have been weathered slake more rapidly than when first taken from the pit.

One method of determining the speed of slaking is to take a ten-gram cube of the clay as it comes from the pit, drying the cube at 100°C and then submerging it in water. The time required for the cube to completely disintegrate was then noted. In order to make sure that the cone of separated particles did not contain an unslaked core a needle point is thrust into the mass as soon as the slaking movement has apparently ceased. The slaking is generally complete by the time the last air bubbles rise to the surface of the water.

Specific Gravity. The specific gravity of a clay is its weight compared with the weight of an equal volume of distilled water at 60°F. The specific gravity of a solid substance may be obtained by weighing it first in air and then in water and dividing its weight in air by its loss of weight in water. The specific gravity of clays usually varies from 1.50 to 2.50, that is, from one and a half to two and a half times the weight of an equal volume of water, but there are some clays whose specific gravity is lower and others whose specific gravity is higher than these limits. Pure kaolin has a specific gravity of from 2 to 2.5. Pure quartz sand has a specific gravity of from 2.5 to 2.8. Where clay is largely a mixture of varying proportions of these two minerals, its specific gravity is not far from 2.5. Clays containing in addition to these minerals, mica and limonite are slightly heavier. The presence of magnetite, however, may greatly increase the specific gravity, while on the other hand the presence of organic matter may decrease it.

Methods of determining specific gravity of clays by investigators are not uniform and different methods may produce different results with the same clay. By the use of the pycnometer the specific gravity of the individual grains is determined and taken as the specific gravity of lumps of clay which have been coated with paraffin to prevent absorption of water is determined. This method considers the pore space a part of the clay. The specific gravity of any clay is less by the latter method.

Porosity. A porous clay is one which contains considerable space not occupied by clay particles. This unoccupied space is called pore space and its volume depends upon the size and shape of the clay particles. The maximum volume of pore space would be attained in a clay containing spherical grains of equal size, but since the shape and size of clay grains are extremely variable this maximum volume is rarely, if ever, attained. The quartz grains in clay are usually well-rounded, water worn particles. The mica grains are little flat crystals with irregular borders. The feldspar grains are either more or less rounded or irregular. The kaolinite crystals are flat with irregular edges. The grains are in contact only at certain points and thus leaving spaces between the particles. These spaces or pores are in connection with other pores and long chains of such connections form irregular tubes. These tubes are of capillary size and the water within the clay passes through them under the influence of capillary attraction.

Porosity is an important property in clays. The amount of water required for tempering the clay depends in a measure upon its porosity. The air shrinkage of the clay is brought about by the loss of this water. The speed of tempering and the speed of drying both depend upon porosity. The larger the pores the more readily water is taken up and given off.

Structure. The structure of a clay refers to its mode of occurrence in the outcrop or pit. A stratified clay is one which occurs in layers. A massive clay is one in which no division planes are visible. A clay which splits readily into thin leaves or irregular blocks is said to be shaly. If the leaves are small, thin and light, the term "chaffy" is applied to it. If in large thin sheets it is called sheety. A slaty clay is one in which the laminae have undergone a considerable degree of induration. Instead of occurring in layers some clays are found in concretionary or pebbly masses. Joint clays are those which are separated into somewhat regular blocks by vertical crevices. This structure is an aid to the mining of many clays. The various structures in clay are the result of deposition, compression, shrinkage and induration. In the process of weathering they are obliterated, and the rapidity of such weathering action is often dependent upon the structure of the clay. The speed with which a mineral producing soluble salts can be removed by weathering will depend upon the structure of the clay. In order that clay may be used in the formation of clay wares, it must be reduced to a structureless mass. For this purpose it is necessary to employ disintegrating or pulverizing machinery. The expense of this process will be determined by the degree of induration which has taken place in the clay structure.

Shrinkage. The amount which a clay contracts in passing from a plastic condition to that of a rigid solid is termed its shrinkage. The water which is added to the clay in order to render it plastic is lost by evaporation, causing a loss of volume. The loss of volume or shrinkage varies greatly in different clays and with different conditions of the same clay. Water added in excess of the amount required for plasticity will cause a greater loss of volume, as will also the presence of air bodies in the clay. Considerable water may exist in the clay without increasing its volume, but whenever the particles of clay are completely enveloped in water the volume and the plasticity will be increased. Water absorbed by a clay exists either interstitial, that is, in the pores, or interparticle, that is, not occupying the pores but separating the particles. It is the latter which increases the volume of a clay. Clays of coarse grain have large interstices and contain large quantities of interstitial water, but less interparticle water than clays of finer grain, therefore, the fine grained clays shrink more than the coarse grained.

Air Shrinkage. The amount of contraction which a clay undergoes when drying in the air is called its air shrinkage. The amount of air shrinkage depends mainly upon two factors: First, the amount of water absorbed, and second, the size of the grain of the clay.

A number of methods of preventing excessive shrinkage is employed. The method more generally in use is that of mixing a sandy clay with the more plastic clay. Under ordinary conditions this is the most economical method but it cannot be employed for all wares. Pure sand and burned clay are used to decrease shrinkage; the use of the latter may extend to more wares than the former.

Fire Shrinkage. The loss of volume which a clay sustains in passing from the raw to the burned condition is termed its fire shrinkage. In the first stages of burning, the clay loses its chemically combined water and the organic matter in the clay is consumed, causing the porosity of the clay to increase. But when the temperature is raised sufficiently high the clay grains soften and run together, destroying the pores and causing a loss of volume. Sandy clays not burned to the point of vitrification may not exhibit any fire shrinkage. Clays containing a high percentage of organic matter where subjected to a rapidly increasing temperature may become viscous on the outside, thus preventing the escape of hydrocarbons formed by the distillation of organic matter and therefore cause the ware to increase slightly in volume. A low fire shrinkage is very desirable in many clay wares. In order to secure the requisite minimum shrinkage it is necessary frequently for the clayworker to use a mixture of clays or grog.

Fineness of Grain. Clay is a mechanical mixture of mineral particles. These particles vary in size from those which are easily detected by the unaided eye to those which may be seen only by the use of a powerful microscope. The mechanical analysis of a clay consists in the separation of these particles into various groups. But because of the extreme degree of gradation in the size of the particles a complete separation is not possible. Fortunately it is not essential for the purposes of the clayworker. Clay grains are very small and in the separation of rock particles in mechanical analysis anything having a diameter of .001 mm. or less is classed as clay. In making a mechanical analysis of clay, sieves of from 200 to 250 meshes to the inch are used for the coarser particles and silk lawn and the elutriator method for the finer particles. The larger percentages of fine clay grains are found among the clays underlying the coals in Indiana and the larger percentages of coarse grains among the residual clays.

Plasticity. A clay is plastic when it can be easily fashioned by the hands into a desired form and when it has the property of retaining that form when so fashioned. Dry clay of any kind is devoid of this property. In order that a clay may become plastic it must be mixed with a certain amount of water. The quantity of water necessary for plasticity varies with the physical condition of the clay. Not all clays become plastic when mixed with water. The white kaolin of Indiana is non-plastic. This fact leads to the conclusion that some clays possess inherent properties which render them plastic upon the addition of a certain proportion of water but that all clays do not possess such properties. Experience demonstrates that plasticity is the result of a combination of conditions rather than the result of a single physical factor. The factors which seem to have the greatest influence upon the plasticity of a clay are the presence of colloidal matter and fineness of grain. It has been demonstrated by various experiments that when colloidal substances are added to clay the plasticity of the clay is increased. It is also a well-known fact that when a plastic clay is subjected to temperatures sufficiently high to destroy the colloidal nature of substances the clay is no longer plastic. It has been demonstrated further that the reduction in the size of the grain of clays has increased their plasticity. But non-plastic substances cannot be rendered as plastic as clay by reduction in size of grain. Other factors which are thought to influence the plasticity of clay are the presence of combined water and the presence of flat and interlocking crystals.

Bonding Power. The bonding power of a clay is its power to hold together particles of non-plastic materials. The bonding power of a clay is dependent in a measure upon the amount of clay substance which the clay contains. It also depends on the size of the grains of the inert matter added to the clay. To illustrate, a larger amount of finely-divided sand may be added to a clay without decreasing its bonding power than of coarse sand. It is often necessary in order to secure the proper shrinkage and drying capacity in a clay ware, to use a mixture of two clays or to add sand or grog to the clay. The quantity of the inert matter which may be added without seriously impairing the strength of the ware will depend on the bonding power of the clay. Bonding power is a very essential property to practically all clay used for ceramic purposes.

Tensile Strength. The amount of resistance which a clay offers to pull is termed its tensile strength. Wet clays possess this property to a slight degree; dry clays to a greater degree and burned clays to a still higher degree. Were it not for this property it would be impossible to handle clay wares because of the ease with which they would be cracked or broken. The tensile strength of a clay is not due to any chemical change, but to the physical cohesion of its particles. It was formerly thought that the tensile strength of a clay was a safe guide to the degree of plasticity but it is no longer so considered for the reason that many very plastic clays have been found to have a very low tensile strength.

In the preparation of clays for the tensile strength test, the clay is crushed to a powder which will pass a 100 mesh sieve. The powdered clay is mixed with sufficient water to render it plastic. It is then wedged into blocks about three inches long by one and one-half in cross-section. These blocks are then clamped in between the halves of a cement mold, the clamps closed and the clay pressed down on the sides until it completely fills the mold. The surplus clay is then stroked off with a putty knife, the surface of which has been moistened. The brickettes are then removed from the mold and placed on edge. After drying in the air they are placed in a desiccating oven and subjected to a temperature of 100°C to remove the hydroscopic moisture.

The brickettes are tested by the use of a Fairbanks cement testing machine. The brickettes are placed in the clips of the machine and subjected to a gradually increasing tension. The increase of tension is secured by the weight of shot discharging into the pail on the lever arm. At the moment of breaking the discharge of shot is stopped automatically. If the brickettes have undergone much shrinkage they will not fit the clips of the machine and it will be necessary to bush them. This may be done by placing cardboard or blotting paper between the clip and the brickette.

The tensile strength is expressed in pounds per square inch of the cross-section of the brickette and the shrinkage must be calculated and taken into account in estimating the tensile strength of the clay. The average of half a dozen tests should be taken.

Fusibility. Matter may exist in three states, viz., solid, liquid or gas. Water, for example, at ordinary temperatures exists as a liquid. At slightly lower temperatures it becomes a solid. At higher temperatures it assumes the form of a gas. When in the solid state, if heat be applied, the solid becomes a liquid. This transformation is termed fusion. The temperature at which the solid becomes liquid is called the fusion point of that substance. The fusion point of any substance is controlled by pressure. All solids, having a definite chemical composition under a fixed pressure, fuse at a certain definite temperature. This definite temperature is called the fusion point.

Ordinary clays, however, are not definite chemical compounds. Clays are composed of a variety of minerals, each having a definite chemical composition and a definite point of fusion. When heat is applied to this aggregate of minerals, the one having the lowest fusion point will be the first to fuse. The molten matter which is free to combine may unite with some other mineral or minerals in the clay and form a compound having a lower fusion point than the original compounds. These when molten may act as fluxes for other minerals and the whole clay be reduced to a molten condition at a temperature considerably lower than the fusion point of its most refractory constituents. The change from the solid to the liquid involves the consumption of heat in raising the temperatures to the fusion point. Some heat is consumed as latent heat, some in chemical reactions.

Three stages are usually recognized in the fusion of a clay, namely, incipient fusion, vitrification and viscosity. In the first stage the more fusible particles become soft and upon cooling cement together the more refractory particles, forming a hard mass. In the second stage the clay particles become soft enough to close up all of the pore spaces so that further shrinkage is impossible. When the mass becomes cool it forms a dense solid body which is glassy on a fractured surface. In the third stage, the clay body becomes so soft as to no longer retain its shape and flows.

The fusibility of a clay depends upon a number of factors but the most important ones, are the amount and kinds of fluxing impurities in the clay and the fineness of grain.

For determining the temperature of kilns and furnaces and the fusion points of different substances, pyrometers of various kinds are used. One of these is the thermo-electric couple which generates an electric current when heated. The intensity of the current increases with the temperature. The current is measured by means of a galvanometer. The thermopile consists of a platinum wire and a wire composed of ninety per cent platinum and ten per cent rhodium. The wires, protected by clay tubes, are inserted into the furnace usually through a small opening in the door.

The fusibility of clays is also determined by the use of Seger cones. These cones are made of a mixture of substances of known fusibility. The cones, together with the clay to be tested, are placed in a furnace or oven and heat applied. The cone which loses its shape at the moment the clay cone does determines the fusion point of the clay. The cones are numbered from 022 to 36. The former fuses at 1,094°F, 590°C and the latter at 3,322°F or 1,850°C.

Many potters use test pieces of the same composition as the body of the ware. The test pieces are made of the same thickness as the ware and pierced with holes for the insertion of hooks when it is desired to withdraw them from the kiln. They are placed near the outer portion of the kiln and also in the interior. From time to time after a definite period of firing, a test piece is withdrawn and its hardness tested. Whenever the test piece reaches the degree of hardness desired for the ware the firing ceases. In a good many potteries the test pieces used are the sections of either round or square tubes. On one side of the tube is placed a small amount of low fusing clay and on the other some feldspar or other substance of about the same fusibility. As the fusion point of these substances are known some idea of the temperature of the kiln at different stages of the firing may be obtained. The tubes also enable the potter to judge of the progress of the firing by the degree of hardness exhibited.

Classes of Indiana Clays.

No deposits of clay of primary origin, that is derived directly from the decomposition of feldspathic rocks in place, occur in Indiana. The clays of the State are of secondary origin that is derived from rocks of secondary origin, such as shales, argillaceous sandstones and shales.

Shales. Shales are clays which have been subjected to compression and to a certain degree of induration. They are not adapted to all ceramic uses and are adaptable to some uses for which clays are not suited. The shales of Indiana are adapted to the manufacture of common and front brick and with proper selection and admixing to sewer pipe and drain tile.

Knobstone Shales. The Knobstone shales of the Mississippian period have been utilized for the manufacture of common and front building brick, paving blocks and wall and drainage tile. Plants utilizing the Knobstone shale in the manufacture of ceramic products are located at New Albany, Martinsville, Brooklyn and Crawfordsville. A plant to be established at Bloomington will use these shales.

For the best results mixtures of two to three of sandy and aluminous varieties of shale are used at Brooklyn and Martinsville. For the outcrop of these and the following shales see the accompanying map.

Chester Shales. The upper portion of the Mississippian in Indiana contains a number of important shale horizons, the Brandy Run, the Elwren, the Indian Springs are the chief horizons. These shales are serviceable in the manufacture of clay products, such as stoneware, paving brick, sewer pipe, drain tile, terra cotta, common and front brick. Chester clays are being used for ceramic purposes on the State Farm near Putnamville.

Pennsylvania Shales. The formations of the Pennsylvanian period in Indiana consists of a series of sandstones, shales, coals, fire clays and limestones. The shale in many parts of the area occupied by the outcrop of these formations are suitable for ceramic products. These may be used in the manufacture of tile, paving and building brick and sewer pipe. Ceramic plants using these shales are located at Evansville, East Montezuma, Brazil, Cayuga, Newport, Hillsdale and Terre Haute and other places.

Clays. The clays underlying the coal beds in the area of the Pennsylvanian rocks are used in a number of ceramic plants. They are used alone or mixed with shales. The clays are used in the manufacture of sewer pipe, hollow blocks, conduits, stone pumps, fire proofing, drain tile, wall coping, culvert pipe, flue linings.

Pleistocene Clays. Glacial clays are used throughout the glaciated area of the State in the manufacture of common brick and drain tile. The clay usually consists of a top yellow sandy clay and an underlying tough bluish clay. The latter in some places attains a thickness of more than one hundred feet.

Residual Clay. Residual clays are used in a number of counties in the non-glaciated region in the manufacture of common brick and drain tile.

BLACK SHALE OR GENESEE SHALE (DEVONIAN).

Analyses of Shale taken from different localities near New Albany.

I.*

Water expelled at 100°C	0.50
Volatile organic matters	14.16
Fixed carbon	9.30
Silica	50.53
*Pyritic iron and alumina	25.30
Calcium oxide	0.09
Magnesium oxide	0.12
	100.00

II.

Water expelled at 100°C, during 4 hours		0.56
Volatile organic matters	14.30	23.60
Fixed organic matters	9.30	
Silicates insoluble in HCl		65.43
Ferric oxide		8.32
Calcium oxide		0.09
Magneisum oxide		0.12
Sulphur		2.08
		100.00

* The amount of pyrite and alumina changes considerably in different layers. This piece had 10.367 per cent iron pyrite and 14.933 per cent alumina.

MISSISSIPPIAN (KNOBSTONE) SHALES

Name	Town	County	Silica	Titanium oxide	Alumina	Water (Combined)	Clay Base and Sand (Total)	Ferric oxide	Ferrous oxide	Lime	Magnesia	Potash	Soda	Fluxes (Total)	Carbon dioxide
Ice House	Attica	Fountain	68.14	1.38	16.03	3.86	89.41	4.54	......	.47	.27	3.80	1.50	10.56	......
Arthur Hadley	Danville	Hendricks	70.84	.18	14.73	3.32	89.07	......	4.82	.38	1.36	3.15	1.42	11.13	.21
Branch & Son	Martinsville	Morgan	70.60	.43	13.89	3.19	88.11	2.33	3.56	.60	.50	2.76	1.60	11.35	.51
Blue Lick	Sec. 6, 5 N. 5 E.	Jackson	59.64	1.05	19.14	4.36	84.19	3.39	4.20	.26	2.31	3.53	.80	14.49	.35
Matthews & Chrisler	Sec. 6, 5 N. 5 E.	Jackson	59.95	1.12	18.65	4.54	84.26	4.87	3.34	.34	2.18	3.57	.78	15.08	.28
D. M. Hughes	Medora	Jackson	64.59	.30	16.37	3.71	84.97	5.37	1.59	.16	1.56	4.24	.97	13.89	.43
Goets Dry Pressed Brick Co.	New Albany	Floyd	63.88	.91	17.85	4.99	87.63	5.38	......	.38	1.47	3.98	1.29	2.50	......
Jas. McLarey	Bloomington	Monroe	58.22	1.15	19.63	6.61	85.61	5.88	......	1.96	2.10	3.93	.68	14.54	......
Wm. Fee	Bloomington	Monroe	74.43	......	6.88	2.24	83.55	6.52	2.10	1.32	1.36	2.40	1.86	15.56	.75
W. H. Gregory	Heltonville	Lawrence	68.90	......	17.20	3.97	90.07	5.05	1.05	......	1.96	.85	1.00	9.90	......

PENNSYLVANIAN SHALES[a]

Name	Town	County	Silica	Titanium oxide	Alumina	Water (Combined)	Clay Base and Sand (Total)	Ferric oxide	Ferrous oxide	Lime	Magnesia	Potash	Soda	Fluxes (Total)	Carbon dioxide
Wabash Clay Co.	Veedersburg	Fountain	59.55	1.00	16.21	5.62	82.38	2.18	7.13	.75	1.58	2.81	.28	14.73	3.15
(2) Mecca Coal & Mining Co.	Mecca	Parke	59.77	.8	20.60	4.53	85.70	2.22	3.70	.64	1.98	3.10	.85	12.49	.90
(9) Mecca Coal & Mining Co.	Mecca	Parke	58.83	.7	22.34	5.22	87.09	5.13	1.44	.49	1.56	4.18	.63	13.43	......
(13) Mecca Coal & Mining Co.	Mecca	Parke	59.02	1.1	20.93	7.59	88.64	4.45	1.56	.51	1.66	2.92	.41	11.51	......
Cayuga Brick & Coal Co.	Cayuga	Vermillion	65.78	1.00	14.79	4.98	86.55	8.03	......	.54	1.42	2.82	.97	13.78	.26
(6) Burns & Hancock Land	Montezuma	Vermillion	46.07	1.19	24.22	9.76	81.24	9.65	.34	.19	1.31	1.66	2.76	15.91	2.87
(11) Burns & Hancock Land	Montezuma	Vermillion	56.32	1.07	24.34	6.33	88.06	5.60	.24	.31	.54	2.85	2.34	11.88	......
Clinton Paving Brick Co.	Clinton	Vermillion	61.46	1.20	16.54	5.09	84.29	3.77	3.71	.66	1.81	3.28	1.09	14.32	1.45
Casto or Thorp Farm	Sugar Creek Tp.	Vigo	61.05	1.20	21.46	6.94	90.65	5.57	.71	.25	.70	2.34	.30	9.87	......
(8) Island Coal Co.—Mine 1	Linton	Greene	55.31	1.15	22.46	7.48	86.40	7.18	.23	.66	.93	2.05	2.06	13.11	1.24
Prospect Hill Mine	Vincennes	Knox	64.05	1.00	16.00	3.79	84.84	.80	5.85	.42	2.00	2.59	1.19	12.94	2.68
Larkin Farm	Loogootee	Martin	59.64	.7	18.93	5.88	85.15	1.62	5.01	.85	1.80	3.66	.81	13.75	2.22
Buckskin Mine	Buckskin	Gibson	57.60	.89	24.74	8.14	91.37	.15	2.41	.70	1.20	3.40	.99	8.85	......
Southern R. R. Shops	Princeton	Gibson	62.04	1.30	18.49	6.50	87.33	7.54	.06	.16	.91	.93	2.04	12.64	......
Evansville Pressed Brick Co.	Evansville	Vanderb'gh	65.87	1.10	14.66	4.59	86.22	6.23	1.37	.39	1.54	2.66	1.31	13.50	......
Cannelton Branch Cut	Lincoln City	Spencer	56.68	.9	20.33	6.54	84.45	4.35	3.69	.57	2.04	3.15	.63	14.43	.47
Amer. Cannel Coal Co.	Cannelton	Perry	53.26	1.05	25.77	7.00	87.08	3.32	3.82	.32	1.90	2.54	.44	12.34	1.10

PENNSYLVANIAN UNDER-CLAYS[11]
(Clays Beneath Coal Beds.)

Name	Town	County	Silica	Titanium oxide	Alumina	Water (Combined)	Clay Base and Sand (Total)	Ferric oxide	Ferrous oxide	Lime	Magnesia	Potash	Soda	Fluxes (Total)	Carbon dioxide
Minnick & Hoagland	Stone Bluff	Fountain	68.46	1.49	16.08	7.04	93.07	1.92	.06	.99	.05	1.31	2.40	6.73	......
Frank Landers	Stone Bluff	Fountain	67.82	1.10	13.60	9.72	92.24	4.04	.45	.57	.44	1.68	1.18	8.36	......
John R. Teegarden	Kingman	Fountain	71.91	.31	17.62	5.37	95.21	2.55	.40	.43	.17	2.00	.15	5.70	......
No. 12 Mecca Coal & Mining Co.	Mecca	Parke	54.46	1.20	25.71	8.50	89.87	5.51	.91	24	.83	2.68	.33	10.50	......
No. 20 Mecca Coal & Mining Co.	Mecca	Parke	63.00	1.10	23.57	6.45	94.12	1.87	.46	.44	.89	2.40	.29	6.35	......
No. 24 Mecca Coal & Mining Co.	Mecca	Parke	67.65	1.01	19.97	5.96	94.59	.72	......	.48	.59	1.75	2.29	5.83	......
No. 27 Mecca Coal & Mining Co.	Mecca	Parke	66.52	1.02	20.12	6.13	93.79	1.03	......	.83	.69	3.05	.57	6.17	......
W. W. Wray	Coxville	Parke	53.91	1.60	26.18	7.99	89.68	3.39	1.25	.30	.95	3.60	.44	9.93	.25
Cayuga Brick & Coal Co.	Cayuga	Vermillion	55.09	1.20	20.76	7.01	84.06	3.00	4.01	1.51	1.18	2.36	.34	12.40	3.04
No. 10 Burns & Hancock	W. Montezuma	Vermillion	83.44	1.29	10.36	3.15	98.24	.27	.28	.36	.14	.03	71	1.79	......
No. 10 Jackson Bros.	Between W. Montezuma and Newport	Vermillion	64.05	1.10	20.55	8.88	94.58	2.61	1.42	.29	.58	.51	.40	5.81	......
PottersClay—Chas. Cooprider	Clay City	Clay	75.57	.85	13.97	4.47	94.86	1.47	......	.96	.71	1.77	.46	5.36	......
Crance Land	Worthington	Greene	73.82	.60	14.46	3.82	92.70	.83	1.95	.30	.58	2.07	.30	6.03	.94
Potters Clay—Bockting Bros.	Huntingburg	Dubois	69.23	1.50	18.97	5.46	95.16	1.57	.55	.12	.36	2.27	.33	5.20	......
Potters Clay—Chris. Fuchs	Huntingburg	Dubois	65.25	......	17.30	5.40	87.95	2.30	......	.50	.20	1.56	.98	5.54	6.50
Amer. Cannel Coal Co.	Cannelton	Perry	57.59	1.10	21.70	6.78	87.15	2.26	4.11	.32	1.12	2.16	.33	10.30	1.73
Ballou Under-clay	Muskingum County	Ohio State	58.78	.94	26.47	9.96	96.15	1.22	......	.59	.32	.99	......	3.12	......

PLEISTOCENE AND RECENT CLAYS.[11]
(Surface Drift, Alluvial, Silty and Loose Clays.)

Name	Town	County	Silica	Titanium oxide	Alumina	Water (Combined)	Clay Base and Sand (Total)	Ferric oxide	Ferrous oxide	Lime	Magnesia	Potash	Soda	Fluxes (Total)	Carbon dioxide
Surface Drift	Rockville	Parke	66.84	......	13.64	4.22	84.70	9.40	......	1.36	Tr.	.64	3.34	14.74	......
Alluvial Clay	Terre Haute	Vigo	66.11	......	13.78	6.34	86.23	5.35	......	1.67	1.78	2.11	1.15	12.06	......
Surface Loess	Princeton	Gibson	71.20	.88	18.56	6.30	96.94	1.34	.15	.14	.52	.32	1.26	3.73	......
Surface Drift	Mooresville	Morgan	73.97	.91	12.33	3.16	90.37	3.87	......	.76	.78	Tr.	4.23	9.64	......
Mrs. Abrilla Merriman	Martinsville	Morgan	49.51	.87	14.73	3.59	68.70	4.65	......	9.48	3.76	.58	2.40	20.87	10.72
Indiana University Campus	Bloomington	Monroe	67.15	......	13.96	3.25	84.36	6.84	......	1.23	.54	1.18	3.33	13.12	3.00
National Stone Co.	Bloomington	Monroe	72.56	.31	10.44	4.54	87.85	7.45	.43	.82	1.09	2.06	.73	12.57	......
Residual Surface Clay above Mitchell Limestone	Bloomington	Monroe	79.99	......	8.66	3.55	92.20	......	3.31	.57	.77	1.93	.83	7.41	.33
Jessup Land	Richmond	Wayne	67.76	.89	13.69	7.10	89.44	5.09	.46	.94	.51	3.26	.61	10.77	......
Silty or Marley Clay	Hobart	Lake	50.56	1.00	13.11	2.76	67.43	2.96	2.32	7.87	5.06	3.74	.70	22.67	9.62
Silty or Marley Clay	Garden City	Porter	50.37	.65	9.93	1.50	62.45	2.10	2.05	10.26	6.26	3.04	.79	24.50	12.50
P. E. Anderson	Chesterton	Porter	53.02	1.30	10.72	2.21	67.25	2.64	2.54	8.38	5.28	3.25	.86	22.53	10.48
Roeske Bros.	Michigan City	Laporte	50.47	1.45	12.77	3.14	67.83	2.44	2.52	8.17	5.22	3.70	.73	22.78	9.80
Silty or Marley Clay	Rochester	Fulton	45.12	.45	6.84	2.43	54.84	5.92	1.38	15.63	4.88	2.28	.74	30.83	14.95
Yellow Drift Clay	Akron	Fulton	71.01	19.02	.13	5.64	96.80	1.35	.62	.52	1.06	.15	.30	3.99	.20

THE CERAMIC INDUSTRY BY COUNTIES.

ADAMS COUNTY

Name of Firm or Individual	Location	Began Operations	Capital Invested	Kind of Product	System of Drying
A. Gottschald	Berne	1886	$10,000	Drain, silo and building tile	Air.
Krik-Tyndall Co.	Decatur	1919	75,000	Drain tile, hollow building blocks	Steam and waste heat.
J. H. Elick	Decatur		6,000	Brick	Sheds, pallets.
Margaret Nayer	Decatur		6,000	Brick	Sheds, pallets.
Lewellen & Smith	Monroe		7,000		

ALLEN COUNTY.

Name of Firm or Individual	Location	Began Operations	Capital Invested	Kind of Product	System of Drying
Fort Wayne Brick Co.	Fort Wayne	1908	$74,000	Solid and hollow building brick	Hot air.
Moellering Brick Co.	Fort Wayne	1910	12,500	Building brick	Air drying.
Woodburn Tile & Brick Co.	Woodburn	1904	18,600	Drain tile, brick and blocks	Air drying.
Wallen Tile & Milling Co.	Wallen	Failed to report in 1919.			
Wm. Miller	Fort Wayne	Failed to report in 1919.			
J. P. Auspach & Son	Edgerton	Failed to report in 1919.			
John C. Braun	Fort Wayne	Failed to report in 1919.			
Chas. W. Gets & Co.	Fort Wayne	Failed to report in 1919.			
J. A. Koehler	Fort Wayne	Failed to report in 1919.			

BARTHOLOMEW COUNTY.

Name of Firm or Individual	Location	Began Operations	Capital Invested	Kind of Product	System of Drying
Chas. D. Glick	Grammer	1900	$12,500	Drain tile	Exhaust steam, live steam at night.
Smith & Henderson	Columbus	1892		Brick	Open air.
W. R. Smith	Columbus				
F. J. Crump	Columbus	Out of business.			
John C. Burns	Elizabethtown	Failed to report in 1919.			
R. D. Stam	Near Hope	Failed to report in 1919.			
Miller & Rominger	Hope	Failed to report.			
Geo. H. Daum	Columbus	Failed to report.			

BENTON COUNTY.

Name of Firm or Individual	Location	Began Operations	Capital Invested	Kind of Product	System of Drying
J. & E. Fowler	Fowlerton	Failed to report in 1919.			
Redkey Brick & Tile Co.	Raub	Failed to report in 1919.			
J. J. Holtam	Earl Park	Failed to report in 1919.			
Wm. Lawson	Otterbein	Failed to report in 1919.			

BLACKFORD COUNTY

Name of Firm or Individual	Location	Began Operations	Capital Invested	Kind of Product	System of Drying
Clark Croninger	Hartford City	Failed to report in 1919.			

BOONE COUNTY.

Name of Firm or Individual	Location	Began Operations	Capital Invested	Kind of Product	System of Drying
Quicks & Allen	Englewood		$1,500	Drain tile	Air and steam.
J. A. Bassett	Lebanon	Out of business.			
Kersey Bros.	Haselrig	Failed to report in 1919.			
Lebanon Brick Works	Lebanon	Failed to report in 1919.			
Wicker Bros.	Lebanon	Failed to report in 1919.			
Saunders & Robinson	Northeast of Lebanon	Failed to report in 1919.			

CARROLL COUNTY.

Name of Firm or Individual	Location	Began Operations	Capital Invested	Kind of Product	System of Drying
H. H. Landis	Bringhurst	Failed to report in 1919.			

THE CERAMIC INDUSTRY BY COUNTIES—Continued.

CASS COUNTY.

Name of Firm or Individual	Location	Began Operations	Capital Invested	Kind of Product	System of Drying
John E. Barnes & Sons..	Logansport....	Failed to report in 1919.			

CLARK COUNTY.

Name of Firm or Individual	Location	Began Operations	Capital Invested	Kind of Product	System of Drying
Jeffersonville Brick Co..	Jeffersonville..	Out of business.			

CLAY COUNTY.

Name of Firm or Individual	Location	Began Operations	Capital Invested	Kind of Product	System of Drying
Ayer, McCarel-Reagon Clay Company........	Brazil..........	1910	$48,000	Silo and building tile	Stream.
American Vitrified Products Co..............	Brazil..........	1892	....	Sewerpipe, siloblocks, flue linings, cement blocks, wall coping.	Steam.
Hydraulic Press Brick Company.............	Brazil..........	1900	300,000	Face brick, silo tile..	Waste heat.
Clay Products Company	Brazil..........	....	1,700,000	Bldg. tile, silo tile, Elec. conduits and shale bldg. tile....	Steam—waste heat.
Chicago Sewer Pipe Co.	Brazil..........	1893	150,000	Sewer pipe and wall coping.............	Steam dryer.
Clyde Griffith..........	Clay City.....	1886	3,000	Stoneware..........	Natural.
Branch of Nat'l Fire Roofing Company.....	Brazil..........	1900	....	Conduit silos and hollow bldg. tile......	Exhaust steam.
Indiana Block Coal Co.	Saline City....	1919	....	Raw clay..........	....
Brazil Clay Co..........	Brazil..........	1915	230,000	Face brick..........	Waste steam.
Excelsior Clay Works...	Brazil..........	Failed to report in 1919.			
Modern Concrete Works.	Clay City.....	Failed to report in 1919.			
Sheridan Brick Works...	Brazil..........	Failed to report in 1919.			
Clay City Brick & Clay Co....................	Clay City.....	Failed to report in 1919.			
Geo. J. Kiser...........	Clay City.....	Failed to report in 1919.			

CLINTON COUNTY.

Name of Firm or Individual	Location	Began Operations	Capital Invested	Kind of Product	System of Drying
Colfax Drain Tile Co....	Colfax.........	1915	$48,500	Drain tile..........	Steam, pallets.
J. B. Lowden...........	Michigantown.	Out of business.			
B. F. Alter & Son.......	Forest.........	Out of business.			
Frankfort Brick and Construction Co..........	Frankfort.....	Out of business.			
M. J. Lee Drain Tile Co..	Colfax.........	Failed to report in 1919.			

DAVIESS COUNTY.

Name of Firm or Individual	Location	Began Operations	Capital Invested	Kind of Product	System of Drying
Joseph Kruts............	Washington....	1868	$10,000	Common brick......	Rack and pallet.
Reister Bros............	Washington ..	Failed to report in 1919.			
Dillon Bros.............	Elnora........	Failed to report in 1919.			
C. E. O'dell............	Odon..........	Failed to report in 1919.			

DEARBORN COUNTY.

Name of Firm or Individual	Location	Began Operations	Capital Invested	Kind of Product	System of Drying
Aurora Brick Works.....	Aurora........	1913	$10,000	Common brick......	Sun.
Suer Bros..............	Lawrenceburg.	1915	....	Common bldg. brick	Sun.
Mitchell Brick Co.......	Lawrenceburg.	Failed to report in 1919.			

DECATUR COUNTY.

Name of Firm or Individual	Location	Began Operations	Capital Invested	Kind of Product	System of Drying
Geo. S. Littell..........	Greensburg....	Failed to report in 1919.			
L. A. Terhune...........	Spring Hill....	Failed to report in 1919.			

THE CERAMIC INDUSTRY BY COUNTIES—Continued.

DEKALB COUNTY.

Name of Firm or Individual	Location	Began Operations	Capital Invested	Kind of Product	System of Drying
Herman Groscop	Altona	Failed to report in 1919.			
Blun & Son	Moore	Failed to report in 1919.			
W. B. Miller	Auburn	Out of business.			
Grogg Bros	Auburn	Failed to report in 1919.			
Fred Groscop	Garrett	Failed to report in 1919.			

DELAWARE COUNTY.

Name of Firm or Individual	Location	Began Operations	Capital Invested	Kind of Product	System of Drying
Gin Clay Pottery Co	Muncie	1883	$150,000	Clay pots for glass factories	
J. D. Mock	Muncie	Out of business.			
Bennett Brick Company	Muncie	Failed to report in 1919.			
Studebaker & Son	Shideler	Failed to report in 1919.			

DUBOIS COUNTY.

Name of Firm or Individual	Location	Began Operations	Capital Invested	Kind of Product	System of Drying
Uhl Pottery Co	Huntingburg	1903	$50,000	Stoneware and flower pots	Steam heat.
Huntingburg Pressed Brick Company	Huntingburg	1892	30,000	Face and common brick	Justice radiated heat dryer.
Wm. Lukemeyer	Huntingburg	Failed to report in 1919.			
Bockting Bros	Huntingburg	Failed to report in 1919.			

ELKHART COUNTY.

Name of Firm or Individual	Location	Began Operations	Capital Invested	Kind of Product	System of Drying
Geo. Bemenderfer	Goshen	Failed to report in 1919.			
Goshen Brick Co	Goshen	Failed to report in 1919.			
Elkhart Brick Co	Elkhart	Failed to report.			
John C. Boss	Elkhart	Failed to report.			

FAYETTE COUNTY.

Name of Firm or Individual	Location	Began Operations	Capital Invested	Kind of Product	System of Drying
A. Fries & Son	Connersville	1892	$50,000	Drain tile	Stream.
C. P. Ariams	Connersville	Failed to report in 1919.			

FLOYD COUNTY.

Name of Firm or Individual	Location	Began Operations	Capital Invested	Kind of Product	System of Drying
Hoosier Brick Company	New Albany			Brick	Waste heat, exhaust steam.
Goetz Bros	New Albany	1903		Wall plaster filler	
Henry Vance	New Albany	Failed to report.			

FOUNTAIN COUNTY.

Name of Firm or Individual	Location	Began Operations	Capital Invested	Kind of Product	System of Drying
Veedersburg Paving Co	Veedersburg	1914	$125,000	Paving brick	Waste heat.
Veedersburg Brick Co	Veedersburg	1902	60,000	Building brick	Waste heat.
Clarence E. Poston	Attica	1907		Wire-cut face and paving brick	Waste heat.
H. H. Prather	Covington	Out of business.			
James A. Furr	Steam Corner	Failed to report in 1919.			
S. P. Cowgill	Kingham	Failed to report in 1919.			
Daniel Carpenter	Mellott	Failed to report in 1919.			

FRANKLIN COUNTY.

Name of Firm or Individual	Location	Began Operations	Capital Invested	Kind of Product	System of Drying
A. Fries & Son	Brookville	Failed to report in 1919.			
Noah B. Waggoner	Blooming grove	Failed to report in 1919.			

THE CERAMIC INDUSTRY BY COUNTIES—Continued.

FULTON COUNTY.

Name of Firm or Individual	Location	Began Operations	Capital Invested	Kind of Product	System of Drying
Swick & Meridith	Fulton	Failed to report in 1919.			
A. A. Gast	Akron	Failed to report in 1919.			
A. A. Gast	Mets	Failed to report in 1919.			

GIBSON COUNTY.

Name of Firm or Individual	Location	Began Operations	Capital Invested	Kind of Product	System of Drying
F. M. Ferguson	Oakland City	Out of business.			
W. M. Mead	Princeton	Failed to report in 1919.			
Jesse Mitchell	Princeton	Failed to report in 1919.			
Stormont Bros	Francisco	Failed to report in 1919.			
Chas. Read	Oakland City	Failed to report in 1919.			
Polk Clay Working Co	Fort Branch	Failed to report in 1919.			

GRANT COUNTY.

Name of Firm or Individual	Location	Began Operations	Capital Invested	Kind of Product	System of Drying
Citizens Brick Co	Jonesboro	Failed to report in 1919.			
Brickner Window Glass Co	Sweetser	Failed to report in 1919.			
John Clifton Sons	Matthews	Failed to report in 1919.			
Herbst Drain Tile Co	Herbst	Out of business.			
L. C. Lillard	Marion	Out of business.			
Bolen Brick Co	Marion	Out of business.			
Diamond Window Glass Co	Gas City	Failed to report in 1919.			
Marion Brick Works	Marion	Failed to report in 1919.			
Sweetser Drain Tile Co.	Sweetser	Failed to report in 1919.			
Fairmount Tile Co	Fairmount	Failed to report in 1919.			
J. & E. Fowler	Fowlerton	Failed to report in 1919.			

GREENE COUNTY.

Name of Firm or Individual	Location	Began Operations	Capital Invested	Kind of Product	System of Drying
C. W. Baughn	Bushrod	Failed to report in 1919.			
Aschman Bros.	Jasonville	Failed to report in 1919.			
Worthington Brick and Tile Company	Worthington	Failed to report in 1919.			
O. Shelton	Switz City	Failed to report in 1919.			
Linton Brick Mfg. Co.	Linton	Failed to report in 1919.			
Smith & Smith	Midland	Failed to report in 1919.			

HAMILTON COUNTY.

Name of Firm or Individual	Location	Began Operations	Capital Invested	Kind of Product	System of Drying
Lacy Seed & Fuel Co.	Noblesville	Failed to report in 1919			
Arcadia Brick Works	Arcadia	Failed to report in 1919.			
Chas. Jessup	Arcadia	Failed to report in 1919.			

HANCOCK COUNTY.

Name of Firm or Individual	Location	Began Operations	Capital Invested	Kind of Product	System of Drying
Wm. Reasoner	Gem	Failed to report in 1919.			
Greenfield Brick Works	Greenfield	Failed to report in 1919.			
Elmer Knight	Maxwell	Failed to report in 1919.			
Madison Brick Co	Shirley	Failed to report in 1919.			

HENDRICKS COUNTY.

Name of Firm or Individual	Location	Began Operations	Capital Invested	Kind of Product	System of Drying
G. B. Pruitt	Coatesville	Out of business.			
J. W. Beck	Danville	Out of business.			
Lingeman Bros	Brownsburg	Failed to report in 1919.			
E. R. Ellis	Coatesville	Failed to report in 1919.			
W. F.-C. S. Boyd	Clayton	1880	$10,000	Drain tile	Open air.
Boyd Tile Works	Hazelwood	Failed to report in 1919.			
Wm. I. Gill	N. Salem	Failed to report in 1919.			

THE CERAMIC INDUSTRY BY COUNTIES—Continued.

HENRY COUNTY.

Name of Firm or Individual	Location	Began Operations	Capital Invested	Kind of Product	System of Drying
Newcastle Brick & Tile Company	Newcastle	Out of business			
Nathaniel Edwards	Grant City	Failed to report in 1919.			

HOWARD COUNTY.

Name of Firm or Individual	Location	Began Operations	Capital Invested	Kind of Product	System of Drying
Kokomo Brick Works	Kokomo	1867		Sand moulding brick	Steam heated tunnels.
Pittsburg Plate Glass Company	Kokomo			Clays used are for glass pot making, etc.	
Richard Cunningham & Son	Russiaville	1885	$3,500	Tile	Dry shed.
Standard Sanitary Mfg. Company	Kokomo	1890		Sanitary earthenware	Air drying.
Kokomo Sanitary Pottery Company	Kokomo	1909		Chinaware	Steam.
J. M. Lynch & Co	Kokomo	Failed to report in 1919.			
Stephen Colescott	Greentown	Failed to report in 1919.			
Richard Cunningham	New London	Failed to report in 1919.			

HUNTINGTON COUNTY.

Name of Firm or Individual	Location	Began Operations	Capital Invested	Kind of Product	System of Drying
Shideler & Mahoney	Majenica	Failed to report in 1919.			
Bippus Drain Tile Co	Bippus	Out of business in 1919.			
Tribolet Bros	S. E. of Huntington	Failed to report in 1919.			
Holsinger Brick Co.	Huntington	Failed to report in 1919.			

JACKSON COUNTY.

Name of Firm or Individual	Location	Began Operations	Capital Invested	Kind of Product	System of Drying
Jackson Brick & Hollow-ware Company	Brownstown	1907	$12,000	Bldg. brick and drain tile	Raidiated heat.
H. H. Koocner	Crothersville	Failed to report in 1919.			
G. M. Ebaugh	Seymour	Failed to report in 1919.			
W. L. Kasting	Seymour	Failed to report in 1919.			
Brownstown Brick and Tile Company	Brownstown	Failed to report in 1919.			
Kurtz Brick & Tile Co.	Kurtz	Failed to report in 1919			

JASPER COUNTY.

Name of Firm or Individual	Location	Began Operations	Capital Invested	Kind of Product	System of Drying
J. I. Miller	Pleasant Grove	Failed to report in 1919.			
Alter & Wolf	Rensselaer	Failed to report in 1919.			
Green & Bowman	Remington	Failed to report in 1919.			
J. F. Erwin & Son	Rensselaer	Failed to report in 1919.			

JAY COUNTY.

Name of Firm or Individual	Location	Began Operations	Capital Invested	Kind of Product	System of Drying
Portland Drain Tile Co.	Portland	1903	$125,000	Tile and blocks	Steam.
Current Tile Co	Redkey	1918		Tile, blocks and bricks	Steam heat and from kilns.
Shaw, Rhodes & Singer	Bryant	Failed to report in 1919.			
F. M. Byrd & Son	New Mount Pleasant	Failed to report in 1919.			
Martin Bros	Greene	Failed to report in 1919.			
Jas. A. Byrd	Boundary	Failed to report in 1919.			
H. L. Huey	Portland	Failed to report in 1919.			
Portland Tile and Hollow Bldg. Block Works	Portland	Failed to report in 1919.			
Redkey Tile & Bldg. Block Company	Redkey	Failed to report in 1919.			

THE CERAMIC INDUSTRY BY COUNTIES—Continued.

Name of Firm or Individual	Location	Began Operations	Capital Invested	Kind of Product	System of Drying
Jefferson County.					
Joseph Heck & Co.	Madison	Failed to report in 1919.			
Ross & Kimmel	Madison	Failed to report in 1919.			
Jennings County.					
Simmons & Son	North Vernon	Failed to report in 1919.			
Harry Harms	North Vernon	Failed to report in 1919.			
Johnson County.					
Land & Britton	Nineveh	Failed to report in 1919.			
J. F. Davis	Franklin	Failed to report in 1919.			
E. C. Halstead	Franklin	Out of business.			
Dickson Bros.	Whiteland	Failed to report in 1919.			
T. J. Spears	Greenwood	Failed to report in 1919.			
H. H. Warner	Franklin	Failed to report in 1919.			
Knox County.					
Wm. F. Kock	Bicknell	Failed to report in 1919.			
Wm. F. Koch	Westphalia	Failed to report in 1919.			
S. Kizmiller	Vincennes	Failed to report in 1919.			
J. P. B. Prullage	Vincennes	Failed to report in 1919.			
Peter Bonewits	Monroe City	Failed to report in 1919.			
Chas. Meyer	Freelandsville	Failed to report in 1919.			
Kosciusko County.					
T. H. Wheeler	Warsaw	Failed to report in 1919.			
Phemel Bros.	Nappanee	Failed to report in 1919.			
M. W. Sellers	Packertown	Failed to report in 1919.			
Thacker Bros.	Silver Lake	Failed to report in 1919.			
Lagrange County.					
B. F. Ditman	Topeka	Failed to report in 1919.			
Brillhart & Miller	Wolcottville	Failed to report in 1919.			
Lake County.					
National Brick Co.	Maynard	1906	$185,579	Common brick	Hot air and steam.
Clark Mfg. Company	Lowell	Failed to report in 1919.			
J. P. Van Kirk	Laporte	Out of business.			
North Indiana Brick Co.	Michigan City	Out of business.			
Roeske Bros.	Michigan City	Out of business.			
Lawrence County.					
Heitger & Winterhaulter	Bedford	Failed to report.			
Madison County.					
National Drain Tile Co.	Summitville	1885	$100,000	Drain tile	Dry sheds, steam heated.
Creston Tile Works	Madison	Failed to report in 1919.			
Cooper Bros.	Anderson	Failed to report in 1919.			
Indiana Brick Co.	Anderson	Failed to report in 1919.			

THE CERAMIC INDUSTRY BY COUNTIES—Continued.

MARION COUNTY.

Name of Firm or Individual	Location	Began Operations	Capital Invested	Kind of Product	System of Drying
U. S. Encaustic Tile Co.	Indianapolis...	1880	$150,000	Clay spur flint.......	Steam.
Cook & Schmidt	Indianapolis...	Failed to report in 1919.			
T. B. Laycock, Son & Co.	Indianapolis..	Out of business.			
Frederick Bremer.......	Indianapolis...	Failed to report in 1919.			
Herman Eilering	Indianapolis...	Failed to report in 1919.			
Herman Luedeman	Indianapolis...	Failed to report in 1919.			
Jas. Mageennis	Indianapolis...	Failed to report in 1919.			
Leonard Neuerburg.....	Indianapolis...	Failed to report in 1919.			
Chas. C. Quack.........	Indianapolis...	Failed to report in 1919.			
John Hohn	Indianapolis...	Failed to report in 1919.			

MARSHALL COUNTY.

Name of Firm or Individual	Location	Began Operations	Capital Invested	Kind of Product	System of Drying
The Bremen Clay Product Company........	Bremen.......	1912	$40,000	Drain tile...........	Open air and steam heat.
J. W. Thomas & Son...	Plymouth.....	Failed to report in 1919.			
Lement & Co..........	Teegarden.....	Failed to report in 1919.			
Sarber & Sarber........	Argos........	Failed to report in 1919.			

MARTIN COUNTY.

Name of Firm or Individual	Location	Began Operations	Capital Invested	Kind of Product	System of Drying
Loogootee Fire Clay Products Co..........	Loogootee....	1918	$50,000	Bldg. and drain tile.	Justice radiated flat dryer.
Moran Bros............	Loogootee...	Out of business.			
Lawhead Bros. Brick & Tile Company	Loogootee....	Failed to report in 1919.			

MIAMI COUNTY.

Name of Firm or Individual	Location	Began Operations	Capital Invested	Kind of Product	System of Drying
F. E. Mock, lessee Peru Electric Sq. Company	Peru.........	1917	$35,000	Elec. porcelains and Wiring devices.....	Steam heat.
Ridgeway & Lamb.....	Amboy........	Failed to report in 1919.			
Brubaker & Kreider....	Perrysburg....	Failed to report in 1919.			

MONROE COUNTY.

Name of Firm or Individual	Location	Began Operations	Capital Invested	Kind of Product	System of Drying
Dolan Brick & Tile Co..	Dolan........	Failed to report in 1919.			

MONTGOMERY COUNTY.

Name of Firm or Individual	Location	Began Operations	Capital Invested	Kind of Product	System of Drying
Poslon Paving Brick Co.	Crawfordsville	1901	$150,000	Face and paving brick	Waste steam.
Armantrout & Childers.	New Market ..	1884	3,000	Drain tile...........	Air drying.
The Standard Brick Co.	Crawfordsville	1895	200,000	Paving, face and common brick....	Waste heat.
R. M. Booker...	New Ross...		4,000	Drain tile...........	Air and sun.
Crawfordsville Shale Brick Co	Crawfordsville	1908	80,000	Bldg. brick	Waste heat.
Wm. Hawkins........	Crawfordsville	Failed to report in 1919.			
Everson & Ferguson ...	Crawfordsville	Failed to report in 1919.			
F. J. Booker...........	New Ross....	Failed to report in 1919.			
H. K. Lee	New Richmond........	Failed to report in 1919.			
T. J. Casey............	Crawfordsville	Failed to report in 1919.			
Robt. Robbins..........	Ladoga.......	Failed to report in 1919.			
Kirk Bros.............	Bowers Station	Failed to report in 1919.			

MORGAN COUNTY.

Name of Firm or Individual	Location	Began Operations	Capital Invested	Kind of Product	System of Drying
Adams Clay Prod. Co..	Martinsville...	1897	$50,000	Face brick..........	Waste heat.
John B. Clark	Martinsville ..	Failed to report in 1919.			
Bradley Brick Co.......	Mooresville...	Failed to report in 1919.			

THE CERAMIC INDUSTRY BY COUNTIES—Continued.

NEWTON COUNTY.

Name of Firm or Individual	Location	Began Operations	Capital Invested	Kind of Product	System of Drying
Brook Terra Cotta Tile and Brick Co.	Brook	1902	$84,000	Drain tile and hollow bldg. tile	Waste heat and steam dryer.
Chas. Foster	Beaver City	Failed to report in 1919.			

NOBLE COUNTY.

Name of Firm or Individual	Location	Began Operations	Capital Invested	Kind of Product	System of Drying
Albion Tile Co.	Albion	1917	$3,000	Drain tile	Weather drying.
Brikkart & Miller	Wolcottville	Failed to report in 1919.			
Frank Landgraff	Albion	Failed to report in 1919.			
Jessie Devoe	Kendallville	Failed to report in 1919.			

PARKE COUNTY.

Name of Firm or Individual	Location	Began Operations	Capital Invested	Kind of Product	System of Drying
Indiana Sewer Pipe Co.	Mecca	1907	$180,000	Sewer pipe, drain tile, wall coping, fire proofing and chimney tops	Steam coils.
Wm. E. Dee Clay Mfg. Co.	Mecca	1904	300,000	Sewer pipe, flue lining and wall coping	....
H. R. Atchins	Annapolis	Failed to report in 1919.			
Marion Brick Works	Montezuma	Failed to report in 1919.			
National Drain Tile Co.	Montezuma	Failed to report in 1919.			
Russell Lee	Bellmore	Failed to report in 1919.			

PERRY COUNTY.

Name of Firm or Individual	Location	Began Operations	Capital Invested	Kind of Product	System of Drying
U. S. Brick Company	Tell City	1908	$65,174 77	Brick and drain tile.	Exhaust steam.
Geo. Dickman	Tell City	Failed to report in 1919.			

PIKE COUNTY

Name of Firm or Individual	Location	Began Operations	Capital Invested	Kind of Product	System of Drying
J. E. Schurz	Petersburg	Out of business.			
Frank D. Read	Petersburg	Failed to report in 1919.			
P. M. Ferguson	Spurgeon	Failed to report in 1919.			
Dempsey & Son	Otwell	Failed to report in 1919.			
Wm. Coldemeyer & Sons	Stendal	Failed to report in 1919.			

PORTER COUNTY.

Name of Firm or Individual	Location	Began Operations	Capital Invested	Kind of Product	System of Drying
Hydraulic Press Brick Co.	Porter	1890	$500,000	Face brick	Waste heat.
P. E. Anderson & Son	Chesterton	Failed to report in 1919.			
R. S. Kenny	Hebron	Failed to report in 1919.			
Chas. E. Lembke & Co.	Valparaiso	Failed to report in 1919.			
Coovert & Clevanger	Valparaiso	Failed to report in 1919.			

POSEY COUNTY.

Name of Firm or Individual	Location	Began Operations	Capital Invested	Kind of Product	System of Drying
Geo. B. Beal	New Harmony	Failed to reoprt in 1919.			
Henry Brinkman	Mount Vernon	Failed to report in 1919.			
Industrial Brick Co	Mount Vernon	Failed to report in 1919.			
Redmond & Co.	Cynthia	Failed to report in 1919.			

PULASKI COUNTY.

Name of Firm or Individual	Location	Began Operations	Capital Invested	Kind of Product	System of Drying
Francesville Clay Products Co.	Francesville	1902	$16,650	Drain tile and brick	....
Clay Tile Factory (Widup)	Winamac	Out of business.			
Orin Sevens	Francesville	Failed to report in 1919.			

THE CERAMIC INDUSTRY BY COUNTIES—Continued.

Name of Firm or Individual	Location	Began Operations	Capital Invested	Kind of Product	System of Drying
PUTNAM COUNTY.					
Wm. C. Rehling	Greencastle	Failed to report in 1919.			
J. A. Rice & Son	Roachdale	Failed to report in 1919.			
David Knoll	Cloverdale	Failed to report in 1919.			
RANDOLPH COUNTY.					
Kelley - Sanders Brick Co.	Winchester	Failed to report in 1919.			
B. H. Strahan	6 mi. N. of Winchester	Failed to report in 1919.			
Geo. Ashley	Stone Station	Failed to report in 1919.			
RUSH COUNTY.					
Arbuckle & Son	Homer	1896	$25,000	Drain tile and brick	Steam dryers.
Rushville Brick & Tile Works	Rushville	Out of business.			
W. M. Bainbridge	Rushville	Failed to report in 1919.			
SCOTT COUNTY.					
D. W. Wyman & Son	Scottsburg	Failed to report in 1919.			
SHELBY COUNTY.					
Thos. Brooks	Shelbyville	Failed to report in 1919.			
Schoelch Bros	Shelbyville	Failed to report in 1919.			
A. C. Bowlby & Son	Shelbyville	Failed to reporr in 1919.			
SPENCER COUNTY.					
Underhill Brick & Tile Company	Rockport	1911	$35,000	Common brick and drain tile.	Radiated heat dryer hot floor.
F. J. Bockting	Mariah Hill	Failed to report in 1919.			
St. Meinrad Brick & Tile Company	St. Meinrad	Failed to report in 1919.			
Palmer Drain Tile Co.	Lake	Failed to report in 1919.			
ST. JOSEPH COUNTY.					
Frank Fisher	South Bend	Failed to report in 1919.			
F. E. Perkins	South Bend	Failed to report in 1919.			
Moehel Bros	Woodland	Failed to report in 1919.			
South Bend Brick Co. Nos. 1 and 2	South Bend	Failed to report in 1919.			
STEUBEN COUNTY.					
Angola Brick & Tile Co.	Angola	1903	$25,000	Drain tile and bldg. brick	Natural dryers.
Chas. A. Bachelor	Angola	Failed to report in 1919.			
SULLIVAN COUNTY.					
Parsons & Heap	Farmersburg	Out of business.			
Jas. Abbott	Hymera	Failed to report in 1919.			
Ziba Howe	Sullivan	Failed to report in 1919.			

THE CERAMIC INDUSTRY BY COUNTIES—Continued.

TIPPECANOE COUNTY.

Name of Firm or Individual	Location	Began Operations	Capital Invested	Kind of Product	System of Drying
Union Brick Co.	Lafayette	Failed to report in 1919.			
Jacob May & Sons	Lafayette	Failed to report in 1919.			
Henry Kneale	Montmorencie	Failed to report.			

TIPTON COUNTY.

Name of Firm or Individual	Location	Began Operations	Capital Invested	Kind of Product	System of Drying
Thos. F. Lindley	Goldsmith	1910	$25,000	Drain tile and bldg. brick	Steam pipes and open slatted floor.
Windfall Mfg. Co.	Windfall	Out of business.			
Tipton Clay Co.	Tipton	Out of business.			
Edw. Henry	Tipton	Failed to report in 1919.			
P. P. Parnell	Goldsmith and Hobbs	Failed to report in 1919.			
Continental Tile & Brick Co.	Curtisville	Failed to report in 1919.			

VANDERBURGH COUNTY.

Name of Firm or Individual	Location	Began Operations	Capital Invested	Kind of Product	System of Drying
Standard Brick Mfg. Co.	Evansville	1903	$150,000	Common face and brick	Waste heat.
First Ave. Brick & Tile Company	Evansville	1895	50,000	Bldg. brick and drain tile	Waste heat.
Sophie Heseman	Evansville	Failed to report in 1919.			
Crown Potteries Co.	Evansville	Failed to report in 1919.			
Ernest Wedeking	Evansville	Failed to report in 1919.			
Standard Brick Mfg. Co.	Evansville	Out of business.			
Henry Alexander	Evansville	Failed to report in 1919.			
Samuel Wellmeier	Evansville	Failed to report in 1919.			
Jn. Waterman	Evansville	Failed to report in 1919.			
Klamer Bros.	Evansville	Failed to report in 1919.			
John Herdink	Evansville	Failed to report in 1919.			

VERMILLION COUNTY.

Name of Firm or Individual	Location	Began Operations	Capital Invested	Kind of Product	System of Drying
Acme Brick Co.	Cayuga	1906	$60,000	Face bldg. brick and floor tile	Steam dryer.
Wm. DeeClay WorksCo.	Newport	1908	113,200	Drain tile and hollow bldg. brick	Kiln heat.
Wm. DeeClay WorksCo.	Cayuga	1919	....	Drain tile and hollow bldg. brick	
Eureka Brick Company	Clinton	Failed to report in 1919.			
Southern Fire & Clay Co.	W. Montezuma	Failed to report in 1919.			
Wm. Martin & Son	Rileysburg	Failed to report in 1919.			
National Drain Tile Co.	Hillsdale	Failed to report in 1919.			
Cayuga Brick & Coal Co.	Cayuga	Failed to report in 1919.			
Newport Brick & Clay Co.	Newport	Failed to report in 1919.			

VIGO COUNTY.

Name of Firm or Individual	Location	Began Operations	Capital Invested	Kind of Product	System of Drying
Vigo American Clay Co.	Terre Haute	1901	$400,000	Hollow bldg. tile	Waste heat.
National Drain Tile Co.	Terre Haute	1902	450,000	Farm drain tile	Steam pipe and hot floors.
Bennett Brick Company	Terre Haute	1916	10,000	Bldg. brick	Sun.
Terre Haute Vitrified Brick Co.	Terre Haute	1894	350,000	Ironstone paving brick	Waste heat.
Vigo American Clay Co.	Terre Haute	1908	2,000,000	Hollow bldg. tile	Waste heat.
Vigo Plant	W. Terre Haute	1902	100,000	Hollow bldg. tile	Waste heat.
Chas. W. Hoff Brick Co.	Terre Haute	1872	50,000	Common brick	Open sheds.
Parle Brick Company	Terre Haute	Failed to report in 1919.			
I. A. Terhune	Spring Hill	Failed to report in 1919.			
C. M. Miller Mining & Mfg. Co.	Terre Haute	Failed to report in 1919.			
Cooperative Brick Co.	Terre Haute	Failed to report in 1919.			
Chas. Smith	Terre Haute	Failed to report in 1919.			
O'Mara Bros.	Terre Haute	Failed to report in 1919.			
W. Bergeman	Terre Haute	Failed to report in 1919.			
Terre Haute Pressed Brick Co.	Terre Haute	Failed to report in 1919.			

THE CERAMIC INDUSTRY BY COUNTIES—Continued.

WABASH COUNTY.

Name of Firm or Individual	Location	Began Operations	Capital Invested	Kind of Product	System of Drying
Manchester Tile Works	N. Manchester	1884	$15,000	Drain tile	Steam.
Edw. Haggerty	Lafontaine	Failed to report in 1919.			
M. Giek	N. Manchester	Failed to report in 1919.			
Ernest H. Carothers	N. W. of Wabash	Failed to report in 1919.			
Albert Shinkel	Wabash	Failed to report in 1919.			

WARREN COUNTY.

Name of Firm or Individual	Location	Began Operations	Capital Invested	Kind of Product	System of Drying
Alex. Hamer	W. Lebanon	Out of business.			
Industrial Brick Works	Boonville	1904	$32,000	Brick, drain tile	Steam heat.

WARRICK COUNTY.

Name of Firm or Individual	Location	Began Operations	Capital Invested	Kind of Product	System of Drying
Warrick Roofing Tile Co.	Newburg	Failed to report in 1919.			
Star Brick & Clay Co.	Newburg	Failed to report in 1919.			
Louis Klostermeier	Boonville	Failed to report in 1919.			
Industrial Brick Works	Boonville	Failed to report in 1919.			
Jarvis & Thene	Elberfeld	Failed to report in 1919.			
John Sertel	Lynville	Failed to report in 1919.			

WASHINGTON COUNTY.

Name of Firm or Individual	Location	Began Operations	Capital Invested	Kind of Product	System of Drying
Kern Shrum	Salem	Failed to report in 1919.			
Franklin Bros.	Little York	Failed to report in 1919.			

WAYNE COUNTY.

Name of Firm or Individual	Location	Began Operations	Capital Invested	Kind of Product	System of Drying
P. Franzman	Cambridge City	Out of business.			
Chas. H. Meyer	Richmond	Out of business.			
W. C. Thistlethwaite	Richmond	Failed to report in 1919.			

WELLS COUNTY.

Name of Firm or Individual	Location	Began Operations	Capital Invested	Kind of Product	System of Drying
Aurora Fire Clay Co.	Bluffton	Out of business.			
Buckner & Irwin	Bluffton	Out of business.			
O. J. Montgomery	Bluffton	Failed to report in 1919.			
J. W. Cook	Poneto	1880	$5,000	Drain tile	Air.
Beaty & Doan	Ossian	Failed to report in 1919.			
J. C. Bell	Craigville	Failed to report in 1919.			

WHITE COUNTY.

Name of Firm or Individual	Location	Began Operations	Capital Invested	Kind of Product	System of Drying
Underhill Brick & Tile Company	Idaville	Failed to report in 1919.			
Hussey & Alford	Monon	Failed to report in 1919.			
J. T. Cuppy	Chalmers	Failed to report in 1919.			
Michael Pine	Idaville	Failed to report in 1919.			

WHITLEY COUNTY.

Name of Firm or Individual	Location	Began Operations	Capital Invested	Kind of Product	System of Drying
W. M. Crabill	Collins	Failed to report.			
S. F. Trembley Co.	Columbia City	Failed to report.			
J. D. Sherwood & Son	Columbia City	Failed to report.			
Erdman & Wynkoop	Columbia City	Failed to report.			
Philip Zuber	Columbia City	Failed to report.			
J. J. Stockert	Churubusco	Failed to report in 1919.			
Cotterly & Pontizius	Columbia City	Failed to report in 1919.			

CHAPTER V.

KAOLIN.

INTRODUCTION.

The deposits of kaolin in southern Indiana have long attracted the attention of the scientist. The occurrence of deposits of kaolin of extraordinary purity and crystalline appearance apparently interstratified with sedimentary rocks was to him an extremely unusual phenomenon.

Less frequently has the attention of the manufacturer been attracted by the commercial possibilities presented through its utilization. Since the discovery in 1874 of a thick bed of kaolin on the property long owned by the late Dr. Joseph Gardner in Lawrence County the kaolin from this deposit has been used intermittently.

In the autumn of 1916 the writer began investigations with a view to determining the origin and of extending the utilization of the koalin. One line of investigation led to the testing of trial mixtures of kaolin and Indiana fire clays for the possible manufacture of refractories. The other line of investigation concerned itself chiefly with a study of the origin of the kaolin. The results of these and other lines of investigations are embodied in this report.

Following the entrance of the United States into the European war, the demand for kaolin suitable for the manufacture of glass pots, white ware, refractories and for purposes connected with chemical warfare became very insistent, for the reason that the European supplies were no longer available because of transoceanic transportation difficulties.

Upon request, the writer secured and made shipments of Indiana kaolin to several ordnance laboratories and to the Bureau of Standards. Shipments of many cars of kaolin were made also from the Gardner mines. This kaolin was used near St. Louis, Mo., in the manufacture of a refractory material called "Malinite."

At the request of Professor H. Reis, who had the investigation of clays and kaolins in charge for the United States Geological Survey, the writer prepared a brief report on the kaolin of Indiana. This report has not been published. It contains a brief discussion of the geological occurrence, the origin and the geographical distribution of the kaolin.

The kaolin of southern Indiana was first mentioned in the geological literature by Leo Lesquereux[1] who suggests that it is a very soft ochrous clay which has resulted from a burning out of a bed of coal. This deposit is near Dover Hill in Martin County.

The occurrence of the white kaolin in Lawrence County is first referred to by E. T. Cox[4] as follows: "One of the most interesting as well as valuable discoveries made during the year is a large bed of White Porcelain Clay in the Carboniferous rocks of Lawrence County." He says that the bed, stratified with the rocks, is from five to six feet thick and may be traced over a large area of land.

"The principal body of clay is on section 21, town 4, range 3. This property has been purchased by Dr. J. Gardner of Bedford, Lawrence County, who has associated with him Messrs. Tempest, Brockmand and Co., the pioneer potters of Cincinnati. This firm has given the clay a thorough practical test and finds that it makes a beautiful white ware equal to the best English ironstone china."

To the kaolin Cox applied the name "Indianaite" and gave a discussion of its origin, which he assigned to the decomposition of a bed of limestone.

Maurice Thompson[8] discusses the mode of origin suggested by Cox and asserts that a large part of the silica composing the kaolin must have been contained in the clay. He thinks the silica was derived from the sandstone by the leaching action of meteoric waters.

W. H. Thompson[6] supports the view of Maurice Thompson and cites examples of the deposition of silica carried in solution by meteoric water.

Geo H. Ashley[3] called attention to the fact that Lesquereux's view of origin has been overlooked and expresses his belief that the latter has hit upon the proper solution to the problem.

W. S. Blatchley[22] says in concluding a discussion of the origin: "While the facts at hand are not sufficient to fully justify either of the conclusions above given as to the origin of the kaolin, that of Lesquereux and Ashley is by far the more plausible. It at least accounts, according to the laws of chemistry, for the presence of the silica and alumina which, with the combined water, made up 98.61 per cent of the deposit."

ACKNOWLEDGMENTS.

The writer desires to acknowledge his indebtedness to the authors of the publications named in the bibliography accompanying this report; to Professor E. R. Cumings who assisted in the microphotographic work and gave many helpful suggestions during the progress of the investigations; to Professor R. E. Lyons for assistance in the chemical investigations; to Professor E. O. Jordan for assistance in the bacterial investigations; to Professor J. A. Badertsher who assisted in securing microphotographs of the bacteria; to Dr. C. A. Malott and other members of the field party of 1919; to Mr. Jacob Papish and Mr. C. C. Beals for field assistance; to Mr. P. B. Stockdale for assistance in the field and in the photographic work; to Mr. J. R. Reeves for assistance in the field and in the preparation of the drawings; to Mr. Willis Richardson and Mr. Howard Legge for assistance in the laboratory; and to Miss Alice O'Connor for stenographic work.

BIBLIOGRAPHY.

[1] Leo Lesquereux, Geological Reconnaissance of Indiana, 1862, p. 320.

[2] H. Reis, Clays, 1914, p. 351.

[3] Geo. H. Ashley, Twenty-third Ann. Rept. Ind. Geol. Surv., 1898, pp. 931, 942, 953.

[4] E. T. Cox, Sixth Ann. Rept. Geol. Sur. Ind., 1874, p. 15.

[5] Maurice Thompson, Fifteenth Ann. Rept. Geol. Sur. Ind, 1886, pp. 34-40.

[6] W. H. Thompson, Sixteenth Ann. Rept. Geol. Sur. Ind., 1888, pp. 77-80.

[7] Zeitschr. f. Geologie, 1910, p. 353.

[8] W. N. Logan, Clays of Mississippi, Miss. Geol. Survey.

[9] W. S. Blatchley, Twentieth Ann. Rept. Geol. Sur. Ind., 1895, p. 103, etc.

[10] John Collett, Seventh Ann. Rept. Geol. Sur. Ind., 1875, pp. 358-9.

[11] John Collett, Eighth, Ninth and Tenth Ann. Repts. Ind. Geol. Surv., 1878, pp. 416-17.

[12] John Collett, Thirteenth Ann. Rept. Ind. Geol. Sur.

[13] John Collett, Fourteenth Ann. Rept. Ind. Geol. Sur., 1884, p. 9.

[14] W. S. Blatchley, Twenty-first Ann. Rept. Geol. Sur. Ind., 1896, p. 19.

[15] G. K. Greene, Second Ann. Rept. Bureau Statistics and Geology Ind., 1880, p. 447.

[16] W. S. Blatchley, Twenty-second Ann. Rept. Sur. Ind., 1897, p. 21.

[17] W. N. Logan, Proc. Ind. Acad. Sci., 1917, p. 227.

[18] W. N. Logan, Proc. Ind. Acad. Sci., 1918, p. 117.

[19] W. N. Logan, Bul. No. —, U. S. Geol. Surv., 1918, p. —, unpublished.

[20] J. F. Newson, 26th Ann. Rept. Geol. Sur. Ind., 1901, p. 285.

[21] W. S. Blatchley, 24th Ann. Report Geol. Sur. Ind., 1897, p. 28.

[22] W. S. Blatchley, 29th Ann. Report. Geol. Sur. Ind., 1904, pp. 55, 221, 231, 273, 297.

[23] W. S. Blatchley, 31st Ann. Report. Geol. Sur. Ind., 1906, pp. 42, 43.

THE PHYSICAL AND CHEMICAL PROPERTIES OF INDIANA KAOLIN.

Definition. The term kaolin is applied by Dana to a rock made up of one or more minerals included in the following group:

Kaolin Minerals.

	Silica	Alumina	Water
Kaolinite, $H_4Al_2(Si_2O_9)$	46.05	39.50	14.00
Meerschaluminite, $2HAl(SiO_4)+aq$.	43.15	41.07	15.78
Halloysite, $H_4Al_2(Si_2O_9)+aq$.......	43.50	36.90	19.60
Newtonite, $H_8Al_2(Si_2O_{11})+aq$......	38.50	32.70	28.80
Cimolite, $H_6Al_4(SiO_3)_9+aq$	63.40	23.90	12.70
Pyrophyllite, $H_2Al_2(SiO_3)_4$.........	66.70	28.30	5.00
Allophane, $Al_2(SiO_5)5H_2O$.........	23.80	40.50	35.70
Collyrite, $Al_4(SiO_8)9H_2O$...........	14.10	47.80	38.00
Schrötterite, $Al_8(SiO_{14})30H_2O$......	11.70	53.10	35.20
Gibbsite, Al_2O_3, $3H_2O$.............		65.40	34.60

Since the Indiana white porcelain-like material seemed to differ from other kaolins, E. T. Cox named it Indianaite. Because of its nearness in chemical composition to halloysite, Dana assigned it as a variety of that mineral species. It is essentially an hydrous aluminium silicate, containing in some places a bluish mineral of the approximate composition of allophane, and so called. The term which is applied to the Indianaite locally is kaolin. Some mineralogists would limit the term kaolin to residual deposits formed from the decomposition of feldspathic rocks. Under such restrictions Indianaite would not be a kaolin nor could it be called a sedimentary kaolin to distinguish it from the residual type, since it is partly of the nature of a replacement mineral, and partly of organic origin. Only a very small portion has been redeposited. The term kaolin as used by Dana, includes such minerals as kaolinite, gibbsite, halloysite and others, and so the term may be properly applied to Indianaite.

Macroscopic Appearance. The purest form of the kaolin is a white substance of porcelain-like appearance. When first taken from the deposit it often exhibits a pale sea-green color and is semi-translucent. After exposure to the air, it loses its green color and becomes opaque-white. The change is probably due to loss of water. The larger masses of the kaolin when exposed to the air break up into small irregular fragments. These fragments do not pass readily into clay. Even when reduced to a fine powder, the kaolin remains non-plastic.

Porosity. The white kaolin is porous, and when dry contains much air. When the kaolin is placed in water the air escapes rapidly from the pores, producing a sound like that accompanying effervescence. The bulk of the air is usually forced out at one or two points, the small spheres rising rapidly toward the surface of the water, like the spurt of a miniature fountain.

Absorption. The white porcelain-like kaolin is porous and readily absorbs water, though the absorbed water does not render it plastic. The amount of water absorbed is greater in the granular kaolin and least in the white porcelain variety. The per cent of absorption by weight varies from 6.6% to 13%.

Fracture. This kaolin has no definite cleavage, but there are flat smooth surfaces which resemble cleavage planes. These planes usually extend for only short distances and may meet other planes or curved surfaces at any. angle. The fracture is distinctly conchoidal in some places and very irregular in others. The fractured surfaces often resemble the surfaces of broken bisque.

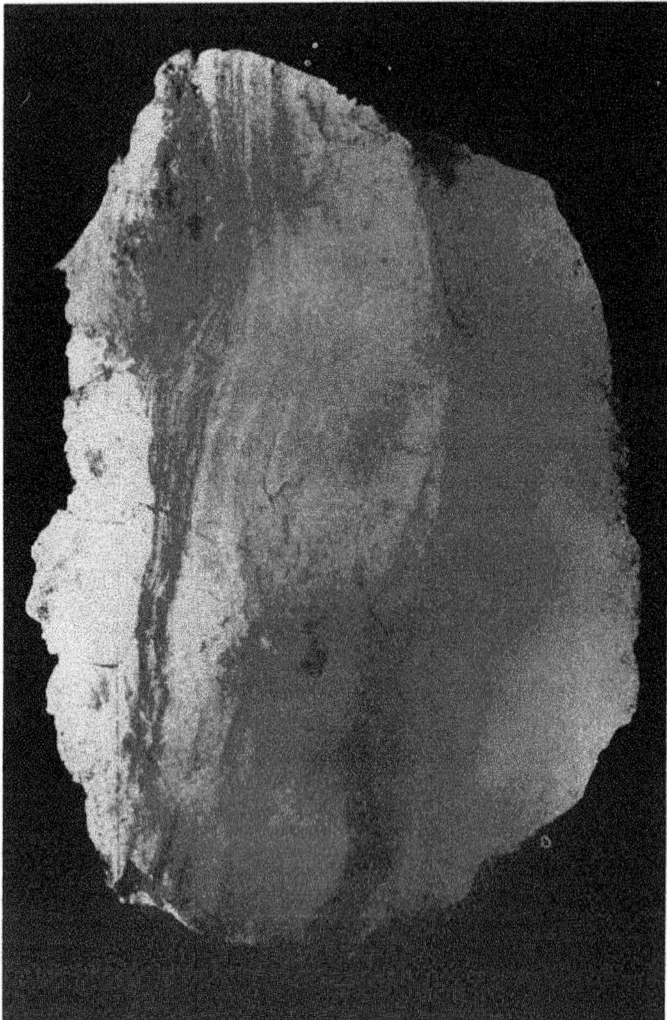

Plate XXIX. Stained kaolin from the Gardner mine. Colors, white, yellow, purple and red.

Hardness. The hardness of the white porcelain-like kaolin is usually above that of gypsum and below that of calcite, on the average it is about 2.5. There is, however, a variety which has a hardness of less than 1, being soft and plastic. This variety becomes hard on exposure to the air. The yellow and purple varieties have a hardness of 1.

Specific Gravity. The specific gravity of the white porcelain kaolin varies from 2.00 to 2.31; white granular, from 1.90 to 1.94; yellow, from 1.94 to 2.01; purple, from 2.26 to 2.33.

Macroscopic Structure. The unstained white material seems devoid of definite structure. (See Plate XXX.) It might be described as being, in gen-

Plate XXX. Masses of white kaolin from the Gardner mine, Lawrence County.

eral, coarsely granular with an occasional plate-like surface. In some places there are cell-like spaces with interior portions covered with botryoidal forms. Again, the kaolin may assume the form of concretionary masses, composed of concentric layers. The stained kaolin may be distinctly laminated, in which case yellow, brown and purple layers alternate with white. In some places the porcelain-like material exhibits a wavy laminated structure with stained lines and surfaces that are covered with projecting spherical masses of white material separating the layers of porcelain-like Indianaite, while in others it is pitted with these small white granules of clay, producing an amygdaloid-like structure. These small granules have about the same degree of hardness as the surrounding Indianaite. In some places the white Indianaite presents an irregular laminated appearance. This lamination seems to be more clearly marked in the yellow clay than in the white. In some cases, however, the laminated appearance is due to the presence of thin layers of yellow clay within the white. Concretionary masses are sometimes found in the white Indianaite. These vary in size from a few inches up to nine or ten inches in diameter. The concentric layers are separated by thin lines of black and yellow stain. The translucent form is, in places, composed of a series of small geode-like masses, arranged in a general horizontal position. The walls are about ⅛ of an inch in thickness and the diameter of the cavities from 1/10 to ⅛ of an inch. The interiors of the cavities are lined with a layer of

Plate XXXI. Masses of stained kaolin from the Gardner mine. Soft when taken from the mine.

milk-colored Indianaite which has a mammillary surface, and the cavities are sometimes wholly or partly filled with round grains of the same material. In one hand specimen five of these cavities are arranged in a line occupying a horizontal distance of two inches. Cavities of various sizes are of common occurrence in the Indianaite.

Color of Indianaite. The purest form is white or greenish-white in color. Much of it has been stained and the most common form of the stained is yellow, due to the presence of hydrated oxide of iron. There are also brown colors, due to larger amounts of the same pigment. In some places there are purple colors and black colors, due to the presence of manganese compounds; however, some of the black stains are due to organic matter. In most cases there is no regularity in the arrangement of colors, but in some cases the arrangement of colors is in bands. The kaolin under such conditions has a distinctly stratified appearance due to alternate layers of white and yellow kaolin. (See Plate XXXI.)

Microscopic Appearance of Indianaite. Under the microscope the massive white kaolin is found to be composed of minute globular granules which are often arranged in a dendritic form. The granules are translucent in appearance and have about the same index of fraction as balsam. Sometimes larger structures occur in the midst of the granules. In cross-section, these are circular bodies made up of concentric rings enclosing a centrally placed nucleus. Radial lines pass through the concentrically arranged rings from the nucleus to the periphery. These circular bodies seem to break up into small granules of the ground mass. Several stages of the process of disintegration are recognizable. There are circles in which the boundary between the periphery and the ground mass is but faintly outlined. These will contain granules but will still show some evidence of concentric and radial structure. In other cases

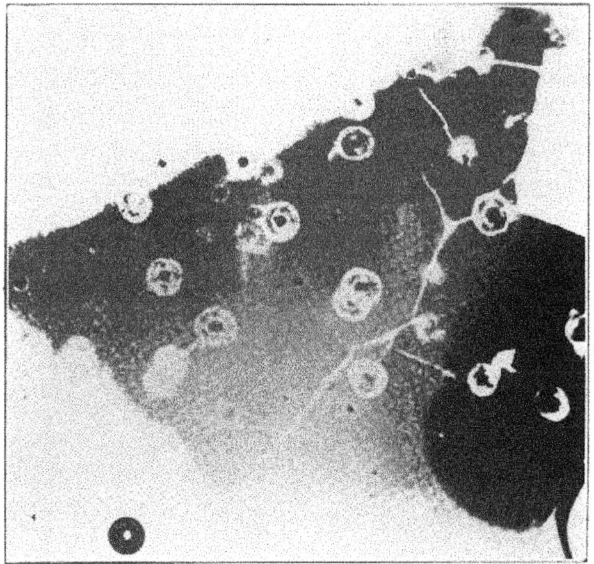

Plate XXXII. Fig. A Micro-photograph of spherules in white kaolin. Enlarged 250 diameters.

all but the nucleus has disintegrated by passing into the granular stage. In some cases the circular body has been completely disintegrated, with the exception of the dendritic or web-like mass of pigment which stained the nucleus. (See Plate XXXII, Fig. B.)

It appears, also, that even this structure may become broken up and its small dark granules be distributed among the translucent ones of the ground mass. In some sections the ground mass and the nucleus are stained yellow, but the concentric rings of the circle remain free from coloring matter. In some places there are rows of circles occupying nearly a straight line. The outer surface of these circles and the intervening ground mass are stained yellow along one side, while the opposite side may remain clear. In some circles the entire peripheral zone is stained. These circular bodies are probably the cross-sections of oölites similar to those produced in the black clay found beneath the white kaolin. Rothpletz[1] believes that "Calcareous oölites with regular zonal and radial structure" are phytogenic; "the product of microscopically small algæ of very low rank, capable of secreting lime." The oölites found in the kaolin, since they possess a zonal and radial structure, may be phytogenic.

In the case of the ferro-aluminium sulphate referred to, spherical bodies were formed on the inside walls of the vessel above the surface of the water. The surfaces of these spheres were covered with acicular projections radiating outward from the spherical surfaces. Sometimes the sulphate would pull away from the surface of the vessel under its own weight. When this occurred and fresh sulphate was deposited, it would form a thin film between the surface of the vessel and the hanging body of sulphate. This film has a convex surface and possesses both radiating and concentric lines. These lines have the same

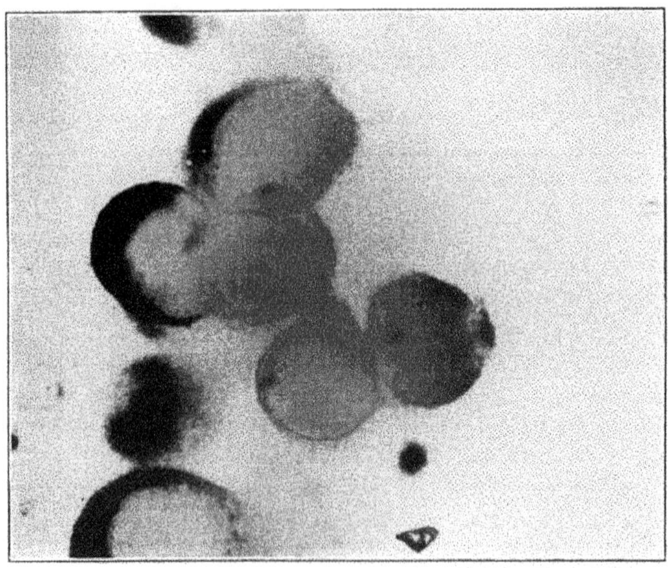

Plate XXXII. Fig. B--Spherules enlarged 800 diameters. Have radial and concentric structure and break up into granules.

appearance as those in the circular bodies. In other parts of the vessel the deposited sulphate had a botryoidal or mammillary appearance. The surfaces of the globular masses were rough. They are fastened to the surface of the vessel by sulphate of a massive structure. In some places plate-like masses were formed with their surfaces parallel with the surface of the vessel. This plate-like structure is also recognizable in the natural kaolin deposits. It would

[1] Botanisches Centralblatt No. 35, 1892, pp. 265-268.

seem that in some instances the sulphate may have been changed into the silicate without change of form.

Composition. The composition of three samples of Indianaite as published in Dana's Mineralogy is as follows:

SiO_2	Al_2O_3	H_2O	H_2O at 100°C	CaO, MgO	Alkalies	Total
39.00	36.00	14	9.50	0.63	.054	99.67
39.35	36.35	22.90		0.40	....	99.00
38.90	37.40	23.60		Undet.	....	99.90

These samples all contain less alumina than is contained in kaolinite which contains 39.5 per cent. of alumina. They approach more closely the composition of halloysite, which contains 43.5 per cent of silica, 36.9 per cent of alumina and 19.6 per cent of water and it is to hallovsite that Dana refers it.

OTHER ANALYSES OF INDIANA KAOLIN.

No.	SiO_2	Al_2O_3	Fe_2O_3	TiO_2	SO_3	CaO	MgO	K_2O	Na_2O	H_2O Comb.	Total
1	41.82	32.65	.29	.04	.085	.40		.12	tr.	18.47	93.875
2	39.00	36.00					.63		.54	9.50 14.00	99.67
3	39.35	36.35				.40				22.90	99.00
4	38.90	37.40								23.60	99.90
5	45.90	40.34				tr.				13.26	99.50
				(MnO_2)							
6	47.05	37.14	tr.	.03			.03			15.55	99.80
7	47.13	36.76	tr.	tr.			.04			15.13	99.06
8	46.00	40.20					.20			12.62	99.02
9	44.75	38.39	.95			.37	.30	.12	.23	15.17	100.28

No. 1. Soft plastic kaolin collected by the writer. L. L. Carrick, analyst.
Nos. 2, 3 and 4. Cox, 8th Annual Rept. Ind. Geol. Sur., pp. 155-156.
Nos. 5, 6 and 7. Cox, 6th Annual Rept. Ind. Geol. Sur., p. 18.
No. 8. M. Thompson, 15th Annual Rept. Ind. Geol. Sur., p. 36.
No. 9. W. A. Noyes, 20th Annual Rept. Ind. Geol. Sur., p. 105.

ANALYSES OF ALLOPHANE.

No.	AiO_2	Al_2O_3	MgO	Na_2O	H_2O Comb.	Total
1	20	40			40.00	100.00
2	15.71	42.74	.59	26.50	14.50	99.54

No. 1. Cox, 6th Ann. Rept. Ind. Geol. Survey, p. 16.
No. 2. Cox, 8th Ann. Rept. Ind. Geol. Survey, p. 156.

The allophane is a bluish colored mineral which occurs in the white kaolin. Sometimes it forms a layer of irregular thickness above sandstone or conglomerate masses in the kaolin. In other places it is a part of concretion-like masses in the kaolin. When it is exposed to the air it breaks down into the granular form of kaolin and loses its bluish tint.

THE GEOLOGICAL CONDITIONS OF OCCURRENCE OF INDIANA KAOLIN AND THE ASSOCIATED ROCKS.

Mode of Occurrence. The kaolin of Indiana lies in beds apparently interstratified with sandstones and shales. The kaolin, however, is more irregular in thickness and distribution than the adjoining beds.

One horizon of the kaolin occurs at the contact between the Chester shales and the Mansfield sandstone of the Pennsylvanian. The surface of the Chester was eroded prior to the Pennsylvanian deposition and the Mansfield was laid down upon this eroded surface. The position of the elevations and depressions in this old Chester surface can still be determined, and it is in connection with the depressions that the best deposits of kaolin have been found. For instance, on the Gardner place the contact of kaolin and Mansfield occur at about 700 feet above sea level, while only one-half mile south the limestone of the Chester occurs at 790 feet, while one mile north the Mansfield contact occurs at 720 feet above sea level.

The kaolin along the contact is closely associated with the Mansfield sandstone and with Chester shales. It always has Mansfield sandstone above and Chester shales below. In some cases the kaolin has passed upward into the sandstone a distance of as much as ten feet, apparently replacing a portion of the sandstone. At one point below the kaolin there is a dark colored clay resting on Chester shales. The dark colored clay contains quartz pebbles similar to those found in the Mansfield, and it would seem that the dark colored clay was of Pennsylvanian age.

The kaolin underlying the Mansfield rests on the Elwren shales at some points, on the Cypress shales at others, on the Golconda shale at others, and on the Hardinsburg shales at other points. In some places the limestone above the shale was removed over a given area and a basin formed by pre-Pennsylvanian erosion. Some one of the Chester shales forms the bottom of the basin and its walls are composed of Chester shales and limestones. Bowlders of limestone, erosion remnants, are scattered over the surface of the basin. This basin was then filled with the Mansfield sediments.

Kaolin in the Chester. The Chester epoch of the Mississippian period in Indiana is represented by a series of shales, sandstones and limestones. All except some of the upper members of the Chester group are represented in the railroad cut east of Huron. (See Plate XXXIII.) Resting on the Mitchell limestone at this place is 4 or 5 feet of Sample shale containing a thin layer of coal. This is followed by 14 feet of Beaver Bend limestone; followed by 24 feet of Brandy Run shale and sandstone; followed by 6 feet of Reelsville limestone; followed by 42 feet of Elwren shales and sandstones; followed by 15 feet of Beech Creek limestone; followed by 6 feet of Cypress sandstone. At Dover Hill in Martin County about 30 feet of Cypress sandstone is followed by 30 feet of Golconda shales and limestones; followed by 20 feet of Hardinsburg sandstone and shales; followed by 25 feet of Glen Dean limestone; followed by sandstone, probably Tar Springs, which is followed by the Mansfield sandstone.

The Chester contains kaolin at several horizons. On the George Cleveland farm in Orange County kaolin was found in a cave lying beneath the Sample sandstone and resting on the Beaver Bend limestone. The kaolin was formed from a thin layer of shale lying between the limestone and the sandstone. In both Lawrence and Monroe counties kaolin occurs beneath the Elwren sand-

Plate XXXIII. Cut east of Huron on B. O. R. R. Sample shale at bottom, Beaver Bend limestone, Brandy Run shale, Reelsville limestone, Elwren shale and Beech Creek limestone in order above.

stone and is formed from the shales of that division of the Chester. In Lawrence and in Martin counties, kaolin occurs below the Cypress sandstone. In a cave in the Beech Creek limestone in Section 1 in Mitchelltree township in Martin County about 18 inches of kaolin rests on the Beech Creek limestone

Plate XXXIV. Masses of kaolin from the Gardner mine. Two of these masses were plastic and soft when removed and shrinkage cracks developed.

and is succeeded above by about 5 feet of shaly sandstone which in turn is succeeded by 25 feet of massive Cypress sandstone.

Conditions of the Outcrops. Occurrences of the pure white kaolin at the outcrop are rare. Small particles of white kaolin are found distributed

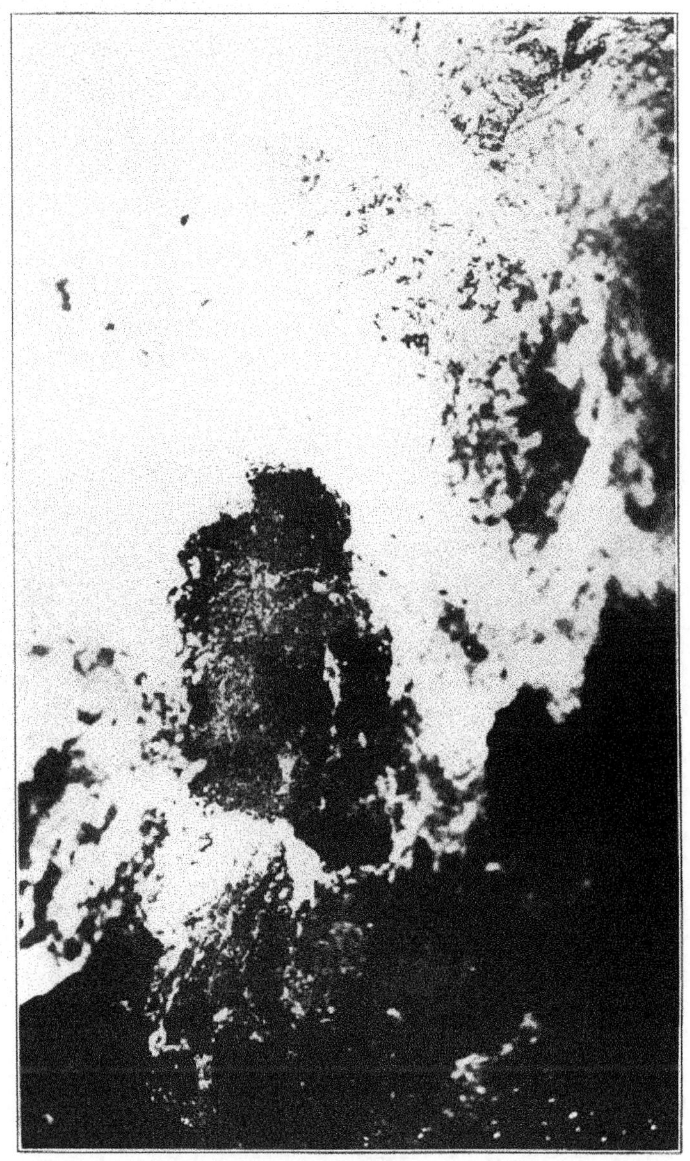

Plate XXXV. Six feet of white kaolin in the Gardner mine. Large mass of imbedded sandstone at the left

through the detritus covering the outcrop, and such occurrences are not uncommon. The weathered portion of the bed at and near the outcrop consists of a mahogany colored clay. This mahogany clay contains small fragments of the white kaolin, especially in the upper portion underneath sandstone ledges that are fairly compact and unfractured. The mahogany clay has doubtless in some instances at least originated from the staining of white Indianaite by oxide of iron. The olive-green shales in places are changed into a maroon colored clay which passes into mahogany colored clay. The following analysis shows the composition of a sample of this mahogany clay:

Analysis of Mahogany Clay.

Silica	33.04%
Alumina	17.33%
Ferric oxide	27.96%
Manganese oxide	2.22%
Calcium oxide	1.52%
Magnesium oxide	1.11%
Titanium oxide	.69%
Alkalies	1.77%
Volatile matter	13.87%
Total	99.51%

Structural Features of the Kaolin. In the mass the kaolin occurs as either coarsely granular material or as concretionary or other masses of porcelain-like appearance. These porcelain-like masses fracture into irregular fragments, exhibiting on some surfaces a conchoidal fracture. In some places the porcelain-like material exhibits a wavy laminated structure with stained lines and surfaces that are covered with projecting spherical masses of white material separating the layers of porcelain-like kaolin. In some places the porcelain-like material is pitted with these small white granules of clay, producing an amygdaloidal structure. These small granules have about the same degree of hardness as the surrounding kaolin. Under the microscope the porcelain-like kaolin exhibits a granular appearance, the grains being apparently spherical in form and in some sections arranged in a dendritic structure. The dendritic structure may be due to the loss of water and consequent shrinkage when the section is heated during its preparation. In some places the white kaolin presents an irregular laminated appearance. This lamination seems to be more clearly marked in the yellow clay than in the white. In some places the clay presents a laminated appearance due to the presence of thin layers of yellow clay within the white. (See Plate XXXIV.) Concretionary masses are sometimes found in the white kaolin. These vary in size from a few inches up to nine or ten inches in diameter. The concentric layers are separated by thin lines of black and yellow stain. The white kaolin is, as a rule, hard, but in some places is soft and very plastic, resembling in appearance a white ball of clay. In other places the plastic clay is yellow in color and is interstratified with yellow or brown sand.

Both varieties of kaolin, granular and massive, occur in a translucent body of greenish color. They also occur in an opaque white form, in which form they resemble the interior surface of broken bisque. The translucent form is, in places, composed of a series of small geode-like masses arranged in a generally horizontal position. The walls are about ⅛ of an inch in thickness and the diameter of the cavities from 1/10 to ⅛ of an inch. The interiors of the cavities are lined with a layer of milk-colored kaolin which has a mammillary surface. The cavities are sometimes filled or partially filled with rounded grains of the same material. In one hand specimen five of these

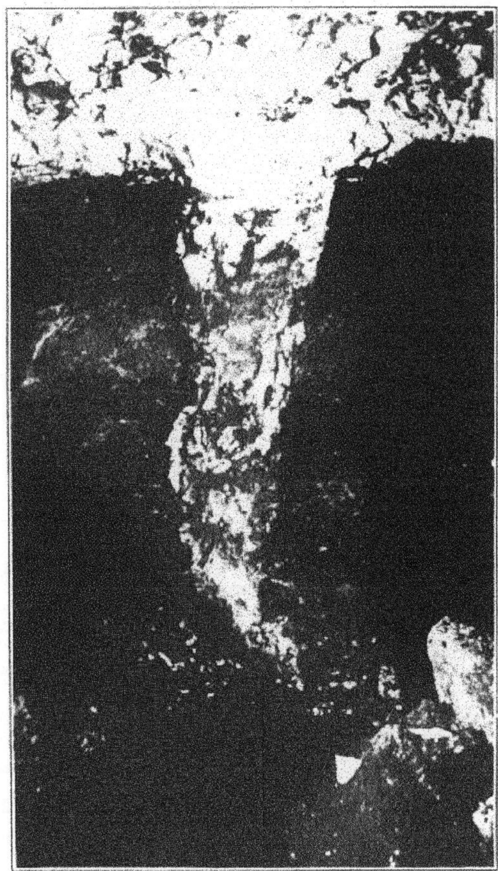

Plate XXXVI. Vein-like mass of white kaolin in sandstone in Gardner mine.

cavities are arranged in a line occupying a horizontal distance of two inches. Cavities of various sizes are of common occurrences in the Indianaite. The porcelain-like form fractures irregularly and also with a conchoidal fracture. Its conchoidal fracture resembles that of flint, and its irregular fracture that of chert. The composition of a sample of the mine-run white clay is as follows:

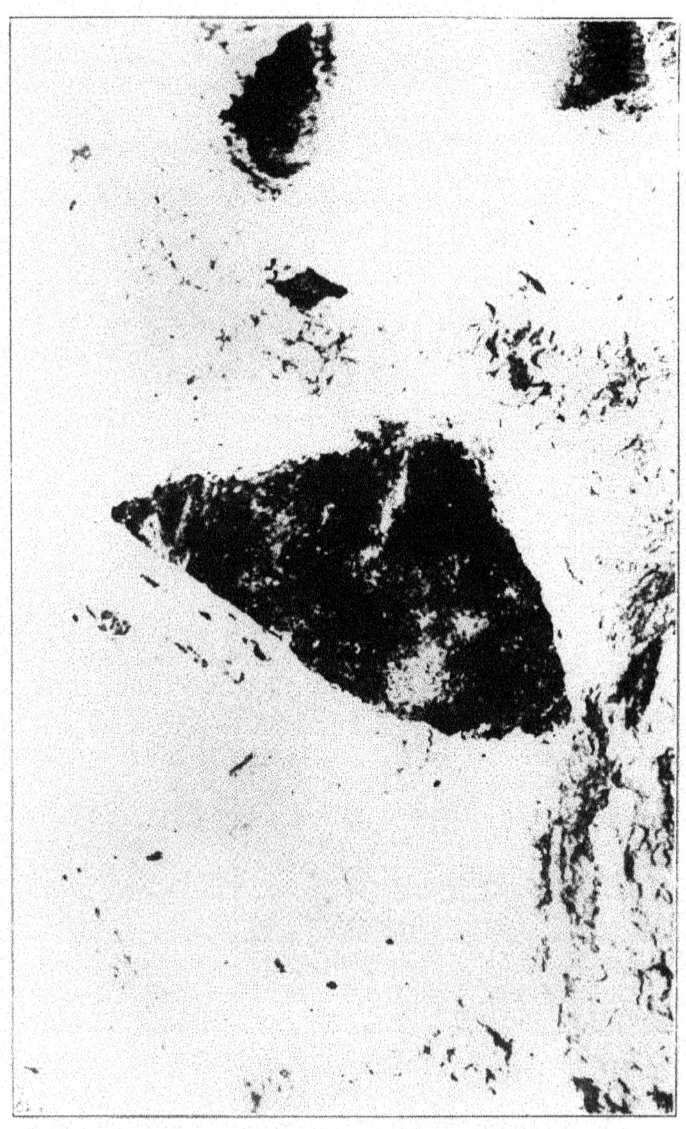

Plate XXXVII. Triangular mass of sandstone in Gardner mine surrounded by white kaolin. Coarse grains of sand cemented with kaolin.

Analysis of Mine-run Sample of White Indianaite.

Silica	40.48
Alumina	39.60
Hygroscopic water	5.01
Combined water	15.03
Ferric oxide	.11
Total	100.23

Layers of coarse or disconnected masses of coarse sand or sandstone, stained with iron oxide and containing small particles of white kaolin, occur in many portions of the bed of kaolin. (See Plate XXXV.) The separated block masses of sandstone are sometimes almost, if not completely, surrounded with white Indianaite. Occasionally a mass of sandstone is penetrated by a vein-like body of kaolin. (See Plate XXXVI.) In one instance a triangular piece of sandstone occurs surrounded on all three sides by white kaolin. (See Plate XXXVII.) In another instance a dark laminated neck-like body of clay extends from a floor of similar material up into the kaolin. This clay contains pellet-like bodies of white clay which are soft and plastic in situ, but hard and granular when dry. This clay also contains white quartz pebbles such as are found in the basal sandstone members of the Pottsville. The clay has a fetid odor similar to that of muck. It also contains concretionary masses of marcasite. At another point in the bed of kaolin next to the sandstone roof there were white and dark layers of kaolin above, then a layer of white, partially laminated kaolin below, then a four-foot layer of sandstone occupying the lower part of the deposit. The sandstone layer was divided near its central portion by a vien of white kaolin extending through the four-foot layer of sandstone. The sandstone layer has a thin fin extending upward obliquely into the white kaolin. This fin is about five feet in length and one foot thick in its thickest portion. Above the lens of sandstone there are several concretion-like masses of kaolin, the porcelain-like layers of which are separated by dark stains. (See Plate XXXVIII.)

Associated Rocks. The rock overlying the kaolin in all cases as far as observations have extended is sandstone. This sandstone is assigned in some instances to the Pennsylvanian, in others to the Mississippian. The character of the layer in immediate contact with the upper surface of the kaolin is variable. In some places the roof is composed of soft pink sandstone; in other places the sandstone is conglomeratic; in other places it is firmly cemented and almost iron ore in composition. The underlying rock is in most places a shale or clay. In one instance a small body of white kaolin is completely surrounded with sandstone. In other instances there are thin layers of sandstone underlying the kaolin. In some places there are lens-like masses of limonite occurring at the base of the kaolin. These masses of limonite contain quartz pebbles, characteristic of the Pottsville formation. Within the mahogany clay irregular eroded masses of limestone are sometimes found. Microscopic sections of this limestone show it to be highly fossiliferous. Within the kaolin bed sandstone, sand, conglomerate and concretionary masses of limonite constitute the principal foreign materials of consequence. A sample of the limonite taken from the base of the kaolin has the following composition:

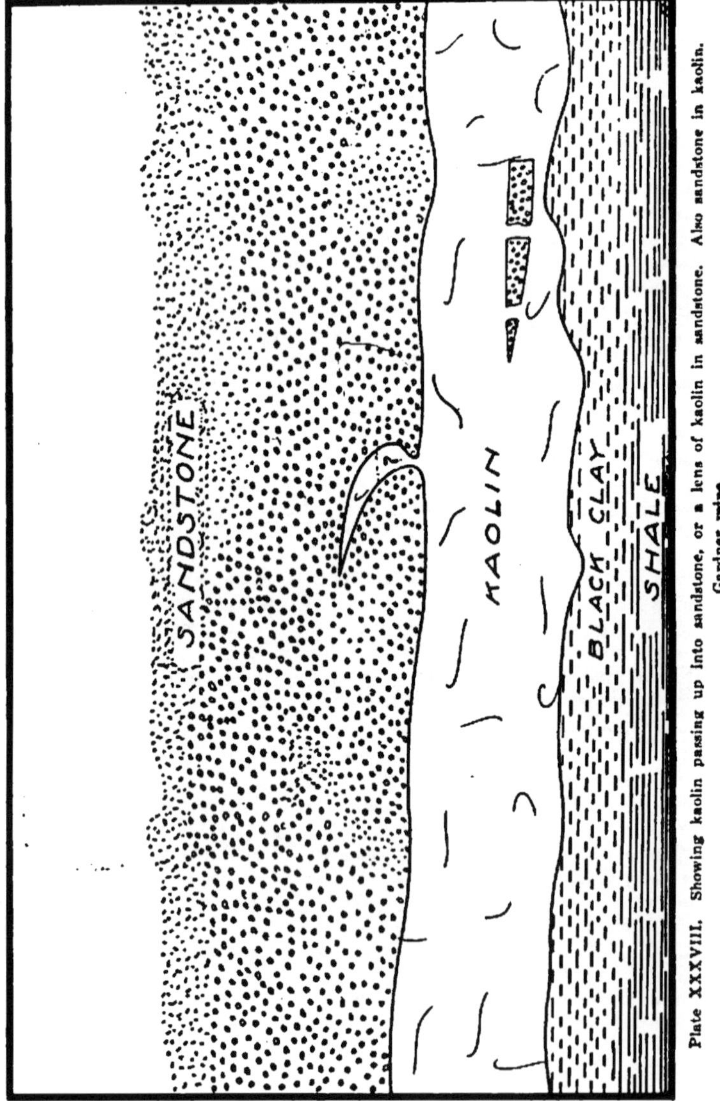

Plate XXXVIII. Showing kaolin passing up into sandstone, or a lens of kaolin in sandstone. Also sandstone in kaolin. Gardner mine.

Analysis of Limonite from Base of Indianaite.

Iron oxide	83.73
Insoluble matter	2.56
Loss on ignition	12.02
Total	98.31

Quantity of Indianaite. A question of absorbing interest is, What is the quantity of white clay? It is a question which under the present state of development is difficult to answer. If one were to base his judgment upon the number of outcrops he would be compelled to say that the quantity was large. If he bases his judgment upon the results of the meager development or upon the visible quantity of mahogany clay he would be forced to a similar conclusion. After going over the greater part of the field, the judgment of the writer is that there is a large quantity of clay of all grades. In fact that it underlies thousands of acres of land in the counties mentioned. How much of it is of the pure white variety cannot be estimated from data which is at hand at present, but that the total amount is large, is probable.

ORIGIN OF INDIANAITE.

The question of the origin of Indianaite has been the source of much speculation and discussion. Two theories of origin have been advanced in the geological literature of Indiana.

The Coal-Ash Theory. Leo Lesquereux[1] suggested that the clay was formed by the burning out of a bed of coal. Speaking of the clay near Dover Hill, he says: "On both sides of the place where this coal is worked there is a bank of very soft ochrous clay, a true powder, as fine as flour, without any trace of coal, though occupying exactly the same horizon. It is overlaid by a clay iron ore, which looks as if it had been roasted. I consider this local formation as the result of the burning of the bank of coal at places where it was exposed along the creek." Assuming, as suggested by Lesquereux, that the clay was formed from the ash of the coal, it would require a bed of coal, the thickness of which would be far beyond any recorded thickness. The coal bed which Lesquereux suggested had been burned out is Coal 1, the ash of which is generally well under 10 per cent. Assuming the ash to be 10 per cent, it would require 110 feet of coal to produce the maximum thickness of kaolin. If we assume an ash of 25 per cent it would still require a bed of coal of a thickness of 44 feet to produce the maximum thickness of kaolin. If it be assumed that kaolin was formed from beds of clay lying adjacent to the coal, it would seem, as suggested by Reis[2], that the burning of the coal would produce dehydration of the clay, evidence of which is lacking. The writer has seen clays which have been dehydrated by the burning out of coal beds in several states, but he never saw anything approaching kaolinization of the dehydrated clay. If a bed of coal had been burned out, it would cause a disturbance of the overlying sandstone, break it up, fracture, fault and possibly brecciate it. No evidence of such disturbance has been detected, although the writer has examined much of the sandstone roof above

[1] See Leo Lesquereux's report of the Geological Reconnaissance of Indiana, 1862, p. 320.
[2] See Reis' Clays, 1914, p. 351.

the kaolin which has already been mined. Moreover, it is probable that the burning of the bed of coal in contact with the sandstone would cause the fusion of the ferruginous sandstones of the latter, at least at the point of contact. No evidence of such fusion exists, or even the slighest evidence of incineration.

The writer examined the outcrop of kaolin at Dover Hill and found that it lies at the contact of the Mansfield sandstone with the Hardinsburg shale. Five hundred feet to the east no kaolin is present, but twenty feet of sandstone lies below the Mansfield and ten feet of limestone is exposed below the sandstone. No coal occurs at the elevation or horizon of the kaolin, but coal occurs on the east side of Dover Hill in the Mansfield at a higher elevation than the kaolin. There is a pink colored, coarse grained sandstone overlying the kaolin on the north side of Dover Hill, but about one-quarter of a mile

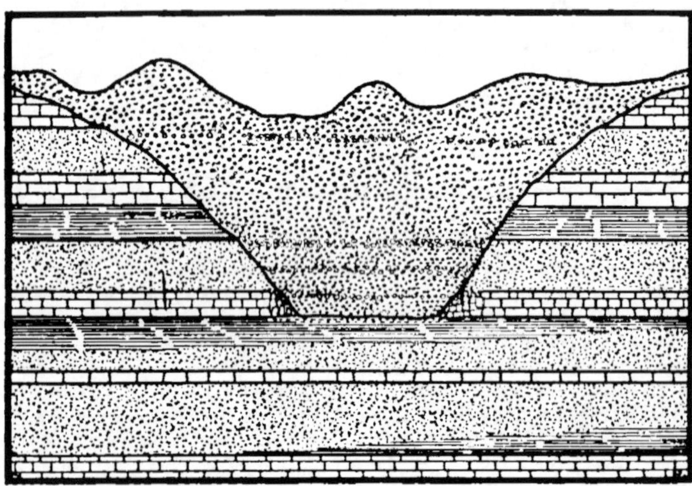

Plate XXXIX. Diagram showing nature of the Chester-Mansfield contact in Martin County. Vertical scale greatly exaggerated. Lowest contact on the Elwren shale in this section.

southwest there is an outcrop of kaolin at about the same elevation underlying sandstone containing limonite. In case of burning out of coal underlying it the limonite would have lost its water of crystallization and been converted into hematite. At this point the kaolin occurs at the Mississippian-Pennsylvanian contact, (see Plate XXXIX) but the coal occurs above the contact in the Pennsylvanian. At Dover Hill the coal is higher than the kaolin when referred to either sea level or stratigraphical position. At no place has the writer seen a coal bed near a bed of kaolin at the same stratigraphical horizon.

The method of making road ballast by burning log heaps, covered with clay, is a familiar one. The material produced is very similar to that of the dehydrated clay or burned-out coal beds. The latter is, if anything, a little more vitrified. Burned clay ballast has been used for years on wet marshy lands and not noticeably hydrated. The point west of Shoals mentioned by Ashley[1]

[1] See Ashley's 23d Annual Report, Geol. Survey of Indiana, p. 931.

is not at the same elevation as the coal. The kaolin is about seven feet higher than the coal.

The kaolin underlies massive conglomeratic sandstone and has associated with it masses of limonite. Chester shales lie below the kaolin, so it is evident that the kaolin lies at the contact between the Mississippian and the Pennsylvanian. The coal occurs about 500 yards west of the kaolin outcrop and lies between beds of shale. This shale seems to belong to the Pennsylvanian. It underlies a ledge of massive sandstone similar in appearance to that under which the kaolin lies. The Mississippian-Pennsylvanian contact must be below this shale. The line of unconformity dips strongly toward the west at this point. The kaolin is deposited on the line of the unconformity and the coal is deposited above the unconformity, but seven feet lower than the kaolin. Though it does not occur at this point it is entirely possible for coal to be deposited on the line of the unconformity at one point and the conditions favorable to the formation of kaolin to occur at another point on the unconformity. While the kaolin of the Pennsylvanian is, in all outcrops observed, connected with the unconformity, in no case has the coal been directly connected with the unconformity, as at least a few feet of shales intervene between the coal and the unconformity. The fact that limonite exists in and above the kaolin at this point increases doubt of the burned-out coal theory.

Kaolin occurs at three different horizons in the Chester, occurring under the Sample sandstone, the Elwren sandstone and the Cypress sandstone. In one place the kaolin under the Elwren has a thickness of six feet. Very little coal occurs in the Chester in Indiana. A few inches of coal occur at widely separated points in the Sample and in the Elwren. It seems an assumption impossible of demonstration that the kaolin of the Chester was formed by the burning out of coal beds.

Residual Limestone Theory. This theory was suggested by E. T. Cox,[1] who in describing a geological section in Lawrence County containing kaolin, says: "It will be seen from the above section that the clay lies immediately beneath the Millstone grit or pebbly conglomerate of the coal measure and here occupies the place of a bed of Archimedes limestone which is seen in situ about two miles southwest of the mine. The overlying sandstone is very ferruginous, and the base, where exposed to the weather, has decomposed and covered the clay in places to a depth of eight or ten feet with ferruginous sand and pebbles. There is a constant oozing of water from this sandstone which has, no doubt, played an important part in the chemistry of the clay and hematite deposit, for though similar in its chemical composition to kaolin, this clay differs physically and owes its origin to an entirely distinct set of causes and effects. While the former is derived from the decomposition of the feldspar of feldspathic rocks, such as granite, porphyry, etc., the porcelain clay of Lawrence County has resulted from the decomposition, by chemical waters of a bed of limestone and the mutual interchange of molecules in the solution, brought about by chemical precipitation and affinity. Where cavities existed in the limestone at the base of the strata, there the chalybeate water found the oxygen to change the carbonate into sesquioxide of iron, which finally filled up the cavity. In places you can trace the passage of the ferruginous water along irregular joints in the clay bed, by the iron-stained

[1] See E. T. Cox, 6th Annual Report of the Geological Survey of Ind., 1874, p. 15.

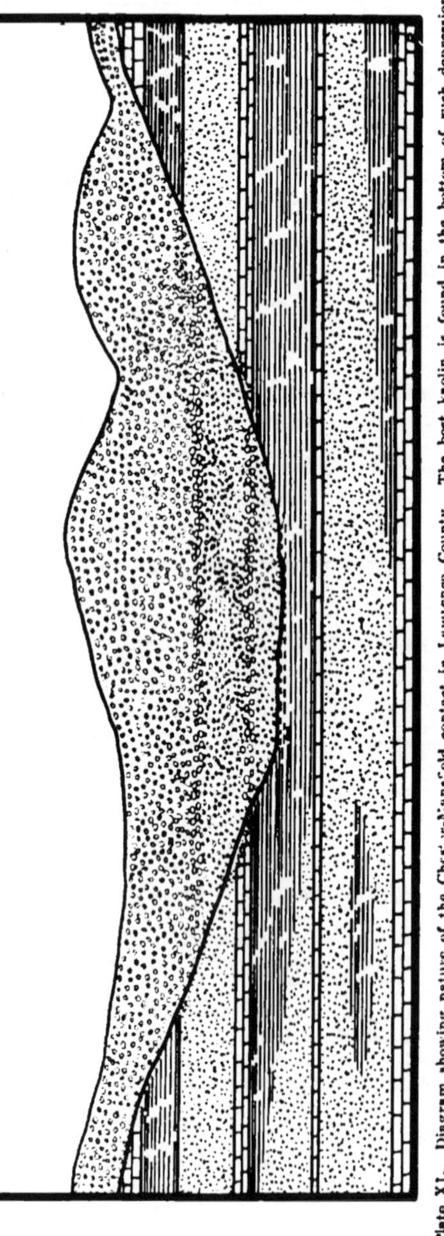

Plate XL. Diagram showing nature of the Chester-Mansfield contact in Lawrence County. The best kaolin is found in the bottom of such depressions. Vertical scale exaggerated.

path which it has left, to the brown hematite ore which lies in a mass at the bottom. The largest beds of hydrated sesquioxide of iron both in Europe and America are found at the base of the Millstone grit and filling up cavities in the cavernous sub-carboniferous limestone."

It appears that the conclusion arrived at by Cox, that the kaolin has been formed by the decomposition of a bed of limestone, "which is seen in situ, 2 miles southeast of the mine," is based on an error. The Archimedes limestone mentioned lies about eighty feet higher than the position of the kaolin because of the unconformity which exists between the Chester and the Pennsylvanian at this point. In the basin in which the kaolin lies more than ninety feet of Chester rocks have been cut out. (See Plate XL.) Although there are in this region numerous occurrences of sandstones overlying beds of limestones and these limestones have been and are being decomposed under the influence of meteoric waters, there exists no evidence of the formation of kaolin, even in its incipient stages under the conditions suggested by Cox. If the kaolin were formed by a dissolving out of beds of limestone, underlying sandstones, we should expect to find evidence of disturbance in the overlying sandstone layer. No evidence of such disturbance exists.

The Mississippian limestones contain a very small amount of insoluble materials, perhaps not more than 10 per cent at the most, and it would require the decomposition of a bed of limestone of the thickness of at least 40 feet, in order to produce the maximum thickness of kaolin. It has been suggested[1] that part of the material composing the kaolin was brought down by meteoric waters from the overlying sandstone. This view seems to me untenable, because there are numerous caves occurring in limestones in this region which have an overburden of sandstone and there are no evidences of these cavities being filled with material brought from above, although there is abundant evidence of percolating waters.

The occasional occurrence of limestone bowlders in connection with kaolin outcrops is a matter requiring explanation. The writer believes them to be erosion remnants left on the surface of the Chester shales during the pre-Pennsylvanian period of erosion. The facts that the kaolin is always found at the contact and overlying Chester shales and that the limestone bowlders belong to Chester limestones, seem to support this view. The conditions are represented in the diagram in Plate XLI. In Section 8, a mile and a half east of Huron on the Wilson farm, there is an outcrop of Beech Creek limestone resting on Elwren shale with a spring at the contact. Above the limestone the Cypress sandstone forms a structural terrace on which the Wilson house stands at an elevation of 720 feet above sea level. No Mansfield is present at this point and no kaolin. One-fourth of a mile northwest of the spring the fifteen feet of Beech Creek limestone and forty feet of Cypress have been cut out and the Mansfield rests on a bed of kaolin which in turn rests on the Elwren shale at an elevation of 660 feet above sea level. A few limestone bowlders have been found in the mahogany clay associated with the kaolin. The limestone undoubtedly gradually thickens toward the spring, where the full thickness is revealed. One mile north of the kaolin the Mansfield rests on the Cypress at an elevation of 695 feet above sea level. It is evident that the kaolin has been formed on the Elwren shale in a pre-Pennsylvanian basin or trough.

[1] See Thompson, 15th Annual Report, Indiana Geological Survey, 1886, p. 37.

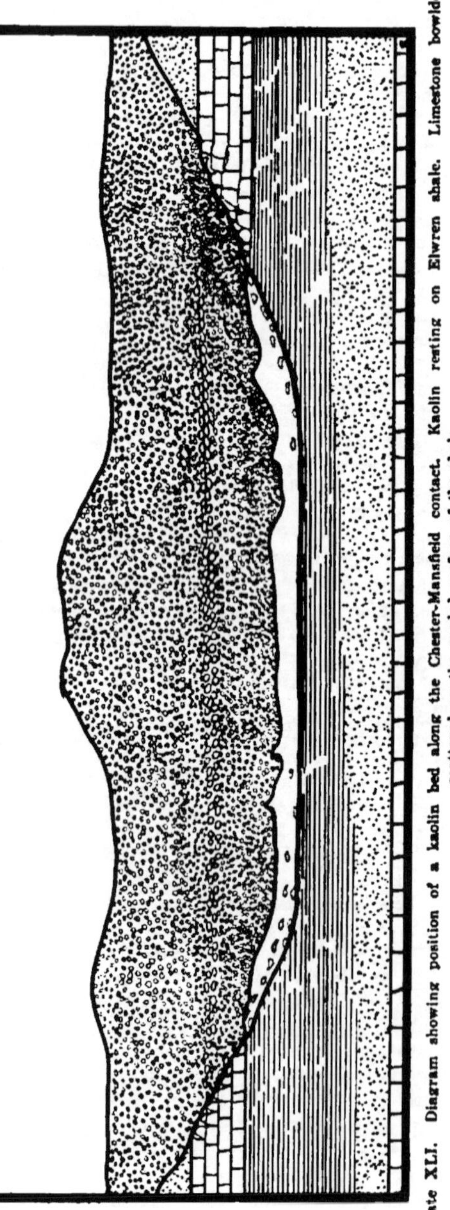

Plate XLI. Diagram showing position of a kaolin bed along the Chester-Mansfield contact. Kaolin resting on Elwren shale. Limestone bowlders scattered on the eroded surface of the shale.

Kaolinization. Were such deposits of white kaolin as those of Indiana to occur in regions of profound crustal disturbance and vulcanism, no time would be lost in assigning their origin to hydrothermal action. But when the geological occurrence does not justify a belief in hydrothermal conditions the problem assumes a more difficult aspect. Upon the discovery of such deposits in practically undisturbed sedimentary rocks one is naturally inclined to assign their origin to weathering, the chemical action of meteoric water or possibly the refinements of selective deposition. So, for Cox to appeal to the weathering of limestone as the origin of the Indiana kaolin was natural, as it was also for the Thompsons to support him in this view. Weiss[1] emphasizes the importance of moor waters as agents of kaolinization, asserting that they contain all the necessary compounds, such as carbon dioxide and organic matter, while oxygen, which is detrimental to the process, is absent.

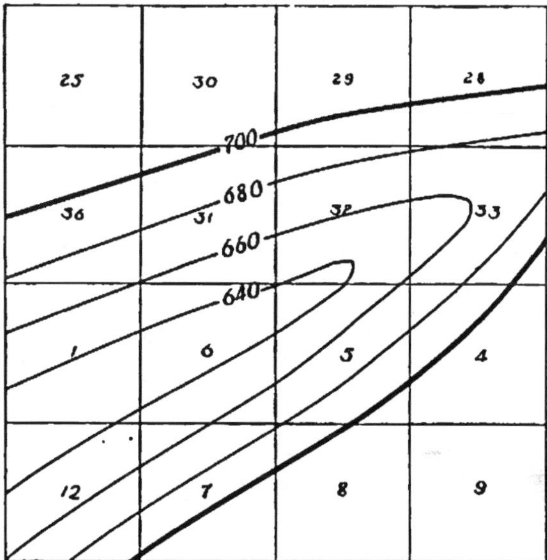

Plate XLIII. Diagram showing contour lines drawn on the Chester-Mansfield contact east of Huron. Contour interval 20 feet. Kaolin occurs in the basin.

To confirm this view experimentally, he proceeds as follows:

He introduced 50 grams each of a greenish and of a yellow clay in separate flasks of 5 liters capacity, to which flasks a mixture of moor water and distilled water was added. He passed carbon dioxide through one flask, and to the other he added grape-sugar and yeast to bring about fermentation, thus generating carbon dioxide. In the later experiments the moor water was replaced by an extract obtained from the treatment of fresh peat with distilled water. The flasks were kept at a temperature of 45°-50° and shaken daily for a period of nine weeks. The sediments were removed, washed with distilled water, and, on analysis, the following results were obtained:

[1] Zeitschr. f. Geologie, 1910, p. 353.

Yellow Clay.

	Original material (dried)	Sample treated with moor water and CO_2	Sample treated with grape-sugar and yeast
SiO_2	(A) 60.90	(B) 62.80	(C) 62.15
Al_2O_3	24.74	30.84	29.84
Fe_2O_3	4.61	1.14	1.64
CaO	3.05	0.60	1.69
MgO	2.43	1.70	1.49
Na_2O	2.05	1.00	0.88
K_2O	3.20	2.40	3.03
	100.98	100.48	100.72

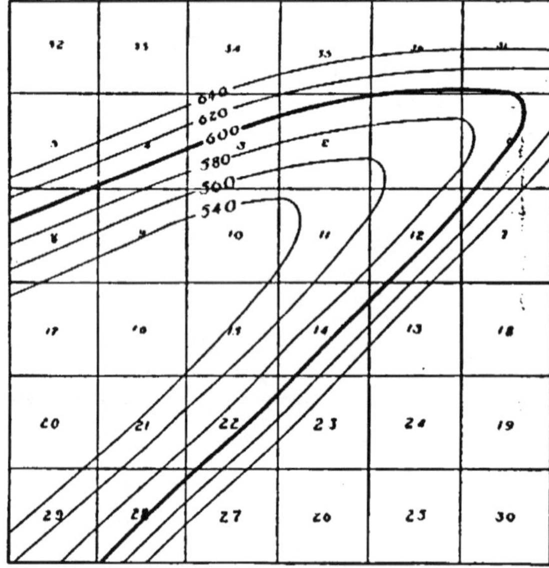

Plate XLII. Diagram showing a pre-Pennsylvanian basin carved in the Chester northeast of Dover Hill in Martin County. Contours drawn on the contact. Contour interval 20 feet.

Greenish Clay.

SiO_2	(D) 56.85	(E) 55.22	(F) 55.10
Al_2O_3	30.22	34.90	35.80
Fe_2O_3	2.82	1.25	1.04
CaO	2.18	2.13	2.01
MgO	2.47	2.05	2.10
Na_2O	2.26	0.97	1.60
K_2O	3.72	3.20	2.50
	100.52	99.72	100.15

The ratio of aluminium oxide to silicon dioxide in the above sample is as follows:

(A)	Al_2O_3	:	SiO_2	40.62	:	100
(B)	Al_2O_3	:	SiO_2	49.11	:	100
(C)	Al_2O_3	:	SiO_2	48.03	:	100
(D)	Al_2O_3	:	SiO_2	53.16	:	100
(E)	Al_2O_3	:	SiO_2	63.20	:	100
(F)	Al_2O_3	:	SiO_2	64.97	:	100

As seen from the preceding analyses, the silica content of the samples decreased on treatment. But the most remarkable thing has been the increase in the proportion of alumina to silica on treatment, thus showing that the clays were approaching the composition of kaolin. The yellow clay, when kaolinized by the process described, approached, at its final stage, the composition of the greenish clay, while the latter, which was naturally nearer kaolin, passed through a higher stage of kaolinization.

It should be noted that the method of procedure of Weiss in these experiments made possible the introduction of bacteria, since they were undoubtedly present in the moor water and the fresh peat, and that these bacteria may have been partly or wholly the agents of kaolinization.

Bio-chemical Theory. From a study of conditions in the field and from investigations conducted by the writer in the laboratory, he is led to suggest another theory of origin for the Indiana kaolin. He is not unaware of the fact that there are conditions existing which are difficult of explanation under the suggested hypothesis, but he believes they are less difficult of explanation than under those which have been proposed previously.

Stratigraphical Conditions. The kaolin deposits of Indiana are confined to the Chester division of the Mississippian period and to the Mansfield division of the Pennsylvanian.

The Chester formations of the Mississippian period in Indiana consist of a series of shales, sandstones and limestones. The following section exhibits the usual stratigraphic conditions and relation to the overlying Pennsylvanian:

Chester-Mansfield Section.

Sandstone, coarse grained, quartz pebbles, iron ore (Mansfield)	100-200′
Unconformity—Kaolin in places	0-11′
Sandstone, fine grained, massive (Tar Springs)	30-45′
Limestone (Glendean)	15-25′
Sandstone, ripple marked, thin bedded (Hardinsburg)	25′
Limestone with shale (Golconda)	50′
Sandstone with shale (Cypress)	20-35′
Limestone breaking into irregular blocks (Beech Creek)	12-16′
Sandstone and shale (Elwren)	30-50′
Limestone weathering red (Reelsville)	4-10′
Sandstones and shale (Brandy Run)	20-50′
Limestone (Beaver Bend)	10-14′
Sandstone and shales (Sample)	25′
Limestone (Mitchell)	

The Mansfield may rest at a given point on any one of the formations of the Chester. The line of the unconformity is very steep in places, being more than one hundred feet to the mile. For instance in section 28, in Spice Valley Township, in Lawrence County, at the north line of the section, the Mansfield rests on the Elwren shales at an elevation of 700 feet above sea level, while only one-half mile south the Golconda limestone occurs at the surface at an elevation of 790 feet above sea level. Kaolin has been found along the contact where the Mansfield rests on the shales of the Golconda, the Hardinsburg, the Cypress, the Elwren and the Sample. (See Plates XL, XLI, XLII and XLIII.)

These shales are of highly aluminous character, as the analyses of the following samples show:

Analyses of Samples of Chester Shale.

Constituents	No. 1	No. 2	No. 3	No. 4	No. 5
Silica	61.24	64.57	58.40	58.68	59.67
Aluminum oxide	18.38	18.67	20.70	21.39	19.75
Iron oxide	4.76	6.22	6.58	6.22	7.32
Calcium oxide	.70	.10	.15	.14	.20
Magnesia	1.94	.65	.86	.82	1.02
Alkalies	3.94	2.62	3.16	2.59	2.42
Loss on ignition	10.04	7.17	8.89	8.90	8.60
Sulphur trioxide		tr.	1.26	1.26	1.02
Total	101.00	100.00	100.00	100.00	100.00

Unweathered, these clays are of an olive-green tint, but under the action of weathering agents they assume a maroon color. The maroon color is probably produced by an iron stain which is formed by the oxidation of pyrite in the shale. Under certain conditions the maroon shale is changed to a deeper red, mahogany colored clay. The mahogany clay often contains small particles of white kaolin.

The composition of a sample of the mahogany clay is given in the following analysis:

Analysis of a Sample of Mahogany Clay.

Constituents	Per Cent
Silica	33.04
Alumina	17.33
Ferric oxide	27.96
Manganese	2.22
Calcium oxide	1.52
Magnesium oxide	1.11
Titanium oxide	.69
Alkalies	1.77
Volatile matter	13.87
Total	99.51

During the erosional period in the Mississippian, large quantities of this mahogany clay and the underlying shale were removed, bleached, somewhat purified and redeposited with organic matter. The same process occurred during the erosional period, existing between the emergence of the Mississippian and the deposition of the Pennsylvanian.

During the early stages of the Pennsylvanian, depositions of a portion of these clays were worked out, sorted and redeposited along with organic matter. These deposits of clay were further purified by a bio-chemical process, mentioned later. The method of separation of the pure Indianaite from the impure clay was revealed in one locality of Lawrence County, where there is a bed of kaolin overlying a bed of dark colored laminated clay, containing small globules of the soft white kaolin. The white clay appears to be composed of minute granules, and to have been secreted by micro-organisms. The method of deposition of a part of the kaolin in this particular instance seems to be somewhat as follows: The porous layer of dark colored clay becomes filled with water from the surface and the secreted kaolin is carried upward and is deposited against the lower surface of the kaolin bed as the water is evaporated. Gradually the layer of white kaolin becomes thick by the constant addition of material to the underside. The porous nature of the kaolin permits the free passage of the water upward, so that the process is not checked even after there has been considerable accumulation of the kaolin. The irregularity of the depositional surface is conspicuous and will assist in explaining the irregular structures occurring in the kaolin.

The white granules of kaolin which occur in the clay may be produced by the action of aluminium sulphate or ferro-aluminium sulphate upon quartz sand and pebbles in the clay, but a large part of it is probably due to bacteriai elaboration, as will be explained later. There is also another source of the kaolin. When during periods of humidity an abundant supply of water is brought in contact with the clay containing the aluminium sulphate and the latter rises, comes in contact with beds of sandstone and is converted into Indianaite.

The ferro-aluminium sulphate is produced by the oxidation of pyrite in the shales or in limestones included in the shales, assisted by the action of bacteria. By the oxidation of pyrite, sulphuric acid and ferrous sulphate, melanterite is formed. The sulphuric acid attacks the aluminium silicate of the clay, forming alunogen, hydrous aluminius sulphate, ($Al_2(SO_4)_3$, $18H_2O$). This compound has the following composition: Suphur trioxide 36 parts, alumina 15.3 parts, water 48.7 parts. The melanterite, hydrous ferrous sulphate, ($FeSO_4$, $7H_2O$) is composed of sulphur trioxide 28.8 parts, iron protoxide 25.9 parts, water 45.3 parts. Both of these minerals are readily soluble in cold water, and it seems that they are absorbed by sulphur bacteria, robbed of part of their sulphur and possibly secreted as a ferro-aluminium compound such as halotrichite ($FeSO_4Al_2(SO_4)_3$, $24H_2O$), the composition of which is sulphur trioxide 34.5 parts, alumina 11 parts, iron protoxide 7.8 parts, water 46.7 parts. (See Plate XLIV.)

Changes to the Silicate. The halotrichite is carried by water until it comes in contact with sand or sandstone, where it is changed into hydrous aluminium silicate. This change may be brought about by the action of the halotrichite on very finely divided silica or it may be produced by the action of silicic acid, which is brought down from the sandstone by percolating waters, or by the elaboration of bacteria.

Experiments with Halotrichite.

Experiment I. Halotrichite was collected from the edge of the pan in which black clay, containing pyrite and bacteria was placed. The halotrichite

was dissolved in water, to the solution, water glass was added, and the greenish white precipitate was obtained. This precipitate was washed with dilute hydrochloric acid and water, and a white granular crystalline powder was obtained. This powder was tested for aluminium with the cobalt nitrate solu-

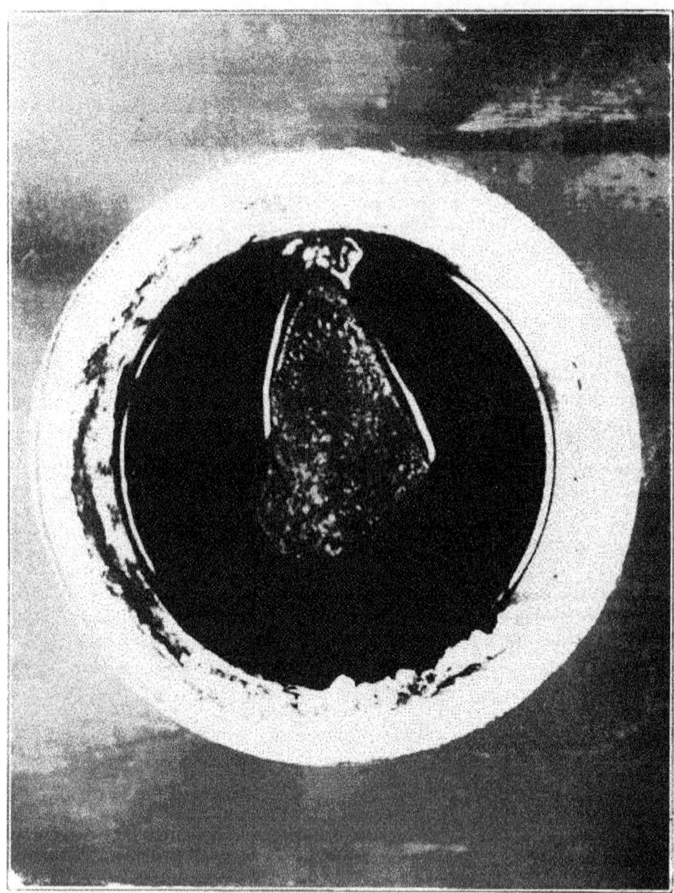

Plate XLIV. Aluminum sulphate from black clay passing through sandstone and collecting on upper surface, also on walls of pan.

tion, and the characteristic blue color was obtained. By the hydrochloric acid test and the borax bead test the presence of silica was determined.

Experiment II. **Halotrichite was dissolved in water and silicic acid was added to the solution. A white precipitate was obtained which gave reaction for both alumina and silica.**

Experiment III. **Halotrichite was dissolved in water in a glass beaker, in which broken fragments of quartz crystals were placed. The solution was**

allowed to stand for three days. At the end of that time a yellow stain was noticeable in the liquid and a slight precipitation occurred in the bottom of the beaker. The precipitate was filtered out and washed with dilute hydrochloric acid and water to remove the yellow iron stain. The test was then made for alumina and silica and both were found to be present. The formation of kaolin was more rapid when bacteria were present.

Experiment IV. Some very fine sand was placed in the bottom of a glass beaker, water was poured over the sand and some halotrichite was dissolved in the water. Bacteria were added after the solution had stood for a few days, a yellowish white precipitate was noticeable on the surface of the sand. The precipitate gave reactions for alumina and silica. From time to time soil water was added and in the course of two months the sand had almost disappeared and kaolin had taken its place.

Experiment V. Some of the black clay which had contained bacteria was placed in a glass tumbler, water was added and a colony of bacteria. A round piece of sandstone was fitted in the upper part of the tumbler above the surface of the water. The halotrichite which was formed passed through the sandstone and formed an incrusting mass on top of it. A white granular substance giving reactions for both alumina and silica was formed on the upper surface of the sandstone after standing many weeks. It was found that the ferro-aluminium sulphate was drawn upward along the surface of glass tubes to a height of more than two feet above the surface of the solution in the vessel.

An experiment was conducted by the writer which demonstrated a probable mode of deposition of a portion of the kaolin. Into a pan of enamelware he placed some of the dark colored clay, taken from beneath the layer of kaolin and covered the clay with distilled water to a depth of an inch. The pan was then placed in a room, the temperature of which was kept at about 70 degrees Fahrenheit. In the course of time a deposit of white and bluish-green colored ferro-aluminium sulphate was deposited in a ring, just above the water, on the inside wall of the pan. The rate of deposition varied at different points, so that the ring of deposited sulphate was very irregular in form, the irregularity extending to all surfaces. The deposition of the sulphate in no way seemed to interfere with the evaporation of the water, which probably passed freely through the sulphate because of its porous nature.

Later, a further experiment was tried. Three holes were drilled in a thin piece of sandstone; into these holes three wooden pegs were inserted. The table thus formed was placed on its pegs in the pan containing the dark colored clay. The lower portion of the sandstone was separated from the clay by the space of one inch. Water was then poured into the pan until its level stood not quite to the top of the sandstone. The pan was placed under the same conditions as before, and in the course of 24 hours sulphate began to make its appearance upon the top of the sandstone. During the next 24 hours it had formed little mounds and pinnacles, one-fourth of an inch high, on the top of the sandstone. (See Plate XLIV.) In the course of a few days it had obtained a thickness of one-half inch. Little globular masses were formed on the upper surface of the sulphate. Surfaces of these globules (See Plate XLV) were covered with spine-like projections. On a portion of the surface of the pan a botryoidal form of the sulphate was deposited. The grape-like masses were not smooth and they were united at their bases to a massive layer of sulphate which was attached to the surface of the pan. When the mass of sulphate collected on the side of the pan became heavy enough to draw away from the pan slightly, and fresh water was added to the clay in the pan, fresh sulphate was deposited in a film, stretching between the old deposit and the surface of the pan. The film consisted of a translucent convex layer of pearly luster with its surface covered with lines, like lines of growth, running longitudinally and transversely.

The sulphate which accumulates when the black clay from beneath the kaolin is placed in water seems in its physical properties to correspond to

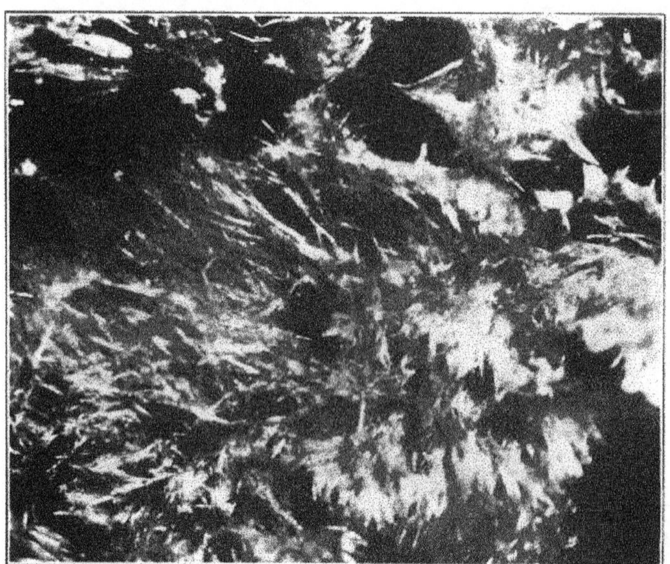

Plate XLV. Micro-photograph of ferro-aluminum sulphate collected from black clay by evaporation.

some of the varieties of halotrichite, though it may be only a mixture of melanterite and alunogen. The accumulated sulphate may vary in composition with the abundance of melanterite and alunogen.

A sample of the sulphate was analyzed by Dr. R. E. Lyons with the following results: "The crushed material heated to 125°C sinters together and changed to chocolate-brown color. Dried at 145°C the mineral suffered a loss of 31.52% (expelled water). Not all of the water is expelled at 145°C. Ignited over a blast, the mineral suffered a loss of 75.61% (water and sulphur trioxide)."

Analysis of Sulphate.

Found	%	Calculated	%
Ferric oxide (Fe_2O_3)	19.35	$FeSO_4$	38.70
Alumina (Al_2O_3)	5.73	$Al_2(SO_4)_3$	19.21
Sulphur trioxide (SO_3)	33.84	Free H_2SO_4	1.23
Water	41.08		
Total	100.00		

The mineral was formed and collected in the air and is lower in per cent of alumina than halotrichite. It corresponds in composition more nearly to the mineral coquimbite when a part of its iron is replaced by aluminium. This mineral is a greenish or yellowish-white mineral which is soluble in cold water but which precipitates iron hydroxide when the solution is heated. The sulphate first collected from the black clay is readily soluble in cold water and does not precipitate iron hydroxide. After pyrite is added to the black clay in order to keep up the deposition of the sulphate the mineral deposited is soluble in cold water and precipitates iron hydroxide on heating.

In the presence of free oxygen the ferrous sulphate, $FeSO_4$, may be oxidized to ferric sulphate, $Fe_2(SO_4)_3$, as follows:

$$6FeSO_4+3O+3H_2O=2Fe_2(SO_4)_3+Fe_2(OH)_6$$

Those portions of the black clay deposits deeply buried may contain iron only in the ferrous state and such compounds as halotrichite may be present. In the presence of free oxygen the ferric compound, coquimbite, might be formed.

Near the outside of the deposit the ferric sulphate, in the presence of water, might produce limonite as follows:

$$Fe_2(SO_4)_3+6H_2O=2Fe(OH)_3+3H_2SO_4$$
$$4Fe(OH)_3=2Fe_2O_3+6H_2O=2Fe_2O_3+3H_2O+3H_2O$$

Limonite is found underneath the white kaolin near the outcrop in the north Gardner mine.

The sulphur which is stored by the bacteria may be converted into sulphuric acid as follows:

$$2S+6Fe_2(SO_4)_3+8H_2O=12FeSO_4+8H_2SO_4$$

This reaction could occur only where there was sufficient free oxygen for the oxidation of ferrous sulphate to ferric sulphate. Whether enough oxygen for this purpose would be set free by the bacteria in the reduction of carbon dioxide is doubtful. The sulphur might be converted into hydrogen sulphide and this oxidized to sulphuric acid.

It may be assumed that the amount of water present fluctuates. With abundance of water it is possible for the ferric sulphate to break up into ferrous sulphate, sulphuric acid and oxygen as follows:

$$Fe_2(SO_4)_3+H_2O=2FeSO_4+H_2SO_4+O$$

Dark Colored Clay Beneath the Kaolin. The clay which lies beneath the kaolin is bluish-black when wet and grayish-black when dry. (See Plate XLVI.) The gray color of the dry clay is due to the evaporation, bringing the soluble salts to the outer surface. In structure it is slightly laminated. The laminated appearance seems due to a change in color rather than to a material change in the size of grain or composition of the clay. The laminæ consist of light and dark bands. The light bands are not always continuous but are made up of lens-like masses with their longer axes in the same general plane. The clay when dried forms hard masses which do not show as many shrinkage cracks as ordinary clay. Generally speaking, there will be one or two large cracks to a mass of clay four inches in diameter. The streak of the clay is gray in color. The surface of the dry clay may be polished to a smooth shiny surface by rubbing the clay on an oilstone without the use of oil or water. The polished surface is black with white irregular particles,

Plate XLVI. Black clay from beneath kaolin in Gardner mine. Rests on Elwren shale and contains pyrite and sulphur bacteria.

irregularly distributed on the smooth surface. The specific gravity is 1.78.

Foreign Matter in the Black Clay. An examination of the black clay underlying the kaolin revealed the presence of considerable foreign material. Both organic and inorganic substances were found.

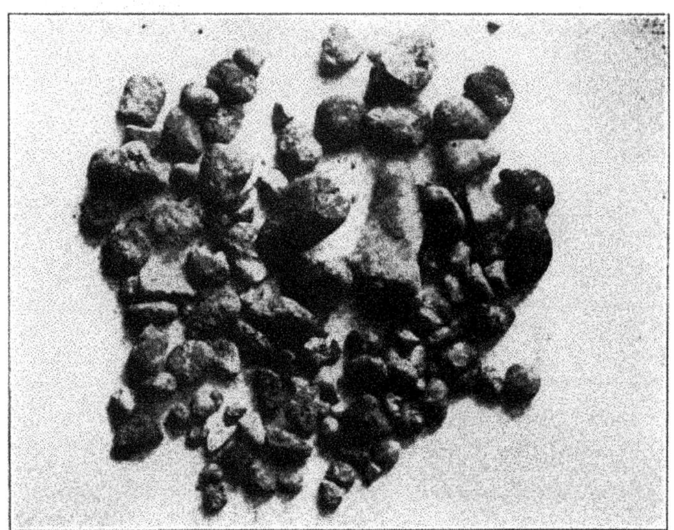

Plate XLVII. Fig. A—Quartz pebbles from black clay. Note etched and corroded surfaces.

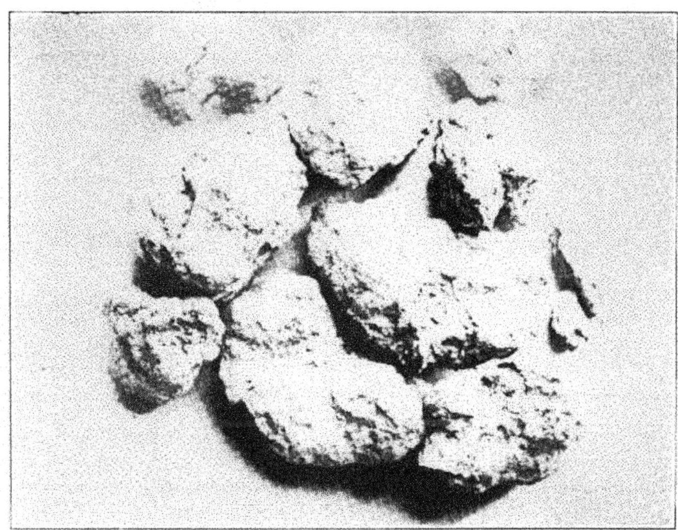

Plate XLVII. Fig. B—Nodules of marcasite from black clay beneath kaolin. Source of the sulphuric acid producing aluminium sulphate.

Quartz Pebbles. The black clay contains pebbles of milk-white quartz. (See Plate XLVII.) The pebbles vary in size from that of a pea to that of a plum pit. The surface of some of the pebbles is etched and corroded, and they are more brittle than the pebbles in other parts of the Mansfield. There seems to be quite a marked difference between the pebbles contained in the Mansfield sandstone which has not been penetrated by solutions from the black clay. The pebbles in size, shape and composition resemble those in the Mansfield sandstone which lies above the black clay. Since no pebbles have been found in the Mississippian, it seems evident that the black clay belongs to the Pennsylvanian, but the shale below the black clay is undoubtedly of Mississippian age, since it contains the same characteristics as the Elwren shale which lies between beds of Mississippian limestone in the same stratigraphical horizon both north and south of this point.

Marcasite Concretions. Irregular concretions of marcasite were found in the clay. They are generally elongate bodies, some are kidney-shaped and some in the form of spherical bodies of small size. All of the concretions exhibit evidence of decomposition. (See Fig. B, Plate XLVII.)

Mica. Small flakes of mica of the muscovite variety appear in some layers of the clay but the quantity is small and is not greater than that usually found in the Chester shales.

Organic Matter. The organic material found in the black clay underlying the kaolin, besides the bacteria, includes lignitic material which is usually in a very finely divided state. This organic matter is probably responsible for the dark color of the clay.

White Clay. Small masses of white clay also occur in masses of black clay. These are generally spherical in form and plastic. As seen under the microscope, the spherical masses are composed of granules. It is difficult to separate the spherical particles of white clay completely from the dark clay matrix. The composition of a number of these white particles, separated as completely as possible from the matrix, is as follows:

Analysis of White Clay Bodies.

	Per Cent
Silica (SiO_2)	41.82
Ferric oxide (Fe_2O_3)	.29
Aluminium oxide (Al_2O_3)	32.65
Titanium oxide (TiO_2)	.04
Calcium oxide (CaO)	.40
Potash (K_2O)	.12
Soda (Na_2O)	trace
Sulphur trioxide (SO_3)	00.085
Loss on ignition	18.47
Total	93.875

White Quartz Sand. A few thin layers of white quartz sand were found in the dark colored clay. Grains of white kaolin occur surrounding the sand grains. Two adjacent layers were examined, each about one-half inch thick. The upper layer contains only a small quantity of kaolin and a larger quantity of quartz sand. In the lower layer the conditions were reversed.

Ferro-Aluminium Sulphate. When placed in water a soluble ferro-aluminium sulphate was derived from the black clay. This compound collected on the walls of the vessel about an inch above the surface of the water. Its behavior is different from alunogen, which may be produced by treating kaolin with sulphuric acid. The latter forms a white incrustation which gradually creeps over the edge of the containing vessel and falls off. The former is greenish-white in color, forms irregular globular bodies and does not advance beyond the edge of the vessel. It is readily soluble in cold water and may be precipitated by the addition of ammonium hydroxide.

Yellow Clay. Not all of the under-clay is black in color. Some of it is yellow and resembles in structure the mahogany clay, but is much lighter, as

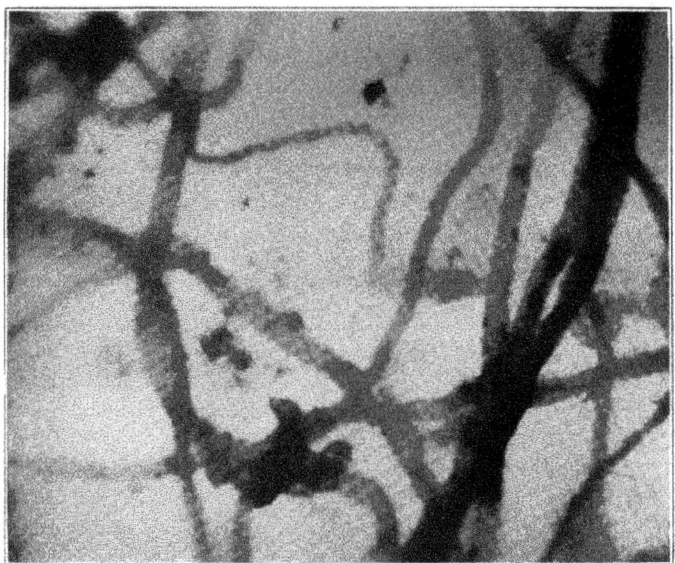

Plate XLVIII. Fig. A Sulphur bacteria taken from black clay beneath white kaolin. Enlarged 1,800 diameters. Mineral matter in cells and along filaments.

though it had been bleached. The yellow clay seems to be an intermediate form between the black clay and the shale which lies below.

Bacteria. The water surface above the clay, after it has been placed in water and allowed to settle, is clear and appears free from sediment or cloudiness of any kind. Later the surface of the water becomes covered with a white film which becomes thick and finally forms a dense white web-like mass, resembling snowflakes on the surface of a pond. The microscopic examination of these web-like bodies shows that they are made up of filaments and small minute, bead-like bodies. Along the side of these filaments there is a collection of mineral matter of a granular nature. Under a microscope of very high power these filaments are found to be very short, irregular cell-like bodies. These organic forms seems to be bacteria of some species closely related to the species of the genus Beggiatoa. These bacteria are capable of living in a strong solution of sulphuric acid. (See Plate XLVIII.)

The bacteria found associated with the black clay underlying the kaolin consist of thread-like filaments. These filaments vary in diameter from 2 to 3 microns. The filaments are segmented. The segments vary in length from 11 to 20 microns. The filaments cluster near the top of the water, forming first a white film and later a thick cotton-like mass. A small colony of bacteria presents the appearance of a pyramid or cone, the apex projecting down into the water and the base lying near the surface of the water, but the whole completely submerged. When separate colonies occur in the water they soon spread out and unite by their edges until the entire surface of the water is covered.

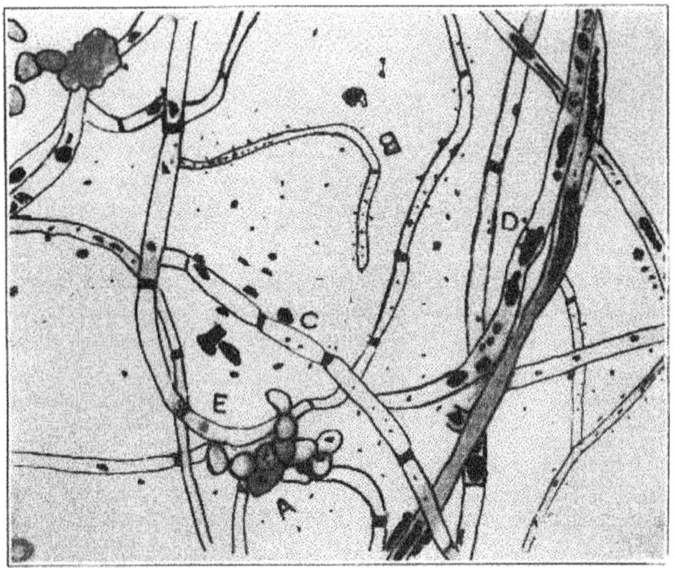

Plate XLVIII. Fig. B—Same as above. Cells outlined. A. Spore cluster. B. Young filament. C. Older form. D. Mineral matter in cell. E. Sulphur? in cell.

The Morphology of the Bacteria. The filaments in their younthful stages are long and slender. They increase in diameter and separate into short segments. These shortened segments subdivide into elongate spore-like cells. Under the low power lens of the microscope these spores appear to be spherical and are arranged in rows which are often sixty or more microns in length. Under a lens of high power these spores are found to be elongated cells, having an average length of 2.8 microns and an average width of 1.5 microns. The shape is somewhat irregular but the ends of the longer diameter are somewhat constricted. When stained with methylene blue, the spores show the presence of small dark bodies within the walls of the spores. These bodies are distinctly irregular in size and number. Not all the spores contain these dark bodies. The filaments also contain what appears to be the same substance. The bodies, however, in the filaments are usually larger than those contained in the spores. These bacteria seem to be closely related to the sulphur bacteria of the genus Beggiatoa. These bacteria accumulate sulphur in their cells, sulphur being derived from hydrogen sulphide. They also convert sulphur into sulphuric acid. They consist of cells without nucleus. The colorless cells are arranged in filaments. The filament has no sheath and is composed of plain rod shaped cells. The filaments move by an undulatory movement of the cell wall. They propagate by the separation of the filaments into segments, the shorter segments by further segmentation forming spores. The cells usually contain stored grains of sulphur. Nathan Zohn found sulphur bacteria which did not oxidize hydrogen sulphide but thio-sulphide and, therefrom, sulphuric acid. Unable to oxidize organic substance, they proclude no carbon dioxide and have no normal respiration. Carbon dioxide is

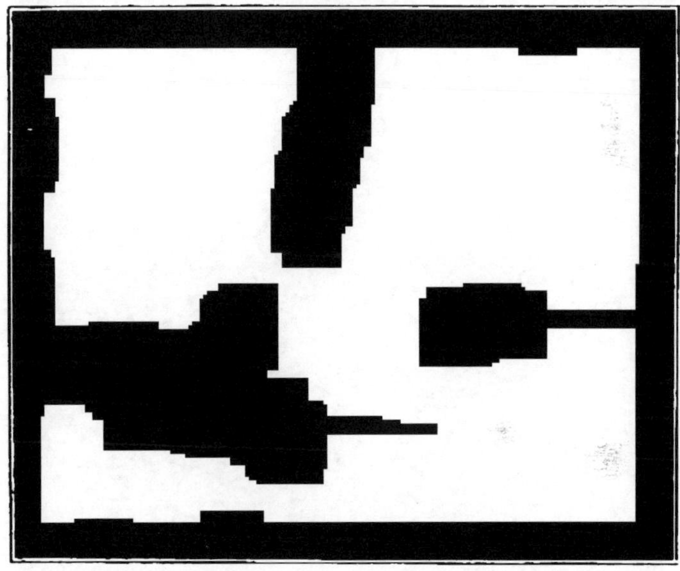

Plate XLIX. Fig. A—Chains of spores. Enlarged 1,800 diameters. Dark spots mineral matter.

a food-stuff and they form organic matter from it. They are organisms which respire inorganic material only.

Composition of the Under-Clay. The dark colored under-clay contains a high per cent of silica, which is due in a large part to the presence of white quartz pebbles and sand. It also contains a high per cent of volatile matter, due no doubt to the presence of pyrite and organic matter.

Acidity. A solution obtained by placing black clay in distilled water is strongly acid. The presence of decomposing pyrite explains the cause and nature of the acidity. Treating the solution with barium chloride, a white precipitate was formed, which was insoluble in hydrochloric acid. This proves the acid to be sulphuric.

Sources of Sulphide of Iron. The sulphide of iron is derived from the Chester shales. Some horizons of these shales contain large quantities of nodular and lens-like masses of sulphide of iron. A part of the sulphide of iron is obtained from the limestones which are embedded in the shales. The decomposition of the limestone sets free the sulphide of iron, which by oxidation forms sulphuric acid. The latter attacks the clay and forms the aluminium sulphate which is acted upon by the bacteria. Some of the limestones of the Chester contain large quantities of pyrite and there is no doubt of their influence in the formation of mahogany clay. Some of the pyrite was also derived from the organic matter deposited with the clay at the base of the Pennsylvanian and the Elwren and other sandstones of the Mississippian.

Bio-chemical Experiments.

Experiment I. Black clay from beneath the white kaolin was placed in a small deep pan and covered with water to the depth of one inch. The solu-

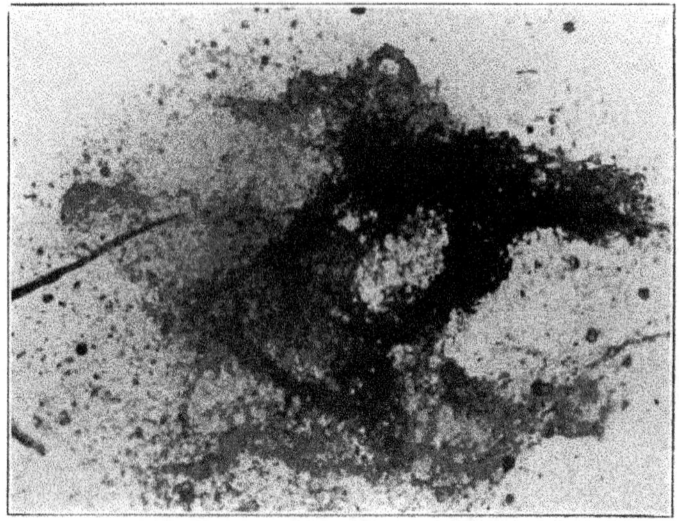

Plate XLIX. Fig. B—Mass of bacteria and secreted mineral. Taken with lens of low power.

tion gave an acid reaction. After standing for awhile, sulphate collected on the edge of the pan. Later, a white film was formed on the surface of the water; this film increased in thickness until it formed a white cottony mass, covering the surface of the water. An examination of these cottony masses proved them to be colonies of bacteria, the meshes of their interlacing fibres containing granular and spherical bodies of mineral matter. The water evaporates and is renewed from time to time. Finally deposition halts, the surface of the water becomes covered with a film of oxide of iron. Some of this iron oxide is carried upward and deposits on the wall of the pan. The colonies of bacteria apparently disappear, but the spores are present in the water. Then a few drops of sulphuric acid are added to the water. The oxide of iron disappears, the water becomes clear, then the colonies of bacteria reappear. The mineral matter found in the cottony masses of bacteria gives the reaction for alumina and silica and its granules are similar to those in the white plastic kaolin in the black clay and to those of the hard white kaolin. This kaolin is undoubtedly a secretion from the cells of the bacteria.

Experiment II. Some black clay from under the kaolin was placed in two test tubes and water was added to each. The solution formed was tested and found to be acid. After standing for a period of three days a small amount of sulphate was deposited on the walls of each test tube. More water was added to each test tube and a small colony of bacteria was placed on the surface of the water in one tube. After standing for one week the number of colonies of bacteria had multiplied, and the deposition of sulphate on the surface of the tube above the water had greatly increased. The uncolonized

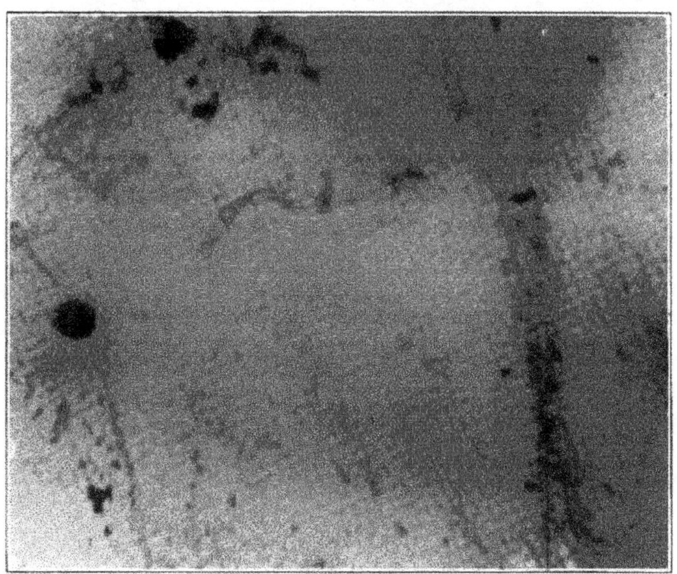

Plate L. Micro-photograph of floating mass of bacterial filaments with mineral matter along filaments.

tube showed only a very slight deposit of sulphate. This experiment seems to afford evidence that the bacteria influenced the formation of the sulphate.

Experiment III. Black clay was placed into jelly glasses and covered with water. Both solutions gave an acid reaction. As the water evaporated a slight deposit of sulphate took place on the walls of the glass, above the water level. Deposition ceased after a few days and the water surfaces were covered with a film of iron oxide. No bacteria were present in either glass. To the water in one of the glasses a colony of bacteria was added; these multiplied in number and the sulphate increased on the walls of the glass. Later, a few drops of sulphuric acid were added and the bacteria showed a greater increase and the deposit of sulphate also increased.

Experiment with Black Clay.

Some of the black clay taken from beneath the white kaolin was placed in a shallow pan of enamelled ware; distilled water was added; in a few days a thin film covered the surface of the water. A portion of the film was examined under the microscope and found to be filamentous micro-organisms, interlaced, and the meshes thus formed either partly or wholly filled with mineral matter. The mineral matter was present in the form of minute granules which were distributed along the walls of the filaments either in single granules or as clusters of granules.

A number of questions presented themselves. Was the mineral a secretion of the micro-organism? It did not take on the form of a crystal; it was granular like secreted masses; it was not present in the solution where the

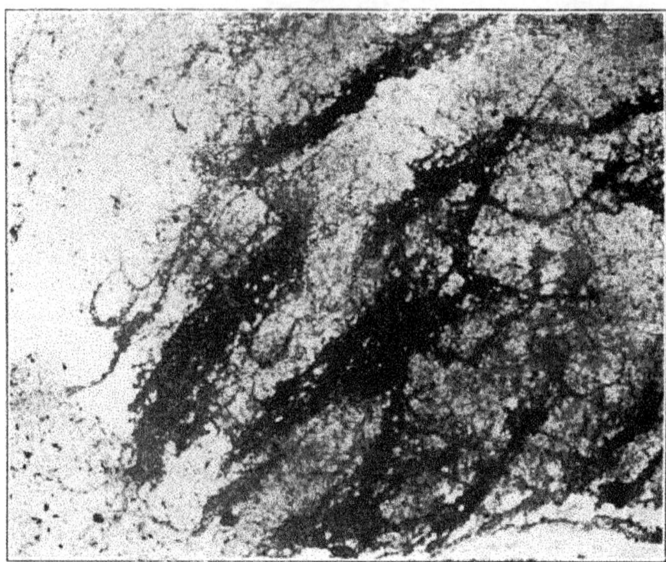

Plate LI. Dense mass of bacteria with enmeshed mineral matter.

micro-organisms did not exist; it was present where even a single filament existed.

Could the mineral be precipitated by the evaporation of water from concentrated portions of the solution in the meshes formed by the filaments of the micro-organism? This did not seem to be a reasonable assumption since no precipitation was taking place along the sides of the vessel, which indicated that the solution was not very concentrated; and then there was the presence of the mineral where no meshes existed but only single filaments.

Was the mineral soluble? The solution was acid. Could a soluble mineral exist in it? It was suggested that it might be soluble but protected by an organic film but the sulphuric acid of the solution would soon break up an organic film. To determine whether the mineral was soluble and precipitated by evaporation of the water the pan was placed under a bell jar and evaporation prevented, but the mineral continued to form, proving that it was not the result of evaporation. The filaments were washed in water and dilute

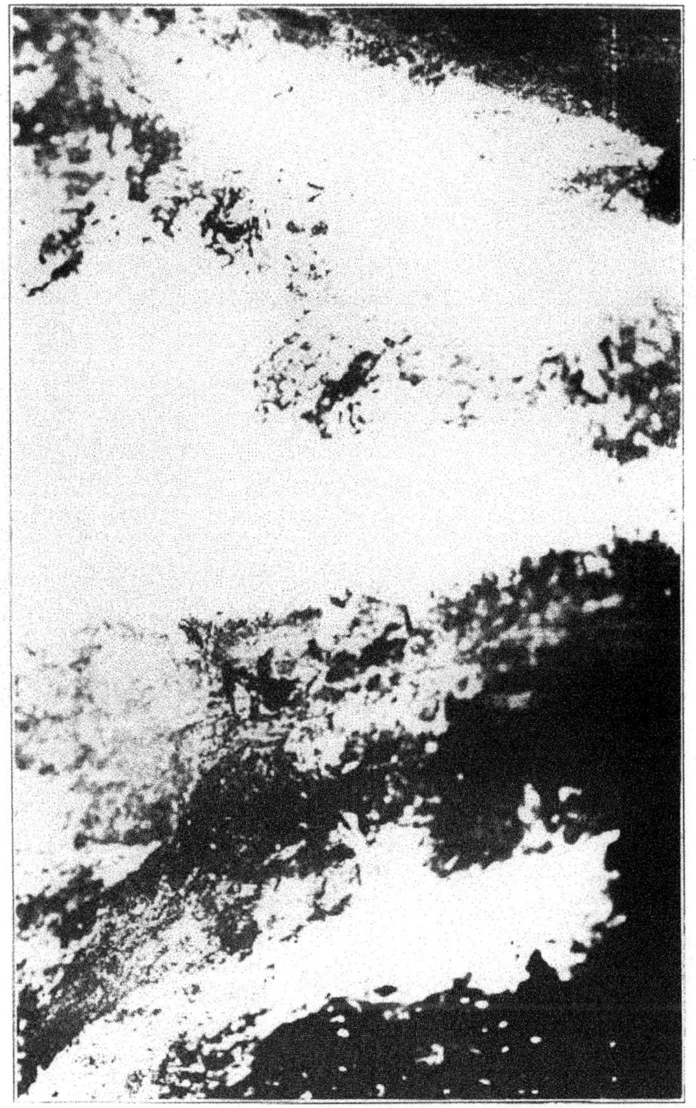

Plate LII. Layer of coarse conglomerate sandstone between layers of white kaolin in Gardner mine. Grains cemented with kaolin.

acids without removing the mineral. Collections of the micro-organisms and the associated mineral matter were dessiccated and gave reactions for alumina and silica. After a period of time granules dropped from the meshes of the filaments to the surface of the clay beneath the water. It was easily distinguished from the clay as it was light and moved back and forth with the

Plate LIII. Aluminum sulphate (Alunogen) collecting on walls of pan and falling over edges. Alunogen formed by treating kaolin with sulphuric acid.

movement of the water. Under the microscope it appeared like the grains in the meshes of the filaments. It took stains in a similar way. The mineral matter in the interior of the bacterial cells was similar in appearance and staining to that along the exterior walls of the cells.

In order to determine the chemical composition of the secretion of the bacteria some of the black clay was placed in a vessel containing water. The vessel was placed under a bell jar. The water above the clay was clear but acidic. After a few days a bacterial growth covered the surface of the water and mineral matter collected among the filaments. The bacteria and enmeshed mineral matter were collected from the surface of the water. The collected mass was analyzed by Professor R. E. Lyons. He first dried the mass at 100° and then for several days over $CaCl_2$ and H_2SO_4. He then found the dessiccated mass to contain:

Silica (SiO_2)	17.34
Alumina (Al_2O_3)	21.56
Ferric oxide (Fe_2O_3)	3.38

The composition of the mass differs from the composition of kaolin, which is to be expected. In the first place, it must be borne in mind that the mass is not a single mineral but contains kaolin, aluminium sulphate and ferrous sulphate.

Therefore, when the volatile matter is driven off the residue is found to be composed of sufficient silica to form about 42.78% of kaolin; an excess of alumina, which is probably derived from the aluminium sulphate, and 3.38% of ferric oxide, which was derived mainly from the ferrous sulphate. Since the kaolin is secreted into a solution of aluminium sulphate and ferrous sulphate which has a leaching action upon it before it is collected, and since some of this solution is collected with it we should not expect this collected matter to have the same composition as pure kaolin. However, in the presence of sand or other forms of silica it would be converted into kaolin, since the excess of free alumina would unite with the silica, and this is what probably occurs when the bacteria are in actual contact with the sand or quartz pebbles in the clay.

It is immaterial whether the alumina unites with the silica within the cells of the micro-organism or unites with it after its secretion. That the union takes place within the cells is probable, because the mineral matter within the cells reacts to stains like that on the outside, its appearance is similar, its solubility is similar and the silica, being soluble, could enter the cells as freely as the aluminium sulphate. The mineral matter within the cells to be visible would of necessity be an insoluble compound.

Experiment with Knobstone Shale.

A sample of aluminous Knobstone shale was selected from an outcrop on the I. C. railroad, east of Bloomington. This shale contained considerable pyrite which was partly decomposed. When the shale was placed in water in a vessel, it gave a strong acid reaction. A colony of bacteria was placed on the surface of the water. In the course of a few days the surface of the water became covered with the bacteria and the secreted mineral matter. Before the bacteria were placed in the solution there was a slight precipitation of sulphates around the margin of the vessel as the water was evaporated. The amount of precipitation increased noticeably with the addition of the bacteria.

Experiment with Chester Shale.

A sample of Elwren shale was placed in a vessel and water and fragments of pyrite were added. In the course of a few weeks ferro-aluminium sulphate collected on the walls of the vessel in small quantities. After standing for three months, the water being renewed from time to time, no evidence of white kaolin was found in the shale.

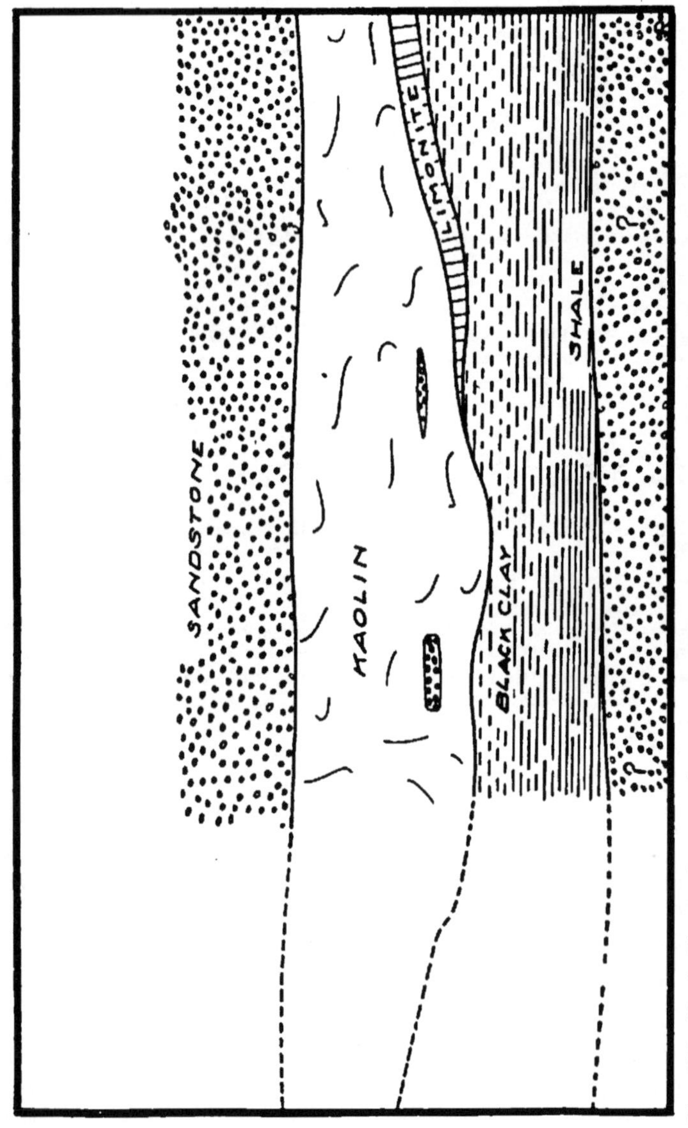

Plate LIV. Fig. A—Diagram showing sandstone layers in kaolin. Limonite layer below at one point, black clay at another. Gardner mine.

In a second vessel an equal amount of the same shale was placed, covered with water and pyrite added as before. At the end of a few weeks, when the sulphates began to appear, a colony of bacteria was placed on the surface of the water. In the course of two weeks the bacteria covered the surface of the

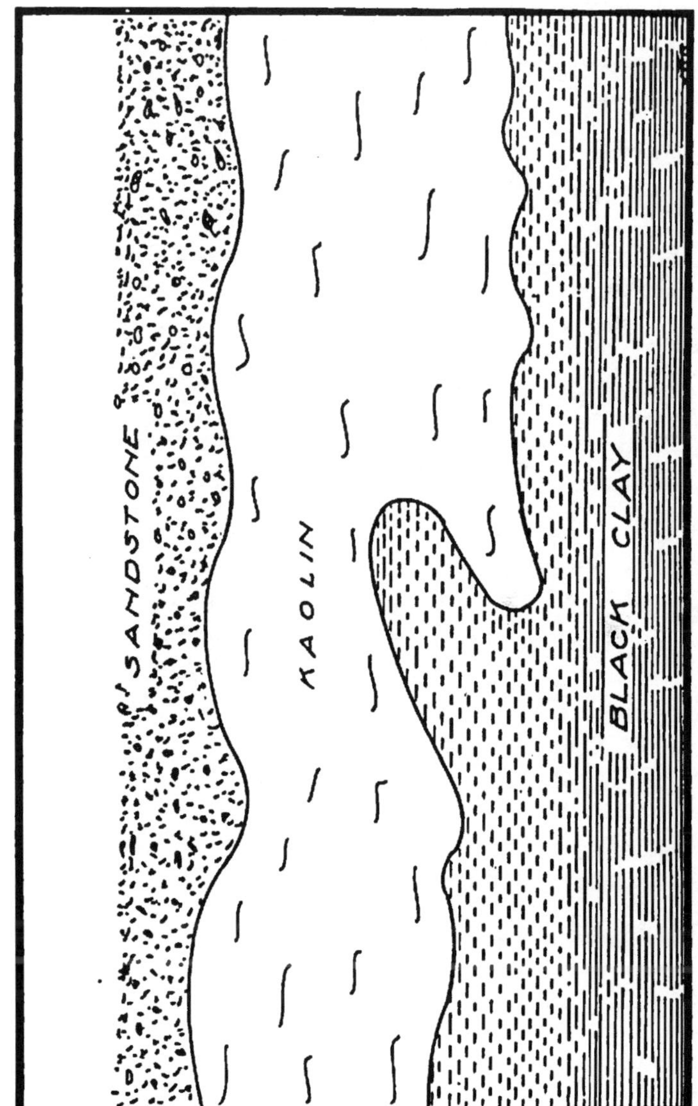

Plate LIV. Fig. B—Diagram showing contact of black clay with kaolin in Gardner mine. At one point neck of clay extends into kaolin.

water and mineral matter was collected in the meshes of the filaments and in a few weeks was dropping down to the surface of the clay.

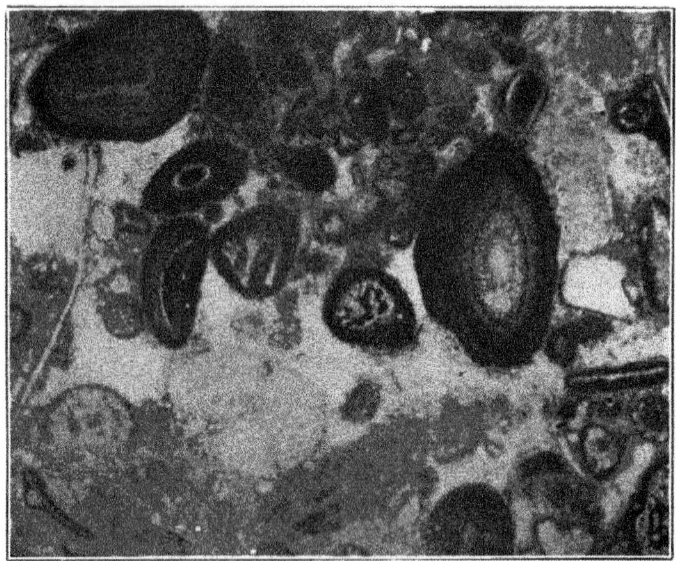

Plate LV. Fig. A—Micro-photograph of a thin section of limestone found in mahogany clay.

Plate LV. Fig. B—Showing contact of white kaolin and black clay, the latter partly converted into kaolin.

Conditions of Accumulation of Kaolin. It is assumed that shale or clay, quartz pebbles and organic matter were deposited in a shallow basin and sand deposited above them. During the decomposition of the organic matter, pyrite is formed. Later deposition was followed by elevation; erosion followed. The edge of the basin was uncovered, but its position was such as to collect the water which percolated through the overlying beds. (See Plate LIV.) The lower portion of the filled basin thereby became saturated with water. The water, carrying oxygen, caused the decomposition of the pyrite, thus forming sulphuric acid and ferrous sulphate. The sulphuric acid attacks the clay and forms aluminium sulphate. Bacteria penetrated the basin. The micro-organisms can live without light or oxygen, so they continue to thrive in the pores of this saturated deposit. They absorb the ferrous sulphate and aluminium sulphate and secrete kaolin and possibly ferro-aluminium sulphate (halotrichite). In the presence of abundant water the sulphate passes upward into the bed of sand and acting on the sand grains, replaces them with kaolin. It has been demonstrated that the sulphate will do this by the experiments already described. After the deposition of the layer of kaolin near the top of the sand, deposition would not be checked, because the porous nature of the kaolin would permit the solution to continue to come through the lower layers of the sandstone, thus a thickness of many feet might be formed. The solution might remove most of the silica from the sand and, if the layer of sand were thin, it is conceivable that it might disappear entirely. As the amount of clay under the sand decreases, it is possible for the sand or sandstone to settle down. The movement being differential, the sandstone may be broken up into blocks. The sand in some of these blocks may be dissolved by the acid solution and the spaces thus formed filled with kaolin. This process would explain the occurence of irregular sandstone blocks which are sometimes found in the beds of kaolin. No caving of the sandstone would take place since only a small area would be affected at one time.

Part Played by the Bacteria. The influence of the bacteria in the formation of Indianaite seems to be positive. They seem to secrete kaolin and their presence seems to be essential to the formation of large quantities of the ferro-aluminium sulphate which attacks the silica so readily. They may also assist in conserving the supply of sulphur which is so essential to continuous formation of aluminium sulphate. There does not appear to be enough pyrite in the shales or in the limestones to supply the necessary amount of sulphuric acid unless the supply is carefully conserved. The sulphur which the bacteria take from the ferrous sulphates would ultimately be converted into sulphuric acid. The latter would attack the clay and a clay-sulphuric acid-alum-bacteria-sulphuric acid cycle would be established. Such a cycle I believe to be essential to the formation of large quantities of kaolin. The amount of pyrite might be small since by the action of the bacteria only a small amount of the sulphur would be lost. By the reduction of the sulphates by the bacteria some oxygen may be set free for the further oxidation of the pyrite or the precipitation of the iron compound. A layer of very pure limonite is sometimes found beneath the kaolin. The bacteria may take part in the formation of colloidal kaolin which is found in the black clay. It seems probable, too, that the granules of plastic white kaolin occurring in the kaolin are secreted by the bacteria. They are composed of rounded grains resembling the oölites of algæ secretions. In the case of lime-secreting algæ "the lime is enclosed in the alga-body in the form of rounded tubercles, which often gather themselves together into larger irregular tubercular bodies." (Rothpletz.)

The similarity of the microscopic grains found within the cells of the bacteria and along the outside walls of the cells to the grains in the soft white kaolin and those in the hard white kaolin is remarkable. A microscopic examination of the filaments of the bacteria reveals the presence of mineral which is insoluble in water entangled in the meshes of the bacterial filaments and distributed along the outside walls of the cells. This compound gives reactions for alumina and silica and contains water of crystallization. Physically, the mineral is made up of tubercles and these are collected into larger masses. These tubercles are similar in size and appearance to those found in the soft plastic kaolin of the black clay and to the granules of the hard white kaolin. They are also similar in appearance and react to stains in the same way as the larger bodies in the cell walls of the bacteria.

It appears that there are here two methods of kaolinization; that in the presence of only a small amount of water the clay or shale is changed in situ by bacterial elaboration, the kaolinization extending downward; that when there is an abundance of water present the soluble ferro-aluminium sulphate is carried upward into beds of sandstone where it is converted into kaolin, a replacement process producing an upward extension of the kaolin deposit.

The conditions for the formation of kaolin by the bio-chemical process are too complex to be met frequently. If the process involved only the decomposition of pyrite in connection with shale, kaolin deposits would be found in most areas of sedimentary rocks because the decomposition of pyrite in shales is of common occurrence. But the conditions for kaolinization demand more than the presence of decomposing pyrite in shale, the essential conditions for the growth of bacteria must be present. There must be present also conditions favorable to the accumulation of kaolin. The ensemble of essential conditions is of rarer occurrence.

Since the activities of micro-organisms are controlled by temperature, the matter of climate enters into the problem. Since sulphur bacteria have been found to be active in the latitude of southern Indiana in the kaolinization process at the present time, are we to infer that the climate of this latitude is to be taken as the favorable one for their activities? If the climatic factor is an important one, it might be assumed that if the southern kaolins were formed by bacterial action, a climate similar to the present climate of southern Indiana must have prevailed in that region some time subsequent to the deposition of the Tuscaloosa and the other formations bearing kaolin. Such climatic condition must have prevailed about the time of the glacial period if not earlier.

The optimum temperature of sulphur bacteria is probably not far from 50°F., since this would be about the uniform temperature of a bog deposit in this latitude and is about the average temperature of the clay at the point where the bacteria were collected.

The question may be asked: Why are not such deposits of kaolin of more widespread distribution? In the first place kaolin deposits of like origin may be more widely distributed than Indianaite. The Tuscaloosa, the Wilcox and the Lafayette formations of the south contain kaolins and white clays that may have a similar origin. These kaolins occur in sedimentary rocks associated with sands and often at great distances from feldspathic rocks. All of these formations contain organic matter in the form of beds of lignite containing pyrite. The composition of samples of kaolin collected by the writer from the Tuscaloosa formation in Mississippi is given in the following table:[1]

Analysis of Tuscaloosa Kaolin from Mississippi.

	No. 1	No. 2	No. 3	No. 4
Moisture (H_2O)	00.87	.48	1.11	......
Volatile matter (CO_2 etc.)	11.96	15.01	13.88	15.17
Silicon dioxide (SiO_2)	38.11	44.23	42.92	44.75
Iron oxide (Fe_2O_3)	11.73	.81	.61	.95
Aluminium oxide (Al_2O_3)	36.42	38.82	41.30	38.39
Calcium oxide (CaO)	0.60	.19	.37	.37
Sulphur trioxide (SO_3)	tr.	.45	.18	alk. .35
Magnesium oxide (MgO)	•.14	.13	.13	.30
Total	99.82	100.12	100.57	100.28

No. 4 is the analysis of a sample of Indianaite made by W. A. Noyes and placed here for comparison.

The presence of such beds of pure white kaolin in sedimentary formations has long been an enigma and a source of speculation to the writer. There appears to the writer no inconsistencies with the view that these southern kaolins may have originated in the same way as the Indianaite. The chemical compositions are similar and the physical properties of certain portions of each are similar.

Origin of Mahogany Clay.

The mahogany clay is found occasionally in connection with limestones occurring in contact with shale. The shale is a highly aluminous formation. The origin of the mahogany clay seems to be in some way connected with the decomposition of the limestone. The bowlder-like masses of limestone are very frequently found in the shales surrounded by mahogany clay. In some of these masses of clay fragments of white kaolin are found. The aluminous shale is decomposed along with the decomposition of the limestone. In some places the limestone contains large quantities of pyrite. In many places the shale also contains pyrite. In the decomposition of pyrite, sulphate of iron and sulphuric acid are formed. The oxidation of the sulphate of iron produces oxides of iron, which give to the mahogany clay its reddish color. The sulphuric acid attacks the aluminium silicate of the clay and forms aluminium sulphate. This aluminium sulphate is converted into kaolin by the action of the sulphur bacteria. During the decomposition of the limestone some calcium sulphate is formed. Crystals of the crystalline form of gypsum selenite are formed in the shale at points near the mahogany clay.

The first change due to weathering is from an olive-green shale to a maroon colored clay, which is very plastic. The maroon clay is changed into the mahogany clay which is less tenacious, and in some outcrops not very plastic.

The mahogany clay is in some instances formed by the staining of beds of white kaolin with iron oxides and by the partial disintegration of the kaolin. This may occur where the sandstone roof above the kaolin has been partly removed so that quantities of surface water may penetrate the bed of kaolin.

[1] Logan Clays of Mississippi, Miss. Geol. Survey, pp. 2 loc. cit.

USES OF INDIANA KAOLIN.

The kaolin of Lawrence and adjoining counties is suitable for use in a number of industries in some of which its value is already well established. Its utilization in quantity has been in the ceramic industry and in the manufacture of aluminium sulphate. The kaolin of Lawrence County was used for many years in the manufacture of this mineral, but cryolite supplanted it and was in turn supplanted by bauxite. Economy of production seems to have been the influencing factor in each substitution.

Alum Cake. The kaolin from Dr. Gardner's place in Lawrence County was used for many years in the manufacture of aluminium sulphate. It is readily soluble in dilute sulphuric acid and is said to be superior to other aluminium silicates for this purpose. Large quantities of alum cake are now used to produce flocculation of clay particles in water intended for domestic use. With an abundance of pyrite for the manufacture of sulphuric acid there is no reason why alum cake should not be manufactured with profit in Indiana.

At the request of the writer, Mr. Jacob Papish offers the following suggestions as to the method of treatment of kaolin in the manufacture of alum cake:

"The plant to be erected for the treatment of kaolin is to consist of a series of tanks, the latter being rectangular or round in shape, built of 2x4 inch lumber and lined with 9 pound lead. A convenient size for a tank is 20x18x6 feet, which is suitable for the treatment of half a carload of kaolin at one time. The boiler and engine should be of a larger capacity than the immediate requirements, so as to take care of future expansion. On the other hand, in order to have the immediate returns as large as possible, it is advisable to procure a cheap kerosene engine. Such an engine will take care of a battery of two tanks at a time, and since stirring and mixing are not a continuous operation with a given batch of material, a cheaper engine would be satisfactory. The cost of erection of one tank and the price of the required accessories are given in the following table:

Tank, 20x18x6	$300 00
Lead lining (9-pound lead)	403 00
Stirrer, lined with lead and shaft	60 00
Belting, pump, spouts, etc.	100 00
Kerosene engine which can be used in connection with several tanks	100 00
Total	$963 00

"The method of treatment, in the main, is as follows: The kaolin should be allowed to weather so as to crumble to pieces. Fifteen tons of this material is to be introduced in a tank and an equal quantity of sulphuric acid added with constant stirring. The stirring should be continued for several hours. After a period of two or three days the cake is ready for shipment. The heat of the action is sufficient to remove surplus water. With proper management, one tank can be made to throw out 90 tons of alum cake in a week, and for one year of 50 weeks this means a production of 4,500 tons.

Cost of Production of 450 tons of Alum Cake. (Crude.)

225 tons of kaolin ($4 per ton)	$900 00
225 tons of sulphuric acid (60°, $16 per ton)	3,600 00
Taxes, insurance (10%)	96 00
Overhead charges, depreciation of plant (20%)	192 00
Labor (20% of selling price)	2,250 00
Total	$7,038 00

"It is seen from the above that the cost of production of one ton of alum cake is $15.64.

"Some municipal water works in Indiana are buying crude aluminum sulphate at $45 per ton. Making an allowance for the difference in active material, the alum cake prepared from kaolin should sell for $25 per ton. The margin of profit offers an excellent inducement to the manufacture. I remember that you made the statement last year that the manufacture of alum cake from kaolin is a paying proposition. Having gone over the different cost data, I am ready to confirm your statement. The main question is the market. The overhead charges and the cost of labor given in my estimate will take care of cost of delivery to railroad cars."

Pottery. The purer varieties of kaolin may be used in the manufacture of white wares. Such wares are manufactured out of a mixture of feldspar, kaolin, quartz and ball clay. Kaolin to be used for this purpose must be very low in percentage of iron to prevent coloring of the ware. The materials used in the manufacture of white ware are reduced to the powdered form, put into a blunger and water added, the portion of each being weighed in order to keep the right proportion in the mixture. After being agitated in the blunger from an hour to an hour and a half, the mixture is put through a sieve, thence through a trough to a vat. In the trough it comes in contact with magnets which remove particles of iron which may have been entered from the grinding machinery. In the vat the mixture is agitated until pumped into the filter press. The leaves of the mixture from the filter press are mellowed and pugged. From the pug mill the clay is taken to the molding room. After being wedged the clay is molded by throwing, jollying or jiggering, pressing or casting. The ware is then finished and fired in the biscuit kiln. It is then brushed, glazed and fired in the glost kiln.

Kaolin is used in the manufacture of the various grades of domestic white ware, sanitary ware, electrical ware and porcelain.

Refractories. Kaolin may be used in the manufacture of refractories such as fire brick, fire proofings, furnace linings, glass pots, saggers and pottery kiln supplies. Since the Indiana kaolin is non-plastic it must be mixed with a plastic clay of high refractoriness in order to meet the requirements of these materials. Indiana possesses fire clays that are plastic and of such a degree of refractoriness as to serve very well as bonding material for kaolin in the manufacture of refractories as the following tests demonstrate:

Results of Kaolin-Fire Clay Tests.

In order to determine the value as a refractory of mixtures of Indiana kaolin and fire clays, these substances were ground, molded into cones and briquettes. The cones and briquettes were marked for the determination of shrinkage. After drying in the air the air shrinkage and loss of water were determined. The test-pieces dried without cracking.

Firing. The cones and briquettes were then dried at 100°C until the weight remained constant. They were then placed in the kiln, the temperature of which was raised to that recorded by the incipient fusion of Cone 30. The results obtained are recorded in the following table:

No. of Fire Clay*	Per Cent of Kaolin	Wet Weight of Cones	Air Dry Weight	Air Shrinkage	Weight Burned Grams	Fire Shrinkage	Absorption	Fired at Cone
18	25	57	44	.04	37	.08	....	30
18	25	46	36	....	30	.12	....	30
18	33⅓	37	29	....	25	.18	.16	30
18	33⅓	46	30	....	32.5	.16	....	30
18	50	49	37	....	30	.14	.066	30
18	50	53	41	....	33	.16	.062	30
20	25	47	36	....	32	.12	....	20
20	25	56	43	.04	40	.12	.027	30
20	33⅓	63	48	....	40	.18	.05	30
20	33⅓	51	39	.016	32	.08	.031	30
20	50	60	46	.04	37	.08	.07	30
20	50	43	33	....	27	.08	.111	30
21	25	60	45.5	.04	....	....	.11	30
21	25	44	33	.016	28	.06	....	30
21	33⅓	41	32	....	26	.16	.11	30
21	33⅓	34	27	.016	22	.06	.045	30
21	50	60	46	.04	37	.08	.08	30
21	50	36	28	.04	23	.12	.30	30
22	25	46	37.5	.04	32.5	12	....	30
22	25	37	29	.10	25	.10	.04	30
22	33⅓	40	28	.06	23	.12	.043	30
22	50	43	29	.08	25	.20	.04	30
22	50	44	30	.04	25	.12	.08	30
22	50	43	29	.08	25	.20	.04	30
23	25	48	38	.04	33	.12	.03	30
23	25	52	41	.04	36	.08	....	30
23	33⅓	42	32	.015	27	.06	.073	30
23	33⅓	36	27	.02	23	.12	.043	30
23	50	50	38	....	32	.08	.12	30
23	50	52	40	.016	33	.10	.15	30

*Fire Clays from Clay County, Indiana.

Refractory Mixtures. In order to determine the value of Indiana kaolin, Chester shales, mahogany clays and fire clays as refractory material mixtures were prepared, molded and fired.

Mode of Preparation. The clays and kaolin were ground separately to pass a screen of 100 meshes to the inch, mixed in proportion by weight of air-dried clay, water added and the mixture molded into cones and briquettes. The cones and briquettes were marked, weighed and the weight marked on each. After drying in the air the loss of water and the air shrinkage was determined.

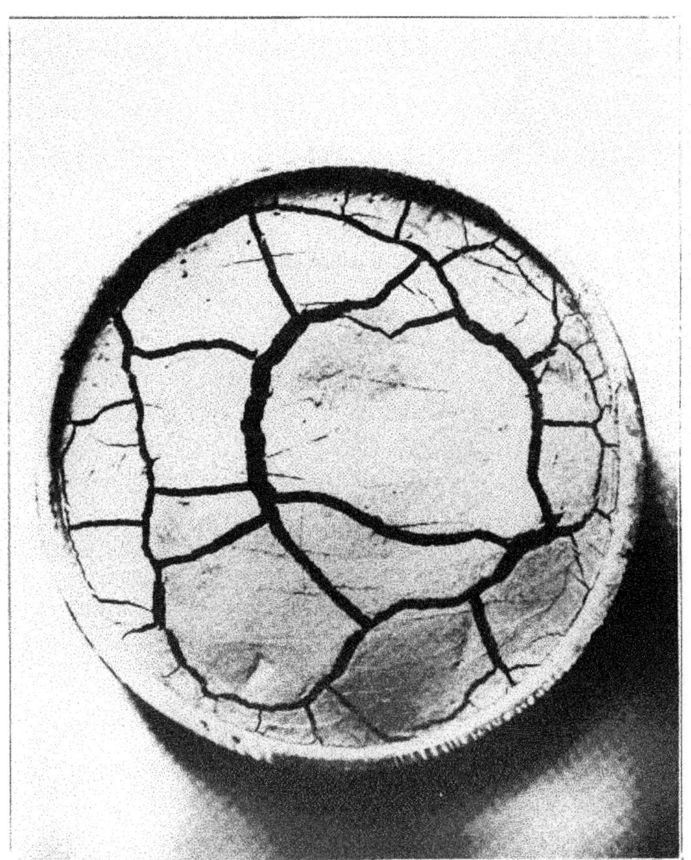

Plate LVII. Chester shale puddled and air-dried. Compare with mahogany clay in Plate LVIII.

Firing. The cones and briquettes were then dried at 100°C until the weight remained constant. They were then placed in the kiln, the temperature of which was raised to that recorded by Cone 30. The results obtained follow the list giving location of the samples.

Mahogany Clay.

1. Near Landreth, Orange County, Orangeville Township.
2. Pearson Place, Lawrence County, Sec. 30, Twp. 4 N., R. 2 W.

3. Wilson Place, Lawrence County, Sec. 8, Twp. 3 N., R. 2 W.
4. Hall Option Place, Monroe County, Twp. 7 N., R. 2 W., Sec. 10.
5. Purlee Place, Orange County, Orangeville Township, Sec. 2.
6. Timberlake Mine, Monroe County, Sec. 28, Twp. 7 N., R. 2 W.
7. Mr. Bridges, S. W. of Fortners, Martin County, Halbert Township,

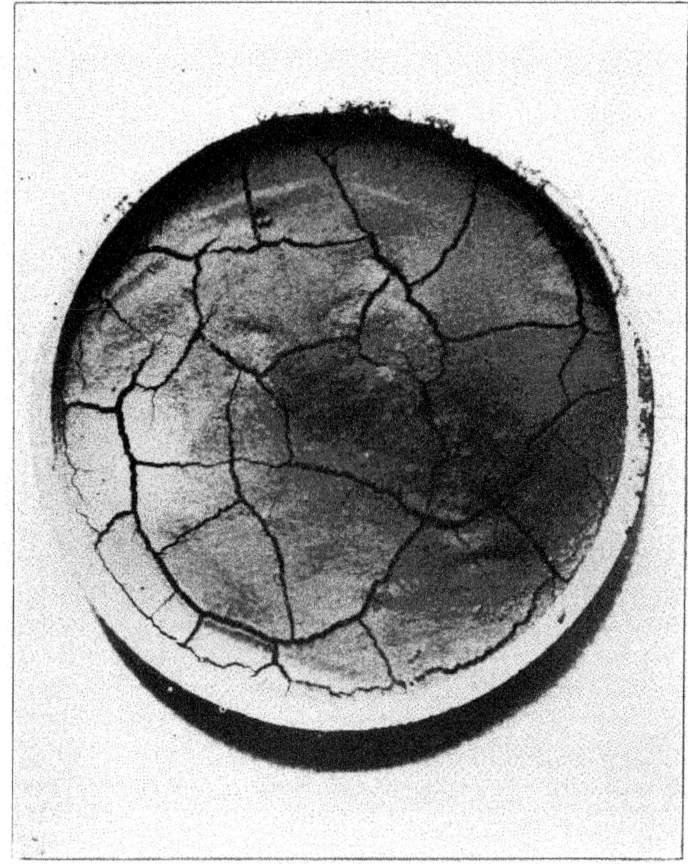

Plate LVIII. Mahogany clay which has been puddled and air-dried. Compare shrinkage cracks with those of the Chester shale, Plate LVII.

Sec. 3.

8. Below Paff house, Monroe County, Sec. 33, Twp. 8 N., R. 2 W.
9. Gardner Place, 2½ miles S. E. of Jones house, Lawrence County, Sec. 34, Twp. 4 N., R. 2 W.
10. South of Landreth House, Lawrence County, Sec. 18, Twp. 3 N., R. 1 W.

58. West of Monon Church, Hendersonville.

59. Edwards, Lawrence County, Sec. 28, Twp. 4 N., R. 2 W.

Shales.

11. From Huron, Lawrence County, Twp. 3 N., R. 2 W., Sec. 5.

12. Near Gardner House, Lawrence County, Twp. 4 N., R. 2 W., Sec. 21.

13. Near Miller Hill, Lawrence County, Twp. 4 N., R. 2 W., Sec. 21.

14. Wilson Place, Lawrence County, Sec. 8, Twp. 3 N., R. 2 W.

15. Miller Hill, Lawrence County, Sec. 31, Twp. 4 N., R. 2 W.

16. Just below Cross Cave where dam washed out, Sec. 21, Twp. 4 N., R. 2 W.

17. One-half mile east of French Lick below Spring, top of mountain, Orange County, French Lick Township.

18. Pilot Knob, above 5 L. S., near Marengo, Crawford County.

19. West end of tunnel, west of Marengo, just above L. S., Crawford County.

20. Lloyd Place, Shoals, Martin County.

21. East of tunnel, west of Marengo, just below first L. S., Crawford County.

22. R. R. cut east of Huron, below first L. S., Lawrence County, Sec. 5, Twp. 3 N., R. 2 W.

23. Isaac Fortner Place, Huron, 50 feet above L. S., Martin County, Hulbert Township.

24. East of tunnel, west of English, below L. S., Crawford County.

25. West of English, near tunnel just below coal measure, Crawford County.

26. South of French Lick, below 3d L. S., Orange County.

27. One and a half miles east of French Lick, below S. S., below L. S., Orange County.

28. East of West Fork, above 3d L. S., Crawford County.

29. One mile south of Grantsburgh, just below 2d L. S., Crawford County.

30. Miller Hill, Huron, Lawrence County, Sec. 21, Twp. 4 N., R. 2 W.

31. Ballard and Clayton farm, 2 miles S. W. of French Lick, red shale, Orange County.

32. Just east of West Forks, above 1st L. S.

33. Sandy layer below mahogany, Timberlake Mine, Monroe County, Sec. 10.

34. One mile west of English, below 1st L. S., Crawford County.

35. West of English, below 4th L. S., Crawford County.

36. North of Mifflin, just above 2d L. S., Crawford County.

37. One and a half miles N. E. of Mifflin, just above 4th L. S., Crawford County.

38. First R. R. cut west of English, just above L. S., Crawford County.

39. One-half mile west of West Forks, above 3d L. S., Crawford County.

40. Timberlake Mine, Monroe County, Sec. 28, Twp. 8 N., R. 2 W.

41. One mile west of English, just above 2d L. S., Crawford County.

42. R. R. cut east of Huron, between 1st and 2d L. S., Lawrence County, Sec. 5, Twp. 3 N., R. 2 W.

43. East of Huron, under 2d L. S., Lawrence County, Sec. 5, Twp. 3 N., R. 2 W.

44. One and a half miles N. E. of Sulphur, just below 4th L. S., Crawford County.

45. N. E. of Huron, Lawrence County, Sec. 32, Twp. 3 N., R. 2 W.

47. Fortner Place in Branch, Martin County, Hulbert Township.

49. Two miles east of French Lick, Geo. Giles Place, below L. S., Orange County.

50. N. E. of Thrasher School House, Monroe County, Sec. 3, Twp. 7 N., R. 2 W.

51. Shoals, by Fred Jones, Martin County.

52. Shoals, Lloyd Place, just below coal, Martin County.

53. Huron Cut, Lawrence County, Sec. 5, Twp. 3 N., R. 2 W.

54. Gardner Place, 2 miles east of Gardner House, just below mahogany, Lawrence County, Sec. 21, Twp. 4 N., R. 2 W.

55. First cut S. W. of French Lick, just below 1st L. S., Orange County.

56. Shale, Gardner Place, south of house, Lawrence County, Sec. 21, Twp. 4 N., R. 2 W.

57. Toward Nail Hill, east of Gardner House, Lawrence County, Sec. 22, Twp. 4 N., R. 2 W.

Mixture of Kaolin and Putnam Fire Clay, State Farm.

1.	$\frac{1}{3}$ K.	$\frac{1}{2}$ K.	$\frac{1}{4}$ K.	$\frac{3}{4}$ K.
11.	$\frac{1}{2}$ K.	$\frac{1}{3}$ K.	$\frac{1}{4}$ K.	

1. White kaolin was used.
11. Mine run kaolin was used.

Results of Shale and Mahogany Tests.

The mahoganies and shales of these tests were treated alike. The material was ground and put through the sixty-mesh screen. Each clay was then molded into two pyramids of about thirty-four grams of weight, the weight was carefully put on them, they were numbered, and a depression exactly an inch long put in by means of a little paddle. These pyramids were allowed to dry in the air for about two weeks. After they were thoroughly air dried, the weight, the shrinkage in weight, and the lateral shrinkage was carefully taken and tabulated. Then one pyramid of each specimen was burned without being previously heated. Many of those broke to pieces because of the violent action of the escaping gases caused by the rapid heating. The other pyramids were put close to the furnace and heated while the first were being burned. They were also heated to a red heat in the muffle before being put in the furnace. The results in the second case showed a marked improvement.

In each case the pyrometric cones used were numbers 022, 03, 10, 30. The temperature at which they fuse is: 022 at 1,094 degrees, 03 at 1,994 degrees, 10 at 2,426 degrees, and 30 at 3,146 degrees, all being given as Fahrenheit.

In the second burning all these cones fused, showing a temperature of at least 3,140°F. or 1,730°C.

After they were burned, the weight, the shrinkage in weight, and the lateral shrinkage was again carefully taken and tabulated. Also the color and the condition was put down. From this data the following reports are made.

These tests were made under the writer's direction by Mr. Willis Richardson.

Data Taken After the Pyramids Had Air Dried.

The first ten and also 58 and 59 are mahoganies.

Number	Weight	Shrinkage in Weight	Lateral Shrinkage
1	17, 18	11, 11	7/60, 7/60
2	24, 22	9, 9	6/60, 6/60
3	24, 24	10, 10	5/60, 5/60
4	24, 24	11, 11	4/60, 4/60
5	24, 23	10, 9½	7/60, 7/60
6	26, 23	11, 10	4/60, 4/60
7	28, 29½	12, 13½	7/60, 7/60
8	23, 24	12, 13½	4/60, 4/60
9	23, 23	13, 14	4/60, 4/60
10	28, 28	9, 9	4/60, 4/60
58	26, 26	14, 15	5/60, 5/60
59	25, 25	11, 12	4/50, 5/60
11	25, 24	9, 8	3/60, 4/60
12	26½, 26	8½, 8	5/60, 5/60
13	22, 25	7, 7	5/60, 5/60
14	31, 26	10, 8	5/60, 5/60
15	26, 25	9, 9	6/60, 5/60
16	25, 26	6½, 7	3/60, 3/60
17	24, 25	7, 6	2/60, 2/60
18	27, 26	8, 8½	4/60, 4/60
19	27, 24	7, 6½	4/60, 3/60
20	28, 29	8, 8	3/60, 3/60
21	27, 27	7, 6	3/60, 2/60
22	28, 29	8, 8	6/60, 4/60
23	23, 23	8, 8	3/60, 3/60
24	32, 32	8, 8	3/60, 3/60
25	23½, 25	9½, 9	3/60, 3/60
26	32, 34	8, 8	4/60, 4/60
27	29, 30	7, 7	1/60, 2/60
28	30, 30	8, 7	4/60, 5/60
29	29, 31	6, 7	2/60, 2/60
30	27, 27	8, 8	5/60, 5/60
31	29, 32	8, 8	5/60, 5/60
32	29, 33	9, 9	5/60, 5/60
33	29, 29	7, 6	2/60, 2/60
34	25, 23	8, 9	6/60, 6/60
35	30, 28	8, 8	5/60, 5/60
36	29, 26	8, 8	2/60, 2/60
37	28, 31	9, 10	5/60, 5/60
38	29, 28	7, 7	3/60, 3/60
39	30, 28	7, 7	3/60, 3/60

Data Taken After the Pyramids Had Air Dried—Continued.

Number	Weight	Shrinkage in Weight	Lateral Shrinkage
40	28, 30	14, 14	4/60, 4/60
41	28, 28	7, 7	3/60, 3/60
42	27, 27	6, 7	4/60, 4/60
43	33, 33	6, 6	2/60, 2/60
44	29, 28	7, 7	4/60, 4/60
45	35, 35	7, 7	2/60, 2/60
47	29, 30	6, 6	2/60, 2/60
49	28, 28	7, 6	4/60, 4/60
50	25, 24	11, 11	1/60, 1/60
51	28, 28	8, 7	2/60, 2/60
52	25, 26	7, 8	2/60, 2/60
53	28, 29	6, 7	4/60, 4/60
54	28, 28	7, 7	3/60, 3/60
55	24, 23	7, 6	3/60, 3/60
56	26, 27	7, 6	5/60, 4/60
57	28, 26	8, 9	4/60, 5/60

Data Taken After the Pyramids Were Fired.

No.	First Weight	Weight After Burning	Lateral Shrinkage After Burning	Remarks
1	26	15	3/16	Incipient fusion.
2	31	20	5/32	Dark brown, hard burned.
3	34	21	5/32	Dark red, hard burned.
4	35	20	2/16	Brownish red, hard burned.
5	32½	20	3/16	Dark red, hard burned.
6	37	22	3/16	Dark red, good, slightly cracked.
7	43	26	1/32	Red, slightly puffed, blubbered.
8	37½	21	1/6	Blue, incipient fusion, fair shape.
9	37	20	3/16	Dark red, very good, hard burned.
10	37	26	1/16	Brown, burned good, hard, firm.
58	41	21	3/16	Gray brown, slightly cracked, good.
59	37	21	3/16	Dark, incipient fusion, vitrified.
11	28	20	....	Dark gray, good shape.
12	34	24	5/60	Dark, incipient fusion, good shape.
13	32	22	4/60	Brown, vitrified, puffed, cracked.
14	34	25	1/16	Gray, vitrified, good shape.
15	34	Broken	in Furnace	Red half gone, good, hard.
16	31	22	3/16	Vitrified, good shape, gray.
17	31	22	7/60	Gray, vitrified, slightly cracked.
18	34½	23	9/60	Red to dark, vitrified, cracked, good.
19	30½	32	4/60	Gray, vitrified. very good.

Data Taken After the Pyramids Were Fired—Continued.

No.	First Weight	Weight After Burning	Lateral Shrinkage After Burning	Remarks
20	37	26	1/16	Gray, very good, red spot.
21	33	25	5/60	Brownish gray, cracked and puffed, slightly vitrified.
22	36	25	9/60	Pink, brown, gray, incipient fusion.
23	37	27	2/32	Brown, incipient fusion.
24	40	30	1/16	Gray, vitrified, slightly cracked.
25	33	14	8/60	Brown, pink, vitrified, half broken away.
26	36	26	6/60	Very good, pinkish-gray color.
27	37	28	8/60	Pink, gray, vitrified, slightly cracked, good shape.
28	31	22	9/60	Blue, gray, incipient fusion at top.
29	37	29	4/60	Pink and gray, slightly cracked, good body.
30	35	24	2/60	Brown, incipient fusion, cracked and puffed out of shape.
31	30	23	8/60	Shows iron vitrified.
32	38	25	....	Expanded, brown, incipent fusion, badly cracked, iron present.
33	35	27	4/60	Pink, gray, slightly cracked, good body.
34	....	....	....	....
35	34	24	5/60	Very good, almost vitrified.
36	37	26	3/60	Brown, gray, vitrified, very good.
37	36	27	2/60	Brown, vitrified, slightly cracked.
38	36	25	....	Brown, vitrified, slightly cracked.
39	37	27	1/60	Cream colored, cracked, vitrified, slightly out of shape.
40	44	24	14/60	Gray, cracked slightly, good shape.
41	35	24	6/60	Pink to brown, out of shape, incipient fusion.
42	34	24	2/60	Slightly cracked, good shape.
43	39	30	1/32	Red, slightly cracked, incipient fusion.
44	36	26	1/60	Well burned, well vitrified, good body.
45	42	31	4/60	Pink, gray, cracked slightly, good body.
47	36	27	1/60	Red, vitrified, incipient fusion, good body.
49	34	24	5/60	Red, puffed, slightly cracked.
50	35	19	2/16	Orange color, slightly cracked, good.
51	35	25	1/32	Cream colored, burned very good.
52	34	23	1/16	Gray, very good, vitrified, pink.
53	36	25	3/32	Red, very good body.
54	35	25	3/60	Gray, incipient fuson, fair shape.
55	29	20	1/16	Red, good shape, burned well.
56	33	24	7/60	Gray, vitirfied, cracked, puffed.
57	36	20	6/60	Pink, cracked badly, broken and partly gone, vitrified.
1	⅓K37	24	8/60	Gray, burned very good, vitrified.
11	⅓K	23	....	Gray, very good shape.
11	¼K41	26	2/16	Gray, very good body.

Remarks on Tests.

Shales showing a great deal of iron: Numbers 12, 18, 28, 54, 30, 32, 13, 25, 14.

1. Average lateral shrinkage: Between 4/60 and 5/60 of an inch.

2. Average shrinkage in weight: About 7 grams. Greatest shrinkage was 8 grams, least 6 grams.

3. Average condition of pyramids: Numbers 20 and 27 are in good condition, while numbers 30, 32, 13 and 25 are badly out of shape, puffed, cracked and broken.

Shales showing less iron: Numbers 21, 41, 27 and 20.

1. Average lateral shrinkage: Between 2/60 and 3/60 of an inch.

2. Average shrinkage in weight: Seven grams. Greatest shrinkage in weight was 8 grams.

3. Average condition of the pyramids: Numbers 20 and 27 are in good condition, while numbers 21 and 41 are badly out of shape, puffed and cracked.

The mahoganies showing a great deal of iron: Numbers 1, 8 and 6.

1. Average lateral shrinkage: About 5/60 of an inch. Number 1 has a shrinkage of 7/60, while numbers 8 and 6 have only 4/60 of an inch.

2. Average shrinkage in weight: 11 grams before burning. 12 grams after burning.

3. Condition of the pyramids: All good and hard, very good shape.

The Mahoganies showing less iron are numbers 2, 3, 5 and 9.

1. Average lateral shrinkage: Between 5/60 and 6/60 of an inch. After burning 5/32 of an inch.

2. Average shrinkage in weight: Air dried, 10 grams. After burning, 12 grams.

3. Condition of the pyramids: All very good shape and hard.

General review of the shales.

1. Average lateral shrinkage:

(a) Air dried, 4/60 of an inch.
(b) After burning, 6/60 of an inch.
(c) Makes 2/60 of an inch of shrinkage due to burning.

2. Average shrinkage in weight:

(a) Air dried, about 8 grams.
(b) After burning, about 10 grams.
(c) Makes a shrinkage of 2 grams due to burning.

General review of the mahoganies.

1. Average lateral shrinkage:

(a) Air dried, 5/60 of an inch.
(b) After burning, about 5/60 of an inch.
(c) There is no difference in shrinkage due to burning.

2. Average shrinkage in weight:

(a) Air dried, 11 grams.
(b) After burning, 13 grams.
(c) Makes an average shrinkage in weight of 2 grams due to burning.

Highest temperature reached in burning was at least 3,146 degrees F. or 1,730 degrees C.

Only three of the pyramids were completely fused. Many of them were partially fused and several showed incipient fusion.

Extra good pyramids are numbers: 51, 44, 52, 47, 53, 55 and 14.

Kaolin and Shale Mixtures. Mixtures of 1 to 3; 1 to 2, and 1 to 1 of some of the Chester shales and kaolins were made and the following data obtained after the pyramids were fired:

19A. Good shape, shows very slightly incipient fusion.
19B. Burned well, shows no fusion. Cream colored.
19C. Same as 19B except it is slightly cracked.
44B. Burned well, porous, buff colored.
44C. Good shape, vitrified, slightly cracked, hard.
47A. Soft and sandy, pink and gray color, very porous.
47B. Same as above, slightly cracked.
47C. Same as above, very porous, not vitrified.
51B. Burned very good, cream colored, no cracks, hard.
51C. Burned well, vitrified, very good.
52A. Burned well, cream colored, rather soft, not vitrified.
52B. Same as above.
52C. Same as above.
53C. Well burned, no cracks, pink colored, rather porous.

All the cones have held their shape well. Most of them could be fired at a higher temperature without fusing them. The kaolin mixture makes a much better fire resisting material than shales above.

Data Collected from Fired Shales and Mixtures.

Number	Vol. C. C.	Difference in Wet and Dry Grams.	Per Cent of Porosity.	Number	Vol. C. C.	Difference in Wet and Dry Grams.	Per Cent of Porosity.	Number	Vol. C. C.	Difference in Wet and Dry Grams.	Per Cent of Porosity.
1	8	1	12.5	24	19	......	......	51	11	1	9.0
2	9	1	11.1	25	22	3	13.6	52	12	......	......
3	9	1	11.1	26	13	3	23	53	11	1	9.0
4	10	4	40	27	14	2	14.2	54	13	2	15.3
5	9	1	11.1	28	15	2	13.3	55	10	1	10
6	9	1	11.1	29	13	1	7.7	56	14	......	......
7	22	......	......	30	20	1	5	......	......	......	......
8	14	1½	10.7	31	10	1	10	1⅓ K	12	1½	12.5
9	9	2	22.2	32	26	1	3.9	11¼ K	13	1	7.7
10	12	2	16.6	33	12	2	16.6	......	......	......	......
58	10	3	30	34	......	......	......	19A	14	2	14.2
59	9	1	11.1	35	13	......	......	19B	14	3	21.4
11	14	1	7.1	36	15	......	......	19C	13	4½	34.6
12	19	......	......	37	30	......	......	44B	12	5	41.6
13	22	......	......	38	19	1	5.2	44C	11	1	9.0
14	14	......	......	39	22	1	4.5	47A	10	3	30
15	......	......	......	40	12	1	8.3	47B	12	4	33.3
16	14	2	14.2	41	16	1	6.2	47C	12	5	50
17	14	1	7.1	42	15	1	6.6	51B	12	3	25
18	12	1	8.3	43	20	......	......	51C	10	1	10
19	10	......	......	44	14	1	7.1	52A	13	4	30.7
20	12	1	8.3	45	26	1	3.8	52B	11	4	36.3
21	14	......	......	47	10	......	......	52C	12	5	41.6
22	18	3	16.6	49	14	1	7.1	53C	12	4	33.3
23	10	1	10	50	9	3	33.3				

Malinite.

A refractory substance called "Malinite" is manufactured out of a mixture of Lawrence County kaolin, from the Dr. Gardner place, and fire clay. Three samples were tested by R. W. Hunt & Co., of Chicago, as follows:

"Three samples of Malinite refractories were selected, one white, one mottled and one brown. Portions of these brick were cut out and shaped into cones and mounted into a pat of Malinite, together with pyrometric cones numbered 30, 32, 34, 35, 36, 37, 38 and 39. After drying, the pat holding the pyrometric cones and test pieces was placed in an electric furnace of Hoskins manufacture, the current turned on and heating continued until the melting of Cone No. 39.

The tests were started about 2:30 p. m., August 21st, and completed at 7:55 p. m. the same date. The atmosphere of the surface was reducing. The tests were conducted in the presence of J. H. Campbell, A. Malinovszky and W. W. Ittner. On August 22d, the pat was taken from the furnace and the specimen of brick examined. There was no indication of fusion on the three samples tested. The temperature reached was 3,542 degrees F. or better. A photograph of the test pat is attached herewith.

Plate LIX. Showing three cones of Malinite after being tested at a high temperature. The numbers mark the positions of the pyrometric cones which were all fused.

Referring to this photograph, the three samples of Malinite refractories will be noted in the center; on each side the numbers show the location of the pyrometic cones. Between cones Nos. 30 and 36 is the white specimen. Between Nos. 34 and 37 is the mottled specimen. Between Nos. 36 and 39 the brown specimen. Examining these specimens with a glass, the complete absence of fusion is apparent.

Load Tests Under Heat.

Full specimens of brick were placed in a furnace and heated up to 1,350 degrees C., about 2,500 degrees F., in a period of 4½ hours, and held at that temperature for 1½ hours. Specimens were allowed to cool in the furnace over night. As tested the brick was set on end and a load of 25 pounds per square inch applied during the heating and cooling, as provided by specification of the American Society of Testing Materials C 16-17 T.

Description of Sample	Mottled	Brown
Dimensions under compression	2.50 in. x 4.56 in.	2.55 in. x 4.50 in.
Height as tested	9.35 in.	9.43 in.
Area under compression	11.40 sq. in.	11.47 sq. in.
Total load applied	285 lbs.	287 lbs.
Load per square inch	25 lbs.	25 lbs.
Height after test	9.34 in.	9.43 in.
Per cent contraction	11/100	None

Neither sample showed signs of checking after the above test.

Cold Crushing Test.

Description of Sample	Mottled		Brown	
Test number	1	2	3	4
Specimen tested	On end	Flat	On end	Flat
Dimensions under compression	2.56"x 4.66"	9.37"x 4.54"	2.45"x 4.59"	9.39"x 4.54"
Area under compression	11.93 sq. in.	42.54 sq. in.	11.25 sq. in.	42.63 sq. in.
Height as tested	9.42 in.	2.49 in.	9.31 in.	2.50 in.
Maximum load	16,750 lbs.	92,290 lbs.	12,430 lbs.	73,680 lbs.
Crushing strength (lbs. per sq. in.)	1,404 lbs.	2,170 lbs.	1,105 lbs.	1,705 lbs.
Failure	Regular	Regular	Regular	Regular

Slagging Test.

Specimens were heated to 1,350 degrees Centrigrade, 2,500 degrees F., in a period of five hours. 12.6 grammes of basic open hearth slag were then placed in the cavity previously prepared in the samples and held at the above temperature for two hours and allowed to cool in the furnace. Size of cavity, 1⅝ inch diameter by 9/16 inch depth.

Description of sample	Mottled	Brown
Slag used	Basic Open Hearth	
Slag penetration sq. in.	0.57	0.48

A test was made on each sample, using powdered silica brick and powdered magnesia brick. These materials showed no penetration under the above test.

The above slagging test is practically in compliance with the specification of the American Society of Testing Materials C 17-17 T, except the weight of slag and diameter of cavity. This was changed in order to permit the placing of three specimens in the furnace under the same conditions at the same time.

Chemical Analysis.

Chemical analysis of portions of the three specimens subjected to fusion test follows:

	White	Mottled	Brown
% Silica	44.90	46.62	47.50
% Iron Oxide	.71	.71	4.60
% Aluminium Oxide	52.39	50.91	46.36
% Calcium Oxide	1.12	1.16	.60
% Magnesium Oxide	.28	None	None
% Sodium Oxide	.04	.24	.26
% Potassium Oxide	None	Trace	.09
% Titanium Oxide	.10	.28	.60

Transverse Test.

A transverse test was made on two samples of each, mottled and brown. The average of these two tests are:

Test	No. 1	No. 2
Description of sample	Mottled	Brown
Average dimensions	4.615x2.50x9.365	4.425x2.465x9.36
Distance between supports, C to C	8 in.	8 in.
Breadth as tested	4.615	4.425
Depth as tested	2.50	2.465
Maximum load sustained	1,140	1,390
Modulus of rupture	476	621

Specific Gravity.

The determination of the specific gravity was made upon two specimens each of mottled and brown, the results of which are:

	Mottled	Brown
Test No.	1	2
Weight used	60 grams	60 grams
Volume	21.1cc	21.2cc
Actual specific gravity	2.84	2.83

Conclusion.

The above tests indicate a refractory with a very high fusion point and one that under temperature maintains its shape when subjected to pressure. It resists slagging action well, and the uniform penetration of the slag indicates a good mixture of the materials before molding. The absence of slagging with the magnesite brick and silica brick suggest that Malinite refractories may be placed in contact with the other refractory materials without slagging action, and that the refractory possesses neutral properties."

Encaustic Tile. The stained kaolins which have been reduced to a moderate degree of plasticity by weathering, the introduction of colloidal iron oxide and possibly to some extent by bacterial action have been used in the manufacture of encaustic tile. It is possible to use the non-plastic kaolin for this purpose by mixing with it some plastic clay to furnish the proper bonding power, though it does not seem desirable to use the better grades of kaolin for this purpose, since the stained kaolin is more abundant and its use more restricted.

Paint Pigment. The pure white kaolin may be used in the manufacture of white paint since it is not readily soluble in acids and would be little affected by atmospheric agents of decomposition. To be used for this purpose the kaolin should be ground very fine in a ball mill and the coarser particles separated by means of the blower process.

Plate LX. Outcrop of mahogany clay in Monroe County.

Kaolin is also used in the manufacture of ultramarine, which was formerly obtained by the grinding of lapis lazuli. The artificial pigment is manufactured from kaolin, charcoal, silica, soda and sulphur, by fusion and roasting, after which it is powdered. It is used in paints and dyes.

Ultramarine Blue Mixtures.

Ultramarine	Pale	Medium	Dark
Kaolin or clay	100	100	100
Soda	9	100	103
Glauber salt	120	0	0
Carbon	25	12	16
Sulphur	16	60	117

The raw materials which are used in the manufacture of Ultramarine are enumerated in the above table. The clay used may be china clay, pottery clay or kaolin of good quality. The clay should be free from excessive amounts of iron or manganese and should be finely divided. The soda used should be a good quality of carbonated soda ash. The glauber salt should be free from acid and iron compounds, and should be reduced to a fine powder. The sulphur should be a good quality of stick sulphur reduced to a finely divided state. The carbon should be a good grade of pine trunk charcoal, containing not

Plate LXI. Outcrop of kaolin in Monroe County.

more than four per cent water and ground in a ball mill to a fine powder. Coal may be used if it is high in carbon and free from sulphur and iron. The silica may be finely ground quartz or Rieselguhr. The tripoli of southern Illinois would probably be admirably adapted to this purpose. Charcoal produced from rice husks contains both carbon and silica and may be used in place of carbon and silica.

Manufacture. Ultramarines rich in silica are made by the direct method. The processes involved are: Mixing, roasting, lixivating, wet-grinding, levigat-

ing, pressing, drying and sifting. The roasting is done in mass ovens or in shaft furnaces. In the indirect method some ultramarines poorer in silica are The furnaces used are muffle or crucible or shaft or cylindrical retort furnaces. manufactured. The preparatory stages are the same as in the direct method.

Paper Manufacture. **Talc, chalk and other substances are used as a filler in the manufacture of paper, and some of minerals so used require expensive**

Plate LXII. Chester limestone surrounded by mahogany maroon and olive green shale. Monroe County.

methods of preparation. The better grade of Indiana kaolin could be used for this purpose and compete, so far as cost is concerned, with other materials.

Other Uses. **It is possible to use Indiana kaolin in the manufacture of fillers, as a filler for varnishes, for fulling cloth, as a catalyzer in the manufacture of poison gas and as an abrasive in buffing.**

THE GEOGRAPHICAL DISTRIBUTION OF KAOLIN IN INDIANA BY COUNTIES.

General Distribution. The kaolin of Indiana occurs along the outcrop of the shales and sandstones of the Chester division of the Mississippian rocks and also along the contact of the Mississippian with the Pennsylvanian rocks. This line of contact extends from Benton County, north of the Wabash River, on the northwest, to Perry County, on the Ohio River, at the south. The line passes through the following counties: Benton, Warren, Fountain, Parke, Montgomery, Putnam, Clay, Owen, Monroe, Greene, Lawrence, Martin, Orange, Dubois, Crawford and Perry. (See maps accompanying this report.) Outcrops of the contact are concealed under a glacial overburden in the northern portion of the area. From Monroe County southward the line of contact lies through the non-glaciated portion of Indiana. In this region outcrops along the line of contact are more abundant. The width of the accessible area of contact is variable, but is about twelve miles in width in the northern portion and about eighteen miles in width in the southern portion. The attenuated margin of the Pennsylvanian has been eroded in this region to such an extent that as a rule its rocks are found only on the tops of the high ridges which form the divides between the present drainage lines. In places these ridges rise 300 or more feet above the adjacent valleys and the line of contact between the Mississippian and Pennsylvanian occurs often 200 or more feet above the valleys. The ridges are decidedly irregular and their margins serrated by the indentation of minor drainage lines with their intervening spurs. The ridges are usually capped with a soft sandstone which crumbles easily, works down over and conceals the outcrops of the kaolin along the contact zone. The mantle of sand is so general along the contact that good outcrops occur only under exceptional conditions. Sometimes a stream cuts in against the line of contact and exposes the kaolin. In other places a spring may discharge its waters from beneath the bed of kaolin and the quantity of water may be sufficient to carry away the detritus and form a perpendicular face at the outcrop.

Kaolin has not been reported from all the counties along the contact zone. It has been found in Owen, Greene, Monroe, Lawrence, Martin and Orange. The only county in which there has been a serious attempt toward development is Lawrence County. Even here actual production has been confined almost entirely to one-half section and to only a small portion of that. Outcrops of white clay have been found in many places and test drifts, pits and drill-holes have revealed the presence of a white clay in many sections in the southwestern part of the county. Explorations have been more extensive in this county than in any other county in the Indianaite belt. For this reason, prospects appear best in this county. Martin County probably offers the next best field for exploration and development and Orange, Greene, Monroe and Owen would perhaps follow in that order.

Monroe County.

General Statement. This county lies largely in the unglaciated portion of Indiana and the bed-rock formations are well exposed. The Monon railroad crosses the county from north to south near the center of the county and the Illinois Central crosses it from east to west through Bloomington and sends a branch from that point south to Victor parallel with the Monon. The kaolin outcrops occur in the southwestern part of the county within from

Plate LXIII. Entrance to the Orchard-Timberlake mine in Monroe County. White kaolin on the dump in front of entrance.

one to three miles of these roads. If enough white kaolin should be found to warrant development on a large scale the transportation situation would present no serious difficulties.

At the present time there are no ceramic industries in this county and no established facilities for making use of the kaolin locally for ceramic purposes.

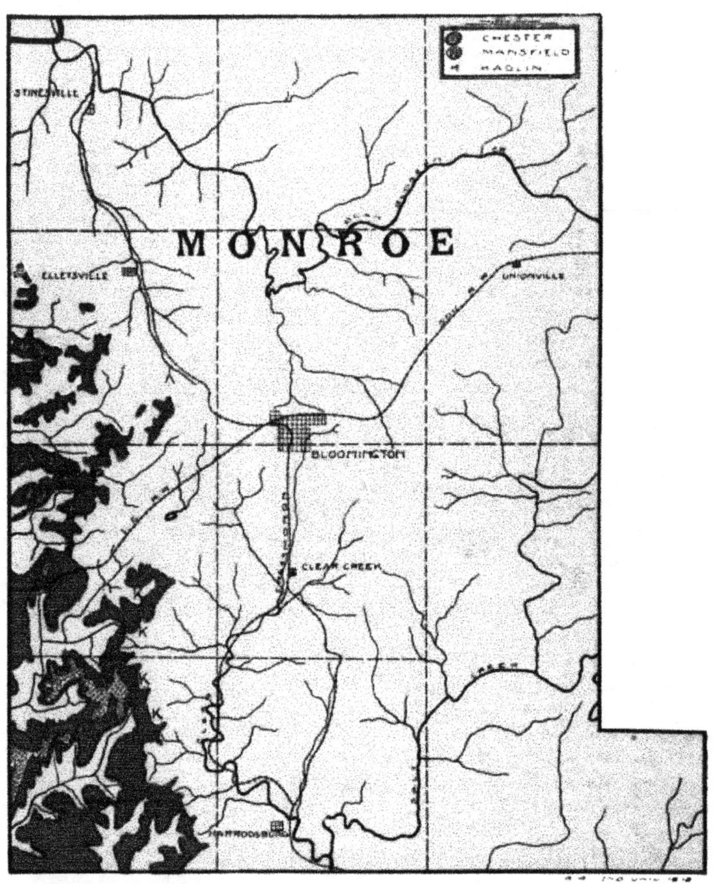

Plate LVI

The greater number of outcrops of kaolin occur below the Elwren sandstone of the Mississippian and only a few at the base of the Mansfield sandstone of the Pennsylvanian.

During field work in 1917 the writer's attention was attracted to an outcrop of reddish colored clay containing fragments of a white clay near the public road in Section 3 of Indian Creek Township. A later examination of the white clay showed it to be kaolin, Indianaite, a variety of halloysite.

In the spring of 1918 Mr. Dick Hall of Bloomington located a number of outcrops of the same kind of clay in the township. One of these outcrops is

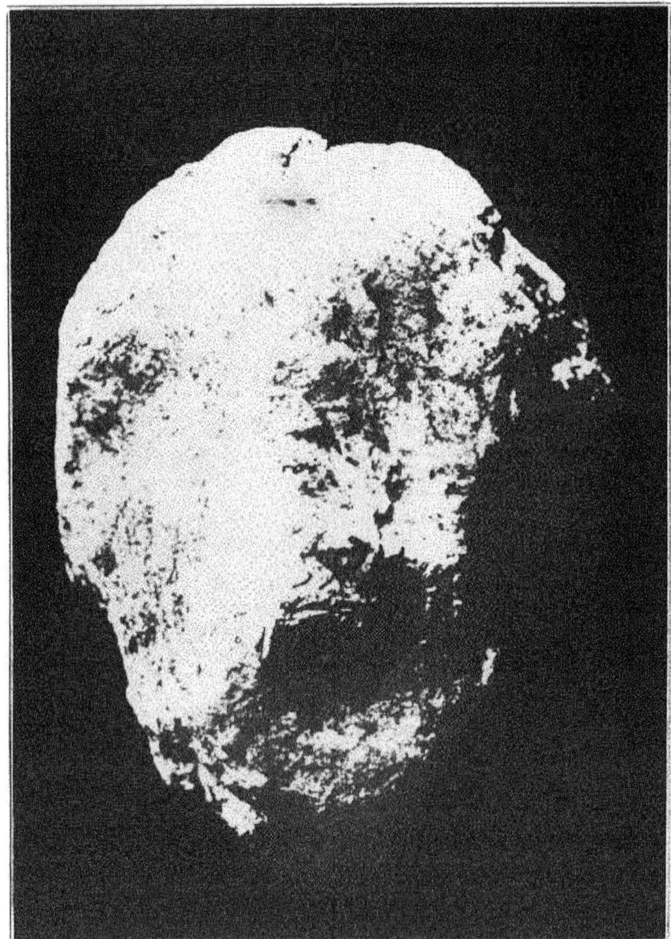

Plate LXIV. Mass of white kaolin from the Orchard-Timberlake mine in Monroe County. Kaolin lies below the Elwren sandstone.

on the public road near the John Koontz place in Section 10. The section exposed consists at the bottom of a shale containing sandy layers near the upper part, overlying this is a layer of mahogany colored clay of a thickness of thirty inches, containing fragments of kaolin, and above is a five-foot layer of sandstone. The kaolin occurs under and in most cases immediately in contact with the sandstone. Where the sandstone is compact and unfissured the Indianaite is more abundant. The thickness of the mahogany clay is

Plate LXV. Outcrop of mahogany clay and white kaolin under Elwren sandstone in Monroe County. See note book and above.

variable, pinching and swelling. In some places it may have a thickness of four feet and pinch down to less than half that amount in less than ten feet.

At one point in Section 28 of Van Buren Township, in a sandstone ledge, there is a thin layer made up of the fragments of kaolin. This occurrence shows that the Indianaite had been formed, eroded and redeposited. Below the sandstone there occurs a layer of mahogany clay which contains small fragments of Indianaite. The mahogany clay rests on a thin bed of sandstone which in turn rests on a bed of greenish colored shales. In the shale there are irregular lens-like masses of limestone. Where exposed at the surface these limestone masses are surrounded with the mahogany clay in which fragments of the white kaolin were found.

Distribution. In Van Buren township kaolin has been found in Sections 27, 28, 33 and 34. The outcrops occur on the slopes of a ridge which rises about 900 feet above sea level and forms a part of the divide between Clear Creek on the east and Indian Creek on the southwest. On the road which connects West pike with the Rockport pike, passing through the center of Section 28, and intersecting the above mentioned ridge, there are a number of outcrops of kaolin. On the northern slope of the ridge at the point where the road crosses it, there is an outcrop of mahogany clay which contains a considerable quantity of kaolin. Underlying the clay and separating it from a bed of shale is a thin layer of sandstone. A bed of sandstone having a thickness of twenty-five feet overlies the clay. The clay has a thickness of four feet at the outcrop but pinches down to about half that in a distance of six feet. The kaolin occurs in hard irregular fragments and also as white plastic streaks in the red colored clay. On the same slope below this outcrop there are some greenish gray shales containing irregular masses of limestone surrounded by mahogany clay. This clay also contains some fragments of the white kaolin.

On the same ridge farther east on the north side there is an outcrop of kaolin six feet thick on the side of a sink hole. On the south side of this ridge in the southeast quarter of Section 28, kaolin occurs under the sandstone, capping the top of the ridge, at about the same elevation as that on the north side. West of the road above mentioned, in Section 33, there is an outcrop of mahogany clay containing considerable Indianaite. The clay occurs between layers of sandstone of very fine grain. The overlying sandstone has a thickness of about thirty feet. The mahogany layer is irregular in thickness, pinching and swelling. Similar outcrops have been found in Section 27 on the southwest side of the ridge and in Section 34 on the east side.

Indian Creek Township. Indications of the presence of kaolin have been found at several places along the ridge which forms the divide between Indian Creek and Clear Creek in this township. In Section 3, outcrops occur in the west half of the section. In Section 10, outcrops of mahogany clay occur at several points, also in Sections 9 and 17. In the northwest corner of Section 10, near the public road, there is an outcrop of a layer of mahogany clay having a thickness of about 30 inches in places but thinning down to about half that in other places. White kaolin occurs in the clay in small irregular fragments which are most abundant under the compact and unfractured portions of the roof of sandstone. The underlying rock is a shale which passes into very sandy shale and lenses of sandstone just below the mahogany clay. The geological section exposed at this point is as follows:

No. 8 (top) Shale	5	feet
No. 7 Sandstone in thin beds....................	5	feet
No. 6 Shale, sandy	6	feet
No. 5 Sandstone	5	feet

No. 4 Shale	20	feet
No. 3 Sandstone, thick layers....................	10	feet
No. 2 Mahogany clay and Indianaite.............	2½	feet
No. 1 (bottom) Shale, sandy toward the top.......	12	feet

This mahogany clay lies near a slight unconformity in the Mississippian system of rocks. The shales above and below the mahogany belong to the Mississippian. The sandstone above is the Elwren sandstone of the Chester group.

State of Development. Small pits have been dug at several places on the outcrop of the mahogany clay but no serious attempt at development has been made. In order to determine whether the kaolin occurs in sufficient quantities to warrant commercial development will require the drilling of wells along the sandstone ridge at some distance from the outcrop. Near the outcrop the clay is nearly always stained with oxides of iron. The number and thickness of the outcrops offer promise of workable beds of the white clay. A tunnel has been driven at one point to a distance of 130 feet. Six feet of fairly white kaolin was found in this tunnel, and the indications are that a marketable quantity exists.

The presence of kaolin in Monroe County was mentioned by G. K. Greene[1] as follows: "Traces of kaolin, mere water-worn fragments, are occasionally found upon the surface. No beds of this deposit are found in Monroe County."

Lawrence County.

General Statement. The kaolin-bearing formations of Lawrence County are confined to the western portion. The greater part of the subsurface is occupied by formations of Mississippian or Lower Carboniferous age. The oldest of these formations forms the bed-rock in the eastern part of the county, and is bordered on the west by the outcrop of the Harrodsburg limestone. The central part of the county is occupied by the Salem (Bedford) and Mitchel. limestones. The Chester (Huron) limestones, sandstones and shales, capped on the higher ridges with the Mansfield (Pottsville) sandstone of the Pennsylvanian or Upper Carboniferous occupy the western portion. In this portion of the county the topography is more rugged and problems of transportation more difficult and it is in this region that the kaolin deposits occur.

The Monon railroad crosses the county from north to south near the central portion and a branch of it traverses the northwestern portion of the county, connecting with the main line at Bedford. The B. & O. S. W. railroad crosses the southern part of the county from east to west, intersecting the Monon at Mitchell. Kaolin deposits occur on both sides of this railroad and near it in the southwestern part of the county. The Southern indiana railroad crosses the county from east to west near the central part and intersects the Monon at Bedford. This road passes near kaolin deposits in the western part of the county. The greater part of the kaolin which has been shipped thus far has been from Huron on the B. & O. S. W. railroad, but smaller shipments have been made from Williams on the Southern Indiana railroad. No industries for the utilization of the kaolin have been established in Lawrence County and all kaolin mined has been shipped long distances.

[1] 2d Annual Report Indiana Geological Survey, p. 447.

Spice Valley Township. In the east half of Section 21 there is a high ridge of sandstone and shales which has upon its slopes a number of exposures of kaolin. These exposures occur near the contact between the Mansfield and the Chester formation, and to a limited extent at least, within the Mansfield. The thickness of the kaolin varies from five to eleven feet, the average thickness being not far from six feet. In appearance the kaolin, in its upper portion is white and comparatively free from discoloration. In the lower portion it contains sandstone masses and is often discolored with oxide

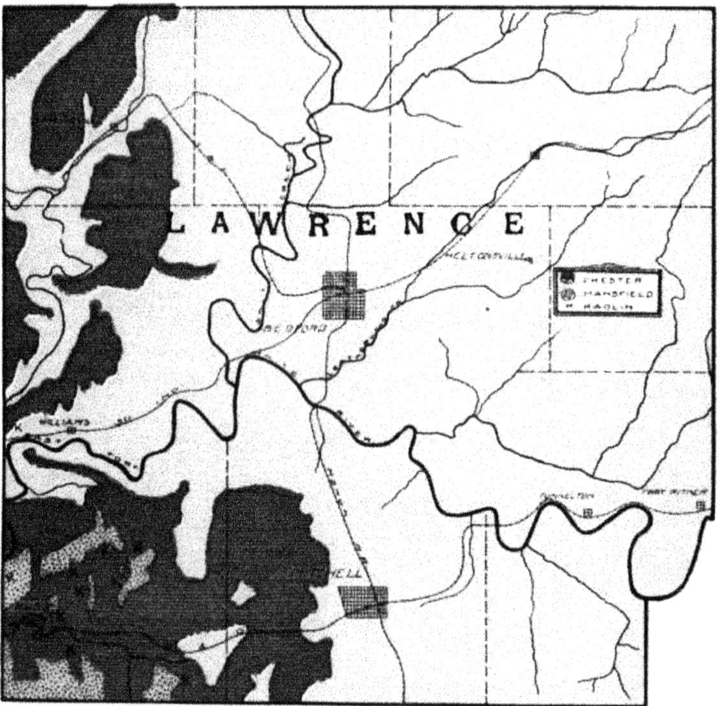

Plate LXVI. Lawrence County.

of iron. Two mines have been opened in this section; the west mine is a drift which is driven on the deposit to a depth of about 200 feet. The room and pillar system of mining was used and the pillars left were forty to fifty feet in diameter. This entry was driven from the west towards the east. The north mine entry is located about three-quarters of a mile north of the west entry and was driven south about 300 feet. Side entries were driven at intervals of about twenty-five feet. The room and pillar system was used in this mine and a large quantity of kaolin has been removed from it. (See Plate LXVIII.) The kaolin rests on the Elwren shale at an elevation of about 700 feet above sea level.

In Section 28 an outcrop of kaolin occurs in the northern part of the northeast quarter and also in the southeast quarter. An outcrop of mahogany clay occurs in the public road in the northwest quarter. In Section 22 indications of kaolin are found in the west half. Mahogany clay and disseminated white fragments occur at one point. There has been no development in this section. In Section 29 there are indications of the presence of kaolin in the east half. There has been no development except a small entry which encountered the mahogany but did not pass through the weathered zone. In Section 30, R. 2 W., drillings were made in the southwest quarter and kaolin was reported from these drillings. Kaolin occurs under the Cypress and under the Mansfield on the Hardinsburg shale at about 690 feet above sea level. In Section 31 the northwest quarter of this section is reported to contain kaolin, the data having been obtained from drill holes. In Section 4, kaolin

Plate LXVII. Six feet of white kaolin in an entry in the Gardner mine.

has been found by explorations made by the American Aluminium Company in both the northeast and northwest quarters. The kaolin lies below the Mansfield and rests on the Elwren shale. Four or five feet of kaolin is revealed in one of the openings. In Section 34 indications of kaolin have been found in the northwest, northeast and southwest quarters. In Section 35 an outcrop of mahogany occurs in the northwest quarter. This outcrop has been explored by a short entry. The other three quarters contain indications of kaolin. In Section 36 kaolin indications have been explored in the northwest and northeast quarters. White kaolin was found in the entries. In Section 8 there is a deposit of kaolin in the northeast quarter which has been explored by drill holes and by entries. The kaolin rests on Elwren shale and underlies Mansfield. Two entries have been driven. In Section 18 indications occur in all four quarters. Some outcrops have been explored by short entries. In Section 13 indications have been found in the northwest quarter. A small entry has been driven in the outcrop and some white clay obtained. In Section 2 indications of kaolin have been found in the northeast quarter and explorations have been made by entry and drilling. The southeast quarter also shows indications. In Section 3 the east half shows indications of the presence of kaolin and small explorations have been made in one place. In the

Plate LXX. Entrance to Gardner mine, Lawrence County. Mansfield sandstone above containing a thin layer of kaolin.

N. E. quarter of the N. E. quarter of Section 30, T. 4 N., R. 2 W., a deposit of mahogany clay and white kaolin occurs at the contact between the Elwren shale and the Mansfield sandstone which has a thickness of 50 or 60 feet. Below the contact the Reelsville limestone occurs at 785 and the Beaver Bend

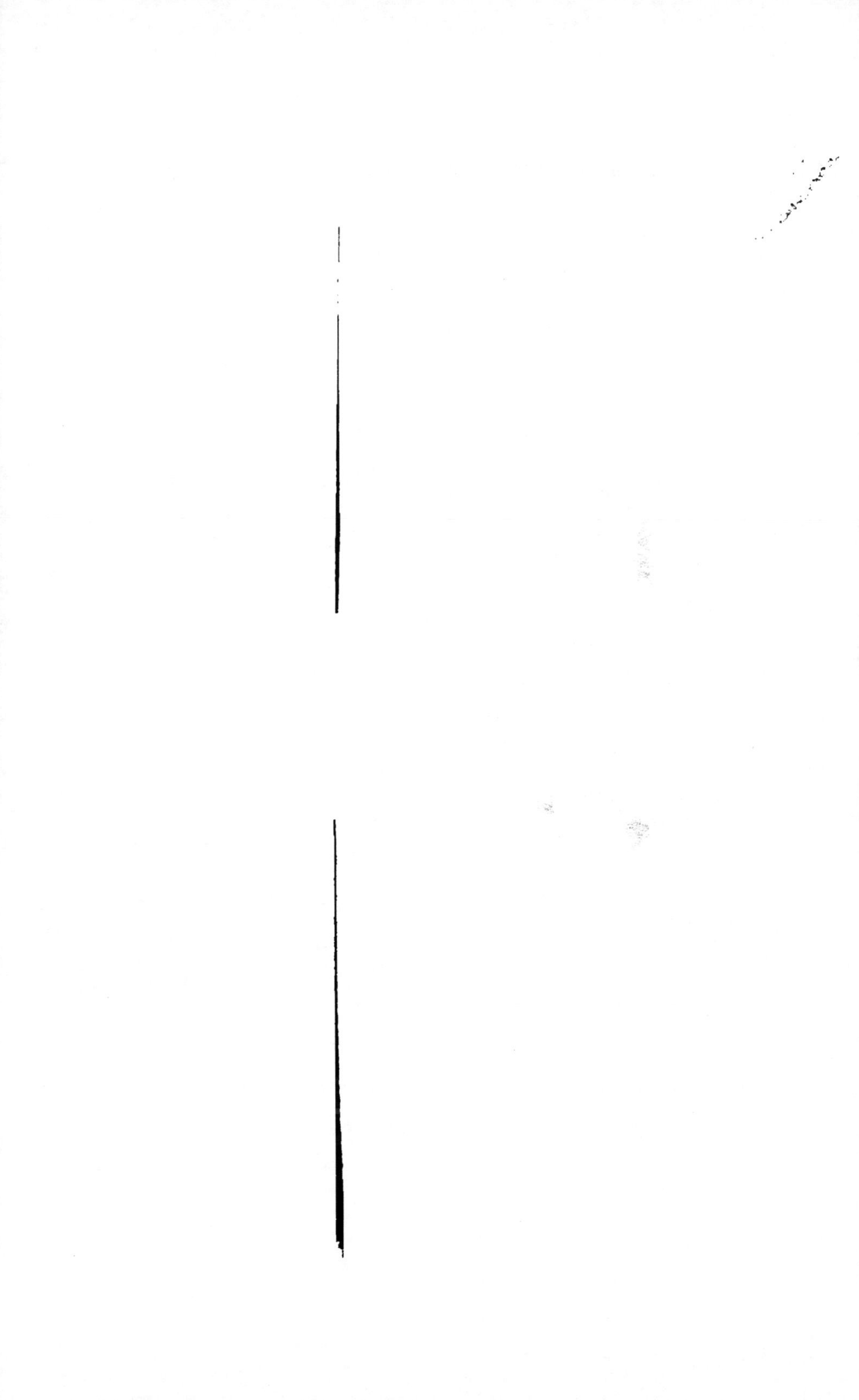

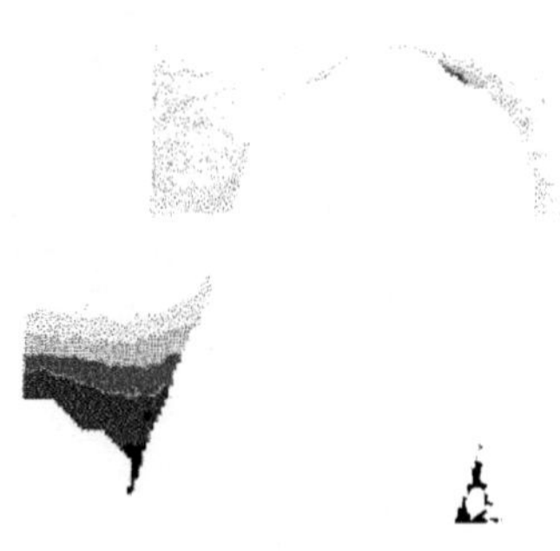

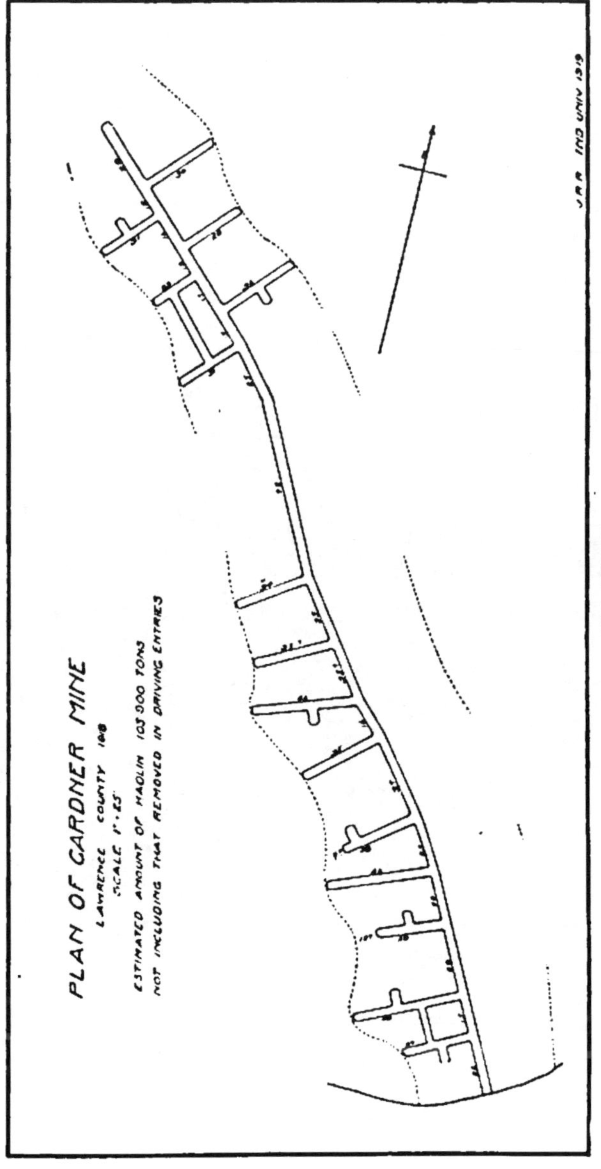

Plate LXVIII.

limestone at 760 feet. They are separated by about 25 feet of Brandy Run shale. Two openings have been made at the kaolin horizon, one in a shaft 18 feet deep, the other a tunnel. Some white kaolin was found in each but neither extended into the deposit far enough to secure a good sandstone roof and consequently the kaolin is much stained. The ridge of Mansfield at this

Plate LXXI. Entrance to Phipps mine east of Huron. Kaolin resting on Elwren shale and containing a few Beech Creek limestone bowlders. Mansfield above, contact rising toward the south.

point extends in a curve toward the northeast for more than one-half mile.

In Section 26, near the S. E. corner, mahogany containing some white kaolin occurs at the contact between the Mansfield and the Sample shale, evidently at the top of the Sample, since a few bowlders of Beaver Bend limestones were found in the mahogany near the entrance of a tunnel being driven into the kaolin deposit by Dr. John Laughlin of Bedford. The Mansfield at this point has a thickness of 180 feet and has been deposited in a depression in the eroded surface of the Chester. Beneath the mahogany which has a thickness of from 4 to 6 feet there is about 20 or 25 feet of shale resting on the top of the Mitchell limestone.

Plate LXXII. White kaolin in mahogany clay in Phipps mine, east of Huron. Limestone bowlder at right. Beech Creek on Elwren shale.

In Section 25, about one-fourth of a mile east of the above outcrop, the Mansfield contact, at approximately the same elevation, exhibits a layer of stained kaolin.

About one-fourth mile north and a little east of the center of Section 36 the basal member of the Mansfield consists largely of uncemented sands which have been eroded into a badland type of topography. The Mansfield rests on

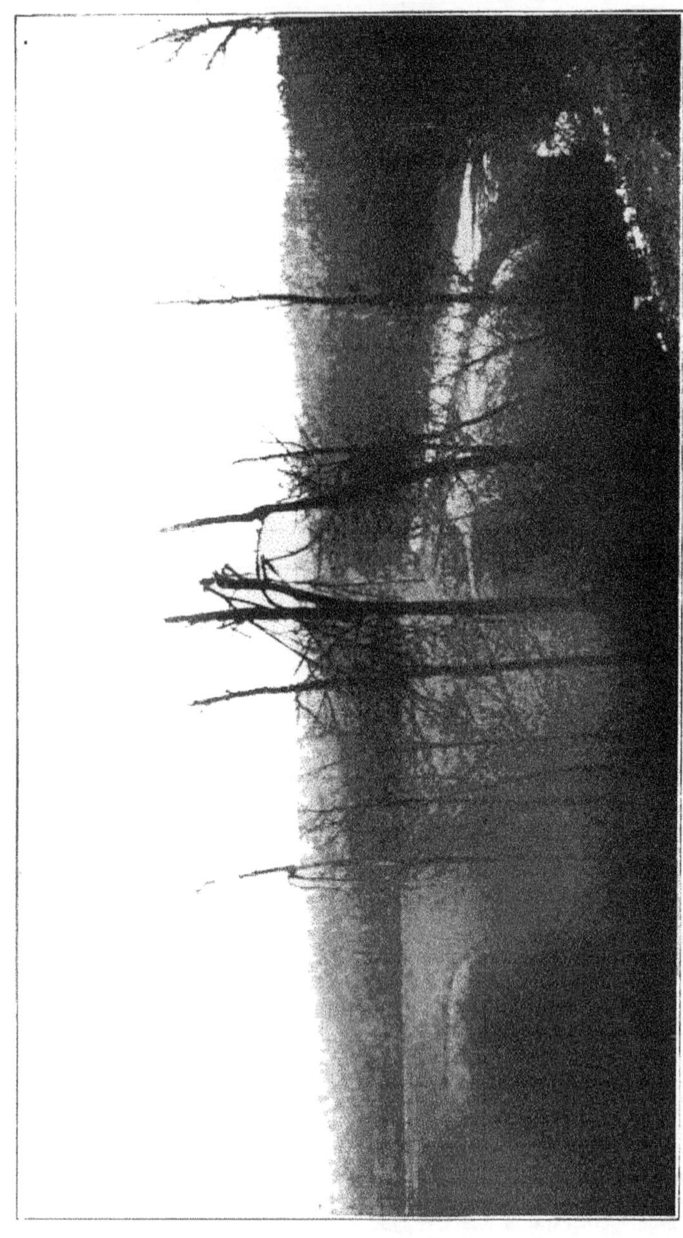

Plate LXXIII. Horseshoe curve in White River in the area mapped in Lawrence County

a bed of kaolin, a part of which is unstained. The full thickness of the kaolin stratum is not revealed but it is several feet thick and rests upon a bed of maroon shale which changes into olive-green below. The elevation of the kaolin at this point is five feet higher than it is in the S. E. corner of Section 26.

Across the ridge in Section 35, southwest of the kaolin outcrop in Section 26, the contact of the Mansfield and Chester is about seventy feet higher than at the latter place. The Mansfield at this point rests on kaolin which has been formed from the Elwren shales. The unconformity has a steep dip, falling seventy feet in about one-half mile.

In the N. E. quarter of Section 27 a kaolin outcrop occurs at the Mansfield contact with the Chester at about 720 feet. The outcrop exhibits four or more feet of mahogany and white kaolin. The thickness of the Mansfield is 120 feet at this point and the prospects for unstained kaolin under the ridge are good. In the southwest quarter of Section 27 a ridge capped with Mansfield sandstone occupies the north part of the quarter. Beneath the Mansfield there is a stratum of stained kaolin which rests on the Elwren shale. The kaolin contains a few much weathered limestone bowlders which belong to the Beech Greek limestone. In Section 5, near the northeast corner, mahogany clay lies underneath the Cypress sandstone. In Section 6, in the northwest quarter, kaolin occurs beneath the Mansfield at an elevation of about 90 feet above the Station at Huron; also below Cypress sandstone.

Indian Creek Township. In Section 6 there is an outcrop of kaolin underlying the Mansfield sandstone in a gravel pit north of the Southern railway. The kaolin rests on the Sample shale at an elevation of 587 feet above sea level.

Marion Township. In the northwest quarter of Section 17 indications of kaolin have been found. In the southeast quarter of this section an outcrop of mahogany clay occurs. This outcrop has been explored by a drift entry and shaft and considerable white material has been found.

Orange County.

General Statement. The geological conditions in the western portion of Orange County are favorable to the occurrence of kaolin. The higher lands are occupied by the Mansfield sandstone while the stream valleys and lowlands contain the outcrops of the Chester formations. The outcrops of kaolin so far encountered have been in the northwest portion of the county and the railroad facilities for this portion are not good. The Monon lies too far east and the B. & O. S. W. runs not quite as far away on the north. The topography is rugged so that transportation over the public roads would be more expensive.

Orangeville Township. In the northeast quarter of Section 10, near the north line of the county, there are several outcrops of kaolin. The geological section exposed in the public road at this point is as follows:

No. 6	(Top) Sandstone (Cypress)	35 feet
No. 5	Mahogany clay with white particles	10 feet
No. 4	Greenish clay and thin bedded sandstone and limestone	88 feet
No. 3	Limestone (Beaver Bend)	23 feet
No. 2	Shale, thin bed of coal and sandstone (Sample)	21 feet
No. 1	Limestone (Mitchell)	18 feet

All quarters of this section contain outcrops of kaolin, some of which exhibit considerable white material. Indications of kaolin have also been found in Section 20. In Section 32, white kaolin occurs in a cave. It lies above the Mitchell limestone and below the Sample sandstone. In Section 2 there is an outcrop of mahogany clay in the public road in the northwest quarter. The mahogany clay rests upon a clay which in turn rests upon

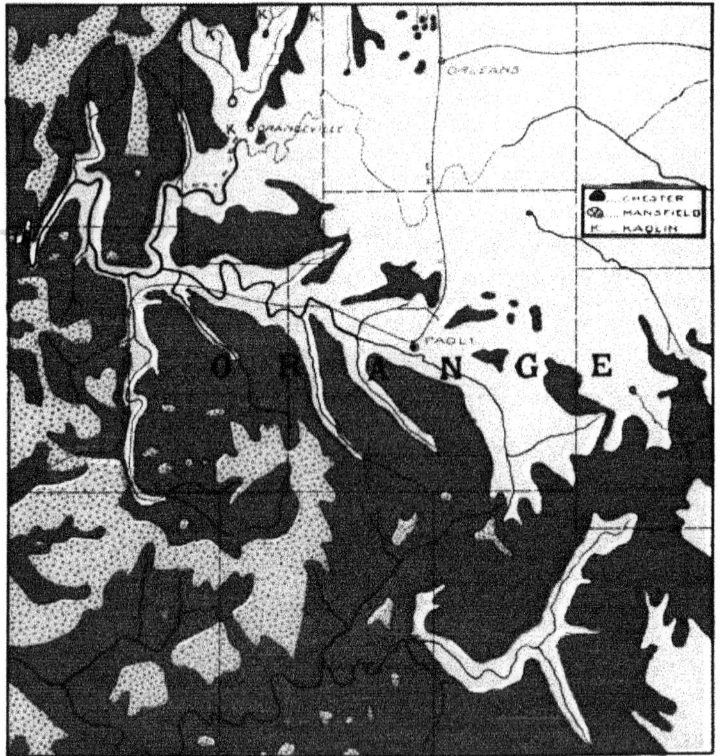

Plate LXXIV. Map of Orange County.

limestone. The overburden is sandstone. In Section 20,[1] T. 3 N., R. 1 W., white kaolin occurs. In Section 7, T. 1 N., R. 2 W., yellow kaolin occurs three feet thick.

Martin County.

General Statement. The Pennsylvanian system of rocks forms the subsurface of a large part of Martin County. The greater part of the county is occupied by the outcrop of the Mansfield sandstone, which contains iron ore

[1] Cox. 7th Annual Report Indiana Geological Survey, 1875, pp. 234-5.

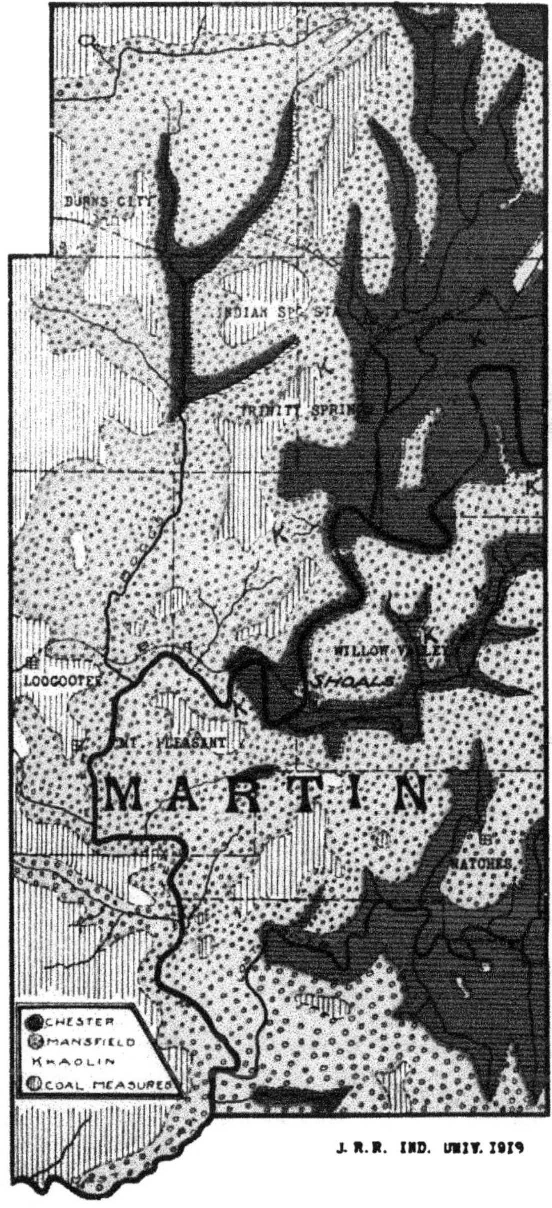

Plate LXXV. Map of Martin County.

which was formerly smelted at Shoals in this county. It was the mining of iron ore at the base of the Mansfield for the smelter at Shoals that led to the discovery of a large deposit of white kaolin in Lawrence County. In the eastern part of Martin County the streams have cut through the Mansfield and exposed the Chester shale, limestones and sandstones.

The Southern Indiana railroad crosses the northern part of Martin County from east to west and passes near the outcrops of kaolin. The B. & O. S. W. railroad crosses the county in an east and west direction, a little south of the center. A few outcrops of kaolin occur near this railroad in the eastern part of the county and just west of Shoals.

A brick plant which was erected at Shoals several years ago is not now in operation. The kaolin in this county is not being utilized at present.

The eastern third of Martin County is occupied by the attenuated border of the Pennsylvanian and exhibits its contact with the underlying Mississippian at many points. Although there are many outcrops of kaolin there has been little serious attempt at development.

Halbert Township. In the southeast quarter of Section 16, where the public road ascends the divide between Beaver Creek and White River, is an exposure of mahogany clay which is several feet thick and underlies sandstones and shales of the Pennsylvanian. The clay contains some white streaks which resemble decomposed kaolin. Limestones of Mississippian age underlie the deposit. No attempt has been made at development. A similar outcrop occurs in the southeast quarter of Section 11, at the side of the public road. Sandstone overlies the mahogany clay at this point and shale underlies it. Several springs mark the line of contact of the sandstone and the shale. An outcrop of white kaolin was found many years ago at the base of a thick bed of sandstone in the northeast quarter of Section 12. It was reported to have been found in a well and below a spring in the southeast quarter of the same section.

West of the public road, in the southwest quarter of Section 2, many fragments of white kaolin were found on a slope which is covered with detritus from a slope which contains sandstone overlying shale. The sandstone is exposed at the top of the ridge and the shale in a small stream bed below the contact. The white fragments were very abundant around the holes made by groundhogs.

In the public road, in the northwest quarter of the same section, there is an outcrop of mahogany clay overlying a bed of bluish shale. Sandstone overlies the clay. The layers of sandstone are thick and the grains are well cemented with oxide of iron. An outcrop of mahogany clay occurs also in Section 1 in the public road near the center of the section. The underlying formation is shale and the overlying rock is sandstone.

Center Township. In R. 4 W., T. 3 N., Section 1, kaolin occurs in the N. W. quarter. It underlies the Mansfield sandstone and rests on the shales of the Hardinsburg formation. The Glendean limestone which outcrops about 375 feet east of the kaolin entry and the twenty feet of Tar Springs sandstone lying above underneath the Mansfield has been cut out by the Mansfield at this point. The Tar Springs sandstone at this point is composed of thin bedded sandstones, some layers of which are capable of being split into very thin laminæ. The thicker layers are suitable for the manufacture of whetstones. The kaolin consists of mahogany clay with fragments of white

Plate LXXVI. Reelsville limestone west of Huron. Brandy Run shale below and Elwren shale above. Contains much pyrite.

Plate LXXVII. Outcrop of Elwren sandstone under which kaolin occurs in Monroe County. Fine of grain and cross-bedded.

Plate LXXVIII. Chester limestone and shale in Monroe County. Limestone where aneroid hangs surrounded by maroon clay.

kaolin at the entrance, but it said that back under the firm sandstone there is a five-foot layer of good white kaolin. This kaolin occurs on the edge of a basin which has been formed in the Chester rocks by pre-Pennsylvanian erosion.

In Section 2, near the southwest corner, there is an outcrop of mahogany clay which contains some small bands of white kaolin. The kaolin underlies the Mansfield sandstone and rests on the shales of the Hardinsburg. The elevation is about five feet lower than in Section 1.

In the northwest quarter of Section 27 kaolin occurs near the bank of White River in a small ravine which enters the river from the south. The Mansfield which occurs in the river bed at Shoals is forty feet above the river.

In Mitchell Tree Township, White River crosses the southeast corner of Section I, producing a high bluff. The base of this bluff contains sixteen feet of Beech Creek limestone, overlying which is about twelve to eighteen inches of kaolin. Above the kaolin thin bedded Cypress sandstone having a thickness of five feet is succeeded by twenty-five feet of massive sandstone belonging to the Cypress. Above the Cypress is a high bluff of Mansfield sixty to seventy-five feet high. On top of the Mansfield are Lafayette gravels consisting of brownish colored gravels such as are found in the Knobstone and geodes such as occur near the Harrodsburg-Knobstone contact.

Along the high ridge between Indian Creek and White River several outcrops of kaolin occur below the Cypress sandstone. One of these occurs in Section 12 at an elevation of 695 feet above sea level.

An outcrop of white kaolin occurs in the public road in the southeast quarter of Section 16. A short tunnel has been dug into this deposit and considerable white kaolin occurs on the dump.

In Columbia Township, in Section 5, there is an outcrop of mahogany clay containing some fragments of white kaolin underlying the Mansfield.

Greene County.

General Statement. The eastern part of Greene County lies within the outcrop of the Chester group of rocks, mainly, though, the Mansfield occupies a part of the region. The remainder of the county has its subsurface occupied by the rocks of the Pennsylvanian system. Outcrops of kaolin occur only in the eastern portion of this county and two railroads intersect it. The Monon branch from Bedford to Bloomfield crosses the southern portion and the Indianapolis branch of the Illinois Central crosses the northern portion.

The eastern part of Greene County lies within the driftless area. The exposures of the Pennsylvanian-Mississippian contact are numerous in the eastern tier of townships. The Mississippian occupies the larger part of the surface, the Pennsylvanian forming irregular areas occupying the higher elevations. Mahogany clay has been found in Center and in Beech Creek townships. White kaolin is reported from the Sullivan, Coombs and Edwards places in these townships. In Center Township mahogany clay outcrops in Sections 1, 2, 10, 11, 12, 22 and 25.

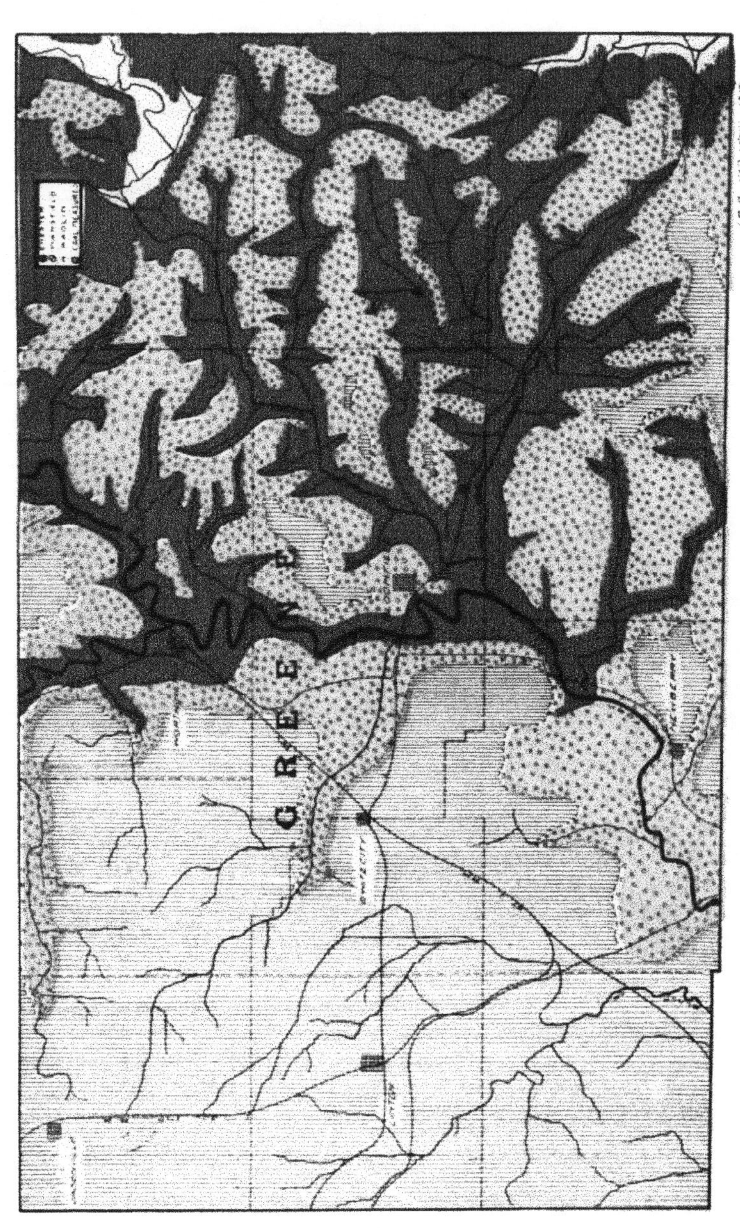

Plate LXXXIX. Map of Greene County.

Owen County.

General Statement. That portion of Owen County adjacent to Monroe and southeast of White River is favorably located for the occurrence of kaolin so far as geological conditions are concerned. Both the Chester and Mansfield are represented by outcrops. Railroad facilities are not good in this

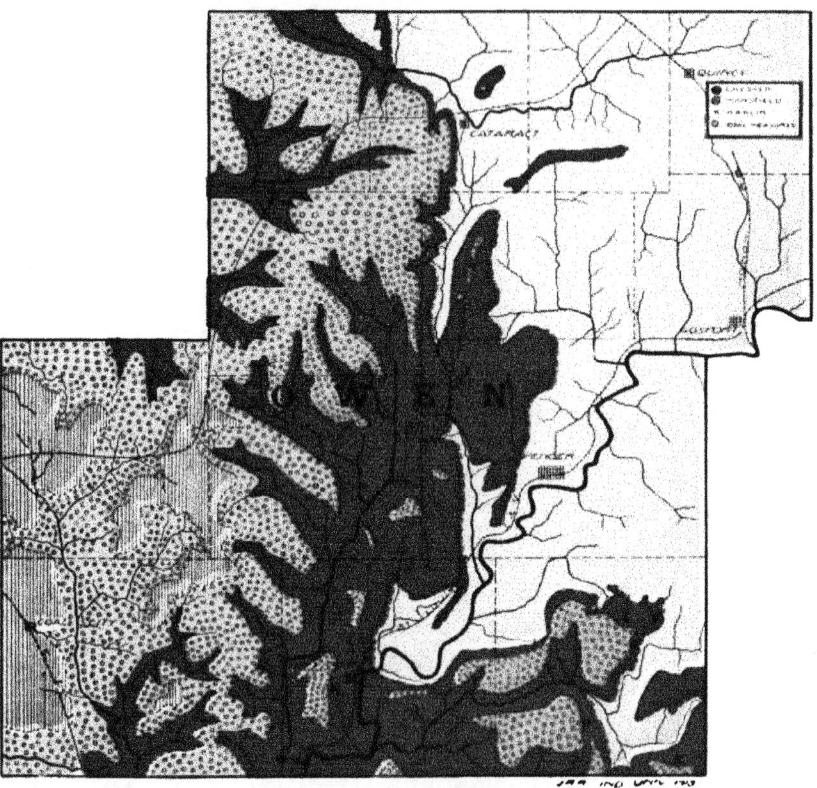

Plate LXXX. Map of Owen County.

part of the county, but the territory lies between the Illinois Central on the south and the Pennsylvania railroad on the north.

Specimens of the white kaolin were reported by Collett[10] from the headwaters of Raccoon and on Jordan and Rattlesnake creeks. Outcrops occur on the northwest quarter of Section 7, T. 9, R. 3, and in Section 27, T. 11, R. 4.

CHAPTER VI.

IRON ORES.

Minerals containing iron are widely distributed in the rocks of the earth. The mantle or surface rocks are rarely ever free from iron compounds. The yellow, brown and red colors of soils are due to the presence of some form of iron. It is estimated that iron constitutes 4.56 per cent of the earth's crust.

Although there are a large number of minerals containing iron there are only a few that may be used for ores from which iron may be extracted. Those minerals commonly used for ores of iron are given below:

Iron Ores.

Magnetite (Fe_3O_4)	72.24 per cent of iron
Hematite (Fe_2O_3)	70.00 per cent of iron
Limonite ($2Fe_2O_33H_2O$)............	59.89 per cent of iron
Siderite ($FeCO_3$)	48.27 per cent of iron

Magnetite is a black mineral, very hard and heavy, and strongly magnetic. It occurs in veins, in lens-like masses and as grains in sands. Aside from an occasional crystal of magnetite found in glacial boulders it does not occur in Indiana except as sand grains. Magnetite bearing sands are not of uncommon occurrence especially in the stream beds of the glacial area when the stream beds lie on shale or clay as they do in the Knobstone area. Some gold flakes are associated with these sands in some of the counties of the area. No deposits of magnetite of economic importance are known in Indiana; nor are they likely to be found since economic deposits of these types are rarely of sedimentary origin.

Hematite. This is a red or steel gray mineral which has a red streak, a slightly lower specific gravity than magnetite. It occurs in many forms and colors. It may be in the form of a powder, granular, nodular, irregular or massive. Red ocher is a mixture of hematite and clay. It occurs in small quantities in many places in Mississippian and Pennsylvanian strata in Indiana.

Hematite also occurs in these rocks in the form of silicious clay, stones and sandstones. The latter is the most abundant form and is found in the Mansfield formation.

Limonite. This ore of iron occurs in a number of forms, as nodular, botryoidal, vesicular, stalactitic, massive or as powder in clay forming yellow or brown ocher. Its color is variable, but usually black brown or yellow. Its specific gravity is low, 3.8. It has a yellow streak which easily distinguishes it from hematite. It also contains water of crystallization and is often associated with other iron ores which are hydrous. This is the most abundant form of iron ore in Indiana. It occurs as concretions in soils, which are poorly drained; as bog ore in swamps and marshes; as lenses, irregular masses, nodules and distinct layers in the rock strata.

Siderite. This is a brown to gray colored ore. It occurs as lenses, concretions or as massive or granular masses. Its principal form of occurrence in Indiana is in nodular masses and lenses in shales and clays.

Impurities. Iron ores as they occur in the earth usually contain one or more mineral impurities. These may be silica, alumina, lime, titanium, phosphorus, sulphur, manganese and small quantities of other elements. Bog ores frequently contain organic matter.

Plate LXXXI. Mansfield sandstone showing cross-bedding and effect of differential weathering. Cementation with iron oxides.

Silica is one of the most common impurities. It is generally present in the form of quartz sand grains. It displaces iron in the ore and is detrimental in all forms of the metal save ferro-silicon. The quantity of lime required for fluxing is increased by the presence of silica. Ores containing as high as forty per cent of silica are used and some foundry iron contains as high as ten per cent of silica. The iron ores in the Mansfield or basal Pennsylvanian of Indiana usually contain a high per cent of silica.

Alumina is present in many iron ores and is generally in the form of a silicate. The silicates are those commonly present in clay and are most abundant in the ochers. Many bog ores contain quantities of clay which is derived from the surrounding land areas. The alumina passes off in the slag which may contain enough to make it valuable in the manufacture of Portland Cement and a great deal of slag is used for this purpose in Indiana.

Lime is present in some iron ores in considerable quantities, especially in ores which have resulted from concentration through the decomposition of beds of limestone containing disseminated iron compounds or in ores deposited with limestone like the Clinton ores. A small quantity of lime is not detrimental but large quantities must be removed by fluxing.

Titanium is a common impurity of magnetite. It is not detrimental to the metal itself but causes loss in smelting by driving large quantities of iron into the slag. It is possible to use titaniferous iron ores with other ores to give the necessary amount of titanium for ferro-titanium. Titanium is not an important element in the iron ores of Indiana.

Phosphorus occurs in iron ores as a calcium phosphate and as a constituent of some silicates such as glauconite. It is detrimental above one-tenth of one per cent in pig iron used for the manufacture of steel by the Bessemer process. Non-Bessemer ores are those containing more than one part of phosphorus to each thousand parts of metallic iron content of the ore. In order to reduce the percentage of phosphorus in some high phosphorus ores, low-phosphorus or non-phosphorus ores are mixed with the former. The iron ores of Indiana range from a trace to 1.03 per cent in phosphorus content.

Sulphur which is present in some ores of iron is usually derived from the iron sulphide, pyrite. Sulphur has a tendency to weaken the metal. It is easily transformed into the gaseous sulphur dioxide which in escaping from the molten metal may leave it porous and render it brittle. Sulphur in the Indiana iron ores ranges from a trace to more than two per cent.

Manganese is found in most limonite ores. It is not detrimental in iron used for certain purposes but is desirable. The amount of manganese in the Indiana ores varies from a trace to nearly one per cent.

Iron Bearing Formations in Indiana.

The geological formations which contain iron ores in Indiana belong to the Mississippian, Pennsylvanian and Post-Pleistocene periods. The ores which have been used in the State have been obtained from formations belonging to the last two periods.

Mississippian Strata. The Knobstone shales contain bands of kidney ore near the contact of the Knobstone with the New Albany shale. The principal outcrops occur in Scott, Clark and Floyd counties. The bands are from two to ten inches thick and from six to ten bands occur in twenty feet of the shale. Important localities are near Henryville in Clark County, and in Finley and Vienna townships in Scott County. The metallic iron content of the ore averages about twenty-eight per cent.

The composition of a sample of the Scott County ore is given by Shannon as follows:[14]

Analysis of Scott County Iron Ore.

Combined water	15.00
Silicic acid	14.00
Protoxide of iron	38.56
Sesquioxide of iron	3.01
Oxide of manganese	4.50
Carbonate of lime	2.02
Carbonate of manganese	.85
Sulphur	.05
Phosphoric acid	.50
Carbonic acid and loss	21.51
Total per cent of iron	32.20

The composition of a sample of ore from north of Henryville, Clark County, is given by the same writer as follows:[14]

Analysis of Henryville Iron Ore.

Moisture dried at 212°	0.500
Insoluble silicates	16.400
Carbonate of iron	49.400
Peroxide of iron	2.171
Manganese	2.500
Alumina	1.500
Carbonate of magnesia	14.000
Carbonate of lime	10.000
Sulphuric acid	0.686
Phosphoric acid	0.779
Loss and undetermined	1.744

A section near Henryville contains the following strata:

Knobstone shale and iron bands	23 feet
Limestone (Rockford)	3 feet
New Albany shale	6 feet

The iron bands are six in number averaging six inches in thckness and are separated by from two to three feet of shale.

Kidney ore also occurs at other horizons in the Knobstone and outcrops containing the ore occur in Washington, Lawrence, Monroe, Brown, Jackson and Morgan counties. The deposits that have been found so far are not of great extent either laterally or vertically. The richness of the ores tested does not exceed the average given above.

The Chester shales which occupy the upper part of the Mississippian underlying the Pennsylvanian contain iron bands and iron bearing concretionary masses. The iron is in the form of limonite and siderite ores. Lenses of the former occur frequently along the Mississippian-Pennsylvanian contact. Some of the Chester limestones, especially the Reelsville, contain iron, in the unweathered portions mostly in the form of pyrite and in weathered portions in the form of limonite. Some ores from the contact have been mined in Lawrence County, but commercial quantities are not abundant.

Pennsylvanian Ores. The Pottsville division of the Pennsylvanian contains the largest deposits of iron ore in Indiana. The basal member of the Pottsville, the Mansfield, which in many places is a conglomerate or coarse sandstone, is in other places a silicious iron stone. The shale which in places lies below the Mansfield contains in many places bands and irregular lenses or masses of pyrite and limonite. The latter is sometimes confined to the weathered zone. Shale lying above the Mansfield sandstone also contains kidney ore, bands and irregular masses. Iron ores occur in the Pottsville in Vermillion, Parke, Vigo, Clay, Owen, Greene, Monroe, Lawrence, Martin, Orange, Crawford and Perry counties. Ores collected from Vermillion, Vigo, Clay, Greene, Monroe, Lawrence and Martin have been used in past years though none of these ores are being used at the present time.

The average composition of some selected samples of limonite ore from the Mansfield-Chester contact in Lawrence County is as follows:

Average Analysis of Lawrence County Limonite.

Hygroscopic water	2.32 per cent
Combined water	8.50
Insoluble silicates	3.50
Sesquioxide of iron	79.50
Sesquioxide of manganese	2.00
Alumina	2.00
Magnesia carbonate	.426
Lime carbonate	.555
Phosphoric acid	.139
Sulphur	trace

A large area east of Bloomfield in Greene County contains iron ores belonging to the Pottsville division. The ores are of two kinds, kidney limonite ore and silicious limonites and hematites. The compositions of a sample of the former is given below:"

Analysis of Kidney Ore, Greene County.

Loss on ignition, water (mostly) and inorganic matter	11.50 per cent
Insoluble silicates	17.00
Sesquioxide of iron, some protoxide, and tr. of Mn.	56.00
Alumina	2.00
Carbonate of lime	10.00
Magnesia	3.50
Total	100.00

Total iron, 39.20 per cent.

The analysis of a sample of the silicious ore is as follows:

Analysis of Silicious Ore, Greene County.

Loss by ignition, water	7.50	per cent
Insoluble silicates	34.00	
Sesquioxide of iron	54.73	
Alumina	2.50	
Manganese	1.14	
Lime	.12	
Magnesia	.03	
Total	100.02	

Total iron, 38.31 per cent.

Some of the best iron ores of Indiana are in the Pottsville of Martin County. The beds of ore range in thickness from a few inches to fifteen or twenty or more feet. The purer ores are seldom more than four or five feet thick. A large area in the eastern part of the county was core drilled in 1905. The analyses of forty-three samples of the ore from this county show the following variation in constituents:

Analyses of Martin County Iron Ores.

	Range per cent
Water	5.25 to 10.50
Silica	6.30 to 56.90
Iron (metallic)	25.02 to 55.44
Alumina	.80 to 15.50
Ferric oxide	35.72 to 79.20
Lime	tr. to 20.00
Phosphorus	tr. to 1.624
Sulphur	tr. to 2.08
Manganese	tr. to 3.06

The southwestern part of Monroe County contains iron ores belonging to the Pottsville division. These ores were used at one time in a furnace located on Indian Creek. The composition of a sample of this ore is as follows:

Metallic iron	52.2	per cent
Insoluble	4.1	
Manganese	1.20	
Phosphorus	.24	

These ores occur in shales in bands and irregular masses. They are often pyritiferous but usually of the limonite type in weathered zones.

There are large quantities of silicious ores in this area connected with the Mansfield formation.

The Post-Pleistocene Iron Ores. Since the glacial period the streams and surface waters of northern Indiana have been draining into the ponds, lakes and marshes of that region and carrying in solution quantities of iron which on being precipitated through the action of organic matter formed the bog ore of that region. The ores are usually of the limonite type and occur in concretionary masses. The ore is found in the marginal deposits of lakes and marshes and in layers in the bottom of peat bogs. These layers are rarely more than two feet thick. It also occurs as masses of variable size in the marsh material, such masses of several tons are occasionally found. The concentration of the iron may be seen in many places in the glaciated area. The water which falls upon the surface passes downward through the porous glacial material until it reaches a bed of impervious till, following along the surface of the till it finally reaches the surface again. In the meantime it has taken the iron into solution, probably in the form of a carbonate, and when exposed to the air the iron compounds are quickly oxidized and precipitated in the form of a yellow hydrated oxide of iron.

The most extensive swamp lands of the region are found along the Kankakee River which controls the drainage for parts of seven counties. Bog ore has been revealed in many places in this area by drainage ditches. There are also a large number of lakes and ponds about the margins of which peat and bog ore have been found. Smaller marshes and peat bogs are distributed throughout the lake region and bog ores are associated with many of them.

Development.[11] During the period from 1830 to 1895 iron ore was smelted in Indiana and native ores were used. During the early part of the period Indiana was one of the leading states in the production of iron. The first smelter was erected in northern Indiana at the present site of Mishawaka in St. Joseph County in 1834. The bog ore from along the marsh lands of St. Joseph River were used. The fuel was charcoal burned in the vicinity. The iron was manufactured into cast iron kettles, pots and other household utensils, and later stoves and railroad iron. A second furnace was started at this place in 1847.

A smelting furnace was established at Rochester in Noble County in 1845. The ore used was bog ore dug from Ore Prairie. Charcoal was used for fuel. About the year 1850 a furnace was located at Lima in Lagrange County. The ore was obtained from the marsh lands along Pigeon River. The ore was smelted into bar iron by the use of charcoal. These plants were all abandoned by the year 1858.

A blast furnace for smelting iron ores in Martin County was established about 1870, at Irontown just east of Shoals. The furnace was destroyed by an explosion due to an accumulation of gases. Some of the iron ores from Martin and Lawrence counties were shipped to Jackson, Ohio, and smelted at that place.

A furnace was established on Richland Creek east of Bloomfield in 1841. Charcoal was used in the production of pig iron from ores mined in the vicinity from the Pottsville division.

The Old Virginia furnace was located on Indian Creek in Monroe County almost directly south of Sanford in 1840. Ores from Greene and Monroe counties were used. Charcoal was the fuel used. A blast furnace was erected in Brazil in 1867. A furnace was also established in Knightsville in 1868. This was a double furnace costing about $200,000. The Masten furnace was established north of Brazil and the Star furnace at Harmony. Native ores were used to a limited extent. Block coal was used as fuel.

The Vigo Blast furnace was established in Terre Haute and did not cease operation until 1895. The native ores of Vigo County were used to a limited extent. They consisted of clay iron bands and kidney ore concretions from the Allegheny division of the Pennsylvanian.

The Old Indiana furnace in Vermillion County made use of the same kinds of ore. It used charcoal for fuel. An attempt to use coal was a failure and the furnace was not operated after 1859.

ANALYSIS OF IRON ORE.

Location	H_2O at 212°	Silica SiO_2	Carb. of Iron $Fe\,CO_3$	Perox of Iron Fe_2O_3	Manganese	Alumina	Carb. of Magnesia $Mg\,CO_3$	Carb. of Lime $Ca\,CO_3$	Sulphuric Acid $S\,O_3$	Phosphoric Acid P_2O_5	Magnesia	Lime	Sulphur	Loss on Ignition	Lime, Mag. and Loss	Iron and Tr. of Manganese	% of Iron	Total
Clark County—North of Henryville	.500	16.400	49.400	2.171	2.500	1.500	14.000	10.000	0.686	0.779				1.744				
Greene County—Near Bloomfield		17.00				2.00		10.00			3 5			11.5		56.00	39.2	100.00
Greene County—Siliceous "black ore"		34.00			1.14	2.5					0.03	0.12				54.73	38.31	100.02
Lawrence County—Near Shoals	13.000	0.90		84.89		Tr.		1.00		0.145						81.5	58.00	
Lawrence County—Clay bank deposit	10.82	3.5				2.00	0.426	0.555		0.139			Tr.					
Martin County—Stevens land, Sec. 1, Tp. 3, R. 3 (a)		27.00		66.4		1.1				Tr.		Tr.	Tr.				44.48	
Martin County—Stevens land, Sec. 1, Tp. 3, R. 3 (b)	1.00	28.60		54.45		7.2				Tr.			Tr.	6.56	1.95			100.00
Martin County—Stevens land, Sec. 1, Tp. 3, R. 3 (c)	1.00	36.80 •		49.95		2.12								8.00				
Martin County—Munson's ridge, Sec. 14-32 (a)	4.00	32.35		53.00										9.11	1.54			
Martin County—Munson's ridge, Sec. 14-32 (b)	1.00	7.00		60.5						Tr.			Tr.	28.00			42.35	
Martin County—Sampson's Hill	1.15	8.00	60.00									6.8		24.05			42.00	100.00
Martin County—Sec. 9-10, Tp. 4, R. 3	1.4	13.00		55.6				5 6					0.9	22.8			38.92	99.3
Martin County—Sec. 9-10, lower portion of middle member	3.00	23.00	59.65			2.7				Tr.		1.15		10.5				
Martin County—Sec. 9-10, upper portion of middle member	3.00	37.75	48.05			1.15				Tr.		2.05		8.00				
Martin County—Sec. 9-10, upper stratum	0.30	8.5												28.5				
Martin County—Huron Bank—Sample 1	5.40	44.70	32.20	46.00	.76	1.60				.33								
Martin County—Huron Bank—Sample 2	5.30	45.65	32.71	46.73	.76	1.20				.33								
Martin County—Huron Bank—Sample 3	5.25	56.90	25.04	35.72		15.50				.22								
Martin County—Huron Bank—Sample 4			39.48	56.40														
Martin County—J. A. Cook Opening—Sample 1	10.50	16.90	45.54	65.07	1.02	4.70				.65		.85						
Martin County—J. A. Cook Opening—Sample 2			32.48	46.40														
Martin County—J. A. Cook Opening—Sample 3			40.88	58.40														

Martin County—Felton S. Bank—Sample 1	9.20	36.50	33.32	47.60	1.28	4.00				54								
Martin County—Felton S. Bank—Sample 2		19.84	39.39	56.27	.378	11.03				60		1.40	.238					
Martin County—Felton S. Bank—Sample 3		10.64	35.75	51.07	.281	4.18				34		5.53	.32					
Martin County—Sarah Horner—Sample 1			55.44	79.20														
Martin County—Sarah Horner—Sample 2	9.20	28.00	38.22	54.60	1.53	4.50				.57		1.00						
Martin County—Sarah Horner—Sample 3		23.14	36.86	52.65	.438	13.00				1.03		4.63	.140					
Martin County—Sarah Horner—Sample 4 (at seam)			40.65	58.07						61								
Martin County—Sarah Horner—Sample 4 (away from seam)			41.66	59.51						.99								
Martin County—Boring, Sample No. 5 (12 ft. 6 in. to 14 ft. 6 in.)		22.86	37.87	54.10	.666	9.80				82		1.60	.016					
Martin County—Boring, Sample No. 5 (43 ft. 6 in. to 45 ft. 6 in.)		9.38	27.32	39.02	.848	3.37				.34		20.00	.067					
Martin County—Stephens (upper ore) Sample 1	10.50	26.30	38.82	55.46	.76	4.80				.46		1.05						
Martin County—Sarah Horner—Sample 1	6.50	11.20	52.22	74.60	3.06	.80				1.00		2.01						
Martin County—Sarah Horner—Sample 2			50.82	72.60														
Scott County—Finley and Vienna Tps.	15.00	14.00		38.56	4.5					.5		2.02	.05	21.51			32.20	

CHAPTER VII.

LIME.

The lime industry in Indiana is an important one but it has not reached that state of development which the abundance of its raw materials and fuel supplies warrant. The quality of the raw material which exists in great abundance and the accessibility of an adequate fuel supply should cause the State to take high rank as a producer of lime.

Definition. Lime, or quick-lime, is an oxide of calcium (CaO), having a high affinity for water under the action of which it produces a new chemical compound accompanied by the generation of heat. It is composed of 71.42 parts of calcium and 28.58 parts of oxygen. Since lime is formed from limestone consisting of carbonate of lime from which only carbon dioxide is eliminated if the limestone be pure calcium carbonate the resulting product will be pure lime. Otherwise the lime will contain whatever impurities are contained in the limestone save such as are volatile. Small amounts of silica,

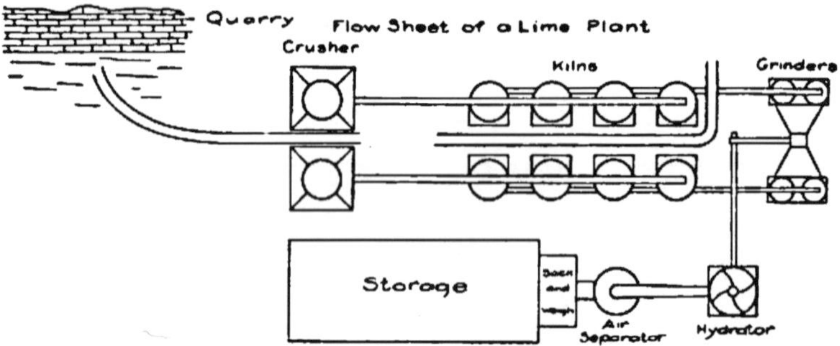

Plate LXXXIII. Flow sheet of a lime plant.

and oxides of iron and alumina are generally present. The essential composition of lime varies with its use. A building lime requires crushing strength, tensile strength and bonding power. These qualities may be improved by the presence of small amounts of silica and alumina but such substances are detrimental to chemical lime. The presence of small quantities of iron compounds in lime used for masonry mortar is permissible but not in finishing mortar.

Properties. Lime in the pure state is a white amorphous solid. The presence of impurities may produce yellow or blue tints. Overburned lime is generally yellow or black in color. Lime is infusible and non-volatile except at high temperatures. It has a strong affinity for water with which it unites and forms hydrated lime. It is porous and slakes readily in water. In an excess of water the hydrated lime particles remain in suspension forming milk of lime.

Classification. Limes are classified for commercial purposes according to their composition as high-calcium lime, magnesian lime, dolomitic lime and super-dolomitic. High-calcium lime is manufactured from limestone containing more than ninety-three per cent of calcium carbonate and contains less than five per cent of magnesia. Magnesian lime is made from stone containing more than seven per cent of magnesium carbonate and contains from five to twenty-five per cent of magnesia. Dolomitic lime contains from five to twenty-five per cent of magnesia and super-dolomitic lime contains over forty-five per cent of magnesia.

Manufacture. Quick-lime is manufactured from limestone. The limestone is burned or calcined in kilns. The first kilns were of a simple type called intermittent kilns. The kiln was built of stone or brick and the firing was continued until the contents of the kiln were calcined when, after cooling, the kiln was emptied. The term "field" kiln has been suggested for this type of kiln. Three types of continuous kilns are recognized. The pot kiln is a vertical shaft kiln. The fuel and limestone are placed in the kiln in alternate layers. The flame kiln is a vertical shaft kiln in which only the flame of the fuel comes in contact with the stone. The other type of continuous kiln is the rotary.

The primitive type of field kiln was constructed of limestone filled with alternate layers of wood and limestone. It was filled one day, burned for two or three days, cooled for two days and then emptied. Objections to the periodic method of burning lie in the loss of time and the impurities arising from the mixing of more or less ash with the lime.

The flame type of kiln is constructed of a cylindrical shell of steel lined with fire brick supported by a back wall of common brick. The kilns are cylindrical in the upper part and conical in the lower portion which is used as a cooling zone. The diameter varies from 6 to 12 feet and the height from 25 to 50. In this type of kiln the fuel does not come in contact with the limestone but is fed into furnaces near the base, the flame rising only through openings into the cupola. Two openings in the fire brick stack opposite each other and placed from three to five feet from the bottom, form the inner ends of the fire boxes. The fire boxes extend outward through the steel casing and have a length of four feet and a cross section of two by three feet. The bottom of the kiln is closed by a conical steel chamber which is six or more feet in diameter at the top, eighteen inches at the bottom and four feet high. At the lower end of this cooling chamber the lime is thrown off through a door. The capacity of the kilns vary from eight to twelve tons per twenty-four hours. Care must be used to keep the lime below the fire boxes and the unburned stone above them.

Another type of continuous kiln is the rotary kiln which consists of a revolving brick-lined steel cylinder into which crushed limestone is placed. The rotary kiln is used where the lime is to be hydrated.

Changes in Calcining. The changes which take place when a pure calcium limestone is burned in a kiln is the conversion of calcium carbonate into calcium oxide and carbon dioxide.

The chemical equation is: $CaCO_3$+heat=CaO (Calcium Oxide)+CO_2 (Carbon dioxide).

One hundred pounds of pure limestone will produce 56 pounds of quick-lime and 44 pounds of carbon dioxide. If the limestone is a mixture of calcium carbonate and magnesium carbonate the products are lime (CaO) and magnesia (MgO). The action of the heat in driving off the carbon dioxide gas renders the lime porous even though the original limestone were very compact.

Plate LXXXIV. Plants and Quarries of the Mitchell Lime Company, Mitchell, Indiana.

The quantity of heat required to drive off the carbon dioxide (decarbonation) varies with the chemical and physical properties of the limestone used. The temperature of decarbonation is between 850°C and 900°C, and the heat may be applied with a high temperature for a short period or with a lower temperature for a longer period. The temperature should not be raised above 1,000° to 1,200°C, as there is danger of overheating the lime. The carbon dioxide should be removed as rapidly as possible, if the draught of the kiln is not sufficient, artificial draught should be employed. It is estimated that it requires 373.5 calories to convert one kilogram of calcium carbonate into lime, or the equivalent of 747 B. T. U.

Uses. To mention all the uses of lime would require the enumeration of a long list of mechanical arts and industries. Broadly speaking, about one-half of the lime produced is used for structural purposes in lime mortars, plasters, cement mortars, concrete and white wash. Both quick-lime and hydrated lime and also calcium and magnesium lime are used as structural materials.

Glass Industry. Lime is used in the manufacture of glass. Many glass manufacturers prefer a high calcium lime, others a magnesium lime. Both kinds are manufactured in Indiana and the glass factories should experience no difficulty in supplying their needs from the home products.

Sugar Refining. Both lime and carbon dioxide are used in the refining of sugar. About one per cent of the total production of lime is used in this industry. The lime is used to remove the excess of organic acids and to coagulate the albumen and mucous. The carbon dioxide is used to remove the calcium which unites with the sugar.

Paper Making. The lime used in the manufacture of paper is about five per cent of the total production. Lime is used in cleansing materials used in paper manufacture. Soft wood pulp is boiled in a solution of sodium carbonate which has been made alkaline by the addition of lime. Resinous wood pulp is boiled in sulphurous acid and milk of lime to remove the tars and oils. In the bleaching of the pulp calcium chloride is used.

Chemical Lime. Lime used in chemical works amounts to about one-seventh of the total production. It is used in the manufacture of calcium compounds, as a desiccator and an absorbent. Common calcium compounds produced are calcium chloride, calcium carbide, calcium cyanide and calcium nitrate.

Tanning. A strong solution of milk of lime is used to remove the hair from hides, and to dissolve the fats and loosen the fibers and render the leather pliable. A depilatory paste is made by mixing sodium sulphite with the lime.

Soap Making. Lime is used to obtain the alkaline hydroxides used in the manufacture of soap. The effect of lime on oils used in soap manufacture is to form organic salts of calcium which is easily replaced by sodium or potassium from some of their compounds, to form the soluble soaps of commerce.

Agricultural Lime. Quick-lime and ground limestone are both applied to soils to correct acidity. Quick-lime should be used only on those soils which carry a high per cent of organic matter. The caustic action of the lime on organic matter in soil causes too rapid depletion unless the organic matter is unusually abundant. On peat or muck soils the application of quick-lime brings quick returns.

Raw Materials. Limestones suitable for the manufacture of lime are widely distributed in Indiana. Limestones suitable for the manufacture of lime are found in Silurian, Devonian, Mississippian, Pleistocene and Recent strata.

Plate LXXXV. Interior view of Mitchell lime plant, near Mitchell, Indiana.

Lime Industry. The amount of lime produced in Indiana is not as large as the quality of the raw material and the quantity and accessibility of fuel would warrant. As long as the amount of lime shipped from other states and consumed in Indiana is greater than the amount of the home product consumed ideal conditions have not been attained. In 1917 118,530 tons of lime were produced in Indiana, 75,444 tons shipped out of the State, 46,772 tons shipped into the State and 43,086 tons of the home product consumed within the State. The production of lime in Indiana since 1903 is given in the following table:

Production of Lime.

Year	Quantity—Tons	Value	Quantity—Rank by States	Plants
1904	100,703	$349,499	....	....
1905	106,408	366,866	....	....
1906	114,819	353,648	....	....
1907	107,964	335,151	....	....
1908	95,988	293,579	....	....
1909	99,325	335,154	14	11
1910	86,811	301,304	14	10
1911	92,229	324,950	11	12
1912	98,086	329,893	13	12
1913	96,359	323,905	11	10
1914	99,185	358,738	10	9
1915	99,916	374,221	11	8
1916	121,306	495,283	10	8
1917	118,530	646,555	11	6

Silurian. Limestones of Silurian age reach the surface in northern and eastern portions of Indiana. (See map for distribution.) These limestones are suitable for the manufacture of lime. The limestones in the northern part of the area contain large percentages of magnesium carbonate. Those in the southern portion contain more calcium and less magnesium. The analyses of samples of the Silurian limestones containing a high per cent of magnesia are given in the following table: (No. 1 Delphi, others from Huntington.)

Analyses of Silurian Limestones High in Magnesia.
Dr. R. E. Lyons, Analyst.

	No. 1	No. 2	No. 3	No. 4
Calcium carbonate ($CaCO_3$)	54.53	53.22	63.03	59.20
Magnesium carbonate ($MgCO_3$)	43.92	44.96	34.15	38.38
Iron oxide and Alumina (Fe_2O_3, Al_2O_3)	.51	.23	2.62	.49
Silica (SiO_2)	.19	.59	.07	.35
Sulphur trioxide (SO_3)	.18	.11		
Moisture			.13	1.80
Total	99.33	99.11	100.00	100.22

The following analyses are of Silurian limestones having a high calcium and a low magnesium content.

	No. 1	No. 2	No. 3	No. 4
Calcium carbonate	74.02	83.00	96.02	93.17
Magnesium carbonate	10.35	6.30	1.04	.25
Iron and Alumina	6.20	2.50	2.00	.93
Silica	5.90	5.30	1.04	5.85
Sulphur trioxide	.90	1.00	(H_2O)	.08
Total	97.37	98.10	100.08	100.28

No. 1 and No. 2 Decatur County, E. T. Cox, Analyst. No. 3 Cass County, Dr. J. N. Hurty, Analyst. No. 4 Madison County, Dr. W. A. Noyes, Analyst.

Devonian Limestones. Limestones belonging to this period outcrop in the southeastern part of Indiana. Only in very limited areas are the Devonian limestones suitable for making lime. They are more useful in the manufacture of natural and Portland cements. The hydraulic limestone used for the manufacture of natural and Portland cement contains from fifty to sixty per cent of calcium carbonate and from sixteen to thirty-five per cent of magnesium carbonate.

Mississippian Limestones. The Mississippian strata of Indiana contain many beds of limestone which are well adapted to the manufacture of lime.

Plate LXXXVI. Quarry and Plant of the Delphi Lime Co., Delphi.

The formations which contain suitable limestones are the Harrodsburg (Warsaw), the Salem (Oölitic), the Mitchell and limestones of the Chester group.

Harrodsburg Limestone. This limestone which lies upon the Knobstone outcrops in an irregular line running through Floyd, Clark, Washington, Jackson, Lawrence, Monroe, Owen, Putnam, Parke, Montgomery and Fountain counties. The width of the outcrop varies from a few feet to several miles. The strata consist of thin irregularly bedded layers among which there are often one or two heavy bedded or massive layers. The thickness of the formation varies from sixty to ninety feet.

Salem (Oölitic) Limestone. This limestone overlies the Harrodsburg and the line of its outcrop extends from the Ohio River in Harrison County into Putnam County. The thickness of the formation varies from a few feet to one hundred feet. It is generally massive, containing few bedding planes. The limestone is granular, the grains consisting of minute shells or fragments of shells which are cemented with calcium carbonate. Some of the shells are covered with concentric layers of lime carbonate and occasionally a true oölite is mingled with the shells. This limestone has a high calcium content and is especially well adapted to the manufacture of high calcium lime. The analyses of several samples of the stone are given in the following table:

Analyses of Salem Limestone.
W. A. Noyes, Analyst.

	No. 1	No. 2	No. 3	No. 4	No. 5
Calcium carbonate ($CaCO_3$)	98.27	98.11	97.90	98.16	98.20
Magnesium carbonate ($MgCO_3$)	.84	.92	.65	.97	.39
Insoluble	.64	.86	1.26	.76	.63
Ferric oxide	.15	.16	.18	.15	.39
	99.90	100.05	99.99	100.04	99.61

No. 1, Bedford. No. 2, Hunter Valley. No. 3, Ramona. No. 4, Twin Creek. No. 5, Bedford. The total lime forming content of these samples averages about ninety-nine per cent.

The Salem limestone has been used successfully in the manufacture of lime at Bedford, Salem and other places. The ground limestone is used by many glass factories in the State.

Mitchell Limestone. This formation consists of a stratum of heavily bedded limestones overlying the Salem and extending along its outcrop from the Ohio River in Harrison County to beyond the northern extension of the Salem into Parke and Montgomery counties. The thickness of the formation varies from 150 to 350 feet. The upper part of the formation contains thin layers of shale and some layers of chert and cherty limestones. The average analysis of four samples of Mitchell limestones is as follows:

Average Analysis of Mitchell Limestones.

Calcium carbonate	98.06
Magnesium carbonate	.91
Iron oxide and alumina	.21
Insoluble (silica, etc.)	.91
Total	99.61

This limestone has been used successfully in the manufacture of lime at Milltown, Mitchell, Putnamville and other places.

Chester Limestones. In the Chester group are a number of beds of limestone which are intercalated with beds of shale and sandstones. The lowermost one which can be definitely separated from the Mitchell is the Beaver Bend limestone which usually attains a thickness of fourteen feet. The Reelsville which lies above the Beaver Bend limestone and is separated from it by twenty to fifty feet of sandstone or shale does not reach a thickness of more than ten feet. The Beech Creek limestone usually attains a thickness of twelve feet and is separated from the Reelsville by from thirty to fifty feet of shales or sandstones. The Golconda limestone attains a thickness of thirty feet and the Glendean a thickness of twenty-five feet. Several of these limestones have been used locally in the production of lime.

Marls. The glacial lakes of Indiana contain quanities of marl which is composed chiefly of carbonate of lime suitable for the manufacture of lime. The average of five analyses of this marl is as follows:

Average Analysis of Indiana Marls.

Calcium carbonate ($CaCO_3$)	88.80
Magnesium carbonate ($MgCO_3$)	3.15
Iron oxide and alumina (Fe_2O_3, Al_2O_3	.35
Insoluble (Silica, etc.)	3.91
Organic matter	3.01
	98.89

See the discussion of marl for its occurrence, properties and distribution.

THE LIME INDUSTRY IN INDIANA.

Huntington. The Kelley Island Lime and Transport Company operates eighteen upright kilns at Huntington. The limestone used is Silurian (Niagara) and contains, according to analyses of the company, 55.32 per cent of calcium carbonate, 43.86 per cent of magnesium carbonate .54 per cent of silica and .28 per cent of iron and alumina.

The composition of the lime produced is given as follows:

	Per cent
Calcium oxide	58.15
Magnesium oxide	38.80
Silica	.53
Iron and alumina	.30

The capacity of the kilns is 125 tons of lump lime or 150 tons of hydrated lime per day.

The lime is used for agricultural, building and chemical purposes. Some of its uses in the arts and trades are as a soil neutralizer, as masonry lime, in sulphite paper pulp manufacture, in glass making as a flux for flint glass.

Salem. At Salem, Washington County, lime has been manufactured for many years. First by the Salem Stone and Lime Company then by the Salem Bedford Stone Company and still later by the Union Cement and Lime Company. The Salem (oölitic) limestone is used for making the lime. Some lime is still being manufactured at this point.

Delphi Lime Company. The Delphi Lime Company is located at Delphi in Carroll County. The lime is manufactured from Silurian (Niagara) limestone. Wood is used for fuel in vertical cylindrical steel kilns. The composition of the lime from an analysis furnished by the company is as follows:

Calcium oxide (CaO)	90.20
Magnesium oxide (MgO)	6.08
Iron and aluminium oxides (Fe_2O_3, Al_2O_3)	1.70
Potassium and sodium oxides (K_2O, Na_2O)	1.54
Silicon oxide (SiO_2)	.15
Moisture (H_2O)	.30
Total	99.97

The accompanying cut is from a photograph supplied by the company.

Silicon oxide (SiO_2)	.15
Moisture (H_2O)	.30
Total	99.97

The accompanying cut is from a photograph supplied by the company.

Harley Bros. Lime Company. The property of this company is located on a switch of the Monon north of Delphi. "The three kilns operated by the company are of stone and are seven feet inside diameter, with an output of 225 bushels each per day. The burning is by the continuous process, coal and a small amount of wood being used as fuel." An analysis of the limestone used was made by Dr. R. E. Lyons with the following result.[1]

Analysis of Stone from Harley Bros. Quarry.

Calcium carbonate ($CaCO_3$)	54.53
Magnesium carbonate ($MgCO_3$)	43.92
Ferric oxide and alumina ($Fe_2O_3+Al_2O_3$)	.51
Insoluble residue (silica)	.19
Sulphuric anhydride (SO_3)	.18
	99.33

Mitchell Lime Company of Chicago. The Mitchell Lime Company of Chicago operate a quarry and a plant for the manufacture of lime at Mitchell. The quarry is located in the Mitchell limestone of the Mississippian. The limestone contains an average of more than ninety-six per cent of calcium carbonate. The thickness of the limestone at the quarry is from eighty to one hundred feet. The limestone is broken from the face of the quarry by blasting. The limestone is hand picked and the calcined product is spread upon a cooling floor and selected.

Two kinds of lime are manufactured: Chemical lime (quick-lime) and hydrated lime. The lump lime is crushed and passed through the hydrator which consists of "six enclosed vertical steel cylinders, one mounted on top of the other. Up through these cylinders rushes a continuous stream of vapor. When the crushed lime reaches the hydrator it is carried by a paddle conveyor. As the lime moves along, numerous fine streams of water play over it from the supply pipe above". After leaving the hydrator the lime passes through an air separator which separates the fine particles of lime from any impurities or grit which it may contain. The Mitchell lime is said to have been used with success in the manufacture of soap, paper, rubber, varnish, insecticides and in oil refining, dehairing hides, sanitation, water purification and a large number of other purposes.

The accompanying illustration of the plant and quarry was furnished by the company.

Milltown.[14] Lime has been manufactured from the Mitchell limestone at this point for many years. The J. B. Speed Company and the Eichel Company have operated quarries at Milltown for the manufacture of lime for many years. No reports of production have been received from these companies recently.

[1] Ind. Geol. Sur. 28th Ann. Rept., p. 230.

CHAPTER VIII.

MARL AND NATURAL ABRASIVES.

Marl.

The lake basins of northern Indiana contain deposits of calcium carbonate mixed with organic matter and slight amounts of other impurities. Such deposits are common in the glaciated area of the United States where lake basins are abundant.

Definition. Marl is chiefly an amorphous form of calcium carbonate containing organic matter and frequently small quantities of sand, clay and other impurities.

Properties. Marl is a soft granular to earthy material of white color usually though the color depends upon the kind and quantity of the impurities contained. Free from impurities it is milky white or greenish white. The porosity of marl is high giving it power to absorb a large quantity of water. The fineness of grain varies from coarse granular to microscopic particles. The mass shrinks markedly leaving large cracks on the surface when it dries. It is readily soluble in hydrochloric acid and easily recognized by its effervescence.

Composition. The principal chemical compound of marl is calcium carbonate but some magnesium carbonate is generally present. Organic matter, iron compounds, alumina and silica are common impurities. The composition of Indiana marl is given in the following table:[11]

Analyses of Indiana Marl.

	No. 1	No. 2	No. 3	No. 4
Calcium carbonate ($CaCO_3$)	91.14	84.00	90.67	84.75
Magnesium carbonate ($MgCO_3$)		6.46	2.42	2.84
Calcium oxide (CaO)				
Magnesia (MgO)	1.31	1.34		
Alumina (Al_2O_3)	.86		.06	.15
Ferric oxide (Fe_2O_3)		1.34*	.26	.35
Loss on ignition		3.68	2.87	5.69
Silica	.85	4.52*	2.48	4.61

* Carbonates and silicates.

No. 1. Average of six samples from Big Turkey Lake, W. R. Oglesby, Analyst. No. 2. Silver Lake, C. R. Dryer, Analyst. No. 3. Tippecanoe Lake, W. A. Noyes, Analyst. No. 4, Little Eagle Lake, W. A. Noyes, Analyst.

Mode of Occurrence. The marl deposits of Indiana occur in existing lake basins, ponds and marshes, and in the former basins of extinct lakes. The marl occurs in all parts of the lakes resting on the floor of the lake sometimes being thicker at the margins and sometimes in the center. After being deposited near the margins it may be moved toward the deeper water by currents. It rests upon beds of clay, sand or gravel and is often covered with peat or muck. The present depth of the water is no indication of the thickness of the marl but thick deposits of marl are indications of greater depths of water.

Size of Marl Deposits. The marl deposits of Indiana vary in area from a few acres to several hundred acres. The majority of the deposits have areas of less than one hundred acres but many have larger areas. The thickness of the deposits ranges from a few inches to forty or more feet.

Origin. The marl is deposited from the lake water. Its original source is from beds of sands and gravel through which ground waters flow on their way to the lake basins. Mixed with the sands, gravels and till are quantities of carbonate of lime from limestones which were ground up by glaciers and mixed with the debris which they left in their final retreat. By the aid of carbon dioxide this calcium carbonate is taken into solution by circulating ground waters and carried into the lakes. If none of the calcium carbonate were removed it is conceivable that the point of saturation might be reached and precipitation occur but the calcium carbonate seems to have been removed long before the point of saturation has been reached. Its removal is accomplished seemingly by organisms. Plants such as Chara are thought to absorb the soluble bicarbonate of lime rob it of carbon dioxide and secrete the monocarbonate form which sinking to the floor of the lake forms the marl. Certain forms of algæ and bacteria are probably other agents of secretion. Perhaps some of the calcium carbonate is precipitated through change in temperature or pressure of the ground water when it enters the lake.

Uses of Marl. Marl when pure may be used for many of the purposes for which limestone is used. It may be used in the manufacture of lime, cement, glass, polishing powder and as a soil addendum. Marl was formerly used in Indiana in the manufacture of lime but with the establishment of the large limestone burning kilns the practice ceased. On account of the water content which must be eliminated before burning it is not probable that marl can compete with limestone in the manufacture of lime.

Marl has been used in Indiana for many years in the manufacture of Portland cement. Some of the plants using marl have suspended operations but there are two, one at Stroh and one at Syracuse that continued to use it. However, the latter is making arrangements to use limestone instead of marl. The marl is used with glacial clay in the manufacture of Portland cement. The principal objection to its use is the large amount of water which must be eliminated at great expense for fuel.

In the manufacture of glass ground marl could be used when the marl is free from impurities which are injurious to glass. The marl would be easier ground than limestone but the absorbed water would be a troublesome factor.

As an ingredient of scouring soaps to furnish the grit marl could be used as it is granular and fine of grain. It can also be used as a polishing and buffing material but it is uncertain whether marl can compete with chalk, ground limestone or other substances used for these purposes.

Marl is as useful as ground limestone for treating soils to remove acidity and to make the growing of leguminous crops more profitable. The nitrogen content of soils may be increased by marling the soils and growing crops of clover and plowing them under. The mechanical condition of the soil can also be improved by the addition of marl which renders a clay soil more porous and a sandy soil more retentive of moisture. Many areas of soil in Indiana are in need of the neutralizing effect of marl, especially soils that have been in cultivation for many years, exposed to the leaching action of agricultural processes, which cause a more rapid removal of lime carbonate than can take place in fallow soils. The use of marl for correction of acidity should appeal strongly to those farmers having lands in the vicinity of the marl deposits. Lime or ground limestones should not be used by them except on the grounds of economy which would be difficult to establish.

ANALYSIS OF INDIANA MARLS.*

ORIGIN OF SAMPLE		Calcium Carbonate (Ca CO_3)	Magnesium Carbonate (Mg CO_3)	Alumina (Al_2O_3)	Ferric Oxide (Fe_2O_3)	Calcium Sulphate (Ca SO_4)	Insoluble Organic (Silica, etc.)	Organic Matter	Total	Authority
LAKE	COUNTY									
Hog Lake	Steuben	90.42	2.88	.14	.28	....	.68	4.13	98.53	W. A. Noyes.
Lime	Steuben	86.00	9.42	....	1.16*	....	1.08	2.32	99.98	C. R. Dryer.
Deep and Shallow	Steuben	93.29	2.67	.04	.12	....	.47	1.56	98.15	W. A. Noyes.
James	Steuben	92.41	2.38	....	.29	.15	1.16	1.97	98.36	W. A. Noyes.
Silver	Steuben	84.00	6.46	....	1.31*	....	4.52	3.68	100.00	C. R. Dryer.
Turkey Lakes	Lagrange	91.11	2.75	.61	.25	....	.85	....	95.60	W. R. Oglesbey.
Loon	Whitley and Noble	82.07	2.63	.41	.42	.22	5.95	6.71	98.41	W. A. Noyes.
Mud	Elkhart	82.89	2.04	.41	.23	....	7.94	3.67	97.18	Osborn Engineering Co.
Cooley	Elkhart	88.21	4.78	.52	.36	....	1.42	2.58	97.87	Osborn Engineering Co.
Syracuse	Kosciusko	88.49	2.71	.90	.31	1.58	1.78	4.23	100.00	S. B. Newberry.
Dewart	Kosciusko	92.35	3.54	.37	.16	....	2.00	2.12	100.54	A. W. Burwell.
Dewart	Kosciusko	84.21	2.85	.18	.30	....	4.52	5.02	97.11	W. A. Noyes.
Tippecanoe (James Basin)	Kosciusko	90.67	2.42	.06	.26	....	2.48	2.87	98.76	W. A. Noyes.
Tippecanoe	Kosciusko	91.02	2.28	....	.29	.05	2.92	2.10	98.66	W. A. Noyes.
Little Eagle	Kosciusko	84.75	2.84	.15	.35	.07	4.61	5.69	98.46	W. A. Noyes.
Manitou	Fulton	87.65	2.60	.19	.30	....	6.39	2.88	100.01	W. A. Noyes.
Maxinkuckee	Marshall	85.02	3.85	.12	.33	.17	5.67	3.21	98.37	W. A. Noyes.
Maxinkuckee	Marshall	85.38	3.50	.05	.33	.17	6.40	3.15	98.98	W. A. Noyes.
Houghton and Moore	Marshall	89.22	2.73	.04	.20	....	2.02	4.15	98.36	W. A. Noyes.
Notre Dame	St. Joseph	91.62	4.02	.05	.07	.14	.19	2.25	98.34	W. A. Noyes.
Chain and Bass	St. Joseph	87.92	2.64	.10	.20	.23	3.10	4.18	98.37	W. A. Noyes.
Kankakee Marsh Deposit	St. Joseph	91.30	2.90	....	.08	.22	.82	3.88	99.20	W. A. Noyes.
North Judson Marsh Deposit	Starke	89.92	2.46	.45	.74	....	1.56	4.51	99.64	W. A. Noyes.
Curran Marsh Deposit	Berrian (Mich.)	89.63	2.12	.37	.25	....	.42	7.55†	100.35	F. S. Kedzie.

*This was given by Dr. Dryer as ferrous carbonate, the form in which the iron probably occurs in the marl.
†This includes undetermined matter, alkalies, etc.

Natural Abrasives.

Definition. The term includes those natural substances or products which are used for grinding or polishing. A great many minerals and rocks are used for such purposes. Among the minerals are corundum, diamond, feldspar, garnet and quartz. Among the rocks are chert, flint, emery, sand, sandstone, quartzite, pebbles, conglomerate, tripoli, pumice, volcanic ash, schist, diatomaceous earth and novaculite.

Uses. Sandstones and quartz conglomerates are used in the manufacture of millstones and buhrstones. Sandstones are used in the manufacture of grindstones and pulpstones. Sandstones and schists in the manufacture of

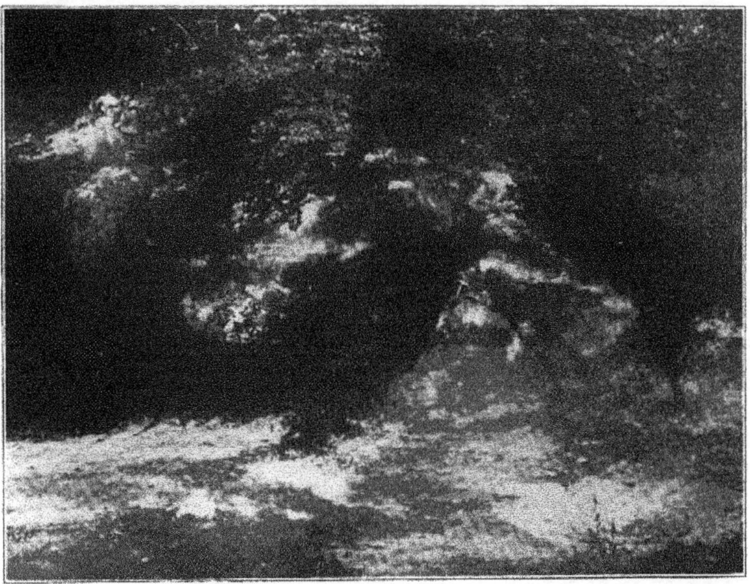

Plate LXXXVII. Outcrop of Mansfield Sandstone, used as an abrasive, Orange County, Indiana.

whetstones and sandstones and novaculite in the manufacture of oil stones. Pumice, volcanic ash, tripoli and diatomaceous earth are used for polishing powders and in scouring soaps. Quartz and garnet are used in the manufacture of sandpaper. Diamonds in glass cutters, diamond drills, diamond saws and to cut and polish diamonds. Corundum is used in the manufacture of corundum wheels and emery in the manufacture of emery wheels. Quartz pebbles are used in pebble mills for grinding raw materials and such products as cement and talc.

Among the minerals named as abrasives only quartz occurs in any abundance in Indiana. Small quantities of all of them occur in the glacial drift but they do not occur in workable quantities. The quartz occurs very rarely as vein quartz. It occurs as crystals in concretionary forms called "geodes". It occurs in the form of flint in limestones and in beds of chert. Quartz pebbles are abundant in glacial drift and in river gravels.

Rocks suitable for the manufacture of certain forms of abrasives are abundant in Indiana. For many years some of these rocks have been used in the manufacture of scythestones, whetstones and oil stones. Some of the sandy shales and sandstones of the Knobstone are suitable for the manufacture of whetstones. As are also some of those from the Chester division of the Mississippian. Fine to coarse grained sandstones occur in the Pennsylvanian rocks which are suitable for abrasives purposes. Rocks of this period occurring in Lawrence, Martin and Orange counties have been used in the manufacture of abrasives.

Plate LXXXIX. A whetstone factory in Orange County near Orangeville. Sandstones of the Mississippian and Pennsylvanian Systems are used in the manufacture of whetstones and oilstones.

A factory located between Georgia and Orangeville has been in successful operation for a number of years. Whetstones, coarse, medium and fine, and oil stones have been manufactured for many years and the product shipped to all parts of this country and to foreign countries. The stone is quarried from quarries near the factory and sawed with steel gang saws fed with sand obtained from a quarry in the Mansfield sandstone near the factory. The stones are ground on rotating disks covered with water carrying sand.

The Hindostan sandstone is used widely for oilstones and merits the attention given it in the abrasive market. A deposit of sandstone suitable for whetstones occurs near Clay City in Clay County. This sandstone lies above the Minshall Coal and is ten to fifteen feet thick at the outcrop.

Sands used for sawing building stone are obtained from the dune area near Michigan City. This sand, because it contains no iron or clay to stain the stone, is well adapted to stone cutting. Some of the Mansfield sands and some of the river sands could be used for feeding gang saws.

Buffing Materials. The raw materials of Indiana which could be used for buffing powders are the white kaolin which may be reduced to an impalpable powder. The ground Salem limestone or the manufactured lime of Indiana may be used for buffing. Another promising material is the stained kaolin known as mahogany clay.

Pebbles for use in pebble mills could be selected from glacial pebble deposits and from river gravels. The lack of uniformity in size, shape and composition of such pebbles would necessitate hand picking, an expensive operation and one which would probably prevent competition with foreign sources.

The flints and cherts of Indiana do not offer great promise of being abrasive sources. The most abundant form of quartz occurs in the geodes at the base of the Harrodsburg limestone. The uses of quartz in such a form would probably be so limited that preparatory processes would be necessary.

Cherts occur in the Mitchell and in some of the Chester limestones and may be found in form and abundance sufficient for use.

Analyses of Chert.

	Loss on Ignition	Silica (SiO_2)	Alumina (Al_2O_3)	Ferric Oxide (Fe_2O_3)	Calcium Oxide (CaO)	Magnesium Oxide (MgO)
Lehigh Portland Cement Co., Mitchell, Lawrence Co....	1.15	93.64	0.76	4.08	0.68	tr.
J. B. Speed & Co., Milltown, Crawford Co..............	14.90	61.28	0.32	3.06	20.36	0.10
The "A. & C." Quarry, East of Greencastle, Ind........	11.71	69.08	0.40	2.90	16.22	tr.

CHAPTER IX.

MINERAL WATERS.

Mineral waters are natural waters which contain in solution either large amounts of common substances or small amounts or rare substances and which are used for remedial purposes. The minerals in such waters have been obtained from the rocks of the crust of the earth through which or over which the waters have passed. In some of the mineral waters the percentage of dissolved salts is sufficiently large to affect the taste, in others the quantity of dissolved salts is so small as not to be recognizable in this way, and yet the water may have a remedial value.

Classification of Mineral Waters. Nearly all classifications of mineral waters are based upon the kinds of dissolved minerals or upon the elements contained in those minerals. For instance, if the water contains a high per cent of common salt it is called saline or of iron compounds it is called chalybeate or ferruginous. Since the elements of the dissolved salts are in a state of ionization the water may be named for the predominant ion or ions, as chloride water where chlorine is the predominating ion, or lithic water where lithium is the principal ion. Probably a large percentage of the ingredients of the mineral waters of Indiana are present as ions. Only in the strongest brines do they occur undissociated.

The positive or base-forming ions which are present in mineral waters are aluminium, ammonium, barium, calcium, iron, lithium, magnesium, manganese, potassium, sodium and strontium.

The negative or acid-forming ions are arsenate, borate, bromide, carbonate, chloride, fluoride, hydrocarbonate, iodide, nitrate, phosphate, silicate, sulphate, and sulphide. In addition some very rare elements and radio-active elements may be present.

According to their chemical reactions waters may be classed as neutral, acid or alkaline. Dilute acids and alkalies have an ion action, due to the presence of acid hydrogen, produce osmotic changes and exert ordinary salt action. They also modify the processes of absorption and digestion. Acids and alkalies are neutralized in the stomach or in the intestines. The system is so constructed that it can take care, for a time at least, of an excess of acid or alkali. This is done by a change in a composition of the urine, so acids and alkalies are excellent diuretics, increasing the ammonia of the urine at the expense of the urea.

Acids in the stomach assist in the action of the pepsin in digestion and increase the flow of the gastric juice. All acids convert proteids into acid-albumins which are soluble in concentrated and very weak acids but insoluble in acids of medium strength.

Alkalies reduce the acidity of the chyme and thus increase the alkalinity of the intestinal fluids even if they are themselves neutralized and absorbed before reaching the duodenum. In this way they may favor the emulsification of fats and the action of the pancreatic ferments, if there is not sufficient alkalies in the intestine. In a normal condition of the system, this action would be of no value, but where there is an excess of mucus, or too great acidity, the alkalies are very useful. On account of the action of undissociated salt, the secretion of the urine is increased.

Probable Therapeutic Action of Ions.

Aluminium is present in some of the mineral waters of Indiana. Little can be said of the action of the aluminium ion, but alums are used locally as astringents, and internally in gastric catarrh, enteralgia, gastralgia and lead colic. Aluminium sulphate and iron sulphates are produced by the oxidation of pyrite in the presence of clay or shale.

Ammonium is present in waters coming in contact with decaying organic matter. Its presence in shallow well waters should be viewed with suspicion as it may be indicative of contamination. It is not harmful in deep waters that are not allowed to stand before using. Ammonium has a marked action on the secretions saliva, mucus and perspiration. It is used as a local expectorant, and as a stimulant of the respiratory centers for cough and asthma.

Barium is not an important constituent of the mineral waters of Indiana. In its ion actions, it resembles the organic groups. Its most important systemic action is a slowing of the heart and a rise in blood pressure. When given in dilute solutions the amount absorbed is small and is deposited in the bones. It may be of use in the treatment of cancerous, scrofulous and other morbid growths.

Calcium is a very common and often a very abundant ion in Indiana mineral waters. Calcium carbonates and calcium sulphates are the two common forms of the undissociated substances. The former are very alkaline in action and in large doses may cause constipation. It is used in chronic diarrhea and in cases of uric-acid gravel and calculi. The latter is one of the constituents which render water "hard". Calcium chloride is said to have a diobstenent effect and to promote the secretion of urine, perspiration and mucus. The use of water containing it is recommended in scrofulous diseases and in chronic eczema empetigo, connected with a lymphatic temperament.

The calcium ion is a universal constituent of protoplasm. It appears to be essential not only to living protoplasm, but also to inorganized ferments.

Iron is a very abundant element in many of the mineral waters of Indiana. Its undissociated forms are commonly carbonates and sulphates. It is present in the hemoglobin of the blood, in the lymph, chyle, gastric juice and other liquids of the body. Chalybeate waters produce a constructive metamorphosis, creating more red blood corpuscles, thereby increasing the specific gravity of the blood and of the bodily weight, reproducing a healthy glow. Chalybeate waters may make up the deficiency of the coloring matter of the blood, observed in anemic states. It matters not though iron be present in small quantities, and few of the carbonated iron waters contain more than five or six grains to the gallon. The blood contains normally about forty-five grains of iron, and this quantity cannot be permanently increased by consuming large quantities. It is probable that the deficiency, no matter how produced never exceeds fifteen or twenty grains.

The tendency of iron is to increase the appetite, promote digestion and relieve a languid or depressed condition of the system. The most common compound is the bicarbonate which is supposed to be the form which most readily enters the circulation.

Lithium is a constituent of some of the mineral waters of the State. It is probably present as a carbonate or bicarbonate, mixed with carbonates or other alkalies. Lithium forms a soluble salt with uric acid and this has led to the extensive use of lithia water in cases of uricemia. It is also useful in the treatment of uric acid, sand, gravel and calculi, and of gout, rheumatoid arthritis and phosphate deposits in the appendix. It has a tendency to increase the excretion of nitrogen.

Magnesium is one of the abundant elements in the mineral waters of the State. In the salt brines it occurs as a chloride. In the waters of meteoric origin it occurs as a carbonate but more frequently as a sulphate. The carbonate has an alkaline action and is useful in acid eructations and pyrosis in sick headaches, when accompanied by constipation, and to check the formation of uric acid gravel and calculi. The chloride is useful to increase the flow of bile and as a mild purgative.

The sulphate promotes the process of endosmosis and exosmosis, and, by extracting the watery elements of the blood, increases the intestinal secretions. Even if the quantity is small, it will tend to promote regularity of the bowels when taken continuously. The best results are observed in disordered conditions of the stomach, liver and bowels with concomitant symptoms of constipation. In sluggish states of the liver, characterized by a sallow countenance, yellowness of the conjunctiva, coating of the tongue and hemorrhoids, the sulphated saline waters are speedily efficacious. They have important and useful functions in eliminating various chronic infections from the system, scrofulous, syphilitic and malarial, as well as in expelling lead, mercury and other metallic poisons. For purgative effects, physicians recommend that the waters be taken before breakfast and followed by a brisk walk in the open air. These waters should not be used in cases of chronic inflammatory condition of the stomach or intestines, or in case of general debility.

It seems probable that the magnesium in water acts as an undissociated salt, as, although soluble, the magnesium ion is incapable of absorption into the blood. Magnesium is practically the only non-absorbable cation which can be used as a cathartic.

Manganese is not an ingredient in the mineral waters of Indiana. When present it is in the form of a bicarbonate or sulphate. It is present in the blood and its function is thought to be reconstructive, to act as a tonic, and to increase the flow of the bile.

Potassium is present in many of the mineral waters of the State, in small quantities. It is rapidly excreted and seems to have no special therapeutic action. As a carbonate, it corrects acidity and acts as a diuretic in connection with other alkalies.

Sodium is one of the most abundant elements in the mineral waters of Indiana. The connate waters, the deeply buried sea-brines, contain large quantities of salt obtained from the original seas. Because of the ready solubility of sodium salts the percolating meteoric waters rapidly strip them from the rocks. Sodium chloride, common salt, is the most abundant sodium or other salt in the connate waters. It is necessary for a healthy growth of the body, as it is a constituent of almost every anatomical element. It is thought to have much to do with the regulations of exudation and absorption, and to assist in maintaining the fluidity of the albuminoids in the blood.

The carbonate and bicarbonate are present in mineral waters. Sodium carbonate is found in the blood, saliva, urine and other fluids of the body and its importance to bodily functions is thus indicated. Waters containing it have an alkaline action. Carbonated alkalies have a marked effect on the mucous membranes and increase the secretions. Sodium carbonate waters are used in catarrhal conditions of the stomach or intestines, especially when accompanied by chronic diarrhea, in cases of excessive acidity in the alimentary canal, to increase the activity of the skin and kidneys.

Strontium is a rare element in mineral waters but occurs as a bicarbonate with similar salts of calcium and magnesium. It has little therapeutic value being an intestinal antiseptic considerable quantities of water containing it might be useful in intestinal disorders. "In dilute solutions only very small amounts are absorbed from the stomach: none from the intestines, since it is converted into phosphates, in which form it is generally deposited in the bones." Arsenate waters do not occur in Indiana but are found in regions where metals such as antimony, bismuth, copper, cobalt and nickel are found. Such waters are recommended in cases of irritative dyspepsia, chronic gastric catarrh, gastralgia and entralgia; also in certain skin diseases such as eczema and psoriasis.

Borates occur in brines and in some alkaline waters. The usual compound present in mineral waters is sodium borate. Strong borate waters occur in lake waters of the west. It is used as a preservative and as a gargle in throat infections and as an eye wash.

Bromides and chlorides occur in brines and are present in some of the connate waters of Indiana. Bromide waters are recommended in cases of lead or mercury poisoning; as sedatives, in cases of epilepsy, in the treatment of scrofulous tumors, ulcerations and chronic cutaneous diseases.

Carbonate waters are those containing carbonates usually of the alkalies. The effects of the alkalies have been discussed. "Bathing in water charged with carbon dioxide causes a prickling sensation, which lasts for some time, and persons in health, on leaving the bath experience a pleasing exhilaration and the inclination to muscular activity is greatly increased."

Nitrates are not present in mineral waters in any quantity. Potassium nitrate is the most common form. Nitrates are toxic if taken in large quantities. The action in the ionic form is large upon the mucous membranes and may result in gastritis, in diuresis and perhaps nephritis, at the place of exit.

Phosphates occur sparingly in mineral waters. Calcium and strontium phosphates are present in the bones of the body. Silicates occur in small quantities in most waters. The soluble silicates occur usually in minute quantities and probably do not have a strong therapeutic action.

Sulphates are common in the mineral waters of Indiana. The most abundant forms are calcium, magnesium, ferrous and sodium sulphates. The effects of the bases has been discussed in former pages and it is difficult to determine the action of the sulphate radical. Sulphates are useful more from their action as undissociated salts than from any action as ions. Sulphuretted hydrogen and free sulphur occur in some mineral waters in Indiana. "Sulphur waters are used in cases of rheumatism, gouty inflammation and chronic joint injuries. They are especially valuable in such cases when used in the form of a hot or mud bath."

Classification of Mineral Waters. The following classification of mineral waters is suggested by Dr. E. H. S. Bailey[31] to whom the writer is indebted for much of the information contained in the discussion of the properties of mineral waters in the foregoing pages.

1. Chloride group, or those in which the chlorine ion (Cl) is the predominant one.

2. Sulphate group, or that in which there is a predominance of the sulphate ion.

3. Chlor-sulphate group, or waters which contain about equal amounts of the sulphate and the chlorine ion.

4. The carbonate group, or those in which the carbonate ion (CO_3) are abundant.

5. The chlor-sulfo-carbonate group, or those containing considerable quantities of each of these ions.

6. The sulfid group, or those waters that give off hydrogen sulphid and are commonly called sulphur waters.

7. The chalybeate or iron group. (This may also contain the few manganese waters.)

8. The special group or those waters containing some special substance like lithium, borax, etc.

9. The soft water group, or those waters that contain only small quantities of any mineral substance.

Use of Mineral Waters. Mineral waters are not cure-alls and should be used only under the advice of a physician. They should be taken at such times and under such conditions as are assigned by him. "The indiscriminate use of mineral waters, either for drinking, or bathing purposes, cannot be too strongly condemned; for while they look bland and harmless, they are potent therapeutic agents which may accomplish much good if judiciously employed, but may also do much harm and may be followed by serious if not fatal results in careless hands."

Geological Occurrence. The mineral waters of Indiana are obtained from a large number of geological formations. The St. Peters sandstone is the oldest water bearing formation. It does not reach the surface in any part of the State and its waters are reached only by deep wells which usually yield strong brines since these waters are largely connate but a sulphuretted magnesium.

The Trenton limestone contains two or more horizons which are porous and contain strong salt waters which are also probably connate and too brackish for medicinal use. These waters are reached in wells less than one thousand feet deep in eastern Indiana and at a depth of more than twenty-three hundred in southwestern Monroe County.

The Silurian (Niagara) limestone yields a potable, slightly saline but strongly sulphuretted water which is being utilized for medicinal purposes at a number of points in the State. A similar water is obtained in places from the Devonian strata.

The Knobstone sandstone produces sulphuretted and soft waters from both springs and deep wells. A slightly saline-sulphuretted water flows from the base of the Harrodsburg limestone in some places.

The Chester sandstones, the Brandy Run, the Elwren, the Cypress, the Hardinsburg and Tar Springs all contain waters which supply wells and springs with waters which are usually soft. The Cypress sandstone rests upon the Beech Creek limestone. At the base of this limestone on the down dip side of the outcrop numerous springs burst forth boldly in clear, cool, sparkling water and rushing forth from the subterranean passages in the limestone comes an ever present breath of cold air.

In the area of outcrop of the Mansfield sandstone are many important springs and many shallow wells draw their supplies from its sands and gravels. Both well and spring waters are obtained from the sandstones of the Coal Measures and from the Merom sandstone.

Water supplies are obtained from the glacial filled valleys lying beyond the glaciated area as well as from the glacial drift.

Artesian Water. The deeper water hearing strata of Indiana presents conditions favorable for artesian water supplies on both sides of the Cincinnati Arch. Such waters have been obtained from the St. Peter sandstone, the Trenton limestone, the Silurian (Niagara) limestone, the Knobstone and some of the Pennsylvanian sandstones. Flowing water has been obtained from the Trenton, the Niagara, the Knobstone, the Chester, the Mansfield and the glacial drift.

Flowing wells from the Trenton are located at Ft. Wayne, Columbus, Logansport, Aurora, Elkhart and other places. Flowing water is derived from the Niagara at Attica, Brownstown, Goshen, Lodi, Martinsville, Gosport, Michigan City, Lafayette, Paoli, Reelsville, Shoals, Spencer, Terre Haute, Worthington and other places. The Devonian probably yields the water at Elkhart, Lambert, Montezuma and Jasper.

Mineral water is obtained from the Mississippian at Orleans, Avoca, Bedford, French Lick, West Baden, Indian Springs, Trinity Springs, Lasalle Springs and elsewhere.

The Pennsylvanian yields mineral waters at De Gonia, Coates, Oakland City, Owensville and elsewhere.

Plate XC. One of Indiana's famous sulphur springs—Trinity Springs. Martin County.

PRODUCTION OF MINERAL WATERS IN INDIANA.

The total value of all the mineral waters used and sold in Indiana annually is about $500,000. The following are or have recently furnished commercial supplies of mineral water.

Artesian Mineral Springs, Martinsville, Morgan County.
Blue Cast Well, Woodburn, Allen County.
Brownson Well, Terre Haute, Vigo County.
Carlson's Mineral Springs, Laporte, Laporte County.
Cartersburg Spring, Cartersburg, Hendricks County.
Colomagna Mineral Springs, Columbus, Bartholomew County.
Coats Springs, Littles, Pike County.
Greenwood Springs, Fort Wayne, Allen County.
Holman Mineral Well, Crawfordsville, Montgomery County.
Hunter Mineral and Mudlava Springs, Kramer, Warren County.
Knotts Mineral Springs, Porter, Porter County.
McCullough Springs, Oakland City, Gibson County.
Martinsville Spring, Martinsville, Morgan County.
Mysenite Well, Silverwood, Fountain County.
Paoli Lithia Springs, Paoli, Orange County.
Paynes Saline Sulphur Wells, Henryville, Henry County.
Pluto, Prosperine and Bowles Springs, French Lick, Orange County.
Reid Mineral Spa Lithia Springs, near Richmond, Wayne County.
Spencer Artesian Wells, Spencer, Owen County.
Trinity Springs, Martin County.
West Baden Mineral Springs, West Baden, Orange County.
White Crane Well, Dillsboro, Dearborn County.

ANALYSES OF INDIANA MINERAL WATERS.

Location				Sodium Chloride (Na Cl)	Magnesium Chloride ($Mg Cl_2$)	Magnesium Sulphate ($Mg SO_4$)	Calcium Sulphate ($Ca SO_4$)	Calcium Carbonate ($Ca CO_3$)	Potassium Bromides (K. Br)	Ferrous Carbonate ($Fe CO_3$)	Silica Alumina Organic Matter	Magnesium Carbonate
Abbott Mineral Well	Fort Wayne	Allen	Alkaline, saline, sulphuretted	2,993.790	148.825	143.283	29.710	597.401	5.469	21.119	43.755	
Columbus Sanitarium	Columbus	Barthol	Saline, sulphuretted	37.947	0.437		3.591	6.786			2.646	8.447
Azalia Mineral Spring	Columbus	Barthol	Chalybeate									
McCarthy's Mineral Spring	Mt. Moriah	Brown	Saline, sulphated	15.28		325.30	8.40			Tr.	55.29	
Nahville Artesian Wells	Nashville	Brown	Saline, sulphuretted									
Delphi Artesian Wells	Delphi	Carroll	Saline, sulphuretted	30.885	1.678		0.004	0.708		0.058		
Logansport Artesian Well	Logansport	Cass	Saline, sulphuretted	792.716	78.570	46.884	104.494	107.109			4.460	
King's Mineral Spring	Dallas	Clark	Saline, sulphated	167.264		406.475	55.937	25.434			0.653	
Paynes Mineral Spring	Blue Lake	Clark	Saline, sulphated	49.698		9.440	184.446	73.471				
Indiana Blue Lick Spring	Blue Lake	Clark	Saline, sulphated									
Samson King Mineral Well	Blue Lake	Clark	Saline, sulphated	238.313		357.97	59.814					
Charlestown Blue Lick Spring	Charlest'n	Clark	Saline, sulphated									
White Sulphur Well	Sulphur	Crawford	Saline, sulphuretted	125.394		43.916	11.449	47.498		1.989		16.793
Tar Springs	English	Crawford	Alkaline, saline			8.95		21.596		3.758		2.078
Haselwood Sulphur Well	English	Crawford	Saline, sulphuretted	23.864		12.296	5.866	20.568				9.459
Cable Mineral Well	Washington	Daviess	Saline, sulphated	1,014.336	88.480		75.712	9.256				
Auroral Artesian Well	Aurora	Dearborn	Saline, sulphuretted									
Cheeks Spring	Aurora	Dearborn	Saline, sulphuretted									
Jasper Artesian Well	Jasper	Dubois	Saline, sulphuretted									
Toussaint Dubois Spring	Jasper	Dubois	Neutral									
Lambert Mineral Well	Elkhart	Elkhart	Saline, carbonated	686.077	30.643			6.370		.035	.479	
Briggs Mineral Well	New Albany	Floyd	Saline, sulphated	18.536		263.280	31.896	29.728		0.141		2.936
Lodi Artesian Well	Silver Wood	Fountain	Saline, sulphuretted	572.464	53.510	3.260	55.553	2.904			1.200	1.104
Attica Artesian Well	Attica	Fountain	Saline, sulphuretted	338.82	14.72		4.10	21.65			.08	
Wallace Mineral Spring	Wallace	Fountain	Chalybeate									
Fleece Mineral Spring	Rochester	Fulton	Chalybeate									
McCullough's Mineral Spring	Oak'd City	Gibson	Alkaline, saline, chalybeate	1.784		87.355		36.339		57.920		
Owensville Artesian Well	Owensville	Gibson	Chalybeate									
Worthington Artesian Well	Worth'gton	Greene	Saline, sulphuretted									
Halsall's Spring	Maxwell	Hancock	Alkaline									
Spring Lake Park Mineral Well	Spring L. P.	Hancock	Alkaline, chalybeate	1.03			1.75	0.40			0.64	12.42
Corydon Sulphur Well	Corydon	Harrison	Saline, sulphuretted									
Cartersburg Mineral Springs	Cartersb'rg	Hendricks	Neutral									
Martha Hadley Min. Well	Amo	Hendricks	Alkaline, saline	381.744				13.94				7.81
Spiceland Mineral Spring	Spiceland	Henry	Chalybeate									
Kokomo Artesian Well	Kokomo	Howard	Saline, sulphuretted, chaly'te									
Seymour Artesian Well	Seymour	Jackson	Saline, sulphuretted									
Rensselaer Mineral Well	Rensselaer	Jasper	Saline, sulphuretted									

ANALYSES OF INDIANA MINERAL WATERS—Continued.

Location				Sodium Chloride (Na Cl)	Magnesium Chloride (Mg Cl₂)	Magnesium Sulphate (Mg SO₄)	Calcium Sulphate (Ca SO₄)	Calcium Carbonate (Ca CO₃)	Potassium Bromides (K. Br)	Ferrous Carbonate (Fe CO₃)	Silica Alumina Organic Matter	Magnesium Carbonate
Austin Mineral Well	N. Madison	Jefferson	Saline									
Greenwood Mineral Well	Greenwood	Johnson	Saline, sulphuretted									
Bradely Mineral Spring	Franklin	Johnson	Chalybeate									
Winona Mineral Spring	Winona L.	Kosciusko	Alkaline	.341			3.649	15.54		.129	.80	8.48
Hammond Artesian Well	Hammond	Lake	Alkaline, saline	20.913			10.003				1.08	9.28
Willowdale Springs	Crown Point	Lake	Alkaline	0.183				19.29			1.664	11.65
East Chicago Artesian Well	E. Chicago	Lake	Chalybeate									
Michigan City Artesian Well	Mich. City	Laporte	Alkaline, saline, sulphure'd	356.68	87.28		134.71	1.98		.37		
Avoca Mineral Spring	Avoca	Lawrence	Saline, sulphuretted									
Feldun Fields Mineral Well	Avoca	Lawrence	Alkaline, saline, sulphure'd	12.387	.64	13.91		16.16		7.48	.92	
Mt. Jackson Sanitarium	Indianap'ls	Marion	Saline, sulphuretted	648.800	140.00		1.3	20.00			.98	24.2
New Haven Well	Lawrence	Marion	Neutral									
Trinity Springs	Trinity Sps.	Martin	Saline, sulphuretted	8.338	1.099	4.165	1.26	5.589			.825	4.107
Indian Springs	Indian S.	Martin	Alkaline, saline, sulphure'd	39.366	.056	30.385		33.10			.439	18.944
La Salle Springs	Mt. Olive	Martin	Alkaline, saline, sulphure'd	1.440	5.616	31.417	46.40	36.68		.232	.728	
Elliott Springs	Willow Val.	Martin	Saline, chalybeate									
Shoals Artesian Wells	Shoals	Martin	Saline, sulphuretted	1,303.44	9.683		203.01	32.525			3.353	
Ketcham's Sulphur Spring	Smithville	Monroe	Saline, sulphuretted									
Orchard's Sulphur Spring	Blooming'n	Monroe	Saline									
Van Clev's Spring	Crawfords'e	Montg'ery	Alkaline, chalybeate	.704		7.320		9.800		.616	.072	3.824
Garland Dell's Min. Springs	Garland D.	Montg'ery	Neutral									
Martinsville Mineral Wells	Martinsv'le	Morgan	Alkaline, saline, sulphur'ed	58.58				16.902			1.217	15.359
Orleans Mineral Well	Orleans	Orange	Saline									
Moore's Mineral Well	Orleans	Orange	Saline	9.5		43.5	67.15				4.09	
Paoli Artesian Well	Paoli	Orange	Saline, sulphuretted	120.433	4.395	52.138	101.12			.251	.840	20.43
French Lick Springs	French L.	Orange	Saline, sulphuretted	118.197	4.246	55.652	13.00	33.470			.549	43.907
Rhodes Mineral Spring	West Baden	Orange	Saline, sulphuretted									
Ryan & Mickler Spring	French L.	Orange	Saline, sulphuretted									
West Baden Springs	West Baden	Orange	Saline, sulphuretted	77.971	11.402	36.14	11.178	41.367			.437	39.154
Lost River Mineral Spring	Orangeville	Orange	Saline, sulphuretted									
Flat Lick Springs	Helix	Orange	Saline, sulphuretted									
Spencer Artesian Well	Spencer	Owen	Saline, sulphuretted	66.813	5.031		.393	12.222			1.073	2.386
Gosport Artesian Well	Gosport	Owen	Saline, sulphuretted									
Montezuma Artesian Well	Montezuma	Parke	Saline, sulphuretted	357.71	9.975		3.553	11.183			.898	17.885
Coats Springs	Littles	Pike	Alkaline, saline, chalybeate									
Sweet Sulphur Springs	Velpen	Pike	Alkaline, saline, sulphure'd	1.693		42.054	18.136	28.347		1.924	1.002	2.154
Porter Artesian Wells	Porter	Porter	Saline, sulphuretted	208.76	38.71			11.14			1.10	
Winamac Artesian Well	Winamac	Pulaski	Saline, sulphuretted									

Mudge Artesian Well	Medaryv'le	Pulaski	Alkaline, sulphuretted	5.419				4.73		.023	.659	2.526
McLean Springs	Greencastle	Putnam	Alkaline, chalybeate	.695		1.036		11.88		2.381	.078	5.335
Mahan's Spring	Greencastle	Putnam	Chalybeate									
Reelsville Artesian Well	Reelsville	Putnam	Saline, sulphuretted									
Snowden Springs	Bainbridge	Putnam	Alkaline, sulp'ed, chalybeate									
Roachdale Mineral Springs	Roachdale	Putnam	Chalybeate									
Johnson Mineral Springs	Versailles	Ripley	Chalybeate									
Clark Artesian Well	Carthage	Rush	Saline	4,101.2	203.2		15.88	1.89		30.85	413.0	
Shelbyville Mineral Well	Shelbyville	Shelby	Alkaline, saline, sulphure'd	696.57				65 1			.86	45.762
Lafayette Artesian Well	Lafayette	Tippecanoe	Saline, sulphure'd carbonate	324.76	29 65		56.01	12.024			.464	.4
Paper Mill Artesian Well	Lafayette	Tippecanoe	Saline, sulphuretted									
Buck Creek Artesian Well	Buck Creek	Tippecanoe	Saline, sulphuretted									
Battle Ground Spring	Battle G'd	Tippecanoe	Chalybeate									
Fritalar Mineral Well	Evansville	Vander'gh	Alkaline, saline, chalybeate	2,227.12			32.70	220.18			9.26	64.81
Seventh Av. Mineral Well	Evansville	Vander'gh	Alkaline, chalybeate sulp'd	3 08			.123	12.96			1.235	2.76
Willard Market Well	Evansville	Vander'gh	Saline, chalybeate	7.88			46.28				.5	7.25
Exchange Mineral Well	Terre Haute	Vigo	Thermal, alkaline, saline, sulphuretted	301.25	11.055		.257	19.92		.035	.759	15.34
Magnetic Mineral Well	Terre Haute	Vigo	Thermal, alkaline, saline, sulphuretted	347.73	13.945		2.078	21.94			.893	16.44
T. H. Gas Co. Artesian Well	Terre Haute	Vigo	Thermal, saline, sulphure'd									
Rose Artesian Well	Terre Haute	Vigo	Alkaline, s'ine, sulphure'd	316.0	6.428		2.325	25.026			1.2	6.42
White's Institute Artesian Well	Wabash	Wabash	Chalybeate									
Indiana Mineral Springs	Attica	Warren	Neutral	.331			1.85	17.696			.964	
Hunter Mineral Spring	Kramer	Warren	Neutral									
Kickapoo Magnetic Spring	Kickapoo	Warren	Alkaline					12.35		.05	.68	5.38
DeGonia Springs	DeGonia	Warrick	Alkaline, saline, chalybeate			56.0	4.0	16.0		4.00	3.00	
Hash Iron Springs	DeGonia	Warrick	Alkaline, saline, chalybeate									
Fairview Springs	Boonville	Warrick	Saline, chalybeate									
Underwood Mineral Well	Salem	Washing'n	Saline									
Beck's Sulphur Spring	Salem	Washing'n	Saline, sulphuretted									
Glenn Miller's Spring	Richmond	Wayne	Chalybeate									
Reid's Spring	Richmond	Wayne	Alkaline, chalybeate									
Hawkes's Spring	Richmond	Wayne	Alkaline, chalybeate	.333		1.599	11.684	9.448		.192	.158	

ANALYSES OF INDIANA MINERAL WATERS—Continued.

Location.				Sodium Sulphate	Potassium Chloride	Calcium Chloride	Calcium Phosphate	Potassium Sulphate	Magnesium Bromide	Alumina Sulphate	Gases	Total Solids
Abbott Mineral Well	Fort Wayne	Allen	Alkaline, saline, sulphure'd								4.677	3,974.355
Columbus Sanitarium	Columbus	Bartholo.	Saline, sulphuretted								4.031	64.9
Azalia Mineral Spring	Columbus	Bartholo.	Chalybeate									
McCarthy's Mineral Spring	Mt. Moriah	Brown	Saline, sulphated	319.92								724.19
Nashville Artesian Wells	Nashville	Brown	Saline, sulphuretted									
Delphi Artesian Wells	Delphi	Carroll	Saline, sulphuretted		0.023						7.76	33.356
Logansport Artesian Well	Logansport	Cass	Saline, sulphuretted									1,134.233
King's Mineral Spring	Dallas	Clark	Saline, sulphated	189.133								914.831
Paynes Mineral Spring	Blue Lake	Clark	Saline, sulphated	303.008								722.455
Indiana Blue Lick Spring	Blue Lake	Clark	Saline, sulphated									
Samson King Mineral Well	Blue Lake	Clark	Saline, sulphated	170.285								826.299
Charlest'n Blue Lick Spring	Charlest'n	Clark	Saline, sulphated									
White Sulphur Well	Sulphur	Crawford	Saline, sulphuretted	10.950				5.397				263.432
Tar Springs	English	Crawford	Alkaline, saline	3.529				1.696				41.652
Haselwood Sulphur Well	English	Crawford	Saline, sulphuretted					1.310				
Cable Mineral Well	Washington	Daviess	Saline, sulphated	438.088					605			1,683.645
Auroral Artesian Well	Aurora	Dearborn	Saline, sulphuretted									
Cheeks Spring	Aurora	Dearborn	Saline, sulphuretted									
Jasper Artesian Well	Jasper	Dubois	Saline, sulphuretted									
Touissaint Dubois Spring	Jasper	Dubois	Neutral									
Lambert Mineral Well	Elkhart	Elkhart	Saline, carbonated		8.050						10,088.00	786.104
Briggs Mineral Well	New Alb'ny	Floyd	Saline, sulphated					4.072				350.589
Lodi Artesian Well	SilverWood	Fountain	Saline, sulphurette l	2.135		47.928		.804			7.94	672.292
Attica Artesian Well	Attica	Fountain	Saline, sulphuretted		Tr.	10.13						390.66
Wallace Mineral Spring	Wallace	Fountain	Chalybeate									
Fleece Mineral Spring	Rochester	Fulton	Chalybeate									
McCullough's Mineral Spring	Oakland C.	Gibson	Alkaline, saline, chalybeate	Tr.						36.407		316.31
Owensville Artesian Well	Owensville	Gibson	Chalybeate									
Worthington Artesian Well	Worthingt'n	Greene	Saline, sulphuretted									
Halsall's Spring	Maxwell	Hancock	Alkaline									
Spring Lake Park Min. Well	Spring L. P.	Hancock	Alkaline, chalybeate									32.480
Corydon Sulphur Well	Corydon	Harrison	Saline, sulphuretted									
Cartersburg Min. Springs	Cartersburg	Hendricks	Neutral									
Martha Hadley Min. Well	Amo	Hendricks	Alkaline, saline		0.992			933				409.336
Spiceland Mineral Spring	Spiceland	Henry	Chalybeate									
Kokomo Artesian Well	Kokomo	Howard	Saline, sulphure'd, chalyb'te									
Seymour Artesian Wells	Seymour	Jackson	Saline, sulphuretted									
Rensselaer Mineral Well	Rensselaer	Jasper	Saline, sulphuretted									

Austin Mineral Well	N. Madison	Jefferson	Saline									
Greenwood Mineral Well	Greenwood	Johnson	Saline, sulphuretted									
Bradely Mineral Spring	Franklin	Johnson	Chalybeate									
Winona Mineral Spring	Winona L.	Kosciusko	Alkaline	.322	.049							29.360
Hammond Artesian Well	Hammond	Lake	Alkaline, saline	29.89								112.741
Willowdale Springs	Crown Point	Lake	Alkaline	.113				.799		.652		35.749
East Chicago Artesian Well	E. Chicago	Lake	Chalybeate									
Michigan City Artesian Well	Mich. City	Laporte	Alkaline, saline, sulphure'd		12.31							
Avoca Mineral Spring	Avoca	Lawrence	Saline, sulphuretted									
Feldun Fields Mineral Well	Avoca	Lawrence	Alkaline, saline, sulphure'd	2.247							22.99	53.743
Mt. Jackson Sanitarium	Indianapolis	Marion	Saline, sulphuretted		1	102.6					4.27	935.800
New Haven Well	Lawrence	Marion	Neutral									
Trinity Springs	Trinity S'gs	Martin	Saline, sulphuretted	.35				.097				26.721
Indian Springs	Indian S'gs.	Martin	Alkaline, saline, sulphure'd	11.828						.829	12.91	163.604
LaSalle Springs	Mt. Olive	Martin	Alkaline, saline, sulphure'd	3.512						.056	10.32	126.328
Elliott Springs	Willow Val.	Martin	Saline, chalybeate									
Shoals Artesian Wells	Shoals	Martin	Saline, sulphuretted		47.3	2.6512			2.237		43.289	1,801.226
Ketcham's Sulphur Spring	Smithville	Monroe	Saline, sulphuretted									
Orchard's Sulphur Spring	Blooming'n	Monroe	Saline									
Van Clev's Spring	Crawf'ds'le	Montg'ry	Alkaline, chalybeate	.200								22.848
Garland Dell's Min. Springs	Garland D'l	Montg'ry	Neutral									
Martinsville Mineral Wells	Martinsville	Morgan	Alkaline, saline, sulphure'd	1.879	1.775						22.1	98.194
Orleans Mineral Wells	Orleans	Orange	Saline									
Moore's Mineral Well	Orleans	Orange	Saline	45.85	7.10							177.17
Paoli Artesian Well	Paoli	Orange	Saline, sulphuretted		2.364						7.505	303.605
French Lick Springs	French Lick	Orange	Saline, sulphuretted	3.391				1.009			11.706	312.692
Rhodes Mineral Springs	West Baden	Orange	Saline, sulphuretted									
Ryan & Mickler Spring	French Lick	Orange	Saline, sulphuretted									
West Baden Springs	West Baden	Orange	Saline, sulphuretted			7.276					10.104	235.734
Lost River Mineral Spring	Orangeville	Orange	Saline, sulphuretted									
Flat Lick Springs	Helix	Orange	Saline, sulphuretted									
Spencer Artesian Well	Spencer	Owen	Saline, sulphuretted	.940	18.11	8.11					13.090	121.702
Gosport Artesian Well	Gosport	Owen	Saline, sulphuretted									
Montezuma Artesian Well	Montezuma	Parke	Saline, sulphuretted		2.68						9.678	422.659
Coats Springs	Littles	Pike	Alkaline, saline, chalybeate									
Sweet Sulphur Springs	Velpen	Pike	Alkaline, saline, sulphure'd		.303						5.987	95.613
Porter Artesian Wells	Porter	Porter	Saline, sulphuretted		13.188	51.93						342.34
Winamac Artesian Well	Winamac	Pulaski	Saline, sulphuretted									
Mudge Artesian Well	Medaryville	Pulaski	Alkaline, sulphuretted								3.31	22.404
McLean Springs	Greencastle	Putnam	Alkaline, chalybeate	.099							3.005	22.157
Mahan's Spring	Greencastle	Putnam	Chalybeate									
Reelsville Artesian Well	Reelsville	Putnam	Saline, sulphuretted									
Snowden Springs	Bainbridge	Putnam	Alkaline, sulphure'd, chaly't									
Roachdale Mineral Spring	Roachdale	Putnam	Chalybeate									
Johnson Mineral Spring	Versailles	Ripley	Chalybeate									
Clark Artesian Well	Carthage	Rush	Saline		19.42	2,188.94						6,974.38
Shelbyville Mineral Well	Shelbyville	Shelby	Alkaline, saline, sulphure'd	48.19	6.02						1.78	874.86
Lafayette Artesian Well	Lafayette	Tippecanoe	Saline, sulphure'd, carbonate			3.72					19.4284	427.54
Paper Mill Artesian Well	Lafayette	Tippecanoe	Saline, sulphuretted									
Buck Creek Artesian Well	Buck Creek	Tippecanoe	Saline, sulphuretted									
Battle Ground Spring	Battle G'd	Tippecanoe	Chalybeate									
Fritzlar Mineral Well	Evansville	Vanderb'h	Alkaline, saline, chalybeate									2,574.43

ANALYSES OF INDIANA MINERAL WATERS—Continued.

Location.				Sodium Sulphate	Potassium Chloride	Calcium Chloride	Calcium Phosphate	Potassium Sulphate	Magnesium Bromide	Alumina Sulphate	Gases	Total Solids
Seventh Av. Mineral Well	Evansville	Vanderb'h	Alkaline, chalyb'te, sulph'd								1.332	28.159
Willard Market Well	Evansville	Vanderb'h	Saline, chalybeate									64.672
Exchange Mineral Well	Terre Haute	Vigo	Thermal, alkaline, saline, sulphuretted		3.625	12.941					12.0017	366.398
Magnetic Mineral Well	Terre Haute	Vigo	Thermal, alkaline, saline, sulphuretted		3.957	16.297					15.259	432.565
T. H. Gas Co. Artesian Well	Terre Haute	Vigo	Thermal, saline, sulphure'd									
Rose Artesian Well	Terre Haute	Vigo	Alkaline, saline, sulphure'd		1.232	4.816						365.067
White's Insti. Artesian Well	Wabash	Wabash	Chalybeate									
Indiana Mineral Spring	Attica	Warren	Neutral								3.836	26.846
Hunter Mineral Spring	Kramer	Warren	Neutral									
Kickapoo Magnetic Spring	Kickapoo	Warren	Alkaline	.99								24.42
DeGonia Springs	DeGonia	Warrick	Alkaline, saline, chalybeate	25.00		4.00	2.00	7.00			8.02	121.00
Hash Iron Spring	DeGonia	Warrick	Alkaline, saline, chalybeate									
Fairview Springs	Boonville	Warrick	Saline, chalybeate									
Underwood Mineral Well	Salem	Washing'n	Saline									
Beck's Sulphur Spring	Salem	Washing'n	Saline, sulphuretted									
Glenn Miller's Spring	Richmond	Wayne	Chalybeate									
Reid's Spring	Richmond	Wayne	Alkaline, chalybeate									
Hawkes's Spring	Richmond	Wayne	Alkaline, chalybeate			.323					4.302	24.907

CHAPTER X.

PART I.

INTRODUCTION.

No industry is more dependent upon science than is the petroleum industry upon the science of geology. The petroleum and natural gas industry of Indiana is of so much importance to the industrial development of the State that it should be given every aid which this science can supply for the solution of its problems. Enormous sums of money have been expended and are still being expended in Indiana in "wild-cat" drilling and the greater part of this form of prospecting is being indulged in without reference to the presence or absence of geological conditions favorable to the accumulation of oil and gas. Very naturally such prospecting leads to enormous losses and few gains.

In the absence of any comprehensive discussion of the subject of petroleum and natural gas in Indiana available for distribution and in response to hundreds of inquiries for information on the subject reaching the office of the Division of Geology, this chapter has been prepared.

The chapter is preliminary to the preparation of a more comprehensive report to be issued later. It was not possible in the limited time and with the limited funds at our disposal to make the chapter more complete. The collected information has not been studied as thoroughly as it should have been and hence conclusions have not been drawn where, perhaps, a more careful study of the evidence would warrant. However, since the industry is changing rapidly through development in parts of the States and decline in others no report can be prepared which will not need revision in a few years. In view of this fact it seems best to present such information as we have been able to bring together with the hope that it may be of immediate assistance to those who have so urgently requested it.

Those who are seeking petroleum in Indiana would do well to bear in mind that the geologist does not use "divining rod" or "witching" methods in the location of oil. He studies the structural conditions of the strata to determine whether such structural conditions are favorable to the accumulation of oil. For the determination of structural conditions he must be able to examine exposures of the bed rock or durolith, the indurated solid portion of the earth underlying the loose mantle of clay, sand and gravel called the regolith.

In certain parts of Indiana the durolith is completely concealed by a thick covering of glacial drift and unless deep well records are available the geologist is without means of determining the structural conditions. The majority of the reported oil seeps from this part of the state are only oil-like films of oxide of iron on water seeping from glacial sands and gravels. Surface indications are of little value in oil prospecting in such a region. To be of value in any region they must be correctly interpreted.

In that portion of Indiana where the glacial covering is attenuated or in the non-glaciated portion the work of the geologist is not so hampered and wherever persistent hard layers of rock are present he is usually able to determine the structural conditions.

For the good of the petroleum industry in the future it is hoped that more money will be expended in securing favorable locations for wells and less expended on the drilling of wells that have been located without reference to the structural conditions. The money expended on one deep well will pay for securing the information and the publication of many thousands of copies of a chapter more comprehensive in its scope than the present one. Mistakes of location are expensive in more ways than one. Aside from the actual pecuniary loss in drilling the well, there is often a loss of confidence in the territory. For example, dry holes in sections one, two and three may condemn good territory in adjacent sections whereas if the structure had first been located the drilling of a single well on the structure might prove the territory.

It is important, therefore, that in all areas of the State where it is possible to determine the structural conditions this be done before any prospecting with the drill takes place.

The oil industry suffers from two classes of individuals, namely, from the purveyor of oil stock of the "blue sky" brand and from the activities of the fake oil expert. The laws of Indiana very wisely provide for the protection of its citizens against the dispenser of inferior foods. No one doubts that the abolition of the food inspection department would result in making the State the dumping ground for all sorts of foods of inferior quality. However, the average consumer of foods has some knowledge of their quality which knowledge is within itself a form of protection. But in the matter of oil stocks, legislation, in some states, affords inadequate protection and few are qualified to judge of the value of oil stocks.

Many states protect their citizens against the unscrupulous dealer in oil stocks of questionable value. States without such protection naturally become the Meccas of jobbers in all sorts of oil stock of the "blue sky" brand. Laws have been provided in Indiana to aid in the protection of the novitious small investor from the machinations of the unscrupulous oil-stock purveyor. Such legislation is not intended to interfere with legitimate attempts at the development of the oil industry in Indiana. It is not intended to prevent the organization of local cooperative companies for the avowed purpose of developing prospective oil properties within the State. Nor should such companies be prevented under proper representations from offering the stock of such companies for sale. For in some parts of the State where it is impossible to determine the structural conditions and the only possible form of prospecting, that with the drill, is extremely hazardous, the expense of such testing should be widely distributed in order that the burden may not fall too heavily upon the few.

The purveyor of all oil stock should be required to furnish to the purchaser of such stock a sworn statement of the location of the oil property, the number of acres under lease, the state of development, and a certified copy of the report of the consulting geologist.

The oil operator, the investor in oil stock, and the general public need protection from the quack, the manipulator of the "divining rod", the witch hazel switch and other devices for the location of oil pools. Novitious oil companies are known to have used the funds secured from the sale of oil stock to small investors to drill a well costing as much as ten or twelve thousand dollars on a location made by the manipulator of a "divining rod".

The success of the competently trained geologist in the location of geological structures favorable to the accumulation of oil and gas has induced a large number of unprepared or illy prepared individuals to assume the role of oil

geologists. Its rewards have also induced many pseudo-scientists to enter the field. Such impostors do not find employment with reputable oil companies of experience, but they gull the public through the mushroom companies of limited experience in the oil industry, and at the same time tend to bring discredit upon the science.

There are two ways of obtaining protection for the public against the activities of such impostors. One is to educate the people to an understanding of the scientific principles of oil geology, a very difficult task. A more immediate and effective method of protection might be secured through legislation which would provide for the licensing of oil experts by the State and measures prohibiting the practice of the profession of oil geologist by persons not possessing the requisite amount of training in the science and practice of geology.

Since this chapter was written much additional information has been collected and a part of it has been included since the preliminary issue of the chapter. All of it could not be prepared at this time because of the necessity for further field work. This information will be presented in a future publication.

ACKNOWLEDGMENTS.

The writer acknowledges his indebtedness to those who have written on the subject of petroleum and natural gas in Indiana. The information contained in the reports of Blatchley and others has been freely drawn upon in the preparation of the county reports. The publications mentioned in the accompanying bibliography have been especially helpful. The reference figures in the text apply to the numerals in this list of publications.

In the field work the writer has had the assistance of the members of the field party of 1919, the names of the members of which are given under Geological Corps.

Especial mention should be made of the assistance and advice of Dr. E. R. Cumings, the field work of Dr. C. A. Malott who, assisted by Mr. P. B. Stockdale, collected data for structural maps of portions of Jennings, Orange and Pike Counties, prepared a structural map of the Bloomington Quadrangle and assisted in other ways. Mr. O. H. Hughes, a member of the field party of 1917 and 1919, collected the data for a structural map of a portion of Jackson County. Dr. S. S. Visher collected data and prepared the report on Sullivan County. Mr. J. R. Reeves, a member of the field party of 1917 and 1919, prepared the maps and charts and assisted in other ways. Mr. B. J. Malott collected data, read manuscript and corrected proof. Miss Alice O'Connor did the stenographic work.

BIBLIOGRAPHY.

1. "The Natural Gas Field of Indiana." Dr. A. J. Phinney, Eleventh Ann. Rept. U. S. G. S. Pt. 1, 1890, pp. 617-742.

2. "Natural Gas and Petroleum." S. S. Gorby, Sixteenth Ann. Rept. Indiana Dept. of Geology and Natural Resources, 1888, pp. 189-301.

3. "Petroleum Industry in Indiana." W. S. Blatchley, Ann. Indiana Dept. of Geology Reports, 1896, pp. 27-96; 1903, pp. 79-210; 1906, 429-558.

4. "The Princeton Petroleum Field of Indiana." R. S. Blatchley, Ann. Rept. of Indiana, Dept. of Geol., 1906, pp. 559-607.

5. Special Report on Oil and Gas, E. Haworth, Uni. Geol. Sur. of Kansas, Vol. IX, 1908.

6. "Oil and Gas Resources of Kansas." Bul. 3, R. C. Moore and W. P. Haynes, State Geol. Survey of Kansas, 1917.

7. "Oil Geology," Dorsey Hager.

8. "Petroleum Technology," Bacon and Hamor.

9. "Oil," etc. Bul. 33, Ill. Geol. Survey, 1915, Kay and Savage.

10. "Petroleum and Natural Gas." Mineral Resources of U. S. for various years.

11. "Oil and Gas in Louisiana." G. D. Harris, Bul. 429, U. S. G. S., 1910.

PART II.

PETROLEUM: ITS PROPERTIES AND ORIGIN.

Definition. Petroleum or crude oil is a mixture of gaseous, liquid and solid hydrocarbons in which the liquid elements predominate, but in which the percentage of each element is not a fixed quantity, but varies in different oils. The solid hydrocarbons are in solution and consist of paraffin or asphaltum or in some oils of both. Those oils with asphaltum in solution are said to have an asphalt base and those containing paraffin to have a paraffin base. The paraffin oils predominate east of the Mississippi River and the asphalt oils west.

Composition. The chemical compounds of which petroleum is a mechanical mixture belong to a number of hydrocarbon series. They include the marsh gas series, C_nH_{2n+2}, ranging from CH_4 to $C_{35}H_{72}$. The first member is gaseous, the middle members liquid, and the last members are solid paraffins. The olefiant series, C_nH_{2n}, is represented by some of its members in small amounts. The acetylene series, C_nH_{2n-2}, is represented in some petroleums. The fourth series is C_nH_{2n-4}. The fifth or benzine series, C_nH_{2n-4}, is represented in nearly all petroleums.

The elementary analyses of various petroleums indicate that the per cent of carbon varies from 83.5 to 86.6; the per cent of hydrogen from 12 to 14.8, and the per cent of oxygen from 0.1 to 6.9. These three elements make up the larger part of the oil, but nitrogen and sulphur occur in minute quantities usually.

Color. The color of petroleum varies with the sand or field. Pennsylvanian oils have a greenish color; the Kansas-Oklahoma oils have a yellowish tint; California oils are black; Indiana oils greenish black; some Kentucky oils are green by reflected light and red by transmitted light. White oils may be produced by filtration.

Odor. The odor of most petroleums is slight, but some oils have an odor resembling some of their products such as gasolene or kerosene. Hydrogen sulphide and pyridine may give distinctive odors.

Density. The specific gravity of petroleum varies from 0.77 in some light oils to 1 in the heavier oils. The average for the American petroleums is about 0.89. The oil from the Lima-Indiana field ranges in specific gravity from 0.816 to 0.86. The Terre Haute oil has a specific gravity of 0.879; the Jasper oil of 0.928.

Boiling Point. The temperature of boiling ranges from 180° F. in Pennsylvanian oils to 338° F. in some German oils. The point of solidification ranges from 82° F. to several degrees below zero.

The Flashing Point. The flashing point of petroleum varies from zero in some Italian oils to 338° F. in some African oils. The fuel value of the oil from the eastern Indiana field is 18,900 B. T. U.

Specific Heat. The specific heat of American petroleums ranges from .3999 to .5000.

Specific Gravity. The specific gravity of a substance is its weight compared with the same volume of water which is assumed to have a specific gravity of 1. Petroleum usually floats on water and has a specific gravity less than that of water. The specific gravity of petroleum may be expressed as a decimal fraction, as .8588, or the Baumé scale may be used for oils lighter than water, in which case it will be expressed in degrees. If the oil has a specific gravity equal to water its specific gravity as expressed on the Baumé scale is 10°.

In the determination of specific gravity of oils the hydrometer is used. This instrument consists of a glass column provided with the Baumé scale graduated in degrees from 10 to 100 and an expanded portion below the scale which contains mercury to sink the hydrometer to the point which registers its specific gravity if the temperature of the fluid is 60° F. For lower or higher temperatures, corrections must be made. The specific gravity may be calculated by adding 130 to the reading on the hydrometer and dividing 140 by the sum, as $\frac{140}{40 + 130} = .8235$ specific gravity. The following table will show the relation between the Baumé scale and specific gravity and weight per gallon:[1]

Degrees Baumé	Specific Gravity	Pounds per Gallon	Degrees Baumé	Specific Gravity	Pounds per Gallon	Degrees Baumé	Specific Gravity	Pounds per Gallon
10	1.0000	8.33	32	.8641	7.20	54	.7608	6.34
11	.9929	8.27	33	.8588	7.15	55	.7567	6.30
12	.9859	8.21	34	.8536	7.11	56	.7526	6.27
13	.9790	8.16	35	.8484	7.07	57	.7486	6.24
14	.9722	8.10	36	.8433	7.03	58	.7446	6.20
15	.9655	8.04	37	.8383	6.98	59	.7407	6.17
16	.9589	7.99	38	.8333	6.94	60	.7368	6.14
17	.9523	7.93	39	.8284	6.90	61	.7329	6.11
18	.9459	7.88	40	.8235	6.86	62	.7290	6.07
19	.9395	7.83	41	.8187	6.82	63	.7253	6.04
20	.9333	7.78	42	.8139	6.78	64	.7216	6.01
21	.9271	7.72	43	.8092	6.74	65	.7179	5.98
22	.9210	7.67	44	.8045	6.70	66	.7142	5.95
23	.9150	7.62	45	.8000	6.66	67	.7106	5.92
24	.9090	7.57	46	.7954	6.63	68	.7070	5.89
25	.9032	7.53	47	.7909	6.59	69	.7035	5.86
26	.8974	7.48	48	.7865	6.55	70	.7000	5.83
27	.8917	7.43	49	.7821	6.52	71	.6829	5.69
28	.8860	7.38	50	.7777	6.48	72	.6666	5.55
29	.8805	7.34	51	.7734	6.44	73	.6511	5.42
30	.8750	7.29	52	.7692	6.41	74	.6363	5.30
31	.8695	7.24	53	.7650	6.37	75	.6222	5.18

[1] American Petroleum Industry, Bacon and Hamor, Vol. 1, p. 95.

Petroleum Products. The various products obtained from crude pertoleum are kerosene, gasolene, benzene, naptha, rhigolene, vaseline, paraffin, lubricating oil, petroleum butter, formolit, asphalt, oil coke, gas carbon, special illuminating oils such as mineral sperm and astral oil.

Origin of Petroleum and Natural Gas.

The close association of petroleum and natural gas points to a common origin. The hydrocarbons which form them are identical or closely related. The gases given off by petroleum are similar to those of natural gas, which may be converted into liquid by increase of pressure at low temperature, as may be the gas given off by petroleum. Natural gas is commonly present in petroleum, and they often exist together, though natural gas may exist alone.

The theories of the origin of oil and gas fall into two classes: the inorganic and the organic.

Inorganic Theories. A chemical theory was suggested by Humboldt and further elaborated by Berthelot[1] and Mendeleeff[2]. This theory assumes that the interior of the earth contains metallic iron and carbides of iron; that the high interior heat of the earth converts water into steam, which attacks the carbides of iron, producing hydrocarbons which are forced toward the surface by the expanding power of steam. According to this theory the hydrocarbons formed should be predominately of the acetylene series, but they are predominately of the methane series; they should be associated with igneous rather than sedimentary rocks.

Another inorganic theory is the volcanic theory of Costé[3] which assumes that oil and gas are the result of volcanic action. Costé asserts that animal remains are not intombed in the rocks and that vegetable remains decompose into carbonaceous matter and further distillation of carbonaceous matter has not taken place in nature; that gaseous, liquid, and solid hydrocarbons are the result of volcanic activity, because oil and gas are under great pressure which must be volcanic; heated oil and gas exists in some fields; oil and gas occur in folded and fissured regions parallel with great orogenic movements; oil and gas and bitumens are never indigenous to the strata in which they are found and that the density of rocks precludes the possibility of anything except volcanic pressure forcing oils and gas through them. Many of these assertions do not accord with the observed facts. The almost complete restriction of oil and gas to sedimentary rocks placed at great distance from volcanic activity and the decrease in pressure in wells are not in harmony with this theory.

Organic Theory. This theory assumes that oil and gas have been generated from animal and vegetable matter by a slow process of distillation. Many accumulated geological facts may be enumerated in support of this theory, such as: The close association of rocks containing organic matter to those containing oil and gas; drops of oil have been found in decaying plant remains; natural gas, a constituent of both oil and gas, is generated from vegetable matter buried in porous beds; it is present in coal as are other hydrocarbons of petroleum; such gases as carbon dioxide, hydrogen, marsh gas and nitrogen are formed during the decay of sea weeds. Hydrocarbons analogous to those

[1] Berthelot, E. M. P. Annales Chem. Phys., Vol. I, 1866, p. 481.
[2] Mendeleeff, D. Der Deutch. Chem. Gesell, 1877, p. 229.
[3] Costé, E. Am. Inst. Min. Eng., Vol. XX, p. 504, 1914.

in natural gas, petroleum and asphalt have been derived from either plant or animal remains. Natural petroleum has optical properties similar to those of organic compounds which inorganically synthesized oil does not possess.

The presence of oil in shales from which as much as twenty-five gallons per ton have been extracted has strengthened the belief that the organic matter of shales is the source of petroleum. It is assumed that the bituminous matter is the form of a solid, organic gum, kerogen, which may be converted into liquid hydrocarbons by the application of heat. McCoy[1] placed an oil shale under pressure and secured liquid hydrocarbons from it and asserts that liquid hydrocarbons can be formed from solid bituminous material at ordinary temperatures and under pressures of 5,000 to 6,000 pounds, such as exist at the depth of oil bearing horizons; and that the only place where such compounds would be formed are in areas of differential movement.

Kemp[2] has recently called attention to the presence of asphaltum in the beach sands of Florida and the possibility of the origin of petroleum from the marine and terrestrial organisms in buried coastal sands.

The optical behavior of petroleum under polarized light is said to be due to the presence of cholesterol, which may be derived from animal fats and phytosterol, which is also a constituent of vegetable oils, facts strongly supporting the organic theory of the origin of petroleum. In fact, the weight of evidence at the present time seems to favor the organic theory. The remains of land plants and animals may have contributed in a minor way to the accumulations of petroleum, but marine organisms were probably the greater contributors of the original compounds from which the petroleum was extracted through long periods of time at possibly only ordinary rock pressures and at moderately low temperatures.

Oil Wells in Indiana.

Counties	1906-1910		1910-1914		Wells Abandoned
	Completed	Dry	Completed	Dry	1915
Adams	112	13	20	1	866
Blackford	158	28	22	5	1,389
Cass	3	2	....	....	....
Daviess	2	....	10	2	....
Delaware	297	75	125	28	1,320
Dubois	5	4	5	1	....
Gibson	96	28	30	2	5
Grant	480	42	11	2	4,141
Hamilton	....	....	....	....	....
Harrison	....	....	2	....	....
Huntington	206	9	....	....	891
Jay	561	112	100	17	554
Knox	4	3	19	17	12
Madison	11	3	3	3	87
Marion	....	....	....	....	15
Martin	2	1	2	1	....
Miami	1	1	3	3	49

[2] Kemp, J. F. Econ. Geol., Vol. XIV, 4 p. 302.
[1] McCoy, Alex. W. Journal of Geol., Vol. XXVII, 4 p. 252.

Oil Wells in Indiana—Continued

Counties	1906-1910 Completed	1906-1910 Dry	1910-1914 Completed	1910-1914 Dry	Wells Abandoned 1915
Pike	280	63	116	25	3
Pulaski	4	3	....	....	..
Randolph	59	18	33	....	213
Shelby	....	....	4	....	....
Sullivan	3	3	758	271	1
Vigo	3	2	7	2	....
Wabash	....	....	1	....	16
Warrick	3	3	1	1	....
Wells	497	71	35	2	3,950
Miscellaneous	128	91	30	15	....

Petroleum Production in Indiana.

(Compiled from Mineral Resources of the United States.)

Date	Barrels	Value
1889	33,375	$31,414
1890	63,496	55,403
1891	136,634	91,545
1892	698,068	388,300
1893	2,335,293	1,494,588
1894	3,688,666	2,654,840
1895	4,386,132	4,780,884
1896	4,680,732	2,954,411
1897	4,122,356	1,880,412
1898	3,730,907	2,214,322
1899	3,848,182	3,363,738
1900	4,874,392	4,693,983
1901	5,757,086	4,822,826
1902	7,880,896	6,526,622
1903	9,186,411	10,474,127
1904	11,339,124	12,235,574
1905	10,964,247	9,404,909
1906	7,673,477	6,770,066
1907	5,128,037	4,536,930
1908	3,283,629	3,203,883
1909	2,296,086	1,997,610
1910	2,159,725	1,568,475
1911	1,695,289	1,228,835
1912	970,009	885,975
1913	956,095	1,279,226
1914	1,335,456	1,548,042
1915	875,758	813,365
1916	769,036	1,207,565
1917	759,432	1,470,548
1918	877,558	2,028,129

PART III.

NATURAL GAS.

Definition. Natural gas is a mixture of hydrocarbons (chiefly) which are gaseous at ordinary atmospheric temperatures. The principal hydrocarbon is marsh gas (CH_4), methane or fire damp. Natural gas also contains small quantities of ethane (C_2H_6), Olefine (C_2H_4), Carbon dioxide (CO_2), Carbon monoxide (CO), Oxygen (O), Nitrogen (N), Hydrogen (H), Helium (He), Neon (Ne) and Hydrogen sulphide (H_2S). However, not all natural gases contain all of these gases.

Physical Properties. Natural gas is colorless and usually odorless, though the presence of such gases as hydrogen sulphide may produce a perceptible odor. It is usually inflammable though some natural gases contain so much nitrogen as to be non-combustile. It burns with a luminous flame and deposits carbon when the flame is brought in contact with objects of lower temperature. It readily mixes with air and forms an explosive mixture.

Gas Pressure. Natural gas as it occurs in the earth is usually under pressure which ranges as high as 2,000 pounds per square inch. This pressure is commonly called "rock pressure" and decreases as the gas becomes exhausted. The pressure is probably due to the expansive force of the confined gas.

Chemical Properties. The maximum amount of the various constituents found in natural gas is: Marsh gas, 98.40%; Ethane, 14.60%; Olifinant, .39%; Carbon dioxide, 1.6%; Carbon monoxide (CO), 2.5%; Oxygen (O), 3.46%; Nitrogen (N), 85.83%; Hydrogen (H), 11.51%; Helium (He), 1.84%, and Hydrogen sulphide (H_2S), .20%.

The composition of natural gases from various fields is given below for comparison with the analysis of a gas from Muncie:

COMPOSITION OF NATURAL GASES.

STATE.	Methane (CH_4)	Ethane (C_2H_6)	Olefine (C_2H_4)	Carbon Dioxide (CO_2)	Carbon Monoxide (CO)	Oxygen	Nitrogen	Hydrogen	Helium	Hydrogen Sulphide (H_2S)	Location
Indiana	92.67	.....	.25	.25	.45	.35	2.53	2.35	......	.15	Muncie
Illinois	73.81	.....	......	.81	......	3.46	21.92	......	......	......	Pittsfield
Ohio	92.61	.....	.30	.26	.50	.34	3.61	2.18	......	.20	Findlay
Kansas	94.40	.....	......	......	......	.23	5.08	......	1.83	......	Iola.
Kansas	14.85	.41	......	......	......	.20	82.70	tr.	1.84	......	Dexter

Origin of Natural Gas. Since natural gas is closely associated with petroleum they are thought to have a common origin. They often occur together, though one may occur without the presence of the other. Nearly all petroleums contain at least small quantities of natural gas. Since natural gas is free to move independent of the movement of water it may accumulate in a different reservoir though having a common origin with petroleum. For instance, it may accumulate, in fact does accumulate, in glacial sands and gravels at a horizon far from its point of origin.

The principal constituent of most natural gases is marsh gas (CH_4). This gas also accumulates in marshes where decaying organic matter is surrounded with porous sands. This gas is also found in coal beds and is one of the constituents of petroleum. These facts argue for an organic origin for natural gas and for a common origin with petroleum.

Production of Natural Gas in Indiana.

(Compiled from Mineral Resources of the United States.)

Year	No. Producers	Value	Wells, Gas	Wells, Dry	Productive Wells
1886	....	$300,000	(Est. Amt. Coal Displaced)		
1887	....	600,000	(Est. Amt. Coal Displaced)		
1888	....	1,320,000	(Est. Amt. Coal Displaced)		
1889	....	2,075,702	(Est. Amt. Coal Displaced)		
1890		2,302,500	....	....	435
1891	93	3,942,500	....	....	305
1892	109	4,716,000	....	....	570
1893	....	5,718,000	....	....	....
1894	....	5,437,000	....	....	....
1895	....	5,203,200	...	....	....
1896	....	5,043,635	....	....	....
1897	452	5,009,208	419	66	2,881
1898	533	5,060,969	706	111	3,325
1899	571	6,680,370	838	109	3,909
1900	670	7,254,539	861	156	4,546
1901	656	6,954,566	985	208	4,572
1902	929	7,081,344	1,331	205	5,820
1903	924	6,098,364	895	242	5,514
1904	846	4,342,409	706	153	4,684
1905	740	3,094,134	252	74	3,650
1906	578	1,750,715	159	46	3,523
1907	687	1,572,605	185	56	3,386
1908	823	1,312,507	187	41	3,223
1909	1,010	1,616,903	190	70	2,938
1910	1,027	1,473,403	69	33	2,955
1911	1,094	1,192,418	110	32	2,744
1912	1,140	1,014,295	96	39	2,547
1913	1,100	843,047	69	24	2,370
1914	1,029	755,407	68	19	2,224
1915	999	695,380	65	11	2,063
1916	995	503,373	43	14	1,967
1917	941	453,000	42	17	1,830
1918	....	899,671	....	....	....

Natural Gas in Indiana.

Counties	Depth of Well	Pressure in Lbs. 1910	Pressure in Lbs. 1914
Adams	1,000-1,050	100 (1912)	0-6
Bartholomew	864- 990	50-250	80-150
Blackford	850-1,100	1- 10	0- 20

Natural Gas in Indiana—Continued

COUNTIES	Depth of Well	Pressure in Lbs. 1910	Pressure in Lbs. 1914
Clark	128- 244	27 (1912)	......
Daviess			25- 40
Martin	300- 600	0- 60	......
Decatur	700-1,200	0-315	5-350
Delaware	728-1,500	0- 70	0- 60
Franklin	728- 730	60 (1913)	......
Grant	830-1,200	2- 50	0- 50
Hamilton	800-1,250	15-180	0-230
Hancock	700-1,100	0-100	6- 80
Harrison	320- 764 (1911)	60-110	0- 50
Henry	800-1,200	0- 90	4-100
Howard	800-1,100	0-220	30-160
Jay	900-1,600	0- 40	0- 40
Jefferson	1,360	10 (1911)	20
Madison	800-1,200	0-190	0-100
Miami	900-1,000	0- 40	......
Marion			
Ripley	880-1,050	40	70-300
Pike	1,000-1,400	125-500	50-225
Randolph	900-1,300	0-180	1-125
Rush	700-1,400	20-325	15-325
Shelby	650-1,020	1-375	20-300
Spencer	1,025	410 (1912)	
Sullivan	698- 795	200	50-185
Tipton	750-1,100	10-230	3-100
Wayne	800-1,150	50-240	45

Gas Depletion.

An examination of the pressure of gas in the wells of Indiana shows that the gas is being rapidly depleted. The pressure recorded in some of the wells in 1910 was 250 pounds per square inch and in 1914 the same wells showed a pressure of only 150 pounds.

The following methods of computing gas depletion are given by the Treasury Department of the United States[1]:

"*Details of production or the performance record of the well* or *property.*—As a general rule the demand on a natural gas property is a variable factor. In certain fields, however, the demand from some wells has from the beginning, or for considerable periods, been greater than the supply, so that the amount of gas marketed per well may, as in the case of oil, show a regular decline, which will be indicative of the total amount that the well may be expected to produce, and also the rate of production. Even where the demand does not greatly exceed the supply, the amount and rate of past production may in certain cases throw light on the future of the well or property.

"*Decline in open-flow capacity.*—Where data are available the decline in open-flow capacity indicates in a general way the rate of exhaustion of the gas field. The relationship is not at all close and varies from field to field

[1] Manual for the Oil and Gas Industry, U. S. Treasury Dept., 1919.

and from well to well. Also for most gas wells accurate data on decline in open-flow capacity are not available. Nevertheless it is probable that for certain properties this method will have value, for with rare exceptions the production of gas from a well leads to a decline in its capacity, and the fraction produced is roughly proportional to the decline.

"*Comparison with life history of similar wells* or *properties, particularly those now exhausted or nearing exhaustion.*—Where no other data are available the rate of depletion of a gas well or property may be approximated by comparison with a neighboring well or property that has reached a later stage in life. Particularly is this applicable in a district where many gas wells have become exhausted. For example, in a region where wells produce from 8 to 12 years, or an average of 10 years, a 10 per cent deduction will be a rough approximation of depletion.

"*Size of reservoir and pressure of gas*, or *the pore-space method.*—For some properties the pore-space method may be best for estimating underground supplies of natural gas and for a good many it will furnish additional evidence of value. The method would be ideal if the average percentage of pore-space, the extent and thickness of the sand, and the pressure of the gas could be accurately ascertained. In computing the reserves of an individual property by this method the migratory character of gas must be considered and the production and behavior of adjacent properties taken into account. The factors that make the method difficult to apply are difficulty of accurately ascertaining the thickness of pay, limits of pool, percentage of pore-space, the effect of encroaching water and oil, and the quantity of gas remaining when commercial production is no longer possible.

"Take, for example, a pool where there is no encroachment by water. Suppose that the pore-space is 25 per cent, the thickness of the pay 20 feet, and the extent of the pool 10 square miles, or roughly 280,000,000 square feet. The volume of the reservoir would be 1,400,000,000 cubic feet, and the amount of gas in the sand could be readily computed by taking into account the closed pressure of the wells.

"*Other indications of depletion.*—Additional evidence of decreasing supply of natural gas in the ground is commonly observable in the behavior of the wells and the provision that must be made for transporting the gas to market. Observations on minute pressure show more or less progressive change as the wells become older and an increasing amount of gas is drawn from the ground. Line pressures and pressures at compressing stations are also likely to show a progressive change in the same direction. The appearance of water or oil in a gas well or in neighboring gas wells may be a very significant symptom of the approaching termination of the life of the well. The clogging of gas wells by paraffin, salt, or other deposits may demand modification of depletion estimates.

Closed-Pressure Method.

"Because of its general applicability, the closed-pressure method is by far the best method of estimating the depletion of gas properties.

"Unfortunately, accurate closed-pressure data have not been kept for all properties or perhaps even for the majority of properties, but the rock pressure in most pools is known or is ascertained with a fair degree of accuracy, and the information drawn from the pressure decline is, with the exception of a few fields, not subject to profound modification, because of factors whose value can not be appraised. The basis of this method is Boyle's law. According to this law of physics, if gas is pumped into a vessel until the pressure is 200 pounds and then is drawn off until the pressure is 100 pounds, the size of the vessel remaining fixed, and ignoring for the moment atmospheric pressure, it may be concluded that one-half of the gas has been drawn out of the vessel. If an underground gas reservoir of fixed dimensions is tapped by wells and the pressure is found to be a thousand pounds, and then if the gas is drawn off through the wells until the gas pressure in the pool is lowered to 100 pounds, we may infer that about nine-tenths of the supply of gas has been exhausted.

"*'Unit Cost' as applied to natural gas.*—Although, as a rule, the number of cubic feet of gas under a tract cannot be satisfactorily estimated and the quantity that will be marketed is even less definite, the 'unit cost method' can be used by regarding pounds of closed pressure as units, for the actual quantity of gas underground commonly varies with the decline in pressure and the relative quantity at the beginning and end of the tax year and at the time of abandonment, is, in the lack of better information, usable for tax purposes.

"*Corrections and refinements of closed-pressure method.*—Several corrections and more or less important refinements are made in applying this method to the computation of depletion, and it should be borne in mind that it does not afford data on the amount of gas originally in the pool or at any later specified time, but only the fraction of the gas that has been removed from its natural reservoir does not remain fixed but becomes smaller as the gas is drawn and water or oil advances into a part of the space formerly occupied by the gas. The pressure is thus prevented from declining at a rate proportionate to the amount of gas drawn from the pool. The correction on account of water or oil encroachment is difficult to make, because of the lack of data to determine the extent of the encroachment. However, in a good many pools, after a study of the distribution of wells that have been 'drowned out' and the history of water troubles in similar nearby pools, it is possible to make allowance for water or oil encroachment which will more or less closely approximate the facts.

"Another refinement applicable to the computation of depletion of natural gas by the closed-pressure method is based upon the fact that even where there is no encroachment of water or oil the depletion is not precisely represented by the gauge readings, though the errors are generally so small that they may be ignored. For example, where the pressure declines from 1,000 to 500 pounds, the gas is not exactly half gone, for the reason the pressures referred to are gauge readings and to each should be added the pressure of the atmosphere—for most fields about 14.4 pounds to the square inch. The fraction remaining in the ground then becomes $\frac{514.4}{1014.4}$.

"Account should also be taken of the pressure at which wells are abandoned in the field or district.

"If wells can not be operated with profit after the pressure has declined to 25 pounds gauge reading (39.4 pounds absolute), then the percentage of recoverable gas remaining when the pressure has declined from 1,000 to 500 pounds gauge reading is not one-half or even the fraction $\frac{514.4}{1014.4}$ but $\frac{475}{975}$. The difference in the fraction where pressures of several hundred pounds are involved is not great and scarcely worth considering in view of the other errors which are certain to affect the result. However, after the pressure has declined to a low figure, the matter of correcting the fraction becomes of considerable importance. Thus, if the pressure of abandonment is 4 pounds gauge reading and during the year the average closed pressure of a pool has declined from 10 to 5 pounds gauge reading, five-sixths instead of one-half of the recoverable gas has been withdrawn.

"Still another refinement that has, as a rule, more theoretical than practical value may be worthy of consideration in certain instances. This arises out of the fact that gases do not expand precisely as the pressure decreases, and that even if the size of the natural reservoir remains fixed the pressure does not decline in exact proportion to the amount of gas removed. The difference amounts to only a few per cent and is greatest for high pressure. In the decline from 1,000 to 500 pounds per square inch the gas expands several per cent more than would be calculated by a strict application of the law and in a decline from 1,500 pounds to 1,000 pounds the departure is still greater. The correction varies from field to field because of the different constitution of the gases, though since most natural gases consist largely of methane the variations on account of differences in gases are not great.

"A fourth detail of refinement arises out of the fact that on the average more gas is marketed from 50 pounds of decline in pressure after the pressure has reach 100 pounds or less than an equal decline while the pressure is high, as, for example, 1,000 pounds per square inch. Also the expense of marketing gas after the pressure has become low is greater than when it was high, largely because of the necessity of installing compressors to push the gas through the pipe lines to the consumers. These two considerations have a tendency to balance each other and, with certain exceptions, will not be of sufficient importance to warrant to apply the corrections.

Method of Gauging.

"In using the closed-pressure method of estimating depletion, the method of gauging is of vital importance and in many fields is not carried out with sufficient care. Care should be taken to make sure that the gauge is accurate, testing it before and after attaching it to the well. If it must be transported far or is subject to much jolting in transportation, a gauge tester should be taken along and used at the well.

"Care should also be taken to empty the well of oil and water by pumping, blowing or siphoning before attaching the gauge, for any liquid in the hole will lower the closed pressure reading.

"The well should be closed long enough to allow pressure to build up to its maximum. The length of time necessary for this purpose varies a great deal from field to field and well to well. The well should remain closed until the pressure will not build up more than 1 per cent in 10 minutes. Ordinarily, 24 hours will be sufficient for this purpose, but for some wells several days or even a longer period will be required, owing to the slowness of equalization of pressure in the sand."

PART IV.

MODE OF ACCUMULATION.

Experience in oil fields has taught that oil may accumulate under certain conditions in either synclines or anticlines. In the absence of water in synclines oil may move downward under the influence of gravitation to the bottom of the syncline. (See Fig. XCIV.) Of course it is not known whether the oil has migrated downward from the limbs of the syncline or the roof of the porous layer or been moved upward by capillarity to the bottom of the syncline from underlying beds of oil bearing shales, though it is doubtful that the latter would produce sufficient concentration. The essential conditions for oil accumulations are: First, a source of the oil which may be a

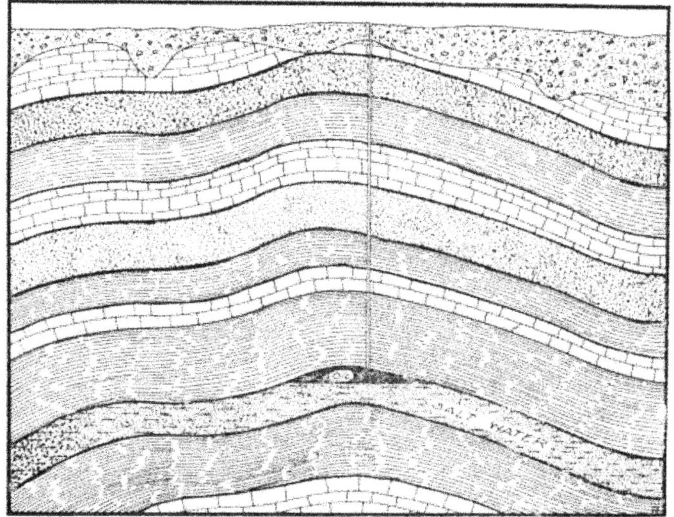

Fig. XCII. A diagrammatic cross section of an anticline showing the mode of occurrence of oil when no gas is present.

bituminous rock probably at no great distance from the point of accumulation. Second, a porous bed of rock which acts as a reservoir. This porous bed must be contained between impervious layers of rock. Third, the presence of flexures in the reservoir. In the absence of water in the reservoir the oil will collect in the downward folds (synclines), but if water is present no oil collects in the synclines but only in the anticlinal or upward folds as the oil advances to the highest point occupied by the water which would be in the upper part of the anticline. From this point it would be impossible for the oil to advance as its progress is checked by the impervious roof layer which dips down below the level occupied by the oil.

Relation of Geological Structure to Oil and Gas Accumulation. Oil and gas are widely distributed in the rocks of the earth as is evident from their presence in rocks, in mines, in seeps, the water of springs and deep wells. But accumulations of oil and gas of economic importance are far less widely distributed since special geological conditions are necessary to the concentration of oil and gas in economic quantities. Oil and gas generated in some bituminous beds, rise under the agencies of migration and reach a porous bed as widely distributed particles, and are, therefore, valueless, from an economic standpoint. The concentration of oil and gas can be brought about if certain geological structures are present in the porous bed containing the oil and the gas. The presence or absence of such concentrating structures may, in most cases, be determined by the geologist so that a knowledge of geology is fundamental to the development of the oil industry.

Oil Sands. The rock in which the oil and gas accumulates is commonly termed the "oil sand" though it is often not a true sand but a porous rock

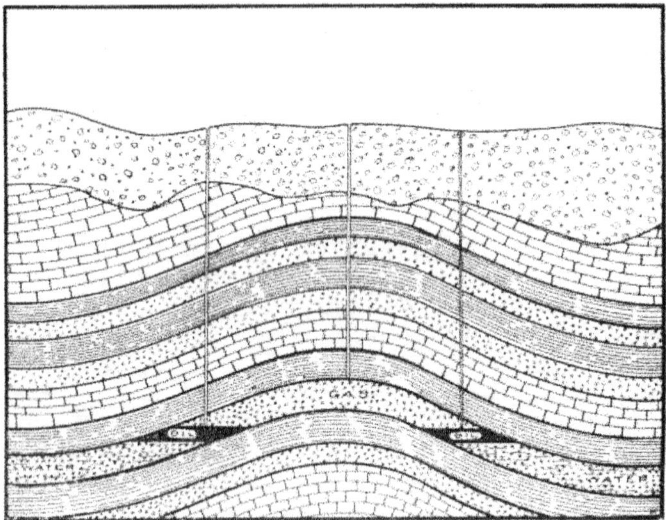

Fig. XCIII. A diagrammatic cross section of an anticline, the most abundant type of oil bearing structure. In this anticline water, oil and gas are present arranged in the order of their specific gravities. The removal of the gas will permit the oil and water to rise higher toward the apex of the structure.

such as limestone. More commonly the oil accumulates in a porous sand, sandstone, or conglomerate, less commonly in porous limestone and very rarely in fissures in shales or in the cavities in igneous rocks. The quantity of oil possible in an oil sand will depend upon the degree of porosity of the sand which in turn depends upon the size and arrangement of the sand grains in the case of a true sand and on the size of the cavities in the case of a porous limestone. The pore space in compacted but uncemented sands ranges as high as 25 per cent, in sandstones to 15 per cent and in conglomerates to as high as 32 per cent. The amount of pore-space produced by the size and the arrangement of the grains may be reduced by deposition of cement in the pores.

Geological Structures Favorable to the Accumulation of Oil. There are certain structural conditions which are favorable to the accumulation of oil and gas. Such conditions may exist without the presence of oil or gas, but so far as is known, accumulations of oil or gas do not occur without the presence of such favorable structural conditions. Among the more favorable structures for the accumulation of oil and gas are: The anticline, monocline, structural terrace, dome, fault, joints, lenses, igneous intrusions and synclines.

The Anticline. The anticline is an upward bend or fold in the rock strata which forms a trap which prevents the escape of the oil or gas when they have once penetrated it. The essential conditions for the accumulation of petroleum in an anticline is the presence in the fold of a porous layer of

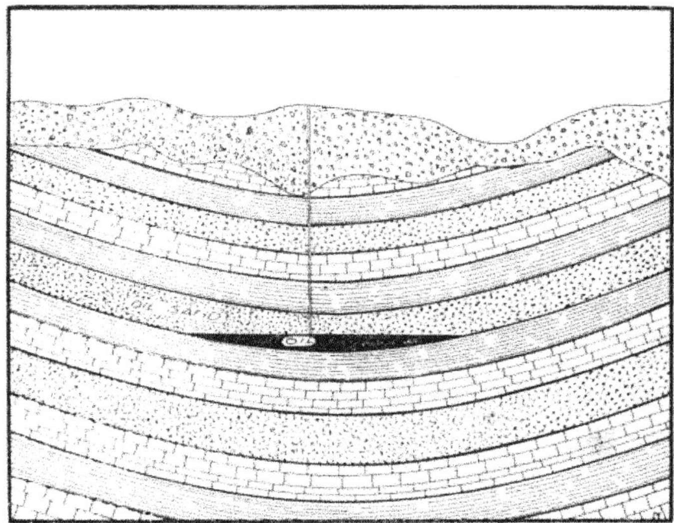

Fig. XCIV. A diagram to show possible mode of accumulation of oil in a syncline. The sand is a dry sand, that is, it does not contain any water. Since the oil is free to move under the action of gravity it will sink to the lowest portion of the porous layer.

rock enclosed between two layers of impervious rock. For example, a layer of porous sandstone between two layers of shale. The presence of water in the porous layer is also essential. If no gas is present, the oil will accumulate in the highest portion of the porous layer. (See figure XCII.) The oil being of lighter specific gravity collects in the upper part of the porous layer above the water. If gas be present, the three will arrange themselves in order of their specific gravities. (See figure XCIII.) The pressure of the gas in this case forces the oil and water to the limbs of the anticline. With the escape of the gas the oil and the water would tend to rise in the porous layer and arrange themselves in order of their specific gravities.

Syncline. The presence of oil in downward folds of rocks called synclines, occurs under certain conditions. (See figure XCIV.) If no water is present in the porous layer, the oil under the influence of gravitation may be carried down to the bottom of the syncline and there remain, held in by impervious layers of rock above and below. Oil is obtained from synclines in Pennsylvania and Ohio. No oil has been obtained from such structures in Indiana. No dry oil reservoirs have been found as yet.

The Dome. The dome or salt dome is an anticlinal structure produced by accumulation of minerals under strata along the plane of a subsurface or a sealed fault. (See figure XCV.) Such structures are common in the Gulf Coastal plain in the states of Louisiana and Texas. According to Harris[1] these domes are produced by water carrying minerals such as salt, gypsum, lime carbonate and magnesium carbonate in solution ascending along a fault plane to a point beneath the surface where the minerals were deposited through

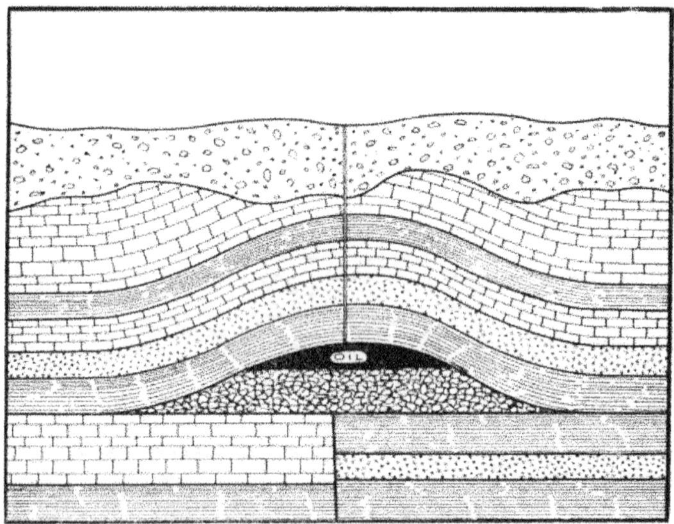

Fig. XCV. A diagrammatic cross section of a salt dome structure favorable to the accumulation of oil. Soluble salts carried by ascending solutions are deposited under strata which are forced upward forming a dome. Oil passing upward along the fault plane accumulates in the porous deposit formed by the salts.

the evaporation of the water. The accumulation of the mineral matter elevates the super-incumbent beds and the oil accumulates in porous beds of limestone or in sands overlying or tilted up against the salt core. Topographically these domes may form conspicuous mounds on the flat prairies of the coast. Continual erosion of the surface of the mound as the salt accumulates may bring deep seated beds of rock 900 feet or more nearer the surface than their normal position for that area. Numerous faults are produced by the doming and the oil and gas pass to the porous beds along these faults. A number of domes may be distributed along a major fault.

The Monocline. Rock strata are often inclined in only one direction and form a monocline. That is they may pass from one horizontal position to

[1] Harris, G. D. Bul. La. Geol. Sur. No. 7, 1908, p. 75 et seq.

another horizontal position or from one inclined position to another inclined position without reversing the direction of dip of the strata. Under certain conditions monoclines afford favorable conditions for the accumulation of oil. (See figure XCVI.) The inclination of the beds is here greatly exaggerated and gives the impression of reversal of dip. Lenses of sand or sandstone enclosed in shales in monoclines furnish favorable conditions for oil concentration.

The Structural Terrace. The structural terrace may be called a flattened monocline. The strata which are inclined pass to a horizontal position or from a greater to less degree of inclination and then back to the same degree of inclination first assumed. (See figure XCVII.) In the horizontal portion the

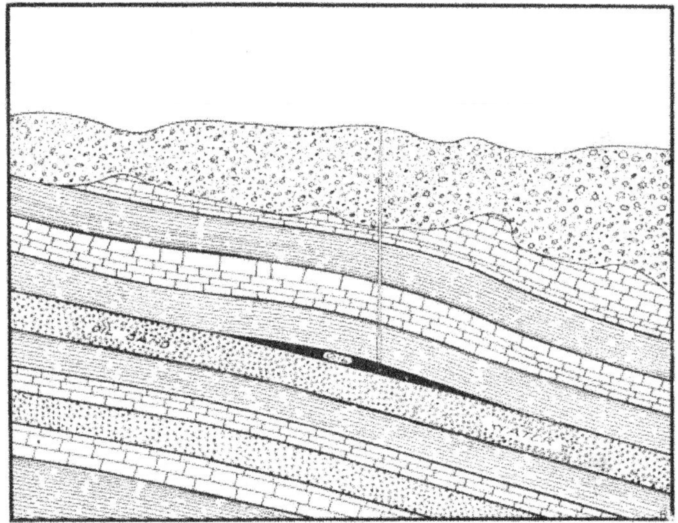

Fig. XCVI. A diagrammatic cross section of a monocline showing a possible mode of oil accumulation. A slight irregularity in the direction of dip in the shale layer above the oil sand produces a condition which is favorable to the accumulation of oil. This irregularity may or may not express itself at the surface.

trap is formed and the oil accumulates if water be present in the porous layer. The structural terrace occurs in the Mississippian area of Indiana in probably more than one locality. There is one at least in Orange County and one in Martin County. Noses and shoulders which are modifications of the terrace occur in Jackson and Jennings counties. In the latter one has produced some gas, though the drilling was not done in the most favorable spot and was done without reference to the structure.

Lens Structure. Lenses of porous sand or sandstones inclosed in bituminous shales may afford conditions favorable to the accumulation of oil and gas. (See figure XCVIII.) The lenses may lie in a horizontal position or be inclined and still furnish the proper conditions for accumulation. Since such structures do not express themselves in any way at the surface and prospecting with the drill is the only method of determining the presence, size, or

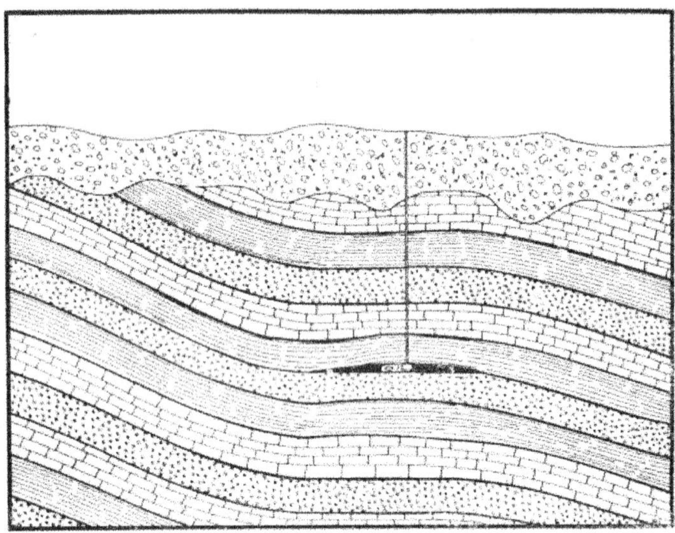

Fig. XCVII. A diagrammatic cross section of a structural terrace. Showing possible mode of accumulation of oil in the flattened portion of the structure when water is present in the oil bearing stratum.

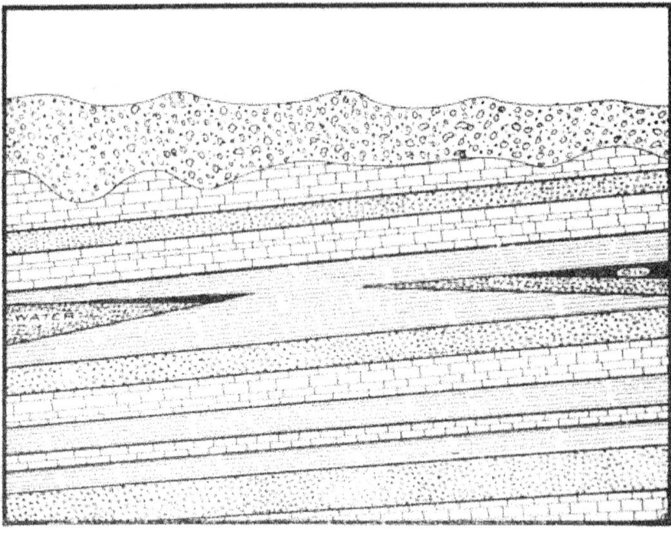

Fig. XCVIII. A diagram showing a possible mode of accumulation of oil in lenses of sand enclosed in beds of shale. Such a structure may exist in the southwestern oil field in Indiana.

shape of the structures, the geologist can locate the position and probable extent of the enclosing shale bed, but cannot indicate the position of the lenses. Sandstones or sands with convex upper surfaces due to unconformable relation with overlying beds or to lenticular shape; or sandstones with higher porosity in some parts than in others furnish adequate conditions for oil and gas accumulation when they are confined in impervious layers of rock. It is probable that such conditions exist in the Mississippian and Pennsylvanian strata of southwestern Indiana and that they are responsible for some of the oil and gas accumulations.

Fault Structure. The occurrence of oil in connection with sealed faults is an established fact. The oil migrates upward along the fault plane until it reaches a porous bed so situated as to form a trap. (Figure XCIX.) Beds of

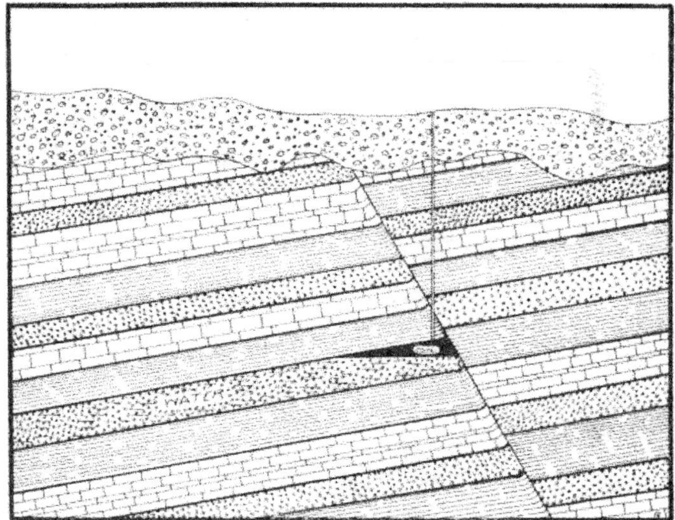

Fig. XCIX. A diagram to show the mode of accumulation of oil on the upthrow side of a fault. A porous layer has been faulted against an impervious layer of shale in such a way as to seal the fault and produce a collecting ground for the oil near the fault line.

bituminous shale and beds of sandstone may be displaced in such a way as to throw shale bed against shale bed, thus sealing the fault. If a porous bed lying between impervious beds is faulted, in such a way as to form a trap, the accumulation of oil may result. In the case of a fault cutting a rising oil and gas bearing sand the fault may seal the sand in such a way as to prevent the upward movement of the oil and the gas and cause it to accumulate. The fault is sealed by bringing the broken end of the sand layer against a shale layer. Since prospecting is more hazardous in connection with faults than anticlinal folds, little testing of the former has taken place. Structures of this type may occur in connection with the Mount Carmel fault in Indiana, but no tests have been made to determine whether they exist and are productive. There is little doubt that the fault is sealed because Knobstone shale

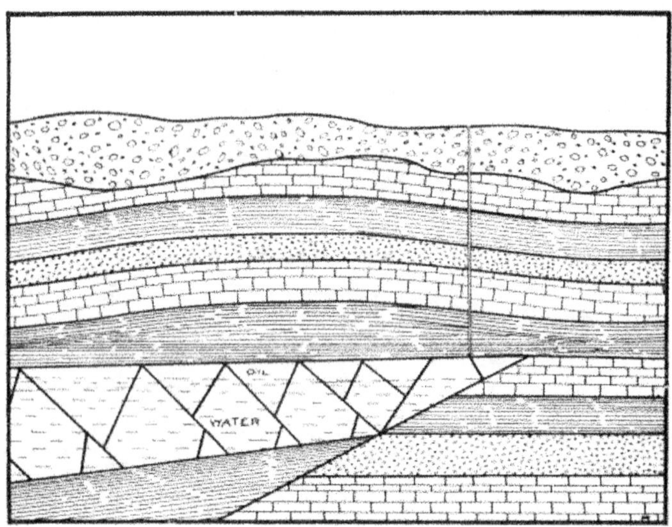

Fig. C. A diagram to show possible method of accumulation of oil in the joints of rocks. This type of structure is not common.

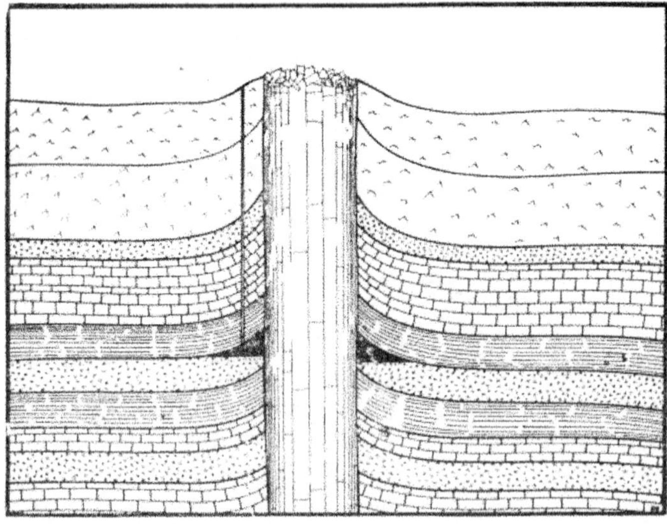

Fig. Cl. A diagram to show possible accumulation of oil in a structure produced by an igneous intrusion. The oil sand in this case may be either of sedimentary or igneous origin. The igneous rock may be either primarily porous like cellular basalt or it may receive its porosity by alteration subsequent to its intrusion.

has been faulted against Knobstone shale and sandstone layers are confined below.

Joints. Oil has been known to accumulate in joints under certain conditions. The conditions are such that the joint virtually acts as the porous layer and must occupy a position between impervious layers and be so situated as to form a trap. (Figure C.) The joint layer of rock in this case forms the reservoir. Such rocks are necessarily hard rocks, unyielding under pressure, and not exposed to the agents of cementation. Oil is found in joint cracks in some fields in California and in Colorado. Structures of this type are not known to occur in Indiana.

Igneous Intrusions.[1] The vertical or nearly vertical intrusion of igneous rocks into sedimentary strata which contain beds of bituminous rocks may result in the accumulation of oil near the intrusion. (Figure CI.) The injection of the igneous rock causes an upturning of the sedimentary beds on the sides of the igneous core. The sealing of the end of the upturned oil and gas reservoir provides conditions favorable to the accumulation of oil and gas.

The sealing may be done by the igneous rock or by hydrothermal action of the porous bed, rendering it impervious. Oil seeps may reach the surface from the oil pools along fault planes produced during the upward bending of the beds. Structures of this type do not occur in Indiana as vulcanism has not expressed itself in the State.

The geological structures favorable to the accumulation of oil and gas which may be encountered in Indiana are anticlinal, monoclinal, terrace, fault and lens structures. Oil bearing synclines are not likely to be present because of the abundance of water in the porous beds of rock. The other types of occurrence are associated with special conditions which do not exist in Indiana.

PART V.

PROSPECTING FOR OIL AND GAS.

The best equipment that an oil prospector can have is thorough training in the science of Geology. He must have a knowledge of the geological conditions of the field in which he is prospecting. This must include a knowledge of the nature of the rocks, not merely at the surface but to a considerable depth. This information he may obtain from surface outcrops, railroad cuts, stream courses, excavations, well records and geological reports.

He will need to have a knowledge of the age of the rocks since the occurrence of oil and gas in the oldest rocks of the earth has not been recorded. He will need to know that oil and gas are not found in igneous and metamorphic rocks, but are confined to sedimentary rocks. He will need to know further that certain kinds of sedimentary rocks are not likely to contain oil and gas. He will learn to look with favor upon rocks containing organic matter or rocks associated with rocks containing organic matter, evidence of which will be found in fossils, lignite and prevailing dark colors. He will look with disfavor upon rocks with prevailing red or yellow color, because the oxidized condition of the iron compounds points to the absence of organic matter.

[1] Clapp, Econ. Geol. VII, 1912, 364.

A knowledge of the structure of the rocks is essential because of its bearing on the accumulation of oil and gas. They accumulate in beds of porous rocks. If the rocks are dry the oil will accumulate in the lower part of the porous bed and the gas in the upper part, if water is present they will be arranged in the order of their specific gravities, with the gas at the top and the water at the bottom. It is obvious that if the porous rock were of uniform

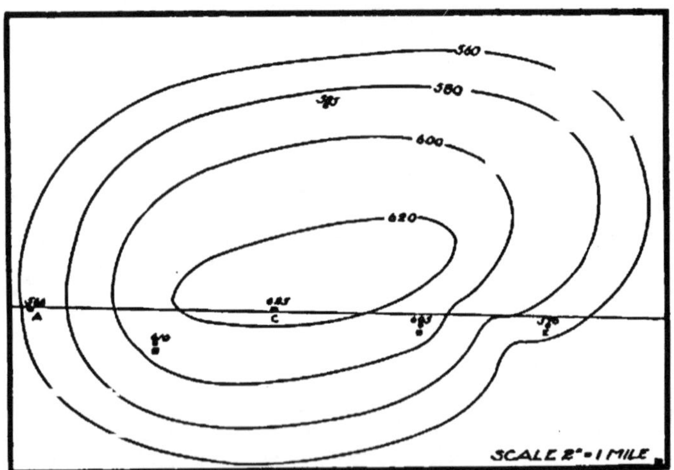

Fig. CII. Diagram of an anticline represented by contours drawn on the surface of a bed of coal. Contour interval twenty feet. Position of the bed of coal determined by well records.

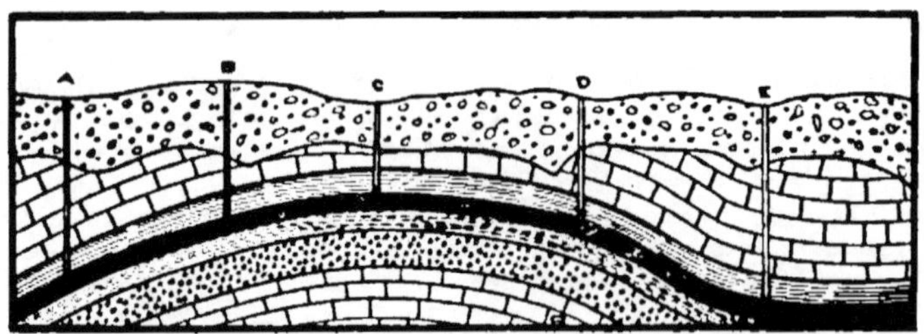

Fig. CIII. A cross section of the above anticline along the line A, B, C, D, E.

thickness and horizontal in position that there would be no concentration of oil and gas. At best there would be only a film of oil on the water. In other words, there must be irregularities of certain kinds either in the bedding or in the structure which will permit the concentration of the oil and gas at one point. And so the prospector must be able to recognize such structures as anticlines, monoclines, synclines, terraces and faults.

If the anticline is small it may be determined frequently by direct observation. If the anticline is broad, or the degree of inclination is slight, other means of determination must be used. In some instances the structure may be determined by locating upon the map the strike and the dip of the strata. The succession of the rocks should be carefully determined then a layer of relatively hard rock which is continuous over a large area should be selected and the strike and the dip of this bed at many points be recorded on a map. By this means reverse dips will be indicated and the nature of the structure determined.

Fig. CIV. Standard derricks. (Ill. Geol. Survey.)

The determination of the structure is often more difficult because of the slight degree of dip or because it may be difficult to find a layer that is continuous over large areas and which may be relied upon as a key formation. In regions where the structure is sufficiently pronounced and where there are established elevations (bench marks) for comparison, the aneroid barometer may be used and the structure be worked upon the key rock. The key rock may be a bed of coal, (figure CII) or a layer of any persistent rock such as limestone or sandstone. The elevations of the key rock above sea level should be determined for the various parts of the area, and upon a map representing this area, the points of equal elevation should be joined. By drawing lines through points of equal elevation for each ten or twenty feet of difference in elevation, the shape and the size of the structure may be exhibited. The elevations of the key rock may be determined at its outcrops by using a plane table and a telescopic alidade and stadia. In the absence of bench marks, they may be set by using plane table and stadia. The outcrops may then be located with aneroid barometer by checking frequently on the established bench marks.

Fig. CV. A steel framed derrick. (Photo by Ill. Survey.)

Exploitation. The development of the oil and gas industry began with the drilling of the first well by Colonel Drake, on Oil Creek in Pennsylvania in 1859. Great progress has been made since that date in both methods and machinery. Haphazard methods by untrained men in small companies having little capital have given way to scientific methods practiced by trained experts in power companies of large capital. No industry responds more readily to careful scientific methods than the oil and gas industry, for this reason the wise company employs trained men in each of the various departments which are a necessary part of the industry. In the absence of a sufficient number of trained engineers some large companies have established apprenticeships for inexperienced men and paid them wages while training them for their positions. In the development of new oil territory much preliminary work must be done before the drilling can be begun.

Locating the Structure. The first work in the new field falls to the Geologist. He is required to locate and to carefully map the geological structure. No wise company starts drilling operations until it has assurance that the geological conditions are favorable for the accumulation of oil. This

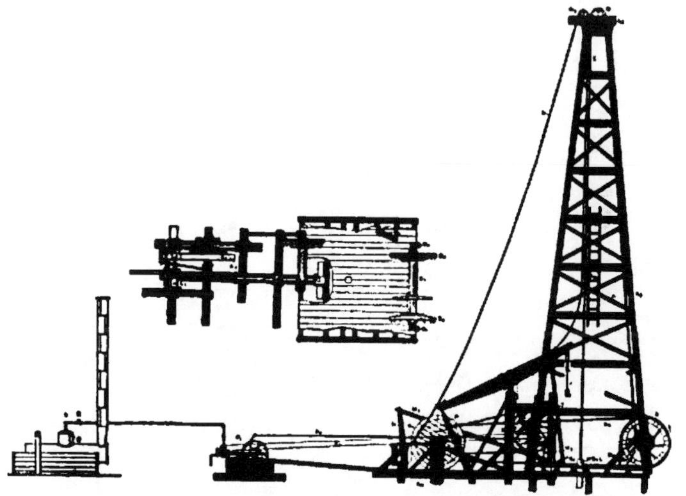

Fig. CVI. Standard drilling outfit, coupled for raising tools. (After Bowman, U. S. Geol. Surv.)

A Derrick foundation posts.
A^2 Mudsils.
A^3 Subsill.
A^4 Main sill.
A^5 Derrick legs.
A^6 Derrick girts.
A^7 Derrick braces.
A^8 Ladder.
A^9 Crown block.
B Crown pulley.
C^1 Drilling cable.
D Bull-wheel shaft.
D^2, D^3 Bull wheels.
D^4 Bull-wheel posts.
D^5 Bull-wheel post brace.
D^6 Bull rope.
D^7 Bull-wheel brake band.
E^1 Calf-wheel.
E^2 Calf-wheel brake lever.
F Sampson post.
G Walking beam.
H Pitman.
J Temper screw.
K^1 Band-wheel.
K^2 Tug pulley.
K^3 Band-wheel crank.
L^1 Sand-reel drum.
L^2 Sand-reel pulley.
M^1 Sand-reel lever.
M^2 Sand-reel reach.
M^3 Sand-reel handle.
N^1 Sand-pump line.
N^2 Sand-pump pulley.
O Calf-wheel posts.
P^1 Throttle-valve wheel.
P^2 Telegraph cord and throttle valve.
P^3 Rod to reverse engine.
Q Globe valve.

assurance can only be given by some one thoroughly trained in the science of geology. There are pseudo-geologists, so-called practical geologists, who can lay small claim to any real knowledge of the science and such men have done much harm to the industry as well as discredit to the science. But so strongly intrenched has the science of geology become in the oil industry that some large companies keep in their employ more than one hundred geologists many of whom have attained high rank in the profession.

Securing Leases. After an oil company has determined the location of favorable geologic structure, leases covering the area are secured as rapidly

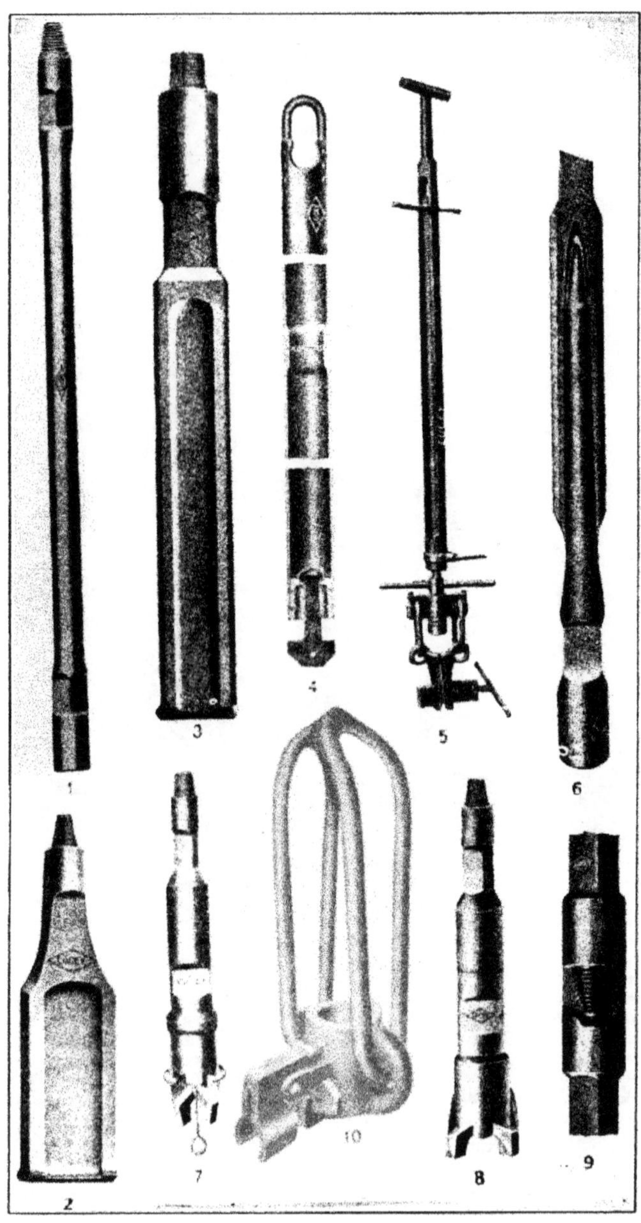

Fig. CVII. Drilling tools. 1. Auger stem; 2. spudding bit; 3. drilling bit; 4. bailer; 5. temper screw; 6. drilling jars; 7-8. underreamer, closed and open; 9. joint; 10. elevator, for lifting casing into derrick. (Lucey.)

as possible. The leases are in the nature of written agreements between the owner of the land and the oil company. The terms of such agreements vary greatly in different states and even in different parts of the same state. The lease gives a description of the land covered by the lease, duration of the lease, and states the compensation to be received by the lessor. The property is usually described by the quarter section, town and range. The time of the duration of the lease may be from one to five years with the option of extending the lease to cover the period of production. The lessor is paid one dollar to make the agreement legally binding. His further compensation may take the form of a fixed rental per acre such as one-fourth of one dollar per acre annually in wild cat territory to many hundreds of dollars in proven territory. The compensation may take the form of a royalty of one-twelfth, one-eighth, or one-sixth of the production. In exceptional good territory an additional bonus of $100 to $300 per acre may be paid.

By the terms of some leases rentals do not begin until after the drilling of the first well which must occur before the expiration of a certain period, say two years. In leases providing for cash yearly rentals no provision is made for the completion of a well; it generally being considered to the advantage of the operator to prove his territory as soon as possible so as to avoid payment of unproductive rentals. Some leases provide for the time of beginning and finishing the first well.

The terms of the lease provide that the lessee shall have access to the land and the use of enough of the surface of the land for the establishing of his equipment and for conducting operations necessary to production. The lessor has the use of all land not necessary to the operations of the lessee. In the event of natural gas instead of oil being found on the property under lease, the owner of the property is protected by a clause in the lease which provides for the payment for the gas based on the number of cubic feet produced. Some leases provide for the payment of from $100 to $150 per year per well to the land holder and free gas for his use.

Locating the Wells. The location of the wells on the structure is a matter of considerable importance. The location of the first well should be chosen with care since a failure tends to condemn the entire structure. When gas and oil are present in an anticlinal structure, as gas, oil and water arrange themselves in the order of their specific gravities, gas may be expected in the highest portion of the porous stratum, oil farther down the dip and water still farther down the structure.

Locations along the crest or apex of the anticline may, under such circumstances yield gas and if oil is desired a location should be made farther down the dip. If gas is not present, oil may occupy the highest part of the porous layer and rest beneath the surface of the apex of the structure.

If the first well is productive, the second well is located near the first following the supposed trend of the structure. The distance between the wells should be governed by the thickness and the porosity of the oil sand. If the oil sand is thin and porous, the wells may be placed further apart, say 1,000 to 1,500 feet. If the oil sand is thick and not very porous, the wells may be placed 500 feet apart or even less. Some operators place one well to every ten acres. In the drilling of deep wells much money is wasted by close placing of wells.

Drilling Methods. Methods of drilling oil wells and the type of drill used varies with the depth of the wells, the character of the rocks penetrated, and other conditions. For moderately shallow wells in soft strata the portable type of drill may be used. Such rigs are easily transported over rough roads and rapidly put down to depths not exceeding 1,200 feet, but wells have been put down to depths of 2,500 feet by the use of such rigs.

The rig most in use for the drilling of deep wells is known as the "Standard" which consists of a derrick, with walking beam, bull wheel, cable with tools attached, and other accessories. (Figure CVI.) The derrick may be either a steel frame (Figure CV) or wood, but consists of four uprights converging toward the top and tied and braced at intervals with cross pieces. The height of the derrick is usually 70 or more feet, about 20 feet wide at the bottom and four feet at the top. The bottom of the derrick rests upon large beams, rocks, or concrete and supports at the top, the crown block bearing the pulleys for the cables attached to the drill and sand pump.

The cable, composed of manilla or wire is wound upon the shaft of the bull wheel, while one end passes over the crown pulley at the top of the derrick and down to the end of the walking beam, to which the temper screw is attached by one end, the other end is clamped to the cable. (Figure CVI.) To the end of the cable is attached the string of tools which consists of the rope socket, sinker bar, jars, auger stem and auger. (Figure CVIII.) The walking beam is pivoted at the middle to an upright post and is attached by a pitman rod to a crank on the band wheel. The motion of the band wheel moves the walking beam up and down alternately lifting and dropping the auger and string of tools in the bore. As the bore is deepened the temper screw (Figure CVIII) is turned until the bore has increased in depth a full screw length, about five feet, when the temper screw is unclamped from the cable, the latter is wound on the bull wheel shaft and the tools are lifted from the well. The well is then bailed by lowering a sand pump or a bailer into the well by a line passing over the sand-reel pulley, allowing it to fill and elevating it to the surface by the same line. The bailer consists of a cylindrical body of galvanized iron with a bail at the top and a stem valve at the bottom. When the stem rests on the bottom of the bore it raises the valve and allows the bailer to fill, but when lifted from the bottom the valve drops into place and the water and drillings are carried to the surface and allowed to escape as the stem of the valve rests on the bottom of the water trough.

An engine and boiler are necessary to furnish power to the drill, the engine being connected to the band wheel by a belt. The fuel used for the boiler may be coal, oil or gas. Water for the boiler may be supplied from wells, springs, streams, or ponds.

Drive Pipe and Casing. Whenever a well is started in loose rock such as glacial drift or forms of mantle rock, a large iron pipe called drive pipe is forced through the mantle rock, following the drill and set on the solid bed rock. This pipe prevents caving of the soft strata and keeps water out of the drill hole. If, during the process of drilling, a porous layer is encountered, containing water under pressure, it may be necessary to lower the string of casing inside the drive pipe and set it on an impervious layer below the water bearing layer in order to shut out the water. If other water bearing layers are encountered, other strings of casings must be lowered. In deep wells it is often necessary to have eight or ten different sizes of casings, starting with an 18-inch casing and ending with a 2-inch.

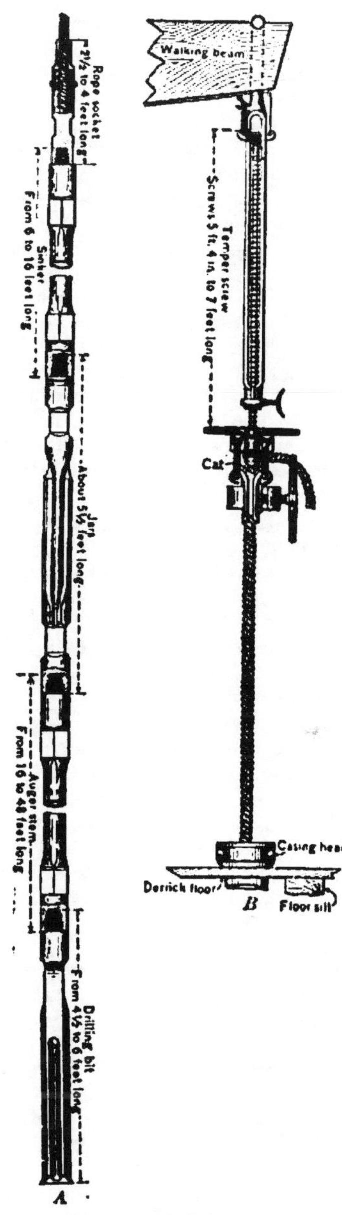

Fig. CVIII. A, string of tools used with standard drilling outfit; B, temper screw. (After Bowman, U. S. G. S.)

Cost of Oil Wells. The cost of an oil well varies with a number of factors, such as depth, character of rock, accessibility to fuel, transportation conditions and others. The cost of work preliminary to the actual drilling is the same regardless of the depth of the well, providing the same type of rig is used for both shallow and deep wells. The cost of actual drilling per foot increases with the depth. The light portable rig which may be used to advantage in Indiana in drilling wells ranging up to 1,200 or 1,500 feet in depth and has been used in wells as deep as 2,400 feet, cost $2,500 to $3,000. The standard rigs because of the construction of the derrick, cost much more. The cost of wells having depths ranging from 800 to 1,000 feet is from $2,000 to $2,500. Wells of twice those depths, cost from $6,000 to $8,000. Drillers usually contract to drill a well to a certain depth at so much per foot for the drilling and installing the casing, which is to be furnished by the owner of the well. The cost of casing varies from $1 per foot for the smaller sizes to $3.50 per foot for the larger sizes. In the glaciated regions of Indiana the largest tubing, the so-called drive pipe, must extend the full thickness of the glacial drift and be set on the solid bed rock. The length of the drive pipe in this region varies from a few feet to more than four hundred feet. In the non-glaciated region except in the alluvial bottoms of rivers the drive pipe rarely exceeds one section of pipe.

A written contract is usually made between the driller and the operator. This contract binds the driller to drill to a certain depth for a certain specified sum per foot; to furnish all necessary equipment; to begin drilling within a certain specified period; to install the casing and to pull it in case of a dry hole. It binds the operator to furnish on the ground the drive pipe, casing, rodding, tubing and other accessories except such as are a part of the drilling equipment; he also allows the driller the use for fuel the oil or gas which exists or may be found in drilling.

Abandoning a Well. If a well is dry or the production too light to be profitable and the well is to be abandoned it must be plugged. The laws of Indiana provide that before the casing can be drawn from a well and abandoned, the nearest State Gas Inspector shall be notified and his presence secured. Under his direction the casing may be drawn and the well plugged.

Shooting Oil Wells. If after an oil well has reached pay sand the oil does not flow freely into the well as it is not likely to do in case of a close-textured rock it becomes necessary to shoot the well. Shooting is accomplished by lowering to the position of the oil sand a charge of nitroglycerine in canisters. The amount of nitroglycerine used will depend upon the texture of the rock, the thickness of the pay sand, danger of flooding and other factors. The amount ordinarily used is from 60 to 100 quarts but the amount may be more or less. The explosive may be exploded by placing a fulminate cap on the charge in the well and dropping a conical iron, the "go-devil" upon it or by dropping a nitroglycerine "jack squib" bearing a fulminate cap upon the charge in the well (Figure CXII). Care must be taken not to get the charge below the pay sand because of the danger of flooding or of getting it above the pay sand in which case the shattered barren rock may interfere with production.

Pumping Oil Wells. When oil exists in the oil sand under great pressure it may be forced to the surface and a flowing well produced. Even a flowing well by decrease of pressure may cease to flow and require pumping. Some wells require pumping from start. Wells may be pumped by separate power units or by central power units. A very common practice is to connect a number of wells, say six, with a central power plant by means of rods

Fig. CIX. Standard pumping jack.

Fig. CX. Steel pumping jack. (Ill. Geol. survey.)

which are attached at the well to pumping "jacks" which transform the horizontal pull of the rods into vertical movement of the pump rods in the well. (Figures CIX and CX.)

Fig. CXI. View of an oil field in Indiana. (Amer. Inst. Min. Engineers.)

Fig. CXII. Broad Ripple oil well after shooting.

Oil Transportation. The most efficient method of oil transportation is by pipe line, pipes laid underground through which oil is pumped. Pipe lines now carry oil from the mid-continental field to the Atlantic Coast. The pipe of the main lines has a diameter of eight inches and the feeders from three to six inches. Pumping stations are distributed at intervals along the main lines. Oil is also transported from the oil field to the refineries by tank cars and tank ships. Some oils, like certain Mexican oils, are too dense to be transported long distances through pipes and such oils are transported in tank cars or tank ships.

Oil Storage. Oil as it is brought from the wells, must be stored in tanks at least temporarily. If the oil field is near the refinery it may be pumped through pipe lines and kept moving from the field thus necessitating only temporary storage. When the field is located at a distance from the refinery and the means of transportation is by tank cars, large storage facilities are a necessity. Storage tanks are built of iron, wood or concrete, in cases of emergency reservoirs of earth, have been made. Tanks may be placed above or below ground. In some of the oil fields concrete tanks placed below ground are being constructed. Less evaporation and greater safety from fires, especially fires caused by lightning, are the claims made for them. The approximate dimensions of tanks of various capacity are given below:

Capacity in Barrels	Height in Feet	Diameter in Feet
5,000	20	40
10,000	30	49 7/12
20,000	30	70
30,000	30	86
55,000	30	115

The gauging tanks range in size from 25 to 100 barrels and the oil is measured in these before being pumped to the storage tanks.

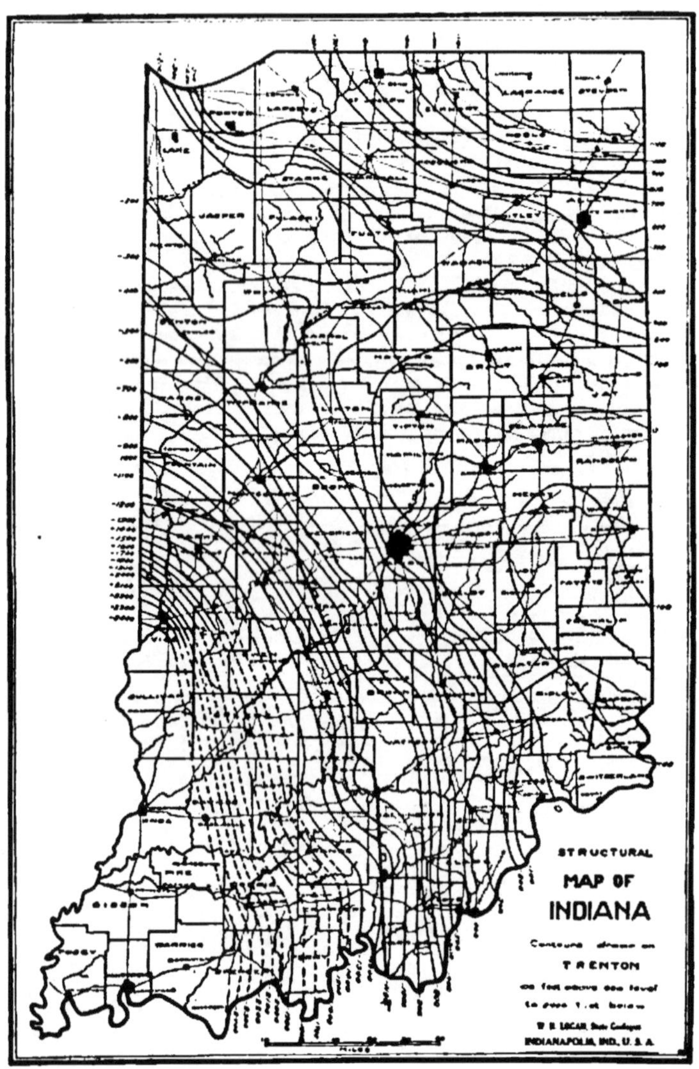

Fig. CXIV. Structural map of Indiana, contours drawn on Trenton.

PART VI.

GENERAL GEOLOGICAL CONDITIONS IN INDIANA.

The general geological conditions of Indiana are not complex. The rocks belong to the sedimentary division. The only rocks of igneous origin known in the State are the bowlders which were carried into the State from the crystalline belt of rock lying far to the north. During a great part of the time that the rocks of Indiana were being deposited the sea occupied the whole or a part of the State. In this sea the fragments of disintegrated rocks of former ages were deposited to contribute to the strata which were later to form the surface of the State. The movement which was to convert the marine Indiana into dry land began on the eastern border and extended across the State northwesterly. Because of this differential uplift the southwestern and the northeastern corners of the State were the last portions to emerge from a gradually retreating sea. Though it is possible the emergence of the northeast corner may have antedated that of the southwest. (See next page for table.)

Potsdam Sandstone. The oldest rock reached by the drill in Indiana is a sandstone which is probably of the age of the Potsdam sandstone of the Cambrian period. Oil or gas has not been found in this formation in this or in the neighboring States. The formation does not outcrop at any point within the State. Wells have penetrated it to a depth of 300 feet without pasing through it.

Lower Magnesian Limestone. Overlying the Potsdam sandstone is a limestone which is thought to be of the age of the Lower Magnesian. No outcrop of the formation occurs within the State. Its thickness as recorded in well records is about 300 feet. It is thought to be equivalent in age to the Calciferous of the New York section.

St. Peter Sandstone. A number of deep wells in Indiana have passed through the Trenton limestone and pierced a stratum of sandstone which has been referred to the St. Peter. The thickness of the sandstone as revealed by well records varies from 150 to 300 feet. It is thought to be equivalent in age to the Chazy of New York.

Trenton Limestone. Overlying the St. Peter sandstone is a limestone which has been the source of the larger part of the oil and gas produced in the State. Portions of the upper part of the limestone have been rendered porous by dolomitization and where the structural conditions of the formation have been favorable oil or gas has been collected in these porous portions. The thickness of the Trenton limestone varies from 470 to 586 feet.

The geological formations which outcrop at the surface or have been revealed in deep wells in Indiana are given in the accompanying table.[1]

[1] For more complete discussions of the subdivisions represented in this table see reports by Ashley, Cumings, Foerste, Newsom, Price, Siebenthal, and others published in the Annual Reports of the Survey. For the subdivisions of the Chester see paper on "The American Bottoms," Indiana Studies, by C. A. Malott. See Part II of this work.

GEOLOGICAL SECTION OF INDIANA.

Area	Period	Epoch	Formation	
Cenozoic	Quaternary	Recent	Alluvium, residual clays.	
		Pleistocene	Glacial drift.	
	Tertiary	Pliocene?	Gravels (Lafayette?).	
Paleozoic	Pennsylvanian	Allegheny	Merom sandstone.	
			Coal measures, coal, shale, etc.	
		Pottsville	Mansfield sandstone.	
		Unconformity		
	Mississippian	Chester	Limestones, sandstones and shales.	
		St. Genevieve	Mitchell limestone.	
		St. Louis		
		Salem	Salem limestone.	
		Osage	Harrodsburg (Warsaw limestone).	
			Knobstone shales.	
			Rockford (Goniatite limestone).	
		Unconformity		
	Devonian	Corniferous	New Albany shale.	
			Sellersburg limestone.	
			(Silver Creek limestone. Beechwood,	
			Geneva (Jeffersonville limestone).	
		Unconformity		
	Silurian		Louisville limestone.	
			Waldron shale.	
			Laurel limestone.	
			Osgood limestone and shale.	
			Brassfield.	
		Unconformity		
	Ordovician	Cincinnatian	Richmond	Elkhorn.
				Whitewater sh. and ls.
				Saluda, sh. and ls.
				Liberty, limestone.
				Waynesville, sh. and ls.
			Maysville	Arnheim, shale.
				Mt. Auburn, limestone.
				Corryville, limestone.
				Bellevue, sh. ls. ss.
				Fairmount, sh. and ls.
				Mt. Hope, sh. and ls.
			Eden	McMicken, sh. and ss
				Southgate, sh. ss. ls
				Economy.
				Fulton.
			Trenton limestone.	
			St. Peter sandstone.	
			Magnesian limestone.	
	Cambrian		Potsdam sandstone	

Geological Section of Indiana.

Cincinnatian. The group of limestones, shales and sandstones overlying the Trenton are usually referred to as the Utica and Hudson River shales in report pertaining to the oil industry of the State. These formations outcrop in the southeastern part of the State and they have been studied and their lithological and paleontological characters determined. The total thickness of the strata of this group is about 700 feet.

Silurian Strata. The formations belonging to the Silurian in Indiana con-

Fig. CXV. Public road in the "Knobstone" formation near New Albany. Photo by Hohenberger.

sist chiefly of limestones with thin layers of calcareous shales. In the records of oil wells they are commonly referred to under the head "Niagara limestone." Over much of the oil and gas territory in the eastern part of Indiana the first stratum of the durolith (bed rock) encountered by the drill is the Silurian limestone. The Silurian strata outcrop in the southeastern portion of the State and in the eastern portion where erosion has removed the glacial drift. The divisions represented in southern Indiana are: Brassfield limestone (Medina), the Osgood limestones and shales, the Laurel limestone, the Waldron shale and the Louisville limestone. The thickness of the Silurian in southern Indiana varies from 95 to 140 feet. The Waterlime is supposed to be represented in northern Indiana and the Schoharie by the Pendleton sandstone.

Devonian Strata. The lower portion of the Devonian consists of the Jeffersonville, the Silver Creek and the Sellersburg limestones. These outcrop in Clark, Jennings and other counties in the southern part of the State where they attain a total thickness of about ninety feet. In the well records these limestones are usually referred to as the Corniferous, though they are probably largely Hamilton. In many places it is sufficiently porous to

Fig. CXVI. An Oölitic (Salem) limestone quarry. The overburden which has been removed is Mitchell limestone. The first cut is being made in the upper surface of the Salem.

allow the accumulation of oil and gas where structural conditions are favorable and some oil and gas production in Indiana is derived from the Corniferous. Above the Devonian limestone lies a black bituminous shale called the "New Albany" which is supposed to be of equivalent age to the Genesee of the New York section.

Mississippian Strata. The lowermost division resting on the New Albany is the Goniatite or Rockford limestone, a thin stratum, often only two feet thick, greenish color on fresh fracture but weathers brown. Overlying the Rockford is the New Providence shale member which is followed by the Knobstone shales and sandstones, containing some lenses of limestone. The term, Riverside sandstone was applied by Foerste to a sandstone in the Knobstone. The Knobstone sandstones frequently contain pockets of gas and there is reason to believe they may form oil reservoirs. The thickness of the Knob-

Fig. CXVII. Cave in Mitchell limestone in Harrison County. Caves and underground water courses are abundant in this limestone in Indiana and Kentucky. (Photo by Hohenberger.)

stone varies from 530 to 650 feet. The Harrodsburg (Warsaw) limestone overlies the Knobstone. The line of contact is marked by a large quantity of quartz geodes. The crystals in the interior of the geodes are usually quartz but in some calcite. This member consists of thin bedded limestone and shales. The limestones are irregularly bedded, very fossiliferous, contain chert, stylolites and coarsely crystalline calcite. Its thickness is from 60 to

90 feet. The Salem furnishes the Indiana oölitic building stone. It occupies in its outcrop, a narrow strip extending from Putnam County to Harrison County, the main quarry district being located in Lawrence and Monroe counties. The limestone occurs in a massive bed usually varying in thickness from 30 to 90 feet. The stone is a fine grained limestone, the grains being composed

Fig. CXVIII. An outcrop of Mansfield sandstone showing differential weathering, the more resistant parts are cemented with iron oxides. This forms one of the oil sands of southwestern Indiana. (Photo by P. B. Stockdale.)

of shells or fragments of shells. It is generally recognized by its massiveness and granular (so-called oölitic) structure.

The Mitchell is composed chiefly of limestone with some thin beds of shale in its upper horizon. It is a harder limestone than the oölitic and is used much for road material. The individual beds of the limestone vary from two to thirty feet in thickness. Some of the layers of the upper portion contain inclusions of chert. Fine grained, lithographic stone is present in some horizons. The thickness of the Mitchell varies from 150 to 200 feet.

The Chester is composed of a series of sandstones, limestones and shales. The sandstones become oil reservoirs in southwestern Indiana, a portion of the oil production of that region being derived from them. Some of the shales of the Chester are oil-bearing though they do not form reservoirs. Some of the limestones are of an oolitic character and some are lithographic. The sandstones are fine grained and are usually distinguishable from the coarser grained Mansfield.

Pennsylvanian Strata. A long period of erosion preceded the deposition of the Pennsylvanian rocks, and the surface of the Mississippian upon which the Pottsville rocks were deposited was very irregular. The Pottsville division is represented by beds of shale, thin beds of coal and a coarse sandstone, the Mansfield. The latter is often conglomeratic and in some places contains irregular masses of limonite. The Mansfield sandstone becomes an oil reservoir in the southwestern part of the State. Many of the shales associated with the coals of the Pottsville are oil bearing. There is an unconformity between the Pottsville and Allegheny divisions in Indiana which in some places is well marked.

Coal Measures. The rocks of the Allegheny division consist of shales, coals, limestones and sandstones. Many of the shales are oil bearing under destructive distillation. The sandstones furnish reservoirs in which oil and gas have accumulated at points where structural conditions are favorable. Many of the productive sands in Gibson and Pike counties belong to the Coal Measures.

Merom Sandstone. This sandstone rests unconformably upon the Coal Measures in some places occupying erosion channels carved in the rocks of the Coal Measures. This sandstone is conglomeratic in its basal portions in some localities.

Tertiary. Some gravel beds which occur in southern Indiana consisting chiefly of chert and flint gravels with geodes probably belong to the Pliocene epoch of the Tertiary Period.

Quaternary. The Pleistocene or glacial deposits cover a large part of the surface of Indiana. There is an area in the southern part of the State lying south of the north line of Monroe County where the two lobes of the Illinoian glacier did not coalesce that was not glaciated. The deposit of glacial drift reaches a thickness of more than 400 feet in places. The presence of the drift has greatly interfered with the development of the oil and gas industry since it concealed the outcrop of the durolith and prevented the determination of structural conditions by direct observation. The strata of the Cincinnati geanticline are buried under the drift and its minor structural irregularities concealed.

The Recent deposits consist of residual clays, loam and soils formed from the decomposition of the durolith, alluvial deposits of the stream valleys, dunes of wind blown sand and marl and peat deposits.

Structural Features of Indiana. The major structural features of Indiana are comprised in the Cincinnati geanticline, the northern basin, the western basin and the Mount Carmel Fault.

The Cincinnati Geanticline which extends northward in Ohio sends off an arm which passes through Indiana in a northwesterly direction. The movement which inaugurated the arching took place during the Ordovician period and continued until the close of the Carboniferous Period but while the movement resulted probably in land condition being produced in southern Ohio, the effect in Indiana was the production of a sub-marine ridge on the slopes and across the top of which the sediments of later periods were deposited. This ridge formed the dividing line between a basin on the north and one on the southwest. The younger rocks dip away from the ridge toward these basins. Sediments of Cambrian and Ordovician age were deposited on the eroded Pre-Cambrian surface before the elevation of the Cincinnati Arch. Through well records we learn that below the Trenton limestone which has a thickness of 500 or more feet there lies a sandstone which probably corresponds to the St. Peter sandstone which outcrops in Wisconsin. Its thickness varies from 150 to 300 feet. That below the sandstone there is a limestone which probably corresponds in age to the Lower Magnesium limestone which has a thickness of about 300 feet and rests on the Potsdam sandstone which has a thickness of more than 300 feet. The Potsdam sandstone belongs to the Cambrian period and is the oldest rock known to occur in situ in Indiana.

The Northern Basin. The center of the northern basin lies north of Indiana about Bay City, Michigan. The southern limit of the basin is the Cincinnati Arch which passes across the State in a northwesterly direction. The sediments deposited in this basin range in age from the Silurian to and including the Coal Measures of the Pennsylvanian. It is very probable that the sediments of these formations were continuous across the arch at one time, but if so, they have been removed by erosion as only the Silurian rocks now rest below the drift and overlie the Ordovician on the top of the Arch. The dip of the strata from the top of the Arch northward is gentle at first not exceeding ten feet to the mile but the dip increases until it reaches thirty or more feet to the mile.

The Southwestern Basin. This basin has its center in southern Illinois toward which the formations laid down on the western and southern flanks of the Cincinnati Arch dip. The dip of the formations varies from thirty to fifty feet to the mile, perhaps in a few places exceeding fifty feet. The total thickness of the sediments deposited in this basin in Indiana on top of the Trenton is probably as much as 3,500 feet.

The Mount Carmel Fault. Early in the fall of 1916 the attention of the writer was attracted to a reversal of dip in some beds of limestone lying in eastern part of Monroe County. In places, this reversal of dip was noticeable in the limestones which overlie the Knobstone shales and and sandstones, in other places in the sandstones of the Knobstone and again in beds of limestone occupying certain horizons in the Knobstone. Upon an investigation of the available geological literature I found in the Report of the State Geologist for 1896, pages 390-91, that Siebenthal discusses the Heltonville Limestone Strip as follows: "Commencing at Limestone Hill, eight miles southeast of Bloomington and extending east of southeast through Heltonville to, and probably beyond Fort Ritner, Lawrence County, is a band of limestone from one-half to one and a half miles in width, bordered sharply, both east and west, by Knobstone, and known in that neighborhood as the Limestone Strip. Isolated patches of similar limestone occur north of this strip and in line with it. The strip is well developed in the vicinity of Heltonvillé, Lawrence County, where it gives exposures of the Harrodsburg, Bedford Oolitic and Mitchell limestones.

"At many points the Knobstone contains intercalated lenticular beds of limestone, and it is possibly conceivable that the conditions which prevailed while these beds were being deposited might have been extended over a narrow territory like the Heltonville strip. However, the fact, first that Knobstone has not been found overlying this limestone, and second, that it shows the lithological facies of the Harrodsburg, the Bedford Oolitic and the Mitchell limestones, and the faunas of these formations, identifies it with them and shows conclusively that it is a narrow band of these formations, occupying a depression in the Knobstone, and not an included member of the Knobstone.

"This depression may have resulted from a double fault or may be an old erosion channel. Some things seem to point to one as the origin and some to the other. The facts at hand incline us to the latter view. The most palpable objection to this view is the fact that no nonconformity exists between the Knobstone and the Harrodsburg limestone at their contact a few miles west of the strip. Another objection is that the bottom of the channel, at present at least, is not all of uniform elevation throughout its length. The principal objections to the view of a double fault are two—at no point was a direct vertical contact of Knobstone and limestone visible, nor was there to be seen any of the tilting, crushing and shattering which usually accompanies faulting. On the other hand, as the vicinity of the contact line is approached the shaly layers of the limestone become more and more argillaceous and apparently pass over into the Knobstone. To determine the exact conditions under which the limestone strip was laid down would require more extended study than is consistent with the scope of this report. What has been done was to trace upon the accompanying maps the outcrop of the Bedford Oolitic and to examine the bed more carefully at places where it is now being quarried, namely at Heltonville and Fort Ritner."

In the proceedings of the Academy of Science of Indiana for 1897, page 262, J. A. Price discusses the boundary of the limestone strip and says in conclusion: "It is not possible, from data in hand, to say surely whether this strip of limestone owes its existence to an unconformity or a fault."

In 1903 J. F. Newsom published a description of a "Geologic Section Across Southern Indiana" as a part of the 26th Annual Report of the State Geologist. On pages 274 and 275 Newsom refers to the structure as a fault in the Knobstone area. He gives its extent as being from near Unionville in Monroe County to a point in the northern part of Washington County.

In referring to the discussions of Siebenthal and Price in the 27th Annual Report of the State Geologist, 1903, on page 90, Ashley says: "It is evident that if the limestone strip north of White River is due to a fault its effects should continue to the south rather than turn and follow the outcrop. A glance at the map in the region north of Campbellsburg is alone sufficient proof of the fault character of the disturbance."

In studying this structure in detail the writer has found that it is much more extensive than Newsom stated; that there is a second fault; that other disturbances were connected with it and that the actual contact which he has found presents some interesting features.

Extent of the Fault. While I have not yet been able to trace the fault to the borders of the State at either of its extremities I have been able to trace it far beyond its mentioned boundaries and feel confident that the particular disturbance under discussion extended from the Ohio to the Wabash along the western border of the Knobstone outcrop and perhaps beyond. Tracing the fault south of Campbellsburg in Washington County is difficult because the area on each side of the rift is occupied by limestone.

Along the northern end of the displacement glacial deposits conceal the bedrock to such an extent as to render observation difficult. Under these circumstances the best that can be done is to trace the disturbance by the reversal of dip of the limestones, as the finding of the rift will be extremely difficult. By such observations as it was possible to make I have traced the disturbance from a point southeast of Campbellsburg in Washington County to a point northwest of Waveland in Montgomery County.

Rift. The actual contact of the rocks along the fault plane is revealed in only a few places. There are numerous places where the harder more resistant stratum of limestone stands forth like a wall on one side of the rift, but the opposite side is occupied by mantle rock which was derived by the weathering of the Knobstone and which conceals the actual rift. Excavations made at such places would doubtless reveal the actual contact of the limestone and the Knobstone.

In a few localities the rift is exposed and the plane of the fault is bordered on the one side with limestone and on the other by shale. One outcrop of the rift zone was found in the bed of the north fork of Leatherwood Creek near Heltonville. At this point the Knobstone occurs on one side of the fault plane and the Harrodsburg limestone on the other. The line of rift is distinct, being marked by a thin bed of breccia. The brecciated zone is composed mainly of fragments of limestone in which small fragments of shale are intermingled. These fragments have been cemented together with calcite and the whole zone more or less marbleized. In a cross-section of the brecciated rock the veins of calcite stand out clearly, as they are whiter than the fragments of limestone and shale which they bind together. Small quantities of other minerals are present in some parts of the brecciated zone, but there is an absence of the more insoluble minerals, such as silica or the silicates. This fact leads to the conclusion that meteoric rather than thermal waters have played the leading role in the concentration of these minerals.

Periods of Movement. The question of whether the displacement took place all at one time or was intermittent is an interesting one. All of my attempts to find an evidence of intermittent movement by an examination of surface features have been unsuccessful. If there were intermittent movements of any considerable extent we would probably find them revealed in hanging valleys on the upthrow side and the rapid broadening of valleys on the downthrow side of the fault. In case there were two stages of movement, and the movement in the last stage an exceedingly slow one, the vertical cutting of the main stream might be as rapid as the uplift, but still the rejuvenation of the tributaries should result in a narrowing of the valleys. In the rift zone there is evidence of two stages of movement though the amount of displacement in the second stage is slight. The time interval between the two movements was of considerable length, since the fragments of the brecciated zone were firmly cemented before the second movement took place. Fragments of shale which were included in the limestone fragments during the first movement were faulted by the second movement. These shale inclusions would not have undergone faulting had they not been held rigidly in place by the cementing material.

Amount of Throw. The amount of throw of the fault varies probably from 200 to 300 feet. Opportunities for measuring the amount of throw are not numerous. It can best be computed by estimating the total amount of eastward dip of the formations along the line of contact between the Harrodsburg and the Knobstone. At a point south of Mt. Carmel the difference in elevation of the contact above sea level is 50 feet in a distance of one-fourth mile. Since the width of the down-thrown block is at least one mile and a half in this locality the throw of the fault is at least 300 feet. The amount of dip of the down-thrown beds in other localities is less than at this point, so much less that the indicated throw is not more than 200 feet.

Age of the Fault. The time at which the dislocation occurred can not be fixed definitely. It is probable that it occurred at the close of the Paleozoic Era when the Appalachian revolution which resulted in the elevation of the eastern part of North America took place. Contemporaneous with or subsequent to that great epeirogenic movement, faulting and minor folding took place in Indiana, Illinois and Iowa, and other States lying as far west as these from the region of maximum disturbance. These faults like the one under discussion have a northwest disturbance.

The Heltonville Fault. About one mile west of the Mt. Carmel fault there is a second fault. This I have named the Heltonville Fault because the rift is exposed a short distance east of Heltonville in the bed of the north fork of Leatherwood Creek, at a point just east of the wagon-crossing under the Southern Indiana railroad. This fault lies approximately parallel with the Mt. Carmel fault. The limestone has been faulted down against the Knobstone. Slickenslides have been produced in the limestone and it has been much fractured. In places the limestone has been thrust backward and fragments of the Knobstone shales have been thrust into the limestone. In places these formations are dovetailed, fingers of limestone projecting into the Knobstone and vice versa as first one and then the other yielded to the pressure. The fragments of limestone containing inclusions of shale have been united by calcite veins.

Though the fault character of the disturbance at this point is incontestable it is not equally clear at other points. The disturbance extends both north and south of this point, but it probably passes into a fold in both directions. In Monroe County near Unionville there is an anticline which occupies about the same position in relation to the Mt. Carmel Fault as the Heltonville fault does. Similar folds have been noted at intervening points and also to the south of Heltonville.

Effect Upon Topography. The general effect upon topographic conditions within the area of disturbance has been to produce a narrow limestone belt extending parallel with the main Knobstone outcrop and bordered on each side by outcrops of Knobstone. In the southern portion of the faulted area the western belt of Knobstone is absent, but its nearness to the surface along the line of the eastward reversal of dip is revealed in the channels of many streams which have carved their valleys at right angles to the line of reversal. Probably the most marked effect is on the drainage. Both surface and underground drainage lines are affected. In the faulted area the ground waters which have found their way through the limestone have a tendency to follow the eastward sloping surface of the Knobstone to the rift, and near this point often come to the surface in a stream valley which lies near the rift and generally parallel with it. This tendency of the underground streams is modified by local dips of the strata north or south.

The surface streams, especially those along the line of the fault plane, have been influenced by the displacement. They have worked off the harder limestones on to the Knobstone in many places. These follow the line of rift until a local north or south dip has caused them to change the direction of their course. Small tributaries of the larger cross-cutting streams have developed, as has been noted again and again, along the line of rift.

The Mount Carmel Fault is one of the most important structural features in Indiana. It extends from near the Ohio River northward to the north part of Putnam County and possibly extends in a westerly and northwesterly direction from that point to the western boundary of the State. The extent of its throw in places exceeds two hundred feet. In a general way it parallels the western limits of the Knobstone outcrop. The downthrow side is west of the fault line. The faulting and the subsequent erosion has resulted in a limestone belt bordered on the east and west by Knobstone, the limestone being on the downthrow side and thus protected from the erosion which caused the removal of the limestone of the same age lying at a higher elevation both east and west. Since the normal dip of the rocks is southwest the downward drop of the block toward the east resulted in a fold lying parallel with the fault plane to the west. As the fault changes its directions in some places north and south components of dip are produced in the fold at such places and conditions favorable for the accumulation of oil and gas produced. One such place occurs in Lawrence County and considerable gas and a showing of oil obtained west of Leesville. Another favorable structure exists near Unionville in Monroe County.

PART VII.

DISCUSSION BY COUNTIES.

The following pages contain a discussion by counties of the oil and gas development and possibilities in each county. The report is only preliminary and the information pertaining to some of the counties is very meager. Much information has been obtained since the preliminary publication of this part of the volume but only a small part of the additional information could be included because of the large quantity of other material to be inserted. The excluded information on petroleum and natural gas will be included in a future publication.

ADAMS COUNTY.

Adams County lies within the glaciated area of Indiana, hence its bed rock (durolith) is covered with a thick over-burden of glacial drift (regolith). The latter varies in thickness from a few feet to eighty or more. It conceals the eroded surface of the Silurian (Niagara limestone). Beneath the Silurian strata lie the shales of the Ordovician which rest upon the Trenton limestone, within porous portions of which oil has been found in this county. The structural conditions cannot be determined in this county by surficial observations. Enough wells have been drilled in the county to furnish sufficient data for

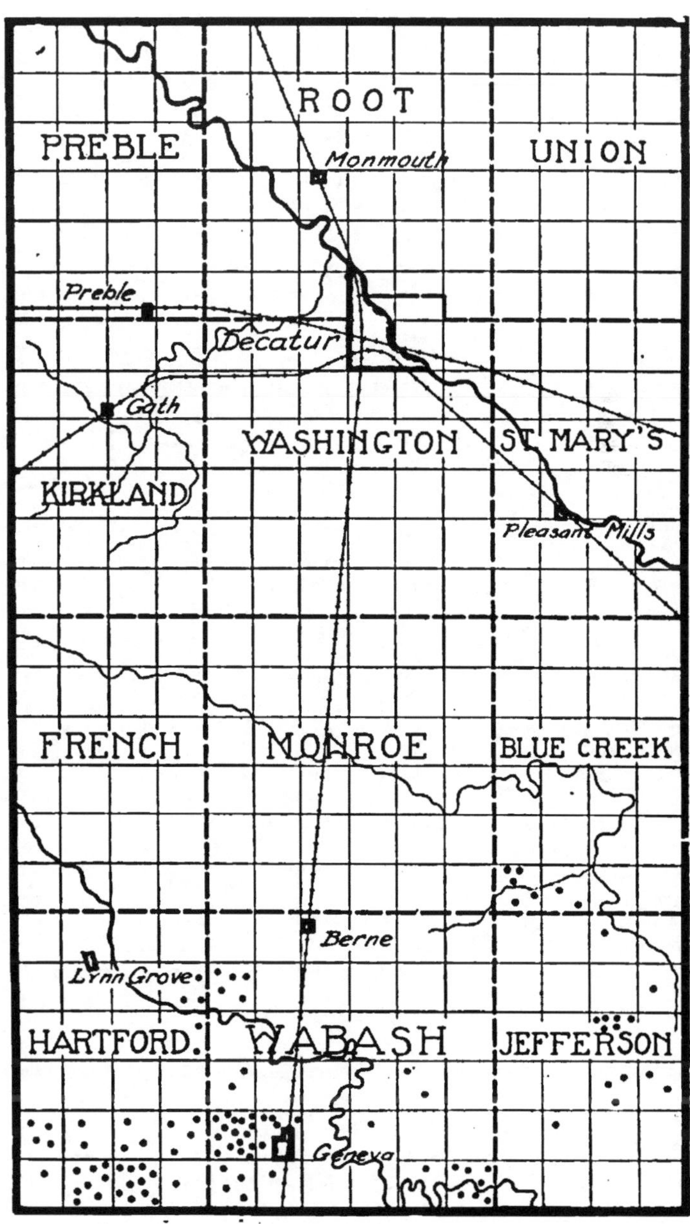

Fig. CXIX. Map of Adams County showing location of wells. The southern tier of townships is in gas and oil territory.

outlining the structural conditions, but unfortunately these records have not been preserved, and so the minor irregularities on the surface of the geanticline cannot be located.

Railroad Elevations.

GRAND RAPIDS AND INDIANA RAILROAD.

Location	Feet above Sea Level	Location	Feet above Sea Level
Geneva	840.5	67th mile post	812.6
55th mile post	849.7	68th mile post	808.7
56th mile post	833.2	69th mile post	803.5
57th mile post	845.2	70th mile post	801.8
58th mile post	847.4	Decatur	799.2
59th mile post	838.7	71st mile post	797.7
60th mile post	841.3	72d mile post	786.4
61st mile post	840.3	73rd mile post	794.0
62d mile post	826.2	Monmouth	789.7
63d mile post	839.2	74th mile post	789.0
64th mile post	825.3	75th mile post	810.0
Monroe	823.8	76th mile post	817.5
65th mile post	823.2	77th mile post	816.1
66th mile post	817.2	Williams	826.2

TOLEDO, ST. LOUIS AND WESTERN RAILROAD.

Location	Feet above Sea Level	Location	Feet above Sea Level
State line	800.7	112th mile post	815.6
101st mile post	795.9	113th mile post	822.0
102d mile post	802.7	Peterson	817.0
Pleasant Mills	799.4	114th mile post	823.8
104th mile post	797.0	115th mile post	829.0
105th mile post	800.1	116th mile post	835.6
106th mile post	804.0	117th mile post	846.3
107th mile post	802.2	118th mile post	851.0
108th mile post	795.6	119th mile post	859.5
Decatur	800.3	120th mile post	855.2
109th mile post	804.9	121st mile post	830.3
110th mile post	809.0	122d mile post	824.0
111th mile post	815.0	123d mile post	814.5

CHICAGO AND ERIE LINE.

Location	Feet above Sea Level	Location	Feet above Sea Level
Bridge No 49	799.0	Bridge No. 53	800.0
Decatur	799.0	Bridge No. 56	809.0
Magley	830.0		

Oil has been produced in the southern tier of townships and in Blue Creek Township. The production was heaviest in Hartford Township.

Washington Township. The following is the record of a well drilled at Decatur as given by Phinney[1]:

Decatur Well.

Drift	47 feet
Limestone	436 feet
Bluish shale	667 feet
Black shale	110 feet
Trenton limestone	40 feet
Total depth	1300 feet
Altitude of well	800 feet

Blue Creek Township. Wells in sections 8, 9, 10, 15, 16, 17, 21, 22, 27, 28, 29, 30, 31, 32, 33 and 34. Light oil production was obtained in 15, 16, 22, 27, 29, 30, 31, 32 and 34. In 1916 five wells were abandoned in section 31 and two in section 32. Dry holes were drilled in sections 8, 9, 10, 15, 17, 21, 28, 29, 30 and 33. Gas was obtained in section 16.

Hartford Township. The most productive territory was found in this township. Oil production was obtained in sections 12 to 36, inclusive. Dry holes were drilled in sections 4, 7, 8, 12, 14, 15, 16, 17, 18, 22 and 23. Some of the wells had an initial production of 180 barrels per day. Thirteen wells were drilled in the northeast quarter of section 25, the average depth of the Trenton being 1004 and the average initial production being one hundred barrels per day[2]. The record of a well drilled on the southwest quarter of section 25 is given by Blatchley[3] as follows:

Record of Well in Section 25.

Drive pipe	110 feet
Casing	230 feet
Trenton struck at	996 feet

Initial production, 150 barrels.
Production in October, 1896, two barrels.

A large number of wells have been abandoned in this township; a partial list is given below:

The wells abandoned in this township are located as follows:

Sec.	Wells	Sec.	Wells	Sec.	Wells
12	3	25	4	34	8
13	2	26	3	35	10
17	1	28	4	36	1
20	1	33	1		

Jefferson Township. Production has been obtained in this township from sections 4, 5, 6, 10, 16, 18, 19, 20, 21, 22, 27, 28, 29, 30, 31, 32 and 34. Dry holes were drilled in sections 3, 7, 8, 10, 15, 16, 17, 18, 22 and 33. Gas was obtained in 16 and 34. The initial production of oil ranged as high as one hundred barrels per day. Abandoned wells are located in section 4, one well; section 10, one well; section 16, seven wells; section 21, three wells; section 22, two wells; section 29, one well.

Wabash Township. Light production was obtained in sections 18, 19, 20, 27, 28, 29, 30, 31, 32 and 36. Dry holes were drilled in sections 1, 2, 3, 5, 6, 7,

8, 9, 10, 11, 12, 13, 14, 15, 16, 17, 18, 19, 20, 21, 22, 24, 25, 32, 33, 35 and 36. A partial list of the wells abandoned is given below:

Sec.	Wells	Sec.	Wells	Sec.	Wells	Sec.	Wells
7	5	27	1	31	4	36	7
19	2	29	5	32	1		
23	2	30	23	35	1		

In Adams County 880 wells have been abandoned, only a partial list of which has been recorded.

ALLEN COUNTY.

The durolith of the northern portion of Allen County is composed of strata of Devonian age, but for the remainder of the county it is composed of Silurian strata. The regolith, which is composed mainly of glacial drift, varies in thickness from one hundred to three hundred feet. Irregularities in the surface of the durolith, irregularities of decomposition and post glacial erosion, account for the difference in thickness of the drift. This county lies largely on the side of the Cincinnati arch dipping toward the north basin. The dip of the surface of the Trenton northward from Ft. Wayne is at the rate of twelve and one-half feet to the mile. At Ft. Wayne the surface of the Trenton lies about 650 feet below sea level, at Stoners near the north line of the county, it is 860 feet and near New Haven it is about 680 feet below sea level. Not enough well records are available to determine accurately the structural conditions and subsurface work is the only possible source of information on account of the concealment of the durolith. The following are some of the railroad elevations in the county:

Railroad Elevations.

Fort Wayne	757.3	Fort Wayne	779.0	New Haven	758.6
State Line	757.8	Dixon	793.5	Gorham	817.8
Fort Wayne	765.1	Carroll	852.6	Washington	811.9
East Yard	802.3	Dawkins	769.1	Maples	790.5
Huntertown	841.6	Hoagland	826.7	Stoner's	837.7
Academie	829.9	Junction	764.4	Wab. Crossing	757.8
Edgerton	758.3	Monroeville	789.6	Adams	790.9
Hadley	840.1	Huntertown	781.1	Wallen	854.6
				Adams	791.9

Wayne Township. Four wells were drilled at Fort Wayne. They range in depth from 1000 to 3000 feet. The records' of Nos. 1 and 2 follow:

Section of Well No. 1, Nov. 18, 1886.

Drift	77 feet
Waterlime	30 feet
Niagara	570 feet
Hudson River and Utica	751 feet
Trenton limestone	15 feet
Total depth	1443 feet
Trenton below sea level	693 feet

Gas with an initial pressure of 160 pounds per square inch was found upon entering the Trenton rock at a depth of 1428 feet; at a depth of 1431 feet a considerable quantity of oil was found.

Section of Well No. 2.

Drift	110 feet
Lower Helderburg	34 feet
Niagara limestone and shale	571 feet
Hudson River limestone and shale	410 feet
Utica shale	312 feet
Trenton limestone	21 feet
Total depth	1458 feet
Trenton below sea level	650 feet

Yielded no gas. Salt water, however, was found in considerable quantities.

Below is given the record of a well drilled in Perry Township[1]:

Section of Well Drilled on Sec. 4, Twp. 32, R. 12.

Surface above sea level	844 feet
Drift	281 feet
Limestone	749 feet
White shale	430 feet
Black shale	240 feet
Trenton limestone	52 feet
Total depth	1752 feet
Trenton below sea level	856 feet

Did not strike gas, oil or salt water. The dip of the surface of the Trenton from Fort Wayne to this point is about twelve and one-half feet to the mile.

Adams Township[1]. N. E. ¼ of section 14 in 1899 made a fair showing of oil, but a second bore resulted in a dry hole. A third bore resulted in a well, described below:

Drive pipe	96 feet
Casing	700 feet
Top of Trenton	1440 feet
Total depth	1496 feet

Several bores were drilled on the farms adjoining the above, but resulted in dry holes.

Jackson Township[1]. Section 3, bore completed on the Amspaugh farm, started with an output of twelve barrels per day. Section 33, a test well was drilled in October, 1903, which resulted in about eighteen barrels.

Monroe Township. A large showing of oil in section 3, also a big supply of gas; caught fire before the drilling was completed. A well in section 3, on the C. K. Dresser property was abandoned in 1919.

BARTHOLOMEW COUNTY.

The glacial drift covering Bartholomew County varies in thickness from five to more than one hundred feet. Underlying the drift in the eastern part of the county are strata of Silurian and lower Devonian age, while in the western part the strata are of the upper Devonian and lower Mississippian age. The Silurian rocks are limestones largely, the Devonian, shales and limestones, and the Mississippian shales and sandstones.

The structural conditions are not easily determined on account of the glacial drift which conceals the outcrop of the bed rock strata. If the proper geological structures exist, it is possible that oil and gas may be found in the Devonian and the Trenton in the western part of the county and from the Trenton in the eastern part of the county. The Trenton lies below the surface at depths ranging from 800 to 1200 feet.

The record of a well drilled at Columbus is given below:

Section of Well No. 1.[2]

Drift	26 feet
Devonian shale	87 feet
Corniferous limestone	32 feet
Niagara limestone	235 feet
Hudson River limestone and shale	440 feet
Utica shale	135 feet
Trenton limestone	155 feet
Total depth	1110 feet

Yielded no gas.

Elevation on Railroads.

Columbus, 627.3; Clifford, 668.3; St. Louis Crossing, 679.5; Wiggs, 615.9; Elizabethtown, 615.8; Waynesville, 601.7.

BENTON COUNTY.

Rock strata belonging to the Devonian, Mississippian and the Pennsylvanian periods underlie the Pleistocene deposits in Benton County. The latter attain a thickness of from 75 to 350 feet. The bed rock strata dip toward the southwest. The Trenton limestone may be reached at a depth of from 800 to 1100 feet, depending upon the surface elevation and location in the county. The structural conditions in the county cannot be determined by surficial methods, and the use of a large number of well records will be necessary in order to gain even a general idea of structural conditions. Without such data, prospecting for oil in this county will be, of necessity, with the drill and attended with exceptional risks.

The following is the reported record of a well at Fowler:

Section of Well No. 1.

Drift	280 feet
Devonian black shale	92 feet
Corniferous limestone	40 feet
Niagara limestone	328 feet
Hudson River and Utica	255 feet
Total depth	995 feet

Railroad Elevations.

Wadena, 800.0; Lochiel, 795; Barce, 808; Swanington, 796; Oxford, 736; State line, 706; Freeland, 720; Atkinson, 712; Gravel Hill, 780; Sheff, 727; Sheldon, 680; Iroquois, 649; Otterbein, 705.3; Vilas, 707; Templeton, 669; Fargo, 771; Chase, 738.3; Boswell, 756.3; Talbot, 763.8; Handy, 743; Ambia, 730.6. The elevations above given used with well records and records of outcrops and an aneroid barometer in the hands of a trained geologist may be the means of determining the structural conditions in this county.

BLACKFORD COUNTY.

The mantle rock in Blackford County is glacial drift varying in thickness from 15 to 150 feet. The drift rests on the Niagara limestone which has been eroded by preglacial streams and varies in thickness with the configuration of that surface. The Silurian (Niagaran) limestone has a thickness of 200 to 350 feet at least. The underlying Ordovician shales (Hudson River and Utican) reach a thickness of 600 feet, while the Trenton limestone has a thickness of about 500 feet.

Licking Township. Producing oil and gas wells have been drilled in this township. The following well records were reported by Gorby[1]:

Hartford City.

	Well No. 1	Well No. 2
Drift	130 feet	82 feet
Niagara limestone	350 feet	280 feet
Hudson River and Utica	473 feet	573 feet
Trenton limestone	30 feet	32 feet
Total depth	983 feet	967 feet
Trenton below sea level	70 feet	40 feet

The first gave a strong flow of gas and the second a very strong flow.

Another well located near the Fort Wayne and Muncie Railroad depot was reported by Phinney[2] as follows:

Drift	125 feet
Limestone	200 feet
Shale	622 feet
Trenton limestone	35 feet
Total depth	982 feet

The elevation of the station is 887.6 feet above sea level.

[1] Gorby, S. S. Ind. Geol. Sur. 1888, p. 247.

[2] Phinney, A. J., 11th Ann. Rept. U. S. G. S., p. 679.

Since the altitude of the mouth of the well is given at 895, the top of the Trenton would lie 52 feet below sea level. This well at first had a flow of gas of 850,000 cubic feet per day; by drilling deeper it was increased to 2,787,000 cubic feet per day. A second well was drilled half a mile southwest of the first and the Trenton reached at 935 feet. This well flowed 7,982,000 cubic feet per day.

A well drilled north of Hartford City reached gas at 980 feet and had a daily capacity of 6,383,000 cubic feet. Gas wells were located in this township

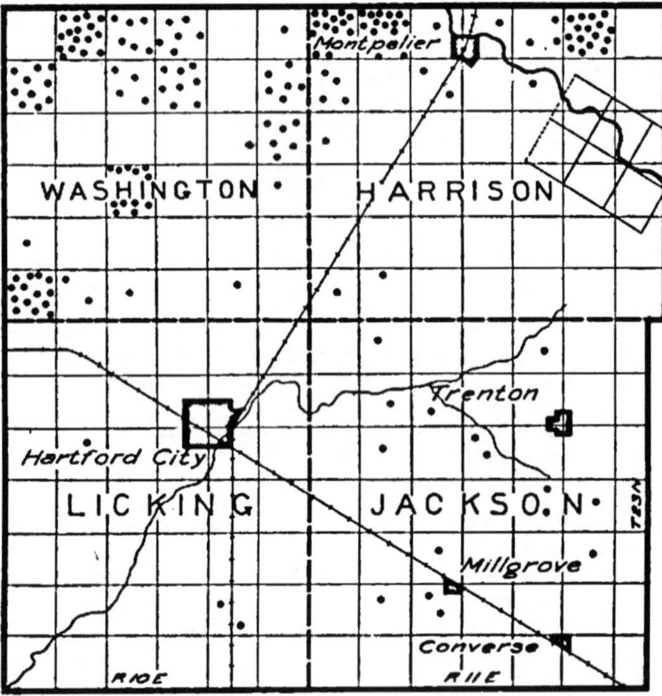

Fig. CXX. Map of Blackford County, showing abandoned wells. Washington and Harrison Townships were oil territory and Licking and Jackson gas territory. A little oil was produced in Licking in the northern part.

in sections 5, 7, 8, 17, 19, 20, 21 and 27. Oil wells were located in sections 1, 2, 3, 4, 5, 6, 7, 9, 10, 11, 14, 16, 18, 22 and 27. The following wells have been plugged: Section 1, one well; section 2, one well; section 6, nine wells; section 8, one well; section 10, four wells; section 11, three wells; section 14, one well; section 15, one well; section 17, one well; section 29, one well; section 35, two wells.

Harrison Township. A well drilled at Montpelier reported by Dr. C. Q. Sholl[1] gives the following section:

[1] Phinney, loc. cit.

1. Drift	16½	feet
2. Gray limestone	23	feet
3. Gravel, 14 ft. and red clay, 16 ft.	30	feet
4. Gray limestone	180	feet
5. Shale (Niagara)	38	feet
6. Bluish limestone	65	feet
7. Bluish shale	35	feet
8. Brownish limestone	35	feet
9. Bluish shale	35	feet
10. Gray limestone	8	feet
11. Bluish green shale	160	feet
12. Brown shale	50	feet
13. Bluish shale	18	feet
14. Black shale	280	feet
15. Trenton limestone	11½	feet
Total depth	975	feet

The elevation of the station at Montpelier is 867 feet above sea level.

The following wells have been plugged in this township: Section 1, 16 wells; section 3, 1 well; section 4, 1 well; sections 5, 14 wells; section 6, 13 wells; section 7, 4 wells; section 9, 2 wells; section 16, 1 well; section 18, 1 well; section 30, 2 wells; section 31, 1 well; section 32, 1 well.

Washington Township. All the sections in the township have produced oil, and gas has been obtained from sections 23, 24, 31, 32, 35 and 36, R. 11 E., oil in 6, 7, 18, 19, 30 and gas in 19 and 30, R. 12 E.

Wells have been plugged as follows: Section 1, 1 well; section 3, 6 wells; section 4, 6 wells; section 5, 23 wells; section 7, 10 wells; section 9, 4 wells; section 10, 8 wells; section 12, 6 wells; section 13, 6 wells; section 21, 12 wells; section 24, 1 well; section 30, 1 well; section 31, 14 wells; section 32, 2 wells; section 33, 1 well; section 35, 1 well.

Jackson Township. Sections 6 and 7 produced oil. Gas was found in sections 5, 17 and 18. Wells have been abandoned in the following sections: Section 2, 1 well; section 5, 1 well; section 8, 1 well; section 9, 1 well; section 15, 2 wells; section 17, 1 well; section 23, 2 wells; section 24, 1 well; section 25, 1 well; section 28, 1 well; section 32, 1 well; section 33, 1 well.

Anna C. Simonton Farm, Sec. 15, Harrison Twp.:

Sand and gravel	134	feet
Limestone	138	feet
Shale	725	feet
Trenton rock	42	feet
Total	1039	feet

Lewis Blount Farm, Sec. 14, Harrison Twp., Blackford County:

Sand and gravel	118	feet
Limestone	162	feet
Shale	700	feet
Trenton rock	47½	feet
Total depth of well	1027½	feet

More than 1398 wells have been abandoned in this county.

BOONE COUNTY.

Strata of Devonian age form the durolith which underlies the eastern and central portions of the surface of Boone County, while strata of Mississippian age underlie the western portion. These strata are concealed by the glacial drift which varies from fifty to one hundred and fifty feet in thickness. The strata of the durolith which are recognizable from the well records are:

Mississippian	Shales and sands—Knobstone
Devonian..................	Shale—New Albany Limestones
Silurian	Shale Limestones—Niagara (?)
Ordovician	Shales—Utica and Hudson River Limestone—Trenton

The structural conditions in Boone County cannot be determined by the use of surficial observations. Deep well records are not sufficiently abundant to furnish the data for subsurface work.

Railroad Elevations.

Zionsville, 842; Whitestown, 928; Hazelrigg, 904; Terhune, 940.8; Max Station, 922; Advance, 928.

A well drilled at Zionsville is reported as follows[1]:

Drift	160 feet
Black shale (trace)	75 feet
Devonian limestone, with sandstone at base........	75 feet
Lower Helderburg and Waterlime.................	50 feet
Niagara limestone	165 feet
Clinton limestone	30 feet
Hudson River and Utica.......................	525 feet
Trenton limestone	33 feet
Total depth	1038 feet
Altitude of well..............................	777 feet

A well drilled at Thorntown has the following record[2]:

Drift	65 feet
Sub-carboniferous limestone and shale............	338 feet
Hamilton shale	87 feet
Corniferous limestone	37 feet
Niagara limestone	405 feet
Hudson River and Utica.......................	373 feet
Trenton limestone	80 feet
Total depth	1287 feet
Trenton below sea level.......................	394 feet

Yielded no gas.

A well was drilled at Lebanon, Indiana, to a total depth of 1800 feet.

Depth to Trenton, 1227 feet. Trenton below sea level, 302 feet. No gas. The record of this well as given by Phinney is as follows:

Drift	210 feet
Blue and black shales	204 feet
Limestones	401 feet
Shale	412 feet
Trenton limestone	373 feet
Total depth	1600 feet
Altitude of well	925 feet

BROWN COUNTY.

The northern part of Brown County lies within the glaciated area but the greater part of the county furnishes good rock exposures as the topography is of a rugged type. Even though the strata are not concealed by glacial drift the determination of the structure is difficult on account of the absence of persistent layers of rock. A lens or perhaps several lenses of limestone occur about one hundred feet below the upper surface of the Knobstone group. These lenses may be used locally as datum for mapping the structure. Sandstone layers occur at many horizons in the Knobstone but they are unreliable because of their lenticular character and cross bedded nature.

Trevlac. A well was drilled on the Bullhimer farm two miles north of Trevlac. The drill passed into the Trenton limestone at 1460 feet. The upper part of the limestone was fossiliferous, porous and contained a showing of gas. The drill passed through the Trenton at 2056 feet, showing 596 feet of limestone.

Johnson Township. A well was drilled in section 7 in Johnson Township and a showing of gas was obtained.

The elevation of the surface on the railroad at Trevlac is 654; Helmsburg, 676; Fruitdale, about 797.

CARROLL COUNTY.

A small area around Delphi, another one in the northern part of the county and another in the eastern portion is occupied by the Niagara limestone as a bed rock formation, the remainder of the county is occupied by the Devonian strata. The bed rock is largely concealed by glacial drift but outcrops occur to a limited extent along the Wabash River and some of its tributaries.

The following are the records of two wells drilled at Delphi[7]:

Niagara limestone	587 feet
Hudson River limestone and shale	220 feet
Utica shale	93 feet
Trenton limestone	12 feet
Total depth	912 feet
Trenton below sea level	334 feet

Yielded no gas.

Section of Well No. 2.

Niagara limestone	565 feet
Hudson River and Utica shale	351 feet
Trenton limestone	434 feet
Potsdam sandstone	12 feet
Total depth	1362 feet
Trenton below sea level	350 feet

Yielded no gas.

A dome-like structure at Delphi has the appearance of a steeply dipping anticline. It has been suggested that the dome is the remnant of a steeply dipping reef. A well drilled upon it failed to secure production.

Railroad Elevations.

Cutter, 722.7; Bringhurst, 18.7; Flora, 699.7; Camden, 659.7; Woodville, 692.7; Pattons, 682.4; Lennox, 663.7; Sleeths, 657.8; Wabash River, 647; N. Delphi, 557; Delphi, 555; Deer Creek, 672; Harley's, 693.5; Ockley, 695; Orvasco, 701.

CASS COUNTY.

Strata of Silurian and Devonian age underlie the surficial deposit of glacial drift in this county. The determination of structural conditions from surficial observations is prevented by the glacial mantle.

A well drilled at Galveston furnishes the following sections:

Sections of Well No. 1[1].

Drift	40 feet
Corniferous and Niagara limestone	410 feet
Hudson River and Utica	480 feet
Trenton limestone	20 feet
Total depth	950 feet

Yielded no gas.

At Logansport a well is reported to have:

Depth to Trenton	995 feet
Trenton below sea level	344 feet

Yielded no gas.

A second record was constructed by Phinney from drillings kept by Dr. J. H. Shultz:

Upper Helderburg limestone	40 feet
Lower Helderburg limestone	30 feet
Water lime	108 feet
Bluish limestone, Niagara	55 feet
Argillaceous limestone	110 feet
White and gray limestone	135 feet
Bluish green shale (Niagara)	2 feet
Clinton limestone steel gray, red grain	53 feet
Hudson River limestone and shales	90 feet

Utica shale	281 feet
Trenton limestone	200 feet
Total depth	1104 feet
Altitude of well	611 feet

Section of well at Royal Center:

Drift	105 feet
Niagara limestone	485 feet
Hudson River	220 feet
Brown shale, Utica shale	110 feet
Trenton limestone	42 feet
Total depth	962 feet

Three wells were drilled in this county in 1909, two were reported dry and the third as showing small production.

CLARK COUNTY.

The greater part of Clark County lies within the unglaciated area but the northeastern part of the county is covered with glacial drift. The strata represented by the outcrops of the county are given in the following table:

System	Group	Formation
Quaternary		Recent: Clays and alluvium
		Pleistocene: Clays, gravels and sand
Mississippian		Mitchell limestone
		Salem limestone
		Harrodsburg limestone
		Knobstone, sandstones and shales
		Rockford limestones
Devonian		New Albany shales
		Sellersburg limestones
		Silver Creek limestone
		Jeffersonville limestone
Silurian		Louisville limestone
		Waldron shale
		Laurel limestone
		Osgood limestone and shale
		Brassfield shale
Ordovician	Richmond	Elkhorn
		Whitewater
		Saluda
		Liberty
		Waynesville
		Arnheim
	Maysville	Mt. Auburn
		Coryville
		Bellevue
		Fairmount
		Mt. Hope

The determination of structural conditions favorable for gas or oil, if such exist, seems possible in this county because of the absence of glacial drift and the rugged condition of the topography which produces many outcrops of the strata. Key formations such as the Louisville limestone, the Sellersburg and the Harrodsburg may be used to advantage in locating structures. In the eastern part of the county the Trenton limestone is a possible source of gas and oil if favorable conditions exist. In the western portion the Trenton, Silurian and the Devonian limestones may furnish oil or gas reservoirs.

The following is the record of a well drilled at Jeffersonville:

Section of Well No. 1.

Alluvium	45 feet
Devonian limestone	40 feet
Niagara limestone	105 feet
Clinton limestone	20 feet
Hudson River limestone and shale	646 feet
Depth to Trenton	856 feet
Trenton below sea level	401 feet

Yielded small flow of gas.

Some gas was obtained from a well north of Jeffersonville.

CLAY COUNTY.

The portion of the Geological column represented by the outcrops in this county is given below:

Quaternary	Recent: Alluvial sands and clays Pleistocene: Glacial sands, gravels and till
Pennsylvanian	Allegheny: Limestones, sandstones, shales and coals Pottsville: Conglomerate, sands, shales and coals
Mississippian	Chester: Shales, sandstones and limestones

On account of the thickness of the mantle of Pleistocene and Recent, outcrops of the bed rock are not numerous but some of the streams have cut through the mantle and revealed the bed rock. Coal strip pits have also uncovered the strata in limited areas. The determination of structural conditions will require the use of sub-surface data, such as the record of wells, coal shafts, etc. Careful discrimination between Pottsville coals and Allegheny coal will be necessary as the use of the latter for key horizons is not always safe, as there is some evidence of a post-Pottsville disturbance.

The following is the record of a well drilled east of Jasonville. These records were obtained from Jesse Liston of Lewis:

Sheets Drill East of Jasonville.

	Feet		Feet
Surface clay	0 to 15	Blue fire clay	38 to 50
Sandstone	15 to 30	Water sandstone	50 to 75
Shale	30 to 36	Shale	75 to 100
Coal	36 to 38	Water sandstone	100 to 130

Sheets Drill East of Jasonville—Continued

	Feet		Feet
Blue shale, soft	130 to 150	Brown limestone	855 to 875
Water sandstone	150 to 170	White limestone, soft	875 to 885
Blue shale, soft	170 to 282	Brown limestone	885 to 925
Coal	282 to 286	White limestone, soft	925 to 932
White shale, soft	286 to 305	Brown limestone	932 to 987
Sandstone	305 to 328	Water sandstone, blue	987 to 1030
Shale	328 to 338	Blue lick water	987
Water sandstone	338 to 352	Brown limestone	1030 to 1052
Shale	352 to 408	Water sandstone	1052 to 1077
Sandstone	408 to 452	Limestone	1077 to 1365
Blue shale	452 to 520	Shale	1365 to 1385
Water sandstone	520 to 580	Shale	1385 to 1550
Hole full of water	550	White shale	1550 to 1685
Shale	580 to 586	Black shale	1685 to 1791
Limestone	586 to 643	Sandstone	1791 to 1836
Shale	643 to 653	Light showing of oil	1836
Blue shale, soft	653 to 668	Sandstone and limestone	1836 to 1858
Limestone	668 to 850	Sandstone, water sand	1836 to 1858
8¼ in. casing at	740	About the same to bottom of hole	1892
White limestone, soft	850 to 855		

Pigg Drill East of Lewis, Indiana.

	Feet		Feet
Coal	75 to 80	Blue sand	1075 to 1100
Broken stuff, soft lime shale and a little broken sand	410	Water	1090
Water sand	410 to 615	Lime	1100 to 1175
Shale	615 to 695	Brown sand	1175 to 1185
Water sand	695 to 725	Lime	1185 to 1290
Black shale	725 to 730	Sandy lime	1290 to 1300
Top of big lime	730	Lime	1300 to 1420
Hard lime	730 to 740	Blue lime	1420 to 1500
White shale	740 to 744	Gray lime	1500 to 1525
Hard lime	744 to 800	Light shale	1525 to 1570
Soft lime	800 to 805	Dark shale	1570 to 1670
Hard lime	805 to 950	Light shale	1670 to 1680
Shale	950 to 955	Riley? sand	1860 to 1684
Hard lime	955 to 1075	Light shale	1684 to 1785
		Dark shale	1785 to 1860

It was drilled some deeper than this, but the record further down was unobtainable.

Merchon Well.

Glacial drift	35 feet	Sandstone and slate	28 feet
Sandstone and shale	35 feet	Gray slate	10 feet
Sandstone	5 feet	Sand and slate	17 feet
Sandy shale	15 feet	Sandstone and slate	50 feet
Slate and stone	6 feet	Limestone	83 feet
Blue shale	20 feet	Slate and sandstone	5 feet
Sandstone	3 feet	Limestone and slate	5 feet
Sandy shale and slate	16 feet	Limestone	30 feet
Black sandstone	13 feet	Black slate	20 feet
Gray slate	75 feet	Blue slate	267 feet
Sandstone	20 feet	Slate	145 feet
Stone and slate	10 feet		
Sandstone	30 feet		979 feet

Measured Line 930		Measured Line 930	
Slate	70 feet	Black shale	19 feet —1208
Blue slate	48 feet	Oil on water, salt water	
Blue shale	46 feet	Limestone	34 feet
Casing set	1094 feet		
Blue shale	95 feet —1189	Total depth	1242 feet

The above is the record of a well drilled about 1⅛ miles southwest of Carbon in the center of Section 12, T. 13 N., R. 7 W. Record secured by Dr. C. A. Malott.

Lewis Township. The Coffee Hill Dome is located principally in Sections 8, 9, 16 and 17. The data was collected and the structure drawn on the Block Coals by Mr. Jesse Liston of Lewis. A shallow well drilled on the structure by the Rescue Oil Co. obtained some oil at 245 feet. A deeper well drilled to the Devonian limestone failed of production. The Trenton was not tested.

CLINTON COUNTY.

With the exception of a small area in the southwestern part of the county which is occupied by strata of the Mississippian age, the entire subsurface of this county is occupied by Devonian strata. The glacial drift overlying the bed rock varies in thickness from 50 to 300 feet. This covering prevents the determination of the structural conditions of the durolith.

The following is the record of one well drilled at Frankfort:

Drift	278 feet
Niagara limestone and shale	380 feet
Limestone	10 feet
Hudson River and Utica	400 feet
Trenton limestone	260 feet
Total depth	1328 feet
Trenton below sea level	327 feet

Yielded no gas.

Railroad Elevations.

Forest, 878.8; Frankfort, 846.7; Colfax, 840.7; Moran, 796.7; Michigantown, 866.2; Jefferson, 859.3; Manson, 857.7; Sedalia, 776.7; Avery, 872; Fickle, 827.3; Kilmore, 829.7; Scircleville, 929.3; Hillsburg, 919.7; Boylston, 903.0; Deniston, 844.1; Mulberry, 772.6.

CRAWFORD COUNTY.

Crawford County lies within the driftless area of Indiana. The strata which outcrop in the county belong to the following divisions and subdivisions:

Quaternary......	Recent: Alluvium Pleistocene: Residuals
Pennsylvanian...	Allegheny: Shales, sandstones, limestones, coal Pottsville: Sandstones, shales and coal
Mississippian....	Chester: Shale, sandstone and limestone Mitchell: Limestone

The structural conditions for a portion of the county can probably be determined by using limestones as the key formations. A portion of Orange County has already been mapped structurally and since the geological conditions are similar, for a part of Crawford County, it may be that the work can be extended to the latter. Possible oil bearing sand may be expected in the Trenton, Devonian and Chester rocks.

Taswell Well.

In 1903 the Highland Investment Company of Chicago, drilled a well in search of gas or oil, near the eastern limits of Taswell, eight miles east of Birdseye, on the Southern Railway. Drilling continued to a depth of 1,690 feet, where they encountered the actual Trenton, and drilled it 100 feet, a total depth of 1,790 feet, and got neither gas nor oil. In the western part of Crawford County there are surface indications of oil that have an extent of five miles in width by ten miles in length. Oil in paying quantities was never found. An oil well and a gas well were drilled in Section 16, Patoka Township, about one mile northwest of Eckerty.

DAVIESS COUNTY.

The mantle of glacial drift varies in thickness in this county from a few feet to more than one hundred feet in the valley of White River. The strata which underlie the drift belong to the Pennsylvanian period. Outcrops of the Pottsville division occur in the east part of the county and of Allegheny in the western portion. Structural conditions of the bed rock cannot be determined by surficial observation so that subsurface work must be resorted to in order to achieve results. Oil has been found in this county south of Cannelburg in Barr Township in the southeastern part of Section 8. One dry hole was drilled in the north; one dry hole in Section 7; one gas well and one dry hole in Section 3; one oil well and one gas well in the northwest quarter of Section 17, and one oil well in Section 30. The productive wells range in

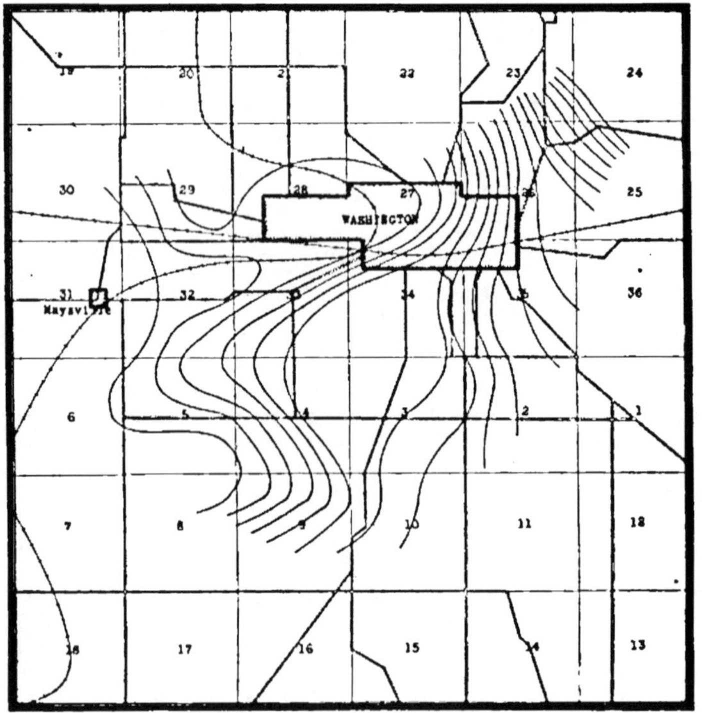

Fig. CXXI. Structural map of an area near Washington, Daviess County. Contours drawn on Coal VII. Data secured by C. A. Malott and P. B. Stockdale of the field party of 1919.

depth from 380 feet to 725 feet. The oil sand probably occurs in the Mansfield and the Chester.

Washington Township. In Section 22 of this Township, on the land of Stanton Barber, a well was drilled and plugged in 1912.

Madison Township. A well was drilled on the land of the Graham Glass Company in Section 34. The well was plugged in 1912.

Reeve Township. A well was drilled on the property of D. A. Brown in Section 10 and plugged in 1910.

Barr Township. The following wells have been plugged in this township: Section 2, Ralph Thompson, 1911. Section 35, Ed Grundy, 1911, and Charles M. Allan in 1913.

Harrison Township. A well drilled in Section 32 on the James Pettigrew property was plugged in 1911 and one in the same section on the property of F. M. Remsel in 1912.

The majority of the wells drilled in this county were drilled from 1910 to 1912.

DEARBORN COUNTY.

The bedrock formations of this county belong to the Ordovician period of geologic time. They are for the most part covered with glacial drift.

It may be possible to determine the structural conditions of this county if enough outcrops and well records can be secured. It lies on the west side of the Cincinnati Arch and the surface of the Trenton may be low enough in the southwest part of the county to be favorable territory. The surface of the Trenton is 158 feet above sea level at Lawrenceburg, where gas was obtained and dips westward where at North Vernon it is 260 feet below sea level.

Lawrenceburg Township. The log of a well drilled at Lawrenceburg in the river valley is as follows:

Alluvium	139 feet
Hudson River limestone and shale	185 feet
Utica shale	25 feet
Trenton limestone	451 feet
Potsdam sandstone	40 feet
Total depth	840 feet
Trenton above sea level	158 feet

The second well was drilled in the fairgrounds and reached the Trenton at 325 feet and showed gas.

Center Township. A well drilled at Aurora has the following log:

Drift	92 feet
Bluish green shale	148 feet
Dark shale, Utica	25 feet
Limestone, Utica, gas	2 feet
Shales and limestone, Utica	18 feet
Shale, Utica	25 feet
Trenton limestone	521 feet
St. Peter	about 170 feet
Total depth	1000 feet
Altitude of well	472 feet

DECATUR COUNTY.

The bed rock formations of this county are of Silurian and Devonian age and are largely concealed by glacial drift which varies in thickness from 10 to 100 feet.

Washington Township. The record of the Greensburg city well as given by Phinney is as follows:

	City Well	No. 1	No. 2	No. 3
Drift	10	7		
Corniferous limestone	4	90		
Niagara limestone	66	90		
Niagara shale	35			
Hudson River and Utica shale	747	823	886	883
Trenton		63		
Total	862	983	886	883
Altitude of well	930	920	925	925

The following wells have been abandoned:

Owner	Section	Date	Wells
Township School	2	1911	1
Aaron Logan	3	1919	1
Wm. Jackson	4	1919	1
City of Greensburg	5	1911	1
S. Logan	10	1919	1

Adams Township. A well drilled on the Chas. White Property was abandoned in 1911.

The surface of the Trenton around Greensburg varies from sea level to 68 feet above sea level. The gas obtained in the wells at Greensburg had a maximum pressure of 350 pounds. In the northwestern portions of the County in Adams Township, at Adams and St. Omer, light flows of gas were obtained.

DEKALB COUNTY.

The subsurface of this county is occupied by strata of the Devonian age which in the region of Auburn seems to have been slightly uplifted. The surface of the eroded Devonian rocks are covered with glacial drift which attains a thickness of more than 300 feet. Deep wells have been drilled at Auburn, Butler, Garrett and Waterloo. The structural conditions of the durolith are determinable only by the use of subsurface data.

The record of one of Auburn wells follows:

Section of Well No. 1.

Drift	280 feet
Black shale	120 feet
Corniferous, water lime and Niagara	963 feet
Hudson River limestone and shale	306 feet
Utica shale	268 feet
Trenton limestone	27 feet
Total depth	1964 feet

The following is the log of the Butler well:

Section of Well No. 1.

Drift	378 feet
Hamilton shale	108 feet
Corniferous, water lime and Niagara	1064 feet
Hudson River and Utica	500 feet
Trenton limestone	89 feet
Total depth	2139 feet

Yielded a small flow of gas, which was found at a depth of 27 feet in the Trenton.

The record of the well at Garrett is given below:

Section of Well No. 1.

Depth to Trenton	1980 feet
Total depth	2160 feet
Trenton below sea level	1098 feet

Yielded a small flow of gas.

DELAWARE COUNTY.

The Niagara limestone occupies the subsurface of this county and is covered with a mantle of glacial drift varying in thickness from 50 to 150 feet. A large number of wells have been drilled in this county and nearly 1,500 have been abandoned. From 1906 to 1914, 422 wells were completed, 103 of which were dry.

Centre Township. Many productive gas wells have been drilled at and near Muncie. The combined records of wells No. 1 and No. 2 are given by Phinney as follows:

Cedarville limestone	90 feet
Bluish limestone (Springfield beds)	135 feet
Niagara shale	40 feet
Hudson River limestone and shale	100 feet
Hudson River shale	340 feet
Utica shale	270 feet
Trenton limestone	481 feet
St. Peter's sandstone	150 feet
Total depth	1606 feet

The records of other wells are given in the table below:

Well	Altitude	Depth to Trenton	Altitude of Trenton
Nut Gas Co., No. 1	936	876	60
Nut Gas Co., No. 2	933	889	44
Reid Well	955	887	68
Highland	949	884	65
Bent Works	946	894	52
West Main St.	930	872	58

Well		Depth to Trenton	Altitude of Trenton
Fay	938	887	51
Winton	936	892	44
Water Works	938	891	47
Anthony	956	891	65
Boycetown No. 1	944	886	58

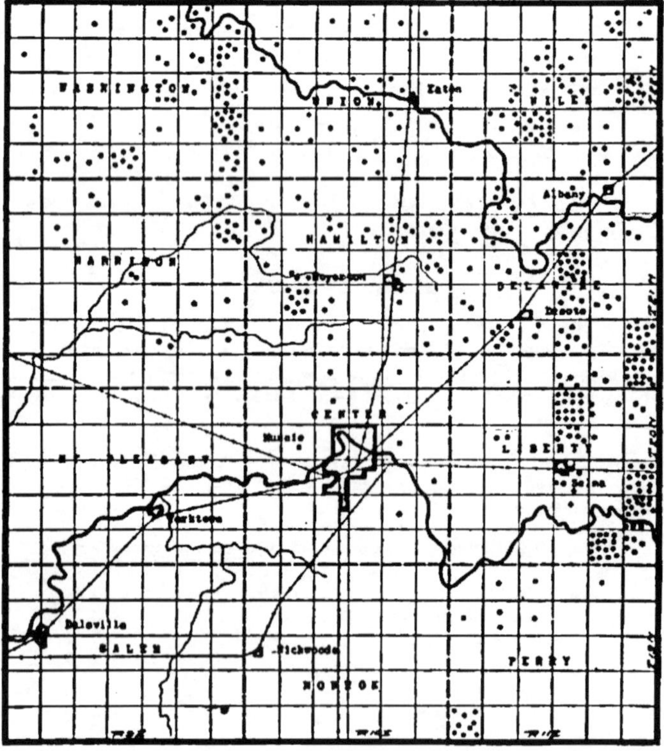

Fig. CXXII. Map of Delaware County showing location of abandoned wells. The eastern tier of townships are oil bearing, the remainder of the county is gas territory.

Two wells were drilled on the J. C. Quick farm in Section 14, one yielded 25 barrel of oil. The records are as follows:

	No. 1	No. 2
Drive pipe	104	108
Casing	348	350
Top of Trenton	898	890
Total depth	1212	1210

In Section 35 a million-foot gas well has the following record:

Drive pipe	47 feet
Casing	350 feet
Top of Trenton	920 feet
Total depth	1015 feet

Abandoned wells are located in the following sections: Section 2, 1 well; Section 4, 1 well; Section 8, 1 well; Section 11, 1 well; Section 13, 1 well; Section 23, 1 well; Section 33, 1 well; Section 34, 1 well.

Delaware Township. A gas well at Albany furnished the following record:

Drift	8 feet
Limestone	200 feet
Shale (Niagara)	68 feet
Hudson River and Utica shale	658 feet
Black shaly limestone	30 feet
Trenton limestone	14 feet
Total depth	978 feet
Altitude of well	940 feet

The table below gives the records of various wells drilled in this township:

		Sec. 16		Sec. 18	Sec. 15	Sec.10
Drive pipe	80	70	70	28	40	27
Casing	380	370	370	294	370	310
Top of Trenton	940	960	965	921	920	921
Total depth	1280	1290	1297	1227	1195	1232

Wells abandoned are located in Section 3, 1 well; Section 5, 6 wells; Section 7, 2 wells; Section 8, 7 wells; Section 9, 1 well; Section 10, 2 wells; Section 11, 2 wells; Section 12, 2 wells; Section 15, 19 wells; Section 16, 1 well; Section 17, 2 wells; Section 18, 1 well; Section 19, 2 wells; Section 22, 4 wells; Section 23, 4 wells; Section 25, 13 wells; Section 27, 7 wells; Section 28, 2 wells; Section 29, 7 wells; Section 30, 1 well; Section 36, 5 wells.

Liberty Township. The record of a well drilled in this township is as follows:

Drift	90 feet
White and buff limestone	85 feet
Soft and Ferruginous	15 feet
Bluish limestone	75 feet
Niagara shale	40 feet
Hudson River	485 feet
Utica shale	210 feet
Trenton limestone	25 feet
Total depth	1025 feet
Altitude of well	1015 feet

As late as 1903, 81 wells were drilled in this township. Fifty-three produced oil and the average initial production was 21 barrels. The records of three wells as given by Blatchley are as follows:

	Section 12	Section 14	
Drive Pipe	85 feet	104	97
Casing	360 feet	350	364
Top of Trenton........	976 feet	984	988
Total depth ....1030 feet		1040	1035

Wells abandoned are located as follows: Section 1, 3 wells; Section 2, 3 wells; Section 3, 16 wells; Section 10, 7 wells; Section 13, 12 wells; Section 14, 1 well; Section 15, 4 wells; Section 17, 2 wells; Section 22, 1 well; Section 24, 26 wells; Section 25, 5 wells; Section 26, 16 wells; Section 34, 4 wells; Section 35, 2 wells; Section 36, 22 wells.

Union Township. A well drilled in Eaton in 1876 produced some gas from Hudson River shale. In September, 1886, the first gas well of importance in Indiana was drilled at this place. The record of the well follows:

Buff limestone	5 feet
Bluish limestone	20 feet
Buff limestone	30 feet
Bluish gray limestone...........................	45 feet
White limestone	35 feet
Shale, bluish green.............................	35 feet
Buff limestone (Clinton)........................	10 feet
Shale, Hudson River and Utica...................	690 feet
Trenton limestone	32 feet
Total depth	922 feet

The Trenton was reached at Shideler at 884 feet. Successful gas wells were drilled at Cowan, Oakville, Yorktown, Royertown and New Corner.

Jas. Dill Farm, Section 26, Township 21 North, Range 11 East.

Top soil	67 feet
Lime	200 feet
Shales	681 feet
Top of sand...................................	948 feet
Into Trenton	325 feet
Salt water struck in Trenton.....................	320 feet

Jefferson Township. An abandoned well is located in Section 21.

Harrison Township. Abandoned wells are located as follows: Section 1, 3 wells; Section 2, 2 wells; Section 5, 2 wells; Section 7, 2 wells; Section 12, 8 wells; Section 16, 2 wells; Section 21, 1 well; Section 23, 1 well; Section 24, 1 well; Section 25, 1 well; Section 27, 2 wells; Section 36, 1 well.

Hamilton Township. Wells were drilled in the following sections: Section 5, 2 wells; Section 7, 2 wells; Section 10, 1 well; Section 11, 4 wells; Section 12, 6 wells; Section 13, 4 wells; Section 16, 2 wells; Section 17, 3 wells; Section 20, 9 wells; Section 21, 2 wells; Section 23, 2 wells; Section 24, 1 well; Section 25, 9 wells; Section 28, 1 well; Section 30, 1 well.

Washington Township. Wells were drilled and abandoned as follows: Section 5, 1 well; Section 10, 2 wells; Section 11, 2 wells; Section 12, 1 well; Section 13, 4 wells; Section 14, 1 well; Section 15, 1 well; Section 22, 2 wells; Section 23, 1 well; Section 24, 4 wells; Section 25, 8 wells; Sections 27, 1 well; Section 31, 1 well; Section 32, 1 well; Section 33, 4 wells; Section 36, 3 wells.

Niles Township. Wells abandoned are as follows. One each in Sections 11, 12, 20, 23, 13, 26, 27, 34, 35 and 36; Section 9, 2 wells; Section 21, 3 wells; Section 22, 5 wells; Section 24, 5 wells; Section 28, 13 wells; Section 15, 2 wells; Section 16, 4 wells, and Section 29, 4 wells.

Union Township. Wells abandoned are located as follows: One each in Sections 9, 10, 11, 12, 18, 25, 27 and 35; 2 in Section 20; 4 in Section 22; 2 in Section 26; 3 in Section 28; 2 in Section 29, and 4 in Section 34.

Perry Township. Abandoned wells are located one each in Sections 2, 4, 7, 9, and 3 in Section 5; 2 in Section 8, and 11 in Section 34.

Wells drilled by Wallace Oil Company, Section 22, Delaware Township. Farm of Marcellius Hitchcock.

Well No. 1

8 inch drive pipe	30 feet
6¼ inch casing	326 feet
Top Trenton	935 feet
Oil (light showing)	1206 feet
Total depth	1209 feet

Shot March 25, 1919. Well pumping.

Well No. 2.

8 inch drive pipe	18 feet
6¼ inch casing	306 feet
Top Trenton	926 feet
Crevice showing light gas	1184 feet
Total depth	1187 feet

Well No. 3.

10 inch drive pipe	28 feet
8 inch drive pipe	140 feet
6¼ inch casing	330 feet
Top Trenton	940 feet
Oil	1215 feet
Total depth	1216 feet

Light oil. Well pumping light.

Well No. 4.

10 inch drive pipe	45 feet
8 inch drive pipe	96 feet
6⅝ inch casing	330 feet
Top Trenton	946 feet
Oil (first)	1220-32 feet
Total depth	1232 feet

Light oil.

Well No. 5.

10 inch drive pipe.............................. 57 feet
8 inch drive pipe.............................. 89 feet
6⅝ inch casing.............................. 315 feet
Top Trenton .... 945 feet
Total depth ....1208 feet

Light oil.

Well No. 1. Farm of Willis Workwell.

8¼ inch drive pipe.............................. 37 feet
6¼ inch casing.............................. 332 feet
Top of Trenton.............................. 935 feet
Oil ....1210 feet
Total depth ....1213½ feet

Light oil.

Well No. 2.

10 inch drive pipe.............................. 20 feet
8 inch drive pipe.............................. 100 feet
6⅝ inch casing.............................. 332 feet
Total depth ....1206 feet
First pay ....1200 feet

Light showing. Well pumping.

Farm of Elmer Ritchie. Well No. 1.

10 inch drive pipe.............................. 39 feet
8 inch drive pipe.............................. 95 feet
6¼ inch casing.............................. 295 feet
Top Trenton .... 927 feet
Total depth ....1206 feet
First oil ....1198 feet
Best oil ....1203 feet

Well No. 2.

8 inch drive pipe.............................. 76 feet
6¼ inch casing.............................. 309 feet
Top Trenton .... 927 feet
Total depth, 1207—oil..............................1200 feet

Monarch Gas Company, Winchester, Indiana. Well No. 1.

S. W. corner of E, ½ of S. W. ¼ of Sec. Delaware Township 21, Range 11. 125 feet drive pipe.

Drive pipe .... 125 feet
Casing .... 325 feet
To sand .... 900 feet
In sand .... 320 feet
Depth of well..............................1230 feet

Well No. 2.

500 feet W. and 500 feet N. of S. E. corner of W. ½ of S. W. ¼ of Sec. Delaware Township 21, Range 11.

Drive pipe	69 feet, 1 inch
Casing	335 feet
To sand	916 feet
Depth of well....................................	1210 feet

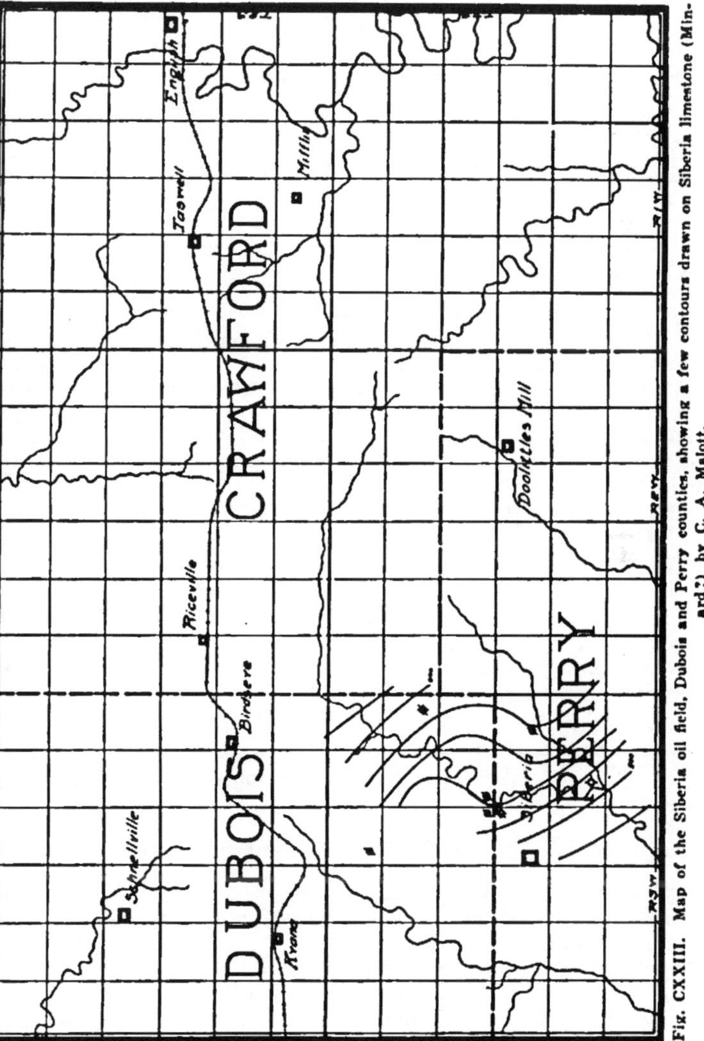

Fig. CXXIII. Map of the Siberia oil field, Dubois and Perry counties, showing a few contours drawn on Siberia limestone (Minard?) by C. A. Malott.

Well No. 3.

650 feet west and 1,000 feet N. of S. E. corner of W. ½ of S. W. ¼ of Sec. Delaware Township 21, Range 11.

Drive pipe	79 feet
Casing	340 feet
To sand	915 feet
In sand	288 feet
Depth of well	1202 feet

DUBOIS COUNTY.

Only the northwestern part of this county lies within the glaciated area. The larger part of its surface is occupied by the outcrops of the strata of the Coal Measures. The divisions and the subdivisions represented by outcrops in the county are given below:

Quaternary	Recent—Alluvium. Pleistocene—Clays, sands and gravels.
Pennsylvanian	Allegheny—Sandstone, limestone shale, coal. Pottsville—Sandstone, shale and coal.
Mississippian	Chester—Shales, limestone and sandstones.

Structural conditions will be difficult to determine in this county because of the absence over a large part of the county of persistent layers which may be used as key horizons. In limited areas it may be possible to use some of the coal beds and in limited areas in the southeastern and the northeastern parts of the county to use some of the limestones of the Chester as key horizons.

Birdseye. A small oil and gas field has been developed about Birdseye. Fourteen wells were drilled in Dubois, Crawford and Perry, oil was obtained in ten, gas in one and three were nonproductive. The depth of the oil bearing sand varied in the various wells from 980 feet to 1,010 feet. The oil sand is probably in the Devonian limestone and occurs about ten feet below the top of the limestone. A black or brownish black shale forty feet thick overlies the limestone.

Patoka Township. A well drilled on the property of J. E. Shertz and Company in Section 36 was plugged in 1911.

About ten wells have been drilled in the county, five of which were dry.

ELKHART COUNTY.

Glacial drift covers the bed rock in this county to a depth of from 50 to 200 feet. The drift overlies the eroded surface of the Devonian and Mississippian strata which dip northward.

Structural conditions of the durolith cannot be determined by direct observation because the outcrop of the durolith is concealed by the drift. Subsurface work is prevented by the absence of sufficient well records.

The record of a well drilled at Elkhart is as follows:[1]

Section of Well No. 1.

Drift	122 feet
Subcarboniferous shale (gray shale)	213 feet
Hamilton black shale	215 feet
Corniferous limestone	65 feet

"At this depth the well was abandoned under the erroneous belief that the drill had passed through the Hudson River and the Utica shales, and that the Corniferous was Trenton limestone."

The record of a well drilled at Goshen is as follows:

Section of Well No. 2.

Drift	165 feet
Shale, sub-carboniferous and Devonian	308 feet
Corniferous limestone	60 feet
Water lime	32 feet
Niagara limestone	728 feet
Hudson River limestone and shale	307 feet
Utica shale	215 feet
Trenton limestone	239 feet
Total depth	2054 feet
Trenton below sea level	1026 feet

Yielded no gas.

Railroad Elevations.

Elkhart, 753; Dunlap, 784.5; Goshen, 797.6; Millersburg, 879.7; Morehouse, 761.4; Bristol, 771.8; Vistula, 794.2; Williams, 845.5; Burns, 894.6; Middlebury, 852.1; Pleasant Valley, 749.9; New Paris, 809.

FAYETTE COUNTY.

The Pleistocene deposits in this county range in thickness from 25 feet to more than 100 feet. The strata underlying the pleistocene belong to the Silurian and Devonian periods. The outcrops of these rocks being concealed by the glacial drift, the determination of the structural conditions favorable to the accumulation of oil and gas is difficult. The surface of the Trenton limestone for the greater part if not all of this county lies above sea level and lies 700 to 900 feet below the surface. At Connersville the following well records have been obtained':

Section of Well No. 1.

Drift, Hudson River and Utica	712 feet
Trenton limestone	522 feet
Potsdam sandstone	12 feet
Total depth	1246 feet
Trenton above sea level	120 feet

Yielded a small flow of gas.

Section of Well No. 2.

Drift	90 feet
Hudson River and Utica	615 feet
Trenton limestone	61 feet
Total depth	766 feet
Trenton above sea level	117 feet

Yielded a small flow of gas.

Harrison Township. A deep well was drilled on the W. H. Wolf property in Section 8 and abandoned in 1912.

Posey Township. A well was drilled on the John Copeland property in Section 3 and abandoned in 1911. Another well was drilled on the property of J. Lambertson in Section 10 and abandoned in 1912.

Gas was obtained in Connersville, Jackson and Posey Townships.

FLOYD COUNTY.

Floyd County lies in the unglaciated area of Indiana. The strata which outcrop in the county belong to the Devonian, Mississippian and Quaternary periods. The subdivisions present are represented in the following table:

Quaternary	Recent—Alluvium. Pleistocene—Residuals.
Mississippian	Mitchell limestone. Salem limestone. Harrodsburg limestone. Knobstone, shales and sandstones. Rockford limestones.
Devonian	New Albany—Shale. Sellersburg—Limestone.

By using the contact of the Knobstone and the Harrodsburg it may be possible to determine the structural conditions of a part of this county. The contact of the New Albany and the Rockford might also be used as a key horizon.

The following is the record of a well drilled at New Albany:

Section of Well No. 1'.

Clay and subcarboniferous shale	80 feet
Devonian shale	104 feet
Corniferous limestone	69 feet
Niagara limestone	209 feet
Hudson River and Utica	545 feet
Trenton limestone	500 feet
Total depth	1507 feet

Yielded no gas or oil.

Railroad Elevations.

New Albany, 498.8; Smith, 565; Floyd, 445.8; Georgetown, 710.

FOUNTAIN COUNTY.

Underlying the glacial drift of this county are strata belonging to the Knobstone, Warsaw, Salem and Chester (?) divisions of the Mississippian and to the Pottsville and the Allegheny divisions of the Pennsylvanian periods. The glacial drift which largely conceals these formations varies in thickness from a few feet to more than one hundred feet. Whether or not structural conditions favorable to the accumulation of oil and gas exist in this county, can be determined only from subsurface data. Surficial methods cannot be used because of the glacial covering which conceals the outcrops. From reliable data collected in the form of well, shaft and outcrop records, it may be possible to determine the structural conditions.

Van Buren Township. Near Veedersburg three wells were drilled to depths of 1,000 feet. In one of them, gas occurred at 610 feet. These wells probably finished in the Devonian.

Two wells were drilled six miles south of Veedersburg to depths of 900 feet.

Cain Township. A well was drilled by the Fountain County Oil and Gas Company, 4½ miles southwest of Hillsboro. The log of the well follows:

Well No. 1.

Drilling log of the David Keller well.

Yellow clay and gravel	0 to 30 feet
Sand, white	30 to 75 feet
Shale, gray	75 to 118 feet
Sand, white	118 to 220 feet
Sand, limey, coarse	220 to 265 feet
Sand, showing oil	265 to 273 feet
Sand, limey, coarse	273 to 290 feet
Sand, very light lime	290 to 305 feet
Lime, very coarse	305 to 315 feet
Shale, gray	315 to 350 feet
Sand	350 to 400 feet
Slate, white	400 to 415 feet
Lime, hard, coarse	415 to 430 feet
Shale, gray	430 to 545 feet
Lime, hard, coarse	545 to 550 feet
Shale, gray	550 to 565 feet
Lime, blue, soft	565 to 595 feet
Lime, hard	595 to 635 feet
Shale, gray	635 to 685 feet
Clay, green	685 to 725 feet
Slate, white	725 to 735 feet
Shale, black	735 to 755 feet
Shale, brown	755 to 800 feet
Slate, white	800 to 810 feet
Shale, brown	810 to 840 feet
Sand, hard, brown	840 to 860 feet
Sand, odor of oil	860 to 890 feet
Niagara lime, containing salt water	890 to 938 feet

Railroad Elevations.

Mellott, 699; Veedersburg, 604.3; Cates, 644.7; Silverwood, 516; Attica, 543; Rob Roy, 634; Aylesworth, 635; Stone Bluff, 622.

FRANKLIN COUNTY.

The subsurface formations of this county belong to the Ordovician and the Silurian periods. They are covered largely by a mantle of glacial drift.

The divisions which are probably represented in the durolith are as follows:

Silurian........	Louisville limestone.
	Waldron shale.
	Laurel limestone.
	Osgood limestone and shale.
	Brassfield.
Ordovician.....	Richmond shales and limestones.
	Maysville shales, limestones and sandstones.
	Eden shales, limestones and sandstones.
	Trenton limestones.
	St. Peter sandstone.

Brookville Township. Seven wells were drilled in the vicinity of Brookville. The log of a well drilled in White Water River Valley is as follows:

Drift	157 feet
Shale	243 feet
Trenton and St. Peter..........................	854 feet
Total depth	1254 feet
Altitude of well............................	575 feet
Salt water at...............................	800 feet

Well No. 2, located in town, reached the Trenton at 550 feet at an altitude of 700 feet. The surface of the Trenton is 150 feet above sea level. A small supply of gas was obtained in this township and in Laurel Township.

Railroad Elevations.

Peoria, 999; Raymond, 1,008; Bath, 1,012.

FULTON COUNTY.

Strata belonging to the Silurian and Devonian periods lie beneath the glacial drift in this county. The drift attains a thickness of more than 300 feet.

The concealment of the strata of the durolith by the drift makes it impossible to determine the structural conditions by surficial methods. The accumulation of the logs of deep wells will greatly aid in such determination. The county lies on the north slope of the Cincinnati Arch, and the Trenton surface which lies nearest sea level near the southern boundary of the county is 351 feet below sea level at Rochester.

The section of a well drilled at Rochester is given below:

Section of Well No. 1.

Drift	245 feet
Niagara limestone	525 feet
Hudson River and Utica	391 feet
Trenton limestone	24 feet
Total depth	1185 feet
Trenton below sea level	351 feet

Yielded no gas.

Record of well drilled at Kewanna:[2]

Section of Well No. 1.

Drift	170 feet
Limestone and shale	879 feet
Trenton limestone	29 feet
Total depth	1078 feet
Trenton below sea level	278 feet

Did not yield gas or oil.

GIBSON COUNTY.

The strata underlying the glacial drift in Gibson County are the coal measures of the Pennsylvanian period. The mantle of drift has been removed in places so that the outcrops of the bed rock occur in the northern and central parts of the county. The extreme southern part of the county was not glaciated. The structural conditions of the bed rock can not be determined from a study of surficial conditions. However, the use of the subsurface data combined with such surficial data as may be obtained from outcrops, may make it possible to locate structures favorable to the accumulation of oil and gas. Several small oil pools have been located in this county but largely, if not wholly, by using the drill. The position of these pools is given in the accompanying map. There are probably three sands in this county from which oil has been obtained. These sands probably all belong to the Pennsylvanian group of rocks or possibly the lower to the Chester.

White River Township. Eight miles northwest of Princeton, gas was obtained at 1,300 feet. At Hazleton three wells were drilled to 2,000 feet. Oil was obtained and this is known as the Hazleton pool.

Patoka Township. Wells in Sections 2, 9, 10 and 32, oil was obtained in two at 871 feet in the Princeton sand.

Center Township. S. E. ¼ of the S. W. ¼ of Section 36, bore drilled came in dry. (Rept., 1908.) Oil is reported in a well drilled near Francisco to a depth of 1,690 feet.

One well in Oakland City[3] field, S. W. ¼ of the N. W. ¼ of Section 13, was reported to have reached a sand lower than the Oakland City sand and to yield an oil of good gravity and of strong sulphur smell. The Oakland City sand in this well was found at 1,228 feet and was eight feet thick. The two leases yielded an output of 150 barrels. The stray sand was found at a lower depth at 1,284 feet and was reported to be 18 feet thick, yielding the sulphur oil.

The No. 1 well on the Montgomery lease, completed in 1907, drilled to a depth of 1,000 feet, sand struck at 845 feet, 87 feet of drive pipe used; dry well. No. 17, drilled to a total depth of 862 feet, sand at 820 feet, drive pipe, eighty feet ten inches; sixty-five barrel well, completed March, 1907. No. 18, completed April, 1907, total depth, 865 feet; sand at 820 feet; 90 feet of drive pipe; 92 barrel well.

Well drilled on the Skinner farm, near Oakland City, total depth, 1,300 feet; encountered salt water, dry well.

H. A. Mauck lease, S. W. ¼ of Section 19; drive pipe (10 inch), 95 feet; casing (8 inch), 130 feet; casing (6¼ inch), 785 feet; top of sand, 918 feet.

Record of a well drilled in Washington Township. L. C. Frederick farm, No. 4.

		Depth	
Surface	30	30	
Sand	20	50	
Shale	20	70	
Broken sand	80	150	Cased with 10"
Black slate	50	200	
Sand, dry	60	260	
Shale	40	300	
Broken sand	20	320	Some water
Black slate	50	370	
Shale	30	400	
Lime shell	10	410	
Shale	90	500	
Black slate	30	530	
Shale	70	600	
Broken lime	20	620	Some water
Black slate	40	660	
Hard lime	10	670	
Coal	2	672	2' coal
Shale	50	722	
Sand	30	752	
Black slate	50	802	
Shale	40	842	
Black slate	20	862	
Sandy shale	60	922	
Black slate	40	962	
Lime	10	970	
Shale	35	1007	
Sand	20	1027	
Shale	5	1032	
Sandy lime	5	1037	
Oil sand	6	1041	
Total depth		1043	

Washington Township, well of McNeece, 2 miles east of Hazelton.

Formation	Depth
Lime and slate	600 to 700 feet
Sand (dry)	700 to 710 feet
Slate	710 to 780 feet
Sand (4 Bailer water)	780 to 830 feet
Slate	830 to 850 feet
Broken sand (dry)	850 to 894 feet
Slate	894 to 920 feet
Sand (soft) strong flow water	920 to 975 feet
Sand (no water)	975 to 980 feet
Slate	980 to 1050 feet
Sand (dry)	1050 to 1080 feet
Slate	1080 to 1170 feet
Lime rock	1170 to 1190 feet
Slate	1190 to 1228 feet
Slate	1228 to 1248 feet
Lime rock	1248 to 1295 feet
Slate	1295 to 1325 feet
Lime rock	1325 to 1327 feet
Slate and shell	1327 to 1335 feet
White sand	1335 to 1340 feet
Slate	1340 to 1405 feet
Lime rock	1405 to 1420 feet
Broken sand and shale	1420 to 1450 feet
Sand (strong flow of water	1450 to 1500 feet
Lime rock	1500 to 1520 feet
Slate	1520 to 1525 feet
Red rock	1525 to 1530 feet
Lime rock	1530 to 1535 feet
Slate	1535 to 1573 feet
Lime rock	1573 to 1590 feet
Slate	1590 to 1595 feet
Sand (dry)	1595 to 1605 feet
Red rock	1605 to 1610 feet
Lime rock	1610 to 1620 feet
White sand	1620 to 1680 feet
Lime rock	1680 to 1690 feet
Slate	1690 to 1718 feet
Lime rock	1718 to 1722 feet
Sand (dry)	1722 to 1740 feet
Lime rock	1740 to 1800 feet
Oil sand (slight show)	1800 to 1806 feet
Lime rock	1806 to 1870 feet
Lime rock	1870 to 2000 feet
Total depth	2000 feet

Log of well No. 5 on the L. C. Frederick farm, Washington Township, Gibson County.

Formation	Depth
Surface	to 30 feet
Sand	to 60 feet
Fire clay	to 65 feet
Coal	to 68 feet
Slate	to 80 feet
Sand	to 125 feet
Dark slate	to 160 feet
Shale	to 195 feet
Sand and water	to 220 feet
Slate	to 235 feet
Lime	to 245 feet
Shale	to 285 feet
Sand (dry)	to 295 feet
Shale	to 320 feet
Broken lime	to 335 feet
Shale	to 355 feet
Sand	to 365 feet
Slate	to 370 feet
Lime	to 378 feet
Shale	to 420 feet
Lime	to 425 feet
Brown shale	to 440 feet
Slate	to 465 feet
Shale	to 493 feet
Coal	to 504 feet
Fire clay	to 508 feet
Sand	to 515 feet
Shale	to 555 feet
Lime	to 562 feet
Shale	to 590 feet
Brown shale	to 620 feet
Sandy lime	to 625 feet
Brown shale	to 675 feet
Broken lime	to 685 feet
Slate	to 700 feet
Lime	to 710 feet

Log of Well No. 5—Continued

Formation	Depth	Formation	Depth
Sand ....to	725 feet	Slate ....to	1245 feet
Slate ....to	730 feet	Shale ....to	1260 feet
Sand ....to	750 feet	Lime ....to	1270 feet
Sandy shale ....to	800 feet	Broken sand (Hoover)..to	1280 feet
Lime ....to	810 feet	Hoover sand ....to	1305 feet
Shale ....to	825 feet	Shale ....to	1310 feet
Slate ....to	875 feet	Broken sand (Gas sand).to	1320 feet
Shale ....to	900 feet	Sand, gas at 1328.......to	1355 feet
Sand ....to	920 feet	Brake ....to	1358 feet
Shale ....to	950 feet	Lime ....to	1380 feet
Water sand ....to	965 feet	Green shale ....to	1405 feet
Shale ....to	1005 feet	Lime ....to	1415 feet
Slate ....to	1025 feet	Shale ....to	1430 feet
Broken sand ....to	1065 feet	Lime ....to	1450 feet
Sand ....to	1165 feet	Shale ....to	1455 feet
Shale ....to	1185 feet	Lime ....to	1455 feet
Lime ....to	1195 feet	Shale ....to	1470 feet
Slate ....to	1220 feet	Broken sand ....to	1485 feet
Lime ....to	1225 feet	Lime ....to	1490 feet
Slate ....to	1235 feet	Sand ....to	1500 feet
Lime shell ....to	1240 feet		

Well No. 3 on the John Zimmerman farm, 200 feet to east line, 665 feet to south line. Section 7, Washington Township, Gibson County.

Formation	Depth	Formation	Depth
Soil ....to	1 foot	Coal ....to	372 feet
Dark clay ....to	20 feet	Gray sand ....to	386 feet
Gravel ....to	22 feet	Dark slate ....to	421 feet
Dark slate ....to	40 feet	Gray sand ....to	431 feet
Dark lime ....to	44 feet	Dark slate ....to	480 feet
Dark slate ....to	98 feet	Brown slate ....to	490 feet
Gray sand ....to	114 feet	Dark slate ....to	525 feet
Dark slate ....to	157 feet	Brown lime ....to	535 feet
Coal ....to	161 feet	Black slate ....to	540 feet
Dark slate ....to	179 feet	Gray lime ....to	546 feet
Gray lime ....to	185 feet	Gray sand ....to	556 feet
Dark slate ....to	186 feet	Light slate ....to	566 feet
Gray lime ....to	191 feet	Dark slate ....to	578 feet
Dark slate ....to	194 feet	Gray lime ....to	583 feet
Gray lime ....to	200 feet	Light slate ....to	610 feet
Gray sand ....to	209 feet	Coal ....to	615 feet
Black slate ....to	240 feet	Dark slate ....to	625 feet
Light slate ....to	245 feet	Gray lime ....to	630 feet
Dark slate ....to	300 feet	Black slate ....to	760 feet
Light sand ....to	320 feet	Gray sand ....to	775 feet
Gray sand, water.......to	365 feet	Dark slate ....to	815 feet
Gray lime ....to	369 feet	Gray sand ....to	825 feet

Well No. 3—Continued

Formation	Depth	Formation	Depth
Dark slate ....to	830 feet	Dark slate ....to	1070 feet
Gray sand ....to	850 feet	Gray sand ....to	1230 feet
Dark slate ....to	865 feet	Dark slate ....to	1245 feet
Gray sand ....to	895 feet	Dark lime ....to	1247 feet
Brown lime ....to	900 feet	Light slate ....to	1254 feet
Black slate ....to	908 feet	Gray sand ....to	1258 feet
Gray lime ....to	912 feet	Dark slate ....to	1265 feet
Black slate ....to	925 feet	Gray sand, gas.........to	1279 feet
Gray sand ....to	939 feet	Dark slate ....to	1295 feet
Dark slate ....to	1001 feet	Gray lime ....to	1306 feet
Black slate ....to	1004 feet	Dark slate ....to	1317 feet
Dark slate ....to	1016 feet	Gray sand, first oil......to	1335 feet
Gray sand ....to	1025 feet	Dark slate ....to	1337 feet
Dark slate ....to	1030 feet	Gray lime ....to	1339 feet
Gray sand ....to	1035 feet	Dark slate ....to	1348 feet

Well No. 4 on the John Zimmerman farm, 200 feet to east line, 200 feet to south line. Section 7, Washington Township, Gibson County.

Formation	Depth	Formation	Depth
Soil ....to	1 foot	Light sand ....to	414 feet
Yellow clay ....to	12 feet	Light shale ....to	462 feet
Light slate ....to	54 feet	Dark slate ....to	500 feet
Dark slate ....to	73 feet	Black slate ....to	506 feet
Dark sand ....to	83 feet	Gray lime ....to	510 feet
Dark slate ....to	108 feet	Coal ....to	512 feet
Dark lime ....to	110 feet	Dark slate ....to	526 feet
Dark slate ....to	113 feet	Dark lime ....to	528 feet
Gray sand ....to	124 feet	Dark slate ....to	536 feet
Light slate ....to	127 feet	Gray lime ....to	540 feet
Gray lime ....to	133 feet	Coal ....to	547 feet
Dark slate ....to	170 feet	Light slate ....to	572 feet
Coal ....to	174 feet	Light sand ....to	578 feet
Dark slate ....to	197 feet	Dark slate ....to	605 feet
Gray lime ....to	208 feet	Light slate ....to	635 feet
Light slate ....to	225 feet	Dark slate ....to	650 feet
Gray sand ....to	233 feet	Gray lime ....to	655 feet
Dark slate ....to	294 feet	Light slate ....to	665 feet
Gray sand ....to	303 feet	Dark slate ....to	776 feet
Dark slate ....to	311 feet	Gray lime ....to	784 feet
Dark sand ....to	314 feet	Light slate ....to	797 feet
Dark slate ....to	325 feet	Gray sand ....to	808 feet
Dark sand ....to	358 feet	Dark slate ....to	838 feet
Dark slate ....to	360 feet	Light sand ....to	862 feet
Light sand ....to	385 feet	Brown lime ....to	865 feet
Black lime ....to	392 feet	Light sand ....to	872 feet
Dark slate ....to	397 feet	Dark slate ....to	874 feet
Light slate ....to	405 feet	Dark sand ....to	910 feet

Well No. 4—Continued

Formation	Depth	Formation	Depth
Gray lime ....to	917 feet	Gray sand ....to	1246 feet
Light slate ....to	930 feet	Dark slate ....to	1250 feet
Dark slate ....to	942 feet	Light slate ....to	1271 feet
Light sand ....to	950 feet	Dark lime ....to	1273 feet
Dark slate ....to	955 feet	Dark slate ....to	1280 feet
Light sand ....to	960 feet	Dark sand ....to	1288 feet
Dark slate ....to	1012 feet	Dark slate ....to	1290 feet
Light sand ....to	1020 feet	Light sand ....to	1295 feet
Dark slate ....to	1025 feet	Light slate ....to	1299 feet
Gray sand ....to	1033 feet	Light sand ....to	1303 feet
Light sand ....to	1042 feet	Dark slate ....to	1314 feet
Dark slate ....to	1050 feet	Gray sand ....to	1316 feet
Gray lime ....to	1052 feet	Brown lime ....to	1324 feet
Light slate ....to	1112 feet	Dark slate ....to	1336 feet
Light sand ....to	1160 feet	Gray sand, first oil at 1338 ....to	1353 feet
Light slate ....to	1175 feet	Dark slate ....to	1361 feet
Light sand ....to	1212 feet		
Light slate ....to	1217 feet		

Well No. 1 on the farm of Mary Shawhan, 200 feet to east line, 700 feet to south line. Section 1, Washington Township, Gibson County.

Formation	Depth	Formation	Depth
Yellow clay ....to	42 feet	Dark slate ....to	520 feet
Light slate ....to	48 feet	Dark slate ....to	545 feet
Gray lime ....to	51 feet	Light slate ....to	555 feet
Light slate ....to	60 feet	Dark slate ....to	588 feet
Dark slate ....to	68 feet	Light slate ....to	602 feet
Light slate ....to	71 feet	Dark slate ....to	615 feet
Light lime, water at 73 ft.to	75 feet	Light slate ....to	630 feet
Light slate ....to	100 feet	Dark lime ....to	632 feet
Coal ....to	104 feet	Dark slate ....to	650 feet
Light slate ....to	176 feet	Light slate ....to	662 feet
Dark lime ....to	180 feet	Dark slate ....to	750 feet
Light slate ....to	223 feet	Light slate ....to	770 feet
Coal ....to	225 feet	Dark slate ....to	775 feet
Light slate ....to	230 feet	Light sand ....to	780 feet
Light sand ....to	240 feet	Dark slate ....to	787 feet
Dark slate ....to	254 feet	Gray lime ....to	790 feet
Light sand ....to	269 feet	Dark slate ....to	835 feet
Gray lime ....to	274 feet	Gray lime ....to	839 feet
Light slate ....to	276 feet	Dark slate ....to	898 feet
Gray lime ....to	279 feet	Light lime ....to	901 feet
Dark slate ....to	305 feet	Dark slate ....to	940 feet
Coal ....to	309 feet	Gray lime ....to	948 feet
Dark slate ....to	417 feet	Dark slate ....to	980 feet
Dark lime ....to	423 feet	Light lime ....to	1015 feet
Light slate ....to	500 feet	Dark slate ....to	1045 feet

Well No. 1—Continued

Formation	Depth	Formation	Depth
Gray sand	to 1075 feet	Gray lime	to 1424 feet
Light sand	to 1103 feet	Gray sand (gas sand)	to 1430 feet
Dark slate	to 1114 feet	White sand, hole full of water at 1436	to 1490 feet
Light sand	to 1151 feet	Gray lime	to 1498 feet
Dark slate	to 1158 feet	Dark slate	to 1504 feet
Gray sand	to 1195 feet	Gray lime	to 1506 feet
Dark slate	to 1200 feet	Dark slate	to 1512 feet
Gray sand	to 1216 feet	Dark lime	to 1529 feet
Dark slate	to 1218 feet	Dark slate	to 1531 feet
Light sand	to 1294 feet	Light sand	to 1533 feet
Dark slate	to 1300 feet	Light sand	to 1537 feet
Brown lime	to 1301 feet	Gray lime	to 1539 feet
Dark slate	to 1338 feet	Light sand	to 1550 feet
Gray sand	to 1343 feet	Dark slate	to 1560 feet
Dark slate	to 1350 feet	Brown lime	to 1569 feet
Gray lime (shows gas)	to 1373 feet	Dark slate	to 1572 feet
Dark slate	to 1384 feet	Gray sand	to 1584 feet
Gray sand (barren)	to 1390 feet	Light sand	to 1589 feet
Dark slate	to 1393 feet	Dark sand	to 1594 feet
Gray lime	to 1400 feet	Light sand	to 1601 feet
Light slate	to 1402 feet	Light sand	to 1612 feet
Gray lime	to 1410 feet	Dark slate	to 1617 feet
Dark slate	to 1423 feet		

Well No. 1, farm of Geo. Colvin, 164 feet to south line, 237 feet to west line. Section 6, Washington Township, Gibson County.

Formation	Depth	Formation	Depth
Yellow clay	to 35 feet	Light slate	to 397 feet
Light slate	to 45 feet	Gray lime	to 400 feet
Gray lime	to 49 feet	Light slate	to 408 feet
Dark slate	to 80 feet	Gray lime	to 418 feet
Light lime	to 100 feet	Light sand	to 425 feet
Light sand	to 145 feet	Dark slate	to 475 feet
Dark slate	to 200 feet	Coal	to 480 feet
Gray lime	to 230 feet	Light slate	to 530 feet
Light slate	to 235 feet	Coal	to 533 feet
Light lime	to 275 feet	Gray lime	to 535 feet
Dark slate	to 285 feet	Dark slate	to 570 feet
Light sand	to 292 feet	Light lime	to 590 feet
Dark slate	to 295 feet	Light slate	to 600 feet
Light sand	to 300 feet	Dark slate	to 690 feet
Dark slate	to 340 feet	Gray sand	to 696 feet
Light lime	to 350 feet	Dark slate	to 715 feet
Gray sand	to 380 feet	Gray lime	to 721 feet
Light slate	to 382 feet	Dark slate	to 725 feet
Gray slate	to 395 feet	Gray lime	to 735 feet

George Colvin Well No. 1—Continued

Formation	Depth	Formation	Depth
Dark slate ....to	770 feet	Dark slate ....to	933 feet
Gray sand ....to	780 feet	Gray sand ....to	947 feet
Dark slate ....to	795 feet	Dark slate ....to	955 feet
Gray sand ....to	800 feet	Light sand ....to	963 feet
Dark slate ....to	820 feet	Dark slate ....to	1010 feet
Gray lime ....to	827 feet	Coal ....to	1014 feet
Dark slate ....to	833 feet	Light sand ....to	1027 feet
Gray sand ....to	835 feet	Dark slate ....to	1040 feet
Dark slate ....to	845 feet	Gray sand ....to	1050 feet
Gray sand ....to	860 feet	Dark slate ....to	1060 feet
Dark slate ....to	907 feet	Brown sand, oil.........to	1062 feet
Gray lime ....to	910 feet	Light sand ....to	1066 feet

Well No. 1 on the Phoebe Hayden farm, 200 feet to north line, 200 feet to east line. Section 7, Washington Township, Gibson County.

Casing Record.

Thirteen feet wood conductor; 1,270 feet 8¼ inch casing; 142 feet 7 inch casing (liner).

Shot Record.

One hundred forty quarts, 1,375 feet to 1,398 feet.

Formation Record.

Formation	Depth	Formation	Depth
Soil ....to	1 foot	Sand and lime..........to	400 feet
Yellow clay ....to	20 feet	Gray lime ....to	409 feet
Quick sand ....to	28 feet	Dark slate ....to	419 feet
Yellow clay ....to	43 feet	Gray lime ....to	423 feet
Dark lime ....to	53 feet	Dark slate ....to	428 feet
Dark slate ....to	73 feet	Light sand ....to	435 feet
Light slate ....to	77 feet	Gray lime ....to	455 feet
Dark lime ....to	79 feet	White lime ....to	465 feet
Light slate ....to	98 feet	White slate ....to	470 feet
Dark lime ....to	104 feet	Light sand ....to	474 feet
Dark slate ....to	155 feet	Light slate ....to	475 feet
Light slate ....to	160 feet	Gray lime ....to	479 feet
Gray lime ....to	220 feet	Light slate ....to	481 feet
Dark slate ....to	232 feet	Dark slate ....to	550 feet
Gray lime ....to	243 feet	Coal ....to	554 feet
Dark slate ....to	248 feet	White slate ....to	565 feet
Gray lime ....to	250 feet	White lime ....to	573 feet
Dark slate ....to	255 feet	Brown slate ....to	580 feet
Gray lime ....to	295 feet	Dark slate ....to	600 feet
Dark slate ....to	325 feet	Gray lime ....to	609 feet
Light slate ....to	335 feet	Dark slate ....to	621 feet
White lime ....to	340 feet	Gray lime ....to	626 feet
Dark slate ....to	380 feet	Dark slate ....to	629 feet

Phoebe Hayden Well No. 1—Continued

Formation	Depth	Fosmation	Depth
Gray lime ....to	641 feet	Gray sand ....to	990 feet
Light slate ....to	648 feet	Dark slate ....to	1012 feet
Dark slate ....to	650 feet	Gray lime ....to	1020 feet
White slate ....to	676 feet	Light sand ....to	1030 feet
Dark slate ....to	705 feet	Dark slate ....to	1128 feet
Gray lime ....to	710 feet	Gray sand ....to	1137 feet
Dark slate ....to	715 feet	Dark slate ....to	1170 feet
White slate ....to	722 feet	Gray sand ....to	1180 feet
Dark slate ....to	770 feet	Dark slate ....to	1190 feet
Gray lime ....to	774 feet	Gray lime ....to	1195 feet
Dark slate ....to	824 feet	Gray sand ....to	1218 feet
Dark slate ....to	830 feet	Dark slate ....to	1240 feet
Gray lime ....to	834 feet	Gray sand ....to	1245 feet
Dark slate ....to	840 feet	Brown lime ....to	1252 feet
Dark sand ....to	845 feet	Dark slate ....to	1260 feet
Dark slate ....to	849 feet	Gray sand ....to	1269 feet
Dark sand ....to	860 feet	Dark slate ....to	1270 feet
Dark slate ....to	865 feet	Brown lime ....to	1272 feet
White lime ....to	869 feet	Dark slate ....to	1286 feet
Brown sand ....to	874 feet	Dark lime ....to	1289 feet
Brown slate ....to	880 feet	Dark slate ....to	1345 feet
Gray lime ....to	884 feet	Light sand ....to	1347 feet
Dark slate ....to	885 feet	Dark slate ....to	1353 feet
Gray lime ....to	889 feet	Gray lime ....to	1360 feet
Dark slate ....to	900 feet	Dark slate ....to	1374 feet
Gray sand ....to	950 feet	Gray sand ....to	1398 feet
White slate ....to	951 feet	Dark slate ....to	1402 feet

First shows oil at 1375 feet.

Well No. 2 on farm of Phoebe Hayden, 200 feet to north line, 200 feet to west line. Washington Township, Section 7, Gibson County.

Shot Record.

One hundred quarts, 1,358 feet to 1,381 feet.

Formation Record.

Formation	Depth	Formation	Depth
Soil ....to	1 foot	Gray sand ....to	140 feet
Yellow clay ....to	23 feet	Light sand ....to	175 feet
Dark slate ....to	37 feet	Light slate ....to	195 feet
Light slate ....to	42 feet	Coal ....to	201 feet
Gray lime ....to	66 feet	Light slate ....to	222 feet
Light slate ....to	73 feet	Gray lime ....to	230 feet
Coal ....to	75 feet	Light slate ....to	240 feet
Dark slate ....to	91 feet	White sand ....to	261 feet
Gray sand ....to	94 feet	Dark slate ....to	295 feet
Light slate ....to	112 feet	Gray lime ....to	298 feet
Dark slate ....to	122 feet	Dark slate ....to	325 feet

Phoebe Hayden Well No. 2—Continued

Formation	Depth	Formation	Depth
Light slate ....to	360 feet	Dark sand ....to	915 feet
Dark sand ....to	373 feet	Dark slate ....to	963 feet
Dark slate ....to	400 feet	Coal ....to	965 feet
Light sand ....to	408 feet	Light sand ....to	985 feet
Dark slate ....to	423 feet	Dark slate ....to	1027 feet
Gray sand ....to	445 feet	Gray lime ....to	1033 feet
Dark slate ....to	460 feet	Dark slate ....to	1048 feet
Light slate ....to	483 feet	Black slate ....to	1052 feet
Dark slate ....to	535 feet	Dark slate ....to	1067 feet
Black slate ....to	550 feet	Light sand ....to	1100 feet
Dark slate ....to	567 feet	Dark slate ....to	1120 feet
Coal ....to	572 feet	Light sand ....to	1200 feet
Gray lime ....to	576 feet	Dark slate ....to	1206 feet
Dark slate ....to	594 feet	Gray sand ....to	1223 feet
Gray lime ....to	603 feet	Dark slate ....to	1225 feet
Light slate ....to	617 feet	Dark lime ....to	1229 feet
Dark slate ....to	624 feet	Dark slate ....to	1243 feet
Gray lime ....to	630 feet	Light sand ....to	1272 feet
Light slate ....to	658 feet	Light slate ....to	1286 feet
Coal ....to	663 feet	Gray lime ....to	1289 feet
Light slate ....to	675 feet	Dark slate ....to	1295 feet
Dark slate ....to	677 feet	Dark sand ....to	1297 feet
Gray lime ....to	680 feet	Dark slate ....to	1303 feet
Dark slate ....to	766 feet	Dark sand ....to	1309 feet
Light sand ....to	770 feet	Light sand ....to	1324 feet
Dark slate ....to	775 feet	Dark slate ....to	1339 feet
Gray sand ....to	785 feet	Gray lime ....to	1342 feet
Dark slate ....to	821 feet	Gray slate ....to	1360
Gray sand ....to	825 feet	Gray sand ....to	1362 feet
Dark slate ....to	835 feet	Dark sand ....to	1366 feet
Gray sand ....to	840 feet	Light sand ....to	1372 feet
Dark slate ....to	885 feet	Brown sand ....to	1381 feet

Top of sand, 1360 feet. First show of oil, 1361 feet.

Well No. 3 on the farm of Phoebe Hayden, 200 feet north to north line, 660 feet to west line. Washington Township, Section 7, Gibson County.

Casing Record.

Fourteen feet wood conductor, 1,298 feet 8¼ inch casing, 134 feet 6⅝ inch casing (liner).

Shot Record.

One hundred twenty quarts, 1,398 to 1,416 feet.

Formation Record.

Formation	Depth	Formation	Depth
Soil ....to	1 foot	Dark slate ....to	65 feet
Yellow clay ....to	11 feet	Coal ....to	67 feet
Yellow sand ....to	60 feet	Dark slate ....to	71 feet

Phoebe Hayden Well No. 3—Continued

Formation	Depth	Formation	Depth
Light lime ....to	73 feet	Gray sand ....to	788 feet
Light slate ....to	110 feet	Dark slate ....to	839 feet
Dark slate ....to	127 feet	Brown lime ....to	845 feet
Gray sand ....to	130 feet	Light slate ....to	855 feet
Dark slate ....to	150 feet	Gray lime ....to	860 feet
Gray lime ....to	205 feet	Gray sand ....to	872 feet
White sand ....to	225 feet	Dark sand ....to	888 feet
Blue slate ....to	250 feet	Dark sand ....to	925 feet
Gray lime ....to	253 feet	Dark slate ....to	930 feet
Dark slate ....to	262 feet	Dark sand ....to	945 feet
Light lime ....to	300 feet	Dark slate ....to	965 feet
Dark slate ....to	320 feet	Dark sand ....to	984 feet
Coal ....to	323 feet	Dark slate ....to	1012 feet
Dark slate ....to	330 feet	White sand ....to	1040 feet
Light slate ....to	400 feet	Dark slate ....to	1088 feet
Gray lime ....to	410 feet	Coal ....to	1090 feet
Light slate ....to	420 feet	Dark slate ....to	1114 feet
Gray sand ....to	440 feet	Light slate ....to	1119 feet
Dark slate ....to	449 feet	Light sand ....to	1135 feet
Light sand ....to	455 feet	Dark slate ....to	1175 feet
Dark lime ....to	463 feet	Light sand ....to	1185 feet
Light sand ....to	480 feet	Dark slate ....to	1190 feet
Light slate ....to	525 feet	Dark lime ....to	1193 feet
Dark slate ....to	560 feet	Dark slate ....to	1197 feet
Black slate ....to	567 feet	Dark sand ....to	1270 feet
Dark slate ....to	577 feet	Dark slate ....to	1272 feet
Light slate ....to	590 feet	Brown lime ....to	1274 feet
Dark slate ....to	597 feet	Light sand ....to	1298 feet
Gray sand ....to	599 feet	Brown lime ....to	1302 feet
Dark slate ....to	603 feet	Light sand ....to	1308 feet
Dark lime ....to	607 feet	Dark slate ....to	1317 feet
Coal ....to	612 feet	Gray lime ....to	1319 feet
Light sand ....to	625 feet	Dark slate ....to	1340 feet
Dark slate ....to	635 feet	Light sand ....to	1359 feet
Dark lime ....to	650 feet	Dark slate ....to	1369 feet
Light slate ....to	665 feet	Gray lime ....to	1383 feet
Brown lime ....to	672 feet	Dark slate ....to	1398 feet
Light slate ....to	685 feet	Dark sand ....to	1403 feet
Dark lime ....to	700 feet	Light sand ....to	1408 feet
Dark slate ....to	720 feet	Gray sand ....to	1415 feet
Dark lime ....to	725 feet	Dark slate ....to	1425 feet
Dark slate ....to	775 feet		

Gas at 1357 feet; 8¼ inch set at 1298 feet; first oil, 1400 feet.

McRoberts Well No. 2. Section 6, Washington Township, Gibson County.

Location		Depth	Location		Depth
Yellow clay	to	25 feet	Sandy lime	to	1003 feet
Shelly slate	to	38 feet	Darke slate, show of oil	to	1042 feet
Light slate	to	116 feet	Gray sand	to	1057 feet
Dark slate	to	218 feet	Dark slate	to	1088 feet
Light lime	to	296 feet	White sand	to	1129 feet
Light sand	to	327 feet	Gray lime	to	1138 feet
Light slate	to	340 feet	White slate	to	1152 feet
Light sand	to	354 feet	Light water sand	to	1207 feet
Broken lime	to	385 feet	Dark slate	to	1240 feet
Light slate	to	407 feet	Gray lime	to	1255 feet
White lime	to	449 feet	Brown slate	to	1286 feet
Light slate	to	502 feet	Brown lime	to	1291 feet
Dark slate	to	536 feet	Brown slate	to	1315 feet
Light slate	to	551 feet	Light sand	to	1318 feet
Dark slate	to	560 feet	Light slate	to	1326 feet
Light lime	to	568 feet	Light lime	to	1331 feet
Light slate	to	621 feet	Light slate	to	1337 feet
Dark slate	to	662 feet	Light sand	to	1342 feet
Light slate	to	748 feet	Light lime	to	1347 feet
Dark slate	to	763 feet	Dark slate	to	1358 feet
Sandy slate	to	783 feet	Light lime	to	1363 feet
Dark slate	to	814 feet	Light sand, water	to	1368 feet
Light slate	to	833 feet	Light sand	to	1382 feet
Gray lime	to	842 feet	Dark slate	to	1448 feet
Dark slate	to	851 feet	Gray lime	to	1453 feet
Light slate	to	859 feet	Dark sand	to	1484 feet
Light slate	to	909 feet	Dark slate	to	1494 feet
Water sand	to	953 feet	Brown lime	to	1501 feet
Dark slate	to	984 feet	Dark slate	to	1516 feet
Brown lime	to	989 feet	White sand	to	1522 feet
Dark slate	to	998 feet	Brown sand	to	1531 feet

Total depth of well, 1,532 feet. Heavy showing of gas. Well completed on September 25, 1919.

Log of L. W. McDonald Well No. 6, located in S. W. corner of the N. E. ¼ of the N. W. ¼ of Section 7, Washington Township, Gibson County.

Top of sand	1323 feet
Oil pay	1324 to 1341 feet
Shelly gray sand	1341 to 1349 feet
Bottom of well	1349 feet

Finished December 4, 1919.

Well No. 1, Ellis Lucas, Section 33, Montgomery Township, Gibson County.

Location	Depth	Location	Depth
Dark soil ....to	10 feet	Dark slate ....to	1360 feet
Light sand ....to	50 feet	Brown sand ....to	1390 feet
White slate ....to	75 feet	Black slate ....to	1425 feet
Broken sand, water.....to	90 feet	Lime shell ....to	1430 feet
White slate ....to	135 feet	Light slate ....to	1440 feet
White sand ....to	145 feet	Light lime ....to	1450 feet
White slate ....to	225 feet	Light slate ....to	1490 feet
Broken sand ....to	235 feet	Dark slate ....to	1500 feet
Gray lime ....to	260 feet	Light slate ....to	1510 feet
Dark slate ....to	430 feet	Dark slate ....to	1520 feet
Lime shell ....to	435 feet	Light slate ....to	1540 feet
White sand ....to	490 feet	Light lime ....to	1565 feet
White slate ....to	555 feet	Light slate ....to	1575 feet
Lime shell ....to	575 feet	Dark lime ....to	1585 feet
Coal ....to	580 feet	Light slate ....to	1600 feet
Black slate ....to	610 feet	Light lime ....to	1605 feet
Gray lime ....to	615 feet	Light slate ....to	1610 feet
Sand ....to	645 feet	Dark lime ....to	1615 feet
White slate ....to	690 feet	Light slate ....to	1625 feet
White lime ....to	695 feet	Sharp sand ....to	1640 feet
Dark slate ....to	790 feet	Black slate ....to	1660 feet
Sand ....to	800 feet	Light lime ....to	1670 feet
Coal ....to	806 feet	Light slate ....to	1690 feet
Lime ....to	811 feet	Sand ....to	1710 feet
White slate ....to	850 feet	Black slate ....to	1718 feet
Broken lime ....to	865 feet	Black lime ....to	1721 feet
White slate ....to	890 feet	Black slate ....to	1735 feet
White lime ....to	910 feet	Gray lime ....to	1745 feet
White slate ....to	920 feet	Slate and lime ....to	1780 feet
White lime ....to	950 feet	Blue slate ....to	1805 feet
Brown slate ....to	990 feet	Light limestone ....to	1815 feet
Brown lime ....to	1005 feet	Blue slate ....to	1875 feet
Dark slate ....to	1007 feet	Lime shell ....to	1880 feet
White sand ....to	1095 feet	Dark slate ....to	1920 feet
White slate ....to	1100 feet	Gray lime ....to	1925 feet
Gray lime ....to	1110 feet	Dark slate ....to	1940 feet
Brown slate ....to	1120 feet	Gray sand ....to	1945 feet
Brown lime ....to	1130 feet	Dark sand ....to	1995 feet
Light slate ....to	1140 feet	Dark lime ....to	2000 feet
Light sand ....to	1215 feet	Dark lime ....to	2010 feet
Dark slate ....to	1235 feet	Dark slate ....to	2040 feet
Dark lime ....to	1245 feet	Light sand ....to	2070 feet
Dark sand ....to	1255 feet	Dark lime ....to	2075 feet
Dark slate ....to	1310 feet	Dark sand ....to	2080 feet
Broken sand ....to	1330 feet		

Well completed December 15, 1919. Dry hole; abandoned.

Since the publication of the report on petroleum and natural gas in Indiana in 1920 much additional material has been obtained on the petrolific conditions of Gibson County, but preparation of other parts of this volume has prevented the working of this material into condition for publication. This will be done in the near future.

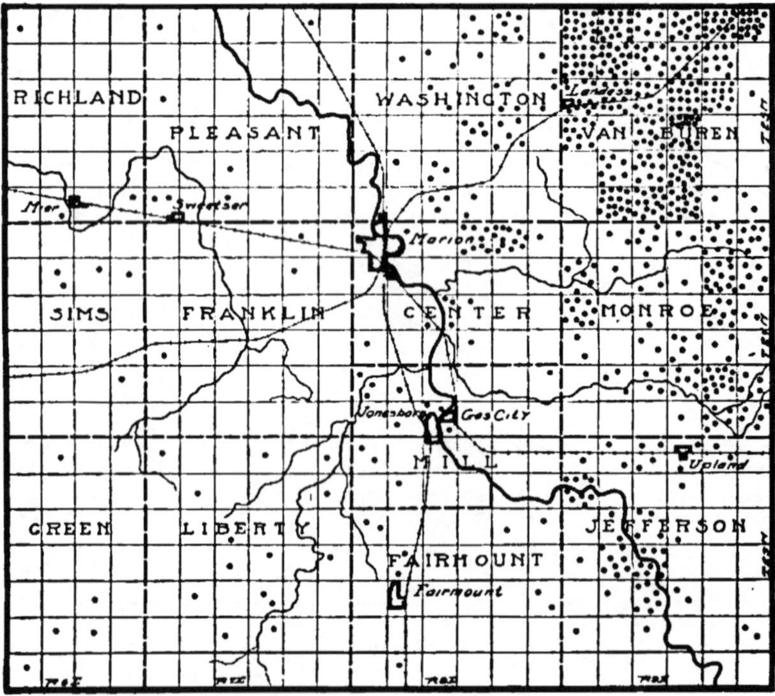

Fig. CXXV. Map of Grant County showing location of recorded abandoned wells. The northeastern part of this county is oil territory and the southeastern part is gas territory.

GRANT COUNTY.

Grant County is covered with glacial drift which varies in thickness from 100 to 425 feet. Except for exposures along the Mississinewa River and Pike Creek the bed rock is completely concealed. The drift rests on the Silurian surface which has been greatly eroded.

Centre Township. The city of Marion in this township was one of the first to prospect for oil and gas. The first well reached the Trenton at a depth of 865 feet, or 60 feet below sea level and produced 350,000 cubic feet of gas daily, but after being deepened produced two million cubic feet daily. The second well reached the Trenton at 880 feet or 83 feet below sea level. At the top of the sand it produced 350,000 cubic feet, 35 feet deeper it produced five million cubic feet daily.

The following table gives the records of some of the wells drilled at Marion:

Record of Marion Wells.

No. of Well	Altitude	Depth of Trenton	Relation to Sea Level	Thickness of Drift	Production in Cu. Ft.
1	840	900	60 feet below	..	2,000,000
2	797	880	83 feet below	..	5,000,000
3	820	878	58 feet below	..	3,500,000
4	802	880	78 feet below	..	2,500,000
5	830	1000	70 feet below	..	4,000,000
6	...	908	....	32	350,000
7	701	904	....	..	3,000,000
8	...	....	....	..	3,000,000
9		....	....	..	1,500,000
10		....	....	..	Oil and gas
11		....	....	..	7,425,000
12		....	....	..	Oil and 350,000
13	.	....	....	..	5,642,000
Soldiers' Home					
2	.	....	....	20	Salt water

Record of Well No. 6.

Drift	32 feet
Limestone	250 feet
Niagara shale	40 feet
Hudson River	336 feet
Utica shale	250 feet
Total depth	908 feet

Wells abandoned in 1911 are located as follows: Section 1, 1 well; Section 2, 12 wells; Section 3, 8 wells; Section 5, 1 well; Section 15, 1 well; Section 16, 4 wells; Section 19, 1 well; Section 22, 1 well.

Mill Township. A well drilled at Jonesboro produced 5,567,000 cubic feet of gas. It was called the "Cyclone" on account of its pressure. A record of the well is given by Phinney as follows:

Drift	162 feet
Limestone	148 feet
Bluish green shale	225 feet
Gray shale	180 feet
Brown shale	197 feet
Trenton	23 feet
Total	935 feet
Altitude at well about	834 feet

The following wells have been abandoned: Section 6, 2 wells, 1911; Section 8, 1 well, 1911; Section 29, 1 well, 1912; Section 30, 3 wells, 1912; Section 32, 1 well, 1913; Section 33, 2 wells, 1912.

Oil has been obtained from nearly all of the sections in this township and gas from many.

Fairmount Township. The first well drilled at Fairmount produced 11,-500,000 cubic feet of gas per day. A second well produced 5,000,000 cubic feet per day.

Record of Fairmount Well No. 1 (Phinney).

Drift	35 feet
Limestone	290 feet
Shale	609 feet
Trenton limestone	31 feet
Total depth	965 feet
Altitude of well	893 feet

A well drilled in Section 25, has a record as follows (Blatchley):

Drive pipe	190 feet
Casing	370 feet
Trenton limestone	950 feet
Oil sand	975 feet
2d oil sand	1030 feet
Total depth	1050 feet
Initial production	50 barrels

Well in Section 2.

Drive pipe	170 feet
Casing	380 feet
Top of Trenton	960 feet
Oil sand	990 feet
Second oil sand	1025 feet
Total depth	1040 feet

Wells have been abandoned in this township as follows: Section 5, 1 well, 1912; Section 13, 1 well, 1912; Section 14, 1 well, 1912; Section 18, 1 well, 1911; Section 20, 1 well, 1911; Section 25, 1 well, 1911.

Jefferson Township. The first well drilled at Upland reached the Trenton at 1,010 feet. The oil sand was 10 feet thick and the total depth of the well 1,040. The drift was 185 feet thick. Sixteen wells were drilled in this township in 1906, 13 were light producers. Forty wells were abandoned the same year. Below is a record of four wells in this township, all of which produce gas, and the first one oil.

	S. E. ¼ of N. W. ¼ of Sec. 28	Sec. 19	Sec. 19	Sec. 17
Drive pipe	100	187	162	100
Casing	420	375	375	365
Top of Trenton	920	933	925	886
Total depth	1020	1035	953	911

Well No. 2, Davis Farm, Jefferson Township, Grant County, Indiana.

Clay, gravel and quicksand	107 feet
Limestone	247 feet
Slate	577 feet
Trenton rock at	931 feet
In Trenton	102 feet
Total depth	1033 feet

Well No. 14, Mary Anderson Farm, Jefferson Township, Grant County, Indiana.

Clay, gravel and quicksand	112 feet
Limestone	268 feet
Slate	563 feet
Trenton rock at	943 feet
In Trenton	107 feet
Total depth	1050 feet

Well No. 1, Highline Farm, Jefferson Township, Granty County, Indiana.

Clay, gravel and quicksand	129 feet
Limestone	227 feet
Slate	580 feet
Trenton rock at	938 feet 6"
In Trenton	102 feet
Total depth	1041 feet

Well No. 2, A. D. Mittank Farm, Jefferson Township, Grant County.

Clay, gravel and quicksand	116 feet
Limestone	249 feet
Slate	549 feet
Trenton rock at	916 feet
In Trenton	100 feet
Total depth	1016 feet

Well No. 3, A. D. Mittank Farm.

Clay, gravel and quicksand	102 feet
Limestone	256 feet
Slate	558 feet
Trenton rock at	916 feet
In Trenton	107 feet
Total depth	1023 feet

Every section in this township has produced either oil or gas or both. The following wells have been abandoned: Section 2, 2 wells; Section 3, 2 wells; Section 4, 4 wells; Section 5, 2 wells; Section 6, 1 well; Section 7, 9 wells; Section 10, 1 well; Section 15, 1 well; Section 16, 4 wells; Section 17, 2 wells; Section 19, 1 well; Section 20, 8 wells; Section 21, 9 wells; Section 27, 2 wells; Section 28, 7 wells; Section 29, 1 well; Section 31, 1 well; Section 33, 1 well; Section 36, 2 wells.

Monroe Township. The records of two wells are given below:

	Sec. 12	Sec. 36
Drive pipe	425 feet	227 feet
Casing	430 feet	403 feet
Top of Trenton	990 feet	995 feet
Gas sand		1030 feet
Water		1049 feet
Total depth	1050 feet	1077 feet

More than one-half of the sections in this township have produced oil or gas or both.

Abandoned wells: Section 1, 8 wells; Section 2, 8 wells; Section 3, 2 wells; Section 4, 5 wells; Section 5, 2 wells; Section 7, 6 wells; Section 8, 1 well; Section 9, 3 wells; Section 10, 1 well; Section 11, 14 wells; Section 12, 6 wells; Section 13, 2 wells; Section 14, 8 wells; Section 15, 1 well; Section 16, 1 well; Section 17, 1 well; Section 18, 10 wells; Section 19, 2 wells; Section 21, 2 wells; Section 22, 6 wells; Section 23, 3 wells; Section 25, 5 wells; Section 26, 14 wells; Section 27, 6 wells; Section 28, 4 wells; Section 29, 2 wells; Section 34, 1 well; Section 35, 6 wells; Section 36, 6 wells.

Pleasant Township. Light producing wells have been found in this township. Two wells were drilled near Jalapa in 1901 and 1903, both light producers. Wells abandoned are located as follows: Section 2, 1 well; Section 6, 1 well; Section 18, 1 well; Section 28, 1 well; Section 31, 2 wells; Section 33, 1 well.

Richland Township. Salt water was obtained in a well drilled about two miles from the north boundary. Wells were abandoned in Section 4, 1 well; Section 34, 1 well; Section 36, 1 well.

Sims Township. A strong gas supply was obtained at Swayzee. Two wells were put down in Section 12, both produced a small supply of oil. Wells abandoned are as follows: Section 2, 1 well, 1913; Section 9, 1 well, 1913; Section 10, 2 wells, 1911; Section 25, 1 well, 1912.

Van Buren Township. The first well drilled contained both oil and gas. A record of the well drilled at Van Buren is given below:

Log of Van Buren Well No. 1.

Drift	91 feet
Limestone	300 feet
Shale	559 feet
Trenton limestone	23 feet
Total depth	973 feet
Altitude of well	843 feet

The following are records of other wells drilled in this township:

	Sec. 2	Sec. 7	Sec. 17
Drive pipe	174	156	412
Casing	460	439	441
Top of Trenton	992	1003	972
Gas	1007	1012	987
Oil (first)	1020	1018	1005
Oil (best)		1038	
Total depth	1046	1085	1032
Initial production	20 bbl.	65 bbl.	30 bbl.

Every quarter section of land in this township has been a producer of oil or gas or both. The following wells have been plugged: Section 1, 15 wells; Section 2, 23 wells; Section 3, 11 wells; Section 4, 7 wells; Section 5, 19 wells; Section 6, 33 wells; Section 7, 22 wells; Section 8, 21 wells; Section 9, 11 wells; Section 10, 11 wells; Section 11, 11 wells; Section 13, 1 well; Section 14, 19 wells; Section 15, 56 wells; Section 16, 6 wells; Section 17, 10 wells; Section 18, 11 wells; Section 19, 11 wells; Section 20, 1 well; Section 21, 12 wells; Section 22, 19 wells; Section 27, 16 wells; Section 28, 10 wells;

Section 29, 8 wells; Section 32, 20 wells; Section 33, 29 wells; Section 34, 38 wells; Section 35, 1 well; Section 36, 2 wells.

Washington Township. A few sections in the northwest corner of the township are the only ones that have not been productive. Wells No. 1 on the N. M. Bradford, and No. 1 on the Ira Bradford farm, in the north half of the southeast ¼ of 16, and No. 11 on the J. T. Bradford in the S. W. ¼ of 16, had the following records:

	No. 1 N. M. B. Feet	No. 2 I. B. Feet	No. 11 J. T. B. Feet
Drive pipe	285	256	341
Casing	509	409	442
Top of Trenton	995	996	994
Gas	1020	1020	1020
Best oil	1040	1030	1055
Total depth	1071	1071	1094
Initial production, bbls.	25	60	15

Section 2, E. J. Hunt Farm, S. W. ¼. An average well on the lease shows the following record:

Drive pipe	300 feet
Casing	500 feet
Top of Trenton	980 feet
Total depth	1055 feet

On Section 3, one mile west of the above farm, a record of bore No. 1 was as follows (the well was a fair producer):

Drive pipe	199 feet
Casing	504 feet
Top of Trenton	1004 feet
Gas struck at	1014 feet
First oil pay	1019-1040 feet
Salt water	1040-1045 feet
Second oil pay	1055-1070 feet
Total depth	1070 feet

A well in the N. E. ¼ of Section 11 had the following record:

Drive pipe	250 feet
Casing	455 feet
Top of Trenton	1014 feet
First pay	1026 feet
Salt water	1073 feet
Total depth	1077 feet

The above well started at 60 barrels.

An average record of ten wells drilled on the Cory lease, west half of the northwest ¼, up to October 1, 1903, is as follows:

Drive pipe	104 feet
Casing	460 feet
Top of Trenton	1001 feet
Total depth	1079 feet

Most of the wells came in with an initial production of 35 to 50 barrels.

A well on the L. W. Smith Farm, Section 16, south half of the N. W. ¼, has the following record:

Drive pipe	220 feet
Casing	470 feet
Top of Trenton	930 feet
Total depth	1000 feet

Section 28, N. E. ¼.

Drive pipe	286 feet
Casing	420 feet
Top of Trenton	987 feet
Struck gas at	1000 feet
Total depth	1074 feet

The well yielded 2,000,000 feet of gas a day for twenty days, with no showing of oil. At the end of that time is was shot with 160 quarts, when a pocket of oil near the bottom of the bore was evidently broken into, as the fluid rose 20 feet above the derrick. The well made 24 barrels the first day and settled down into a fair producer.

The Hawkins lease, on the N. W. ¼ of Section 34, has 7 or 8 fair producers. The record of No. 7 being as follows:

Drive pipe	173 feet
Casing	440 feet
Top of Trenton	997 feet
First oil pay	1027 feet
Second oil pay	1054 feet
Total depth	1070 feet

Abandoned wells are located as follows: Section 2, 8 wells, 1913; Section 3, 3 wells, 1913; Section 9, 2 wells, 1912; Section 10, 1 well, 1916; Section 12, 2 wells, 1912; Section 13, 3 wells, 1912; Section 14, 4 wells, 1912; Section 15, 1 well, 1913; Section 16, 2 wells, 1912; Section 22, 5 wells, 1913; Section 23, 5 wells, 1913; Section 28, 3 wells, 1913; Section 33, 3 wells, 1912; Section 34, 2 wells, 1912.

Green Township. Abandoned wells are located as follows:

Owner	Date	Section	Range	Wells
E. Pennington	1912	3	6E	1
Joe Hoe	1913	4	6E	1
J. J. Johnson		16	6E	1
G. M. Kilgore	1912	26	6E	1
N. J. Lacure	1912	34	6E	1
Chas. Lear	1913	35	6E	1

Liberty Township. A list of the abandoned wells is given below:

Owner	Date	Section	Range	Wells
A. W. Jay	1912	1	7E	1
Henry Daugherty	1913	3	7E	1
A. Gimmell	1912	6	7E	1

P. and N. Muchmore........	1912	8	7E	1
Thos. Shady	1912	12	7E	1
F. A. Stewart..............	1912	16	7E	1
W. W. Elliott..............	1912	21	7E	1
John Harold	1912	22	7E	1
Jessie Haisley	1912	24	7E	1
Frank Mason & Webb Winslow	1912	27	7E	2
Woodie Clark		29	7E	1
Thos. Shady		33	7E	1
Wm. Harvey		34	7E	1

Franklin Township. Wells were drilled and abandoned as follows:

Owner	Date	Section	Range	Wells
H. J. Paulus................	1912	2	7E	1
B. D. Tharp................	1911	11	7E	1
Mat Sheffield	1911	19	7E	1

GREENE COUNTY.

The mantle of glacial drift covering this county is light, varying from five to fifty feet in thickness except in the White River Valley where it may exceed one hundred feet. The rock strata underlying the drift belongs to the Mississippian and the Pennsylvanian periods. In the eastern part of the county the structure may be determined by locating elevations on the surface of some of the Chester limestones which may be used as datum planes for drawing structural contours. In the western part of the County where the coal measures outcrop the coal beds may be used, with proper methods of discrimination, for a like purpose. The surface of the Trenton limestone probably lies from 2,000 to 2,500 feet below the surface in this county. The Devonian, which may be oil bearing where the structure is favorable, may be reached at depths ranging from 1,500 to 1,800 feet.

Jefferson Township. A well drilled at Worthington reached water in the Niagara limestone at 1,430 feet. The well was completed at 1,445 feet.

Taylor Township. A well drilled in Taylor Township has the following record:

Well No. 1 on Section 31.

Surface to	15 feet—Soil, drift and mud.	
15 to	20 feet—Quicksand	5 feet
20 to	40 feet—Soft mud	20 feet
40 to	45 feet—Limeshell	5 feet
45 to	72 feet—Shale and water..............	27 feet
72 to	80 feet—Limeshell	8 feet
80 to	100 feet—Shale and water..............	20 feet
100 to	120 feet—Lime	20 feet
120 to	125 feet—Broken shale	5 feet
125 to	250 feet—Limestone full of water.	
250 to	300 feet—Soft black mud.	
300 to	310 feet—Limeshell	10 feet
310 to	610 feet—Hard limestone	300 feet
610 to	615 feet—Soft lime	5 feet

At 610 feet lime got soft and brown, with a smell of gas and a rainbow the color of oil was just noticeable.

615 to 710 feet—Brown limestone.
710 to 800 feet—Brown lime full of water.
800 to 1250 feet—Black shale.
1250 to 1285 feet—Limeshell.
1285 to 1290 feet—Very hard lime.
1290 to 1400 feet—Dark shale.
1400 to 1487 feet—Brown shale.
1487 to 1642 feet—Niagara rock.

Total depth of well..............................1642 feet

Washington Township. A small gas and oil field was located at Lyons. The production was never very large. Wells were abandoned in Section 4, Section 6, Section 9, Section 11, Section 15 and Section 16. The following is a record of the Kaufman well:

Drift	26 feet
Sandy lime	60 feet
Coal	4 feet
Sand and water	86 feet
Slate	20 feet
White lime	30 feet
Red rock	35 feet
Sandy slate	10 feet
Dark slate	55 feet
Bedford lime	8 feet
Dark shaley lime	342 feet
Shell and lime	100 feet
Brown slate and water	10 feet
Black lime	40 feet
Hard white lime	50 feet
Slate and shale	60 feet
White lime	40 feet
Black slate	250 feet
Brown sand	50 feet
White slate	238 feet
Trenton rock	221 feet

Still in Trenton when finished at 1,959. Big water at 1,950. Filled up to 1900. This well probably finished in the Niagara rather than the Trenton. Casing record: 10 inch, 209 feet; 8 inch, 620 feet; 6⅝ inch, 1,188 feet.

Stafford Township. Two wells were drilled in this township, one on the property of J. L. Morgan and in Glenns Valley.

HAMILTON COUNTY.

The bed rock formations of this county belong to the Silurian and Devonian periods of geologic time. These formations are largely concealed by glacial drift varying in thickness from 50 to 300 feet. The surface of the Trenton lies from 800 to 1,200 feet below the surface and for the greater part of the county is above sea level. The dip of the strata is southwest.

Noblesville Township. A well drilled at Noblesville gave the following log:

Drift	140 feet
Limestone	286 feet
Shale	410 feet
Trenton limestone	7 feet
Total	843 feet
Altitude of well	750 feet

Many gas wells are located in this Township. Abandoned wells are located in Section 11, 1 well; Section 17, 1 well; Section 18, 2 wells.

Delaware Township. Gas wells were located at New Britton and Fishers.

Fall Creek Township. Oil wells were located in Sections 1, 2, 36 and others.

Logs of Wells in Section 2'.

	No. 1	No. 2	No. 3
Drive pipe	56	54	54
Casing	380	384	381
Top of Trenton	886	889	885
Best oil at	914	918	914
Total depth	926	955	935
Initial output, bbls.	65	2	50

Jackson Township. Oil wells were located in Sections 5, 6, 31, 33, 36 and others. Three abandoned wells are located in 28 and one in 23. Logs of some of the wells are as follows:

	Sec. 6	Sec. 5	Sec. 36
Drive pipe	203	240	70
Casing	525	545	
Top of Trenton	1003	1010	916
Total depth	1063	1064	927

The record of a well drilled at Cicero is given below:

Drift	161 feet
Niagara limestone and shale	300 feet
Hudson River and Utica	490 feet
Trenton limestone	32 feet

Phinney gives the following record of a well drilled at Arcadia:

Drift	130 feet
Limestone	120 feet
Blue limestone	130 feet
Shale	581 feet
Trenton limestone	13 feet
Total depth	974 feet
Altitude of well	868 feet

Adams Township. At Sheridan gas was obtained at 1,076 feet and the top of the Trenton at 1,069 feet.

Washington Township. At Westfield the top of the Trenton was reached at 1,040 feet and salt water at 1,080 feet. Blatchley gives the records of five wells in this township, the first three are in the S. W. ¼ of Section 13, and the last two in the east half of Section 20.

	No. 1	No. 2	No. 3	No. 4	No. 5
Drive pipe	305	231	234	160	161
Casing	560	500	500	515	515
Top of Trenton	1024	1020	1022	1005	1000
Total depth	1042	1037	1050	1032	1019

An abandoned well is located in Section 26, on the Allen Stalker property.

Clay Township. At Carmel gas was obtained.

Wayne Township. Abandoned wells are located in this territory as follows: One well each in Sections 3, 5, 9, 10, 17 and 20 and two in Section 9.

White River Township. Abandoned wells are located as follows: One each in Sections 3, 9, 10, 17 and 34.

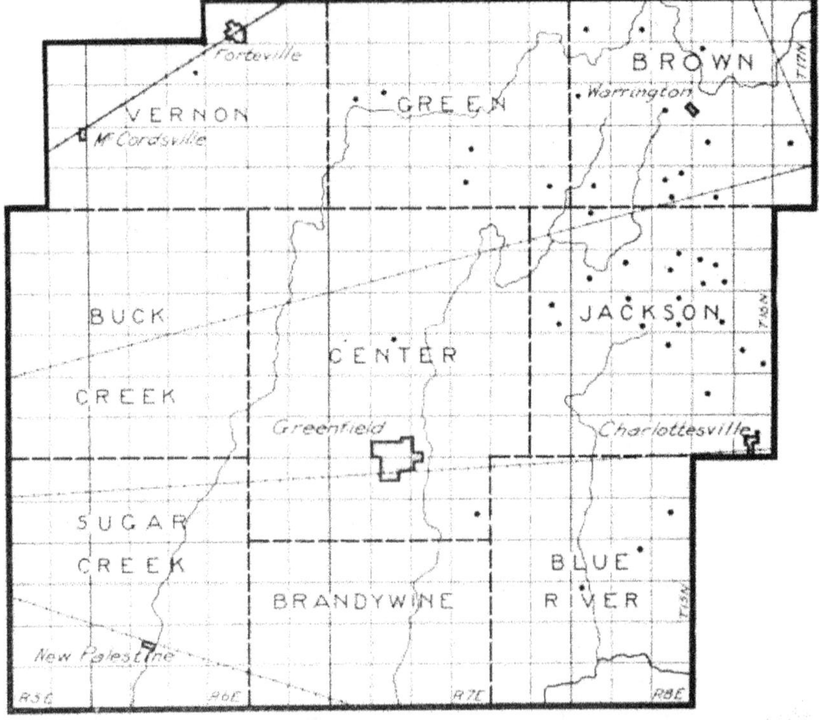

Fig. CXXVI. Map of Hancock County showing abandoned wells. Gas areas occur in the following townships: Vernon, Buck Creek, Brandywine and Blue River.

HANCOCK COUNTY.

This county is covered with glacial drift varying in thickness from 50 to 250 feet. The durolith formations belong to the Silurian and the Devonian periods. In the greater part of the county the surface of the Trenton is above sea level; in the southwest corner of the county it lies below sea level.

Centre Township. Productive gas wells were drilled at Greenfield. A record of Well No. 1 as given by Phinney is below:

Drift	215	feet
Corniferous limestone	65	feet
Shale (Upper Niagara)	17	feet
Limestone (Niagara)	68	feet
Shale	135	feet
Black shale	45	feet
Bluish green shale	138	feet
Limestone	2	feet
Brown shale	300	feet
Trenton limestone	14½	feet
Total depth	999½	feet
Altitude of well	902	feet

Wells drilled on the property of Joe Docman and Max Franks were abandoned in 1912 and one on the property of Joe Branny in Section 20 in 1913.

Sugar Creek Township. The record of a well drilled at Palestine is as follows:

Drift	285	feet
Limestone	122	feet
Shale	593	feet
Trenton limestone	60	feet
Total	1060	feet
Altitude of well	839	feet
Salt water at	1003	feet

Vernon Township. Gas was obtained in wells at Fortville and McCordsville, Vernon Township. The following wells were abandoned, one each on the property of Wm. Fort and J. Lindamood and one on the property of Nelson Fort in Section 16, all in 1913.

Greene Township. Wells abandoned in this township are located:

Owner	Section	Date	Wells
Sarah Martin	19	1912	1
Mark O'Mailey	20	1913	1
S. E. Stubbs	27	1916	1
David Jones	34	1916	1
Ora Peacock	36	1919	1

Brown Township. Abandoned wells are located as follows:

Owner	Section	Date	Wells
Harry Davies	7	1916	1
H. Cook	8	1919	1
J. W. Hedrick	14	1911	1
Madison Brooks	19	1913	1
Frank Burgis	21	1912	1
Hayes	25	1913	1

Owner	Section	Date	Wells
Joe Van Matre	27	1913	1
Carwood	33	1913	1
W. Keck	33	1913	1
Joe Van Matre	33	1913	1

Jackson Township. Gas was obtained at Charlottsville in Section 35, and in many other sections. The following wells have been abandoned: Section 6, 1 well; Section 7, 1 well; Section 8, 1 well; Section 9, 2 wells; Section 10, 4 wells; Section 13, 2 wells; Section 15, 1 well; Section 16, 2 wells; Section 17, 2 wells; Section 21, 1 well; Section 23, 2 wells; Section 27, 1 well; Section 35, 3 wells.

Blue River Township. The following wells have been abandoned in this Township: One each in Sections 9, 10, 17 and 19.

HARRISON COUNTY.

Harrison County lies wholly within the unglaciated area of the State. The greater part of its surface is occupied by the Mitchell peneplain through the surface of which the major streams have cut to the underlying formations. The strata represented by outcrops in the county belong to the following divisions:

Quaternary	Recent—Residual clays and alluvium. Pleistocene—Possible residuals.
Mississippian	Chester sandstones, limestones and shale. Mitchell limestone. Salem limestone. Harrodsburg limestone. Knobstone shales and sandstones.

In that portion of the county occupied by the Mitchell limestone the determination of structural conditions will be difficult because of the absence of definite and persistent horizons in the Mitchell. Where numerous outcrops of the Knobstone-Harrodsburg contact can be found, this may be used as a key horizon. If structural conditions are favorable, oil and gas reservoirs may be found in the Trenton, Silurian and Devonian limestones. Gas has been obtained at Tobacco Landing from the Devonian. A record of one of the wells follows:

Section of Well No. 1.

Keokuk limestone	15 feet
Knobstone	390 feet
Depth to Devonian shale	405 feet

A good flow of gas was found in the Devonian shale. The gas pressure in 1911 was from 60 to 110 pounds. In 1914 it was only 50 pounds. Gas and oil wells range in depth from sixty to nine hundred feet. Six oil wells range in depth from 135 feet to 700 feet. The initial production was from five to thirty barrels per day.

HENDRICKS COUNTY.

The strata underlying the glacial drift in this county belong to the Devonian and Mississippian periods. The New Albany shale occupies the subsurface in the eastern part of the county and the Knobstone in the western portion. The glacial drift conceals the bed rock almost completely and reaches a thickness of two hundred feet.

A well was drilled at Plainfield at an altitude of 742 feet. The total depth was 1,386 feet and a slight flow of gas was obtained at a depth of 350 feet.

The surface of the Trenton in all parts of this county is below sea level, probably 400 to 600 feet. If oil or gas in quantity is obtained in this county it will probably be in terraces or spurs or small domes connected with the Cincinnati geanticline. The position of such structures, if they exist, cannot be determined by surficial observations because the outcrop of the strata is concealed largely by the drift. Not enough well records have been secured to enable one to secure sufficient data for subsurface work.

HENRY COUNTY.

The surface of the durolith of Henry County is formed by Silurian strata (Niagara limestone), which is covered with glacial drift varying in thickness

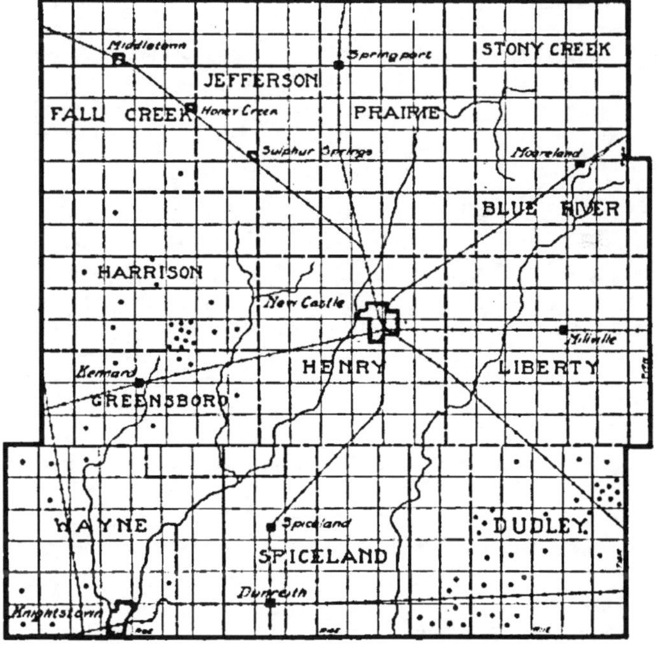

Fig. CXXVII. Map of Henry County showing location of abandoned wells. The northern tier of townships is in gas territory.

from 25 to 500 feet. The surface of the Trenton lies from 500 to 1,200 feet below the surface of the county and for a large part of the county is above sea level.

Henry Township. Well No. 1 at New Castle has the following log:

Drift	333 feet
Hudson River shales	200 feet
Utica shales	343 feet
Trenton limestone	421 feet
Total depth	1297 feet
Trenton above sea level	104 feet

The surface of the Trenton about New Castle varies in height above sea level from 104 to 137 feet, the average is about 125 feet.

Prairie Township. A well drilled at Mt. Summit gave the following log:

Drift	230 feet
Limestone	50 feet
Shale	736 feet
Trenton limestone	66 feet
Total depth	1082 feet
Altitude of well	1110 feet

Two wells were drilled at Springport, the record of the second follows:

Drift	156 feet
Limestone	90 feet
Bluish green shale	600 feet
Black shale	111 feet
Trenton limestone	63 feet
Total depth	1020 feet
Altitude of well	1004 feet

Spiceland Township. At Spiceland a well was drilled which has the following log:

Drift	151 feet
Hard cherty limestone	62 feet
Limestone	67 feet
Shale	10 feet
Bluish green and brown shale	710 feet
Trenton limestone	62 feet
Total depth	1002 feet
Altitude of well	1023 feet

Wayne Township. A well drilled at Knightstown has the following log:

Drift	64 feet
Niagara limestone	200 feet
Hudson River limestone and shale	360 feet
Utica shale	199 feet
Trenton limestone	213 feet
Total depth	1036 feet
Trenton above sea level	113 feet

At this point the surface of the Trenton varies from 112 to 121 feet above sea level. Three wells were drilled in Knightstown recently. The records of these wells are given below. No. 1 was drilled on the lot of Mrs. Walter Garrison; No. 2 on lot of James Oakerson; No. 3 on lot of L. P. Wenly.

	No. 1	No. 2	No. 3
Drift to lime rock	57	63	60 feet
Thickness of lime rock	200	200	200 feet
Thickness of slate to shale	560	555	560 feet
To Trenton	817	818	820 feet
Drilled in Trenton	8	10	10 feet
Total depth	825	828	830 feet

HOWARD COUNTY.

A mantle of drift covers the bed rock of this county to a depth of 40 to 100 or more feet. Underlying the drift are the limestones of the Silurian period. This county was among the first to drill for gas and as early as 1886 brought in a well of 2,000,000 cubic feet capacity. The depth to the surface of the Trenton varies from 800 to 1,100 feet and the surface of the Trenton is from 50 to 350 feet below sea level.

Center Township. The township has produced much gas. The first well was drilled in 1886. The following is a list of 14 wells drilled in or near Kokomo[1]:

No. of Well	Depth to Trenton, feet	Depth to Gas, feet	Altitude feet	Trenton Below Sea Level	Thickness of Drift	Capacity in cu. ft., per day
1	912	922	825	87	....	2,000,000
2	913	922	....	....	....	1,117,000
3	905	910	830	75	65	810,000
4	936	944	....	....	....	1,500,000
5	895	901	....	....	....	4,462,000
6	889	893	....	....	....	1,555,000
7	908	912	....	....	....	3,015,000
8	904	914	....	....	....	1,072,000
9	900	....	....	....	....	2,500,000
10	902	....	....	....	40	2,800,000
11	932	....	....	....	90	2,000,000
12	....	....	....	....	....	3,650,000
13	903	....	....	....	....	3,727,000
14	905	....	....	....	....	2,330,000

Record of Well No. 4 (Wm. Moore).

Drift	65 feet
Water lime	10 feet
Bluish limestone	80 feet
White shaly limestone	15 feet
Bluish limestone	65 feet
Niagara shale (calcareous)	35 feet
Gray limestone	75 feet
Hudson River shale	255 feet
Utica shale	256 feet
Trenton limestone	22 feet
Total depth	958 feet

At Tarkington the Trenton was reached at 965 feet and the drift has a thickness of 140 feet. In Section 19, 4 wells; Section 20, 8 wells, and in Section 24, 1 well, were abandoned from 1911 to 1913.

Jackson Township. Gas wells were produced at Sycamore in this township. Wells abandoned are located as follows: Section 7, 2 wells; Section 12, 3 wells; Section 13, 3 wells; Section 17, 2 wells; Section 18, 1 well; Section 20, 1 well; Section 23, 1 well; Section 24, 5 wells; Section 26, 1 well; Section 31, 1 well; Section 32, 1 well.

Liberty Township. At Greentown a strong flow of gas was obtained. The depth of the Trenton is 936 feet, gas obtained at 965 feet, and the drift was 79 feet. Wells have been abandoned as follows: Section 4, 1 well; Section 6, 1 well; Section 7, 1 well; Section 19, 2 wells; Section 27, 2 wells.

Union Township. The Trenton was reached at 934 feet, gas at 959, and the thickness of the drift was 107 feet. Wells abandoned are located in Section 6, 1 well; Section 7, 1 well; Section 15, 1 well; Section 17, 1 well; Section 20, 2 wells; Section 21, 1 well; Section 23, 1 well; Section 29, 1 well.

Taylor Township. The Fairfield well reached the Trenton at 937, drift 55 feet, McNeal well went through 32 feet of drift and reached the Trenton at 925 feet. Wells abandoned are located in Section 12, 2 wells; Section 15, 1 well; Section 18, 1 well; Section 26, 1 well; Section 30, 3 wells.

Howard Township. The Templin well passed through 80 feet of drift and reached the Trenton at 921 feet. The Weaver well passed through 100 feet of drift and reached the Trenton at 921 feet. A well drilled on the Underwood place in Section 15 was abandoned in 1913.

Harrison Township. A well located on the property of Jackson Morrow in Section 13 and one on the property of Mary A. Frances were abandoned in 1912 and 1913.

Through the courtesy of the Indiana Natural Gas and Oil Company of Kokomo the writer obtained the records and elevations of two hundred thirty-nine wells located in Howard and Tipton counties. By the aid of these records I was able to determine the structural conditions for a considerable area in these two counties. I found that the main arch (geanticline) trends, in that region, north of west and that a series of minor folds cross the surface of the arch north of east in nearly parallel lines. It was found that the productive gas territory was confined to these minor folds. No production was recorded for the wells which had been drilled in the synclines between the minor anticlines, though some of these were high up on the main arch. The fact that approximately ten per cent of the wells on the anticlines were dry, that is produced only a show or very little gas, is explained on the ground of non-porous conditions of the rock at those points. Such lack of porosity being due to a lack of dolomitization of the Trenton limestone.

The fact, however, which has been demonstrated, is that structural conditions and not porosity of sand are the main controlling factors in the accumulation of oil and gas in the Trenton field in Indiana.

If accurate records of all the wells that have been drilled in this territory were obtainable the structural conditions of almost the entire field could be determined.

HUNTINGTON COUNTY.

The Niagara limestone underlies the glacial drift in this county. The surface of the limestone has been deeply eroded and the drift varies much in thickness. Outcrops of the limestone occur along the banks of the Salamonie River. The southern part of this county has been good oil territory in the past and the field has been extended slightly recently. The county lies on the north side of the arch and the strata dip toward the north. Structural conditions can be determined only by subsurface work.

The following are the records of some of the wells that have been drilled in this county:

Jefferson Township. Sections 7, 8, 17 and 18 were all productive territory in 1905. The following wells have been abandoned: Section 7, 2 wells; Section 13, 2 wells; Section 19, 3 wells; Section 21, 6 wells; Section 24, 1 well; Section 28, 7 wells; Section 31, 11 wells; Section 33, 27 wells; Section 34, 10 wells; Section 35, 13 wells; Section 36, 8 wells.

Salamonie Township. Twenty-five new wells were drilled in this township in 1905, which was formerly known as salt water territory. All were good producing wells.

An average record of the bores on the S. E. ¼ showed:

Drive pipe	58 feet
Casing	385 feet
Top of Trenton	1007 feet
Total depth	1087 feet

March Petroleum Co. Mill Lot Well No. 1. Located S. E. ¼ of Section 20, Salamonie Township:

Drive pipe	77 feet
Casing	406 feet
Top of sand	978½ feet
Drilled in sand	29 feet
Total depth	1007½ feet

Shot with 100 quarts April 4, 1919.

Pumped 50 barrels oil first 24 hours.

J. L. Priddy Lease No. 9. N. W. ¼ of Section 20, Salamonie Township:

Drive pipe .... 52 feet 10"
Casing .... 428 feet
Top of sand ....1007 feet
Drilled in sand .... 30 feet
Total depth ....1037 feet

Not shot, plugged April 11, 1919.

Calvin Perdue Lease No. 3, S. E. ¼ of Section 29:

Drive pipe .... 72 feet
Casing .... 425 feet
Top of sand .... 960 feet
Drilled in sand .... 22 feet
Total depth .... 982 feet

Shot with 60 quarts April 11, 1919.

Pumped 125 barrels first 24 hours.

Calvin Perdue Lease No. 4. S. E. ¼ of Section 29, Salamonie Township:

Drive pipe .... 61½ feet
Casing .... 432 feet
Top of sand .... 975 feet
Drilled in sand .... 32 feet
Total depth ....1007 feet

Shot with 100 quarts. Pumped first 24 hours, 50 barrels.

Calvin Perdue Lease No. 5. S. E. ¼ of Section 29:

Drive pipe .... 89 feet
Casing .... 432 feet
Top of sand .... 956 feet
Drilled in sand .... 25½ feet
Total depth .... 981½ feet

Shot 80 quarts. Pumped 70 barrels first 24 hours.

Calvin Perdue Lease No. 6. S. E. ¼ of Section 29:

Drive pipe .... 91 feet
Casing .... 400 feet
Top of sand .... 968 feet
Drilled in sand .... 24 feet
Total depth .... 992 feet

Shot with 80 quarts. Pumped 55 barrels first 24 hours.

Calvin Perdue Lease No. 7. S. W. ¼ of Section 29:

Drive pipe .... 32 feet
Casing .... 417 feet
Top of sand.................................. 952 feet
Drilled in sand .... 32 feet
Total depth .... 975 feet

Shot with 80 quarts. Pumped 150 barrels 24 hours.

Calvin Perdue Lease No. 8. N. E. ¼ of Section 29:

Drive pipe	44 feet
Casing	407 feet
Top of sand	976 feet
Drilled in sand	25 feet
Total depth	1001 feet

Shot with 120 quarts July 25, 1919. Pumped 180 barrels first 24 hours.

Calvin Perdue Lease No. 9. N. E. ¼ of Section 29:

Drive pipe	52 feet 8"
Casing	508 feet
Top of sand	975 feet
Drilled in sand	28 feet
Total depth	1003 feet

Shot with 140 quarts August 6, 1919. Pumped 90 barrels first 24 hours.

Calvin Perdue Lease No. 10. S. E. ¼ of Section 29:

Drive pipe	32 feet
Casing	420 feet
Top of sand	958½ feet
Drilled in sand	20 feet
Total depth	978½ feet

Shot with 60 quarts August 9, 1919.

Calvin Perdue Lease No. 11. N. E. ¼ of Section 29:

Drive pipe	44 feet 2"
Casing	400 feet
Top of sand	968 feet
Drilled in sand	222 feet
Total depth	990 feet

Shot with 100 quarts. Pumped 45 barrels first 24 hours.

L. S. Jones Lease No. 16. S. E. ¼ of Section 20:

Drive pipe	16½ feet
Casing	412 feet
Top of sand	968 feet
Drilled in sand	25 feet
Total depth	993 feet

Shot with 80 quarts May 2, 1919. Pumped 50 barrels first 24 hours.

Frank Malott Lease No. 1. N. E. ¼ of Section 29:

Drive pipe	33 feet
Casing	428½ feet
Top of sand	965 feet
Drilled in sand	27 feet 4"
Total depth	992 feet

Shot with 100 quarts. Pumped 100 barrels first 24 hours.

Frank Malott Lease No. 2, N. E. ¼ of Section 29.

Drive pipe	73 feet
Casing	415 feet
Top of sand	977 feet
Drilled in sand	26 feet
Total depth	1003 feet

Shot with 120 quarts. Pumped first 24 hours, 90 barrels.

Frank Malott Lease No. 3, N. W. ¼ of Section 29:

Drive pipe	69 feet
Casing	394 feet
Top of sand	986 feet
Drilled in sand	18 feet
Total depth	1004 feet

Pumped salt water first 24 hours.

Calvin Perdue Lease No. 12, N. E. ¼ of Section 29.

39 feet to limestone.
376 feet through limestone.
572 feet of shale.
22 feet of Trenton rock.

Pumped 20 barrels first 24 hours.

Calvin Perdue Lease No. 13, S. E. ¼ of Section 29.

39 feet to limestone.
363 feet through limestone.
566 feet of shale.
23 feet of Trenton rock.

Pumped 40 barrels first 24 hours. Finished August 30, 1919.

Calvin Perdue Lease No. 14, N. E. ¼ of Section 29.

28 feet to limestone.
367 feet through limestone.
573 feet of shale.
24½ feet of Trenton limestone.

Pumped 140 barrels first 24 hours.

Calvin Perdue Lease No. 15, N. E. ¼ of Section 29.

31 feet to limestone.
369 feet through limestone.
586 feet of shale.
25 feet of Trenton limestone.

Pumped 25 barrels first 24 hours.

Calvin Perdue Lease Well No. 16, N. E. ¼ of Section 29.

46 feet 3 inches to limestone.
384 feet through limestone.
550 feet of shale.
24 feet of Trenton.

Pumped 110 barrels first 24 hours.

Calvin Perdue Lease Well No. 17, S. W. ¼ of Section 29.

28 feet to limestone.
287 feet through limestone.
542 feet of shale.
20 feet of Trenton.

Pumped 35 barrels first 24 hours. Finished September 19, 1919.

Calvin Perdue Lease Well No. 18, N. E. ¼ of Section 29.

34 feet to limestone.
371 feet through limestone.
575 feet of shale.
25 feet of Trenton.

Water, 24 hours.

Calvin Perdue Lease Well No. 19, S. W. ¼ of Section 29.

33 feet to limestone.
384 feet through limestone.
542 feet of shale.
19 feet of Trenton.

Pumped 15 barrels first 24 hours. Finished October 18, 1919.

Calvin Perdue Lease Well No. 20, N. E. ¼ of Section.29.

78 feet to limestone.
336 feet through limestone.
563 feet of shale.
20 feet of Trenton.

First 24 hours, 25 barrels. Finished October 10, 1919.

Calvin Perdue Lease Well No. 21, N. E. ¼ of Section 29.

56 feet to limestone.
357 feet through limestone.
578 feet of shale.
19 feet of Trenton.

First 24 hours, 20 barrels. Finished October 18, 1919.

Frank Malott Lease Well No. 4, N. W. ¼ of Section 29.

58 feet to limestone.
422 feet through limestone.
503 feet of shale.
20½ feet of Trenton.

First 24 hours, 20 barrels. Finished October 3, 1919.

Wayne Township. Ten wells were drilled in the west half of Section 36, in 1904 and 1905, all of which started at about 100 barrels. Well No. 5, on the Hamilton lease, S. W. ¼ of Section 25, finished August, 1905, may, except in production, be taken as an average for this territory, its record being as follows:

Drive pipe	221 feet
Casing	512 feet
Top of Trenton	1001 feet
Total depth	1064 feet
Initial production (barrels)	100

Well No. 6 on the Pinkerton Lease, N. E. ¼ of Section 13, Jefferson Township, had the following record:

Drive pipe	170 feet
Casing	520 feet
Top of Trenton	971 feet
Total depth	1023½ feet
Initial production	143 barrels

Wells have been abandoned in this township as follows: Section 3, 1 well; Section 11, 1 well; Section 12, 8 wells; Section 13, 9 wells; Section 22, 1 well; Section 23, 5 wells; Section 24, 15 wells; Section 27, 1 well; Section 31, 2 wells; Section 34, 10 wells; Section 35, 15 wells; Section 36, 1 well.

Monroe Wyley Lease No. 1, S. E. ¼ of Section 12:

Drive pipe	137 feet
Casing	427 feet
Top of sand	1001 feet
Drilled in sand	30 feet
Total depth	1031 feet

No showing of oil. July, 1919.

Chas. H. Freck Lease No. 1, S. W. ¼ of Section 13.

28 feet to limestone.
417 feet through limestone.
556 feet of shale.
17 feet of Trenton.

Finished August 20, 1919. Slight showing of oil, but not enough to shoot. August 20, 1919.

Geo. Good Lease No. 1, N. E. ¼ of Section 32:

33 feet to limestone.
357 feet through limestone.
580 feet of shale.
19 feet of Trenton.
4 barrels first 24 hours.

Geo. Good Lease No. 2, N. E. ¼ of Section 32:

54 feet to limestone.
374 feet through limestone.
570 feet of shale.
38 feet of Trenton.

Finished October 13, 1919. No showing of oil.

Well of Grant Myres, No. 1:

Surface	0	to	42	feet
Gravel	42	to	215	feet
Red rock	215	to	235	feet
Slate	235	to	259	feet
Lime	259	to	350	feet
Slate	350	to	370	feet
Lime	370	to	390	feet
Slate	390	to	510	feet

Shale	510	to	600 feet
Brown shale	600	to	680 feet
Light shale	680	to	750 feet
Brown shale	750	to	900 feet
Slate	900	to	992 feet
Trenton rock	992	to	1002 feet

Very hard, light showing of oil. Water found at 1,002 feet. Total depth 1,002 feet. Drilled by Blosser, Phipps and others.

Section 17. Well No. 1. Ed Mossburg, S. W. ¼ of S. W. ¼: 8-inch drive pipe, 52 feet; 5⅝-inch casing, 437 feet. Top of sand (Trenton) 1013 feet. Salt water at 1,032 feet. Total, 1041 feet. Plugged January 29, 1919. Elevation of mouth, 821 feet. Trenton, 192 feet.

Well No. 2. S. E. ¼ of S. E. ¼ of Section 17: Drive pipe, 35 feet. Top of Trenton, 1027 feet. Elevation, 831 feet. Trenton, 196 feet.

Well No. 1, Martha A. Raugh: S. E. ¼ of Section 17: 8-inch drive pipe, 32 feet; 5⅝-inch casing, 395 feet. Top of sand, 1027 feet. Big dose salt water at 1050 feet. No showing of oil. Drilled June, 1918. Plugged June, 1918.

Section 20. S. E. ¼, Old Home Well No. 1: Top of Trenton, 965. Elevation, 816. Trenton, 149.

No. 2, 10 rods east. Top of Trenton, 979. Elevation, 926. Trenton, 153.

No. 3, 500 feet north of No. 2. Top of Trenton, 986 feet. Elevation, 826. Trenton, 160.

No. 4, S. E. of No. 2 500 feet. Top of Trenton, 985. Elevation, 827. Trenton, 158.

No. 5, S. E. of No. 4 500 feet. Top of Trenton, 983. Elevation, 826. Trenton, 157.

No. 6, S. E. of No. 5 500 feet. Top of Trenton, 972. Elevation, 816. Trenton, 156.

No. 7, S. E. of No. 6 500 feet. Top of Trenton, 979. Elevation, 827. Trenton, 152.

No. 8, north of No. 7 500 feet. Top of Trenton, 982. Elevation, 828. Trenton, 154.

No. 9, north of tanks near No. 1, not drilled. Elevation, 827.

Well No. 14. L. S. Jones, south half of N. E. ¼, Section 20: 8-inch drive pipe, 58 feet; 5⅝-inch casing, 412 feet. Top of sand, 990 feet. Total depth, 1015 feet. Drilled 25 feet in sand. Shot 80 quarts. First 24 hours 30 barrels. Drilled February 21, 1919.

Well No. 7. J. L. Priddy. S. E. corner of N. W. ¼, Section 20: 8-inch drive pipe, 28 feet; 5⅝-inch casing, 412 feet. Top of sand, 988 feet. Total depth, 1007 feet. In sand 19½ feet. Shot 100 quarts. Production first 24 hours, 24 barrels. Drilled January, 1919.

Well No. 8. J. L. Priddy. S. E. corner of N. W. ¼ of Section 20: 8-inch drive pipe, 64 feet; 5⅝-inch casing, 400 feet. Top of sand, 990 feet. Total depth, 1015 feet. Drilled 25 feet in sand. Shot 80 quarts. First 24 hours, 30 barrels. Drilled February 21, 1919.

Well No. 15, L. S. Jones. S. ½ of N. E. ¼ of Section 20: 8-inch drive pipe, 72 feet; 5⅝-inch casing, 400 feet. Top of sand, 983 feet. Total depth, 1003 feet. 20 feet in sand. Drilled February, 1919. Production first 23 hours, 40 barrels. Elevation of mouth, 837 feet.

Well No. 3. J. L. Priddy. S. ½ of N. W. ¼: 8-inch drive pipe, 52 feet; 5⅝-inch casing, 424 feet. Top of sand, 1007 feet. First pay, 10 feet. Total depth, 1029 feet. Drilled September, 1918.

Well No. 8. L. S. Jones. S. ½ of N. E. ¼ of Section 20: 8-inch drive pipe, 62 feet; 5⅝-inch casing, 425 feet. Sand at 1007 feet. Total depth, 1027 feet. Drilled August, 1918. Production 24 hours, 20 barrels.

Section 20. Well No. 10. L. S. Jones. S. ½ of N. E. ¼ of section 20: 8-inch drive pipe, 75 feet 10 inches; 5⅝-inch casing, 415 feet. Top of sand, 1006 feet. Bottom of sand, 1027 feet. Drilled October 3, 1918. Production first 24 hours, 22 barrels.

Well No. 4. J. L. Priddy. S. ½ of N. W. ¼: 8-inch drive pipe, 62 feet; 5⅝ casing, 425 feet. Top of sand, 987 feet. Total depth, 1017 feet. Showing of oil 7 feet in sand. Second pay 22 feet in. Drilled October, 1918. Produced 45 barrels first 24 hours.

Well No. 11. L. S. Jones. N. E. ¼: 8-inch drive pipe, 72 feet; 5⅝ casing, 400 feet. Top of sand, 997 feet. Total depth, 1034 feet. In sand 27 feet. Shot October 21, 1918.

Well No. 12. L. S. Jones. N. E. ¼: 8-inch drive pipe, 71 feet; 5⅝ casing, 425 feet. Top of sand, 1007 feet. Total depth, 1029 feet. In sand 22 feet. Production first 24 hours, 12 barrels.

Well No. 13. L. S. Jones. N. E. ¼ of Section 20: 8-inch drive pipe 58 feet; 5⅝ casing, 415 feet. Top of sand, 995 feet; 22 feet in sand. Drilled November 30, 1918.

Well No. 5. J. L. Priddy, S. ½ of N. E. ¼: 8-inch drive pipe, 9. feet; 5⅝ casing, 405 feet. Top of sand, 986 feet. Pay, 14 feet in sand. Depth, 1017 feet. Drilled 31 feet in sand. Shot 80 quarts. Production first 24 hours, 60 barrels.

Well No. 6. J. L. Priddy. S. ½ of N. E. ¼: 8-inch drive pipe, 117 feet 6 inches; 5⅝ casing, 401 feet. Top of sand, 982 feet. Total depth, 1007 feet. Drilled 25 feet in sand.

Well No. 6. L. S. Jones. S. W. ¼ Section 20: 8-inch drive pipe, 56 feet; 5⅝ casing, 425 feet. Top of sand, 991 feet. Total depth, 1018 feet. In sand 27 feet. Showing of oil 10 feet in. Pay at 24 feet in. Shot July 17, 1918.

Well No. 1. J. L. Priddy. N. ½ of N. W. ¼: Drive pipe, 68 feet; 5⅝ casing, 418 feet. Top of sand, 989 feet. Bottom, 1022 feet. First pay 10 feet in. Second pay 28 feet in sand. Shot July 22, 1918. Production first 24hours, 80 barrels.

Well No. 7. L. S. Jones. S. E. ¼: Drilled August 9, 1918. Top of sand, 994 feet. Drilled 27 feet in sand. Total depth, 1021 feet; 8-inch drive pipe, 57 feet; 5⅝ casing, 425 feet. Production first 24 hours, 20 barrels.

Well No. 2. J. L. Priddy. S. ½ of N. W. ¼: 8-inch drive pipe, 70 feet; 5⅝ casing, 425 feet. Top of sand, 991 feet. Total depth, 1024 feet; first pay 8 feet in. All pay. Shot August 23, 1918. 100 quarts. Production first 24 hours, 145 barrels.

Well No. 9. L. S. Jones. S. E. ¼: 8-inch drive pipe, 70 feet 3 inches;

5⅝ casing, 423 feet. Top of sand, 991 feet. Total depth, 1019 feet. Pay at 19 feet. In sand 28 feet. Completed September 6, 1918. Production first 24 hours, 50 barrels.

Well No. 1. L. S. Jones. N. E. ¼ Section 20: 8-inch drive pipe, 25 feet 5 inches; 6¼ casing, 404 feet 1 inch. Top of sand, 987 feet. First pay at 991 feet. Total depth, 1005 feet.

Well No. 2. L. S. Jones. S. E. ¼: 8-inch drive pipe, 29 feet; 6¼ casing, 402 feet. Top of sand, 978 feet. Pay sand 14 feet. Total depth, 1006 feet. Elevation of mouth, 831 feet.

Well No. 3. L. S. Jones. S. E. ¼: 8-inch drive pipe, 47 feet; 5⅝ casing, 415 feet. Top of sand, 990 feet. First pay 4 feet in sand. Total depth, 1016 feet. Elevation of mouth, 841 feet.

Well No. 4. L. S. Jones. S. E. ¼: 8-inch drive pipe, 58 feet; 5⅝ casing, 415 feet. Top of sand, 990 feet. First pay 2 feet in. Total depth, 1010 feet.

Well No. 5. L. S. Jones. S. E. ¼: 8-inch drive pipe, 47 feet; 5⅝ casing, 414 feet. Top of sand, 988 feet. First pay at 12 feet in sand. Show of oil at 16 feet. Salt water at 18 feet. Total depth, 1031 feet. Elevation, 831 feet. Drilled 41 feet in sand. Production, salt water. Plugged June 17, 1918.

Section 21. Well No. 1. Eliza P. Thompson. N. E. corner of S. W. ¼: 8-inch drive pipe, 67 feet; 5⅝ casing, 425 feet. Top of sand, 1007 feet; 28 feet in sand. No showing of oil. Plugged November, 1918.

Section 27. Well No. 1. Raper Holmes. N. E. corner of the S. W. ¼: 8-inch drive pipe, 87 feet; 5⅝ casing, 420 feet. Top of sand, 1023 feet. Bottom of sand, 1067 feet. Show of oil 26 feet in sand, not shot. Production all salt water. Plugged June 21, 1918.

Section 28. Well No. 1. Louisa Beard. S. E. ¼: 8-inch drive pipe, 32 feet; 5⅝ casing, 410 feet. Top of sand, 1001 feet. First pay 10 feet in. Total depth, 1021 feet.

Well No. 1. A. J. Gephart. S. E. corner N. E. ¼: 8-inch drive pipe, 70 feet; 5⅝ casing, 425 feet. Top of sand, 1007 feet. Total depth, 1028 feet. In sand 21 feet. Production first 24 hours, 1 barrel.

Section 29. Well No. 1. Catherine Beard. W. ½ S. W. ¼: 8-inch drive pipe, 28 feet 4 inches; 6¼ casing, 412 feet 10 inches. Top of sand, 999 feet. Total depth, 1038 feet. Pay all way along; water at 1038 feet.

Section 29. Well No. 1. Calvin Perdue. N. ½ of S. E. ¼: 8-inch drive pipe, 22 feet; 5⅝ casing, 405 feet. Top of sand, 968 feet. In sand 25 feet. Total depth, 993 feet. Drilled January, 1919. Production first 24 hours, 15 barrels.

Well No. 2. Calvin Perdue. N. E. corner S. E. ¼: 8-inch drive pipe, 23 feet; 5⅝ casing, 404 feet. Top of sand, 995 feet. In sand 32 feet. Total depth, 1027 feet. No showing of oil. Drilled March, 1919. Elevation of mouth above sea level, 836 feet.

Wells abandoned in this township are as follows: Section 3, 1 well; Section 4, 1 well; Section 12, 1 well; Section 20, 1 well; Section 24, 2 wells; Section 25, 5 wells; Section 26, 2 wells; Section 29, 1 well; Section 31, 13 wells; Section 34, 9 wells; Section 35, 4 wells; Section 36, 19 wells.

JACKSON COUNTY.

The bed rock in the eastern part of Jackson County belongs to the New Albany shale division of the Devonian; the remainder of the county is occupied by the Knobstone division of the Mississippian. The northwest portion lies within the unglaciated region and the remainder of the county is covered with drift varying in thickness from a few feet to more than one hundred feet. In the region not covered with glacial drift the study of structural conditions is difficult because of the absence of persistent layers of rock in the Knobstone.

Fig. CXXVIII. Outline of a structure in Jackson County, Owen Township. Constructed from data secured by O. H. Hughes. Probably not a definite anticline but a shoulder. Key formation, a lens of limestone in the Knobstone.

In the region west of Brownstown there is a layer of limestone, a ledge in the Knobstone, and an accompanying bed of sandstone, which may be used for a datum plane for the registering of the structure. Using this limestone and the sandstone, Mr. O. H. Hughes located a small terrace or shoulder which is represented on the accompanying map. It is possible that under the proper structures oil or gas may be found in the Devonian or in the Trenton in this county. The Trenton lies below the surface in the county at a depth of from 1,200 to 1,500 feet.

The following is the record of a well drilled at Brownstown:

Section of Well No. I.

Drift	43 feet
Knobstone shale	275 feet
Devonian shale	147 feet
Corniferous and Niagara limestone	225 feet
Hudson River and Utica	658 feet
Trenton limestone	100 feet
Total depth	1448 feet

Yielded no gas but at a depth of 1371 feet a slight flow of oil was obtained.

The following is the record of a well drilled at Seymour:

Section of Well No. 1.

Drift	75 feet
Sub-carboniferous sandstone	15 feet
Devonian shale	115 feet
Corniferous limestone	20 feet
Niagara limestone	190 feet
Hudson River limestone and shale	520 feet
Utica shale	165 feet
Trenton limestone	94 feet
Total depth	1194 feet

JASPER COUNTY.

The northwestern extension of the Cincinnati Arch passes through this county and the strata in the southern part of the county dip in the opposite direction to those of the northern portion of the county. Differential movements in the arch have produced structures favorable to the accumulation of oil and gas. These structures occur for the most part on the north side of the arch. Since the bed rock is covered with a mantle of glacial drift ranging in thickness from five to more than one hundred feet, these structures cannot be located by surface examinations. For this reason prospecting operations have been confined to the drill. Such prospecting has not been so expensive in this county on account of the oil sand being found at shallow depths. The geological formations underlying the drift belong to the silurian, Devonian, Mississippian, and the Pennsylvanian periods. Several small oil pools occur in this county, the oil being drawn from the Devonian strata at shallow depths. The map shows the location of these oil fields. Many of the wells indicated as producing wells have been abandoned since the map was prepared or prior to it.

The following is the section of a well drilled at Remington[2]:

Section of Well No. 1.

Drift	5 feet
Devonian shale	85 feet
Corniferous limestone	50 feet
Niagara limestone	260 feet

Hudson River and Utica	570 feet
Trenton limestone	295 feet
Total depth	1265 feet

Yielded no gas.

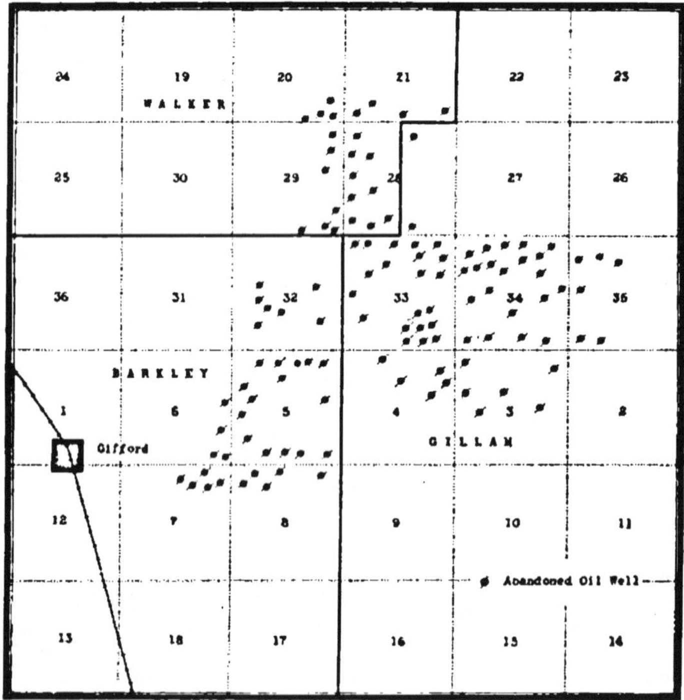

Fig. CXXIX. Map showing location of oil wells in the Jasper County field near Gifford.

JAY COUNTY.

The Silurian forms the bed rock in this county and outcrops along the Wabash River near the north line and on the Salamonie near Portland. The bed rock is largely concealed by the glacial drift which has a thickness of 25 to 125 feet. The general geologic conditions as represented in a well drilled at Portland are given below:

Wayne Township.

Section of Portland Well No. 1.

Drift	58 feet
Niagara limestone	192 feet
Shales	740 feet
Trenton limestone	500 feet
St. Peter	20 feet
Total depth	1510 feet

A small flow of gas and oil yielding 25 barrels a day was obtained. Five wells drilled near Portland reached the Trenton at 17, 63, 62, 67, and 71 feet below sea level. Oil was obtained from sections 5, 6, 10, 21, and 26, and gas from 5, 6, 10, 17, 21, and 22.

Richland Township. At Red Key the Trenton was reached at 900 feet and a flow of gas a few feet below the top of the Trenton resulted. At Dunkirk the Trenton was reached at 925 feet and a flow of 5,000,000 cubic feet of gas obtained. A second well reached the Trenton at 930 feet and produced a strong flow of gas at 955 feet. A section of this well is given below:

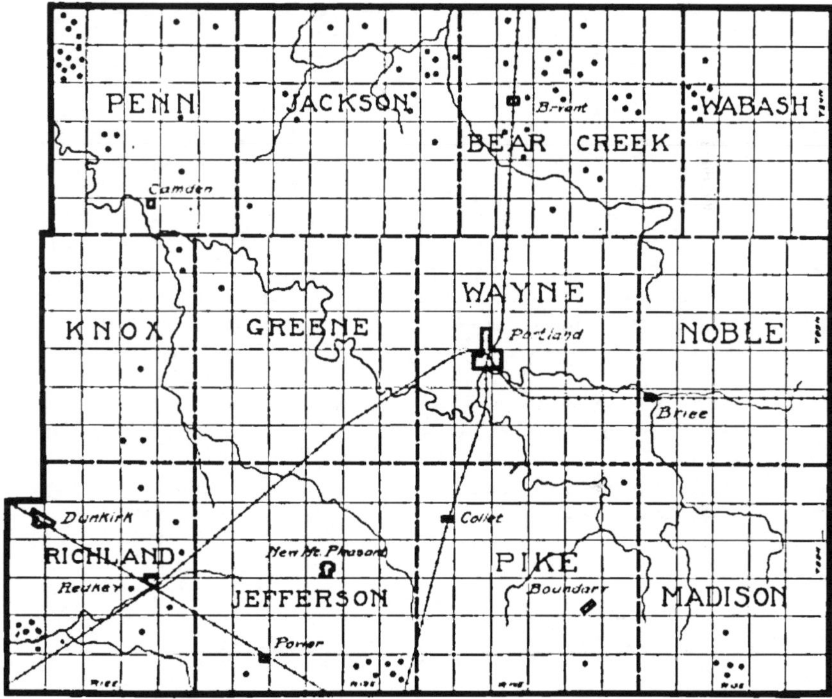

Fig. CXXX. Map of Jay County showing location of abandoned wells. Oil territory in the northern tier of townships. Gas in the central part and in western part of Knox Township. See large map.

Section of Dunkirk Well.

Drift	60 feet
Niagara limestone	230 feet
Hudson River and Utica	640 feet
Trenton limestone	25 feet
Total depth	955 feet

Oil was obtained in this township in sections 13, 16, 24, 25, 28, and 36, and gas in 9, 24, and 26. Wells have been abandoned as follows: Section 2,

1 well; section 12, 1 well; section 13, 1 well; section 23, 1 well; section 24, 1 well; section 26, 1 well; section 29, 5 wells.

Penn Township. At Camden the Trenton is reached at 935 feet and gas at 963 feet. The average depth of the drift at Camden is 35 feet and the average depth of the Trenton 925 feet. Nearly all the sections in this township have produced oil or gas or both. Wells have been abandoned in section 1, 1 well; section 2, 1 well; section 5, 4 wells; section 8, 9 wells; section 14, 1 well; section 21, 3 wells; section 26, 1 well.

Jefferson Township. Gas is reported to have been found at Coneo in this township. The following wells were drilled and plugged: Section 32, 1 well; section 35, 3 wells; section 36, 4 wells.

Greene Township. Oil was obtained in sections 8, 17, 20, and 24, and gas in 4, 5, 6, 7, 18, 19, 20, 23, 26, 28, 31, 32, 34, and 35. Wells were abandoned in section 7, 1 well.

Jackson Township. Oil was obtained in this township from sections 1, 2, 3, 4, 5, 6, 7, 8, 9, 10, 11, 12, 13, 14, 15, 16, 17, 19, 22, 23, 24, 25, 26, 27, 28, 30, 31, 32, 33, 35, and 36. Gas was obtained in sections 7, 18, 19, 20, 21, 25, 29, 30, and 32. Wells have been plugged in section 3, 1 well; section 4, 1 well; section 11, 1 well; section 12, 5 wells; section 14, 2 wells; section 17, 2 wells; section 24, 1 well; section 31, 1 well.

Knox Township. Oil was found in sections 1, 4, 11, and gas in sections 1, 2, 25, and 36. Wells have been abandoned in section 1, 2 wells.

Pike Township. Oil was found in sections 7, 8, and 34. Wells have been abandoned in section 23, 1 well; section 35, 2 wells. Recently wells were drilled in this township as follows:

James Tharp No. 1. N. E. ¼, N. W. ¼, Section 29, Township 22 N., Range 14 E.:

Mud, sand and gravel	189 feet
Limestone	91 feet
Slate and shale	800 feet
In Trenton limestone	20 feet
Total depth	1100 feet

Grant Whitenack, No. 2. S. E. ¼, N. W. ¼, Section 28, Township 22 N., Range 14 E.:

Mud, sand and gravel	137 feet
Limestone	203 feet
Slate and shale	690 feet
In Trenton limestone	20 feet
Total depth	1050 feet

Cornelius Whitenack No. 2. N. E. ¼, N. W. ¼, Section 28, Township 22 N., Range 14 E.:

Mud, sand and gravel	140 feet
Limestone	690 feet
Slate and shale	691 feet
In Trenton limestone	32 feet
Total depth	1553 feet

Wells drilled by Union Heat, Light and Power Company.

Noble Township. Oil occurred in sections 3, 4, 5, 17, and 27. Gas in sections 8 and 17.

Bear Creek Township. At Bryant the top of the Trenton is 1,020 feet or 160 feet below sea level. Oil was obtained 30 feet below the top of the Trenton. The following are the records of two wells drilled on the Kuhn lease, in the southwest quarter of section 28:

	Well No. 7	Well No. 2
Drive pipe (drift)	78 feet	104 feet
Casing	245 feet	238 feet
Top of Trenton	1004 feet	997 feet
Total depth	1050 feet	1048 feet

The record of a well drilled by W. J. Heeter in section 3 is given as follows:

Drift	73 feet
White limestone	131 feet
White slate	10 feet
White lime	20 feet
Slate (shale)	30 feet
Limestone	15 feet
Slate	40 feet
Blue lime	5 feet
White slate	75 feet
Blue lime	10 feet
White slate	305 feet
Brown shale	300 feet
Black slate	12 feet
Trenton rock	50 feet
Total depth	1076 feet

Showing of oil at 20 feet in Trenton. Salt water, strong flow.

Oil was obtained in sections 1, 2, 3, 4, 5, 6, 7, 8, 9, 10, 11, 12, 13, 16, 17, 18, 21, 24, 30, and 31, and gas in sections 14, 18, 19, 22, 26, 27, 29, 30, 31, and 34. Wells have been plugged in section 3, 1 well; section 5, 1 well; section 8, 1 well; section 9, 4 wells; section 10, 2 wells; section 14, 4 wells; section 16, 2 wells; section 17, 1 well; section 20, 3 wells; section 26, 2 wells; section 27, 2 wells; section 33, 1 well.

Wabash Township. Oil was found in sections 3, 4, 5, 6, 7, 8, 17, 18, 19, and 32; and gas in 19. Wells have been plugged in section 7, 1 well; section 18, 5 wells.

Well No. 2, Bon Macy Farm.

Gravel, sand and mud	50 feet
Limestone	200 feet
Slate and limestone	300 feet
Slate	300 feet
Brown shale	150 feet

Gray shale .... 25 feet

Top of Trenton..............................1025 feet
Into Trenton .... 45 feet

Madison Township. Wells have been drilled at various points in this township. Four wells were drilled and abandoned in section 33 and 1 in section 28.

JEFFERSON COUNTY.

The strata which outcrop in Jefferson County belong to the Ordovician, Silurian, Devonian and Quaternary periods. The subdivisions as given by Cummings, Siebenthal and others are given in the following outlines:

Period	Formation	Subdivisions
Quaternary	Recent—clays and alluvium	
	Pleistocene—sand gravel and till	
Devonian	New Albany—shales	
	Sellersburg—limestone	
	Silver Creek—limestone	
	Jeffersonville—limestone	
Silurian	Louisville—limestone	
	Waldron—shale	
	Laurel—limestone	
	Osgood—limestone and shale	
	Brassfield—shales	
Ordovician	Richmond—shales	Elkhorn Whitewater Saluda Liberty Waynesville Arnheim
	Marysville—shales	Mt. Auburn Corryville Bellevue Fairmount Mt. Hope
	Eden—shales	McMicken Southgate Economy Fulton

The Quaternary covering in this county varies in thickness from a few feet to fifty feet. Sufficient outcrops of the bed rock may be obtained to determine the structure. Probably the best key horizon for the west part of the county will be the contact between the Sellersburg limestone and the New Albany shale. Farther east the Laurel or the Louisville limestone might be used. Some gas has been obtained from near Foltz in the Niagara limestone. These wells were reported to have a pressure of 20 pounds in 1914.

Middle Fork Well.

In 1920 a well was drilled about one and one half miles southeast of Middle Fork and one fourth mile east of the Pennsylvania railroad, about Section 5, R. 10 E., T. 4 N. The elevation of the well is about 795 feet above sea level. Water was encountered at 90, 1,470, 1,635, 1,835, 1,965, 2,060, and 2,595 feet. The record of the well is as follows:

Soil and clay	7 feet
Hard lime (Devonian)	143 feet
Shale	25 feet
Lime (Louisville)	50 feet
Broken lime	25 feet
Shale (Waldron)	50 feet
Lime (Laurel)	50 feet
Broken lime and shale	50 feet
Shale	40 feet
Broken lime	60 feet
Shale	200 feet
Limy shale	50 feet
Shale	150 feet
Lime (Trenton)	520 feet
Sand (St. Peter)	30 feet
Sandy lime (St. Peter)	65 feet
Sand (St. Peter)	115 feet
Lime (Lower Magnesian)	250 feet
Sand (Lower Magnesian)	45 feet
Lime	225 feet
Sandy lime	50 feet
Lime	800 feet
Total depth	3000 feet

The surface of the Trenton is about 155 feet below sea level.

JENNINGS COUNTY.

The strata which outcrop in Jennings County are given in the table below:

Quaternary	Recent—residual clays and alluvium Pleistocene—gravels, sand and till
Devonian	New Albany—shales Sellersburg—limestone Silver Creek—limestone Jeffersonville—limestone
Silurian	Louisville—limestone

The Devonian and the Silurian strata are largely concealed by the surficial deposits of drift and alluvium, but enough outcrops have been obtained to enable the construction of a structural map covering a large part of the county. See page 156. The field work in the preparation of the map was done by Dr. C. A. Malott and P. B. Stockdale, members of the field party of 1919.

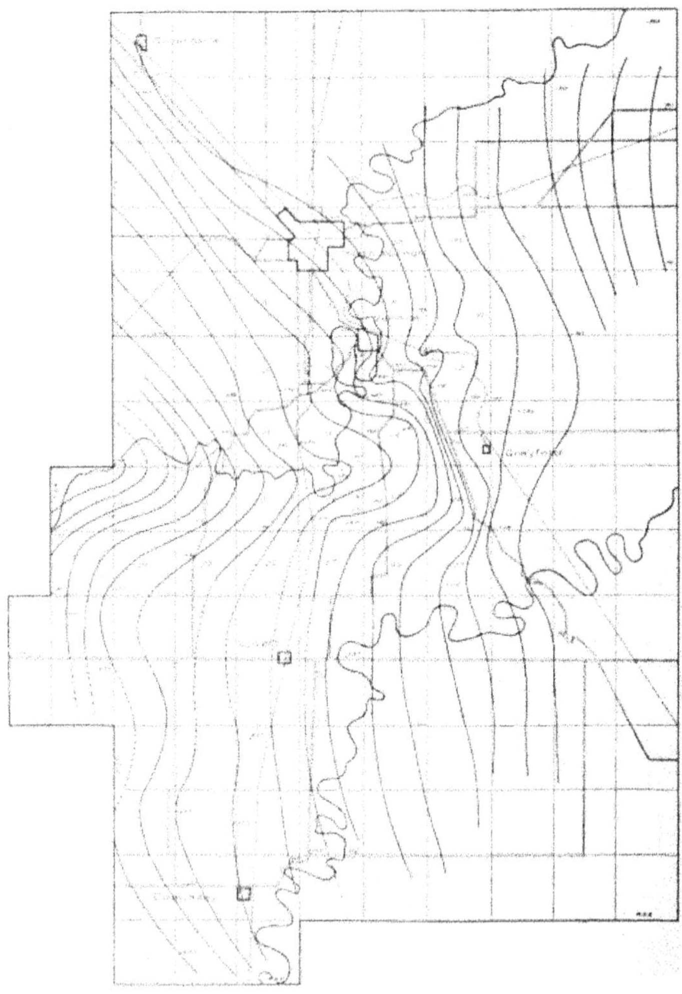

Fig. CXXXI. **Map of a portion of Jennings and Jefferson counties showing structural contours drawn on the contact of the Sellersburg limestone and New Albany shale. Data collected by C. A. Malott and P. B. Stockdale.**

Gas has been obtained at North Vernon in wells drilled on structure, though perhaps not on the best part of it. A record of one of the wells is given below[1]:

Section of Well No. 1.

Surface clay	11 feet
Corniferous limestone	28 feet
Niagara limestone	252 feet
Clinton (?) limestone	29 feet
Hudson River limestone	440 feet
Utica shale	220 feet
Trenton limestone	470 feet
Total depth	1450 feet
Trenton below sea level	253 feet

Yielded medium flow of gas.

JOHNSON COUNTY.

The subsurface rocks of Johnson County consist of the New Albany black shale, which occupies the eastern portion of the county and the Knobstone group occupying the western part. The surface is covered with glacial drift.

A well drilled in Nineveh Township about nine miles south of Franklin reached the Trenton at 1,273 feet; the first 60 feet of the Trenton was porous and contained a showing of oil. It was a wildcat well drilled without any reference to structure. The Trenton was passed through at 1,820 feet, showing 547 feet of Trenton at this point.

The Trenton limestone was reached in the southeastern part of the county at 987 feet; in the central part at 1,042 feet, and in the north central part at 1,220 feet.

The following are the records of some of the wells drilled in the county:

Vandivin Well No. 1, Section 9, Nineveh Township.

Drift	to 16 feet
Sandy lime	to 33 feet
Gray shale	to 285 feet
Slate	to 295 feet
Red rock	to 327 feet
Sandy lime	to 335 feet
Brown shale	to 425 feet
Jeffersonville lime	to 702 feet
Gray shale	to 704 feet
Brown lime	to 745 feet
Gray shale	to 750 feet
Gray lime	to 795 feet
Slate	to 953 feet
Dark brown lime	to 977 feet
Slate	to 892 feet
Gray lime	to 1017 feet

Slate	to 1022 feet
Gray lime	to 1072 feet
Slate	to 1077 feet
Gray lime	to 1082 feet
Slate	to 1097 feet
Brown lime	to 1105 feet
Gray lime	to 1107 feet
Slate	to 1273 feet
Trenton rock	to 1830 feet

This well showed some oil in the first 15 feet of Trenton rock.

Mullidore Well No. 1, Section 3, Nineveh Township.

Drift	to 34 feet
Gravel	to 37 feet
Hardpan	to 65 feet
Gray shale	to 100 feet
Gray lime	to 105 feet
Gray shale	to 160 feet
Lime	to 165 feet
Brown shale	to 261 feet
Jeffersonville lime	to 440 feet
Gray shale	to 445 feet
Gray lime	to 482 feet
Gray shale	to 487 feet
Gray lime	to 531 feet
Gray shale	to 765 feet
Slate	to 843 feet
Gray lime	to 848 feet
Utica shale, dark gray	to 1129 feet
Trenton rock	to 1164 feet

Drilled 28 feet in the Trenton. Small showing of oil in the first five feet of rock.

A well drilled at Franklin was reported by Dr. D. A. Owen as follows:

Drift	170 feet
Black shale	34 feet
Blue and gray limestone	71 feet
Sandstone	27 feet
Blue shale (upper Niagara)	23 feet
Gray and white limestone	120 feet
Greenish blue shale varying to black	597 feet
Trenton limestone	71 feet
Total depth	1113 feet
Altitude of well	736 feet

A well drilled in the southeastern part of the county at Edinburg is reported as follows:

Drift	115 feet
Shale	20 feet
Limestone	220 feet
Shale	622 feet
Trenton limestone	500 feet
Shale	83 feet
White sandstone	10 feet
Total depth	1580 feet
Altitude of well	670 feet

A well was drilled at Greenwood the record of which is as follows:

Drift	210 feet
Black shale	90 feet
Limestone	280 feet
White shale	40 feet
Gray shale	300 feet
Dark Utica shale	300 feet
Trenton limestone	55 feet
St. Peter sandstone	200 feet
Lower Magnesian limestone.	

KNOX COUNTY.

This county lies in the area occupied by the strata of the Pennsylvania division, but these bed rock strata are covered with a mantle of glacial drift and alluvium which varies in thickness from twenty-five to more than one hundred feet, so that the determination of structure by direct observational methods is not possible. Subsurface work will depend upon the amount of data secured from well records. To secure a sufficient number of such records will require a large amount of wild cat drilling. A well drilled about eight miles south of Vincennes has produced some oil and the prospects of the extension of favorable structures north of the Gibson County line are encouraging. It may be possible, by using data from coal mines, wells, etc., to outline the structure on some of the coals.

Washington Township. A well drilled in the southeast quarter of section 30 reached a dry sand at 1,252 feet.

Decker Township. A well was drilled on the property of J. Cunningham in section 12 and plugged in 1912. No record of the well has been obtained.

Record of Bore Northeast of Vincennes.

Drive pipe to bed rock	to	45 feet
Yellow standstone	to	80 feet
Slate and shale	to	195 feet
Sandstone, limestone and shale	to	335 feet
Coal	to	340 feet
Blue limestone	to	350 feet
Light shale	to	360 feet
Soapstone	to	390 feet
Limestone	to	425 feet
Light shale	to	435 feet
Sandstone	to	465 feet
Slate and shale	to	485 feet
Fire clay	to	505 feet
Blue shale	to	520 feet
Limestone	to	525 feet
Blue slate	to	545 feet
Black shale	to	565 feet
Sandstone	to	580 feet
Soapstone	to	590 feet

Slate	to 625 feet
Limestone and slates	to 640 feet
White sandstone and salt water	to 670 feet
Slate and shale	to 700 feet
Blue limestone	to 702 feet
Soapstone and shale	to 785 feet
White sandstone and salt water	to 800 feet
Sandstone	to 815 feet
Sandstone and shale alternately	to 940 feet
Limestone	to 950 feet
Black slate	to 980 feet
Sandstone	to 1000 feet
Slate	to 1020 feet
Streaks of slate and limestone	to 1130 feet
Sandstone	to 1180 feet
Shale	to 1200 feet
Sandstone	to 1292 feet
Shale	to 1298 feet
Gray limestone	to 1310 feet
Shale	to 1315 feet
Soapstone	to 1325 feet
Shale	to 1335 feet
Blue limestone	to 1340 feet
White sandstone	to 1365 feet
Shale	to 1375 feet
Blue limestone	to 1385 feet
Slate	to 1400 feet
Red rock	to 1410 feet
Sandstone and salt water	to 1430 feet
Shale (cased)	to 1535 feet
Gray limestone	to 1655 feet
Shale	to 1660 feet
Blue limestone	to 1665 feet
Slate and shale	to 1690 feet
Sandstone and sulphur water	to 1740 feet
Slate	to 1750 feet
Shale	to 1755 feet
Gray limestone	to 1765 feet
Shale and gray limestone	to 1820 feet
Red rock	to 1825 feet
Hard gray limestone	to 1840 feet
Soapstone	to 1845 feet
Gray limestone	to 1850 feet
Soapstone	to 1860 feet

Vincennes Artesian Salt Well.

Sand and gravel	80 feet
Sandstone	18 feet
Soapstone	100 feet
Hard pebble rock	10 feet
Sandy shale	15 feet
Soapstone	32 feet
Blue sandstone	35 feet
Sandy shale	20 feet
Soapstone	10 feet
Coal	3 feet
Soapstone	18 feet
Coal	5 feet
Soapstone	18 feet
Black shale	41 feet
Soapstone	138 feet
Coal	5 feet
Limestone	10 feet
Blue shale	27 feet
Black slate	30 feet
Soapstone and shale	80 feet
Sandstone	15 feet
Slate and soapstone	75 feet
Sandstone and salt water	25 feet
Slate and shale	95 feet
Sandstone	175 feet
Shale and black slate	140 feet
Sandstone	96 feet
Total depth	1336 feet

Well No. 1 on the Geo. Ryan farm, 200 feet N., 200 feet to west line. Section 36, Twp. 2 N., R. 11 W. Knox County. Oct. 8, 1919. Well plugged and abandoned.

Soil	to 6 feet
Gravel	to 10 feet
Slate, white	to 175 feet
White lime	to 179 feet
Slate, black	to 185 feet
White lime	to 195 feet
Slate	to 310 feet
Lime	to 315 feet
Slate	to 340 feet
Lime	to 343 feet

Slate	to 360 feet	Lime, hard	to 1442 feet
Sand	to 480 feet	Slate	to 1460 feet
Slate	to 500 feet	Sand	to 1466 feet
Sand	to 540 feet	Slate	to 1485 feet
Black slate	to 600 feet	Sand	to 1545 feet
Sand	to 624 feet	Lime	to 1551 feet
Black slate	to 635 feet	Slate	to 1558 feet
White slate	to 655 feet	Lime	to 1564 feet
Black slate	to 673 feet	Slate	to 1600 feet
Lime	to 675 feet	Lime	to 1615 feet
White slate	to 710 feet	Slate	to 1635 feet
Black slate	to 716 feet	Red rock	to 1643 feet
Lime	to 746 feet	Lime	to 1648 feet
Black slate	to 775 feet	Sand	to 1653 feet
Lime	to 781 feet	Slate	to 1675 feet
White slate	to 840 feet	Sand, hole full of water	to 1727 feet
Sand	to 865 feet	Slate	to 1735 feet
Slate, black, soft	to 900 feet	Lime	to 1753 feet
Lime	to 904 feet	Slate	to 1759 feet
Slate	to 960 feet	Sand	to 1772 feet
Sand	to 1060 feet	Slate	to 1777 feet
Slate	to 1080 feet	Brown lime	to 1808 feet
Lime	to 1100 feet	Lime	to 1814 feet
Slate	to 1120 feet	Sandy lime, oil	to 1818 feet
Sand, hole full of water	to 1145 feet	Sand	to 1824 feet
Black slate	to 1195 feet	Red Rock	to 1827 feet
Sand	to 1215 feet	Lime	to 1838 feet
White slate	to 1220 feet	Sand	to 1850 feet
Lime, hard	to 1222 feet	Lime	to 1868 feet
Slate	to 1310 feet	Dark lime	to 1882 feet
Sand	to 1350 feet	Lime shell, volites	to 1894 feet
Lime, hard	to 1365 feet	White lime	to 1897 feet
Slate	to 1370 feet	Lime, brown, hard	to 1920 feet
Sand, hard	to 1395 feet	Lime, soft	to 1930 feet
Slate	to 1400 feet	Lime	to 2004 feet
Lime	to 1405 feet		
Sand	to 1440 feet	Total depth	to 2004 feet

KOSCIUSKO COUNTY.

Underlying the glacial drift which covers the surface of this county are strata of Devonian age consisting of a series of limestones and shales. The strata dip northward away from the arm of the Cincinnati Arch, which passes through Indiana. The drift attains a thickness of over two hundred fifty feet in this county.

Warsaw. The record of a well drilled at Warsaw is given below:

Drift	248 feet
Limestone (Silurian and Devonian)	652 feet
Shale (Ordovician)	487 feet
Trenton limestone	50 feet
Total depth	1437 feet
Altitude of well	815 feet

Syracuse. The record of a well drilled at Syracuse on the property of the Sandusky Cement Company was furnished the writer by Mr. S. B. Newberry, president of the company. The record of the well shows sixty-three feet of New Albany (Devonian) shale underlying the drift. The well probably ended in the Jeffersonville limestone of the Devonian. By consulting the Warsaw well record above, it will be seen that the total thickness of the Devonian and the Silurian limestone is recorded as being 652 feet. In the Elkhart well the New Albany shale has a thickness of 215 feet, which is to be expected as it is down in the basin of the arch. The well stopped in limestone at sixty-five feet. From the evidence of these wells the Devonian limestone is thicker here than in the southern part of Indiana.

Sand, gravel, clay and boulders	278 feet
Gray and dark shale	63 feet
Gray argillaceous limestone	42 feet
Crystalline gray and white limestone showing oil	20 feet
Total	403 feet

Depth	Constituent	Per cent
341-351 feet	Carbonate of lime	51.40
	Carbonate of magnesia	11.93
	Insoluble	32.40
351-361 feet	Carbonate of lime	72.00
	Carbonate of magnesia	7.73
	Insoluble	17.50
361-371 feet	Carbonate of lime	66.60
	Carbonate of magnesia	9.24
	Insoluble	21.24
371-381 feet	Carbonate of lime	48.60
	Carbonate of magnesia	8.57
	Insoluble	37.87

This appears to be similar to the cement rock of southeastern Indiana, but of much greater thickness than recorded in that region.

Depth	Constituent	Per cent
381-390 feet	Carbonate of lime	71.60
	Carbonate of magnesia	23.52
	Insoluble	3.36
392-403 feet	Carbonate of lime	75.40
	Carbonate of magnesia	19.32
	Insoluble	2.00

LAGRANGE COUNTY.

Glacial drift occupies the surface of this county to a depth probably varying in thickness from 100 to 200 feet. The bed rock formations consist of strata belonging to the Devonian and the Mississippian periods. As these formations lie to the north of the Indiana extension of the Cincinnati arch they dip toward the north.

On account of the covering of the glacial drift the structural conditions of the bed rock cannot be determined by surficial observation. The possibility of oil and gas accumulations are connected with the possible occurrences of terraces, or small anticlines in the strata of the northward dipping formations. These can be located by means of well records only.

LAKE COUNTY.

Silurian and Devonian strata underlie the glacial drift in this county. Because of the overlying mantle of drift stratigraphical and structural conditions of the bed rock are difficult to determine. At Crown Point the Tren-

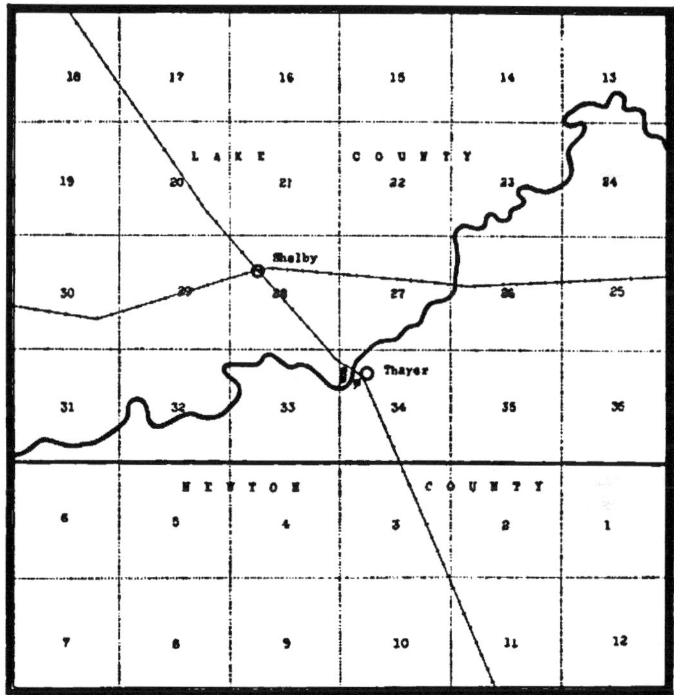

Fig. CXXXII. Map of a portion of Lake and Newton counties showing the location of the Thayer oil field.

ton lies 919 feet below the surface, south of this point it should be encountered nearer the surface for points of the same or less elevation than Crown Point. At the north it will be found to lie deeper as the strata dip to the north.

Center Township. The following is the record of the well at Crown Point:

Drift	176 feet
Black shale	76 feet
Limestone	433 feet
Bluish green shale	55 feet

Clinton limestone	37 feet
Bluish green Hudson River shale	122 feet
Trenton limestone	342 feet
White limestone (sandy)	89 feet
Limestone	15 feet
Total depth	1365 feet
Altitude of well	736 feet
Trenton below sea level	183 feet

West Township. The following is the record of a well drilled on the farm of Marion Driscoll, Section 23, T. 33 N., R. 9 E., Lake County:

Drift	73 feet	
Gray limestone	73 to	598 feet
Red shale	598 to	607 feet
Green-gray slate	607 to	640 feet
Shelly limestone	640 to	705 feet
Limestone	705 to	715 feet
Limestone with salt water	715 to	735 feet
Dark gray limestone	735 to	795 feet
Slate	795 to	850 feet
Dark gray limestone	850 to	870 feet
Hard white limestone	870 to	890 feet
Gray limestone	890 to	1025 feet
Trace of oil		905 feet
Good showing of oil		925 feet
Total depth		1025 feet

Well plugged August 18, 1914.

LA PORTE COUNTY.

Underlying the glacial drift in this county are strata of Devonian age. The drift attains a thickness of three hundred feet or more. The dip of the bed rock is toward the north. The drift at Laporte has a thickness of 295 feet and overlies black shale. At Michigan City the drift is 250 feet thick and overlies limestone.

Michigan Township. The drift varies from 170 to 250 feet and overlies black shale and limestone at Michigan City.

Center Township. A deep well drilled at Laporte contained the following section:

Drift	295 feet
Black shale	125 feet
Shale and limestone	460 feet
Limestone	500 feet
Trenton limestone	520 feet
St. Peter and Low. Magnesian	600 feet
Potsdam sandstone	323 feet
Total depth	2823 feet

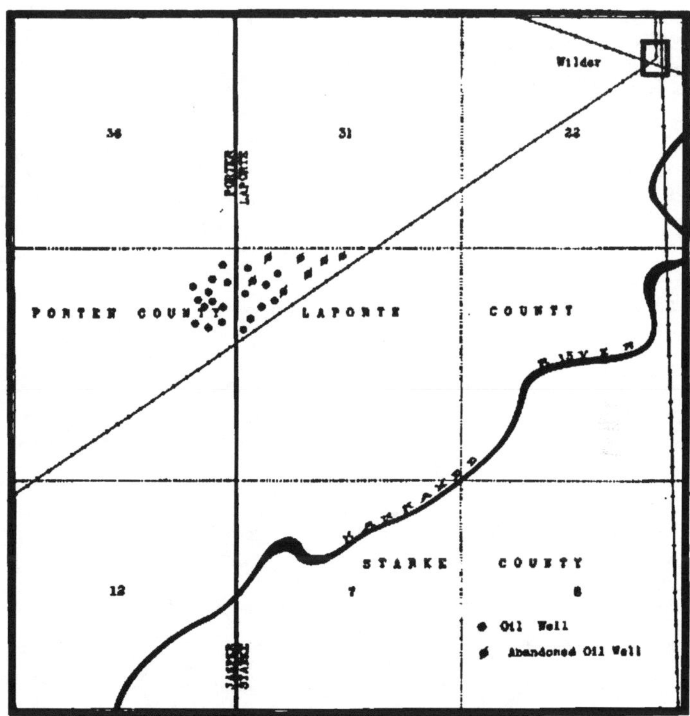

Fig. CXXXIII. Map showing the location of oil wells in the Wilder oil field on the border between Laporte and Porter counties.

Galena Township. A deep well was drilled on the property of O. L. Sutherland two miles east of Reason in section 2. No record was obtained of this well. It was plugged in 1911.

LAWRENCE COUNTY.

Geology. A small portion of the surface in the eastern part of the county is occupied by the Knobstone, the remainder of the Harrodsburg limestone, the central portion by the Salem and the Mitchell limestone and the western portion by the Chester formations and the Pottsville.

Structure. The presence of the Mount Carmel fault and the Heltonville fault in the eastern part of the county produce a fold extending in a general north and south direction parallel to these faults. A change in direction of the Mt. Carmel fault at Leesville produces an anticlinal area southwest of Leesville which has been productive of gas and has to date produced a showing of oil in the Corniferous. The wells which have been drilled have not gone to the Trenton. In the neighborhood of Heltonville three wells have been drilled and a small amount of gas and oil obtained. These wells are located near the fold produced by the down throw of the strata, but the structure seems not to have been considered.

Heltonville Well. In 1913 the Bedford Oil and Gas Company drilled three wells near Heltonville. One of these wells was drilled to a depth of 1707 feet, entered the Trenton at 1,633 feet and encountered a showing of oil at about 1,675 feet.

Record of Heltonville Well.

	Thickness	Depth
Surface soil, etc.	15 feet	
Shale (Knobstone)	85 feet	100 feet
Limestone (lens in Knobstone)	60 feet	160 feet
Sand (7 feet of oil sand)	20 feet	180 feet
Shale	10 feet	190 feet

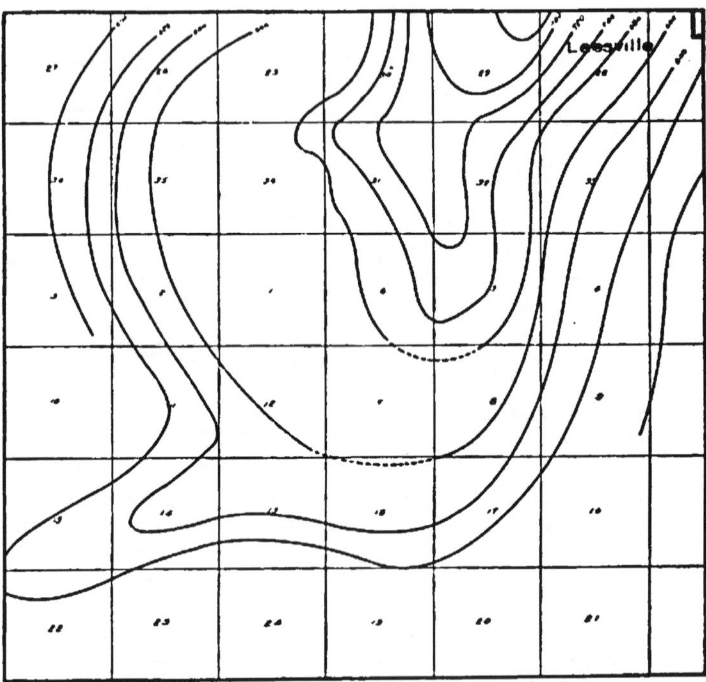

Fig. CXXXIV. Structural map of a portion of Lawrence County.

Shale (white)	310 feet	500 feet
Shale	100 feet	600 feet
Shale	40 feet	640 feet
Sand, gas and oil bearing	10 feet	650 feet
Shale	50 feet	700 feet
Limestone	15 feet	715 feet
Shale	38 feet	753 feet
Oil sand	3 feet	756 feet
Limestone (water)	334 feet	1090 feet
Shale	543 feet	1633 feet
Trenton limestone	74 feet	1707 feet

The following is a record of Easton Well No. 1 drilled in the same township:

Easton No. 1 Well. Pleasant Run Township.

Drift	20 feet	
Gravel	5 feet	
Lime	5 feet	
Shale	75 feet	
White mud	150 feet	
Lime	200 feet	
Black shale	5 feet	
White slate	95 feet	
Brown shale	40 feet	
Lime	70 feet	
Brown sand	15 feet	Mineral water.
Lime	15 feet	
Gray sand	5 feet	Some gas.
White sandy lime	5 feet	
Blue lime	90 feet	
Gray sand	35 feet	Mineral water.
Lime	100 feet	
White slate	150 feet	
Lime	5 feet	
Brown shale	100 feet	
Broken shale with lime	50 feet	
Brown shale	240 feet	
Trenton at	1540 feet	
Gray sand at	1620 feet	No. oil, 15 foot sand
Light brown sand at	1715 feet	5 foot sand
Finished at	1750 feet	

The second well was drilled near the first to a depth of 1,100 feet and encountered a moderate flow of gas at 1,090 feet. A third well was drilled about a mile south of the first two and resulted in a dry hole.

Flinn Township. Gas has been obtained from the Corniferous in this township in sections 3, 4, 5, and 28. Four of these wells were drilled by Mr. W. H. Wheitknecht and associates, and the fifth by Mr. Claude Malott. The following are brief records of the Wheitknecht wells: No. 1 is located in section 3, No. 2 in section 4, and Nos. 3 and 4 in section 28. No. 5 is in section 5.

	No. 1	No. 2	No. 3	No. 4	No. 5
Elevation above sea	587	566	709	608	570
Top of Corniferous	597	616	683	600	512
Water	635	655	714	636	550

These wells were all drilled on the east side of the structure where the strata are dipping toward the fault line. A showing of oil was found in two of the wells. These wells all started in the Knobstone and passed through four feet of Rockford Goníatite limestone and one hundred and twenty-five feet of New Albany shale and about thirty-eight feet of Devonian limestone before reaching water. A slightly different interpretation of the well records might modify the outline of the structure shown on the structural map. The elevation taken on the contact by the use of the aneroid barometer may vary slightly from the true elevations but probably not enough to make a serious change in the structural map.

MADISON COUNTY.

The eroded surface of the Niagara limestone underlies the glacial drift in this county and may be reached at from five to one hundred and fifty feet. Gas has been produced in every township and oil in some parts of the county. The oil sand is reached at from 800 to 1,200 feet. The surface of the Trenton lies between 100 feet above and 100 feet below sea level.

Anderson Township. The record of a well drilled at Anderson is given belw:

Drift	114 feet
Niagara limestone and shale	186 feet
Clinton (?)	20 feet
Hudson River and Utica	494 feet
Trenton limestone	24 feet
Total depth	838 feet
Trenton above sea level	66 feet

A great many wells were drilled in this township, most of which produced gas.

Boone Township. Wells were drilled and abandoned in section 11, 2 wells; section 19, 1 well.

Monroe Township. A well drilled at Alexandria has the following log:

Drift	20 feet
Niagara limestone	261 feet
Hudson River and Utica	611 feet
Trenton limestone	5 feet
Total depth	897 feet

Well No. 3, B. Markle, Monroe Township.

Clay, gravel and quicksand	84 feet
Limestone	246 feet
Slate	593 feet
Trenton rock at	923 feet
In Trenton	77 feet
Total depth	1000 feet

Many wells were drilled and much gas and oil were obtained from this township. Wells have been abandoned in section 2, 3 wells; section 4, 2 wells; section 8, 1 well; section 10, 1 well; section 12, 2 wells; section 13, 1 well; section 15, 1 well; section 19, 1 well, section 24, 2 wells; section 27, 1 well; section 32, 1 well; section 33, 1 well; section 34, 1 well.

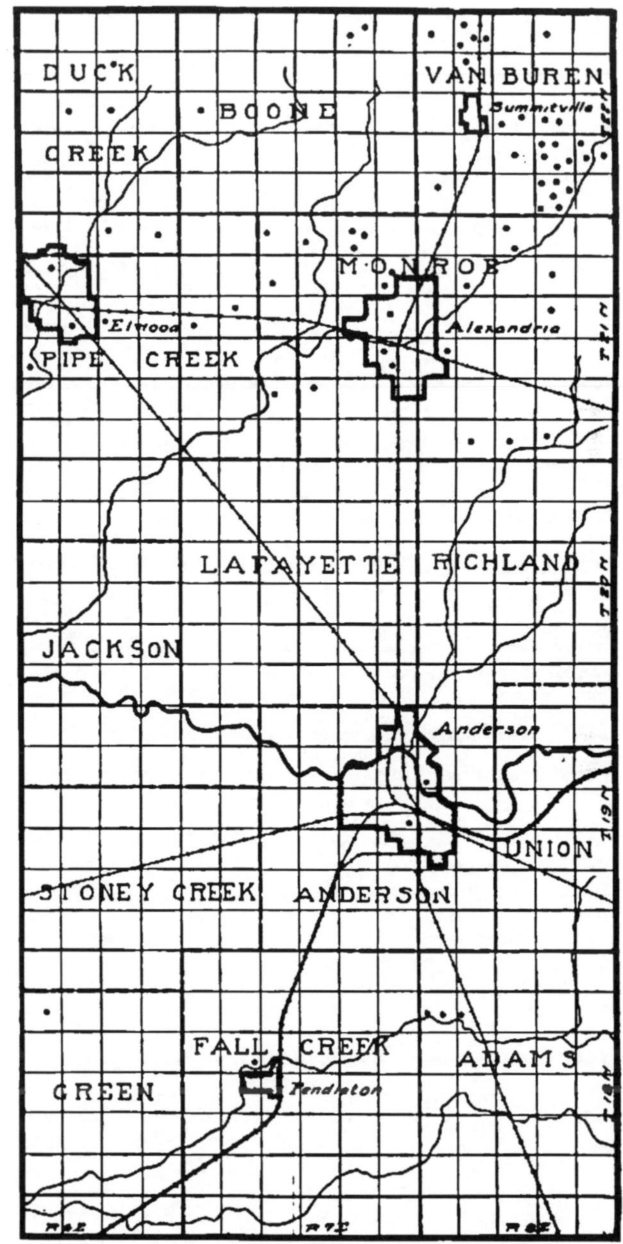

Fig. CXXXV. Map of Madison County showing location of abandoned wells. Gas territory exists in the eastern tier and in Fall Creek Township, oil in Richland and Monroe.

Van Buren Township. At Summit, the strata encountered are:

Drift	98 feet
White limestone	45 feet
Blue limestone	195 feet
Soft bluish green shale	388 feet
Black shale	200 feet
Trenton limestone	45 feet
Total depth	971 feet

An oil well in the eastern limits of Summit yielded 120 barrels per day and has the following log:

Drive pipe	120 feet
Casing	440 feet
Top of Trenton	940 feet
Total depth	1042 feet

Wells have been abandoned in the following sections: Section 7, 1 well; section 8, 1 well; section 10, 1 well; section 17, 1 well; section 21, 2 wells; section 22, 2 wells; section 26, 3 wells; section 27, 5 wells; section 28, 1 well; section 31, 1 well; section 34, 6 wells.

Duck Creek Township. A well drilled on the William Shafer farm in section 24 has the following record:

Drive pipe	36 feet
Casing	228 feet
Top of Trenton	938 feet
Total depth	1238 feet

One well has been plugged in section 14, and one in section 15.

Pipe Creek Township. The following strata were encountered in a well at Elwood:

Drift	54 feet
Niagara limestone and shale	270 feet
Hudson river limestone	260 feet
Utica shale	340 feet
Trenton limestone	16 feet
Total depth	940 feet
Trenton below sea level	66 feet

At Frankton the strata pierced by a well are as follows:

Drift	88 feet
Niagara limestone and shale	272 feet
Hudson River and Utica	480 feet
Trenton limestone	22 feet
Total depth	862 feet

Two wells have been plugged, one in section 15, the other in section 28.

Fall Creek Township. At Pendleton, the first well drilled passed through the following:

Drift	5 feet
Corniferous limestone	2 feet
Sandstone	14 feet
Upper Niagara shale	20 feet
Limestone	200 feet
Shale (Lower Niagara)	5 feet
Limestone	4 feet
Shale (green and brown)	610 feet
Trenton limestone	87 feet
Total depth	947 feet
Altitude of well	841 feet

Wells were plugged in 1913 in section 1 and one in section 16, in 1916.

Adams Township. Gas wells were drilled in and near Markleville. A few wells are still supplying gas (1919).

Wells were plugged in Green Township in sections 6 and 21 in 1911, and 1913, and in Lafayette Township in section 18 in 1916.

MARION COUNTY.

The bed rock of this county consists of limestone of the Silurian age and limestones and shales of the Devonian and the Mississippian ages. These formations are concealed by glacial drift which varies in thickness from 25 feet to 200 feet. The surface of the Trenton lies from 100 feet above to 200 feet below sea level and the depth to the Trenton is from 800 to 1,100 feet.

Washington Township. At Broad Ripple a number of oil wells have been brought in recently. The record of one of the wells is given below:

Drift	55 feet
Corniferous limestone	48 feet
Niagara limestone	257 feet
Hudson River and Utica	504 feet
Trenton limestone	24 feet
Total depth	888 feet
Trenton below sea level	109 feet

Centre Township. A well at Brightwood passed through 199 feet of drift and reached the Trenton at 951 feet, below this a little gas and oil were obtained and salt water reached at 1,181 feet. Eight producing gas wells were obtained northeast of Brightwood.

Lawrence Township. At Lawrence a number of wells were drilled. One was reported to have reached the Trenton at 1,010 feet and salt water at 1,015 feet.

Warren Township. A well drilled at Irvington reached the Trenton at 966 feet and salt water at 990 feet. At Cumberland the Trenton was reached at 1,039 feet.

Wayne Township. The log of a well reported by Judge E. B. Martindale at Bridgeport is as follows:

Drift	160 feet
Black shale	140 feet
Limestone	360 feet
Shale	490 feet
Trenton limestone	50 feet
Total depth	1200 feet
Altitude of well about	750 feet

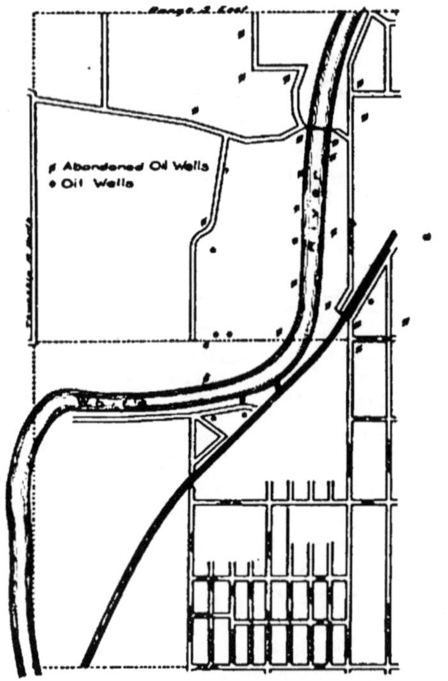

Fig. CXXXVI. Map showing location of Broad Ripple oil field in Marion County. Position of abandoned and pumping wells shown.

The record of a well drilled one and one-half miles northwest of Bridgeport is as follows:

Drift (clay and gravel)	170 feet
Soapstone (Knobstone shale)	85 feet
Black and brown Genesee shale	125 feet
Corniferous limestone	140 feet
Niagara shale	50 feet
Niagara limestone	100 feet
Total depth	670 feet

The following are the records of wells drilled on the farm of D. H. Wiggins, Broad Ripple, in 1918-1919:

	No. 1.	No. 2.	No. 4.	No. 7.	No. 8.	
Drive pipe	26	35	24	31	40	feet
Casing	368	365	360	365	340	feet
Sand at	858	854	855	848¼	860	feet
Total depth	878	862¼	868	860	875	feet

Two wells drilled on the farm of Mr. Britton of Broad Ripple in 1919 are as follows:

	No. 3.	No. 5.	
Drive pipe	26	39	feet
Casing	365	380	feet
Sand at	864	867	feet
Total depth	883	883	feet

The following are the logs of two wells drilled on the Wheeler farm, in Broad Ripple, in 1919:

	No. 1.	No. 2.	
Drive pipe	51	72½	feet
Casing	340	315	feet
Sand at	853	847.4	feet
Total depth	871	859	feet

The following well was drilled on the Carter farm in 1919, in the Broad Ripple field:

Drive pipe	35	feet
Casing	360	feet
Sand at	851	feet
Total depth	866½	feet

MARSHALL COUNTY.

Shales and limestones of Devonian age underlie the glacial drift in this county. The dip of the strata is toward the north, so for points of equal elevation above sea level, the Trenton is nearer the surface in the southern part of the county than in the northern part. The glacial drift which lies on the eroded surface of the bed rock has a thickness of from one hundred to two hundred and fifty feet. Plymouth has a number of flowing artesian wells which are forty to fifty feet deep and draw their supply from the glacial drift. The total thickness of the glacial drift at this point is 242 feet. In a deep well drilled at Plymouth, the Trenton was reached at 1,368 feet. The altitude of the well is 783 feet, and the surface of the Trenton is 585 feet below sea level.

Minor folds may exist in the Trenton underlying the county, but the structural conditions of the strata cannot be determined by direct observation because the outcrops of the durolith are concealed by the glacial drift. Well records and other subsurface data are not of sufficient abundance to warrant the mapping of structural conditions.

MARTIN COUNTY.

Martin County lies within the area of outcrop of strata of Pennsylvanian and Mississippian age. Except for some filled-in valleys, the bed rock has been little affected by glacial deposition. The accessibility of the strata renders stratigraphical and structural work possible though the pronounced unconformity between the rocks of the ages mentioned above somewhat adds to the difficulties of correct interpretation. A general section of the rocks exposed in this county would include formations from the top of the Mitchell to and including a small part of the Allegheny. A generalized section is as follows:

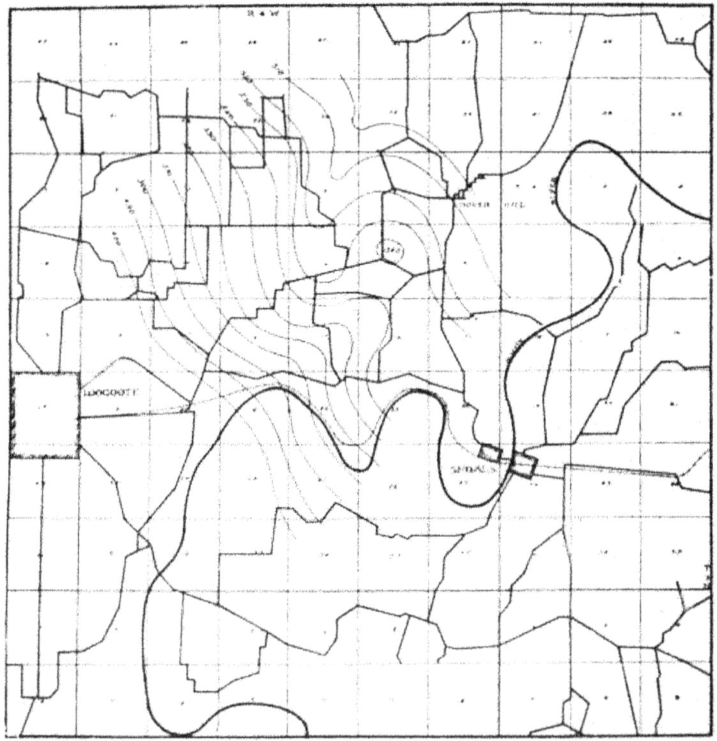

Fig. CXXXVII. A structural map of a portion of Martin County showing presence of a terrace. Contours drawn on limestone of Chester series.

Shales and sandstones containing coal (Coal Measures)	100 feet
Conglomeratic sandstone, iron ore, shales and coal (Mansfield)	200 feet
Shales, sandstones and limestones, Chester (Mississippi)	200 feet

One of the best datum planes for use in drawing structural contours is the contact between the Beech Creek limestone and the Cypress sandstone which lies above. The extreme regularity in thickness of the Beech Creek, the presence of bold springs below, the massive character of the sandstone in connection with its position immediately overlying the limestone render the contact easy of recognition and materially lessens the possibility of its being confused with other limestone contacts of frequent occurrence in the Chester.

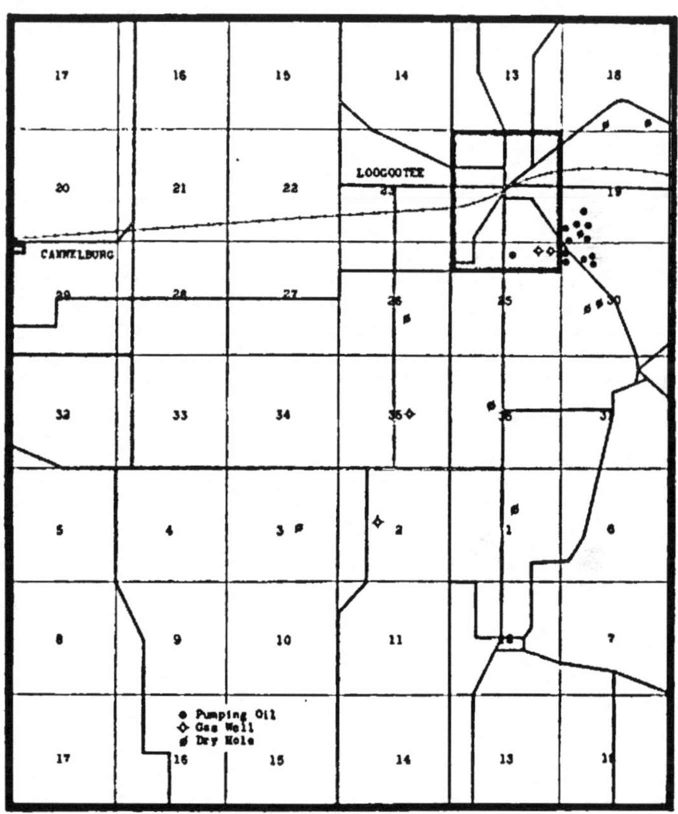

Fig. CXXXVIII. Map of Loogootee oil and gas field showing location of oil, gas and dry wells, Martin County. Data collected by field party of 1919.

A structural map of a portion of Martin County has been constructed from data collected by the writer, Dr. C. A. Malott and other members of the field party of 1919. This map shows the presence of a terrace or possibly a low anticline in the area southwest of Dover Hill. Since the Loogootee field is so near, this may prove productive territory.

A deep well was drilled to a depth of 2,200 feet southwest of Shoals and it is said that a small amount of gas was obtained.

A well drilled west of Shoals in section 26 reached oil at 1,400 feet in the Corniferous. A well drilled in White River valley in the eastern part of Shoals reached salt water at 960 feet. This well was probably finished in the Knobstone.

Perry Township. A small oil field is located in the southwest part of section 19, the northwest part of section 30, and the southeast part of section 24. Dry holes were drilled in sections 19, 36, and 1. See map.

Rutherford Township. Two dry holes were drilled in section 1 on the property of Jno. D. Allen and D. E. Elliott.

MIAMI COUNTY.

The eroded surface of the Silurian and the Devonian strata underlie the glacial drift in the county. The drift varies in thickness from a few feet to as much as 325 feet. Outcrops of the bed rock occur along the bed of Big Pipe Creek between Bunker Hill and the western boundary of the county. The rocks of Devonian age consist of limestone. Outcrops of Silurian rocks occur along the bed of Little Pipe Creek, the Wabash and the Mississinewa Rivers. Gas has been found in this county at Peru, Bunker Hill, Amboy and Xenia. The surface of the Trenton dips northward from Bunker Hill to Peru at the rate of 9 feet per mile.

The records of wells drilled at these points as given by Gorby and others are as follows:

Xenia Well[2].

Soil	4 feet
Gravel	46 feet
Water lime	31 feet
Niagara	238 feet
Hudson River and Utica	587 feet
Trenton limestone	31 feet
Total depth	937 feet
Altitude of well	815 feet
Trenton below sea level	91 feet

Record of well drilled at Bunker Hill:

Section of Well No. 1.

Drift	58 feet
Corniferous and Niagara limestone	503 feet
Hudson River and Utica	431 feet
Trenton limestone	12 feet
Total depth	1004 feet
Trenton below sea level	155 feet

Record of well drilled at Peru:

Section of Well No. 4.

Drift	36 feet
Niagara (and Clinton) limestone	325 feet

Hudson River and Utica	454 feet
Trenton limestone	30 feet
Total depth	905 feet
Trenton below sea level	218 feet

A small quantity of oil was found at a depth of 808 feet. Salt water occurred at 900 feet. This well was drilled in the northern part of the city.

Section of Well No. 2.

Drift	10 feet
Water-lime and Niagara limestone	455 feet
Clinton (?) limestone	15 feet
Hudson River and Utica	449 feet
Trenton limestone	27 feet
Total depth	956 feet
Trenton below sea level	229 feet

Yielded a small quantity of oil and gas, but not sufficient for use. This well was bored a little south of the city limits, about 1¼ miles from well No. 1.

Section of Well No. 3.

Drift	70 feet
Niagara limestone	490 feet
Hudson River and Utica	400 feet
Trenton limestone	42 feet
Total depth	1002 feet

A light flow of gas was obtained from this well. The above well was situated on the Younce farm, seven miles southeast of Peru.

Section of Well No. 4.

Drift	324 feet
Niagara limestone	276 feet
Hudson River and Utica	407 feet
Trenton limestone	35 feet
Total depth	1042 feet

Yielded no gas.

Record of wells drilled in sections 16 and 28, S. E. ¼ of the N. E. ¼ of section 28:

Alluvium—river drift	36 feet
Niagara limestone	385 feet
Hudson River and Utica	454 feet
Top of Trenton	875 feet
Total depth	905 feet
Surface above sea level	657 feet
Top of Trenton below sea level	218 feet

S. W. ¼ of section 16 (27 N. 4 E.):

Drift	324 feet
Niagara limestone	379 feet
Hudson River and Utica shale	307 feet
Top of Trenton	1010 feet
Total depth	1041 feet
Surface above sea level	757 feet
Top of Trenton below sea level	253 feet

Hospital Hill.

Drift	20 feet
Niagara limestone	375 feet
Hudson River shales and limestone	255 feet
Utica shale	248 feet
Top of Trenton at	898 feet
Total depth	933 feet

This well was drilled in October, 1897, and produced 400 barrels of oil a day for four days. The production gradually dropped to 300 barrels when three weeks old.

Jackson Township. Record of well drilled at Amboy:

Section of Well No. 1.

Drift	35 feet
Niagara limestone and shale	350 feet
Hudson River and Utica	522 feet
Trenton limestone	33 feet
Total depth	940 feet

Yielded a strong flow of gas.

The following is a record of wells abandoned in this township:

Owner.	Date	Sec.	Town	Range	Wells
C. C. Hull	1911	14	25	5E	1
E. L. Daniels	1913	20	25	6E	1
Chas. Friemal	1913	20	25	6E	1
E. L. Daniels	1913	24	25	5E	1
E. Hooper	1913	29	25	6E	1
E. L. Carter	1913	30	25	6E	1
E. Gross	1913	32	25	6E	1

MONROE COUNTY.

Geology. The eastern portion of the county lies within the area occupied by the Knobstone, the central portion is occupied by the Harrodsburg, Salem and Mitchell limestones, the western portion by the Chester shales, limestones and sandstones while the highlands in the extreme western portion are occupied by the Pottsville.

Structure. The Mount Carmel fault crosses the eastern part of the county and near Unionville makes a change in direction which makes conditions favorable to anticlinal folds. The fault itself with its downthrow toward the east produces an anticlinal fold extending parallel to the fault but not a closed structure except at such places as cross flexures are produced.

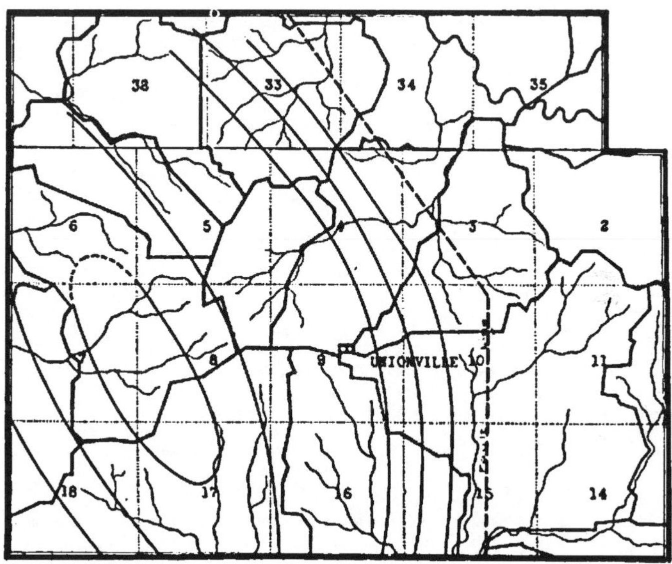

FIG. CXXXIX. A structural map of a portion of Monroe County.

Bloomington Well. A deep well was drilled in the courthouse yard in 1885 to a depth of 2,730 feet. A generalized record of the well follows:

Surface loam	6 feet
Mississippian limestone and shales	749 feet
Devonian shales and limestones	170 feet
Niagara limestone	240 feet
Hudson River limestone	485 feet
Utica shale	180 feet
Trenton limestone	626 feet
Potsdam sandstone (?) St. Peter	274 feet
Total depth	2730 feet
Altitude of well	770 feet

No oil or gas was found in this well, which was drilled for an artesian water supply. The complete record is given below:

Earth	6 feet
St. Louis limestone, water	30 feet
Keokuk limestone	89 feet

Knobstone	630 feet
Red shale	20 feet
Blue limestone	5 feet
Brown shale, gas	10 feet
Black slate, Devonian	120 feet
Gray limestone, Portland cement	15 feet
Brown limestone, Niagara	240 feet
Shaly limestone	15 feet
Light brown limestone	130 feet
Flinty limestone	30 feet
Light colored limestone	100 feet
Brown limestone	70 feet
Blue shale	40 feet
Blue limestone	40 feet
Blue shale, streaks of limestone	60 feet
Blue shale	180 feet
Gray limestone, some shale	586 feet
Blue shale	40 feet
Hard, white sandstone	4 feet
Shaly limestone and sandstone	20 feet
Gray limestone and sandstone	20 feet
Shaly limestone, sandstone quartzite	98 feet
White and yellow, hard sandstone, iron	22 feet
White sandstone, softer	20 feet
White sandstone, soft	40 feet
Gray limestone and sandstone, mixed	42 feet
Gray limestone, sulphur-water increasing rapidly	8 feet
Total	2730 feet
Trenton below sea level about	1060 feet

Well east of Coleman House, west of Thrasher Schoolhouse:

Oolitic limestone at	130 feet
Soil	6 feet
Sandstone and iron ore	7 feet
White sandstone	5 feet
Iron stone	5 feet 6 inches
Brown sandstone	34 feet
Coal	6 feet
Blue sandstone	22 feet 6 inches
Blue sand	17 feet
Iron stone	27 feet
Limestone	3 feet
Total depth	133 feet

Well southeast of Thrasher Schoolhouse:

	Thickness	Total Depth
Drift	10 feet	10 feet
Iron stone	5¼ feet	15¼ feet
Shale	6 feet	21¼ feet

Iron stone	4½	feet	26	feet
Blue sandstone	15	feet	41	feet
Kaolin	5½	feet	46½	feet
Blue sandstone	22	feet	68½	feet
Coal	4½	feet	73	feet
Blue shale	26	feet	99	feet
Blue sandstone	19	feet	118	feet
Iron stone	22	feet	140	feet
White limestone, water	140	feet	280	feet
Shale	60	feet	340	feet
Brown limestone	45	feet	385	feet
Shale	5	feet	390	feet
Brown limestone	20	feet	410	feet
Blue limestone, water	130	feet	540	feet
Quartz	20	feet	560	feet
White sandy shale	100	feet	660	feet
Blue shale	360	feet	1020	feet
Sand	25	feet	1045	feet
Blue shale	127	feet	1172	feet
Red rock	7	feet	1179	feet
Shale	18	feet	1197	feet
Limestone	7	feet	1204	feet
White sandstone	6	feet	1210	feet
Dark shale	35	feet	1245	feet
Iron pyrite	10	feet	1255	feet
Brown shale and iron	33	feet	1288	feet
Black shale, hard	30	feet	1318	feet
White limestone	24	feet	1342	feet
Limestone	9	feet	1351	feet
White gray limestone	85	feet	1436	feet
Brown limestone	25	feet	1461	feet
Gray limestone	10	feet	1471	feet
Brown limestone	25	feet	1496	feet
White limestone	16½	feet	1511	feet
Brown and gray limestone	28	feet	1539	feet
Gray limestone	17	feet	1556	feet
Black shale	22	feet	1578	feet
Limestone (Niagara)	6	feet	1584	feet
Pure white limestone	92	feet	1676	feet
Black limestone	24	feet	1700	feet
Gray limestone	25	feet	1725	feet
Gray limestone and water	11	feet	1736	feet
Coarse limestone and gas	9	feet	1745	feet
Gray limestone	35	feet	1780	feet
Brown limestone	23	feet	1803	feet
Gray limestone	10	feet	1813	feet
Blue limestone	37	feet	1850	feet
Blue shale	15	feet	1865	feet
Blue shale	15	feet	1880	feet
Blue limestone	9	feet	1889	feet

Blue shale	11	feet	1900	feet
Shale (Utica)	50	feet	1950	feet
Shale (Utica)	15	feet	1965	feet
Black shale	30	feet	2000	feet
Blue and black shale	274	feet	2274	feet
Trenton limestone	301	feet		
Total depth			2575	feet

Top of Trenton at 2272 feet.
Oil sand at 2301 feet. Light initial production.
Altitude at mouth of the well 975 feet.

There is a dip of thirty-five feet to the mile for the Trenton limestone between the deep well at Bloomington and the Koontz well. In the former the Trenton is 1,060 feet below sea level and in the latter 1,300 feet.

MONTGOMERY COUNTY.

A small area of the bed rock in the western portion of this county is occupied by Pennsylvanian strata, but the greater part of the subsurface of the county is occupied by the strata of the Mississippian age. The covering of the glacial drift in a large measure prevents the determination of strue-tural conditions of the strata. The surface of the Trenton lies from 1,200 to 1,600 feet below the surface of the county. The dip of the strata is toward the southwest, dipping away from the Cincinnati arch, which lies to the north. The surface of the Trenton lies from 400 to 800 feet below sea level.

The following is the record of well No. 1 drilled at Crawfordsville[9]:

Drift	140	feet
Sub-Carboniferous rocks	410	feet
Devonian shale	80	feet
Corniferous limestones	55	feet
Niagara limestone	380	feet
Hudson River and Utica	365	feet
Trenton limestone	69	feet
Total depth	1499	feet
Trenton below sea level	664	feet

Yielded no gas.

Railroad Elevations.

Linden, 787; Cherry Grove, 797.5; Manchester, 753.4; Crawfordsville, 738.5; Whitesville, 871; Ladoga, 822.5; New Ross, 877; Pawnee, 846; Lapland, 840; Penobscot, 859; Waveland, 744; Sand Creek, 582.

MORGAN COUNTY.

The glacial mantle covering the bed rock in this county varies from a few feet to ninety feet. The Knobstone division of the Mississippian underlies the drift over a large part of the county. Outcrops of the Knobstone occur, but they are not sufficiently abundant to be of much service in locating favorable structural conditions. Even if a sufficient number of outcrops could be found the absence of sufficient number of persistent hardy layers of rock would render the determination of structural conditions exceedingly difficult. In the presence of favorable conditions, oil and gas sands may be found in the Devonian and Trenton strata. The Trenton will be found below the surface at a depth ranging from 1,400 to 1,600 feet.

Two wells were drilled south of Hall, in 1916. The first one was drilled to a depth of about 860 feet and had a showing of oil in the Corniferous limestone. The well was shot, but did not increase the show of oil.

Section of Well No. 1, Martinsville, Ind.

Drift	85 feet
Sub-carboniferous rocks	323 feet
Hamilton shale	120 feet
Corniferous limestone	62 feet
Niagara limestone	236 feet
Hudson River and Utica	571 feet
Trenton limestone	51 feet
Total depth	1448 feet
Trenton below sea level	780 feet

Yielded no gas.

Jackson Township. A well was drilled on the Donald Stewart property in Section 1 in 1911 and another on the Emory Hilderman property in Section 36 in 1912. Both were non-productive.

NEWTON COUNTY.

The subsurface of Newton County is occupied by the strata of the Silurian in the central portion and northern portion of the county and by the Devonian strata in the southern portion of the county. The strata of the northern portion dip north and those of the southern portion toward the south. Slight variations in the uplift of the arch formed has resulted in the creation of at least one minor fold favorable to the accumulation of oil. This occurs in the boundary between Newton and Lake Counties near the town of Thayer.

The following formations will be encountered in this county between the surface of the glacial drift and the surface of the Trenton:

	Thickness.
Glacial drift	100 to 150 feet
Devonian (in south part)	50 to 145 feet
Silurian	280 to 300 feet
Hudson River	300 feet
Utica	210 feet

On the north boundary at Thayer the Trenton is encountered at 846 feet where the surface elevation is 650 feet. At Kentland in the south part of the county at an elevation of 680 feet the Trenton is encountered at 1,060 feet. The dip of the Trenton surface is more than 57 feet to the mile toward the south.

On account of the covering of glacial drift which attains a thickness of more than one hundred feet, the geological structures favorable to the accumulation of oil cannot be determined or located by the use of surficial methods. The oil which has been found is probably in the Trenton limestone. The following is a log of well No. 2 drilled on the Grant farm west of Thayer by the Thayer Oil and Gas Co., Lincoln Township:

Oil sand	at 615 feet
Thickness of gas sand	20 feet
Salt water	at 675 feet
Trenton rock	at 846 feet
Oil	at 850 feet
Total depth	862 feet

This well was plugged in 1919, as was a well on the Rebecca Spitter property.

Well No. 3.

Drift	73 feet
Niagara limestone	283 feet
Hudson River limestone	300 feet
Utica shale	190 feet
Trenton limestone	6 feet
Total depth	852 feet

Record of well drilled at Kentland:

Section of Well No. 1.

Drift	100 feet
Black shale (New Albany)	100 feet
Corniferous	45 feet
Niagara limestone	305 feet
Hudson River limestone	300 feet
Utica shale	210 feet
Trenton limestone	60 feet
Total depth	1120 feet
Trenton below sea level	379 feet

Yielded no gas.

NOBLE COUNTY.

Noble County probably lies wholly within the area occupied by the Devonian strata, though its bed rock is concealed by a heavy mantle of glacial drift. A well record at Albion shows a thickness of 375 feet and at Kendallville of 485 feet of drift. The well at Kendallville reached the Trenton at 1,920 feet.

A well drilled at Albion furnished the following log[1]:

Section of Well No. 1.

Drift	375 feet
Devonian shale	65 feet
Devonian limestone	65 feet

Sandstone	5 feet
Hydraulic limestone	30 feet
Niagara and Clinton (?) limestone and shale	815 feet
Hudson River limestone and shale	285 feet
Utica shale	250 feet
Trenton limestone	24 feet
Total depth	1914 feet
Trenton below sea level	1161 feet

Yielded small flow of gas.

The surface of Trenton dips northward through this county at the rate of from thirty-five to thirty-eight feet to the mile. If there are structures developed in these northward dipping strata they are not visible at the surface because of the thick over-burden of drift, which prevents the detection of reverse dips.

Railroad Elevations.

LaOtto, 872.9; Swan, 872; Avilla, 962.9; Kendallville, 974.7; Rome City, 920.3; Grismore, 868.2; Ligonier, 893.8; Wawaka, 952.1.

OHIO COUNTY.

The Cincinnatian Division of the Ordovician including the Eden (Utica) Maysville, (Lorraine) and Richmond from the strata underlying the Pleistocene and Recent deposits of this county. The Pleistocene deposits vary in thickness from a few to fifty feet. The Ordovician sediments that are revealed consist of a series of shales and limestones. The Trenton limestone lies below these formations. The number and abundance of outcrops will probably make it possible to determine the structural conditions existing in this county, but careful detailed work will be required. The table below gives the subdivisions which are represented in the county.

- Quaternary
  - Recent, clays and alluvium
  - Pleistocene, gravel, sands and till
- Ordovician
  - Cincinnati
    - Richmond
      - Elkhorn
      - Whitewater
      - Saluda
      - Waynesville
      - Arnheim
    - Maysville
      - Mt. Auburn
      - Coryville
      - Bellevue
      - Fairmont
      - Mt. Hope
    - Eden
      - McMicken
      - Southgate
      - Economy
      - Fulton
    - Trenton

ORANGE COUNTY.

This county lies within the unglaciated area and the structural conditions of the rocks may be determined for the greater part of the county by surficial observations. The eastern part of the county contains the Salem and the Mitchell limestones of the Mississippian. The western part of the county contains the shales, sandstones and the limestones of the Chester division of the Mississippian and the conglomeratic sandstones of the Pottsville division of the Pennsylvanian. Where the geologic conditions are favorable there is a probability of the accumulation of oil and gas in the Devonian strata (Corniferous limestone) which may be reached in the western part of the county

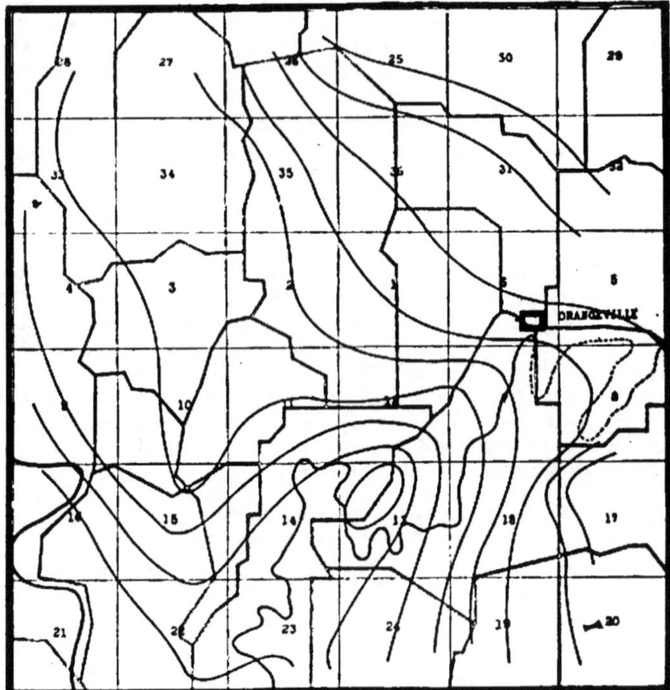

Fig. CXLI. Map of a portion of Orange County showing structural conditions near Orangeville. Contour lines drawn on Chester limestone. Data secured by C. A. Malott and P. B. Stockdale of field party of 1919.

at a depth of from 1,100 to 1,400 feet. There is also a probability of oil and gas accumulating under such structures in the Trenton though the Trenton limestone may lack porosity due to the lack of dolomitization.

A general geological section in this county would include:

Reddish conglomeratic sandstone with iron ore (Pottsville)	200 feet
Fine grained massive sandstone (Tar Springs, Chester)	45 feet
Limestone, gray (glendeane)	10 feet
Sandstone and sandy shales (Hardinsburg, Chester)	50 feet

Limestone, thin bedded (Golconda)	16 feet
Sandstone, massive (passing to shale, Cypress)	35 feet
Limestone, massive (Beech Creek)	12 feet
Shales and sandstones (Elwren)	32 feet
Limestone, pyritiprous, reddish (Reelsville)	4 feet
Sandstone and shales (Brandy Run)	13 feet
Limestone, massive ledges (Beaver Bend), Top of Mitchell	10 feet
Limestone (Mitchell, Salem, Harrodsburg)	400 feet

Two wells were drilled at Paoli, one to a depth of 1,000 feet, the other to a depth of 1,130 feet. In the first mineral water was found at 250 feet and in a blue shale at 1,000 feet. The bottom of this well is probably in the Silurian shale. Its altitude is about 580 feet. The second well encountered mineral water in a limestone at 1,130 feet and probably was completed in the Silurian limestone. These wells were drilled for oil or gas and were drilled without reference to structure. By consulting the structure map accompanying this report it will be evident that no favorable structure is present. A well drilled in Section 8 southwest of Paoli reached a depth of over 1,200 feet before being abandoned. This well was drilled on a slight shoulder or terrace as will be seen by consulting the structure map. The field work necessary to the preparation of this map was done by Dr. C. A. Malott and Mr. P. B. Stockdale.

OWEN COUNTY.

The geological formation represented by the outcrops in this county are found in the following section:

Quaternary	Recent—River alluvium. Pleistocene—Glacial gravels, sands and clays.
Pennsylvanian	Coal measures—coal beds, sandstones, shales and limestones. Mansfield (Pottsville) sandstone shales and coal.
Mississippian	Chester, shales, limestones, and sandstones. Mitchell, limestones and shales. Salem, limestone. Warsaw, limestone. Knobstone, shales and sandstones.

The Pleistocene deposits mantle the surface in all places except along the courses of streams, where it has been removed by postglacial erosion. The number of outcrops may be sufficient in some places in the county to enable the structure of the bed rock to be determined.

Washington Township. Three wells were drilled in Spencer to the Niagara limestone from which a supply of sulphur-saline water was obtained. A well was also drilled south of Spencer and a showing of oil obtained at a depth of 800 feet. This well was drilled deeper, but did not strike production.

A well was drilled on the Tanner property in Section 20 west of Spencer in 1913. No record of this well has been obtained.

PARKE COUNTY.

The strata of the Pennsylvanian period underlie the glacial drift in Parke County. Outcrops of the bed rock occur along the beds of some of the streams, but the structural condition cannot be determined from surficial observations.

The Trenton limestone lies from 2,000 to 2,500 feet below the surface in this county. The following is the record of a well drilled at Rockville:

Section of Well No. 1.

Drift	96 feet	Limestone	118 feet
Gray sandstones	44 feet	Brown sandstones	46 feet
Brown shale	25 feet	White limestones	135 feet
White sandstones	110 feet	Crystallized limestone	85 feet
Black shale	25 feet	White shale, like kaolin	48 feet
Black shale	105 feet	Limestone	108 feet
White sandstone	50 feet	Dark shale (Utica)	324 feet
Limestone	170 feet		——
Gray shale	305 feet	Total depth to Trenton	2100 feet
Sandstone	100 feet	Altitude of well	688 feet
White shale	114 feet	Trenton below sea level	1412 feet
Black shale	102 feet	Yielded no gas.	

In 1908 a bore was sunk to a depth of 1,200 feet near Diamond, in Parke County but was dry.

Where structural conditions are favorable oil may be found in the Devonian in this county.

PERRY COUNTY.

As Perry County occupies a part of the unglaciated area of Indiana the outcrop of its strata is unconcealed. The formations of the county belong to the following divisions:

Quaternary	Recent—alluvium and residuals. Pleistocene—residuals.
Pennsylvanian	Allegheny—shales, sandstones, limestones, coals. Pottsville—shales, sandstones and coal.
Mississippian	Chester—limestones, sandstones and shales.

No structural map of this county has been attempted, but it seems possible to determine the structural conditions for a large part of the county by using the limestones of the Chester as key formations.

Some oil was found in two wells in Section 19 near Uniontown, also in Sections 24 and 26. The records of these wells are given below:

Wells drilled in Clark Township east of Siberia, near Anderson River six miles south of Birdseye.

Well in Southwest ¼ of Section 24.

Drive pipe	40 feet
Casing	595 feet
Top of pay	1010 feet
Total depth	1030 feet

Well in Southeast ¼ of Section 26 (3S. 3W.)

Drive pipe	10 feet
Casing	735 feet
Total depth	1280 feet

The above well came in as a salt water without a showing of oil.

Northeast ¼ of the Southwest ¼ of Section 19.

Drive pipe	60 feet
Casing	600 feet
Total depth	1040 feet

Better producer than No. 1.

Record of Deep Well at Cannelton.

	Thickness.	Depth.
Sand	47 feet	47 feet
Shale	110 feet	157 feet
White sand	63 feet	220 feet
Shale	9 feet	229 feet
Limestone	41 feet	270 feet
Shale	5 feet	275 feet
Hard limestone, white	55 feet	330 feet
Shale	16 feet	346 feet
Limestone	6 feet	352 feet
White sand	5 feet	357 feet
Shale	3 feet	360 feet
Sand	13 feet	373 feet
Shale	23 feet	396 feet
Black limestone	10 feet	406 feet
Gray shale	30 feet	436 feet
White limestone	9 feet	445 feet
Gray shale	15 feet	460 feet
White shale salt water at 480	51 feet	511 feet
Shale	7 feet	518 feet
White limestone salt water at 733	218 feet	736 feet
Limestone salt water at 744	204 feet	940 feet
Dark sandy shale	87 feet	1027 feet
Dark brown limestone	81 feet	1108 feet
Limestone	572 feet	1780 feet
Shale (Utica)	120 feet	1900 feet
Limestone (Trenton)	633 feet	2533 feet

Tell City Well Record.

	Thickness.	Depth.
Soil	25 feet	25 feet
Gray shale	10 feet	35 feet
White sand	40 feet	75 feet
Brown sand	80 feet	155 feet
White limestone	30 feet	185 feet
Dark gray shale	30 feet	215 feet
Shally lime	10 feet	225 feet

Limestone	5 feet	230 feet
Greenville shale	45 feet	275 feet
Limestone	71 feet	346 feet
Gray sand	6 feet	352 feet
Gray limestone and shale	43 feet	395 feet
Sand	15 feet	410 feet
Variegated shales	116 feet	526 feet
Limestone	33 feet	559 feet
Gray shale	36 feet	595 feet
Gray sand	20 feet	615 feet
Limestone and shale	3 feet	618 feet
Limestone	17 feet	635 feet
Brown shale	13 feet	648 feet
Gray sand	27 feet	675 feet
Brown shale	5 feet	680 feet
Sand stone	62 feet	742 feet
No record	10 feet	752 feet
Gray limestone	168 feet	920 feet
Light limestone	245 feet	1165 feet

PIKE COUNTY.

The strata of Pike County belong to the coal measures with the exception of a mantle of glacial drift in the northern portion, of glacial lake deposits in the central portion and recent residuals covering the southern portion and overlying the coal measures. As many as eight distinct veins of coal occur in the county. Three or four of these are workable over considerable area. For the determination of structural conditions it is possible that some use may be made of the Coal Measures. Oil fields have been developed northeast of Petersburg, southwest and southeast, in Washington Township, Madison, Monroe, Patoka and Logan Townships. Some of the structures in this county were outlined on the Petersburg coal and published in the Ditney folio[1].

Madison Township. Oil sands range in depth from 960 to 1,340. Five sands are reported.

Well No. 3, D. & R. Snyder farm. Section 35, Madison Township:

Soil	to	5 feet	Blue sand	to	453 feet
Mud	to	45 feet	Dark lime	to	463 feet
Quick sand	to	55 feet	Slate	to	523 feet
White sand	to	95 feet	Lime	to	540 feet
Slate	to	100 feet	Shale	to	690 feet
Coal	to	105 feet	Sand	to	710 feet
Slate	to	165 feet	Water sand	to	800 feet
Sand	to	175 feet	Slate	to	845 feet
Slate	to	375 feet	Lime	to	850 feet
Shale	to	391 feet	Slate	to	870 feet
Lime	to	420 feet	Sharp sand	to	935 feet
Coal	to	423 feet	Gray slate	to	945 feet
Sand	to	433 feet	Lime	to	960 feet

[1] See Ditney Folio, U. S. G. S.

Slate ....to 1005 feet
Sand ....to 1015 feet
Shale ....to 1035 feet
Water sand.............to 1140 feet
Slate ....to 1145 feet
Little lime..............to 1163 feet
Slate ....to 1193 feet
Dark lime..............to 1200 feet
Shale ....to 1205 feet
Lime ....to 1215 feet
Slate ....to 1245 feet
Dark sand..............to 1253 feet
Big lime................to 1275 feet
Slate ....to 1285 feet
Sand ....to 1295 feet
Shale ....to 1300 feet
Gas sand...............to 1303 feet
Slate ....to 1304 feet
Snyder sand............to 1313 feet
Slate ....to 1323 feet
Brown shell............to 1330 feet
Slate ....to 1341 feet
Dark sand..............to 1347 feet
Brown sand ....to 1348 feet

Total depth.........to 1348 feet

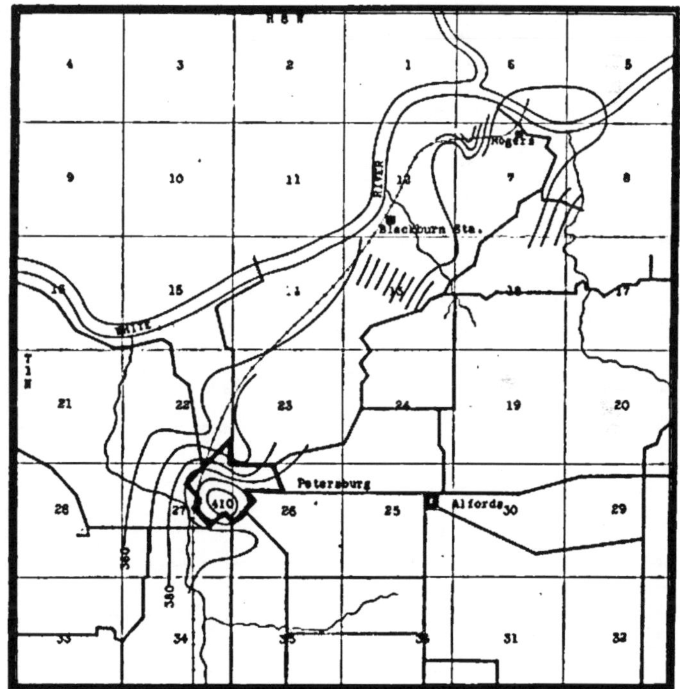

Fig. CXLII. Map of portion of Pike County showing outline of structure at Petersburg. Contour lines drawn on Coal V by C. A. Malott and P. B. Stockdale, field party, 1919.

Casing Record.

12½ in.	71 feet
10 in.	392 feet
9¼ in.	945 feet
6½ in.	1210 feet

Well No. 5, L. Johnson Farm. Madison Township, Pike County, October 6, 1919:

Clay	61 feet	Slate	38 feet
Slate	79 feet	Water sand	110 feet
Coal	1 foot	Lime	8 feet
Slate	319 feet	Slate	7 feet
Lime	6 feet	Sand	35 feet
Slate	414 feet	Slate	35 feet
Sand	55 feet	Big lime	27 feet
Slate	75 feet	Slate	29 feet
Oil sand	15 feet	Oil sand	24 feet
Lime	7 feet		
		Total depth	1346 feet

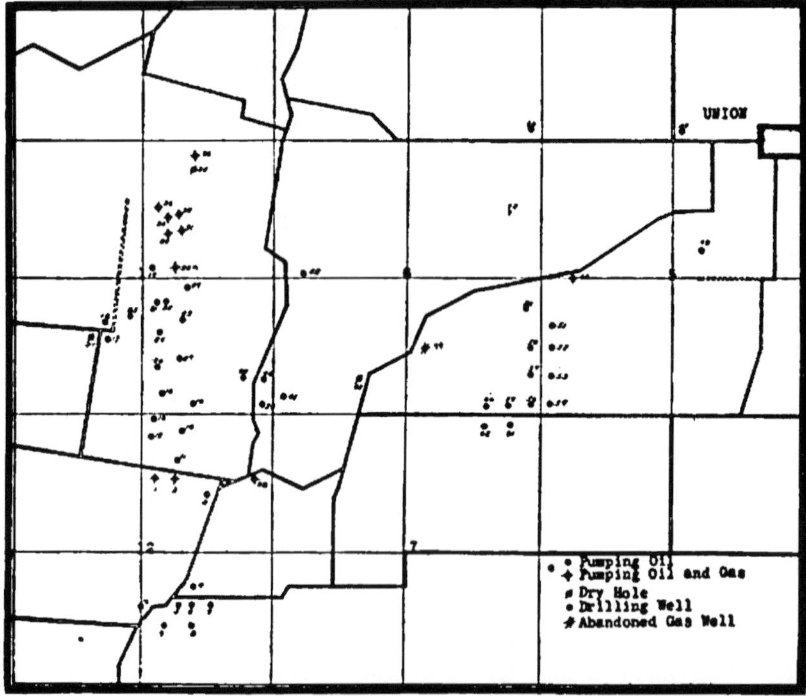

Fig. CXLIII. Map of the Union oil field showing oil and gas wells and dry holes. Data collected by C. A. Malott and P. B. Stockdale of field party of 1919.

Well No. 5, M. E. Sutton farm, Madison Township, Pike County:

Surface	to 1 foot	Shale	to 58 feet
Quick sand	to 16 feet	Lime	to 95 feet
Shale	to 30 feet	Shale	to 115 feet
Coal	to 55 feet	Lime	to 125 feet

Shale	to 130 feet	Shale	to 600 feet
Lime	to 180 feet	Lime	to 720 feet
Shale	to 200 feet	Shale	to 725 feet
Sand	to 285 feet	Sand	to 800 feet
Shale	to 315 feet	Shale	to 885 feet
Lime	to 420 feet	Lime	to 925 feet
Shale	to 428 feet	Sand oil	to 942 feet
Sand	to 580 feet		——
		Total depth	963 feet

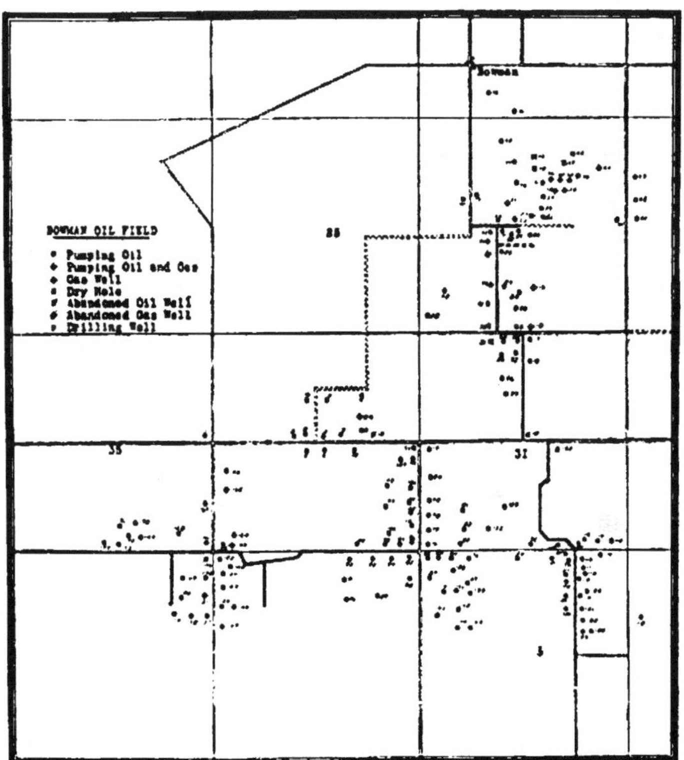

Fig. CXLIV. Map of the Bowman oil field of Pike County, showing location of oil wells, dry holes and gas wells. Data secured by C. A. Malott and P. B. Stockdale, field party of 1919.

Record of Dan Snyder No. 2, Section 36, Madison Township:

12¼-inch casing	92 feet	6¼-inch casing	1240 feet
10-inch casing	210 feet	4¾-inch liner	80 feet
8¼-inch casing	915 feet		

Soil		35 feet	Lime shell	95 feet to 130 feet
Quick sand	35 feet to	45 feet	Blue mud	130 feet to 155 feet
Blue mud	45 feet to	95 feet	Brown shale	155 feet to 200 feet

Coal	200 feet	to 205 feet
Gray mud	205 feet	to 230 feet
White mud ...	230 feet	to 265 feet
Lime	265 feet	to 295 feet
Sand	295 feet	to 340 feet
Blue mud	340 feet	to 370 feet
White mud ...	370 feet	to 435 feet
Lime	435 feet	to 440 feet
Blue mud	440 feet	to 490 feet
Lime	490 feet	to 500 feet
Gray mud	500 feet	to 535 feet
Shale	535 feet	to 585 feet
Brown mud ...	585 feet	to 640 feet
Water sand ...	640 feet	to 660 feet
Slate and shale	660 feet	to 720 feet
Light shale ...	720 feet	to 780 feet
Slate	780 feet	to 850 feet
Shale	850 feet	to 875 feet
Water sand ...	875 feet	to 910 feet
Brown mud ...	910 feet	to 930 feet
Slate	930 feet	to 994 feet
***Rumble sand..**	**994 feet**	**to 1011 feet**
Slate	**1011 feet**	**to 1016 feet**
Brown mud...	**1016 feet**	**to 1025 feet**
Sand	**1025 feet**	**to 1060 feet**
Sandy shale...	**1060 feet**	**to 1095 feet**
Water sand...	**1095 feet**	**to 1125 feet**
Slate	**1125 feet**	**to 1135 feet**
Shale	**1135 feet**	**to 1158 feet**
Gray mud	**1158 feet**	**to 1170 feet**
Lime	**1170 feet**	**to 1190 feet**
White mud ...	**1190 feet**	**to 1195 feet**
Hard lime	**1195 feet**	**to 1210 feet**
Blue mud	**1210 feet**	**to 1234 feet**
Big lime	**1234 feet**	**to 1249 feet**
Slate	**1249 feet**	**to 1264 feet**
Red rock	**1264 feet**	**to 1273 feet**
Shale	**1273 feet**	**to 1283 feet**
Snyder sand ..	**1283 feet**	**to 1302 feet**
Total depth		**1302 feet**

*No oil.

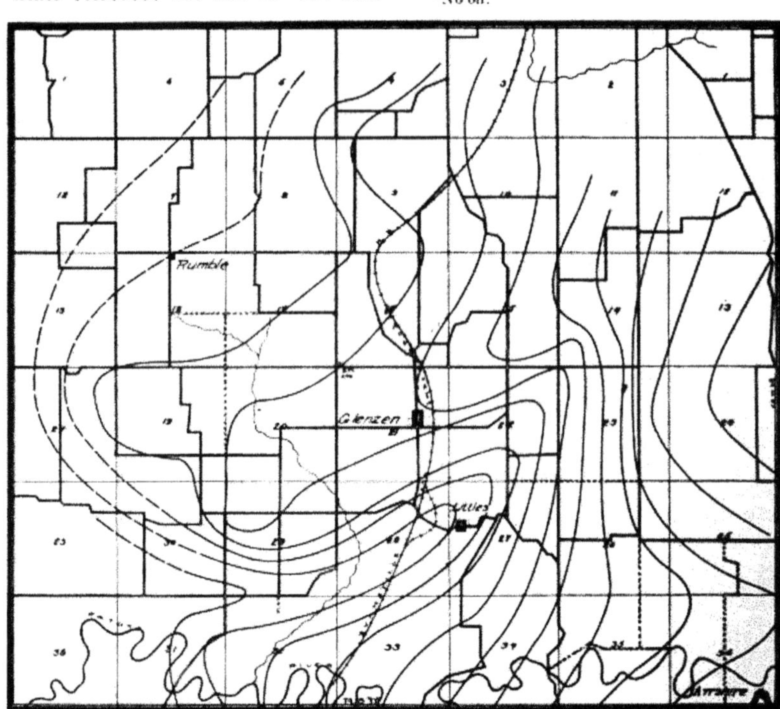

Fig. CXLV. Map of the Glenzen terrace in Pike County. Structure lines drawn on Coal V. Data secured by C. A. Malott and P. B. Stockdale, field party of 1919.

Central Refining Co.

Well No. 9. Section 35, Madison Township:

Clay ....to	18 feet	Gray slate ....to	160 feet
Sand ....to	70 feet	Lime ....to	165 feet
Brown shale ....to	75 feet	Gray slate ....to	180 feet
Coal ....to	77 feet	Brown slate ....to	205 feet
Brown shale ....to	135 feet	Coal ....to	208 feet
Lime ....to	148 feet	Lime ....to	211 feet

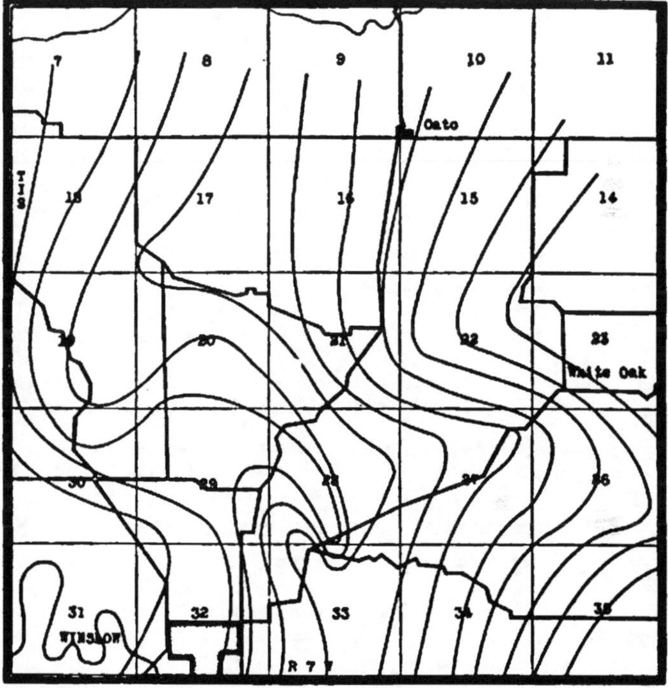

Fig. CXLVI. Map of structural conditions near Winslow. Contours drawn on Coal V. Data secured by C. A. Malott and P. B. Stockdale of the field party of 1919.

Brown slate ....to	215 feet	Brown slate ....to	440 feet
Lime ....to	222 feet	Lime ....to	443 feet
Gray slate ....to	250 feet	Brown slate ....to	450 feet
Brown slate ....to	260 feet	Gray slate ....to	490 feet
Lime ....to	288 feet	Brown slate ....to	488 feet
Brown slate ....to	290 feet	Lime ....to	500 feet
Gray slate ....to	310 feet	Coal ....to	503 feet
Brown slate ....to	350 feet	Gray slate ....to	515 feet
Gray slate ....to	390 feet	Lime ....to	522 feet
Brown slate ....to	430 feet	Brown slate ....to	560 feet
Sand ....to	438 feet	Gray slate ....to	600 feet

Brown slate	to 650 feet	Gray slate	to 1200 feet
Gray slate	to 690 feet	Lime	to 1210 feet
Sand	to 696 feet	Gray slate	to 1220 feet
Brown slate	to 715 feet	Lime	to 1225 feet
Gray slate	to 755 feet	Gray slate	to 1230 feet
Sand	to 760 feet	Big lime	to 1238 feet
Brown slate	to 800 feet	Gray slate	to 1250 feet
Gray slate	to 830 feet	Lime	to 1258 feet
Brown slate	to 858 feet	Gray slate	to 1261 feet
Sand	to 872 feet	Lime	to 1265 feet
Brown slate	to 880 feet	Brown slate	to 1275 feet
Sand	to 940 feet	Red rock	to 1277 feet
Gray slate	to 946 feet	Brown slate	to 1313 feet
Lime	to 950 feet	Sand	to 1319 feet
Gray slate	to 955 feet	Brown shell	to 1325 feet
Lime	to 965 feet	Slate	to 1326 feet
Gray slate	to 1050 feet	Lime	to 1328 feet
Sand	to 1060 feet	Slate	to 1340 feet
Gray slate	to 1080 feet	Gray sand	to 1345 feet
Sand	to 1160 feet	Brown sand	to 1348 feet
Little lime	to 1172 feet		
		Total depth	1348 feet

Casing Record.

12½-inch	70 feet	8¼-inch	948 feet
10-inch	442 feet	6¼-inch	1231 feet

Wells abandoned in this township are located in Section 1, 1 well; Section 2, 2 wells; Section 6, 2 wells; Section 25, 1 well; Section 35, 2 wells; Section 36, 2 wells.

Log of M. F. Snyder well. Located in Section 2, Madison Township:

Yellow clay	to 10 feet	Lime	to 277 feet
Gray slate	to 30 feet	Coal	to 280 feet
Sand	to 47 feet	Gray slate	to 295 feet
Gray slate	to 85 feet	Lime	to 310 feet
Sand	to 93 feet	Sand	to 362 feet
Gray slate	to 95 feet	Brown slate	to 372 feet
Coal	to 97 feet	Gray slate	to 380 feet
Lime	to 103 feet	Sand	to 384 feet
Gray slate	to 135 feet	Gray slate	to 405 feet
Brown slate	to 150 feet	Brown slate	to 407 feet
Lime	to 165 feet	Lime	to 442 feet
Gray slate	to 170 feet	Gray slate	to 460 feet
Sand	to 180 feet	Brown slate	to 463 feet
Brown slate	to 185 feet	Coal	to 467 feet
Sand	to 210 feet	Gray slate	to 483 feet
Gray slate	to 240 feet	Lime	to 485 feet
Sand	to 265 feet	Brown slate	to 495 feet
Brown slate	to 268 feet	Lime	to 513 feet
Coal	to 272 feet	Gray slate	to 525 feet

Brown slate	to 528 feet	Brown slate	to 1055 feet
Coal	to 530 feet	Sand	to 1078 feet
Brown slate	to 558 feet	Gray slate	to 1160 feet
Lime	to 560 feet	Lime	to 1178 feet
Brown slate	to 580 feet	Gray slate	to 1188 feet
Gray slate	to 590 feet	Sand	to 1192 feet
Brown slate	to 625 feet	Lime	to 1212 feet
Sand	to 635 feet	Gray slate	to 1217 feet
Sand	to 675 feet	Sand	to 1232 feet
Brown slate	to 700 feet	Gray slate	to 1237 feet
Gray slate	to 735 feet	Lime	to 1245 feet
Brown slate	to 790 feet	Gray slate	to 1260 feet
Sand	to 820 feet	Lime	to 1270 feet
Brown slate	to 832 feet	Brown slate	to 1280 feet
Sand	to 838 feet	Red rock	to 1283 feet
Gray slate	to 842 feet	Gray slate	to 1298 feet
Sand	to 869 feet	Sand	to 1307 feet
Coal	to 872 feet	Gray slate	to 1312 feet
Gray slate	to 875 feet	Sand	to 1327 feet
Sand	to 930 feet	Brown lime	to 1335 feet
Sand	to 965 feet	Gray slate	to 1340 feet
Gray slate	to 994 feet	Sand	to 1345 feet
Lime	to 997 feet	Oil sand	to 1347 feet
Gray slate	to 1010 feet		
Lime	to 1033 feet	Total depth	1347 feet

Estate of Michael Murphy (deceased) Oil Co. Well No. 5, S. T. Rumble farm, Madison Township. Finished July 7, 1919. Dry.

Lime shell	80 to	85 feet	
Coal	85 to	88 feet	
Sandy lime	88 to	110 feet	
Slate	110 to	130 feet	
Lime	130 to	145 feet	
Coal	145 to	150 feet	Water
White slate	150 to	210 feet	
Sandy lime	210 to	325 feet	Water, 2 bbls. per hour
Dark slate	325 to	420 feet	
Lime	420 to	425 feet	
White slate	425 to	460 feet	
Lime	460 to	470 feet	
Broken lime	470 to	550 feet	
White slate	550 to	625 feet	
Dark slate	625 to	715 feet	
Sand	715 to	750 feet	More water
Slate	750 to	840 feet	
Sandy lime	840 to	920 feet	
Water sand	920 to	940 feet	Salt water
Dark slate	940 to	1050 feet	
Lime cave	1050 to	1075 feet	
Water sand	1075 to	1110 feet	

Lime	1110 to 1130 feet	
Dark slate	1130 to 1160 feet	
Sand	1160 to 1180 feet	
Little lime	1180 to 1200 feet	
Slate	1200 to 1220 feet	
Lime and sand	1220 to 1250 feet	
Dark slate	1250 to 1270 feet	
Big lime	1270 to 1292 feet	
Slate	1292 to 1302 feet	
Red rock	1302 to 1310 feet	
Slate	1310 to 1322 feet	
Oil sand	1322 to 1332 feet	Dry—Snyder sand
Lime	1332 to 1345 feet	Oakland City sand
Sand and lime	1345 to 1384 feet	Water—brown

Casing Record.

12½-inch	21 feet	8¼-inch	840 feet
10-inch	150 feet	8¼-inch	1265 feet

In Madison Township a well on the Thomas farm, Section 30, pumped 50 barrels from a depth of 1280 feet. The Bement Oil and Gas Co.'s No. 1 well on the L. C. Thomas farm, in the S. W. ¼ of the S. W. ¼ of Section 32, is estimated at 50 to 100 barrels. The depth is 1170 feet.

Log of M. F. Snyder well No. 9, located in Section 35, Madison Township:

Yellow clay	to 18 feet	Lime	to 443 feet
Sand	to 70 feet	Brown slate	to 450 feet
Brown slate	to 75 feet	Gray slate	to 490 feet
Coal	to 77 feet	Brown slate	to 498 feet
Brown slate	to 135 feet	Lime	to 500 feet
Lime	to 148 feet	Coal	to 503 feet
Gray slate	to 160 feet	Gray slate	to 515 feet
Lime	to 165 feet	Lime	to 522 feet
Gray slate	to 180 feet	Brown slate	to 560 feet
Brown slate	to 205 feet	Gray slate	to 600 feet
Coal	to 208 feet	Brown slate	to 650 feet
Lime	to 211 feet	Gray slate	to 690 feet
Brown slate	to 215 feet	Sand	to 696 feet
Lime	to 222 feet	Brown slate	to 715 feet
Gray slate	to 250 feet	Gray slate	to 755 feet
Brown slate	to 285 feet	Sand	to 760 feet
Lime	to 288 feet	Brown slate	to 800 feet
Brown slate	to 290 feet	Gray slate	to 820 feet
Gray slate	to 310 feet	Sand	to 830 feet
Brown slate	to 350 feet	Brown slate	to 858 feet
Gray slate	to 390 feet	Sand	to 872 feet
Brown slate	to 430 feet	Brown slate	to 880 feet
Sand	to 438 feet	Sand	to 940 feet
Brown slate	to 440 feet	Gray slate	to 948 feet

Lime	to 950 feet
Gray slate	to 955 feet
Lime	to 965 feet
Gray slate	to 1050 feet
Sand	to 1060 feet
Gray slate	to 1080 feet
Sand	to 1160 feet
Lime	to 1172 feet
Gray slate	to 1200 feet
Lime	to 1210 feet
Gray slate	to 1220 feet
Lime	to 1225 feet
Gray slate	to 1230 feet
Lime	to 1238 feet
Gray slate	to 1258 feet
Lime	to 1261 feet
Gray slate	to 1265 feet
Black slate	to 1275 feet
Red rock	to 1277 feet
Black slate	to 1313 feet
Gas sand	to 1319 feet
Brown shell	to 1325 feet
Black slate	to 1340 feet
Black lime	to 1342 feet
Sand	to 1348 feet
Total depth	to 1348 feet

Casing Record.

12½-inch	70 feet
10-inch	442 feet
8-inch	948 feet
6-inch	1231 feet
4⅞-inch	1325 feet

Well No. 1 on the F. P. Robling farm, 200 feet to South line, 200 feet to West line, Section 35, Madison Township, Pike County:

Soil	to 1 foot
Clay	to 11 feet
White sand	to 51 feet
Blue slate	to 90 feet
White lime	to 100 feet
Blue slate	to 140 feet
Black slate	to 170 feet
Gray slate	to 190 feet
White sand	to 200 feet
Coal	to 202 feet
Fire clay	to 210 feet
Gray shale	to 255 feet
White sand	to 325 feet
Gray shale	to 330 feet
White sand	to 355 feet
Black slate	to 375 feet
White sand	to 390 feet
White shale	to 395 feet
White lime	to 430 feet
Black shale	to 450 feet
Light slate	to 515 feet
Coal	to 518 feet
Light slate	to 563 feet
Dark slate	to 598 feet
Coal	to 600 feet
Light slate	to 645 feet
Light sand	to 655 feet
Dark slate	to 660 feet
Brown lime	to 665 feet
Light sand (gas)	to 675 feet
Dark slate	to 780 feet
Gray sand	to 797 feet
Gray slate	to 815 feet
White sand	to 890 feet
Dark lime	to 897 feet
Dark slate	to 960 feet
White shale	to 962 feet
Brown sand (oil)	to 967 feet
Gray sand	to 977 feet
Brown sand (best pay)	to 992 feet
Dark slate	to 1006 feet
Gray sand	to 1022 feet
Light sand	to 1028 feet
Light sand	to 1119 feet
Dark lime	to 1131 feet
Blue mud	to 1134 feet
Dark sand	to 1147 feet
Dark lime	to 1152 feet
White sand	to 1167 feet
Brown lime	to 1176 feet
Red rock	to 1179 feet
Dark sand	to 1191 feet
Dark slate	to 1197 feet
White lime	to 1233 feet
Dark slate	to 1270 feet
Dark sand (show of oil)	to 1285 feet
Dark sand	to 1292 feet
Dark lime	to 1294 feet

Light lime ....to	1297 feet	Dark lime ....to	1316 feet
Black slate ....to	1302 feet	Gray sand ....to	1322 feet
Light sand ....to	1304 feet	Top of pay.............to	1316 feet
Dark slate ....to	1313 feet		

Well No. 2, on the F. P. Robling farm, 333 feet to West line; 200 feet to North line, Section 35, Madison Township, Pike County:

Black soil ....to	1 foot	Light slate ....to	1021 feet
Yellow clay ....to	6 feet	Gray sand (sand dry)...to	1029 feet
Brown mud ....to	100 feet	Dark slate ....to	1035 feet
Coal ....to	104 feet	Gray lime..............to	1039 feet
Gray lime ....to	109 feet	Same ....to	1054 feet
Blue mud ....to	129 feet	Dark lime..............to	1059 feet
Gray shale ....to	189 feet	Whie sand ....to	1149 feet
Blue mud ....to	204 feet	Gray lime..............to	1164 feet
Gray lime ....to	208 feet	Blue slate..............to	1169 feet
White mud ....to	253 feet	Gray lime ....to	1177 feet
Brown shale ....to	353 feet	Blue slate ....to	1181 feet
White shale ....to	403 feet	Gray lime..............to	1185 feet
White lime ....to	445 feet	Blue slate..............to	1190 feet
Brown mud ....to	545 feet	Gray lime..............to	1202 feet
Gray lime ....to	550 feet	Blue slate..............to	1208 feet
White mud ....to	640 feet	Gray lime..............to	1220 feet
White lime ....to	648 feet	Blue slate..............to	1237 feet
Gray mud ....to	678 feet	White lime, top of big lime ....to	1240 feet
Brown shale ....to	703 feet	Dark lime,.............to	1245 feet
Gray lime shell.........to	710 feet	White lime..............to	1250 feet
White mud ....to	735 feet	Dark gray lime.........to	1275 feet
Brown shale ....to	755 feet	Red rock ....to	1280 feet
White sandy shell.......to	760 feet	Dark slate ....to	1292 feet
Gray slate ....to	810 feet	Gray lime ....to	1302 feet
Brown shale ....to	885 feet	Dark lime ....to	1317 feet
Blue shale ....to	893 feet	Black lime ....to	1323 feet
White sand ....to	903 feet	Brown shell ....to	1326 feet
Blue slate ....to	909 feet	Dark slate ....to	1330 feet
White sand ....to	926 feet	Gray sand ....to	1335 feet
Gray lime ....to	928 feet	Dark slate ....to	1343 feet
Gray sand ....to	943 feet	Gray sand ....to	1348 feet
Black slate ....to	950 feet	Top of pay sand........to	1343 feet
Gray lime ....to	952 feet		
Dark slate ....to	1003 feet		

Well No. 3, on the farm of F. P. Robling, 200 feet to East line, 400 feet S. E. Well No. 2, Section 35, Madison Township, Pike County:

Soil ....to	1 foot	Dark slate ....to	209 feet
Yellow clay ....to	6 feet	Gray lime ....to	213 feet
Dark slate ....to	100 feet	Light slate ....to	258 feet
Coal ....to	104 feet	Dark slate ....to	308 feet
Gray lime ....to	109 feet	Gray sand ....to	323 feet
Dark slate ....to	134 feet	Light slate ....to	343 feet
White lime ....to	154 feet	Dark lime ....to	353 feet

Dark slate ....to 393 feet
Black slate ....to 398 feet
Light slate ....to 435 feet
Gray lime ....to 450 feet
Light slate ....to 465 feet
Gray slate ....to 485 feet
Light slate ....to 495 feet
Gray lime ....to 500 feet
Light slate ....to 505 feet
White lime ....to 525 feet
Light slate ....to 540 feet
Coal ....to 545 feet
Dark slate ....to 560 feet
Brown lime ....to 565 feet
Dark slate ....to 595 feet
Light slate, 8¼ in. set at 635 ft. ....to 645 feet
Brown lime ....to 650 feet
Dark slate ....to 700 feet
White sand ....to 715 feet
Dark slate ....to 720 feet
White lime ....to 750 feet
Dark slate ....to 790 feet
Brown slate ....to 850 feet
Light slate ....to 890 feet
White sand, hole full of water ....to 923 feet
Light slate ....to 926 feet
White sand ....to 931 feet
Dark slate ....to 953 feet
Brown lime ....to 958 feet
Dark slate ....to 1015 feet
Brown sand ....to 1030 feet
Dark slate ....to 1045 feet
Dark slate ....to 1050 feet
Brown lime ....to 1055 feet
Light sand, hole full of water 1060............to 1160 feet
Gray lime ....to 1168 feet
Dark slate ....to 1173 feet
White lime ....to 1180 feet
Dark slate ....to 1185 feet
Gray lime ....to 1205 feet
Light slate ....to 1212 feet
White lime ....to 1216 feet
Dark slate ....to 1220 feet
White lime ....to 1223 feet
Dark slate ....to 1234 feet
Brown lime ....to 1259 feet
Light slate ....to 1279 feet
Light mud ....to 1289 feet
Dark lime ....to 1305 feet
Dark slate ....to 1315 feet
Dark sand ....to 1320 feet
Dark lime ....to 1327 feet
Dark slate ....to 1340 feet
Gray lime ....to 1342 feet
Gray sand ....to 1346 feet
Pay sand ....1342 to 1346 feet

Well Record No. 4, C. Burkhart farm, Section 35, Madison Township:

Soil ....to 8 feet
Mud ....to 16 feet
Sand ....to 28 feet
Water sand ....to 40 feet
Slate ....to 70 feet
Black slate ....to 90 feet
Coal ....to 95 feet
Lime ....to 105 feet
Slate ....to 110 feet
Mud ....to 190 feet
Slate ....to 220 feet
Coal ....to 227 feet
Slate ....to 280 feet
Sand ....to 300 feet
Slate ....to 305 feet
Sand ....to 335 feet
Lime ....to 340 feet
Slate ....to 385 feet
Sandy lime............to 405 feet
White slate ....to 410 feet
Black slate ....to 435 feet
White slate ....to 440 feet
Lime ....to 445 feet
Black slate ....to 450 feet
White slate ....to 460 feet
Lime ....to 465 feet
Black slate ....to 480 feet
White slate ....to 560 feet
Black slate ....to 580 feet
Lime ....to 585 feet
White slate ....to 615 feet
Water sand ....to 650 feet
Black slate ....to 680 feet
Water sand ....to 705 feet
Black slate ....to 735 feet
White slate ....to 750 feet
Black slate ....to 775 feet
Sand ....to 790 feet

Black slate ....to	820 feet
Sand ....to	870 feet
Sandy lime ....to	880 feet
Water sand ....to	900 feet
Black slate ....to	920 feet
White slate ....to	930 feet
Black slate ....to	975 feet
Lime ....to	979 feet
Oil sand ....to	1007 feet
Total depth ....to	1024 feet

Washington Township. Two oil sands are reported from this township at depths ranging from 1110 to 1226 feet. The following is the log of a well completed Sept. 27, 1919, on the J. R. Chew farm, Section 32, Pike County:

Surface ....to	45 feet
Sand rock ....to	55 feet
Slate ....to	105 feet
Lime shell ....to	110 feet
Coal ....to	112 feet
Slate ....to	170 feet
Sand ....to	180 feet
Slate ....to	240 feet
Lime ....to	245 feet
Coal ....to	247 feet
Slate ....to	317 feet
White mud ....to	367 feet
Slate ....to	383 feet
Sandy shale ....to	470 feet
Sand ....to	525 feet
Slate ....to	550 feet
Sand ....to	590 feet
Slate ....to	630 feet
Water sand ....to	680 feet
Dark slate, mud........to	710 feet
Sand ....to	725 feet
Dark slate ....to	770 feet
Lime shell ....to	815 feet
Sand ....to	868 feet
Sandy lime ....to	876 feet
Broken slate ....to	877 feet
Little lime ....to	887 feet
Light slate ....to	920 feet
Light lime ....to	935 feet
Sand ....to	959 feet
Lime ....to	962 feet
Slate ....to	965 feet
Big lime ....to	1001 feet
Slate ....to	1036 feet
Shell ....to	1039 feet
Slate ....to	1041 feet
Oakland City sand......to	1050 feet
Slate ....to	1053 feet
Oakland City sand......to	1061 feet
Brown lime ....to	1071 feet
Slate ....to	1081 feet
Oil sand ....to	1094 feet
Hard shell ....to	1095 feet
Broken sand	1107 feet
Brown oil sand..........to	1109 feet
Total depth ....to	1109 feet

Log of Rogers well, Rogers Station, E. & I. R. R.:

Common top sand......to	25 feet
Shale and limestone shells to	90 feet
Streak of soft sand......to	115 feet
Soft muddy shale........to	140 feet
Coal and black shale.. to	150 feet
White sand and black shale ....to	160 feet
Streaks of very sharp sand ....to	187 feet
White sand ....to	200 feet
White and limestone shells ....to	220 feet
Shale ....to	230 feet
Shells ....to	247 feet
Coal ....to	250 feet
Caving slate and shale..to	260 feet
Sand, small flow of gas on top ....to	290 feet
Black shale ....to	320 feet
Limestone and shale....to	360 feet
Shale ....to	375 feet
Limestone, shells and slate ....to	460 feet
Sand shells ....to	470 feet
Sand ....to	505 feet
Limestone, shells and slate ....to	560 feet
Sand, shells and slate...to	600 feet
Straight salt sand.......to	692 feet
Straight limestone ......to	885 feet
Limestone ....to	900 feet
Sandstone and slate.....to	920 feet

Sand and limestone......to	945 feet	Slate and sand oil......to	1027 feet
Sand with small streaks of slate..............to	992 feet	Sand ....to	1057 feet
Streak of red marl......to	994 feet	Slate ....to	1075 feet
Case brick penal cave....to	1008 feet	Sand and limestone......to	1161 feet
		Limestone ....to	1185 feet

Report on Oil.

24 Degrees gravity
20 Degrees cold test
300 Degrees fire test
504 Vis. at 70 degrees

Section 28, Washington Township:

13 inch drive pipe	57 feet
10 inch dripe pipe	124 feet
8¼ inch casing pipe	791 feet
6¼ inch casing	1075 feet
Top of gas sand	1162 feet

Drilled in three feet. Tested 3,162,000 cubic feet capacity. Completed March 24, 1909. One well has been abandoned in Section 19, 3 in Section 27, 3 in Section 28 and 1 in Section 30.

A. B. Bement's No. 10, on the L. C. Thomas farm, Section 32, Washington Township, pumped 20 barrels from the brown sand. The top of the sand was struck at 1,123 feet and drilled to a total depth of 1,138 feet.

Monroe Township. Record of the Yeager No. 1 well, N. E. ¼ of the S. W. ¼ of Section 26, Monroe Township:

Surface, mud, shale, loam and quick sand..........	52 feet	Limestone	40 feet
Coal measures, shale, coal etc.	408 feet	Shale	10 feet
Sandstones (Mansfield and Huron)	410 feet	Limestone	70 feet
Limestone	30 feet	Shale	5 feet
Shale	15 feet	Limestone	54 feet
		Shale	46 feet
		Limestone and shale........	41 feet
		Total depth	1181 feet

The following wells have been abandoned: Section 21, 6 wells; Section 22, 1 well; Section 23, 9 wells; Section 24, 1 well; Section 26, 6 wells; Section 27, 2 wells; Section 28, 3 wells; Section 30, 1 well; Section 35, 4 wells.

Logan Township. Two oil pools are located in this township, the Union and the Oatsville. The following is the record of a well from the Oatsville pool. A second well drilled on this lease reported oil at 1320 feet. Drilled July 17, 1919.

Well No. 1, John Cornelius Farm, Section 27.

Surface clay ....to	25 feet	Lime ....to	145 feet
Blue slate ....to	50 feet	Black slate ....to	155 feet
Shell, first water........to	55 feet	Sand ....to	175 feet
Slate ....to	80 feet	Sandy shale............to	200 feet
Sand ....to	135 feet	Lime and coal..........to	210 feet

White lime	to	220 feet
White slate	to	265 feet
Black slate	to	285 feet
Sandy slate	to	320 feet
Sandy slate	to	330 feet
White slate	to	360 feet
Coal	to	366 feet
Slate	to	415 feet
Lime	to	420 feet
Slate	to	485 feet
Lime	to	489 feet
Slate	to	540 feet
Sandy lime	to	600 feet
White slate	to	630 feet
Black slate	to	675 feet
Sandy lime	to	745 feet
Slate	to	795 feet
Salt water sand	to	900 feet
White slate	to	935 feet
Sand hard	to	949 feet
Sandy slate	to	956 feet
Black slate	to	999 feet
Coal	to	1000 feet
Black slate	to	1010 feet
Sand	to	1116 feet
Blue slate	to	1126 feet
Gray lime	to	1146 feet
Blue slate	to	1177 feet
Lime	to	1232 feet
Blue slate	to	1239 feet
Gray lime	to	1244 feet
Slate broken	to	1269 feet
Sand top	to	1269 feet
First oil	to	1275 feet
Coarse brown sand	to	1281 feet
Fine white sand	to	1292 feet

Show water in last foot.

One well in Section 27 and another in Section 35 were abandoned.

In the Union field oil sands are reported at depths ranging from 1,070 to 1,774 feet.

Patoka Township. A large number of wells have been drilled in this township. Wells have been abandoned as follows: Section 11, 7 wells; Section 13, 2 wells; Section 14, 18 wells; Section 15, 1 well; Section 18, 1 well.

Lockhart Township. One well was drilled in Section 5 and one in Section 21.

Clay Township. One well was abandoned in Section 3 and one in Section 32.

Jefferson Township. Wells were drilled in Sections 4, 8 and 31.

Much additional information has been obtained in reference to the petrolific conditions of Pike County since the publication of the report on petroleum and natural gas in Indiana in 1920. Only a brief summary can be given at this time.

Davidson Field. Location, Sections 26 and 35, T. 1 N., R. 9 W. Elevation of surface from 440 to 491 feet above sea level. Depth to oil sand from 934 to 1,100 feet. Initial production from 5 to 150 barrels.

Oatsville Field. Location, Sections 14, 15, 21, 22, 23, and 27, T. 1, S. R. 9 W. Elevation of surface from 417 to 541 feet above sea level. Five sands were encountered, the McDaniel, Hightower, Crecelius or Gladdish, brown, and McCloskey. The first at 1,000 feet and the last at 1,650. The initial production was from 10 to 200 barrels.

Bowman Field. Location, Sections 25, 29, 31, 35, 36, 6, 1, and 2 in T. 1 N. and 1 S., and R. 6 and 7 W. Surface elevations range from 440 to 560 feet above sea level. Four oil sands, Rumble, Barker, Oakland City and brown are encountered. The wells range in depth from 935 to 1,300 feet. The initial production ranges from a few barrels to 150 barrels.

Other fields in this county are the Oakland City, Union and Alford.

PORTER COUNTY.

Devonian strata probably underlie the whole of Porter County, though it is possible that preglacial streams may have been cut through to the Silurian. The eroded surface of the Devonian is covered with glacial drift which attains a thickness of 200 feet or more. The record of a well drilled at Valparaiso is given by Phinney[1] as follows:

Drift	125 feet
Black shale	65 feet
Corniferous, lower Helderburg and water	230 feet
Niagara limestone	380 feet
Niagara shale	5 feet
Clinton limestone	55 feet
Bluish-green Hudson River shales	160 feet
Chocolate-brown limestone (Galena)	256 feet
Trenton limestone	68 feet
Total depth	1344 feet
Altitude of well	715 feet

Another well reported by Gorby[1] for the same place is recorded below:

Section of Well No. 1.

Drift	125 feet
Hamilton shale	65 feet
Corniferous limestone	55 feet
Niagara limestone	565 feet
Clinton (?) limestone	10 feet
Hudson River limestone and shale	185 feet
Utica shale	295 feet
Trenton limestone	144 feet
Total depth	1444 feet
Trenton below sea level	602 feet

The surface of the Trenton appears to dip northward through this county at the rate of about twenty feet to the mile.

POSEY COUNTY.

Posey County lies within the area of outcrop of strata of the Pennsylvanian age. As it lies between the Wabash and Ohio Rivers, a goodly portion of its area is covered with alluvium. A somewhat larger area is mantled with glacial drift, though a portion of the county is unglaciated. With the exception of the river valleys, outcrops are not wanting in many parts of the county. Careful detailed work will probably reveal the structural conditions favorable to the accumulation of oil and gas if such exist. The coal beds and beds of limestone will probably be the most useful keys for unlocking structure.

The following is the record of a well drilled at Mt. Vernon[1]:

Yellow clay	27 feet
Brown soapstone	44 feet

[1] Ashley, Coal Report, 1898, p. 1416.

White sandstone (Merom)	32 feet
Coal	2 inches
Limestone with streaks of clay	4 feet
Blue shale	7 feet
Coal	1 foot
Fire clay	5 feet
Sulphur mixed with fire clay	3 feet
Soapstone	3 feet
Dark blue shale	25 feet
Limestone	7 feet
Coal	2 inches
Dark shale	25 feet
Sandstone	6 inches
Soapstone	22 feet, 6 inches
Sandstone	5 feet, 6 inches
Sandstone and shale, about every alternate foot	19 feet
Coal	6 feet
Shale streaked with sandstone	5 feet, 6 inches
Soapstone	10 feet
Dark shale	17 feet, 6 inches
Black coal shale	3 feet
Coal	4 inches
Blue fire clay	12 feet
Dark fire clay	13 feet
Sandstone	3 feet
Shale streaked with sand	4 feet, 6 inches
Blue shale with small white streaks	46 feet
Soft dark blue shales	46 feet, 6 inches
Black shale	1 foot
Bastard shale	1 foot, 6 inches
Rock	6 inches
Coal	1 foot, 3 inches
Fire clay	7 feet, 3 inches
Soapstone	7 feet, 3 inches
Total depth	407 feet

Point Township. A deep well was drilled in Section 2 on the property of W. E. Hastings, and was plugged in 1913. No record of this well was obtained.

PULASKI COUNTY.

Silurian and Devonian strata underlie the glacial drift in Pulaski County. Gas was found in some wells drilled at Francesville. The record of the first well drilled is given below:

Section of Well No. 1.

Drift	8 feet
Niagara limestone	542 feet
Hudson River limestone and shale	235 feet

Utica shale	100 feet
Trenton limestone	10 feet
Total depth	895 feet
Trenton below sea level......................	200 feet

Yielded a small quantity of gas.

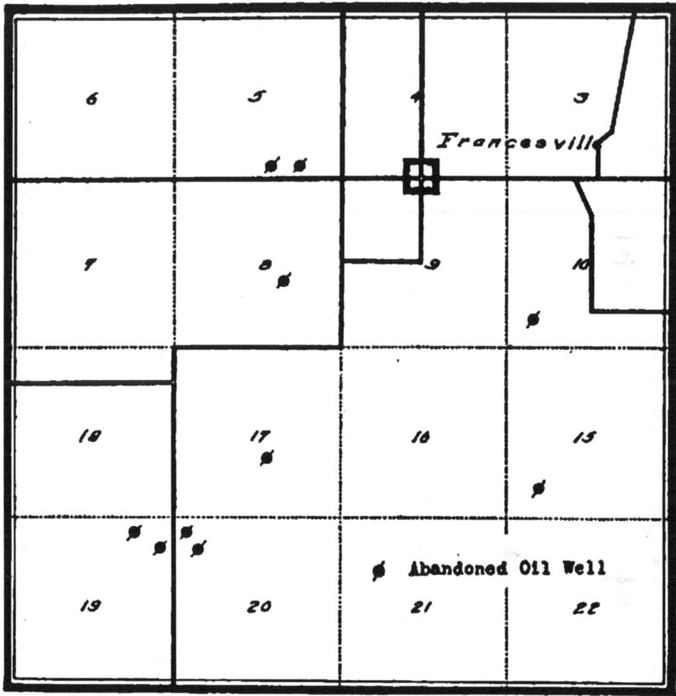

Fig. CXLVII. Map showing the location of oil wells in the Francesville oil field in Pulaski County.

The concealment of the bedrock strata by the glacial drift prevents the determination of the structural conditions so that it is impossible to say whether structures favorable to the accumulation of oil and gas exist in other parts of the county or not. The surface of the Trenton in the southern part of the county is about 200 feet below sea level, and the depth increases to more than 400 feet in the northern portion of the county.

Railroad Elevations.

Boone, 725.1; Thornhope, 710.8; Star City, 697.7; Winamac, 700.3; Monterey, 714; Francesville, 680; Medaryville, 688.1; Clarks, 705.4; Anthonys, 706.6; Lawton, 713; Beardstown, 713.

PUTNAM COUNTY.

The glacial drift in Putnam County is thin so that the bedrock is exposed in many places. The drift is of greater thickness in the northern part of the county than in the southern part, and consequently the outcrops of the bedrock are more numerous in the latter. The rocks underlying the drift belong to the Knobstone, Warsaw, Salem, Mitchell and Chester divisions of the Mississippian

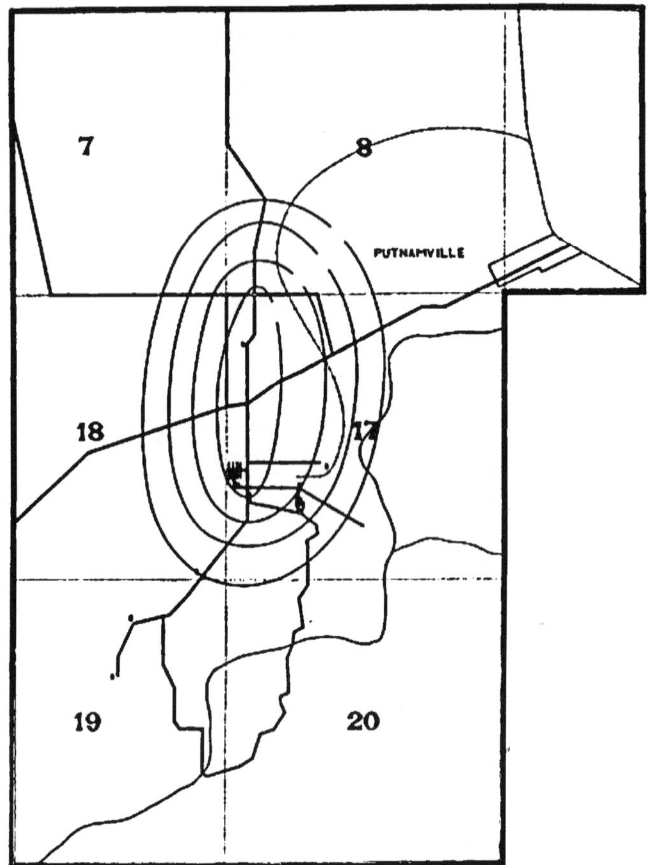

Fig. CXLVIII. Map showing outline of small anticline on the State Farm near Putnamville. Contours on the surface of limestone.

and Pottsville (Mansfield) and coal measures (Allegheny) divisions of the Pennsylvanian. In the southern part of the county in the region occupied by the Chester and the Pennsylvanian divisions the structural conditions may be determined. A small structure has been outlined by the writer on the State Farm, and others may exist in the county.

A well was drilled in Section 28 of Russell Township to a depth of 800 feet. It probably encountered the Corniferous limestone in the Devonian, at which point a strong flow of salt water and a slight showing of gas were encountered.

A well at Reelsville, in Washington Township, at an elevation of 600 feet above sea level, penetrated the Niagara limestone at 1240 feet and secured an artesian supply of salt water.

Bainbridge. A well was drilled on the Miller farm, one and one-half miles west of Bainbridge, to a depth of 1647 feet. A little oil was obtained at 1450 feet. This was evidently in the Trenton, the surface of which must be about 1400 feet or a little below.

Several wells have been drilled around Greencastle, but no records have been obtained.

RANDOLPH COUNTY.

This county lies within the glaciated area, where the drift is from 25 to 150 feet thick. The drift rests upon the eroded surface of the Niagara. The concealment of the bedrock prevents the determination of the structural conditions favorable to the accumulation of oil.

White River Township. The first well drilled in Winchester passed through the following strata[1]:

Drift	147 feet
Niagara limestone	110 feet
Niagara shale	40 feet
Hudson River	430 feet
Utica shale	330 feet
Trenton limestone	20 feet
Total depth	1077 feet

Trenton below sea level, 24 feet. A feeble flow of gas and a few barrels of oil were obtained. A second well drilled one mile north of No. 1 found the Trenton 38 feet higher, well shot, only a feeble flow of gas. No. 3 was drilled one-fourth mile northeast of No. 2, and the Trenton found 72 feet above sea level, well shot, flow feeble. No. 4, located west of No. 1, yielded a little gas and oil, as did No. 5, located east of No. 3. No. 6, located three-quarters of a mile northeast of Winchester, reached the Trenton at 1044 and yielded gas at 1056 to 1060. No. 7, located sixty rods southeast of No. 6, reached the Trenton at 1036 feet and gas between 1060 and 1071. No. 8, located one-half mile northeast of No. 7, was dry. No. 9, located forty rods north of No. 7, gave a good flow. No. 10, located south of No. 7, produced 1,500,000 cubic feet per day.

Wells drilled on the Prickett farm in Section 23, southeast of Winchester, produced 20 barrels of oil per day.

Record of Prickett Wells.

Drive pipe	85 feet
Casing	226 feet
Top of Trenton	1091 feet
Total depth	1156 feet

Wells drilled on the Eliza Goodrich farm, near Winchester, produced a small amount of oil and gas. The records of two of these wells are as follows:

	No. 7	No. 8
Drift	102 feet	70 feet
Niagara limestone	85 feet	110 feet
Hudson River	549 feet	520 feet
Utica shale	300 feet	332½ feet
Trenton limestone	49 feet	51½ feet
Total depth	1085 feet	1084 feet

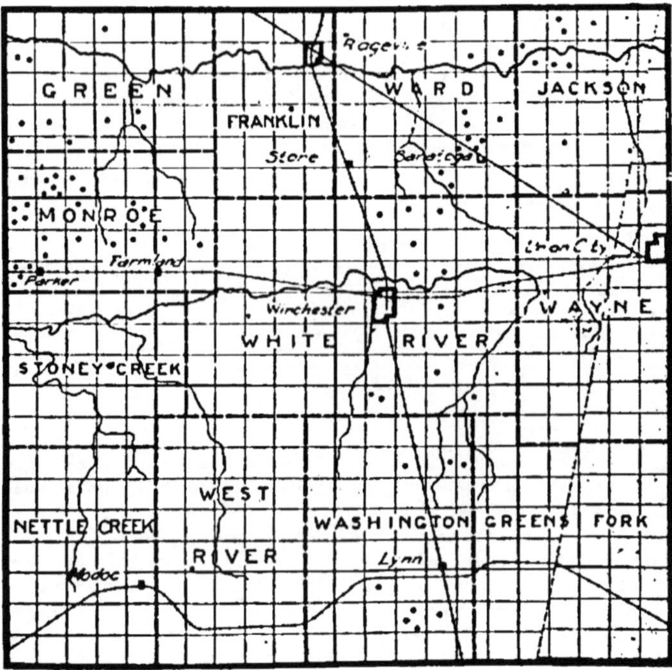

Fig. CXLIX. Map of Randolph County showing location of recorded abandoned wells. The western part of Green and Monroe townships is oil territory and the western part of Stony Creek and Nettle Creek is gas territory.

Wells have been plugged in this township as follows: Section 2, 1 well; Section 3, 1 well; Section 4, 1 well; Section 5, 1 well; Section 9, 2 wells; Section 15, 1 well; Section 16, 1 well; Section 22, 1 well; Section 27, 1 well; Section 32, 1 well; Section 35, 1 well.

Monroe Township. Seven wells were drilled at Farmland. Four produced some gas. A section of well No. 1 is given below:

Farmland Well No. 1.

Drift	55 feet
Niagara limestone	160 feet
Hudson River	585 feet
Utica shale	185 feet
Trenton limestone	32 feet
Total depth	1017 feet
Trenton above sea level	55 feet

Oil was found in this township in Section 3, 4, 5, 8, 9, 10, 11, 15, 16, 17, 21, gas in 9 and 27. Wells abandoned are located in: Section 1, 3 wells; Section 3, 1 well; Section 5, 4 wells; Section 7, 1 well; Section 8, 1 well; Section 9, 2 wells; Section 10, 1 well; Section 12, 1 well; Section 13, 1 well; Section 15, 1 well; Section 17, 4 wells; Section 32, 1 well; Section 33, 5 wells; Section 34, 2 wells.

Stony Creek Township. Oil was obtained from Sections 19 and 30. The record of a dry hole in Section 32 is given below:

Drive pipe	64 feet
Casing	320 feet
Top of Trenton	956 feet
Total depth	307 feet

Greene Township. Oil was found in Sections 20, 28 and 29. Wells have been plugged in Sections 2, 1 well; Section 6, 1 well; Section 8, 2 wells; Section 20, 1 well; Section 21, 1 well; Section 23, 1 well; Section 24, 2 wells; Section 27, 1 well; Section 29, 1 well; Section 35, 1 well.

Section 1, R. 12 E., Greene Township.

Top soil	42 feet
Lime	220 feet
Shale	694 feet

Drilled 446 feet into Trenton.

Total depth	1402 feet
Dry hole. Showing of oil very good at	1006 feet
Showing of sand favorable for oil	1250 feet

Well shot. No good results.

Section 26, T. 21 N., R. 11 E., J. W. Bartlett Farm.

Top soil	46 feet
Lime	280 feet
Shales	632 feet
Trenton at	958 feet
Into Trenton	145 feet
Total depth	1103 feet
Oil showing at	975 feet
Good oil showing at	1103 feet

Franklin Township. At Ridgeville three dry holes were drilled. The Trenton was reached at 981, 2 feet above sea level. A well was drilled and abandoned in Section 23, on the J. M. Addington property, in 1919.

Wayne Township. A little gas was found at Union City. Four wells were drilled. In well No. 4 north of the city, the Trenton was reached at 1093 and is 83 feet below sea level. The record of the first well kept by A. Jaqua is as follows:

Union City Well No. 1.

Drift	98 feet
White limestone (Niagara)	72 feet
Dark gray limestone	62 feet
Bluish limestone	38 feet
Niagara shale	40 feet
Clinton (?) limestone	15 feet
Bluish-green shale	400 feet
Gray shale	175 feet
Brown shale	175 feet
Black shale	80 feet
Trenton limestone	525 feet
Gray sandstone (St. Peter)	100 feet
Total depth	1780 feet
Altitude of well	1079 feet

Another well at Union City yielded traces of gas between depths 1155 and 1162 feet. The record of this well follows:

Union City Well.

Drift	98 feet
Niagara limestone	250 feet
Hudson River and Utica	800 feet
Trenton limestone	540 feet
Total depth	1688 feet
Trenton below sea level	40 feet

Nettle Creek Township. A well was drilled at Losantville and after passing through 173 feet of drift and 821 feet of rock the Trenton was reached at 994 feet. The total depth was 1105 feet. No oil, gas or water was found in the Trenton. The top of the Trenton is 146 feet above sea level.

Washington Township. Abandoned wells are located in this township as follows: Section 5, 1 well; Section 9, 3 wells; Section 10, 1 well; Section 14, 1 well; Section 15, 1 well; Section 16, 1 well.

Jackson Township. Wells have been abandoned in this township as follows: Section 4, 1 well; Section 5, 2 wells; Section 7, 2 wells; Section 8, 2 wells; Section 29, 1 well.

Ward Township. Abandoned wells are located as follows: Section 11, 1 well; Section 12, 1 well; Section 23, 2 wells; Section 24, 1 well; Section 26, 4 wells; Section 34, 2 wells.

RIPLEY COUNTY.

The geological formations represented by outcrops in this county are given below:

Quaternary........	Recent, Alluvial sands and clays. Pleistocene, glacial gravel, sands and till.
Devonian	Limestones.
Silurian	Limestones and shales.
Ordovician	Shales and limestones.

The Pleistocene covering the bed rock varies from a few feet to fifty feet in thickness and rests on the eroded surface of the bed rock. The latter outcrops at many points so that it may be possible to determine the structural conditions by surficial observations.

The surface of the Trenton is about sea level, in the eastern part of the county, and lies probably as much as 150 feet below in the western part of the county. If the structural conditions are favorable there is a possibility of oil or gas accumulations in the Devonian and the Trenton strata.

Railroad Elevations.

Milan, 1,007; Pierceville, 1,007; Osgood, 990; Dabney, 966; Holton, 923; Sunman, 1,016.6; Morris, 997.5.

RUSH COUNTY.

Rocks belonging to the Devonian and Silurian periods form the bedrocks of this county. The glacial drift lies upon the surface of these formations to a depth of from 50 to 100 feet and prevents the determination of structural conditions.

Rushville Township. At Rushville three wells obtained gas. The record of well No. 1 is given below:

Drift	60 feet
Chert and cherty limestone (Corniferous).........	40 feet
Niagara limestone and shale.....................	200 feet
Hudson River limestone and shale................	200 feet
Utica shale	360 feet
Total to Trenton................................	860 feet
Trenton above sea level.........................	124 feet

From drillings preserved by G. W. Clark, Phinney constructed the following record of one of the wells:

Drift	48 feet
White limestone	42 feet
Blue limestone	30 feet
Gray limestone (Clinton)........................	5 feet
Hudson River limestone and shale................	420 feet
Utica shale	262 feet
Gray limestone	25 feet

Brown limestone, Trenton	35 feet
White limestone	30 feet
Total	922 feet
Alitude of well	996 feet

Union Township. At Glenwood the top of the Trenton was reached at 950 feet, or 166 feet above sea level.

A well at Milroy in Anderson Township was unproductive.

Ripley Township. A log of a well drilled at Carthage is as follows:

Drift	50 feet
Limestone	100 feet
Shale	670 feet
Trenton limestone	20 feet
Total depth	840 feet

Wells drilled and abandoned in this township are as follows:

Owner.	Section.	Date.	Wells.
J. Phares	10	1912	1
F. K. Mull	15	1912	1
W. P. Stanley	18	1912	1
V. Robertson	20	1913	1
Benton Henley	25	1912	1
Noah Moore	27	1916	1
J. Vasbinder	6	1911	1
Jabez Reddick	28	1912	1
J. H. Powers	34	1911	1
John Swain	34	1912	1
William Dille	35	1913	1

Washington Township. A large number of wells were drilled in this township. The following have been abandoned: Section 1, 3 wells; Section 3, 1 well; Section 14, 1 well; Section 33, 1 well; Section 4, 3 wells; Section 5, 2 wells; Section 16, 2 wells; Section 34, 1 well; Section 7, 6 wells; Section 8, 3 wells; Section 22, 1 well; Section 9, 2 wells; Section 11, 5 wells; Section 26, 2 wells; Section 12, 2 wells; Section 13, 1 well; Section 32, 1 well.

Jackson Township. Wells were drilled and abandoned in the following Sections: 5, 6, 10 and 20, one well each.

Posey Township. A well drilled in Section 4 was abandoned in 1911, on J. Piper property.

Walker Township. A well on the Tillie Trees property in Section 15 was abandoned in 1913.

SCOTT COUNTY.

The geological formations outcropping in this county belong to the Devonian, Mississippian and Quaternary periods. The divisions represented are given below:

Quaternary .... { Recent—Clays and alluvium.
Pleistocene—Sands, gravels and till.

Mississippian	Knobstone—Shales and sandstones. Rockford—Limestones.
Devonian	New Albany—Shales. Sellersburg—Limestones. Silver Creek—Limestones. Jeffersonville—Limestones.

Because of the removal of much of the regolith, outcrops of the durolith are perhaps numerous enough to permit the determination of the structural conditions for the greater part of the county. The best key horizon will be the contact between the Sellersburg limestone and the New Albany shales for the eastern part of the county and the Rockford limestone for the western part.

Three deep wells were drilled in this county in search of oil, but no production was obtained and the records of the wells were not obtained.

Railroad Elevations.

Blocher, 677; Lexington, 620.

SHELBY COUNTY.

The drift in Shelby County varies in thickness from 50 to 150 feet and overlies Devonian limestones and shales. There are a few outcrops of Silurian rocks in the southeastern part of the county.

Addison Township. According to Phinney[1], five wells were drilled in the vicinity of Shelbyville. He gave the following general section:

Drift	45 feet
Limestone	265 feet
Shale	527 feet
Trenton limestone	100 feet
Total depth	937 feet
Altitude of well..............................	772 feet

Logs of the first two wells drilled at Shelbyville are given below:

	No. 1	No. 2
Drift	48 feet	80 feet
Corniferous limestone	30 feet	...
Niagara limestone	102 feet	769 feet
Hudson River limestone and shale.......	657 feet	...
Trenton limestone	86 feet	...
Total depth	923 feet	849 feet
Trenton below sea level.................	79 feet	...

Hanover Township. At Morristown a well drilled on the Charles F. Muth farm was reported by Phinney[1] as in No. 1 below and by Gorby[2] as in No. 2.

	No. 1	No. 2
Drift	140 feet	140 feet
Limestone	20 feet	...
Niagara	120 feet	130 feet
Hudson River and Utica shale........	638½ feet	628 feet

Two wells were abandoned in Section 1, three in Section 17 and one in Section 18. The Trenton was reached at St. Paul in Noble Township at 820 feet. The thickness of the drift is 90 feet and the altitude is 844 feet.

Marion Township. A well drilled on the S. A. Haven property in Section 6 was abandoned in 1911.

Union Township. A well drilled on the property of H. W. and J. W. Moore was abandoned in 1911, and one on the property of Charles Brown in Section 17 in 1913.

Van Buren Township. Wells drilled on the property of Walter Hadley and Elias Miller in Section 17 were abandoned in 1913.

SPENCER COUNTY.

The strata occupying almost the whole of the surface of Spencer County belong to the Allegheny division of the Pennsylvanian, though some outcrops of the Pottsville probably occur on the banks and in the bed of the Anderson River, which forms the eastern boundary. The rocks are sandstones, shales and limestones with intercalated beds of coal. Three divisions of coal occur in the county. It is possible that coal and some of the associated limestones may prove valuable as key formations by the use of which the structure may be determined. Gas has been found in the county in Jackson Township, near Graysville.

The following is a record of the well drilled on the Fred Frakes farm, Section 3, R. 6 W., Jackson Township, near Gentryville, Spencer County:

10-inch drive pipe	80 feet
8-inch drive pipe	400 feet
Showing of oil	720 feet
6¼-inch casing	900 feet
Gas sand	990 feet
Finished	1025 feet

Capacity of first twenty-four hours, 1,000,000 cubic feet. A well was drilled in Section 1 of this township in 1913, three miles east of Graysville. Two dry holes were drilled in 1916.

Harrison Township. A well was drilled north of St. Meinrad in Section 12 in 1913 without securing production.

Railroad Elevations.

Dale, 432.0; Lincoln City, 459.0; Gentryville, 413.0; Pigeon, 403.0; Lincolnville, 459; Buffaloville, 427; Lamars, 411; Evanston, 413; Bradleys, 460; Chrisney, 447; Millers, 423; Ritchies, 409; Rock Hill, 400; Rockport, 380.

STARKE COUNTY.

This county lies on the north side of the extension of the Cincinnati arch passing through Indiana. Its bed rock strata consists of limestones and shales of Devonian age. On the eroded surface of these rocks there has been deposited an overburden of glacial drift which attains a thickness of several hundred feet. Because of the covering of glacial drift the structural conditions existing in the bed rock of this county can not be determined by direct observation. If a sufficient number of deep well records could be obtained, the structures might be determined. Until such records are available the location of structures favorable to the accumulation of oil and gas can not be located if such exist in the county.

The surface of the Trenton lies between 250 and 500 feet below sea level in this county, being nearer sea level in the southern part of the county.

Railroad Elevations.

Hamlet, 702; Knox, 702; Toto, 703; North Judson, 697; San Pierre, 704; Grovertown, 719.8; Davis, 681.7; Ora, 718; Bass Lake Jct., 711; Aldine, 715.

ST. JOSEPH COUNTY.

The strata of the Devonian age underlie the glacial drift of this county. The glacial drift reaches a thickness of one hundred and fifty or more feet. The dip of the bed rock is toward the north.

The section of a well in South Bend constructed from drillings furnished Phinney[1] by J. D. Oliver is as follows:

Drift sand and gravel	137 feet
Waverly shale (bluish green, calcareous)	143 feet
Black shale	70 feet
Brown shale	25 feet
Gray limestone upper (Helderburg)	60 feet
Blue limestone	20 feet
Lower Helderburg, with gypsum	170 feet
Water lime	55 feet
Niagara limestone (gray, buff and white)	470 feet
Buff Clinton limestone	30 feet
Hudson River limestone and shale	220 feet
Utica shale	183 feet
Trenton limestone (chocolate colored)	85 feet
Total	1670 feet
Altitude of well	725 feet

Salt water was encountered at 375, 610 and 1,670 feet.

The record of a well drilled on the Studebaker farm follows[2]:

Drift	160 feet
Sub-Carboniferous and Hamilton shale	220 feet
Corniferous limestone	60 feet
Lower Helderburg limestone	40 feet
Niagara limestone	640 feet
Clinton (?) limestone	60 feet
Hudson River and Utica	420 feet
Trenton limestone	427 feet
Total depth	2027 feet
Trenton below sea level	855 feet

Yielded no gas or oil.

The structural conditions of the durolith in this county cannot be determined by direct observation because of the glacial drift which conceals the outcrop of the strata. The surface of the Trenton lies from 600 to 1,000 feet below sea level.

STEUBEN COUNTY.

The strata underlying the glacial drift in Steuben County belong to the Mississippian and the Devonian periods of geological times. The bed rock formation consists of shales and limestones. The outcrops of these rocks are concealed by a thick mantle of glacial drift which was deposited on their eroded surface and attains a total thickness of several hundred feet. The dip of the bed rock is toward the north away from the westward extension of the Cincinnati arch through Indiana. Because of the glacial drift the structural and the stratigraphical conditions of the bed rock can not be determined by surficial methods of observation. Deep well records are not at present available for the determination of the structure by the use of subsurface data. Prospecting for oil and gas in this county, for the above reasons, will prove extremely hazardous.

The surface of the Trenton probably lies between 1,500 and 2,000 feet below sea level in this county, being nearer the surface in the southern part.

Railroad Elevations.

Hamilton, 926; Ashley, 999; Fredrick, 972.2; Helmer, 986; Stubenville, 991; Pleasant Lake, 976.1; Angola, 1,055.3; Fremont, 1,058.1; Ray, 1,077.8.

SULLIVAN COUNTY.

(By Dr. S. S. Visher.)

Location. There are seven major pools or oil fields producing at present in Sullivan County. These pools are about 30 miles south of Terre Haute, in the Wabash Valley. They are within a few miles of Sullivan, northwest, west and southwest. Their combined area is about twelve square miles. The location of the pools is shown on the accompanying map on which the elevation of numerous points is also shown.

Production. The present production is about 380 barrels per day. Widespread production commenced in August, 1913; it became considerable in 1914, reaching 3,000 barrels a day by June 1; increased somewhat in 1915 and reached a maximum in that year. Since 1915 it has declined somewhat steadily, in spite of the opening of two new pools and the bringing in of a number of producers in the older pools. The daily production, when greatest, was about 3,500 barrels per day, or nearly three times the present production.

Number of Wells. October 1, 1919, about 480 wells were being pumped. More than 1,000 wells have been sunk for oil in the county. (Four hundred between April 1, 1913 to June 1, 1914, of which 225 were producers, according to Barrett.) Probably more non-producing wells have been drilled in the county than producers. Every month a few wells formerly pumped are abandoned, because it no longer pays to pump them. Two outfits are at present engaged in drilling new wells. Before the war, several outfits were kept busy thus. New producers are added to the total of producers every month, but more wells are abandoned than added, so that the number of producing wells is decreasing, and has been for the last two or three years. The decrease in production is greater than the decrease in the number of wells, however, the declining yield of the existing wells, being the cause.

The average production per well is already distinctly less than a barrel per day. Many wells yielded 20 barrels their first day, and some yielded 100 to 150 and a few somewhat more. At present, many wells yield as little as one-fourth barrel. With the present high price of oil, a producer is not abandoned until it yields less than that, unless it needs recasing.

The presence of 480 wells in an area of 12 square miles, means that on the average there are 40 wells per square mile. In the better parts of the 12 square miles, the wells are drilled only 400 or 460 feet apart, 9 on each 40 acres; wells being drilled 200 feet from the outside lines of the 40 and on a

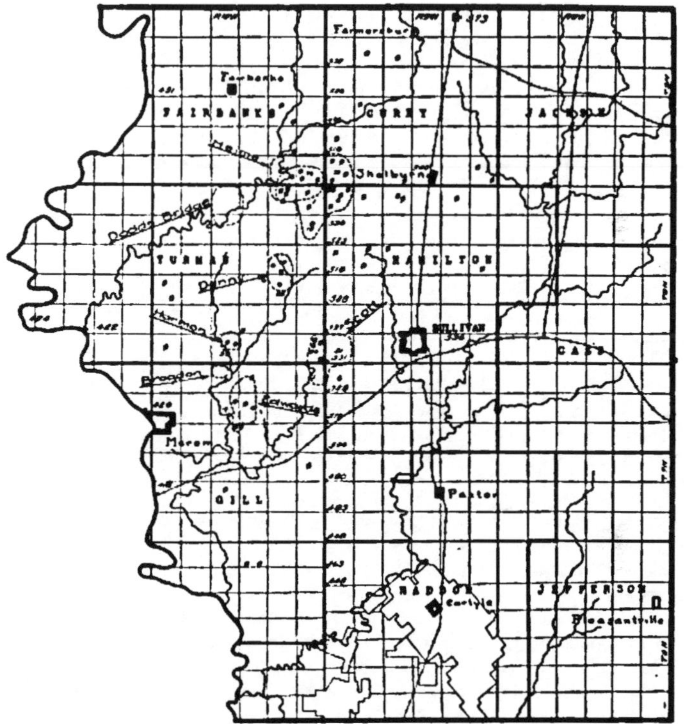

Fig. CL. Map of Sullivan County showing the location of the oil fields.

central row where each well is 460 feet from another of the tract. Nine wells on a 40, is at the rate of 144 per square mile.

"Wild-catting" is the only method known in this area to discover new pools. Wells are drilled at increasing, or irregular distances from the original producing area. If a pool extends that far, production is obtained; if it does not extend that far, a "dry hole" results, unless a new pool is entered.

Similarity of the Pools. The pools of Sullivan County are similar in several respects: (1) the oil is of similar quality, a good, light oil, for the most part (that in the Bragdon pool is the heaviest; that in the Shelburn or Heim pool, the lightest. All the oil is pumped together to the refinery.

The refinery for Illinois Pipe Line Company is at Marshall, Ill.) (2) The oil all comes from "oil sands." (3) The depth to corresponding rock formations is approximately the same in all the pools because the surface slopes to the southwest at approximately the same rate that the rock formations dip in that direction. The region has slight relief. (4) In all the pools, all the four oil sands are present. In one of them, all the four sands are productive. Each of the four sands is the chief productive sand in one or more of the pools. (5) The pools are all small, the largest, the Shelburn or Heims has considerable production from only three square miles. The smallest, the Bragdon is only 40 acres. (6) The production per well averaged approximately the same in each pool when it was opened up. (7) The decline per well in yield has been at a somewhat similar rate in each of the pools. (8) Most of the producing wells yield a little gas, more when new than later, however. (Five strong gas wells have been struck in the county, but none in a pool. Four are just southeast of the Scott pool, near Sullivan. Their gas is piped to the city. (9) In none of the pools is the main gas supply associated with the oil sands.

Production, Etc., of the Sullivan County Oil Pools.

NAME	No. of Producing Wells	Daily Production in Bbls.	When Opened	Average Elevation of Surface Above Sea Level	Average Depth to Sand	Productive Sand	Depth of Sand Below Sea Level
Heims or Shelburn	260	231	1913	520	615–645	1, 2, 3 and 4	100–130
Dodds' Bridge	72	51	1915	580	635–683	1 and 2	130–185
Denny	14	14	1914–18	520	800–810	4	180–280
Harmon or Raley	19	10	1914–15	480	770–780	3 and 4	290–300
Bradgen	6	3	1917	460	....	4	....
Edwards or Buff	32	44	1911	480	740–760	2 and 3	260–280
Scott or Jamison	79	27	1913–18	530	730–775	2	200–250

Geology of the Pools. Production is obtained in Sullivan County from four oil sands. The highest of these is quite certainly along the unconformity between the Allegheny and the Pottsville Divisions of the Coal Measures. It occurs below Coal III and in most places above the level of the lower Minshall coal. As is to be expected on an erosion surface, this sand is higher at some points than at others. It is about 90 feet below Coal III in many places, elsewhere it is only 40 feet. In some places it is found below the level at which the upper Minshall coal occurs in not distant wells. In none of the logs however was it found actually below that coal though in some composite logs it is necessarily so shown. Erosion removed both the upper and lower Minshall coals at some points. The sand deposited along such an erosion valley might be below the level of these coals where they occur in intervalley areas. The existence of such valleys is indicated by a number of the well records.

The second and third oil sands are a short distance below the lower block coal and are thus in the Mansfield sandstone of the Pottsville Division of the Pennsylvanian Formation ("The Coal Measures"). The lowest, fourth, oil sand is probably also in the Mansfield, but it may be barely possible that it is in the uppermost Mississippian Formation, the Chester.

The correlation of the coals upon which the above conclusions, as to the ages of the oil sands depends in part was by means of (1) Ashley's identifications of the higher coals in the mines just east of the pools, at Shelburn,

Sullivan, Farmersburg and Curry. (2) Upon the spacings of the coals and their thickness as compared with the conditions stated by Ashley in the 1898 and 1908 reports of the State Geologist, to be characteristic of these horizons where they are penetrated by many mines in the eastern half of this county. (3) A few logs are sufficiently detailed so far as the rocks overlying and underlying the coals are concerned so that some of the coals may be identified by characteristic roof or floor rock. (4) Coal IV contains more gas in this area than does the other coals. In some of the logs mention is made of this gas at this horizon and hence has aided in the correlation. It is of course recognized that there may be mistakes in the numbering of the coals in the following logs. The determination of the age of the oil sands does not, however, depend solely upon the correlation of the coals. The clear evidence of the erosion surface occupied by the first (Heims or Shelburn pool) oil sand is independent of the correlation of the coals. The existence of three coals below this oil sand proves that it is not Mansfield in age, as it has been considered. The fact that no coal has been found below the lower oil sands in the several wells which have gone deeper proves that these sands are below the Block coal. The fact that the second and third sands are within a hundred feet or so of the lowest coals proves that they are Pennsylvanian in age, rather than older.

The existence of more than two productive sands has not previously been clearly recognized in this oil field. Many operators have assumed indeed, that there is only one, in spite of indisputable evidence to the contrary long available. Some few operators recognized that two sands are productive, and one operator suspected that three are. A study of the more than 100 well records upon which this study is largely based, shows that a failure to appreciate that more than one oil sand is productive, has reduced production greatly. Many well records show that drilling was terminated only a few feet above the horizon where, in not distant wells valuable production was obtained. In not a few cases, a small amount of oil was found in one of the higher sands. After pumping the oil out of this sand, the well should have been deepened to the next sand, instead of being abandoned, as it has been in nearly every case. Of the four sands which yield oil in paying quantities in one or more wells in this field, the top sand is productive in at least two pools. It is entered at from 610 to 660 feet depending upon the topography of the surface and the location of the well. Most of the production in the chief pool, the Heims or Shelburn, is from this level. Much of the production from the Dodds' Bridge pool is also from this level. The second sand is productive in at least three pools. It yields most of the oil in the Scott and the Edwards pools and much of that in the Dodds' Bridge pool.* The third sand is productive in at least four pools, the Edwards, Harmon, Scott and Heims. The fourth sand is productive in at least four of the pools, the Bragdon, Harmon, Denny and Heims. The second sand occurs at approximately 660 to 700 feet varying with the pools and the surface. The third sand is at about 730 to 775 and the fourth sand at 800 feet or so. The depth to the sands is less in the Heims pool than in the pools to the west or south because the rock formations dip southwest at a little greater angle than the surface slopes in that direction.

None of the oil sands are uniformly productive. Even only a few hundred feet from a productive well, the corresponding sand in another well may yield no oil. Commonly such a non-productive condition is due to the sand not being porous. That is, it is clayey. In other cases the sand is so thin as to yield little oil. In still other cases it is filled with water. Some of the water is salty. Before abandoning a well, where the sand is filled with water it might pay to pump the water a while. Sometimes oil is obtained after the water has been removed.

The productive sand is from 20 to 30 feet thick in most of the producing wells. Considerable production is obtained in some wells, however, from sands less than 10 feet thick.

The variation in the thickness of the sand in nearby wells, and its presence at some points and absence nearby indicates that the sand deposits are lenticular or along channels. There commonly is a conspicuous thinning of the sand outward from the center of the pool. In most dry holes, no oil sand, or sand of any kind at that horizon is penetrated. This thinning is not only at the edge. Many dry holes have been drilled within pools. In most of them the sand is so impure as not to be porous, however. In some, it is lacking.

The pools are not known to be related to any local folding or doming. Much oil elsewhere has been proven to have accumulated in paying quantities along sandy channels of ancient streams. The evidence at hand does not warrant a dogmatic statement in regard to the reasons for the pools of Sullivan County being where they are. The indications are, however, that the several pools represent lenses of sand along the valley of an aggrading stream or streams.

The fact, established by a number of well logs, that the depth to the sand is often less near the central part of the pool than near its periphery probably is to be explained by the lenticular shape of the deposits of sand rather than as being due to doming. The depths to the overlying coals do not clearly indicate dynamic doming. The fact that some coals are higher at one point than nearby often shows that the coals themselves were not laid down horizontally, because often one coal in a well will be higher than normal and another will be lower, deeper, than normal.

Glacial drift of considerable thickness overlies most of the area. In some wells it is penetrated for nearly 100 feet, in others it is very thin. It has been removed by erosion along some of the valleys in some of the pools. In Dodds' Bridge pool and in Scott pool for example, a seam of coal is exposed in the valley side only a few rods from some oil wells and only a few feet lower.

SPECIAL PROBLEMS.

1. The present cost of a completed well is about $2,200. When most of the wells were drilled, the average cost was between $1,600 and $1,700. At the present price of oil, a well yielding less than one-half barrel a day will ordinarily not pay for itself, even if located most favorably, in respect to other wells. Rather than abandon such a well, however, it pays to pump it if it can be connected up to a nearby pump. It will bring good interest on the casing and pay the cost of pumping, but not the cost of drilling.

2. Salt water occurs just beneath the lowest oil sand at many points. If the well is drilled too deep, salt water may enter, making the well valueless, in many cases. Many of the dry holes near the pools and elsewhere stop in a salt sand, because of the conviction that when that sand is struck, there is no further hope for oil. This belief is supported by experience, as many wells have gone deeper. However, salt water occurs in some wells at levels far above the lowest oil sand. Thence the striking of salt sand is a proper

occasion for the abandonment of the hole only when it is struck at about 800 feet.

3. Where the numerous coal seams are penetrated, the casing is etched, probably by sulphuric acid developed from the sulphur in the coal. The pipe becomes bright within a few weeks. Many wells have to be recased or abandoned after only a few months. (If the well does not yield more than one-half barrel, it is not recased.)

CONCLUSIONS.

Sullivan County has several pools now yielding oil. The oil comes from four oil sands in the lower Coal Measures partly just above the Pottsville, and partly from the Mansfield horizon of the Pottsville. Undoubtedly other pools will be discovered for in the past the existence of the four oil sands has not been clearly recognized. Many wells have been abandoned before the underlying sands have been tested. The deepest oil sand is only about 800 feet beneath the surface[1].

Composite log for Heims Pool. Based on 50 logs to first oil sand. Average elevation of surface about 523 feet. Surface relief in area about 20 feet:

Coal 8, (average thickness 3 ft.) top at	60	to	70	feet
Coal 7, (4 ft.)	110	to	150	feet
Coal 6a (5 ft.)	170	to	180	feet
Coal 6 (6 ft.)	220	to	240	feet
Coal 5a (3 ft.)	260	to	280	feet
Coal 5 (5 ft.)	305	to	320	feet
Coal 4a (rare) 5 ft.	340	to	...	feet
Coal 4 (5 ft.)	425	to	440	feet
Coal 3a (5 ft.)	480	to	500	feet
Coal 3 (4 ft.)	520	to	560	feet

Gas pockets present in coals 3 and 4.

Ist (main) oil sand	615	to	645	feet
Coal 2 (locally) (2 ft.)	608	to	...	feet
Minshall (4 ft.)	636	to	640	feet
Upper block?	655	to	661	feet
2nd oil sand	660	to	680	feet
Lower block coal?	690	to	694	feet
3rd oil sand	705	to	708	feet
Salt sand or oil sand (Osborn pool)	775	to	815	feet

Composite log for Section 35, Fairbanks Township in N. W. part Heims' Pool. Based on 4 logs for coals and on 6 for oil. Elevation of surface about 500 feet:

Coal 8, top at	..	to	70	feet
Coal 6a	170	to	175	feet
Coal 6	220	to	225	feet
Coal 4	435+		...	feet

[1] The author received much information from L. H. Crews, Shelburn, the local manager of the Ohio Oil Co., the dominant company in this area, and from John Kerens, Sullivan, the local gager for the Illinois Pipe Line Co. Some of the logs studied are given in the 38th (1913) Report of the State Geologist.

Coal 3a ....492 to ... feet
Coal 3 ....550 to ... feet
Best oil (2 wells) ....615 to 620 feet
Best oil (2 wells) ....645 to ... feet
Best oil (2 wells) ....658 to 669 feet

Composite log from Section 36, Fairbanks Township, in north central part of Heims' Pool. Based on 5 logs for coals and in 16 logs for sand. Elevation of surface about 520 feet:

Coal 6a top at............................175 to 205 feet
Coal 6 top at............................215 to 226 feet
Coal 5 top at............................310 to 320 feet.
Coal 4 top at............................425 to 450 feet
Coal 3a top at............................485 to 495 feet
Coal 3 top at............................540 to 560 feet
Best oil (7 wells)..........................615 to 620 feet
Best oil (4 wells)..........................661 to 681 feet
Best oil (1 well)...........................709 to ... feet

Composite log, Section 1, Turman Township, in central part of Heims' Pool. Based on 11 logs for coals and 18 logs for sand. Elevation of surface about 530 feet:

Coal 8 top at............................ .. to 60 feet
Coal 7 top at............................110 to 150 feet
Coal 6a top at............................150 to 205 feet
Coal 6 top at............................225 to ... feet
Coal 5a top at............................260 to 280 feet
Coal 5 top at............................305 to 315 feet
Coal 4 top at............................445 to 470 feet
Coal 3a top at............................480 to ... feet
Coal 3 top at............................510 to 565 feet
Best oil (13 wells)..........................622 to 645 feet
Best oil (4 wells)..........................666 to 675 feet

Some gas found in coal 3 and above it and in coal 4.

Log for N. ½, N. W. ¼, N. E. ¼, Section 12, Turman Township, south edge of Heims' Pool. Elevation about 525 feet:

Coal 6a ....173 to 178 feet
Coal 6 ....238 to 240 feet
Coal 4a ....340 to 345 feet
Coal 4 ....470 to 474 feet
Sand (oil) ....640 to 655 feet

Went to 685 but found no more sand.

Log of well 1 mile west of Heims' Pool, in Section 3, Turman Township (T. 8 N., R. 10 W.) High ground about 500 feet:

Coal 6 ....200 to 207 feet
Coal 5a ....250 to 253 feet
Coal 3a ....490 to 495 feet
Oil sand ....619 to 634 feet
Upper block coal or Minshall....... ......655 to ... feet

Log of Emery Smith well No. 2, S. E. ¼, Section 4, T. 8 N., R. 9 W., southwest of Shelburn, about 2 miles east of Heims' Pool. Elevation of top of well, 540 feet:

Gravel and quick sand....to	43 feet	Coal 4a ....to	405 feet
Hard lime shell..........to	49 feet	Brown slate ....to	420 feet
Gray sandstone ....to	90 feet	Sandstone ....to	433 feet
Gray shale ....to	120 feet	Coal 4 ....to	440 feet
White slate ....to	130 feet	Gray shale ....to	445 feet
Gray shale ....to	165 feet	Brown shale ....to	455 feet
Light sand...............to	175 feet	Lime shell ....to	460 feet
Black slate ....to	179 feet	Sandy slate ....to	465 feet
Coal 6a ....to	184 feet	Brown shale ....to	475 feet
Fire clay ....to	202 feet	Coal 3a and black slate...to	485 feet
Brown shale ....to	225 feet	Gray shale ....to	495 feet
Sandstone ....to	230 feet	White slate ....to	546 feet
Brown slate ....to	239 feet	Brown slate ....to	551 feet
Hard lime shell..........to	244 feet	Coal 3 ....to	556 feet
White slate..............to	250 feet	Brown shale ....to	580 feet
Sandy shale ....to	265 feet	Lime shell ....to	583 feet
Gray shale...............to	310 feet	Light slate ....to	600 feet
White slate ....to	335 feet	White chocolate sand ....to	615 feet
Coal 5 ....to	350 feet	Black slate ....to	622 feet
Black slate ....to	355 feet	Dark hard oil sand.......to	633 feet
White slate ....to	365 feet	Black slate ....to	643 feet
Brown slate ....to	380 feet	Total depth ....to	643 feet
Black slate ....to	400 feet	Well abandoned July 5, 1919.	

Record of William Scott well No. 1, N. W. ¼ of S. E. ¼ of Section 33, Township 9 N., R. 9 W., 2 miles east of Heims' Pool. Elevation of surface about 540 feet:

Gravel and sand ....to	28 feet	Black slate ....to	380 feet
Pink rock ....to	30 feet	Coal 4a ....to	384 feet
Gray sandstone ....to	45 feet	Brown shale ....to	440 feet
Gray slate ....to	70 feet	Light slate ....to	460 feet
Black slate ....to	95 feet	Coal 4 ....to	466 feet
Fire clay ....to	105 feet	Brown shale ....to	490 feet
Sandy shale ....to	140 feet	Black shale ....to	505 feet
Sandy shale ....to	172 feet	Sandy shale ....to	545 feet
Coal 6a ....to	176 feet	Gray shale ....to	570 feet
Fire clay ....to	186 feet	Black slate ....to	600 feet
Gray sandstone ....to	206 feet	Light slate ....to	620 feet
Black slate ....to	216 feet	Gray shale ....to	635 feet
Coal 6 ....to	222 feet	Hard dark sand lime.....to	665 feet
Fire clay ....to	226 feet	Light slate ....to	675 feet
Dark slate ....to	240 feet	Gray slate ....to	690 feet
Black slate ....to	260 feet	Black slate ....to	700 feet
Gray shale ....to	270 feet	Light brown sand........to	710 feet
Hard lime shell..........to	275 feet	Dark sandy shale........to	728 feet
Coal 5a ....to	277 feet	Salt, sand and water.....to	730 feet
Black slate ....to	320 feet	Total depth ....to	730 feet
Light slate ....to	360 feet	Well abandoned.	

Record of Smith well No. 1, located in the S. E. ¼ of the N. E. ¼ of Section 4, Township 8 N., R. 9 W., Curry Township, 5 miles north of Heims' Pool, 2 miles west of Farmersburg. Elevation above sea level at top of well about 530 feet:

Drift	to 44 feet
Soft sand	to 50 feet
Hard shell	to 55 feet
Red rock	to 65 feet
Slate	to 85 feet
Sandstone and water	to 90 feet
Gray slate	to 130 feet
Brown slate	to 141 feet
Coal No. 7	to 146 feet
Black slate	to 160 feet
Sandstone	to 170 feet
Brown slate	to 180 feet
Coal 6a	to 186 feet
White slate	to 195 feet
White slate	to 205 feet
Black slate, some gas	to 210 feet
White slate	to 230 feet
Sandstone	to 247 feet
Hard shell	to 250 feet
Hard lime shell	to 257 feet
Black slate	to 261 feet
Coal 6	to 266 feet
White slate	to 290 feet
Brown slate	to 352 feet
Coal 5	to 358 feet
Black slate	to 404 feet
Coal 4a	to 408 feet
Light slate	to 415 feet
Sand and water	to 430 feet
Water and sand	to 435 feet
Black slate	to 443 feet
Coal 4	to 450 feet
Brown slate	to 460 feet
White slate	to 470 feet
Gray slate	to 484 feet
Hard shell	to 486 feet
Coal 3a	to 488 feet
Black slate	to 500 feet
White slate	to 520 feet
White sand	to 535 feet
Brown slate	to 560 feet
Coal 3	to 565 feet
Brown slate	to 575 feet
Brown slate	to 580 feet
Black slate or shale	to 589 feet
Gray shale	to 595 feet
Gray shale and coal	to 602 feet
Lime shell	to 608 feet
Gray shale and lime	to 614 feet
Gray shale and lime	to 620 feet
Gray slate	to 626 feet
Gray sandy shale	to 630 feet
Oil sand, no production	to 635 feet

Composite log for Dodds' Bridge Pool. Based on 8 logs in Sections 3, 4 and 9, Turman Township. Elevation about 500 feet. Relief about 30 feet:

Coal 6a top at	170 to 183 feet
Coal 6	... to 200 feet
Coal 5a	250 to 260 feet
Coal 5	305 to 320 feet
Coal 3a	460 to 490 feet
Coal 3	... to 550 feet
Oil sand (slight production)	... to 619 feet
Coal Minshall (where present)	620 to 636 feet
Main oil sand (Minshall absent)	634 to 656 feet
Upper block	634 to 660 feet
Oil sand	656 to 683 feet
Lower block	... to 690 feet
Third oil sand	... to 730 feet

Composite log for Harmon pool, Sections 28 and 33, Turman Township. Based on 6 logs. Elevation of surface about 480.

Coal 8 (average 5 feet) top at	85 feet
Coal 7 (3 feet) top at	160 feet
Coal 6 (5 feet) top at	270 feet
Coal 5a (4 feet) top at	330 feet
Coal 5 (5 feet) 3 wells	360 feet
Coal 4a (4 feet)	380 to 385 feet
Coal 4 (5 feet)	440 feet
Coal 3a (5 feet)	580 feet
Coal 3 (5 feet) 2 wells	600 feet
Slate sand (first oil sand)	620 feet
Minshall coal, 6 feet	629 to 642 feet
2nd oil sand (95 barrel well)	740 feet
Main oil sand	767 to 780 feet
4th oil sand (75 barrel well)	818 feet

Composite log for Scott pool, based on 10 logs in Sections 31 and 36, Turman Township. Average elevation of surface about 530. Relief in pool about 20 feet.

Coal 8 top at	80 to 90 feet
Coal 7 (5 feet) top at	95 to 115 feet
Coal 6a (4 feet) top at	170 to 172 feet
Coal 6 (5 feet) top at	255 to 290 feet
Coal 5 (6 feet) top at	320 to 348 feet
Coal 4 (5 feet) top at	440 to ... feet
Coal 3 (5 feet) top at	560 to 585 feet
Upper Block (4 feet)	680 feet
Lower Block (3 feet)	700 feet
Oil sand, top of	730 to 775 feet

Composite log for Edwards pool and vicinity, in Sections 3, 9, 10 and 16. Gill Township. Based on 7 fairly detailed logs. Elevation of surface about 480 feet.

Coal 8 (3 feet) top at	80 to 90 feet
Coal 7 (2 feet) top at	107 to 110 feet
Coal 6a (5 feet) top at (if present)	210 feet
Coal 6 (4 feet) top at	230 to 245 feet
Coal 5a (5 feet) top at (if present)	285 feet
Coal 5 (4 feet) top at	330 to 340 feet
Gas sand, top at	375 to 385 feet
Coal 4a (1 foot)	421 feet
Gas sand (coal 4 level)	460 feet
Coal 3 (5 feet)	585 feet
Coal 2 (5 feet) where present	605 feet
Heims' pool oil sand top at	630 to 640 feet
Minshall (5 feet)	660 to 685 feet
Upper Block (5 feet)	715 feet
Scott pool oil sand	740 to 756 feet
3rd oil sand	770 to 785 feet
4th oil sand	820 to 822 feet
Salt sand	884 to 900 feet

Log of wells 9 and 10, G. W. Buff farm, N. W. ¼, S. W. ¼, Section 3, Gill Township, near N. E. corner of Edwards pool.

Clay and shale	... to 80 feet
Coal 8	80 to 82 feet
Shale	82 to 240 feet
Hard shell	240 to 245 feet
Coal 6	245 to 251 feet
Shale	251 to 330 feet
Coal 5, with some gas	330 to 336 feet
Shale	336 to 400 feet
Limestone	400 to 408 feet
Shale and mud	408 to 500 feet
Sand with water	500 to 522 feet
Shale and mud	522 to 640 feet
First salt sand, some oil	640 to 675 feet
Limestone	675 to 685 feet
Coal, Minshall	685 to 690 feet
Dark shale	690 to 715 feet
Coal, upper block	715 to 720 feet
Dark shale	720 to 756 feet
3rd oil sand	756 to 770 feet
Dark shale	770 to 785 feet
4th lower oil sand	785 to 805 feet
Dark shale	805 to 817 feet

Note: Well No. 9 got 10 barrels production at 756 sand. Well No. 10 found merely a show of oil there, but got production at 785. Well No. 10 is 444 feet east of well No. 9.

Township 8 North, Range 10 West.

Chastine No. 2.

10-inch drive pipe	61 feet
Salt sand	300 to 350 feet
Coal	465 to 470 feet
8-inch casing	550 feet
Salt sand	560 to 600 feet
6¼-inch casing	750 feet
Oil sand	786 feet
Total depth	800 feet

Bell No. 4.

10-inch pipe	42 feet
Coal	446 to 450 feet
8-inch casing	350 feet
6¼-inch casing	629 feet
Oil sand	786 feet
Total depth	797 feet

McClure No. 4.

10-inch drive pipe	32 feet
Coal	465 to 471 feet

8-inch casing		540 feet
Salt sand	560 to	740 feet
6¼-inch casing		775 feet
Oil sand		796 feet
Total depth		806 feet

Well No. 2 on Oscar Hunt farm, Sullivan County:

From top of surface, red clay	to 42 feet
Dark mud	to 80 feet
Coal	to 84 feet
Mud and shale	to 120 feet
Sand and some water	to 180 feet
Hard limestone shell	to 190 feet
Dark mud	to 220 feet
Sandy and hard material	to 260 feet
Coal	to 265 feet
White and black mud	to 340 feet
Coal	to 345 feet
White mud	to 350 feet
Hard shell	to 355 feet
Dark and white mud	to 465 feet
Limestone shell—hard	to 475 feet
White and dark mud	to 525 feet
Coal	to 530 feet
Dark shale	to 560 feet
Sand with some water and nice showing of oil	to 600 feet
Coal with plenty of water	to 605 feet
Dark shale	to 625 feet
Sand—hard	to 635 feet
Dark shale	to 685 feet
Coal and gas water flowing out of hole	to 690 feet
Dark shale	to 730 feet
Top of oil sand	to 752 feet
Broken sand and shale	to 775 feet
Oil sand	to 780 feet
Dark shale	to 800 feet
Total depth of well	to 800 feet
10-inch pipe	to 42 feet
8¼-inch pipe	to 355 feet
6¼-inch pipe	to 721 feet

SWITZERLAND COUNTY.

The following strata outcrop in Switzerland County:

Quaternary	Recent—Alluvial sands and clays. Pleistocene—Glacial gravel, sand and till.
Silurian	Limestones and shales.
Ordovician	Shales and limestones.

The glacial deposits vary in thickness from a few feet to fifty. Many outcrops of the bed rock occur. It is possible that the structural conditions may be determined by surficial observations. The outcrop of the Trenton in the eastern part of the county precludes the possibility of securing oil from that formation in that locality, but in the western part of the county, where the thickness of the overlying formations is adequate, oil may be present in the Trenton if the proper structural conditions exist.

The following is the record of a well drilled at Vevay:

Record of Well Drilled Near Vevay.

Surface, soil and clay	60 feet
Limestone shell and shale, 6 inches thick alternating	105 feet
Limestone	75 feet
Layers of shale and limestone 5 feet thick alternating	60 feet
Dark hard limestone	22 feet
Shale, soft	1 foot
Limestone, very hard and full of salt water	32 feet
Total depth	355 feet

TIPPECANOE COUNTY.

Beneath the Pleistocene and recent deposits of this county lie the strata of the New Albany division of the Devonian, which occupies the northeast portion of the county and the Knobstone division of the Mississippian. The contact between the two formations is revealed between the Wabash River and West Lafayette by an outcrop of Goniatite limestone, which lies at the base of the Knobstone, just above the unconformity between the Devonian and the Mississippian. Small outcrops of the Warsaw occur near Montmorenci, but the number is too small to be of much service in determining structural conditions. The Pleistocene deposits vary in thickness from a few feet to more than one hundred feet. The mantle of glacial drift is everywhere so complete that little can be learned of stratigraphical or structural features of the bed rock. If oil structures are present in the county, they can be outlined only by the use of subsurface data derived from the records of deep wells, and to be of value the wells should be located within less than a mile of each other, as the structures will probably be small.

A well drilled at Lafayette reached the top of the Niagara limestone at 235 feet. The top of the Trenton should be reached at about 1100 feet.

Railroad Elevations.

Clark's Hill, 818.6; Stockwell, 810; Crane, 736; Altamount, 645; Lafayette, 542 (Monon Station); Dayton, 647.1; Summit, 608; Balls, 697; Montmorenci, 692.

TIPTON COUNTY.

Tipton County lies within the glaciated area and is covered with glacial drift, varying in thickness from 50 to 150 feet. The drift rests on the Silurian and Devonian limestones. The surface of the Trenton lies from about sea level to 150 feet below.

Cicero Township. Three wells were drilled at Tipton, and the record of No. 1 is given below:

Drift	139 feet
Limestone	326 feet
Shale	532 feet
Trenton limestone, gas 11 ft., oil 3 ft., water 19 ft..	33 feet
Total	1030 feet
Altitude of well	868 feet

Well on the R. H. Foster farm, N. E. corner of the S. ½ of S. ½ of N. W. ¼ of Section 30, Township 22, R. 4 E., Cicero Township.

Top of sand	1002 feet
Drilled in sand	14 feet
Total depth	1016 feet
Casing used	503 feet
Drive pipe	147 feet

Dry hole.

Wells drilled in Sections 20 and 28 were abandoned in 1911.

Madison Township. At Hobbs gas was obtained and the first well has the following log:

Drift	134 feet
Limestone	330 feet
Shale and limestone	529½ feet
Trenton limestone	13½ feet
Total depth	1007 feet
Altitude of well	875 feet

A well drilled in Section 19 was abandoned in 1911.

Wild Cat Township. At Windfall the Trenton was reached at 937 feet and salt water at 1002 feet. Wells drilled in Sections 8, 17, 18, 20 and 31 were abandoned in 1911 and 1919.

Liberty Township. At Sharpsville gas was obtained from wells in which the following strata were encountered:

Drift	70 feet
Limestone	460 feet
Shale	432 feet
Trenton limestone	8 feet
Total	970 feet

Well drilled N. of the S. W. corner of Section 19, T. 22 N., R. 4 E., on the S. J. Smith farm:

Top of sand	1008 feet
Depth drilled in sand	18 feet
Total depth of well	1026 feet

Dry hole.

Abandoned wells occur in this township as follows: Section 1, 1 well; Section 5, 2 wells; Section 13, 1 well; Section 18, 2 wells; Section 22, 1 well; Section 23, 1 well; Section 31, 1 well; Section 35, 1 well; Section 36, 1 well.

Jefferson Township. At Kempton the upper surface of the Trenton is 93 feet below sea level. The log of the Kempton well follows:

Drift	306 feet
Limestone	293 feet
Shale	424 feet
Trenton limestone	12 feet
Total depth	1035 feet
Altitude of well	930 feet

A well drilled in Section 9 was abandoned in 1913 and one in Section 20 in 1912.

Prairie Township. Wells have been abandoned in this township as follows: Section 2, 2 wells; Section 10, 1 well; Section 15, 1 well; Section 16, 1 well; Section 22, 1 well; Section 23, 1 well; Section 24, 2 wells; Section 26, 1 well; Section 28, 1 well; Section 32, 1 well; Section 33, 1 well; Section 34, 3 wells.

The gas field discussed under Howard County extends down into the northern part of Tipton County. The gas in this county has accumulated in minor folds on the larger arch, a continuation of the Cincinnati geanticline.

UNION COUNTY.

Strata of Ordovician and Silurian age form the bed rock of this county. The Silurian rocks have been removed from all except the northeastern part of the county. The thickness of the Ordovician rocks is about 800 feet. The overlying drift has a thickness of from twenty-five to seventy-five feet. Since the drift is not as thick as in other counties and outcrops of the bed rock are more numerous, it may be possible by detailed work to determine the structural conditions in this county.

Since the Trenton is high in this region the probabilities of obtaining gas are much better than those of obtaining oil. If possible the minor structures on the surface of the geanticline should be located before drilling is undertaken for even then the risk is sufficiently great.

The record of a well drilled at Liberty is given by Phinney as follows:

Drift	70 feet
Limestone (Hudson River)	15 feet
Grayish shale	450 feet
Dark shale	356 feet
Gray Trenton limestone	25 feet
Blue Trenton limestone	55 feet
Total depth	971 feet
Altitude of well	965 feet

Gas was reported in small quantities from the Hudson River shale, but none in the Trenton.

The surface of the Trenton is probably more than 100 feet above sea level in the southwestern part of the county and descends to sea level in the northeastern part.

Railroad Elevations.

Cottage Grove, 1,039; Kitchell, 1,096; Wilts, 1,119; Loties, 1,039; Liberty, 980; Brownsville, 793.

VANDERBURGH COUNTY.

Vanderburgh County lies within the unglaciated area of the State. The strata which outcrop in the county belong to the Pennsylvanian period. The rocks consist of sandstones, shales and limestones, with intercalated beds of coal. The southern part of the county is occupied by the alluvium of the Ohio River valley and outcrops of the bed rock are not found. It is doubtful whether a sufficient number of outcrops of persistent layers can be found to determine structural conditions. It may be possible to use well records, mine shaft records and outcrops and thus determine the structural conditions of the strata. Care should be exercised in using the dip of the rocks of the coal measures to discriminate between purely local dips which are so abundant and dips of regional extent.

The following is the record of a well drilled on the east bank of Pidgeon Creek, near Evansville:

Section in Crescent City Artesian Well.

Soapstone	31 feet	
Gray sandstone	2 feet	6 in.
Soapstone and shale	37 feet	6 in.
Very hard gray sandstone	1 foot	
Slaty coal	1 foot	6 in.
Shale	6 feet	
Gray shale or sandstone	44 feet	6 in.
Soft shale	11 feet	
Soft gray sandstone	18 feet	
Hard dark sandstone	5 feet	
Gray flint	2 feet	
Dark gray sandstone	62 feet	
Salt water		
Hard black shale (coal?)	73 feet	
Gray sandstone	65 feet	
Flint	6 feet	
Hard gray shale	5 feet	
Hard argillacious sandstone	34 feet	
Gray shales (soapstone)	55 feet	
Coal (L?)	1 foot	6 in.
Gray shale and sandstone	134 feet	
Dark sandstone with salt water flowing seven gallons per minute, 3 degrees Baume	5 feet	
Hard, pure sandstone, conglomerate	50 feet	
Coal and slate		6 in.
Soapstone	10 feet	

Coal (A?) and slate	1 foot	6 in.
Fire clay		6 in.
	682 feet	
Surface	17 feet	
Total	709 feet	

Section in Avondale Bore.

Surface	9 feet	6 in.
Blue clay	30 feet	6 in.
Gray sand	2 feet	6 in.
Blue mud, quicksand	22 feet	3 in.
Gravel, sand and shells	6 feet	
Fire clay and sand	28 feet	3 in.
Gravel and sand	1 foot	
Sandstone	2 feet	
Fire clay	2 feet	9 in.
Sandstone	11 feet	
Fire clay	7 feet	9 in.
Sandstone	7 feet	
Fire clay with pebbles	2 feet	8 in.
Silicious clay	1 foot	
Sandstone with iron balls	72 feet	
Concretion	1 foot	10 in.
Sandstone	36 feet	10 in.
Rock slate	6 feet	
Black slate	2 feet	10 in.
Coal	4 feet	
Total	255 feet	9 in.

Section of Inglefield Bore.

Surface clay	10 feet	
Red Merom sandstone	36 feet	
Carbonaceous parting, coal		4 in.
Hard, flinty limestone	4 feet	
Clay parting, second rash coal	1 foot	8 in.
Flinty gray limestone	6 feet	
Light gray sandstone	20 feet	
Soft white limestone	8 feet	
Soapstone, first rash coal	16 feet	3 in.
Shale	20 feet	
Gray, flinty limestone	3 feet	2 in.
Soapstone	26 feet	
White limestone	30 feet	
Gray shale	20 feet	
Fire clay	10 feet	
Coal (N?)	1 foot	6 in.
Fire clay	4 feet	

Gray shale	10 feet
Soapstone	28 feet
Sandstone	3 feet
Black slate	2 feet
Sandstone	17 feet
Total	276 feet 5 in.

Scott Township. A well was drilled on the John M. Hart farm in 1913. It resulted in no production. A well drilled on the R. Cutter farm in 1918 was non-productive. Records of the wells could not be obtained.

VERMILLION COUNTY.

Vermillion County lies wholly within the area occupied by the Pennsylvanian strata, the outcrop of which is covered by the Pleistocene and recent deposits. These deposits of mantle rock attain a thickness of more than one hundred feet. This regolith has been largely removed along the courses of the streams and outcrops of the durolith occur. It may be possible that by using these outcrops in connection with coal openings and the records of wells to determine the structural conditions of the bed rock, though careful work will be necessary and much time required.

The surface of the Trenton is probably 1600 or more feet below the level of the sea. If structures are present, oil may be found in Trenton, Devonian or Pennsylvanian strata.

Railroad Elevations.

Cayuga, 522; State Line (T., St. L. & W.), 626; Rileysburg, 646; Gessie, 616; Perrysville, 582; Dickason, 526; Malone, 507; Walnut Grove, 528; Newport, 496; Dorner, 510; Worthy, 489; Mt. Silica, 492; West Montezuma, 488; Hillsdale, 488; Logan, 496; Summit Grove, 520; Norton Crossing, 493; Jackson, 495; Clinton, 494.

VIGO COUNTY.

Strata belonging to the Pennsylvanian period occupy the sub-surface in Vigo County. The rocks are sandstones, shales and limestones, with intercalated beds of coal. A covering of glacial drift largely conceals the outcrop of the durolith, the thickness of the latter varying from a few feet to more than one hundred feet. The structural conditions of the durolith can probably be determined by using coals IV and V as key horizons and relying on data secured from well records and coal outcrops for the position of these beds.

Harrison Township. Oil has been produced from a single well in Terre Haute for more than thirty years. The following is the record of a well drilled on the bank of the river at Terre Haute in 1869:[1]

Record of Terre Haute Well.

	Feet	Inches	Feet	Inches
1. Sand and gravel	100		100	
2. Soapstone	64	6	164	6

[1] Report of Indiana State Geological Survey for 1870.

Surfa
Blue
Gray
Blue r
Grave
Fire c
Grave
Sands
Fire c
Sands
Fire
Sands
Fire c
Silicio
Sands
Concr
Sands
Rock
Black
Coal

T

Surfa
Red M
Carbo
Hard,
Clay
Flinty
Light
Soft
Soapst
Shale
Gray,
Soapst
White
Gray
Fire c
Coal
Fire c

	Feet	Inches	Feet	Inches
52. White limestone	21		915	
53. Gray limestone	5		920	
54. Limestone and soapstone	5		925	
55. Gray limestone	5		930	
56. White limestone	15		945	
57. Fine blue limestone	2		947	
58. Dark gray limestone and flint	73		1020	
59. Light gray limestone	7		1027	
60. Blue gray limestone	7		1034	
61. Soapstone (fire clay)	26		1060	
62. Gray limestone	24		1084	
63. Gray sandstone	3		1087	
64. Soapstone (fire clay)	5		1092	
65. Quartz and shale mixed	166		1258	
66. Quartz, slate and soapstone	3		1261	
67. Slate rock	21		1282	
68. Soapstone	33		1315	
69. Slate rock	7		1322	
70. Soapstone	235		1557	
71. Soapstone and sandstone	10		1567	
72. Fine sandstone	15		1582	
73. Blue soapstone	40		1622	
74. Black shale	15		1637	
75. Red shale	5		1642	
76. Black shale	15		1657	
77. Lime rock	5		1662	
78. Black shale	5		1667	
79. Gray lime rock, oil near top	149		1816	
80. Gray sand rock	23		1839	
81. Lime rock	73	4	1912	4

In discussing the geology of Vigo County in the annual report of the Indiana Survey for 1896, Dr. J. T. Scovell publishes the following well records:

Swan Street Well on Banks of Wabash.

Sand, gravel sandstone, shale and limestone	1110 feet	1110 feet
Limestone	450 feet	1560 feet
Shale	50 feet	1610 feet
Limestone	3 feet	1613 feet

Oil Sand and Oil.

Limestone	967 feet	2580 feet
Shale	100 feet	2680 feet
Limestone (perhaps Trenton)	250 feet	2930 feet

	Feet	Inches	Feet	Inches
3. Coal	6	2	170	8
4. Hard sandstone	2	3	172	11
5. Soapstone	10		182	11
6. Coal	3		185	11
7. Soapstone	4	3	190	2
8. Gray sandstone	5	10	196	
9. Blue soapstone		10	196	10
10. Gray sandstone		6	197	4
11. Blue soapstone	12	.9	210	1
12. Soft black shale	6		216	1
13. Coal		9	216	10
14. Soapstone	7	7	224	5
15. White sandstone (conglomerate)	30	3	254	8
16. Blue shale	7	2	261	10
17. Coal	2	3	264	1
18. Black shale	10		274	1
19. White soapstone	3		277	1
20. Black shale	15		292	1
21. White soapstone	8		300	1
22. Black shale	3	3	303	4
23. Coal	3		306	4
24. Soapstone	17	8	324	
25. Sand rock	3		327	
26. Soapstone	20		347	
27. Sand rock	10		357	
28. Blue shale	22		379	
29. Limestone	2		381	
30. Blue shale	31		412	
31. Light shale	5		417	
32. Blue shale	60		477	
33. Sandstone	7		484	
34. Blue shale	24		508	
35. Sandstone	3		511	
36. White shale	10		521	
37. Blue shale	147		668	
38. Hard, gritty slate rock	11	7	679	7
39. Hard gray sandstone	14	5	694	
40. Hard limestone	11		705	
41. White limestone	24		729	
42. Gray limestone	2		731	
43. Limestone	14		745	
44. White limestone	82		827	
45. Soapstone	3		830	
46. Brown limestone	35		865	
47. Soapstone	5		870	
48. Lime rock	9		879	
49. Soapstone	6		885	
50. White limestone	7		892	
51. Soapstone or Gypsum?	2		894	

Coal at 22 feet below the top of shale.

Limestone	5 feet	145 feet
Shale	95 feet	240 feet
Limestone	10 feet	250 feet
Shale	40 feet	290 feet
Limestone	20 feet	310 feet
Shale	210 feet	520 feet
Limestone	23 feet	543 feet
Shale	10 feet	553 feet
Limestone, hard and flinty	82 feet	635 feet
Shale	5 feet	640 feet
Limestone	160 feet	800 feet
Limestone with sand	70 feet	870 feet
Sandstone	30 feet	900 feet
Limestone	25 feet	925 feet
Sandstone	65 feet	990 feet
Limestone	30 feet	1020 feet
Shale	180 feet	1200 feet
Sandstone, white	50 feet	1250 feet
Sandstone and shale	50 feet	1300 feet
Sandstone, white	150 feet	1450 feet
Shale	122 feet	1572 feet
Limestone—oil rock	11 feet	1583 feet

Show of oil at 1575 and sulphur at 1578 feet.

Alden Well.

On northwest quarter of Section 23-12-9:

Sand and gravel	130 feet	80 feet
Shale	110 feet	190 feet
Limestone	20 feet	210 feet
Shale	300 feet	510 feet
Sandstone	10 feet	520 feet
Shale	30 feet	550 feet
Sandstone	160 feet	710 feet
Limestone	300 feet	1010 feet
Sandstone	90 feet	1100 feet
Shale with sand	132 feet	1232 feet

Salt water at 525 feet and between 600 and 700 feet.

Section of the Elliott Well.

Near west line of Section 23 and Wabash avenue, Terre Haute.

Sand and gravel	128 feet	78 feet
Shale	260 feet	338 feet
Sandstone	35 feet	373 feet
Limestone	40 feet	413 feet
Sandstone	98 feet	511 feet
Limestone	23 feet	534 feet
Sandstone	179 feet	713 feet
Shale	110 feet	823 feet

Section of Kinser Well.

Located between Fourteenth and Fifteenth streets, just east of the center of Section 22-12-9, near Liberty avenue.

Soil, gravel and sand	80 feet	80 feet
Shale or soapstone	70 feet	150 feet
Sandstone	10 feet	160 feet
Shale	90 feet	250 feet
Sandstone	70 feet	320 feet
Shale or slate	130 feet	450 feet
Sandstone	140 feet	590 feet
Limestone	360 feet	950 feet
Limestone with some shale	185 feet	1135 feet
Limestone with quartz	85 feet	1220 feet
Shale	25 feet	1245 feet
Limestone with shale	225 feet	1470 feet
Shale or soapstone	5 feet	1475 feet
Sandstone or limestone	15 feet	1490 feet
Shale or soapstone	138 feet	1628 feet
Limestone or oil rock	20 feet	1648 feet

A little oil was present near the surface of the limestone. To reduce these records and the following to the level of the river fifty feet was deducted from the thickness of the first stratum.

Section of the Big Four Well.

Located in the northeast corner of the northwest quarter of Section 23-12-9.

Soil	6 feet	6 feet
Gravel	10 feet	16 feet
Sand	102 feet	118 feet
Shale	117 feet	185 feet
Sandstone or limestone	2 feet	187 feet
Shale	207 feet	394 feet
Salt water at 78 below the top of shale.		
Limestone or sandstone	41 feet	435 feet
Shale or slate	50 feet	485 feet
Limestone or sandstone	12 feet	497 feet
Shale or slate	53 feet	550 feet
Sandstone	50 feet	600 feet
Limestone	600 feet	1200 feet
Shale with some limestone	190 feet	1390 feet
Shale or slate	210 feet	1600 feet
Limestone, oil rock, sulphur water	18 feet	1618 feet

Section of Exchange Well.

Situated a little west of the center of Section 22-12-9:

Soil and coarse gravel	80 feet	30 feet
Sand fine	45 feet	75 feet
Shale and slate	65 feet	140 feet

Coal at 22 feet below the top of shale.

Limestone	5 feet	145 feet
Shale	95 feet	240 feet
Limestone	10 feet	250 feet
Shale	40 feet	290 feet
Limestone	20 feet	310 feet
Shale	210 feet	520 feet
Limestone	23 feet	543 feet
Shale	10 feet	553 feet
Limestone, hard and flinty	82 feet	635 feet
Shale	5 feet	640 feet
Limestone	160 feet	800 feet
Limestone with sand	70 feet	870 feet
Sandstone	30 feet	900 feet
Limestone	25 feet	925 feet
Sandstone	65 feet	990 feet
Limestone	30 feet	1020 feet
Shale	180 feet	1200 feet
Sandstone, white	50 feet	1250 feet
Sandstone and shale	50 feet	1300 feet
Sandstone, white	150 feet	1450 feet
Shale	122 feet	1572 feet
Limestone—oil rock	11 feet	1583 feet

Show of oil at 1575 and sulphur at 1578 feet.

Alden Well.

On northwest quarter of Section 23-12-9:

Sand and gravel	130 feet	80 feet
Shale	110 feet	190 feet
Limestone	20 feet	210 feet
Shale	300 feet	510 feet
Sandstone	10 feet	520 feet
Shale	30 feet	550 feet
Sandstone	160 feet	710 feet
Limestone	300 feet	1010 feet
Sandstone	90 feet	1100 feet
Shale with sand	132 feet	1232 feet

Salt water at 525 feet and between 600 and 700 feet.

Section of the Elliott Well.

Near west line of Section 23 and Wabash avenue, Terre Haute.

Sand and gravel	128 feet	78 feet
Shale	260 feet	338 feet
Sandstone	35 feet	373 feet
Limestone	40 feet	413 feet
Sandstone	98 feet	511 feet
Limestone	23 feet	534 feet
Sandstone	179 feet	713 feet
Shale	110 feet	823 feet

The Smith well, drilled near the southwest corner of Wabash avenue and Tenth street, southwest of the southwest section 22-12-9, reached the oil-bearing limestone at 1632 feet.

The Guarantee No. 3, between Eighth and Ninth streets, near Wabash avenue, reached oil rock at 1569 feet.

The Guarantee No. 4, between Wabash avenue and Chestnut street, on Tenth-Half street, reached sulphur water at 1590 feet.

The Guarantee No. 5, near southwest corner South Fifth and Farrington streets, southeast of the northeast section 28-12-9, reached oil sand at 1700 feet.

Section of Guarantee Well No. 6.

Northeast corner Third and Mulberry streets, northwest ¼ of the southeast ¼, section 21-12-9.

Soil, gravel and sand	128 feet	78 feet
Shale	44 feet	122 feet
Coal	5 feet	127 feet
Shales and sandstone	308 feet	435 feet
Limestone	40 feet	475 feet
Shale, blue and black	90 feet	565 feet
Limestones	415 feet	980 feet
Limestone, coarse	25 feet	1005 feet
Shale with some limestone	55 feet	1060 feet
Shale with some limestone	40 feet	1100 feet
Limestone with some shale	320 feet	1420 feet
Shale	25 feet	1445 feet
Limestone	9 feet	1454 feet
Shale	43 feet	1497 feet
Black shale, lime shell	72 feet	1569 feet
Coarse shale	9 feet	1578 feet
Limestone, black	20 feet	1598 feet

Salt water at 800 feet, gas at 925, 160 and 1100 feet, sulphur water at 1598 feet.

Guarantee No. 1 (Diall well) located on the alley between Chestnut and Eagle streets and between Ninth and Tenth, was drilled to oil on May 8, 1888. Oil rose fifty feet above the surface, "flowed out over the whole region into the sewer and down to the river and its villainous odor filled the air for squares."

The Phenix well was drilled 300 feet south between Eagle and Mulberry streets and became a good producer.

Guarantee No. 3, near Wabash avenue, between Eighth and Ninth streets, also produced some oil. The productive area is very small. Wells were drilled in all directions from the productive wells, but yielded water only.

Riley Township. The Riley oil field is located southeast of the town of Riley in sections 23 and 24. Oil has been produced from about twenty-five wells. The largest initial production is about twenty-five barrels per day. The locations of the producing wells on the accompanying map were made by Dr. C. A. Malott.

Joslin Well Record.

A well was completed October 7, 1912, on the Charles N. Joslin farm,

Section 23, Township 11 North, Range 8 West, Riley Township, Vigo County, Indiana, by Bill Brothers. The following is a complete log of the well:

Clay	12 feet
Sand rock to 21 feet	9 feet
Lime to 40 feet	19 feet
Slate to 76 feet	36 feet
10-inch pipe	76 feet
Lime to 85 feet	9 feet
Brown shale to 120 feet	35 feet
Sand rock to 180 feet	60 feet
Coal to 182 feet	2 feet
Brown shale to 196 feet	14 feet
Lime to 210 feet	14 feet
Slate to 240 feet	30 feet
Lime to 248 feet	8 feet
Slate to 275 feet	27 feet
White sand (water) to 290 feet	15 feet
Slate to 340 feet	50 feet
Lime to 355 feet	15 feet
Slate to 390 feet	35 feet
Salt sand (more water) to 420 feet	30 feet
Slate to 450 feet	30 feet
Lime to 465 feet	15 feet
Slate to 490 feet	25 feet
White sand to 560 feet	70 feet
Slate to 620 feet	80 feet
Lime to 625 feet	5 feet
Salt sand to 645 feet	20 feet
Lime to 660 feet	15 feet
Hard lime to 710 feet	50 feet
8¼-inch casing	663 feet
White lime with small break 750 feet	40 feet
Hard lime to 820 feet	70 feet
White lime to 990 feet	170 feet
Slate and shells to 1060 feet	70 feet
Slate to 1100 feet	40 feet
Lime to 1115 feet	15 feet
Slate to 1160 feet	45 feet
Lime to 1170 feet	10 feet
Slate to 1220 feet	50 feet
Lime to 1230 feet	10 feet
Slate to 1250 feet	20 feet
Black slate to 1290 feet	40 feet
Lime to 1310 feet	20 feet
Slate to 1370 feet	60 feet
Lime to 1380 feet	10 feet
Slate to 1440 feet	60 feet
Lime to 1445 feet	5 feet
Slate to 1455 feet	10 feet
Lime shell to 1458 feet	3 feet
Slate to 1507 feet	49 feet
6⅝-inch casing	1507 feet
Lime to 1520 feet	13 feet
Slate and shells to 1555 feet	35 feet
Brown shale to 1615 feet	60 feet
Lime to 1617 feet	2 feet
Slate to 1619 feet	2 feet
Sand or cap rock to 1621 feet	2 feet
First oil to 1623 feet	2 feet
Light brown shale to 1625 feet	2 feet
Dark brown sand to 1629 feet	4 feet
Light and lime sand to 1631.	2 feet
Gray shelly sand to 1637 feet	6 feet
Light shelly sand to 1641 feet	4 feet

Oil only in one place, 1621 to 1625.

Linton Township. A deep well was drilled in this township just west of Pimento in section 14. No production was obtained. A well was also drilled in section 1 of this township without favorable results. Many wells in this township have been drilled to coal V, which is penetrated at depths ranging from 320 feet to 500 feet.

Sugar Creek Township. The record of a well drilled at St. Mary's-in-the Wood, on the northeastern quarter, southwestern quarter, Section 6-12-9, is given by Scovell as follows:

	Feet	Total feet
Surface soil and yellow clay	20	...
Blue clay	55	...
Blue clay and quicksand (low water)	25	...

	Feet	Total feet
White shale	25	25
Coal, probably coal "N"	5	30
White shale—fire clay and shale	65	95
Coal, probably coal "M"	6	101
White shale—fire clay and shale	90	191
Coal, probably "L," the big vein	10	201
Fire clay and white shale	50	251
White sand rock	40	291
White shale	229	520
Sandstone	80	600
Limestone	490	1090
Fresh water at 730 feet.		
Shale	50	1140
Brown sandstone	20	1160
White shale	250	1410
Limestone and sandstone	180	1590
Brown shale	115	1705
Limestone	250	1955

Sulphur water at 1905 feet, but no show of oil or gas reported.

WABASH COUNTY.

The bed rock strata in this county belong to the Silurian period. The drift overlying varies from 25 to 300 feet, and conceals the bed rock strata to such an extent that stratigraphical and structural conditions are difficult to determine. The surface of the Trenton lies from 100 to 400 feet below sea level. The total thickness of the Niagara in this county is probably about 450 feet. The following are records of wells drilled in Wabash County.[1]

Drift	36 feet
Bluish limestone	54 feet
White limestone	20 feet
Bluish limestone varying to green	140 feet
Whitish limestone	30 feet
Bluish limestone	60 feet
Bluish green Niagara shale	35 feet
Bluish gray limestone (Clinton)	20 feet
Hudson River limestones and shales	205 feet
Utica shale	280 feet
Trenton limestone (salt water)	7 feet
Total depth	887 feet
Altitude of well	680 feet

[1] Record of a well drilled at North Manchester.

Section of Well No. 2.

Drift	28 feet
Niagara limestone and shale	525 feet
Hudson River and Utica	325 feet
Trenton limestone	54 feet
Total depth	932 feet
Trenton below sea level	198 feet

Did not yield gas nor oil.

The following is the log of a well drilled in S. W. ¼ of Section 34, Liberty Township. Drilled in 1903:

Drive pipe	202 feet
Casing	470 feet
Top of Trenton	945 feet
Total depth	965 feet

A log of a well drilled at LaFontaine is given below:

Section of Well No. 1.

Drift	300 feet
Niagara limestone	225 feet
Hudson River limestone and shale	175 feet
Utica shale	200 feet
Trenton limestone	23 feet
Total depth	923 feet
Trenton below sea level	6 feet

Yielded strong flow of gas.

Section of Well No. 1.

Drift	274 feet
Niagara limestone and shales	300 feet
Hudson River limestone and shales	250 feet
Utica shale	306 feet
Trenton limestone	50 feet
Total depth	1180 feet
Trenton below sea level	365 feet

Yielded no gas.

WARREN COUNTY.

The bed rock formations, which have been recognized by direct observation, belong to the Knobstone, Harrodsburg (Warsaw), Salem, Mitchell and Chester Divisions of the Mississippian and the Mansfield (Pottsville) and coal measures (Allegheny) divisions of the Pennsylvanian. Overlying these formations are Pleistocene and recent deposits of sand clay and gravel. The mantle rock or drift attains a thickness of more than two hundred feet. The Pennsylvanian rocks attain a thickness of about 225 feet, the Mississippian of about 110 feet, and the Devonian of about 525 feet. Devonian and Trenton strata, which may be productive of oil and gas, if the proper geological structures exist, lie below the formations mentioned above. The surface of the Trenton lies probably from 1500 to 1800 feet below the surface of the county. The outcrops of the bedrock are not sufficiently numerous to make it possible to determine the structural conditions under which the formations exist. By the aid of well records, coal shaft records and outcrops, it may be possible to determine the structural conditions favorable to the accumulation of oil and gas.

A deep well was drilled at Williamsport which struck salt water at 1200 feet. It is not probable that this well reached the Trenton limestone; it more probably reached the upper part of the Silurian.

Railroad Elevations.

Pine Village, 702; Chatterton, 714; Winthrop, 677; Kickapoo, 546; Independence, 521; State Line, 694 (C. & E. I.); Pence, 700; Finney, 719; Judyville, 771.

WARRICK COUNTY.

This is another one of the counties lying wholly within the unglaciated area of the state, and the outcrops of the strata, where concealed, are only by alluvium and residual deposits of glacial and post-glacial age. The rocks of the Pennsylvanian period outcrop in the county. The structural conditions of the county are difficult to study because of the absence of outcrops of persistent layers in sufficient numbers. In the region of coal mines some of the coal beds may be used as key formations in determining the structures. The Petersburg coal, for instance, is an important and persistent bed of coal from the line of its outcrop to the western line of the county and might be used if a sufficient number of shafts or drill holes reached it. Structural lines were drawn on the surface of this coal for a part of this county and published in the Ditney Folio.

Not many well records are available for this county. The following have been reported:

Ohio Township. A well was drilled to a depth of 1450 feet in section 15, but no production was obtained.

Lane Township. A well was drilled in section 29 on the Elisha Burr property and plugged in 1911.

Record of dry hole on the John N. Miller lease, S. E. ¼ of the N. W. ¼ of Section 19, Boone Township.

Surface loam and shale	to	40	feet	Fire clay and shale	to	322	feet
Shale	to	60	feet	Shale and shells	to	333	feet
Lime and shale	to	85	feet	Limestone	to	336	feet
Shale	to	105	feet	Coal	to	341½	feet
Fire clay	to	120	feet	Shale and shells	to	390	feet
Black shale (cave)	to	130	feet	Limestone and shells	to	416	feet
Black shale	to	143	feet	Brown shale	to	465	feet
Coal	to	149	feet	White shale	to	567	feet
Hard shale	to	152	feet	Brown shale	to	617	feet
White shale	to	202	feet	Shale and shells	to	717	feet
Black shale	to	222	feet	Black shale	to	767	feet

Lime shells ....to	787	feet	Shale ....to	1265	feet
Gray shale ....to	827	feet	Brown lime ....to	1280	feet
Black shale ....to	837	feet	Black shale ....to	1292	feet
White sand (full of salt water) ....to	907	feet	Red cave ....to	1300	feet
White shale ....to	947	feet	Soft black shale.......to	1323	feet
Brown shale ....to	1047	feet	Salt sand, yielding salt water ....to	1383	feet

The second dry hole in Warrick County was on the Barkely lease in the S. E. ¼ of the N. E. ¼ of Section 21, Hart Township. Its record showed a total depth of 1310 feet. A very slight showing of oil occurred at 1220 feet.

WASHINGTON COUNTY.

Washington County lies largely within the unglaciated area of the State, only a small area in the northwestern part of the county is covered with glacial drift. The rocks which appear at the surface of this county belong to the Quaternary and the Mississippian periods. The sub-divisions are given in the table below:

Quaternary	Recent—Sands, clays and alluvium. Pleistocene—Sands and gravels.
Mississippian	Mitchell limestone. Salem limestone. Harrodsburg limestone. Knobstone, shales and sandstones.

A large part of the surface of the county is included in the Mitchell plain on which there are few outcrops that can be used in determining structures favorable for the accumulation of oil. The best key formation is the contact between the Knobstone and the Harrodsburg (Warsaw). Some gas was obtained at Salem from the Devonian limestone, but the structural conditions existing there have not been determined. The following is the record of a well drilled at that point:

Section of Well No. 1.

Soil	7 feet
Keokuk limestone	53 feet
Sub-carboniferous sandstone	567 feet
Hamilton shale	103 feet
Devonian limestone	40 feet
Niagara limestone	215 feet
Clinton (?) limestone	30 feet
Hudson River limestone and shale...............	535 feet
Utica shale	180 feet
Trenton limestone	45 feet
Total depth	1775 feet
Trenton below sea level........................	1000 feet

Yielded good flow of gas. The gas was found in the limestone underlying the Devonian shale.

WAYNE COUNTY.

Rocks of Ordovician and Silurian age occupy the sub-surface of this county, but are exposed at few places, being covered with glacial drift, which attains a thickness of more than two hundred feet.

Wayne Township. At Richmond a well was drilled the log of which was recorded as follows by Gorby:[3]

Hudson River limestone and shale	500 feet
Utica shale	380 feet
Trenton limestone	510 feet
St. Peter's sandstone	10 feet
Total depth	1400 feet
Trenton above sea level	79 feet

Another well reached the Trenton at 945 feet, another at 886 feet and another at 972 feet.

Jefferson Township. At Hagerstown gas was found in a number of wells. One of the wells passed through 100 feet of drift, reached the Trenton at 846 feet, 167 feet above sea level.

Jackson Township. Two wells drilled at Cambridge City gave the following sections:

Drift	96 feet
Niagara limestone	2 feet
Hudson River and Utica	668 feet
Trenton limestone	134 feet
Total depth	900 feet
Trenton above sea level	174 feet

No. 2 passed through 100 feet of drift and reached the Trenton at 847 feet.

The records of other wells drilled in the county as given by Phinney are as follows:

	Dublin	Dalton	Washington	Russell	Fountain
Drift	300	275	212	...	185
Depth of Trenton	868	960	976	909	1025
Altitude of surface	1066	...	1100	1029	1011
Altitude of Trenton	198	...	124	120	86

WELLS COUNTY.

This county lies within the area occupied by the Silurian strata, which is covered with glacial drift. The stratigraphical and the structural conditions can be determined by the study of well records. This county has produced oil and the old field has recently been extended in the western part of the county. The records of some of the wells are given below:

Chester Township. A large number of wells were drilled in this township. Two wells drilled in 1908, started at 80 and 85 barrels each. The abandoned wells are: Section 2, 1 well; Section 5, 6 wells; Section 6, 5 wells; Section 7,

9 wells; Section 8, 19 wells; Section 9, 1 well; Section 10, 4 wells; Section 14, 13 wells; Section 15, 37 wells; Section 16, 3 wells; Section 17, 18 wells; Section 18, 7 wells; Section 22, 4 wells; Section 23, 19 wells; Section 27, 1 well; Section 30, 21 wells; Section 31, 8 wells; Section 32, 16 wells; Section 33, 2 wells; Section 34, 11 wells.

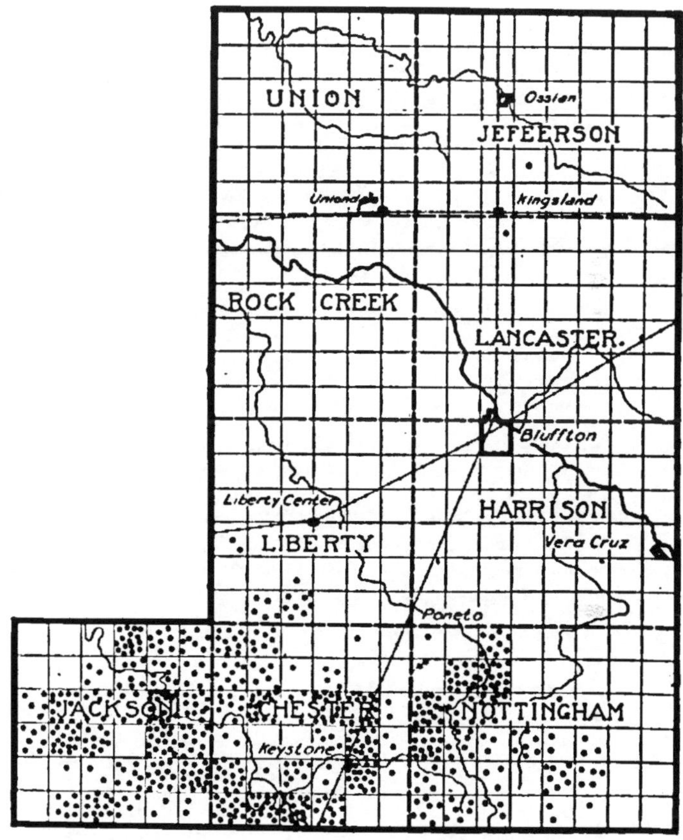

Fig. CLII. Map of Wells County showing location of recorded abandoned wells. The southern tier of townships is oil territory. Some extension has been made recently in Liberty Township.

Jackson Township. A well was drilled in 1908 in Section 12, S. E. ¼ and yielded 110 barrels the first day. The following is the average record of the wells in the N. W. ¼ of Section 20:

Drive pipe	153 feet
Casing	385 feet
Top of Trenton	989 feet
Total depth	1045 feet

A bore on the Palmer lease, east half of the N. W. ¼ of Section 31, had the following record:

Drive pipe	130 feet
Casing	340 feet
Top of Trenton	985 feet
Total depth	1045 feet

The abandoned wells are as follows: Section 1, 9 wells; Section 2, 8 wells; Section 3, 9 wells; Section 9, 3 wells; Section 10, 8 wells; Section 11, 8 wells; Section 12, 9 wells; Section 13, 16 wells; Section 14, 26 wells; Section 15, 8 wells; Section 16, 13 wells; Section 17, 11 wells; Section 18, 5 wells; Section 19, 10 wells; Section 21, 27 wells; Section 22, 1 well; Section 23, 28 wells; Section 24, 15 wells; Section 25, 40 wells; Section 26, 12 wells; Section 27, 8 wells; Section 28, 3 wells; Section 29, 1 well; Section 32, 7 wells; Section 33, 14 wells; Section 34, 7 wells; Section 35, 2 wells; Section 36, 7 wells.

Nottingham Township. A well drilled on the Dickinson tract, in the N. E. ¼ of Section 28, has the following record:

Drive pipe	38 feet
Casing	332 feet
Top of Trenton	1005 feet
Total depth	1050 feet
Initial output	30 bbls.

Abandoned wells are as follows: Section 4, 9 wells; Section 6, 1 well; Section 9, 15 wells; Section 8, 15 wells; Section 14, 1 well; Section 16, 8 wells; Section 17, 6 wells; Section 18, 21 wells; Section 19, 29 wells; Section 20, 7 wells; Section 21, 3 wells; Section 22, 7 wells; Section 23, 2 wells; Section 24, 2 wells; Section 25, 3 wells; Section 26, 8 wells; Section 28, 3 wells; Section 29, 1 well; Section 30, 5 wells; Section 31, 7 wells; Section 32, 6 wells; Section 33, 3 wells; Section 35, 2 wells; Section 36, 1 well.

Harrison Township. Section of well No. 1, Bluffton, Indiana:

Drift	12 feet
Niagara limestone and shale	413 feet
Hudson River limestone and shale	340 feet
Utica shale	285 feet
Trenton limestone	150 feet
Total depth	1200 feet
Trenton below sea level	213 feet

Yielded no gas.

Section of well No. 2, Bluffton, Indiana:

Drift	15 feet
Water lime	30 feet
Niagara limestone	479 feet
Hudson River limestone and shale	340 feet
Utica shale	175 feet
Trenton limestone	31 feet
Total depth	1106 feet
Trenton below sea level	238 feet

Liberty Township. A large number of wells were drilled in this township. The following have been abandoned: Section 19, 2 wells; Section 28, 1 well; Section 32, 5 wells; Section 33, 8 wells.

Lancaster Township. A well was abandoned in Section 4 on the property of H. Rupright in 1919.

Jefferson Township. A well drilled on the property of Grover Gibson in Section 27 was abandoned in 1919.

WHITE COUNTY.

Strata of the Mississippian age occupy the subsurface of the southwestern portion of this county; Devonian strata, the central portion; and Silurian strata the eastern portion. A mantle of glacial drift largely conceals these strata and attains a thickness of from 200 to 300 feet. The structural condition of the strata of the durolith cannot be determined by direct observation because of the overlying drift.

The record of a well drilled at Monticello is given below:

Section of Well No. 1.

Drift	205 feet
Niagara limestone	515 feet
Hudson River limestone and shale	120 feet
Utica shale	170 feet
Trenton limestone	63 feet
Total depth	1073 feet
Trenton below sea level	338 feet
Yielded no gas.	

A well drilled at Monon is reported as follows:

Limestone	530 feet
Shale	30 feet
Petroliferous limestone (Clinton?)	25 feet
Shale	285 feet
Trenton limestone	50 feet
Total depth	920 feet
Altitude of well	664 feet

The surface of the Trenton lies from 250 to 400 feet below sea level in this county.

Railroad Elevations.

Burnettsville, 711.2; Idaville, 709.7; Monticello, 677.9; Reynolds, 691.2; Seafield, 697.7; Walcott, 714.1; Lee, 671; Monon, 672.2; Wheelers, 690.7; Chalmers, 708.9.

WHITLEY COUNTY.

The strata which form the bed rock for this county belong to the Silurian and Devonian periods. The strata dip northward. They are concealed by an overburden of glacial drift which attains a thickness of more than three hundred feet. At Columbia City a deep well was drilled and salt water was encountered at 900 and at 1,375 feet. A bed of salt 25 feet thick was reported at a depth of 872 feet. The record of the well follows:

Drift	224 feet
Limestone	350 feet
Shale	776 feet
Trenton limestone	25 feet
Total depth	1375 feet
Altitude of well	816 feet

Gorby gave the following log of a well at Columbia City:

Section of Well No. 1.

Drift	224 feet
Niagara limestone and shale	526 feet
Hudson River limestone and shale	400 feet
Utica shale	218 feet
Trenton limestone	39 feet
Total depth	1407 feet
Trenton below sea level	545 feet
Yielded no gas.	

Another well drilled at Larwill, northwest of Columbia City, has the following log:

Drift	365 feet
Blue limestone	300 feet
Whitish limestone	200 feet
Bluish limestone	22 feet
Niagara shale	43 feet
Clinton limestone (salt water)	14 feet
Shale	43 feet
Limestone, salt water	43 feet
Bluish green shale	212 feet
Black shale	300 feet
Trenton limestone	51 feet
Total depth	1593 feet
Altitude of well	950 feet

The structural conditions of the durolith are not determinable by the direct observations on account of the glacial covering. Subsurface work will depend upon data secured from deep wells.

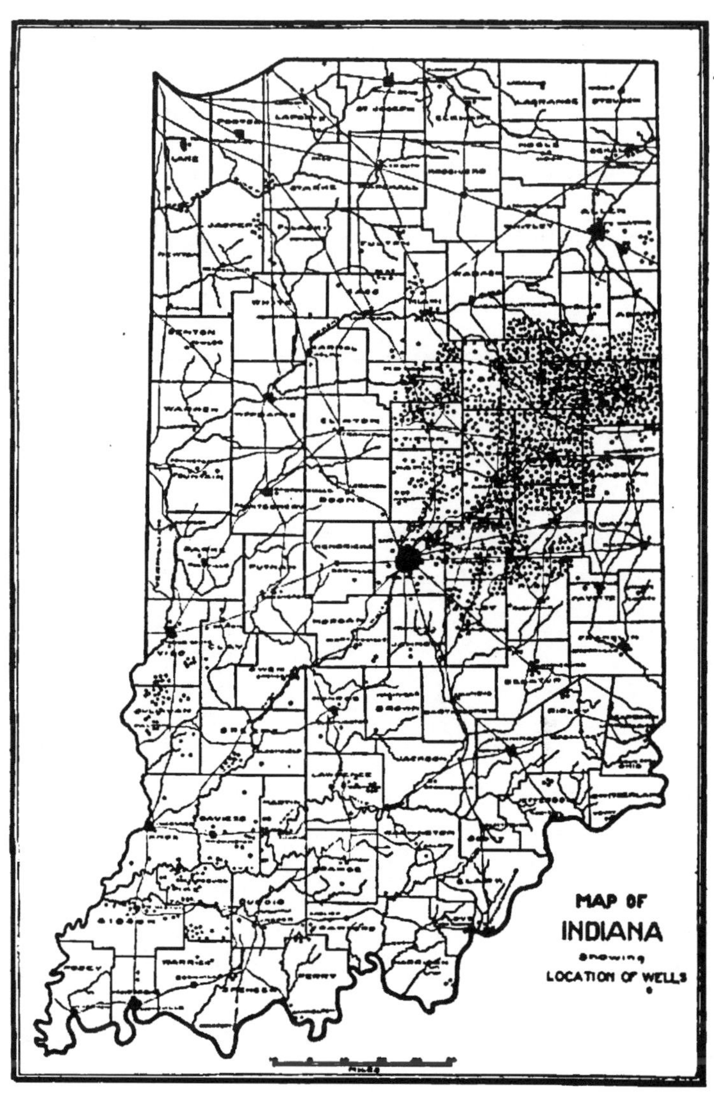

Fig. CLIII. Map showing distribution of oil, gas and dry wells drilled in Indiana. Space does not permit the location of all wells drilled in the oil and gas producing areas.

ADDITIONAL LOGS AND LOCATION OF WELLS.

Logs and Locations of Wells in Adams County.

Township	Section	Depth, Feet	Oil Sands, Feet
Wabash	28	1,057	1,001

Blackford County.

Township	Section	Depth, Feet	Oil Sands, Feet
Harrison	14	1,027½	980
Harrison	15	1,039	997
Jackson	27	1,287	925

Carroll County.

Township	Section	Depth, Feet	Oil Sands, Feet
Deer Creek	20	919	895

Delaware County.

Township	Section	Depth, Feet	Oil Sands, Feet
Delaware	....	1,206	1,197
Delaware	....	1,213	1,211
Delaware	....	1,240	1,238
Delaware	....	1,199	1,197
Delaware	28	1,208	926
Delaware	28	1,207	923
Delaware	27	1,255	971
Delaware	27	1,227	948
Delaware	11	1,196	915
Delaware	26	1,273	948
Delaware	....	1,313(dry)	972
Delaware	....	1,262	1,260
Delaware	22	1,232	946
Delaware	22	1,187	926
Delaware	22	1,216	940
Delaware	22	1,208	945
Delaware	22	1,213	935
Delaware	22	1,212	945
Delaware	22	1,206	1,200
Delaware	22	1,206	927
Delaware	22	1,207	927
Delaware	....	1,175	888
Delaware	....	1,239	930
Delaware	22	1,209	935
Hamilton	....	1,220	932 to 1,205
Hamilton	....	1,212	932 to 1,240

Delaware County—Continued.

Township	Section	Depth, Feet	Oil Sands, Feet
Hamilton	....	1,222	....
Hamilton	17	1,232	....
Hamilton	....	1,237	939
Liberty	1	1,208	941
Liberty	....	1,242	975
Niles	21	1,250	950
Union	....	1,250	947 to 1,177
?	....	1,279	1,220
....	....	1,223	1,221
?	....	1,196	1,191
?	....	1,202	1,197

Dubois County.

Township	Section	Depth, Feet	Oil Sands, Feet
Jefferson	10	1,360	1,163

Gibson County.

Township	Section	Depth, Feet	Oil Sands, Feet
Center	9	1,617	?........
Montgomery	33	2,090	?........
Patoka	32	1,705	....
Washington	1	1,617	1,584
Washington	6.	1,066	1,062
Washington	6	1,532	....
Washington	7	1,349	1,323
Washington	7	1,402	1,375
Washington	7	1,381	1,360
Washington	7	1,425	1,400
Washington	7	1,361	1,338
Washington	....	1,500	....
Washington	7	1,348	1,320
Washington	6	1,355	....
Washington	....	2,210	....
Washington	....	1,375	....
(Geo. Sturgis farm)	....	....	....

Grant County.

Township	Section	Depth, Feet	Oil Sands, Feet
25	5	965	913
25 N. E. ¼	5	....	857 top of sand
25 N. E. ¼	5	942	887½ top of sand
25 N. E. ¼	9	918	866 top of sand
25 N.	29	921	873
25	5	980	928
Jefferson	....	1,033	931
Jefferson	....	1,023	916
Jefferson	....	1,016	916
Jefferson	....	1,041	938
Jefferson	....	1,050	943
Liberty	....	....	920 to Trenton
Pleasant	9 N. ½	994	939½
Pleasant	....	955	897½
Pleasant	4	922	884
Pleasant	....	990	938
Pleasant	4	927	867
Pleasant	9	926½	926½
Pleasant	4	920	853
Pleasant	....	980	909
Pleasant	4	940	869
Pleasant	....	989	881
Pleasant	....	941	879
Pleasant	....	....	917
Pleasant	....	....	920 top sand
Pleasant	....	960	922
Pleasant	....	1,102	936
Pleasant	9	978½	927
Pleasant	9	910	866
Pleasant	9	976	902
Pleasant	9	976	902
Pleasant	9	934	872
Pleasant	9	979	907
Pleasant	9	975½	902½
Pleasant	9	988	931
Pleasant	9	988	926
Pleasant	9	982	912
Pleasant	5	942	887
Pleasant	....	965	913 to Trenton
Pleasant	....	977	920 to Trenton
Pleasant	....	922	....
Pleasant	....	986	....
Richland	....	....	916 to Trenton
Richland	....	....	873 to Trenton
Richland	....	....	939 to Trenton

Hamilton County.

Township	Section	Depth, Feet	Oil Sands, Feet
Adams	2	1,049	1,038 to Trenton
Washington	....	....	1,034to Trenton
20 N. R. 3 W	7	1,049	1,038
19 N	18	1,048	1,034

Hancock County.

17 N. R. 6 E	30	....	1,050
Ewing farm	....	1,017	1,002

Henry County.

Wayne	Knightstown.	830	820
Wayne	Knightstown. town lot...	825	817
Wayne	....	828	818

Huntington County.

Jefferson	32	1,078	1,054
Jefferson	12	1,031	1,001
Salamonie	29	1,053½	1,034
Salamonie	29	1,006	981
Salamonie	29	987	964
Salamonie	29	1,170	1,131
Salamonie	29	1,040½	1,028
Salamonie	29	1,043	1,021
Salamonie	....	1,009	....
Salamonie	29	990	968
Salamonie	29	978½	958½
Salamonie	29	1,003	975
Salamonie	29	1,001	976
Salamonie	29	975	952
Salamonie	29	992	968
Salamonie	29	982	956
Salamonie	29	1,007	975
Salamonie	29	982	960
Salamonie	29	1,060	1,044
Salamonie	29	1,072	1,054½
Salamonie	29	1,052	1,029

Huntington County—Continued.

Township	Section	Depth, Feet	Oil Sands, Feet
Salamonie	29	1,029	1,011
Salamonie	29	1,004	986
Salamonie	29	1,003	977
Salamonie	29	992	965
Salamonie	20	1,036	1,018
Salamonie	20	1,050	1,033
Salamonie	20	1,013	991
Salamonie	30	1,014	986
Salamonie	20	993	968
Salamonie	20	1,037	1,007
Salamonie	20	1,007	978½
Salamonie	20	1,049	1,031
Salamonie	20	1,045	1,030
Salamonie	23	1,061	1,036
Salamonie	28	499	481
Salamonie	20	977	956½
Salamonie	20	995	968 Tr.
Salamonie	20	982½	962 Tr.
		1,009	987
		1,011	986
		1,000	973

Jay County.

Township	Section	Depth, Feet	Oil Sands, Feet
Penn	34	1,068	927
Penn	34	1,010	932
Wayne	2	1,108	
Wayne	2	1,106	
Wayne	2	1,100	
Wayne	2	1,110	
		1,070	

Jennings County.

Township	Section	Depth, Feet	Oil Sands, Feet
6 N. R. 8 E	27 N. W. ¼	1,234	

Johnson County.

Township	Section	Depth, Feet	Oil Sands, Feet
Nineveh		1,830	1,273
Nineveh		1,164	1,129

Knox County.

Township	Section	Depth, Feet	Oil Sands, Feet
Johnson	36	2,004	....

Madison County.

Township	Section	Depth, Feet	Oil Sands, Feet
Monroe	....	1,000	923

Pike County.

Township	Section	Depth, Feet	Oil Sands, Feet
Logan	22	1,236½	995
Logan	27	1,274	1,274
Logan	27	1,575	....
Logan	27	1,292	1,269
Madison	30	1,242	....
Madison	....	1,441	....
Madison	....	1,325	....
Madison	36	1,350	1,287
Madison	36	1,296	1,270
Madison	36	1,350	1,005
Madison	30	1,312	....
Madison	....	1,344	1,002
Madison	36	1,307	1,221
Madison	....	1,314	1,293
Madison	1	1,186½	....
Madison	35	1,316	1,005½
Madison	2	1,347	1,347
Madison	35	1,024	....
Madison	35	1,331	....
Madison	35	1,348	....
Madison	35	1,348	....
Madison	35	1,322	1,316
Madison	35	1,348	1,343
Madison	35	1,346	1,342
Madison	35	1,005½	975
Madison	36	1,302	....
Madison	36	1,350	1,287
Madison	36	1,334	....
Madison	....	1,332	1,301
Madison	....	1,346	....
Madison	....	963	....
Madison	....	1,356	1,050
Washington	32	1,140	....
....	22	1,339	1,331
....	22	2,050	1,576
....	27	1,650	....

Pulaski County.

Township	Section	Depth, Feet	Oil Sands, Feet
....	....	970	905

Randolph County.

Township	Section	Depth, Feet	Oil Sands, Feet
Greene	1	1,402	956
21 N. R. 11 E	....	1,230	1,100
21 N	....	1,103	975
R. 11 E	26	1,103	958
R. 12 E	1	1,402	1,008
....	....	1,264	974

Sullivan County.

Township	Section	Depth, Feet	Oil Sands, Feet
9 N. 9 W	33	730	....
....	....	800	....
Curry	4	643	....
Curry	4	635	630
Gill	....	....	....
Gill	3	817	....
Gill	32	780	....
Hamilton	....	952	750

Tipton County.

Township	Section	Depth, Feet	Oil Sands, Feet
Cicero	30	1,016	1,002
Cicero	33	1,045	1,021
Liberty	19	1,026	1,008

Vigo County.

Township	Section	Depth, Feet	Oil Sands, Feet
Linton	15	2,150	....
Linton	16	1,290	....
Linton	22	860	....

Wabash County.

Township	Section	Depth, Feet	Oil Sands, Feet
Liberty	1	921 to Tr.	....
Liberty	26	931	879
26 N	N. W. ¼ 36	985½	921½
26	....	....	870

Wells County.

Township	Section	Depth, Feet	Oil Sands, Feet
25	9 S W. ¼	1,027	878

County	Location: Township, Range and Section	Depth to Pay Sand, Feet	Total Depth, Feet
Clay	Twp. 9 N. R. 7 W., Sec. 16	311	311
Clay	Twp. 9 N. R. 7 W., Sec. 16	209	259
Clay	Twp. 9 N. R. 7 W., Sec. 15		308
Clay	Twp. 9 N. R. 7 W., Sec. 16	1,847	1,900
Grant	Twp. 25 N. R. 7 E., Sec. 9	833½	935
Grant	Twp. 25 N. R. 7 E., Sec. 9 W	862	915
Grant	Twp. 25 N. R. 7 E., Sec. 9 W	915	967½
Grant	Twp. 25 N. R. 7 E., Sec. 9 W	933	976
Grant	Twp. 25 N. R. 7 E., Sec. 9 W	927	975
Grant	Twp. 25 N. R. 7 E., Sec. 9 W	861	910
Grant	Twp. 25 N. R. 7 E., Sec. 9 W	927	972
Grant	Twp. 25 N. R. 7 E., Sec. 9 W	935	976
Grant	Twp. 25 N. R. 7 E., Sec. 9 W	931	975
Grant	Twp. 25 N. R. 7 E., Sec. 9 W	858	905
Grant	Twp. 25 N. R. 7 E., Sec. 9 W	851	902
Grant	Twp. 25 N. R. 7 E., Sec. 9 W	935	988
Grant	Twp. 25 N. R. 7 E., Sec. 9 W	863	919
Grant	Twp. 25 N. R. 7 E., Sec. 9 W	932	983
Grant	Twp. 25 N. R. 7 E., Sec. 9 W	890	941
Grant	Twp. 25 N. R. 6 E., Sec. 1 N. W.	939	993
Grant	Twp. 25 N. R. 7 E., Sec. 15	891	947½
Grant	Twp. 25 N. R. 7 E., Sec. 4	907	955
Grant	Twp. 25 N. R. 7 E., Sec. 4	884	937
Grant	Twp. 25 N. R. 7 E., Sec. 4	915	968
Grant	Twp. 25 N R. 7 E., Sec. 3	906	948½
Grant	Twp. 25 N. R. 7 E., Sec. 3	872	874
Grant	Twp. Pleasant, R. 7 E., Sec. 9	940	986
Grant	Twp. Richland, R. 6 E., Sec. 2	948	1,003
Gibson	Twp. Jefferson, Sec. 7		1,671
Gibson	Twp. Patoka	875	875
Gibson	Twp. Patoka		877
Gibson	Twp. Washington, Sec. 6	806	806
Gibson	Twp. Washington, Sec. 6	755	755
Gibson	Twp. Washington, Sec. 6		798
Gibson	Twp. Washington, Sec.	744	803
Gibson	Twp. Washington, Sec.	720	823
Gibson	Twp. Washington, Sec.	725	810

County	Location: Township, Range and Section	Depth to Pay Sand, Feet	Total Depth, Feet
Gibson	Twp. Washington, Sec. 6	1,350	1,609
Gibson	Twp. Washington, Sec. 20	....	1,301
Gibson	Twp. Washington, Sec. 20	1,300	1,592
Gibson	Twp. Washington, Sec. 20	1,296	1,573
Gibson	Twp. Washington, Sec. 20	1,300	1,574
Gibson	Twp. Washington, Sec. 20	1,295	1,312
Gibson	Twp. Washington, Sec. 20	1,286	1,300
Gibson	Twp. Washington, Sec. 20	....	1,798
Gibson	Twp. Washington, Sec. 28	....	1,650
Gibson	Twp. White River, Sec. 36	....	1,282
Gibson	Twp. Stockton, Sec. 7	....	1,610
Gibson	Twp. 1, South, Sec. 31	1,671	1,750
Hamilton	Twp. 19, Sec. 7	1,035 Tr.	1,044
Hamilton	Twp. 19 N., Sec.	1,041	1,066
Hamilton	Twp. Washington, Sec. 7	1,034	1,048
Hamilton	Twp. Washington, Sec. 7	1,038	1,049
Hancock	Twp. Buck Creek, Sec. 12	309	1,328
Jackson	Sec. 30	1,350 Tr.	1,752
Jay	Twp. Bear Creek, Sec. 18	....	1,027
Jay	Twp. Bear Creek, Sec. 18	....	1,034
Jay	Twp. Bear Creek, Sec. 18	....	1,035
Jay	Twp. Bear Creek, Sec. 18	....	1,037
Jay	Twp. Bear Creek, Sec. 18	....	1,078
Jay	Twp. Bear Creek, Sec. 18	....	1,047
Pike	Twp. Clay, Sec. 5	722	812
Pike	Twp. Clay, Sec. 5	....	812
Pike	Twp. Madison, Sec. 6	1,110	1,110
Pike	Twp. Madison, Sec. 6	923	923
Pike	Twp. Madison, Sec. 25	940	1,290
Pike	Twp. Madison, Sec. 6	909	911
Pike	Twp. Madison, Sec. 5	1,121	1,319
Pike	Twp. Logan, Sec. 22	....	996
Pike	Twp. Logan, Sec. 22	982	982
Pike	Twp. Logan, Sec. 22	964	987
Pike	Twp. Logan, Sec. 22	987	1,006
Pike	Twp. Logan, Sec. 22	974	1,008
Pike	Twp. Logan, Sec. 22	999	1,021
Pike	Twp. Logan, Sec. 22	990	1,006
Pike	Twp. Logan, Sec. 22	984	1,011
Pike	Twp. Logan, Sec. 22	....	1,024
Pike	Twp. Logan, Sec. 22	1,005	1,023
Pike	Twp. Logan, Sec. 22	1,026	1,026
Pike	Twp. Logan, Sec. 22	1,020½	1,022

surface is covered. Thus gradually the basin becomes filled and the pond or lake has become extinct and a peat or quaking bog has taken its place.

CHAPTER XI.

PEAT AND PYRITE.

Peat.

Peat is an organic substance which forms the chief source of fuel for some countries. In appearance it varies from a light brown spongy mass in which vegetable fibers are clearly seen to a dark carbonized mass in which are marks of occasional forms of vegetation. In the presence of water it is generally soft, somewhat gelatinous and easily molded, on drying it becomes hard and develops shrinkage cracks. Under the microscope peat is seen to be composed largely of vegetable matter with small quantities of inorganic substances such as clay and sand.

Distribution of Peat in Indiana.[*] The peat deposits of Indiana are closely associated with the Wisconsin drift sheet which extended to the central part of the state. However, since the greatest lake development in Indiana is found in the three northern tiers of counties, it is there that largest deposits of peat are found. The workable peat beds cover approximately 36,000 acres and contain nearly three billions cubic feet. The distribution is shown on the accompanying map.

Origin. Peat originates from the accumulation of vegetable matter in lakes, ponds and marshes. The basins in which the vegetation accumulates were formed in part as "kettle hole" or morainic basins, in part as depressions among sand dunes and in part as meander basins caused by the shifting of stream courses. The largest deposits are connected with the glacial lake basins. There are two varieties of peat, distinguished from one another by the character of vegetation entering into their composition. The moss peat is made largely of mosses of which the species, Spagnum Cymbifolium forms the largest part. This form of peat has the highest fuel value because its accumulation is in localities freer from terrestrial contamination. The other variety is formed from the accumulation of various forms of vegetation, such as rushes, reeds, and sedges. These accumulate in shallower waters, where land derived materials are more likely to be deposited.

The essential conditions for the accumulation of peat are an abundant growth of vegetation and the presence of sufficient water to prevent oxidization and the destructive action of bacteria. As long as peat deposits remain below the level of ground water they are preserved, but as soon as they are brought above the water table decomposition takes place and it rapidly loses its fuel value.

In a peat deposit which is being formed of spagnum moss three zones are recognized. At the top is a layer of greenish colored moss and below a layer of light brown spongy dead moss. Below the latter is a layer of dark brown more highly compacted and carbonized. This layer is the one having the highest fuel value but is not as valuable for some other purposes as the middle layer.

The growth of vegetation which forms a peat deposit starts in the shallow water around the margin of the basin and as the old vegetation sinks to the bottom new vegetation takes its place. As the filling continues the mosses reach farther and farther out on the surface of the water until the entire surface is covered. Thus gradually the basin becomes filled and the pond or lake has become extinct and a peat or quaking bog has taken its place.

Composition. The composition of peat compares in a general way with coal, that is, it contains the same elements but less carbon and hydrogen, and more oxygen and ash. As a rule it contains less sulphur and has a lower fuel value. Air dried peat has a higher fuel value than air-dried wood and kiln-dried peats than kiln-dried wood.

The analyses of the following five samples of peat were made by Dr. R. E. Lyons and published in the thirty-first Annual Report of the Indiana Geological Survey:

Analyses of Samples of Indiana Peat.

	No. 1	No. 2	No. 3	No. 4	No. 5
Moisture (105°C)	17.16	12.24	11.40	8.99	10.20
Volatile (air dried)	73.31	70.21	65.52	70.97	66.43
Fixed carbon (air dried)	22.53	23.45	20.65	19.08	24.30
Coke (air dried)	26.67	29.78	34.47	29.09	37.55
Ash (air dried)	4.14	6.33	13.82	10.01	13.25
Nitrogen (air dried)	2.56	2.22	3.31	3.91	2.96
Sulphur (oven dried)	.74	.87	1.33	.83	.96
Phosphoric acid in ash	1.90	1.51	1.17	1.26	1.53
Potash in ash	1.56	1.35	.96	.96	.82

No. 1. Dekalb, Sec. 9, 33 N. 12 E. *No. 2.* St. Joseph, Secs. 28, 33 and 34, 36 N. 2 E. *No. 3.* St. Joseph, Sec. 3, 36 N. 1 E. *No. 4.* Marshall, Secs. 10 and 11, 33 N. 1 E. *No. 5.* Starke, Sec. 10, 33 N. 3 E.

Uses. Peat may be used in many ways, namely, as fuel, as bedding, as packing, as an absorbent, deodorizing, filtering, as a filler for fertilizer, for making ethyl alcohol, in the manufacture of paper and cloth, and artificial boards, and for medicinal baths.

Fuel. Because of the past abundance of wood and the present abundance of coal in Indiana the utilization of peat for fuel has and is receiving little attention. With the complete exhaustion of our timber and a greatly restricted coal supply will come a greater demand for peat as fuel. Peat, as it exists in the earth, contains a large amount of water. Before it can be used for fuel this water must be removed. A part may be evaporated in the open air, but oven drying is essential to the use of peat on a large scale. Peat may be used for fuel in the form in which it is taken from the pit or it may be molded into brickettes while still moist and the brickettes dried either in the open air or in kilns. The peat may be dried and molded into brickettes either with or without the use of a binder. Perhaps the most economical and modern method of using peat for fuel would be to reduce it to the powdered form and feed it to the firebox with air under pressure.

Peat coke is manufactured successfully in European countries in retort ovens. The coke has a higher heating power than either high grade bituminous coal or anthracite, since temperatures of 4,700° F. have been obtained by use of the former and only 4,000° F. and 4,500° F. with the latter. By this method of coking, valuable by-products, acetic acid, alcohols, ethyl and methyl, ammonia, analine colors, benzol, naphtha, lubricating oils and other substances may be obtained.

Peat may be used successfully in the manufacture of producer gas. In large industrial plants this would probably be the most economical way of using it for fuel.

Fertilizer. As will be seen from the analyses given on a previous page peat contains three elements, nitrogen, potassium and phosphorus, which are used by plants and it is therefore a fertilizer. However, it is used on account of its light weight, as a filler for commercial fertilizer carrying salts of nitrogen and phosphorus. Some of the fertilizer companies in Indiana use it for that purpose. Since it is necessary to use some form of filler to prevent too strong an application of the mineral salts at one place, a better one than peat could not be selected.

Bedding. The use of peat as a litter in barns for stock is desirable because of its absorptive power for odors and liquids and because of the fact that when the bedding is later spread upon the land, as it should be, for fertilizing purposes, additional fertilizing elements have been gained from the peat.

Packing. Peat makes excellent packing material and can be used for all purposes for which excelsior is used. There is nothing superior to peat for packing around the roots or stems of live plants because of its absorptive powers.

Surgical Dressings. Mull, which is the finer matter separated from moss peat by screening, is used for deodorizing and disinfecting purposes. During the period of the World War, when it became difficult to obtain absorbent cotton, mull was substituted in the preparation of surgical dressings, for which it was thought to be in some respects, superior to cotton. Its power of absorption is as high as ten times its own weight. The water from peat has strong antiseptic qualities and has been used for wound dressing and in the preparation of medicinal baths.

Paper and Lumber. The use of peat in the manufacture of certain grades of paper and cardboard has been successful. But its use for this purpose has not been extensive on account of the cheapness of other materials which are used for that purpose. The present high price of paper of all grades and the consequent advance in the price of raw materials should open the market for peat. The high price of lumber should also make possible the successful competition of peat lumber with the cheaper grades of the natural product.

Alcohol. Experiments have demonstrated that alcohol can be manufactured from peat. It would be reasonable to suppose that peat could easily compete with other substances used in the manufacture of alcohol in so far as the price of the raw material is a factor. The quantity of alcohol obtained from peat varies with the composition of the peat but is ordinarily from eight to ten gallons per ton.

Miscellaneous. Peat may be used as a decolorizer, as a filter, as a non-conductor of heat, in the manufacture of charcoal for gunpowder, in the manufacture of dyes and in the manufacture of fireworks.

Development. Little attempt has been made to develop the peat beds of Indiana. In widely scattered localities it has been used to a very limited extent as fuel or as a fertilizer. Fertilizer companies have used it in small quantities. It has also been used by one company in the manufacture of mull. This deposit is located about one mile south of Garrett in the northeast quarter of Section 9, in Butler Township, Dekalb County. It was worked for many years by Baker & Company. The deposit covers about sixty acres, and has a maximum thickness of about forty feet, with an average of much less. The overburden is from three to eighteen inches and the peat is of the spagnum variety, a good quality of medium to dark chocolate brown color.

Peat was used for fuel for several years by Mr. C. F. Brown of Tyner City, Marshall County. The peat was air dried then stacked in a shed. Peat was also used for fuel at one time from the deposits in the Kankakee Valley, in the vicinity of South Bend in St. Joseph County. It has been used locally in a number of counties for various purposes and has been used by some fertilizer factories.

Analyses of Peat.

COUNTY, TOWNSHIP, RANGE AND SECTION	Moisture, 105° C.	Volatile, Air Dried	Fixed Carbon Air Dried	Coke, Air Dried	Ash, Air Dried	Nitrogen, Air Dried	Sulphur, Oven Dried	Per Cent. of P_2O_5 in Ash	Per Cent. of K_2O in Ash
Dekalb, Sec. 9 (33 N. 12 E.)	17.16	73.31	22.53	26.67	4.14	2.56	0.74	1.90	1.56
Marshall, Secs. 10, 11 (33 N. 1 E.)	8.99	70.97	19.08	29.09	10.01	3.91	0.83	1.26	1.53
Starke, Sec. 10 (33 N. 3 E.)	10.20	62.43	24.30	37.55	13.25	2.96	0.96	0.96	0.82
St. Joseph, Secs. 28, 33, 34 (36 N. 2 E.)	12.24	70.21	23.45	29.78	6.33	2.22	0.87	1.51	1.35
St. Joseph, Sec. 3 (36 N. 1 E.)	11.40	65.52	20.65	34.47	13.82	3.31	1.33	1.17	0.96

Peat Analysis.

County, Township, Range and Section	B. T. U. Oven Dried 105°C.	Calories, Oven Dried 105°C.	Evaporative Effect (Oven Dried)
Dekalb, Sec. 9 (33 N. 12 E.)	10,232.77	5,684.8	10.6
Elkhart, Sec. 4 (36 N. 5 E.)	8,637.89	4,799.4	8.9
Elkhart, Secs. 10-11 (36 N. 6 E.)	7,211.22	4,006.0	7.4
Elkhart, Secs. 26-27 (35 N. 5 E.)	7,613.06	4,229.4	7.8
Elkhart, Sec. 18 (38 N. 6 E.)	9,628.78	5,349.3	9.9
Jasper, Secs. 12-13-14 (30 N. 6 W.)	8,273.44	4,596.3	8.0
Kosciusko, Secs. 11-12-13 (31 N. 6 E.)	9,715.68	5,397.6	10.0
Kosicusko, Secs. 31-33 (33 N. 6 E.)	6,129.32	3,405.1	6.3
Lake, Secs. 34-35-36 (35 N. 9 W.)	8,731.34	4,850.7	9.0
Lagrange, Secs. 2-11-12 (36 N. 8 E.)	8,513.29	4,729.5	8.8
Lagrange, Secs. 4-9 (37 N. 9 E.)	8,924.47	4,958.0	9.2
Marshall, Secs. 10-11 (33 N. 1 E.)	9,946.19	5,525.6	10.3
Marshall, Sec. 1 (34 N. 2 E.)	8,497.72	4,720.9	8.8
Marshall, Sec. 10 (34 N. 1 E.)	10,466.40	5,814.6	10.8
Newton, Secs. 32-33 (31 N. 8 W.)	9,033.50	5,018.9	9.3
Noble, Secs. 28-29 (33 N. 9 E.)	10,335.57	5,741.9	10.7
Noble, Sec. 18 (33 N. 11 E.)	9,217.28	5,120.7	9.5
Pulaski, Sec. 9 (31 N. 1 W.)	9,774.87	5,430 4	10.2
Pulaski, Secs. 7-8-9 (31 N. 3 W.)	9,064.65	5,035.9	9.3
Pulaski, Secs. 3-9-10-11 (31 N. 4 W.)	8,472.80	4,707.1	8.7
Porter, Secs. 1-2-3 (37 N. 5 W.)	5,635.03	3,130.5	5.8
Starke, Sec. 10 (32 N. 3 E.)	9,905.70	5,503.1	10.2
Steuben, Sec. 34 (37 N. 12 E.)	9,422.87	5,234.3	9.7
St. Joseph, Secs. 28-33-34 (36 N. 2 E.)	9,840.28	5,466.8	10.1
St. Joseph, Sec. 3 (36 N. 1 E.)	9,024.15	5,013.4	9.3
St. Joseph, Secs. 11-12 (37 N. 1 E.)	8,503.95	4,724.4	8.8
St. Joseph, Sec. 16 (37 N. 2 E.)	8,236.06	4,584.4	8.5
St. Joseph, Sec. 20 (37 N. 2 E.)	8,491.40	4,717.5	8.8
Whitley, Sec. 30 (31 N. 10 E.)	4,541.67	2,523.1	4.7

Pyrite.

Pyrite (FeS_2) is a mineral compound of sulphur and iron which is widely disseminated through the rocks of the earth's exterior. When chemically pure it consists of 46.6 per cent of iron and 53.4 per cent of sulphur. In its crystal habit it belongs to the isometric system, forming cubes or octohedrons, etc. Its color is brassy, silvery or bronze and its luster is metallic. Marcasite is a mineral having the same composition but lighter color and crystallizes in the orthorhombic system forming tabular crystals. The marcasite decomposes more rapidly under the influence of oxygen.

Oxidation. In the presence of moisture the oxygen of the air attacks the pyrite, producing a number of compounds. Ferrous sulphate is the most common mineral formed by the oxidization of pyrite. It commonly occurs as a white powder coating the oxidizing mineral. Often there is a little free sulphur formed by the reduction of the ferrous sulphate, or hydrogen sulphide and sulphuric acid may result. The reduction of the ferrous sulphate may result in the formation of hematite or, when moisture is present, of limonite.

Composition. Indiana pyrite contains sulphur varying in amount from forty to forty-seven per cent. In the coals of Indiana, sulphur ranges from .89 in Coal IV to 5.14 per cent in Coal III as mined. This does not include the amount discarded in mining operations. In some places the latter would bring the sulphur content up to ten per cent of the coal mined.

Mode of Occurrence. The pyrite of Indiana occurs in shales, in clays, in limestones, in sandstones and in coals. As a rule commercial quantities occur in coals and in shales or clays associated with coal beds, but occasionally in limestones and shales not so associated. Pyrite occurs in limestones in grains or crystals and in lenses; in shales in balls, crystals, lenses and irregular masses; in clay as nodules; in coal and in the shale associated with coal in bands, the purest pyrite generally occurs in this way. In lenses, which are generally impure, due to the presence of foreign matter. In balls which vary in diameter from a few inches to several feet, frequently perfect spheres weighing from one to four thousand pounds. The interior of these contain a high per cent of calcium carbonate and the pyrite is confined to the outer few inches. While the per cent of sulphur is low in the larger ones it may range as high as forty per cent in the small ones. Pyrite may occur in coal in joint veins varying from the fraction of an inch to three inches in thickness. It may occur also as petrifactions forming the casts of limbs and trunks of trees and as irregular masses of various shapes and sizes.

In the sheety black shale, which occurs in the roof of Coal V, there are large spherical boulders of pyritiferous material, weighing from a few to several thousand pounds. The pyrite is usually confined to a few inches of the outer zone of the ball. The recoverable pyrite is found in bands and joint fillings in the coal.

Distribution. Pyrite occurs in small quantities in nearly every geological formation in the state. It is sent in to the office from all parts of the state, usually under the impression that it is a mineral of great value. Though very widely distributed in the state, rarely does it occur in commercial quantities."

Devonian Pyrite. A band of pyrite often occurs at the line of contact between the New Albany shale and the Devonian limestones which lie below;

whether it assumes proportions of commercial importance or not has not been determined.

Mississippian Pyrite. Pyrite occurs in grains and nodular masses in the Knobstone shales, but it has not been found in commercial quantities. It also occurs in Chester shales and limestones. The Reelsville limestone, which forms a persistent layer of from four to ten feet in thickness, contains in all places enough pyrite so that when it is weathered the color of the stone is changed to a rusty iron-yellow color. There is no reason to doubt that in places it contains pyrite of commercial value. The line of outcrop of this limestone extends from Reelsville, in Putnam County, in a generally southward direction to the Ohio River.

Pennsylvania Pyrite. Both the Pottsville and the Allegheny divisions of the Pennsylvanian in Indiana contain pyrite. There are lenses and nodular masses of pyrite associated with the Pottsville shales in Monroe, Martin, Putnam, Greene and other counties.

Among the Allegheny coals, No. III contains the largest amount of pyrite, the recoverable percentage running as high as six per cent of the tonnage mined, which does not include the larger masses of pyrite which are discarded in the mining process, which may bring the total several per cent higher. Samples of pyrite collected from this coal range in per cent of sulphur from 42.27 to 46.07. Coal No. III is mined in the vicinity of Burnett, Clinton, Jasonville, Midland, Staunton, Seelyville and other points.

Coal No. VI, also contains pyrite of good quality. The pyrite occurs in the coal and also in the shale which lies above the coal. Samples of the pyrite contain 46.05 per cent of sulphur. This bed of coal is mined most extensively in Sullivan County. Representative mines occur near Cass and Shelburn.

Coal No. V contains pyrite in the form of bands, balls and lenses. It extends the entire length of the Indiana coal field and is mined from north to south in Vigo, Vermillion, Sullivan, Knox, Daviess, Pike, Gibson, Warrick and Vanderburgh counties. About sixty per cent of the coal mined in Indiana comes from this bed. The pyritiferous layers in the coal may run as high as three per cent of the coal. These layers are usually discarded in the mine.

Coal beds IV, VII and VIII contain small quantities of pyrite but no recoverable amounts. No. IV is mined extensively but contains little pyrite, the other two are not mined extensively.

Production. Only a small amount of pyrite has been produced in Indiana. This small production has come as a by-product in the mining of coal. The total production amounts to a few hundred tons, which have been used in producing sulphuric acid for the treatment of fertilizers. The greater part of the recovery has been from the coal as it passed through the tipple and from gob or culm piles. Factors which have discouraged production in the Indiana coal field are: The low price paid for the pyrite because of its impurities; the lack of equipment at the mine for washing and hauling the pyrite. The only form of cleaning which has been used is hand work, which is too expensive to be profitable at the usual prices.

With the installation of machinery for washing and handling the pyrite, and fairly stable market conditions to insure steady production there is reason to believe that pyrite might become a valuable by-product in the mining of coal in Indiana. It is estimated that 200,000 or 250,000 tons might be recovered each year.

Uses. The pyrite recovered from coal is used in the manufacture of sulphuric acid. The sulphuric acid obtained from pyrite contained in coal is used largely in the manufacture of fertilizers. The total amount of sulphuric acid used annually in the United States for this purpose is 2,400,000 tons, which is nearly three times as much as that used for all other purposes.

CHAPTER XII.

ROAD MATERIALS.

Rocks occurring in Indiana which may be used for road materials belong to the three great divisions of rocks, namely, sedimentary, igneous and metamorphic. The last two divisions are not as important as the first in Indiana, occurring only in the glacial drift.

Igneous Rocks. No outcrops of igneous rock occur in the state, and bed rock of this type is deeply buried under sedimentary rocks. The only rocks of the igneous type found in the state are in the glacial debris of the drift-covered portion of the state. The igneous rock is in the form of boulders, which have been transported from the Crystalline belt of rocks lying principally north of our national boundary. Several kinds of igneous rock are represented but they are generally classed under the general term "granite," by road builders. These rocks, when unweathered, as they often occur in the later drift, are very serviceable for pavement blocks. They also are desirable in bottom courses of macadam roads, but lack the necessary cementing qualities for top courses. The irregularity and often attenuated condition of their distribution prevents their widespread use for road materials.

Metamorphic Rocks. Metamorphic rocks are derived from both igneous and sedimentary rocks. Boulders of gneiss and shist, which are derived from igneous rocks, are found in the glacial drift in Indiana. Gneiss is generally classed with granite as a road building material. Shist is not of sufficient strength to be valuable road material. Slate and marble both occur in the drift, but not in sufficient quantities to be of importance. Quartzite is one of the most abundant and the most important metamorphic rocks for road material in the glacial drift. They may be used for pavement blocks and for lower courses, but lack of cementing qualities prevents their use for upper courses. The statements made with reference to the distribution and use of igneous rocks apply also to metamorphic rocks in Indiana.

Sedimentary Rocks. These furnish the main source of the road building materials in Indiana. These materials consist of sands, gravels, shales, limestones, sandstones, and conglomerates. The geological formations which yield road materials include all the divisions represented in the state from the Ordovician to the Recent period.

Glacial Gravels and Sand.

The glacial drift covers the bed rock to such a depth in the large part of northern Indiana that the bed rock is not exposed even by the larger streams. In this region the only road material available consists of boulders, sands and gravels from the glacial drift. The coarser materials of the drift are

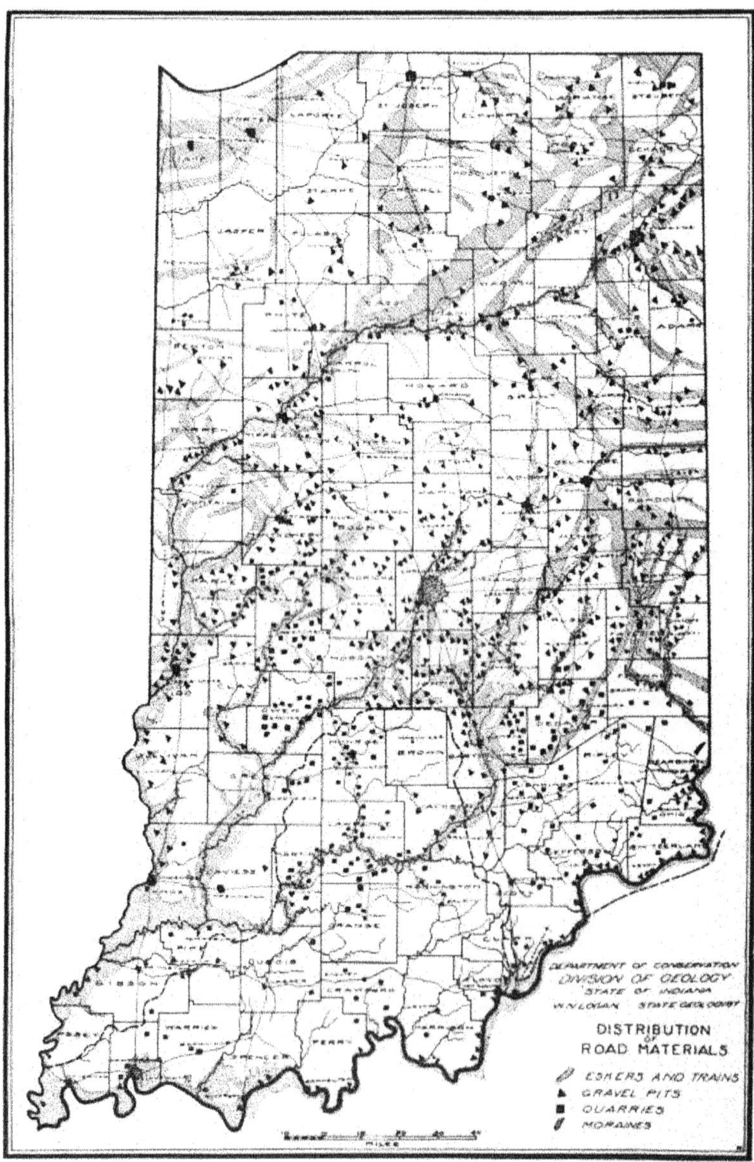

Plate CLVI. Map showing distribution of road materials in Indiana.

concentrated in the glacial deposits known as moraines, drumlins, and eskers. Good examples of morainic hills are found in the vicinity of Valparaiso, South Bend, Bristol, Otterbein, Fowler and other places. Eskers occur in southern Tippecanoe, in Montgomery and in Madison counties. The best known esker is the Anderson esker, which lies partly in the city of Anderson and extends northward to Grant County. The distribution of the glacial gravels is indicated on the accompanying map.

The glacial gravels have been used in many parts of the glaciated area in the building of gravel-surfaced roads. Clean glacial gravels are not suitable for top courses as they lack cementing properties. In some places they contain sufficient clay for bonding purposes or, perhaps, oxide of iron. If pebbles of limestone are present, and these are placed in the top courses, they will be pulverized by the traffic and may form the proper bond for the gravel. The boulders and coarser gravels may be used for the lower courses of macadam roads, for which no more durable material can be obtained.

In the building of concrete roads or of brick-surfaced roads, with a concrete base, the glacial gravels are very valuable. When used for this purpose the gravel should be free from clay and other materials, except clean sand.

Glacial sands may be used in the building of sand-clay roads. If the sand does not contain enough clay or oxide of iron for bonding the clay must be added. Sand-clay roads, if well cared for, are very satisfactory.

Fluviatile Sands and Gravels. During the period of glaciation in Indiana large quantities of sand and gravel were carried from the glaciated region by the waters of the melting glaciers and concentrated in stream valleys, many of which were almost completely filled. Such fluviatile deposits lie not only within the glaciated area, but extend beyond it into the non-glaciated portion of the state. The sands and gravels of such deposits form useful materials in the construction of roads. Fluviatile deposits occur along the valleys of the Wabash River, the White River and other streams passing through the glaciated area.

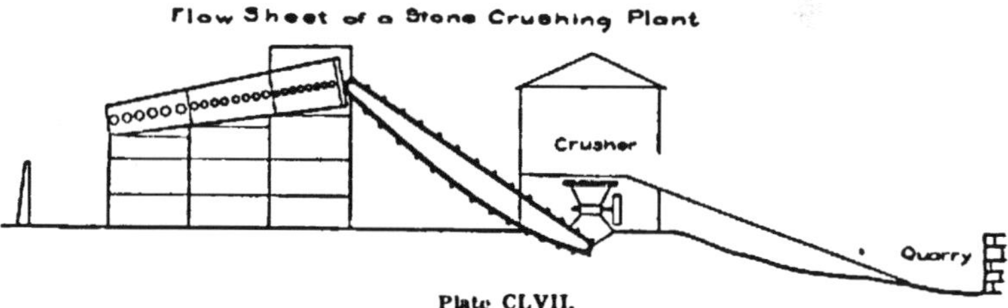

Plate CLVII.

Residual Gravels of the Non-glaciated Area. A portion of the State of Indiana lying south of the northern boundary of Monroe County was not covered with glacial ice and its topography unaffected by glacial erosion. The weathering of the durolith of this area produced gravels of the more resistant portions of the rocks. The Knobstone shales and sandstones yield gravels produced from sandstone, silicious and ferruginous concretions. They have been used for the building of roads in the area of outcrop of that formation.

The Harrodsburg limestone contains silicious geodes which, in some places, contribute beds of gravel which may be used in road construction, but their use for this purpose is extremely limited.

The Mitchell limestone contains chert which adds to the residual gravels of its area of outcrop. The Chester shales contribute concretions, the limestones, chert and the sandstones, concretions.

The Mansfield sandstone yields beds of gravel. Added to these are some beds of gravel of probable Pliocene age.

Durolith Road Materials.

The durolith or bed rock of Indiana is used principally in road construction, for the manufacture of crushed stone, for macadam and concrete.

The Rocks of the Ordovician, Silurian, Devonian, Mississippian and Pennsylvanian.

Ordovician Rocks. The Cincinnatian division of the Ordovician is represented in southeastern Indiana by the outcrop of a series of shales and limestones. The limestones are thin bedded, varying in thickness from two to ten inches. They are hard and of a bluish-gray color. These rocks are exposed along the stream courses where the glacial drift has been removed in Decatur, Dearborn, Ohio, Switzerland and the eastern part of Jefferson, Fayette, Franklin, Ripley, Union, and Wayne counties. These limestones have been used in the manufacture of crushed stone for macadam in several of the counties mentioned.

The chemical composition of samples of the Ordovician limestones is recorded in the following table:[m]

Composition of Ordovician Limestones.

Constituent.	No. 1	No. 2	No. 3	No. 4	No. 5
Alumina (Al_2O_3)	.80	.76	.73	.33	.35
Iron oxide (Fe_2O_3)	.51	.25	.25	.25	.51
Lime (CaO)	52.65	52.20	53.15	54.50	55.00
Magnesia (MgO)	.50	.50	.63	tr.	tr.
Phosphoric acid	.10	......	.38	.62	.54
Insoluble in HCl	5.03	5.45	2.92	2.28	2.15
Loss on ignition	40.69	40.36	41.90	42.14	41.33
Total	100.28	99.52	99.96	100.12	99.88

1. John Wiesahan, Weisburg, Dearborn County.
2. Thomas Croxton, Dillsboro, Dearborn County.
3. Henry Shroeder, Randolph Township, Ohio County.
4. Samuel Locke, Vevay, Switzerland County.

Analyses by Bureau of Public Roads, Washington, D. C. Thirtieth Annual Report Indiana Geological Survey.

The physical analyses of samples of these limestones, as made by the Bureau, are given below. No. 6 is the average of 192 samples of limestone taken as a standard for comparison. No. 5 is from the land of Albert Stein, Richmond, Wayne County. The other samples are as indicated above.

Physical Analyses of Ordovician Limestones.

	No. 1	No. 2	No. 3	No. 4	No. 5	No. 6
Specific gravity	2.7	2.7	2.7	2.7	2.7	2.7
Weight per Cu. Ft. in lbs.	168.4	168.4	168.4	168.4	168.4	165.7
Water absorb. Cu. Ft. in lbs.	.62	.78	.46	.57	.89	1.39
Per cent. of wear	4.68	4.8	5.5	5.4	4.5	5.70
French coeff. of wear	8.3	8.3	7.3	7.5	8.9	8.7
Hardness	2.5	9.8	4.7	4	9.6	3.4
Toughness	8	7	9	8	6	11.0
Cementing Val. dry	44.0	67	15	25	15	32.0
Cementing Val. wet	59	100	98	46	28	59

No. 1. "Below the average in hardness and toughness for limestone, and about the average in resistance to wear, with good cementing value. Excellent for highway and country road traffic."

No. 2. "A rather hard limestone, somewhat low in toughness; average resistance to wear and excellent cementing value. Should give excellent results under highway and country road traffic."

No. 3. "A soft limestone of average toughness, with a low resistance to wear, but with a good cementing value. Suitable for light traffic or as a binder in connection with harder material.

No. 4. "Somewhat above the average in hardness for limestone, but below in toughness and resistance to wear; cementing value good. Best suited for light highway and country road traffic."

Silurian Limestones. The Silurian limestones of Indiana vary much in physical and chemical composition. The color ranges from nearly white to blue. In some places it is composed of massive hard layers, in other places it is shaly and very soft. As may be seen by consulting the areal geological map accompanying this report, the Silurian limestones are more widespread than the Ordovician, and occupy a large area in the northern and eastern part of Indiana. In the northern part of its area it is deeply buried under glacial drift, except for outcrops along some of the stream courses. Such outcrops occur at Bluffton, Delphi, Huntington, Kentland, Logansport, Monon, Portland, Wabash, Warren and other places. In the southern portion of the area in Decatur, Ripley, Jennings, Jefferson and Clark counties, where the glacial drift is thin, outcrops are more aboundant. The Silurian limestones have been used in many places in the manufacture of crushed stone for macadam and concrete. Its chemical and physical properties are revealed in the following tables, which include the chemical analyses and the physical tests of a large number of samples."

Composition of Silurian Limestones.

	No. 1	No. 2	No. 3	No. 4	No. 5	No. 6	No. 7
Alumina (Al_2O_3)	.63	1.14	.60	1.60	.75	.25	1.10
Iron oxide (Fe_2O_3)	.51	tr.	tr.	.25	......	.24	......
Lime (CaO)	49.55	52.00	49.75	44.85	48.65	53.10	41.10
Magnesia (MgO)	3.46	2.30	2.98	7.24	3.49	.61	3.48
Phosphoric acid (P_2O_5)	.26	.16	......	......	.10	......	......
Insol. in HCl	6.24	1.77	4.27	2.95	5.13	4.00	18.65
Loss on ignition	39.28	42.40	42.17	43.21	41.64	41.65	36.00
Total	99.93	99.77	99.77	100.10	99.76	99.85	100.33

No. 1. George Hotchkiss, Soapville, Switzerland County.
No. 2. Richard Johnson, North Madison, Jefferson County.
No. 3. Greensburg Limestone Company, Decatur County.
No. 4. Big Four Stone Company, New Point, Decatur County.
No. 5. Goff Quarry, Kentland, Newton County.
No. 6. Casparis Stone Company, Logansport, Cass County.
No. 7. Wabash Stone Company, Wabash, Wabash County.

The results of the physical tests on these and other samples, as made by the United States Bureau of Public Roads, are given in the following table, which is taken from the Thirtieth Annual Report of the Indiana Geological Survey.[26]

ORIGIN OF SAMPLE	Specific Gravity	Wt. per Cu. Ft., lbs.	Water Absorbed, per Cu. Ft., lbs.	Per Cent. of Wear	French Coefficient of Wear	Hardness	Toughness	Cement Value, Dry	Cement Value, Wet
Goff Stone Co., Kentland, Newton County	2.7	168.4	.33	4.1	9.7	12	12	19	33
Caparis Stone Co., Logansport, Cass County	2.65	165	1.31	......	......	......	......	11	47
Edw. Heby, Monon, White County	2.70	168	1.04	......	......	7.5	9	29	43
Delphi Stone Co., Delphi, Carroll County	2.77	171.5	.75	5.09	7.86	10	14	14	42
F. Showalter, Washington Twp., Carroll County	2.73	171.5	.50	8.73	4.58	12	21	7	45
Chaffin & Co., Kokomo, Howard County	2.44	152.8	6.50	4.13	9.69	5.3	17	28	42
J. M. Leach & Co., Kokomo, Howard County	2.45	153	6.21	4.20	9.60	......	8	46	63
Bridges & Son, Wabash, Wabash County	2.56	159	2.43	4.00	8.70	......	......	41	......
Bridges & Son, Wabash, Wabash County	2.80	171.5	.43	2.30	12.1	11.5	9	17	21
Wabash Stone Co., Wabash, Wabash County	2.60	162.2	3.29	3.80	10.5	9.3	10	20	27
Keefer & Bailey, Huntington, Huntington County	2.75	172.0	1.21	......	......	......	......	18	36
E. Woods & Co., Pleasant Mills, Adams County	2.56	159.0	3.56	10.23	3.91	......	......	10	22
J. S. Bowers, Decatur, Adams County	2.70	168.0	1.11	......	......	......	......	14	40
Meyers & Co., Bluffton, Wells County	2.70	168.4	1.65	6.3	6.3	10.8	13	9	37
Quarry near Rockford, Wells County	2.70	165.3	3.29	5.1	7.8	13	14	22	37
Shoemaker Bros., Bluffton, Wells County	2.70	171.5	.55	5.8	6.9	13.5	8	14	29
Baltes Stone Co., Montpelier, Blackford County	2.70	165.3	2.97	3.7	10.8	9.3	......	35	52
Marion Stone Co., Marion, Grant County	2.70	159.0	4.72	5.2	7.8	2.5	15	44	56
Ingalls, Madison County	2.60	168.4	.29	5.6	7.2	8	5	16	50
N. J. Nicoson, Alexandria, Madison County	2.70	165.3	1.41	4.3	9.3	1.8	7	26	28
Dan'l Abbott, Frankton, Madison County	2.7	165.3	1.57	4.4	9.2	7.8	7	26	29
Armfield & Cartwright, Ridgeville, Randolph County	2.65	165.3	.77	7.2	5.6	7	6	11	19
Portland Lime & Stone Co., Portland, Jay County	2.7	168.4	.69	7	5.7	4.5	8	19	45
Griffin & Co., Noblesville, Hamilton County	2.8	171.5	1.08	4.9	8.1	13.8	9	15	26
Jas. Ochiltree, Rushville, Rush County	2.7	168.4	2.18	2.7	14.8	5.3	6	34	65
Frank Moore, New Salem, Rush County	2.6	162.2	2.32	12.7	3.1	4	5	13	24
Big Four Stone Co., New Point, Decatur County	2.7	168.4	.93	3.5	11.3	9.3	10	12	34
Greensburg Lime Co., Greensburg, Decatur County	2.7	168.4	1.27	4.5	9	12	8	51	56
Geo. Hotchkiss, Soapville, Switzerland County	2.7	168.4	.80	4	9.9	3	9	10	20
Richard Johnson, North Madison, Jefferson County	2.6	165.3	3.21	5.2	7.8	3	12	36	53

Devonian Limestones. Three divisions of the Devonian limestones of Indiana are recognized: the Jeffersonville, the Silver Creek and the Sellersburg, all geographical names of southern Indiana, where the best exposures of Devonian strata occur. The Silver Creek is an hydraulic limestone, used in the manufacture of natural cement, but containing too much clay and not sufficiently indurated for good road metal. The Jeffersonville is a bluish to gray limestone, fairly well indurated and useful as building stone and road metal, it occurs in massive layers and also in thin layers. It ranges in thickness from fifteen to fifty feet. In many exposures the Silver Creek limestone, and in many places the Sellersburg also forms the overburden, which must be removed

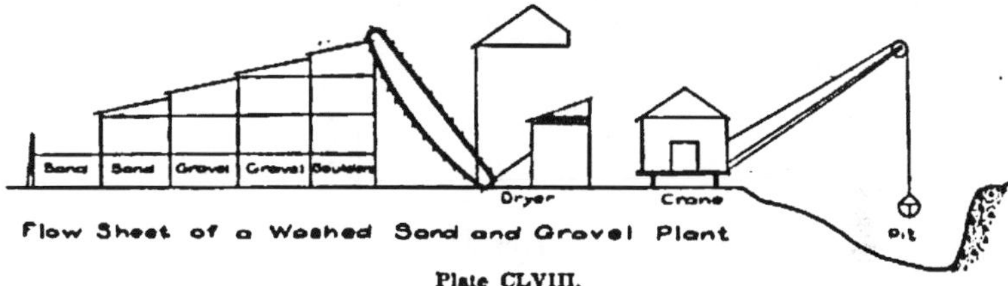

Flow Sheet of a Washed Sand and Gravel Plant

Plate CLVIII.

in order to quarry the Jeffersonville. The Sellersburg is a white or gray crinoidal limestone which is serviceable as road metal. It lies above the Silver Creek limestone, and the New Albany shale forms the overburden in many of its exposures. Its thickness ranges from a few to about twelve feet.

The Devonian limestones are best exposed in southeastern Indiana, where the glacial drift is thin. A few outcrops occur in the northern part in Carroll, Miami and White counties, near the Wabash and Tippecanoe rivers.

The chemical and physical properties of the Devonian limestones are given by the Bureau of Public Roads in the following tables:*

Composition of Devonian Limestones.

	No. 1	No. 2	No. 3	No. 4	No. 5
Alumina (Al_2O_3)	.59	1.40	.18	.41	.70
Iron oxide (Fe_2O_3)	.51	.23	.25	.24	tr.
Lime (CaO)	30.00	31.90	47.80	46.35	34.90
Magnesia (MgO)	19.15	15.18	3.44	5.67	17.00
Phosphoric acid (P_2O_5)	tr.	.25	.52	......	......
Insoluble in HCl	5.13	7.88	6.40	5.00	1.78
Loss on ignition	44.37	42.75	41.04	42.71	45.86
Total	99.75	99.59	99.63	100.38	100.24

	No. 6	No. 7	No. 8	No. 9	No. 10
Alumina (Al_2O_3)	......	.11	.71	.70	.25
Iron oxide (Fe_2O_3)	tr.	.51	.25	.25	.25
Lime (CaO)	55.70	52.70	53.35	44.85	53.45
Magnesia (MgO)	.23	.51	.22	5.43	.80
Phosphoric acid (P_2O_5)	tr.	1.08	.54	.25	......
Insoluble in HCl	.86	3.51	2.65	7.39	3.22
Loss on ignition	43.32	41.68	42.10	40.85	41.98
Total	100.11	100.10	99.82	99.72	99.95

No. 1. Stephen Lewis, Hanover, Jefferson County.
No. 2. David Robertson, Deputy, Jefferson County.
No. 3. William S. Baker, Spencer Township, Jennings County.
No. 4. I. B. Stearns, Sand Creek Township, Jennings County.
No. 5. L. C. Bunker, Greensburg, Decatur County.
No. 6. Lewis Solomon, Hope, Bartholomew County.
No. 7. D. M. Walker, Burnsville, Bartholomew County.
No. 8. M. A. Rainey, Grammer, Bartholomew County.
No. 9. W. E. English, Lexington Township, Scott County.
No. 10. Quarry near Charlestown, Clark County.

PHYSICAL TESTS OF DEVONIAN LIMESTONES.

ORIGIN OF SAMPLE	Specific Gravity	Wt. per Cu. Ft., lbs.	Water Absorbed, per Cu. Ft., lbs.	Per Cent. of Wear	French Coefficient of Wear	Hardness	Toughness	Cement Value, Dry	Cement Value, Wet
Lewis Solomon, Hope, Bartholomew County	2.7	171.5	1.71	5.9	6.7	—1	5	20	41
D. M. Walker, Burnsville, Bartholomew County	2.7	168.4	.96	6	6.7	—8.3	5	18	25
M. A. Rainey, Grammer, Bartholomew County	2.5	159	1.37	6.2	6.5	8	8	38	49
L. C. Bunker, Greensburg, Decatur County	2.6	159	2.99	4.7	8.5	.	9	8	38
Wm. S. Baker, Hayden, Jennings County	2.7	168.4	.93	3.2	12.6	8.5	9	49	97
I. B. Stearns, Brewersville, Jennings County	2.7	168	.57	3.7	10.9	6.3	8	28	94
Thos. Croxton, Dillsboro, Dearborn County	2.7	168.4	.78	4.8	8.3	9.8	7	67	100
David Robertson, Deputy, Jefferson County	2.7	168.4	.43	4.7	8.5	—3	7	53	81
Stephen Lewis, Hanover, Jefferson County	2.5	159	3.58	12.3	3.2	—16	7	23	42
W. E. English, Lexington, Scott county	2.65	165	2.27	3.3	12.3	2.7	9	28	103
City Quarry, Charlestown, Clark County	2.6	162.2	1.41	5.6	7.1	3.5	8	17	52
B. L. Burt, Jeffersonville, Clark County	2.65	165	1.23	3.4	11.7	3	7	31	91

Mississippian Limestones. The Mississippian strata of Indiana contain some of the best limestones of the State for road metal. The Harrodsburg, the Salem, the Mitchell, the Beaver Bend, the Reelsville, the Beech Creek, the Golconda and the Glen Dean are all limestones of the Mississipian period occurring in Indiana. Nearly all of these are suitable for the manufacture of crushed stone to be used in macadam and concrete.

Harrodsburg Limestone. This limestone rests upon the Knobstone and underlies the Salem. It is a thin and somewhat irregularly bedded limestone, containing many fossils, particularly crinoids. It attains a thickness of ninety feet in the central part of the area of its outcrop and thins rapidly northward. The extent of its outcrop is indicated on the areal geological map accompanying this report. It has been quarried most extensively in Putnam, Owen, Monroe, Lawrence and Washington counties.

The chemical and physical properties of the Harrodsburg limestone are set forth in the following tables, United States Bureau of Public Roads, reference as above.*

Composition of Harrodsburg Limestone.

	No. 1	No. 2
Alumina and iron oxide ($Al_2O_3+Fe_2O_3$)	0.65	0.63
Lime (CaO)	52.95	50.50
Magnesia (MgO)	.20	2.50
Phosphoric acid (P_2O_5)	....	.32
Insoluble in HCl	3.76	4.53
Loss on ignition	42.04	41.65
Total	99.60	100.13

No. 1. Upper Harrodsburg limestone, Indiana University, Monroe County.
No. 2. Lower Harrodsburg limestone, North Pike quarry, Monroe County.

Plate CLV. A gravel pit in the Mansfield, from which gravel is being taken for road surfacing.

Physical Tests of No. 1.

Specific gravity	2.7	Hardness	14.5
Weight per cu. ft., lbs.	168.4	Toughness	6.
Water absorbed per cu. ft., lbs.	3.32	Cementing value, dry	19.
Per cent of wear	10.60	Cementing value, wet	49.
French coefficient of wear	3.8		

"A very soft limestone, low in toughness, with very low resistance to wear. Develops good cementing value. Best suited for country road traffic."

66—20642

Physical Tests of No. 2.

Specific gravity	2.6	Hardness	3.
Weight per cu. ft., lbs.	165.3	Toughness	7.
Water absorbed cu. ft., lbs.	.97	Cementing value, dry	23.
Per cent of wear	5.3	Cementing value, wet	62.
French coefficient of wear	7.5		

"A soft and brittle limestone, with rather low resistance to wear, but good cementing value. Best suited for highway and country road traffic."

Salem Limestone (Indiana Oölitic). This limestone is one of the best known building materials of the United States and occupies an enviable position among structural materials. It is a soft stone with poor wearing qualities and is not desirable for road building, save where traffic is of the lightest. The waste from the quarries could be used with economy on such roads. The planer dust from the mills can be used for top dressing and cementing material for harder stone.

The nearness of the harder Harrodsburg and the still harder and more desirable Mitchell limestone render the use of the Salem for road metal unnecessary.

Composition of Salem Limestone from Hunter Valley Quarries, Monroe County.

Alumina (Al_2O_3)	tr.	Insoluble in HCl	1.12
Lime (CaO)	56.10	Loss on ignition	42.50
Magnesia (MgO)	tr.		
		Total	99.72

Bureau Public Roads. Reference as above.[36]

Physical Tests of Hunter Valley Sample.

Specific gravity	2.4	Hardness	4.75
Weight per cu. ft., lbs.	149.7	Toughness	4.
Water absorbed per cu. ft., lbs.	4.44	Cementing value, dry	18.
Per cent of wear	10.8	Cementing value, wet	63.
French coefficient of wear	3.7		

"A very soft limestone, very low in toughness and resistance to wear, developing a good cementing value. Best suited for country road traffic."

Physical Tests of Salem from Edwardsville, Floyd County.

Specific gravity	2.65	Hardness	4.6
Weight per cu. ft., lbs.	165.	Toughness	9.
Water absorbed per cu. ft., lbs.	1.32	Cementing value, dry	25.
Per cent of wear	4.4	Cementing value, wet	114.
French coefficient of wear	9.		

"A soft but fairly tough limestone, with a rather low resistance to wear, which develops an excellent cementing value. Suitable for all but very heavy traffic and as a binding material."

Mitchell Limestone. This limestone is one of the best limestones of Indiana for macadam purposes. It is a hard, tough limestone, having good cementing qualities. It is formed of compact, massive layers of dark blue to gray limestone, the upper layers of which contain chert. Quarries and crushing plants are located in Putnam, Owen, Monroe, Lawrence, Orange, Crawford and other counties which prepare Mitchell limestone for concrete, macadam, ballast and other purposes. The distribution of the Mitchell limestone is exhibited on the accompanying geological map. The chemical and physical properties are given in the following tables:[33]

Composition of Mitchell Limestone.

	No. 1	No. 2	No. 3	No. 4	No. 5	No. 6
Alumina (Al_2O_3).. } iron oxide (Fe_2O_3) }	.80	tr.	1.50	.85	.50	.50
Lime (CaO)....................	53.50	55.25	29.70	50.25	55.00	47.00
Magnesia (MgO)................	......	tr.	7.00	tr.	tr.	5.29
Phosphoric acid (P_2O_5)..........	......	.54	.10	......	tr.	......
Insoluble in HCl................	2.77	.85	30.43	9.49	1.84	5.44
Loss on ignition................	43.00	43.31	31.38	39.61	42.69	42.03
Total...................	100.07	99.95	99.91	100.20	100.03	100.26

No. 1. J. D. Torr quarry, Putnam County.
No. 2. McGaughey quarry, Putnam County.
No. 3. Quarry Northwest of Spencer, Owen County.
No. 4. Spencer Stone Company quarry, Spencer, Owen County.
No. 5. Upper Mitchell, Blair quarry, Bloomington, Monroe County.
No. 6. Lower Mitchell, Robinson quarry, Bloomington, Monroe County.

The above analyses and the physical tests given in the following table were furnished by the United States Bureau of Public Roads and published in the Thirtieth Annual Report of the Geological Survey of Indiana:

PHYSICAL TESTS OF MITCHELL LIMESTONE.

Origin of Sample	Specific Gravity	Wt. per Cu. Ft., lbs.	Water Absorbed, per Cu. Ft., lbs.	Per Cent. of Wear	French Coefficient of Wear	Hardness	Toughness	Cement Value, Dry	Cement Value, Wet
Jerry Clifford, Russellville, Putnam County..........	2.67	165.3	.99	4.44	9.01	6	9	16	29
J. D. Torr, Oakalla, Putnam County...	2.69	168.4	.79	3.87	10.34	14	11	8	24
J. B. Hillis, Greencastle, Putnam County..........	2.7	168.4	.43	4.34	9.22	4.3	7	13	22
Simpson McGaughey, Greencastle, Putnam County..........	2.63	165.3	1.53	4.5	8.89	10	8	16	20
J. E. Oldshoe, Waveland, Montgomery County..........	2.7	165.3	1.35	3.9	10.2	—5.2	5	23	43
Spencer Stone Co., Spencer, Owen County..........	2.68	168.4	.88	3.68	10.87	6.3	10	8	21
Roadside Quarry, N. W. of Spencer, Owen County..........	2.6	162.2	3	3.8	11.8	—5.5	9	20	45
Jas. Blair, Bloomington, Monroe County	2.6	165.2	1.7	3	13.4	9.3	13	30	41
F. M. Robinson, Bloomington, Monroe County..........	2.6	162.3	2.16	3.5	11.5	10.5	12	14	74
Southern Ind. R. R. Co., Williams, Lawrence County..........	2.7	168	.7	4.1	9.7	12	8	34	141
Mitchell Lime Co., Mitchell, Lawrence County..........	2.6	162.2	1.66	4.6	8.8	4	9	28	26
O. P. Turley, Orleans, Orange County	2.5	155.9	3	4.4	9.1	8	5	11	37
Jas. McKinster, Corydon, Harrison County..........	2.5	156	1.97	4.5	8.8	13.6	11	24	131
J. B. Speed & Co., Milltown, Crawford County..........	2.66	165.3	.46	3.4	11.6	10	8	18	95
Marengo Mfg. Co., Marengo, Crawford County..........	2.65	165	1.61	3.7	10.8	8.9	9	16	53

Chester Limestones. The Chester division of the Mississippian contains a number of limestones of good quality for road metal. Many of these have been used locally for road building.

The Chester limestones, taken in order from the lowermost, are the Beaver Bend, about twelve feet thick; the Reelsville about five feet thick; the Beech Creek about fourteen feet thick; the Golconda about twenty feet thick, and the Glen Dean about ten feet thick. These limestones are separated by beds of shale or sandstones.

These limestones are composed largely of calcium and magnesium carbonates, with minute quantities of iron oxide, alumina and silicia. The outcrop of the Chester is indicated on the geological map accompanying this report.

Composition of Chester Limestones.[23]

	No. 1	No. 2
Alumina (Al_2O_3)	.25	1.30
Iron oxide (Fe_2O_3)	.75	.75
Lime (CaO)	50.40	51.80
Magnesia (MgO)	....	2.10
Insoluble in HCl	7.17	2.74
Loss on ignition	40.92	41.36
Total	99.49	100.20

No. 1 is from a sample of Chester limestone from the land of George Cox, near the crossing of the Indianapolis and Southern Railway and Richland Creek, near Bloomfield, Greene County.

No. 2 is from a sample of Chester (Beech Creek) limestone from the land of John Scott, three miles southwest of Dover Hill, in Martin County.

Physical Tests of Chester Limestones.[24]

	No. 1	No. 2
Specific gravity	2.7	2.6
Weight per cubic foot (lbs.)	168.4	162.3
Water absorbed per cubic feet (lbs.)	1.15	2.
Per cent of wear	3.	3.2
French coefficient of wear	13.2	12.4
Hardness	13.	9.
Toughness	10.	9.
Cementing value, dry	21.	18.
Cementing value, wet	66.	67.

No. 1. "A very hard and fairly tough limestone, with good resistance to wear and cementing value. Suitable for suburban and highway traffic."

No. 2. "A rock of fair hardness, toughness and resistance to wear and good cementing value. Suited to suburban and highway traffic."

These samples were tested by the United States Bureau of Public Roads, and the results published as above in the Thirtieth Annual Report of the Geological Survey of Indiana.

Pennsylvanian Limestones. Thin beds of limestone are interstratified with the sandstones, shales and coals of the Pennsylvanian in Indiana. These limestones are being used to a limited extent as a source of road metal. The beds are never very thick, rarely exceeding five feet and more often carrying an over-burden than not. The limestones are generally impure, carrying considerable clay and quartz sand.

Composition of Pennsylvanian Limestones.

	No. 1	No. 2	No. 3
Alumina (Al_2O_3)	1.36	2.45	1.90
Iron oxide (Fe_2O_3)	2.44	.50	.50
Lime (CaO)	39.00	9.60	45.40
Magnesia (MgO)	7.61	2.71	tr.
Phosphoric acid (P_2O_5)	.15	.16	tr.
Insoluble in HCl	8.75	72.06	15.32
Loss on ignition	40.47	12.14	36.93
Total	99.78	99.61	100.05

No. 1. Kurtz heirs, Princeton, Gibson County.
No. 2. L. George, Petersburg, Pike County.
No. 3. Louis Meyer, Boonville, Warrick County.

Physical Tests of Pennsylvanian Limestones.

Origin of Sample	Specific Gravity	Wt. per Cu. Ft. (lbs.)	Water Absorbed per Cu. Ft. (lbs.)	Per Cent of Wear	French Coefficient of Wear	Hardness	Toughness	Cement Val.—dry	Cement Val.—wet
Kurtz Heirs, Princeton, Gibson County	2.7	168	1.13	......	......	......	......	24	191
L. George, Petersburg, Pike County	2.75	172	.81	4.9	8.1	–22.5	9	24	71
Louis Meyer, Boonville, Warrick Co.	2.7	168.4	.97	3.3	12.3	14	11	15	126
C. S. Lambscher, Evansville, Vanderburgh County	2.7	168	1.38	4	10	11	8	17	52
W. M. Williams, Mt. Vernon, Posey Co.	2.7	168	.82	1.8	21.7	9.8	8	43	56
Wyatt H. Williams, Mt. Vernon, Posey County	2.7	168	1.67	3.5	11.3	11.5	15	22	108

Tests by Bureau of Public Roads, reference as above.

Sandstones for Road Metal.

Sandstones occur in the Mississippian and in the Pennsylvanian systems of

Indiana, which may serve at least a local use as road building material. The most valuable of such materials are to be found in the latter system.

The Knobstone Sandstones. This formation is largely composed of shales, but there are layers of sandstone at many horizons. Some of these sandstones are soft and poorly cemented. These disintegrate readily under even light traffic, and there is not enough cementing material to bind the sand grains formed. Other ledges, however, exhibit a certain degree of induration being cemented with iron oxide. Such sandstones form a fair quality of road metal, and are used to a limited extent.

Chester Sandstones. The Chester division of the Mississippian system in Indiana contains a large number of sandstones. These include the Sample, the Brandy Run, the Elwren, the Cypress, the Hardinsburg and the Tar Springs. These sandstones are not as a rule sufficiently indurated to be useful as road metal. They are fine-grained and when they disintegrate they form beds of fine sand. These sands may be utilized in the building of sand clay roads.

Pennsylvanian Sandstones. A sandstone, the Mansfield, belonging to the Pottsville division of the Pennsylvanian system in Indiana, supplies an important road building material. The Mansfield is usually a coarse grained sandstone not very well indurated, but in some places it is cemented with oxide of iron into hard ledges. These ledges furnish suitable road metal. But the best material is supplied by the gravelly or conglomerate layers of the Mansfield. The gravel surface roads formed by the use of the Mansfield are among the most durable roads in the State.

Sandstones also occur in the Allegheny division of the Pennsylvanian. These are for the most part fine grained and only moderately cemented. Some of them are suitable for roads where the traffic is light and some of them are being used for such roads.

CHAPTER XIII.

SANDS—BUILDING, FOUNDRY, GLASS.

FOUNDRY SANDS.

Foundry sand is used for making molds in which metals are cast and for making the cores which fill the hollow spaces in the castings. The sands used for these two purposes differ in texture. The core sands are coarse, the sands used for the mold are finer and more loamy. Sands of fine grain are selected for making molds for small castings and coarse sands for the molds of large castings. The molds are often faced with graphite, which is highly refractory. Core sands necessarily contain little clay and require an artificial binder.

Essential Properties. The essential properties of a molding sand include cohesiveness, refractoriness, texture, porosity and durability. In order that the molds may be formed the grains must cohere, and the coherence is produced by the clay present. The sand must be refractory enough to withstand the high temperature of the molten metal in order to prevent fusion of the mold or penetration and union with the metal. Its texture must be adapted to the purpose for which he sand is intended. Porosity is essential, so that the gases from the metal may permeate the sand of the mold and escape, for if confined they would produce flaws in the metal. The sand must have durability so that it may be used more than once.

Chemical Composition. The chemical analyses of a large number of molding sands show that the silica content ranges from 58 to 87 per cent alumina from 3 to 16 per cent; ferric oxide from .80 to 8 per cent; lime from a trace to 11 per cent; magnesia from 0 to 6 per cent, and alkalies from 0 to 3 per cent.

Mechanical Composition. The mechanical analysis, which consists of separating the particles of the sand into various sizes, shows that in a large number of molding sands from .04 to 42 per cent of the particles were retained on a sieve having 20 meshes to the inch; from .06 to 16 per cent on a 100 mesh sieve; from 2.6 to 88 per cent on a 200 mesh sieve, and that from 1.06 to 42 per cent pass the 250 mesh sieve, or aré classed as clay particles.

Distribution. Foundry sands are widely distributed in Indiana. Molding sands have been reported from thirty-six counties in the State, but only about one-third of them furnish quantity production. The following counties contain deposits of molding sands: Allen, Fort Wayne and New Haven, and used in Fort Wayne; Bartholomew, near Columbus and used there; Cass, core sand from Lake Cicott; Clark, near New Albany, used there; Clay, near Brazil; Elkhart, Goshen; Grant, Marion; Henry, Newcastle; Jackson, Brownstown and Seymour; Jefferson, near Madison; Knox, Vincennes; Lagrange, Wolcottville; Lake, East Gary and Hammond; Laporte, Michigan City, dune sand, core sand; Marion, Indianapolis; Monroe, Bloomington; Morgan, Centerton, Landersdale, Paragon; Owen, Gosport; Porter, Valparaiso; St. Joseph, Mishawaka and South Bend; Spencer, Richland Junction and Rockport; Starke, North Judson; Steuben, Pleasant Lake; Tippecanoe, Lafayette; Vanderburgh, Evansville; Vermillion, Hillsdale, and Vigo, Terre Haute.

Geological Occurrence. Foundry sands generally belong to that class of materials known as mantle rock. They may consist largely of residual materials or largely of transported materials. The transportation of the individual particles may be accomplished through the agency of ice, water or wind, of any two or all. The core sand obtained from the dune area of Lake Michigan was first transported by ice and later by winds. The molding sands of the glacial portion of the State are probably partly glacial in origin and partly residual. Practically all the foundry sands in Indiana are of Pleistocene or Post-Pleistocene age.

Production. In 1918 Indiana used 160,851 tons of foundry sand, which cost $310,478. Only 55,116 tons of a value of $72,326 were produced in the State, while 105,735 tons, valued at $238,152, were supplied from outside sources.

GLASS SAND.

The principal ingredient of the mixture used in the manufacture of glass is some form of silica. The silica may be derived from quartzose sands, sandstones or quartzites. The silica used in glass making in Indiana, which is produced within the State, is derived from quartzose sands and sandstones.

Essential Properties. One of the essential properties of a sand used in the manufacture of glass is size of grain. In order that the sand may be fused without burning, the majority of the grains should not pass a sieve of 120 meshes to the inch. In order to fuse properly the grains of sand should all be small enough to pass a 20 mesh sieve. The shape of the grain is probably not important. Some glass makers prefer the angular or sub-angular grains, but experiments tend to show that round-grained sands may be used as successfully. In selecting a sand for the manufacture of glass it is essential that both the chemical and mechanical properties of the sand be taken into consideration.

Chemical Composition. The sand used in the manufacture of glass must of necessity have a high silica content, and it rarely falls below 98 per cent in the sands used for glass making in the United States. The impurities found in glass sands are generally compounds of calcium, magnesium, iron, titanium, aluminium and organic matter. The calcium is usually in the form of the carbonate and is not detrimental, since ground limestone is added as a flux. The small amount of magnesia is not detrimental. Iron oxide when present in small amounts, even colors the glass green. This is detrimental only in the manufacture of clear glass. Arsenic may be used to mask the iron, or a part of the iron may be removed from the sand by washing. The alumina is present in clay which clouds the glass.

One of the purest glass sands in the United States contains 99.45 per cent of silica, .30 per cent of alumina and .13 per cent of lime. One of the best Indiana glass sands contains 99.99 per cent of silica, .008 per cent of alumina, only a trace of iron oxide and no lime. This sand is from Wolcott in White County.

Mechanical Composition. It is essential that glass sands should be composed of quartz grains, which are neither very large or very small. If the grains are large they do not fuse readily; if they are small, the batch may burn. The largest grains should pass a 20 mesh sieve and should be retained on a 120 mesh sieve. One of the best glass sands in the United States has grains 100 per cent of which pass a 40 mesh sieve, 92 per cent pass a 60 mesh and only 25 per cent a 100 mesh sieve. Of a dozen glass sands in the United States on the average 99.5 per cent pass a 20 mesh sieve, 66 per cent pass a 40 mesh, 29 per cent pass a 60 mesh and 5 per cent pass a 100 mesh sieve.

The mineral substances used in the manufacture of glass vary with the kind of glass and the use it is to serve. Silica is the major constituent of all forms, and constitutes from 60 to 75 per cent of the mass of the finished product. In addition to the silica some substances are added to the glass mixture as fluxes to increase the fusibility of the silica. Some substances are added to decrease coloring, others to produce coloring and some to make the glass harder. The following table shows the proportion by weight of the constituents of various kinds of glass:

Proportions by Weight of Glass Constituents.

	Plate	Window	Green bottle	Dead flint	Lime flint
Silica (SiO_2) sand	100	100	100	100	100
Salt cake (Na_2SO_4)	..	42	38	...	...
Soda Ash (Na_2CO_3)	36	...	...	...	36
Limestone ($CaCO_3$)	24	40	34	...	...
Carbon (C)	.75	6	5	...	...
Arsenic (As_2O_3)	1	2	...	.15	.02
Hydrated lime ($Ca(OH)_2$)	..	...	...	...	12
Potash (K_2CO_3)	..	...	...	34	...

Red lead ($2PbO+PbO_2$)	...	...	48	...
Niter ($NaNO_3$)	...	...	6	1
Manganese (MnO_2)	...	...	.06	6.66
Antimony (Sb)	...	...	.02	.23

Distribution. One of the largest areas of glass sand in Indiana is the dune region in Lake, Porter and Laporte counties. Here large areas are covered with dunes of pure white sand of large size. The sands were deposited in the glacial lakes, Chicago and Kankakee, and later heaped into dunes by wind action. The dune area, bordering on Lake Michigan, is principally between Gary and Michigan City, a strip from one-half to one and one-half miles wide, in which the dunes vary from fifty to one hundred and ninety feet high. Some of the higher dunes cover areas of one to two hundred acres. This area contains an inexhaustible supply of glass sand. A sand dune area lying south of the Kankakee River covers an area of about two thousand square miles. A sample of sand from Michigan City contains 91.98 per cent of silica, 4.44 per cent of alumina, .56 per cent of iron oxide, 2.20 per cent of lime. These impurities may be partly if not completely removed by washing.

Mansfield Sandstone. The Mansfield sandstone of the Pottsville division of the Pennsylvanian in Indiana has a maximum thickness of two hundred feet and an average probably not far from one hundred feet. In some places it is found suitable for glass sand. In many places it contains large quantities of pebbles; in others it contains inclusions of clay, and iron oxides in nearly all places. Glass sands have been obtained from it near Attica in Fountain County along the Wabash River and its tributaries. This sand has been used by the Attica Glass Sand Company. It contains 98.84 per cent of silica, .38 per cent of alumina, .10 per cent of oxide of iron and .03 per cent of magnesia.

At Fern, in Putnam County, on Big Walnut Creek, the Mansfield has a thickness of at least fifty feet. The sandstone is quarried and prepared by the Root Glass Company, and is used in the manufacture of glass in their bottle factory in Terre Haute.

The Mansfield sandstone is quarried from a deposit which is probably more than one hundred feet thick at a point on the Baltimore and Ohio Southern Railroad about one and one-half miles east of Loogootee in Martin County. After being blasted from the face of the quarry the sandstone is crushed, washed and dried. The product is shipped to points within and without the State. The prepared sand at this point contains 97.78 per cent of silica, 1.13 per cent of alumina, .10 per cent of ferric oxide and .06 per cent of lime.

There is an outlier of Mansfield sandstone near Wolcott, in the western part of White County. Sand was obtained from a quarry in the outlier by the American Window Glass Company. The average of four analyses of samples of the sand made by the company show a silica content of 99.79 per cent, .237 per cent of alumina and .024 per cent of iron oxide. The best grade of the sand is suitable for the manufacture of the highest grades of glassware and flint glass, and the poorest grade for the manufacture of window glass.

The Mansfield sandstone is a promising formation for the production of glass sand in a large number of counties. In Martin County it attains a thickness of two hundred feet, and presents thick exposures in many places. In Clay, Dubois, Orange, Parke and other counties there are promising exposures of the sandstone.

At Coxville, in Parke County, there is a deposit of sandstone filling an old valley which was carved in the Pottsville rocks of the Pennsylvanian. The deposit lies at the unconformity between the Pottsville and the Allegheny and belongs in age to the latter. The sandstone is quarried and used by the Acme Glass Sand Company. The sandstone ledge has a thickness of forty feet. The sandstone shows little induration and is composed of medium-sized, angular grains. The color of the sandstone ranges from white to yellow. The sandstone is blasted down, passed through a crusher, through corrugated and smooth rolls and then dried in a rotary drier. The sand, as it comes from the quarry, may be used in making colored glass. It must be washed before being used for flint and window glass. The sand contains 98.61 per cent of silica, .74 per cent of alumina, .22 per cent of iron oxide and .12 per cent of lime.

The Ohio River sands of Tertiary age, which occur near DePauw, in Harrison County, have been used in the past in the manufacture of glass. Similar sands in the neighborhood of New Albany have also been used. The New Albany sand contains 92 per cent of silica, 1.50 of alumina and 5 per cent of iron oxide.

Building Sands.

Sands for building purposes are well distributed in Indiana. White finishing sands may be obtained from the dune region south of Lake Michigan. Michigan City is an important distributing point. Finishing sands may be obtained from some of the ridges of sand and gravel of the glacial area and from the beds and valleys of the streams which head in or pass through the glacial area.

Ordinary mortar sands are widely distributed in the glacial drift and in the filled valleys of the streams of the unglaciated area, which originate in the glacial area. Sands may be obtained also from the deltas built into the lakes of the glacial region.

In the unglaciated area building sand may be obtained from the stream beds which cross the Knobstone area, from the Mansfield, the Elwren, the Cypress and other sandstones, from the streams which cross their area of outcrop and from the dunes formed along their valleys. Indiana produces about two million tons of building sand per year.

Production of Sand in Indiana.

Glass sand	40,000 tons
Molding sand	300,000 tons
Building sand	2,000,000 tons
Abrasive sand	50,000 tons
Engine sand	60,000 tons
Paving sand	260,000 tons
Railroad ballast sand	100,000 tons
Other sands	140,000 tons
Total	2,950,000 tons

Total amount of gravel produced is about five million tons.

ANALYSES OF GLASS SANDS OF INDIANA.

Location of Deposit	Owner	Authority	Silica (SO_2)	Alumina (Al_2O_3)	Ferric Oxide (Fe_2O_3)	Calcium Oxide (CaO)	Magnesium Oxide (Mg O)	Loss by Ignition (Water—Organic Matter)	Other Minerals	Total	Remarks
Clark County Jeffersonville	Newport (Ky.) Sand Bank Co.	Laboratory, Cin'ti College of Phar., Cincinnati, Ohio	94.00	5.00	1.00					100.00	Fine-grained molding sand.
Floyd County New Albany	Newport (Ky.) Sand Bank Co.	Laboratory, Cin'ti College of Phar., Cincinnati, Ohio	92.00	1.50	5.00	Tr.				98.50	Fine light-colored molding sand.
Fountain County Attica	Western Silica Co., Danville, Ill.	State Geol. of Indiana	98.84	.38	.10		.03	.32	Titanium oxide, tr.	99.67	
Hancock County Maxwell	E. L. Dobbins, (Greenfield)		85.00			.50	13.50		Loam, 1 per cent.	100.00	
Jackson County Brownstown	Newport Sand Bank Co., Newport, Ky.	Cin. College of Phar., Cincinnati, Ohio	92.00	6.50	1.00		.50			100.00	Coarse red molding sand.
Laporte County Michigan City	Pinkston Sand Co.		91.98	4.44	.56	2.20	Tr.	.72	Loss 10 per cent.	100.00	Molding sand, grinding sand.
Martin County Loogootee	Loogootee Glass Sand Co.	J. F. Elson, New Albany	97.78	1.13	.10	.06				99.07	Glass sand, selected sample.
Loogootee	Loogootee Glass Sand Co.	Operator—Rept. of State Geol.	96.26	2.50	.92		.16		Potassium and sodium oxides (K_2O) and (Na_2O) 0.13%.	99.97	Glass sand, crude.
Parke County Coxville (near Rosedale)	Acme Glass Sand Co., Terre Haute	W. A. Noyes, Rose Polytechnic	98.61	.74	.22	.12	Tr.	.32		100.01	Glass sand, unwashed.
Clay County Brazil	Dr. Sourwine	W. M. Blanchard, Chem. DePauw Uni. Greencastle	96.52	2.60	.18	.19	.06	.70		100.34	
Brazil	Dr. Sourwine	W. M. Blanchard, Chem. DePauw Uni., Greencastle	95.41	2.79	.40	.19	.17	.80		99.76	
Brazil	Dr. Sourwine	W. M. Blanchard, Chem. DePauw Uni., Greencastle	94.88	3.05	.66	.32	.07	.93		99.91	
Brazil	Dr. Sourwine	W. M. Blanchard, Chem. DePauw Uni., Greencastle	98.00	.61	.32	.10	.08	.36	Sulphuric anhydride .13%.	99.60	
Brazil	Dr. Sourwine	W. M. Blanchard, Chem. DePauw Uni., Greencastle	93.69	2.68	1.05	.51	.08	1.05	Sulphuric anhydride .13%, soda 50%.	99.70	

CHAPTER XIV.

FERTILIZERS, GYPSUM, GOLD, HYDRAULIC LIMESTONES, LITHOGRAPHIC LIMESTONES, MANGANESE, DIATOMACEOUS EARTH, MINERAL PAINTS, PRECIOUS STONES, SALT AND SULPHUR.

Fertilizers.

A fertilizer is any substance capable of increasing the fertility of the soil. They may be classed as inorganic or mineral and organic. Substances which contain the essential plant foods, phosphorus, potassium and nitrogen and the correctives of acidity in soils, marl and limestone, may be included as fertilizers. Phosphoric acid is a minor constituent of some of Indiana limestones, but it has not been found in important amounts. It is a constituent of the bones and shells of animals and thus is found in many limestones, but economically important deposits occur only where concentration has taken place.

Potash occurs in some of the brines of the state and is a constituent of some clays and shales. The New Providence shale at the base of Knobstone contains in places between four and five per cent of potash. Experiments are now being carried on to see whether this potash can be extracted in commercial quantities and profitably.

Nitrogen occurs in calcium nitrate in some of the caves of Indiana, but whether it occurs in commercial quantities is not known. The writer collected some of the crystals from a cave south of Georgia in Lawrence County. These crystals occurred in the surface of an earthy deposit on the floor of the cave. This earthy deposit was in places four feet thick.

The fertilizing properties of limestone, marl and peat have been discussed under those heads.

Gypsum.

Gypsum is an hydrated form of calcium sulphate. It contains 46.6 per cent of sulphur trioxide, 32.5 per cent of lime and 20.9 per cent of water when pure. It occurs in transparent crystals, in which form it is called selenite; as fibrous masses, called satin spar; as massive white, called alabaster, and as an earthy variety, called gypsite. It is a soft mineral easily scratched with the thumb-nail. Its color is usually white, though other colors occur.

Gypsum occurs sometimes in thick beds, which have probably been deposited as a chemical precipitate in an enclosed sea, and as crystals in shales, having been formed through the action of sulphuric acid, produced by the oxidation of pyrite, on calcium carbonate.

No thick beds of gypsum are known in Indiana. Selenite crystals are common in many formations. A layer of selenite a few inches thick occurs in the railroad cut east of Huron in Lawrence County. This layer rests between the Beaver Bend limestone and the Brandy Run shale. The oxidation of pyrite formed sulphuric acid which, coming in contact with the limestone through ground water circulating in about three feet of sandy shales, produced the selenite. Crystals of selenite are often found filling minute crevices in shales and sandstone in the Mississippian and Pennsylvanian rocks of the state. Some of the well waters which penetrate these rocks are rendered hard by the abundance of dissolved gypsum. Gypsum is used in the manufacture of cement and of plaster of paris, which have many uses in the arts.

Gold.

Small quantities of gold have been found in many counties of Indiana. These counties are located within the glaciated area or in counties adjacent which receive a part of their drainage from it. No gold-bearing veins occur in the state. The gold found is the placer type, occurring in sands and gravels. These sands and gravels have been derived from the glacial drift. The particles of native gold are usually free, but occasionally a few flakes are found in small quartz pebbles.

From all the evidence collected it seems that the gold of Indiana was brought down by glaciers from some point lying north of the northern boun-

Plate CLIX. Entrance to Wyandotte, one of Indiana's most noted caves.

dary of the United States. Vein stone carrying the gold was removed from the outcrop of the veins by the erosive action of glaciers and transported to its present resting place. The particles of gold were set free from the vein stone by the disintegrating action of the weathering agents. Under the action of gravity the particles of gold settled to the surface of the bed rock and was concentrated along the courses of streams.

A few men "panned" gold for the greater part of a lifetime in Brown and Morgan counties and are said to have realized as high as $1.75 per day for their labor. The most of the panning is done on the bars of the small streams. The gold obtained varies in size from microscopic flakes or grains to small nuggets weighing as much as one-fourth ounce Troy.

The counties in which gold has been found or reported are: Brown, Cass,

Clark, Dearborn, Franklin, Greene, Jackson, Jennings, Montgomery, Morgan, Ohio, Putnam, Vanderburgh and Warren.

Gold in Indiana has been referred to in the following publications: Frazer, Jour. Franklin Institute, June, 1850; Collett, Geol. Sur. Indiana, 1873, p. 244; 1874, pp. 77-110 and 1875 p. 294; Owen, Geol. Reconnaisance of Ind., 1862, p. 119; Cox, Geol. Sur. Ind., 1878, p. 106; Sutton, Proc. Assoc. Adv. Sci., XXX, 1881, pp. 77-185; Haymond, Geol. Sur. Ind., 1869, pp. 185-190; Brown, 13th Ann. Rept. Geol. Sur. Ind., 1903, pp. 11-38; 31st Ann. Rept., pp. 68-72.

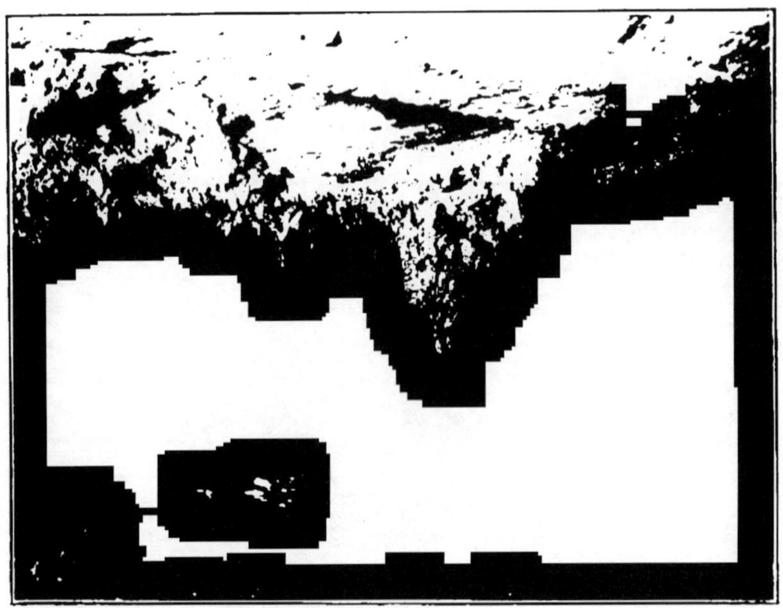

Plate CLX. Inside Wyandotte Cave. View showing deposition of stalactites and stalagmites.

Hydraulic Limestones.

Among the limestones of Devonian age in Indiana is the Silver Creek hydraulic limestone. This limestone attains its greatest thickness, sixteen feet, in Clark County, and thins rapidly toward the north. The Silver Creek limestone lies between the Jeffersonville limestone below and the Sellersburg limestone above, both of Devonian age. The limestone is a homogeneous fine-grained limestone, usually massive, but in some is divided into two or three ledges by bedding planes. The color on freely exposed surfaces is usually blue or bluish gray and on weathered surfaces buff. The chief impurities are silica, clay and iron compounds. It is composed of calcium carbonate and magnesium carbonate, with the impurities mentioned.

The Silver Creek limestone has been used for many years in the manufacture of natural cement. At one time a large number of kilns were in operation in Clark County, at the present time only one firm is manufacturing natural cement. The composition of the raw material and the manufactured product is given under the chapter on Cement in the discussion of the Speed plant.

Lithographic Stone.

In the making of lithographs a fine-grained homogeneous limestone is often used to receive the impression from which the lithograph is made. A limestone to be useful for this purpose should be porous enough to absorb the oil of the printer's ink. It should be soft enough to be easily carved, but hard enough to resist ordinary wear and not crumble under light blows. It should be of uniform texture and free from fossils, cavities, veins and other irregu-

Plate CLXI. Hydro-electric plant on east fork of White River at Williams.

larities. Lithographic limestone usually has a conchoidal fracture and is not easily squared with a hammer but is better shaped by sawing.

Lithographic limestone occurs in Indiana among the Mississippian rocks. The Salem, the upper Mitchell and the Golconda divisions all contain lithographic stone. Small slabs free from imperfections are not difficult to obtain, but the larger slabs are not often obtainable because of the presence of fossils. No extensive use has been made of any of the Indiana lithographic limestones. They are distributed through the Mississippian belt in Monroe, Lawrence, Orange and Crawford counties and probably occur in some others.

Manganese.

The principal ores of manganese are the oxides, pyrolusite (MnO_2), psilomelane (MnO_2, H_2O), and wad, which is an earthy variety of the oxides. Limonite often contains oxides of manganese. Small quantities of the oxides of manganese have been found in the Kaolin deposits of Indiana. Some of the geode-like concretions in the Mansfield iron ores contain cavities lined with oxides of manganese. The iron ores of the lower Knobstone also contain oxides of manganese. Manganese ores have been reported as occurring in deep wells in the southeastern part of the state, but no deposits of economic importance have been found.

In the iron ores at the base of the Knobstone as much as five per cent of manganese oxide and carbonate has been found.

Diatomaceous Earth.

This is an earthy substance composed largely of the shells of diatoms, with an admixture of clay and sand. It is used as a polishing powder, in the manufacture of scouring soaps, as a non-conductor for steam pipes and boilers, and for other purposes. It is reported from Dubois, Lawrence, Jackson, Sullivan and Washington counties.

Mineral Paints.

A large number of minerals and rocks are used in the manufacture of paints. Some of these are used as they are taken from the mine or quarry, others must be ground, separated or washed. One of the common sources of mineral pigments is from ochers, which are an admixture of iron oxides and clay. The common varieties are red, yellow, umber and sienna. Red ocher is a mixture of clay and the red oxide of iron, hematite. The yellow ocher consists of a mixture of clay and the hydrated oxide of iron. The yellow ocher may be changed to red by heating it and driving off the water of crystallization. The brown ocher or umber is a mixture of clay manganese oxide and iron oxide. Sienna is a yellowish-brown variety of umber. The percentage of iron oxide in the ochers ranges from ten to seventy-five.

The iron ores, hematite, limonite and siderite, are also used for pigments. Other substances which are used for pigments are clay, shale, slate, gypsum, barite, asbestos, graphite, chalk, whiting and other substances.

Clays suitable for paints are found in Lawrence and Martin counties. Shale occurs in Vigo and other counties which is suitable for paints.

Iron ores occur in Martin, Greene, Monroe, Orange and other counties which can be used for paints. Ochers also occur in these counties in some places. Yellow ocher of good quality exists in Owen, Greene, and Orange counties. The limstones of Indiana furnish a good quality of calcium carbonate which may be used as a pigment.

Precious Stones.

Under this head are classed all minerals and rocks which are used for gems or ornaments. The greatest variety of such rocks and minerals found in Indiana are obtained from the glacial drift. They were transported by glaciers from the crystalline rocks lying beyond our national boundary. In some cases, when found, they are imbedded in fragments of crystalline rocks, but more often the gem stones are single crystals or fragments of crystals occurring in loose sand or gravel. The writer has identified crystals of quartz, rutile, topaz, garnet, tourmaline (brown and black) feldspar, amethyst, and diamond. Also in pebbles he has recognized agate, opal, jasper and chalcedony.

Diamonds have been found in the older glacial drift (Pre-Wisconsin) in Morgan and Brown counties. More than a score have been found in these two counties, largely by people who have washed the stream sands and gravels for gold. The largest one reported weighed nearly five carats, but the majority were under two carts. In Brown County, diamonds have been found on Indian Creek, Goss Creek, Lick Creek and Salt Creek, and in Morgan County on Gold Creek, Little Indian Creek, Sycamore Creek and Possum Creek.

Transparent Quartz crystals line many of the geodes found near the Knobstone-Harrodsburg contact. Many of them are clear and transparent and suitable for ornamental purposes.

Pearls are obtained from fresh-water clams or mussels in the streams of the state. Clam fishing is carried on in the Ohio, White, Wabash and other rivers. Button form factories are located at Shoals and Petersburg, where the forms for buttons are cut from the clam shells and sent to button factories to be completed. The fishermen often find pearls in the clams, but it is only occasionally that one of economic importance is found. A few are reported to have sold for more than one hundred dollars.

Sulphur.

Free sulphur is not known to occur in Indiana in deposits of commercial importance. A little free sulphur is found in coal and in shales associated with coal. This sulphur has been set free by the oxidation of pyrite. Hydrogen sulphide is a common constituent of some of the mineral waters of Indiana. The principal source of sulphur in Indiana is from pyrite which is discussed under that head.

Salt.

Common salt, sodium chloride (NaCl), is present in Indiana in brines, which occur in deeply buried sedimentary rocks which were deposited under the sea and from which it is probable that the salt has never been completely removed. The St. Peter's sandstone, the Ordovician, Silurian and Devonian limestones are rocks which contain these saliferous waters from which salt could be extracted by evaporation. Rock salt is not known in the state.

Bibliography.

1. Building Stones: Indiana Geological Survey, Annual Reports I, II, III, V, VI, VII, VIII, XII, XIII, XVII, XX, XXI, XXVI, XXVIII, XXX and XXXII.
2. Bedford Oolitic Limestone, T. C. Hopkins and C. E. Seibenthal, XXI, p. 291, et. seq.
3. Indiana Oolitic in 1900, C. E. Seibenthal, p. 331, et. seq.
4. Indiana Oolitic in 1907, T. C. Hopkins, C. E. Seibenthal and R. S. Blatchley, XXXII, p. 301, et. seq.
5. Cement: Ind. Geol. Sur. Ann. Repts III, V, VIII, XVII, XVIII, XXII, XXV and XXXI.
6. Silver Creek Hydraulic Limestones of So. East. Ind., C. E. Seibenthal, XXV, p. 331.
7. Cement, Resources, W. S. Blatchley, XXXI, p. 50.
8. Coal: Ann. Reports II, III, V, VII, XI, XIII, XV, XVII, XVIII, XXI, XXII, XXIII, XXVII, XXVIII, XXIX, XXX, XXXI, XXXII, XXXIII and XXXIV.

9. Coal Deposits of Indiana, Geo. H. Ashley, XXIII, 1898, and XXXIII, 1908.
10. Clays: Annual Reports I, II, III, V, VI, VII, XII, XV, XX, XXII, XXIII, XXVI, XXIX, XXX, XXXI.
11. Clays and Clay Industries of Indiana, W. S. Blatchley, XXIX, 1904.
12. Kaolin—see text page.
13. Iron Ores: I, II, III, V, VI, VII, XXXI, Annual Reports Indiana Survey.
14. Iron Ore Deposits in Indiana, C. W. Shannon, XXXI, p. 299, et. seq.
15. Lime: Annual Reports II, III, V, XI, XIII, XVII, XXVIII, XXXI.
16. Lime Industry in Indiana, W. S. Blatchley, XXVIII, p. 211, et. seq.
17. Marl: Annual Reports VII, XV, XXV, XXVIII, XXIX, XXXI.
18. Lakes and Marls of Northern Indiana, W. S. Blatchley and Geo. H. Ashley, XXV, p. 33, et. seq.
19. Abrasives: Indiana Reports II, VII, XX, XXVIII, XXXI.
20. Whetstone and Grindstone Rocks of Indiana, E. M. Kindle, XX, p. 329, et. seq.
21. Mineral Waters of Kansas, E. H. S. Bailey, Vol. VII, Uni. Geol. Sur.
22. Mineral Waters of Indiana, W. S. Blatchley
23. Petroleum and Natural Gas: Ind. Geol. Reports II, III, VII, VIII, XV, XVI, XVII, XVIII, XIX, XXI, XXII, XXVIII, XXVII, XXIX, XXX, XXXI, XXXIII, XXXIV, XXXIX, XL, XLI, XLII.
24. See Oil Report—Phinney.
25. Peat: Indiana Reports III, V, XXII, XXXI, XXXII.
26. Peat Deposits of Northern Indiana, E. A. Taylor, XXXI, p. 73, et. seq.
27. Pyrite in Indiana, L. P. Dove, XLIII, Y. B., p. 21, et. seq.
28. Road Materials of Indiana, W. S. Blatchley and assistants, XXX, p. 1, et seq.
29. Sands: Indiana Reports V, XII, XXVIII, XXX, XXXI, XXXVI, XLIII.
30. Molding Sands of Indiana, Allen D. Hole, XLIII, p. 196.

PART VI

Preliminary Report on the Oil Shales of Indiana

By

John R. Reeves

CONTENTS

LIST OF ILLUSTRATIONS

Preliminary Report on the Oil Shales of Indiana

INTRODUCTION

The New Albany "black" shale of southeastern Indiana has been a well known formation to geologists for many years. As an oil shale it has not been looked upon with much interest, excepting the work done by Duden in 1896, and the reference and analyses made by Ashley in 1916. The author believed a more detailed study of the formation would be profitable to the general public and those interested in oil shales. It was with this in view that the present work was begun. In addition to the New Albany shale some results of experiments and a brief discussion of the bituminous shales associated with the coals of Indiana are given.

The writer wishes to acknowledge his indebtedness to the authors of the publications named in the bibliography below; to Professor W. N. Logan, State Geologist, who has made the report possible and who gave valuable help and criticism to Professor E. R. Cumings; to Dr. C. A. Malott; to Messrs. K. W. Ray, Fraze, and Quinn for a part of the analytical work and suggestions.

The maps of this report, prepared by the author, were compiled from field notes and sketches and other geological maps of the Indiana Geological Survey.

Author's Note: This report was completed in May, 1921. A large amount of data since collected modifies to some extent several statements made herein. Wherever possible, footnotes have been added, bringing information up to date. Oil shale experimentation is now being conducted (1922) at the Oil Shale Experiment Station, Indiana University, Bloomington, Indiana, in co-operation with the U. S. Bureau of Mines and the Division of Geology, Department of Conservation.

BIBLIOGRAPHY.

Indiana Geological Survey, 1874, p. 124, p. 157.
Indiana Geological Survey, 1875, p. 169.
Indiana Geological Survey, 1891, 17th Rept., p. 180.
Indiana Geological Survey, 1896, 21st Rept., p. 108.
Indiana Geological Survey, 1897, 22nd Rept., p. 20.
Indiana Geological Survey, 1900, 25th Rept., p. 338 and 529.
Kentucky Geological Survey, Jefferson County, Charles Butts, p. 130, 193.
Indiana Geological Survey, 1915, 40th Rept., p. 226.
United States Geological Survey, G. H. Ashley, Bull. 641, p. 319.
Russell D. George, Chem. Age, Vol. 28, No. 12.
M. B. Blacker, Shale Review, Vol. II, No. 12.
L. C. Karrick, Reports of Investigations, Serial No. 2229, March, 1921. U. S. Bureau of Mines.

CHAPTER I

GENERAL CONDITIONS AND THE OIL SHALE INDUSTRY

During the year 1919 the United States Geological Survey estimated the total petroleum content of the United States to be 11,300,000,000 barrels, of which some 4,600,000,000 barrels had been extracted. This leaves as the surplus 6,700,000,000 barrels. These figures are fairly accurate, especially the ones on production, and show clearly our present status in regard to a petroleum supply for the future. During the year 1919 the United States produced 350,131,000 barrels of petroleum. The production in the first ten months of 1920 amounted to 369,790,000 barrels and it seems likely that the total for the year of 1920 will be something like 441,000,000 barrels. At the rate petroleum was consumed in the United States in September, 1920, it is safe to say the consumption in 1921 will amount to 606,000,000 barrels, which doesn't include the possibility of an increase in the consumption during this year, which surely will exist. In case the above expected figures on the consumption for 1920-1921 prove to be correct it seems more than likely that half of our total petroleum resources will be exhausted at the end of the year.

The tabulated production of petroleum in the United States is as follows:

Year	United States
Prior to 1908	1,806,600,000
1908	178,500,000
1912	222,900,000
1916	300,760,000
1920*	441,000,000

This table clearly shows the remarkable increase in production of petroleum. The increase in consumption of petroleum is even more remarkable. During the year 1920 the railroads doubled their 1919 consumption figure. The gasoline engine consumed nearly one-third more gasoline in 1920 than in 1919. In 1920 there was a large increase over the 1919 consumption of crude oil for marine purposes with a strong probability of the 1921 increase being even greater. The United States Shipping Board alone contemplates (1920) the consumption of 66,000,000 barrels within two years. The use of fuel oil for other purposes also is rapidly increasing.

During the year 1919 the United States produced 61.11% of the world's petroleum and Russia produced 24.66%. North America and Russia combined produced over 90% of the world's total petroleum production. The United States imports about 70% of Mexico's production, which came next to Russia in the production of petroleum. Mexico's production now exceeds that of Russia.

It may be seen, not only by the few general facts above mentioned, but many others, that imminent danger of a petroleum shortage exists. To make up for such a shortage recourse may be had to several substitutes. There is a possibility of new fields being discovered in other countries. The use of new and improved methods for the production of petroleum, especially in old fields, will no doubt bring up the total, as well as new and improved methods in the refining of petroleum. The development of engines that will more efficiently utilize fuel, such as the Diesel engine, and internal combustion engines that use less and lower grade fuels, as kerosene. The substitution of alcohol

*Expected production.

for motor fuel has been considered to replace gasoline. A larger development and use of water power can displace to some extent the use of fuel-oil on land. Distillation of coals producing benzol and other coal-tar products already is, and in the future probably will to a greater extent, take the place of petroleum. However, all these substitutes have their limitations and the consumers of motor spirits, fuel, and lubricating oils are looking forward with an increasingly greater interest to the oil produced from oil shales as a substitute for petroleum. The amount of oil that can be educed from oil shales known to exist is unlimited. As long as petroleum was cheaply produced in quanities large enough to meet the demand, little heed was given to oil shales. Now that the demand has grown so large that production and resources meet it with difficulty, it is necessary to turn to this new and inexhaustible source for oil.

An oil shale does not contain oil as such, but it contains a substance which when the shale is subjected to destructive distillation yields gas, oil, water, and nitrogen compounds, such as ammonia, as well as, in some cases, other products.

Oil shales vary in characteristics. In color they are usually brown to black. The main body of most oil shales is composed largely of sand well cemented. In hardness they vary from the curly paper shales which are tough and soft to the massive shale as hard as a medium limestone. They occur much the same as an ordinary shale or fine grained sandstone with the exception of the oil-producing matter.

This matter in oil shales probably originated from plant life growing at the time the shale was deposited, the plant matter in a finely divided condition for the most part, being included in the sediments. Animal matter in some cases has also added largely to this matter. A more detailed discussion of the origin of the oil producing matter of oil shales occurs in a later chapter. In some cases this matter may be dissolved by such solvents as carbon bisulphide, ether, and benzol, and in some cases it cannot. However, both types, containing either soluble or insoluble matter, yield oil upon distillation. To that insoluble material or matter "which is neither bitumen, petroleum, or resin", Cum Brown has given the name "kerogen". The writer is inclined to believe oil-producing matter of oil shales as an oil shale is defined above, is bituminous matter much the same as coal in its various stages older than peat. A shale which is impregnated with petroleum is an oil shale, but is not generally classed as such.

Oil shales occur in many countries of the world as well as in several states of this country. Large deposits exist in the United States, Canada, Nova Scotia, New South Wales, Australia, New Zealand, France, Scotland, Tasmania, Brazil, Spain, Turkey, Austria, South Africa, and Sweden. In this country large deposits exist in Colorado, Wyoming, Utah, California, Montana, Indiana, and Kentucky, besides minor deposits in other states not mentioned. The Colorado and adjacent shales are Tertiary in age. The Indiana shale and the Kentucky shale is for the most part Devonian in age and partly Mississippian. Oil shales also occur in the Pennsylvanian rocks associated with the coals.

Obtaining oil by distillation is an old process. As long ago as 1839 the French were distilling bituminous material for oil on a commercial scale by a process perfected by a chemist, Seligue. In 1850, James Young, after long investigations, obtained a patent for the distillation of bituminous material, and the industry began to flourish in England and Scotland. The bituminous material first used was cannel coal and the distillation was carried on at a low temperature in order to obtain the greatest amount of oil. It was discovered this same year that hoghead coal upon distillation yielded more oil than the cannel coal and it was from then on used exclusively. The extensive demand raised the price of this material, but within sixteen years it had become exhausted.

In 1859 experiments were made in the distillation of the Scottish oil shales at Broxburn. About the time the hoghead coal became exhausted several plants were in operation using this shale. From this time on to the present the industry in Scotland has been one of more or less success. In 1860 there were six plants in operation, in 1870 ninety, in 1880 twenty-six, in 1890 fourteen, in 1900 nine, and at the present time four.

America has not been without her share of the industry. During the years from 1850 to 1860 several plants in the eastern states were distilling the imported boghead coal from England. As many as 55 companies in Pennsylvania were distilling the coals of that region principally for gas and illuminating oils. Canadian companies were also distilling Albertite.

In 1864 the introduction of American petroleum killed the industry in America and sadly crippled it in England and Scotland. Improvements on the process of distillation allowed greater yields of oil and by-products and graually the industry became normal or stabilized again.

France and New South Wales also have produced some oil by distillation from oil shales. In 1905 there were 74 plants in operation in France. Very few plants are operating at present in France. Only 17,425 tons of shale were retorted in New South Wales in 1916 producing some 11,000 barrels of oil.

Inasmuch as the Scotch were the first to successfully produce shale oil and are at present profitably operating shale oil plants it is well to give a brief outline of their plant operations, which with modifications were used in France, and are now beginning to be used in America. A plant consists of quarrying or mining machinery, crushers, retorts, condensers, and ammonia and oil scrubbers.

The shale broken up into small pieces is passed into iron retorts varying in size but usually of two tons per twenty-four hour per day capacity. The shale is fed continuously into the retort as well as discharged continuously. The temperatures of retorting varies from 400 degrees C. to 700 degrees C. depending upon the kind of shale used and the primary product desired. The temperature control of the retort is very important, for upon this depends largely the quality of the oil produced and the amount of ammonia. Superheated steam is passed into the retort near the base, this increasing the amount of ammonia and oil. The gases from the retort are led to a condenser. The uncondensed or permanent gas is drawn from the condenser through water and oil scrubbers to remove ammonia and light oil. The scrubbed gas which is inflammable is returned and used to heat the retort. The ammonia liquor is distilled, the ammonia converted to ammonium sulphate by passing into sulphuric acid. The ammonium sulphate is an important item in the Scotch industry, being sold and used as fertilizer.

If the plant has a refinery the crude oil from the condenser goes there to be treated. The products from refining are motor spirits, burning oil, fuel oil, lubricating oil, paraffin, and coke. Burning oil and fuel oil constitute nearly one-half of the total crude oil.

CHAPTER II

GEOLOGY OF THE NEW ALBANY SHALE

The New Albany shale was named by Borden after the city of New Albany, because of the excellent exposures found along the banks of the Ohio River in that vicinity. For a long time the shale was thought to be Devonian in age but of late question has arisen as to this, which will be discussed in a later paragraph. In Indiana the shale is overlain by a Missis-

Plate I. The New Albany-Rockford-New Providence contact in Falling Run Creek under the electric railway bridge between New Albany and Silver Hills. Note the thinning out of the Rockford to the left. a—New Albany. b—Rockford. c—New Providence.

sippian limestone, the Rockford Goniatite and is underlain by the Beechwood limestone, which is Devonian in age. In the absence of the Rockford limestone in Jefferson County, Kentucky, across the Ohio River from New Albany, Charles Butts says that an unconformity exists between the New Albany and the New Providence which lies in Indiana, immediately above the Rockford.

Along Falling Run Creek in Floyd County, Indiana, which is across the river from Jefferson County, the Rockford is present and shows evidence of unconformity in that it thins out and thickens from none to two feet. The New Albany-Rockford-New Providence contact may be clearly seen in Falling Run Creek under the electric railway bridge that connects Silver Hills with New Albany. (See Plate I.)

The Rockford farther north in Indiana is persistent, having been found at Rockford, Jackson County, Henryville, Clark County, as well as several other places between Rockford and the Ohio River. According to Butts the Rockford is absent in Jefferson County. Taking it for granted that the Rockford is correctly correlated with the Kinderhook group of Missouri and Illinois as has been generally done, and that no such representative exists between the New Albany and the New Providence, he measures there a hiatus equal to the Kinderhook group found elsewhere. Butts further states that if the upper

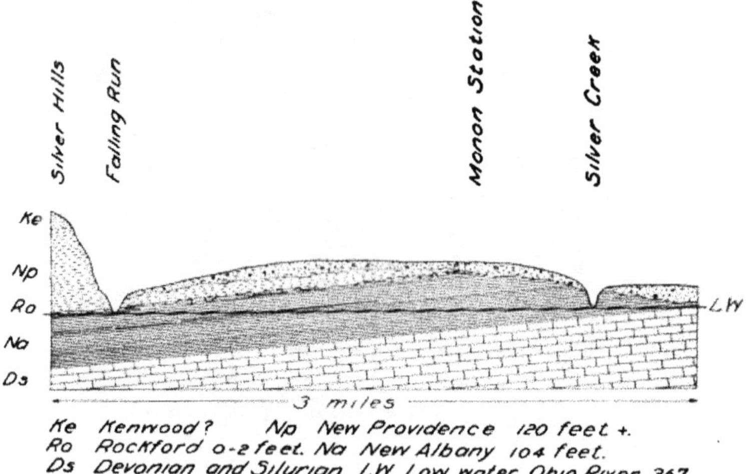

Fig. I. Cross-section east and west through the city of New Albany, showing the dip of the rocks and the relation of the New Albany to the water level of the Ohio River.

New Albany is found to be Mississippian in age instead of Devonian, the hiatus would be considerably reduced and the upper part of the formation correlated with the Kinderhook.

In thickness the shale varies from 104 feet at New Albany at the foot of the Knobstone escarpment to as much as 140 feet in other localities. (See list of well records at the end of this chapter showing depth and thickness of the formation at various places in the state.) In the district of its outcrop the entire thickness is not always represented, as part of it has been removed by erosion. The northern three-fourths of the outcrop (Fig. II) is thickly covered with glacial drift except along the Wabash River near Delphi where the shale may be seen in extensive outcrops. From Bartholomew County south to the Ohio River the drift is thin or not present at all and outcrops occur in large numbers. In the drift area a full thickness of the shale is rarely found, due to erosion subsequent to the glacial drift deposition. West of the outcrop area where the shale is deeper and covered with overlying rocks the full thickness is shown by well records. In the area of the outcrop the shale is thickest at the western side and thinner on the eastern side, due to the dip of the rocks in this region to the west and southwest.

The shale has, on a fresh surface, a brown-black velvet color when dry and densely black when wet. In places throughout the formation there are

Plate II. A close view of the shale, taken near New Albany.

thin variations of blue and green shale and light colored streaks of sand. These thin sandstone layers are found usually within fifteen feet of the base of the formation. The following section taken one-half mile south and one mile east of Henryville shows the presence of these blue shale layers.

Black shale	15 feet 00 inches
Blue-green shale	8 inches
Black shale	4 inches

Green shale	10 inches
Black shale	6 inches
Blue shale	12 inches
Black shale	10 feet 00 inches

When freshly dug the shale is compact and homogeneous with a subconchoidal fracture but extensive weathering reveals fissility and the shale splits into sheets. It is fairly hard, being about 3, and is difficult to dig into with a miner's pick. Weathering softens the shale somewhat and causes its color to change to blue and then gray.

The shale as shown at New Albany is highly jointed (see Plate VI). Here the joints parallel with the river and about five feet apart, form a series of steps to the water's edge. These perfect joints extend vertically through the shale as much as 70 feet, and perhaps more although greater than this has never been measured. In localities cross jointing allows the shale to be wedged out in large rectangles or rhomboids. This jointing is typical of the formation wherever it is found in Indiana.

The body of the shale is composed of fine sand made up of grains of quartz, mica, and kaolinite, some argillaceous matter, and in some parts, as the lower fifteen feet at New Albany and North Vernon, contains a small amount of calcium carbonate. Pyrite is well distributed throughout the shale, in nodular form, ranging from microscopic size to two inches in the longest measurement. Pyrite is also found in lenses. These pyrite concretions occur usually in certain well defined bedding planes. Large limestone concretions, elliptical in shape, are abundant. Where these have formed, the shale about them has been bent and folded by the force exerted in their expansive growth. The following is an analysis of one of these concretions:

CO_2, Volatile, Organic	32.45%
Insoluble in HCl	18.73
Fe and Al	24.25
CaO	21.26
SO_3	2.38
MgO	.55
P_2O_5	.06
	99.68%

Thin slabs and slivers of the shale will, when placed in a fire or lighted by a match, burn quite freely, with a yellow flame giving off a sooty smoke and leaving apparently as much ash as the original bulk.

The following analyses were made by Duden (21st Annual Report, Indiana Geological Survey), the samples having been collected from different localities in the vicinity of New Albany.

Water expelled at 100 degrees C.	0.50
Volatile organic matter	14.16
Fixed carbon	9.30
Silica	50.53
Pyritic iron and alumina	25.30
Calcium oxide	0.09
Magnesium oxide	0.12
	100.00

Water expelled at 100 degrees C. for 4 hours		0.56
Volatile organic matter	14.30	23.60
Fixed organic matter	9.30	
Silicates insoluble in HCl		65.43
Ferric oxide		8.32
Calcium oxide		0.09
Magnesium oxide		0.12
Sulphur		2.08
		100.00

Sea water in which this shale was deposited contains, among other salts in solution, small amounts of calcium sulphate and magnesia which react with the iron oxide present in the water and the mud of the sea bottom, forming ferrous sulphate. This is then reduced to ferric sulphide or pyrite. This is probably the origin of the pyrite found in the shale. Water circulating through the shale after its deposition, took into solution some of this ferric sulphide, depositing it in favorable places in the form of nodules of pyrite.

When exposed to the influences and agencies of weathering, water and oxygen combine with ferric sulphide oxidizing it to ferrous sulphate, water, and free sulphuric acid. The sulphuric acid reacts with the sand, forming aluminum sulphate. This and the ferrous sulphate are water soluble and are leached out, forming white deposits on the weathered surfaces of shale.

The following analysis (by Ray) was made of the shale, sample dried at 110 degrees C.

SiO_2	51.09%
CaO	1.06
Fe_2O_3	10.98
Al_2O_3	9.84
MgO	.98
P_2O_5	.07
SO_3	.18
TiO_2	.88
Na_2O	.81
K_2O	.58
Li_2O	.47
Loss on ignition	24.06
	99.94%

The New Albany shale seems to have been deposited in a shallow sea bordering lowlands upon which grew abundant vegetation. This vegetation may have been both terrestrial and aquatic, the aquatic being either fresh or salt, but more apt to be the latter. As the sediment forming the main body of the shale was washed into the sea, it carried with it much of the vegetable matter in a partly decomposed condition, at least in a fine state of division. In some cases leaves and stems of plants were included.

Microscopic examination of thin sections of the shale show the presence of spores or parts of spores very similar to those found in coal. Surrounding these spores is the black sedimentary part of the shale, the black color being due to disseminated bituminous matter and carbon. It is thought the oil obtained by distillation is derived from this bituminous matter of the shale. When the original sediment enclosing the vegetable matter, began to harden and solidify, a process similar to coalification was begun upon the vegetable matter and continued to a more or less degree to the present, thus forming of the vegetable matter a bituminous substance.

Plate III. A view from a quarry showing the Jointing and the depth to which weathering effects the shale.

Plant fossils are not abundant in the New Albany but Duden found, among four, the *sporangites*, a fossil spore which he believes to be the spore of the large fucoid Parenchymophycus.*

For a long time the New Albany was treated as a single unit of the same age and correlated with the Genesee shale of New York and the Huron shale of Ohio. The age of the formation is a question unsettled. Charles

* The author has since found two fossil tree trunks of at least two feet in diameter. A number of others have been reported.

Butts, in his Geology of Jefferson County, Kentucky, has summed up briefly as follows: "Until recent years the New Albany shale in its entirety has been correlated with the Genesee shale of New York and the Huron shale of Ohio. Of late that correlation has been questioned by Ulrich, who maintains that only the lower few feet containing Schizobolus concentricus, Lei orhynchus quadricostacus, etc., known elsewhere only from the Genesee shale, is of Genesee age. The upper 85 to 90 feet of the New Albany, including as a

Plate IV. The lower fifteen feet of the formation in the B. & O. railroad cut in the east edge of North Vernon. a—Gray shale. b—Black shale. c—Sandstone.

basal bed the calcareous sandstone zone, Ulrich assigns to the Carboniferous system on evidence that cannot be fully stated here. One important item, however, is the presence of Lingula *melie*, which is a common form of the Sunbury shale in the Carboniferous of Ohio. He also argues from considerations of a more general character such as widespread movement of elevation and a long period of dry land between the deposition of the lower

68—20642

part of the shale ascribed to the Genesee and the deposition of the upper 90 feet which is regarded as Carboniferous. If the latter supposition is correct, there is a stratigraphic hiatus in the New Albany about ten feet from the bottom measured by over 5,000 feet of shale and sandstone of Portage and Chemung age in central Pennsylvania, and this, too, without any irregularity or discordance of bedding to mark the contact of the Genesee and the much younger Carboniferous parts of the New Albany, if the upper part is indeed Carboniferous." The New Albany is the same as the black shale found bordering the Jessamine anticline of Kentucky, an extension of the Cincinnati Arch. It is also found in Tennessee, Missouri, and Illinois.

Plate V. A view of the gray shale in the cut shown in Plate IV. The slabs in the lower center are from the thin sandstone above. See text. Note the thin limestone in the upper center.

The records given below by counties show the thickness of the New Albany shale with the thickness of the overlying rocks or drift. The records in some cases do not show the entire thickness of the shale, because a portion of it has been removed by erosion. This is true in the drift region and the region of areal outcrop. These records have been taken from reports of the Indiana Geological Survey as well as the author's notes.

Jackson County.

Well at Seymour.

Drift	75 feet
Knobstone	15 feet
New Albany shale	115 feet

Well at Brownstown.

Drift	43 feet
Knobstone	275 feet
New Albany shale	147 feet

Bartholomew County.

Well at Columbus.

Drift	26 feet
New Albany shale	87 feet

Johnson County.

Well at Franklin.

Drift	170 feet
New Albany shale	34 feet

Well at Edinburg.

Drift	115 feet
New Albany shale	20 feet

Well at Greenwood.

Drift	210 feet
New Albany shale	90 feet

Morgan County.

Well at Martinsville.

Drift	85 feet
Knobstone	323 feet
New Albany shale	120 feet

Monroe County.

Well at Bloomington.

Soil	6 feet
Mississippian	749 feet
Devonian	170 feet

Well at Thrasher School.

Soil	10 feet
Mississippian	749 feet
New Albany shale	108 feet

Marion County.

Well at Bridgeport.

Drift	160 feet
New Albany shale	140 feet

Plate VI. Upper--A shale quarry near Marysville. Center—A seventy-foot bluff of shale on Silver Creek near New Albany. Lower—Outcrop of the shale of the banks of the Ohio River at New Albany, under the K. & I. bridge.

Boone County.

Well at Thorntown.

Drift	65 feet
Knobstone	339 feet
New Albany shale	87 feet

Plate VII. An outcrop of the shale, taken one mile north of Henryville, showing the development of the jointing.

Jasper County.

Well at Remington.

Drift	5 feet
New Albany shale	85 feet

Newton County.

Well at Kentland.

Drift	100 feet
New Albany shale	100 feet

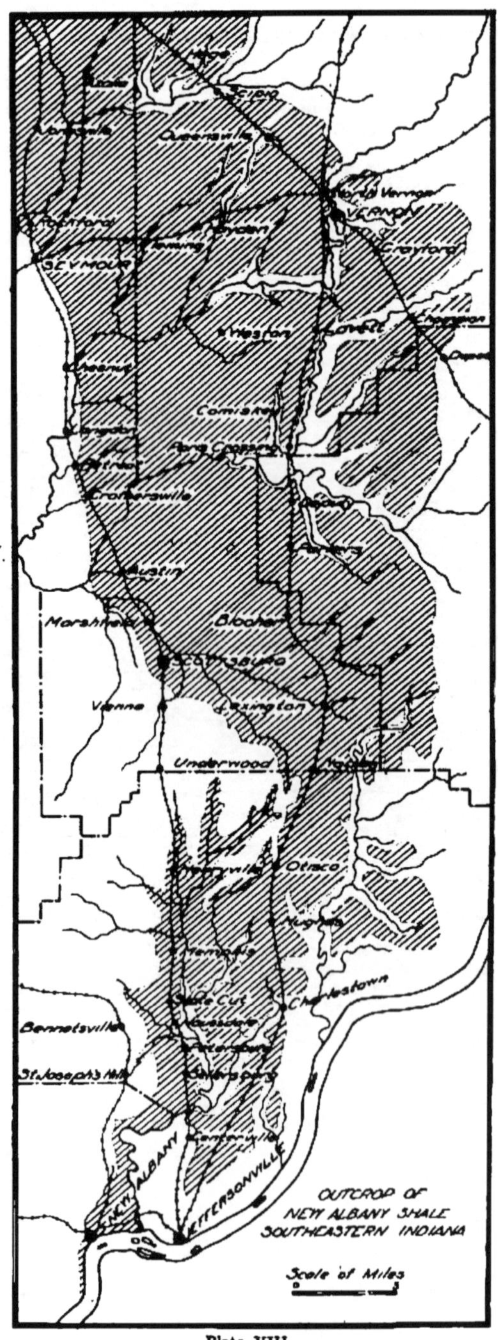

North Vernon
Crothersville
Comiskey
Austin
Marshfield
Blocher
Vienna
Lexington
Underwood
Otisco
Charlestown
Bennetsville
St.Joseph's Hill
NEW ALBANY
JEFFERSONVILLE
OUTCROP OF
NEW ALBANY SHALE
SOUTHEASTERN INDIANA
Scale of Miles

Plate VIII.

Laporte County.

Well at Laporte.

Drift	295 feet
New Albany shale	125 feet

Lake County.

Well at Crown Point.

Drift	176 feet
New Albany shale	76 feet

Montgomery County.

Well at Crawfordsville.

Drift	140 feet
Knobstone	410 feet
New Albany shale	80 feet

Putnam County.

Well south of Russellville.

New Albany shale	100 feet

Washington County.

Well at Salem.

Soil	7 feet
Keokuk limestone	53 feet
Knobstone	567 feet
New Albany shale	103 feet

CHAPTER III.

AREAL DISTRIBUTION OF THE NEW ALBANY SHALE.

The shale outcrop covers a strip of land from five to twenty miles wide lying in a northwest by southeast direction from Monticello in White County, through Marion County, to New Albany on the Ohio River in Floyd County. North of Jennings County the outcrop is covered over deeply with glacial drift and cannot be seen except in a very few places. Lying under this glacial drift the shale does not actually outcrop; however, no consolidated rocks overlie it. Under this drift the thickness and extension of the formation can be told only by well records that penetrate it. Near Delphi, Carroll County, the Wabash River has carried away the glacial material, cutting down several feet into the New Albany shale. The shale is also exposed along the Tippecanoe River near Monticello. South of Bartholomew County the glacial drift and wash is present on the highlands and outcrops are found in abundance, especially along the streams that have cut down into the shale. Actually the shale outcrops only in Jennings, Jefferson, Scott, Jackson, Clark, and Floyd Counties, besides those mentioned above. A scattered outcrop or two has been found in other counties which will be mentioned. It is in the above mentioned five counties that the shale has been studied and samples collected. A fuller discussion of the distribution is given below.

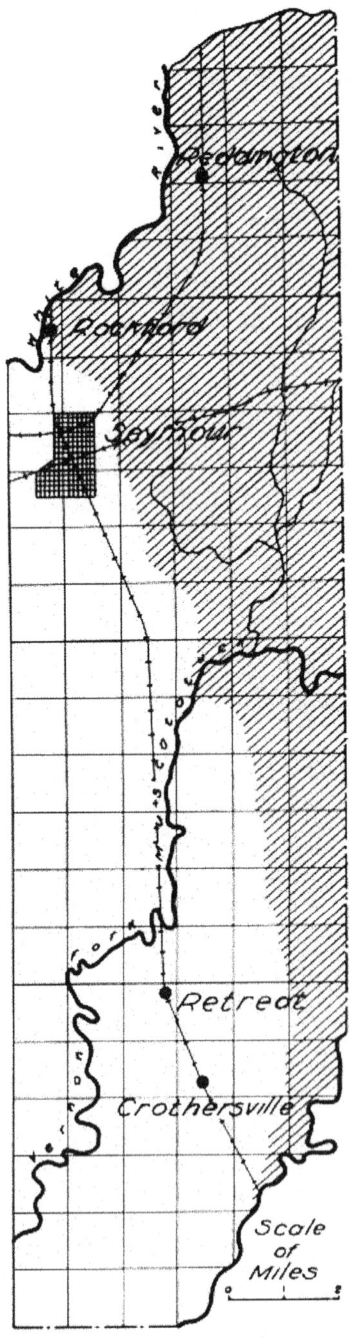

Plate IX. Map of the eastern portion of Jackson County, showing the outcrop of the New Albany shale.

Johnson County. One known outcrop of shale occurs in this county and it is found in the bed of Sugar Creek in Blue River Township, Sections 9 and 17. The shale outcrops along the banks and in the bed of the stream.

Bartholomew County. An outcrop of shale is found in the west center of Section 18, T. 8 N., R. 6 E., about five miles south of Columbus. The outcrop is very small in extent and above it lies the Rockford Limestone, showing this to be the top of the formation. Other outcrops probably occur in this county as most of its surface is underlain by the formation.

Jackson County. This county lies on the western side of the outcrop, where the shale dips under the Knobstone rocks. Only the eastern part of the county has any outcrops. In most places the shale is covered over to a considerable depth with glacial drift. The filled valleys of White and the Muscatatuck Rivers have also obscured the formation. Twenty-five feet or more the shale outcrops along White River one mile north of Rockford and also in the river bed at Rockford. The shale in the river bed at Rockford can be seen only at low water.* The following section was taken one mile north of Rockford:

Drift	70 feet
New Albany shale	25 feet plus

Another outcrop is found along the Muscatatuck River near Langdon. South of Seymour near Chestnut the shale is also found outcropping along the banks of the Muscatatuck River. From here south to Crothersville the land is quite level and a fairly thick layer of glacial drift and river fill overlies the shale. Two miles south of Crothersville, in the bed of the Muscatatuck River, several feet of the shal is found outcropping.

Jennings County. The eastern outcrop of the shale extends in a northwest by southeast direction across the county from Scipio in the northwest to Champion in the southeast. From this line west the shale is found covering the entire county except along Sand Creek, Storm's Creek, and the Muscatatuck River, which have cut through the shale, more so in the eastern part, to the underlying limestone below. At Scipio about five feet or more of the Hamilton limestones outcrop and only a few feet of the shale is found on the hills to either side. A short distance to the west of Scipio the shale becomes much thicker on the hills. From Queensville south to Hayden in the bed of Storm's Creek several feet of the limestones are found outcropping, the contact between the shale and the limestone being above the level of the creek some few feet. At Queensville the shale is about twenty feet thick, but one-half mile east of Hayden the hills are capped with ninety feet of shale, the bottom of the formation being found in the bed of the creek under the railroad bridge east of the town. At the county line three miles west of Hayden the shale is 125 feet thick. In the vicinity of North Vernon and Vernon, the Muscatatuck River has cut from thirty to forty feet below the base of the New Albany shale. The following is a section taken one mile east of Vernon at Tunnel Mill:

Soil	10-20 feet
New Albany shale	10 feet
Limestone	4 feet

* The raising of the water level by a dam has obscured these outcrops.

In the town of North Vernon the shale is about twenty feet thick, outcropping along the B. and O. railroad both east and west of the station in the city limits. About one-half mile from the station east on the B. and O. is a cut in which the lower part of the shale is exposed. The base of the

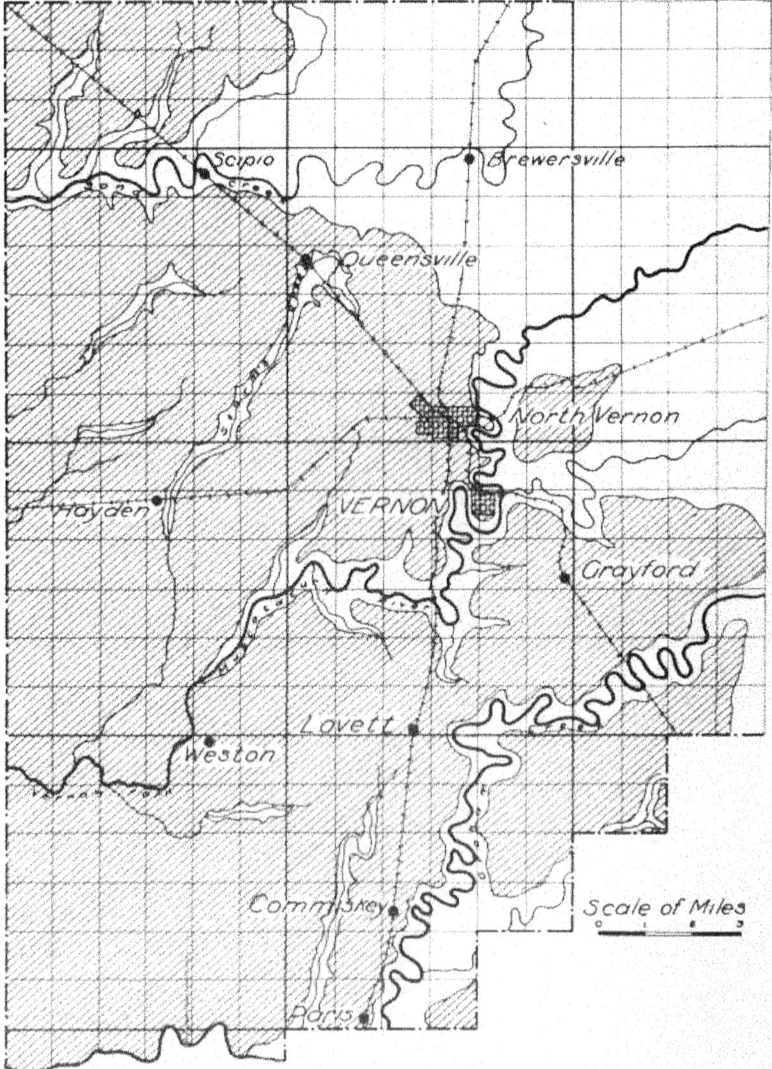

Plate X. Map of a portion of Jennings County, showing the outcrop of the New Albany shale.

cut is about fifteen feet above the base of the formation. The following is a section taken at this place:

6 feet 0 inches, black shale.
0 feet 1 inches, rectangular green sandstone. Pyritic.
0 feet 1 inches, black shale.
0 feet 1 inches, rectangular green sandstone.
0 feet 2 inches, black shale.
0 feet 1 inches, rectangular green sandstone.
2 feet 0 inches, black shale.
2 feet 0 inches, blue-green shale.
0 feet 1 inches, blue-gray hard limestone. Pyritic nodules.
4 feet 0 inches, gray shale.

Two miles east of North Vernon across the river, as much as twenty feet of shale is exposed in a B. and O. railroad cut. The highest hills about here are capped with shale, usually less than twenty feet in thickness. The following section was taken on Jordan's Branch near the old Kuchner quarry:

Soil	15 feet
New Albany shale	42 feet
Limestone	..

The high land southeast of Vernon about Grayford is capped with shale up to a thickness of thirty feet, but more often less. The Muscatatuck River flows across Jennings County on the Devonian limestones to within three miles of the west side of the county, where due to the dip of the rocks being greater than the stream gradient, the stream flows onto the shale. Six miles southwest of Hayden along the Muscatatuck is found a forty-foot bluff of shale which was at one time burning, and now bears the name "Burning Hill." The shale at this place has been quarried extensively and used for road purposes. Between North Vernon and Hayden along the railroad there is from ten to fifteen feet of drift overlying the shale. Several cuts between these two towns show good exposures of the shale. Near Lovett the shale is thin and is slightly covered with drift. East of Lovett there are no thick and extensive outcrops. Coffee Creek, one mile west of Commiskey, has cut through the shale. The thickness of the shale from Commiskey to Coffey Creek is from thirty to forty feet. A short distance west of Paris Crossing in Section 30 the following section was taken along Coffee Creek:

Soil	10 feet
New Albany shale	40 feet
Limestone	4 feet

East of Paris the shale is very thin, capping only the highest hills. From Paris to Deputy the shale has all been cut away by Graham Creek.

Jefferson County. Only in the western part of this county is the shale found, and being on the eastern side of the outcrop does not exceed forty feet in maximum thickness. From Deputy south along the railroad an occasional outcrop is found, although the shale is present, but mostly covered with drift. The shale is found outcropping along Middle Fork Creek in Lancaster Township and on Smock's Creek three miles west of Hanover. It is reported the shale attains a thickness of forty feet along Middle Fork Creek. The outcrop three miles west of Hanover is the easternmost outcrop of the shale known to the author in this state.

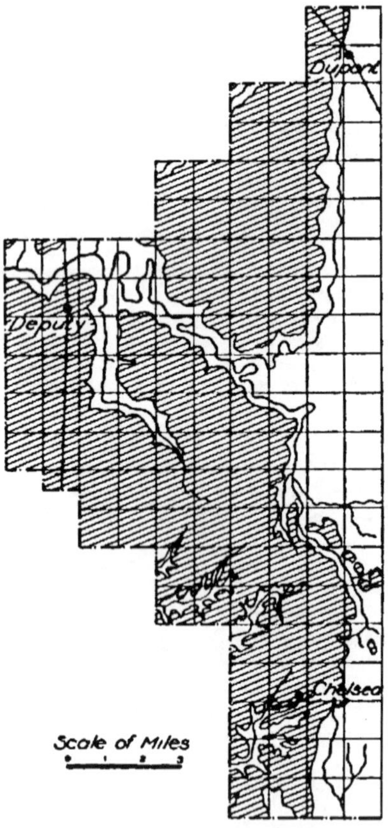

Plate XI. Map of the western part of Jefferson County, showing the outcrop of the New Albany shale.

Scott County. The western outcrop of the shale in this county runs in a general line from one mile east of Austin, south through Scottsburg, to one mile east of Nabbs. The county east of this line is underlain by New Albany shale. At Blocher the shale is forty feet thick, becoming thicker to the west, where, at Marshfield, it probably attains a thickness of 100 feet. At Lexington the underlying limestones are found outcropping in the bed of the stream that flows to the west of the town. The following is a section taken just west of Lexington:

Soil	15 feet
New Albany shale	40 feet
Limestone	..

Two miles northeast and one mile west of Lexington the shale in the hills is some sixty feet thick. East of Lexington the shale is thin, occurring only on the tops of the highest hills. From Lexington west for five miles there are several outcrops. About two miles west of the town is a quarry and another is found on Kimberlin's Creek five miles west, each showing the shale in

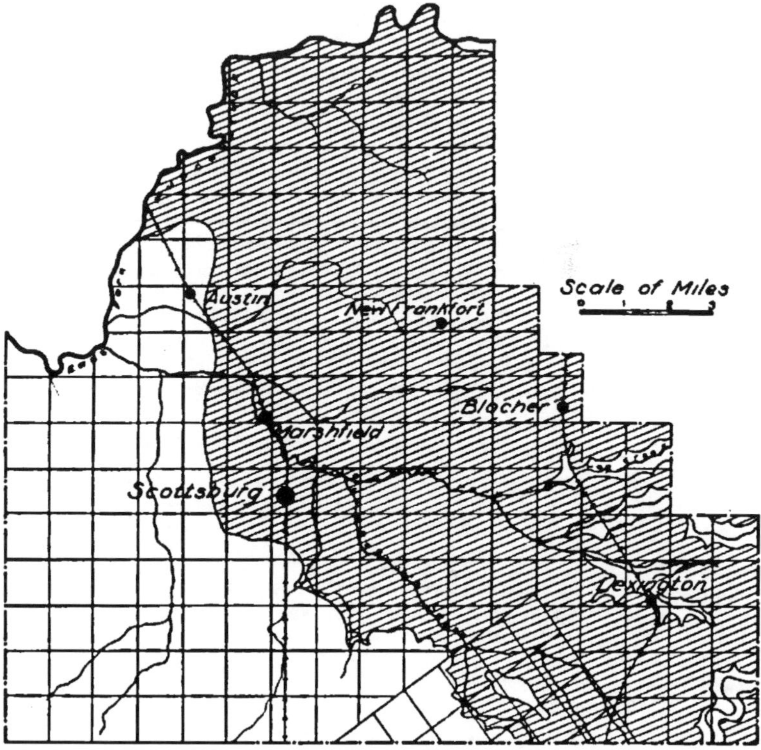

Plate XII. Map of a portion of Scott County, showing the outcrop of the New Albany shale.

good exposures. The shale at Kimberlin's Creek quarry is eighty feet thick. The following is the log of a well drilled one-half mile east of Scottsburg:

Soil	11 feet
Rockford	1 foot
New Albany	118 feet
Limestone	2 feet

One mile east of Scottsburg is a large quarry along the creek showing a twenty-foot exposure and five miles due north is found another good outcrop. From Scottsburg north to New Frankfort there are several outcrops. Two miles north of New Frankfort is another large quarry showing a twenty-foot exposure. The shale here in its entirety is probably 100 feet thick, as it is near the western edge of the outcrop. The main road from Lexington to Nabbs goes over several hills of shale, each from forty to sixty feet high, showing the shale to be at least that thick in this vicinity. The shale north of Scottsburg to the Muscatatuck River probably averages seventy-five feet or more in thickness. There are many outcrops between these two points. East of Nabbs toward the eastern border of the outcrop district there is very little shale.

Clark County. The outcrop in this county lies roughly between Fourteen Mile Creek in the east to Silver Creek in the west. On the road east through Marysville for two miles there are several outcrops and one quarry. The eastern limit of the outcrop does not extend beyond Fourteen Mile Creek in this vicinity. Northwest of Marysville one and one-half miles in Section 264, at Split Stump Quarry, twenty to thirty feet of shale is exposed. A well seventy-six feet deep here had not gone through the shale. One-fourth mile south of this village an outcrop is found in the bed of a small stream and two miles south one of the largest quarries is located, showing a thirty-foot exposure. The shale is present but thin on the high land between Otisco and Fourteen Mile Creek. Between Otisco and Henryville there is a good thickness of shale and several extensive outcrops. From Henryville north for two miles along the creek bed there are extensive outcrops of shale. The shale outcropping along this stream is of the upper twenty feet of the formation. At the State Forest Reserve, one mile north of Henryville, the following well record was obtained:

Clay	22 feet
Soapstone	40 feet
Slate	134 feet 6 inches
Limestone	5 feet
Slate	7 feet

The forty feet of soapstone is the New Providence shale of the Knobstone group. The total thickness of the shale is shown here to be at least 134 feet 6 inches. The writer doubts the presence of five feet of limestone at the base of the formation and believes this to be an error. In Henryville, which is somewhat lower than the location of this well, the shale is seventy-five feet thick as disclosed by several wells which have been drilled through it. South of Henryville, about one-fourth mile, a forty-foot exposure of shale is found at the top of of which is the Rockford limestone. The following section was taken one-half mile south and one mile east of Henryville:

Brown chocolate-colored shale	15 feet
Blue-green shale	.. feet 8 inches
Black shale	.. feet 4 inches
Blue-green shale	.. feet 4 inches
Black shale	.. feet 6 inches
Blue-green shale	.. feet 12 inches
Black shale	10 feet

The blue-green shale shown here is not uncommon in the formation and rarely attains a thickness greater than one foot, except in the northern portion of the state near Delphi. Between Henryville and Memphis outcrops are abundant in the stream valleys. Two miles west of Memphis the New Providence

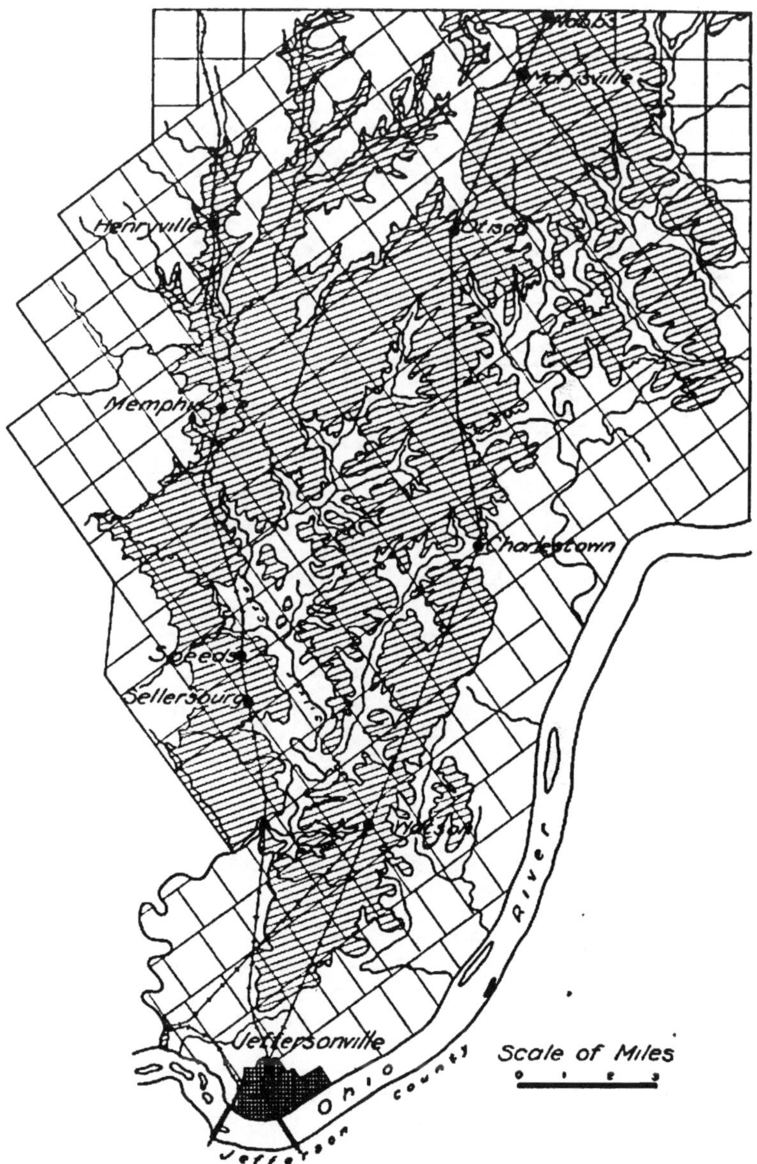

Plate XIII. Map of a portion of Clark County, showing the outcrop of the New Albany shale.

shale is found capping the hills. In Section 203, one-fourth mile northwest of Memphis, a well record shows the shale to be eighty-eight feet thick with no overburden. From Otisco south to Charlestown the shale at its thickest is only forty feet, while at Charlestown and along the stream flowing southwest to Silver Creek, the underlying limestones are found outcropping. There is but little shale east of Charlestown. Between Charlestown and Sellersburg the hill tops are capped with shale, the maximum thickness being about fifty feet. Around Sellersburg and Speeds in the bed of Silver Creek the extensive outcrop of the Silver Creek limestone is made use of for cement. The limestones outcrop along Silver Creek nearly as far north as Memphis and south to where the Pennsylvania railroad bridge crosses Silver Creek in Section 48 about three miles south of Sellersburg. Along the electric railway, north from Speeds about two miles, is an outcrop showing the contact between the under-

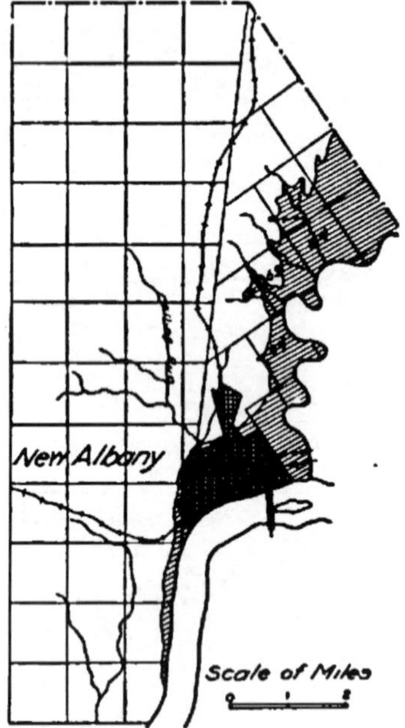

Plate XIV. Map of the eastern part of Floyd County, showing the outcrop of the New Albany shale.

lying limestones and the New Albany. There are also several other outcrops in Sections 149 and 167.

Floyd County and the Ohio River District. Outcrops occur all along the Ohio River and inflowing streams from less than one mile above the mouth of Silver Creek for five miles down the river. It is well exposed in many places, especially in Silver Creek and along the banks of the Ohio from the K. and I. bridge to the mouth of Silver Creek. Under the K. and I bridge twenty-five feet is exposed on the river bank at low water and from fifteen to thirty feet along Silver Creek near its mouth which at this place has cut a gorge into the shale. In Falling Run Creek, which flows at the foot of the Knobstone escarpment at the western edge of the city of New Albany, six feet of the top of the formation is exposed above low water. Above this lies 175 feet of the bluff-forming Knobstone rocks. The total thickness of the shale at New Albany at the foot of the escarpment is 104 feet. (Dr. Clapp, Ind. Geol. Sur., 1873.) In the city of New Albany there is a thickness of about fifty feet of shale above low water level. North of New Albany on the main road to Sellersburg, in Sections 44, 63, and 64, there are several excellent outcrops where the small streams tributary to Silver Creek have cut deeply into the shale. In Section 44, on the west side of Silver Creek, seventy feet of shale occurs as a bluff. The place is known as the "Devil's Step off."

Carroll County. The shale in this county is found outcropping along the Wabash River at Delphi and from there ten miles east along the river. The formation in this northern part of the state apparently contains more blue and green shale than in the southern part of the state. The following section will give some idea as to this:

Drift	7 feet	
Bluish black shale, sheety and tough.....	45 feet	
Drab grayish, slightly sandy............	4 feet	6 inches
Band of gray colored concretions........	6 feet	4 inches
Drab colored sandy shale...............	10 feet	6 inches
Bluish gray sandstone	4 feet	10 inches
Drab colored sandy shale...............	5 feet	6 inches
Covered	5 feet	6 inches
Devonian limestone	8 feet	

The shale also outcrops at Monticello in White County north of this region. At Rockfield thirty feet of black shale is found outcropping along a creek that flows just north of the village. The following section was taken here:

Black shale	20 feet	
Blue shale	.. feet	4 inches
Black shale	5 feet	.. inches
Blue shale	.. feet	6 inches
Black shale	6 feet	.. inches
Blue limey clay with limestone concretions	2 feet	.. inches

The upper section was taken from Kindle, Ind., Dept. of Geol. and Nat. Resources, 25th Rep., p. 533.

CHAPTER IV.

RESULTS OF EXPERIMENTS ON THE NEW ALBANY SHALE.

RETORTING.

Kinds of retorts used. Different kinds of retorts have been used in the laboratory distillation of the shales mentioned in this report. A retort constructed of a six-inch section of a three-inch malleable iron pipe, capped at both ends, with one cap removable, was first used. Dufficulty was had keeping this retort from leaking and it was discarded. An iron mercury retort of one-half pint capacity was then employed. The cap of this retort is machined to fit the top and is fastened by a clamp. The delivery tube leads from the top of the lid. This is a stock retort carried by most laboratory supply companies. This type of retort was used by the United States Geological Survey in its field work on the western oil shales. A complete description of the retort and its use may be found in Bulletin 641, p. 147. The retort is capable of taking a charge of about 240 grams or eight ounces of shale. The retort is heated in the field by gasoline burners or by Meeker or Fisher burners in the laboratory. It is difficult to get the retort heated to as high a temperature as sometimes desired, but by proper insulation this may be done.

In order to employ steam in the distillation, a one-half pint iron retort, as described above, was secured and a small hole drilled in the side near its base. Into this hole, after it was threaded, was screwed a one-eighth by six inches brass tube so that the inner end of the tube came to the middle of the inside of the retort. To the other end of the brass tube was connected a rubber tube or steam line which in turn was connected to a steamer. To super-heat the steam a Bunsen burner with a wing top was placed under the brass tube up near to the retort. This retort so connected with the steamer gave very good results.

An iron mercury retort of about two gallons capacity and similar in shape to that mentioned above, capable of taking a charge of at least four pounds of shale was also used. The lid of this retort was machined to fit the top of the bowl and fastened down by several bolts. The outlet for the vapors was a two-inch pipe fitted in the top of the lid. This outlet pipe reduced from two inches to three-fourths inch to fit the condenser tube. Through the top of the lid, standing vertically, was a three-eighths inch by three feet iron rod enclosed in a one-inch pipe, the lower end of which had two arms, the upper end of which was fitted with a wheel. This rod was revolved by a motor connected with the wheel by a belt. This fixture was used to stir the mass of shale in the retort during distillation. Two thermometers, encased in metal jackets and slotted to permit reading, were also placed in the top of the retort. One thermometer extended down into the mass of the shale and the other extended into the mouth of the outlet. Thus the temperature of the shale mass and of the outgoing vapors was had.

For the small retorts an ordinary twenty-inch glass laboratory condenser was used, supplied with running water from the faucet. For the large retort a three-fourths inch copper tube six feet long wound in spirals and placed in a fifteen-gallon can of water was used. Both of these condensers gave satisfactory results.

The retort recently described by L. C. Karrick of the Bureau of Mines was also used. This retort has been adopted as standard. It is essentially the same as the above mentioned retorts with the exception of a reflux condenser and the delivery tube which fits close to the retort. The capacity of this retort is one pint.

Temperature. In destructive distillation of the New Albany shale without steam more than one-half of the oil obtained comes off between the temperatures of 750 degrees and 850 degrees F. The oil usually begins to come off at about 600 degrees, while usually most of the water comes off at a lower temperature. Above 850 degrees the remainder of the oil comes off and is of a slightly higher specific gravity. By allowing super-heated steam to pass into the retort, a larger yield of oil may be had at a lower temperature. At no temperature below red heat has this shale shown any tendencies to coke or intumesce.

Time of distillation. In using the small one-half pint retort, at the end of one and one-half hours if the temperature is up to 1,000 degrees F., ninety per cent or more of the oil has been extracted. Better results have been had by heating up slowly until the oil begins to come, then more rapidly. At no time should the heat be applied so fast that a white gas forms. The greater the quantity of this gas the smaller the quantity of oil. For complete distillation, obtaining the total yield of oil, gas, and ammonia, the retort should be heated slowly at first until the oil begins to come off, then more rapidly until the temperature has arisen to about 1,100 degrees F., after which the heat may be applied as fast as desired, for at this temperature all of the oil should have been extracted. It will usually take from one and one-half to two hours for a distillation to obtain oil alone and from three to four hours to get all the ammonia and gas.

Rates of distillation. The rate of distillation depends upon the amount of heat available for the retort, the material and thickness of the sides of the retort, and the size of the charge and its distribution within the retort. It is well not to try to hasten a distillation by applying the heat too rapidly at first or before the first oil has appeared. The best guide is the formation of excessive gas. Too rapid application of heat will cause more gas to be formed and less oil. A little experience will enable one to judge nicely the rate of distillation.*

Size of particles. The size of any piece of shale in the retort should be small enough that the temperatures of all parts of it shall reach the maximum temperature of the retort. A piece of shale may be so large that its interior will not be as hot as its exterior due to the insulation afforded by this exterior part. It is obvious that if the inner part of a piece of shale is not brought to a temperature high enough to volatilize the bituminous matter in the interior the yield of oil will not be as high as it should. The thermal conductivity of the New Albany shale is about the same as that of limestone or fire-clay brick, which is relatively low. Theoretically the finer the division of the shale the more complete the distillation. This holds true within certain limits. The table below tends to show that the shale needs not be crushed finer than one-fourth inch. For this experiment a sample of fresh shale was crushed and quartered. The pieces of one-half inch size were hand picked at random from one portion. The remaining portions were crushed and ground to the sizes to be used. The experiment was conducted by using the standard apparatus of the Bureau of Mines.

Size of mesh	Amount used	Gallons of oil
½ inch	8½ ounces	12.00
Through ¼ inch on 14	8½ ounces	13.00
Through 14 inches on 20	8½ ounces	13.00
Finer than 60	8½ ounces	13.25

* Experimental work has since shown the amount of oil produced from a given sample can be made to vary as much as 8% by varying the rate of retorting. The largest yield is obtained when the retorting time is about 1¼ hours.

When the shale is ground so that a quantity of it is finer than 100 mesh the vapor from the retort, when steam is used, carries over with it into the condenser some of this fine material. The carrying over of fine material must be guarded against in commercial practice. The size of pieces of New Albany shale to be used in commercial retorts will vary with the retort, but it will best be determined by experimentation after the retort is in use. It is the writer's opinion that the sizes for a commercial retort should not be larger than one-fourth inch.

OIL FROM THE NEW ALBANY SHALE.

How produced. When the shale in the retort is subjected to heat the bituminous matter volatilizes. The volatile vapors are conducted from the retort into a condenser where some of the vapors are condensed to oil, some go on through the condenser as inflammable gas, and some go through as ammonia. The vapors that do not condense are called permanent gases. The amount of oil, its composition, and the amount of gas and its composition, and the amount of ammonia all vary according to the manner in which the shale is retorted.*

Quantity. The New Albany shale yields from six to fourteen gallons of oil per ton. These figures are verified by many experiments on samples from many different parts of the shale area. Samples collected from outcrops such as bluffs, ditches, and the beds of streams, are not usually entirely free from weathering, having lost some of the bituminous matter, and consequently yield less oil. Weathering affects the shale on most outcrops to a depth of from two to three feet and a sample taken from depths less than three feet cannot be said to be fresh sample and representative.

The following table gives the results of analyses of samples collected at New Albany. The first two are from Falling Run Creek and the last two from the K. and I. bridge. Sample 4 was cut only six inches deep and the others probably not deep enough to get the freshest shale. The samples were collected by Geo. H. Ashley and Chas. Butts, analyses made by D. T. Day. (U. S. Geological Survey, Bull. 641, p. 319.)

Sample No.	Amount Used	Oil Gallons	Water Gallons	Gas Cu. ft.	Ammonia Pounds
1	6 oz.	3.5	5.6	719	.08
2	6 oz.	9.1	4.1	958	0.
3	6 oz.	4.9	8.4	1097	0.
5	6 oz.	11.9	6.3	1197	0.

The writer does not believe these samples to be representative of the New Albany shale as a whole.

The following analyses made by Ray by the steam distillation are of samples collected on outcrops, but fairly fresh. The amounts of ammonium sulphate are somewhat large as they include pyridine and other alkali compounds possibly present.

* It has since been shown by McKee and Lyder that the first product of decomposition is a heavy semi-solid bitumen, soluble in benzene and carbon bisulfide. The formation of the oil is a result of the cracking of this bitumen.

Oil Gallons	Gas Cu. ft.	Ammonium Sulphate Pounds	% Fixed Carbon	% Volatile	% Ash
10.00	997	28.5	6.25	12.84	80.91
12.75	912	22.2	8.50	13.14	78.36
18.00	1836	20.4	6.90	22.81	70.29
9.50	570	38.4	8.00	10.90	81.10
9.50	886	35.5	4.00	9.50	86.50
16.50	1350	22.4	7.90	18.00	76.40
12.00	989	24.0	8.30	10.40	81.30

The following analyses were made by the author, using the standard apparatus of the Bureau of Mines. Samples 7, 8, and 9 were weathered.

Sample Number	Amount Used	Oil in Gallons	Water in Gallons
1	8½ oz.	13.0	9.0
2	8½ oz.	14.0	
3	8½ oz.	11.0	7.0
4	8½ oz.	12.5	6.0
5	8½ oz.	12.0	5.5
6	8½ ox.	13.0	7.0
7	8½ oz.	9.0	6.5
8	8½ oz.	7.5	8.0
9	8½ oz.	7.5	7.0
10	8½ oz.	13.5	6.0
11	8½ oz.	10.0	3.0

The samples came from the following places: No. 1, near Crothersville. No. 2, same as No. 1. No. 3, Section 44, Floyd County, two miles northeast of New Albany. No. 4, Blocher. No. 5, Lexington. No. 6, Vinegar Mill, North Vernon. No. 7, two and one-fourth miles west of Commiskey. No. 8, one and three-fourths miles north of Blocher. No. 9, Blue River Township, Bartholomew County. No. 10, "Burning Hill," six miles south of Hayden. No. 11, Rockfield, Carroll County.

Composition of the oil. The oil from this shale, as said before, varies according to the manner in which it is retorted. An oil formed by fast and high heating is apt to have a larger percentage of unsaturated hydrocarbons than one formed by applying the heat more slowly. It is thought that the composition of the oil is dissimilar to the primary compounds in the shale, this difference being due to reactions between certain compounds during or after volatilization. The average specific gravity of the oil is 25 degrees Baumê, or .9032. It will often vary, the specific gravity ranging from .8900 to .9100.

The following is the result of the fractionation of a sample of oil from the New Albany shale:

Began to boil at 65 degrees C.
150 degrees C.................................. 18.43%
200 degrees C.................................. 13.12%
250 degrees C.................................. 15.31%
300 degrees C.................................. 35.31%
Tar and loss.................................. 17.38%

The amount of unsaturated hydrocarbons varies somewhat but averages 40 per cent in the fractions up to 300 degrees C. According to quantitative analyses, made by Mr. Ray, the oil contains from 2 per cent to 5 per cent pyridine and pyridine bases. Most of the pyridines are low boiling. Creosote in the oil varies from 1 per cent to 2 per cent and phenol from none to ½ per cent. Benzene has been found in traces in the low-boiling fractions. No anthracene has been found, although very delicate tests have been made for it. The oil contains a fairly large per cent of sulphur and sulphur compounds. One sample analyzed contained 2.34 per cent sulphur. The heating value of one sample was 19,200 B.T.U.s per pound. It is thought that this sulphur comes from organic compounds, as the temperature in the retort is not high enough to affect the iron pyrite of the shale. The amount of sulphur in the shale runs from 2 per cent to 4 per cent, mostly in the form of pyrite.

GAS.

The gas given off in the distillation comes from the non-condensable portion of the vapors formed by the volatilization of the bituminous matter. The following is an average analysis of the gas, the sample being taken in three parts; at a low temperature, middle temperature, and high temperature (by Ray:)

Hydrogen sulphide	38.1 per cent
Carbon dioxide	.5 per cent
Sulphur dioxide	1.1 per cent
Carbon monoxide	1.2 per cent
Heavy hydrocarbons, ethene, acetylene, etc.	5.5 per cent
Hydrogen	7.6 per cent
Methane, etc.	46.0 per cent

In the gas taken off at low temperature about 50 per cent is hydrogen sulphide, while at the end of the distillation little hydrogen sulphide is given off. It is probable that at a higher temperature, more carbon monoxide would be formed.

The following is an analysis of the gas made by passing steam through the retort when at high temperature:

Hydrogen sulphide	36.0 per cent
Carbon dioxide	1.8 per cent
Sulphur dioxide	1.5 per cent
Carbon monoxide	2.8 per cent
Heavy hydrocarbons, ethane, acetylene, etc.	8.4 per cent
Hydrogen	10.0 per cent
Methane, etc.	39.0 per cent

The amount and composition of the gas produced varies with the rate of retorting and with the final temperature reached. The amount varies from 500 to 1,800 cubic feet. Its net heating value averages 750 B.T.U.s per cubic foot.

Duden, in his study of the shale, gives the following results: (21st An. Rept. Dept. of Geol. and Nat. Resources, 1896.)

5 pounds Pittsburg coal	105 gallons of gas
8.5 pounds black slate	45 gallons of gas

8.5 pounds black slate, Ohio banks..................... 50 gallons of gas
8.5 pounds black slate, Falling Run banks.............. 65 gallons of gas
15.0 pounds freshly broken slate........................ 105 gallons of gas
15.0 pounds of same after exposure to air for 14 days.... 100 gallons of gas

Duden also quotes from a letter describing an experiment in the production of gas at the New Albany Gas Light & Coke Company's plant: "I carbonized three tons of the New Albany black slate and obtained a yield of 2.20 cubic feet per pound of 22 candlepower gas. Ordinary unenriched coal gas is about 18 candlepower. The quality of the gas is therefore better and the yeld 45 per cent of that obtained from Pittsburg coal. . . . With the arrangement we have for making gas, it would not pay us to use the slate, even if we could obtain it for nothing. The slate was obtained from near the exposed surface of a creek bottom, and I am sure that if a sample was gotten at a greater depth, a much better yield of gas would be obtained."

AMMONIA.

From the results of several analyses made of coal and cannel coal, Clarke, in his Data of Geochemistry (U.S.G.S. Bull. 616), infers that plant remains have contributed but a small part of the nitrogen contained in coals and cannel coals. He then concludes that the obvious source of the nitrogen is animal matter, and points out in support of this the presence of fish remains in cannel coal. However, large amounts of nitrogen have been found in coal, cannels, and certain shales in which no evidence of animal life is found. This does not prove though that no animal remains contributed to the nitrogen of coal or shale.

The nitrogen exists in the New Albany shale probably in compound with organic substances. When these compounds are heated they break down, part of the nitrogen forming ammonia and part engaging itself with other secondary hydrocarbons. It is thought that when hot steam is let into the still hotter mass of shale, some of the hydrogen of decomposed water is set free and unites with the nitrogen, thus forming a larger yield of ammonia.

When the ammonia, in mixture with the other gases, has been scrubbed it is fixed as ammonium sulphate by sulphuric acid. This is done by passing the ammonia into the precipitating tank which holds the acid.

Only a few analyses of the nitrogen of the New Albany shale have been made. By Kjeldahl's method, Mr. Ray obtained from three somewhat weathered samples, total nitrogen in the following quantities: .34 per cent, .47 per cent, and .46 per cent. Two analyses made by the author of fresh samples gave 1.04 per cent and 1.20 per cent. These amounts calculated as ammonium sulphate will run from 30 to 60 pounds per ton. However, in commercial practice probably not more than one-half of the total nitrogen will be recoverable, which would reduce the above figures considerably.

OTHER BY-PRODUCTS.

Potash. No extensive attempts have been made to determine the potash content of the New Albany shale. The only potash that could be profitably extracted is that in the form of water soluble salts. One analysis of the shale, after having been retorted at a high temperature, gave 1.3 per cent of water soluble K_2O (potassium oxide). This is thought to be a little high. Two other analyses of the same kind gave less than 1 per cent water soluble K_2O. Extensive claims have been made by a company for potash in the same formation at Clay City, Kentucky, the yields by analysis running as high as 50 pounds per ton.

The potash, if it exists in quantities to make it profitable to extract, will be obtained by passing the hot shale from the retort into water. The water taking the soluble salts of potash in solution is then concentrated and the salts precipitated. The potash thus obtained is probably in the form of potassium chlorate or sulphate in the presence of other water soluble salts as calcium and lithium. The whole mass of salt is then sold on a K_2O basis.

Other analysis of the total potash and water soluble potash are being made, but at this time are incomplete.*

Other by-products have been suggested from the shale and shale ash, such as use of ash for cement, paint, and road material. The use of the retorted ash for road material seems most feasible and would help to dispose of it. The shale has been and is quarried and used for road building in the shale area and makes a fairly good road bed. The ash may be used for cement, but the expense of handling and erection of a plant makes it impracticable. The finely-ground shale, mixed with heavy, quick-drying oils might be put to use as a rough paint.

The following is an analysis of the ash dried at 110 degrees C.:

SiO_2	65.03 per cent	TiO_2	1.02 per cent
CaO	1.26 per cent	Na_2O	.91 per cent
Fe_2O_3	14.84 per cent	K_2O	.72 per cent
Al_2O_3	13.96 per cent	Li_2O	.58 per cent
MgO	1.32 per cent	Loss on ignition	1.81 per cent
P_2O_5	.08 per cent		
SO_3	.22 per cent		99.75 per cent

CHAPTER V.

ASPECTS OF THE NEW ALBANY SHALE AS AN OIL PRODUCER.

There are problems not mentioned before in this report which must be solved before oil from the New Albany shale can be profitably extracted, refined, and marketed. It must be considered in the beginning that all oil shales are not similar, and that certain methods of distillation applicable to one shale are not necessarily to another. A retort, distilling Colorado shales, may have a high efficiency, but by distilling New Albany shale the efficiency of the same retort may be quite low, because most Colorado shales are dissimilar to the New Albany shale. Oil produced from different shales or from the same shale by different types of retorts are very apt to vary in composition, thus requiring different treatment in refining. It must also be borne in mind that methods and types of retorts will differ according to the primary product desired from the shale or oil. It is generally conceded that an oil containing the largest amount of gasoline and lubricants is desired, but this may not always be the case. Those expecting to establish the industry in Indiana can not afford to disregard the experiences of the Scotch. It is quite possible that

* Analyses made since have shown the potash content of the shale to run from 4 to 5%, occurring as a silicate form which is not water soluble.

the United States will go far in advance of the Scotch methods, but such advance will come quicker if heed is given to the common essentials of shale retorting which the Scotch have largely perfected. Intelligent and scientific construction and management will play the larger part in a successful oil shale enterprise.

There are two main steps in the making of a marketable shale oil and its products, retorting and refining. The first of these steps may be divided into four parts: the quantity of oil, the quality of oil, the by-products, and the cost.

In giving some of the detailed problems below the author has drawn freely from all publications available besides his own knowledge. To secure the greatest quantity of oil from the shale the following factors are to be regarded, some of which have proven to be or may prove to be practical:

1. Size of the pieces of shale.
2. Construction of the retort.
3. Rate and temperature of heating.
4. Use of steam.
5. Thickness of the shale layer in the retort.
6. The forming of emulsions.
7. The forming of undesirable compounds.
8. The cracking of vapors while still in the retort.
9. Thermal conductivity of the shale.
10. Use of vacuums.
11. Use of pressures.
12. Catalizers.
13. Time required for complete distillation.

To secure an oil of a quality amenable to refining with least loss the following factors are to be regarded:

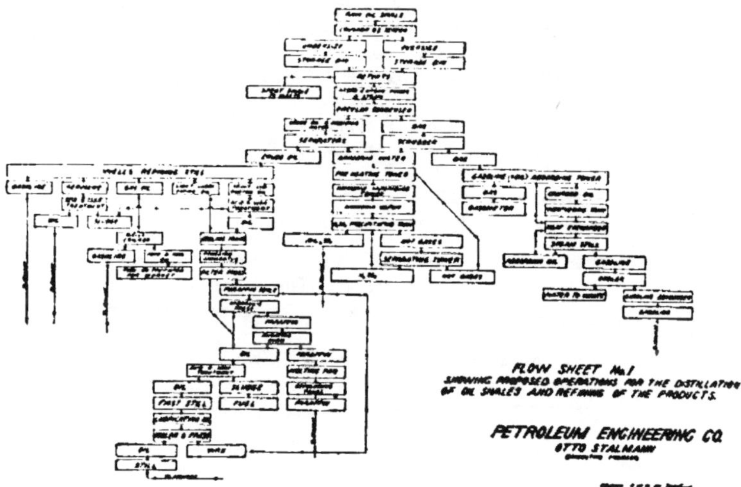

Plate XV.

1. Temperature control.
2. Use of steam.
3. Rapid withdrawal of vapors from the retort.
4. Forming of undesirable compounds such as olefines.
5. Cracking of vapors.

To secure from the shale the largest amount of by-products without decreasing the quality and quantity of oil the following factors may be regarded as important:

For Ammonia.

1. Use of steam.
2. Maximum temperature.

In addition to the ammonia formed in the retort such factors as its fixation and recovery from ammonia liquor are important. The recovery of potash is only possible from the water soluble salts that exist in the shale ash. Leaching is the most important factor in this operation. There are many other by-products or so called by-products that have been advocated, but they are for the most part speculative.

To operate an oil shale plant most efficiently the following factors should be considered:

1. Capital cost of retort per ton of shale throughput per 24 hours.
2. Labor cost per ton of shale throughput per 24 hours.
3. Fuel cost per ton of shale throughput per 24 hours.
4. Power and steam cost per ton of shale throughput per 24 hours.
5. Repair of retort.
6. Depreciation.
7. Initial cost of shale land.

The second main step in the making of a marketable shale oil is its refining. The large petroleum refineries are used to handling a crude petroleum containing usually about ten per cent unsaturated hydrocarbons, the bulk of the crude petroleum being paraffins. It will be necessary to refine shale oil with its 50 per cent to 60 per cent unsaturated hydrocarbons without losing these valuable compounds. There is a considerable percentage of pyridines to be had in the New Albany shale oil, which is a valuable product. Shale oil gasoline will have to be made with a good and permanent color besides a sweet odor. There has been some difficulty in removing the brownish purple color of shale gasoline and eliminating an unpleasant odor.

The main problems of the oil shale industry, which are applicable to all shales, have been for the most part solved. There are many details such as mentioned above to be tried or modified to each shale retorted. It is doubtful whether laboratory experimentation will prove or disprove all these factors or whether laboratory results should be depended upon too literally, but such results obtained from conscientious work should be used as a guide. No plant of commercial size has made continuous runs on the New Albany shale, and until one does, the most desirable results will not be known.

CHAPTER VI.

THE COAL MEASURES SHALES.

There are certain thin bedded black shales or oil shales which occur associated with the coals of Indiana. During the summer of 1920 a number of samples were collected and taken to the laboratory for distillation, where it was found they yielded considerable oil.

Stratigraphically these beds occur immediately overlying the Minshall Coal, overlying a thin coal under Coal III, always overlying Coal IIIa, in localities overlying Coal III, always above Coal IVa, Coal V, and as far as is known, above Coal VIII. Occasionally a bone coal occurs under Coal VI, above the lower Block Coal, and Coal IV. The persistence of these beds makes them a characteristic feature in determining the stratigraphic position of the coals

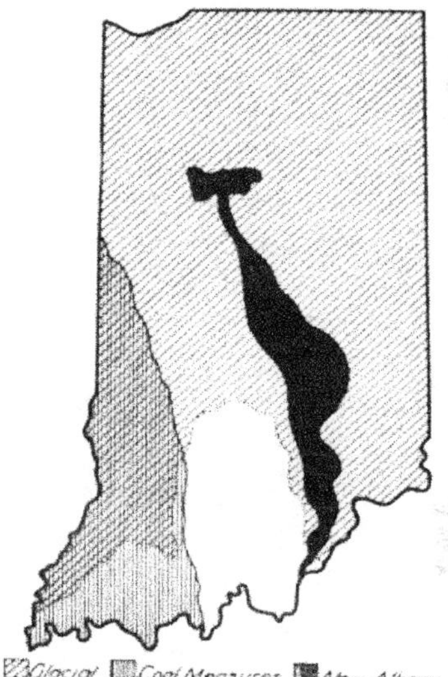

Fig. II. Outline map of Indiana showing the distribution of the New Albany shale and the Coal Measures and their relation to glacial drift.

with which they are associated. These beds of shale also have associated with them limestones, which are for the most part diagnostically fossiliferous.

The black shale beds may be divided into two classes: first, the "sheety" shales, and second, the massive shales tending to be bone coal. The distinct horizontal planes developed in the first class of these shales has given rise to the name "sheety" shale. Where freshly dug the sheets are from one-eighth to one-fourth inch thick. The shale readily splits along the planes, more so when slightly weathered, and small crystals of calcite and selenite may be seen on the surfaces. Vertical joints are, in localities, well developed. The second class of these shales, which are for the most part bone coal, black jack, or cannel coal, have no joints, planes, or definite fracture that are easily noticeable.

In color the shales are dull carbon black. The sheety shales sometimes have thin streaks of gray limestone and black shiny streaks of coal or anthraxylon, which are the coalified remains of a limb or twig. The shales of the second class have more of a lustre than the sheety shales except the one found above Coal V, which is in part, gray-black and hard. Upon weathering, minor bedding planes become visible in the sheety shales, giving rise to the name, "paper" shales. Extensive weathering reduces the color of the shales to gray-white.

The hardness of these shales varies from 2 to 3. The softest are tough and may be cut easily with a knife while the harder ones can be scratched only with a knife. The gray-black shale, above Coal V, is the hardest, due to the probable presence of considerable calcium carbonate.

Plate XVI. Outcrop of Coal V at Alum Cave. Note the warping of the sheety shale about the concretions, one of which has fallen down. In the lower center are slabs of sheety shale on the ground. a Coal V. b Sheety shale. c Hard gray shale. d Limestone.

The shales, when placed in a fire, burn readily with a bright yellow flame and give off much smoke.

All these shales contain fossiliferous fauna except that above the Minshall Coal.

The shale above Coal V contains large limestone concretions, which in growing have caused the tough shale to bend, fault, and give way to the expansive growth. Small pyrite nodules are not uncommon in this shale. Weathered outcrops show pure sulphur and iron sulphate, a result of the oxidation of the pyrite within the shale. The shale above Coal IIIa is also quite pyritic, the pyrite occurring in both nodules and lenses.

Minshall Coal Shale. The shale above the Minshall Coal varies in thickness from one to four feet, but averages nearly four feet. It is of the sheety type and non-fossiliferous.

Coal IIIa Shale. Black shale is found above Coal IIIa. In localities as at Staunton, Clay County, the entire interval between Coal III and Coal IIIa is occupied by this black shale. The interval of five beet in this locality is subnormal. Coal III usually has an overlying sandstone. The shale above Coal IIIa is from one to five feet thick. The shale is most often sheety but in places it is bone coal. A small lens of non-fossiliferous limestone, compact and very hard, is found within this shale and another highly fossiliferous limestone occurs above this.

Coal IVa Shale. The shale above Coal IVa is from two to five feet thick and is as often bone coal as it is sheety. In certain localities it is pure cannel coal. A gray-black limestone, fine grained and compact, from two to six inches thick is found within this shale.

Coal V Shale. This shale always comes above Coal V and in mines serves as a very good roof. It is from three to fifteen feet thick. With it are associated two limestones, the first coming immediately on top of the shale and the second a few feet above the first, separated usually by an interval of shale. The second, a brecciated limestone, is not always present. The lower limestone contains diagnostic fossils. The lower two to four feet of this shale is usually sheety and the upper part, gray-black and hard. In localities the entire thickness is of the gray-black variety and in other localities it is entirely sheety. The sheety shale frequently contains pyritized fossils. The limestone concretions found in this shale are very hard and form "horses" and "nigger-heads" in the mine roof.

Coal VIII Shale. This shale, from one to four feet thick, lies immediately above Coal VIII and is always of the sheety variety. In places, as at Vigo Post Office, Vigo County, two inches of gray shale separate the coal from the black shale. A limestone comes in above this shale also.

Post-Allegheny	Merom	Merom
	Shelburn	Sommerville Limestone
		Coal VIII
Allegheny	Petersburg	Coal VII (Millersburg)
		Coal VI
		Coal Va
		Coal V (Petersburg)
		Coal IVa
	Staunton	Coal IV (Linton)
		Coal IIIa
		Coal III (Seelyville)
Pottsville	Brazil	Coal II (Upper Minshall)
		Minshall
		Upper Block
		Lower Block
	Mansfield	Coal I

The Pennsylvanian of Indiana, showing the position of the coals.

Origin of the shales. It is thought that these shales were deposited in the same basin and after the deposition of the coal forming materials. The black color is due to carbonization of the plant matter deposited with the sediments, such as spores, leaves, and small twigs and limbs, as well as vegetable matter partly decomposed. The bituminous matter in the shales amounts to from ten to forty per cent of the shale by weight. The body of the shale is composed largely of fine grained sand but in some cases there is present a considerable amount of calcium carbonate.

Distribution. These shales, as named above, occur persistently associated with the coals throughout Clay, Vigo, Greene, and Sullivan counties. However, in the northern and southern part of the coal field in Indiana their occurrence and relative position in reference to the coals is somewhat different. Generally these shales occur at the above mentioned horizons over the whole state, if not entirely over the whole Indiana-Illinois coal basin. The coal field in Indiana covers the southwestern part of the state and contains approximately 7,000 square miles. All the samples collected were from the counties mentioned in this paragraph.

Results of distillation. In order to determine the amount of oil contained in these shales, the samples were tested by dry destructive distillation in an iron retort. No attempt was made to determine the amount of ammonia given off from the distillation or to measure the gas, although it is known that the amounts are unusually large, depending upon the sample and the manner of retorting. The samples were collected from mines, stripping pits and outcrops. Some of the samples had been weathered and some had not. It is interesting to note that the sheety shales, after having been weathered to the paper state, gave a larger yield of oil than the same shales unweathered. This is due probably to the fact that a finer division of the shale mass allows a greater heating surface and more complete distillation. The oil obtained does not exist in the shales as such, but is bituminous matter, which upon the application of heat, volatilizes and condenses to oil with some of the gases going off as permanent gas and ammonia. The table below gives the sample number, the yield of oil, and the yield of ammonia liquor. Several tests were made from the same shale at the same place and the results given as in samples Number 1, 33, etc.

	Oil in gallons per ton	Ammonia liquor gallons per ton
1--Under Coal III	18-24-22	13
2—Minshall Coal Shale	25	..
3—Below Coal IV. Bone Coal	54	20
4—Above Coal V	25	..
5—Above Coal IVa	32	18
6—Minshall Coal Shale	20	10
7—Above Coal VIII	20	22
8—Above Coal VIII	20	..
9—Above Coal VIII	22	20
10—Above Coal IVa	22	20
11—Above Coal IVa	24	12
12—Above Coal IVa	10	30
13—Above Coal IVa	16	20
14—Above Coal V	23	10

	Oil in gallons per ton	Ammonia liquor gallons per ton
15—Above Coal V	18	16
16—Above Coal V	18	16
17—Above Coal V	14	14
18—Above Coal IIIa	12	36
19—Above Coal IIIa	21	20
20—Above Coal IVa	29	16
21—Above Coal V	18	18
22—Above Coal IVa	27	18
23—Above Coal V	12	20
24—Above Coal VIII	15	20
25—Above Coal V	24	14
26—Above Coal IVa	12	14
27—Above Coal IIIa	30	16
28—Above Coal V	22	12
29—Above Coal V	7	14
30—Above Coal IIIa	32	12
31—Above Coal V	8	16
32—Below Coal IIIa	26	..
33—Above Coal V	15.4-14	
34—Above Coal IIIa	20-24-20-15	..
35—Above Minshall Coal	29-27	
36—Above Coal IIIa	27-27	
37—Above Coal V	22-22	
38—Above Coal IVa	22	

Number of samples and the locality from which it was taken:

1. Middlebury Hill south of Clay City.
2. One mile south of Middlebury Hill.
3. Twin Mines east of Jasonville.
4. Three miles south of Riley. McNabney Mine.
5. One mile southeast of Cory.
6. One mile south of Middlebury Hill. Four feet thick. Unweathered.
7. Three and one-half miles southwest of Pimento. Unweathered.
8. Hutton Post Offiee.
9. One and one-half miles east of Hutton. Five feet thick.
10. One-half mile south of Coffey Hill. Weathered. Three feet thick.
11. One mile southwest of Coffey Hill. Weathered. Three feet thick.
12. One and one-half miles southwest of Coffey Hill.
13. One and one-half miles southwest of Coffey Hill.
14. Alum Cave north of Coalmont.
15. Alum Cave north of Coalmont.
16. Alum Cave north of Coalmont.
17. One-half mile east of Hymera.
18. Two miles east of Jasonville.
19. Two miles east of Jasonville.
20. Two miles south of Jasonville. Five feet thick. Weathered.
21. Southeast of Jasonville.
22. One and one-half miles southeast of Jasonville. Weathered. Bone.
23. Six miles west of Jasonville.

24. Merom. Four feet thick.
25. Two and one-half miles southeast of Dugger. Unweathered.
26. One mile northwest of Staunton. Unweathered.
27. One mile northwest of Staunton. Unweathered.
28. Two and one-half miles west of Linton. One foot thick.
29. One mile east of Dugger. Five feet thick.
30. S. E. ¼ Sec. 17, T. 12 N., R. 7 W.
31. One-half mile east of Kellar.
32. Section 11, T. 8 N., R. 7 W.
33. Boonville. (From G. H. Ashley. Bull. 641, U. S. G. S.)
34. Old Hill. One foot thick.
35. Fairview Church. Near Coffey Hill.
36. Unknown.
37. Alum Cave. North of Coalmont.
38. Unknown.

The specific gravity of the oil obtained from the sheety shales averages about 20 degrees Baumé. It is thin, has a red-brown color, and has an offensive odor, somewhat sulphurous. The first oil to come off in the distillation of the bone coals is light, with a specific gravity of about 20 degrees Baumé, but as the retort becomes hotter the oil given off is more like coal tar, this latter distillate being heavier than water. The specific gravity of the oil obtained and the oil itself varies greatly according to the application of the heat in the distillation. The following is the result of fractioning the oil from a compound sample:

Began to boil at 55 degrees C.

Temperature in C.	
150	17.75%
200	11.75%
250	17.75%
300	16.00%
Above 300	8.50%
Coke and tar	28.25%

The oil showed a fairly large per cent of pyridine. The shales do not gum on being distilled, but at a high heat slightly coke. By the injection of steam into the retort during the distillation a larger yield of oil and ammonia is obtained. One sample tested by dry distillation yielded 12 gallons of oil, but by using steam 18 gallons were obtained.

Results of the distillation of a four-pound sample of shale:

Temperature F.	Water in ounces	Oil in ounces
630	1.75	begins
700	2.00	.75
758	2.00	2.50
980	2.25	5.00
1000	2.50	5.25

Total amount of oil 20.47 gallons per ton.

Specific gravity, 20 degrees Baume.

From this it may be seen that most of the oil comes off between the temperatures of 758 and 980. It is possible that a little more oil could be obtained by running the temperature higher.

The following analyses are taken from G. H. Ashley's report on black shales of eastern United States. The analyses were made by David T. Day. The samples were collected at Boonville from above Coal V. (U. S. G. S. Bull. 641, page 319.)

Amount used Ounces	Oil Gallons	Water Gallons	Gas Cubic Feet	Ammonia pounds
6	14	11.2	2,522	.97
6	none	9.8	479	none
6	15.4	12.6	2,922	.61

Sample No. 2 is not a true oil shale, but was cut to determine whether the dark shales yielded any oil.

70—20642

GENERAL INDEX